Lecture Notes in Computer Science 3149

Commenced Publication in 1973
Founding and Former Series Editors:
Gerhard Goos, Juris Hartmanis, and Jan van Leeuwen

Editorial Board

David Hutchison
  Lancaster University, UK
Takeo Kanade
  Carnegie Mellon University, Pittsburgh, PA, USA
Josef Kittler
  University of Surrey, Guildford, UK
Jon M. Kleinberg
  Cornell University, Ithaca, NY, USA
Friedemann Mattern
  ETH Zurich, Switzerland
John C. Mitchell
  Stanford University, CA, USA
Moni Naor
  Weizmann Institute of Science, Rehovot, Israel
Oscar Nierstrasz
  University of Bern, Switzerland
C. Pandu Rangan
  Indian Institute of Technology, Madras, India
Bernhard Steffen
  University of Dortmund, Germany
Madhu Sudan
  Massachusetts Institute of Technology, MA, USA
Demetri Terzopoulos
  New York University, NY, USA
Doug Tygar
  University of California, Berkeley, CA, USA
Moshe Y. Vardi
  Rice University, Houston, TX, USA
Gerhard Weikum
  Max-Planck Institute of Computer Science, Saarbruecken, Germany

Marco Danelutto Domenico Laforenza
Marco Vanneschi (Eds.)

Euro-Par 2004 Parallel Processing

10th International Euro-Par Conference
Pisa, Italy, August 31 - September 3, 2004
Proceedings

Volume Editors

Marco Danelutto, Marco Vanneschi
University of Pisa, Department of Computer Science
Via F. Buonarroti 2, 56127 Pisa, Italy
E-mail:{marcod,vannesch}@di.unipi.it

Domenico Laforenza
High Performance Computing Laboratory, ISTI, CNR
Via G. Moruzzi 1, 56126 Pisa, Italy
E-mail: Domenico.Laforenza@isti.cnr.it

Library of Congress Control Number: 2004110451

CR Subject Classification (1998): C.1-4, D.1-4, F.1-3, G.1-2, H.2

ISSN 0302-9743
ISBN 3-540-22924-8 Springer Berlin Heidelberg New York

This work is subject to copyright. All rights are reserved, whether the whole or part of the material is concerned, specifically the rights of translation, reprinting, re-use of illustrations, recitation, broadcasting, reproduction on microfilms or in any other way, and storage in data banks. Duplication of this publication or parts thereof is permitted only under the provisions of the German Copyright Law of September 9, 1965, in its current version, and permission for use must always be obtained from Springer. Violations are liable to prosecution under the German Copyright Law.

Springer is a part of Springer Science+Business Media

springeronline.com

© Springer-Verlag Berlin Heidelberg 2004
Printed in Germany

Typesetting: Camera-ready by author, data conversion by Olgun Computergrafik
Printed on acid-free paper   SPIN: 11306603   06/3142   5 4 3 2 1 0

Preface

Euro-Par Conference Series

Euro-Par is an annual series of international conferences dedicated to the promotion and advancement of all aspects of parallel computing. The major themes can be divided into the broad categories of hardware, software, algorithms and applications for parallel computing. The objective of Euro-Par is to provide a forum within which to promote the development of parallel computing both as an industrial technique and an academic discipline, extending the frontier of both the state of the art and the state of the practice. This is particularly important at a time when parallel computing is undergoing strong and sustained development and experiencing real industrial take-up. The main audience for, and participants at, Euro-Par are seen as researchers in academic departments, government laboratories and industrial organizations. Euro-Par's objective is to be the primary choice of such professionals for the presentation of new results in their specific areas. Euro-Par also targets applications demonstrating the effectiveness of parallelism. This year's Euro-Par conference was the tenth in the conference series. The previous Euro-Par conferences took place in Stockholm, Lyon, Passau, Southampton, Toulouse, Munich, Manchester, Paderborn and Klagenfurt. Next year the conference will take place in Lisbon. Euro-Par has a permanent Web site hosting the aims, the organization structure details as well as all the conference history: http://www.europar.org. Currently, the Euro-Par conference series is sponsored by the Association for Computing Machinery (ACM) in cooperation with SIGACT, SIGARCH and SIGMM, by the International Federation for Information Processing (IFIP) in cooperation with IFIP WG 10.3, and by the Institute of Electrical and Electronics Engineers (IEEE) in cooperation with TCPP.

Euro-Par 2004 at Pisa, Italy

Euro-Par 2004 was organized by the University of Pisa, Department of Computer Science and by the National Research Council Institute of Information Science and Technologies (ISTI), also located in Pisa. The conference took place at the "Palazzo dei Congressi" (the Conference Center of Pisa University). A number of tutorials and invited talks extended the regular scientific program. The tutorials and the invited talks covered major themes in the current high-performance parallel computing scenarios:

- The full-day tutorial "Open Source Middleware for the Grid: Distributed Objects and Components in ProActive", by Denis Caromel (OASIS, INRIA Sophia Antipolis, France) covered major aspects in the implementation of middleware supporting the development of grid applications through suitable, open-source, component-based middleware.

- The half-day tutorial "Achieving Usability and Efficiency in Large-Scale Parallel Computing Systems" by Fabrizio Petrini and Kei Davis (Los Alamos National Laboratory, USA) went through the techniques used to set up high-performance clusters exploiting the most recent and advanced interconnection technology.
- The half-day tutorial "Grid Resource Management and Scheduling", by Ramin Yahyapour (University of Dortmund, Germany) presented an overview of the state of the art and experiences in Grid resource management and scheduling.
- The invited talk by Dennis Gannon (Indiana University, USA) "Building Grid Applications and Portals: An Approach Based on Components, Web Services and Workflow Tools" dealt with advanced techniques for building grid applications.
- The invited talk by Manuel Hermenegildo (Technical University of Madrid and University of New Mexico) "Some Techniques for Automated, Resource-Aware Distributed and Mobile Computing in a Multi-paradigm Programming System" dealt with advanced techniques for distributed and mobile computing.
- The invited talk by Mateo Valero (DAP-UPC Barcelona, Spain) "Kilo-instruction Processors" was about innovative processor technology aimed at reducing the problems raised by memory access latency.
- Last but not least, the invited talk by Murray Cole (School of Informatics, Edinburgh, UK) "Why Structured Parallel Programming Matters" was focused on the relevance and effectiveness of structured parallel programming with respect to performance, correctness and productivity.

The tutorials all took place on Tuesday, 31st of August. Two invited talks were scheduled for the beginning of the regular scientific program of the conference, on Wednesday, 1st of September, and two were scheduled for the end, on Friday, 3rd of September.

Euro-Par 2004 Statistics

Euro-Par 2004 followed the classic Euro-Par conference format and consisted of a number of topics. Each topic was supervised by a committee of four persons: a global chair, a local chair and two vice chairs. This year the conference ran 19 different topics. In addition to classic Euro-Par topics, the High-Performance Bioinformatics topic was introduced for the first time. A Demo Session topic included in the conference program after the positive experience of Euro-Par 2003 received a very small number of submissions. It was eventually canceled from the conference program and the accepted submissions were moved to another, related topic. The overall call for papers was closed in early February 2004 and attracted a total of 352 submissions; 124 papers were accepted after a refering process that collected an average 3.6 referee reports per paper, for a grand total of more than 1300 referee reports that involved more than 600 different reviewers; 105 papers were accepted as regular papers, 15 as research notes, and 4 as distinguished papers. Eventually, 2 papers were withdrawn by the authors, and therefore 122 papers are actually included in these proceedings. Papers were submitted from 39 different countries, according to the corresponding author country. Papers were accepted from 27 different countries. The major number of accepted papers came from the USA (22 papers), France and Spain (15 papers each), and Germany and Italy (13 papers each).

Acknowledgments

Different institutions and a number of individuals contributed to Euro-Par 2004 in different ways. The University of Pisa, its Computer Science Department, the National Research Council (CNR) and the ISTI (Institute of Information Science and Technologies) were the main conference institutional sponsors. A number of different industrial sponsors gave their contributions and/or participated in organizing the industrial booths at the conference location. Their names and logos are all collected on the Euro-Par 2004 conference Web site: http://www.di.unipi.it/europar04. A special thanks is due to the authors of the submitted papers as well as to the topic committee members and to the reviewers who overall contributed to making this conference possible. Also, the organizing committee thanks the Euro-Par steering committee for the continuous support during the conference organization. Special support was provided by Christian Lengauer, chair of the steering committee and by Harald Kosch, co-organizer of the 2003 Euro-Par conference. A special thanks is for the local organization team, Sonia Campa, Nicola Tonellotto, Marco Aldinucci, Massimo Coppola, Tiziano Fagni, Alessandro Paccosi, Antonio Panciatici, Diego Puppin, and Fabrizio Silvestri, and for the NORCI team, which solved many problems related to the conference organization and set up.

We all hope that the attendees had a pleasant stay in Pisa at the Euro-Par 2004 conference!

Pisa, June 2004
Marco Danelutto
Domenico Laforenza
Marco Vanneschi

Euro-Par Steering Committee

Chair
Christian Lengauer   University of Passau, Germany
Vice Chair
Luc Bougé   ENS Cachan, France
European Representatives
Marco Danelutto   University of Pisa, Italy
Rainer Feldmann   University of Paderborn, Germany
Christos Kaklamanis   Computer Technology Institute, Greece
Harald Kosch   University of Klagenfurt, Austria
Paul Kelly   Imperial College, United Kingdom
Thomas Ludwig   University of Heidelberg, Germany
Emilio Luque   Universitat Autònoma of Barcelona, Spain
Luc Moreau   University of Southampton, United Kingdom
Rizos Sakellariou   University of Manchester, United Kingdom
Henk Sips   Technical University Delft, The Netherlands
Non-European Representatives
Jack Dongarra   University of Tennessee at Knoxville, USA
Shinji Tomita   Kyoto University, Japan
Honorary Members
Ron Perrott   Queen's University Belfast, United Kingdom
Karl Dieter Reinartz   University of Erlangen-Nuremberg, Germany
Observers
José Cunha   New University of Lisboa, Portugal

Euro-Par 2004 Local Organization

Euro-Par 2004 was organized by the University of Pisa, Department of Computer Science and by the Italian National Research Council Institute of Information Science and Technologies.

Conference Chair
  Marco Danelutto
  Domenico Laforenza
  Marco Vanneschi
Committee
  Sonia Campa   Marco Aldinucci   Massimo Coppola
  Tiziano Fagni   Alessandro Paccosi   Antonio Panciatici
  Diego Puppin   Fabrizio Silvestri   Nicola Tonellotto
  Lita Sabini

Euro-Par 2004 Programme Committee

Topic 1: Support Tools and Environments
Global Chair
José C. Cunha — Faculdade de Ciências e Tecnologia Universidade Nova de Lisboa, Portugal

Local Chair
Dieter Kranzlmüller — Institut für Technische Informatik und Telematik Johannes Kepler Universität Linz, Austria

Vice Chairs
Allen Malony — Department of Computer and Information Science University of Oregon, USA

Arndt Bode — Institut für Informatik, Technische Universität München, Germany

Topic 2: Performance Evaluation
Global Chair
Wolfgang E. Nagel — Center for High Performance Computing, Technical University Dresden, Germany

Local Chair
Lorenzo Donatiello — Department of Computer Science, University of Bologna, Italy

Vice Chairs
Thomas Ludwig — Institute of Computer Science, Ruprecht-Karls-Universität Heidelberg, Germany

Jeffrey Vetter — Oak Ridge National Laboratory, Oak Ridge, USA

Topic 3: Scheduling and Load Balancing
Global Chair
Emilio Luque — Dept. Computer Science, Universitat Autònoma de Barcelona, Spain

Local Chair
Raffaele Perego — ISTI-CNR Pisa, Italy

Vice Chairs
Evangelos Markatos — Institute of Computer Science (ICS), Crete, Greece

José G. Castanos — IBM Thomas J. Watson Research Center, New York, USA

Topic 4: Compilers for High Performance

Global Chair

Hans P. Zima — Institute for Software Science, University of Vienna, Austria, and Jet Propulsion Laboratory, California Institute of Technology, Pasadena, California, USA

Local Chair

Beniamino Di Martino — Department of Information Engineering, Second University of Naples, Italy

Vice Chairs

Siegfried Benkner — Institute for Software Science, University of Vienna, Austria

Michael O'Boyle — University of Edinburgh, UK

Topic 5: Parallel and Distributed Databases, Data Mining and Knowledge Discovery

Global Chair

David Skillicorn — School of Computing, Queen's University, Kingston, Ontario, Canada

Local Chair

Salvatore Orlando — Department of Computer Science, University of Venice, Italy

Vice Chairs

Abdelkader Hameurlain — IRIT, Université Paul Sabatier, Toulouse, France

Paul Watson — School of Computing Science, University of Newcastle upon Tyne, United Kingdom

Topic 6: Grid and Cluster Computing

Global Chair

Thierry Priol — IRISA/INRIA, Rennes, France

Local Chair

Diego Puppin — ISTI/CNR, Pisa, Italy

Vice Chairs

Craig Lee — Aerospace Corporation, El Segundo, USA

Uwe Schwiegelshohn — University of Dortmund, Germany

Topic 7: Applications on High-Performance Computers

Global Chair
Jack Dongarra — Computer Science Department, University of Tennessee, USA

Local Chair
Rolf Hempel — German Aerospace Center DLR, Cologne, Germany

Vice Chairs
David W. Walker — Department of Computer Science, University of Wales, Cardiff, UK

Peter Arbenz — Institute of Computational Science, Zürich, Switzerland

Topic 8: Parallel Computer Architecture and Instruction-Level Parallelism

Global Chair
Kemal Ebcioglu — IBM T.J. Watson Research Center, New York, USA

Local Chair
Marco Aldinucci — ISTI-CNR, Pisa, Italy

Vice Chairs
Wolfgang Karl — Institut für Rechnerentwurf und Fehlertoleranz, Universität Karlsruhe (TH), Germany

André Seznec — IRISA, Campus de Beaulieu, Rennes, France

Topic 9: Distributed Algorithms

Global Chair
Henri Bal — Dept. Computer Science, Vrije Universiteit, Amsterdam, The Netherlands

Local Chair
Giuseppe Prencipe — Department of Computer Science, University of Pisa, Italy

Vice Chairs
Andrzej M. Goscinski — School of Information Technology, Deakin University, Victoria, Australia

Eric Jul — Dept. of Computer Science, University of Copenhagen, Denmark

Topic 10: Parallel Programming: Models, Methods and Programming Languages

Global Chair
Paul H.J. Kelly — Department of Computing, Imperial College of Science, Technology and Medicine, London, UK

Local Chair
Sergei Gorlatch — Institut für Informatik, Universität Münster, Germany

Vice Chairs
Daniel Quinlan — Center for Applied Scientific Computing, Lawrence Livermore National Laboratory, Livermore, USA

Christoph Kessler — Institutionen för datavetenskap, Linköpings Universitet, Sweden

Topic 11: Numerical Algorithms

Global Chair
Emilio L. Zapata — Dept. of Computer Architecture, University of Malaga, Spain

Local Chair
Pasqua D'Ambra — ICAR-CNR, Naples, Italy

Vice Chairs
Oscar Plata — Dept. of Computer Architecture, University of Malaga, Spain

David Keyes — Dept. of Appl. Physics and Appl. Mathematics, Columbia University, New York, USA

Topic 12: High-Performance Multimedia

Global Chair
Odej Kao — Department of Computer Science, University of Paderborn, Germany

Local Chair
Harald Kosch — Institute of Information Technology, University of Klagenfurt, Austria

Vice Chairs
Anreas Uhl — Department of Scientific Computing, University of Salzburg, Austria

Subramania Sudharsanan — Queen's University, Kingston, Ontario, Canada

Topic 13: Theory and Algorithms for Parallel Computation

Global Chair
Christos Kaklamanis — Computer Technology Institute and Department of Computer Engineering & Informatics, University of Patras, Greece

Local Chair
Andrea Pietracaprina — Department of Information Engineering, University of Padova, Italy

Vice Chairs
Nancy Amato — Department of Computer Science, Texas A&M University, USA

Danny Krizanc — Mathematics Department, Wesleyan University, Middletown, Connecticut, USA

Topic 14: Routing and Communication in Interconnection Networks

Global Chair
José Duato — Technical University of Valencia, Spain

Local Chair
Fabrizio Petrini — Los Alamos National Laboratory, USA

Vice Chairs
Olav Lysne — Simula Research Lab and University of Oslo, Norway

Angelos Bilas — University of Crete, Greece

Topic 15: Mobile Computing

Global Chair
Sajal K. Das — Center for Research in Wireless Mobility and Networking (Crewman), University of Texas at Arlington, Texas, USA

Local Chair
Gianluigi Ferrari — Department of Computer Science, University of Pisa, Italy

Vice Chairs
Jiannong Cao — Hong Kong Polytechnic University, Hong Kong

Jie Wu — Florida Atlantic University, USA

Topic 16: Integrated Problem-Solving Environments

Global Chair
Domenico Talia — DEIS, University of Calabria, Rende (CS), Italy

Local Chair
Daniela Di Serafino — Department of Mathematics, Second University of Naples, Italy

Vice Chairs
Peter Sloot — Faculty of Science, University of Amsterdam, The Netherlands

Elias Houstis — Department of Computer and Communications Engineering, University of Thessaly, Volos, Greece, and Department of Computer Sciences Purdue University, West Lafayette, IN, USA

Topic 17: High-Performance Bioinformatics

Global Chair
Mohammed J. Zaki — Department of Computer Science, Rensselaer Polytechnic Institute, Troy, NY, USA

Local Chair
Concettina Guerra — Department of Computer Engineering, University of Padova, Italy

Vice Chairs
David A. Bader — Electrical and Computer Engineering Department, University of New Mexico, Albuquerque, USA

Johan Montagnat — Creatis, INSA, Villeurbanne, France

Topic 18: Peer-to-Peer Computing

Global Chair
Seif Haridi — Department of Computer Science, National University of Singapore, Singapore

Local Chair
Michele Colajanni — Department of Engineering Science, University of Modena and Reggio Emilia, Modena, Italy

Vice Chairs
Karl Aberer — Institute for Core Computing Science (IIF), School for Computer and Communication Science (I&C), Lausanne, Switzerland

Peter Van Roy — Department of Computing Science and Engineering, Catholic University of Louvain, Louvain-la-Neuve, Belgium

Topic 19: Demo Session

Global Chair
    Henk J. Sips   Department of Computer Science, Delft University of Technology, The Netherlands

Local Chair
    Massimo Coppola   Department of Computer Science, University of Pisa, Italy

Vice Chairs
    Luc Moreau   University of Southampton, UK

    Bernd Mohr   Research Center Juelich, Juelich, Germany

Euro-Par 2004 Referees

(not including members of the programme or organization committees)

Abella, Jaume
Acher, Georg
Agrawal, Gagan
Aida, Kento
Aldinucci, Marco
Almasi, George
Alt, Martin
Aluru, Srinivas
Amamiya, Satoshi
Amherst, Umass
Andreolini, Mauro
Antoniu, Gabriel
Apostolopoulos, George
Arbenz, Peter
Arioli, Mario
Asenjo, Rafael
Ashby, Tom
Attanassio, Dick
Avis, Nick J.
Badia, Rosa M.
Baggio, Aline
Bagherzadeh, Nader
Baiardi, Fabrizio
Baker, Mark
Bal, Henri
Banâtre, Jean-Pierre
Bandera, G.
Banerjee, Nilanjan
Basermann, Achim
Beck, Micah
Beckmann, Olav
Beivide, Ramon
Bell, Robert
Benkner, Siegfried
Beraldi, Roberto
Bern, Marshall
Bertoldo, Alberto
Bilas, Angelos
Bischof, Holger
Bisseling, Rob H.
Bode, Arndt
Bodin, Francois
Bokhari, Shahid
Bononi, Luciano
Borgeest, Rolf
Brandes, Thomas
Brandic, Ivona
Bravetti, Mario
Breton, Vincent
Bretschneider, Timo
Brezany, Peter
Browne, Jim
Bruguera, Javier D.
Brunst, Holger
Bucchignani, Edoardo
Buchty, Rainer
Budé, Porfidio Hernández
Buyya, Rajkumar
Cabri, Giacomo
Caceres, Edson Norberto
Cafaro, Massimo
Campa, Sonia
Camus, Herve
Canali, Claudia
Cao, Chun
Cao, Jiannong
Cappello, Franck
Caragiannis, Ioannis
Cardei, Mihaela
Cardellini, Valeria
Carissimi, Alexandre
Caruso, Antonio
Casado, L.G.
Casanova, Henri
Cascaval, Calin
Casola, Valentina
Castanos, Jose
Ceccanti, Andrea
Ceccarelli, Michele
Cela, Jose M.
Chakravarty, Manuel
Chalabine, Mikhail

Chapman, Barbara
Chassin de Kergommeaux, Jacques
Chatzigiannakis, Ioannis
Chbeir, Richard
Cheng, Perry
Chessa, Stefano
Choi, Tae-Uk
Choi, Wook
Chung, I-Hsin
Ciriani, Valentina
Clematis, Andrea
Codina, Josep M.
Colajanni, Michele
Collet, Raphael
Congiusta, Antonio
Coppola, Massimo
Corsaro, Stefania
Cortes, Ana
Courtiat, Jean-Pierre
Cremet, Vincent
Cunha, Jose C.
Cuppens, Frédéric
Czajkowski, Karl
Czech, Zbigniew J.
D'Agostino, Daniele
D'Ambra, Pasqua
D'Angelo, Gabriele
Dahlgren, Tamara
Dahm, Ingo
Dail, Holly
Dalgic, Ismail
Danelutto, Marco
Das, Sajal
Dash, Manoranjan
Davis, Kei
Davoli, Renzo
Daydé, Michel
de Sola, Francesc Giné
de Supinski, Bronis R.
Deb, Kalyanmoy
Deelman, Ewa
Delmastro, Franca
Dew, Robert
di Serafino, Daniela
Dias, Artur M.

Dimakopoulos, Vassilios V.
Doallo, R.
Dogan, Atakan
Dolbeau, Romain
Drozdowski, Maciej
Duato, José
Duennweber, Jan
Eisenbach, Susan
El-Ansary, Sameh
Elbern, Hendrik
Engelbrecht, Gerhard
Epema, Dick
Epperly, Tom
Ernemann, Carsten
Erway, Charles
Esener, Sadik
Evripidou, Skevos
Faber, Peter
Fagni, Tiziano
Fernandez, J.J.
Ferrari, Gianluigi
Ferretti, Stefano
Ferrini, Renato
Figueiredo, Renato
Filippone, Salvatore
Flocchini, Paola
Folino, Gianluigi
Fraboulet, Antony
Franco, Daniel
Franke, Bjoern
Franke, Hubertus
Friebe, Lars
Frolov, Pavel
Fürlinger, Karl
Fursin, Grigori
Galizia, Antonella
Gannon, Dennis
Garcia, Inmaculada
Garzon, E.M.
Gazzaniga, Giovanna
Geisler, Stefan
Gennaro, Claudio
Gerlach, Jens
Gerndt, Michael
Gervasi, Vincenzo

Getov, Vladimir
Ghini, Vittorio
Ghodsi, Ali
Gil-García, Reynaldo
Girdzijauskas, Sarunas
Glynn, Kevin
Gómez, María Engracia
Gorlatch, Sergei
Goscinski, Andrzej M.
Gottschling, Peter
Grauer, Manfred
Griebl, Martin
Grigoras, Romulus
Grimstead, Ian J.
Grolaux, Donatien
Grove, David
Gschwind, Michael
Guarracino, Mario
Guerra, Concettina
Gupta, Anshul
Gusat, Mitchell
Gutierrez, Eladio
Hameurlain, Abdelkader
Hansen, Audun Fosselie
Haridi, Seif
Hartstein, Allan
Harz, Wolf-Dietrich
Hellwagner, Hermann
Hempel, Rolf
Herrmann, Christoph
Heydemann, Karine
Heymann, Elisa
Hobbs, Michael
Hochberger, Christian
Hofstedt, Petra
Hollingsworth, Jeff
Homburg, Philip
Horn, Geir
Houstis, Elias
Huet, Fabrice
Inostroza, Mario
Jefferson, David
Jégou, Yvon
Jeitner, Jürgen
Jesshope, Chris
Ji, Yongchang
Jin, Hai
Johnsen, Einar Broch
Johnson, Chris
Jones, Tim
Jorda, Jacques
Joshi, Anupam
Kao, Odej
Karakostas, Georgios
Karczmarek, Michal
Karonis, Nicholas
Katevenis, Manolis
Keane, John
Kelly, Paul
Kenyon, Chris
Kerbyson, Darren
Kereku, Edmond
Kersken, Hans-Peter
Kessler, Christoph
Keyes
Kielmann, Thilo
Klasing, Ralf
Kleinjung, Jens
Knijnenburg, Peter
Knoop, Jens
Kosch, Harald
Koziris, Nectarios
Kranzlmüller, Dieter
Krechel, Arnold
Krizanc, Danny
Kuchen, Herbert
Kudriavtsev, Alexei
Kumar, Raj
Kumar, Sanjeev
Kumfert, Gary
Kyriacou, Costas
Laccetti, Giuliano
Lacour, Sébastien
Lafage, Thierry
Laforenza, Domenico
Lakamraju, Vijay
Lamont, Gary
Lancellotti, Riccardo
Lanteri, Stéphane
Lauria, Mario

Lecussan, Bernard
Lee, Craig
Lee, Jack
Lee, Pete
Legrand, Arnaud
Leong, Hong-Va
Leshchinskiy, Roman
Li, Kai
Li, Tianchao
Li, Vivian
Li, Yaohang
Lipasti, Mikko
Lodi, Giorgia
Lombardo, Saverio
Lonetti, Francesca
Lopes, Paulo
Lopez, Pedro
Lottiaux, Renaud
Lou, Wei
Loulergue, Frederic
Lowe, Welf
Lucchese, Claudio
Ludwig, Thomas
Luque, Emilio
Lysne, Olav
Maassen, Jason
Machì, Alberto
MacLaren, Jon
Madria, Sanjay
Maehle, Eric
Magklis, Gregory
Magoutis, Kostas
Maheswaran, M.
Mairandres, Martin
Maloney, Andrew
Malony, Allen D.
Mandal, Anirban
Marazakis, Manolis
Margalef, Tomas
Marinescu, Dan
Markatos, Evangelos
Martin, Maria J.
Martoglia, Riccardo
Martorell, Xavier
Maselli, Gaia

Masuzawa, Toshimitsu
Mattsson, Hakan
Matyska, Ludek
May, John
Medeiros, Pedro
Mehammed, Hamza
Mehofer, Eduard
Mehra, Pankaj
Melhem, Rami
Mellia, Marco
Mendes, Celso L.
Merzky, Andre
Mesaros, Valentin
Mezzetti, Nicola
Michaud, Pierre
Miguel-Alonso, Jose
Mihajlovic, Milan
Miller, Patrick
Mix, Hartmut
Mohr, Bernd
Monnet, Sébastien
Montagnat, Johan
Montella, Raffaele
Montresor, Alberto
Moreau, Luc
Moreira, Jose
Moscato, Francesco
Moshovos, Andreas
Moulitsas, Irene
Mucci, Phil
Mullai, Srinivasan
Müller-Pfefferkorn, Ralph
Mutlu, Onur
Nagel, Wolfgang E.
Nakada, Hidemoto
Nakhimovski, Iakov
Namyst, Raymond
Naroska, Edwin
Nenadic, Goran
Networks, Redback
Nguyen, Thu D.
Ni, Yang
Niculescu, Virginia
Nigel, Thomas
Nolte, Joerg

O'Boyle, Michael
Obertelli, Graziano
Oklobdzija, Vojin
Oliker, Leonid
Omang, Knut
Overeinder, Benno
Ozguner, Fusun
Paccosi, Alessandro
Pagani, Elena
Pagli, Linda
Paker, Yakup
Pakin, Scott
Palazzari, Paolo
Palazzi, Claudio
Panda, D.K.
Papadopoulos, Constantinos
Papaioannou, Evi
Parashar, Manish
Park, Seungjin
Park, Y.W.
Parthasarathy, Srinivasan
Pazat, Jean-Louis
Pelagatti, Susanna
Perego, Raffaele
Pérez, Christian
Perla, Francesca
Petrini, Fabrizio
Pflüger, Stefan
Pierre, Guillaume
Pierson, Jean-Marc
Pietracaprina, Andrea
Pinotti, Cristina
Pirsch, Peter
Plale, Beth
Plank, James S.
Plata, Oscar
Pnevmatikatos, Dionisios
Podlipnig, Stefan
Pokam, Gilles
Pozniansky, Eli
Prencipe, Giuseppe
Prieto, Manuel
Primet, Pascale Vicat-Blanc
Priol, Thierry
Pucci, Geppino

Pugliese, Andrea
Puppin, Diego
Quinlan, Dan
Rajic, Hrabri L.
Ralha, Rui
Ramakrishnan, Naren
Rana, Omer F.
Rastello, Fabrice
Rathmayer, Sabine
Rauber, Thomas
Raynal, Michel
Reinemo, Sven-Arne
Ricci, Laura
Ripoll, Ana
Rivera, Francisco F.
Robert, Yves
Robles, Antonio
Roccetti, Marco
Roig, Concepció
Romero, Felipe
Rossi, Davide
Rothkrantz, Leon
Roussopoulos, Mema
Ruenger, Gudula
Ryan, Stein Jørgen
Sainrat, Pascal
Sakellariou, Rizos
Sánchez, Jesús
Sancho, José Carlos
Sander, Volker
Sanders, Peter
Sane, Aamod
Santos, Luis Paulo
Sathaye, Sumedh
Scheuermann, Peter
Schimpf, Joachim
Schintke, Florian
Schmermbeck, Stefan
Schmidt, Andreas C.
Schmidt, Bertil
Schoettner, Michael
Schojer, Peter
Schordan, Markus
Schramm, Peter
Schueller, Anton

Schulz, Martin
Schütt, Thorsten
Schwiegelshohn, Uwe
Senar, Miquel A.
Sergeyev, Yaroslav D.
Serot, Jocelyn
Serpanos, Dimitrios
Sevcik, Ken
Seymour, Keith
Seznec, André
Shan, Hongzhang
Shen, Kai
Shende, Sameer
Shestak, Vladimir
Sie, H.J.
Silcock, Jackie
Silla, Federico
Silva, Fernando
Silva, Joao Gabriel
Silva, Luis Moura
Silvestri, Claudio
Silvestri, Fabrizio
Simmel, Derek
Singer, Daniel
Singh, Arjun
Sips, Henk J.
Skeie, Tor
Skillicorn, David
Smith, Jim
Sobe, Peter
Solihin, Yan
Solsona, Francesc
Song, S.W.
Spezzano, Giandomenico
Spielman, Daniel
Stiliadis, Dimitrios
Stockinger, Heinz
Streit, Achim
Strohmaier, Erich
Stunkel, Craig
Su, Alan
Sudharsanan, S.
Sugavanam, Prasanna V.
Talia, Domenico
Tao, Jie

Taylor, Ian J.
Terpstra, Dan
Theiss, Ingebjorg
Theys, Mitchell D.
Thiyagalingam, Jeyarajan
Tierney, Brian
Tikir, Mustaka
Tjortjis, Christos
Tonellotto, Nicola
Toraldo, Gerardo
Touati, Sid
Toulouse, Michel
Trabado, Guillermo P.
Trenkler, Bernd
Trinitis, Carsten
Trunfio, Paolo
Truong, Hong-Linh
Trystram, Denis
Tseng, Chau-Wen
Tulone, Daniela
Tuosto, Emilio
Turek, Stefan
Turi, Giovanni
Ubik, Sven
Uhl, Andreas
Ungerer, Theo
Ungor, Alper
Urpi, Alessandro
van den Berghe, Sven
van Nieuwpoort, Rob
van Reeuwijk, Kees
van Renesse, Robbert
Van Roy, Peter
Vanneschi, Marco
Venticinque, Salvatore
Verdoscia, Lorenzo
Vetter, Jeffrey
Villalba, Julio
Vogels, Werner
Volpato, Alessandra
von Laszewski, Gregor
von Praun, Christoph
Vukovic, Jaksa
Wagner, Gustavo
Walker, David W.

Watson, Paul
Weerawarana, Sanjiva
Weidendorfer, Josef
White, Brian
White, James B., III
Williams, Chadd
Williams, Tiffani L.
Winkler, Manuela
Wismüller, Roland
Wiszniewski, Bogdan
Wolf, Wayne
Wolinski, Krisztof
Wong, Adam K.L.
Wook, Choi
Wu, Jie
Wu, Jiesheng
Xue, Jingling
Yahyapour, Ramin
Yang, Zhang
Yates, Kim
Yeung, Kwok Cheung
Yi, Qing
Yoo, Andy
Zahorjan, John
Zaki,
Zanichelli, Francesco
Zhang, Liang
Zhang, Yongbing
Zhou, Xiaobo (Joe)
Ziegler, Wolfgang
Zilles, Craig
Zima, Hans
Zimeo, Eugenio
Zlatev, Zahari
Zoccolo, Corrado
Zomaya, Albert

Table of Contents

Invited Talks

Building Grid Applications and Portals:
An Approach Based on Components, Web Services and Workflow Tools .. 1
    *D. Gannon, L. Fang, G. Kandaswamy, D. Kodeboyina, S. Krishnan,
B. Plale, and A. Slominski*

Maintaining Thousands of In-Flight Instructions 9
    A. Cristal, O.J. Santana, and M. Valero

Some Techniques for Automated, Resource-Aware Distributed
and Mobile Computing in a Multi-paradigm Programming System 21
    M. Hermenegildo, E. Albert, P. López-García, and G. Puebla

Why Structured Parallel Programming Matters 37
    M. Cole

Topic 1: Support Tools and Environments

Topic 1: Support Tools and Environments 38
    J.C. Cunha, A.D. Malony, A. Bode, and D. Kranzlmueller

Profiling and Tracing OpenMP Applications
with POMP Based Monitoring Libraries 39
    L. DeRose, B. Mohr, and S. Seelam

Efficient Pattern Search in Large Traces Through Successive Refinement . 47
    F. Wolf, B. Mohr, J. Dongarra, and S. Moore

DRAC: Adaptive Control System with Hardware Performance Counters . 55
    M. Pillon, O. Richard, and G. Da Costa

Reproducible Network Benchmarks with CONCEPTUAL 64
    S. Pakin

Optimizing Cache Access: A Tool for Source-to-Source Transformations
and Real-Life Compiler Tests 72
    R. Müller-Pfefferkorn, W.E. Nagel, and B. Trenkler

Detecting Data Races in Sequential Programs with *DIOTA* 82
    M. Ronsse, J. Maebe, and K. De Bosschere

A Time-Coherent Model for the Steering of Parallel Simulations 90
    A. Esnard, M. Dussere, and O. Coulaud

MATE: Dynamic Performance Tuning Environment 98
    A. Morajko, O. Morajko, T. Margalef, and E. Luque

Imprecise Exceptions in Distributed Parallel Components 108
    K. Damevski and S. Parker

Topic 2: Performance Evaluation

Topic 2: Performance Evaluation 117
    W.E. Nagel, T. Ludwig, J. Vetter, and L. Donatiello

Overhead Compensation in Performance Profiling 119
    A.D. Malony and S.S. Shende

A Data Structure Oriented Monitoring Environment
for Fortran OpenMP Programs 133
    E. Kereku, T. Li, M. Gerndt, and J. Weidendorfer

An Approach for Symbolic Mapping of Memory References 141
    L. DeRose, K. Ekanadham, and S. Sbaraglia

Impact of Cache Coherence Models
on Performance of OpenMP Applications 149
    J. Tao and W. Karl

Evaluating OpenMP Performance Analysis Tools
with the APART Test Suite 155
    M. Gerndt, B. Mohr, and J.L. Träff

Collective Communication Performance Analysis
Within the Communication System 163
    L.A. Bongo, O.J. Anshus, and J.M. Bjørndalen

Understanding the Behavior and Performance
of Non-blocking Communications in MPI 173
    T. Saif and M. Parashar

Generation of Simple Analytical Models
for Message Passing Applications 183
    G. Rodriguez, R.M. Badia, and J. Labarta

Parallel Hybrid Particle Simulations Using MPI and OpenMP 189
    M. Hipp and W. Rosenstiel

Distributed Shared Memory: To Relax or Not to Relax? 198
    V. Iosevich and A. Schuster

Design-Time Data-Access Analysis for Parallel Java Programs
with Shared-Memory Communication Model 206
    R. Stahl, F. Catthoor, R. Lauwereins, and D. Verkest

Parallel PEPS Tool Performance Analysis
Using Stochastic Automata Networks................................. 214
    L. Baldo, L.G. Fernandes, P. Roisenberg, P. Velho, and T. Webber

Topic 3: Scheduling and Load Balancing

Topic 3: Scheduling and Load Balancing 220
    E. Luque, J.G. Castaños, E. Markatos, and R. Perego

Scheduling Under Conditions of Uncertainty: A Bayesian Approach 222
    L.P. Santos and A. Proenca

From Heterogeneous Task Scheduling
to Heterogeneous Mixed Parallel Scheduling 230
    F. Suter, F. Desprez, and H. Casanova

Scheduling of MPI Applications: Self-co-scheduling 238
    G. Utrera, J. Corbalán, and J. Labarta

Scheduling Tasks Sharing Files from Distributed Repositories........... 246
    A. Giersch, Y. Robert, and F. Vivien

A Scheduling Algorithm
for Running Bag-of-Tasks Data Mining Applications on the Grid 254
    F.A.B. da Silva, S. Carvalho, and E.R. Hruschka

Lookahead Scheduling for Reconfigurable GRID Systems 263
    J. Andersson, M. Ericsson, W. Löwe, and W. Zimmermann

Topic 4: Compilers for High Performance

Topic 4: Compilers for High Performance 271
    H.P. Zima, S. Benkner, M.F.P. O'Boyle, and B. Di Martino

More Legal Transformations for Locality 272
    C. Bastoul and P. Feautrier

Link-Time Optimization of IA64 Binaries 284
    *B. Anckaert, F. Vandeputte, B. De Bus, B. De Sutter,
    and K. De Bosschere*

A Polyhedral Approach to Ease the Composition
of Program Transformations.. 292
    A. Cohen, S. Girbal, and O. Temam

Compiler-Guided Code Restructuring
for Improving Instruction TLB Energy Behavior 304
  I. Kadayif, M. Kandemir, and I. Demirkiran

Using Data Compression to Increase Energy Savings
in Multi-bank Memories 310
  M. Kandemir, O. Ozturk, M.J. Irwin, and I. Kolcu

Automatic Customization of Embedded Applications for Enhanced
Performance and Reduced Power Using Optimizing Compiler Techniques. 318
  E. Özer, A.P. Nisbet, and D. Gregg

Architecture-Independent Meta-optimization
by Aggressive Tail Splitting 328
  M. Rock and A. Koch

If-Conversion in SSA Form 336
  A. Stoutchinin and G. Gao

Topic 5: Parallel and Distributed Databases, Data Mining and Knowledge Discovery

Topic 5: Parallel and Distributed Databases, Data Mining
and Knowledge Discovery 346
  D. Skillicorn, A. Hameurlain, P. Watson, and S. Orlando

A Highly Scalable Parallel Caching System
for Web Search Engine Results 347
  T. Fagni, R. Perego, and F. Silvestri

OLAP Query Processing in a Database Cluster 355
  A.A.B. Lima, M. Mattoso, and P. Valduriez

Efficient Parallel Hierarchical Clustering 363
  M. Dash, S. Petrutiu, and P. Scheuermann

Parallelizing EM Clustering Algorithm on a Cluster of SMPs 372
  L. Glimcher and G. Agrawal

A Parallel Knowledge Discovery System for Customer Profiling 381
  M. Coppola, P. Pesciullesi, R. Ravazzolo, and C. Zoccolo

A Large-Scale Digital Library System to Integrate Heterogeneous Data
of Distributed Databases 391
  M. Di Giacomo, M. Martinez, and J. Scott

Topic 6: Grid and Cluster Computing

Topic 6: Grid and Cluster Computing 398
    T. Priol, C. Lee, U. Schwiegelshosh, and D. Puppin

Agreement-Based Interactions for Experimental Science 399
    K. Keahey, T. Araki, and P. Lane

A Double Auction Economic Model for Grid Services 409
    L. Joita, O.F. Rana, W.A. Gray, and J. Miles

Towards a Monitoring Framework
for Worldwide Grid Information Services 417
    S. Zanikolas and R. Sakellariou

A Model for a Component Based Grid-Aware Scientific Library Service .. 423
    S. Lombardo and A. Machì

Developing Grid-Aware Applications
with DRMAA on Globus-Based Grids 429
    J. Herrera, E. Huedo, R.S. Montero, and I.M. Llorente

Exposing MPI Applications as Grid Services 436
    E. Floros and Y. Cotronis

Globus-Based Grid Computing Simulations
of Action Potential Propagation on Cardiac Tissues 444
    J.M. Alonso, V. Hernández, and G. Moltó

Profiling Grid Data Transfer Protocols and Servers 452
    G. Kola, T. Kosar, and M. Livny

Storage Power Management for Cluster Servers
Using Remote Disk Access 460
    J.H. Choi and H. Franke

A Path Selection Based Algorithm for Maximizing Self-satisfiability
of Requests in Real-Time Grid Applications 468
    M. Eltayeb, A. Doğan, and F. Özgüner

Topic 7: Applications on High Performance Computers

Topic 7: Applications on High Performance Computers 476
    P. Arbenz, R. Hempel, and D. Walker

Using a Structured Programming Environment
for Parallel Remote Visualization 477
    P. Ammirati, A. Clematis, D. D'Agostino, and V. Gianuzzi

Parallel and Grid Computing in 3D Analysis
of Large Dimmension Structural Systems 487
    J.M. Alonso, C. de Alfonso, G. García, and V. Hernández

FlowVR: A Middleware for Large Scale Virtual Reality Applications..... 497
    J. Allard, V. Gouranton, L. Lecointre, S. Limet, E. Melin,
    B. Raffin, and S. Robert

Topic 8: Parallel Computer Architecture and ILP

Topic 8: Parallel Computer Architecture
and Instruction-Level Parallelism 506
    K. Ebcioğlu, W. Karl, A. Seznec, and M. Aldinucci

Exploring the Possibility of Operating in the Compressed Domain 507
    V. De La Luz, M. Kandemir, A. Sivasubramaniam, and M.J. Irwin

Contents Management in First-Level Multibanked Data Caches 516
    E.F. Torres, P. Ibañez, V. Viñals, and J.M. Llabería

A Mechanism for Verifying Data Speculation 525
    E. Morancho, J.M. Llabería, and À. Olivé

Feasibility of QoS for SMT 535
    F.J. Cazorla, P.M.W. Knijnenburg, R. Sakellariou, E. Fernandez,
    A. Ramirez, and M. Valero

Improving Data Cache Performance via Address Correlation:
An Upper Bound Study.. 541
    P.-f. Chuang, R. Sendag, and D.J. Lilja

Exploiting Spatial Store Locality
Through Permission Caching in Software DSMs...................... 551
    H. Zeffer, Z. Radović, O. Grenholm, and E. Hagersten

CacheFlow: A Short-Term Optimal Cache Management Policy
for Data Driven Multithreading..................................... 561
    C. Kyriacou, P. Evripidou, and P. Trancoso

SCISM vs IA-64 Tagging: Differences/Code Density Effects............. 571
    G. Gaydadjiev and S. Vassiliadis

Topic 9: Distributed Systems and Algorithms

Topic 9: Distributed Systems and Algorithms......................... 578
    H.E. Bal, A.M. Goscinski, E. Jul, and G. Prencipe

Exploiting Differentiated Tuple Distribution in Shared Data Spaces 579
   G. Russello, M. Chaudron, and M. van Steen

Enhancing Efficiency of Byzantine-Tolerant Coordination Protocols
via Hash Functions... 587
   D. Tulone

Accelerating Apache Farms
Through Ad-HOC Distributed Scalable Object Repository 596
   M. Aldinucci and M. Torquati

A Proxy Server-Network for Real-Time Computer Games 606
   J. Müller, S. Fischer, S. Gorlatch, and M. Mauve

Topic 10: Parallel Programming: Models, Methods and Languages

Topic 10: Parallel Programming:
Models, Methods and Programming Languages 614
   P.H.J. Kelly, S. Gorlatch, C. Kessler, and D. Quinlan

Comparison of Parallelization Models
for Structured Adaptive Mesh Refinement 615
   J. Rantakokko

Task-Queue Based Hybrid Parallelism: A Case Study 624
   K. Fürlinger, O. Schenk, and M. Hagemann

Co-array Python: A Parallel Extension to the Python Language 632
   C.E. Rasmussen, M.J. Sottile, J. Nieplocha, R.W. Numrich,
   and E. Jones

Targeting Heterogeneous Architectures in ASSIST: Experimental Results . 638
   M. Aldinucci, S. Campa, M. Coppola, S. Magini, P. Pesciullesi,
   L. Potiti, R. Ravazzolo, M. Torquati, and C. Zoccolo

A Fusion-Embedded Skeleton Library................................. 644
   K. Matsuzaki, K. Kakehi, H. Iwasaki, Z. Hu, and Y. Akashi

Cross Component Optimisation
in a High Level Category-Based Language........................... 654
   T.J. Ashby, A.D. Kennedy, and M.F.P. O'Boyle

Concurrent Table Accesses in Parallel Tabled Logic Programs 662
   R. Rocha, F. Silva, and V. Santos Costa

Handling Exceptions Between Parallel Objects 671
   C. Pérez, A. Ribes, and T. Priol

Parallel Software Interoperability by Means of CORBA
in the ASSIST Programming Environment 679
  S. Magini, P. Pesciullesi, and C. Zoccolo

Designing Parallel Operating Systems via Parallel Programming 689
  E. Frachtenberg, K. Davis, F. Petrini, J. Fernandez, and J.C. Sancho

Topic 11: Numerical Algorithms

Topic 11: Numerical Algorithms 697
  E.L. Zapata, O. Plata, D. Keyes, and P. D'Ambra

Reducing the Deteriorating Effect of Old History
in Asynchronous Iterations 698
  Y. Yalçınkaya and T. Steihaug

Execution Schemes for Parallel Adams Methods 708
  T. Rauber and G. Rünger

A Data Management and Communication Layer
for Adaptive, Hexahedral FEM 718
  J. Hippold and G. Rünger

A Parallel PSPG Finite Element Method
for Direct Simulation of Incompressible Flow 726
  J. Stiller, K. Fraňa, R. Grundmann, U. Fladrich, and W.E. Nagel

Load and Memory Balanced Mesh Partitioning
for a Parallel Envelope Method 734
  O. Medek, P. Tvrdík, and J. Kruis

Combining Explicit and Recursive Blocking for Solving Triangular
Sylvester-Type Matrix Equations on Distributed Memory Platforms 742
  R. Granat, I. Jonsson, and B. Kågström

Topic 12: High Performance Multimedia

Topic 12: High Performance Multimedia 751
  O. Kao, H. Kosch, S. Sudharsanan, and A. Uhl

Towards User Transparent Data and Task Parallel Image
and Video Processing: An Overview of the Parallel-Horus Project 752
  F.J. Seinstra, D. Koelma, and A.D. Bagdanov

Visual Data Rectangular Memory 760
  G. Kuzmanov, G. Gaydadjiev, and S. Vassiliadis

Dynamic Granularity Switching
in Parallel Block-Matching Motion Compensation 768
   F. Tischler and A. Uhl

Placement of Nodes in an Adaptive Distributed Multimedia Server 776
   B. Goldschmidt, T. Szkaliczki, and L. Böszörményi

A Practical Performance Analysis of Stream Reuse Techniques
in Peer-to-Peer VoD Systems 784
   L.B. de Pinho and C.L. de Amorim

Supporting Caching and Mirroring
in Distributed Video-on-Demand Architectures 792
   X.Y. Yang, F. Cores, A. Ripoll, P. Hernández, B. Qazzaz, R. Suppi, and E. Luque

An Efficient Fair Queuing Algorithm for Guaranteed Disk Bandwidth ... 799
   Y.J. Nam and C. Park

Topic 13: Theory and Algorithms for Parallel Computation

Topic 13: Theory and Algorithms for Parallel Computation 803
   K. Christos, N. Amato, D. Krizanc, and A. Pietracaprina

Distributed Largest-First Algorithm for Graph Coloring 804
   J. Hansen, M. Kubale, Ł. Kuszner, and A. Nadolski

Parallel Delaunay Refinement with Off-Centers 812
   D.A. Spielman, S.-h. Teng, and A. Üngör

Near-Optimal Hot-Potato Routing on Trees 820
   C. Busch, M. Magdon-Ismail, M. Mavronicolas, and R. Wattenhofer

A Coarse-Grained Parallel Algorithm
for Spanning Tree and Connected Components 828
   E.N. Cáceres, F. Dehne, H. Mongelli, S.W. Song, and J.L. Szwarcfiter

Topic 14: Routing and Communication in Interconnection Networks

Topic 14: Routing and Communication in Interconnection Networks 832
   J. Duato, F. Petrini, O. Lysne, and A. Bilas

Implementing MPI on the BlueGene/L Supercomputer 833
   G. Almási, C. Archer, J.G. Castaños, C.C. Erway, P. Heidelberger, X. Martorell, J.E. Moreira, K. Pinnow, J. Ratterman, N. Smeds, B. Steinmacher-burow, W. Gropp, and B. Toonen

Modular On-Chip Multiprocessor for Routing Applications 846
    S. Berrayana, E. Faure, D. Genius, and F. Pétrot

OMULT: An Optical Interconnection System for Parallel Computing 856
    B.P. Sinha and S. Bandyopadhyay

Distributing InfiniBand Forwarding Tables 864
    A. Bermúdez, R. Casado, and F.J. Quiles

Tuning Buffer Size in InfiniBand to Guarantee QoS 873
    F.J. Alfaro and J.L. Sánchez

Reducing Power Consumption in Interconnection Networks
by Dynamically Adjusting Link Width............................. 882
    M. Alonso, J.M. Martínez, V. Santonja, and P. López

A Methodology to Evaluate the Effectiveness
of Traffic Balancing Algorithms 891
    J.E. Villalobos, J.L. Sánchez, J.A. Gámez, J.C. Sancho, and A. Robles

Load Unbalance in k-ary n-Cube Networks 900
    J. Miguel-Alonso, J.A. Gregorio, V. Puente, F. Vallejo, and R. Beivide

Understanding Buffer Management for Cut-Through 1D Rings 908
    C. Izu and R. Beivide

Topic 15: Mobile Computing

Topic 15: Mobile Computing 916
    S.K. Das, J. Cao, J. Wu, and G. Ferrari

Coverage and Connectivity in Networks with Directional Sensors 917
    E. Kranakis, D. Krizanc, and J. Urrutia

Local Route Recovery Algorithms for Improving Multihop TCP
Performance in Ad Hoc Wireless Networks 925
    Z. Li and Y.-K. Kwok

Cluster Based Distributed Mutual Exclusion Algorithms
for Mobile Networks 933
    K. Erciyes

An Improved Handoff Technique for a Seamless Multimedia Services 941
    S.-W. Lee, H.-O. Hong, S.-H. Park, Y.-W. Park, and K.-D. Chung

An Improved Approximation Algorithm
for the Minimum Energy Consumption Broadcast Subgraph 949
    V. Bilò and G. Melideo

UVOD: A Jini-Based Ubiquitous VOD Service Architecture 957
    T.U. Choi and K.-D. Chung

Topic 16: Integrated Problem Solving Environments

Topic 16: Integrated Problem Solving Environments 962
    D. di Serafino, E. Houstis, P. Sloot, and D. Talia

Pattern/Operator Based Problem Solving Environments 964
    C. Gomes, O.F. Rana, and J. Cunha

Databases, Workflows and the Grid in a Service Oriented Environment .. 972
    Z. Jiao, J. Wason, W. Song, F. Xu, H. Eres, A.J. Keane, and S.J. Cox

A Parallel Programming Tool for SAR Processors 980
    M. Lettere, D. Guerri, and R. Fontanelli

Topic 17: High Performance Bioinformatics

Topic 17: High Performance Bioinformatics........................... 988
    M.J. Zaki, D.A. Bader, J. Montagnat, and C. Guerra

A Generic Parallel Pattern-Based System for Bioinformatics 989
    W. Liu and B. Schmidt

Parallel Inference of a 10.000-Taxon Phylogeny
with Maximum Likelihood 997
    A. Stamatakis, T. Ludwig, and H. Meier

Cache-Based Parallelization of Multiple Sequence Alignment Problem ... 1005
    G. Parmentier, D. Trystram, and J. Zola

Topic 18: Peer-to-Peer and Web Computing

Topic 18: Peer-to-Peer and Web Computing 1013
    S. Haridi, K. Aberer, P. Van Roy, and M. Colajanni

The ncast Primitive for Peer-to-Peer Networks 1014
    J. Pascoe and V. Sunderam

A P2P Grid Services-Based Protocol: Design and Evaluation 1022
    D. Talia and P. Trunfio

A Synchronization and Session Control Based Protocol
for 3D Multi-user Virtual Environments in a Peer-to-Peer Networks 1032
    A. Boukerche, R. Rabêlo, M. Laffranchi, and R. Araujo

Large-Scale Deployment in P2P Experiments
Using the JXTA Distributed Framework 1038
  G. Antoniu, L. Bougé, M. Jan, and S. Monnet

Experimental Study of Multi-criteria Scheduling Heuristics
for GridRPC Systems 1048
  Y. Caniou and E. Jeannot

Using Similarity Groups to Increase Performance of P2P Computing 1056
  J. Bourgeois, J.B. Ernst-Desmulier, F. Spies, and J. Verbeke

Improving the Scalability
of Logarithmic-Degree DHT-Based Peer-to-Peer Networks 1060
  B. Carton and V. Mesaros

A Small World Overlay Network for Resource Discovery 1068
  K. Vanthournout, G. Deconinck, and R. Belmans

Author Index 1077

Building Grid Applications and Portals: An Approach Based on Components, Web Services and Workflow Tools

D. Gannon, L. Fang, G. Kandaswamy, D. Kodeboyina, S. Krishnan, B. Plale, and A. Slominski

Department of Computer Science, Indiana University
Bloomington, Indiana, USA
gannon@cs.indiana.edu
http://www.extreme.indiana.edu

Abstract. Large scale Grid applications are often composed a distributed collection of parallel simulation codes, instrument monitors, data miners, rendering and visualization tools. For example, consider a severe storm prediction system driven by a grid of weather sensors. Typically these applications are very complex to build, so users interact with them through a Grid portal front end. This talk outlines an approach based on a web service component architecture for building these applications and portal interfaces. We illustrate how the traditional parallel application can be wrapped by a web service factory and integrated into complex workflows. Additional issues that are addressed include: grid security, web service tools and workflow composition tools. The talk will try to outline several important classes of unsolved problems and possible new research directions for building grid applications.

1 Introduction

This talk will consider the basic problem of building Grid applications by composing web services and CCA components in a dynamic, distributed environment. There are three basic issues that we address. First we consider the way the user accesses the Grid applications. It is our experience that this is best accomplished with a web portal that can provide the user with a secure framework for accessing all his or her applications and services. The design of such a portal is non-trivial and plays a central role in the overall security architecture of the Grid. The second important problem is to provide a way for the user to encapsulate legacy applications as Grid services that can be securely accessed by remote, authorized clients. Finally we consider the problem of integrating dynamic workflow with the static composition model of component frameworks like the Common Component Architecture. In this extended abstract we briefly consider each of these problems.

2 The Portal

There is a four layer model that can best illustrate the role of the portal in its relation to the user and Grid/web services. At the bottom layer we have the physical resources that comprise the Grid fabric. At the layer above that we have a core service framework like WSRF or OGSI, which provides the common language to talk about the real and virtual resources of the Grid. Above that we have the Open Grid Service Architecture , which is s set of specifications for the common Grid services that applications use. Finally the Portal Server provides the gateway for the user. The portal server provides the environment in which the user 's "grid context" is established. This server is based on emerging standards [11] and is available from OGCE [15] or the GridSphere [14] project.

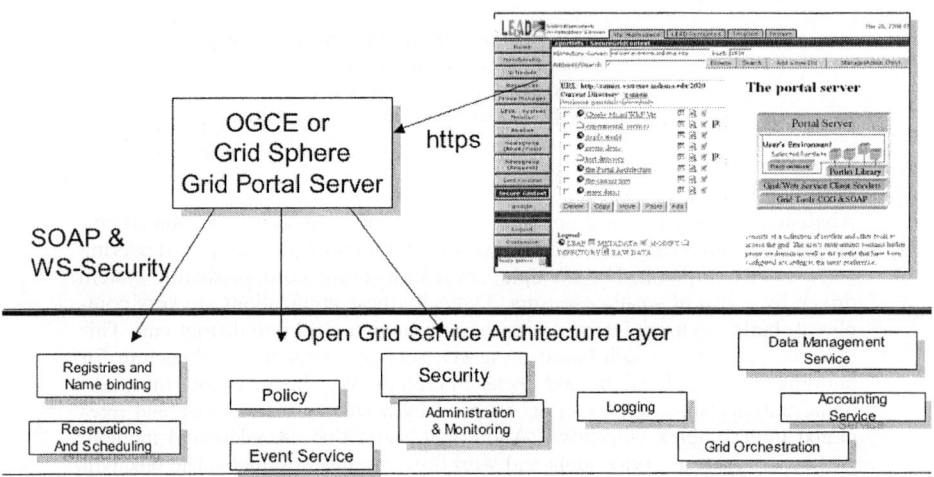

Fig. 1. Users access the portal via a standard web browsers. The portal server becomes the proxy for the user in his or her interactions with the Grid.

The OGCE and GridSphere portal servers are each based on the concept of "portlets", which are portal components that each provide a specialized capability, such as access to some service. For example, the OGSA Data Access and Integration service is a Grid service to access databases and metadata services. Figure 2 illustrates the portlet that is presented to the user from the portal server.

Users access the portal via a web browser over https, so they can be assured their interaction with the portal server are secure. The portal server fetches and hold the user's proxy credentials , so that it can act on the user behalf for the duration of the session. An important special case where the portal acts as the agent for the user while interacting with the Grid, is when a Grid service must project a user interface to the user. The way this is accomplished is illustrated in Figure 3.

The user first makes a request to the portal server to load the interface for the protected Grid Service. The portal server then calls the service in question using the user's credentials and asks for the service interface. The Grid service stores this interface as a standard service resource. It may be an HTML form document or it may be an applet. The portal delivers this interface back to the user. The user uses the interface to formulate a query for the remote service and sends that back to the portal server which tunnels it to the service.

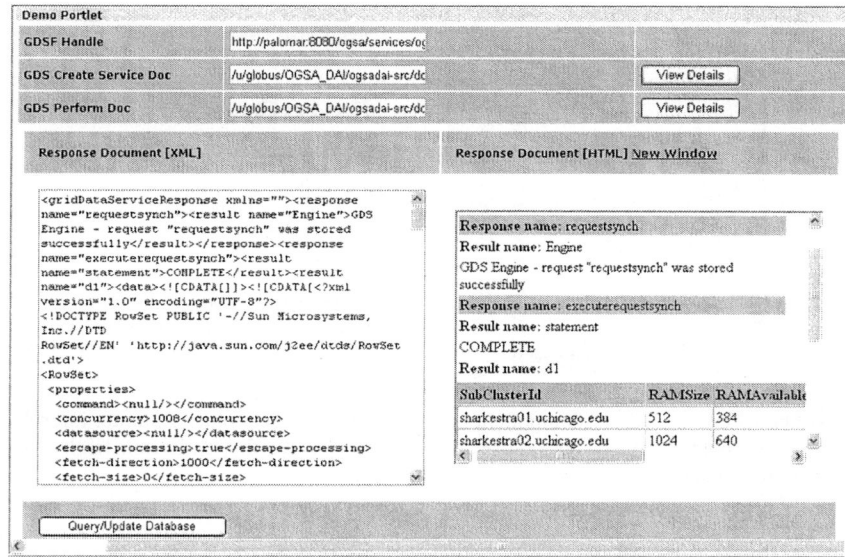

Fig. 2. The OGSA DAI portlet allows user to pose queries to Grid data services.

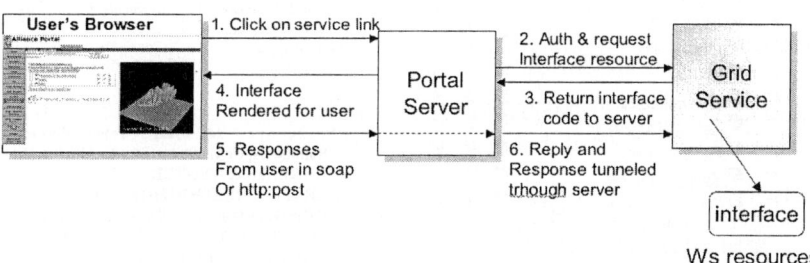

Fig. 3. Using the portal server to tunnel service requests to the back-end Grid services.

It should be noted that this model is very similar to WSRP [12], but it places the emphasis on security and using the built-in capability of WSRF or OGSI. WSRP is an OASIS standard for building Remote Portlets for Web Services. Our approach, which is based on dynamically loading interfaces into the portal server has some advantages. First, it allows the service to define the interface in the same way that a web site defines the way it presents itself to users. But, most important, loading the interface first into the portal server, allows that interface to share the users portal context information. This context information includes security credentials, such as proxy certificates and capability tokens.

It is also possible for the portal to generate an interface for simple web services automatically. Octav Chipara and Aleksander Slominski developed Xydra Ontobrew [21], (see figure 4 and 5) which can dynamically load a WSDL file that defines a web service and generate an interactive interface for it. While it does not work in cases where message have extremely complex types, it works in those cases where the service may be designed for human interaction.

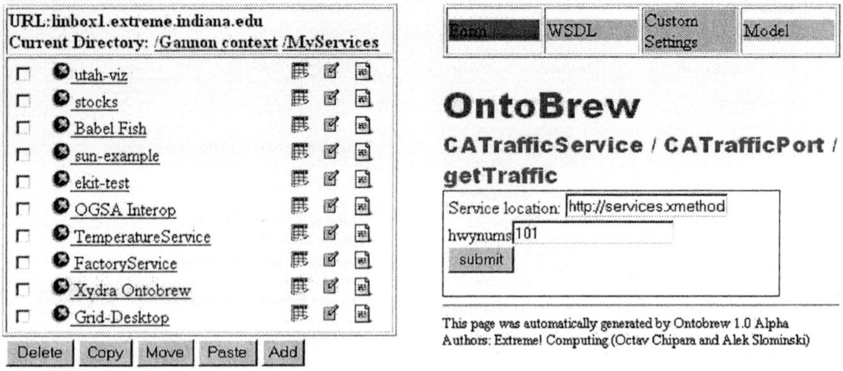

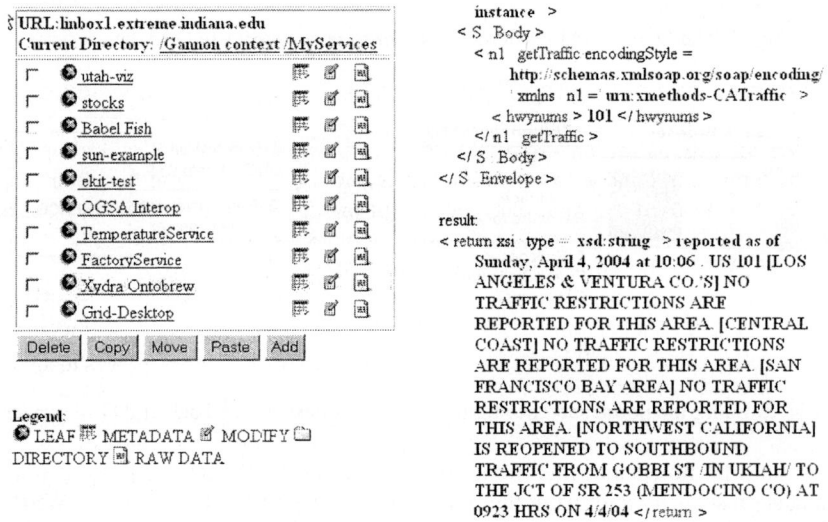

Fig. 4. OntoBrew is a portal component that automatically generates a human interface for a web service, given its WSDL definition. In this case we see the California Highway Traffic Information Service.

Fig. 5. The result of the query to the traffic service for information about Highway 101.

We find the portal layer is one of the most important parts of the Grid system because it is where the user meets the Grid services. If it is not designed well, the users will not use the Grid at all. There must be a clear advantage to using the portal over dealing with the Grid in terms of basic distributed computing primitives. In fact, for scientific users, they will use the portal only if it has something that can be accessed with three or fewer mouse clicks and is something that they can access no other way.

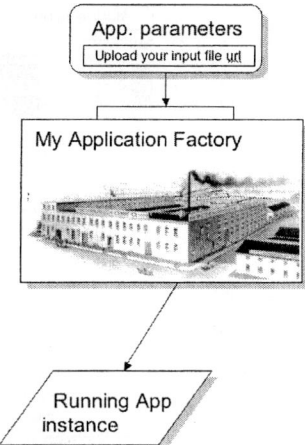

Fig. 6. The application factory model. The factory presents a user interface to the user through the portal that allows the user to supply any configuration parameters needed by the application prior to its execution.

3 Incorporating Legacy Applications Using a Factory Service

Legacy applications are an important part of almost all large scale Grid application. The challenge for legacies is to incorporate them as components into larger workflows and composed applications. One solution is to provide a Factory service, as illustrated in Figure 6, that is capable of launching instances of the application based on parameters supplied to the factory by the user.

There are two significant issues that are important here. The first is that it is most application scientists are not trained in writing web services and a mechanism is needed to help them do this. Fortunately, it is possible to write a factory generator that can live within the portal server and can be used to create a factory on behalf of the application programmer. What the programmer must provide is a simple xml document that describes the application and its configuration parameters and how it is launched. From this information we can automatically generate a factory service for the application.

The second issue is solving the problem of how to authorize other users to use a factory service. Unlike most services, which run as "root", the factory service runs with the identity of the application program that created it. The solution we use is to allow the owner of the factory service to create capability tokens which allow certain users, or groups of users access to the factory. These tokens are simple xml documents that are signed by the owner of the service stating that the intended user shall have the authorization to interact with the owner's factory.

As shown in Figure 7, these capabilities can be stored in a separate capability store and a user's capabilities can be fetched by the portal each time the user logs into the portal. The ability to encapsulate an application, and provide secure access and authorization to it, is very similar in spirit to the original Netsolve and Ninf projects [4, 13], which pioneered the concept of viewing an application as a service.

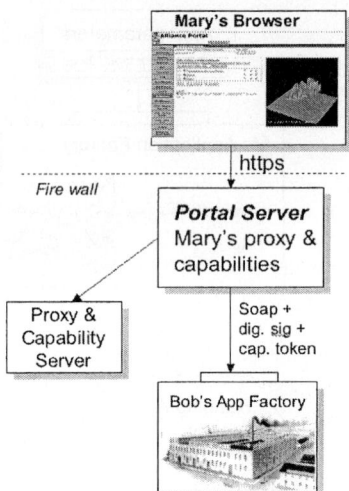

Fig. 7. Users access the portal via a standard web browsers. The portal server becomes the proxy for the user in his or her interactions with the Grid.

4 Workflow and Component Composition

There are a host of good workflow tools for scientific Grid applications including Dagman [5], ICENI [8], Triana [18], Keppler and Ptolemy II [17], Taverna and GridAnt [19]. On the industrial side there is BPEL4WS [3]. In addition there are component based software systems like CCA [2, 16, 23]. In the presentation we will focus on two aspects of doing workflow in Grid systems. First we will illustrate a large and complex example motivated by mesoscale storm simulation [6] and then we will discuss the interesting challenges of compiling very high level graphical specifications into one or more of the languages listed above.

There are a host of open problems for research in Grid Programming in general. (See [22] for a good overview). In the area of component based systems, there are two general models: dataflow based composition (Triana and Ptolemy are good examples of this), in which as stream of data is pushed from a source port to a destination port and remote procedure call (RPC) based systems like CCA a source invokes a function on the remote side and waits for a response. In both cases, the underlying assumption is that a source and destination port are both live partners in a data exchange. Workflow systems sequence activities and the connection between component services may only be a causal dependence. However this is not always the case. Modern web services allows us to send a message to one service and ask that the reply be sent to different service. This is possible because most web services operate on a message level rather than RPC. However, CCA can be implemented using web services (see XCAT3 [23]) and we can build workflows that incorporate both models.

Most of the tools for Grid workflow assume that the user wants to build the workflow as a simple graph of service component. However, this compact graph model is not always appropriate. This may be the case when the workflow is simply too large and complex to effectively "program" it as a graph. (However, some tools allow one to turn a Graph into a new first-class component or service, which can then be included as a node in another graph (a workflow of workflows). This technique allows graphs of arbitrary complexity to be constructed). But size complexity is not the only problem. It may be the case that the workflow is driven by events that determine its structure. It may also be the case that the workflow structure is defined by processes that are negotiated at runtime. For example, suppose one component service must be replaced by another and that new service requires other new services to be connected into the picture. These new services may also require a different organization of the upstream components of the workflow. Or, for example, a workflow may be able to dynamically optimize its structure when it sees that a component service is not needed. While applications of workflows of this complexity may appear to be science fiction at the present time, the application of Grids to autonomic systems may require this level of intelligence and adaptability.

Finally, the static graph model may fail as a means to characterize the workflow in the case that the graph is implicit as in the case when the workflow is expressed as a set of desired outcomes that can be satisfied by a number of different workflow enactments. For example, a data query may be satisfied moving a large data set across country or it may be cheaper to recreate the data in a local environment. This is a simple example of the Griphyn virtual data concept.

For larger systems, or for workflows of most complex dynamically structured type, something that resembles a complete programming language is required. BPEL4WS is the leading candidate for a "complete programming language" for workflows. In fact, BPEL provides two usage patterns for expressing workflow behaviors. The first usage pattern is based on the concept of an abstract process which represents a role, such as "buyer" or "seller" in a business process and the "graph" based language to describe their interaction. These interactions are defined by partner links and BPEL provides the language to describe the public aspects of the protocol used in the interaction between these partners. The other usage pattern of BPEL involves the "algebraic" language needed to define the logic and state of an executable process in terms of web service resources and XML data, and tools to deal with exceptions and failures.

There are a number of significant general issues for workflow authoring and enactment in Grid systems. Many of these point to general research problems for the future.

One area that has received little attention in the current Grid workflow projects is the issue of security. It is dangerous to trust a complex workflow enactment to assure that resources are protected. To what extent is a single user's identity/authority sufficient to authroize a complex workflow enactment that may require a large set of capabilities to complete a set of distributed tasks?

Another interesting area of research involves the way in which we can use a workflow document as part of the scientific provenance of a computational experiment? Under what conditions can we publish a workflow script as a document that can be used by others to verify a scientific claim? If the workflow was triggered by sequences of external events, can the monitoring of the workflow capture these events well enough so that the enactment can be repeated?

There are many other interesting questions. Can we build systems that allow workflows to automatically do incremental self-optimization? Can we automatically discover new properties and services of a Grid that enable such self-optimization? If there is time, we will address additional issues in the final presentation.

References

1. Agarwal, M., and Parashar, M. . Enabling Autonomic Compositions in Grid Environments.*Proceedings of the 4th International Workshop on Grid Computing (Grid 2003)*, Phoenix, AZ, USA, IEEE Computer Society Press, pp 34 - 41, November 2003.
2. Armstrong, R., Gannon, D., Geist, A., Keahey, K., Kohn, S. McInnes, L., Parker, S., and Smolinski, B.. Towards a common component architecture for high performance scientific computing. In *Proceedings of the The Eighth IEEE International Symposium on High Performance Distributed Computing*, 1998.
3. Business Process Execution Language for Web Services Version 1.1. http://www-106.ibm.com/developerworks/library/ws-bpel/
4. Casanova, H. and Dongarra, J, NetSolve: a network server for solving computational science problems. Proceedings SC 96.
5. Condor Dagman, http://www.cs.wisc.edu/condor/dagman/
6. Droegemeier, K.K., V. Chandrasekar, R. Clark, D. Gannon, S. Graves, E. Joseph, M. Ramamurthy, R. Wilhelmson, K. Brewster, B. Domenico, T. Leyton, V. Morris, D. Murray, P. Plale, R. Ramachandran, D. Reed, J. Rushing, D. Weber, A. Wilson, M. Xue, and S. Yalda, 2004: Linked environments for atmospheric discovery (LEAD): A cyberinfrastructure for mesoscale meteorology research and education. Preprints, 20th. Conf. on Interactive Info. Processing Systems for Meteor, Oceanography, and Hydrology, Seattle, WA, Amer. Meteor. Soc.
7. Foster, I., Kesselman, C., Nick, J., Tuecke, S., The Physiology of the Grid An Open Grid Services Architecture for Distributed Systems Integration, www.globus.org/research/papers/ ogsa.pdf
8. Anthony Mayer, Steve McGough, Nathalie Furmento, Jeremy Cohen, Murtaza Gulamali, Laurie Young, Ali Afzal, Steven Newhouse, John Darlington, ICENI: An Integrated Grid Middleware to support e-Science. Workshop on Component Models and Systems for Grid Applications, Saint-Malo, June 26, 2004.
9. Open Grid Computing Environment (OGCE), http://www.ogce.org.
10. GridLab, The GridSphere Portal http://www.gridsphere.org
11. JSR-168 Portlet Specification. http://www.jcp.org/aboutJava/communityprocess/final /jsr168/
12. Kropp, A., Leue, C., Thompson, R., Web Services for Remote Portlets (WSRP), OASIS http://www.oasis-open.org
13. Matsuoka, et. al., Ninf: A Global Computing Infrastructure, http://ninf.apgrid.org/ welcome.shtml
14. Navotny, J. Developing grid portlets using the GridSphere portal framework, http://www-106.ibm.com/developerworks/grid/library/gr-portlets/
15. The Open Grid Services Infrastructure Working Group. http://www.gridforum.org/ogsi-wg, 2003.
16. S.G. Parker and C.R. Johnson. SCIRun: A scientific programming environment for computational steering. In *Supercomputing '95*. IEEE Press, 1995.
17. "Ptolemy II," Web Page. [Online]. Available: http://ptolemy.eecs.berkeley.edu/ptolemyII/
18. "Triana Workflow," Web Page. [Online]. Available: http://www.triana.co.uk
19. Gregor von Laszewski, Kaizar Amin, Mihael Hategan, Nestor J. Zaluzec, Shawn Hampton, Albert Rossi, GridAnt: A Client-Controllable GridWorkflow System, proceedings 37th Hawai'i International Conference on System Science, Jan 5-8, 2004.
20. WS-Resource Framework. http://www.globus.org/wsrf
21. O. Chipara, A. Slominski, Xydra OntoBrew, http://www.extreme.indiana.edu/xgws/xydra/
22. D. Laforenza, Grid programming: some indications where we are headed, Parallel Computing, Vol. 28, no. 12, 2002. pp. 1733-1752.
23. M. Govindaraju, S. Krishnan, K. Chiu, A. Slominski, D. Gannon, and R. Bramley. Merging the CCA Component Model with the OGSI Framework. In Proceedings of CCGrid2003, 3rd International Symposium on Cluster Computing and the Grid, Tokyo, Japan, pages 182-189, 2003.

Maintaining Thousands of In-Flight Instructions

Adrian Cristal, Oliverio J. Santana, and Mateo Valero

Departament d'Arquitectura de Computadors
Universitat Politècnica de Catalunya
Barcelona, Spain
{adrian,osantana,mateo}@ac.upc.es

Abstract. Superscalar processors tolerate long-latency memory operations by maintaining a high number of in-flight instructions. Since the gap between processor and memory speed continues increasing every year, the number of in-flight instructions needed to support the large memory access latencies expected in the future should be higher and higher. However, scaling-up the structures required by current processors to support such a high number of in-flight instructions is impractical due to area, power consumption, and cycle time constraints.
The kilo-instruction processor is an affordable architecture able to tolerate the memory access latency by supporting thousands of in-flight instructions. Instead of simply up-sizing the processor structures, the kilo-instruction architecture relies on an efficient multi-checkpointing mechanism. Multi-checkpointing leverages a set of techniques like multi-level instruction queues, late register allocation, and early register release. These techniques emphasize the intelligent use of the available resources, avoiding scalability problems in the design of the critical processor structures. Furthermore, the kilo-instruction architecture is orthogonal to other architectures, like multi-processors and vector processors, which can be combined to boost the overall processor performance.

1 Introduction

A lot of research effort is devoted to design new architectural techniques able to take advantage of the continuous improvement in microprocessor technology. The current trend leads to processors with longer pipelines, which combines with the faster technology to allow an important increase in the processor clock frequency every year.

However, this trend generates two important problems that processor designers should face up. On the one hand, long pipelines increase the distance between branch prediction and branch resolution. The longer a pipeline is, the higher amount of speculative work that should be discarded in case of a branch misprediction. In this context, an accurate branch predictor becomes a critical processor component. On the other hand, the higher clock frequency increases the main memory access latency. Since the DRAM technology improves at a speed much lower than the microprocessor technology, each increase in the processor clock frequency causes that a higher number of processor cycles are required to access the main memory, degrading the potential performance achievable with the clock frequency improvement.

If the main memory access latency increase continues, it will be a harmful problem for future microprocessor technologies. Therefore, dealing with the gap between the processor and the memory speed is vital in order to allow high-frequency microprocessors to achieve all their potential performance. A plethora of well-known techniques has been proposed to overcome the main memory latency, like cache hierarchies or data prefetching, but they do not completely solve the problem. A different approach to tolerate the main memory access latency is to dramatically increase the number of in-flight instructions that can be maintained by the processor.

A processor able to maintain thousands in-flight instructions can overlap the latency of a load instruction that access to the main memory with the execution of subsequent independent instructions, that is, the processor can hide the main memory access latency by executing useful work. However, maintaining a high number of in-flight instructions requires scaling-up critical processor structures like the reorder buffer, the physical register file, the instruction queues, and the load/store queue. This is not affordable due to cycle time limitations. The challenge is to develop an architecture able to support a high number of in-flight instructions while avoiding the scalability problems involved by such a big amount of in-flight instructions.

2 Increasing the Number of In-Flight Instructions

Figure 1 shows an example of the impact of increasing the maximum number of in-flight instructions supported by a four instruction wide out-of-order superscalar processor. The main memory access latency is varied from 100 to 1000 cycles. Data is provided for both the SPECint2000 integer applications and the SPECfp2000 floating point applications. A first observation from this figure is that the increase in the main memory latency causes enormous performance degradation. In a processor able to support 128 in-flight instructions, the integer applications suffer from an average 45% performance reduction. The degradation is even higher for floating point applications, whose average performance is reduced by 65%. Nevertheless, a higher number of in-flight instructions mitigates this effect. Increasing the number of in-flight instructions in a processor having 1000-cycle memory latency causes an average 50% performance improvement for the integer programs, while the floating point programs achieve a much higher 250% improvement.

These results show that increasing the number of in-flight instructions is an effective way of tolerating large memory access latencies. On average, executing floating point programs, a processor able to maintain up to 4096 in-flight instructions having 1000-cycle memory latency performs 22% better than a processor having 100-cycle latency but only being able to maintain up to 128 in-flight instructions. Executing integer programs, the processor supporting 4096 in-flight instructions cannot achieve a better performance than the processor support-

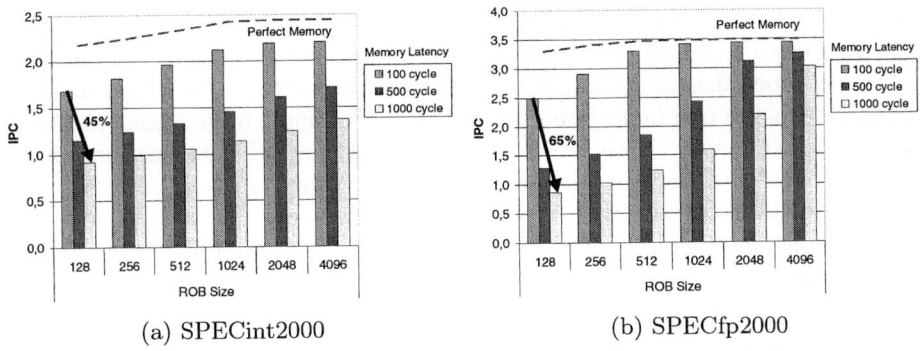

Fig. 1. Average performance of a 4-wide out-of-order superscalar processor executing both the SPEC2000 integer and floating point programs. The maximum number of in-flight instructions supported is varied from 128 to 4096, and the main memory access latency is varied from 100 to 1000 cycles.

ing 128 in-flight instructions. However, it is only 18% slower despite having a memory latency 10 times larger. The improvement achieved in floating point applications is due to the higher instruction-level parallelism available in these programs. The amount of correct-path instructions that can be used to overlap a long-latency memory operation in integer programs is limited by the presence of chains of instructions dependent on long-latency loads, especially other long-latency loads (i.e. chasing pointers), as well as the higher branch misprediction rate. Nevertheless, the continuous research effort devoted to improve branch prediction suggests that integer programs can achieve an improvement as high as floating point programs in a near future. Moreover, the higher number of in-flight instructions also enables different optimizations, like instruction reuse and value prediction, which can provide even better performance.

Therefore, high-frequency microprocessors will be able to tolerate large memory access latencies by maintaining thousands of in-flight instructions. The simplest way of supporting so much in-flight instructions is to scale all the processor resources involved, that is, the reorder buffer (ROB), the physical register file, the general purpose instruction queues (integer and floating point ones), and the load/store queue. Every decoded instruction requires an entry in the ROB, which is not released until the instruction commits. Every decoded instruction also requires an entry in the corresponding instruction queue, but it is released when the instruction issues for execution. The load/store queue has a different behavior. An entry in this queue is not released until it can be assured that no previous instruction will modify the contents of the memory address that will be accessed by the load or store instruction. In addition, store instructions cannot release their entries until commit because the memory hierarchy must not be updated until the execution correctness is guaranteed. Finally, every renamed instruction that generates a result requires a physical register, which is not released until it can be assured that the register value will not be used by a later instruction, that is, when the commit stage is reached by a new instruction that overwrites the contents of the registers.

However, scaling-up the number of entries in these structures is impractical, not only due to area and power consumption constraints, but also because these structures often determine the processor cycle time [15]. This is an exciting challenge. On the one hand, a higher number of in-flight instructions allows to tolerate large memory access latencies and thus provide a high performance. On the other hand, supporting such a high number of in-flight instructions involves a difficult scalability problem for the processor design. Our approach to overcome this scalability problem, while supporting thousands of in-flight instructions, is the kilo-instruction processor.

3 The Kilo-Instruction Processor

In essence, the kilo-instruction processor [4] is an out-of-order processor that keeps thousands of in-flight instructions. The main feature of our architecture is that its implementation is affordable. In order to support thousands of in-flight instructions, the kilo-instruction architecture relies on an intelligent use of the processor resources, avoiding the scalability problems caused by an excessive increase in the size of the main processor structures. Our design deals with the problems of each of these structures in an orthogonal way, that is, we apply particular solutions for each structure. These solutions are described in the following sections.

3.1 Multi-checkpointing

Checkpointing is a well established and used technique [8]. The main idea is to create a checkpoint at specific instructions of the program being executed. A checkpoint can be though of as a snapshot of the state of the processor, which contains all the information required to recover the architectural state and restart execution at that point. Several recent proposals are based on checkpointing. Cherry [10] uses a single checkpoint outside the ROB. The ROB is divided into a region occupied by speculative instructions, which can use the checkpoint for recovery, and a region occupied by non-speculative instructions, which depend on the ROB for recovery. Runahead execution [14] also creates a single checkpoint when the head of the ROB is reached by a load that has missed in the second level cache, allowing to speculatively execute the following instructions in order to perform data and instruction prefetches.

Instead of using a single checkpoint, the kilo-instruction architecture relies on a multi-checkpointing mechanism. Figure 2 shows an example of our checkpointing process [6]. First of all, it is important to state that there always exists at least one checkpoint in the processor (timeline A). The processor will fetch and issue instructions, taking new checkpoints at particular ones. If an instruction is miss-speculated or an exception occurs (timeline B), the processor rolls back to the previous checkpoint and resumes execution from there. When all instructions between two checkpoints are executed (timeline C), the older checkpoint among the two is removed (timeline D).

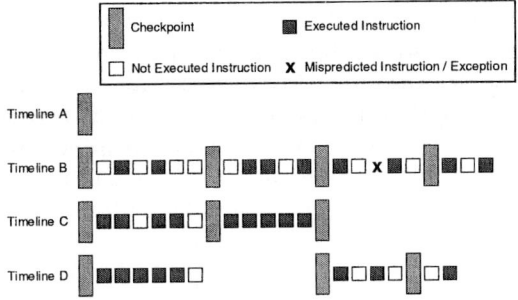

Fig. 2. The checkpointing process performed by the kilo-instruction processor.

Our multi-checkpointing mechanism has been designed as an efficient way to control and manage the use of critical resources inside the processor [3]. The novelty of this mechanism is that the kilo-instruction architecture uses checkpointing to allow an early release of resources. The multi-checkpointing mechanism enables the possibility of committing instructions out-of-order, allowing to early release ROB entries, which can lead to an architecture where the classical ROB is essentially unnecessary [6]. Early release can also be applied to physical registers, reducing the number of required registers. In addition, multi-checkpointing makes it possible to early manage the entries of the instruction queues, implementing a two-level structure. In summary, the multi-checkpointing mechanism is the key technique that makes our kilo-instruction architecture affordable.

3.2 Out-of-Order Commit

In a superscalar out-of-order processor, all instructions are inserted in the reorder buffer (ROB) after they are fetched and decoded. The ROB keeps a history window of all in-flight instructions, allowing for the precise recovery of the program state at any point. Instructions are removed from the ROB when they commit, that is, when they finish executing and update the architectural state of the processor. However, to assure that precise recovery is possible, the instructions should be committed in the program order.

In-order commit is a serious problem in the presence of large memory access latencies. Let us suppose that a processor has a 128-entry ROB and 500-cycle memory access latency. If a load instruction does not find a data in the cache hierarchy, it accesses the main memory, and thus it cannot be committed until its execution finishes 500 cycles later. When the load becomes the older instruction in the ROB, it blocks the in-order commit, and no later instruction will commit until the load finishes. Part of these cycles can be devoted to do useful work, but the ROB will become full soon, stalling the processor during several hundreds cycles. To avoid this, a larger ROB is required, that is, the processor requires a higher number of in-flight instructions to overlap the load access latency with the execution of following instructions. However, scaling-up the number of ROB entries is impractical, mainly due to cycle time limitations.

The kilo-instruction architecture solves this problem by using the multi-checkpointing mechanism for enabling out-of-order commit [3, 6]. The presence of a previous checkpoint causes that in-order commit is not required for preserving correctness and exception preciseness. If an exception occurs, the processor returns to the previous checkpoint and restarts execution. Indeed, the ROB itself becomes unnecessary. In spite of this, we keep a ROB-like structure in our kilo-instruction processor design. This structure, which we call pseudo-ROB [6], has the same functionality of a ROB. However, the instructions that reach the head of the pseudo-ROB are removed at a fixed rate, not depending on their state. Since the processor state can be recovered from the pseudo-ROB, generating a checkpoint is only necessary when the instructions leave the pseudo-ROB. Delaying checkpoint generation is beneficial to reduce the impact of branch mispredictions. Over 90% of mispredictions are caused by branches that are still inside the pseudo-ROB. This means that most branch mispredictions do not need to roll back to the previous checkpoint for recovering the correct state, minimizing the misprediction penalty. In this way, the combination of the multi-checkpointing and the pseudo-ROB allows to implement the functionality of a large ROB without requiring an unimplementable centralized structure with thousands of entries.

3.3 Instruction Queues

At the same time that instructions are inserted in the ROB, they are also inserted in their corresponding instruction queues. Each instruction should wait in an instruction queue until it is issued for execution. Figure 3 shows the accumulative distribution of allocated entries in the integer queue (for SPECint2000 programs) and in the floating point queue (for SPECfp2000 programs) with respect to the amount of total in-flight instructions. The main observation is that, in a processor able to support up to 2048 in-flight instructions, the instruction queues need a high number of entries. To cope with over 90% of the scenarios the processor is going to face, the integer queue requires 300 entries and the floating point queue requires 500 entries, which is definitely going to affect the cycle time [15].

Fortunately, not all instructions behave in the same way. Instructions are divided in two groups: blocked-short instructions when they are waiting for a functional unit or for results from short-latency operations, and blocked-long instructions when they are waiting for some long-latency instruction to complete, like a load instruction that misses in the second level cache. Blocked-long instructions represent by far the largest fraction of entries allocated in the instruction queues. Since these instructions take a very long time to even get issued for execution, maintaining them in the instruction queues just takes away issue slots from other instructions that will be executed more quickly. Multilevel queues can be used to track this type of instructions, delegating their handling to slower, but larger and less complex structures. Some previous studies have proposed such multilevel queues [9, 2], but they require a wake-up and select logic which might be on the critical path, thus potentially affecting the cycle time.

The kilo-instruction processor also takes advantage of the different waiting times of the instructions in the queues, but doing it in an affordable way. First

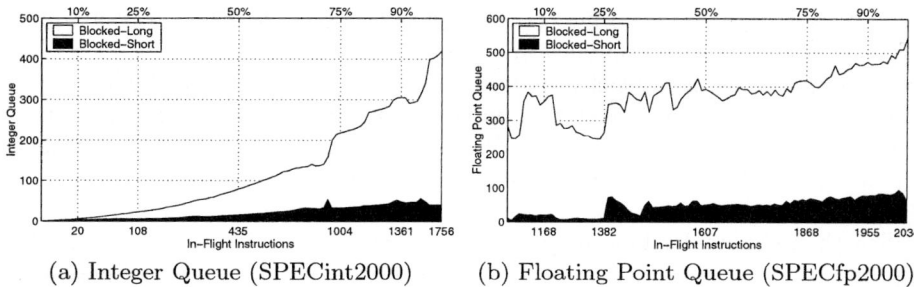

Fig. 3. Accumulative distribution of allocated entries in the integer queue (using the SPECint2000 programs) and in the floating point queue (using the SPECfp2000 programs) with respect to the amount of total in-flight instructions. For example, in floating point applications, 50% of the time there are 1600 or less in-flight instructions, requiring 400 floating point queue entries. Data presented corresponds to a processor able to maintain up to 2048 in-flight instructions and having 500-cycle memory access latency.

of all, our mechanism detects those instructions that will take a very long time to get issued for execution. The presence of the pseudo-ROB is beneficial for this detection process, since it allows to delay the decision of which instructions will require a long execution time until it can be effectively known. This not only increases the accuracy of long-latency instruction detection, but also greatly reduces the complexity of the logic required.

Long-latency instructions are removed from the general purpose instruction queues and stored in-order in a secondary buffer, which allows to free entries from the instruction queues that can be used by short-latency operations. The secondary buffer is a simple FIFO-like structure that we call Slow Lane Instruction Queue (SLIQ) [6]. The instructions stored in the SLIQ will wait until there is any need for them to return to the respective instruction queue. When the long-latency operation that blocked the instructions finishes, they are removed from the SLIQ and inserted back into their corresponding instruction queue, where they can be issued for execution. This mechanism allows to effectively implement the functionality of a large instruction queue while requiring a reduced number of entries, and thus it makes it possible to support a high number of in-flight instructions without scaling-up the instruction queues.

3.4 Load/Store Queue

Load and store instructions are inserted in the load/store queue at the same time they are inserted in the ROB. The main objective of this queue is to guarantee that all load and store instructions reach the memory system in the correct program order. For each load, it should be checked if an earlier store has been issued to the same physical address, and thus use the value produced by the store. For each store, it should be checked if a later load to the same physical address has been previously issued, and thus take corrective actions. Maintaining a high number of in-flight instructions involves an increase in the number of loads and stores that should be taken into account, which can make the load/store queue a true bottleneck both in latency and power.

Some solutions have been proposed for this problem. In [17], the load/store queue scalability is improved by applying approximate hardware hashing. A Bloom filter predictor is used to avoid unnecessary associative searches for memory operations that do not match other memory operations. This predictor is also used to separate load/store queue partitions, reducing the number of partitions that should be looked up when a queue search is necessary. Another approach is using multilevel structures [1, 16]. These works propose different filtering schemes that use two-level structures for storing most or all instructions in a big structure, while a smaller structure is used to easily check the dependencies.

3.5 Physical Register File

A great amount of physical registers is required to maintain thousands of in-flight instructions. Figure 4 shows the accumulative distribution of allocated integer registers (for SPECint2000 programs) and floating point registers (for SPECfp2000 programs) with respect to the amount of total in-flight instructions. To cope with over 90% of the scenarios the processor is going to face, almost 800 registers are required for the integer programs and almost 1200 are required for the floating point programs. Such a large register file will be impractical not only due to area and power limitations, but also because it requires a high access time, which will surely involve an increase in the processor cycle time.

In order to reduce the number of physical registers needed, the kilo-instruction processor relies on the different behaviors observed in the instructions that use a physical register. Registers are classified in four categories. Live registers contain values currently in use. Blocked-short and blocked-long registers have been allocated during rename, but are blocked because the corresponding instructions are waiting for the execution of predecessor instructions. Blocked-short registers are waiting for instructions that will issue shortly, while blocked-long registers are waiting for long-latency instructions. Finally, dead registers are no longer in use, but they are still allocated because the corresponding instructions have not yet committed.

It is clear that blocked-long and dead registers constitute the largest fraction of allocated registers. In order to avoid blocked-long registers, the assignment of physical registers can be delayed using virtual tags [12]. These virtual register mapping keeps track of the rename dependencies, making unnecessary the assignment of a physical register to an instruction until it starts execution. Dead registers can also be eliminated by using mechanisms for early register recycling [13]. These mechanisms release a physical register when it is possible to guarantee that it will not be used again, regardless the corresponding instruction has committed or not.

The kilo-instruction architecture combines these two techniques with the multi-checkpointing mechanism, leading to an aggressive register recycling mechanism that we call ephemeral registers [5, 11]. This is the first proposal that integrates both a mechanism for delayed register allocation and early register

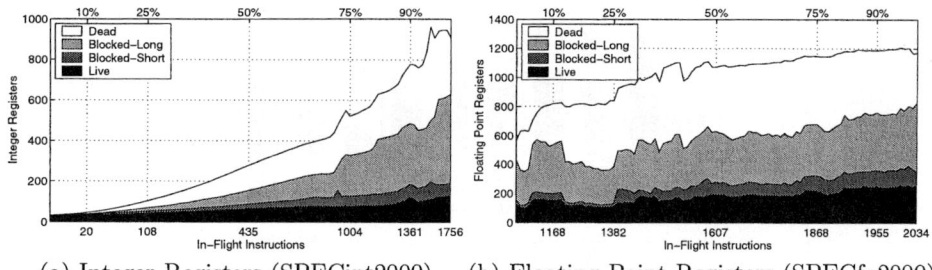

(a) Integer Registers (SPECint2000)   (b) Floating Point Registers (SPECfp2000)

Fig. 4. Accumulative distribution of allocated integer registers (using the SPECint2000 programs) and floating point registers (using the SPECfp2000 programs) with respect to the amount of total in-flight instructions. Data presented corresponds to a processor able to maintain up to 2048 in-flight instructions and having 500-cycle memory access latency.

release and analyzes the synergy between them. The combination of these two techniques with checkpointing allows the processor to non-conservatively deallocate registers, making it possible to support thousands of in-flight instructions without requiring an excessive number of registers.

4 Real Performance

Figure 5 provides some insight about the performance achievable by the kilo-instruction processor. It shows the average performance of a 4-wide processor executing the SPEC2000 floating point applications. The kilo-instruction processor modeled is able to support up to 2048 in-flight instructions, but it uses just 128-entry instruction queues. It also uses 32KB separate instruction and data caches as well as an unified 1MB second level cache. The figure is divided into three zones, each of them comprising the results for 100, 500, and 1000 cycles of main memory access latency. Each zone is composed of three groups of two bars, corresponding to 512, 1024, and 2048 virtual registers or tags [12]. The two bars of each group represent the performance using 256 or 512 physical registers.

In addition, each zone of the figure has two lines which represent the performance obtained by a baseline superscalar processor, able to support up to 128 in-flight instructions, and a limit unfeasible microarchitecture where all the resources have been up-sized to allow up to 4096 in-flight instructions. The main observation is that the kilo-instruction processor provides important performance improvements over the baseline superscalar processor. Using 2048 virtual tags, the kilo-instruction processor is more than twice faster than the baseline when the memory access latency is 500 cycles or higher. Moreover, a kilo-instruction processor having 1000 cycles memory access latency is only a 5% slower than the baseline processor having a memory access latency 10 times lower.

These results show that the kilo-instruction processor is an effective way of approaching the unimplementable limit machine in an affordable way. However,

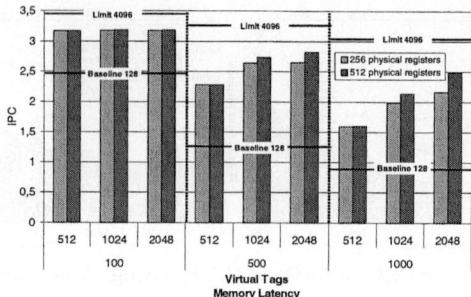

Fig. 5. Average performance results of the kilo-instruction processor executing the SPECfp2000 programs with respect to the amount of virtual registers, the memory latency, and the amount of physical registers.

there is still room for improvement. The distance between the kilo-instruction processor performance and the limit machine is higher for larger memory access latencies. This causes that, although the performance results for more aggressive setups nearly saturate for a memory access latency of 100 or 500 cycles, the growing trend is far from saturating when the memory access latency is 1000 cycles. This trend suggests that a more aggressive machine, able to support a higher number of in-flight instructions, will provide even a better performance.

5 Current Research Lines

The kilo-instruction architecture is not only an efficient technique for tolerating the memory access latency. It is also a flexible paradigm able to cross-pollinate with other techniques, that is, it can be combined with other mechanisms to improve its capabilities and boost the processor performance.

The kilo-instruction architecture can be combined with multiprocessors or simultaneous multithreaded processors. In particular, the kilo-instruction multiprocessor architecture [7] uses kilo-instruction processors as computing nodes for building small-scale CC-NUMA multiprocessors. This kind of multiprocessor system provide good performance results, showing two interesting behaviors. First, the great amount of in-flight instructions makes it possible for the system not just to hide the latencies coming from local and remote memory accesses but also the inherent communication latencies involved in preserving data coherence. Second, the significant pressure imposed by many in-flight instructions translates into a very high contention for the interconnection network. This fact remarks the need for better routers capable of managing high traffic levels, dictating a possible way for building future shared-memory multiprocessor systems.

Other interesting research line is the combination of the kilo-instruction architecture with vector processors. Vector architectures require high bandwidth memory systems to feed the vector functional units. The ability of bringing a high amount of data from memory reduces the total number of memory accesses, also reducing the impact of large memory access latencies. However, this improvement is limited to the vector part of a program. The performance of the scalar part will still be degraded due to the presence of a large memory access latency. Our objective is to use a kilo-instruction processor to execute the scalar part of vector programs, removing this limitation and thus providing a great improvement in the performance of vector processors.

We are also analyzing the combination of kilo-instruction processors with other techniques, which were previously proposed for processors with small instruction windows. Combining value prediction with kilo-instruction processors makes it possible to predict the results of long-latency memory operations. This technique allows to break the dependencies between long-latency loads and other instructions, increasing the available instruction-level parallelism, and thus improving the kilo-instruction processor ability of tolerating the memory access latency. Kilo-instruction processors can also be combined with mechanisms for detecting the reconvergence point after branch instructions. Correctly identifying the control independent instructions will enable the possibility of reusing independent instructions after branch mispredictions, alleviating the misprediction penalty. A different approach for avoiding the branch misprediction penalty is combining the kilo-instruction processor with techniques for multi-path execution. In general, the possibility of combining the kilo-instruction architecture with other techniques in an orthogonal way creates a great amount of new appealing ideas for future research.

6 Conclusions

Tolerating large memory access latencies is a key topic in the design of future processors. Maintaining a high amount of in-flight instructions is an effective mean for overcoming this problem. However, increasing the number of in-flight instructions requires up-sizing several processor structures, which is impractical due to power consumption, area, and cycle time limitations. The kilo-instruction processor is an affordable architecture able to support thousands of in-flight instructions. Our architecture relies on an intelligent use of the processor resources, avoiding the scalability problems caused by an excessive increase in the size of the critical processor structures. The ability of maintaining a high number of in-flight instructions makes the kilo-instruction processor an efficient architecture for dealing with future memory latencies, being able to achieve a high performance even in the presence of large memory access latencies.

Acknowledgements

This research has been supported by CICYT grant TIC-2001-0995-C02-01, the European Network of Excellence on High-Performance Embedded Architecture and Compilation (HIPEAC), and CEPBA. O. J. Santana is also supported by Generalitat de Catalunya grant 2001FI-00724-APTIND. Special thanks go to Francisco Cazorla, Ayose Falcón, Marco Galluzzi, Josep Llosa, José F. Martínez, Daniel Ortega, and Tanausú Ramírez for their contribution to the kilo-instruction processors.

References

1. H. Akkary, R. Rajwar, and S. T. Srinivasan. Checkpoint processing and recovery: towards scalable large instruction window processors. *Procs. of the 36th Intl. Symp. on Microarchitecture*, 2003.
2. E. Brekelbaum, J. Rupley, C. Wilkerson, and B. Black. Hierarchical scheduling windows. *Procs. of the 35th Intl. Symp. on Microarchitecture*, 2002.
3. A. Cristal, M. Valero, A. Gonzalez, and J. Llosa. Large virtual ROBs by processor checkpointing. *Technical Report UPC-DAC-2002-39*, Departament d'Arquitectura de Computadors, Universitat Politècnica de Catalunya, 2002.
4. A. Cristal, D. Ortega, J. Llosa, and M. Valero. Kilo-instruction processors. *Procs. of the 5th Intl. Symp. on High Performance Computing*, 2003.
5. A. Cristal, J. F. Martinez, J. Llosa, and M. Valero. Ephemeral registers with multicheckpointing. *Technical Report UPC-DAC-2003-51*, Departament d'Arquitectura de Computadors, Universitat Politècnica de Catalunya, 2003.
6. A. Cristal, D. Ortega, J. Llosa, and M. Valero. Out-of-order commit processors. *Procs. of the 10th Intl. Symp. on High-Performance Computer Architecture*, 2004.
7. M. Galluzzi, V. Puente, A. Cristal, R. Beivide, J. A. Gregorio, and M. Valero. A first glance at kilo-instruction based multiprocessors. *Procs. of the 1st Conf. on Computing Frontiers*, 2004.
8. W. M. Hwu and Y. N. Patt. Checkpoint repair for out-of-order execution machines. *Procs. of the 14th Intl. Symp. on Computer Architecture*, 1987.
9. A. Lebeck, J. Koppanalil, T. Li, J. Patwardhan, and E. Rotenberg. A large, fast instruction window for tolerating cache misses. *Procs. of the 29th Intl. Symp. on Computer Architecture*, 2002.
10. J. F. Martinez, J. Renau, M. Huang, M. Prvulovic, and J. Torrellas. Cherry: checkpointed early resource recycling in out-of-order microprocessors. *Procs. of the 35th Intl. Symp. on Microarchitecture*, 2002.
11. J. F. Martinez, A. Cristal, M. Valero, and J. Llosa. Ephemeral registers. *Technical Report CSL-TR-2003-1035*, Cornell Computer Systems Lab, 2003.
12. T. Monreal, A. Gonzalez, M. Valero, J. Gonzalez, and V. Viñals. Delaying physical register allocation through virtual-physical registers. *Procs. of the 32nd Intl. Symp. on Microarchitecture*, 1999.
13. M. Moudgill, K. Pingali, and S. Vassiliadis. Register renaming and dynamic speculation: an alternative approach. *Procs. of the 26th Intl. Symp. on Microarchitecture*, 1993.
14. O. Mutlu, J. Stark, C. Wilkerson, and Y. N. Patt. Runahead execution: an alternative to very large instruction windows for out-of-order processors. *Procs. of the 9th Intl. Symp. on High-Performance Computer Architecture*, 2003.
15. S. Palacharla, N. P. Jouppi, and J. E. Smith. Complexity-effective superscalar processors. *Procs. of the 24th Intl. Symp. on Computer Architecture*, 1997.
16. I. Park, C. Ooi, and T. Vijaykumar. Reducing design complexity of the load/store queue. *Procs. of the 36th Intl. Symp. on Microarchitecture*, 2003.
17. S. Sethumadhavan, R. Desikan, D. Burger, C. Moore, and S. Keckler. Scalable hardware memory disambiguation for high ILP processors. *Procs. of the 36th Intl. Symp. on Microarchitecture*, 2003.

Some Techniques for Automated, Resource-Aware Distributed and Mobile Computing in a Multi-paradigm Programming System

M. Hermenegildo[1,2], E. Albert[3], P. López-García[1], and G. Puebla[1]

[1] School of Comp. Sci., Technical U. of Madrid
{herme,pedro,german}@fi.upm.es
[2] Depts. of Comp. Sci. and Elec. and Comp. Eng., U. of New Mexico (UNM)
herme@unm.edu
[3] School of Comp. Sci., Complutense U. of Madrid
elvira@sip.ucm.es

Abstract. Distributed parallel execution systems speed up applications by splitting tasks into processes whose execution is assigned to different receiving nodes in a high-bandwidth network. On the distributing side, a fundamental problem is grouping and scheduling such tasks such that each one involves sufficient computational cost when compared to the task creation and communication costs and other such practical overheads. On the receiving side, an important issue is to have some assurance of the correctness and characteristics of the code received and also of the kind of load the particular task is going to pose, which can be specified by means of *certificates*. In this paper we present in a tutorial way a number of general solutions to these problems, and illustrate them through their implementation in the Ciao multi-paradigm language and program development environment. This system includes facilities for parallel and distributed execution, an assertion language for specifying complex programs properties (including safety and resource-related properties), and compile-time and run-time tools for performing automated parallelization and resource control, as well as certification of programs with resource consumption assurances and efficient checking of such certificates.

Keywords: resource awareness, granularity control, mobile code certification, distributed execution, GRIDs.

1 Introduction

Distributed parallel execution systems speed up applications by splitting tasks into processes whose execution is assigned to different nodes in a high-bandwidth network. GRID systems [12] in particular attempt to use for this purpose widely distributed sets of machines, often crossing several administrative domain boundaries. Many interesting challenges arise in this context.

A number of now classical problems have to be solved when this process is viewed from the *producer side*, i.e., from the point of view of the machine in charge of starting and monitoring a particular execution of a given application (or a part of such an application) by splitting the tasks into processes whose execution is assigned to different nodes (i.e., *consumers*) on receiving sides of the network. A fundamental problem involved in this process is detecting which tasks composing the application are independent and can thus be executed in parallel. Much work has been done in the areas of parallelizing compilers and parallel languages in order to address this problem. While obviously interesting, herein we will concentrate instead on other issues.

In this sense, a second fundamental problem, and which has also received considerable attention (even if less than the previous one), is the problem of grouping and scheduling such tasks, i.e., assigning tasks to remote processors, and very specially the particular issue of ensuring that the tasks involve sufficient computational cost when compared to the task creation and communication costs and other such practical overheads. Due to these overheads, and if the *granularity* of parallel tasks (i.e., the work necessary for their complete execution) is too small, it may happen that the costs are larger than the benefits of their parallel execution. Of course, the concept of small granularity is relative: it depends on the concrete system or set of systems where parallel programs are running. Thus, a *resource-aware* method has to be devised whereby the granularity of parallel tasks and their number can be controlled. We will call this the *task scheduling and granularity control* problem. In order to ensure that effective speedup can be obtained from remote execution it is obviously desirable to devise a solution where load and task distribution decisions are made automatically, specially in the context of non-embarrassingly parallel and/or irregular computations in which hand-coded approaches are difficult and tedious to apply.

Interestingly, when viewed from the *consumer side*, and in an open setting such as that of the GRID and other similar overlay computing systems, additional and novel challenges arise. In more traditional distributed parallelism situations (e.g., on clusters) receivers are assumed to be either dedicated and/or to trust and simply accept (or take, in the case of work-stealing schedulers) available tasks. In a more general setting, the administrative domain of the receiver can be completely different from that of the producer. Moreover, the receiver is possibly being used for other purposes (e.g., as a general-purpose workstation) in addition to being a party to the distributed computation. In this environment, interesting security- and resource-related issues arise. In particular, in order to accept some code and a particular task to be performed, the receiver must have some assurance of the *correctness and characteristics of the code received* and also of *the kind of load the particular task is going to pose*. A receiver should be free to reject code that does not adhere to a particular *safety policy* involving more traditional safety issues (e.g., that it will not write on specific areas of the disk) or *resource*-related issues (e.g., that it will not compute for more than a given amount of time, or that it will not take up an amount of memory or other resources above a certain threshold). Although it is obviously possible to interrupt a task after a certain time or if it starts taking too much memory this will be wasteful of resources and require recovery measures. It is clearly more desirable to be able to detect these situations a priori.

Recent approaches to mobile code safety involve associating safety information in the form of a *certificate* to programs [28, 21, 26, 1]. The certificate (or proof) is created at compile time, and packaged along with the untrusted code. The consumer who receives or downloads the code+certificate package can then run a *verifier* which by a straightforward inspection of the code and the certificate, can verify the validity of the certificate and thus compliance with the safety policy. It appears interesting to devise means for certifying security by enhancing mobile code with certificates which guarantee that the execution of the (in principle untrusted) code received from another node in the network is *safe* but also, as mentioned above, *efficient*, according to a predefined safety policy *which includes properties related to resource consumption.*

In this paper we present in a tutorial way a number of general solutions to these problems, and illustrate them through their implementation in the context of a multi-paradigm language and program development environment that we have developed, Ciao [3]. This system includes facilities for parallel and distributed execution, an assertion language for specifying complex programs properties (including safety and resource-related properties), and compile-time and run-time tools for performing automated parallelization and resource control, as well as certification of programs and efficient checking of such certificates.

Our system allows coding complex programs combining the styles of logic, constraint, functional, and a particular version of object-oriented programming. Programs which include logic and constraint programming (CLP) constructs have been shown to offer a particularly interesting case for studying the issues that we are interested in [14]. These programming paradigms pose significant challenges to parallelization and task distribution, which relate closely to the more difficult problems faced in traditional parallelization. This includes the presence of highly irregular computations and dynamic control flow, non-trivial notions of independence, the presence of dynamically allocated, complex data structures containing pointers, etc. In addition, the advanced state of program analysis technology and the expressiveness of existing abstract analysis domains used in the analysis of these paradigms has become very useful for defining, manipulating, and inferring a wide range of properties including independence, bounds on data structure sizes, computational cost, etc.

After first reviewing our approach to solving the granularity control problem using program analysis and transformation techniques, we propose a technique for resource-aware security in mobile code based on safety certificates which express properties related to resource usage. Intuitively, we use the granularity information (computed by the cost analysis carried out to decide the distribution of tasks on the producer side) in order to generate so-called *cost certificates* which are packaged along with the untrusted code. The idea is that the receiving side can reject code which brings cost certificates (which it cannot validate or) which have too large cost requirements in terms of computing resources (in time and/or space) and accept mobile code which meets the established requirements.

The rest of the paper proceeds as follows. After briefly presenting in Section 2 the basic techniques used for inferring complex properties in our approach, including upper and lower bounds on resource usage, Section 3 reviews our approach to the use of bounds on data structure sizes and computational cost to perform automatic granularity control. Section 4 then discusses our approach to resource-aware mobile code certification. Section 5 finally presents our conclusions.

2 Inferring Complex Properties Including Term Sizes and Costs

In order to illustrate our approach in a concrete setting, we will use CiaoPP [15] throughout the paper. CiaoPP is a component of the Ciao programming environment which performs several tasks including *automated parallelization and resource control*, as well as *certification* of programs, and efficient *checking* of such certificates. CiaoPP uses throughout the now well-established technique of *abstract interpretation* [5]. This technique has allowed the development of very sophisticated global static program analyses which are at the same time automatic, provably correct, and practical. The basic idea of abstract interpretation is to infer information on programs by interpreting ("running") them using abstract values rather than concrete ones, thus obtaining safe approximations of program behavior. The technique allows inferring much richer information than, for example, traditional types. The fact that at the heart of Ciao lies an efficient logic programming-based kernel language allows the use in CiaoPP of the very large body of approximation domains, inference techniques and tools for abstract interpretation-based semantic analysis which have been developed to a powerful and mature level in this area (see, e.g., [2, 27, 6, 16] and their references) and which are integrated in CiaoPP. As a result of this, CiaoPP can infer at compile-time, always safely, and with a significance degree of precision, a wide range of *properties* such as data structure shape (including pointer sharing), bounds on data structure sizes, determinacy, termination, non-failure, bounds on resource consumption (time or space cost), etc.

All this information is expressed by the compiler using *assertions*: syntactic objects which allow expressing "abstract" – i.e. symbolic – properties over different abstract domains. In particular, we use the high-level assertion language of [29], which actually implements a two-way communication with the system: it allows providing information to the analyzer as well as representing its results.

As a very simple example, consider the following procedure inc_all/2, which increments all elements of a list by adding one to each of them (we use functional notation for conciseness):

```
inc_all([])    := [].
inc_all([H|T]) := [ H+1 | inc_all(T)].
```

Assume that analysis of the rest of the program has determined that this procedure will be called providing a list of numbers as input. The output from CiaoPP for this program then includes the following assertion:

```
:- true pred inc_all(A,B)
        : ( list(A,num), var(B) )
       => ( list(A,num), list(B,num), size_lb(B,length(A))
        + ( not_fails, is_det, steps_lb(2*length(A)+1)).
```

Such "true pred" assertions specify in a combined way properties of both: ":" the entry (i.e., upon calling) and "=>" the exit (i.e., upon success) points of all calls to the procedure, as well as some global properties of its execution. The assertion expresses that procedure inc_all will produce as output a list of numbers B, whose length is at least (size_lb) equal to the length of the input list, that the procedure will never fail (i.e., an output value will be computed for any possible input), that it is deterministic (only one solution will be produced as output for any input), and that a lower bound on its computational cost (steps_lb) is $2\ length(A) + 1$ execution steps (where the cost measure used in the example is the number of procedure calls, but it can be any other arbitrary measure). This simple example illustrates type inference, non-failure and determinism analyses, as well as lower-bound argument size and computational cost inference. The same cost and size results are actually obtained from the upper bounds analyses (indicating that in this case the results are exact, rather than approximations). Note that obtaining a non-infinite upper bound on cost also implies proving *termination* of the procedure.

As can be seen from the example, in our approach cost bounds (upper or lower) are expressed as functions on the sizes of the input arguments and yield bounds on the number of execution steps required by the computation. Various measures are used for the "size" of an input, such as list-length, term-size, term-depth, integer-value, etc. Types, modes, and size measures are first automatically inferred by the analyzers and then used in the size and cost analysis.

While it is beyond the scope of this paper to fully explain all the (generally abstract interpretation-based) techniques involved in this process (see, e.g., [15, 10, 11] and their references), we illustrate through a simple example the fundamental intuition behind our lower bound cost estimation technique.

Consider again the simple inc_all procedure above and the assumption that type and mode inference has determined that it will be called providing a list of numbers as input. Assume again that the cost unit is the number of procedure calls. In a first approximation, and for simplicity, we also assume that the cost of performing an addition is the same as that of a procedure call. With these assumptions the exact cost function of procedure inc_all is $\text{Cost}_{inc\_all}(n) = 2\ n + 1$, where n is the size (length) of the input list.

In order to obtain a lower bound approximation of the previous cost function, CiaoPP's analyses first determine, based on the mode and type information inferred, that the argument size metric to be used is list length. An interesting problem with estimating lower bounds is that in general it is necessary to account for the possibility of failure of a call to the procedure (because of, e.g., an inadmissible argument) leading to a trivial lower bound of 0. For this reason, the lower bound cost analyzer uses information inferred by non-failure analysis [9], which can detect procedures and goals that can be guaranteed not to fail, i.e., to produce at least one solution (which would indeed be the case for inc_all) or not terminate.

In general, in order to determine the work done by (recursive) clauses, it is necessary to be able to estimate the size of input arguments in the procedure calls in the body of the procedure, relative to the sizes of the input arguments. For this, we use an abstraction of procedure definitions called a data dependency graph. Our approach to cost analysis consists of the following steps:

1. Use data dependency graphs to determine the relative sizes of variable bindings at different program points.
2. Use the size information to set up difference equations representing the computational cost of procedures
3. Compute lower/upper bounds to the solutions of these difference equations to obtain estimates of task granularities.

The size of an output argument in a procedure call depends, in general, on the size of the input arguments in that call. For this reason, for each output argument we use an expression which yields its size as a function of the input data sizes. For the inc_all procedure let $\text{Size}^2_{\text{inc\_all}}(n)$ denote the size of the output argument (the second) as a function of the size of its first (input) argument n. Once we have determined that the size measure to use is list length, and the size relationship which says that the size of the input list to the recursive call is the size of the input list of the procedure head minus one, the following difference equation can be set up for inc_all/2:

$\text{Size}^2_{\text{inc\_all}}(0) = 0$ (boundary condition from base case),

$\text{Size}^2_{\text{inc\_all}}(n) = 1 + \text{Size}^2_{\text{inc\_all}}(n-1)$.

The solution of this difference equation obtained is $\text{Size}^2_{\text{inc\_all}}(n) = n$.

Let $\text{Cost}^L_p(n)$ denote a lower bound on the cost (number of resolution steps) of a call to procedure p with an input of size n. Given all the assumptions above, and the size relations obtained, the following difference equation can be set up for the cost of inc_all/2:

$\text{Cost}^L_{\text{inc\_all}}(0) = 1$ (boundary condition from base case),

$\text{Cost}^L_{\text{inc\_all}}(n) = 1 + \text{Cost}^L_{\text{inc\_all}}(n-1)$.

The solution obtained for this difference equation is $\text{Cost}^L_{\text{inc\_all}}(n) = 2\,n+1$. In this case, the lower bound inferred is the exact cost (the upper bound cost analysis also infers the same function). In our approach, sometimes the solutions of the difference equations need to be in fact approximated by a lower bound (a safe approximation) when the exact solution cannot be found. The upper bound cost estimation case is very similar to the lower bound one, although simpler, since we do not have to account for the possibility of failure.

3 Controlling Granularity in Distributed Computing

As mentioned in Section 1, and in view of the techniques introduced in Section 2, we now discuss the task scheduling and granularity control problem, assuming that the program is already parallelized[1]. The aim of such distributed granularity control is to replace parallel execution with sequential execution or vice-versa based on some conditions related to task size and overheads. The benefits from controlling parallel task size will obviously be greater for systems with greater parallel execution overheads. In fact, in many architectures (e.g. distributed memory multiprocessors, workstation "farms", GRID systems, etc.) such overheads can be very significant and in them automatic parallelization cannot in general be done realistically without granularity control. In some other architectures where the overheads for spawning goals in parallel are small (e.g. in small shared memory multiprocessors) granularity control is not essential but it can also achieve important improvements in speedup.

Granularity control has been studied in the context of traditional programming [20, 25], functional programming [17, 18], and also logic programming [19, 7, 30, 8, 23, 24]. In [24] we proposed a general granularity control model and reported on its application to the case of logic programs. This model proposes (efficient) conditions based on the use of information available on task granularity in order to choose between parallel and sequential execution. The problems to be solved in order to perform granularity control following this approach include, on one hand, estimating the cost of tasks, of the overheads associated with their parallel execution, and of the granularity control technique itself. On the other hand there is also the problem of devising, given that information, efficient compile-time and run-time granularity control techniques.

Performing accurate granularity control at compile-time is difficult because some of the information needed to evaluate communication and computational costs, as for example input data size, is only known at run-time. A useful strategy is to do as much work as possible at compile-time, and postpone some final decisions to run-time. This can be achieved by generating at compile-time cost functions which estimate task costs as a function of input data size, which are then evaluated at run-time when such size is known. Then, after comparing costs of sequential and parallel executions (including all overheads), it is possible to determine which type of execution is profitable.

The approximation of these cost functions can be based either on some heuristics (e.g., profiling) or on a *safe* approximation (i.e. an upper or lower bound). We were able to show that if upper or lower bounds on task costs are available, under a given set of assumptions, it is possible to ensure that some parallel, distributed executions will always produce speedup (and also that some others are best executed sequentially). Because of these results, we will in general require

---

[1] In the past two decades, quite significant progress has been made in the area of automatically parallelizing programs in the context of logic and constraint programs, and some of the challenges have been tackled quite effectively there – see, for example, [13, 14, 4] for an overview of this area.

the cost information to be not just an approximation, but rather a well-defined bound on the actual execution cost. In particular, we will use the techniques for inferring upper- and lower-bound cost functions outlined in the previous section.

Assuming that such functions or similar techniques for determining task costs and overheads are given, the remainder of the granularity control task is to devise a way to actually compute such costs and then dynamically control task creation and scheduling using such information. Again the approach of doing as much of the work as possible at compile-time seems advantageous. In our approach, a transformation of the program is performed at compile time such that the cost computations and spawning decisions are encoded in the program itself, and in the most efficient way possible. The idea is to perform any remaining computations and decisions at run-time when the parameters missing at compile-time, such as data sizes or node load are available. In particular, the transformed programs will perform (generally) the following tasks: computing the sizes of data that appear in cost functions; evaluating the cost functions of the tasks to be executed in parallel using those data sizes; safely approximating the spawning and scheduling overheads (often also a function of data sizes); comparing these quantities to decide whether to schedule tasks in parallel or sequentially; deciding whether granularity control should be continued or not; etc.

As an example, consider the inc_all procedure of Section 2 and the program expression:

..., Y = inc_all(X) & M = r(Z), ...

which indicates that the procedure call inc_all(X) is to be made available for execution in parallel with the call to r(Z) (we assume that analysis has determined that inc_all(X) and r(Z) are independent, by, e.g., ensuring that there are no pointers between the data structures pointed to by X,Y and Z,M. From Section 2 we know that the cost function inferred for inc_all is $Cost^L_{inc\_all}(n) = 2\,n + 1$. Assume also that the cost of scheduling a task is constant and equal to 100 computation steps. The previous goal would then be transformed into the following one:

..., (2*length(X)+1 > 100 -> Y = inc_all(X) & M = r(Z)
      ; Y = inc_all(X), M = r(Z)), ...

where (*if* -> *then* ; *else*) is syntax for an if-then-else and "," denotes sequential execution as usual. Thus, when $2 * length(X) + 1$ (i.e., the lower bound on the cost of inc_all(X)) is greater than the threshold, the task is made available for parallel execution and not otherwise. Many optimizations are possible. In this particular case, the program expression can be simplified to:

..., (length(X) > 50 -> Y = inc_all(X) & M = r(Z)
      ; Y = inc_all(X), M = r(Z)), ...

and, assuming that length_gt(L,N) succeeds if the length of L is greater than N (its implementation obviously only requires to traverse at most the n first elements of list), it can be expressed as:

..., (length_gt(LX,50) -> Y = inc_all(X) & M = r(Z)
      ; Y = inc_all(X), M = r(Z)), ...

```
:- module(qsort, [qsort/2], [assertions]).

qsort([X|L],R) :-
        partition(L,X,L1,L2),
        qsort(L2,R2), qsort(L1,R1),
        append(R1,[X|R2],R).
qsort([],[]).

partition([],_B,[],[]).
partition([E|R],C,[E|Left1],Right):-
        E < C, partition(R,C,Left1,Right).
partition([E|R],C,Left,[E|Right1]):-
        E >= C, partition(R,C,Left,Right1).

append([],Ys,Ys).
append([X|Xs],Ys,[X|Zs]):- append(Xs,Ys,Zs).
```

Fig. 1. A qsort program.

As mentioned before, scheduling costs are often also a function of data sizes (e.g., communication costs). For example, assume that the cost of executing remotely Y = inc_all(X) is 0.1 $(length(X) + length(Y))$, where $length(Y)$ is the size of the result, an upper bound on which (actually, exact size) we know to be $length(X)$. Thus, our comparison would now be:

$$2\ length(X) + 1 > 0.1\ (length(X) + length(Y)) \equiv$$
$$2\ length(X) + 1 > 0.1\ (length(X) + length(X)) \equiv$$
$$2\ length(X) + 1 > 0.2\ length(X) \cong$$
$$2\ length(X) > 0.2\ length(X) \equiv$$
$$2 > 0.2$$

Which essentially means that the task can be scheduled for parallel execution *for any input size*. Conversely, with a communication cost greater than $0.5(length(X) + length(Y))$ the conclusion would be that it would never be profitable to run in parallel.

These ideas have been implemented and integrated in the CiaoPP system, which uses the information produced by its analyzers to perform combined compile–time/run–time resource control. The more realistic example in Figure 1 (a quick-sort program coded using logic programming) illustrates additional optimizations performed by CiaoPP in addition to cost function simplification, which include improved term size computation and stopping performing granularity control below certain thresholds. The concrete transformation produced by CiaoPP adds a clause: "qsort(X1,X2) :- g_qsort(X1,X2)." (to preserve the original entry point) and produces g_qsort/2, the version of qsort/2 that performs granularity control (where s_qsort/2 is the sequential version) is shown in Figure 2.

Note that if the lengths of the two input lists to the recursive calls to qsort are greater than a threshold (a list length of 7 in this case) then versions which

```
g_qsort([X|L],R) :-
        partition_o3_4(L,X,L1,L2,S1,S2),
        ( S2>7 -> (S1>7 -> g_qsort(L2,R2) & g_qsort(L1,R1)
                         ; g_qsort(L2,R2), s_qsort(L1,R1))
                ; (S1>7 -> s_qsort(L2,R2), g_qsort(L1,R1)
                         ; s_qsort(L2,R2), s_qsort(L1,R1))),
        append(R1,[X|R2],R).
g_qsort([],[]).
```

Fig. 2. The qsort program transformed for granularity control.

continue performing granularity control are executed in parallel. Otherwise, the two recursive calls are executed sequentially. The executed version of each such call depends on its grain size: if the length of its input list is not greater than the threshold then a sequential version which does not perform granularity control is executed. This is based on the detection of a recursive invariant: in subsequent recursions this goal will not produce tasks with input sizes greater than the threshold, and thus, for all of them, execution should be performed sequentially and, obviously, no granularity control is needed. Procedure `partition_o3_4/6`:

```
partition_o3_4([],_B,[],[],0,0).
partition_o3_4([E|R],C,[E|Left1],Right,S1,S2) :-
        E<C, partition_o3_4(R,C,Left1,Right,S3,S2), S1 is S3+1.
partition_o3_4([E|R],C,Left,[E|Right1],S1,S2) :-
        E>=C, partition_o3_4(R,C,Left,Right1,S1,S3), S2 is S3+1.
```

is the transformed version of `partition/4`, which "on the fly" computes the sizes of its third and fourth arguments (the automatically generated variables S1 and S2 represent these sizes respectively) [22].

4 Resource-Aware Mobile Computing

Having reviewed the issue of granularity control, and following the classification of issues of Section 1 we now turn our attention to some resource-related issues on the receiver side. In an open setting, such as that of the GRID and other similar overlay computing systems, receivers must have some assurance that the received code is safe to run, i.e., that it adheres to some conditions (the *safety policy*) regarding what it will do. We follow current approaches to mobile code safety, based on the technique of *Proof-Carrying Code* (PCC) [28], which as mentioned in Section 1 associate safety *certificates* to programs. A certificate (or proof) is created by the code supplier for each task at compile time, and packaged along with the untrusted mobile code sent to (or taken by) other nodes in the network. The consumer node who receives or takes the code+certificate package (plus a given task to do within that code) can then run a *checker* which by a straightforward inspection of the code and the certificate can verify the validity of the certificate and thus compliance with the safety policy. The key benefit of this approach is that the consumer is given by the supplier the capacity of ensuring compliance with the desired safety policy in a simple and efficient way.

Indeed the (proof) checker used at the receiving side performs a task that should be much simpler, efficient, and automatic than generating the original certificate. For instance, in the first PCC system [28], the certificate is originally a proof in first-order logic of certain *verification conditions* and the checking process involves ensuring that the certificate is indeed a valid first-order proof.

The main practical difficulty of PCC techniques is in generating safety certificates which at the same time:

- allow expressing interesting safety *properties*,
- can be generated *automatically* and,
- are easy and *efficient* to check.

Our approach to mobile code safety [1] directly addresses these problems. It uses approximation techniques, generally based on abstract interpretation, and it has been implemented using the facilities available in CiaoPP and discussed in the previous sections. These techniques offer a number of advantages for dealing with the aforementioned issues. The expressiveness of the properties that can be handled by the available abstract domains (and which can be used in a wide variety of assertions) will be implicitly available to define a wide range of safety conditions covering issues like independence, types, freeness from side effects, access patterns, bounds on data structure sizes, bounds on cost, etc. Furthermore, the approach inherits the inference power of the abstract interpretation engines to automatically generate and validate the certificates. In the following, we review our standard mobile code certification process and discuss the application in parallel distributed execution.

Certification in the Supplier: The certification process starts from an initial program and a set of assertions provided by the user on the producer side, which encode the safety policy that the program should meet, and which are to be verified. Consider for example the following (naive) reverse program (where append is assumed to be defined as in Figure 1):

```
:- entry reverse/2 : list * var.
reverse( [] )    := [].
reverse( [H|L] ) := ~append( reverse(L), [H] ).
```

Let us assume also that we know that the consumer will only accept purely computational tasks, i.e., tasks that have no side effects, and only those of polynomial (actually, at most quadratic) complexity. This safety policy can be expressed at the producer for this particular program using the following assertions:

```
:- check comp reverse(A,B)
         + sideff(free).
:- check comp reverse(A,B)
         : list * var
         + steps_ub( o(exp(length(A),2)) ).
```

The first (computational –comp) assertion states that it should be verified that the computation is pure in the sense that it does not produce any side effects (such as opening a file, etc.). The second (also computational) assertion states that it should be verified that there is an upper bound for the cost of this predicate in $O(n^2)$, i.e., quadratic in n, where n is the length of the first list (represented as length(A)). Implicitly, we are assuming that the code will be accepted at the receiving end, provided all assertions can be checked, i.e., the intended semantics expressed in the above assertions determines the safety condition. This can be a policy agreed a priori or exchanged dynamically.

Note that, unlike traditional safety properties such as, e.g., type correctness, which can be regarded as platform independent, resource-related properties should take into account issues such as load and available computing resources in each particular system. Thus, for resource-related properties different nodes may impose different policies for the acceptance of tasks (mobile code).

Generation of the Certificate: In our approach, given the previous assertions defining the safety policy, the certificate is automatically generated by an *analysis engine* (which in the particular case of CiaoPP is based on the *goal dependent*, i.e., context-sensitive, analyzer of [16]). This analysis algorithm receives as input a set of entries (included in the program like the entry assertion of the example above) which define the base, boundary assumptions on the input data. These base assumptions can be checked at run-time on the actual input data (in our example the type of the input is stated to be a list). The computation of the analysis process terminates when a fixpoint of a set of equations is reached. Thus, the results of analysis are often called the *analysis fixpoint*.

Due to space limitations, and given that it is now well understood, we do not describe here the analysis algorithm (details can be found in, e.g., [2, 16]). The important point to note is that the certification process is based on the idea that the role of certificate can be played by a *particular and small subset of the analysis results* (i.e., of the analysis fixpoint) computed by abstract interpretation-based analyses.

For instance, the analyzers available in CiaoPP infer, among others, the following information for the above program and entry:

```
:- true pred reverse(A,B)
        : ( list(A), var(B) )
       => ( list(A), list(B))
        + ( not_fails, is_det, sideff(free),
            steps_ub( 0.5*exp(length(A),2)+1.5*length(A)+1 )).
```

stating that the output is also a list, that the procedure is deterministic and will not fail, that it does not contain side-effects, and that calls to this procedure take at most $0.5\ (length(A))^2 + 1.5\ length(A) + 1$ resolution steps. In addition, given this information, the output shows that the "status" of the three check assertions has become checked, which means that they have been validated and thus the program is safe to run (according to the intended meaning):

```
:- checked comp reverse(A,B)
        + sideff(free).
```

```
:- checked comp reverse(A,B)
       : list * var
       + steps_ub( o(exp(length(A),2)) ).
```

Thus, we have verified that the safety condition is met and that the code is indeed safe to run (for now on the producer side). The analysis results above can themselves be used as the *cost and safety certificate* to attest a safe and efficient use of procedure `reverse` on the receiving side.

In general the verification process requires first generating a *verification condition* [1] that encodes the information in the check assertions to be verified and then checking this condition against the information available from analysis. This validation may yield three different possible status: i) the verification condition is indeed checked and the fixpoint is considered a *valid certificate*, ii) it is disproved, and thus the certificate is not valid and the code is definitely not safe to run (we should obviously correct the program before continuing the process); and iii) it cannot be proved nor disproved. Case iii) occurs because the most interesting properties are in general undecidable. The analysis process in order to always terminate is based on approximations, and may not be able to infer precise enough information to verify the conditions. The user can then provide a more refined description of initial entries or choose a different, finer-grained, abstract domain. However, despite the inevitable theoretical limitations, the analysis algorithms and abstract domains have been proved very effective in practice. In both the ii) and iii) cases, the certification process needs to be restarted until achieving a verification condition which meets i). If it succeeds, the fixpoint constitutes a valid certificate and can be sent to the receiving side together with the program.

Validation in the Consumer: The *validation* process performed by the consumer node is similar to the above certification process except that the analysis engine is replaced by an *analysis checker*. The definition of the analysis checker is centered around the observation that the checking algorithm can be defined as a very simplified "one-pass" analyzer. Intuitively since the certification process already provides the fixpoint result as certificate, an additional analysis pass over it cannot change the result. Thus, as long as the fixpoint is valid, one single execution of the abstract interpreter validates the certificate.

As it became apparent in the above example, the interesting point to note is that abstract interpretation-based techniques are able to reason about computational properties which can be useful for controlling efficiency issues in a mobile computing environment and in distributed parallelism platforms. We consider the case of the receiver of a task in a parallel distributed system such as a GRID. This receiver (the code consumer) could use this method to reject code which does not adhere to some specification, including usage of computing resources (in time and/or space). Reconsider for example the previous reverse program and assume that a node with very limited computing resources is assigned to perform a computation using this code. Then, the following "check" assertion can be used for such particular node:

```
:- check comp reverse(A,B)
        : ( list(A, term), var(B) )
        + steps_ub( length(A) + 1 ).
```

which expresses that the consumer node will not accept an implementation of `reverse` with complexity bigger than linear. In order to guarantee that the cost assertion holds, the certificate should contain upper bounds on computational cost. Then, the code receiver proceeds to validate the certificate. The task of checking that a given expression is an upper bound is definitely simpler than that of obtaining the most accurate possible upper bound. If the certificate is not valid, the code is discarded. If it is valid, the code will be accepted only if the upper bound in the certificate is lower or equal than that stated in the assertion. In our example, the certificate contains the (valid) information that `reverse` will take at most $0.5\ (length(A))^2 + 1.5\ length(A) + 1$ resolution steps. However, the assertion requires the cost to be at most $length(A) + 1$ resolution steps. A comparison between these cost functions does not allow proving that the code received by the consumer satisfies the efficiency requirements imposed (i.e. the assertion cannot be proved)[2]. This means that the consumer will reject the code. Similar results would be obtained if the worst case complexity property `steps_ub( o(length(A)) )` was used in the above check assertion, instead of `steps_ub( length(A) + 1 )`.

Finally, and interestingly, note that the certificate can also be used to approximate the *actual costs of execution* and make decisions accordingly. Since the code receiver knows the data sizes, it can easily apply them to the cost functions (once they are verified) and obtain values that safely predict the time and space that the task received will consume.

5 Conclusions

We have presented an abstract interpretation-based approach to resource-aware distributed and mobile computing and discussed their implementation in the context of a multi-paradigm programming system. Our framework uses modular, incremental, abstract interpretation as a fundamental tool to infer resource and safety information about programs. We have shown this information, including lower bounds on cost and upper bounds on data sizes, can be used to perform high-level optimizations such as resource-aware task granularity control. Moreover, cost information and, in particular, upper bounds, inferred during the previous process are relevant to certifying and validating mobile programs which may have constraints in terms of computing resources (in time and/or space). In essence, we believe that our proposals can contribute to bringing increased flexibility, expressiveness and automation of important resource-awareness aspects in the area of mobile and distributed computing.

---

[2] Indeed, the lower bound cost analysis in fact disproves the assertion, which is clearly invalid.

Acknowledgments

This work has been supported in part by the European Union IST program under contracts IST-2001-38059 "ASAP" and "GridCoord", by MCYT project TIC 2002-0055 "CUBICO" and FEDER infrastructure UNPM-E012, and by the Prince of Asturias Chair in Information Science and Technology at the University of New Mexico.

References

1. E. Albert, G. Puebla, and M. Hermenegildo. An Abstract Interpretation-based Approach to Mobile Code Safety. In *Proc. of Compiler Optimization meets Compiler Verification (COCV'04)*, April 2004.
2. M. Bruynooghe. A Practical Framework for the Abstract Interpretation of Logic Programs. *Journal of Logic Programming*, 10:91–124, 1991.
3. F. Bueno, D. Cabeza, M. Carro, M. Hermenegildo, P. López-García, and G. Puebla (Eds.). The Ciao System. Reference Manual (v1.10). The ciao system documentation series–TR, School of Computer Science, Technical University of Madrid (UPM), June 2002. System and on-line version of the manual available at http://clip.dia.fi.upm.es/Software/Ciao/.
4. J. Chassin and P. Codognet. Parallel Logic Programming Systems. *Computing Surveys*, 26(3):295–336, September 1994.
5. P. Cousot and R. Cousot. Abstract Interpretation: a Unified Lattice Model for Static Analysis of Programs by Construction or Approximation of Fixpoints. In *Proc. of POPL'77*, pages 238–252, 1977.
6. Patrick Cousot and Radhia Cousot. Abstract interpretation and application to logic programs. *The Journal of Logic Programming*, 13(2 and 3):103–179, 1992.
7. S.K. Debray, N.-W. Lin, and M. Hermenegildo. Task Granularity Analysis in Logic Programs. In *Proc. of the 1990 ACM Conf. on Programming Language Design and Implementation*, pages 174–188. ACM Press, June 1990.
8. S.K. Debray and N.W. Lin. Cost analysis of logic programs. *ACM Transactions on Programming Languages and Systems*, 15(5):826–875, November 1993.
9. S.K. Debray, P. López-García, and M. Hermenegildo. Non-Failure Analysis for Logic Programs. In *1997 International Conference on Logic Programming*, pages 48–62, Cambridge, MA, June 1997. MIT Press, Cambridge, MA.
10. S.K. Debray, P. López-García, M. Hermenegildo, and N.-W. Lin. Estimating the Computational Cost of Logic Programs. In *Static Analysis Symposium, SAS'94*, number 864 in LNCS, pages 255–265, Namur, Belgium, September 1994. Springer-Verlag.
11. S.K. Debray, P. López-García, M. Hermenegildo, and N.-W. Lin. Lower Bound Cost Estimation for Logic Programs. In *1997 International Logic Programming Symposium*, pages 291–305. MIT Press, Cambridge, MA, October 1997.
12. I. Foster, C. Kesselman, J. Nick, and S. Tuecke, editors. *The Grid: Blueprint for a New Computing Infrastructure*. Morgan-Kaufmann, 1999.
13. G. Gupta, E. Pontelli, K. Ali, M. Carlsson, and M. Hermenegildo. Parallel Execution of Prolog Programs: a Survey. *ACM Transactions on Programming Languages and Systems*, 23(4):472–602, July 2001.
14. M. Hermenegildo. Automatic Parallelization of Irregular and Pointer-Based Computations: Perspectives from Logic and Constraint Programming. In *Proceedings of EUROPAR'97*, volume 1300 of *LNCS*, pages 31–46. Springer-Verlag, August 1997.

15. M. Hermenegildo, G. Puebla, F. Bueno, and P. López-García. Program Development Using Abstract Interpretation (and The Ciao System Preprocessor). In *10th International Static Analysis Symposium (SAS'03)*, number 2694 in LNCS, pages 127–152. Springer-Verlag, June 2003.
16. M. Hermenegildo, G. Puebla, K. Marriott, and P. Stuckey. Incremental Analysis of Constraint Logic Programs. *ACM Transactions on Programming Languages and Systems*, 22(2):187–223, March 2000.
17. L. Huelsbergen. Dynamic Language Parallelization. Technical Report 1178, Computer Science Dept. Univ. of Wisconsin, September 1993.
18. L. Huelsbergen, J. R. Larus, and A. Aiken. Using Run-Time List Sizes to Guide Parallel Thread Creation. In *Proc. ACM Conf. on Lisp and Functional Programming*, June 1994.
19. S. Kaplan. Algorithmic Complexity of Logic Programs. In *Logic Programming, Proc. Fifth International Conference and Symposium, (Seattle, Washington)*, pages 780–793, 1988.
20. B. Kruatrachue and T. Lewis. Grain Size Determination for Parallel Processing. *IEEE Software*, January 1988.
21. T. Lindholm and F. Yellin. *The Java Virtual Machine Specification*. Addison-Wesley, 1997.
22. P. López-García and M. Hermenegildo. Efficient Term Size Computation for Granularity Control. In *International Conference on Logic Programming*, pages 647–661, Cambridge, MA, June 1995. MIT Press, Cambridge, MA.
23. P. López-García, M. Hermenegildo, and S.K. Debray. Towards Granularity Based Control of Parallelism in Logic Programs. In Hoon Hong, editor, *Proc. of First International Symposium on Parallel Symbolic Computation, PASCO'94*, pages 133–144. World Scientific, September 1994.
24. P. López-García, M. Hermenegildo, and S.K. Debray. A Methodology for Granularity Based Control of Parallelism in Logic Programs. *Journal of Symbolic Computation, Special Issue on Parallel Symbolic Computation*, 22:715–734, 1996.
25. C. McGreary and H. Gill. Automatic Determination of Grain Size for Efficient Parallel Processing. *Communications of the ACM*, 32, 1989.
26. G. Morrisett, D. Walker, K. Crary, and N. Glew. From system F to typed assembly language. *ACM Transactions on Programming Languages and Systems*, 21(3):527–568, 1999.
27. K. Muthukumar and M. Hermenegildo. Compile-time Derivation of Variable Dependency Using Abstract Interpretation. *Journal of Logic Programming*, 13(1, 2, 3 and 4):315–347, 1992.
28. G. Necula. Proof-Carrying Code. In *Proc. of POPL'97*, pages 106–119. ACM Press, 1997.
29. G. Puebla, F. Bueno, and M. Hermenegildo. An Assertion Language for Constraint Logic Programs. In *Analysis and Visualization Tools for Constraint Programming*, pages 23–61. Springer LNCS 1870, 2000.
30. X. Zhong, E. Tick, S. Duvvuru, L. Hansen, A.V.S. Sastry, and R. Sundararajan. Towards an Efficient Compile-Time Granularity Analysis Algorithm. In *Proc. of the 1992 International Conference on Fifth Generation Computer Systems*, pages 809–816. Institute for New Generation Computer Technology (ICOT), June 1992.

Why Structured Parallel Programming Matters

Murray Cole

School of Informatics, University of Edinburgh,
King's Buildings, Mayfield Road, Edinburgh, EH9 3JZ, Scotland
mic@inf.ed.ac.uk

Abstract. Simple parallel programming frameworks such as Pthreads, or the six function core of MPI, are universal in the sense that they support the expression of arbitrarily complex patterns of computation and interaction between concurrent activities. Pragmatically, their descriptive power is constrained only by the programmer's creativity and capacity for attention to detail.

Meanwhile, as our understanding of the structure of parallel algorithms develops, it has become clear that many parallel applications can be characterized and classified by their adherence to one or more of a number of generic patterns. For example, many diverse applications share the underlying control and data flow of the pipeline paradigm, whether expressed in terms of message passing, or by constrained access to shared data.

A number of research programs, using terms such as *skeleton*, *template*, *archetype* and *pattern*, have sought to exploit this phenomenon by allowing the programmer to explicitly express such meta-knowledge in the program source, through the use of new libraries, annotations and control constructs, rather than leaving it implicit in the interplay of more primitive universal mechanisms.

While early work stressed productivity and portability (the programmer is no longer required to repeatedly "reinvent the wheel") we argue that the true significance of this approach lies in the capture of complex algorithmic knowledge which would be impossible to determine by static examination of an equivalent unstructured source. This enables developments in a number of areas. With respect to *low-level performance*, it allows the run-time system, library code or compiler to make clever optimizations based on detailed foreknowledge of the evolving computation. With respect to *high-level performance*, it enables a methodology of improvement through powerful restructuring transformations. Similarly, with respect to *program correctness*, it allows arguments to be pursued at a much coarser, more tractable grain than would otherwise be possible.

Topic 1
Support Tools and Environments

José C. Cunha, Allen D. Malony, Arndt Bode, and Dieter Kranzlmueller

Topic Chairs

Due to the diversity of parallel and distributed computing infrastructures and programming models, and the complexity of issues involved in the design and development of parallel programs, the creation of tools and environments to support the broad range of parallel system and software functionality has been widely recognized as a difficult challenge.

Current research in this topic continues to address individual tools for supporting correctness and performance issues in parallel program development. However, standalone tools are sometimes insufficient to cover the rich diversity of tasks found in the design, implementation and production phases of the parallel software life-cycle. This has motivated interest in interoperable tools, as well as solutions to ease their integration into unified development and execution environments.

Modern cluster and grid computing systems with possibly larger numbers of nodes and several degrees of heterogeneity are posing new challenges to the tool designers, developers, and users. Such environments require increased flexibility in the individual tools, and in the support for tool interaction and composition. Concurrently, the dynamic nature of modern computing applications requires new solutions for dynamic and adaptive tools and environments. Capabilities such as computational steering and dynamic performance tuning are growing in interest and necessity.

This year 27 papers were submitted to this topic area. Overall, they illustrate the current importance of tools and environments for supporting the development of parallel programs, as well as the continued efforts in the search for improved concepts and solutions. Among the submissions, 9 papers were accepted as full papers for the conference (33%). These papers address a diversity of issues and approaches, including instrumentation and monitoring support techniques, and management of large traces, performance evaluation and optimization issues, correctness support tools for concurrent programs, and integrated environments for dynamic performance tuning, and for steering of parallel simulations.

Profiling and Tracing OpenMP Applications with POMP Based Monitoring Libraries

Luiz DeRose[1], Bernd Mohr[2], and Seetharami Seelam[3]

[1] Cray Inc.
Mendota Heights, MN, USA
ldr@cray.com
[2] Forschungszentrum Jülich, ZAM,
Jülich, Germany
b.mohr@fz-juelich.de
[3] University of Texas at El Paso
El Paso, TX, USA
seelam@cs.utep.edu

Abstract. In this paper we present a collection of tools that are based on the POMP performance monitoring interface for analysis of OpenMP applications. These POMP compliant libraries, POMPROF and the KOJAK POMP library, provide respectively the functionality for profiling and tracing of OpenMP applications. In addition, we describe a new approach to compute temporal overhead due to scheduling (load-imbalance), synchronization (barrier time), and the runtime system. Finally, we exemplify the use of these libraries with performance measurement and visualization of the ASCI sPPM benchmark code. Our examples show that the information provided by both tools is consistent, and provides data that is helpful for users to understand the source of performance problems.

1 Introduction

OpenMP is today's de facto standard for shared memory parallel programming of scientific applications. It provides a higher level specification for users to write threaded programs that are portable across most shared memory multiprocessors. However, application developers still face application performance problems such as load imbalance and excessive barriers overhead, when using OpenMP. Moreover, these problems are difficult to detect without the help of performance tools.

The MPI specification defines a standard monitoring interface (PMPI), which facilitates the development of performance monitoring tools. Similarly, a Java Virtual Machine Profiler Interface (JVMPI) is an experimental feature in the Java 2 SDK, which is intended for users and tools vendors to develop profilers that work in conjunction with the Java virtual machine implementation. Unfortunately, OpenMP does not provide yet a standardized performance monitoring interface, which would simplify the design and implementation of portable OpenMP performance tools. Mohr et. al. [9] proposed POMP, a performance monitoring interface for OpenMP. This proposal extends experiences of previous implementations of monitoring interfaces for OpenMP [1, 7, 10]. POMP describes an API to be called by "*probes*" inserted into the application by a compiler, a pre-processor, or via a binary or dynamic instrumentation mechanism. With such performance monitoring interface, users and tools builders can then define their own POMP compliant libraries for performance measurement of OpenMP applications.

In [6], we presented DPOMP, a POMP instrumentation tool based on binary modification, which takes as input a performance monitoring library that conforms to the POMP API and an OpenMP application binary. DPOMP instruments the binary of the application with dynamic probes containing POMP calls defined in the library. In this paper we present two POMP compliant libraries, POMPROF and the KOJAK POMP library, which provide respectively the functionality for profiling and tracing of OpenMP applications. In addition, we exemplify the use of these libraries with performance measurement and visualization of the ASCI sPPM benchmark code.

The remainder of this paper is organized as follows: In Section 2 we briefly describe the main features of DPOMP. In Section 3 we describe our POMP compliant library for profiling of OpenMP applications. In Section 4 we describe the KOJAK POMP library for tracing of OpenMP programs. In Section 5 we present examples of utilization of these two libraries. Finally, we present our conclusions in Section 6.

2 A POMP Instrumentation Tool Based on Binary Modification

DPOMP was implemented using DPCL [5], an object-based C++ class library and runtime infrastructure based on the Dyninst Application Programming Interface (API) [3]. Using DPCL, a performance tool can insert code patches at function entry and exit points, as well as before and after call sites. Since the IBM compiler translates OpenMP constructs into functions that call *"outlined"* functions containing the body of the construct, DPOMP can insert calls to functions of a POMP compliant monitoring library for each OpenMP construct in the target application[1]. The main advantage of this approach lies in its ability to modify the binary with performance instrumentation with no special preparations, like re-compiling or re-linking. In addition, since it relies only on the binary, DPOMP works independently of the programming language used for the OpenMP program.

DPOMP takes as input an OpenMP application binary and a POMP compliant performance monitoring library. It reads the application binary, as well as the binary of the POMP library and instruments the application binary, so that, at locations which represent events in the POMP execution model the corresponding POMP monitoring routines are called. From the user's point of view, the amount of instrumentation can be controlled through environment variables which describe the level of instrumentation for each group of OpenMP events as proposed by the POMP specification. From the tools builder point of view, instrumentation can also be controlled by the set of POMP routines provided by the library, i.e., instrumentation is only applied to those events that have a corresponding POMP routine in the library. This means that tool builders only need to implement the routines which are necessary for their tool and that instrumentation is minimal (i.e., no calls to unnecessary dummy routines which don't do anything). By

---
[1] Notice that DPOMP has a dependence on the naming convention used by the IBM XL compiler. However, it can be easily extended to any other OpenMP compiler that uses the "function outline" approach.

default, DPOMP instruments all OpenMP constructs for which there is a corresponding POMP function in the library. It also instruments all user functions called from the main program provided that there is a definition in the library for performance monitoring of user functions. In addition, for MPI applications, by default DPOMP instruments all MPI calls in the program. Once instrumentation is finished, the modified program is executed.

3 The POMP Profiler Library

We implemented a POMP compliant monitoring library (POMPROF) that generates a detailed profile describing various overheads and the amount of time spent by each thread in three key regions of the parallel application: parallel regions, OpenMP loops inside a parallel region, and user defined functions. In addition, POMPROF provides information for analysis of the slowdown caused by sequential regions, as well as measurements of execution time from user functions and OpenMP API calls such as set lock. The profile data is presented in the form of an XML file that can be visualized by a graphical interface as shown in Figure 3.

POMPROF performs overhead analysis, which is a key incremental method for performance tuning of parallel applications [4]. Historically, both temporal and spatial overheads have been defined as a function of two execution times of an application: namely, the execution time of the sequential code, and the execution time of the parallel code [4, 12]. One of the problems of this approach is that it requires an application programmer to have two versions of a program, a serial and a parallel. Moreover, often overhead definitions for OpenMP do not include the property of measurability of basic blocks of code, reason being that it is not always possible to measure all overhead sources at the OpenMP construct level.

Since we can use our binary instrumentation infrastructure to monitor blocks of OpenMP code, which are packaged into outlined functions, we compute temporal overheads with measurable times. Measurable temporal overhead often results from an imbalanced loop, a barrier for synchronization, or from the runtime code. So, POMPROF focuses on temporal overhead that is due to scheduling (load-imbalance), synchronization (barrier time), and the overhead due to the runtime system.

The main goal of our profiler library is to help application programmers tune their parallel codes, by presenting a "thread-centered" temporal overhead information. Hence, we define temporal overhead in terms of the total amount of time in the OpenMP constructs by each of the "N" threads in a parallel region. We assume that each of the "N" threads runs on a different processor. To illustrate our definition and measurement of the three components of overhead, we use a simple triangular loop inside an OpenMP parallel region with an implicit barrier at the end, shown in Figure 1(a).

The code in Figure 1(a) leads to an execution pattern similar to the one depicted in Figure 1(b). Let us assume that N threads are created by the parallel region. We define for each thread i, where $(1 \leq i \leq N)$, δ_i^b as the runtime system overhead before the thread begins executing the work; δ_i^e as the runtime system overhead after the thread ends the execution of the work; β_i as the time spent on barrier synchronization; γ_i as the time spent executing the allocated work; and finally ϕ_i as the exit overhead,

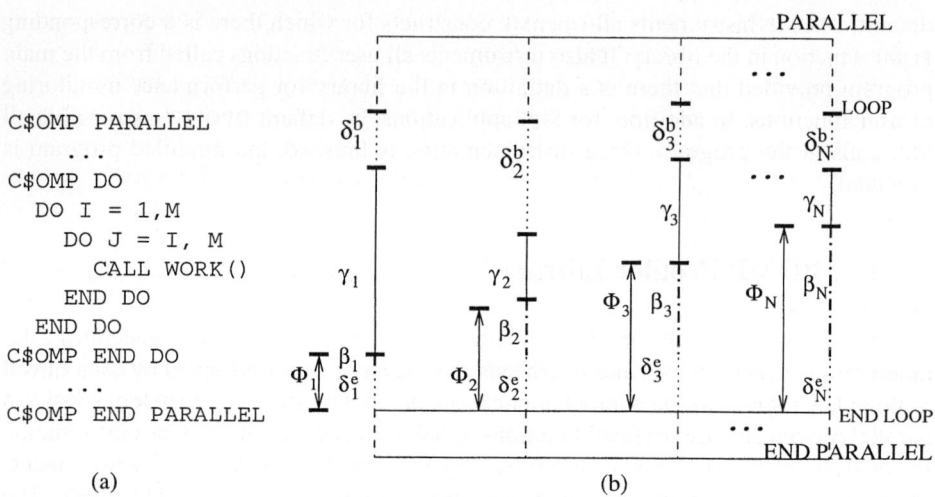

Fig. 1. (a) Pseudo code for a triangular loop inside an OpenMP parallel region, and (b) possible execution pattern

which is the sum of the barrier overhead and the exit overhead of the runtime system, i.e., $\phi_i = \beta_i + \delta_i^e$. The total execution time (τ_i) of each thread i is the sum of these components, i.e., $\tau_i = \delta_i^b + \delta_i^e + \beta_i + \gamma_i$.

Load imbalance basically occurs because of un-equal amount of work is being distributed to the threads. In [2] load imbalance is defined as the difference between the time taken by the slowest thread and the mean thread time, which essentially is a constant for all threads. As we can see from Figure 1, each thread i may have a different amount of execution time $\gamma_i \neq 0$ leading to different amounts of load imbalance. Hence we need a thread-centered definition of load imbalance.

For performance enhancement reasons (e.g., selecting an appropriate scheduling technique), in our approach to define load imbalance we focus on how much worse each thread i is performing with respect to the thread j that takes the minimum amount of time. Hence our definition of load imbalance is the percentage of extra time spent by each thread in computation normalized with respect to the time of the fastest thread. Thus, the load imbalance of each thread i, expressed in percentage, is computed as

$$l_i = \frac{(\gamma_i - \min(\gamma_i))}{\min(\gamma_i)} \times 100.$$

Barrier overhead is the amount of time spent by each thread i while waiting for synchronization with other threads, after its work share is completed. In OpenMP there are two types of barriers: explicit and implicit. Explicit barriers are often visible at the binary level; hence, they can be instrumented and measured for each thread. On the other hand, implicit barriers, which are normally used by OpenMP loop constructs, such as the one shown in our example, are executed inside of the runtime library, and are not visible in the application binary. Hence, they cannot be instrumented with DPOMP [6].

In order to estimate the implicit barrier overhead (β_i) for implicit barriers, we use the measured total exit-overhead ϕ_i, as follows:

We consider that the last thread to join the barrier (say thread j) incurs the least amount of exit overhead and has zero barrier time. Hence, since $\phi_j = \beta_j + \delta_j^e$, with $\beta_j = 0$, we can assume that the runtime exit overhead $\delta_j^e = \phi_j$. The runtime exit system overhead δ_i^e of each thread i is a constant ε for all practical purposes, because all threads executes the same runtime system code. Hence, we can compute the barrier time of each thread i as $\beta_i = \phi_i - \varepsilon$, where $\varepsilon = min(\delta_i^e)$ for $(1 \le i \le N)$.

In case of a "NO WAIT" clause at the end of the loop, the threads incur only an exit runtime overhead, and the barrier overhead is considered zero.

Runtime overhead is the amount of time taken by each thread i to execute the runtime system code. For each thread i, the runtime overhead δ_i is computed as the sum of the overheads at the beginning of the loop (δ_i^b) and at the end of the loop (δ_i^e), i.e., $\delta_i = \delta_i^b + \delta_i^e$. In case of an explicit barrier at the end of the loop, both δ_i^b and δ_i^e can be instrumented and measured. However, for implicit barriers, only δ_i^b is measured, while δ_i^e is estimated as described in the barrier overhead.

4 The KOJAK POMP Library

We implemented a POMP monitoring library which generates EPILOG event traces. EPILOG is an open-source event trace format used by the KOJAK performance analysis tool framework [13]. Besides defining OpenMP related events, it provides a thread-safe implementation of the event reading, writing, and processing routines. In addition, it supports storing hardware counter and source code information and uses a (machine, node, process, thread) tuple to describe locations. This makes it especially well suited for monitoring OpenMP or mixed MPI/OpenMP applications on today's clustered SMP architectures. EPILOG event traces can either be processed by KOJAK's automatic event trace analyzer EXPERT or be converted to the VTF3 format used by the commercial Vampir event trace visualization tool [11] (not shown here due to space limitation).

Figure 2 shows a screen-dump of the resulting display of the EXPERT automatic event trace analyzer. Using the color scale shown on the bottom, the severity of performance problems found (left pane) and their distribution over the program's call tree (middle pane) and machine locations (right pane) is displayed. The severity is expressed in percentage of execution time lost due to this problem. By expanding or collapsing nodes in each of the three trees, the analysis can be performed on different levels of granularity. We refer to [13] for a detailed description of KOJAK and EXPERT.

If a more detailed (manual) analysis is needed, EPILOG traces can be converted to VTF3 format suitable for Vampir. The conversion maps OpenMP constructs into Vampir symbols and activities, as well as OpenMP barriers into a Vampir collective operation. This allows users to investigate the dynamic behavior of an OpenMP application using a Vampir time-line diagram as well as to use Vampir's powerful filter and selection capabilities to generate all kind of execution statistics for any phase of the OpenMP application. In addition, all source code information contained in a trace is preserved

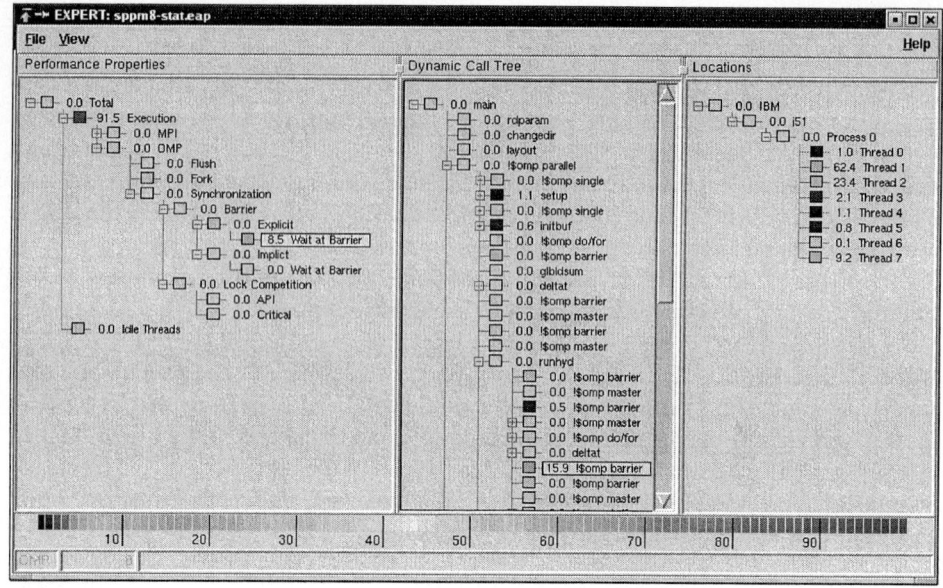

Fig. 2. Result display of EXPERT automatic trace analyzer.

during conversion; allowing the display of the corresponding source code simply by clicking on the desired activity.

5 Examples of Use

In this section we exemplify the use of both libraries with performance measurements and visualization of the ASCI SPPM benchmark code [8], which solves a 3D gas dynamics problem on a uniform cartesian mesh using a simplified version of the Piecewise parabolic method, with nearest neighbor-communication. The SPPM benchmark is a hybrid code (OpenMP and MPI), written in Fortran 77 with some C routines. Figure 3 shows the summary view for the program and the detailed view for one of the loops (loop 1125) from the profile data obtained with POMPROF, when using "static" scheduling for the OpenMP loops, running 8 threads on an IBM p690+. In the summary view, which displays the highest value for each metric, we observe very high values for "% Imbalance" on all the loops. The detailed view for the loop in line 1125 confirms this imbalance. When replacing the static scheduling by dynamic scheduling, we observe a much better behavior of the code with respect to "% Imbalance", as shown in Figure 4.

These findings are also confirmed by the automatic trace analysis of EXPERT. As shown in Figure 2, the imbalance caused by the static scheduling of OpenMP loop iterations results in a total of 8.5% of waiting time at explicit barriers. For the selected barrier, the uneven distribution of this waiting time can be seen in the right pane. For dynamic scheduling (not shown here), waiting time only amounts to 0.5%. By selecting the property "Execution" in the left pane and one of the "!$omp do/for" loops in the middle pane, the right pane of EXPERT would show the (im)balance in execution time.

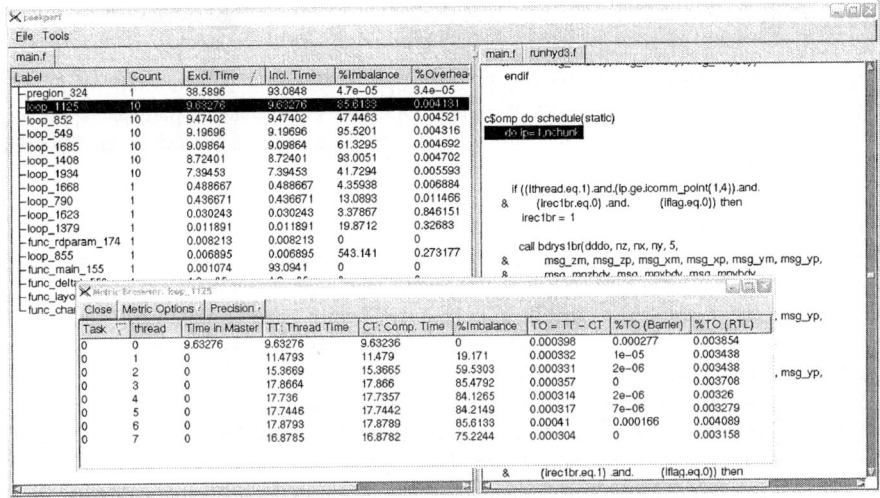

Fig. 3. Visualization of the POMPROF data from SPPM and detailed view of Loop 1125 implemented with static scheduling

Task	thread	Time in Master	TT: Thread Time	CT: Comp. Time	%Imbalance	TO = TT − CT	%TO (Barrier)	%TO (RTL)
0	0	8.86989	8.86989	8.86223	0	0.007664	7.3e-05	0.086337
0	1	0	9.02216	9.01806	1.75842	0.004096	1.1e-05	0.046166
0	2	0	9.03998	9.03598	1.96059	0.004007	1.1e-05	0.045168
0	3	0	9.03129	9.02727	1.86237	0.004019	8e-06	0.0453
0	4	0	9.03619	9.03217	1.9176	0.004021	0	0.04533
0	5	0	9.03915	9.03514	1.95119	0.004004	8e-06	0.045131
0	6	0	9.00262	8.99855	1.53825	0.004072	8.1e-05	0.045824
0	7	0	9.04206	9.03798	1.98319	0.004081	1.1e-05	0.046002

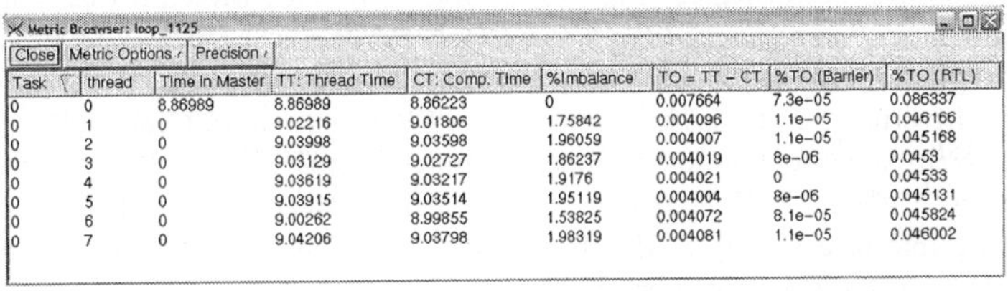

Fig. 4. Detailed view of Loop 1125, modified to use dynamic scheduling

6 Conclusion

We presented a collection of tools for the analysis of OpenMP applications based on the DPOMP instrumentation infrastructure. The use of a standard monitoring interface like POMP allows the utilization of a variety of measurement methods and provides the flexibility for development of tools ranging from profilers to tracers for performance analysis, debugging, and tuning. We presented the POMPROF profiling library and the KOJAK performance analysis framework which not only includes the EXPERT automatic trace analyzer but is also integrated with the Vampir trace visualization system. Other tools can be easily integrated or developed from scratch by implementing a POMP compliant monitoring library. In addition, we exemplified the use of these libraries with performance measurement and visualization of the ASCI sPPM benchmark code. Through the use of appropriate PMPI wrapper libraries which, if needed, record MPI point-to-point and collective communication performance data, the DPOMP infrastructure can also be used to monitor hybrid OpenMP / MPI applications. However, this requires an extra relinking step before execution. We are currently working on extending DPOMP to avoid this problem.

References

1. E. Ayguadé, M. Brorsson, H. Brunst, H.-C. Hoppe, S. Karlsson, X. Martorell, W. E. Nagel, F. Schlimbach, G. Utrera, and M. Winkler. OpenMP Performance Analysis Approach in the INTONE Project. In *Proceedings of the Third European Workshop on OpenMP - EWOMP'01*, September 2001.
2. M. K. Bane and G. D. Riley. Automatic overheads profilers for openmp codes. In *Second European Workshop on OpenMP (EWOMP 2000), Edinburgh, Scotland*, September 2000.
3. B. R. Buck and J. K. Hollingsworth. An API for Runtime Code Patching. *Journal of High Performance Computing Applications*, 14(4):317–329, Winter 2000.
4. J. M. Bull. A hierarchical classification of overheads in parallel programs. In *First IFIP TC10 International Workshop on Software Engineering for Parallel and Distributed Systems, Chapman Hall*, pages 208–219, 1996.
5. Luiz DeRose, Ted Hoover Jr., and Jeffrey K. Hollingsworth. The Dynamic Probe Class Library - An Infrastructure for Developing Instrumentation for Performance Tools. In *Proceedings of the International Parallel and Distributed Processing Symposium*, April 2001.
6. Luiz DeRose, Bernd Mohr, and Seetharami Seelam. An Implementation of the POMP Performance Monitoring Interface for OpenMP Based on Dynamic Probes. In *Proceedings of the fifth European Workshop on OpenMP - EWOMP'03*, September 2003.
7. Seon Wook Kim, Bob Kuhn, Michael Voss, Hans-Christian Hoppe, and Wolfgang Nagel. VGV: Supporting Performance Analysis of Object-Oriented Mixed MPI/OpenMP Parallel Applications. In *Proceedings of the International Parallel and Distributed Processing Symposium*, April 2002.
8. Lawrence Livermode National Laboratory. *the sPPM Benchmark Code*, 2002. http://www.llnl.gov/asci/purple/benchmarks/limited/sppm/.
9. B. Mohr, A. Mallony, H-C. Hoppe, F. Schlimbach, G. Haab, and S. Shah. A Performance Monitoring Interface for OpenMP. In *Proceedings of the fourth European Workshop on OpenMP - EWOMP'02*, September 2002.
10. Bernd Mohr, Allen Malony, Sameer Shende, and Felix Wolf. Towards a Performance Tool Interface for OpenMP: An Approach Based on Directive Rewriting. In *Proceedings of the Third European Workshop on OpenMP - EWOMP'01*, September 2001.
11. W. Nagel, A. Arnold, M. Weber, H-C. Hoppe, and K. Solchenbach. Vampir: Visualization and Analysis of MPI Resources. *Supercomputer*, 12:69–80, January 1996.
12. G. D. Riley, J. M. Bull, and J. R. Gurd. Performance improvement through overhead analysis: A case study in molecular dynamics. In *International Conference on Supercomputing*, pages 36–43, 1997.
13. Felix Wolf and Bernd Mohr. Automatic performance analysis of hybrid mpi/openmp applications. *Journal of Systems Architecture, Special Issue 'Evolutions in parallel distributed and network-based processing'*, 49(10–11):421–439, November 2003.

Efficient Pattern Search in Large Traces Through Successive Refinement*

Felix Wolf[1], Bernd Mohr[2], Jack Dongarra[1], and Shirley Moore[1]

[1] University of Tennessee, ICL
1122 Volunteer Blvd Suite 413
Knoxville, TN 37996-3450, USA
{fwolf,dongarra,shirley}@cs.utk.edu

[2] Forschungszentrum Jülich, ZAM
52425 Jülich, Germany
b.mohr@fz-juelich.de

Abstract. Event tracing is a well-accepted technique for post-mortem performance analysis of parallel applications. The EXPERT tool supports the analysis of large traces by automatically searching them for execution patterns that indicate inefficient behavior. However, the current search algorithm works with independent pattern specifications and ignores the specialization hierarchy existing between them, resulting in a long analysis time caused by repeated matching attempts as well as in replicated code. This article describes an optimized design taking advantage of specialization relationships and leading to a significant runtime improvement as well as to more compact pattern specifications.

1 Introduction

Event tracing is a well-accepted technique for post-mortem performance analysis of parallel applications. Time-stamped events, such as entering a function or sending a message, are recorded at runtime and analyzed afterward with the help of software tools. For example, graphical trace browsers, such as VAMPIR [1], allow the fine-grained investigation of parallel performance behavior using a zoomable time-line display. However, in view of the large amounts of data usually generated, automatic analysis of event traces can provide the user with the desired information more quickly by automatically transforming the data into a more compact representation on a higher level of abstraction.

The EXPERT performance tool [9] supports the performance analysis of MPI and/or OpenMP applications by automatically searching traces for execution patterns that indicate inefficient behavior. The performance problems addressed include inefficient use of the parallel programming model and low CPU and memory performance. EXPERT is implemented in Python and its architecture consists

---

* This work was supported in part by the U.S. Department of Energy under Grants DoE DE-FG02-01ER25510 and DoE DE-FC02-01ER25490 and is embedded in the European IST working group APART under Contract No. IST-2000-28077.

of two parts: a set of pattern specifications and an analysis unit that tries to match instances of the specified patterns while it reads the event trace once from the beginning to the end. Each pattern specification represents a different performance problem and consists of a Python class with methods to identify instances in the event stream. Although all pattern classes are organized in a specialization hierarchy, they are specified independently from each other resulting in replicated code and prolonged execution time whenever a pattern implementation reappears as part of another more specialized version.

This article describes an optimized design and search strategy leading to a significant speed improvement and more compact pattern specifications by taking advantage of specialization relationships. The design shares information among different patterns by looking in each step for more general patterns first and then successively propagating successful matches to more specialized patterns for refinement. We evaluate two implementations of the new design, one in Python and one in C++.

The article is outlined as follows: Section 2 describes related work. In Section 3, we outline EXPERT's overall architecture together with the current search strategy in more detail. After that, we explain the successive-refinement strategy in Section 4. Section 5 presents experimental results, followed by our conclusion in Section 6.

2 Related Work

The principle of successive refinement has also been used in the KAPPA-PI [3] post-mortem trace-analysis tool. KAPPA-PI first generates a list of idle times from the raw trace file using a simple metric. Then, based on this list, a recursive inference process continuously deduces new facts on an increasing level of abstraction.

Efficient search along a hierarchy is also a common technique in online performance tools: To increase accuracy and efficiency of its online bottleneck search, Paradyn stepwise refines its instrumentation along resource hierarchies, for example, by climbing down the call graph from callers to callees [2]. Fürlinger et al. [6] propose a strategy for online analysis based on a hierarchy of agents transforming lower-level information stepwise into higher-level information.

Ideas based on successive refinement can also be found in performance-problem specification languages, such as ASL [4] and JavaPSL [5], which is basically a Java version of ASL. Both use the concept of *metaproperties* to describe new performance problems based on existing ones.

3 Overall Architecture

EXPERT is part of the KOJAK trace-analysis environment [8, 9] which also includes tools for instrumentation and result presentation (Figure 1). Depending on the platform and the availability of tools, such as built-in profiling interfaces of compilers, the application is automatically instrumented and then executed.

During execution, the program generates a trace file in the EPILOG format to be automatically postprocessed by EXPERT. EPILOG is able to represent region entry and exit events, MPI point-to-point and collective communications, as well as OpenMP fork-join, synchronization, and work-sharing operations.

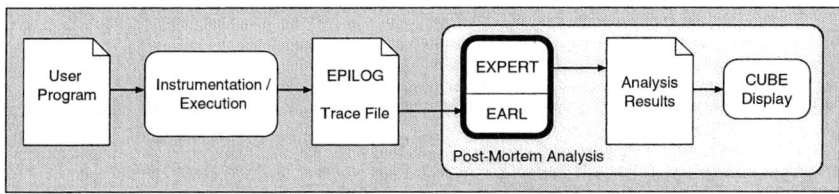

Fig. 1. The KOJAK trace-analysis environment.

To simplify pattern specification, EXPERT uses the EARL library [8] to access individual events. EARL provides a high-level interface to the event trace with random-access capabilities. Events are identified by their relative position and are delivered as a set of attribute values, such as time and location. In addition to providing random access to single events, EARL simplifies analysis by establishing links between related events, such as a link pointing from a region exit event to its corresponding enter event, and identifying event sets that describe an application's execution state at a given moment, such as the set of send events of messages currently in transit. EARL is implemented in C++ and provides both a C++ and a Python interface. It can be used independently of EXPERT for a large variety of trace-analysis tasks.

EXPERT transforms event traces into a compact representation of performance behavior, which is essentially a mapping of tuples (performance problem, call path, location) onto the time spent on a particular performance problem while the program was executing in a particular call path at a particular location. Depending on the programming model, a location can be either a process or a thread. After the analysis has been finished, the mapping is written to a file and can be viewed using CUBE [7], which provides an interactive display of the three-dimensional performance space based on three coupled tree browsers.

3.1 Current Search Strategy

There are two classes of search patterns, those that collect simple profiling information, such as communication time, and those that identify complex inefficiency situations, such as a receiver waiting for a message. The former are usually described by pairs of enter and exit events, whereas the latter are described by more complex compound events usually involving more than two events. All patterns are arranged in a hierarchy. The hierarchy is an inclusion hierarchy with respect to the inclusion of execution-time interval sets exhibiting the performance behavior specified by the pattern.

EXPERT reads the trace file once from the beginning to the end. Whenever an event of a certain type is reached, a callback method of every pattern class that has registered for this event type is invoked. The callback method itself then might access additional events by following links or retrieving state information to identify a *compound event* representing the inefficient behavior. A compound event is a set of events hold together by links and state-set boundaries and satisfying certain constraints.

3.2 Example

The example of a process waiting for a message as a result of accepting messages in the wrong order illustrates the current search strategy. The situation is depicted in Figure 2. Process A waits for a message from process B that is sent much later than the receive operation has been started. Therefore, most of the time consumed by the receive operation of process A is actually idle time that could be used better. This pattern in isolation is called *late-sender* and is enclosed in the spotted rectangle. EXPERT recognizes this pattern by waiting for a receive event to appear in the event stream and then following links computed by EARL (dashed lines in Figure 2) to the enter events of the two communication operations to determine the temporal displacement between these two events (idle time in Figure 2).

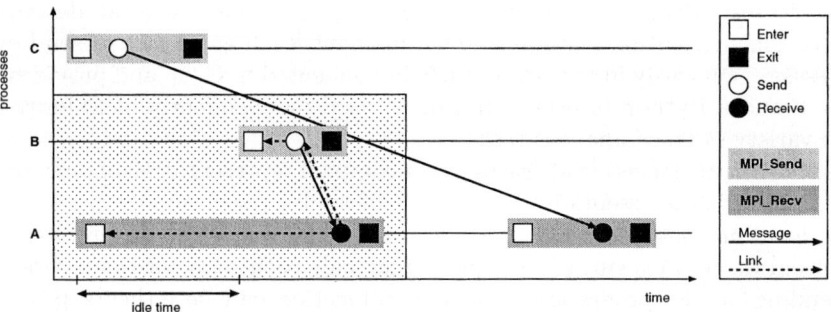

Fig. 2. Idle time as a result of receiving messages in the wrong order.

Looking at this situation in the context of the other message sent from process C to A allows the conclusion that the late-sender pattern could have been avoided or at least alleviated by reversing the acceptance order of these two messages. Because the message from C is sent earlier than that from B, it will in all probability have reached process A earlier. So instead of waiting for the message from B, A could have used the time better by accepting the message from C first. The late-sender pattern in this context is called *late-sender / wrong-order*. EXPERT recognizes this situation by examining the execution state computed by EARL at the moment when A receives the message from B. It inspects the queue of messages (i.e., their send events) sent to A and checks whether there are older messages than the one just received. In the figure, the queue would contain the event of sending the message from C to A.

This combined situation is a specialization of the simple late-sender. However, in spite of this relationship, both of them are currently computed independently. Whenever a receive event is reached, the late-sender callback follows links to see whether there is a receiver waiting unnecessarily. In addition, the callback method responsible for detecting the combined late-sender / wrong-order situation first checks the message queue for older messages and, if positive, tries to identify a late-sender situation as well – resulting in a part of the work done twice.

3.3 Profiling Patterns

A similar problem occurs with patterns collecting profiling information. Here, every pattern filters out matching pairs of enter and exit events that form region instances satisfying certain criteria, such as invocations of synchronization routines, to accumulate the instances' durations. Here also, a large fraction of the work is done more than once, when for example, one pattern accumulates communication time while a specialization of it accumulates collective-communication time.

4 Successive Refinement

The new search strategy is based on the principle of successive refinement. The basic idea is to pass a compound-event instance, once it has been detected by a more general pattern, on to a more specialized pattern, where it can be reused, refined, and/or prepared for further reuse. In the previous version, patterns have only been able to register for primitive events, that is, events as they appear in the event stream, as opposed to compound events consisting of multiple primitive events. The new design allows patterns also to publish compound events that they have detected as well as to register for compound events detected by others.

Figure 3 shows three pattern classes in EXPERT's pattern hierarchy. The hierarchical relationships between the three classes are expressed by their `parent()` methods. The `register()` methods register a callback method with the analysis unit to be invoked either upon a primitive event or a compound event. Descriptions of valid compound-event types for which a callback can be registered are stored in a developer map and can be looked up there manually. At runtime, the callback methods may return either a valid compound-event instance or `None`. If a valid instance is returned, the analysis unit invokes all callbacks subscribing for this type of compound event and a data structure holding the instance is supplied as an argument to theses callbacks.

The class `P2P` registers the method `recv()` as callback for a primitive receive event. `recv()` returns a compound event representing the entire receive operation (`recv_op`) including the receive event and the events of entering and leaving the MPI call. This corresponds to the left gray bar on the time line of process A in Figure 2. The `LateSender` class registers the method `recv_op()` for the compound event returned by `recv()` and tries there to identify the remaining

```
01  class P2P(Pattern):
02      [...]
03      def register(self, analyzer):
04          analyzer.subscribe('RECV', self.recv)
05      def recv(self, recv):
06          [...]
07          return recv_op
08
09  class LateSender(Pattern):
10      [...]
11      def parent(self):
12          return "P2P"
13      def register(self, analyzer):
14          analyzer.subscribe('RECV_OP', self.recv_op)
15      def recv_op(self, recv_op):
16          if [...]
17              return ls
18          else:
19              return None
20
21  class LateSendWrOrd(Pattern):
22      [...]
23      def parent(self):
24          return "LateSender"
25      def register(self, analyzer):
26          analyzer.subscribe('LATE_SENDER', self.late_sender)
27      def late_sender(self, ls):
28          pos      = ls['RECV']['pos']
29          dest_id  = ls['RECV']['loc_id']
30          queue    = self._trace.queue(pos, -1, dest_id)
31          if queue and queue[0] < ls['SEND']['pos']:
32              loc_id   = ls['ENTER_RECV']['loc_id']
33              cnode_id = ls['ENTER_RECV']['cnodeptr']
34              self._severity.add(cnode_id, loc_id, ls['IDLE_TIME'])
35          return None
```

Fig. 3. Late-sender / wrong-order pattern based on successive refinement.

parts of the late-sender situation, as depicted in the spotted rectangle in Figure 2. In the case of a positive result, the successfully matched instance is returned and passed on to the LateSendWrOrd class. It is supplied as the ls argument to the late_sender() method, which has previously been registered for this type of compound event. ls is a Python dictionary containing the various constituents of the compound event, such as the receive event and the enter event of the receive operation, plus calculated values, such as the idle time lost by the receiver. The method examines the queue to see whether there are older messages that could have been received before the late message. If positive, the idle time is accumulated.

The difference from the old version (not shown here) is that the results are shared among different patterns so that situations that appear again as part of others are not computed more than once. The sharing works in two ways. First, if a compound event is successfully matched, it is passed along a path in the pattern hierarchy and is refined from a common to a more specialized situation by adding new constituents and constraints. Second, since subscribers on a deeper level of the hierarchy are not invoked if the match was already unsuccessful on a higher level, negative results are shared as well, which increases the search efficiency even further by suppressing matching attempts predicted to be unsuccessful. In contrast, the old version might, for example, try to match the late-sender situation twice: the first time as part of the late-sender pattern and the second time as part of the late-sender / wrong-order pattern even if the simple late-sender pattern was already unsuccessful. In addition to being more efficient, the new design is also more compact since replicated code has been eliminated.

The profiling patterns mentioned in Section 3.3 offered a similar opportunity for optimization. Previously, every profiling pattern had to do both: accumulating time and hardware-counter values and then filtering based on call-path properties. The new design performs accumulation centrally by calculating a (call path, location) matrix for the execution time and every hardware counter recorded in the trace. After the last event has been reached, the matrices are distributed to all profiling-pattern classes where the filtering takes place. Because now the accumulation is done once for all patterns and because filtering is done only once per call path as opposed to once per call-path instance, the new version is again much more efficient.

5 Evaluation

We evaluated our new strategy using five traces from realistic applications, three from pure MPI codes and two from hybrid OpenMP/MPI codes (Table 1). For one of the applications, the SWEEP3D ASCI benchmark, we ran the original MPI version monitoring also cache-miss values and a hybrid version without hardware counter measurements. The second and the third rows of the table contain the number of CPUs used to generate the traces as well as the trace-file sizes in millions of events. Please refer to [8] for further details about the codes.

Table 1. Execution times of the old and the new implementation.

		TRACE	CX3D	SWEEP3D-HW	SWEEP3D	REMO
Type		MPI	MPI	MPI	hybrid	hybrid
No. of CPUs		16	8	16	16	16
Events	$[10^6]$	19.7	1.9	0.4	4.7	11.1
Old Python	[min]	332.0	30.9	4.8	60.3	285.1
New Python	[min]	42.0	12.5	1.7	14.6	22.0
New C++	[min]	44.0	13.9	2.3	6.8	2.5
New Python	[speedup]	7.9	2.5	2.8	4.1	13.0
New C++	[speedup]	7.6	2.2	2.1	8.9	114.0

We compared two implementations of our new strategy, one in Python and one in C++, against the old Python implementation. The test platform was an IBM AIX system with a Power4+ 1.7 GHz processor. The first three rows below the event numbers list the plain execution times in minutes, the last two rows give speedup factors in relation to the old version for a more convenient comparison.

In all examples the new strategy implemented in Python achieved a speedup of at least a factor of 2.5 and in many cases the speedup was even much higher (13.0 maximum). Although the C++ version obtained significant additional speedup for hybrid traces (e.g., 114.0 vs. 13.0 for REMO), it was surprisingly unable to deliver any additional performance in the analysis of pure MPI traces, which we hopefully can improve in the near future.

6 Conclusion

The benefit of our new design is twofold: a significant runtime improvement by avoiding repetition of detection work on the one hand and less redundant and therefore more compact pattern specifications on the other hand. In particular the latter achievement will allow us to extend the set of patterns more easily in the future. An integration of the underlying concepts into the ASL [4] specification language might help share these results with a broader community. The KOJAK software is available at http://www.fz-juelich.de/zam/kojak/.

References

1. A. Arnold, U. Detert, and W. E. Nagel. Performance Optimization of Parallel Programs: Tracing, Zooming, Understanding. In R. Winget and K. Winget, editors, *Proc. of Cray User Group Meeting*, pages 252–258, Denver, CO, March 1995.
2. H. W. Cain, B. P. Miller, and B. J. N. Wylie. A Callgraph-Based Search Strategy for Automated Performance Diagnosis. In *Proc. of the 6th International Euro-Par Conference*, volume 1999 of *Lecture Notes in Computer Science*, Munich, Germany, August/September 2000. Springer.
3. A. Espinosa. *Automatic Performance Analysis of Parallel Programs*. PhD thesis, Universitat Autonoma de Barcelona, September 2000.
4. T. Fahringer, M. Gerndt, B. Mohr, G. Riley, J. L. Träff, and F. Wolf. Knowledge Specification for Automatic Performance Analysis. Technical Report FZJ-ZAM-IB-2001-08, ESPRIT IV Working Group APART, Forschungszentrum Jülich, August 2001. Revised version.
5. T. Fahringer and C. Seragiotto Júnior. Modelling and Detecting Performance Problems for Distributed and Parallel Programs with JavaPSL. In *Proc. of the Conference on Supercomputers (SC2001)*, Denver, Colorado, November 2001.
6. K. Fürlinger and M. Gerndt. Distributed Application Monitoring for Clustered SMP Architectures. In *Proc. of the 9th International Euro-Par Conference*, Klagenfurt, Austria, August 2003.
7. F. Song and F. Wolf. CUBE User Manual. Technical Report ICL-UT-04-01, University of Tennessee, Innovative Computing Laboratory, Knoxville, TN, 2004.
8. F. Wolf. *Automatic Performance Analysis on Parallel Computers with SMP Nodes*. PhD thesis, RWTH Aachen, Forschungszentrum Jülich, February 2003. ISBN 3-00-010003-2.
9. F. Wolf and B. Mohr. Automatic performance analysis of hybrid MPI/OpenMP applications. *Journal of Systems Architecture*, 49(10-11):421–439, 2003. Special Issue "Evolutions in parallel distributed and network-based processing".

DRAC: Adaptive Control System with Hardware Performance Counters

Mauricio Pillon*, Olivier Richard, and Georges Da Costa

ID-IMAG Laboratory / Project APACHE
Grenoble, France
{Mauricio.Pillon,Olivier.Richard,Georges.Da-Costa}@imag.fr

Abstract. The memory hierarchy becomes the bottleneck for multiprocessors systems as its evolution does not keep pace with processor technology. This study intends to identify the relationship between performance slow-down and memory pressure, using hardware performance counters. Based on this relationship, we propose an adaptive control system that improves the efficiency of load balancing among the computer resources. The DRAC system, our adaptive control system, observes the access requests on the memory bus. It then adapts its user-level scheduling strategy to maximize the resource utilization. We describe the DRAC system and its mathematical model. We show experimental results that prove the DRAC system is nearly optimal with our model.

1 Introduction

The impact of memory hierarchy on the multiprocessor performance is well known [5]. As the technological development of processors is much faster than the one of memories, it has become a widely studied bottleneck. A strong relation between the slow-down on multiprocessor architecture and memory pressure [9, 3] has been identified.

Adaptive control system approach can be used to prevent apparition of bottlenecks, like Paradyn [8] and Autopilot [11] which are based on user defined sensors and actors that implement the mechanism of control. However those systems provide frameworks for developing applications, not improving the concurrent execution of a set of applications. Another system, the Polychronopoulos's project [9] proposes to monitor memory swap requests to adapt its scheduling strategy at the kernel-level. This system tries to avoid using swap.

These systems need low-level information to adapt their scheduling strategy. Most of them use hardware counters that make it possible to observe a large number of low level events like cache misses or memory bus transactions.

The first generation of tools provided raw data from hardware counters to the users. Tools like TAU [1] and HPM Toolkit [4] propose processing of higher level: statistic tools, automatic instrumentation and visualization tools. The most important problem of these counters is their heterogeneity among processors. The counter type and access modes differ from one processor family to another. Two projects, PCL [2] and PAPI [6], address this problem and propose an application programming interface (API) to

---

* Supported by scholarship 200 242/00-8 from CNPq - Brazil.

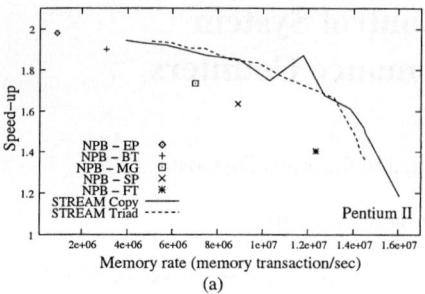 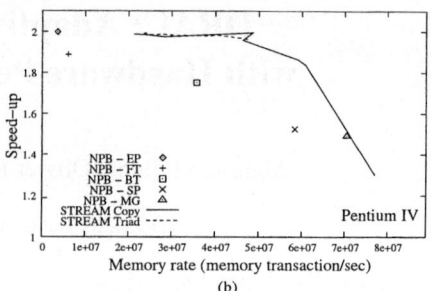

Fig. 1. Evolution of the speed-up according to the increase of memory rate (Dual Pentium II and IV Processor).

abstract the architecture of many processors. The PAPI project provides a simple interface for all counters and all events in many architectures.

Based on the hardware counters, we propose the DRAC project (*aDaptive contRol system with hArdware performance Counters*). Its aim is to maximize the efficiency of memory hierarchy utilization. The current DRAC system was developed in order to optimize memory management. It identifies memory pressure during runtime and reacts to reduce the amount of memory hierarchy saturation. It has a robust dynamic control of processes.

This paper is organized as follows. Section 2 presents the bases on which DRAC is build. In Section 3 we show the DRAC System, in Section 4 we explain the mathematical model used in our work, and in Section 5 we show an experimental evaluation of the DRAC system and its comparison with our model. In the last section we conclude and give directions for future works.

2 Links Between Application Behavior and Memory Activity

An efficient adaptive control system of memory hierarchy based on hardware counters needs to identify the relationship between speed-up and event counters, and needs applications with foreseeable behavior to provide robustness.

We have observed the evolution of the memory bandwidth utilization over the time for several SPEC2000 Benchmarks [12] (11 SPEC INT programs and 9 SPEC FP programs). We identified three different categories: *stable*, *unstable* and *multi-regions*. The first one has a steady memory utilization during the execution (70% of the test programs are following this behavior). For the second one, the memory access rate changes for each probe (around 10% of the programs). Finally, the last one has different stable areas over the time (the last 20%).

Thus, as the memory behavior of applications is foreseeable, the DRAC system can use previous information on memory behavior of applications to improve the scheduling strategy. In the following, we will introduce the relationship between speed-up and memory bus rate. This relationship might be used to detect the memory utilization efficiency. It might detect available memory bandwidth as well as memory bottlenecks.

The test platform is composed of two computers. The first one is a dual PII processor of 450 MHz, its memory bus runs at 66MHz, and it has 512 KB of memory cache. The second one is a dual PIV processor of 1.7 GHz, its memory bus runs at 400 MHz and it has 256KB of memory cache.

Figure 1 presents the speed-up according to the memory rate on P6 and PIV family computers. In our tests we used memory benchmarks like STREAM [7], and computing benchmarks like NPB-3.0 [13] (OpenMP version). We used the performance counters to measure the memory rate in memory hierarchy (memory bus transactions) with the *perfctr*'s [10] library (a PAPI [6] library component). Even if this library uses low-level hardware mechanisms and sensible event counters, the standard deviation of the measures remains small and is not represented.

Figure 1(a) (resp. Figure 1(b)) shows the tests on a Pentium II dual processor computer (resp. PIV). We used the events that count the number of memory bus transactions to measure the memory bus activity. For NAS Parallel Benchmarks [13] we have one dot in the graph for each program (EP, BT, MG, SP and FT). We used the STREAM benchmarks [7] (Triad, Copy, Add and Scale) to measure the speed-up for different memory bus rates. To change the memory bus rate of the STREAM benchmark, we modified the main loop. We added assembler instructions (*nop*) in the source code to change the memory rate evolution. We developed another program to measure the memory bus activity: *Stride*. It copies an array and accesses different memory pages in order not to use the cache.

We observe on Figure 1 that when the memory bus rate is low, the speed-up is optimal (two). When the memory bus rate is reaching the full capability of the bus, the speed-up falls to one. The NPB-3.0 EP speed-up is close to the optimal and has a low memory access rate. Other NPB Benchmark programs reach the full capacity of the bus and their speed-up fall. We do not show all STREAM benchmarks nor *Stride* results because all these programs have the same behavior than *Copy* and *Triad*. They have a gradual increase of memory bus rate utilization and the same characteristics of speed-up and memory rate.

As a shared resource, the memory can be a bottleneck. It can lead to slow-down in the execution of applications. We showed that for some areas, it is possible to link memory behavior and speed-up. For example, on Intel PII, applications with memory rate under around 5.10^6 memory transactions per second have a high speed-up, whereas applications with 12.10^6 memory transactions per second have a low one. Moreover as most applications have a steady memory utilization, the DRAC system can use this information to adapt its scheduling and optimize memory utilization.

3 DRAC: Adaptive Control System

In this section we present the first implementation of the DRAC system, currently an adaptive control system that reacts to memory bottleneck. It schedules processes to optimize memory management.

The DRAC system monitors the memory bus and detects slow-down possibility (*cf.* section 2). Then it schedules processes according to their memory consumption to obtain a better memory utilization. The goal of DRAC is to avoid the cases where requests exceed the memory bus capacity.

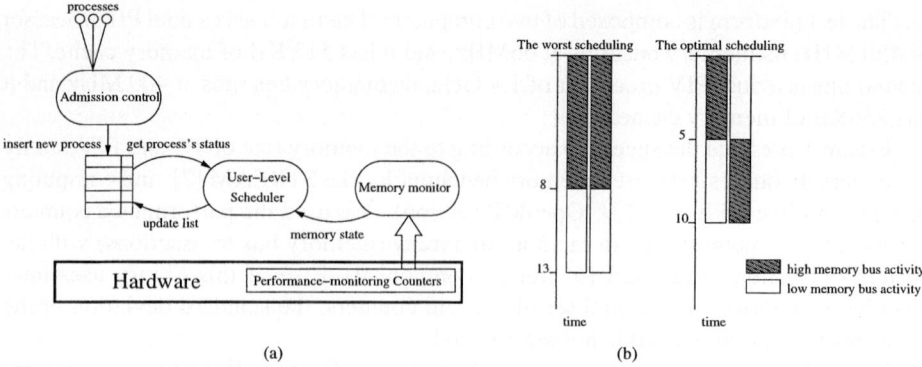

Fig. 2. (a): The DRAC System architecture. The simple scheduler uses a list of running processes and the memory bus activity to take a decision. The admission control waits for a new process. (b): An hypothetic scenario with the worst and the optimal scheduling.

We can define the DRAC system as a *dynamic*, *preventive*, *adaptive* and *non-intrusive* system as described in [9]. It is dynamic, since the DRAC base identifies the processes behavior at runtime. It is preventive, because this system reacts to situations which may lead to memory bus saturation. It is adaptive as it performs online scheduling changes. It is non-intrusive, because it does not adversely affect the program execution.

3.1 System Architecture

DRAC is composed of two different layers: the *hardware layer* and the system one. The hardware layer manages the register counters. The *system layer* takes care of the user-level scheduling. Our system observes the memory hierarchy activity through the *perfctr* [10] low-level library. We developed a new simple scheduler that bases its decisions on this information.

Figure 2 (a) shows this architecture. The user-level scheduler takes a list of processes and may change its scheduling according to the memory hierarchy activity provided by the memory monitor. The list of processes is shared between the scheduler and the admission control. This list contains all the process information. The admission control receives the requests for new processes. Then it checks if the system has free memory available and the number of running processes. If both values are under limits, a new process is executed, else it is suspended until the end of another process.

3.2 Implementation

Thanks to the low performance intrusion of the hardware counters, the DRAC system can estimate at runtime the speed-up of applications. Indeed, the memory access rate can be measured with the hardware counters and we showed in section 2 the relationship between the speed-up and this rate.

The overhead of context switching misuses resources during the runtime, so we need a strong decision system to avoid unnecessary ones. We choose to use an hysteresis control on account of its robustness with disturbed data. We take the decision according to two limits: the first one *max_limit* represents the maximum wanted utilization of the bus. When this limit is reached one considers that the bus might become saturated. The second one *min_limit* is the threshold under which resources are considered under-used. These limits are fixed manually for each processor.

When the scheduler identifies a memory bus saturation, it stops the most memory bus consuming process. It substitutes this process with another one that does less memory access. The same policy decision was done in Polychronopoulos's system [9] but in another memory hierarchy level. We proceed according to the memory bus activity, whereas the Polychronopoulos's system focuses on swap requests. The DRAC system observes the process behavior at runtime and saves the information within the list of processes. When the system needs to know how much memory bus a process uses, it checks the list of processes. If a process has never been started, there are no information on it in the list. If no process has the right characteristics, the system takes a random process.

The Figure 2 (b) shows two examples of scheduling: the optimal one, and the worst one. We consider two kinds of processes: the grey process that does many access on the memory bus (memory bus pressure) and the white one that does nearly no access (no memory bus pressure). The sequential time of the grey process is 5 units of time. If two grey processes are executed during the same time, they take a delay. So, we observe that with an optimal scheduler, the final time is 10 units of time whereas with a worst one it is 13 units.

The DRAC system follows the hypothesis of the model presented in the next section. As the DRAC system decision algorithm depends on previous memory access rate values, it needs applications with a steady memory access rate. Hopefully, most applications have such behavior (*cf.* section 2).

4 Mathematical Model

The previous sections showed that memory load influences the application execution time and our adaptive control system. But there is still to know how this time is affected by different scheduling strategies. We describe the mathematical model used to formally approach the problem. Changes in the execution environment lead to several behaviors. In this study we choose to evaluate the influences of the memory management on a finite set of jobs. This model can be used to evaluate the quality of DRAC scheduling as well as its overhead.

4.1 Slow-Down According to the Memory System Load

We will model here the impact of the memory system load on the runtime as it can be experimentally seen in the previous sections.

Hypothesis:
1. The memory model is simple: a memory bus shared by all processors.
2. Dependencies between the n running processes is limited to bus memory access (it implies that processes do not use the swap).

3. The instruction flows of the processes are the same when they run alone or with other parallel processes.
4. The memory access rate demand of the processes is constant over time[1].

Definitions: Let n be the number of jobs, running on a p processor computer. Let $d_{seq}(i)$ be the memory rate of processes i for $i \in \{1 \ldots n\}$ and d_{max} is the full capacity of the memory bus. We define the following load:

- The load of the process i: $\alpha(i) = \frac{d_{seq}(i)}{d_{max}}$, where α is the proportion of the memory bus the process i uses when running alone. If $\alpha(i) = 0$ the process does not use the memory bus, if $\alpha(i) = 1$ the process saturates the memory bus.
- Theoretical load of the computer: $\frac{\sum_{i=1}^{n} d_{seq}(i)}{d_{max}}$. The theoretical load can be greater than 1 if the sum of the memory rates is higher than the maximum capacity of the bus.
- Real load of the computer: $min(1, \frac{\sum_{i=1}^{n} d_{seq}(i)}{d_{max}})$, the theoretical load of the computer or 1 when the former exceeds 1.

The total time of the program execution is: $\boxed{T_{all} = \frac{t_{seq}}{p} * max(1, \frac{\sum_{i=1}^{n} d_{seq}(i)}{d_{max}})}$.
where the sequential time t_{seq} is the processor time spent by n processes one after another ($t_{seq} = \sum_{i=1}^{n} t_i$).

4.2 Case-Study: Dual Processor Architectures

In this section, we emphasize the dual processor case to show an application of this mathematic model. We will use three schedulings: T_{min} the Best Scheduling, T_{max} the Worst Scheduling and T_{rand} the Random Scheduling. For those three cases, we will evaluate the parallel execution times T_{all}.

The scenario is composed of two types of processes P_0 and P_1. P_0 does not access memory ($\alpha(P_0) = 0$), whereas P_1 does ($\alpha(P_1) \in [0, 1]$). The speed-up of a process is t_{seq}/t_{all} when running two processes P. The speed-up of P_0 is two. The speed-up of P_1 is δ ($\delta = \frac{2}{max(1, 2\alpha)}$). β is the ratio of processes P_1, so $\beta = \frac{n_{P_1}}{N}$. N is the total number of processes: $N = n_{P_0} + n_{P_1}$. The parallel execution time of P_i and P_j is $t_p(P_i, P_j)$ where $i \in \{0 \ldots 1\}$ and $j \in \{0 \ldots 1\}$. We suppose that when P_0 and P_1 are running alone, their execution times are the same.

The following formula is the total execution time of the two set of processes. As P_0 does not use the memory bus, running P_0 in parallel with P_1 does not slow down the execution. Running twice P_1 may slow down the execution with a factor of $2/\delta$ by definition of δ. $T_{P_i P_j}$ is the total execution time of all sequences where P_i is in parallel with P_j during the runtime and $t_{th}(P_i P_j)$ is the execution time with no memory congestion.

$$T_{all} = t_{th}(P_0 P_0) + t_{th}(P_1 P_0) + t_{th}(P_0 P_1) + \frac{2}{\delta} t_{th}(P_1 P_1)$$

For *the Random Scheduling* we will calculate the time expectancy of a parallel execution. To calculate this expectancy we need the probability of each different execution:

---
[1] If this rate varies too much, it becomes impossible to predict the behavior of the process.

$Prob(P_0P_0) = (1-\beta)^2$, $Prob(P_1P_0) = Prob(P_0P_1) = \beta(1-\beta)$, $Prob(P_1P_1) = \beta^2$.
The expected time of a random execution is: $T_{rand} = \frac{N}{2}(1 + \beta^2(\frac{2}{\delta} - 1))$.

The Best Scheduling is obtained by minimizing the time when P_1 is running in parallel with itself:

- If $0\% \leq \beta \leq 50\%$, so $n_{P_0} \geq n_{P_1}$: $T_{min} = \frac{N}{2}$
- If $50\% \leq \beta \leq 100\%$, so $n_{P_0} \leq n_{P_1}$: $T_{min} = N(1 - \beta + \frac{1}{\delta} \cdot (2 \cdot \beta - 1))$.

The Worst Scheduling is obtained by maximizing the time when P_1 is running in parallel with itself: $T_{max} = \frac{N}{2}(1 + \beta(\frac{2}{\delta} - 1))$

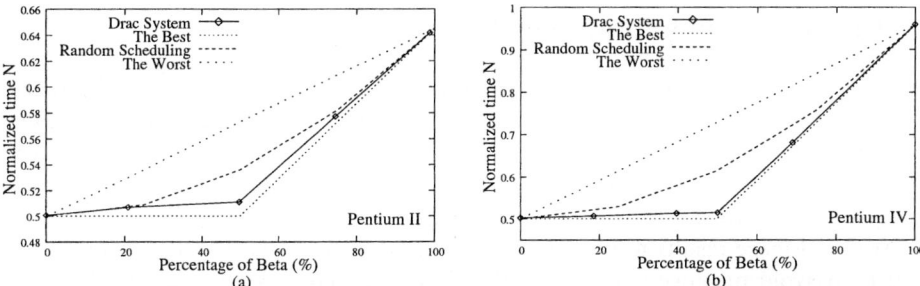

Fig. 3. Normalized execution time of four different scheduling strategies according to the percentage of processes with high memory bus activity on the dual Pentium II and IV processor.

The relationship (*cf.* Section 2) between the memory load and the slow-down has been compared with the experimentations. We used this as the base of the model. An optimal memory management can be achieved. It can be proved that the maximum time variation according to the scheduling for dual processors is when $\beta = 50\%$. In this case the best scheduling is 50% faster than the worst one. These model can be generalised to *p-processor*, and keeps the same properties with $p > 2$. As the model provides the best and random scheduling, it will help to validate the DRAC scheduling.

5 Experimental Results

In this section, we present the experimental validation of our model hypothesis and some experimental evaluations of the DRAC system. For all these experiments, we used the GNU/Linux operating system with kernel 2.4.18 and the GNU compiler for C language (gcc-3.2 with -O3 optimization). These results are obtained with the DRAC system (*cf.* section 3). We used a set of jobs with two kinds of processes, one with high memory bus activity (SPEC FP ART) and another one with low memory bus activity (SPEC FP MESA). We performed tests on two different dual processors, one PII, and one PIV (*cf.* section 2). Our experiments used four different schedulers that restrict the total number of running processes to the number of processors: *the Worst Scheduling*, *the Best Scheduling*, *the Random Scheduling* which are obtained with the mathematical model described in Section 4, and the *DRAC System*.

The experimental results on Pentium II (Figure 3(a)) show that the DRAC system is 2% slower than *the Best Scheduling* and its worst result over *the Random Scheduling* is when $\beta < 25\%$ where DRAC is as fast as it. In the case where $\beta = 50\%$, our system runs 5% faster than *the Random Scheduling* and 12% faster than *the Worst Scheduling*.

The results on Pentium IV (Figure 3(b)) show that the DRAC system achieves nearly the same time as *the Best Scheduling* strategy (less than 2% slower). When β is 50%, the DRAC system is 20% faster than *the Random Scheduling* and 41% faster than *the Worst Scheduling*.

These results show that the DRAC system provides an efficient scheduling according to the memory bus activity of processes, without a large overhead (less than 2%).

6 Conclusion and Future Works

We presented a study of the impact of the memory hierarchy utilization on multiprocessor system. We verified that the memory bus bottleneck slows-down processes. Then we identified the relationship between the memory access rate and this slow-down. Moreover, with the model, we proved that it is possible to obtain a speed-up of 50% compared to the worst scheduling in some cases.

We used these results to propose DRAC, an adaptive control system whose main goal is to avoid memory bottleneck. DRAC is supposed to work with steady memory utilization processes on account of its scheduling strategy. Some NAS, STREAM and SPEC2000 benchmarks are following our hypotheses, we showed that DRAC can achieve a significant improvement on real applications. The DRAC system has been tested with several kind of processes. Its performance has been compared with the model and it has been verified that its scheduling was nearly optimal (less than 2% slower) for dual processor. The same tests have been done with four processor and the results seem at least as good.

Future works go in the direction of testing the DRAC system with other real applications. As soon as possible we will test the memory behavior of the Intel Itanium II multiprocessor. We have already started to study the hardware performance counters in this architecture. Another part of this future work is to broaden this study to the case of shared memory systems following the OpenMP specifications, and to take into account workloads induced by other resources like the network or the disks on the memory hierarchy.

References

1. R. Bell, A. D. Malony, and S. Shende. Paraprof: A portable, extensible, and scalable tool for parallel performance profile analysis. volume LNCS 2790, pages 17–26. Euro-Par 2003, Springer, Berlin, 2003.
2. R. Berrendorf and H. Ziegler. Pcl – the performance counter library: A common interface to access hardware performance counters on microprocessors. In *Internal Report FZJ-ZAM-IB-9816, Forschungszentrum Julich.*, 1998.
3. F. Cappello, O. Richard, and D. Etiemble. Investigating the performance of two programming models for clusters of smp pcs. In *Proc. of the 6th Int. Symposium on High Performance Computer Architecture Conference, Toulouse, France*, pages 349–359, 2000.

4. L. DeRose. The hardware performance monitor toolkit. In *Procedings of Euro-Par, Manchester, United Kingdom.*, pages 122–131, August 2001.
5. R. Jin and G. Agrawal. Performance prediction for random write reductions: a case study in modeling shared memory programs. In *Proc. of the 2002 ACM SIGMETRICS int. conf. on Measurement and modeling of computer systems, New York, USA*, pages 117–128, 2002.
6. K. London, J. Dongarra, S. Moore, P. Mucci, K. Seymour, and T. Spencer. End-user tools for application performance analysis, using hardware counters. International Conference on Parallel and Distributed Computing Systems, August 2001.
7. J. D. McCalpin. A survey of memory bandwidth and machine balance in current high performance computers. In *IEEE Computer Society Technical Committee on Computer Architecture (TCCA) Newsletter.*, December 1995.
8. B. P. Miller and all. The paradyn parallel performance measurement tool. In *IEEE Computer.*, volume 28, pages 37–46, 1995.
9. D. S. Nikolopoulos and C. D. Polychronopoulos. Adaptive scheduling under memory pressure on multiprogrammed clusters. In *Proc. of the Second IEEE/ACM Int. Symp. on Cluster Computing and the Grid (CCGrid 2002), Berlin, Germany, (Best Paper Award)*, May 2002.
10. M. Pattersson. Linux x86 performance-monitoring counters driver. http://user.it.uu.se/~mikpe/linux/perfctr/.
11. R. L. Ribler, H. Simitci, and D. A. Reed. The autopilot performance-directed adaptive control system. In *Future Generation Computer Systems.*, volume 18, pages 175–187, 2001.
12. SPEC2000. Standard performance evaluation corporation. spec cpu2000 benchmarks. December 1999. http://www.specbench.org/osg/cpu2000.
13. R. F. Van-Der-Wijngaart and P. Wong. Nas parallel benchmarks i/o version 2.4. In *Technical Report NAS-03-002, NASA Ames Research Center*, 2003.

Reproducible Network Benchmarks with coNCePTuaL

Scott Pakin

Los Alamos National Laboratory, Los Alamos, NM 87545, USA,
pakin@lanl.gov
http://www.c3.lanl.gov/~pakin

Abstract. A cornerstone of scientific progress is the ability to reproduce experimental results. However, in the context of network benchmarking, system complexity impedes a researcher's attempts to record all of the information needed to exactly reconstruct a network-benchmarking experiment. Without this information, results may be misinterpreted and are unlikely to be reproducible.

This paper presents a tool called coNCePTuaL which simplifies most aspects of recording and presenting network performance data. coNCePTuaL includes two core components: (1) a compiler for a high-level, domain-specific programming language that makes it possible to specify arbitrary communication patterns tersely but precisely and (2) a complementary run-time library that obviates the need for writing (and debugging!) all of the mundane but necessary routines needed for benchmarking, such as those that calibrate timers, compute statistics, or output log files. The result is that coNCePTuaL makes it easy to present network-performance data in a form that promotes reproducibility.

1 Introduction

Network and messaging-layer performance measurements are used for a variety of purposes, such as explaining or predicting system and application performance, procuring large-scale systems, and monitoring improvements made during system deployment or messaging-layer development. Unfortunately, following a truly scientific approach to measuring network performance is not easy. In the absence of clear but precise experimental descriptions, the consumers of network performance data may draw incorrect conclusions, leading to dire consequences.

Consider a standard ping-pong latency test, which reports the time needed to send a message of a given size from one node to another by calculating half of the measured round-trip time. Fig. 1(a) shows the result of running a latency test atop two messaging layers and networks: p4 [1] (based on TCP) over Gigabit Ethernet [2, Sect. 3] and Tports over Quadrics Elan 3 [3]. The latency test is implemented using MPI [4] and both the p4 and Tports layers are integrated as MPICH channel devices [5]. The latency program was compiled with GCC 2.96 using the -O3 and -g flags. All experiments were performed across the same – otherwise idle – pair of nodes, each containing two 1 GHz Itanium 2 processors

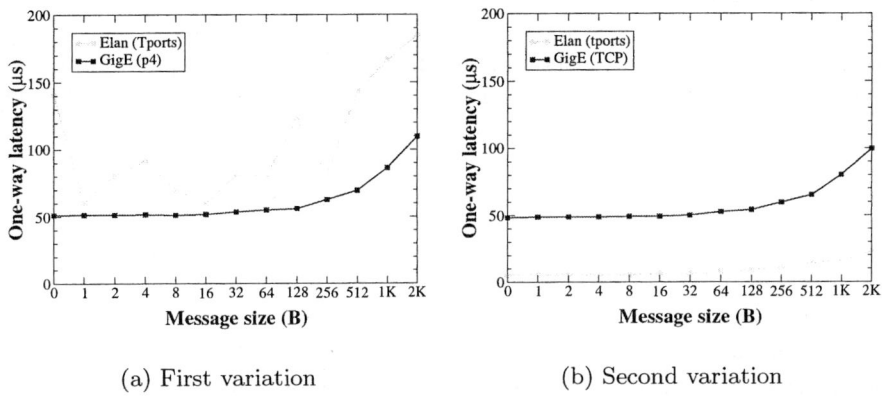

(a) First variation (b) Second variation

Fig. 1. Two variations of a latency test.

(one unused). Each data point represents the arithmetic mean of 11 executions of the latency test. As Fig. 1(a) shows, the p4/GigE version of the latency test exhibits lower latency than the Tports/Elan 3 version on all message sizes from 0 bytes to 2 kilobytes. Furthermore, the p4/GigE latency increases smoothly while the Tports/Elan 3 latency varies erratically.

Figure 1(b) also shows the result of running an MPI-based latency test. This test was run atop the same two messaging layers and networks as the previous test. The same compiler was used and the experiments were performed across the same nodes of the same cluster. As before, each data point represents the arithmetic mean of 11 sequential executions of the latency test and, as before, nothing else was running on the system. However, Fig. 1(b) delivers the opposite message from that delivered by Fig. 1(a): Fig. 1(b) shows that Tports/Elan 3 is significantly *faster* than p4/GigE. Also, the Tports/Elan 3 curve shown in Fig. 1(b) is smooth, unlike the erratic curve presented in Fig. 1(a).

Naturally, something is different between the experiment/experimental setup used in Fig. 1(a) from that used in Fig. 1(b) – but what? Although we defer the answer to Sect. 2 the point is that even with all of the experimental setup described above, the performance results are untrustworthy; some critical piece of information is missing. This simple exercise demonstrates the problem with the status quo of network benchmarking: performance data that lacks a complete and precise specification is subject to misinterpretation. This paper proposes a solution in the form of a programming environment called coNCePTuaL which was designed specifically to simplify the implementation of reproducible network-performance tests. The rest of this paper is structured as follows. Sect. 2 motivates and describes coNCePTuaL and showcases some sample output. Sect. 3 places coNCePTuaL in context, discussing the types of tasks for which coNCePTuaL is best suited. Finally, Sect. 4 draws some conclusions about the implications of using coNCePTuaL for network performance testing.

2 coNCePTuaL

Fig. 1 on the preceding page shows how subtle differences in experimental setup can lead to radically different performance results. coNCePTuaL (= "Network Correctness and Performance Testing Language") is a programming environment designed to help eliminate the ambiguities that can limit the usefulness of performance results. It centers around a high-level, domain-specific language created for the express purpose of writing network benchmarks. The design decision to introduce a new language instead of merely creating a performance library stemmed from the desire to make coNCePTuaL programs more readable than a jumble of function calls and control structures. Although a library can certainly encapsulate all of the functionality needed for the scientific acquisition and reporting of data and a textual or pseudocode description of a benchmark can convey the basic idea, coNCePTuaL combines the best features of both approaches:

1. Like pseudocode or prose but unlike the combination of a general-purpose programming language and a library, coNCePTuaL programs are English-like and can largely be read and understood even by someone unfamiliar with the language.
2. Like the combination of a general-purpose programming language and a library but unlike pseudocode or prose, coNCePTuaL programs precisely describe all aspects of a benchmark, most importantly the implementation subtleties that may be omitted from a description yet have a strong impact on performance (as demonstrated by the previous section's description of Fig. 1).

The coNCePTuaL compiler, written in Python with the SPARK compiler framework [6], sports a modular design that enables generated code to target any number of lower-level languages and messaging layers[1]. Hence, the same high-level coNCePTuaL program can be used, for example, to compare the performance of multiple messaging layers, even semantically disparate ones such as MPI and OpenMP. The generated code links with a run-time library that takes care of most of the mundane aspects of proper benchmarking, such as calculating statistics, calibrating timers, parsing command-line options, and logging a wealth of information about the experimental setup to a file.

The coNCePTuaL language provides too much functionality to describe in a short paper such as this; the reader is referred to the coNCePTuaL user's manual [7] (available online) for coverage of the language's syntax and semantics. In lieu of a thorough description of the language, we now present a few trivial code samples with some accompanying explanation to convey a basic feeling for coNCePTuaL. Listing 1 presents the complete coNCePTuaL source code which produced the data for Fig. 1(a) and Listing 2 presents the complete coNCePTuaL source code that produced the data for Fig. 1(b). One thing

---
[1] Currently, the only backends implemented are C + MPI and C + Unix-domain datagram sockets; more backends are under development.

Listing 1. Source code that produced Fig. 1(a)

```
1  msgsize is "Message size (bytes)" and comes from "--bytes" or "-b"
2  with default 0.
3
4  All tasks synchronize then
5  task 0 resets its counters then
6  task 0 sends a msgsize byte message to task 1 then
7  task 1 sends a msgsize byte message to task 0 then
8  task 0 logs elapsed_usecs/2 as "1/2 RTT".
```

Listing 2. Source code that produced Fig. 1(b)

```
1   msgsize is "Message size (bytes)" and comes from "--bytes" or "-b"
2   with default 0.
3
4   All tasks synchronize then
5   for 100 repetitions {
6     task 0 resets its counters then
7     task 0 sends a msgsize byte message to task 1 then
8     task 1 sends a msgsize byte message to task 0 then
9     task 0 logs the median of elapsed_usecs/2 as "1/2 RTT"
10  }
```

to note is that the language is very English-like; coNCePTuaL programs are intended to read much like a human would describe a benchmark to another human. Although one can probably grasp the gist of Listings 1–2 without further explanation, the following details may be slightly unintuitive:

- The first statement in Listings 1 and 2 parses the command line, assigning the argument of --bytes (or simply -b) to a variable called msgsize.
- To simplify the common case, **sends** is synchronous and implies a matching synchronous **receives**. Either or both operations can be asynchronous. coN-CePTuaL also supports data verification and arbitrary buffer alignment.
- The coNCePTuaL run-time library automatically maintains the elapsed_usecs ("elapsed time in microseconds") counter and many other counters, as well.

If not yet obvious, the explanation of the performance discrepancy shown in Fig. 1 is that the code shown in Listing 1 measures only a single ping-pong while the code shown in Listing 2 reports the median of 100 ping-pong iterations. Unlike p4, which uses TCP and therefore goes through the operating system for every message, Tports is a user-level messaging layer that communicates directly with the Elan. However, the Elan, which can transfer data from arbitrary addresses in an application's virtual-memory space, must pin (i.e., prevent the paging of) message buffers before beginning a DMA operation. Because pinning requires both operating-system intervention and a sequence of costly I/O-bus crossings, a large startup overhead is incurred the first time a message buffer is utilized. The code shown in Listing 1 does not amortize that overhead while the code shown in Listing 2 does. Neither latency test is unreasonable; although codes like Listing 2 are more common in the literature, codes like Listing 1 are used when they more accurately represent an application's usage pattern. For example, two of the three execution modes of Worley's COMMTEST benchmark, which is used to help tune climate and shallow-water modeling applications [8], specify that each message be sent only once.

Although Listings 1–2 state simply "**task** 0 **logs** ⟨*something*⟩", the CONCEPTUAL run-time library takes this as a cue to write a set of highly detailed log files, one per task. Each log file contains information about the execution environment, a list of all environment variables and their values, the complete program source code, the program-specific measurement data, and a trailer describing the resources used during the program's execution.

A sample log file is shown below. This particular file – selected arbitrarily from those used in the preparation of this paper – corresponds to one of the values averaged to make the upper-right data point of Fig. 1(a), although the file represents a later run than was used in the figure.

```
 1  #########################################################################
 2  # ====================
 3  # coNCePTuaL log file
 4  # ====================
 5  # coNCePTuaL version: 0.5.1
 6  # coNCePTuaL backend: c_mpi (C + MPI)
 7  # Executable name: /home/pakin/src/coNCePTuaL/latency1-elan
 8  # Working directory: /home/pakin/src/coNCePTuaL
 9  # Command line: ./latency1-elan --bytes=2048 --logfile=latency1-elan-2048-run4-%d.log
10  # Number of tasks: 2
11  # Processor (0-1): 0
12  # Host name: a11
13  # Operating system: Linux 2.4.21-3.5qsnet #2 SMP Thu Aug 7 10:51:04 MDT 2003
14  # CPU vendor: GenuineIntel
15  # CPU architecture: ia64
16  # CPU model: 1
17  # CPU count: 2
18  # CPU frequency: 1300000000 Hz (1.3 GHz)
19  # Cycle-counter frequency: 1300000000 Hz (1.3 GHz)
20  # OS page size: 16384 bytes
21  # Physical memory: 2047901696 bytes (1.9 GB)
22  # Library compiler+linker: /usr/bin/gcc
23  # Library compiler options: -Wall -W -g -O3
24  # Library linker options: -lpapi -lm -lpopt
25  # Library compiler mode: LP64
26  # Dynamic libraries used: /usr/local/lib/libpapi.so /lib/libm-2.2.4.so /usr/lib/libpopt.so.0.0.0 /usr/
27  # Average microsecond timer overhead [inline assembly code]: <1 microsecond
28  # Microsecond timer increment: 1.00887 +/- 0.256529 microseconds (ideal: 1 +/- 0)
29  # Process CPU-time increment [getrusage()]: 976.57 +/- 0.49757 microseconds (ideal: 1 +/- 0)
30  # WARNING: Process timer exhibits poor granularity (not a serious problem).
31  # Log file template: latency1-elan-2048-run4-%d.log
32  # Message size (bytes): 2048
33  # Number of minutes after which to kill the job (-1=never): -1
34  # List of signals that should not be trapped: 14
35  # Compilation command line: /usr/lib/mpi/mpi_gnu/bin/mpicc -I/tmp/ncptl/include -I/usr/local/include
36  # Log creator: Scott Pakin
37  # Log creation time: Thu May 27 18:45:43 2004
38  #
39  # Environment variables
40  # ---------------------
41  # HOME: /home/pakin
42  # PATH: /home/pakin/bin:/usr/local/bin:/usr/bin:/usr/sbin:/bin:/sbin:.
43  # PWD: /home/pakin/src/coNCePTuaL
44  # RMS_JOBID: 16588
45  # RMS_MACHINE: a
46  # RMS_NNODES: 2
47  # RMS_NODEID: 0
48  # RMS_NPROCS: 2
49  # RMS_PROCID: 0
50  # RMS_RANK: 0
51  # RMS_RESOURCEID: parallel.17079
52  # RMS_STOPONELANINIT: 0
53  # SHELL: /bin/tcsh
```

```
54   # USER: pakin
55   #
56   # coNCePTuaL source code
57   # ---------------------
58   #     msgsize is "Message size (bytes)" and comes from "--bytes" or "-b"
59   #     with default 0.
60   #
61   #     All tasks synchronize then
62   #     task 0 resets its counters then
63   #     task 0 sends a msgsize byte message to task 1 then
64   #     task 1 sends a msgsize byte message to task 0 then
65   #     task 0 logs elapsed_usecs/2 as "1/2 RTT".
66   #
67   ##########################################################################
68   "1/2 RTT"
69   "(all data)"
70   207
71   ##########################################################################
72   # Program exited normally.
73   # Log completion time: Thu May 27 18:45:43 2004
74   # Elapsed time: 0 seconds
75   # Process CPU usage (user+system): 0 seconds
76   # Task IDs assigned to processor 0: 0
77   # Processors assigned to task ID 0: 0
78   ##########################################################################
```

The key point is that a CONCEPTUAL log file contains not just performance measurements but also a detailed description of how they were produced, which helps third parties understand the performance results. Such complete log files are also of use to the people who generated them: How often does a researcher struggle to use old data in a new paper or presentation, not remembering if results.dat was produced with the environment variable MPI_BUFFER_MAX set to 2048 or 1048576; whether the experiment was run with the older 2.8 GHz processors or the newer 3.2 GHz processors; or, even if the test transmitted messages synchronously or asynchronously? Furthermore, aspects of the execution environment that cannot be determined automatically (e.g., characteristics of the network fabric) can be inserted manually into a log file with a command-line option to the benchmark program. In short, with CONCEPTUAL, log files present a complete picture of an experiment, making them far more valuable than measurement data alone.

3 Discussion

The programs presented in Sect. 2 are simple to express in CONCEPTUAL but – apart from the creation of such content-rich log files – would be almost as simple to express in any other language. In general, CONCEPTUAL's usefulness increases with the complexity of the communication pattern being tested. For example, the 4×4 synchronous-pipe pattern described in a MITRE report [9] requires 248 lines of LSE, a terse but low-level language for describing communication benchmarks. Because CONCEPTUAL is a high-level language, the same code (in fact, a more general $M \times N$ synchronous pipe) can be expressed in only 26 lines of CONCEPTUAL – far less than the LSE version and not significantly more than what would be needed for a textual description of the communication pattern. It is not merely short code lengths that make CONCEPTUAL useful; the increased comprehensibility of a CONCEPTUAL program over the equivalent program written in a lower-level language and the increased precision of a CONCEPTUAL program over a prose description make CONCEPTUAL a useful tool for any sort of network performance testing. The language even supports a hybrid coding style in which lower-level language code can be inlined and executed from a coNCePTuaL program, thereby ensuring that no functionality is lost by programming in coNCePTuaL instead of a lower-level language.

coNCePTuaL is not intended to be a replacement for existing communication-benchmark suites such as the Pallas MPI Benchmarks [10] or SKaMPI [11]. Rather, its real strengths lie in its ability to rapidly produce customized tests of network and messaging-layer performance:

- A coNCePTuaL mock-up of an application may make "what if" scenarios more easy to evaluate than would rewriting the original application. For instance, a user can evaluate how altering the communication pattern (caused, for example, by a different data decomposition) should affect overall application performance. A coNCePTuaL mock-up of Sweep3D [12] is currently under development.
- System and application performance problems can be diagnosed by generating a simple but representative communication benchmark and successively refining it to hone in on the source of the problem. (This methodology was recently used to nearly double the performance of an application running on ASCI Q [13].)
- Network-performance tests unique to a particular domain or otherwise unfamiliar to a target audience can be presented in a precise, easily understood manner.

The coNCePTuaL source code will soon be available from http://www.c3.lanl.gov/~pakin/software/. Making the software open-source enables researchers to scrutinize the code so that coNCePTuaL can be used as a trustworthy replacement for C as a network-benchmarking language.

4 Conclusions

In the domain of network benchmarks, recorded performance data cannot blindly be trusted. As demonstrated in Sect. 1, subtle variations in experimental setup – even for a benchmark as trivial as a latency test – can lead to grossly varying performance curves, even leading to different conclusions being drawn about relative performance. The problem is that the complexity of current computer systems makes it difficult (not to mention tedious) to store a sufficiently thorough depiction of an experiment that was run and the experimental conditions under which it ran. As a consequence, performance tests can rarely be reproduced or validated in a scientific manner. Even unpublished performance data used locally suffers from lack of reproducibility; a researcher may unearth old measurements but have no record of what benchmark produced them or what parameters were utilized in the process.

This paper proposes the coNCePTuaL programming environment as a solution to the problem of irreproducible network performance results. coNCePTuaL tries to codify the best practices in network and messaging-layer performance testing into a high-level domain-specific language and accompanying run-time library. coNCePTuaL was designed specifically to support and facilitate all aspects of network and messaging-layer performance testing, from expressing complex communication patterns tersely yet unambiguously through storing in self-contained log files everything needed to reproduce an experiment. Using coNCePTuaL, a researcher can easily present in a paper or report a benchmark's actual source code – not pseudocode, which may inadvertently omit critical details. Although it will always be possible to misrepresent network performance, coNCePTuaL makes it much easier to be meticulous.

References

1. Butler, R.M., Lusk, E.L.: Monitors, messages, and clusters: The p4 parallel programming system. Parallel Computing **20** (1994) 547–564
2. LAN/MAN Standards Committee: Carrier sense multiple access with collision detection (CSMA/CD) access method and physical layer specifications. IEEE Standard 802.3, IEEE Computer Society, Technical Committee on Computer Communications, New York, New York (2002)
3. Petrini, F., Feng, W., Hoisie, A., Coll, S., Frachtenberg, E.: The Quadrics network: High-performance clustering technology. IEEE Micro **22** (2002) 46–57
4. Message Passing Interface Forum: MPI: A Message-Passing Interface Standard. (1995)
5. Gropp, W., Lusk, E., Doss, N., Skjellum, A.: A high-performance, portable implementation of the MPI message passing interface standard. Parallel Computing **22** (1996) 789–828
6. Aycock, J.: Compiling little languages in Python. In: Proceedings of the Seventh International Python Conference, Houston, Texas (1998) 69–77
7. Pakin, S.: coNCePTuaL user's guide. Los Alamos Unclassified Report 03-7356, Los Alamos National Laboratory, Los Alamos, New Mexico (2003) Available from http://www.c3.lanl.gov/~pakin/papers/conceptual.pdf.
8. Drake, J.B., Hammond, S., James, R., Worley, P.H.: Performance tuning and evaluation of a parallel community climate model. In: Proceedings of SC'99, Portland, Oregon (1999)
9. Monk, L., Games, R., Ramsdell, J., Kanevsky, A., Brown, C., Lee, P.: Real-time communications scheduling: Final report. Technical Report MTR 97B0000069, The MITRE Corporation, Bedford, Massachusetts (1997)
10. Pallas, GmbH: Pallas MPI Benchmarks—PMB, Part MPI-1. (2000)
11. Reussner, R., Sanders, P., Prechelt, L., Müller, M.: SKaMPI: A detailed, accurate MPI benchmark. In Alexandrov, V., Dongarra, J., eds.: Recent Advances in Parallel Virtual Machine and Message Passing Interface: Proceedings of the 5th European PVM/MPI Users' Group Meeting (EuroPVM/MPI'98). Volume 1497 of Lecture Notes in Computer Science., Liverpool, United Kingdom, Springer-Verlag (1998) 52–62
12. Hoisie, A., Lubeck, O., Wasserman, H.: Performance and scalability analysis of teraflop-scale parallel architectures using multidimensional wavefront applications. The International Journal of High Performance Computing Applications **14** (2000)
13. Petrini, F., Kerbyson, D.J., Pakin, S.: The case of the missing supercomputer performance: Achieving optimal performance on the 8,192 processors of ASCI Q. In: Proceedings of SC2003, Phoenix, Arizona (2003)

Optimizing Cache Access:
A Tool for Source-to-Source Transformations and Real-Life Compiler Tests

Ralph Müller-Pfefferkorn, Wolfgang E. Nagel, and Bernd Trenkler

Center for High Performance Computing (ZHR)
Dresden University of Technology
D-01062 Dresden, Germany
{mueller-pfefferkorn,nagel,trenkler}@zhr.tu-dresden.de

Abstract. Loop transformations are well known to be a very useful tool for performance improvements by optimizing cache access. Nevertheless, the automatic application is a complex and challenging task especially for parallel codes. Since the end of the 1980's it has been promised by most compiler vendors that these features will be implemented - in the next release. We tested current FORTRAN90 compilers (on IBM, Intel and SGI hardware) for their capabilities in this field. This paper shows the results of our analysis. Motivated by this experience we have developed the optimization environment *Goofi* to assist programmers in applying loop transformations to their code thus gaining better performance for parallel codes even today.

1 Introduction

Usually, a developer focuses on implementing a correct program which solves the problem underneath. Applications which do not take into account the cache hierarchy of modern microprocessors, most times achieve only a small fraction of the theoretical peak performance. Tuning a program for better cache utilization has become an expensive and time consuming part of the development cycle.

The EP-CACHE project[1] [1, 2] is developing new methods and tools to improve the analysis and the optimization of programs for cache architectures. The work presented here is part of this research activity and focuses on the optimization of the source code.

One way to optimize the cache usage of applications are source-to-source transformations of loops. There are a number of transformations known that improve data locality and therefore the reuse of the data in the cache, like loop interchange, blocking, unrolling etc. (see Fig. 1).

Modern compilers claim to use loop transformations in code optimization. We have tested three FORTRAN90 compilers (Intel ifc 7.1 [3], SGI MIPSpro 7.3 [4], and IBM xlf for AIX V8.1.1 [5]) for loop transformations. On one hand, an

---

[1] Funded by the German Federal Ministry of Education and Research (BMBF) under contract 01IRB04.

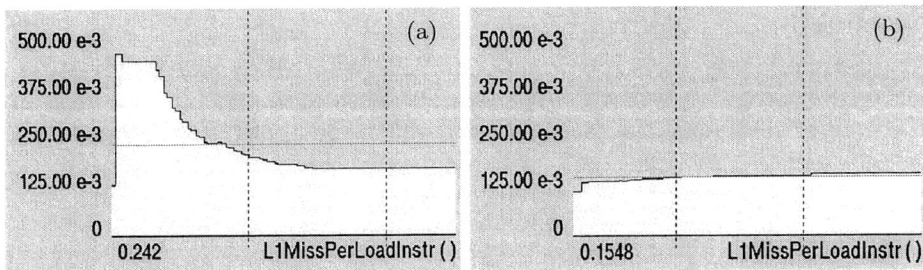

Fig. 1. Visualization of the measured L1 misses per load instruction as a function of time in Vampir [6] of a nested loop (a) before and (b) after the "unrolling" transformation.

example program was compiled with optimizations switched on. On the other hand, the example source code was optimized manually. In Sect. 2 the tests and their results are outlined in more detail.

Restructuring source code (e.g. applying loop transformations) by hand is a complicated and error-prone task. Therefore, we developed a tool to assist developers in optimizing their FORTRAN applications: loop transformations are done automatically on user request. Sect. 3 gives and overview of this tool. Finally, in Sect. 4 our future work and intentions are pointed out.

2 Current Compilers and Their Optimization Capabilities

2.1 Compiler Tests

In order to test the optimization capabilities of current compilers in terms of cache access, we compared the code generated by compilers with code optimized by hand. An example program was written in Fortran, which solves a system of linear equations based on the Gaussian algorithm. Step by step the original code of the algorithm was optimized manually and executed, both sequential and in parallel. OpenMP was used for the realization of the parallel processing.

The original source code did not utilize any optimizations. The elements of the coefficient matrix are accessed line by line. As this is contrary to the internal storage order of the matrix elements, a large number of cache-misses are produced. In the compilation process the maximum optimization level (-O5 for IBM's xlf, -O3 for the other compilers) was used with all tested compilers.

A first manual optimization was realized by implementing a loop interchange. This adapts the data access to the storage order on FORTRAN, increasing temporal locality in the cache and therefore decreasing the cache miss rate.

Further optimizations were tested, both for the sequential and parallel case. The following changes were applied: loop fission; replacing of multiple divisions with one division and multiple multiplications; loading of loop invariant matrix and vector elements into temporary variables; use of different names for loop indices.

2.2 Results

Figure 2 shows the results on SGI Origin 3800 with MIPS R12000 processors (400 MHz) using the MIPSpro 7.3 compiler. With optimizations switched on, the compiler recognizes the cache access problems and optimizes the sequential code ("sequential: -O3" in Fig. 2). Further manual transformations provide only very few improvements in the runtime of the sequential code ("sequential: loop interchange" and "sequential: additional optimizations").

The parallel programs are not optimized by the compiler automatically ("4 threads: -O3"). This can be concluded from the fact, that the manually optimized code yields much better runtimes ("4 threads: loop interchange" and "4 threads: additional optimizations"). The runtime even increases in parallel mode without manual changes compared to the sequential case, which is probably caused by the very good optimization results of the sequential code.

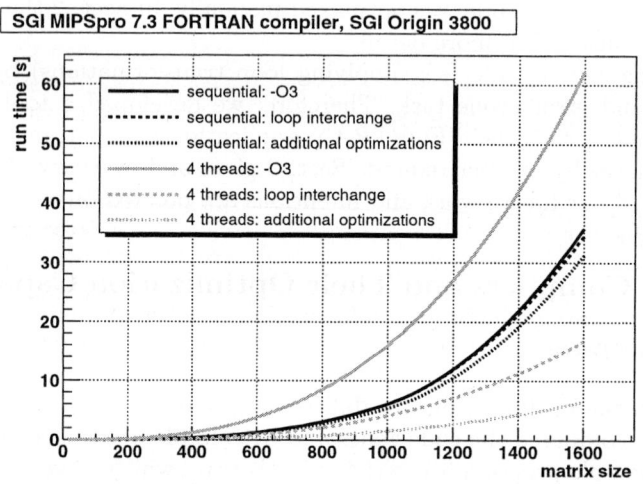

Fig. 2. Runtime as function of the matrix dimension on the SGI Origin 3800 with the MIPSpro 7.3 FORTRAN compiler.

In Figs. 3 and 4 the results of the measurements on Fujitsu Siemens Celsius 670 and IBM Regatta p690 are illustrated. The Celsius machine consists of two Intel Xeon 2.8 GHz processors running in hyperthreading mode (thus having only 4 logical CPUs in contrast to the other machines with 4 physical CPUs). As compiler Intel's ifc Version 7.1 was used. On the Regatta with its Power4 1.7GHz processors, the code was compiled with IBM's xlf FORTRAN compiler for AIX V8.1.1. For either compiler and both in sequential and parallel processing the findings are similar: the improvement in runtime due to manual optimizations is significant and larger than on the SGI Origin. The speedup is in the order of about 10. In contrast to the MIPSpro7 compiler on the Origin, the appliance of the additional optimizations does not result in any improvement. This implies, that comparable optimizations were already done by the compilers.

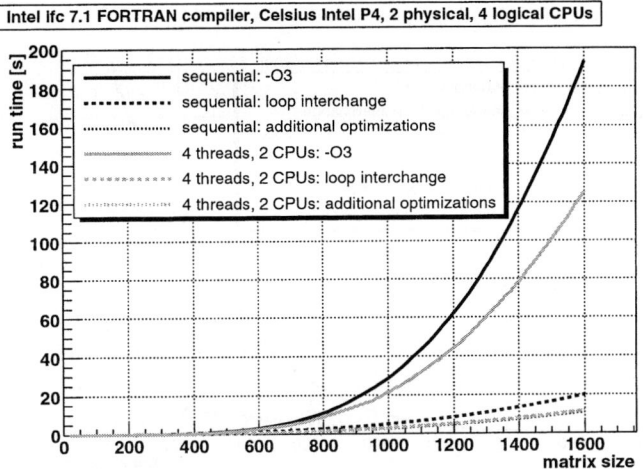

Fig. 3. Runtime as function of the matrix dimension on a Fujitsu Siemens Celsius 670 (2 CPUs in hyperthreading mode) with Intel's FORTRAN compiler 7.1; the measurement curves of the two manually optimized parallel codes and of the code with additional optimizations are on top of each other.

As an example the runtimes for a matrix size of 1600 are listed in Table 1.

Table 1. Runtime in seconds for all tested compilers and cases (matrix size: 1600) Remark: In the parallel case the Intel machine uses Hyperthreading mode. This means that the 4 threads run on 2 physical CPUs only.

	SGI MIPSpro 7.3	Intel ifc 7.1	IBM xlf V8.1.1
sequential			
-O3	35.8	190.5	82.4
loop interchange	34.5	19.9	8.7
additional optimizations	31.2	11.4	7.8
parallel, 4 threads			
-O3	62.4	125.7	28.5
loop interchange	16.7	11.2	2.6
additional optimizations	6.4	11.2	3.0

2.3 First Summary

Our measurements demonstrate, that the capabilities of the three tested FORTRAN compilers to optimize cache behaviour vary. Only MIPSpro7 is able to automatically optimize sequential code in such a way, that the resulting speedup is comparable with a manual optimization.

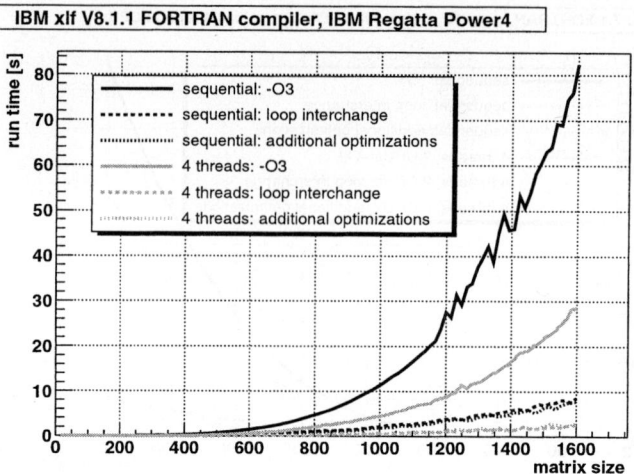

Fig. 4. Runtime as function of the matrix dimension on a IBM Regatta p690 system with IBM's xlf for AIX FORTRAN V8.1.1 compiler; the measurement curves of the two manually optimized parallel codes are on top of each other.

In the case of parallel OpenMP processing, none of the compilers can improve the original source code. Currently, the only way to improve cache access problems in FORTRAN programs seem to be the manual optimizations like loop transformations.

3 Assisting the Developer: *Goofi*

3.1 Goals

There are three drawbacks in a manual optimization of source code: it is time consuming, error-prone and can become quite complicated. Therefore, we developed the tool *Goofi* (Graphical Optimization Of Fortran Implementations) to support cache optimizations for FORTRAN applications. The goals are to provide the user with

- a graphical interface,
- an easy way to request and specify transformations,
- automatic transformations by one mouse click and
- the possibility to easily compare original and transformed source code.

3.2 Doing Loop Transformations with *Goofi*

Making loop transformations with *Goofi* is done in two steps: Insert one or a chain of transformation request directives (called "trafo directives") and secondly, ask *Goofi* to carry out the transformations. A "trafo directive" is composed of the name of the transformation requested and their parameters. It is inserted as a FORTRAN comment beginning with a special character sequence. This allows the user to save the directives with the source code for later reuse without interfering with compilation.

The following directive e.g. requests a blocking of size 8 of the loop immediately following the directive (I loop) with the loop nested one level below (J loop):

```
                            DO I_1=1,N,8
                            I_UPPER_1 = N
!TRA$ BLOCKING 8 1          DO J_1=1,N,8
DO I=1,N           ⟹        J_UPPER_1 = N
  DO J=1,N                  DO I=I_1,MIN(I_1+7,I_UPPER_1),1
                              DO J=J_1,MIN(J_1+7,J_UPPER_1),1
```

It is also possible to specify a chain of directives at once like "normalize a loop and then merge it with another one".

Directives can be inserted either by hand or in a more comfortable way by using *Goofi*. For the latter, the user loads the FORTRAN file he/she wants to optimize into the *Goofi* environment. The original source code will appear on the left side of a window splitted into two parts (see Fig. 5). Now, the user can simply insert directives by right clicking into the source code at the places where he/she wants to apply them. After selecting a special transformation, a popup window will appear, where the transformation parameters can be filled in or adjusted.

Finally, the user requests *Goofi* to do the transformations either by a button click or by a selection from a menu. The resulting transformed source file will show up in the right window of the split screen, making direct visual comparison easily possible. It is also possible to edit the source code directly in *Goofi*, which is supported by syntax highlighting.

Currently, the following loop transformations are implemented:

- optimizing loop transformations
    - Index Set Splitting: split a loop into several loops
    - Fission/Distribution: split a loop into several loops distributing the loop body → increase temporal locality by reusing cached data that were overwritten before
    - Fusion: merge two loops → increase temporal locality by reusing data distributed in several loops before
    - Interchange: exchange two nested loops → increases temporal locality
    - Blocking: decompose loops over arrays into blocks → improves cache line usage and data locality
    - Unrolling: replicate the body of a loop → minimizes loop overhead and increases register locality
    - Outer Loop Unrolling: replicate the body of a outer loop in a loop nest → minimizes loop overhead and increases register locality

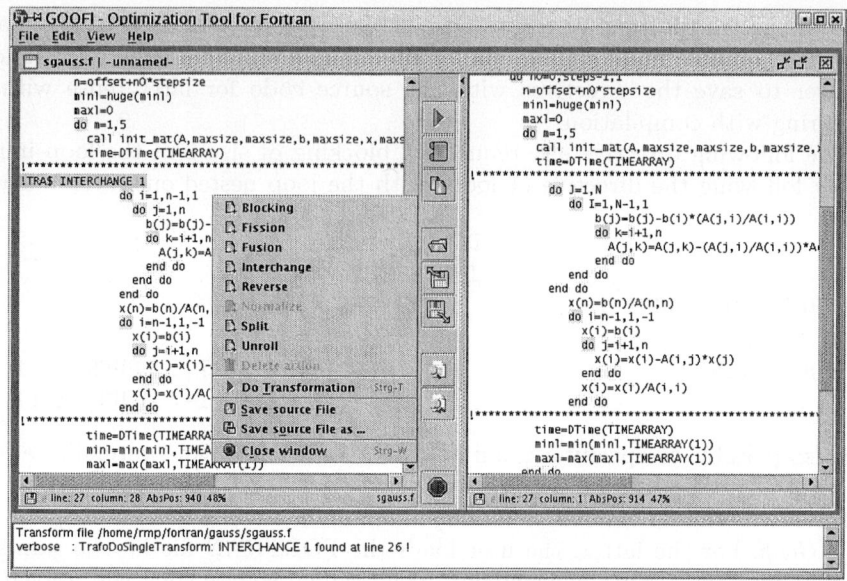

Fig. 5. Screenshot of *Goofi* with original and transformed source files and the transformation selection window.

- preparing loop transformations
  - Normalization: bring the loop in the normalized form
  - Reversal: reverse the iteration order

The last group of transformations does not directly improve cache utilization. However, it can be used to prepare loops for further transformations. E.g. sometimes a loop can be fused with another one after the normalization step only.

When using OpenMP, *Goofi* is capable to take into account changes needed in OpenMP contructs that are affected by a transformation. As an example, the change of the variable in the "private" clause of the outer parallized loop after an interchange transformation can be mentioned.

Applying transformations can improve the runtime behaviour of applications as cache access due to a prior cache miss is costly. Though *Goofi* is not directly aimed to provide improvements especially for parallel codes the above statement is true both for sequential and parallel applications. A distributed loop still experiences pretty much the same problems, bottlenecks or advantages in the cache behaviour as a sequential one.

3.3 Basics of *Goofi*

Goofi is based on a client-server architecture (see Fig. 6). The clients graphical user interface is written in Java to make it portable to different computing

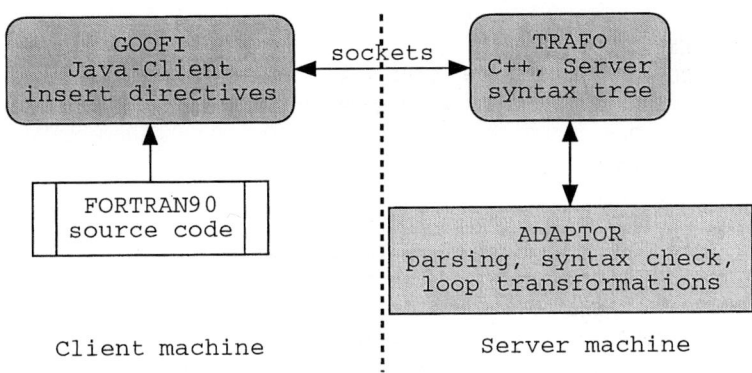

Fig. 6. Basic architectural principles of *Goofi*. The client is just a user interface, the server analyses the FORTRAN code and applies the transformations.

platforms. This allows to bring the tool to the users source code on the platform the code runs on. *Goofi* is dynamically configurable with XML-files. This includes the GUI (like menu structures) and the language preferences. Currently, the english and german language preferences are implemented. There is a strict separation between the server and the client regarding their knowledge on FORTRAN. The client is only the GUI and editor sending the FORTRAN source code to the server for analysis and transformation. Server and client communicate via sockets and use their own protocol. The server is implemented in C++ and C using the front end of the ADAPTOR [7] compilation system to analyse the FORTRAN code. After scanning and parsing the source code, definition and syntax check, a syntax tree is constructed (see Fig. 7). The "trafo directives" requested by the user are inserted into this tree. The transformations are done on the tree first. This includes restructuring the loops and statements (e.g. in an interchange transformation) or inserting new syntax elements (e.g. new lines with upcounted indices in an unrolling transformation). Finally, the transformed source code is generated by unparsing the tree and is sent back to the client.

Additional care has to be taken if there are dependencies in the code. For the syntax analysis, all elements (variables, functions etc.) have to be known. In FORTRAN, there are two possibilities to introduce such dependencies: include files and the use of modules. Dependencies due to include-files are resolved by simply inserting the include-files during parsing. Thus, in the client the user has to specify the directories where *Goofi* can look for them. Then, they have to be sent to the server.

For the usage of modules their syntax trees have to be created before opening files depending on such a module. These syntax trees are saved in files on the server side. If a "USE" statement appears the server can load the syntax tree from file. On the client side, the user only has to specify the directories containing the source code of the modules. By one further click the modules are transfered to the server and their syntax trees are created.

```
         Elem                   =    ACF_DO
            Line                =    10
            DO_ID               =    LOOP_VAR
                LOOP_VARNAME    =    VAR_OBJ
                    Pos         =    10
                    Ident       =    I
            DO_RANGE            =    SLICE_EXP
                FIRST           =    CONST_EXP
                    C           =    INT_CONSTANT
                        value       =  1
                        kind        =  4
                STOP            =    VAR_EXP
                    V           =    USED_VAR
                        VARNAME     =  VAR_OBJ
                            Pos     =  10
                            Ident   =  N
                INC             =    ...
            DO_BODY             =    ...
            TRANSFO             =    TINTERCHANGE_STMT
                depth           =    1
         ...
```

Fig. 7. Example of the syntax tree of a DO-loop. The TRANSFO element marks the transformation to be performed on this loop.

4 Future Work

Our next steps are to evaluate and validate the performance improvements possible with *Goofi* on large applications. Two representative parallel simulation applications will be used: the local model of the DWD (Deutscher Wetterdienst, German National Meteorological Service) and the geophysical package GeoFEM of RIST, Tokio.

Having a tool like *Goofi* to assist a developer in applying optimizations is just one part in an optimization process, actually it is even the second step. The first major part is the identification and the understanding of bottlenecks in a program due to cache access problems. General information about cache misses are not very useful, as they give the user only the information that something goes wrong, but not where and why.

The EP-CACHE project [2] is also intended to overcome this problem. We are working on the exploration of hardware monitoring and monitor control techniques (TU München, [8–10]) that can help the user to gain more precise and detailed information about the cache behaviour of a program. Combined with the useful performance visualization VAMPIR [6,11,12] and further optimization tools (SCAI Fraunhofer Gesellschaft, Sankt Augustin) the user will be enabled to easily speedup his or her application.

Acknowledgement

First of all, we want to thank Thomas Brandes (SCAI Fraunhofer Gesellschaft, St. Augustin) for providing the compiler tool ADAPTOR and for all his help. We thank the students of the Java practical 2002 at the TU Dresden for their initial work. By providing access to their IBM p690 (JUMP) the John-von-Neuman Institute at the Research Center Juelich made a part of the measurements possible.

References

1. EP-CACHE: Tools for Efficient Parallel Programming of Cache Architectures. WWW Documentation (2002) http://www.scai.fhg.de/292.0.html?&L=1.
2. Brandes, T., et al.: Werkzeuge für die effiziente parallele Programmierung von Cache-Architekturen. In: 19.PARS-Workshop, Basel (2003)
3. Intel Corporation: Intel®Fortran Compiler for Linux Systems. (2003)
4. Silicon Graphics Inc.: MIPSpro Fortran 90. (2003)
5. IBM: XL Fortran for AIX V8.1.1. (2003)
6. Brunst, H., Nagel, W.E., Hoppe, H.C.: Group Based Performance Analysis for Multithreaded SMP Cluster Applications. In: Proceedings of Euro-Par2001. Volume 2150 of Lecture Notes in Computer Science., Manchester, UK, Springer-Verlag Berlin Heidelberg New York (2001) 148ff
7. Brandes, T.: ADAPTOR - High Performance Fortran Compilation System. Institute for Algorithms and Scientific Computing (SCAI FhG), Sankt Augustin. (2000) http://www.scai.fhg.de/index.php?id=291&L=1.
8. Schulz, M., Tao, J., Jeitner, J., Karl, W.: A Proposal for a New Hardware Cache Monitoring Architecture. In: Proceedings of the ACM/SIGPLAN workshop on Memory System Performance, Berlin, Germany (2002) 76–85
9. Tao, J., Karl, W., Schulz, M.: Using Simulation to Understand the Data Layout of Programs. In: Proceedings of the IASTED International Conference on Applied Simulation and Modeling (ASM 2001), Marbella, Spain (2001) 349–354
10. Tao, J., Brandes, T., Gerndt, M.: A Cache Simulation Environment for OpenMP. In: Proceedings of the Fifth European Workshop on OpenMP (EWOMP '03), Aachen, Germany (2003) 137–146
11. Brunst, H., Hoppe, H.C., Nagel, W.E., Winkler, M.: Performance Optimization for Large Scale Computing: The Scalable VAMPIR Approach. In: Proceedings of ICCS2001. Volume 2074 of Lecture Notes in Computer Science., San Francisco, USA, Springer-Verlag Berlin Heidelberg New York (2001) 751ff
12. Brunst, H., Nagel, W.E., Malony, A.D.: A distributed performance analysis architecture for clusters. In: IEEE International Conference on Cluster Computing, Cluster 2003, Hong Kong, China, IEEE Computer Society (2003) 73–81

Detecting Data Races in Sequential Programs with *DIOTA*

Michiel Ronsse, Jonas Maebe, and Koen De Bosschere

Department ELIS, Ghent University, Belgium
{michiel.ronsse,jonas.maebe,koen.debosschere}@UGent.be
http://www.elis.UGent.be/diota/

Abstract. In this paper we show that data races, a type of bug that generally only causes havoc in parallel programs, can also occur in sequential programs that use signal handlers. Fortunately, it turns out that adapting existing data race detectors to detect such bugs for sequential programs is straightforward. We present such a tool, and we describe the modifications that were necessary to detect data races in sequential programs. The experimental evaluation revealed a number of data races in some widely used programs.

1 Introduction

Developing parallel programs is encumbered by the fact that proper synchronisation must be used to access shared variables. A lack of synchronisation will lead to a data race [1]: two or more parallel executing threads access the same shared memory location in an unsynchronised way, and at least one of the threads modifies the contents of the location. Note that the prerequisite *parallel execution* does not imply that the application should be executed on a parallel computer with multiple processors on which true physical parallel execution is possible. A data race is also possible on computers with only one processor, as the scheduler simulates logical parallellism between the threads.

As data races are (most of the time) considered bugs, they should be removed. Fortunately, automatic data race detection is possible and a number of data race detectors have been build in the past [2, 3]. Basically, two types of information are needed to perform data race detection: a list of all load/store operations (with information about the memory address used, the thread that executed the instruction, the type of operation (load, store or modify)) and information about their concurrency. The latter information can be deduced by intercepting the synchronisation operations (e.g. thread creation, mutex operations, semaphore operations, ...).

Developers of sequential programs do not have to worry about data races, except if the program uses signal handlers. We'll show that signal handlers introduce *logical* parallellism and as such data races can show up.

In the remainder of this paper, we start with a short description of UNIX type signals and signal handlers. Next, we'll show that signal handlers indeed introduce parallellism, even in sequential programs, and we'll describe how an existing data race detector for IA32 binaries running on Linux has been adapted to detect them. We end with an experimental evaluation. To the best of our knowledge, there are no other data race detectors that deal with signal handlers.

2 Signals

A signal is an important communication mechanism that was already supported by the very first UNIX-implementations [4]. A signal is basically a message (signal number, identity of the sender, context information, ...) that can be sent to a process. The signal can either be sent by a process (including the recipient) using the `kill` system call, or the kernel can trigger one upon the occurrence of a certain event (e.g. a processor trap). In both cases, the kernel will notify the application by sending the appropriate signal (read: setting a bit in the process context). The next time the application is scheduled, the application will execute the corresponding signal handler. Such a signal handler is a function that is registered with the kernel by means of the `signal` or `sigaction` system call. E.g. Figure 1 shows a program that will print You pressed ^C! each time ^C is pressed.

Typical signals are SIGSEGV (the application tries to access an invalid memory location), SIGILL (an illegal instruction was executed), SIGFPE (a floating point exception occurred), SIGPIPE (broken pipe), SIGALRM (normally sent by an application to itself in order to execute a function at a certain time in the future), SIGINT (interrupt signal sent by the shell when you press ^C), SIGUSR1 (a general purpose signal), SIGWINCH (sent by the window manager when the size of the window changes), ...

As mentioned above, when a signal reaches[1] an application, the normal program execution is interrupted and a *signal handler* is executed instead. As the program execution can be interrupted at any time, esp. if a signal was sent by another application, a data race can occur if one does not pay attention to this fact. Figure 2 shows an example: in `main()` the variable `global` is incremented and decremented a (large) number of times, resulting in a final value of 0 for `global`. However, each time ^C is pressed, a signal handler gives `global` an additional increment, resulting in a positive final value for `global`. This should be considered to be a data race: two unsynchronised write operations to the same variable. Of course, this is an artificial example, but similar problems can occur in real life applications, e.g. a graphical application that receives a SIGWINCH signal should update its width and height with care.

In [5], a number of Linux applications were examined, and the author states that "80 to 90% of signal handlers we have examined were written in an insecure manner". The manual inspection revealed races in such high-profile applications as `sendmail`, `WU-FTP` and `screen`.

---

[1] Two signals (SIGKILL and SIGSTOP) never reach an application, but are directly handled by the kernel.

```
#include <signal.h>

void sigint(){puts("You pressed ^C!");}

main(){
  signal(SIGINT, sigint);
  while(1){}
}
```

Fig. 1. A simple signal handler example.

```
#include <signal.h>

#define N 1<<24
unsigned global=0;

void sigint(){global++;}

main(){
  unsigned i;
  signal(SIGINT, sigint);
  i=N; while (i--) global++;
  i=N; while (i--) global--;
}
```

Fig. 2. A small program exhibiting a data race involving `global`.

3 Dealing with Signals in the Context of Data Race Detection

In order to detect data races in signal handlers, one has to trace the memory operations and gather information about their concurrency. The latter requires some attention: although signal handlers normally do not use synchronisation operations in order to prevent data races, not all signal handler memory operations are parallel with all memory operations executed by the application itself.

Attention should be paid to the following points:

- the data race detector should take into account that it is impossible to execute a signal handler before the signal handler is registered with the kernel. E.g. Figure 3 shows a program in which `global` is modified in `main()` and in the `sigint()` signal handler. As the signal is installed after the modification of `global` in `main()`, the accesses to `global` will never be executed in parallel, and hence no data race can occur.
- the above also applies to a signal that was sent by the application itself, e.g. using the `kill()` or `alarm()` function. Figure 4 shows an application that sends an alarm signal to itself. In this case, no data race occurs as the

```
#include <signal.h>

unsigned global=0;

void sigint(){global++;}

main(){
  global++;
  signal(SIGINT, sigint);
  sleep(10);
}
```

Fig. 3. This program does not exhibit a data race involving `global` as `main` accesses `global` before the signal handler is installed.

```
#include <signal.h>

unsigned global=0;

void sigalarm(){ ❹
  global+;+
} ❸❺

main(){
  signal(SIGALRM, sigalarm); ❶
  global+;+
  alarm(10); //send a SIGALRM to ourself in 10 seconds ❷
}
```

Fig. 4. This program does not exhibit a data race involving `global` as `main` accesses `global` before sending the signal. However, a data race will occur if another process also sends a `SIGALRM` to the application. The numbers refer to the modifications that were applied to our data race detector, see section 4.3.

> `alarm()` call always precedes the execution of the `sigalarm()` handler. Of course, as the `SIGALRM` can also be sent by another application (resulting in a data race), the signal context should be checked; if the sender is the process itself, the data race detector should update the concurrency information.
> – the easiest way to prevent data races in signal handlers is blocking the arrival of signals. Signals can be blocked automatically each time a certain signal handler is executed or explicitly using the `sigprocmask` system call. For both methods, a mask with one bit per signal is registered in the kernel. A data race detector should take this blocking into account.
> – during the execution of a signal handler, the thread executing the signal handler should be assigned a new thread number. This also applies to a signal handler preempting the execution of another (or the same) signal handler.

4 Implementation in *DIOTA*

We have implemented data race detection for sequential programs in the data race detector that is part of our *DIOTA* (Dynamic Instrumentation, Optimisation and Transformation of Applications) framework for Linux running on IA32 processors [6].

4.1 Description of *DIOTA*

DIOTA is implemented as a shared library for the Linux/80x86 platform. It instruments programs at the machine code level, so it is completely compiler- and language-agnostic and can also cope with hand-written assembler code. It has support for extensions to the 80x86 ISA such as MMX, 3DNow! and SSE and is written in an extensible and modular way so that adding support for new instructions and new types of instrumentation is easy.

An environment variable is used to tell the dynamic linker to load the *DIOTA* library whenever a dynamically linked application is started[2]. An init routine allows *DIOTA* to be activated before the main program is started, after which it can gain full control and start instrumenting.

The instrumentation happens gradually as more code of the program (and the libraries that it uses) is executed, so there is no need for complex analysis of the machine code to construct control-flow graphs or to detect code-in-data. The instrumented version of the code is placed in a separate memory region (called the "clone"), so the original program code is not touched and as such neither data-in-code nor the variable-length property of 80x86 instructions pose a problem.

The behaviour of *DIOTA* can be influenced by using so-called backends. These are dynamic libraries that link against *DIOTA* and tell it what kind of instrumentation should be performed. They can ask for any dynamically linked routine to be intercepted and replaced with a routine of their own, ask to be notified of each memory access, of each basic block that is executed and of each system call that is performed (both before and after their execution, so their behaviour can be modified as well as analysed).

4.2 Description of the Data Race Backend of *DIOTA*

The data race detection backend [7] is implemented as a shared library that uses the services provided by *DIOTA*. The backend requests *DIOTA* to instrument all memory operations and to intercept all pthread synchronisation operations and provides functions that should be called whenever a memory or synchronisation operations occurs. The data race detector works as follows:

- For all parallel pieces of code, the memory operations are collected and compared. This is based on the fact that all memory operations between

---
[2] A dynamically linked helper program is used to instrument statically linked binaries.

two successive synchronisation operations (called segments) satisfy the same concurrency relation: they are either all parallel or not parallel with a given operation and therefore with the segment containing the latter operation. Given the sets $L(i)$ and $S(i)$ containing the addresses used by the load and store operations in segment i, the parallel segments i and j contain racing operations if and only if

$$\Big((L(i) \cup S(i)) \cap S(j)\Big) \cup \Big((L(j) \cup S(j)) \cap S(i)\Big) \neq \emptyset$$

Therefore, data race detection basically boils down to collecting the sets $L(i)$ and $S(i)$ for all segments executed and comparing parallel segments.
– In order to detect parallel segments, a vector timestamp [8,9] is attached to each segment. As vector clocks are able to represent the *happened-before* [10] relation (they are strongly consistent), two vector clocks that are not ordered must belong to parallel segments. This gives us an easy way to detect parallel segments.

4.3 Modifications to Support Data Race Detection in Signal Handlers

In order to detect data races in sequential programs, the backend was adapted as follows (the numbers refer to the annotated Figure 4):

❶ the data race detector is informed when the application installs a signal handler. The current segment is ended, a new segment is started and the vector timestamp is saved (as `VC_install`).
❷ the data race detector is informed when the application sends a signal. If the target of the signal is the application itself, the current segment is ended, a new segment is started and the vector timestamp is saved (as `VC_signal`).
❸ each time signals become unblocked (by exiting a signal handler, or by using `sigprocmask`), a new segment is started and the vector timestamp is saved (as `VC_install`) for each signal that becomes unblocked.
❹,❺ the data race detector is informed at the begin and at the end of the execution of a signal handler. The backend then uses a new thread number for all memory and synchronisation operations executed by this signal handler. The initial vector timestamp for the signal handler is
  – if the signal was sent by the application itself (can be detected by checking the signal context), the signal handler gets the vector clock that was saved when the signal was sent (`VC_install`).
  – if the signal was sent by another application, the signal handler gets the vector clock that was saved when the signal handler was registered with the kernel (`VC_signal`).

The additional space required is therefore the space needed to store two vector timestamps for each signal handler (there are 64 possible signals in Linux). The actual size of such a vector timestamp can become quite large as each execution of a signal handler gets a new thread number. The data race backend can cope with arbitrarily large vectors as as they are enlarged when a new thread starts. The potentially large memory consumption could be limited by using *accordeon clocks* [11] that can shrink as threads exit or by re-using thread numbers.

5 Experimental Evaluation

In order to test our implementation, we used a number of widely used Linux applications. Using the data race backend is easy: after compiling *DIOTA*, type `diota-dr` and the data race detector will attach itself to applications you start in the same shell. As the data race detector intercepts all memory operations, a huge slowdown should be expected (e.g. the mozilla browser incurs a slowdown of 63.4× [12]).

Our test revealed a number of applications with data races:

vim: (an editor) resizing the VIM window results in a `SIGWINCH` handler setting the variable `do_resize` to `TRUE`. This variable is then checked in the main event loop of VIM, where appropriate action is taken. Although our data race backend flags this as a data race, this is actually no real data race. This technique (setting a boolean in the signal handler and dealing with the signal in the main loop) is used in a lot of applications (e.g. the pine e-mail client, the Apache web server,...). The reason is that, although the signal handler is executed on behalf of the application, a signal handler is restricted in its capability to execute kernel calls. Delaying the actual signal handling so that the main program can handle it at a later point helps overcome this problem.

links: (a text-only web browser): resizing the window or pressing ^C causes `queue_event()` to enter this event in a global queue (`ditrm`). As such, handling queued events while resizing causes havoc.

lynx: (another text-only web browser): global variables `LYlines` and `LYcols` are used for the width and height of the window. The `SIGWINCH` handler changes these variables. The `highlight()` function highlights a link and uses `LYcols` to check the available space.

Although we only tested a small number of applications, we were surprised that we found data races in signal handlers, which clearly shows that developers don't pay much attention to this kind of problem.

6 Conclusion

In this paper, we have shown that data races can show up in sequential programs. Fortunately, extending data race detectors for sequential programs is fairly straightforward: use a temporary thread number during the execution of an asynchronous signal handler and use additional rules for updating the vector clocks. The experimental evaluation revealed a number of data races in widely used programs. *DIOTA* is released to the public under the GPL and can be downloaded from `http://www.elis.UGent.be/diota/`.

Acknowledgements

Jonas Maebe is funded by the Institute for the Promotion of Innovation by Science and Technology in Flanders (IWT). The research was funded by Ghent University, and by the Fund for Scientific Research-Flanders (FWO-Vlaanderen).

References

1. Netzer, R.H., Miller, B.P.: What are race conditions? some issues and formalizations. ACM Letters on Programming Languages and Systems (1992)
2. Savage, S., Burrows, M., Nelson, G., Sobalvarro, P., Anderson, T.: Eraser: A dynamic data race detector for multithreaded programs. ACM Transactions on Computer Systems **15** (1997) 391–411
3. Ronsse, M., De Bosschere, K.: An on-the-fly Data Race Detector for RECPLAY, a Record/Replay System for Parallel Programs. In: Proceedings of the 16th ACM Symposium on Operating Systems Principles, Saint-Malo (1997) (on CD)
4. Stevens, W.R.: Advanced Programming in the UNIX Environment. Addison Wesley (1993)
5. Zalewski, M.: Delivering signals for fun and profit. http://razor.bindview.com/publish/papers/signals.txt (2001) RAZOR, BindView Corporation.
6. Maebe, J., Ronsse, M., De Bosschere, K.: DIOTA: Dynamic instrumentation, optimization and transformation of applications. In Charney, M., Kaeli, D., eds.: Compendium of Workshops and Tutorials Held in conjunction with PACT'02: Intl. Conference on Parallel Architectures and Compilation Techniques, Charlottesville, VA (2002)
7. Ronsse, M., De Bosschere, K.: Non-intrusive detection of synchronization errors using execution replay. Automated Software Engineering **9** (2002) 95–121
8. Mattern, F.: Virtual time and global states of distributed systems. In Cosnard, Quinton, Raynal, Roberts, eds.: Proceedings of the Intl. Workshop on Parallel and Distributed Algorithms. Elsevier Science Publishers B.V., North-Holland (1989) 215–226
9. Fidge, C.J.: Logical time in distributed computing systems. In: IEEE Computer. Volume 24. (1991) 28–33
10. Lamport, L.: Time, clocks, and the ordering of events in a distributed system. Communications of the ACM **21** (1978) 558–565
11. Christiaens, M., De Bosschere, K.: Accordion clocks: Logical clocks for data race detection. In Sakellariou, R., Gurd, J., Freeman, L., eds.: Proceedings of the 7th International Euro-Par Conference, Manchester, Springer (2001) 494–503
12. Ronsse, M., Stougie, B., Maebe, J., De Bosschere, K.: An efficient data race detector backend for diota. In: Proceedings of the International Conference ParCo2003. (2004) To be published.

A Time-Coherent Model for the Steering of Parallel Simulations

Aurélien Esnard, Michaël Dussere, and Olivier Coulaud

ScAlApplix Project, INRIA Futurs and LaBRI UMR CNRS 5800,
351, cours de la Libération, F-33405 Talence, France

Abstract. The on-line visualization and the computational steering of parallel simulations come up against a serious coherence problem. Indeed, data distributed over parallel processes must be accessed carefully to ensure they are presented to the visualization system in a meaningful way. In this paper, we present a solution to the coherence problem for structured parallel simulations. We introduce a hierarchical task model that allows to better grasp the complexity of simulations, too often considered as "single-loop" applications. Thanks to this representation, we can schedule in parallel the request treatments on the simulation processes and satisfy the temporal coherence.

1 Introduction

Thanks to the constant evolution of computational capacity, numerical simulations are becoming more and more complex; it is not uncommon to couple different models in different distributed codes running on heterogeneous networks of parallel computers (e.g. multi-physics simulations). For years, the scientific computing community has expressed the need for new computational steering tools to better grasp the complex internal structure of large-scale scientific applications.

The computational steering is an effort to make the simulations more interactive. To reach this goal, we focus on three key functionalities: the *control* which allows the end-user to precisely follow the simulation execution and to suspend its execution; the *on-line visualization* of intermediate results which requires to efficiently access the simulation data at run-time; the *interactions* which allow the end-user to modify steering parameters on-the-fly or to remotely fire steering actions at particular points of the simulation. The complex parallel simulations commonly used in scientific computing raise the crucial issue of the temporal coherence for the steering operations. For instance, an action must occur at the very same moment for all the simulation processes. In the same way, the distributed data collection requires that all the pieces come from the same timestep to remain meaningful.

The different approaches proposed to address the temporal coherence can be compared by considering both the time management and the simulation model. Magellan [1] represents a simulation with an unstructured collection of instrumentation points. To notify the time evolution, a timestamp is explicitly specified by the end-user on each instrumentation point but nothing ensures their coherence over the parallel processes. CUMULVS [2] considers parallel simulations as single-loop programs with a unique instrumentation point. Thanks to this simplification, it implements synchronization mechanisms relying on the internal loop counter and provides time-coherent accesses to distributed data. Only VASE [3] proposes a high-level model of simulations based on a control-flow graph (CFG) and provides a data-flow coupling model. However, this environment is only intended for sequential simulations. These environments are representative of the state-of-the-art: efforts have to be done to achieve an efficient steering environment which combines both the temporal coherence and a precise model of parallel simulations.

In this paper, we focus on the steering of parallel simulations like SPMD or simple MPMD applications (MPI, PVM, etc.). We first present the basis and the architecture of *EPSN*[1], our computational steering environment. Then, we introduce the high-level model used in *EPSN* to specify the steering of parallel simulations. Finally, we describe a strategy of scheduled request treatments to guarantee the temporal coherence.

2 *EPSN* Steering Environment

EPSN is a distributed environment based on a client/server relationship between user interfaces (clients) and simulations (servers). Both simulations and user interfaces must be instrumented with the *EPSN* API. The clients are not tightly coupled with the simulation; actually, they can interact on-the-fly through asynchronous and concurrent requests. These characteristics make *EPSN* environment very flexible and dynamic.

EPSN uses an internal communication infrastructure based on CORBA which provides the interoperability between applications running on different architectures. We previously discussed this option and other aspects of the architecture in [4]. As shown in Fig. 1, *EPSN* environment runs a permanent thread attached to each process of the simulation. These threads are in charge of all the steering treatments and each one contains a CORBA server waiting for the client requests. In addition, an independent application, called *proxy*, provides an unified view of the simulation and acts as a relay for the requests between the clients and all the *EPSN* threads. Once a request for data is received, the thread accesses the simulation process memory and transfers the data directly to the remote client. This transfer occurs concurrently to the simulation execution, in respect of the access areas placed in the simulation source code. As both the simulation and the client can be parallel applications, we use a MxN redistribution library, called *RedSYM*, with a CORBA communication layer.

Once a simulation has been instrumented with the *EPSN* library, it can be deployed as usually. The clients locate it on the network using the CORBA naming service and then connect it through the proxy. To follow the time evolution

---

[1] *EPSN* project (http://www.labri.fr/epsn) is supported by the French ACI-GRID initiative.

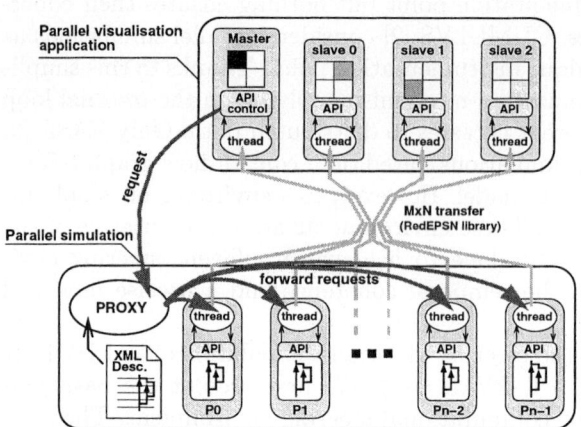 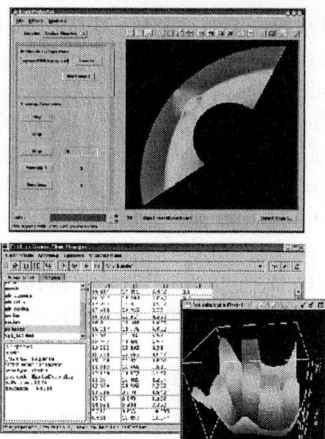

Fig. 1. Architecture of *EPSN* environment.   **Fig. 2.** Visualization clients.

of data, one can use a high-level visualization system based on *AVS/Express* (top of Fig. 2) or *VTK*. In addition, one can monitor a simulation and basically interacts thanks to our *Generic Client* (bottom of Fig. 2).

3 High-Level Model of Steerable Simulations

In order to closely interact with any parallel simulations, we introduce a high-level description model. Traditionally, numerical simulations are described as a collection of steerable items (data, breakpoint, loop, etc.) which are remotely manipulated by a user-interface. In the *EPSN* environment, we organize these items in a hierarchical task model that reflects the essential structure of SPMD and simple MPMD applications. To describe a simulation according to this model, we use an XML file parsed at the initialization of *EPSN* (*epsn_init*). As illustrated in Fig. 3, the XML is divided into three parts, the accessible data for clients (section *data*), the task program structure (section *htg*) and the possible interactions (section *interaction*).

3.1 Data Model

Data currently supported in *EPSN* are *scalars* and *fields* (dense multi-dimensional arrays). Ongoing works will extend the data model for *particles* and *meshes*. All accessible data must be declared, in the XML file (Fig. 3, lines 4–8), with a unique ID, a type (short, long, float, etc.), the number of dimensions for fields, and the distribution mode. Scalars and fields can be either *located* on a single process or *replicated* among all processes. Fields can also be *distributed* over the processes with any generic rectilinear decomposition such as the classical block-cyclic distribution. Finally, in order to make data accessible for clients

```xml
1  <?xml version="1.0" encoding="UTF-8" ?>
2  <!DOCTYPE simulation SYSTEM "epsn.dtd">
3  <simulation id="sample">
4    <data>
5      <scalar id="initial-temp" type="double" distribution="replicated"/>
6      <scalar id="mean-temp" type="double" distribution="located"/>
7      <field id="heat" type="double" dimension="2" distribution="distributed"/>
8    </data>
9    <htg access="protected">
10     <task id="init" access="protected"/>
11     <loop id="main-loop" access="accessible">
12       <task id="init-loop">
13         <data-context ref="mean-temp" context="out"/>
14       </task>
15       <loop id="sub-loop">
16         <data-context ref="heat" context="out"/>
17       </loop>
18       <breakpoint id="bp0" state="up"/>
19     </loop>
20     <task id="ending"/>
21   </htg>
22   <interaction>
23     <writing data="initial-temp">
24       <point node="init" position="end"/>
25     </writing>
26     <action id="restart">
27       <point node="main-loop" position="iterate"/>
28     </action>
29   </interaction>
30 </simulation>
```

Fig. 3. Example of a simulation description in XML.

(*publish*), the user must complete its description with the *EPSN* API and must precise the data storage and the field distribution. Moreover, correlated data may be logically associated in a *group*, and thus *EPSN* will ensure the temporal coherence among the group members.

3.2 Hierarchical Task Model

The basis of our hierarchical task model consists in representing SPMD simulations through a hierarchy of tasks (Fig. 3, lines 9–21). In this paper, we extend the definition of *hierarchical task graphs* [5] or HTGs for the purpose of computational steering. Basically, an HTG is a directed acyclic graph, which captures the control flow between blocks of code and hierarchical nodes (condition, loop).

Here, we consider three types of nodes: (1) *basic tasks* encapsulate logical blocks (e.g. program routine); (2) *loop tasks* represent the iterative structures (for, while); (3) *conditional tasks* represent the conditional structures (if-then-else, switch-case). All these nodes are hierarchical in nature, as they can contain a serie of other nodes. We also introduce *breakpoint* nodes, that are used to explicitly control the execution flow and have no sub-nodes. Initially, a main task encapsulates all the hierarchy which is delimited in the source code by the *epsn_init* and the *epsn_exit* calls. In practice, the program structure can be described at different levels of details and the user can consider only the relevant tasks for the steering. The instrumentation is achieved by manually annotating the begin and the end of each task with the *EPSN* API (*epsn_task_begin*, *epsn_task_end*, etc.). Figure 4 illustrates, on a simple "two-loop" program, that the instrumentation exactly reflects the HTG structure and its XML description. As a result, the instrumentation process can be partially automated.

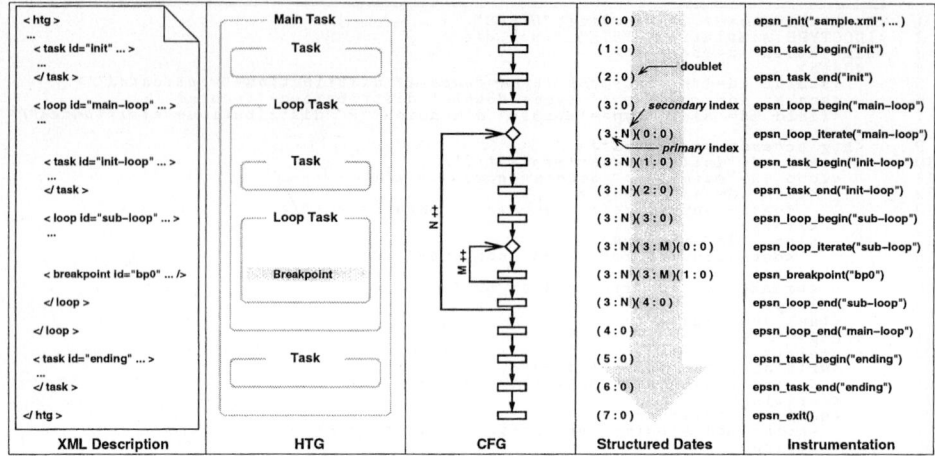

Fig. 4. Comparison of the different representation modes.

In addition, a task may have input and output data, explicitly specified in the XML through a *context*. The input data are purely formal information, whereas the output data are those modified by a task. *EPSN* prevents any client access to output data during all the task execution and automatically releases them at the end. Furthermore, a task may contain a specific access context, specifying the protected and accessible data. Such access areas are fundamental in *EPSN* as they exhibit where and when a data can be safely and asynchronously retrieved.

In simple MPMD cases that we consider, the processes are divided into very few groups (*process-partition*) involved in different programs (e.g. ocean atmosphere coupled models) and running on the same cluster. We assume that the HTGs of the different groups have their upper hierarchy level in common, typically, they share a synchronized main loop. Consequently, we can describe the whole simulation as a unified HTG whose lower levels are distributed in sub-HTGs among the groups (*distributed-sections*). The user must also complete the XML description to precise over which process group the data are replicated or distributed (*location* attribute). The unified HTG provides the end-user with a single representation of the simulation, which can be steered either at an upper level addressing all processes or at a lower level addressing just a group.

3.3 Steering Interactions

As shown in Fig. 3 (lines 22–29), *EPSN* proposes two kinds of steering interactions: the modification of simulation data (*writing*) and the callback of user-defined functions (*action*). As we carefully consider such interactions, we only enable them to occur on instrumentation points, identified in the XML description. Only the actions require to be instrumented in the source code by associating a callback function with the action ID.

4 Temporal Coherence

The efficiency of a steering system for parallel simulations depends on its capability to execute time-coherent operations over all the processes. Different strategies can be used to maintain the temporal coherence, but they all rely on the ability to compare the current positions in the execution of parallel processes.

4.1 Structured Dates

We consider the date of a process as its position in the execution flow. The succession of the dates related to the instrumentation points define a discrete time system. As we deal with structured applications typically involving a hierarchy of loops, a simple timestamp, incremented at each instrumentation point, would be ambiguous. As a result, we define a structured date as a collection of doublets (e.g. $(F_0 : A_0)(F_1 : A_1)...(F_n : A_n)$). As shown in Fig. 4, each doublet corresponds to a level of the hierarchy and it is composed by a *primary* and an *secondary* index. When the execution flow traverses a level of the HTG, the corresponding *primary* index is simply incremented at each instrumentation point. When it enters a sub-level of the HTG (loop, switch), the *secondary* index marks the current iteration or the chosen branch; then, a new doublet is added to follow the progression in the sub-level.

The use of structured dates presents several advantages for a steering environment. It first allows a client to clearly identify the simulation progress. Structured dates are also relevant for elaborating precise requests. Finally, they constitute a robust discrete time system: thanks to the inherent hierarchy of the structured dates, a desynchronization in a sub-level will not propagate into the upper ones.

4.2 Coherence Strategies

The toolkits and libraries developed for the steering of parallel and distributed applications present different strategies to maintain the temporal coherence. In the strategy of *post-collection ordering*, the environment first collect on a central server the informations of the simulation, then these informations are interpreted together. This strategy is mainly used for the ordering of monitoring events like in Falcon [6] or PathFinder [7]. Considering that it requires a systematic centralized data collection, it is poorly adapted to the on-line visualization of large amount of data. In the *centralized request management* strategy, at each step of the simulation, the processes refer to a central server which indicates them if they have to do a treatment. This strategy is used by client-server-client architectures like DAQV [8], but it imposes a very strong synchronization and the simulation processes advance at the server's pace. For *EPSN* environment, we have chosen a third strategy which consists in scheduling the treatments independently on each process of the simulation. The *scheduled treatment* strategy is the one used in CUMULVS [2] to plan data transfers between a simulation and its clients. This strategy does not imply a synchronization with a server and if there is no request,

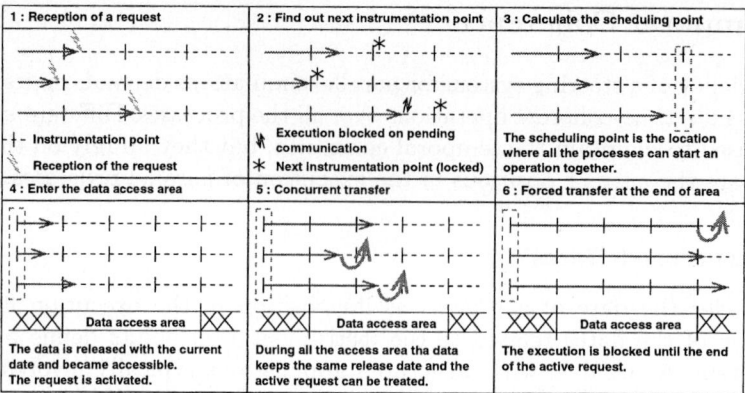

Fig. 5. Scheduled treatment of parallel steering requests.

the simulation is not disturbed. However, to ensure the temporal coherence, the requests must be scheduled at the same date, even asynchronously, in all the processes and these processes must follow the same evolution.

4.3 Scheduled Treatments

We call *scheduling point* the date when the simulation processes schedule the treatment of a request. It is also the earliest instrumentation point that no simulation process has already passed when it get the request. The steps 1 to 3 of Fig. 5 illustrate the choice of the scheduling point. When the instrumentation threads of the processes receive a request, they lock the instrumentation points, they exchange their current dates to determine the scheduling point and then they unlock the instrumentation points. Even if a process is between two instrumentation points, whether it is blocked on pending communications or just computing, its instrumentation thread can communicate its current date to the other threads. As a result, the research of the scheduling point can not lead to a deadlock. When a simulation process goes past the scheduling point, the request is activated independently of the other processes, and it will be treated as soon as its associated constraints are satisfied. For control requests and actions, the constraints are based on the current date. For data access requests, they are based on the data *release*: the date when the data has been lately modified. The release remains unchanged during all the access area after the data modification, and the data can be treated at any time in this area. The constraints based on the release are consistent because the data evolution is linked to the control-flow through the task description and consequently it is the same on all the processes.

The steps 4 to 6 of Fig. 5 illustrate the use of the data release in the case of a simple on-the-fly data access request. When a simulation process enters the access area, the data is released and the request constraint is satisfied. During all the access area, the request can be asynchronously treated because the simulation description ensures that the data will not be modified. At the end of the access area, the request treatment is forced (flushed) before a process enters in a new modifying area. Figure 5 exhibits all the mechanisms that ensure the temporal coherence. However, in the best cases, the simulation would be synchronized enough for the requests to happen at the same moment of the simulation, and for the scheduling point to be chosen without blocking any process. In the same way, if the access area is long enough, the data transfer would be completely overlapped by the simulation.

5 Conclusion and Future Works

In this paper we address problems raised by the computational steering of parallel simulations. We have introduced a high-level model to better grasp the complexity of these simulations and to closely follow their evolution. In this context, the use of *hierarchical task graphs* has appeared to be very promising. In order to allow the precise positioning in these graphs, we have defined the structured dates which are the basis for our scheduling strategy of the steering treatments. This temporal coherence strategy involves very little synchronization between processes and allows the overlapping of data transfers.

The developments of *EPSN* are now oriented on the support of massively parallel simulations and distributed applications, involving a hierarchy of proxies. Ongoing works also aim at the specification of a generic data model for particles and meshes which would be suitable for the MxN redistribution.

References

1. Vetter, J.: Experiences using computational steering on existing scientific applications. In: Ninth SIAM Conf. Parallel Processing. (1999)
2. Papadopoulos, P., Kohl, J., Semeraro, B.: CUMULVS: Extending a generic steering and visualization middleware for application fault-tolerance. In: Proceedings of the 31st Hawaii International Conference on System Sciences (HICSS-31). (1998)
3. Jablonowski, D., Bruner, J., Bliss, B., , Haber, R.: VASE: The visualization and application steering environment. In: Proceedings of Supercomputing'93. (1993) 560–569
4. Coulaud, O., Dussère, M., Esnard, A.: Toward a computational steering environment based on CORBA. In Joubert, G., Nagel, W., Peters, F., Walter, W., eds.: Parallel Computing: Environments and Tools for Parallel Scientific Computing. Advances in Parallel Computing, Elsevier (2004)
5. Girkar, M., Polychronopoulos, C.: The hierarchical task graph as a universal intermediate representation. International Journal of Parallel Programming **22** (1994) 519 – 551
6. Gu, W., Eisenhauer, G., Schwan, K., J.Vetter: Falcon: On-line monitoring for steering parallel programs. Concurrency: Practice and Experience **10** (1998) 699–736
7. Miller, D., Guo, J., Kraemer, E., Xiong, Y.: On-the-fly calculation and verification of consistent steering transactions. In: ACM/IEEE SC2001 Conference. (2001)
8. Hackstadt, S., Harrop, C., Malony, A.: A framework for interacting with distributed programs and data. In: HPDC. (1998) 206–214

MATE: Dynamic Performance Tuning Environment*

Anna Morajko, Oleg Morajko, Tomàs Margalef, and Emilio Luque

Computer Science Department. Universitat Autònoma de Barcelona
08193 Bellaterra, Spain
ania@aomail.uab.es, olegm@aia.ptv.es,
{tomas.margalef,emilio.luque}@uab.es

Abstract. Performance is a key issue in the development of parallel/distributed applications. The main goal of these applications is to solve the considered problem as fast as possible utilizing a certain minimum of parallel system capacities. Therefore, developers must optimize these applications if they are to fulfill the promise of high performance computation. To improve performance, programmers search for bottlenecks by analyzing application behavior, finding problems and solving them by changing the source code. These tasks are especially difficult for non-expert programmers. Current approaches require developers to perform optimizations manually and to have a high degree of experience. Moreover, applications may be executed in dynamic environments. Therefore, it is necessary to provide tools that automatically carry out the optimization process by adapting application execution to changing conditions. This paper presents the dynamic tuning approach that addresses these issues. We also describe an environment called MATE (Monitoring, Analysis and Tuning Environment), which provides dynamic tuning of applications.

1 Introduction

Parallel/distributed systems offer high computing capabilities that are used in many scientific research fields. They facilitate the determination of the human genome, computing atomic interactions or simulating the evolution of the universe. So biologists, chemists, physicists and other researchers have become intensive users of applications with high performance computing characteristics. They submit applications to powerful systems to get their results as fast as possible. In this context, performance becomes a key issue. To satisfy user requirements, applications must reach high performance standards. An application is inefficient and useless when its performance is below an acceptable limit. Therefore, applications must not only be systematically tested from the functional point of view to guarantee correctness, but must also be optimized to ensure that there are no performance bottlenecks.

The optimization (or tuning) process requires a developer to go through the application performance analysis and the modification of critical application parameters. First, performance measurements must be taken to provide data about the application's behavior. This phase is known as monitoring and collects data related to the

---

* This work has been supported by the MCyT (Spain) under contract TIC2001-2592, by the European Commission under contract IST- 2000-28077 (APART 2) and has been partially supported by the *Generalitat de Catalunya* – GRC 2001SGR-00218.

execution of the application. Then, the performance analysis of this information is carried out. It finds performance bottlenecks, deduces their causes and determines the actions to be taken to eliminate these bottlenecks. Finally, appropriate changes must be applied to the application code to overcome problems and improve performance. However, all these tasks are somewhat complicated, especially for non-expert users.

The classical approach to performance analysis is based on the visualization of program execution. Tools that support this approach show the execution of the application in graphical and numerical views [1, 2, 3]. Then, users must analyze generated views recognizing the most problematic performance regions, determining the causes of the bottlenecks and finally changing the application source code. To reduce developer effort and relieve them of such duties as analysis of graphical information and determination of performance problems, an automatic analysis has been proposed. Tools using this type of analysis are based on the knowledge of well-known performance problems. Such tools are able to identify critical bottlenecks and help in optimizing applications by providing suggestions to developers [4, 5, 6].

All mentioned tools involve developers changing a source code, re-compiling, re-linking and restarting the program. They require a certain degree of knowledge and experience of parallel/distributed applications and hence are appropriate for developers rather than for such users as biologists, chemists, physicists or other scientists. To tackle these problems, it is necessary to provide tools that automatically perform program optimizations. A good, reliable and simple optimization tool performing automatic improvement could be profitable for non-expert users as well as for developers. Because of the complexity of the solution, there are not many tools that support automatic application optimization during run time [7, 8]. Moreover, they lean towards automated tuning, which requires certain changes to the application.

This paper addresses the problem of automatic and dynamic tuning of parallel/distributed applications. Section 2 presents this approach showing its fundamental concepts. Section 3 describes a dynamic tuning environment called MATE. Section 4 shows a catalog of tuning techniques that we investigated as part of our study. Finally, section 5 presents the conclusions of this study.

2 Dynamic Performance Tuning

The main goal of dynamic automatic tuning is to improve application performance by modifying its execution without recompiling or rerunning it. In this approach, the following steps can be distinguished: application monitoring, performance analysis and modifications of the running program. All of these must be performed automatically, dynamically, and continuously during application execution. The dynamic analysis and introduced modifications enables adaptation of the application behavior to changing conditions in the application itself or in the environment. Dynamic tuning appears as a promising technique that exempts non-experts or programmers from some of the performance-related duties. The most useful dynamic tuning is that which can be used to successfully optimize a broad range of different applications. It would be desirable to tune any application even though its source code and application-specific knowledge is not available. However due to incomplete information this kind of tuning is highly challenging and at the same time the most limited. The key question is: *what can be tuned in an "unknown" application?*

2.1 Tuning Layers

The answer to this key question can be found by investigating how an application is built. Each application consists of several layers: application-specific code, standard and custom libraries, operating system libraries, and hardware. An application is based on services provided by an operating system (OS). OS offers a set of libraries so that system users do not need to worry about low-level hardware details. The application uses the system calls to perform hardware/system-specific operations. Besides that, applications use standard libraries that support them with a variety of functions, e.g. higher level I/O, mathematical, string manipulation functions and so on. Additionally, applications may use custom libraries that provide domain-specific functionality, e.g. communication libraries, numerical methods, programming frameworks. These libraries insulate programmers from low level details as they offer a higher level of abstraction. Finally, each application contains application-specific implementation and consists of modules that solve a particular problem.

Considering OS and library layers, the tuning process is based on well-known features for them. By investigating particular OS and libraries it is possible to find their potential drawbacks and hence determine problems common to many applications. For each drawback, a tuning procedure can be identified. Optimizing the application code is the most complex and less reusable, due to the lack of application-specific knowledge. Each application implementation can be totally different and there may be no common parts, even though they may provide the same functionality. An application can be tuned if there is knowledge of its internal structure. Therefore, to optimize the application layer, dynamic tuning should be supported in some way with certain information about the application.

2.2 Approaches to Tuning

Considering the available knowledge, we have defined two main approaches to tuning: automatic and cooperative. In the automatic approach, an application is treated as a black-box, because no application-specific knowledge is provided by the programmer. This approach attempts to tune any application and does not require the developer to prepare it for tuning (the source code does not need to be adapted) and, therefore, it is suitable for tuning such layers as the operating system and libraries. We can find many general tuning procedures common to many applications. For each particular problem, all the necessary information, such as what should be measured, how it should analyzed, and what should be changed and when, can be provided automatically. The cooperative approach assumes that the application is tunable and adaptable. This means that developers must prepare the application for the possible changes. Moreover, developers must define an application-specific knowledge that describes what should be measured in the application, what model should be used to evaluate the performance, and finally what can be changed to obtain better performance. The cooperative approach is suitable for the application tuning layer.

2.3 Performance Analysis

Performance analysis examines application behavior based on the collected measurements, identifies performance bottlenecks, and provides specific solutions that overcome these problems. Application behavior can be characterized by an analytical performance model. Such a model may help to determine a minimal execution time of the application or predict the application performance. Such a model can contain formulas and/or conditions that recognize a bottleneck and facilitate determination of the optimal behavior. As input, a model needs the measurements extracted from the application execution. Based on these and applying adequate formulas, the performance model can estimate the desired application behavior, e.g. the optimal value of some parameter. Finally, the application can be tuned by changing the value of that parameter.

To make the presented approaches to tuning (automatic and cooperative) homogeneous and to make optimization on the fly possible and effective, we concluded that the application knowledge should be described as the following terms:

- measure point – a location where the instrumentation must be inserted
- performance model – determines an optimal application execution time
- tuning point – the code element that may be changed
- tuning action – the action to be performed on a tuning point
- synchronization – policy determining when the tuning action can be invoked.

2.4 Dynamic Modifications of an Application

All phases of improving the application performance must be done "on the fly". To instrument the application without accessing the source code, the code insertion must be deferred till the application is launched. Modifications cannot require source code recompilation or restart. The technique that fulfills these requirements is called dynamic instrumentation. It permits insertions of a piece of code into a running program. Dynamic instrumentation was used in Paradyn [6] to build an automatic analysis tool. The Paradyn group developed a library called DynInst [9].

Considering DynInst's possibilities and our definition of application knowledge, we determined tuning actions that can be applied on the fly to a tuning point. A tuning point can be any point found by DynInst in the application executable (e.g. function entry, function exit). We consider the following to be tuning actions:

- function replacement – function calls are replaced with a call to another function
- function invocation – an additional function call is inserted at a specified point
- one-time function invocation – a specified function is invoked just once
- function call elimination – a specified function call is eliminated
- function parameter changes – the value of an input parameter is modified
- variable changes – the value of a particular variable is modified.

All modifications must be performed carefully to ensure that the application continues its execution correctly and does not crash. Therefore, each tuning action defines the synchronization policy that specifies when the action can be invoked in a safe manner. E.g. to avoid reentrancy problems, race hazards or other unexpected behavior, a breakpoint can be inserted into an application at a specific location. When the execution reaches the breakpoint, the actual tuning action is performed.

3 MATE

To provide dynamic automatic tuning of parallel/distributed applications we have developed a prototype environment called MATE (Monitoring, Analysis and Tuning Environment). For the purpose of our work we have made the assumption of targeting our tuning system to C/C++ parallel/distributed PVM [10] applications running on a UNIX platform. MATE performs dynamic tuning in three basic and continuous phases: monitoring, performance analysis and modifications. This environment dynamically and automatically instruments a running application to gather information about the application's behavior. The analysis phase receives events, searches for bottlenecks, detects their causes and gives solutions on how to overcome them. Finally, the application is dynamically tuned by applying a given solution. MATE consists of the following main components that cooperate among themselves, controlling and trying to improve the application execution:

- Application Controller (AC) – a daemon-like process that controls the application execution on a given host (management of tasks and machines). It also provides the management of task instrumentation and modification.
- Dynamic monitoring library (DMLib) – a shared library that is dynamically loaded by AC into application tasks to facilitate instrumentation and data collection. The library contains functions that are responsible for registration of events with all required attributes and for delivering them for analysis.
- Analyzer – a process that carries out the application performance analysis, it automatically detects existing performance problems "on the fly" and requests appropriate changes to improve the application performance.

Figure 1 presents the MATE architecture in a sample PVM scenario. In this example the PVM application consists of 3 tasks distributed on 2 different machines. When the Analyzer has been started, it distributes the AC to all machines where the application is running to control all the tasks. Once the AC is distributed the Analyzer receives from it information about the configuration of a virtual machine. The performance analysis is based on tunlets. Each tunlet is a shared library that implements the analysis logic for one particular performance problem. A tunlet uses the Analyzer's Dynamic Tuning API (DTAPI) to perform the performance monitoring and tuning of a program. Tunlets are passive components that drive the analysis by responding to a set of incoming events. The Analyzer provides a container that is responsible for managing a set of tunlets simultaneously. Tunlets provide the Analyzer with an initial set of measure points that are forwarded to all ACs. Next, the Analyzer asks the AC to start the application. The AC loads the shared monitoring library (DMLib) to the task memory that enables its instrumentation.

During execution, the ACs manage the instrumentation of each task. The shared monitoring libraries are responsible for delivering registered events directly to the Analyzer. When an event record is received, the Analyzer notifies the corresponding tunlet and this tunlet in turn finds bottlenecks and determines their solutions. By examining the set of incoming event records, the tunlet extracts measurements and then uses the built-in performance model to determine the actual and optimal performance. If the tunlet detects a performance bottleneck, it decides whether the actual performance can be improved in existing conditions. If this is the case, it then asks the Analyzer to apply the corresponding tuning actions. A request determines what should be

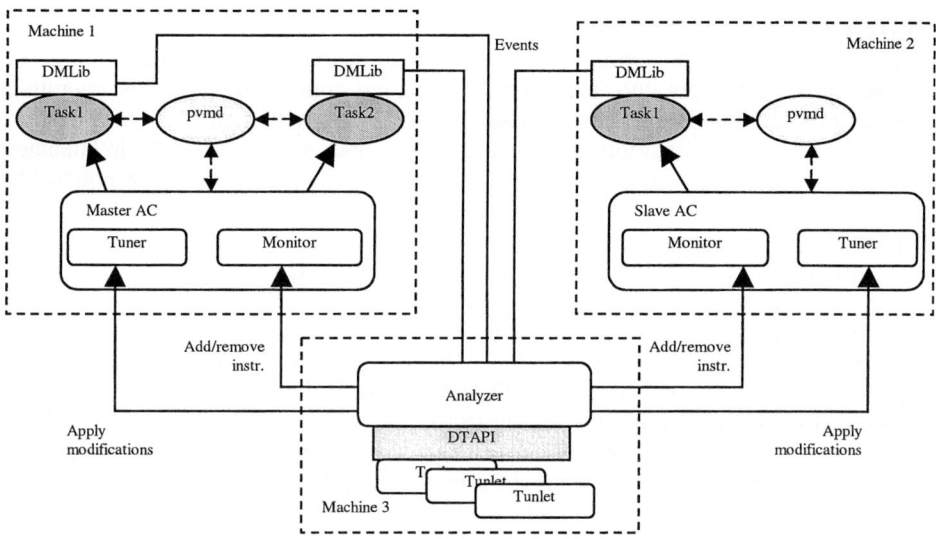

Fig. 1. Architecture of the MATE dynamic tuning for PVM.

changed (tuning point/action/synchronization) and it is sent to the appropriate instance of AC.

4 Tuning Techniques

We conducted a number of experiments on parallel/distributed applications to study how our approach works in practice. We proved that running applications under the control of MATE may be effective, profitable, and the adaptation of the application behavior to the existing conditions results in performance improvements. All required information related to one particular performance problem is what we call a tuning technique. Each tuning technique describes a complete optimization scenario:

- it specifies a potential performance problem of a parallel/distributed application
- it determines what should be measured to detect the problem (measure points)
- it determines how to detect the problem (performance model)
- it provides a solution for how to overcome the problem (tuning point/action/sync).

All experiments were conducted in a cluster of Sun UltraSPARC II workstations connected by LAN network. To investigate the profitability of dynamic tuning we used the following applications: (1) a set of synthetic master-worker programs; (2) Xfire – Forest Fire Propagation application [11], a computation-intensive program that simulates fireline propagation; (3) IS – NAS Integer Sort benchmark [12], a communication-intensive application that ranks integers using a bucket sort method.

We organized tuning techniques into a catalog in accordance with the tuning approach and the layer at which a modification occurs. Each tuning technique is implemented in MATE as a tunlet. Currently, we are focused on investigating tuning techniques separately. The catalog available in MATE is the following:

Operating system layer (Automatic approach):
- Message aggregation – minimizes communication overhead by grouping sets of consecutive messages into large ones. The tuning consists of replacing operating system function calls that transmit data, e.g. `write()`, with their optimized custom version with an aggregation mechanism. In networks with non-ignorable latencies and applications that send small messages, the technique can produce noticeable time savings (up to 91% in the case of synthetic applications).
- TCP/IP buffers – maximize the network transmission performance across high-performance networks using TCP/IP-based protocol. This is done by setting the send and receive socket buffers to an optimal value that can be estimated calculating bandwidth delay product. The tuning action includes one time system call invocation `setsockopt()` using `SO_SNDBUF` and `SO_RCVBUF` socket options. [13] reports improvements ranging from 17% up to 500% for FTP transmissions.

Standard library layer (Automatic approach):
- Memory allocation – improves performance by optimizing memory allocations. Programs that make intensive use of memory may benefit from optimized pool-based allocators if they perform a large number of small object allocations and de-allocations. The tuning action replaces the standard allocator with the optimized pool allocator. The specialized pool allocators perform much better, giving up to 60-70% experimenting with synthetic applications.

Custom library layer (Automatic approach):
- PVM communication mode – minimizes the PVM communication overhead by switching the messaging to point-to-point mode. The tuning action includes one-time function call `pvm_setopt(PvmRoute,PvmRouteDirect)`. Changing the communication mode resulted in faster communication, up to 50% in synthetic applications and 17% in IS benchmark. The measured intrusion did not exceed 3,5% of the improved execution time.
- PVM encoding mode – minimizes the PVM encoding overhead by skipping data encoding/decoding phase. The tuning action includes input parameter modification of `pvm_initsend()` that changes the encoding mode from default XDR to data raw. We observed important benefits from data raw encoding mode (up to 74% in synthetic applications and up to 47% in IS benchmark). The intrusion reached up to 2,8% of the total application execution time.
- PVM message fragment size – selects the optimal size of message fragments to minimize the PVM communication time. The tuning action includes one-time function call `pvm_setopt(PvmFragSize,OptFragSize)`. To calculate the optimal fragment size, we used the experimentally deduced formula. This technique gave up to 55% profit in synthetic applications and up to 28% in IS benchmark. The intrusion reached 4,9%.

The PVM experiments that we conducted are described in more detail in [14].

Application layer (Cooperative approach):
- Workload balancing – balances the amount of work that is distributed by the master to workers considering the capacities and load of the machines where the application is running. During program execution, the simulation algorithm estimates the optimal workload. The tuning action changes the work factor by updating the variable value in the master that adapts the work assignment. This technique produced

up to 50% profit in synthetic applications and 48% in Xfire. The overhead was small – about 2% of the improved execution time.
- Number of workers – optimizes the number of workers assigned to perform a specified amount of work in the master/worker application. To calculate the optimal number of workers we used the performance model presented in [15]. The tuning action changes the number of workers by updating the variable value in the master process. We observed that changes to the number of workers could produce profits ranging from 20% to 300% in synthetic applications.

5 Related Work

The Autopilot [8] project bases on closed loop and allows parallel applications to be adapted in an automated way. It contains a library of runtime components needed to build an adaptive application. Autopilot provides a set of distributed performance sensors, decision procedures and policy actuators. The programmer can decide what sensors/actuators are necessary and then manually inserts them in the application source code. The toolkit includes fuzzy logic engine that accepts performance sensor inputs and selects resource management policies based on observed application behavior. Autopilot is similar to the MATE cooperative approach. However, it differs from the black-box approach where necessary measure and tuning points are decided and inserted dynamically and automatically. The Autopilot uses fuzzy logic to automate the decision-making process, while MATE is based on simple, conventional rules and performance models. Moreover, MATE is based on the dynamic instrumentation where measure and tuning points are inserted on the fly while in Autopilot it is done manually.

Active Harmony [7] is a framework that allows an application for dynamic adaptation to network and resource capacities. In particular, Active Harmony permits automatic adaptation of algorithms, data distribution, and load based on the observed performance. The application must contain a set of libraries with tunable parameters to be changed. Moreover, it must be Harmony-aware, that is, to use the API provided by the system. Active Harmony manages the values of the different tunable parameters and changes them for better performance. The project focuses on the selection of the most appropriate algorithm. This is conceptually similar to the cooperative approach of MATE. However, it differs from the automatic method that treats applications as black-boxes and does not require them to be prepared for tuning. Active Harmony automatically determines good values for tunable parameters by searching the parameter value space using heuristic algorithm. Better performance is represented by a smaller value of the performance function, and the goal of the system is to minimize the function. MATE uses a distinct approach where performance models provide conditions and formulas that describe the application behavior and allow the system to find the optimal values.

The AppLeS [16] project has developed an application-level scheduling approach. This project combines dynamic system performance information with application-specific models and user specified parameters to provide better schedules. A programmer is supplied information about the computing environment and is given a library to facilitate reactions to changes in available resources. Each application then selects the resources and determines an efficient schedule, trying to improve its own performance without considering other applications. MATE is similar to AppLeS in that it tries to maximize the performance of a single application. However, it focuses on the efficiency of resource utilization rather than on resource scheduling.

6 Conclusions

Parallel/distributed programming offers high computing capabilities to users in many scientific research fields. The performance of applications written for such environments is one of the crucial issues. It is therefore necessary to provide good, reliable and simple tools that automatically carry out tasks involving performance analysis of parallel/distributed programs and behavior optimization. Our goal was to investigate and prove that dynamic tuning works, is applicable and may be effective. We also wanted to demonstrate that it is possible to support a user with a specific functioning environment for automatic dynamic tuning. Our work concluded with the prototype environment called MATE. It includes the monitoring, analysis and modifications of an application on the fly without stopping, recompiling or rerunning the application. The MATE environment tries to adapt the application to dynamic behavior. The conclusion of this work is that although dynamic tuning is a complicated task, it is not only possible, but also provides real improvements in application performance. This methodology seems to be a promising technique for accomplishing the successful performance of applications with dynamic behavior.

References

1. Heath, M.T., Etheridge, J.A. "Visualizing the performance of parallel programs". IEEE Computer, vol. 28, pp. 21-28. November, 1995.
2. Nagel, W.E., Arnold, A., Weber, M., Hoppe, H.C., Solchenbach, K. "Vampir: Visualization and Analysis of MPI Resources". Supercomputer, vol. 12, pp. 69-80. 1996.
3. DeRose, L., Reed, D.A. "SvPablo: A Multi-Language Architecture-Independent Performance Analysis System". Proceedings of the ICPP 99, pp. 311-318. Japan, September, 1999.
4. Espinosa, A., Margalef, T., Luque, E. "Automatic Performance Analysis of PVM applications". EuroPVM/MPI 2000, LNCS 1908, pp. 47-55. 2000.
5. Wolf, F., Mohr, B. "Automatic Performance Analysis of Hybrid MPI/OpenMP Applications". Euro PDP 2003, pp. 13-22. Italy, February, 2003.
6. Miller, B.P., Callaghan, M.D., Cargille, J.M. Hollingswoth, J.K., Irvin, R.B., Karavanic, K.L., Kunchithapadam K., Newhall, T. "The Paradyn Parallel Performance Measurement Tool". IEEE Computer vol. 28. pp. 37-46. November, 1995.
7. Tapus, C., Chung, I-H., Hollingsworth, J.K. "Active Harmony: Towards Automated Performance Tuning". SC'02. November, 2002.
8. Vetter, J.S., Reed, D.A. "Real-time Performance Monitoring, Adaptive Control, and Interactive Steering of Computational Grids". IJHPCA, Vol. 14, No. 4, pp. 357-366. 2000.
9. Buck, B., Hollingsworth, J.K. "An API for Runtime Code Patching". University of Maryland, Computer Science Department, Journal of High Performance Computing Applications. 2000.
10. Geist, A., Beguelin, A., Dongarra, J., Jiang, W., Manchek, R., Sunderam, V. "PVM: Parallel Virtual Machine, A User's Guide and Tutorial for Network Parallel Computing". MIT Press, Cambridge, MA, 1994.

11. Jorba, J., Margalef, T., Luque, E., Andre, J, Viegas, D.X. "Application of Parallel Computing to the Simulation of Forest Fire Propagation", Proc. 3rd International Conference in Forest Fire Propagation, Vol. 1, pp. 891-900. Portugal, November, 1998.
12. Bailey, D.H., Harris, T., Saphir, W., Wijngaart, R., Woo, A., Yarrow, M. "The NAS Parallel Benchmarks 2.0", Report NAS-95-020, December, 1995.
13. "Automatic TCP Window Tuning and Applications". Issue of NLANR Packets, http://www.nlanr.net/NLANRPackets/v2.1/autotcpwindowtuning.html August, 2000.
14. Morajko, A., Morajko, O., Jorba, J., Margalef, T., Luque, E. "Dynamic Performance Tuning of Distributed Programming Libraries". LNCS, 2660, pp. 191-200. 2003.
15. César, E., Mesa, J.G., Sorribes, J., Luque, E. "Modeling Master-Worker Applications in POETRIES". IEEE 9th International Workshop HIPS 2004, IPDPS, pp. 22-30. April, 2004.
16. Berman, F., Wolski, R., Casanova, H., Cirne, W, Dail, H., Faerman, M., Figueira, S., Hayes, J., Obertelli, G., Schopf, J., and Shao, G., Smallen, S., Spring, N., Su, A., Zagorodnov, D. "Adaptive Computing on the Grid Using AppLeS". IEEE Transactions on Parallel and Distributed Systems, Vol. 14, No. 4, pp 369-382. April, 2003.

Imprecise Exceptions
in Distributed Parallel Components

Kostadin Damevski and Steven Parker

School of Computing, University of Utah, Salt Lake City UT 84112, USA

Abstract. Modern microprocessors have sacrificed the exactness of exceptions for improved performance long ago. This is a side effect of reordering instructions so that the microprocessor can execute instructions which were not to be executed due to an exception. By throwing more circuits at the problem, microprocessors are designed so that they are able to roll back to the instruction causing the exception. However, some microprocessors, like the HP Alpha, do not roll back and impose a paradigm of inaccurate exceptions. This decision can reduce circuit complexity and increase speed. We propose a similar method of handling exceptions in a component environment that achieves high performance by sacrificing exception accuracy when dealing with parallel Single Program Multiple Data (SPMD) components. The particular domain this design is intended for is high performance computing, which requires maximum resource use and efficiency. A performance-centric way to handle exceptions is explained as well as additional methodology to enforce exception strictness if required.

1 Introduction

The component software abstraction is one that has been leveraged in multiple computing environments. One such environment is that of high performance scientific computing (see [1]). High performance computing is an environment that, more so than others, emphasizes optimizing the performance of the application. In order to achieve the maximum possible performance and leverage distributed computing potential, the execution of an application may be performed in parallel by multiple parallel computing threads. Typical parallel algorithms divide the problem space in some fashion among the available computing threads, which work in concert to solve the problem, sometimes communicating part of the developing solution to each other. These algorithms often employ the Single Program Multiple Data (SPMD) programming paradigm.

In a component environment, multiple computing threads can be coupled to form one component called an SPMD parallel component. This collection of computing threads is intended to work collaboratively to solve some problem. In order to provide interaction between SPMD parallel components in a distributed environment, collective method invocations are used from a proxy[1] to a parallel component. Collective invocations are defined as ones which require the involvement of each computing thread. That is, all callee computing threads receive an invocation for the same method and all caller threads make an invocation. Therefore, a method invocation launches all computing threads that proceed to execute the same method on separate pieces of data (as the SPMD paradigm suggests). The movement and separation of the data are problems that have been discussed in literature [3, 4, 6, 2].

Parallel components have already been introduced by multiple component frameworks intended for high performance applications. However, these frameworks differ slightly in the way parallel component interaction is defined. For instance, the Parallel Remote Method Invocation (PRMI) mechanism of the PARDIS framework is indirect so that a parallel component's invocation to another parallel component is serialized before it is sent over the wire and deserialized afterwards. That is, the data between components is transferred through a single communication link [Figure 1 (left)]. The authors of PARDIS later acknowledged the performance penalty of the serialization [7] and provided direct, multiport component interaction [Figure 1 (right)]. The improved performance of the direct method, in this case, was to be expected as this method provides better utilization of the network and imposes less synchronization on the computing threads. Our system, SCIRun2, leveraged the previous work in this field and was built with collective invocations that provide direct communication between each parallel computing thread (as it is needed).

Another important aspect of collective invocations is the level of synchronization imposed in their implementation by the underlying system. In order to provide a guarantee of collectiveness, systems often enforce a barrier on the collective invocation. This unarguably provides better error reporting, but at a very significant cost of execution speed, which is unacceptable for many high performance computing applications. Because of this, SCIRun2 and other equivalent systems extend unsynchronized behavior as much as possible. The imprecise exceptions described here, only make sense in a direct and mostly unsynchronized component framework. In fact, their main goal is to loosen the synchronization reigns that regular implementation of exceptions would require.

2 Problem Description and Related Work

This direct mode of invocation and communication introduces our nemesis when dealing with exceptions. Specifically, what if only one computing thread on the invoked (callee) component throws an exception? This exception will be propagated quickly to one of its proxies on the invoking (caller) component, which would cause one of the invoking threads to throw an exception. The rest of the

---

[1] Proxies are objects used to represent other objects. All requests to the proxy are forwarded to the represented object. They are used frequently in distributed middleware in order to invoke a method on a component. In this context, proxies are synonymous to caller components.

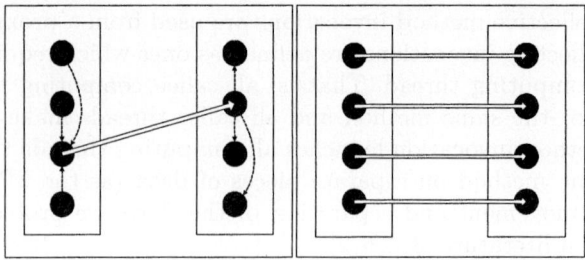

Fig. 1. Serialized and multiport (direct) collective invocation of parallel components.

invoking threads are not aware of the exception as they do not participate in the invocation on the excepting thread. The result is that the invoking component has an incoherent state, possibly destroying the collective behavior of the caller. The user is not in control of which invoking thread has received the exception and is left with a mess to clean up. There is little useful that could come out of this situation. One possibility is to do away with exceptions altogether. Another is to provide the user with a necessary framework to propagate the exception to all the processes of the invoking component. The latter approach has an additional problem: due to the communication lag and the general unsynchronized nature of the processes making the invocation, the exceptions may not arrive exactly during the method invocation. We call these imprecise exceptions. Appropriate thread synchronization can be in place to mitigate this problem; however synchronization will always hurt the performance of the application. Instead we introduce imprecise exceptions, allowing the programmer to trade off accuracy of the exception and performance.

One category of exceptions that CORBA supports are those that represent various network and system failures. These are termed system exceptions. Conversely, user purpose exceptions are explicitly defined in the IDL via a throws clause in each method. While CORBA's user exceptions occur only on the server side, system exceptions can occur in both client and server. An exception hierarchy of classes also exists. The base classes impose methods on the derived classes which report useful information about the exception, such as whether or not method completed executing or not. However, exception inheritance of any sort is not allowed within the CORBA IDL.

Imprecise exceptions have been suggested for the Haskell functional programming language [5]. This is a language that achieves high performance and flexibility through lazy evaluation of expressions. Adding imprecise exceptions to Haskell provides the stack unwinding error reporting mechanism of exceptions, without sacrificing rich transformations or other strengths of the language. Some of our ideas are similar to the ideas presented by Peyton-Jones et al. [5], although the domain is quite different.

3 Design

This work will provide the utility of component exceptions in a distributed SPMD component setting, while attempting not to sacrifice performance. In doing this we ought to emphasize that very little additional overhead is imposed in cases in which an exception does not occur. If an exception does occur, it is thrown and the application incurs a performance penalty (as it should).

Let us consider an example that enables us to propose a design for imprecise parallel component exceptions. In discussing this design, we will use the following interface (written in Scientific Interface Definition Language (SIDL) [8]):

```
interface SomeThrowSomeNo {
    void exceptionOnemethod() throws exceptionOne;
    void exceptionTwomethod() throws exceptionTwo;
    void noexceptionmethod();
}
```

The SomeThrowSomeNo interface contains two methods that throw exceptions and one method that does not. The *noexceptionmethod* method has no implementation. The two exception-throwing methods have the following implementation:

```
void exceptionOnemethod() {
    if (rank == 1) throw exceptionOne;
}
void exceptionTwomethod() {
    if (rank == 2) throw exceptionTwo;
}
```

For the purposes of this discussion, we consider two SPMD parallel components of three parallel threads each; one component will be the caller and the other the callee. We assume that a proxy has already been created by the caller component. Each thread of the caller component has a parallel proxy and is able to invoke a method on any callee thread. However, the collective interaction dictates that all callee threads should be invoked only once. Therefore, we have a one-to-one correspondence in the collective invocation between the threads of the parallel component. For the sake of clarity let's assume that the caller thread whose rank is zero invokes the method on the zero ranked callee thread, rank one invokes on rank one, etc.

3.1 Reporting Imprecise Exceptions

Upon the execution of this SPMD code:

```
stsn-proxy->exceptionOnemethod();
stsn-proxy->noexceptionmethod();
```

The caller thread ranked one receives an exception message back from its invocation. Threads zero and two have continued execution to some other part of the code. We are concerned that threads zero and two eventually receive information about the exception that occurred. Thread one has the responsibility of informing the other threads of the exception in *exceptionOnemethod*. It does so through an asynchronous message to each of the other threads. After this, thread one can throw the C++ equivalent of *exceptionOne* and goes into its appropriate catch block (if it exists).

All remote method invocation stubs contain code to check for the existence of asynchronous messages notifying of exceptions in other threads. In the scenario we discussed, we assume the threads are perfectly synchronized (i.e. they execute the same instruction at the same time) and an asynchronous message travels at lightning speed. Upon the invocation of *noexceptionmethod*, threads zero and two will instantaneously detect the existence of *exceptionOne* and will throw it within their context. At this time all threads would have excepted and reached a steady state.

Asynchronous messages provide the appropriate behavior since they do not impose synchronization of the processes in order to communicate the exception and because reads for these messages come by at an extremely small cost. Performing a check for an exception message within all remote invocations may be unnecessary at times. At other times, it would be overzealous. Overall, we feel that providing the check through each proxy use strikes a perfect balance in most cases. The cost of checking for an asynchronous message is extremely small (typically checking a global flag) and negligible even if there are many frequent uses of the proxy. However, if the opposite is true and proxy use is very infrequent it could decrease the granularity of the exception checks to an unacceptable limit. We discuss this problem in the next section. The extreme example of this case is when there are no proxy uses after an exception and no opportunity to report this exception. This problem is considered in Section 5.

3.2 getException Method

We realize that it is unfair to assume lightning speed message delivery as we did before. Therefore, if the threads are not synchronized, we should be prepared to see *exceptionOne* thrown sometime later than the invocation of the *noexceptionmethod*. In many cases this is not acceptable. Executing *noexceptionmethod* may depend on the successful finish of *exceptionOnemethod*.

If the programmer is unable to count on exceptions to occur within a specific invocation, an exception can crop up too late or too early and provide the wrong behavior. In order to allow the programmer to control where exceptions occur in the code, we provide a *getException* method which is invoked on the proxy (similar to Peyton-Jones et al.'s use of the same method). Once we invoke *getException* we guarantee that all exceptions that have occurred have been propagated to this thread. The only real way to ensure this in a parallel and distributed environment is to synchronize the caller threads. That is, this method requires a barrier and as such imposes a performance penalty. Despite this, it is useful for it provides the programmer with a wider range of uses for exceptions.

4 Corner Cases

Although the design presented above is straightforward, the complex interaction between two parallel components requires that a number of non-trivial scenarios be analyzed.

Consider what would happen if the implementation of *exceptionOnemethod* were something like this:

```
void exceptionOnemethod() { throw exceptionOne; }
```

Collective invocation to this method would result with each caller thread receiving a separate exception. When this occurs, threads should ignore the asynchronous exception messages reporting exceptionOne in order not to throw the same exception multiple times. To do this, we keep an exception log which makes sure that we are aware of exception we have already produced. In doing this we have to make sure that we keep enough information to help us determine the right course of action. We use a combination of tags to uniquely identify each proxy invocation. The guarantee that all collective methods are invoked by all computing threads can be relied on to uniquely tag each collective invocation without imposing additional communication.

Let us consider the following example where method invocations are executed by the caller component using the SomeThrowSomeNo interface:

```
stsn-proxy->exceptionOnemethod();
stsn-proxy->exceptionTwomethod();
```

Both *exceptionOne* and *exceptionTwo* have been thrown in this scenario. Caller thread one and two respectively are the first ones to receive each exception. Let us suppose that they both contact thread zero at about the same time so that thread zero has received messages about two distinct exceptions. Now thread zero has a dilemma about which exception to report: *exceptionOne* or *exceptionTwo*. Here are some possible ways to resolve this problem:

- Allow non-determinism. This allows for either exception to be thrown. It is probably an improvement over no exceptions being thrown, however it is definitely not a clean solution as the collective nature of the caller is not preserved, and the application may except differently in each run. This may confuse the programmer and result with unwanted behavior.
- Always report the user exception before the system or vice versa. This may prove to be effective in some situations. However, this choice would work poorly when the confusion is between two user and two system exceptions.
- Report the first method's exception first. Using the information tags that we collect per each proxy invocation, we can easily make this determination. This would provide consistent behavior and error reporting which would be of use to a programmer. We opt for this solution.

This scenario also requires that thread rank two comes to its senses and throws *exceptionOne* instead of *exceptionTwo*. In order to provide this behavior, we should synchronize and determine the correct exception between all the exceptions that have been witnessed among all of the threads. This includes each thread that comes in direct contact with an exception. It should not rush and throw an exception before it makes sure that it is the correct exception to throw. In order to perform this, our implementation modifies the tournament barrier algorithm to determine the best exception to be thrown and propagate the result to all the threads. This algorithm is also applied in the implementation of the *getException* method, since similar exception election is needed there as well. Detailed discussion of the tournament barrier and other barrier types can be found in [9].

One situation not captured by our running example is one where a collective invocation of the same method yields two different exceptions caught by separate caller threads. This is a situation that is not likely to occur often. We propose a random yet consistent choice of which exception to throw. The exception reporting mechanism will use a consistent way of choosing to throw the same exception across different runs or compilations and in all of the computing threads. In our system exceptions are related to exception IDs assigned by the compiler roughly representing the order of the exceptions in the throws clause. The lower exception ID is chosen to resolve this confusion. This provides consistent behavior to all runs of the code compiled by our IDL compiler. As every compiler and parallel component architecture could make this determination differently (we do not propose that everyone uses exception IDs), we fully expect that a different exception may be raised in the same situation in a different environment. This is something that the parallel component programmer should be prepared to face. Another possibility would be to prefer the exception type thrown by the lowest (or highest) rank.

The exception reporting mechanism relies on the checks performed by the proxy upon each method invocation. If no method invocation existed in the executing code we could fail to report an exception. In order to remedy this, we could impose a *getException* call upon the proxy destructor. This would guarantee that exceptions are reported and is a viable solution. However, if proxies are created and deleted often, we would significantly constrain the performance of the application by synchronizing the computing threads upon each invocation to *getException*. Another possibility is to save the actual deletion of the proxy until the very end of the program. Synchronization at the end of a program will not impact performance. Our suggestion is to perform a *getException* synchronization within the Object Request Broker[2] (ORB) destructor. The ORB's deallocation by the destructor typically occurs when main finishes on a particular machine.

An imprecise exception can come before, during, or after the excepting method invocation in caller threads that do not directly receive an exception. So far, we have mostly discussed exceptions arriving after the collective method invocation to the exception method is finished. We have not yet considered the case of them arriving before. Exceptions arriving before the invocation are very awkward to face. In essence, we have been given a glimpse of the future and

---

[2] Object Request Broker exists on each node and manages requests among components. An ORB should be deallocated when all component interactions on one node cease.

we respond to this sight even though we have not yet reached this point of execution. The best alternative is to rule this case out altogether using the proxy infrastructure we have already described. This comes at almost no extra cost and does not impact performance. We can hold off and ignore the existence of a specific exception from an asynchronous message until we encounter the method which has produced it. When we reach this method we do not invoke it but throw the exception at that time.

5 Conclusions

We have presented a paradigm of imprecise exceptions for parallel distributed components. Based on the fact that less synchronization improves performance, we claim that this approach saves us from a drastic execution speed penalty with exceptions. We examined the difficult scenarios that are possible with imprecise exceptions. We also provided a way to synchronize imprecise exceptions so that they provide the same behavior as regular exceptions. This imprecise exception mechanism is targeted at a component-based distributed Problem-Solving Environment, but could be applied to a variety of scenarios involving multiple communicating SPMD parallel programs.

6 Results

The performance improvement of our design is somewhat self-validating. We provide an exception handling method that allows a greater degree of asynchrony, which in turn betters performance. In spite of this, we implemented a simple example and provided some runtimes for it in order to show that our concepts work and to exemplify how much of a performance increase we can achieve.

The use scenario we chose involves a parallel query tool sending queries to a parallel data server. We implemented these as parallel components and tested various modes of exception use: total imprecise exceptions, use of the *getException* method after each query, and use of regular (precise) exceptions. Assuming the time to process the query is close to zero, here are the results of this test (between two Pentium4 class machines connected by Gigabit Ethernet):

Table 1. Average time for a single query (averaged for 10000 queries).

Instances of Query Tool	Imprecise Exceptions	getException()	Precise Exceptions
1	658.39	675.89	672.91
2	978.55	1487.46	1595.33
3	1924.08	3482.97	4238.06
4	1705.58	6882.35	8899.77

Times are in microseconds.

The results were largely what we expected. The performance increase of the imprecise exceptions scales as the degree of parallelism increases. It is important to note that the imprecise exceptions we measured do not make use of the *getException* method. This total asynchrony is unrealistic in practice; therefore the difference in performance would be slightly less exaggerated and peaking at the numbers we have shown above.

Acknowledgements

The authors would like to acknowledge Matthew Flatt for initially pointing us to Haskell exceptions. This work was funded by the Department of Energy Center for Component Technology for Terascale Simulation Software (CCTTSS) and by NSF ACI-0113829.

References

1. R. Armstrong, D. Gannon, A. Geist, K. Keahey, S. Kohn, L. McInnes, S. Parker, and B. Smolinski. Toward a Common Component Architecture for High-Performance Scientific Computing. In *Proceedings of the 8th IEEE International Symposium on High Performance Distributed Computing*, 1999.
2. F. Bertrand, Y. Yuan, K. Chiu, and R. Bramley. An approach to parallel MxN communication. In *Proceedings of the 3rd Los Alamos Computer Science Institute (LACSI) Symposium*, October 2003.
3. K. Damevski and S. Parker. Parallel remote method invocation and m-by-n data redistribution. In *Proceedings of the 3rd Los Alamos Computer Science Institute (LACSI) Symposium*, October 2003.
4. G. A. Geist, J. A. Kohl, and P. M. Papadopoulos. CUMULVS: Providing fault-tolerance, visualization and steering of parallel applications. In *Environment and Tools for Parallel Scientific Computing Workshop*, Domaine de Faverges-de-la-Tour, Lyon, France, August 1996.
5. S.L. Peyton Jones, A. Reid, F. Henderson, C.A.R. Hoare, and S. Marlow. A semantics for imprecise exceptions. In *Proceedings of the SIGPLAN Conference on Programming Language Design and Implementation*, pages 25–36, 1999.
6. K. Keahey, P. K. Fasel, and S. M. Mniszewski. PAWS: Collective invocations and data transfers. In *Proceedings of the 10th IEEE International Symposium on High Performance Distributed Computation*, July 2001.
7. K. Keahey and D. Gannon. Developing and evaluating abstractions for distributed supercomputing. *Journal of Cluster Computing, special issue on High Performance Distributed Computing*, 1(1), 1998.
8. S. Kohn, G. Kumfert, J. Painter, and C. Ribbens. Divorcing language dependencies from a scientific software library. In *Proceedings of the 10th SIAM Conference on Parallel Processing*, Portsmouth, VA, March 2001.
9. J.M. Mellor-Crummey and M.L. Scott. Algorithms for scalable synchronization on shared-memory multiprocessors. *ACM Transactions on Computer Systems*, 9(1):21–65, 1991.

Topic 2
Performance Evaluation

Wolfgang E. Nagel, Thomas Ludwig, Jeffrey Vetter, and Lorenzo Donatiello

Topic Chairs

Besides excellent performance results for quite some applications for large scale systems, measurements very often reveal performance limitations for parallel programs. They result either from the architecture of the parallel systems (memory organization, communication hardware), from system software (parallel compilers, message passing implementations, operating systems), or sometimes even from implementation aspects of the given application. However, one of the main observations still is that on a single PE, many, if not most applications which are relevant to the technical field do not benefit adequately from clock rate improvement. The reason for this is memory accesses: most data read and write operations have to access memory which is relatively slow compared to processor speed. With several levels of caches we now have architectures which, in principal, provide plenty of options to keep the data as close as possible to the processor. Nevertheless, the resulting chip architectures are quite complex, and the compiler development progresses more slowly than anticipated.

In the near future, we will see chip architectures which integrate more than one core on a chip. A simple pocket calculator analysis shows that in most cases this will limit memory bandwidth even further. After the delivery of powerful SMP nodes from most vendors, we have learned that coupling these kinds of chips to a parallel shared memory computer and integrating these new SMP basic blocks to larger SMP-Clusters is even more challenging. It does not really matter whether some kind of message passing (like MPI), some kind of shared memory (like OpenMP) or may be a heterogeneous combination of both is used as the parallelization concept. The announced system peak performance numbers are quite impressive, based on the simple multiplication of single PE peak performance numbers. In contrast to the new opportunities, a large number of application programmers are still struggling with many complex features of todays modern computer architectures. Quite often, it takes several hundreds of clock ticks to get one element out of the memory. In future, it is expected that the numbers get worse. This will limit the success of a technology which could give a boost to new application areas like virtual products (engineering), new medication (biology field) or even better insight to complex simulation problems (nanotechnologies).

Therefore, the techniques and methods developed in our field have a high potential to address a couple of system bottlenecks, to improve their localization and understanding, and to provide solutions. Our workshop brings together people working in the different fields of performance modeling, evaluation, prediction, measurement, benchmarking, and visualization for parallel and distributed applications and architectures. It covers aspects of techniques, implementations, tools, standardization efforts, and performance-oriented development of distributed and parallel applications. This year, 24 papers were submitted to this workshop, one was selected as distinguished paper, eight were selected as regular papers, and three as short papers.

The removal of measurement overhead and intrusion from the measurement process - one of the more general and very important problems in our field - is the topic of the distinguished paper from Malony and Shende. New algorithms for quantifying and eliminating the overhead on-the-fly for profiling mechanisms have been developed and implemented. The tests with the NAS parallel benchmarks have shown that they are effective at reducing the error in estimated performance. Another very interesting approach is the detection of performance bottlenecks. The paper from Gerndt et. al. focuses on the capability of two OpenMP performance analysis tools.

We already have stressed the aspect of performance limitation through memory access. The contribution from Kereku et.al. is focused on a monitoring environment where the access to data structures is analyzed by using a combination of a hardware monitor and a software simulator. A quite similar approach is going to be presented by Luiz DeRos et. al. A data-centric component analyzing the memory behaviour requires the mapping of memory references to the corresponding symbolic names of the data structure. The paper presents the algorithms implemented in the SIGMA environment. The third paper in that area - from Tao and Karl - investigates cache effects. Based on simulator results, these studies concentrate on several different cache coherence protocols.

Message passing is still the dominant parallelization paradigm for large scale computations. Therefore, performance limitations in such implementations have a large impact on the application performance. Based on a monitoring system, the contribution from Bongo et.al. investigates the performance of collective operations. For the "allreduce" operation, they have identified parameters which have significant impact on the load balance of this operation. In the paper from Saif and Parashar, several MPI implementations for non-blocking communication operations have been analyzed. By using the results from this analysis in a test application, performance improvements between 25% and about 50% have been reported. The third contribution from Hipp and Rosenstiel describes a heterogeneous parallelization approach for a particle simulation code.

The first paper from the last part of the workshop from Iosevich and Schuster compares the performance of two different multithreaded memory coherence protocols. While the coherence protocol always is the critical decision when a DSM system is realized, the runtime results have shown no dramatic differences between a simple and a highly optimized variant. The second paper from Stahl et.al. describes techniques to exploit task-level parallelism in sequential object-oriented programs to obtai parallel Java. The third paper from Rodriguez et. al. presents a method to derive accurate and simple analytical models which are able to describe the performance of parallel applications. Finally, the last paper from Fernandes et.al. presents a theoretical performance analysis of a tool (PEPS) used to analyze Stochastic Automata Networks.

Overhead Compensation in Performance Profiling

Allen D. Malony and Sameer S. Shende

Performance Research Laboratory, Department of Computer and Information Science
University of Oregon, Eugene, OR, USA
{malony,sameer}@cs.uoregon.edu

Abstract. Measurement-based profiling introduces intrusion in program execution. Intrusion effects can be mitigated by compensating for measurement overhead. Techniques for compensation analysis in performance profiling are presented and their implementation in the TAU performance system described. Experimental results on the NAS parallel benchmarks demonstrate that overhead compensation can be effective in improving the accuracy of performance profiling.

Keywords: Performance measurement and analysis, parallel computing, profiling, intrusion, overhead compensation.

1 Introduction

Profiling and tracing are the two main approaches for empirical parallel performance analysis. Parallel tracing is most often implemented as a measurement-based technique. Here, the application code is instrumented to observe *events* which are recorded in a trace buffer on each occurrence during execution [5, 28]. In contrast, performance profiling [16] can be implemented in one of two ways: 1) *in vivo*, with measurement code inserted in the program (e.g., see [6, 9, 21, 23, 24]), or 2) *ex vivo*, by periodically interrupting the program to assign performance metrics to code regions identified by the halted program counter (e.g., see [10, 13, 14, 22, 26]). The first technique is commonly referred to as *measurement-based profiling* (or simply *measured profiling*) and is an *active* technique. The second technique is called *sample-based profiling* (also known as *statistical profiling*) and is a *passive* technique since it requires little or no modification to program.

There are significant differences of opinion among performance tool researchers with regards to the merits of measured versus statistical profiling. The issues debated include instrumentation, robustness, portability, compiler optimizations, and intrusion. Ultimately, the profiling methods must be *accurate*, else the differences of opinion really do not matter. Herein lies an interesting performance analysis conundrum. How do we evaluate analysis accuracy when the "true" performance is unknown? Some technique must be used to observe performance, and profiling tools will always have limitations on what performance phenomena can and cannot be observed [17]. Is a tool inaccurate if it does not provide information about particular performance behavior at a sufficient level of detail? Furthermore, no tool is entirely passive, and any degree of execution intrusion can result in performance perturbation [17]. Should not all profiling tools be considered inaccurate in this case? Parallel performance analysis is no different from experimental methods in other sciences. "Truth" always lies just beyond the reach of observation. As long as this is the case, accuracy will be a relative assessment.

Until there is a systematic basis for judging the accuracy of profiling tools, it is more productive to focus on those challenges that a profiling method faces to improve its accuracy. Our work advocates measured profiling as a method of choice for performance analysis [21]. Unfortunately, measured profiling suffers from direct intrusion on program execution. This intrusion is often reported as a percentage slowdown of total execution time, but the intrusion effects will be distributed throughout the profile results. The question we pose in this paper is whether it is possible to compensate for these effects by quantifying and removing the overhead from profile measurements.

Section §2 describes the problem of profiling intrusion and outlines our compensation objectives. The algorithms for overhead removal are described in Section §3. We tested these on a series of case studies using the NAS parallel benchmarks [1]. In Section §4, we report our findings with an emphasis on evaluating relative accuracy. Our approach can improve intrusion errors, but it does not fully solve the overhead compensation problem. Section §5 discusses the thorny issues that remain. Conclusions and future work are given in Section §6.

2 Measured Profiling and Intrusion

Profiling associates performance metrics with aspects of a program's execution. Normally, it is the program's components being profiled, such as its routines and code blocks, and profile results show how performance data are distributed across these program parts. Execution time is the most common metric, but any source of performance data is valid (e.g., hardware counters [4, 6, 23, 30]). *Flat profiles* show performance data distributed onto the static program structure, while *path profiles* [11] map performance data to dynamic program behavior, most often represented as program execution paths (e.g., *routine calling paths* [7, 11]). Profiling both measures performance data and calculates performance statistics at runtime.

Measured profiling requires direct instrumentation of a program with code to obtain requisite performance data and compute performance statistics. Generally, we are interested in observing *events* that have *entry / exit* semantics. Instrumentation is done both at the event "entry" point (e.g., routine begin) and the event "exit" point (e.g., routine return). Profile statistics report performance metrics observed between event entry and exit. We typically speak of *inclusive* performance, which includes the performance of other descendant events that occur between the entry and exit of a particular event, as well as of *exclusive* performance, which does not include descendant performance.

Unfortunately, the measurement code alters the program's execution, primarily in execution time dilation, but it also affects hardware operation. Intrusion will be reflected in the profile results unless techniques can compensate for it. Other research work has sought to characterize measurement overhead as a way to bound intrusion effects or to control the degree of intrusion during execution. For instance, the work by Kranzlmüller [15] quantifies the overhead of MPI monitors using the benchmarking suite SKaMPI, and Fagot's work [8] assesses systematically the overhead of parallel program tracing. The work by Hollingsworth and Miller [12] demonstrates the use of measurement cost models, both predicted and observed, to control intrusion at runtime via measurement throttling and instrumentation disabling.

Our interest is to compensate for measurement intrusion. Thus, we need to understand the intrusion effects are manifested in performance profiles. Let us first consider inclusive execution time profiles, using routines as our generic events. Two timing measurements are necessary to determine inclusive time for a routine every time it is called: one to get the current clock value at time of entry and one to get the clock value at exit. The difference between the clock samples is the inclusive time spent in this call. Let Δ_i represent the overhead to measure the inclusive time. If the routine A is executed N times, its inclusive time is increased by $N * \Delta_i$ measurement overhead.

However, A's inclusive time is also increased by the overhead incurred in the inclusive time measurement of descendant routines invoked after A is called and before it exits. If M routines are called while A is active (M covers all calls to A), the inclusive time is increased additionally by $M * \Delta_i$ because each descendant routine profiled will incur inclusive overhead.

To calculate A's exclusive time, the inclusive time spent in each direct descendant is subtracted from A's inclusive time. Since this occurs at A's exit, the inclusive times for all direct descendant calls must be summed. If Δ_e is the measurement overhead to do the summation upon each direct descendant exit, the total overhead that gets reflected in A's exclusive time is $L * \Delta_e$ for L direct descendant calls. (The subtraction from A's inclusive time is small and will be ignored for sake of discussion.) It is important to observe that all overhead accumulated in the inclusive times of A's direct descendants gets cancelled in the exclusive time computation. Unfortunately, if we calculate the inclusive and exclusive times simultaneously, as is normally the case, the $L * \Delta_e$ overhead must be added to A's inclusive time.

To summarize, the total performance profile of routine A will show an increase in inclusive time due to measurement overhead of $N * \Delta_i + M * \Delta_i + L * \Delta_e$, where N is the total number of times A is called, M is the number of times descendant routines of A are called, and L is the number of times A's direct descendant routines are called. The performance profile of routine A will also show an increase in exclusive time of $L * \Delta_e$ due to overhead.

Standard profiling practice does not report the measurement overheads included in per event inclusive and exclusive profiles. Typically, overhead is reported as a percetage slowdown in total execution time, with the implicit assumption that measurement overhead is equally distributed across all measured events. Clearly, this is not the case. Events with a large number of descendant event occurrences will assume a greater proportion of measurement overhead. If we cannot compensate for this overhead, the performance profile will be distorted as a result. Furthermore, if processes of a parallel application exhibit different execution behavior, the parallel profiles will show skewed performance results.

3 Overhead Analysis and Compensation

The above formulation allows us to quantify the measurement overheads that occur in inclusive and exclusive times as a result of measured parallel profiling. If we were to actually compute these overheads, it must be done at runtime because the overheads depend on the dynamic calling behavior. However, our goal is to additionally *remove* the measurement overhead incurred in parallel profiling. Thus, the profiling system must both track the overhead and compensate for it dynamically in profile computations.

As a first step, we must determine the values Δ_i and Δ_e. These overhead units will certainly depend on the machine and compilers used, but could also be influenced by the application code, the languages and libraries, how the application is linked, and so on. Also, it should not be assumed that Δ_i and Δ_e are constant. In fact, they may change from run to run, or even within a run. This means that we must measure the overhead values at the time the application executes, and even then the values will be approximate. The approach we propose is to conduct overhead experiments at application startup.

3.1 Overhead Analysis

To ascertain the overhead values Δ_i and Δ_e, we must measure the instrumentation code that calculates inclusive and exclusive performance. However, the Δ_i and Δ_e values may be at the same scale as the measurement precision. Thus, we must construct a test that guarantees statistical accuracy in our estimation of Δ_i and Δ_e.

Our approach is to conduct a two-level experiment where the inclusive measurement code is executed k number of times (in a tight loop) and the performance data being profiled is measured and stored. This procedure is then repeated j times and the minimum performance value across the j experiments is retained. We use that value divided by k to compute Δ_i. While this does not entirely insure against anomalous timing artifacts, its statistical safety can be improved by increasing j. The two-level experiment is then repeated, this time with the exclusive measurement code included, resulting in an approximation of $\Delta_i + \Delta_e$. We can substract our Δ_i approximation to find Δ_e. It is important to note that we need to do this overhead evaluation at the beginning of the application execution for the specific set of profile performance data being measured[1].

---

[1] In this regard, Δ_i and Δ_e are really vectors of overhead values.

3.2 Overhead Compensation

Given the approximations for Δ_i and Δ_e, we are ready to apply overhead compensation. There are several ways to go about it. One way is to use the formulas above and remove the overhead at the end of the execution. To do so, however, we must determine the variables N, M, and L for every event. Without going into details, calculating N, M, and L amounts to maintaining a dynamic call graph for every currently active event as the root of its own call tree. Knowing this allows us to consider a second way that removes inclusive overhead on-the-fly with every event exit. In this case, we need to only determine m_i and l_i values for each ith event occurrence ($M = \sum m_i$ and $L = \sum l_i$, for $1 \leq i \leq N$). Since we must calculate profile values at event exit, it is reasonable to have these be compensated calculations.

Using compensated inclusive calculations at event exit, we now have a choice for calculating compensated exclusive profile values. Without loss of generality with respect to other performance metrics, consider only execution time. The exclusive time for an event A is the difference between A's inclusive time and the sum of the inclusive times of all of A's direct descendants. Regardless of compensated or uncompensated inclusive times, we have to accummulate the inclusive times of A's direct descendants. However, for uncompensated inclusive times, an additional subtraction at A's exit of $l_i * \delta_e$ is needed to calculate the exclusive profile time for this invocation of A. Thus, the scheme we advocate for on-the-fly overhead compensation of exclusive time is to subtract the compensated inclusive times of A's direct descendants (as they exit) from A's running exclusive time, and add A's compensated inclusive time back in when A finally returns.

4 Experiments with Compensation Analysis in TAU

To evaluate the efficacy of overhead compensation, we must implement the methods described above in a real parallel profiling system and demonstrate their ability to improve application profiling results. To this end, we have implemented the overhead compensation techniques in the TAU parallel performance system [21]. TAU uses measured profiling to generate both flat profiles and callpath profiles. Inclusive and exclusive overhead compensation is implemented in TAU for both profiling modes.

How will we know if our overhead compensation works successfully? The standard measure for intrusion error is usually given as a percentage slowdown in total execution time. Thus, one can test the ability of a compensation-enabled profiling tool to accurately recover total execution time from profile measurements with varying levels of instrumentation. As the instrumentation increases, so likely will the intrusion and the overhead compensation techniques will be more stressed. However, it is important to understand that accurate performance profiling will also depend on the precision of measurement, in particular, the ability to observe small performance phenomena. Overhead compensation can improve measurement accuracy, but it cannot remove measurement uncertainty for small events.

For any level of instrumentation, it is reasonable to expect that less instrumentation leads to more accurate profiling results than more instrumentation. Thus, if the total execution time is accurate, we might assume the rest of the profile statistics are also. However, there are two issues to keep in mind. First, performance variability due to environmental factors can arise even in un-instrumented applications. Second, the success of overhead compensation on profile statistics is difficult to assess given that the "real" profile values are not known. The best we can do then is to compare profiling results at one level of instrumentation with results from using less instrumentation, under the assumption that the execution is relatively stable and the results from less instrumented runs are more reliable and accurate.

4.1 Experimental Methodology

The experimental methodology we use to evaluate overhead compensation characterizes the profiling measurement for an application with respect to levels of instrumentation and sequential versus parallel execution. For the experiments we report here, we used three levels of instrumentation. The *main only (MO)* instrumentation is used to determine the total execution time for the "main" routine. This will serve as our standard estimate for overall performance using as little instrumentation as possible. The *profile all (PA)* instrumentation generates profile measurements for every source-level routine of the program. The *callpath all (CA)* instrumentation uses TAU's callpath profiling capabilities to generate profile measurements for routine callpaths of the program. Obviously, this *CA* instrumentation is significantly greater than *PA* and will further stress overhead compensation.

Five experiments are run for an application using the three levels of instrumentation. The *MO* experiment gives us a measure of total execution time. For parallel SPMD applications, we profile the "main" routine of the individual processes, using the maximum as the program's total execution time. The per process times can also be used for evaluation under the assumption the program's behavior is well-behaved. The *PA* experiment returns profiling measurements without compensation. We let *PA-comp* represent a *PA*-instrumented run with compensation enabled. Similarly, a *CA* experiment returns callpath profiling measurements without compensation and a *CA-comp* experiment returns callpath profile results after overhead compensation.

We can compare the "main" profile values from *PA*, *PA-comp*, *CA*, and *CA-comp* runs to the *MO* run to evaluate the benefit of overhead compensation. However, we can also look at other indirect evidence of compensation effectiveness. Assuming the *PA-comp* run delivers accurate profile results, we can compare the associated statistics from the *CA-comp* profile to see how closely they matched. This can also be done for the *PA* and *PA-comp* runs with different levels of instrumentation. Per process values can be used in all parallel cases for comparison under SPMD assumptions.

Ten trials are executed for each experiment. We have a choice of using profile results with the minimum "main" values or the average "main" values in the evaluation. Our preference is to use the profiles reporting minimums. The reason is that these runs are likely to have less artifacts in the execution (i.e., anomalies not directly attributed to the program) and, thus, represent "best case" performance. On the other hand, an argument can be made to take the average profile values, since artifacts may be related to the instrumentation. We report both values in our results below. However, it is important to note that calculating average profiles may not be reliable for programs that do not behave in a deterministic manner.

Following the experimental methodology above, we tested overhead compensation on all NAS parallel benchmark applications [1]. As the application codes vary in their structure and number of events, we expected differences in the effectiveness of compensation. We ran the ten experiments for each application sequentially and on 16 processors. Problems sizes were chosen mainly to achieve runtimes of reasonable durations. The parallel system used in our study was a Dell Linux cluster[2]. In the following sections, we report on six of the NAS benchmarks: SP, BT, LU, CG, IS, and FT.

4.2 Sequential Experiments

Table 1 shows the total sequential execution time of "main" in microseconds from the different profiles for the different applications. The minimum and mean values are reported. We also calculate the percentage error (using minimum and mean values) in approximating the *MO* time for "main." The dataset size (A or W) used in the experiments is indicated.

An important observation is that the TAU measurement overhead per event is already very small, on the order of 500 nanoseconds for flat profiling on a 2.8 GHz Pentium Xeon processor. This can be easily seen in the TAU profile results (not shown) where the overhead estimation is given as an event in the profile. Of course, the slowdown seen in the *PA* and *CA* runs depends on the benchmark and the number of events instrumented and generated during execution. Because more events are created for callpath profiling, we expect to see more slowdown for the *CA* runs.

The results show that overhead compensation is better at approximating the total execution time, both for flat profiles and for callpath profiles. This is generally true for all of the NAS benchmarks we tested. In the case of IS-A, the flat profile compensation (*PA-comp*) shows remarkable improvement, from a 193% error in the *PA* measurement to within 2.1% of the "main" execution time. The improvements in compensated callpath profiles for SP-W to less than 1% error are also impressive.

To be clear, we are instrumenting *every* routine in the program as well as every depth of callpath. If, as a result, we instrument a small routine that gets

---

[2] Hardware: 16 dual-processor 2.8 GHz Intel® Pentium 4 Xeon™ CPUs, 512 KB cache, 4 GB memory per node, gigabit ethernet interconnect, hyperthreading enabled. OS: Red Hat Linux 2.4.20-20.8smp kernel.

Table 1. Overhead Compensation Results for NAS Benchmarks on Linux Cluster - Sequential

Experiment		MO μsecs	PA μsecs	PA-comp μsecs	CA μsecs	CA-comp μsecs
SP-A	min	387588657	397602281	392833924	405226516	399405895
	mean	388540699	398360423	394245841	407233889	401650317
%error (min:mean)			2.5 : 2.5	1.3 : 1.4	4.5 : 4.8	3.0 : 3.3
SP-W	min	65427051	67942093	66404006	71812623	65517453
	mean	66178471	69254426	67104562	73659688	66687843
%error (min:mean)			3.8 : 4.6	1.4 : 1.3	9.7 : 11.3	0.1 : 0.7
BT-A	min	522765488	549063282	542479898	553178345	532736660
	mean	524248915	552617635	545409236	555959945	536680190
%error (min:mean)			4.6 : 5.2	3.4 : 3.8	5.8 : 6.0	1.9 : 2.3
LU-W	min	297366632	300993317	302786082	306287598	303405699
	mean	299395075	302941264	305796049	307849925	306172285
%error (min:mean)			1.4 : 3.3	0.0 : -0.6	10.2 : 8.9	3.4 : 2.6
CG-A	min	5368659	5733951	5740469	6824800	6536302
	mean	5560969	5758157	5764569	6916842	6628535
%error (min:mean)			6.8 : 3.5	6.9 : 3.6	27.1 : 24.3	21.7 : 19.1
IS-A	min	5967910	17540614	6094620	35457776	2632054
	mean	5987002	17667114	6215288	36008102	4441510
%error (min:mean)			193.9 : 195.0	2.1 : 3.8	494.1 : 501.4	-55.8 : -25.8
FT-A	min	24593893	25418103	25296244	29104159	28754736
	mean	25215853	25549141	25557557	29470907	28918045
%error (min:mean)			3.3 : 1.3	2.8 : 1.3	18.3 : 16.9	16.9 : 14.6

called many times, overheads can accumulate significantly. For callpath profiling with instrumentation including a small event, overheads will be effectively multiplied by the number of callpaths containing the small routine. This is what is happening in IS-A. Flat profile compensation can deal with the error, but callpath compensation cannot. It is interesting that the reason can be attributed to the small differences in overhead unit estimation, ranging in this case from 957 nanoseconds (minimum) to 1045 (maximum). This seemingly minor 90 nanoseconds difference is enough in IS-A callpath profiling to cause major compensation errors. Certainly, the proper course of action is to remove the small routine from instrumentation.

4.3 Parallel Experiments

Table 2 reports the results for parallel execution of the six NAS benchmarks on the Linux Xeon cluster. All of the applications execute as SPMD programs using MPI message passing for parallelization across 16 processors. For evaluation purposes, we compare minimum "main" values for each process to those for the MO run. Each process will complete its execution separately, resulting in different "main" execution times. We show the range of minimum values (labeled "high"

Table 2. Overhead Compensation Results for NAS Benchmarks on Linux Cluster - Parallel

Experiment		MO μsecs	PA μsecs	PA-comp μsecs	CA μsecs	CA-comp μsecs
SP-A	min (high)	67369049	67519758	67758618	72968801	73416350
	min (low)	64346890	64834412	64963104	67047549	67124742
%error (mean:high)			0.6 : 0.2	0.8 : 0.5	4.3 : 8.3	4.3 : 8.9
SP-W	min (high)	13874506	14217942	14257427	15336991	13985473
	min (low)	11306714	11602819	11628739	12539279	11064565
%error (mean:high)			2.5 : 2.4	2.5 : 2.7	9.9 : 10.5	-1.5 : 0.7
BT-A	min (high)	76799427	77454300	77839767	85876074	85835820
	min (low)	74182308	74696115	74937243	78018235	77721303
%error (mean:high)			0.6 : 0.8	1.0 : 1.3	5.5 : 11.8	5.4 : 11.7
LU-A	min (high)	36966517	37783314	37629343	52540729	52395303
	min (low)	34399415	35194131	35099696	43787261	43176436
%error (mean:high)			2.2 : 2.2	1.8 : 1.7	27.5 : 42.1	25.7 : 41.7
CG-A	min (high)	4353851	4612676	4525479	8677331	8291439
	min (low)	1848843	2076113	1950485	4252990	3691704
%error (mean:high)			7.4 : 5.9	3.8 : 3.9	84.1 : 99.3	65.6 : 90.4
IS-A	min (high)	5420444	5973752	5836727	8301860	5585069
	min (low)	2772617	3490618	3080709	5789329	1634756
%error (mean:high)			17.9 : 10.2	11.1 : 7.6	76.9 : 53.1	1.4 : 3.0
FT-A	min (high)	8085574	8195461	8088853	9620210	9366497
	min (low)	5422766	5518819	5485972	6021030	6029058
%error (mean:high)			0.8 : 1.3	-0.1 : 0.0	8.9 : 18.9	8.6 : 15.8

and "low" in the table), the mean error over this range (comparing process-by-process with the minimum results from the *MO* run), and the error of the mean and "high" values (effectively execution time of Node 0's "main").

Overall, the results show that compensation techniques improve performance estimates, except in a few cases where the differences are negligible. This is encouraging. However, we also notice that the results display a variety of interesting characteristics, including differences from the sequential results.

For instance, the *PA* values for SP-A, BT-A, and FT-A are practically equivalent to "main" only results, yet the sequential profiles show slowdowns. The *PA-comp* values are within less than 1% in these cases. We believe this suggests that the instrumentation intrusion is being effectively reduced due to parallelization, resulting in fewer events being measured on each process. We tend to characterize the SP-W flat profile experiments in the same way, since the errors are reasonably small and the minimum ranges are tight.

Other benchmarks show differences in their range of minimum "main" execution times. IS-A is one of these. It also has the greatest error for flat profile compensation. Compared to the sequential case, there is a significant reduction in *PA* error (193.9% to 10.2%) due to intrusion reduction, but the compensated values are off by 11.1% on average per process and 7.6% for Node 0's "main" time (compared to 2.1% minimum error in the sequential case). This suggests a possible correlation of greater range in benchmark execution time with poorer compensation, although it does not explain why.

CG-A also has a significant difference in its "high" and "low" range, but its *PA-comp* errors are lower than IS-A. However, as more events are profiled with callpath instrumentation, the *CA* and *CA-comp* errors increase significantly. Compared to the sequential *CA* and *CA-comp* runs, we also see a slowdown in execution time compared to the sequential case. This is odd. Why, if we assume the measurement intrusion is being reduced by parallelization, do we see an execution slowdown? Certainly, the number of events is affecting compensation performance, as was the case in the sequential execution, but the increase in execution times beyond the sequential results suggests some kind of intrusion interdependency. In addition, we see the execution time range is widening.

Looking for other examples of widening execution time range with increased number of events, we find additional evidence in the callgraph runs (*CA* and *CA-comp*) for SP-A, BT-A, LU-A, and FT-A. The effect for LU-A is particularly pronounced. Together with the observations above, these findings imply a more insidious problem that may limit the effectiveness of our compensation algorithms. We discuss these problems below.

5 Discussion

The experiments we conducted were stress tests for the overhead compensation algorithms. We profiled all routines in the application source code for flat profiles and we profiled all routine calling paths for callpath profiles[3]. While the results show the overhead compensation strategies implemented in TAU are generally effective, we emphasize the need to have an integrated approach to performance measurement that takes into account the limits of measurement precision and judicious choice of events. It should not be expected that performance phenomena occurring at the same granularity as the measurement overhead can be observed accurately. Such small events should be eliminated from instrumentation consideration, improving measurement accuracy overall. In a similar vein, there is little reason to instrument events of minor importance to the desired performance analysis.

TAU implements a dynamic profiling approach where events to be profiled are created on-the-fly. This is in contrast with static profiling techniques where all events must be known beforehand [7]. Static approaches can be more efficient in the sense that the event identifiers and profile data structures can be allocated *a priori*, but these approaches do not work for usage scenarios where events occur dynamically. While TAU's approach is more general, modeling the overhead is more complicated. For instance, we do not currently track event creation overhead, which can occur at any time. Future TAU releases will include this estimate in the overhead calculation. The good news is that we made significant

---

[3] We did eliminate a few very small routines: BINVCRHS, MATMUL_SUB, and MATVEC_SUB, from the BT benchmark, and ICNVRT from CG.

improvements in the efficiency of TAU's profiling system in the course of this research. Our callpath profiling overheads were improved ten times by switching to a more efficient callpath calculation and profile data lookup mechanism.

Callpath profiling is more sensitive to recovery of accurate performance statistics for two reasons. First, there are more callpath events than in a flat profile and each callpath event is proportionally smaller in size. Second, we only estimate the flat profiling overhead at this time in TAU. The overhead for profiling measurements of callpaths is greater because the callpath must be determined on-the-fly with each event entry and the profiling data structures used must be mapped dynamically at runtime (flat profile data structures are directly linked). Nevertheless, it is encouraging how well compensation works with callpath profiling using less exact (smaller) overhead values. Also, TAU's implementation of callpath profiling allows the depth of callpath to be controlled at execution time. A callpath depth of 0 results in a flat profile. Setting the callpath depth to d results in events for all callpaths of length $\leq d$ being profiled. This callpath depth control can be used to limit intrusion.

The most important result from the research work is the insight gained on overhead compensation in parallel performance profiling. Our present techniques are necessary for compensating measurement intrusion in parallel computations, but they are not sufficient. Depending on the application's parallel execution behavior, it is possible, even likely, that intrusion effects seen on different processes are interdependent. Consider the following scenario. A master process sends work to worker processes and then waits for their results. The worker processes do the work and send their results back to the master. A performance profile measurement is made with overhead compensation analysis. The workers see some percentage intrusion with the last worker to report seeing a 30% slowdown. The compensated profile analysis works well and accurately approximates the workers "actual" performance. The master measurement generates very little overhead because there are few events. However, because the master must wait for the worker results, it will be delayed until the last worker reports. Thus, its execution time will include the last worker's 30% intrusion! Our compensated estimate of the master's execution time will be unable to eliminate this error because it is unaware of the worker's overhead. We believe a very similar situation is occurring in some, if not all, of the parallel experiments reported here.

Figure 1 depicts the above scenario. Figure 1(a) shows the measured execution with time overhead indicated by rectangles and termination times by triangles. The large triangle marks where the program ends. The overhead for the master is assumed to be negligible. The arrows depict message communication. Figure 1(b) shows the execution with all the overhead bunched up at the end as a way to locate when the messages returning results from the workers (dashed arrows) would have been sent and the workers would have finished (small shaded triangles), if measurements had not been made. Profile analysis would correctly remove the overhead in worker performance profiles under these circumstances. However, the master knows nothing of the worker overheads and, thus, our current compensation algorithms cannot compensate for it. The master

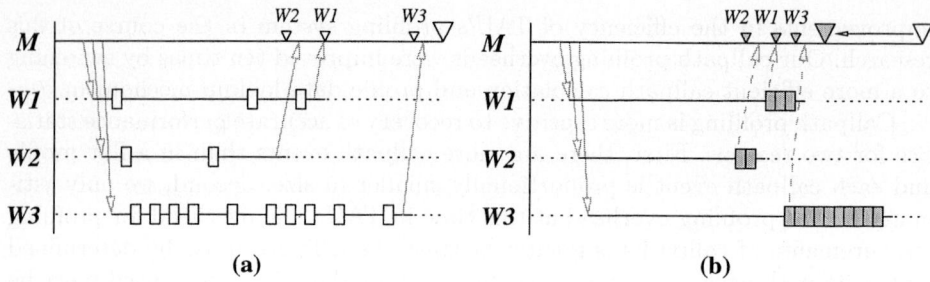

Fig. 1. Parallel Execution Measurement Scenario.

profile will still reflect the master finishing at the same time point, even though its "actual" termination point is much earlier.

Unfortunately, parallel overhead compensation is a more complex problem to solve. This is not entirely unexpected, given our past research on performance perturbation analysis [18–20]. However, in contrast with that work, we do not want to resort to a parallel trace analysis to solve it. The problem of overhead compensation in parallel profiling using only profile measurements (not tracing) has not been addressed before, save in a restricted form in Cray's MPP Apprentice system [29]. We are currently developing algorithms to do on-the-fly compensation analysis based on those used in trace-based perturbation analysis [25], but their utility will be constrained to deterministic parallel execution only, for the same reasons discussed in [17, 25]. Implementation of these algorithms also will require techniques similar to those used in PHOTON [27] and CCIFT [2, 3] to embed overhead information in MPI messages. While Photon extends the MPI header in the underlying MPICH implementation to transmit additional information, the MPI wrapper layer in the CCIFT application level checkpointing software allows this information to piggyback on each message.

6 Conclusion

Measured profiling has proven to be an important tool for performance analysis of scientific applications. We believe it has significant advantages over statistical profiling methods, but there are important issues of intrusion that must be addressed. In this paper, we focus on the removal of measurement overhead from measured profiling statistics. The algorithms we describe for quantifying the overhead and eliminating it on-the-fly have been implemented in the TAU performance system. Testing these algorithms using TAU on the NAS parallel benchmarks shows that they are effective at reducing the error in estimated performance. This is demonstrated for both flat and callpath profiling. In general, the overhead compensation techniques can be applied to any set of performance metrics that can be profiled using TAU.

However, there are still concerns and problems to address. We need to validate the compensation analysis approach on other platforms where different factors will influence observed overhead. We need to better understand the limits of overhead compensation and its proper use in an integrated instrumentation and measurement strategy. In particular, the problems with overhead compensation analysis in parallel profiling require further study. While the current methods do reduce intrusion error in parallel profiling, they are unable to account for interdependent intrusion effects. We will address this problem in future research.

References

1. D. Bailey, T. Harris, W. Saphir, R. van der Wijngaart, A. Woo, M. Yarrow, "The NAS Parallel Benchmarks 2.0," Technical Report NAS-95-020, NASA Ames Research Center, 1995.
2. G. Bronevetsky, D. Marques, K. Pingali, and P. Stodghill, "Collective Operations in an Application-level Fault Tolerant MPI System," *International Conference on Supercomputing (ICS)*, 2003.
3. G. Bronevetsky, D. Marques, K. Pingali, and P. Stodghill, "Automated Application-level Checkpointing of MPI Programs," *Principles and Practice of Parallel Programming (PPoPP)*, 2003.
4. S. Browne, J. Dongarra, N. Garner, G. Ho, and P. Mucci, "A Portable Programming Interface for Performance Evaluation on Modern Processors," *International Journal of High Performance Computing Applications*, 14(3):189–204, Fall 2000.
5. H. Brunst, M. Winkler, W. Nagel, H.-C. Hoppe, "Performance Optimization for Large Scale Computing: The Scalable VAMPIR Approach," In V. Alexandrov, J. Dongarra, B. Juliano, R. Renner, K. Tan, (eds.), *International Conference on Computational Science*, Part II, LNCS 2074, Springer, pp. 751–760, 2001.
6. L. De Rose, "The Hardware Performance Monitor Toolkit," *Euro-Par Conference*, 2001.
7. L. De Rose and F. Wolf "CATCH - A Call-Graph Based Automatic Tool for Capture of Hardware Performance Metrics for MPI and OpenMP Applications," *Euro-Par Conference*, LNCS 2400, Springer, pp. 167–176, 2002.
8. Alain Fagot and Jacques Chassin de Kergommeaux, "Systems Assessment of the Overhead of Tracing Parallel Programs," *Euromicro Workshop on Parallel and Distributed Processing*, pp. 179–186, 1996.
9. T. Fahringer and C. Seragiotto, "Experience with Aksum: A Semi-Automatic Multi-Experiment Performance Analysis Tool for Parallel and Distributed Applications," *Workshop on Performance Analysis and Distributed Computing*, 2002.
10. S. Graham, P. Kessler, and M. McKusick, "gprof: A Call Graph Execution Profiler," *SIGPLAN Symposium on Compiler Construction*, pp. 120–126, June 1982.
11. R. Hall, "Call Path Profiling," *International Conference on Software Engineering*, pp. 296–306, 1992.
12. J. Hollingsworth and B. Miller, "An Adaptive Cost System for Parallel Program Instrumentation," *Euro-Par Conference*, Volume I, pp. 88–97, August 1996.
13. IBM, "Profiling Parallel Programs with Xprofiler," IBM Parallel Environment for AIX: Operation and Use, Volume 2.
14. C. Janssen, "The Visual Profiler," http://aros.ca.sandia.gov/~cljanss/perf/vprof/.
15. D. Kranzlmüller, R. Reussner, and C. Schaubschläger, "Monitor Overhead Measurement with SKaMPI," *EuroPVM/MPI Conference*, LNCS 1697, pp. 43–50, 1999.

16. D. Knuth, "An Empirical Study of FORTRAN Programs," *Software Practice and Experience*, **1**:105–133, 1971.
17. A. Malony, "Performance Observability," Ph.D. thesis, University of Illinois, Urbana-Champaign, 1991.
18. A. Malony, D. Reed, and H. Wijshoff, "Performance Measurement Intrusion and Perturbation Analysis," *IEEE Transactions on Parallel and Distributed Systems*, **3**(4):433–450, July 1992.
19. A. Malony and D. Reed, "Models for Performance Perturbation Analysis," *ACM/ONR Workshop on Parallel and Distributed Debugging*, pp. 1–12, May 1991.
20. A. Malony "Event Based Performance Perturbation: A Case Study," *Principles and Practices of Parallel Programming (PPoPP)*, pp. 201–212, April 1991.
21. A. Malony, S. Shende, "Performance Technology for Complex Parallel and Distributed Systems," In G. Kotsis, P. Kacsuk (eds.), *Distributed and Parallel Systems, From Instruction Parallelism to Cluster Computing, Third Workshop on Distributed and Parallel Systems (DAPSYS 2000)*, Kluwer, pp. 37–46, 2000.
22. J. Mellor-Crummey, R. Fowler, and G. Marin, "HPCView: A Tool for Top-down Analysis of Node Performance," *Journal of Supercomputing*, **23**:81–104, 2002.
23. P. Mucci, "Dynaprof," http://www.cs.utk.edu/~mucci/dynaprof
24. D. Reed, L. DeRose, and Y. Zhang, "SvPablo: A Multi-Language Performance Analysis System," *International Conference on Performance Tools*, pp. 352–355, September 1998.
25. S. Sarukkai and A. Malony, "Perturbation Analysis of High-Level Instrumentation for SPMD Programs," *Principles and Practices of Parallel Programming (PPoPP)*, pp. 44–53, May 1993.
26. Unix Programmer's Manual, "prof command," Section 1, Bell Laboratories, Murray Hill, NJ, January 1979.
27. J. Vetter, "Dynamic Statistical Profiling of Communication Activity in Distributed Applications," *ACM SIGMETRICS Joint International Conference on Measurement and Modeling of Computer Systems*, ACM, 2002.
28. F. Wolf and B. Mohr, "Automatic Performance Analysis of SMP Cluster Applications," Technical Report IB 2001-05, Research Centre Juelich, 2001.
29. W. Williams, T. Hoel, and D. Pase, "The MPP Apprentice Performance Tool: Delivering the Performance of the Cray T3D," *Programming Environments for Massively Parallel Distributed Systems*, North-Holland, 1994.
30. M. Zagha, B. Larson, S. Turner, and M. Itzkowitz, "Performance Analysis Using the MIPS R10000 Performance Counters," *Supercomputing Conference*, November 1996.

A Data Structure Oriented Monitoring Environment for Fortran OpenMP Programs[*]

Edmond Kereku, Tianchao Li, Michael Gerndt, and Josef Weidendorfer

Institut für Informatik, Technische Universität München,
Boltzmannstr. 3, D-85748 Garching bei München, Germany
{kereku,lit,gerndt,weidendo}@in.tum.de

Abstract. This paper describes a monitoring environment that enables the analysis of memory access behavior of applications in a selective way with a potentially very high degree of detail. It is based on a novel hardware monitor design that employs an associative counter array to measure data structure related information at runtime. A simulator for this hardware monitor is implemented, providing the capability of on-the-fly simulation targeting shared memory systems. Layers of software are constructed to operate and utilize the underlying hardware monitor, thus forming a complete monitoring environment. This environment is useful to help users to reason about optimizations based on data reorganization as well as on standard loop transformations.

1 Introduction

This paper describes a novel approach for analyzing the memory access behavior of OpenMP applications developed in the German project EP-Cache. It is based on a hardware monitor designed to be integrated into cache controllers which provides counters that can be configured to measure events for certain address ranges.

This hardware monitor enables nonintrusive analysis of access overhead to data structures in the program. While state of the art hardware counters in modern CPUs can only be used to measure access behavior for program regions or program lines, this hardware monitor is capable to give, for example, information about the number of cache misses for a specific array in a loop of the program.

This information can then be used in the identification of transformations for optimizing the cache behavior of applications. Especially for data structure transformations, such as padding, it will be very useful.

To exploit the abilities of the new hardware monitor, a software infrastructure is required to collect and map the measured information back to the symbols in the program. This paper describes this software infrastructure as well as a simulator for the hardware monitor which enables us to investigate the advantages

---

[*] The work presented in this paper is mainly performed in the context of the EP-Cache Project, funded by the German Federal Ministry of Education and Research (BMBF).

obtained from having this data structure-related information in the optimization process.

Section 2 gives an overview of the monitoring infrastructure. Section 3 describes the individual components of the infrastructure. Section 4 presents a monitoring scenario and Section 5 the analysis and optimization of Gauss Elimination.

2 An Overview of the Monitoring Infrastructure

Monitoring data structures in a selective way is a demanding task. It requires knowledge about the involved data structures, the state of application execution, as well as the state of different memory levels present in the system - all of these have to be taken in account in a coordinated way. We designed our monitoring system to meet those requirements in the first place, but taken others into consideration as well, such as portability and extendibility. The primary goal is to give more detailed information about memory access behavior in Fortran OpenMP programs targeting specific data structures. Programs written in other programming languages as well as MPI programs can also be analyzed with our environment provided that the language-specific instrumenters for program regions and data structures are available.

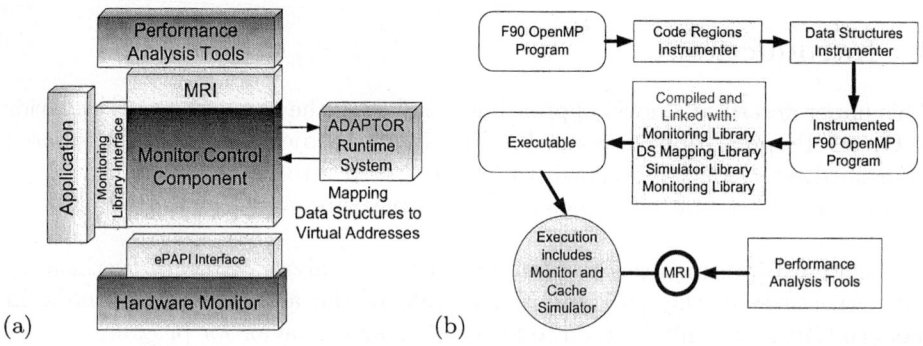

Fig. 1. (a) The central component is the Monitoring Control Component(MCC), which controls the application's execution, and manages different hardware and software monitoring resources. (b) An application must go through several preprocessing procedures before it can be monitored.

Figure 1 depicts the building blocks of our monitoring system, focusing on the *Monitor Control Component (MCC)*. The MCC provides two APIs: *Monitoring Request Interface (MRI)*[4] and the monitoring library interface.

Via MRI, performance analysis tools[1] can specify monitoring requests and retrieve runtime data. The MCC converts MRI requests into many low level

---
[1] In the rest of the paper we interchangeably use the terms "Tool" and "Performance Tool" always refereeing to a Performance Analysis Tool.

requests which are addressed to different sensors[2]. Those sensors can be hardware counters, e.g. counting cache misses, or software sensors, e.g. information about the current state of program execution. The information provided by the sensors is subsequently processed by the MCC - for example, the information is aggregated through region instances or threads upon request from the tools, producing in this way the required profile or trace data. The monitored application calls the monitoring library at instrumented regions. This library implements parts of the MCC.

Some preprocessing procedures are necessary before we can start monitoring an application (Figure 1). First, a *Code Region Instrumenter* (see Section 3.3) inserts calls to the monitoring library(Section 3.5) at the entry and exit of regions in a selective way. A second instrumentation is done by the *Data Structure Instrumenter* based on ADAPTOR [7] to generate information about the application's data structures, such as their virtual address range. In the next step, the application is compiled and linked with libraries, including the monitoring library, the simulator libraries and the ADAPTOR runtime system library, producing an executable. Finally the program is executed which includes a simulation of the cache hierarchy and the new hardware monitor. During the execution, performance tools request and access performance data via the MRI.

The whole process can be automated. We provide a simple script which takes source files of the application as input and generates the executable.

3 Resources for Data Structure Monitoring

3.1 The Hardware Monitor

The proposed hardware monitor [6] enables non-intrusive performance monitoring of access overhead for specific data structures. It can be configured into two working modes. The *static mode* allows to count predefined events or accesses to specific memory regions of interest. This can be used to monitor given data structures or parts of arrays, in order to get such information as L1 cache misses for array A, L2 caches hits for array B etc. The *dynamic mode* enables a fine-grained monitoring of memory accesses to selected address ranges. It provides a histogram, e.g., for the cache misses of array A. The histogram's granularity can be configured as multiples of cache lines.

3.2 Hardware Monitor Simulator

While the hardware monitor is still only a concept, currently, a simulation of the monitor is used instead. Though this environment can only catch memory accesses happening in user space of target processes, this enables the development of performance tools and optimization techniques.

---

[2] Our Hardware Monitor plays in the Figure 1 the role of sensor. MCC accesses it through ePAPI, a PAPI [10] alike interface with the extension of histogram support.

The simulation and monitoring environment is capable of providing detailed information about the runtime cache access behavior, and is composed of the following modules: a runtime instrumentation module which instruments the program's load/store operations while the program is executing, a cache simulation module which simulates a hierarchy of caches on processors with shard memory, and a monitor simulation module which simulates the hardware monitor exactly as it was described in the last section.

The runtime instrumentation module provides the capability of on-the-fly cache simulation for OpenMP programs, which is enabled by the Valgrind [9] runtime instrumentation framework. This allows catching all memory references of an IA-32 binary, including SIMD instructions introduced in modern Intel processors (MMX/SSE/SSE2). While the instrumentation is done inside of the Valgrind CPU emulation layer, the cache simulator and monitor simulator are linked to the target binary, and are notified via a callback mechanism each time there is a memory access.

3.3 Code Region Instrumenter

To be able to measure performance data for regions in the program code, the application has to be instrumented. We developed a Fortran 95 instrumenter based on the NAG compiler frontend. It instruments sequential regions, taking into account multiple exits from regions, as well as OpenMP regions [5]. For OpenMP we follow the work of Bernd Mohr et. al. in Opari [3] and POMP [1].

The programmer can select which regions are instrumented by using appropriate command line switches. This is especially important when instrumenting sequential loops. Sometimes it might be beneficial to instrument nested loops, although the measurement overhead can be quite high.

The instrumenter generates information about the instrumented region in an XML-Format that was designed in the APART working group. It is called the Standard Intermediate Program Representation (SIR). Besides the file and the lines covered by a region and the region type, we also collect information on the data structures accessed in the region. For arrays, the generated XML-representation specifies the data type as well as the array size if it is statically known. This information can be used to help the user and automatic tools to select appropriate data structures for measurements.

3.4 Data Structure Instrumenter

To be able to measure, for example, the L1 cache misses of a data structure with the hardware monitor, we need to know the virtual address range of the data structure at runtime. Our current implementation uses some components of the ADAPTOR compilation system. ADAPTOR inserts for all arrays code which handles array descriptors at runtime. These store, besides other information, the address range to which the array is mapped. For scalar variables this information can easily be obtained without special support. The ADAPTOR runtime system also provides an interface through which our monitor can retrieve the current virtual address range for a specific data structure.

This kind of implementation restricts our monitoring to data structures in code regions that are already allocated when the region is entered.

3.5 Lightweight Monitoring Library

The monitoring library implements the function calls that are inserted into the application by the Program Region Instrumenter. Although the instrumentation is done only for selected regions, most of the instrumented regions will actually not be measured. The tools can and should request only information required for the analysis via the MRI.

Thus, most of the calls will actually be empty calls and should have as little overhead as possible. Due to the numbering scheme of code regions, which is based on a file number and the regions first line number, we had to implement the *Configuration Table* for storing MRI requests as a hash table (see Figure 2). Each table entry represents an instrumented region. It has a flag that is only set if there is an MRI request appended to the region. This way, the minimal weight of a library call is basically reduced to the access time for the hash table.

3.6 Monitor Control Component

MCC (ref. Figure 2) is the central component that glues all the resources described in the previous sections together and communicates with the performance analysis tool. It is responsible for the initialization and configuration of resources, for handling MRI requests, and for postprocessing and delivery of the runtime information. For handling MRI requests for specific data structures, the MCC translates the variables names into virtual addresses and vice versa.

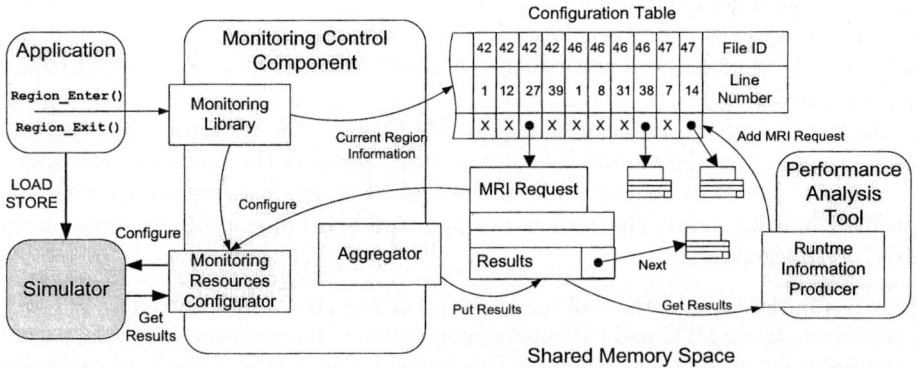

Fig. 2. A more detailed view of our system revealing some internal functionality and implementation details.

4 Monitoring Scenario

Our monitoring system is structured into two processes, the application process and the tool process. These processes communicate via a System V shared memory segment. Accordingly, the system is implemented as two libraries. The first library, linked to the application, contains the monitoring library, the MCC functionality and the simulator. The other library, linked to the analysis tool, implements the *Runtime Information Producer (RIP)* responsible for MRI processing at the client side (Figure 2).

At runtime, the first statement executed by the application is the initialization of the MCC. The application is blocked during initialization until the analysis tool specified MRI requests and released the application. The MCC initializes all monitoring sensors, retrieves the region structure of the application from SIR, and builds up a configuration table in the shared segment. The configuration table is implemented as a hash table, where the couple of file ID and region first line number serves as the key.

When the tool is started, it will first initialize the RIP and wait for a READY signal of the MCC. It will then specify MRI requests which are validated and attached to the appropriate region entries in the configuration table by the RIP. The requests specify the requested runtime information (e.g. the number of L1 cache hits), information about the target code region and eventual data structures, and any desired aggregation. The data structure provides also space for the measurement results.

Once the tool finished its initial requests[3], it specifies where the program should stop (usually the end of a region) and the MCC is notified to start the application's execution.

At the enter and exit point of each instrumented region, the region instrumenter inserted a call to the monitoring library. When the control flow enters the library, MCC looks up the configuration table for an MRI request that is appended to the current region. If such a request exists, symbolic data structure information will be translated into virtual addresses if necessary and the monitor will be configured accordingly. After that, the control is returned to the application.

At the end of the monitored region, MCC stops the monitor, retrieves the results, aggregates the results if required, and transfers the results to the space reserved by the MRI request. When the end of a region corresponds with the specified halting point, the tool is notified and the application, together with MCC, is blocked again.

---

[3] We say initial because the tool can make other requests at any time. The synchronization between MCC and RIP allows interruption of the program for getting partial results or for making new requests. This is particulary useful if the Tool can make new decisions (followed by new requests) starting from the data gathered until the present state of application's execution.

5 Analysis and Optimization of Gauss Elimination

In order to illustrate the usefulness of the monitoring system, we use a Fortran 90 program for solving linear equations, $Ax = b$, using Gauss elimination without pivoting. Variable A, a two dimensional real array (100x100), is chosen as the target data structure of monitoring.

Figure 3 presents two access histograms for L1 cache, including events of read hits, read misses, write hits and write misses. The x-axis shows the relative position of the access, with the whole memory range for A evenly divided into 25 portions. The y-axis represents the number of events. Both measurements target code regions that constitute the numerical kernel, excluding the initializations etc.

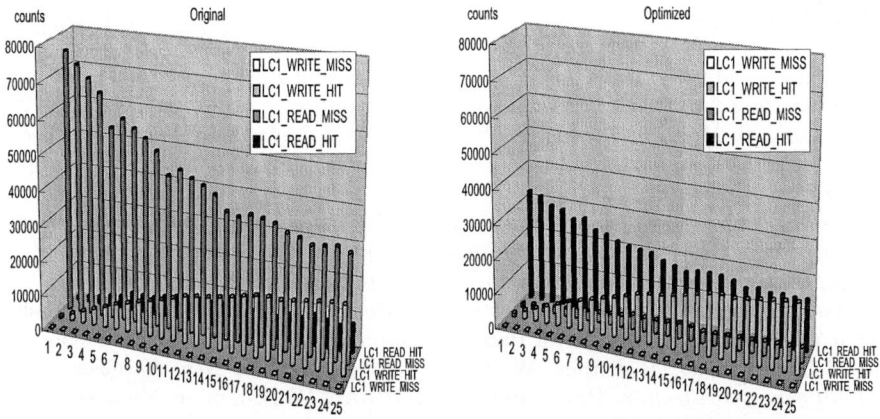

Fig. 3. Cache access histograms for the original and optimized Gauss program.

The left part of Figure 3 illustrates the access histogram of the original Gauss elimination program. The most significant feature of this histogram is the high value of read misses and very low level of read hits. This encouraged us to optimize the original program with different optimization techniques, such as loop unrolling etc, and we finally derived an optimized version of the Gauss program with better utilization of the hardware. The right part of Figure 3 presents the access histogram of the optimized program. It can be clearly seen from this diagram, that the read misses are significantly reduced.

6 Future Work

This paper presents the current status of the development of a monitoring system for measuring the memory access behavior for specific data structures. It is based on a novel hardware monitor design.

In the near future we plan to enhance the automation scripts driving the preprocessing and to make a full implementation of the monitor available. We will develop a version of the monitor that works with the PAPI interface and will thus be able to access the current hardware counters of modern microprocessors. Another effort will be made to port our ePAPI interface to Itanium architectures taking advantage from this processor's support on monitoring predefined data address spaces. We are also working on mapping variables to virtual address based on debug information.

Within the EP-Cache project we will also develop an automated performance analysis tool that does an incremental online analysis as described in the monitoring scenario on top of our monitor. The tool will be based on a formalization of performance property with the APART Specification Language (ASL).

References

1. B. Mohr, A. Malony, S. Shende, F. Wolf: *Design and Prototype of a Performance Tool Interface for OpenMP*, Journal of Supercomputing, Vol. 23, pp. 105 - 128, 2002
2. A. Malony, B. Mohr, S. Shende, F. Wolf: *Towards a Performance Tool Interface for OpenMP: An Approach Based on Directive Rewriting*, EWOMP 01, Third European Workshop on OpenMP, 2001
3. *OpenMP Pragma and Region Instrumentor*, www.fz-juelich.de/zam/kojak/opari
4. M. Gerndt, E. Kereku: *Monitoring Request Interface Version 1.0*, http://wwwbode.in.tum.de/~kereku/epcache/pub/MRI.pdf
5. M. Gerndt, E. Kereku: *Selective Instrumentation and Monitoring*, to be published: 11th Workshop on Compilers for Parallel Computers (CPC 04), Kloster Seeon, 2004
6. M. Schulz, J. Tao, J. Jeitner, W. Karl: *A Proposal for a New Hardware Cache Monitoring Architecture*, Proceedings of SIGPLAN Workshop on Memory System Performance (MSP 2002), Berlin, Germany. June 2002
7. *ADAPTOR (Automatic DAta Parallelism TranslaTOR)* http://www.scai.fraunhofer.de/291.0.html
8. A-T. Nguyen, M. Michael, A. Sharma, J. Torrellas: *The Augmint Multiprocessor Simulation Toolkit for Intel x86 Architectures*, Proceedings of 1996 International Conference on Computer Design. October 1996
9. N. Nethercote and J. Seward: *Valgrind: A Program Supervision Framework*, Proceedings of the Third Workshop on Runtime Verification (RV'03), Boulder, Colorado, USA. July 2003
10. S. Browne, J. Dongarra, N. Garner, G. Ho, P. Mucci: *A Portable Programming Interface for Performance Evaluation on Modern Processors*, The International Journal of High Performance Computing Applications, 14(3), Fall 2000. Pp. 189–204.

An Approach for Symbolic Mapping of Memory References

Luiz DeRose[1], K. Ekanadham[2], and Simone Sbaraglia[2]

[1] Cray Inc.
Mendota Heights, MN, USA
ldr@cray.com
[2] IBM Research,
Yorktown Heights, NY, USA
(eknath,sbaragli)@us.ibm.com

Abstract. Understanding and tuning memory system performance is a critical issue for most scientific programs to achieve reasonable performance on current high performance systems. Users need a *data-centric* performance measurement infrastructure that can help them understand the precise memory references in their program that are causing poor utilization of the memory subsystem. However, a *data-centric* performance tool requires the mapping of memory references to the corresponding symbolic names of the data structure, which is non trivial, especially for mapping local variables and dynamically allocated data structures. In this paper we describe with examples the algorithms and extensions implemented in the SIGMA environment for symbolic mapping of memory references to data structures.

1 Introduction

A variety of performance measurement, analysis, and visualization tools have been created to help programmers tune and optimize their applications. These tools range from source code profilers, tracers, and binary analysis tools (e.g., SvPablo [4], TAU [8], KOJAK [10], VampirGuideView [5], HPCView [6], and Paradyn [7]) to libraries and utilities to access hardware performance counters built into microprocessors (e.g., Perfctr [9], PAPI [1], and the HPM Toolkit [2]).

In general, performance measurement and visualization tools tend to be *control-centric*, since they focus on the control structure of the programs (e.g., loops and functions), where traditionally application programmers concentrate when searching for performance bottlenecks. However, due to the advances in microprocessors and computer systems designs, there has been a shift in the performance characteristics of scientific programs from being computation bounded to being memory/data-access bound. Hence, understanding and tuning memory system performance is today a critical issue for most scientific programs to achieve reasonable performance on current high performance systems. Thus, users also need a *data-centric* performance measurement infrastructure that can help them understand the precise memory references in their program that are causing poor utilization of the memory hierarchy. Fine-grained information such as this is useful for tuning loop kernels, understanding the cache behavior of new algorithms, and to investigate how different parts of a program and its data structures compete for and interact within the memory subsystem.

In [3], we presented SIGMA (Software Infrastructure to Guide Memory Analysis), a new data collection framework and a family of cache analysis tools. The goal of the SIGMA environment is to provide detailed cache information by gathering memory reference data using software-based instrumentation. This is used to provide feedback to programmers to help them apply program transformations that improve cache performance. An important characteristic of the SIGMA infrastructure is that it is both control-centric and data-centric; it presents performance metrics expressed in terms of functions, as well as data structures defined in the source code.

In order to provide such relation to the source program, whenever a memory reference is made, SIGMA gathers the address of the instruction that made the reference and the address of the referenced data. It then associates the instruction address to the source line in the program that made the reference, and the data address to the symbolic name of the data structure that corresponds to the reference. While the transformation of the instruction address into the corresponding source line can be performed easily by analyzing the line number and symbol table stored in the executable, mapping the data address to the corresponding symbolic name of the data structure requires considerably more work. This is especially true if we want to support local variables and dynamically allocated data structures, which were not supported in the work presented in [3]. The main problem in supporting the mapping of local (automatic) variables and dynamically allocated variables is that the virtual address cannot be known statically at link time.

The remainder of this paper is devoted to describing the algorithms implemented in SIGMA that map each memory reference to the corresponding data structure. Section 2 presents a brief overview of the SIGMA infrastructure. In Section 3, we describe with examples our algorithms for mapping memory references to variable names. Finally, in Section 4, we present our conclusions.

2 The SIGMA Approach

The SIGMA environment consists of three main components: a pre-execution instrumentation utility that reads the binary file to locate and instrument all instructions that refer to memory locations; a runtime data collection engine that collects the streams of instructions and memory references generated by the instrumentation, performs the symbolic mapping, and a highly efficient lossless trace compression. It also provides a number of simulation and analysis tools that process the compressed memory reference trace to provide programmers with tuning information.

SIGMA uses a binary instrumentation approach so that it can gather data about the actual memory references generated by optimizing compilers rather than using source instrumentation which would gather data about the user specified array references. The simulation and analysis tools include a TLB simulator, a data cache simulator, a data prefetcher simulator, and a query mechanism that allows users to obtain performance metrics and memory usage statistics for current and hypothetical architectures. For more details on the SIGMA environment, please refer to [3].

The runtime symbolic conversion engine, which is the subject of this paper, performs the transformation:

$$(instructionAddress, dataAddress) \rightsquigarrow (sourceLine, varName, arrayElement) \quad (1)$$

and maintains a table where each column represents an *active* data structure and each row represents a function in the program. Each entry in the table contains counters for the memory events of interest, such as cache accesses, hits, and misses. Once the symbolic transformation (1) is completed the module updates the counters for the entry corresponding to the source line *sourceLine* and the data structure *varName*. If the variable is an array the precise element references is also available.

3 Mapping Memory References to Data Structures

In order to perform the transformation: $dataAddress \rightsquigarrow varName$ the runtime engine builds and maintains a linked list shown in Figure 1, where each entry corresponds to an allocated virtual address range and carries the information about the symbolic name of the data structure that corresponds to each address range. When an address "a" is accessed, the runtime engine searches the list for an entry "i" such that $a_i \leq a < b_i$, in order to match the reference to the data structure x_i.

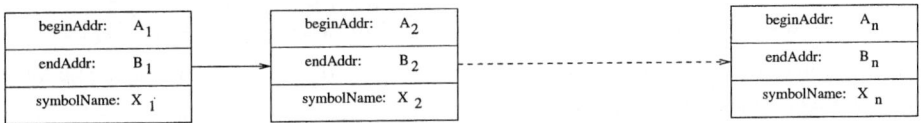

Fig. 1. Table to map memory addresses to symbolic names.

Clearly, since data structures can be allocated and deallocated dynamically, this list has to be dynamically updated at runtime. Moreover, the instrumentation engine must capture the information about allocations and deallocations. A further difficulty is presented by the fact that stack variables are not identified by a global address range, but by an offset within the stack pointer of the function where they are defined.

In this section we describe our approach to support the mapping of each data address to the corresponding symbolic name of the data structure. We divide the description in three parts: the algorithm for mapping of global variables, the algorithm for mapping of dynamically allocated variables, and the algorithm for mapping of stack variables. Our examples are based on the IBM compilers for AIX and their binary representation (XCOFF), but the ideas can be extended to other environments.

3.1 Mapping Global Variables

Global variables are allocated in the *Data Segment* of the binary, where the information about the virtual address assigned to each variable and its size is completely known at link time. If the program is compiled with the debugger flag, this information is stored by the compiler in the executable in the form of tables. Hence, by analyzing these tables we can build the linked list of Figure 1 statically, before the application starts executing. For example, let us consider the following C pseudocode:

```
double var1;
double var2 = 1;
int array1[100];
int main(int argc, char *argv[]) { ... }
```

By analyzing the symbol table of the compiled executable, we find the following entries that refer to the data structure `var1`:

```
Symbol  Class = C_EXT   Value = 0x20000660  Name = var1
Stab    Class = C_GSYM  Value = 0x00000000  Name = var1:G-13
```

The first entry classifies the symbol `var1` as a variable statically allocated at the virtual address $0x20000660$. The second entry identifies the variable as a global variable (symbol G) of type -13, which is the internal code for *double*. We can therefore infer that the variable `var1` will be attributed the address range $[0x20000660, 0x20000668)$ and we can build an entry in the linked list. Similarly, the array `array1` and the variable `var2` are represented as:

```
Symbol  Class = C_EXT   Value = 0x200001e0  Name = var2
Stab    Class = C_GSYM  Value = 0x00000000  Name = var2:G-13
Symbol  Class = C_EXT   Value = 0x20000668  Name = array1
Stab    Class = C_GSYM  Value = 0x00000000  Name = array1:G6
Stab    Class = C_DECL  Value = 0x00000000  Name = :t6=ar0;0;99;-1
```

where the last Stab entry defines the type 6 as an array $0, \ldots, 99$ of integers. The table for this binary is shown in Figure 2.

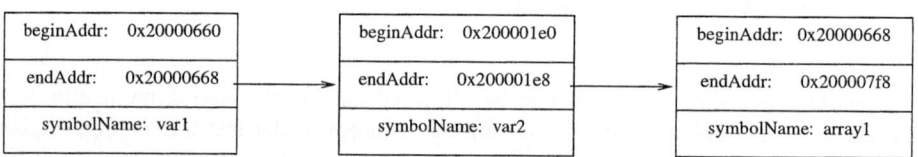

Fig. 2. Table to map memory addresses to symbolic names.

3.2 Mapping Dynamically Allocated Variables

Normally, the sizes of dynamic data structures depend on user input and are often unknown at compile time. Moreover, a data structure size may need to be changed during the program execution. Hence, an address range of a dynamically allocated data structure is assigned to the variable at runtime, and cannot be deduced uniquely from the executable tables. Furthermore, allocated variables can be "released" when they are no more needed, and the same address range (or a subset of it) can be re-assigned to some other dynamic variable.

In order to account for such situations, we needed to expand our instrumentation utility to capture the allocation and deallocation requests, the address range allocated or freed, and the symbolic data structure that is bound from time to time to the address range. To illustrate the algorithm, let us consider the following C example and its corresponding entries in the symbol table:

```
int *A;
A = (int *)malloc(n);

Stab    Class = C_DECL   Value = 0x00000000  Name = :t4=*-1
Symbol  Class = C_EXT    Value = 0x2000055c  Name = A
Stab    Class = C_GSYM   Value = 0x00000000  Name = A:G4
```

The symbol table of the executable contains an entry for the global variable A (of type 4, i.e., pointer to integer, as specified in the first Stab entry), which is associated to the virtual address $0x2000055c$. When the malloc function is called, it returns the address x of the newly allocated address range. By intercepting at runtime this return address and the corresponding argument passed to malloc, we can infer that a new address range $[x, x+n)$ has been allocated. Unfortunately, the malloc call does not directly provide any information about the symbolic name that the new address range is associated with. However, the compiler usually stores the new address x into the memory location identified by A. The code generated for the malloc call is usually of the following type, where R3 indicates the register that is used to pass the first argument to a function and to collect its return code:

```
store n into R3
call malloc (which returns the address x in the register R3)
store R3 into 0x2000055c
```

An optimizing compiler might avoid storing the address x and just keep it in a register. In this case, we are unable to bind the address range with the symbolic name, and we classify such references to belong to a dynamically allocated area whose name is unknown. In practice, however, this situation only occurs when the allocated memory is used for a very limited amount of time and then released, and is not encountered often in real applications where the allocated memory is heavily reused (for example in a loop) before being released.

The algorithm to track dynamically allocated memory is the following: first, by analyzing the executable tables we create an entry for A, as shown in Figure 3. Second, from the malloc instrumentation, we obtain the parameter passed to malloc (size n) and the returning value (address x), and instantiate an entry, as shown in Figure 4, where the name is still undefined. Then, when a store instruction: store x into a is executed, we search the list for an entry whose name is undefined and x is in the begin address of its range. If we find such an entry, we infer that the address range is now associated with the data structure whose virtual address is a. In the example above, when store R3 into 0x2000055c is executed, we change the list by assigning the name A to the allocated variable, as shown in Figure 5.

Finally, each time *free* is called, we capture the address passed to it and search the table for an entry that corresponds to it. We then remove the entry from the table, since it is no more "active".

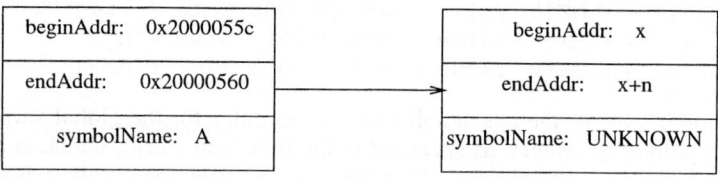

Fig. 3. Table entry for a pointer declaration.

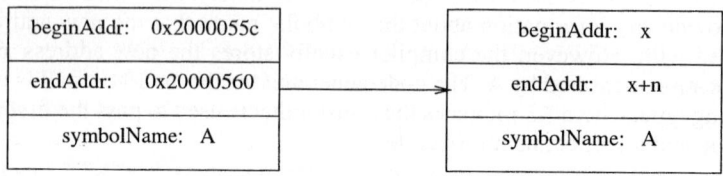

Fig. 4. Table entry after a dynamic allocation at address x of size n.

beginAddr:	0x2000055c
endAddr:	0x20000560
symbolName:	A

beginAddr:	x
endAddr:	x+n
symbolName:	A

Fig. 5. Table entry of a dynamically allocated variable after symbolic mapping.

3.3 Mapping Stack Variables

When a stack variable is declared in the source code, such as an automatic variable in C or a local variable in Fortran 90, the virtual address that will be assigned to the variable will depend on the position of the stack pointer when the function is called, and is therefore unknown at compile time. However, the compiler stores in the executable the offset of each stack variable in the stack frame. For instance, the pseudocode

```
int foo(void) {
   int var1;
   int array1[100];
   ...
}
```

would carry in the executable the following symbols:

```
Stab     Class = C_FUN    Value = 0x00000000  Name = foo:F-1
Symbol   Class = C_FCN    Value = 0x100003d4  Name = .bf
Stab     Class = C_LSYM   value = 0x00000040  Name = var1:-1
Stab     Class = C_LSYM   Value = 0x00000048  Name = array1:3
Stab     Class = C_DECL   Value = 0x00000000  Name = :t3=ar0;0;99;-1
Symbol   Class = C_FCN    Value = 0x100003f0  Name = .ef
```

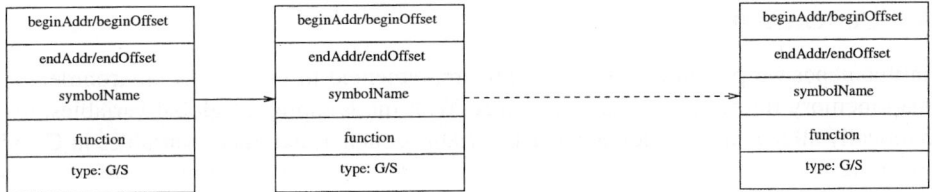

Fig. 6. Table to map memory addresses to symbolic names.

which identify `var1` as a symbol defined locally in the function `foo`. `var1` is a symbol of type *int*, allocated at offset $0x40$ in the stack frame, and `array1` is an array $0, \ldots, 99$ of integers, allocated at offset $0x48$. The `.bf` and `.ef` symbols denote the begin and end of a function variable definition.

A local variable is therefore defined by a pair *(function, offset)*, where *function* indicates the function where the variable is defined and *offset* indicates the stack offset where the variable will be allocated. In order to map stack memory references to the symbol associated with them, we need to be able to identify, for each stack reference, the stack offset of the reference and the function whose stack frame we are accessing. The approach is the following: first, the table in Figure 1 was extended to accommodate the information about local variables, as shown in Figure 6. The *beginAddr* and *endAddr* fields are interpreted as absolute addresses or offsets, depending whether the variable is a global or local symbol (G or S in the type field). The field *function* indicates the function where the variable was declared and is empty for global variables. Second, we extended the instrumentation utility to fetch the value of the stack pointer. Each time the stack pointer changes, we record the event as a couple *(stkPointer, function)* where *stkPointer* is the new value of the stack pointer and *function* is the function that is currently executing.

During execution, we maintain an internal stack structure called *stkList*. Each time the stack pointer changes we search the *stkList* from the top and if we find an entry that matches the current stack pointer, we make that entry the top of the *stkList*. Otherwise, we add the current couple *(stkPointer, function)* to the top of the *stkList*. For instance, let us consider the following call and return sequence:

```
f1() --> f2() --> f3() --> f2() --> f1()
```

and let us further assume, for the sake of simplicity, that there are no stack pointer changes other than the ones involved in the function calls and returns. When `f1` is called, a new stack pointer s_1 is allocated. We capture the stack pointer change and create an entry $(f1, s_1)$ in the *stkList*. Then, when `f2` is invoked, and therefore a new stack pointer s_2 is allocated, we add it to our stack: $(f1, s_1) \rightarrow (f2, s_2)$. Similarly for `f3` we update the table as $(f1, s_1) \rightarrow (f2, s_2) \rightarrow (f3, s_3)$. When `f3` returns, its stack frame is poped and the new value of the stack pointer becomes s_2. We then delete the entry from the list: $(f1, s_1) \rightarrow (f2, s_2)$ and so on. In this way, if the function `f2` accesses a variable whose address x is, say, in the range $[s_1, s_2]$ we can immediately identify it as a reference to a variable declared in `f1`. We will then search our symbolic linked list for an entry whose function field equals `f1` and such that *beginOffset* $\leq x \leq$ *endOffset*.

4 Conclusions

In this paper we presented the algorithms implemented in the SIGMA environment to map memory references to data structures. With the mapping of global variables, dynamically allocated variables, and stack variables, such as automatic variables in C and local variables in Fortran 90, SIGMA provides *data-centric* performance information that is useful for tuning loop kernels, understanding the cache behavior of new algorithms, and to investigate how different parts of a program and its data structure compete for and interact within the memory subsystem in current and hypothetical systems.

In addition we described the extensions to SIGMA that were needed to support the symbolic memory mapping. These extensions include the binary instrumentation of the functions for allocation (`malloc`) and deallocation (`free`) of variables, needed for symbolic mapping of dynamic data structures, and the simulation of a stack frame in the runtime data collection engine, needed for symbolic mapping of local variables.

References

1. S. Browne, J. Dongarra, N. Garner, G. Ho, and P. Mucci. A Portable Programming Interface for Performance Evaluation on Modern Processors. *The International Journal of High Performance Computing Applications*, 14(3):189–204, Fall 2000.
2. Luiz DeRose. The Hardware Performance Monitor Toolkit. In *Proceedings of Euro-Par*, pages 122–131, August 2001.
3. Luiz DeRose, K. Ekanadham, Jeffrey K. Hollingsworth, and Simone Sbaraglia. SIGMA: A Simulator Infrastructure to Guide Memory Analysis. In *Proceedings of SC2002*, Baltimore, Maryland, November 2002.
4. Luiz DeRose and Daniel Reed. Svpablo: A Multi-Language Architecture-Independent Performance Analysis System. In *Proceedings of the International Conference on Parallel Processing*, pages 311–318, August 1999.
5. Seon Wook Kim, Bob Kuhn, Michael Voss, Hans-Christian Hoppe, and Wolfgang Nagel. VGV: Supporting Performance Analysis of Object-Oriented Mixed MPI/OpenMP Parallel Applications. In *Proceedings of the International Parallel and Distributed Processing Symposium*, April 2002.
6. John Mellor-Crummey, Robert Fowler, Gabriel Marin, and Nathan Tallent. HPCView: A tool for top-down analysis of node performance. *The Journal of Supercomputing*, (23):81–101, April 2002.
7. Barton P. Miller, Mark D. Callaghan, Jonathan M. Cargille, Jeffrey K. Hollingsworth, R. Bruce Irvin, Karen L. Karavanic, Krishna Kunchithapadam, and Tia Newhall. The Paradyn Parallel Performance Measurement Tools. *IEEE Computer*, 28(11):37–46, November 1995.
8. Bernd Mohr, Allen Malony, and Janice Cuny. TAU Tuning and Analysis Utilities for Portable Parallel Programming. In G. Wilson, editor, *Parallel Programming using C++*. M.I.T. Press, 1996.
9. Mikael Pettersson. *Linux X86 Performance-Monitoring Counters Driver.* http://user.it.uu.se/~mikpe/linux/perfctr/.
   Computing Science Department; Uppsala University - Sweden, 2002.
10. Felix Wolf and Bernd Mohr. Automatic performance analysis of hybrid mpi/openmp applications. *Journal of Systems Architecture, Special Issue 'Evolutions in parallel distributed and network-based processing'*, 49(10–11):421–439, November 2003.

Impact of Cache Coherence Models on Performance of OpenMP Applications

Jie Tao and Wolfgang Karl

Institut für Rechnerentwurf und Fehlertoleranz
Universität Karlsruhe (TH)
76128 Karlsruhe, Germany
{tao,karl}@ira.uka.de

Abstract. OpenMP is becoming an important shared memory programming model due to its portability, scalability, and flexibility. However, as it is a fact with any programming paradigms, cache access behavior significantly influences the performance of OpenMP applications. Improving cache performance in order to reduce misses therfore becomes a critical issue for High Performance Computing. This can be achieved by optimizing the source code, but also gained through adequate coherence schemes.

This work studies the behavior of various cache coherence protocols, including both hardware based mechanisms and software based relaxed models. The goal is to examine how well individual schemes perform with different architectures and applications, in order to find general ways to support the cache design in shared memory systems. The study is based on a simulation environment capable of modeling the parallel execution of OpenMP programs. First experimental results show that relaxed models are scalable and can be used as efficient alternative for those hardware coherence mechanisms.

1 Introduction

OpenMP [3] is becoming increasingly popular in both research and commercial areas. In contrast to the message passing models, like PVM and MPI [8], OpenMP has the advantages of being capable of releasing the programmers from the burden of explicitly specifying the control and communication between processes and threads. OpenMP also enforces a nested structure in the parallel program allowing to exploit the parallelism to the full extent. In addition, OpenMP is portable, flexible, and scalable.

Traditionally, OpenMP is developed to program Symmetric Multiprocessor (SMP) machines that deploy a physically shared memory. With the endeavor of compiler developers, however, OpenMP applications can currently also run on top of cluster environment with software Distributed Shared Memory (DSM). Examples are the Nanos Compiler [4] and the Polaris parallelizing compiler [2]. These compilers enable to maintain the architectural properties, e.g. scalability and cost-effectiveness, of modern cluster systems, while at the same time use OpenMP for developing parallel programs.

A problem with parallel programming is that applications usually can not achieve expected performance. Among those various reasons cache utility is rather critical. Since data stored in caches can often not be reused, applications still suffer from large amount of cache misses and the thereby resulted long access time. This issue is specially critical for OpenMP programs on cluster architectures since OpenMP exploits fine-grained parallelism, which introduces more data transfers between processors.

A primary source causing cache misses is cache line invalidation. On a multiprocessor system, each processor cache can hold a copy of the same data and all these copies have to be kept consistent. Tightly coupled systems usually use hardware based consistency mechanisms, and clusters with software DSM often deploy relaxed consistency models. These coherence protocols vary significantly in techniques for selecting cache lines to invalidate and hence probably result in rather different behavior of invalidation.

This work investigates various cache coherence protocols, including those existing schemes and several novel ones. This study is based on an OpenMP cache simulator that models shared memory architectures and the parallel execution of OpenMP programs. First experimental results with standard benchmark applications show that hardware consistency mechanisms perform well on small systems, while relaxed models often behave better on larger machines.

The remainder of this paper is organized as follows. Section 2 briefly describes common used cache coherence protocols and several new ones. In Section 3 the OpenMP simulation environment is introduced. This is followed by the first experimental result in Section 4 that shows the different validation behavior of the investigated coherence models. The paper concludes in Section 5 with a short summary and a few future directions.

2 Cache Consistency Model

Shared memory systems are typically constructed from commodity processors with on-chip caches or cache hierarchies. Replications of data in caches have to be consistent giving the user a unified memory abstraction. This kind of consistency is maintained by cache coherence protocols.

For bus-based multiprocessors, like SMPs, usually a snooping protocol is deployed. A snoopy coherence protocol uses an invalidation strategy with this concept: when a processor writes to its cache, it sends the address in a special invalidation message through the bus to other processors; cache copies in these processors are correspondingly invalidated. This strategy does not result in much traffic due to the shorter invalidation message. Snooping protocols often combine a cache tag with different states and uses a simple state machine to adjust the status of cache tags. A well known representative is MESI [1], which uses four states: modified, exclusive modified, shared, and invalid, to identify the cache tags.

Another group of hardware coherence protocols are directory based mechanisms. In contrast to the snooping protocols, the directory based protocols maintain a distributed directory for information about which caches contain which memory entries. When an entry is modified, the directory invalidates the copies in other caches with that entry. Since no broadcast is required during the invalidation process, directory based protocols are usually applied on multiprocessors with hardware DSM like NUMAchine [5] and SGI Origin O2000/3000 [7].

However, both groups of coherence mechanisms can not be used on clusters with software DSM since they rely on hardware support. On these machines the consistency behavior is controlled in software and usually a relaxed coherence protocol is deployed.

Relaxed consistency models [12] formalize the use of inconsistent memory hence giving the programmer a consistent memory abstraction. They are generally combined with synchronization primitives, like locks and barriers, which are contained in shared memory codes. These primitives force the caches consistent by invalidating cache entries with an *Acquire* access. Existing systems in this area often deploy a full invalidation mechanism that invalidates the whole cache by a synchronization operation. This is easy to implement, but introduces significant overheads and cache line misses.

Actually, full invalidation is not necessary because often only partial entries in the caches are updated before a synchronization operation is required. Therefore, we propose two novel schemes based on partial invalidations: Selective I and Selective II. The former only invalidates cache lines holding remote memory contents, while the latter invalidates only those remote cache lines which require to be invalidated. In contrast to the former scheme, the second one is capable of leading to less invalidations. For example, locks usually protect specific regions. Cache lines that would need to be invalidated under a lock might not be updated outside the critical regions and need not to be invalidated during the next barrier operation.

3 Simulation Platform

The cache coherence mechanisms described above have been evaluated using an OpenMP simulation platform [10] based on Valgrind [11]. Valgrind is a trace-driven memory simulator and debugger for x86-GNU/Linux. It is capable of supervising the execution of a program and recording all memory reads and writes. More specially, Valgrind models the POSIX threads supporting therefore any shared memory programming models based on this thread library. Additionally, Valgrind uses a runtime instrumentation technique to mark each memory access instruction within the executables. This enables to build an independent OpenMP simulator allowing the use of any compiler including the commercial ones.

Valgrind itself contains a cache simulator, which supports two level cache hierarchy with fixed write back behavior and LRU replacement policy. However, this simulator does not model multiprocessors, but uniprocessor systems. In order to enable the study of the memory system on shared memory machines, we have integrated into Valgrind a backend, while using Valgrind as a frontend for establishing the basic simulation infrastructure and generating memory references.

The backend [9] is a self-developed package that models the target architectures. As we focus on the research work on the memory system, the backend contains mainly mechanisms for modeling the complete memory hierarchy in detail. This includes a flexible cache simulator which models caches of arbitrary levels and various cache coherence protocols, a memory control simulator which models the management of shared and distributed shared memories and a set of data allocation schemes, and a network mechanism simulating the interconnection traffic.

The cache simulator models a multilevel cache hierarchy and allows each cache to be organized as either write-through or write-back depending on the target architectures. All relevant parameters including cache size, cache line size, and associativity can be specified by the user. In addition to simulating the caches themselves, the cache simulator also models a set of cache coherence protocols. This includes those mechanisms described in Section 2 and an optimal scheme in order to evaluate them. All schemes maintain the caches on the system consistent in the following way:

- *MESI*: invalidates all valid cache copies on other processors at every shared write.
- *Optimal*: within this scheme, invalidations are not done by write operations, rather by read accesses. At each shared read, the accessed cache line is invalidated only when the content has been updated by other processors since the last loading. This scheme hence performs invalidations only when necessary.
- *Full Invalidation*: invalidates all valid cache lines on other processors whenever an application initiates a lock or a barrier.
- *Selective I*: similar to the *Full Invalidation*, but invalidates only those cache lines holding remote memory contents.
- *Selective II*: similar to *Selective I*, but invalidates only those remote cache lines which require to be invalidated.

4 Experimental Results

Using this simulation platform described above, we could study the influence of various cache coherence policies on the cache performance of OpenMP applications. All applications for this study are chosen from the NAS parallel benchmark [6]. This benchmark is designed for comparing the performance of parallel computers and is widely recognized as a standard indicator of computer performance. Selected applications are FT, LU, MG, and CG.

FT contains the computational kernel of a three-dimensional FFT-based spectral method. It performs three one-dimensional fast Fourier transformations, one for each dimension. LU is a simulated CFD application that uses symmetric successive overrelaxation (SSOR) to solve a block lower triangular and block upper triangular system of equations resulting from an unfactored finite-difference discretization of the Navier-Stokes equations in three dimensions. MG uses a multigrid algorithm to compute the solution of the three-dimensional scalar Poisson equation. CG tests unstructured grid computation and communication using a matrix with randomly generated locations of entries. For the following experiment, the working set size is chosen: $64 \times 64 \times 32$ for FFT, $32 \times 32 \times 32$ for LU and MG, and 1400 for CG.

Firstly, we examine the invalidation behavior of these coherence protocols on DSM systems in order to give an initial comparison. For this, all tested codes are simulated on a loosely-coupled multiprocessor with 4 processor nodes, each deploying a 16KB, 2-way L1 cache and a 512KB, 4-way L2 cache. Access latency on memory locations is chosen with 1 cycle for L1, 5 cycles for L2, 50 cycles for the main memory, and 2000 cycles for inter-node communication.

Impact of Cache Coherence Models on Performance of OpenMP Applications

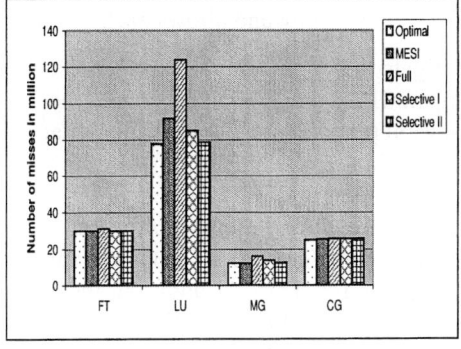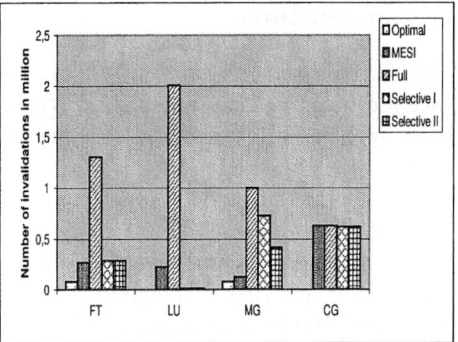

Fig. 1. Cache misses (right) and invalidations (left) on a 4-node system.

Figure 1 shows the experimental results with the left presenting the number of invalidations and the right the number of cache misses caused by different coherence protocols. It can be observed that applications vary significantly.

For FT similar behavior can be seen with all protocols in terms of miss number. A surprised case is the Full scheme, where even 1.3 million invalidations have been caused, which are nearly 5 factors more than the MESI protocol, the execution behavior of FT is not clearly influenced. This can be explained by the fact that the invalidated cache lines do not contain reused data and hence those additional invalidations result in no more cache misses.

The LU code, however, shows a quite different behavior. As expected, the Optimal protocol behaves better than MESI due to its ability of only performing invalidations as necessary. MESI behaves better than Full. This is caused by the full cache invalidation of the latter. Nevertheless, both Selective schemes performs well with even a better performance than MESI. This can be explained by the fact that LU contains many barriers and locks, where Full and the Selective protocols perform invalidations. Since both Selective schemes only invalidate partial cache lines, rather than the whole cache as Full does, frequently reused data can be maintained in the caches.

For the MG code, it can be observed that all relaxed consistency models result in more cache misses than MESI. This poor behavior is caused by the larger number of invalidations which can be seen in the right figure. While only one barrier is used in this code, both selective schemes can not show their advantages.

The last code CG shows a close behavior with FT in terms of cache misses. While Optimal presents a slight improvement, the others behave similarly with only a 0.06% gain for MESI in comparison to Full and the Selective schemes.

Overall, the experimental results show that the relaxed consistency models are generally not worse than MESI on small systems. This leads us to further examine the invalidation behavior on larger system and using larger working set size. The experimental results show that relaxed consistency models are beneficial on larger systems and larger working set, and can provide an efficient alternative for hardware coherence mechanisms.

5 Conclusions

This work uses a simulation platform to investigate the impact of various cache coherence protocols on the execution behavior of OpenMP applications. The studied mechanisms include hardware based schemes and several relaxed models. It is found that the former, like the MESI protocol, works well on machines with small number of processors, while the latter often outperforms on larger systems and working set.

In the next step of this research work, we will examine more applications with complex access patterns, especially those realistic ones. The goal is to classify various applications and detect adequate coherence schemes for them. In addition, the same work would be done on actual cluster systems.

References

1. J. Archibald. A Cache Coherence Approach for Large Multiprocessor Systems. In *Proceedings of the International Conference on Supercomputing*, pages 337–345, Vovember 1988.
2. A. Basumallik, S.-J. Min, and R. Eigenmann. Towards OpenMP Execution on Software Distributed Shared Memory Systems. In *Proceedings of the 4th International Symposium on High Performance Computing (ISHPC 2002)*, pages 457–468, 2002.
3. L. Dagum and R. Menon. OpenMP: An Industry-Standard API for Shared-Memory Programming. *IEEE Computational Science & Engineering*, 5(1):46–55, January 1998.
4. Marc Gonzàlez, Eduard Ayguadé, Xavier Martorell, Jesús Labarta, Nacho Navarro, and José Oliver. NanosCompiler: Supporting Flexible Multilevel Parallelism in OpenMP. *Concurrency: Practice and Experience*, 12(12):1205–1218, 2000.
5. T. S. Grbic, S. Brown, S. Caranci, G. Grindley, M. Gusat, G. Lemieux, K. Loveless, N. Manjikian, S. Srbljic, M. Stumm, Z. Vranesic, and Z. Zilic. Design and Implementation of the NUMAchine Multiprocessor. In *Proceedings of the 1998 Conference on Design Automation*, pages 66–69, Los Alamitos, CA, June 1998.
6. H. Jin, M. Frumkin, and J. Yan. The OpenMP Implementation of NAS Parallel Benchmarks and Its Performance. Technical Report NAS-99-011, NASA Ames Research Center, October 1999.
7. J. Laudon and D. Lenoski. The SGI Origin: A ccNUMA Highly Scalable Server. In *Proceedings of the 24th International Symposium on Computer Architecture*, pages 241–251, May 1997.
8. I. Pramanick. MPI and PVM Programming. In R. Buyya, editor, *High Performance Cluster Computing*, volume 2, Programming and Applications, chapter 3, pages 48–86. Prentice Hall PTR, 1999.
9. J. Tao, M. Schulz, and W. Karl. A Simulation Tool for Evaluating Shared Memory Systems. In *Proceedings of the 36th Annual Simulation Symposium*, pages 335–342, Orlando, Florida, April 2003.
10. J. Tao and J. Weidendorfer. Cache Simulation Based on Runtime Instrumentation for OpenMP Applications. In *Proceedings of the 37th Annual Simulation Symposium*, Arlington, VA, April 2004. to appear.
11. WWW. Valgrind, an open-source memory debugger for x86-GNU/Linux, 1999. http://developer.kde.org/~sewardj/
12. Y. Zhou, L. Iftode, J. P. Singh, K. Li, B. R. Toonen, I. Schoinas, M. D. Hill, and D. A. Wood. Relaxed Consistency and Coherence Granularity in DSM Systems: A Performance Evaluation. In *Proceedings of the Sixth ACM SIGPLAN Symposium on Principles and Practice of Parallel Programming*, pages 193–205, June 1997.

Evaluating OpenMP Performance Analysis Tools with the APART Test Suite*

Michael Gerndt[1], Bernd Mohr[2], and Jesper Larsson Träff[3]

[1] Institut für Informatik, Technische Universität München, Germany
gerndt@in.tum.de
[2] Forschungszentrum Jülich GmbH, ZAM, Jülich, Germany
b.mohr@fz-juelich.de
[3] C&C Research Labs, NEC Europe Ltd., St. Augustin, Germany
traff@ccrl-nece.de

Abstract. We outline the design of ATS, the *APART Test Suite*, for evaluating (automatic) performance analysis tools with respect to their *correctness* and *effectiveness* in detecting actual performance problems, with focus on the ATS test programs related to OpenMP. We report on results from applying two OpenMP performance analysis tools to the test cases generated from ATS.

1 Introduction

Achieving high performance on parallel computers most often requires performance tuning. The programmer identifies performance problems with the help of manual or automatic performance analysis tools, and transforms the code to improve its performance. The members of the European IST APART working group are developing automatic performance analysis tools for parallel and grid environments. The APART group defined the APART Specification Language (ASL) for writing portable specifications of typical performance problems [2]. Recently, an extensive set of performance properties for hybrid parallel programs combining MPI and OpenMP has been collected [4].

Automatic performance analysis tools, such as those implemented by APART members [1,5,7,8], must be tested with respect to *correctness* and *effectiveness* in detection of actual performance problems. For an automatic performance analysis tool *(positive) correctness* means that the tool is able to detect manifested performance problems in a given application; negative correctness means that the tool does not falsely report performance problems where none exist. To aid the correctness testing and to provide a "standardized" testbed that can be applied to different tools, we are developing the *APART Test Suite (ATS)* framework which allows for easy construction of synthetic positive and negative

---

* Part of this work is funded by the European Commission via the working group on Automatic Performance Analysis: Real Tools (APART),
http://www.fz-juelich.de/apart/

Table 1. OpenMP performance properties related to synchronization.

`critical_section_locking`: critical section overhead without competing threads
`critical_section_contention`: critical section overhead with competing threads
`serialization_due_to_critical_section`: all work in parallel loop in critical section
`frequent_atomic`: excessive time spent in simple atomic operation
`setting_lock`: overhead for setting a lock without competition
`lock_testing`: overhead for lock testing
`lock_waiting`: overhead for waiting for a lock
`all_threads_lock_contention`: locking overhead due to contention
`pairwise_lock_contention`: locking overhead by pairs of threads
---

test programs. The current version includes test cases for MPI and OpenMP performance properties in C and Fortran [6].

This paper gives an overview of typical OpenMP performance properties (Section 2), explains the basic structure of ATS (Section 3) and reports on first findings of an evaluation study of two OpenMP performance tools (Section 4).

2 A Performance Property Hierarchy for OpenMP

To describe performance properties of parallel programs an object-oriented, functional formalism called ASL (APART Specification Language) has been developed [2]. In ASL terminology a *performance property* characterizes a particular performance-related behavior of a program based on available or required *performance data*.

Performance data is primarily *dynamic information* collected during one or more sample runs of the program, and can be either trace or summary information. Performance data, however, also includes *static information* about the program (block structure, program and data flow information, loop scheduling

Table 2. OpenMP load imbalance performance properties.

`imbalance_in_parallel_region`: different amount of work per thread
`imbalance_at_barrier`: different arrival time at explicit barrier
`imbalance_in_parallel_loop`: different amount of work in iterations
`imbalance_in_parallel_loop_nowait`: imbalanced parallel loop without implicit barrier
`imbalance_in_parallel_section`: different amount of work in parallel sections
`imbalance_due_to_uneven_section_distribution`: more sections than threads, some threads executed multiple sections
`imbalance_due_to_not_enough_sections`: less sections than threads
`unparallelized_in_master_region`: idle threads due to OpenMP master region
`unparallelized_in_single_region`: idle threads due to OpenMP single region
`unparallelized_in_ordered_loop`: thread serialization
`imbalance_in_ordered_loop`: different amounts of work in ordered region

Table 3. OpenMP performance properties related to control of parallelism.

`dynamic_scheduling_overhead`: scheduling overhead due to dynamic scheduling
`scheduling_overhead_in_parallelized_inner_loop`: inner loop with few iterations was parallelized even though outer loop has much more iterations
`insufficient_work_in_parallel_loop`: loop overhead dominates execution
`firstprivate_initialization`: overhead for initialization of firstprivate variables
`lastprivate_overhead`: initialization overhead of lastprivate variables
`reduction_handling`: overhead for reduction operation handling

Table 4. OpenMP performance properties related to inefficient serial execution.

`false_sharing_in_parallel_region`: overhead for access to different array elements on same cache line

---

information etc.) and the programming model/paradigm. Examples of performance properties are load imbalance, abundant or mis-scheduled communication, and cache misses.

A performance property is described by a boolean *condition*, and has an associated *severity* for expressing the relative importance of the property. A performance property is a *performance problem* if it is present and its severity exceeds a preset threshold. A *performance bottleneck* is a most severe performance problem. In this framework performance engineering consists in locating and eliminating performance bottlenecks.

Using the ASL formalism, the APART group has compiled hierarchically structured specifications of typical performance properties for the programming paradigms MPI, OpenMP, and HPF [2]. Specifications for OpenMP can be found in [3], and considerably more detailed in [4]. We briefly summarize the properties recorded in [4], since these are the properties that are closely mirrored in ATS. The specification is divided into four categories: (i) *synchronization*, (ii) *load imbalance*, (iii) *control of parallelism*, and (iv) *inefficient serial execution*. Concrete properties in the four categories are listed in Tables 1 to 4. For hybrid OpenMP/MPI programming, additional categories contain properties related to MPI communication and to parallel I/O.

3 APART Test Suite Design

We briefly describe the design of the APART Test Suite, especially as pertaining to the OpenMP properties listed in the previous section. The first version of ATS covers the "standard" parallel programming paradigms MPI and OpenMP, but the modular structure of the design easily allows to add modules for other programming paradigms like HPF, PVM, or POSIX threads.

The main idea of our design is a collection of functions with a standardized interface, hierarchically organized into modules that can be easily combined to produce a program exhibiting desired performance properties. Thus, functions

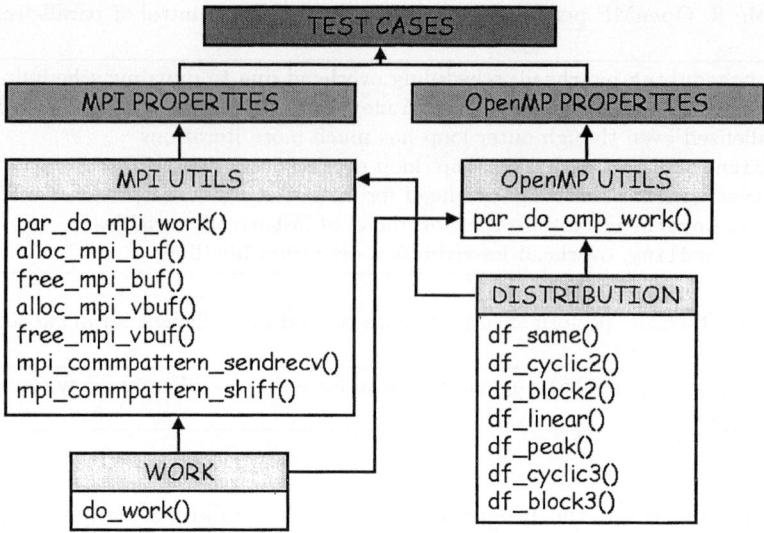

Fig. 1. Basic Structure of the ATS framework.

from the modules should have as little context as possible, and whatever context is necessary is provided through standardized parameters. Furthermore, since (automatic) performance analysis tools have different thresholds/sensitivities, it is important that the test suite is parametrized so that the relative severity of the properties can be controlled by the user.

Figure 1 shows the basic structure of the ATS framework for the MPI and OpenMP programming paradigms. The boxes with shaded titles represent the basic modules. Arrows indicate *used-by* relationships. For the lower levels, the functions provided by those module are listed.

The lowest two modules, work and distribution, provide basic functionality to specify the amount of generic work to be executed by the individual threads or processes of a parallel program. The next level provides generic support for the two main parallel programming paradigms MPI and OpenMP. The third level implements *property functions* which when executed exhibit one specific performance property. For OpenMP this means that we implemented one such function for each property listed in Section 2. Finally, there are several ways of calling the property functions so that different aspects of performance correctness testing can be addressed. For a full description of the ATS framework, see [6].

4 Evaluation of OpenMP Performance Tools

We used the ATS to evaluate four OpenMP performance analysis tools, namely the Hitachi Profiling Tool pmfunc specifically for Hitachi parallel supercomputers, Intel's performance tool Vtune, the platform-independent tools EXPERT and Vampir. Vtune, Vampir and the Hitachi Profiling Tool are manual, while EXPERT is an automatic tool. Due to limited space, we only discuss pmfunc and EXPERT below.

4.1 Hitachi Profiling Tool

The Hitachi SR8000 supercomputer is a clustered SMP design. Each node consists of eight processors that can be used by applications. Applications are developed in the hybrid programming model, MPI across nodes and OpenMP or COMPASS, a Hitachi proprietary shared memory programming API, within nodes.

The Fortran, C and C++ compilers on the Hitachi SR8000 can automatically instrument program regions. A compiler switch pmfunc directs the compiler to instrument user functions. The switch pmpar instruments all COMPAS parallel regions, independent of whether they are generated by automatic parallelization or by manual transformation. In OpenMP programs, the switch instruments only OMP PARALLEL REGION, OMP PARALLEL DO, and OMP PARALLEL SECTION. It does not instrument work-sharing constructs within parallel regions. The compiler switches not only insert calls to the monitoring routines, but also link a performance monitoring library to the instrumented code. This library measures for each instrumented region, that is, user function or parallel region, among other values execution time, cache misses, load/store instructions, floating point operations, and number of executions. See [9] for details.

For each node a separate information file is generated that can be inspected with the pmpr command. This command displays the information contained in the output files in a humanly readable form (Figure 2).

The information measured for a function or parallel region is presented separately for each thread. This makes it possible to investigate differences among

```
imbalance_due_to_uneven_section_distribution[2](omp_pattern.c+560)
      CPU time  FLOP      Inst     LD/ST    D-cache   MFLOPS     MIPS    Times
      ---------------------------------------------------------------------------
IP0   4.492<    16>   76903k<   36190k<    2272k     0.000>    17.120>    4>
IP1   4.492     16>   76903k<   36190k<    2273k>    0.000     17.120     4>
IP2   4.493>    16>   76903k<   36190k<    2272k     0.000     17.116     4>
IP3   4.492     16>   76903k<   36190k<    2272k     0.000     17.119     4>
IP4   4.493      8<   38452k<   18095k<    1136k<    0.000<     8.559<    4>
IP5   4.492      8<   38452k<   18095k<    1137k     0.000      8.559     4>
IP6   4.493      8<   38452k<   18095k<    1137k     0.000      8.559     4>
IP7   4.493      8<   38452k<   18095k<    1137k     0.000      8.559     4>
      ---------------------------------------------------------------------------
TOTAL 35.940    96   461419k   217138k   13635k     0.000    102.693    32
      ---------------------------------------------------------------------------
Element parallelizing rate : (TOTAL)/(Max * IPs)
     CPU time :   99.98[%] =   35.940/(4.493168*8)
     FLOP     :   75.00[%] =   96/(16*8)
```

Fig. 2. Example information file showing load imbalance due to uneven section distribution. The code had 12 sections, so the first four threads got two sections. The evaluation program pmpr marks the largest and smallest values in each column with '<' and '>' respectively.

the threads, for example, resulting from load imbalance. For parallel regions the values can be compared directly. For functions, the data have to be interpreted more carefully. If a function is started on the master processor but includes parallel regions, the data of the other threads are accumulated in the master processor. If, on the other hand, a function is called in a parallel region, the execution information is reported for each thread individually.

Since execution time and instruction counts are given on a per thread basis for parallel regions, load imbalance properties could be identified. The execution time did show the imbalance only in imbalance_in_parallel_loop_nowait since in all other cases the implicit barrier ensures equal execution times. In those cases, the imbalance was detected from the differences in the instruction counts.

The difference between unparallelized_ordered_loop and imbalance_in_- ordered_loop was not shown since ordered loops are executed by the Hitachi compilers as sequential loops.

The only test case based on cache misses, false_sharing_in_parallel_- region, could be detected from a very high cache miss rate (about 75%). The tool did not give any indication that the misses resulted from false sharing.

Properties related to synchronization bottlenecks could not be identified since the tool gives no information about synchronization operations. The same is true for properties checking parallelism overhead.

4.2 EXPERT

The EXPERT automatic event trace analyzer [8] is part of the KOJAK project (Kit for Objective Judgment and Knowledge-based Detection of Performance Bottlenecks), whose aim is a generic automatic performance analysis environment for parallel programs. Performance problems are specified in terms of *execution patterns* that represent situations of inefficient behavior. These are input to an analysis process that recognizes and quantifies inefficient behavior in event traces. The pattern specification in EXPERT is different from ASL, as it allows to specify how performance metrics are calculated out of basic event attributes.

The KOJAK analysis process is composed of two parts: a semi-automatic multi-level instrumentation of the user application followed by an automatic analysis of the generated performance data. Running an instrumented executable generates a trace file in the EPILOG format. After program termination, the trace file is fed into the EXPERT (Extensible Performance Tool) analyzer. The analyzer generates an analysis report, which serves as input for the EXPERT presenter. A screen dump is shown in Figure 3. Using the color scale shown on the bottom, the severity of performance problems found (left pane) and their distribution over the program's call tree (middle pane) and machine locations (right pane) is displayed. By expanding or collapsing nodes in each of the three trees, the analysis can be performed on different levels of granularity.

The experiments were performed on a 4 CPU Intel IA32 Linux system. As can be seen in the left part of Figure 3, for OpenMP, EXPERT is currently able to identify performance problems related to Flush, thread startup (Fork), barrier and locking overhead. It also shows that EXPERT could detect all properties

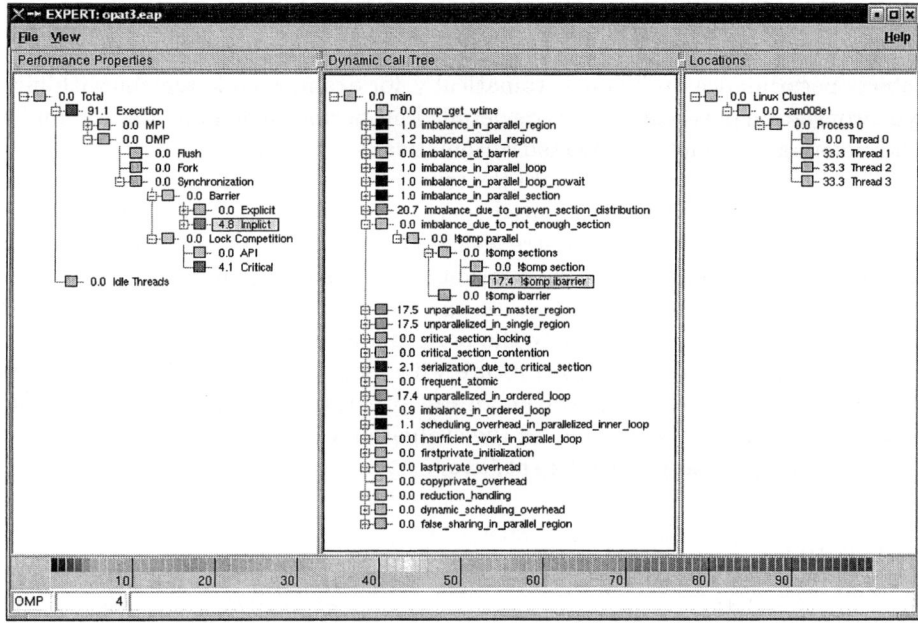

Fig. 3. Expert Presenter result display. Left pane shows performance problems, middle pane their distribution, and right pane their machine location.

related to load imbalance, as indicated by darker boxes in the middle pane. The numbers shown are the percentage of the total execution time lost because of the identified performance property. The load imbalance problems are detected due to high barrier overhead. Similar properties could be defined to distinguish lock contention and frequent locking. By selecting specific call tree nodes the distribution of the problem over the machine, processes, and threads can easily be investigated in the right pane (shown for the property function imbalance_-due_to_uneven_section_distribution). EXPERT was also able to identify performance problems related to locks and critical regions (not visible in Figure 3).

5 Conclusion and Future Work

We listed the current set of OpenMP performance property functions in the APART Test Suite (ATS), a framework which can be used to generate test cases for the evaluation of (automatic) performance analysis tools. We ran the full set of OpenMP functions with a semi-automatic vendor performance analysis tool and the automatic EXPERT tool.

The effectiveness of the tools depends highly on the information provided by the runtime monitor. With the Hitachi tool no synchronization information is available. Thus, even simple load imbalance problems cannot be easily detected. The hardware counter information can be used instead to get hints to load imbalances and have proved very useful to identify false sharing. The Hitachi profiling tool provides only summary tables in ASCII form. The EXPERT tool detects performance problems automatically, for example code regions with high synchronization overhead, but more detailed properties explaining the reason for load imbalance cannot be detected automatically.

The ATS enabled us to evaluate the four tools (results for Vtune and Vampir are not discussed here due to limited space). Due to the well defined semantics of the property functions, the strength and weaknesses of the different tools can be easily identified. However, a formal comparison or ranking of the tools is quite difficult since, except for EXPERT, the user has to interpret the information provided by the tools and identify the performance properties manually.

We plan to extend ATS in the future with more performance properties and to work on the automatic generation of test programs combining the individual modules. ATS is freely available for the evaluation of other performance tools at http://www.fz-juelich.de/apart/ats/.

References

1. A. Espinosa. *Automatic Performance Analysis of Parallel Programs*. PhD thesis, Universitat Autonoma de Barcelona, 2000.
2. T. Fahringer, M. Gerndt, B. Mohr, F. Wolf, G. Riley, and J. L. Träff. Knowledge specification for automatic performance analysis. Technical Report FZJ-ZAM-IB-2001-08, Forschungszentrum Jülich, 2001.
3. T. Fahringer, M. Gerndt, G. Riley, and J. L. Träff. Formalizing OpenMP performance properties with ASL. In *Workshop on OpenMP: Experience and Implementations (WOMPEI), Intl. Symposium on High Performance Computing (ISHPC2K)*, LNCS 1940, pp. 428–439, 2000.
4. M. Gerndt. *Specification of Performance Properties of Hybrid Programs on Hitachi SR8000*. Peridot Technical Report, TU München, 2002
5. B. P. Miller, M. D. Callaghan, J. M. Cargille, J. K. Hollingsworth, R. B. Irvine, K. L. Karavanic, K. Kunchithapadam, T. Newhall. The Paradyn Parallel Performance Measurement Tool. *IEEE Computer*, 28(11):37–46, 1995.
6. B. Mohr and J. L. Träff. Initial Design of a Test Suite for (Automatic) Performance Analysis Tools. In *8th Intl. Workshop on High-Level Parallel Programming Models and Supportive Environments (HIPS 2003)*, pp. 77–86, 2003.
7. H.-L. Truong, T. Fahringer. SCALEA: A Performance Analysis Tool for Distributed and Parallel Programs. In *Euro-Par 2002*, LNCS 2400, pp. 75–85, 2002.
8. F. Wolf, B. Mohr. Automatic Performance Analysis of Hybrid MPI/OpenMP Applications. *Journal of Systems Architecture*, 49(10-11):421–439, 2003. Special Issue "Evolutions in parallel distributed and network-based processing".
9. www.lrz-muenchen.de/services/compute/hlrb/manuals

Collective Communication Performance Analysis Within the Communication System

Lars Ailo Bongo, Otto J. Anshus, and John Markus Bjørndalen

Department of Computer Science, University of Tromsø, Norway
{larsab,otto,johnm}@cs.uit.no

Abstract. We describe an approach and tools for optimizing collective operation spanning tree performance. The allreduce operation is analyzed using performance data collected at a lower level than by traditional monitoring systems. We calculate latencies and wait times to detect load balance problems, find subtrees with similar behavior, do cost breakdown, and compare the performance of two spanning tree configurations. We evaluate the performance of different configurations and mappings of allreduce run on clusters of different size and with different number of CPUs per host. We achieve a speedup of up to 1.49 for allreduce. Monitoring overhead is low, and the analysis is simplified since many subtrees have similar behavior. However, the calculated values have large variations, and reconfiguration may affect unchanged parts.

1 Introduction

Clusters are becoming an increasingly important platform for scientific computing. Many parallel applications run on clusters use a communication library, such as MPI [9], which provides collective operations to simplify the development of parallel applications. Of the eight scalable scientific applications investigated in [16], most would benefit from improvements to MPI's collective operations.

The communication structure of a collective operation can be organized as a spanning tree, with threads as leafs. Communication proceeds along the arcs of the tree and a partial operation is done in each non-leaf node. Essential for the performance of a collective operation is the shape of the tree, and the mapping of the tree to the clusters in use [7, 12–14].

We present a methodology and provide insight into performance analysis within the communication system. We demonstrate and evaluate the methodology by comparing and optimizing the performance of different allreduce configurations.

Performance monitoring tools for MPI programs [8] generally treat the communication system as a black box and collect data at the MPI profiling layer (a layer between the application and the communication system). To understand why a specific tree and mapping have better performance than others it is necessary to collect data for analysis inside the communication system. We describe our experiences about what type of data is needed, how to do the analysis, and what the challenges are for collective communication performance analysis.

Usually, MPI implementations only allow the communication structure to be implicitly changed either by using the MPI topology mechanism or by setting attributes of communicators. To experiment with different collective communication configurations, we use the PATHS system [2], since it allows to inspect, configure and map the collective communication tree to the resources in use.

For a given tree configuration and mapping, our analysis approach and visualizations allows us to find performance problems within the communication system, and to compare the performance of several configurations. This allows us to do a more fine grained optimization of the configuration than approaches that only use the time per collective operation (as in [14]).

In addition to remapping trees, collective operation performance can be improved by taking advantage of architecture specific optimizations [12, 13], or by using a lower-level network protocol [6, 13]. However, the advantage of these optimizations depends on the message size. For example, for small message sizes, as used by most collective operations, point-to-point based communication was faster than Ethernet based broadcast in [6], and a shared memory buffer implementation for SMPs in [12]. Our approach allows comparing advantages of changing the communication protocol, or synchronization primitives for different message sizes.

On SMPs, collective communication performance can also significantly be reduced, by interference caused by system daemons [10]. Reducing the interference will make the performance analysis within the communication system even more important.

Mathematical models can be used to analyze the performance of different spanning trees (as in [1]), but these do not take into account the overlap and variation in the communication that occurs in collective operations [14].

For each thread, our monitoring system traces messages through a *path* in the communication system. We calculate latencies and wait times, and use these to detect load balance problems, find subtrees with similar behavior, do cost breakdown for subtrees, and compare the performance of two configurations.

Monitoring overhead is low, from nearly 0 to 3%. Analysis is simplified since many subtrees have similar behavior. Monitoring overhead is low, and the analysis is simplified since many subtrees have similar behavior. However, the calculated values have large variation, reconfiguration may affect unchanged parts, and predicting the effect of reconfigurations is difficult. Despite these problems we achieved a speedup of up to 1.49 for an allreduce benchmark using our tools.

The rest of this paper proceeds as follows. PATHS is described in section 2. Our monitoring tool and analysis approach are described in section 3 and demonstrated in section 4. In section 5 we discuss our results, and finally, in section 6 we conclude and outline future work.

2 Reconfigurable Collective Operations

We use the PATHS system [2] to experiment with different collective operation spanning tree configurations and mappings of the tree to the clusters in use. In

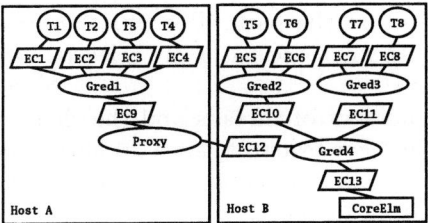

Fig. 1. An allreduce tree used by threads T1–T8 instrumented with event collectors (EC1–EC13). The result is stored in a PastSet element (CoreElm).

allreduce, each thread has data that is reduced using an associative operation, followed by a broadcast of the reduced value. The semantics differs from MPI in that the reduced value is stored in the PastSet structured shared memory [17].

Using PATHS we create a spanning tree with all threads participating in the allreduce as leafs (figure 1). For each thread we specify a *path* through the communication system to the root of the tree (the same path is used for reduce and broadcast). On each path, several *wrappers* can be added. Each wrapper has code that is applied as data is moved down the path (reduce) and up the path (broadcast). Wrappers are used to store data in PastSet and to implement communication between cluster hosts. Also, some wrappers, such as allreduce wrappers, join paths and handle the necessary synchronization.

The PATHS/PastSet runtime system is implemented as a library that is linked with the application. The application is usually multi-threaded. The PATHS server consists of several threads that service remote clients. The service threads are run in the context of the application. Also, PastSet elements are hosted by the PATHS server. Each path has its own TCP/IP connection (thus there are several TCP/IP connections between PATHS servers). The client-side stub is implemented by a *proxy* wrapper. Wrappers are run in the context of the calling threads, until a wrapper on another host is called. These wrappers are run in the context of the threads serving the connection.

In allreduce, threads send data down the path by invoking the wrappers on their path. The allreduce wrappers block all but the latest arriving thread, which is the only thread continuing down the path. The final reduced tuple is stored in the PastSet element before it is broadcasted by awakening blocked threads that return with a copy of the tuple. For improved portability, synchronization is implemented using Pthread condition variables.

3 Monitoring and Analysis

To collect performance data we use the EventSpace system [3]. The paths in a collective operation tree are instrumented by inserting *event collectors*, implemented as PATHS wrappers, before and after each wrapper. In figure 1, an allreduce tree used by threads T1–T8 is instrumented, by event collectors EC1–EC13. For each allreduce operation, each event collector records a timestamp when moving down, and up the path. The timestamps are stored in memory and written to trace files when the paths are released. In this paper analysis is done post-mortem.

Depending on the number of threads and the shape of the tree, there can be many event collectors. For example, for a 30 host, dual CPU cluster, a tree has 148 event collectors collecting 5328 bytes of data for each call (36 bytes per event collector). The overhead of each event collector is low (0.5 μs on a 1.4 GHz Pentium 4) compared to the hundreds of microseconds per collective operation. Most event collectors are not on the slowest path, thus most data collecting is done outside the critical path.

There are three types of wrappers in an allreduce spanning tree: gred (partial allreduce), proxy (network) and core (PastSet). For core wrappers the two timestamps collected by the event collector above it (EC13 in figure 1) are used to calculate the *store latency*; the time to store the result in PastSet. For proxy, we calculate the two-way TCP/IP latency by $(t_4-t_1)-(t_3-t_2)$, where t_1 (down) and t_4 (up) are collected by the event collector above the proxy in a path, and t_2 (down) and t_3 (up) are collected by the event collector below. To achieve the needed clock synchronization accuracy for calculating one-way latencies (tens of μs) special hardware is needed [11].

Gred wrappers have multiple children that contribute with a value to be reduced. The contributor can be a thread or data from another gred wrapper (in figure 1 threads T5–T6 contribute to gred2, while gred1–3 contribute to gred4). There is one event collector on the path to the parent that collects timestamps t_2 and t_3, while the paths from the P parents each have an event collector collecting timestamps $t_{1,i}$, and $t_{4,i}$. We define the *down latency* for a gred wrapper to be $t_2 - t_{1,l}$, the down latency for the last arrival l. The *up latency* is $t_{4,f} - t_3$, the up latency for the first departurer f.

For a given number of collective operations we calculate for each participant the *arrival order distribution* and the *departure order distribution*; that is the number of times the contributor arrived and departed at the gred wrapper as the first, second, and so on. In addition we calculate: *arrival wait time* $t_{1,l}-t_{1,i}$; the amount of time the contributor i had to wait for the last contributor l to arrive, and *departure wait time* $t_{4,i}-t_{4,f}$; elapsed time since the first contributor f departed from the gred wrapper, until contributor i departed.

For the analysis we often divide the path from a thread to a PastSet element into several stages consisting of the latencies and wait times described above. To calculate the time a thread spent in a specific part of the tree (or a path), we add together the time at each stage in the tree (or path). Usually mean times are used. Similarly, we can do a hotspot analysis of a tree by comparing the mean times.

For the performance analysis we (i) detect load balance problems, (ii) find paths with similar behavior, (iii) select representative paths for further analysis, (iv) find hotspots by breaking down the cost of a path into several stages, (v) reconfigure the path, and (vi) compare the performance of the new and old configuration.

For applications with load balance problems, optimizing collective operations will not significantly improve performance since most of the time will be spent waiting for slower threads. To detect load balance problems, we use the arrival order at each gred wrapper to create a weighted graph with gred wrappers and threads as nodes, and the number of last arrivals as weights on the edges. The part of the tree (or thread) causing a load imbalance can be found by searching for the longest path (i.e. with most last arrivals).

Usually, many threads have similar paths, with similar performance behavior. Thus, we can simplify the analysis and optimization by selecting a representative path for each type of behavior. Also, fast paths can be excluded since the allreduce operation time is determined by the slow paths. We have experimented with, but not succeeded in finding analysis and visualization approaches that allows for a detailed analysis of multiple paths at the same time.

When analyzing a representative path we break the cost of an allreduce operation into subtrees that can be optimized independently such as the subtree for an SMP host, a cluster, or TCP/IP latencies within a cluster and on a WAN. As for a gred wrapper, we calculate for a subtree the down latency, up latency, and departure wait time (using for the subtree on host B in figure 1 event collectors EC5–8 and EC13). A large down latency implies that the computation in the operation takes a long time. A large up latency indicates performance problems in the gred implementation, while a large departure wait time implies scalability problems of the synchronization variables, leading to unnecessary serialization.

4 Experiments

In this section we analyze the performance of different allreduce configurations for a blade cluster with ten uni-processor blades (Blade), a cluster of thirty two-way hosts (NOW), a cluster of eight four-way hosts (4W), and a cluster of four eight-way hosts (8W). The clusters are connected through 100 Mbps Ethernet, and the operating system is Linux. We designed experiments to allow us to measure the performance of different shapes and mappings of the spanning trees used to implement allreduce. A more detailed analysis can be found in [4].

We use a microbenchmark, Gsum, which measures the time it takes T threads to do N allreduce operations. The allreduce computes a global sum. The number of values to sum is equal to the number of threads, T. Threads alternate between using two identical allreduce trees to avoid two allreduce calls to interfere with each other.

We also use an application kernel, SOR, a red-black checker pointing version of successive over-relaxation. In each iteration, black and red points are computed and exchanged using point to point communication, before a test for converge which is implemented using allreduce. A communication intensive problem size was used (57% of the execution time was spent communicating). The computation and point-to-point communication results in a more complex interaction with the underlying system than in Gsum.

Gsum was run ten times for 25000 iterations on each cluster, but only the results from one execution are used in the analysis. The Gsum execution time has a small standard deviation (less than 1%). Also, the slowdown due to data collection is small (from no slowdown up to 3%). For SOR the execution time variation is similarly low, and the monitoring overhead is lower since SOR has relatively less communication than Gsum.

The standard deviation for the allreduce operation time is large. On the NOW cluster the mean is about $1000\,\mu s$ and the standard deviation is about $200\,\mu s$. Also the variation is large for the computed latencies and wait times. For many stages both the mean and the standard deviation are about $10-25\,\mu s$. The only values with low standard deviation are the TCP/IP, and store latencies. Despite the large variation, using mean in the analysis gives usable results.

The large standard deviation for gred up, and down latency is caused by queuing in the synchronization variables, while the departure wait time distribution is a combination of several distributions since the wait time depends on the departure order (which, in our implementation, depends on the arrival order). We expect most implementation to have similar large variations for these stages. For Gsum, arrival wait time variation is caused by variations on the up-path, while for SOR and other applications there are additional noise sources such as system daemons [10].

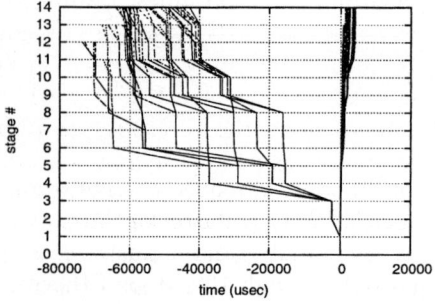
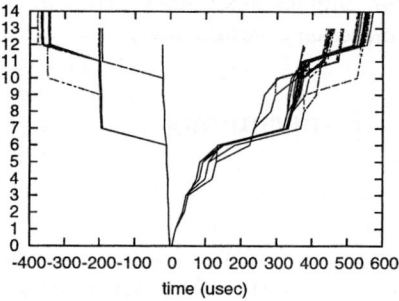

(a) SOR configuration with arrival wait times.

(b) Gsum configuration without arrival wait times.

Fig. 2. Timemap visualizations for NOW cluster spanning tree configurations.

For well balanced applications, such as Gsum and SOR, there is not one thread causing a load imbalance. SOR has a load imbalance when run on the 4W and 8W clusters caused by differences in point-to-point communication latency since two threads on each host communicate with a neighbor on a different host. Due to the load imbalance performance will not significantly be improved by optimizing allreduce as shown in figure 2a, where the down paths ($x < 0$) are dominated by arrival wait time (i.e. load imbalance). Based on our experiences most of the optimizations can be done on the up paths ($x > 0$). SOR can be reimplemented to hide the point-to-point communication latency, and thereby reducing the load imbalance.

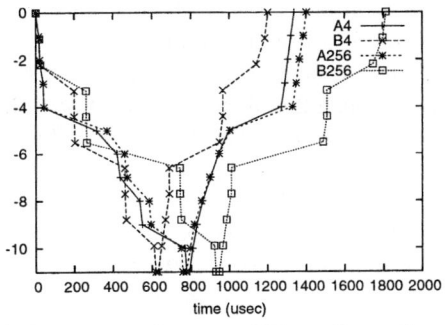 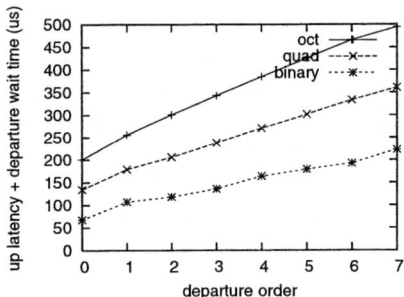

(a) Timemaps for two 4W configurations when using 4 and 256 byte messages.

(b) Serialization introduced by 8W host subtree configurations.

Fig. 3. Visualizations for comparing the performance of different configurations.

Differences in network latency also cause a load imbalance within the communication system when Gsum is run on a multi-cluster. However, the problem is caused by the spanning tree and can be improved by reconfiguring the tree. For the remaining analysis we ignore the arrival wait times, since these hide the difference between fast and slow paths.

To get an overview of the communication behavior of the different threads we use a *timemap* visualization that shows the mean time spent (x-axis) in each stage of the path (y-axis) when moving down and up the tree. $X = 0$ is when the threads enter the bottommost wrapper. For more details we use tables with statistics for each thread, and for each stage. The timemap in figure 2b shows that the 60 threads run on the NOW cluster have similar behavior. Arrival wait times are not shown, and the up-path has variations due to the arrival-departure order dependency. The threads can roughly be divided into classes according to the number of TCP/IP connections in their paths. An optimized configuration for 2W has a more irregular shape complicating the analysis due to rather large variation for most stages.

For the mostly used allreduce message sizes (below 256 bytes [15]), a cost breakdown shows that broadcast is more expensive than reduce. The up, and down gred-latency are only a few μs, hence the time to do the reduce operation is insignificant. For SMPs the departure wait time can be large, but for the best configuration the TCP/IP stages dominate the execution time. Also, the time spent storing the result in PastSet is insignificant. For some single-CPU cluster configurations we have a few outliers in the proxy stage that can significantly reduce performance. For one Blade Gsum configuration they caused a slowdown of 54.2.

When tuning the performance of a configuration it is important to find the right balance between load on *root* hosts and number of network links. Load on root hosts can be reduced by moving gred wrappers to other hosts. This will also improve potential parallelism, but it will introduce another TCP/IP connection to some threads paths. For SMPs, the performance of a host subtree can be improved by adding or removing an additional level in the spanning tree.

Reconfiguration improved the performance of Gsum up to a factor of 1.49. However, the effects of a reconfiguration can be difficult to predict. For a 4W Gsum configuration we doubled the number of TCP/IP connections on a threads path, but the time spent in these stages only increased by 1.55 due to the TCP/IP latency being dependent on load on the communicating hosts. A reconfiguration can also have a negative performance effect on unchanged subtrees. In addition, the best configuration for a cluster is dependent on CPU speed on hosts, LAN latency, number of hosts in cluster, and the message size used in the collective operation.

An overview of the differences between two configurations is provided by a timemap visualization that shows several configurations for one thread. Since paths in two configurations can have unequal length, the y-coordinates are scaled such that both have the same y_0 and y_{max}. Figure 3a shows that by moving a gred wrapper to another 4W host (B4 and B256), gives better performance for 4 byte messages, but worse for 256 byte messages, due to a trade-off between increased TCP/IP latencies, and single host subtree performance.

Figure 3b shows how adding additional levels to a subtree improves performance on an 8-way host. The figure shows the introduced latency (x=0), and the amount of serialization (slope of the curve, flatter is better). On the x-axis the order of departure is shown, and on the y-axis the up latency + departure wait time is shown. Notice that the up-latency is the same for all departures, and that for the first departurer (x=0) has zero departure wait time. The optimal height of the tree depends on the load on the hosts.

To find the fastest configuration, it is usually enough to compare the paths that are on the average slowest, but not always. A 4W Gsum configuration A, is slower than configuration B even if the 12 slowest threads are faster, because the remaining 20 threads in A are slower, and the large variation causes A to have the slowest thread for most allreduce calls.

5 Discussion

The timestamps collected by EventSpace allows us to analyze the performance of spanning trees and their mapping to the clusters in use. However, information from within the operating system is needed to understand why a synchronization operation, or a TCP/IP connection is slow. Hence, information from within the operating system should be collected and used in the analysis.

For ease of prototyping we used our own parallel programming system (PATHS). Our analysis approach should be applicable for MPI runtime systems provided that we can collect timestamps that can be correlated to a given MPI collective operation call. Once monitoring tools using the proposed MPI PERUSE interface[5] are available these may be used to collect the data necessary for our analysis. However, other runtime systems will have performance problems not treated by this work, for example with regards to buffering.

The calculated values used in the analysis have large variation, and outliers can have a significant effect on performance. A complete trace of all messages sent within a given time period provides enough samples for statistical analysis, and can be used to detect any periodical performances faults (e.g. caused by system daemons [10]). EventSpace allows to collect such traces with a small overhead and memory usage (1 MB of memory can store 29.127 EventSpace events).

We only studied one application; SOR. Since MPI defines the semantics of collective operations the communication pattern of SOR is general and frequent. Since SOR has a load imbalance a better application for the study would have been one of the applications described in [10]. Also, only the allreduce operation has been analyzed. We believe the analysis will be similar for other collective operations with small message size such as reduce and barrier. Message arrival order, synchronization points, and network latencies are also important for the performance of operations with larger messages, such as alltoall, allgather, and the MPI-IO collective operations. For MPI-IO we can wrap the I/O operations using PATHS wrappers.

6 Conclusion and Future Work

We have described systems for monitoring and tuning the performance of collective operation within the communication system. For each thread we trace the messages through a *path* in the communication system. We demonstrated an analysis approach and visualizations by evaluating and optimizing different spanning-tree configurations and cluster mappings of the allreduce operation.

The monitoring overhead is low, from nearly 0 to 3%, and the analysis is simplified since many paths have similar behavior. However, the computed latencies and wait times have large variation, reconfiguration may affect unchanged parts, and it is difficult to predict the effect of some changes.

As future work, we will use the EventSpace system [3] for run-time analysis. Also, we will examine how data collected inside the operating system can be used in the analysis, and if some load balance problems can be avoided by reconfiguring the collective operation spanning trees. Finally, our long-term goal is to build a communication system where collective communication is analyzed, and adapted at run-time.

References

1. BERNASCHI, M., AND IANNELLO, G. Collective communication operations: Experimental results vs.theory. *Concurrency: Practice and Experience 10*, 5 (1998).
2. BJØRNDALEN, J. M. *Improving the Speedup of Parallel and Distributed Applications on Clusters and Multi-Clusters.* PhD thesis, Tromsø University, 2003.
3. BONGO, L. A., ANSHUS, O., AND BJØRNDALEN, J. M. EventSpace - Exposing and observing communication behavior of parallel cluster applications. In *Euro-Par* (2003), vol. 2790 of *Lecture Notes in Computer Science*, Springer, pp. 47–56.

4. BONGO, L. A., ANSHUS, O., AND BJØRNDALEN, J. M. Evaluating the performance of the allreduce collective operation on clusters: Approach and results, 2004. Technical Report 2004-48. Dep.of Computer Science, University of Tromsø.
5. JONES, T. Personal communication. 2003.
6. KARWANDE, A., YUAN, X., AND LOWENTHAL, D. K. CC-MPI: a compiled communication capable MPI prototype for Ethernet switched clusters. In *Proc. of the ninth ACM SIGPLAN symposium on Principles and practice of parallel programming* (2003), ACM Press, pp. 95–106.
7. KIELMANN, T., HOFMAN, R. F. H., BAL, H. E., PLAAT, A., AND BHOEDJANG, R. A. F. Magpie: Mpi's collective communication operations for clustered wide area systems. In *Proceedings of the seventh ACM SIGPLAN symposium on Principles and practice of parallel programming* (1999), ACM Press, pp. 131–140.
8. MOORE, S., D.CRONK, LONDON, K., AND J.DONGARRA. Review of performance analysis tools for MPI parallel programs. In *8th European PVM/MPI Users' Group Meeting, Lecture Notes in Computer Science 2131* (2001), Springer Verlag.
9. MPI: A Message-Passing Interface Standard. *Message Passing Interface Forum* (Mar. 1994).
10. PETRINI, F., KERBYSON, D. J., AND PAKIN, S. The case of the missing supercomputer performance: Achieving optimal performance on the 8,192 processors of ASCI Q. In *Proc. of the 2003 ACM/IEEE conference on Supercomputing* (2003).
11. PASZTOR, A., AND VEITCH, D. Pc based precision timing without gps. In *Proceedings of the 2002 ACM SIGMETRICS international conference on Measurement and modeling of computer systems* (2002), ACM Press, pp. 1–10.
12. SISTARE, S., VANDEVAART, R., AND LOH, E. Optimization of mpi collectives on clusters of large-scale smp's. In *Proceedings of the 1999 ACM/IEEE conference on Supercomputing* (1999), ACM Press.
13. TIPPARAJU, V., NIEPLOCHA, J., AND PANDA, D. Fast collective operations using shared and remote memory access protocols on clusters. In *17th Intl. Parallel and Distributed Processing Symp.* (May 2003).
14. VADHIYAR, S. S., FAGG, G. E., AND DONGARRA, J. Automatically tuned collective communications. In *Proceedings of the 2000 ACM/IEEE conference on Supercomputing* (2000).
15. VETTER, J., AND MUELLER, F. Communication characteristics of large-scale scientific applications for contemporary cluster architectures. In *16th Intl. Parallel and Distributed Processing Symp.* (May 2002).
16. VETTER, J. S., AND YOO, A. An empirical performance evaluation of scalable scientific applications. In *Proceedings of the 2002 ACM/IEEE conference on Supercomputing* (2002), IEEE Computer Society Press.
17. VINTER, B. *PastSet a Structured Distributed Shared Memory System.* PhD thesis, Tromsø University, 1999.

Understanding the Behavior and Performance of Non-blocking Communications in MPI*

Taher Saif and Manish Parashar

The Applied Software Systems Laboratory
Department of Electrical and Computer Engineering
Rutgers University, Piscataway, NJ 08854, USA
{taher,parashar}@caip.rutgers.edu

Abstract. The behavior and performance of MPI non-blocking message passing operations are sensitive to implementation specifics as they are heavily dependant on available system level buffers. In this paper we investigate the behavior of non-blocking communication primitives provided by popular MPI implementations and propose strategies for these primitives than can reduce processor synchronization overheads. We also demonstrate the improvements in the performance of a parallel Structured Adaptive Mesh Refinement (SAMR) application using these strategies.

1 Introduction

The Message Passing Interface (MPI) [1] has evolved as the de-facto message passing standard for supporting portable parallel applications - with commercial as well as public-domain implementations available for most existing platforms including general purpose clusters to high-performance systems such as IBM SP.

An important design goal of the MPI standard is to allow implementations on machines with varying characteristics. For example, rather than specifying how operations take place, the MPI standard only specifies what operations do logically. Consequently, MPI can be easily implemented on systems that buffer messages at the sender, receiver, or do no buffering at all. It is typically left to the vendors to implement MPI operations in the most efficient way as long as their behavior conforms to the standards. As a result of this, MPI implementations on different machines often have varying performance characteristics that are highly dependant on factors such as implementation design, available hardware/operating system support and the sizes of the system buffers used.

The behavior and performance of MPI non-blocking message passing operations are particularly sensitive to implementation specifics as they are heavily dependant on available system level buffers and other resources. As a result,

---

* The work presented here was supported in part by the National Science Foundation via grants numbers ACI 9984357 (CAREERS), EIA 0103674 (NGS) and EIA 0120934 (ITR), and by DOE ASCI/ASAP (Caltech) via grant number PC295251.

naive use of these operations without an understanding of the underlying implementation can result in serious performance degradations, often producing synchronous behaviors.

We believe that an efficient and scalable use of MPI non-blocking communication primitives requires an understanding of their implementation and its implication on application performance. This paper has two key objectives: (1) To investigate and understand the behavior of non-blocking communication primitives provided by two popular MPI implementations: the public domain MPICH [2] implementation on a Linux cluster, and the proprietary IBM implementation on an IBM SP2 [3]. (2) To propose and evaluate usage strategies for these primitives that the parallel programmer can implement to reduce processor synchronization and optimize application performance. We use the proposed strategies to optimize the performance of parallel implementations of scientific/engineering simulations that use finite difference methods on structured adaptive meshes [4].

2 Non-blocking MPI: Behavior and Performance

The generic operation of a non-blocking MPI communication is a "three step" process in which the implementation (of the non-blocking communication) decouples the send and receive operations by using system and/or application buffers at the sender and receiver processes, allowing computation and communication to be overlapped. However, this decoupling is strictly limited by the size of the buffers available to copy the message. MPI implementations typically switch to a synchronous communication mode when the message size exceeds the available buffer size, where the sender waits for an acknowledgement from the receive side before sending out the data.

In this section we experimentally investigate the behavior and performance of non-blocking MPI communications in two popular MPI implementations: MPICH on a Beowulf cluster, and IBM MPI on the IBM SP2.

The test kernel used for these experiments - as illustrated in Figure 1 - is a typical non-blocking communication implementation between two processes in which one sends and the other receives. In this kernel the sending process (process 0) issues MPI_Isend (IS) at time-step T0 to initiate a non-blocking send operation while the receiving process (process 1) posts a matching MPI_Irecv (IR) call. Both processes then execute unrelated computation before executing an MPI_Wait call at T3 to wait for completion of the communication. In the following discussion we denote MPI_Wait posted on the send side as Ws and the MPI_Wait posted on the receive side as Wr. The processes synchronize at the beginning of the kernel and use deterministic offsets to vary values of T0, T1, T2 and T3 at each process. For each configuration (value of T0, T1, T2 and T3 at each process) we conducted a number of experiments varying the message size, system buffer size and number of messages exchanged. The objectives of these experiments included determining thresholds at which the non-blocking calls synchronize, the semantics of synchronization once this threshold is reached, and possibility of deadlocks.

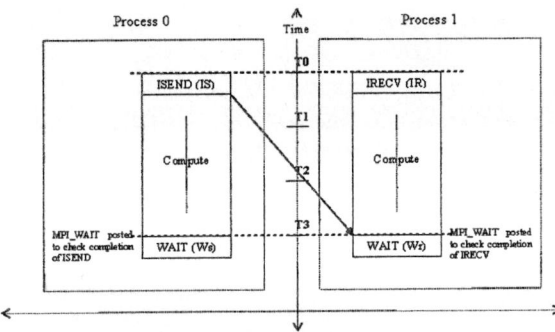

Fig. 1. Operation of the test kernel used in the experimental investigation.

2.1 MPICH on a Linux Beowulf Cluster

The first MPI implementation analyzed is MPICH version 1.2.5, release date January 6, 2003 [3] on Frea, a 64 node Linux Beowulf SMP cluster at Rutgers University. Each node of cluster has a 1.7 GHz Pentium 4 processor with 512 MB main memory. The MPICH profiling tool Upshot [14] is used for the profiles and timing graphs presented below.

Our first experiment investigates the effect of message size on non-blocking communication semantics. In this experiment the value of T0 - T3 are approximately the same on the two processes, and the message size was varied. The system buffer size was maintained at the default value of 16K. For smaller message sizes (1KB), we observe that *IS* and *IR* return without blocking (Figure 2). Furthermore, *Ws* and *Wr*, posted after local computations, return almost immediately, indicating complete overlap. However, for message sizes greater than or equal to 60 KB, *IS* blocks and returns only when the receiver process posts *Wr* (Figure 3). We can further see from the Figure that *Wr* blocks until the message delivery completes. This threshold is dependent on the system buffer size as discussed below.

To further understand the synchronizing behavior of MPI_Isend for large message sizes, we modified our experiment to post a matching MPI_Test (a non-blocking variant of MPI_Wait) on the receiver side (i.e. process 1) in the middle of the computation phase. As shown in Figure 4, in this case MPI_Isend returns as soon as MPI_Test is posted. It was also seen that the *Wr* posted after com-

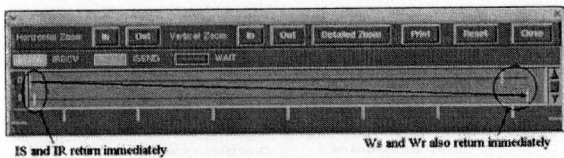

Fig. 2. Profile of the test on MPICH where process 0 (top) sends a 1 KB message to process 1 (bottom).

Fig. 3. Profile of the test on MPICH where process 0 (top) sends a 60 KB message to process 1 (bottom).

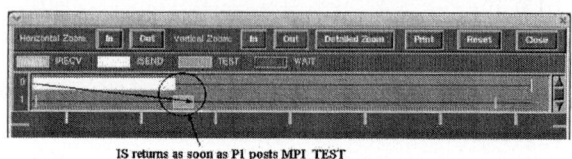

Fig. 4. Profile of the test on MPICH in which process 1 (bottom) posts an intermediate MPI_Test.

putation returns almost immediately, indicating that the message was already delivered during the computation. This indicates that MPI_Isend blocks for large messages until the completion of the corresponding MPI_Irecv is checked using either blocking (MPI_Wait) or non-blocking (MPI_Test). Note that, as the MPI implementation optimizes the number of messages sent to a single destination, the message size threshold is cumulative. That is, in the above case MPI_Isend switches to blocking semantics when the cumulative size of outstanding messages to a particular process is 60KB. For example, when we repeated the test using 3 sends of size 20KB each (instead of one of size 60KB), the same non-blocking behavior was observed.

The experiment plotted in Figure 5 evaluates potential deadlocks if two processes simultaneously send large messages to each other using MPI_Isend and block. The Figure shows that process 1 initially blocks but then returns after a certain time instead of waiting for process 0 to post a *Wr*. Process 0 however blocks until *Wr* is posted on process 1. This behavior seems to indicate a non-deterministic time out mechanism to ensure progress.

In the case of the Frea Beowulf cluster, default TCP socket buffer size is 16KB. This can be increased by either using the environment variable P4_SOCKBUFSIZE or using the command line option -p4sctrl bufsize=<size>.

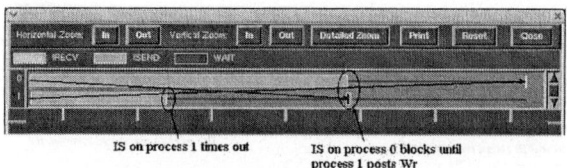

Fig. 5. Profile of the test on MPICH in which both processes post MPI_Isend. Message size is 60Kb.

We repeated the experiments using messages of size 60KB but increased the TCP socket buffer size. We observed that for a TCP socket buffer size of 64KB, IS did not block.

Analysis and Discussion. To try to understand the behavior of the MPICH non-blocking communication presented above let us consider its implementation. MPICH has a layered software architecture [5] consisting of (a) a high level MPI API layer (b) a middle Abstract Device Interface (ADI) layer and, (c) a Dlevice layer. The Device layer defines three protocols to send messages based on the message size: short, eager (long) and rendezvous (very long). In the case of relatively short messages, for example 1KB, MPICH can copy the message directly into the system socket buffer and thus send the message out onto the network using the eager protocol, allowing MPI_Isend to return immediately.

In the case of larger messages (e.g. 60KB), the system socket buffer is not large enough to accommodate the message and MPICH cannot directly copy the message into the buffer. Instead, it switches to the rendezvous protocol, which requires the sending process to synchronize with the receiving process before the message is sent out. As a result MPI_Isend, which should return immediately irrespective of the completion mode, now has to wait for the corresponding Wr for an acknowledgement(Figure 3). Similarly, when a matching MPI_Test is posted at the receiver process, it essentially sends an acknowledgement back to the sender which caused the blocked MPI_Isend to return. When the TCP/IP socket buffer size is increased to 64 KB, MPICH can copy the 60 KB message directly into the socket buffer and use the eager protocol allowing the MPI_Isend call to return without blocking.

Finally, due to MPICH optimizations, the blocking behavior of MPI_Isend depends on the system socket buffer and the cumulative size of the outstanding messages rather than the actual number of messages sent. Consequently, reducing the message size by breaking up the total message into smaller messages will not yield any performance improvement.

Optimization Strategies. Based on the analysis presented above we identify two strategies to address the blocking behavior of MPI_Isend in MPICH. The first strategy is obvious, increase the TCP socket buffer size. However this option is not scalable since the total buffer space grows with the number of processes. Further, every system imposes a hard limit on the total socket buffer size. As a result this option has only limited benefits and any further optimization must be achieved at the applications level.

It is clear from the analysis presented above that the only way to prevent MPI_Isend from blocking is for the receiving process to return an acknowledgement using a (blocking or non-blocking) test for completion call. Our second strategy is to use calls to the non-blocking test for completion (MPI_Test or its variant) on the receive side to release a blocked sender.

To illustrate this consider the code snippet (Figure 6) for a typical loose-synchronous application, for example, a finite-difference PDE solver using ghost

```
for m=1 to number_of_messages_to_receive {       for m=1 to number_of_messages_to_receive{
        MPI_IRECV(m, recv_msgid_m)                       MPI_IRECV(m, recv_msgid_m)
}                                                }
***COMPUTE***                                    ***COMPUTE***
for n=1 to number_of_messages_to_send{           for n=1 to number_of_messages_to_send{
        MPI_ISEND(n, send_msgid_n)                       MPI_ISEND(n, send_msgid_n)
        MPI_WAIT(send_msgid_n)                           MPI_WAIT(send_msgid_n)
}                                                }
***COMPUTE***                                    MPI_TESTALL(recv_msgid_*)
MPI_WAITALL(recv_msgid_*)                        ***COMPUTE***
                                                 MPI_WAITALL(recv_msgid_*)
```

Fig. 6. MPICH: Unoptimized algorithm.   **Fig. 7.** MPICH: Optimized algorithm.

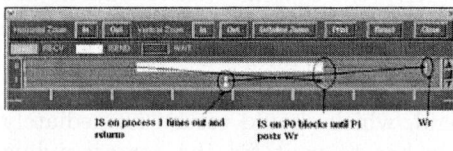
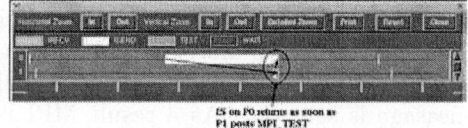

Fig. 8. MPICH: Unoptimized algorithm.   **Fig. 9.** MPICH: Optimized algorithm.

communications. In this pseudo-code, each process posts non-blocking receive calls before computing on its local region of the grid. After finishing computation, it then sends its data to update the ghost regions of its neighboring processors using the MPI_Isend/MPI_Wait pair. The process may do some further local computation and then finally waits to update it own ghost regions, possibly using an MPI_Waitall. In this case, if the message size is greater than 60KB the MPI_Isend will block until the corresponding MPI_Waitall is called on the receiving process as shown in Figure 8.

If we now insert an intermediate MPI_Testall call as shown in Figure 7, MPI_Isend returns as soon as the receiver posts the test (Figure 9). While the MPI_Testall call does have a cost, this cost is small compared to the performance gain.

2.2 IBM MPI on the SP2

The second MPI implementation analyzed is the IBM native implementation (version 3 release 2) [3] on the IBM SP2, BlueHorizon, a teraflop-scale Power3 based clustered SMP system at the San Diego Supercomputing Center. The machine consists of 1152 processors, each having 512 GB of main memory. Once again, our first experiment investigates the effect of message size on non-blocking communication semantics. For smaller message sizes (1KB), we observe the expected non-blocking semantics. This is also true for larger messages sizes (greater than 100 KB) as shown in Figure 10.

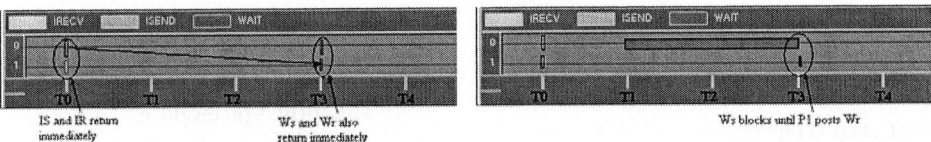

Fig. 10. SP2: Ws and Wr posted at the same time-step.

Fig. 11. SP2: Ws and Wr posted at different time-steps.

To further understand the effect of increasing message size on the behavior of non-blocking communications in the IBM MPI, we moved Ws to T1, i.e. directly after the send to simulate the situation where one might want to reuse the send buffer. Wr remained at T3. In this case, for message sizes greater than or equal to 100KB, Ws blocked until Wr was posted by the receiver at T3 (Figure 11).

In an experiment where both processes exchange messages, IS and IR are posted at T0, process 0 posts Ws at T1 while process 1 posts Ws at T2, and both processes post Wr at T3. The message size is maintained at 100KB. In this case deadlock is avoided in an interesting manner since Ws, posted at T1 and blocks on process 0, returns as soon as process 1 posts Ws at T2, rather than waiting for the corresponding Wr on T3.

Analysis and Discussion. The SP2 parallel environment imposes a limit (called the eager limit) on the total message size that can be sent out asynchronously. When message sizes exceed this limit, the IBM MPI implementation switches to a synchronous mode. However, in this case, it is the Ws call that blocks until an acknowledgement is received from the receiver process. Consequently in the experiment above, Ws blocks until Wr is posted at the receiving process. The analysis above also shows that the synchronization call on the receive side need not be a matching wait. In fact the receiver may post any call to MPI_Wait (or any of its variants) to complete the required synchronization.

Optimizations Strategies. The POE users' guide [3] specifies the environment variable, MP_EAGER_LIMIT, which defines the size of MPI messages that can be sent asynchronously. However, as the number of processes increase, trying to increase MP_EAGER_LIMIT simply reduces the amount of memory available to the application.

A more scalable strategy is to address this at the application level by appropriately positioning IS, IR, Ws and Wr calls. The basic strategy consists of delaying Ws until after Wr and is illustrated in Figures 12 and 13.

To illustrate the strategy, consider a scenario in which two processes exchange a sequence of messages and the execution sequence is split into steps T0-T3. Both processes post MPI_Irecv (IR) calls at T0 and Wall denotes a MPI_Waitall call. Assume that, due to load imbalance, process 0 performs computation until T2 while process 1 computes only till t1. Ws posted on process 1 at T1 will block until process 0 posts Ws at T2. For a large number of messages, this delay can

```
for n=1 to number_of_messages_to_receive{
        MPI_IRECV(n, msgid_n)
}
***COMPUTE***
for n=1 to number_of_messages_to_send{
        MPI_ISEND(n, send_msgid_n)
        MPI_WAIT(send_msgid_n)
}
MPI_WAITALL(recv_msgid_*)
```

Fig. 12. SP2: Unoptimized algorithm.

```
for n=1 to number_of_messages_to_receive{
        MPI_IRECV(n, msgid_n)
}
***COMPUTE***
for n=1 to number_of_messages_to_send{
        MPI_ISEND(n, send_msgid_n)
}
MPI_WAITALL(recv_msgid_*+send_msgid*)
```

Fig. 13. SP2: Optimized algorithm.

become quite significant. Consequently, to minimize the blocking overhead due to *Ws* on process 1, it must be moved as close to T2 as possible. Now, if *Ws* is removed from the send loop and a collective MPI_Waitall is posted as shown in Figure 13, it is observed that process reaches T2, it has already posted *IS* for all of its messages and is waiting on Wall, thus reducing synchronization delays.

3 Evaluation of Communication Performance in SAMR

Dynamic Structured Adaptive Mesh Refinement (SAMR) techniques [4] for solving partial differential equations provide a means for concentrating computational effort to appropriate regions in the computational domain. These methods (based on finite differences) start with a base coarse grid with minimum acceptable resolution that covers the entire computational domain. As the solution progresses, regions in the domain requiring additional resolution are recursively tagged and finer grids are laid over these tagged regions of the coarse grid [4].

Parallel implementations of hierarchical SAMR applications typically partition the adaptive heterogeneous grid hierarchy across available processors, and each processor operates on its local portions of this domain in parallel [7]. Due to their irregular load distributions and communication requirements across levels of the grid hierarchy, parallel SAMR applications make extensive use of non-blocking MPI primitives so as to overlap intra-level communications with computations on the interior region.

A typical implementation of intra-level communications in parallel SAMR applications is similar to the ghost communication associated with parallel finite difference PDE solvers as described in Section 2. Clearly, the optimizations proposed by us in Section 2 can be applied here to reduce the synchronization costs.

Evaluation Using the RM3D Kernel. To evaluate the impact of the proposed optimization strategies on application performance we used the 3-D version of the compressible turbulence application kernel (RM3D) which uses SAMR

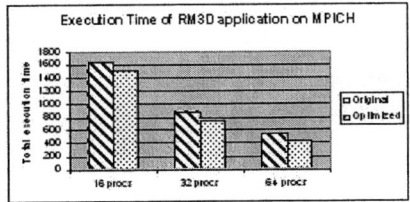
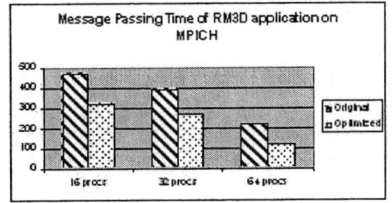

Fig. 14. Comparison of execution and communication times on Frea (MPICH).

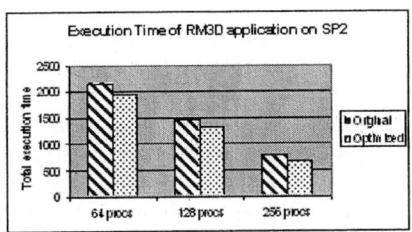
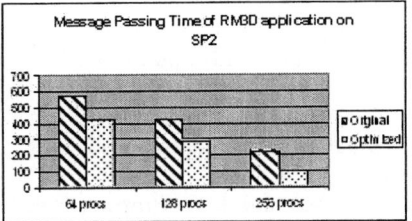

Fig. 15. Comparison of execution and communication times on SP2 (IBM POE).

techniques to solve the Richtmyer-Meshkov instability [9]. The experiments consist of measuring the message passing and application execution times for the RM3D application kernel before and after incorporating our optimizations strategies outlined in this paper, on both Frea and Blue Horizon. Except for the optimizations in the message passing algorithm, all other application-specific and refinement-specific parameters are kept constant. The results of the evaluation for MPICH on Frea for 16, 32 and 64 processors are shown in Figure 14. These runs used a base grid size of 128*32*32 and executed 100 iterations. We observe that the reduction in communication time is approximately 27%.

On the SP2 the evaluation run used a base grid size of 256*64*64 and executed 100 iterations. Figure 15 shows the comparisons of the execution times and communication times respectively for 64, 128 and 256 processors. In this case we observe that the reduction in communication time is approximately 44%.

4 Summary and Conclusions

In this paper we experimentally analyzed the behavior and performance of non-blocking communication provided by two popular MPI implementations. It is important to note that the blocking behavior described by us is not a bug in the message passing softwares. Rather it is due to inherent limitations in the underlying hardware architectures.

We used the strategies proposed in this paper to optimize the performance of the SAMR-based Richtmyer-Meshkov compressible turbulence kernel. Our evaluation shows that the proposed strategies improved the performance by an average of approximately 27% for MPICH and 44% for IBM MPI.

References

1. The MPI Forum. The MPI Message-Passing Interface Standard 2.0. http://www.mpi-forum.org/docs/mpi-20-html/mpi2-report.html, September 2001.
2. MPICH - A Portable MPI Implementation. http://www-unix.mcs.anl.gov/mpi/mpich/
3. Parallel Environment (PE) for AIX V3R2.0: Operation and Use, Vol. 1. December, 2001.
4. M. Berger and J. Oliger, "Adaptive Mesh Refinement for Hyperbolic partial Differential Equations", *Journal of Computational Physics*, Vol. 53, pp. 484-512, 1984.
5. W. Gropp, E. Lusk, A. Skjellum and N. Doss, "MPICH: A High-Performance, Portable Implementation for MPI Message-Passing Interface", *Parallel Computing*, 22, 1996, pp. 789-828.
6. W. Gropp and E. Lusk, "MPICH Working Note: The Second-Generation ADI for the MPICH Implementation of MPI", http://www-unix.mcs.anl.gov/mpi/mpich/, 1996.
7. M. Parashar and J. C. Browne, "On Partitioning Dynamic Adaptive Grid Hierarchies", Proceedings of the *29th Annual Hawaii International Conference on System Sciences*, Maui, Hawaii, IEEE Computer Society Press, pp. 604-613, January 1996.
8. V. Herrarte and E. Lusk, "Studying Parallel Program Behavior with upshot", Technical Report ANL-91/15, Argonne National Laboratory, 1991.
9. J. Cummings, M. Aivazis, R. Samtaney, R. Radovitzky, S. Mauch, and D. Meiron, "A virtual test facility for the simulation of dynamic response in materials" *Journal of Supercomputing*, 23:39-50, 2002.

Generation of Simple Analytical Models for Message Passing Applications*

German Rodriguez, Rosa M. Badia, and Jesús Labarta

CEPBA-IBM Research Institute,
Technical University of Catalonia (UPC),
Campus Nord, Mòdul D6, Jordi Girona, 1-3, 08034 Barcelona, Spain
grodrigu@ac.upc.es, {rosab,jesus}@cepba.upc.es

Abstract. We present a methodology which allows to derive accurate and simple models which are able to describe the performance of parallel applications without looking at the source code. A trace is obtained and linear models are derived by fitting the outcome of a set of simulations varying the influential parameters, such as: processor speed, network latency or bandwidth.

The simplicity of the linear models allows for natural derivation of interpretations for the corresponding factors of the model, allowing for both prediction accuracy and interpretability to be maintained.

We explain how we plan to extend this approach to extrapolate from these models to be apply it to predict for processor counts different to the one of the given traces.

1 Motivation and Goal

Obtaining performance models of parallel applications is extremely useful for a broad range of purposes: determining if a platform achieves its expected performance, identification of the source of performance problems[16], scheduling, and many others.

Models based on simulations are used to explore the parameter space design, but are time consuming and lack the abstraction and possibility of interpretation that analytic models provide [8–10].

In this work we want present, extend and validate a methodology that we are developing [18] to extend the analysis capabilities of the tools DIMEMAS[1] and PARAVER[2] developed at CEPBA.

We want to derive simple analytical models including the number of processors as a parameter starting from post-mortem trace-files of the application under study. The simplest model presents non-linearities and we show how we treat them being methodological while gaining insight of how the application is affected by the architectural parameters.

---

* This work has been partially funded by the Ministry of Science and Technology of Spain under CICYT TIC2001-0995-CO2-01 and by a Ministry of Education of Spain *FPU* grant.

2 Methodology

We apply the following methodology: for each application that we want to analyze, we obtain traces from its real execution varying the number of processors. The dynamic instrumentation technology of MPIDtrace[11] (based on DPCL[12]) allows us to obtain traces of production binaries without needing access to its source code. This traces contain information of communication requests of all processes as well as the CPU demands between those requests. The usefulness of MPIDtrace has been shown in [13–15]. Other previously proposed methods needed the source code and knowledge of the program structure to be able to instrument the application [17].

This traces feed the DIMEMAS[1] simulator. DIMEMAS implements a simple abstract model of a parallel platform in which the MPI software layers are taken into account. It allows to simulate for varying Latencies, Bandwidths, and CPU ratios (that models the CPU performance ratio between the source and target machine). Given a trace for a particular number of processors, DIMEMAS is not able to perform a simulation of the trace under a different processor count.

STORM[3] performs DIMEMAS simulations of a particular trace file, randomly varying the influential architectural parameters.

From these results, we fit a linear model from these data of the elapsed time of the application against the architectural parameters. The coefficients characterize the application.

2.1 Brief Review of Previous Results and Methodology

We have already addressed the problem of finding an accurate model, simple and at the same time general enough, to explain and predict the behavior of the same set of programs [18], using a formula like: $T_P(L, BW) = \alpha_P + \beta_P \cdot L + \frac{\gamma_P}{BW}$.[1]

We obtained the α_P, β_P, γ_P through linear fitting of DIMEMAS simulations varying L and BW. P was not a parameter.

A further step was the unification of these formulas to a single one, including the processor count P as one additional parameter in the form of:

$$T(P, L, BW) = \alpha(P) + \beta(P) \cdot L + \frac{\gamma(P)}{BW} \quad (1)$$

$\alpha(P)$ represents the time of the critical path, $\beta(P)$ the number of non-overlapped latencies and $\gamma(P)$ the number of bytes exchanged between processors not overlapped with computation. By knowing these functions of P we would be able to predict for a different processor count of that of the original traces.

It would be simple to model $\alpha(P)$ as: $\alpha(P) = \alpha_0 + \frac{\alpha_1}{P}$ (following Amdahl's law). However, in a parallel program, this is not necessarily true. We observe a non-linearity on P for the "critical path" ($\alpha(P)$) when the distribution of computation improves locality and therefore results in a better memory hierarchy performance.

---

[1] L stands for Latency, BW, Bandwidth, and P for the number of processors.

2.2 Decoupling Non-linearities

Such a non-linearity is found in the $\alpha(P)$ parameter for the NAS BT benchmark. Locality plays and important role, and executions for increasing processor counts account for smaller total amounts of computation[2], as would be expected following Amdahl's law.

We use the term *Locality Factor* to quantitatively express the difference between the expected amount of work using Amdahl's law, and the real one. This *Locality Factor* is computed as the relation between the total CPU time for all processors of the trace for the minimum processor count (*reference P*) and the *actual* amount of time accounted for a particular processor count P:
$LocalityFactor(P) = \frac{TotalCPUtime(P)}{TotalCPUtime(P_{\text{reference}})}$.

The *Locality factors* of the NAS BT executions do not vary linearly with the number of processors. Amdahl's law would predict a constant behavior (Locality factor being always 1). However the BT total sequential execution times decrease as much as 55% from 4 to 49 processors (non-linearly).

In order to apply our methodology, this non-linear effect should be decoupled, and therefore we propose the following formula:

$$T(P, L, BW) = (\alpha'(P)) \cdot LocalityFactor(P) + \beta'(P) \cdot L + \frac{\gamma'(P)}{BW}, \quad (2)$$

Where α', β' and γ' are obtained having previously corrected the effect of increased locality. We simply scale all the CPU bursts of the simulations by setting the DIMEMAS CPU efficiency scaling parameter, CPU_ratio, to the value of the *Locality factor*. The same methodology explained in the two first paragraphs of Subsect. 2.1 is applied to the trace of the execution of the application for *each* processor count P with the corresponding CPU efficiency scaling factor applied in the simulation to obtain the α'_P.

Given these α'_P we then apply a linear regression (this time with no non-linearities involved) to obtain an estimate of $\alpha'(P)$ of the form:

$$\alpha'(P) = \alpha'_0 + \frac{\alpha'_1}{P} \quad (3)$$

We checked (Sect. 3) that β'_P and γ'_P are statistically identical to β_P and γ_P and independent of the *Locality factor* for a sensible range of correcting factors.

We expect to find linear behaviors in $\beta(P)$ and $\gamma(P)$, as they roughly represent the non–overlapped latencies and communication bandwidth (see [18]).

The model obtained predicts with very high accuracy the DIMEMAS simulations obtained with the Locality Factor correction. However, this model is not predicting the real application, but an ideal one where the memory effects are not taken into account. To model the real application, the corresponding $\alpha'(P)$ parameter should be again multiplied by the *Locality factor* corresponding to the processor count to obtain what would be the real $\alpha(P)$ (critical path's time). We checked this approach in Sect. 3 and we show that for the applications considered, this process does not change the communication pattern.

---

[2] In DIMEMAS terms, *Total CPU time*: sum of all CPU bursts of all processors.

3 Example of Methodology Use

To validate this methodology we have used IBM SP2 traces, from 4 to 49 processors, of the following applications: NAS BT[4], Sweep3D[6], RNAfold[5] and POP[7]. We show how we applied the process and show the relative errors obtained.

The last column of Table 1 shows the α_P factors obtained through regression of the set of simulations with no Locality Factor correction; the second column shows the $\alpha'(P)$ parameters[3] obtained using (4), in which its coefficients where obtained by fitting the data obtained of a set of simulations using the corresponding *Locality Factor* (shown in the third column), and therefore decoupling the non-linearity. Finally, the correctness of the model is validated by reconstruction of the α_P parameter as the product $\alpha'(P) \cdot LocalityFactor(P)$, shown in the fourth column. In this table, the $\alpha'(P)$ factors are calculated as:

$$\alpha'(P) = \alpha'_0 + \frac{\alpha'_1}{P} \quad (4)$$

where α'_0 and α'_1 were calculated by doing a linear regression on the α'_P obtained having previously applied the *Locality Factor* correction. The α'_P followed a linear behavior on the inverse of the number of processors, whereas the original α_P, with non-corrected Locality Factors, did not.

Table 1. Going back to the α_P factors from the linear model built using the α'_P parameters.

Procs.	NAS BT			
	$\alpha'(P)$	$Loc.Factor(P)$	$\alpha'(P) \cdot Loc.Factor(P)$	α_P
4	14.748	1.000	14.748	14.718
9	6.589	0.986	6.496	6.585
16	3.734	0.900	3.361	3.351
25	2.412	0.672	1.620	1.598
36	1.694	0.575	0.974	0.964
49	1.261	0.553	0.697	0.699

In Table 1 we show that we could eliminate the non-linear behavior of the α_P parameter through a simple transformation, and then go back to the original one using a function $LocalityFactor(P)$ for which we understand its meaning and could derive a model. But we also have to show that this transformation does not affect the other factors: β_P and γ_P.

Table 2 shows that the β_P parameters are independent from the scaling of the CPU bursts according to the $LocalityFactor(P)$.

Given these facts, it was possible to use the $\alpha'(P)$, $\beta'(P)$ and $\gamma'(P)$ to go back to (2), recovering what the $\alpha(P)$, $\beta(P)$ and $\gamma(P)$ parameters would be and validate the model and the explained transformations. Table 3 shows the values obtained for the linear models used to fit the three factors.

---
[3] i.e., the $\hat{\alpha}'(P)$ parameters.

Table 2. β_P and γ_P factors against β'_P and γ'_P factors.

	NAS BT			
# P	β_P	β'_P	γ_P	γ'_P
9	320.37	321.96	0.46	0.44
25	527.92	521.80	0.50	0.55
36	675.39	677.55	0.27	0.23

Table 3. Parameters: N.S. means "Non Significant" (regression analysis).

Benchmark	α'_0	α'_1	β'_0	β'_1	γ'_0	γ'_1
NAS	0.062	58.743	200.814	12.240	0.453	N.S.
POP	1.366	337.444	14478.395	N.S.	2.739	N.S.
SWEEP	2.258	105.007	1850.302	21.488	N.S.	0.569
RNAfold	0.411	91.778	3528.737	N.S.	0.202	N.S.

Table 4. Maximum Relative Errors of validation against equation 2.

Processors	NAS	POP	SWEEP	RNAfold
4	0.24%	0.35%	—	—
8	—	—	0.47%	0.05%
9	1.39%	0.93%	—	—
12	—	—	—	0.43%
16	0.27%	2.48%	1.15%	—
25	1.37%	2.83%	—	—
28	—	—	—	0.88%
32	—	—	3.50%	—
36	1.13%	9.09%	—	—
49	0.28%	—	5.16%	—

We validated our results with more than 200 randomly selected bandwidths and latencies for each of the traces for varying P. In Table 4 we show the Maximum Relative Errors using (2) (where the parameters were obtained simulating with the corrected *Locality Factors*). These relative errors compare the prediction of model (2) and the actual times calculated by DIMEMAS.

4 Conclusions and Future Work

We have analyzed the possibility of easily deriving simple models that accurately characterize the behavior of a set of representative parallel benchmarks, from which one of them is a real application: POP[7].

The methodology under development allows to understand the underlying factors that influence the performance of a program. We analyzed the meaning of the parameters in a previous work [18] and are extending that analysis.

We have overcome the difficulties imposed by non-linearities and we are now testing this methodology to extrapolate the results to any number of processors.

We plan to extrapolate the $\alpha(P)$ model using similar techniques to the ones in [15] (PMACS), as well as determining the regions of linearity in a more or less automatic way.

References

1. DIMEMAS: http://www.cepba.upc.es/dimemas/
2. PARAVER: http://www.cepba.upc.es/paraver/
3. STORM: a tool for stochastic analysis, http://www.easi.de/storm
4. Bailey, D., Harris, T., Saphir, W., van der Wijngaart, R., Woo, A., Yarrow, M.: "The NAS Parallel Benchmarks 2.0". The International Journal of Supercomputer Applications, 1995.
5. Hofacker, I.L., Fontana, W., Bonhoeffer, L. S., Tacker, M., Schuster, P.: "Vienna RNA Package", http://www.tbi.univie.ac.at/ ivo/RNA, October 2002.
6. "The ASCI sweep3d Benchmark Code", http://www.llnl.gov/asci_benchmarks/asci/limited/sweep3d/asci_sweep3d.html
7. "Parallel Ocean Program": http://climate.lanl.gov/Models/POP/
8. Culler, D., Karp, R., Patterson, D., Sahay, A., Schauser, K. E., Santos, E., von Eicken, T.: "LogP: Towards a Realistic Model of Parallel Computation". Proc. of the 4th ACM SIGPLAN Symposium on Principles and Practice of Parallel Programming, May 1993.
9. Mathis, M. M., Kerbyson, D. J., Hoisie, A.: "A Performance Model of non-Deterministic Particle Transport on Large-Scale Systems". Proc. Int. Conf. on Computational Science (ICCS), Melbourne, Australia, Jun 2003.
10. Jacquet, A., Janot, V., Leung, C., Gao, G. R., Govindarajan, R., Sterling, T. L.: "An Executable Analytical Performance Evaluation Approach for Early Performance Prediction". Proc. of IPDPS 2003.
11. MPIDtrace manual, http://www.cepba.upc.es/dimemas/manual_i.htm
12. DeRose, L.: "The dynamic probe class library: an infrastructure for developing instrumentation for performance tools". International Parallel and Distributed Processing Symposium, April 2001.
13. Girona, S., Labarta, J.: "Sensitivity of Performance Prediction of Message Passing Programs". International Conference on Parallel and Distributed Processing Techniques and Applications (PDPTA'99), Monte Carlo Resort, Las Vegas, Nevada, USA, July 1999.
14. Girona, S., Labarta, J., Badia, R. M.: "Validation of Dimemas communication model for MPI collective operations". EuroPVM/MPI'2000, Balatonfüred, Lake Balaton,Hungary, September 2000.
15. Snavely, A., Carrington, L., Wolter, N., Labarta, J., Badia, R. M., Purkayastha, A.: "A framework for performance modeling and prediction". SC 2002.
16. Crovella, M. E., LeBlanc, J. L.: "The Search for Lost Cycles: A New Approach to Parallel Program Performance Evaluation (1993)". Tech. Rep. 479, Computer Science Department, University of Rochester, Dec., 1993.
17. Mehra, P., Schulbach, C., Yan, J.C.: "A comparison of two model-based performance-prediction techniques for message-passing parallel programs". Sigmetrics'94, pgs. 181-190, May 1994.
18. Badia, R. M., Rodriguez G., Labarta, J.: "Deriving analytical models from a limited number of runs". ParCo 2003 Proceedings

Parallel Hybrid Particle Simulations Using MPI and OpenMP*

M. Hipp and W. Rosenstiel

Wilhelm-Schickard-Institut für Informatik
Department of Computer Engineering, Universität Tübingen
Sand 13, D-72076 Tübingen, Germany
{hippm,rosen}@informatik.uni-tuebingen.de

Abstract. We present a library for the parallel computation of particle simulations called ParaSPH. It is portable and performs well on a variety of parallel architectures with shared and distributed memory. We give details of the parallelization for hybrid architectures (clustered SMPs) using MPI and OpenMP and discuss implementation issues, performance results and memory consumption of the code on two parallel architectures, a Linux Cluster and a Hitachi SR8000-F1. We show the advantage of hybrid parallelization over pure message-passing especially for large node numbers for which we gain a maximum speedup of about 350 for hybrid parallelization compared to 120 for message-passing.

1 Introduction

The Collaborative Research Center (CRC) 382 works in the field of Computational Physics with a main focus on astrophysical simulations. Our group is responsible for the parallel computing. Some methods used in the CRC 382 are particle based and thus we develop efficient parallel particle libraries to support these applications. [6]

There is always a need for larger simulations needing more memory and computing power. A parallelization combining threads and message passing is a promising way to reduce the parallel overhead in respect of memory and performance on the increasing number of hybrid architectures over pure message-passing parallelization.

2 Hybrid Architectures

For development and performance analysis we are working on two parallel systems. A Hitachi SR8000 installed at the HLRB in Munich and a Linux Cluster installed locally at Tübingen University.

---

* This project is funded by the DFG within CRC 382: *Verfahren und Algorithmen zur Simulation physikalischer Prozesse auf Höchstleistungsrechnern* (Methods and algorithms to simulate physical processes on supercomputers).

The Hitachi SR8000 consists of 8-way SMP nodes coupled by a fast communication network. The installation in Munich has 168 nodes with a total FPU performance of 2 TFlops.

The Linux cluster is built of commodity hardware and has two partitions. One partition consists of 96 2-way SMP Intel Pentium 3 nodes with 650 MHz processor speed and the second partition consists of 32 2-way AMD Athlon nodes with 1667 MHz processor speed. All nodes from both partitions are connected with a switched full-bisection-bandwidth Myrinet network. The peak FPU performance of the Linux cluster is 338 GFlops.

3 Motivation

The majority of parallel machines in the TOP500 lists are so called hybrid parallel architectures, a combination of N-way shared memory nodes with message passing communication between the nodes.

Hybrid parallelization is the combination of a thread based programming model for parallelization on shared memory nodes together with message-passing based parallelization between the nodes. The standard library for message passing is MPI[8] OpenMP[9] has become the standard for thread based programming for scientific applications.

Obviously, the share of common data structures on shared memory nodes can reduce the amount of required memory. More important is the reduction of the communication. Every parallel implementation has a maximum speedup limited by its serial parts. In our non trivial particle codes the major serial part is the (often collective) communication. A hybrid implementation reduces the amount of transferred data because the communication of shared data on the SMP nodes is implicit.

But the communication itself is faster, too. A simple test shows this for the MPI-*Allgather* call. We distribute a 200 MByte data array on the Hitachi SR8000 between all nodes. K is the number of nodes.

In the first test (pure-MPI) a MPI process runs on every SR8000 processor. The $K \times 8$ processors send and receive $200/(K \times 8)$ MByte to/from each *processor*. In the second test (Hybrid) the same amount of data is just sent between the master threads of each node and thus each call sends and receives $200/K$ MByte to/from each *node* (other non-master threads are idle).

The numbers in the table are the time for 50 MPI-*Allgather* calls in seconds on the SR8000-F1 at the HLRB in Munich.

	K=1	K=2	K=4	K=8	K=16	K=32
Pure-MPI	16	51	97	126	172	200
Hybrid	10	15	20	21	22	21

One can see a big difference between the hybrid communication (comparable to the hybrid programming model with implicit intra-node communication) and the *pure-MPI* communication model with explicit MPI intra-node communication. Because the inter-node communication should be independent of the parallelization strategy for a good implemented *Allgather* call, for $K > 1$ one would expect for $K > 1$ the same 16 to 10 ratio of the $K = 1$ case (showing the time of the intra-node communication and the overhead for 8 times more messages). Instead, for the $K = 32$ run, the hybrid programming model is 8 times faster.

3.1 OpenMP

We chose OpenMP to keep the implementation portable. Compilers for OpenMP are available on all important hybrid platforms including Hitachi SR8000, NEC SX5/6 and Linux (Intel C++ or Portland Group compilers). OpenMP has advantages over explicit thread programming, for example POSIX threads. First, it annotates sequential code with compiler directives (pragmas) allowing an incremental and portable parallelization. Non-OpenMP compilers ignore the directives. Second, POSIX threads require to implement parallel sections in separate functions and functions with more than one argument need wrappers, since the POSIX threads API supports only one argument. OpenMP allows joining and forking threads at arbitrary positions in the source code.

3.2 HPF

A complete different approach is using a HPF compiler such as the Vienna Fortran Compiler, which is able to generate hybrid code. If a code is parallelized with HPF and performs well on a message-passing architecture this can be an interessting option. Since our code-base was written in C and there was a MPI parallelisation, this was no option for us.

4 The SPH Method

An important particle method used in the CRC 382 is Smoothed Particle Hydrodynamics (SPH). SPH is a grid-free numerical Lagrangian method for solving the system of hydrodynamic equations for compressible and viscous fluids. It was introduced in 1977 by Gingold [3] and Lucy[7]. It has become a widely used numerical method for astrophysical simulations. Later, the method was also applied to other problems such as the simulation of a fluid jet to model the primary break-up of a diesel jet as it is injected into the cylinder of an engine[2].

Rather than being solved on a grid, the equations are solved at the positions of the pseudo particles, representing a mass element with certain physical quantities while moving according to the equations of motion. Due to the mesh-less nature of the method, SPH is well suited for free-boundary problems and can handle large density gradients.

Each particle interacts with a limited number of particles in its neighborhood. Because the particle positions and therefore the neighbor interactions change after each time-step one cannot find a perfect load balancing and domain decomposition for the parallelization in advance. Instead, a reasonable fast and communication optimized domain decomposition, which must be applied after every time-step is crucial. The general parallelization strategies are explained in more detail in [5].

5 Hybrid Implementation

The hybrid implementation is an extension of the ParaSPH library for particle simulations. ParaSPH is written in C and parallelized with MPI. The library separates the parallelization from the physics and numeric code. The interface between the library and the application is optimized for particle simulations. The library provides an iterator concept to step through all particles and their neighbors and later communicates the results.

In parallel mode, the library transparently distributes the work amongst all processors. Every local iterator processes only a subset of all particles. The code performs well on machines with a fast message passing network. We tested the code on Cray T3E, Hitachi SR8000, IBM SP and a Linux cluster.

For the hybrid implementation, OpenMP is used for the inner intra-node parallelization. MPI is still used for inter-node communication. To achieve a better portability, because not all MPI implementations are thread safe, only the master thread calls the MPI library. The performance penalty is small for our applications, since the expensive communication calls are collective operations.

Experiments showed, that it is necessary to optimize the load balancing. Therefore, we introduced two different balancing strategies. The standard load balancer for distributed memory is used only for a coarse load balancing between the nodes. For the fine balancing on the node, the user can choose between two new balancers, a fixed load balancing and a dynamic master-worker load balancing. The master-worker algorithm promises the best load balancing for inhomogeneous problems, because of its inherent load-steeling. The disadvantage is the worse cache utilization, because the data is not bound to a specific CPU for successive runs over the particle list. For SPH simulations with a fixed number of neighbour interactions, the static load balancing is faster. One may consider the dynamic load balancer for computations with a variable number of interactions and large density gradients.

From OpenMP, we used 15 *parallel* pragmas, one *barrier* pragma and one *threadprivate* pragma. We also need two additional locks to protect internal structures.

The first version, ParaSPH frequently called sections with a *critical* region and showed a bad performance on the SR8000. Explicit locking instead of critical regions improved the performance only a little. Lock-free implementations of the static load balancer and iterator fixed the problem. We setup up an independent particle list with its own iterator for each thread and omit shared counters and pointers.

Another problem was a compiler flaw in the Hitachi OpenMP implementation, if the program uses a local constant value like

```
{ /* begin local code section */
    double const aValue = 2.0/3.0;
    ... some code using aValue ...
}
```

The compiler generated a shared variable together with an initialization for every run through the local code section instead of using an immediate value or a register. The cache trashing leads to a two times slower code.

On the Linux platform, we used Intel's C++ compiler. We had to replace the *threadprivate* directives by explicit memory allocation for every thread, because the *threadprivate* directive triggers a compiler bug.

No additional code change was necessary to instrument 95% of the parallel code with OpenMP. But 5% remaining serial code limits the speedup to about 6 on one Hitachi node with its 8 CPUs per node. So, we redesigned parts of the code to increase the parallelization ratio. The strategy was to first identify the hot spots in the remaining 5% and find a lock-free implementation. Most parts need minor changes to omit necessary locks or heavy usage of shared data structures. For the remaining parts with no lock-free alternative, we tried to optimize the serial code itself to reduce the run time. With these changes we gain a parallelization ratio of about 98% compared to the *pure-MPI* version.

6 Results

Our standard application – a typical astrophysical problem of an accretion disk – is a 2-dimensional SPH simulation with 300 000 particles and 80 interaction partners per particle. This medium sized simulation requires about 900 MBytes of memory on one node. One integration step needs about 74 seconds on one Hitachi node (8 processors, *pure-MPI* mode). Large production runs use more particles and are often calculated in three dimensions. They usually need 1 000 or more integration steps resulting in several days of computation time on eight CPUs and weeks on one CPU.

6.1 Hitachi SR8000 Single Node Performance

First we compared the performance of the hybrid and *pure-MPI* code on one SR8000-F1 node. We used the hybrid version for this test, although there is no message-passing. The application was about 25% slower (95 sec. compared to 74 sec.).

We found three main reasons for the performance impact of the hybrid version.

Serial Parts. The code is not fully annotated with OpenMP instructions. There is a small serial part, which is not present in the pure MPI version. This causes a performance decrease especially on machines with a great number of processors per node like the SR8000. In hybrid mode, the serial part is a fixed overhead over *pure-MPI* parallelization. For the SR8000, the remaining 2% serial code in ParaSPH result in a performance decrease of about 14%.

OpenMP Overhead. Similar to the serial overhead is the OpenMP overhead itself (thread creation, locking of critical sections) together with some additional parallelization work (for example the fine load balancing described above). Flaws in the OpenMP compiler may additionally decrease performance and are difficult to find. Usually, the generated assembler code has to be verified after identifying the problem with a profiler.

Cache. The parts computing the physical quantities have a near perfect hybrid parallelization (no serial code) together with a near perfect load balancing (about 99%) But unfortunately, the total computing speed of the physical part is much slower in hybrid mode. When we monitor the number of data load/stores and the data cache-misses we get a near equal number of load/stores but 3 to 4 times higher data cache-misses. The reason is not yet investigated. There is no simple reason, since the intra-node load balancer tries to schedule the same load to the same thread for successive runs over the particle list and concurrent memory access of the same data exists only for reading (causing no cache-trashing). Only data writes from different threads may fall into the same cache line resulting in a higher cache miss rate.

6.2 Linux Cluster Single Node Performance

We used gcc for the *pure-MPI* version and Intel icc for the OpenMP version. The Intel icc is about 4% slower than the gcc in *pure-MPI* mode.

So, we expect only a little difference between the *pure-MPI* and the hybrid version for the Linux Cluster with only two processors per node. The reality was different. The hybrid version is 30% slower on one node than *pure-MPI*.

The reason may be a again a much worse cache utilization discussed in the section above. Additionally, the Linux kernel may frequently reschedule the threads on different processors while MPI processes are better stuck to one CPU.

6.3 Parallel Speedup

Figure 1 shows the speedup comparison of the *pure-MPI* and the hybrid parallelization for different processor numbers. Now the inter-node communication becomes important. On the SR8000 with 8 processors per node the hybrid speedup for large node numbers is much better. There is only little difference between the two strategies on the Linux cluster. "Physics" is the computation of the physical quantities including their communication. It shows a linear hybrid speedup until 256 CPUs. Since it contains communication, there is no linear speedup for larger node numbers. For *pure-MPI*, the "Physics" is the dominant part while for hybrid the parallelization overhead (without communication) is dominant. On the Linux cluster one can see a super-linear speedup for the "Physics" curve. The reason is the increasing cache efficiency. The smaller per-processor problem sizes for large processor numbers result in a better cache reuse. The effect is smaller for *pure-MPI*, since one node cache efficiency is higher.

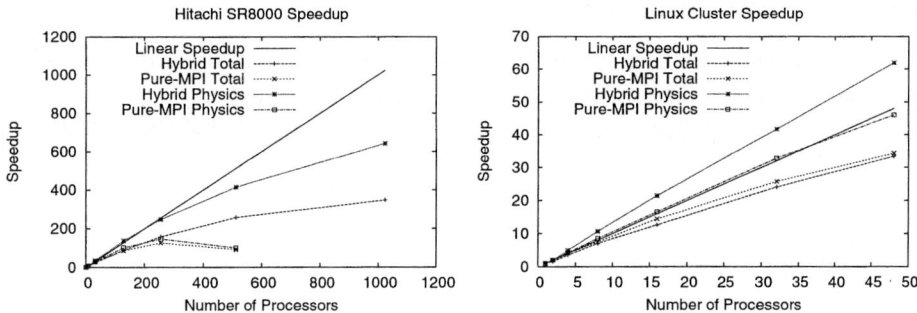

Fig. 1. Speedups using different parallelization strategies on Hitachi SR8000 (left) and the Linux Cluster (right).

The reduced communication is the reason why the hybrid version is faster than the *pure-MPI* version for large node numbers (see 2). The time consuming call in the communication part is a MPI_Allgather and (after a code change) several calls to MPI_Allgatherv together with some post processing to reorganize the received data. In the Allgather-test in section 3 we sent the same amount of data independent of the number of nodes. For the SPH method, the amount of data sent to the neighbors even increase with the number of nodes, because the interaction area between the domains increases compared to the domain size. On the SR8000 the hybrid communication time is near constant for large node numbers, while the *pure-MPI* communication significantly increases for more than 128 processors. On the Linux Cluster, there is only little difference between hybrid and *pure-MPI* parallelization.

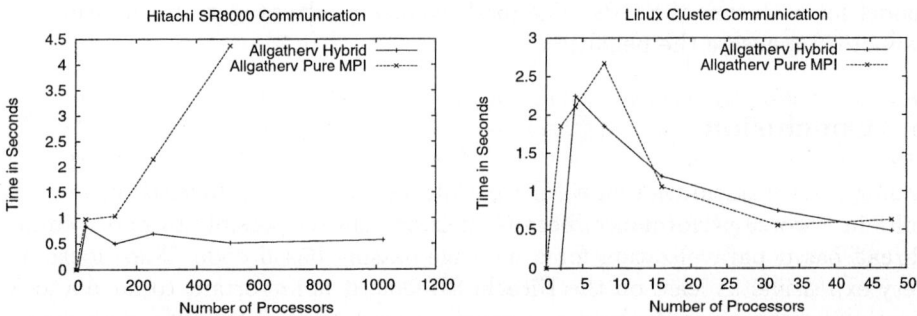

Fig. 2. The communication is the reason for the limited speedup in *pure-MPI* mode for large node numbers.

6.4 Memory Consumption

The SPH method with its neighbour interactions shares a lot of data between processor domains. The advantage of reducing the number of domains for the memory consumption is obviously. While there is only little thread overhead, a *pure-MPI* run on one SR8000 node takes about the same amount of memory as a hybrid job on 8 nodes. The exact reduction depends on several simulation parameters. A good approximation is that we can double our problem size by switching to hybrid mode.

7 Related Works

The comparison of a *pure-MPI* with a hybrid OpenMP/MPI programming model is done in [1] for the NAS benchmark on a IBM SP. To achive a good hybrid implementation the NAS benchmark is profiled to find loops for OpenMP parallelization. Comparing only computation time, the *pure-MPI* version is near always faster. There are only advantages for the hybrid version, if communication time dominates the benchmark.

In [10] there is an extensive comparison of different hybrid programming models. The author compares hybrid-masteronly, hybrid-multiple and *pure-MPI* communication on several parallel platforms. The author concludes, that on some platforms (for example IBM SP) only a hybrid-multiple programming style can achive full inter-node bandwith because a single CPU cannot saturate the network unless the MPI library itself uses a thread based model to optimize the communication.

Henty [4] presents the results of a hybrid implementation of a Discrete Element Modelling (DEM) code. The author made similar experiences, that a scalable hybrid implementation is not achieved without effort. Worse cache utilization and locking in a thread based parallelization reduces overall performance. Therefor, the *pure-MPI* parallelization of the DEM code is better than a hybrid model for a cluster of SMPs. The load balancing advantages of a hybrid code cannot compensate the penalties.

8 Conclusion

A simple instrumentation of a MPI parallelized code with OpenMP usually results in a worse performance, because it is not always possible to find a simple thread based parallelization for a message-passing based code. Since locks are very expensive, at least on the Hitachi SR8000, it is important to have a lock-free implementation. We gain a parallelization rate of about 98% with simple OpenMP instrumentation together with minor code rewrites and explicit optimization of serial parts.

With these enhancements, a hybrid parallelization is possible. It has serveral advantages over *pure-MPI* such as reduced memory consumption and can increased performance for larger node numbers on machines with bigger SMP nodes such as the 8-way Hitachi SR8000, while there is no performance advantage or even a disadvantage over *pure-MPI* parallelization for small node numbers or machines with only 2-way SMP nodes. The reasons are a worse cache efficiency, thread scheduling problems together with compiler flaws and inefficient implementations of some OpenMP features.

For large node numbers the parallelization profits from the reduced communication overhead in hybrid mode. For the SPH method, one can achieve a speedup of about 257 on 512 CPUs and 346 on 1024 CPUs while the *pure-MPI* version is limited to a speedup of about 120 on 256 CPUs on the Hitachi SR8000.

The future work is to further investigate the reason for the reduced cache utilization of the hybrid parallelization and to provide a hybrid version with good performance for small node numbers, too.

References

1. Franck Cappello and Daniel Etiemble. MPI versus MPI+OpenMP on the IBM SP for the NAS Benchmarks. In *Proc. Supercomputing '00, Dallas, TX*, 2000.
2. S. Ganzenmüller, M. Hipp, S. Kunze, S. Pinkenburg, M. Ritt, W. Rosenstiel, H. Ruder, and C. Schäfer. Efficient and object oriented libraries for particle simulations. In E. Krause, W. Jäger, and M. Resch, editors, *High Performance Computing in Science and Engineering 2003*, pages 441–453. Springer-Verlag, 2003.
3. R. A. Gingold and J. J. Monaghan. Smoothed particle hydrodynamics: Theory and application to non-spherical stars. *Monthly Notices of the Royal Astronomical Society*, 181:375–389, 1977.
4. D. S. Henty. Performance of hybrid message-passing and shared-memory parallelism for discrete element modeling. In *Proc. Supercomputing '00, Dallas, TX*, 2000.
5. M. Hipp, S. Kunze, M. Ritt, W. Rosenstiel, and H. Ruder. Fast parallel particle simulations on distributed memory architectures. In E. Krause and W. Jäger, editors, *High Performance Computing in Science and Engineering 2001*, pages 485–499. Springer-Verlag, 2001.
6. S. Hüttemann, M. Hipp, M. Ritt, and W. Rosenstiel. Object oriented concepts for parallel smoothed particle hydrodynamics simulations. In *Proc. of the Workshop on Parallel/High-Performance Object-Oriented Scientific Computing (POOSC'99)*, 1999.
7. L. B. Lucy. A numerical approach to the testing of the fission hypothesis. *The Astronomical Journal*, 82(12):1013–1024, 1977.
8. Message Passing Interface Forum. MPI: A message passing interface. In *Proc. Supercomputing '93*, pages 878–883. IEEE Computer Society, 1993.
9. OpenMP Architecture Review Board. OpenMP C and C++ Application Program Interface, March 2002. http://www.openmp.org.
10. Rolf Rabenseifner. Hybrid Parallel Programming on HPC Platforms. In *Proc. European Workshop on OpenMP '03*, 2003.

Distributed Shared Memory: To Relax or Not to Relax?

Vadim Iosevich and Assaf Schuster

Technion–Israel Institute of Technology
Computer Science Dept.
{vadim_ds,assaf}@cs.technion.ac.il

Abstract. Choosing a memory consistency model is one of the main decisions in designing a Distributed Shared Memory (DSM) system. While Sequential Consistency (SC) provides a simple and intuitive programming model, relaxed consistency models allow memory accesses to be parallelized, improving runtime performance. In this article we compare the performance of two multithreaded memory coherence protocols. The first protocol implements Home-based Lazy Release Consistency (HLRC) memory semantics and the second one implements SC semantics using a MULTIVIEW (MV) memory mapping technique. This technique enables fine-grain access to shared memory while using the virtual memory hardware to track memory accesses. We perform an "apple-to-apple" comparison on the same testbed environment and benchmark suite, and investigate the effectiveness and scalability of both these protocols.

1 Introduction

A *Distributed Shared Memory* (DSM) system provides the distributed application with an abstraction of the shared address space in such a way that all data stored in this space is shared between all nodes in the cluster. Generally, each node uses its local virtual memory as a cache of the shared memory, often identifying the presence of data in the local cache by utilizing the virtual memory hardware. If the data is located on a remote node, the DSM system is responsible for fetching it, while maintaining the correctness of the shared memory. This concept was first proposed by Li and implemented in the first software DSM system, named IVY [1]. In order to keep the cache in a coherent state, the shared data is grained to atomic segments, like lines in a real cache. These segments are called *coherency units*.

A formal specification of how memory operations appear to execute to the programmer is called a *memory consistency model*. Since the introduction of Lamport's now-canonical sequential consistency (SC) model [2], various consistency models have been proposed by researchers [3, 4]. The idea of these models is to postpone the propagation of coherence information until synchronization points are reached. There are two types of synchronization operations, ACQUIRE and RELEASE, used respectively to obtain and yield exclusive access to shared data.

Lazy Release Consistency (LRC) [4] is a refinement of the *Release Consistency* (RC) model [3]. The RC model requires that shared memory accesses be performed globally upon a **RELEASE** operation only. The idea of LRC is to make those accesses visible only to the processor that acquires a lock rather than perform all operations globally. False-sharing is alleviated by allowing different processes to access the same page simultaneously if these operations are not synchronized. A home-based implementation of LRC (HLRC) was proposed by Iftode [5]. In this implementation each shared page has an assigned home node. This home node always hosts the most updated contents of the page, which can then be fetched by a non-home node that needs an updated version.

2 Contribution

This work compares the runtime performance of two memory coherence protocols: a multithreaded implementation of the HLRC model and an efficient multithreaded implementation of the SC model that uses a MULTIVIEW [6] memory mapping technique. **Both coherence protocols are implemented within the same DSM system, where all code that is not consistency-specific is used by both protocols.** We use the same benchmark suite, where all applications produce the same output for both tested protocols. We show that an efficient implementation of the SC memory model can match the performance of the HLRC coherency protocol for the majority of tested applications. We analyze the impact of multithreading on DSM performance and show that the majority of tested applications improve their performance in the multithreaded mode.

3 Implementation

Our implementation is based on a MILLIPEDE DSM system [6]. MILLIPEDE implements a technique called MULTIVIEW, which allows an efficient implementation of SC and fine-grain access to the shared memory. MULTIVIEW can completely eliminate false sharing, treating each shared variable as a coherency unit. If false sharing is to be eliminated, the DSM system must be aware of the size of each shared variable, and this information is naturally supplied by an application via allocation requests. That means that each variable must be allocated separately, and generally requires only a small change to an application source code. A small size of a coherency unit can result in a large number of faults. An application's data set can be allocated by chunks of few variables as a trade-off between false sharing and data prefetching. Generally, there is an optimal allocation pattern for each application that results in a minimal execution time.

Niv and Shuster [7] proposed a mechanism that automatically changes the shared memory granularity during runtime. This mechanism, called the *dynamic granularity* was proven to be a successful technique for improving MULTIVIEW's performance. It is based on a history of shared memory accesses and aggregates variables in larger coherency units when the application accesses them coarsely.

When different nodes start to request different parts of this large coherency unit, it is disassembled to separate variables.

Our complementary work [8] details the efficient implementation of the HLRC memory coherence protocol that supports preemptive multithreading. Previous HLRC implementations proposed non-preemptive multithreading [9] or creating a process for each CPU in an SMP node [10, 11]. The only HLRC implementation that supports preemptive threads is mentioned by Antoniu and Bougé in [12].

The protocol is implemented over the *Virtual Interface Architecture* (VIA) [13] – a standard architecture for high-speed networking. The implementation details are not provided here for lack of space. To make a real "apple-to-apple" comparison, we port a previous version of MILLIPEDE to this communication layer in order to evaluate both protocols on the common substrate.

4 Performance Evaluation

In this section we compare and analyze the performance of two multithreaded shared memory coherence protocols: SC implemented with MULTIVIEW (further denoted as SC/MV) and HLRC. Our testbed environment is a cluster of twelve Compaq Professional Workstations AP550. Each node is an SMP PC with two 733MHZ Pentium-III processors, a 512KB L2 cache, a 512MB physical memory and a 32-bit/33MHz PCI bus. All nodes run the Win2000 operating system. The cluster is interconnected by the ServerNet-II [14] VIA-based network.

We also investigate the effect of chunk allocation in order to estimate the potential of the SC/MV technique and discover the best static granularity for the particular application. In addition, we try to estimate how the dynamic granularity change can boost the performance of the MULTIVIEW technique. Our system does not support the dynamic granularity protocol, but we try to estimate the runtime performance for the dynamic granularity on the basis of results presented in [7]. We estimate the performance gain achieved with dynamic granularity versus fixed granularity and the presented results are only an approximation.

4.1 Benchmark Application Suite

Our benchmark suite consists of two microbenchmarks, NBodyW and NBody; eight applications from the SPLASH-2 [15] benchmark suite (Barnes, Volrend, LU, Water-nsq, Water-sp, FFT, Radix and Ocean); and TSP and SOR from the TreadMarks [16] benchmark applications.

NBodyW is a microbenchmark that imitates a kernel of n-body applications. The program operates with a large set of 64-byte bodies and performs three phases as follows: (1) Each of the P application's threads reads the entire set of bodies. (2) Each of the P application's threads processes and updates $1/P$ of the bodies. The processing of a body is simulated by a constant-length busy loop. (3) A single thread updates all the bodies (sequential phase). NBody is a shortened modification of NBodyW that contains only the first two phases. Hence, this application contains one coarse phase and one fine phase.

Table 1. Benchmark characteristics. B stands for barriers, L stands for locks.

Application	Input data set	Shared memory	Sharing granularity	Synch	Allocation pattern
Water-nsq	8000 molecules	5.35MB	a molecule (672B)	B, L	fine
Water-sp	8000 molecules	10.15MB	a molecule (680B)	B, L	fine
LU	3072×3072	72.10MB	block (coarse)	B	coarse
FFT	2^{20} numbers	48.25MB	a row segment	B	coarse
TSP	A graph of 32 cities	27.86MB	a tour (276B)	L	fine
SOR	2066×10240	80.73MB	a row (coarse)	B	coarse
Barnes	32768 bodies	41.21MB	body fields (4–32B)	B, L	fine
Radix	10240000 keys	82.73MB	an integer (4B)	B, L	coarse
Volrend	a file "head.den"	29.34MB	a 4×4 box (4B)	B, L	fine
Ocean	a 514×514 grid	94.75MB	grid point (8B)	B, L	coarse
NBody	32768 bodies	2.00MB	a body (64B)	B	fine
NBodyW	32768 bodies	2.00MB	a body (64B)	B	fine

A detailed description of the other benchmarks can be found in related papers and is not provided here for lack of space. Table 1 summarizes input data sets and memory sharing characteristics of all tested benchmarks.

4.2 Performance Analysis and Comparison

Fig. 1 summarizes the speedups obtained for all applications, with one thread running on each node. We can divide the tested benchmarks into three groups, according to their performance with each of the two memory coherence protocols mentioned above.

The first group contains the majority of tested applications, for which the SC/MV protocol matches HLRC's performance, or for which the gap between the two can be eliminated by using the proper granularity level or dynamic granularity protocol. These applications are: NBody, NBodyW, Volrend, TSP, Water-nsq, Water-sp, Radix and FFT.

The second group comprises three applications: SOR, LU and Ocean, which achieve a better speedup with the SC/MV protocol. HLRC's poor performance for these applications is due to the excess coherence operations that are required by the consistency model but not justified in this specific case. Generally, these operations are performed on barriers, when all threads are notified about modifications performed by all other threads. Nevertheless, each thread processes its own part of a data set and is not interested in the modifications of other threads.

To the third group we ascribe Barnes only. For this application, the HLRC coherence protocol significantly outperforms the SC/MV, and this gap cannot be bridged by optimal or dynamic chunking levels. Dynamic granularity does improve the SC/MV protocol's performance, but it is still a long way from that of HLRC.

Investigating the impact of multithreading on DSM performance, we found that while generally beneficial, multithreading can also be detrimental in some

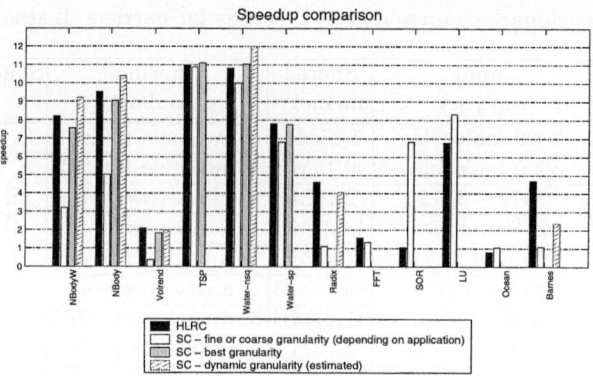

Fig. 1. Speedup comparison for all tested applications. The speedup is measured as the relation between the serial execution time and the minimal execution time on a 12-node cluster, where each node runs only one application thread.

cases. First of all, there are sequential phases performed by a single thread (as in NBodyW, Volrend and LU) or phases where the same task is performed by all threads regardless of their location, as in the first phase of NBody(W). Furthermore, multithreading does not always provide computation-communication overlap and can lead to increased contention in the shared memory protocol and communication layer. Consequently, multithreading affects all applications to a different extent, as can be observed from Fig. 2. In summary, the HLRC protocol benefits from multithreading, with an average performance improvement of 35.5%. The SC/MV protocol without optimized allocation improves its performance by 21.9% on average. With the optimized allocation pattern, the SC/MV protocol gains 30.7%.

5 Conclusions

The coherency state information that must be kept by the SC/MV protocol is quite simple. Only the presence of all page replicas must be tracked. In contrast, the HLRC protocol is much more memory-consuming and cumbersome. Tracking the causality relation between memory accesses requires complex data structures that must be referenced by different threads. This requires a very accurate mutual exclusion mechanism to keep the state data valid.

If data has to be fetched from a remote node as the result of a fault, this will always take one round trip in HLRC – a request message is sent to a home node that answers with the requested content. In the SC/MV protocol, however, this operation generally takes three messages. The first message is sent to a manager (if the faulting node is not the manager itself), and the second is forwarded by the manager to the page owner (if the manager is not the owner itself), which, in turn, replies to a requester. This results in a more costly fault in the SC/MV implementation. It should be noted that although chunk allocation increases the

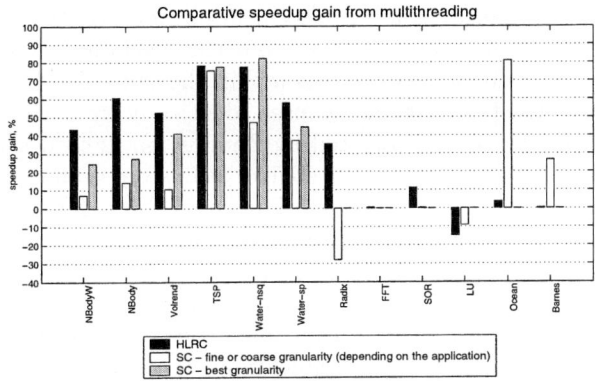

Fig. 2. Speedup gain from multithreading. The presented measure is the difference in the speedup achieved on a 12-node cluster with each node running one vs. two application threads.

average cost of a fault (due to the increased portion of data that must be sent to the requester), the overall number of faults is reduced, alleviating page fault overhead.

We found that the average speedup of the HLRC protocol is 5.97, while the average speedup of the SC/MV protocol with non-optimized allocation is 4.5. This means that HLRC exhibits a 32.7% performance advantage over a non-optimized SC/MV. Considering the best possible allocation pattern for each application, we raise the average speedup of the SC/MV protocol to 5.6, which is very close to the average HLRC speedup. This decreases the HLRC performance advantage to only 6.6%. Taking into account the speedup achieved by the SC/MV protocol if the granularity is changed dynamically at runtime, we get an average SC/MV speedup of 6.2.

6 Related Works

Dwarkadas et al. [17] compared the Shasta [18] (SC) and Cashmere [19] (RC) DSM systems, both of which were tuned to run on a cluster of four-CPU 400 MHz multiprocessors connected via a Memory Channel network. The authors concluded that for the eight applications that were written and tuned for hardware DSM systems, Shasta performed 6.1 times better than Cashmere. For the five programs that were written or tuned for page-based DSM, Cashmere performed 1.3 times better than Shasta. When all the tested applications were optimized separately for both coherence protocols, Cashmere performed 1.15 times better than Shasta. Both of the aforementioned protocols differ from those investigated in our study, and we ran our DSM on nodes with more powerful CPUs. Nevertheless, the results confirm our conclusion that the SC protocol, if implemented efficiently, performs no worse than relaxed memory consistency models. In addition, Shasta and Cashmere are two completely different DSM systems: they differ not only in coherence protocols but also in implementing services such as synchronization primitives, communication, and the tracking of memory accesses. We implement both coherence protocols within the same DSM system where all code that is not consistency-specific is used identically by both protocols.

Additional research was conducted by Zhou et al. [20] on a noncommodity hardware system that supports access control at custom granularity. This allows the use of a uniform access control mechanism for both fine-grain and coarse-grain protocols. The authors tried to find the best combination of granularity and consistency protocols for different classes of applications. Three consistency protocols with four sizes of coherence granularity were tested: SC, single-writer LRC, and HLRC. The results show that no single combination of protocol and granularity performs best for all the applications. A combination of the SC protocol and fine granularity works well with 7 of the 12 applications. The combination of a HLRC protocol and page granularity works well with 8 of the 12 applications. It should be noted that the testbed environment used in [20] consisted of 66MHz processors, while our processors are an order of magnitude faster. In contrast to this study, we use standard hardware to implement DSM protocols.

Keleher [21] compared single- and multiple-writer versions of the LRC protocol, and the SC protocol, all implemented in the CVM software DSM system. Keleher found that the multiple writer LRC protocol performs an average of 34% better than the SC protocol. While not specified, the SC protocol used in this study is probably a page-based protocol and therefore cannot perform well for applications that use fine-grain data sharing. Nevertheless, it performs well for coarse-grain applications like FFT and LU. The results show that the SC protocol performs like multiple-writer LRC for FFT and outperforms it for LU. This confirms our results for these applications. The use of slow (66MHz) processors make it difficult to compare this study to our research.

References

1. Li, K.: Ivy: A shared virtual memory system for parallel computing. In: Proc. of the Int'l Conf. on Parallel Processing (ICPP'88). Volume 2. (1988) 94–101
2. Lamport, L.: How to make a multiprocessor computer that correctly executes multiprocess programs. IEEE Transactions on Computers **28** (1979) 690–691
3. Gharachorloo, K., Lenoski, D., Laudon, J., Gibbons, P.B., Gupta, A., Hennessy, J.L.: Memory consistency and event ordering in scalable shared-memory multiprocessors. In: 25 Years ISCA: Retrospectives and Reprints. (1998) 376–387
4. Keleher, P., Cox, A.L., Zwaenepoel, W.: Lazy release consistency for software distributed shared memory. In: Proc. of the 19th Annual Int'l. Symp. on Computer Architecture (ISCA'92). (1992) 13–21
5. Iftode, L.: Home-based shared virtual memory (thesis). Technical Report TR-583-98, Princeton University, Computer Science Department (1998)
6. Itzkovitz, A., Schuster, A.: MultiView and Millipage–fine-grain sharing in page-based DSMs. In: Proc. of the 3rd Symp. on Operating Systems Design and Implementation (OSDI-99), Berkeley, CA (1999) 215–228
7. Niv, N., Schuster, A.: Transparent adaptation of sharing granularity in MultiView-based DSM systems. Software Practice and Experience **31** (2001) 1439–1459

8. Iosevich, V., Schuster, A.: Multithreaded home-based lazy release consistency over VIA. In: Proc. of the 18th Int'l. Parallel and Distributed Processing Symp. (IPDPS'04). (2004)
9. Rangarajan, M., Divakaran, S., Nguyen, T.D., Iftode, L.: Multi-threaded home-based LRC distributed shared memory. In: The 8th Workshop of Scalable Shared Memory Multiprocessors (held in conjunction with ISCA). (1999)
10. Stets, R., Dwarkadas, S., Hardavellas, N., Hunt, G.C., Kontothanassis, L.I., Parthasarathy, S., Scott, M.L.: Cashmere-2L: Software coherent shared memory on a clustered remote-write network. In: Symp. on Operating Systems Principles. (1997) 170–183
11. Bilas, A.: Improving the Performance of Shared Virtual Memory on System Area Networks. PhD thesis, Dept. of Computer Science, Princeton University (1998)
12. Antoniu, G., Bougé, L.: DSM-PM2: A portable implementation platform for multi-threaded DSM consistency protocols. In: Proc. 6th Int'l. Workshop on High-Level Parallel Programming Models and Supportive Environments (HIPS '01). (2001)
13. Compaq, Intel and Microsoft Corporations: Virtual Interface Architecture Specification. Version 1.0. http://www.viarch.org (1997)
14. Heirich, A., Garcia, D., Knowles, M., Horst, R.: ServerNet-II: A reliable interconnect for scalable high performance cluster computing. Technical report, Compaq Computer Corporation, Tandem Division (1998)
15. Woo, S.C., Ohara, M., Torrie, E., Singh, J.P., Gupta, A.: The SPLASH-2 programs: Characterization and methodological considerations. In: Proc. of the 22th Int'l Symp. on Computer Architecture. (1995) 24–36
16. Keleher, P., Dwarkadas, S., Cox, A.L., Zwaenepoel, W.: Treadmarks: Distributed shared memory on standard workstations and operating systems. In: Proc. of the Winter 1994 USENIX Conf. (1994) 115–131
17. Dwarkadas, S., Gharachorloo, K., Kontothanassis, L., Scales, D.J., Scott, M.L., Stets, R.: Comparative evaluation of fine- and coarse-grain approaches for software distributed shared memory. In: Proc. of the 5th IEEE Symp. on High-Performance Computer Architecture (HPCA-5). (1999) 260–269
18. Scales, D.J., Gharachorloo, K., Thekkath, C.A.: Shasta: A low overhead, software-only approach for supporting fine-grain shared memory. In: Proc. of the 7th Symp. on Architectural Support for Programming Languages and Operating Systems (ASPLOSVII). (1996) 174–185
19. Kontothanassis, L.I., Hunt, G., Stets, R., Hardavellas, N., Cierniak, M., Parthasarathy, S., Meira, Jr., W., Dwarkadas, S., Scott, M.L.: VM-based shared memory on low-latency, remote-memory-access networks. In: Proc. of the 24th Annual Int'l. Symp. on Computer Architecture (ISCA'97). (1997) 157–169
20. Zhou, Y., Iftode, L., Li, K., Singh, J.P., Toonen, B.R., Schoinas, I., Hill, M.D., Wood, D.A.: Relaxed consistency and coherence granularity in DSM systems: A performance evaluation. In: Proc. of the Sixth ACM SIGPLAN Symp. on Principles and Practice of Parallel Programming (PPOPP'97). (1997) 193–205
21. Keleher, P.: The relative importance of concurrent writers and weak consistency models. In: Proc. of the 16th Int'l. Conf. on Distributed Computing Systems (ICDCS-16). (1996) 91–98

Design-Time Data-Access Analysis for Parallel Java Programs with Shared-Memory Communication Model

R. Stahl, F. Catthoor, R. Lauwereins, and D. Verkest

IMEC vzw, Kapeldreef 75, B-3001 Leuven, Belgium
richard.stahl@imec.be

Abstract. In the era of future embedded systems the designer is confronted with multi-processor architectures both for performance and energy reasons. Exploiting (sub)task-level parallelism is becoming crucial because the instruction-level parallelism alone is insufficient.

The challenge is to build compiler tools that support the exploration of the task-level parallelism in the programs. To achieve this goal, we have designed an analysis framework to estimate the potential parallelism from sequential object-oriented programs in Java.

Data-access analysis is one of the crucial techniques for estimation of the transformation effects. We have implemented support for platform-independent data-access analysis and profiling of Java programs. Herein, we focus on the technique for design-time data-access analysis. It complements our earlier work on parallel performance analysis. We demonstrate the feasibility and effectiveness of our approach on a number of Java applications.

1 Context and Related Work

Data-access and communication analysis for parallel programs is an important topic, as motivated in the abstract. We can identify two main categories: data-access analysis for single-processor platforms with memory hierarchy and communication analysis for parallel programs on multi-processor platforms.

We have been partially inspired by work of Ding and Zhong [4], who introduce platform-dependent, run-time monitoring of data accesses. This approach is based on compiler-directed instrumentation of single-threaded C programs. A similar approach for data-access analysis has been introduced by Bormans et al. [5] They use design-time data-access analysis to identify all possible data-accesses in the sequential C programs. Afterwards, the executable specification is profiled and the data-access traces generated. Leeman et al. [6] introduce a technique for data-access profiling for power estimation. They use method-level data-access summaries, which are inserted into the program code at design-time so that the run-time system can gather the data-access traces for arbitrary data types. We distinguish from these approaches in the following way: first, we have introduced the concept of parallel execution [15], which allows the designer to perform parallel program analysis without the previous mapping to the target platform, and second, we have introduced the concept of parallel communicating tasks [16] for which we analyse the computation as well as communication cost.

In the area of parallel systems, the research focus has been mainly on communication analysis and optimisation. These approaches usually require explicit communication between the tasks. Miller et al. [10] have introduced Paradyn - parallel performance measurement tools. It focuses on the profiling and post-processing of profile information for long-running large-scale programs written in high-level data-parallel languages. Haake et al. [11] have introduced a similar approach, but they have implemented profiling support for the Split-C programs with Active Messages. It is based on fine grain communication profiling while the program traces are post-processed off-line. Another approach, implemented by Vetter [9] analyses the performance of parallel programs with message passing communication. The main contribution of this work is in the classification of communication inefficiencies, i.e., it is a post-processing phase of performance analysis that gives the designer concise and interpreted performance measures. Chakrabarti, et al. [7] introduce communication analysis and optimisation techniques for High-Performance Fortran programs. Even though the approach includes performance analysis, the main focus is on the optimisation of the global program communication. We distinguish from the previous approaches by introducing automated data-access analysis support for high-level programming languages. Additionally, these approaches are intended for a platform-specific performance analysis for particular machines, as opposite to our platform-independent analysis.

We believe that the approach introduced by Tseng [8] is one of the closest to our work. The technique focuses on communication analysis for machine-independent High-Performance Fortran programs, and provides application-oriented analysis of the communication in the parallel programs. We, on the other hand, introduce design-time data-access analysis for high-level concurrent object-oriented programs. Moreover, we introduce the above mentioned concept of parallel-execution environment with support for performance and data-access profiling.

2 Parallel-Performance Analysis Framework for Java

The proposed performance analysis tool is based on a concept of parallel-execution time [15, 16], which allows one to abstract specific architectural features of the platform. The concept is used to simulate parallel execution of program tasks while the program is actually executed on the underlying platform, which does not need to be the final target platform. The tools work as follows. Firstly, the program is automatically transformed based on designer's input constraints. This phase consists of two complementary transformations: parallel performance analysis and data-access analysis. Secondly, the parallel program execution is simulated and profiled. Finally, the profiling information is analysed and interpreted to provide the designer with a more convenient form of profiling output.

Herein, we focus on the design-time data-access analysis. The analysis identifies all potential data-accesses which are then profiled at run time. For this purpose, we define a data-access model (Figure 1) that consists of main execution thread, number of separate threads and shared data. The implementation of the transformation passes as well as intra and inter-procedural program analysis is based on the existing transformations in the SOOT optimisation framework [1].

3 Design-Time Data-Access Analysis

The data-access model (Figure 1) is used as a representation for modelling the accesses to the data shared between different program tasks present in a sequential or parallel program. Therefore, it serves as the conceptual base for the design-time data-access analysis. The data-access model consists of the following components: main-program thread, separate threads/methods and shared data. No cache memories are included yet this extention is possible. All shared data belong to main method and they are stored in the shared-data section. The separate methods require an amount of data to be read from the shared-data section before they can proceed with execution. On the other hand, the methods generate and write an amount of data to be stored back to the shared-data section.

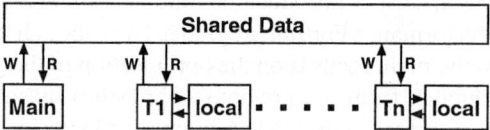

Fig. 1. Data-access model: a shared memory communication model of an abstract architecture, used in the data-access analysis to separate local and non-local accesses of separate threads T_i.

The designer specifies a set of separate methods, and the analysis identifies all the accesses performed on the shared data. To achieve this, it traverses the corresponding parts of the program representation based on a method-call graph (Figure 2). In the case of a polymorphic method call the analysis resolves all method call candidates while only one of them is selected and profiled at run-time. The analysis identifies all the data created outside the scope of a separate method, and accessed within its scope. Based on the Java programming language specification [2], access to this shared data can be performed only via method parameters (P_M - list of parameter for method M), class members (F_M - list of class fields) and return statements (R_M). We can conceptually split the data-access analysis into two parts: data-read and data-write analysis:

$$\forall M \in SeparateMethods \rightarrow analyseR(M); analyseW(M)$$

The data-access analysis algorithms are implemented recursively because of their iterative nature without manifest bounds on the exploration depth. The upper bound of computation complexity of the algorithms is $O(n \times m_R \times k_R)$ for forward pass and $O(n \times m_W \times k_W)$ for backward pass, where n is number of methods selected by designer, m_R (m_W) is number of methods accessed from selected methods for data read (resp. write) and k_R (k_W) is number of shared data which are read (resp. written) within the scope of selected methods. However, we have not observed this worst-case behaviour in analysis of tested programs.

The data-access analysis identifies and distinguishes three groups of data types:

- primitive type $T_P = char, short, integer, long, float, double,$
- reference type $T_R = ObjectReference, ArrayReference,$
- array type $T_A = ArrayReference$ [1].

---

[1] The Java arrays implement distinct concept within the Java language so we consider them a special case of the reference type.

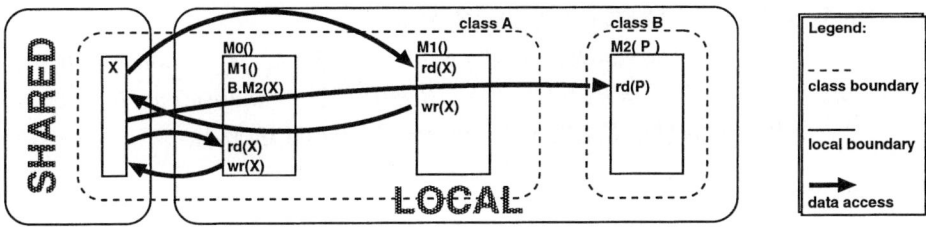

Fig. 2. Example: the data-access analysis identifies all possible read and write accesses from thread's local scope to shared-data section, where M0() is the separate (or thread's main) method.

3.1 Forward Data-Read Analysis

The data-read analysis identifies the read accesses to the shared data resulting in purely forward recursive traversal of the program representation. The potential candidates for the data-read access are only method parameters and class members. The analysis consists of the following steps: analysis of method calls for own class methods ($ClassMethods$), analysis of method parameters (P_M) and analysis of class members (F_M) accessed within each separate method.

$ana\ lyseR(M)$:
  if $(M \notin SeparateMethods) \rightarrow M_c = clone(M)$
  $\forall\ M' : isInvokedIn(M, M') \wedge isClassMethods(M, M') \rightarrow analyseR(M')$
  $\forall\ P_M : isReadIn(M, P_M) \rightarrow analyseTypeR(P_M)$
  $\forall\ F_M, isReadIn(M, F_M) \rightarrow analyseTypeR(F_M); FieldList.identify(F_M, M_c)$

Method $clone(M)$ creates a clone M_c of given method M with unique thread identification tag. Each top-level method and sub-graph of the method-call graph accessed from this method is identified by an identification tag. Method M' is analysed if it is invoked with M method body ($isInvokedIn(M, M')$) and it is a method implemented with the same class ($M' \in ClassMethods$).

Method $FieldList.identify(Field, Method)$ identifies and resolves given class member within global program scope. The analysis uses this global information to resolve accesses to all members of given classes. Thus, it removes aliasing of the members which is introduces by assignments in different methods of the program. This alias removal results in minimal and realistic data-access patterns. As shown below, the same method is used for the data-write analysis that results in accurate information on read and write accesses to all the class members.

The algorithm enters $analyseTypeR(P_M)$ method for each of the method parameters P_M and class members F_M used within M method body ($isReadIn(M, P_M)$). The method analyses the given argument based on its type (primitive T_P, reference T_R or array T_A) as follows.

$ana\ lyse\ TypeR(P_M)$:
  if $(ass\ ignedTo(P'_M, P_M)) \rightarrow analyseTypeR(P'_M)$
  if $(P_M \in T_P) \rightarrow addRdType(M_c, P_M, T_P)$

$\quad if\ (P_M \in T_R) \rightarrow addRdType(M_c, P_M, T_R)$
$\quad \forall\quad M' : isClassMethods(P_M, M') \wedge isInvokedIn(M, M') \rightarrow analyseR(M')$
$\quad if\ (P_M \in T_A) \rightarrow \forall E_M = P_M[idx] : E_M.isReadIn(M) \rightarrow analyseTypeR(E_M);$
$\quad\quad addRdType(M_c, E_M, T_A); ArrayList.identify(P_M, M_c);$

Call to $addRdType(Method, Parameter, Type)$ does the actual code annotation for later use in other phases of program transformation. The method adds parameter-specific annotation to the code segment of cloned method (M_c). The annotation consists of parameter name (P_M) and type (T_P, T_R or T_A). Each data (parameter or member) is assigned a unique global identification tag that it is identifiable within the global program scope, i.e., for all the separate methods.

In the case of array data type (T_A), the analysis annotates all array elements (E_M) accessed withing the method body ($isReadIn(M, E_M)$). Thus, the annotation made by method $addRdType(M_c, E_M, T_A)$ includes also the actual array index, e.g., if the annotated program is later profiled the actual index value is identified, which results in a detailed data-access trace for the arrays.

Method $ArrayList.identify(Array, Method)$ has similar usage as $FieldList$ method yet it resolves even method-local array data. This way, the analysis keeps global information on the accesses to individual array elements that results in alias removal on the level of array elements.

3.2 Backward Data-Write Analysis

The data-write analysis traverses the representation forward until it finds any write access to a shared data then it starts a backward traversal to identify the origin of the data assignment. Furthermore, it inspects all potential accesses to this data. Thus, it uses the forward analysis to resolve all read accesses to this newly identified data which is eventually written into shared-data section. The potential candidates are non-primitive method parameters, class members, and return statements.

$ana\ lyseW(M)$:
$\quad if\ (M \notin SeparateMethods) \rightarrow M_c = clone(M)$
$\quad \forall\ M' : isInvokedIn(M, M') \wedge M' \in ClassMethods \rightarrow analyseW(M')$
$\quad \forall\ P_M : isWrittenIn(M, P_M) \rightarrow analyseTypeW(P_M)$
$\quad \forall\ F_M : isWrittenIn(M, F_M) \rightarrow analyseTypeW(F_M); FieldList.identify(F_M, M_c)$
$\quad if\ (hasReturn(M)) \rightarrow R_M = returnedFrom(M); analyseTypeW(R_M)$

The method $analyseTypeW(P_M)$ is defined as follows.
$ana\ lyse\ TypeW(P_M)$:
$\quad if\ (P_M \in T_P) \rightarrow DataWriteType(M_c, P_M, T_P)$
$\quad if\ (P_M \in T_R) \rightarrow$
$\quad\quad \forall\ M' : isClassMethods(P_M, M') \wedge isInvokedIn(M, M') \rightarrow analyseW(M')$
$\quad\quad if\ (assignedFrom(P'_M, P_M) \wedge findOrigin(P'_M)) \rightarrow addWrType(M_c, P_M, T_R)$
$\quad if\ (P_M \in T_A) \rightarrow$

$if\ (assignedFrom(P'_M, P_M) \land findOrigin(P'_M)) \rightarrow addWrType(M_c, P_M, T_A)$
$\forall\ E_M = P_M[idx] : asignedFrom(E'_M, E_M) \land findOrigin(E'_M) :$
$addWrType(M_c, E_M, T_A); ArrayList.identify(P_M, M_c);$
$analyseTypeR(E_M); analyseTypeW(E_M)$

Method $findOrigin(P_M)$ is called for each assignment $(assignedFrom(P'_M, P_M))$ of a reference, array and array element . Based in the result of its analysis an appropriate method annotation is made $(addWrType(M_c, P_M, Type))$.

$findOrigin(P_M) :$
$if\ (isDefLocally(P_M)) \rightarrow return(true)$
$if\ (isDefByMember(P_M) \land assignedFrom(P'_M, P_M)) \rightarrow$
$\quad analyseTypeR(P'_M); analyseTypeW(P_M); return(false)$
$if\ (isDefByParameter(P_M)) \rightarrow$
$\quad if(M \in SeparateMethods) \rightarrow return(false)$
$\quad M' = getCaller(M); P'_M = getParam(P_M, M'); return(findOrigin(P'_M))$

Method $findOrigin(P_M)$ traverses the program representation backwards to search for the original assignment of the given data (P_M). It returns true only in case of true local assignment, i.e., if the given data was created locally within the scope of given separate method.

In the case of definition by parameter ($isDefByParameter(P_M)$), if the analysis operates in the representation of a particular separate method ($SeparateMethods$) the returned value is false and no further analysis is needed. This situation corresponds to assignment from input parameter of the given separate method to its output data, which means that no actual read and/or write access is performed.

Design-time data-access analysis traverses all potential method calls in the subgraph of the method-call graph while performing alias checks to isolate given data structures and thus provide accurate information on potential data accesses in the program.

4 Experimental Results

In experiments accomplished we have focused on the usability of the proposed framework [16]. We have used the host platform and corresponding memory model[3] to obtain the absolute timing information for program execution and data communication while the main interest of our platform-independent program characterisation is in the relative comparison between them. The data communication timing is calculated as a product of the number of memory accesses and data-access timing for the above referenced memory model.

For the evaluation of the performance analysis framework we have used a 3D application [13], an MPEG video player [12] and the following set of applications from the Java Grande Forum Thread Benchmark Suite [14]: JGFCrypt, JGFSparseMatMult,

Table 1. Interpreted analysis results: $program_{Xca/sa}$ (X - number of threads, ca/sa - concurrent or sequential data-access mode), communication time, total-execution time and speedup for zero/page/random-mode communication.

program	$T_{C,pg}[ms]$	$T_{C,rnd}[ms]$	$T_{T,pg}[ms]$	$T_{T,rnd}[ms]$	S_0	S_{pg}	S_{rnd}
$3D_{4ca}$	3.44	30.9	1519	1546	2.18	2.18	2.15
$3D_{4sa}$	9.28	83.6	1525	1600	2.18	2.18	2.07
$MPEG_2$	962	8653	19679	27370	1.97	**1.88**	**1.35**
$Crypt_2$	204	1836	2979	4611	1.82	1.70	1.10
$Crypt_{4ca}$	102	918	1726	2542	3.12	2.93	1.99
$Crypt_{4sa}$	306	2754	1930	4378	3.12	2.62	1.16
$Matrix_2$	1800	1980	11947	29954	1.98	1.68	0.67
$Matrix_{4ca}$	800	7208	6007	12415	3.85	**3.34**	1.62
$Matrix_{4sa}$	2400	21624	7607	26831	3.85	2.63	**0.75**
$RayT_2$	925	8320	55391	62786	1.99	1.96	1.73
$RayT_{4ca}$	465	4183	27863	31581	3.97	3.90	3.44
$RayT_{4sa}$	1385	12466	28783	39864	3.97	3.77	2.73
$MCarlo_2$	81	729	27359	28007	1.85	1.84	1.80
$MCarlo_{4ca}$	40.5	365	16193	16517	3.12	**3.11**	3.05
$MCarlo_{4sa}$	120	1095	16272	17247	3.12	3.09	**2.92**

JGFRayTracer and JGFMonteCarlo. The results of the performance analysis can be interpreted as follows (Table 1): data-communication time (T_C) is calculated separately for concurrent and sequential data accesses to the shared-data section ($program_{Xca/sa}$). The total execution time (T_T) is a sum of execution time and data-communication time. Thus, the achievable speed-up ranges from the speedup for random data-access model (S_{rnd}) to the speedup for page-mode data-access model (S_{pg}). For comparison reasons we present also speedup for an ideal reference (S_0 - no data-communication overhead).

An example of application with heavy data communication is the MPEG video player. Based on the interpretation of the analysis results (Table 1, $MPEG_2$), we see that in case of random data-access mode, the communication corresponds to 31% of the total execution time. Thus, compared to the ideal reference ($S_0 = 1.97$), the realistic speed-up is considerably degraded ($S_{rnd} = 1.35$). Another example of such an application is JGFSpareMatMult (Table 1, $Matrix_4$). On the other hand, an example of less data dominated application is the JGFMonteCarlo benchmark programs. Even though, the program complexity is similar the communication time ranges from 0.3 - 6.3%. Thus, the final speedup (ranging from 2.92 to 3.11, with reference speedup of 3.12) does not depend on the data-communication as heavily as in the previous example.

The execution time of the data-access analysis tool ranges from 245 ms for JGFSparseMatMult program (2 analysed methods and 11 identified data accesses) to 33798 ms for MPEG video player (41 methods and 773 data accesses), executed using Sun J2SDK1.3.1 on desktop PC with Pentium-4 1.6GHz, 640MB RAM. Typical execution time is in the range of 1 to 4 seconds (programs with 20 to 60 methods and 10 to 140 identified accesses).

5 Conclusions

We have introduced the design-time data-access analysis which is a crucial part of our transformation framework for exploration of task-level parallelism in sequential object-oriented programs. The main difference of our approach compared to related work is the introduction of the design-time data-access analysis for the programs with shared-memory communication model. To increase the usability of our technique we have implemented automatic tool that performs the data-access analysis on Java programs.

References

1. Vallee-Rai, R., Hendren, L., Sundaresan, V., Lam, P., Gagnon, E., Co, P.: Soot - a Java Optimization Framework, Proc. of CASCON, 1999
2. Gosling, J., Joy, B., Steele, G. and Bracha, G.: The Java Language Specification, Second Edition Addison-Wesley, 2000
3. Micron: Calculating Memory System Power For DDR, www.micron.com, TN-46-03
4. Ding, C., Zhong, Y.: Compiler-Directed Run-time Monitoring of Program Data Access, Proceedings of the workshop on Memory system performance, Berlin, Germany, pp.1-12, 2003
5. Bormans, J., Denolf, K., Wuytack, S., Nachtergaele L. and Bolsens, I.: Integrating System-Level Low Power Methodologies into a Real-Life Design Flow, Proceeding of IEEE Workshop on Power and Timing Modeling, Optimization and Simulation (PATMOS), Kos, Greece, pp.19-28, 1999
6. Leeman, M. et al.: Power Estimation Approach of Dynamic Data Storage on a Hardware Software Boundary Level, Proceeding of IEEE Workshop on Power and Timing Modeling, Optimization and Simulation (PATMOS), Torino, Italy, pp.289-298, 2003
7. Chakrabarti, S., Gupta, M., Choi, J.D: Global Communication Analysis and Optimisation, Proceedings of Conference on Programming Language Design and Implementation, pp.68-78, 1996
8. Tseng, C-W.: Communication Analysis for Shared and Distributed Memory Machines, Proceedings of the Workshop on Compiler Optimizations on Distributed Memory Systems, 1995
9. Vetter, J.: Performance Analysis of Distributed Applications using Automatic Classification of Communication Inefficiencies, Proceeding of ACM International Conference on Supercomputing, Santa Fe, USA, 2000
10. Miller, B.P., et al.: The Paradyn Parallel Performance Measurement Tool, Journal IEEE Computer, vol.28, num.11, pp.27-46, 1995
11. Haake, B., Schauser, K.E., Scheiman, C.: Profiling a parallel language based on fine-grained communication, Proceedings of the ACM/IEEE conference on Supercomputing, Pittsburgh, USA, 1996
12. Anders, J.: MPEG-1 player in Java, http://rnvs.informatik.tu-chemnitz.de/ jan/MPEG/MPEG_Play.html
13. Walser, P.: IDX 3D engine, http://www2.active.ch/ proxima
14. Java Grande Forum Benchmarks, http://www.epcc.ed.ac.uk/javagrande/javag.html
15. R.Stahl et al.: Performance Analysis for Identification of (Sub)task-Level Parallelism in Java, Proceedings of SCOPES'03, Austria, 2003
16. R.Stahl et al.: High-Level Data-Access Analysis for Characterisation of (Sub)task-Level Parallelism in Java, to be published in Proceedings of HIPS'04 workshop, Santa Fe, USA, April 2004

Parallel PEPS Tool Performance Analysis Using Stochastic Automata Networks*

Lucas Baldo[1], Luiz Gustavo Fernandes[1], Paulo Roisenberg[2], Pedro Velho[1], and Thais Webber[1]

[1] Faculdade de Informática, PUCRS
Avenida Ipiranga, 6681 Prédio 16 - Porto Alegre, Brazil
{lucas_baldo,gustavo,pedro,twebber}@inf.pucrs.br
[2] HP Brazil
Avenida Ipiranga, 6681 Prédio 91A - TecnoPuc - Porto Alegre, Brazil
paulo.roisenberg@hp.com

Abstract. This paper presents a theoretical performance analysis of a parallel implementation of a tool called Performance Evaluation for Parallel Systems (PEPS). This software tool is used to analyze Stochastic Automata Networks (SAN) models. In its sequential version, the execution time becomes impracticable when analyzing large SAN models. A parallel version of PEPS using distributed memory is proposed and modelled with SAN formalism. After, the sequential PEPS itself is applied to predict the performance of this model.

1 Introduction

In recent years, the effort of many authors has confirmed the importance of performance prediction of parallel implementations. Some authors have presented generic studies offering options to the performance prediction of parallel implementations [1, 2]. The research community classifies the different approaches in three quite distinct groups: monitoring, simulation and analytical modelling. This last approach (analytical modelling), compared to the two first ones, is more rarely employed to achieve parallel programs performance prediction. This happens due to a frequent misconception: the most known formalisms to analytical modelling, *e.g.*, Markov chains [3] and queueing networks [4], are not very suitable to represent parallelism and synchronization. In this paper, we adopted the analytical modelling approach using a formalism called Stochastic Automata Networks (SAN). The reader interested in a formal description of the formalism can consult previous publications [5]. The SAN formalism describes a complete system as a collection of subsystems that interact with each other. Each subsystem is described as a stochastic automaton, *i.e.*, an automaton in which the transitions are labelled with probabilistic and timing information. The analysis of the theoretical SAN models is performed by a software package called Performance Evaluation for Parallel Systems (PEPS) [6]. Although PEPS has proven its usability during the past years, it presents an important drawback: the execution time to analyze SAN models with too many states is very often impracticable. Thus, the main contribution of this paper is to

---

* This work was developed in collaboration with HP Brazil R&D.

verify the feasibility of a parallel implementation of the PEPS tool using a SAN model, identifying the requirements and advantages of such approach.

2 PEPS Implementation Analysis

The input of PEPS is a SAN model described in a predefined grammar. The current application loads this grammar and builds an equation system using the events rates. The SAN models allow a better management of the needs for space memory than Markov Chains, because they are described in a more compact and efficient way. This optimization can be carried out due to the use of tensorial algebra [5]. Thus, a model is no more described by a unique matrix, but instead, by a new structure called Markovian Descriptor. This structure is the kernel of the PEPS tool and it has a significant impact over its execution time. Considering a network with N automata and E synchronizing events, the Markovian Descriptor is given by:

$$Q = \bigoplus_{i=1}^{N} Q_l^{(i)} + \sum_{e=1}^{E} \left(\bigotimes_{i=1}^{N} Q_{e+}^{(i)} + \bigotimes_{i=1}^{N} Q_{e-}^{(i)} \right) \quad (1)$$

In this equation, there are N matrices $Q_l^{(i)}$ representing the local events and $2E$ matrices $Q_{e_k}^{(i)}$ representing the synchronizing events, which result in a total of $N + 2E$ stored matrices. The basic operation to solve a SAN model in numeric iterative methods (like the Power Method, GMRES, etc.) is to multiply a probability vector π by a Q matrix stored in the Markovian Descriptor form. This probability vector assigns a probability π_i ($i \in \{1, 2, ..., n\}$) to each one of the n states of the Markov Chain equivalent to the SAN model. Each element of Q may represent a local event or a synchronizing event. The local events are given by a sum of normal factors. On the other hand, synchronizing events are given by a product of normal factors [5]. Hence, the most important point in this descriptor-vector multiplication is to know how one can multiply normal factors. In spite of the Markovian Descriptor optimization, the PEPS application still suffers from a performance decline at the same time that the complexity[1] of the model grows (more details can be found in [6]).

3 PEPS Parallel Version

In order to improve the PEPS software tool performance, this paper proposes a parallel version for this tool based on the features of a specific kind of high performance architecture. To reinforce the usability of this new version, the hardware environment should be based on an usual and not very expensive (compared to a supercomputer) one. Following this scenario, the new version is designed to run over a cluster architecture which has several processors connected by a dedicated fast network.

As seen in section 2, PEPS solves a SAN model using numeric iterative methods. In order to represent a SAN model, many matrices are created, describing local and

---

[1] Complexity here is related to synchronizing events and states amount.

synchronized events rates. Another feature of PEPS is that the number of iterations necessary to make the model converge is different from one input model to another. Due to this, it is not possible to deduce how many iterations are necessary to determinate the convergence of a model. The solution proposed here is based on a synchronized master-slave model. Considering that a SAN model is described as a set of tensorial matrices (from the Markovian Descriptor, equation 1) and each iteration represents the multiplication of a probability vector by these matrices, the main idea is to distribute a set of matrices to each slave and execute the multiplications concurrently. In each iteration, the master node sends, in broadcast, the vector from the ith iteration to the slaves. Each slave takes some time processing its own task, and returns to the master a piece of the next iteration vector. For each iteration, the master must wait for the results from all slaves (reduce operation) to combine them into the next iteration vector. Finally, the master sends this new probability vector in broadcast to the slaves, starting a new iteration. This procedure will continue until the convergence criteria of the numerical method is matched.

4 SAN Model for Parallel PEPS

Taking on a parallel point of view, the main relevant states to be modelled using SAN are those who are focused on processes data exchange and computing time. That happens because the trade-off between these two features is the key to determine the success of the parallel implementation. The SAN model which describes the PEPS parallel implementation explained in section 3 is presented in Fig. 1. The model contains one automaton *Master* and P automata $Slave^{(i)}$ ($i = 1..P$).

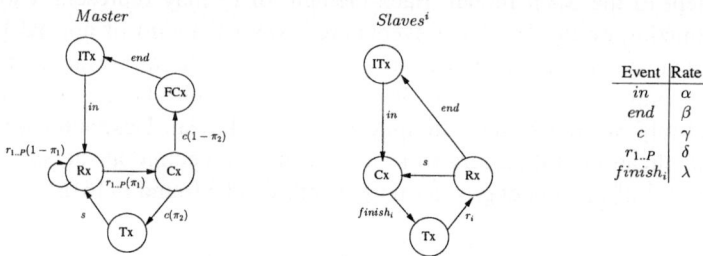

Fig. 1. The PEPS SAN model.

The automaton *Master* has five states *ITx*, *Tx*, *Rx*, *Cx* and *FCx*. It is responsible over the distribution of iteration vectors to the slaves. The states *ITx*, *Tx* and *Rx* mean, respectively, the initial transmission of the matrices and the first vector to the slaves, transmission of new iteration vectors to slaves, and the reception of the resulting vectors evaluated by the slaves. The states *Cx* and *FCx* represent, respectively, the time spent to sum all slaves evaluated vectors and write on file the asked results. The occurrence of the synchronizing event *in* broadcasts the matrices and the first vector to the slaves. On the other hand, the occurrence of the event *end* finalizes the communication between master and slaves. The synchronizing event s broadcasts the vector of the ith iteration to slaves. The synchronizing event r_i receives one resulting vector of the slave i. The occurrence of events $r_{1..P}$ can change the state of master automaton or not, depending on the probability π_1. The master will change to Cx state when the last slave sends its results and goes to Rx state, *i.e.*, the master must wait until all slaves send their results of the ith iteration. The *Master* automaton initializes new iterations or finalizes the descriptor-vector multiplication by the occurrence of the local event c. In 99% of the times (represented by the probability π_2) the event c will initialize new iterations. On the remaining time $(1 - \pi_2)$, the event c will finalize the descriptor-vector multiplications.

Automaton $Slave^{(i)}$ represents the slave of index i, where $i = 1..P$. It has four states: I (Idle), Cx (Computing), Tx (Transmitting) and Rx (Receiving). All slaves start their tasks when synchronizing event in occurs. Slave i stops processing a task by the occurrence of the local event $finish_i$. The same slave sends its resulting vector to the master with the synchronizing event r_i. When the event s occurs, all slaves start to compute a new step. The occurrence of the event end finalizes the execution of the slaves processes.

In order to complete the model representation it is necessary to assign occurrence rates to its events. The rate of one event is the inverse of the time spent in its associated state. The occurrence rate of the event end is extremely high because the time spent to finish a model is insignificant. The occurrence rates of events s, in and c are inversely dependent by the probability vector size and also by the number of slaves nodes. On the other hand the rate of the events $r_{1..p}$ are only inversely dependent by the probability vector size. The rate of the events $finish_{1..p}$ are inversely dependent by the number of multiplications each slave will perform. This number of multiplications is directly dependent by the probability vector size and inversely dependent by the number of slaves nodes. All these rates were obtained through sample programs performed over the target architecture[2], *e.g.*, the time spent by one slave to compute a probability vector was obtained through a program that simulates the number of multiplications performed by one slave).

5 Results

The results given by PEPS are steady state probabilities. For instance, the time that the system stays in state Tx of the automaton *Master* is equal to the probability of the system to be in this state. These probabilities are important information to indicate the amount of time the system will stay in a given state. For a parallel implementation this information will help to verify if the proposed parallel solution has bottlenecks which compromise its performance.

Three case studies will be presented next. Each one uses an hypothetical different input SAN model which differs in the number of states (directly related to the size of the probability vector). The behavior of the PEPS parallel version will be analyzed for each input model. Readers must pay attention that the sequential version of PEPS is used to analyze the SAN model that represents the proposed parallel version of PEPS, which needs some input SAN models to solve. Another important remark is that for

---

[2] The target architecture is a COW (Cluster of Workstations).

Table 3. Results for the case study 16,384.

slaves	Tx_{Master}	Rx_{Master}	Cx_{Master}	Cx_{Slave}
2	0.3690	0.1834	0.0025	0.0585
3	0.3790	0.1887	0.0011	0.0369
4	0.3810	0.1910	0.0005	0.0251
5	0.3811	0.1939	0.0003	0.0177
6	0.3808	0.1961	0.0002	0.0128
7	0.3805	0.1975	0.0001	0.0092

Table 4. Results for the case study 65,536.

slaves	Tx_{Master}	Rx_{Master}	Cx_{Master}	Cx_{Slave}
2	0.3687	0.2258	0.0007	0.0797
3	0.3788	0.2034	0.0002	0.0449
4	0.3808	0.1987	0.0001	0.0289
5	0.3809	0.1984	0.0001	0.0197
6	0.3806	0.1988	3.79e-5	0.0139
7	0.3804	0.1992	2.31e-5	0.0099

Table 5. Results for the case study 327,680.

slaves	Tx_{Master}	Rx_{Master}	Cx_{Master}	Cx_{Slave}
2	0.3628	0.2422	6.95e-5	0.0926
3	0.3764	0.2098	2.45e-5	0.0490
4	0.3796	0.2018	1.14e-5	0.0306
5	0.3802	0.2000	6.18e-6	0.0206
6	0.3802	0.1999	3.58e-6	0.0144
7	0.3802	0.1999	2.18e-6	0.0102

Table 6. Heterogeneous results - 16,384.

degree	Cx_{S1}	Cx_{S2}	Cx_{S3}	Cx_{S4}	Cx_{S5}	Cx_{S6}
none	0.0128	0.0128	0.0128	0.0128	0.0128	0.0128
low	0.0120	0.0144	0.0120	0.0144	0.0120	0.0114
medium	0.0161	0.0161	0.0051	0.0161	0.0161	0.0016
high	0.0218	0.0190	0.0087	0.0087	0.0044	0.0044

each new slave added, the SAN model of the PEPS parallel version changes because a new automaton Slave must be added to the model. Thus, the sequential PEPS execution time to solve the parallel PEPS model strongly increases and that is the reason why we have carried out experiments with only until seven slaves nodes for each case study.

Table 3, Tab. 4 and Tab. 5 show the state stay probabilities for each case study. The first important remark is that the probabilities of slave nodes be computing (Cx_{slave}) decline as the number of slaves grows, indicating that more work is done in parallel. Another interesting result can be observed in the probability of master to be receiving data from slaves(Rx_{master}): each case study presents a different behavior. In the 16,384 states case study, this probability increases as the number of slaves grows. That happens because slaves have progressively less work to compute, thus increasing the frequency they send their results to the master. Therefore, for this size of input model, more than two slaves seems to compromise the master efficiency. Looking at the same probability for the others input models (65,536 and 327,680 states) and applying the same analysis, it is possible to identify the best number of slaves (*i.e.*, the point the master becomes the bottleneck of the system) at five for the second case study and probably seven for the third case study. Finally, looking comparatively at the probability of slaves to be computing (Cx_{slave}) in all three case studies, one can see that, as the number of states of the input model grows, the time each slave spends computing gets higher. This is a coherent result, because if the system has a higher workload, it is expected that the nodes spend more time computing.

Until this point, all results presented were based on an homogeneous environment, *i.e.* all slaves have exactly the same computational power and workload (represented by the matrices computational cost). In order to verify the importance of the workload balance for the parallel PEPS implementation, heterogeneity in the slaves behavior was introduced in the SAN model for the parallel version of PEPS. That was done by using different computing rates for each slave. Table 6 presents the results of this experiment, where using the input model with 16,384 states and fixing the number of slaves on 6, four different configurations were analyzed: none, low, medium or high heterogeneity. The results show that as heterogeneity grows, the degree of discrepancy among the slaves to be in state Cx increases. Through this analysis, the model seems to be capable of identify the impact heterogeneity in the slaves behavior could represent over a parallel implementation.

6 Conclusion

The main contribution of this paper is to point out the advantages and problems of the analytical modelling approach applied to predict the performance of a parallel implementation. According to authors best knowledge, some others efforts to predict performance of parallel systems were not so formal. The recent work of Gemund [7] exploits a formal definition used for simulations to automatically generate a formal description of a parallel implementation. Gelenbe et al. [4] employs a more conservative formalism (queueing networks) in which the synchronization processes are not easy to describe. The SAN formalism, instead, was quite adequate to describe the behavior of a master-slave implementation. The definition of the events was very intuitive; their rates and probabilities were a little bit more hard to obtain but still intuitive. The stationary solution of the model provided enough prediction information about the behavior of the parallel version of the PEPS tool. The master and slaves nodes behaviors were clearly understood, and an analysis about the workload balance was also possible. The natural future work for this paper is to verify the accuracy of the proposed model by comparison with the real parallel implementation, in order to validate the correctness of the modelling choices made.

References

1. Hu, L., Gorton, I.: Performance Evaluation for Parallel Systems: A Survey. Technical Report 9707, University of NSW, Sydney, Australia (1997)
2. Nicol, D.: Utility Analisys of Parallel Simulation. In: Proceedings of the 17th Workshop on Parallel and Distributed Simulation (PADS'03), San Diego, California, USA, ACM (2003) 123–132
3. Stewart, W.J.: Introduction to the numerical solution of Markov chains. Princeton University Press (1994)
4. Gelenbe, E., Lent, R., Montuoria, A., Xu, Z.: Cognitive Packet Network: QoS and Performance. In: 10th IEEE International Symposium on Modeling, Analysis, and Simulation of Computer and Telecommunications Systems (MASCOTS'02), Fort Worth, Texas, USA (2002)
5. Fernandes, P., Plateau, B., Stewart, W.J.: Efficient descriptor - Vector multiplication in Stochastic Automata Networks. Journal of the ACM **45** (1998) 381–414
6. Benoit, A., Brenner, L., Fernandes, P., Plateau, B., Stewart, W.J.: The PEPS Software Tool. In: Proceedings of the 13th International Conference on Modelling Techniques and Tools for Computer Performance Evaluation, Urbana and Monticello, Illinois, USA (2003)
7. van Gemund, A.J.C.: Symbolic Performance Modeling of Parallel Systems. IEEE Transactions on Parallel and Distributed Systems **14** (2003) 154–165

Topic 3
Scheduling and Load Balancing

Emilio Luque, Jose G. Castaños, Evangelos Markatos, and Raffaele Perego

Topic Chairs

Scheduling and Load Balancing techniques are key issues for the performance of applications executed in parallel and distributed environments, and for the efficient utilization of these computational resources. Research in this field has a long history and is well consolidated. Nevertheless, the evolution of parallel and distributed systems toward clusters, computational grids, and global computing environments, introduces new challenging problems that require a new generation of scheduling and load balancing algorithms. Topic 3 in Euro-Par 2004 covers all aspects related to scheduling and load balancing from application and system levels, to theoretical foundations and practical tools. All these aspects are addressed by contributed papers.

A total of 29 papers were submitted to Topic 3, from where our reviewers selected 7 full-length papers to be included in the Conference program. We divided their presentation into two sections of three papers each, while the remaining paper will be presented in conjunction with Topic 6 (Grid and Cluster Computing).

Two papers present new theoretical results for specific scheduling problems. The paper *Scheduling Under Conditions of Uncertainty: a Bayesian Approach*, by L. P. Santos and A. Proenca, proposes the use of Bayesian decision networks to dynamically schedule a parallel application among the nodes of a distributed system. J. Andersson, M. Ericsson, W. Löwe, and W. Zimmermann in *Lookahead Scheduling for Reconfigurable GRID Systems* discuss the problem of scheduling simple, stateless data parallel applications in an evolving grid environment. The remaining papers propose and evaluate scheduling and load balancing algorithms that experimentally exhibit performance improvements. H. Casanova, F. Desprez, and F. Suter in *From Heterogeneous Task Scheduling to Heterogeneous Mixed Parallel Scheduling* present techniques for scheduling mixed data-parallel and task-parallel codes in heterogeneous clusters of homogeneous machines. In *Scheduling Tasks Sharing Files from Distributed Repositories*, A. Giersch, Y. Robert, and F. Vivien focus on the problem of scheduling a large collection of independent tasks which share files stored in distributed repositories. The goal of this work is to minimize the total execution time by finding an appropriate task allocation and communication schedule. The paper *A Scheduling Algorithm for Running Bag-of-Tasks Data Mining Applications on the Grid*, by F. A. Barbosa da Silva, S. Carvalho, and E. R. Hruschka, presents a new scheduling approach devoted to improve the performance and the scalability of Bag of Tasks Data Mining applications running on the grid. The solution is based on a extension of the MyGrid middle-ware. In *Scheduling of MPI Applications: Self Co-Scheduling*, G. Utrera, J. Corbalàn, and J. Labarta present a new technique aimed at scheduling MPI parallel jobs on SMP architectures. The proposed approach combines co-scheduling and space-sharing. Its performance is assessed either with artificial workloads composed of jobs having all the same number of processes and the same arrival time, and with realistic applications. The last paper *Experimental Study of Multi-Criteria Scheduling Heuristics for GridRPC Systems*, by Y. Caniou, and E. Jeannot presents a study of scheduling heuristics for GridRPC. The authors focus on NetSolve, an instantiation of GridRPC, and compare the standard scheduling heuristic of NetSolve (MCT) , with three proposed enhancements. This paper extensively measures the performance of these alternatives.

We would sincerely like to thank all of the contributing authors for their work, as well as to express our gratitude to the numerous experts in the field for their assistance in the reviewing process. Their hard work produced an interesing and stimulating trak.

Scheduling Under Conditions of Uncertainty: A Bayesian Approach[*]

Luis Paulo Santos and Alberto Proenca

Departamento de Informatica
Universidade do Minho
Campus de Gualtar
4710-057 Braga
Portugal
{psantos,aproenca}@di.uminho.pt

Abstract. The efficient execution of irregular parallel applications on shared distributed systems requires novel approaches to scheduling, since both the application requirements and the system resources exhibit an unpredictable behavior. This paper proposes Bayesian decision networks as the paradigm to handle the uncertainty a scheduler has about the environment's current and future states. Experiments performed with a parallel ray tracer show promising performance improvements over a deterministic approach of identical complexity. These improvements grow as the level of system sharing and the application's workload irregularity increase, suggesting that the effectiveness of decision network based schedulers grows with the complexity of the environment being managed.

1 Introduction

Shared parallel and cluster computing, coupled to the exploitation of unused cycle times on interconnected workstations, is pushing the computing paradigms towards new levels of expectations. To assure that parallel applications get a performance close to the theoretical available processing power, the workload must be effectively distributed over the available resources. The role of the scheduler is to ensure this adequate correspondence between the workload structure – both code and data – and the distributed system resources.

The scheduling problem complexity increases when applications exhibit unpredictable computing and communication requirements, and the available set of computing resources is dynamically shared among several users and applications: the scheduler must still correctly distribute the workload on the shared set of resources, but exact prediction of the environment behavior is impossible due to its dynamic nature [1].

Dynamic scheduling strategies that regularly measure the environment state seem appropriate, since they enable the scheduler to react to fluctuations on the environment behavior. However, the data gathered through the scheduler's

---

[*] This work was partially supported by grant PRAXIS 2/2.1/TIT/1557/95.

sensors, about the environment's current state, quantifies only those quantities that were included on the scheduler's simplified model of the world, may suffer from noise and will age with time. Furthermore, the future state of the environment and the outcome of the scheduler selected actions can not be accurately predicted, because other processes, independent of the scheduler, are acting on the shared computing resources. The scheduler is required to decide and act under conditions of uncertainty about past, present and future system states and workload profiles.

The problem of handling uncertainty on computational models of real world problems can be solved by using decision theory. Decision theory combines probabilities and utilities to evaluate alternative actions. One of the tools proposed by decision theory for rational decision making under conditions of uncertainty are decision networks [2–4]. Decision networks allow the inclusion, on the reasoning process, of the uncertainty about the environment current state and about the consequences of the scheduling agent's selected actions, i.e., the environment next state after each scheduling event. The hypothesis put forward by this paper is that a scheduler is more effective if it uses decision making mechanisms that deal explicitly with this uncertainty, and that Bayesian decision networks are such a mechanism.

The use of decision networks to dynamically schedule a parallel application among the nodes of a distributed shared system is the main contribution of this paper; the authors have no knowledge of any related work where decision networks have been applied to solve this problem. Previous works modelled stochastically either only the environment's current state (e.g., Bayesian probabilistic networks [5], stochastic structural execution models [6, 7], handling information aging [8, 9]) or only the suitability of each possible action, given the environment's current state, such as with stochastic learning automata [10, 11]. Bayesian decision networks allow the integration of the uncertainty on both these factors on a single paradigm, and provide an automated process for computing the expected utility of each action, enabling rational decision making by selecting the action which maximizes the expected utility.

2 Handling Uncertainty

A dynamic scheduler collects information about the system state and workload profile and creates an internal image of the environment current state. Using this information and its execution model of the world, it generates estimates about the environment next state for each possible action a_i. The scheduler's decision making mechanism must select the action that leads the environment to the most desirable next state [12]. Uncertainty, however, derives from four main sources and may hinder the scheduler from meeting its performance requirements:

1. it is too expensive, or even impossible, to get exact and accurate information about the current environment state, i.e., the environment is not totally measurable;

2. the image the scheduler has about the environment state gets obsolete with time – information aging;
3. the environment complexity requires that some simplifications be included in the execution model, either by neglecting or summarizing some of the environment characteristics; therefore, the model can not provide exact and accurate predictions of the environment near future behavior;
4. the workload profile and system behavior can be unpredictable, both due to the application characteristics and to the background workload imposed upon the distributed shared system by other users and/or applications.

2.1 Bayesian Decision Networks

Due to uncertainty the scheduler's knowledge is, at best, a degree of belief on the environment's most relevant aspects. Probability theory provides a tool to process and combine these beliefs [3, 13]. A probabilistic model consists of a set of stochastic variables that represent some important aspect of the world. The joint distribution assigns probabilities to all possible states of the world. Its size grows exponentially with the number of variables and these probabilities are seldom easy to assess.

High–dimensional probabilistic models can only be computationally tractable if modularity is introduced [3, 14]. This may be achieved by introducing the concept of conditional independence among variables and by representing these local interactions. Each stochastic variable X_i is directly influenced by at most k other variables, referred to as $Parents(X_i)$. The remaining variables carry no informational relevance to X_i once the relevant ones are known. The probabilistic model may be structured in terms of direct influences among the variables, quantified by conditional probability tables (CPT) $\mathbf{P}(X_i|Parents(X_i))$. By integrating the notion of conditional independence, the required number of independent numeric values is drastically reduced and, if the model is causally constructed, the expert task of assessing each variable CPT is simplified, since the resulting model represents the expert natural way of thinking about the problem.

A Bayesian belief network is a directed acyclic graph, whose nodes are the model variables and whose links represent local causal dependencies. The network topology can be thought of as an abstract knowledge base that holds independently of the numerical assignment of conditional probabilities [2–4, 15–18]. A fully specified Bayesian network can be used as a probabilistic inference engine, which computes the posterior probability distribution for a set of query variables given the probability distribution for some evidence variables.

Decision networks [2–4], combine belief networks with two additional nodes: a decision node, which represents the choices available to the agent and has its value imposed to represent actions, and a utility node, which represents the utility function to be optimized. Evaluating an action amounts to impose the value of the decision node, which, on a non–deterministic environment, may have several different outcomes. This setting alters the probability distribution of a set of stochastic variables in the network, resulting on the probability distribution over all possible next states of the environment for that action. To be able to select among different actions, an agent assigns utilities to all the different possible states. Utility theory states that the agent will prefer states with higher utility. The expected utility of each possible action can be computed by weighing the utility of each of the various possible outcomes of that action with the probabilities that these outcomes may occur. An agent exhibits rational behavior if it selects the action that yields the highest expected utility.

Since the numerical parameters required to quantify all the CPTs can be substantial and hard for an expert to assess, the initial assessments of probabilities for some stochastic variables U_i can be updated whenever new data is available. At each inference step n the agent observes the environment state as perceived by its sensors. The observed variables are represented by the evidence vector **E**. The inference algorithm updates the belief given **E** on all stochastic variables $\mathbf{P}(U_i|\mathbf{E})$. These values can be used to update the probabilities $\mathbf{P}(U_i)$ using a process known as sequential updating [13, 19, 20]:

$$\mathbf{P}_{n+1}(U_i) = \mathbf{P}_n(U_i) + \alpha * [\mathbf{P}_n(U_i|\mathbf{E}) - \mathbf{P}_n(U_i)] \ . \tag{1}$$

This is a crucial capability, allowing the development of systems that can overcome errors in their initial model and adapt to changes in the dynamics of the environment being modelled.

3 Applying Decision Networks to a Dynamic Scheduler

On a decision network two different subsets may be identified: one related to modelling the current state, the other, referred to as the transition model, related to the outcome of actions and the environment's next state.

The Environment's Current State. The scheduling agent acquires information about the environment through its sensors and infers its belief distribution on the state of the world. Imperfect information about the stochastic variable X, representing some metric of the world, can only be obtained through perfect information about the sensor readings E and the sensor model, $P(E|X)$ – which is a causal relationship: X determines the sensor readings, not the other way around. The inference process uses Bayes' theorem to compute the probability distribution over X, $\mathbf{P}(X|E)$. The possibility of noisy or incorrect sensor readings are accounted for on the sensor model. Information age, i.e.the time elapsed since the sensor was actually read, T, can be included on the model and influences the belief distribution over the sensor's readings, $\mathbf{P}(E|X,T)$. The prior probabilities associated with the metrics, $\mathbf{P}(X)$, are ideal candidates for sequential updating, since these are the quantities that actually describe the environment state. Higher level variables can be used to model abstract concepts, such as the distributed system degree of load balancing or the communication network availability (Fig. 1).

The Environment's Next State. The outcome of a scheduler's action depends not only on the current state and on the selected action, but also on the dynamics of the environment itself. These uncertainties are modelled by the CPTs related to the variables that represent the next state N_s, given the current state C_s and the selected action a, $\mathbf{P}(N_s|C_s,a)$. This set of CPTs is referred to as the state transition model.

Generic Structure. Figure 1 presents a generic structure for a scheduling decision network, whose topology complies with causal relationships: the sensors readings are an effect of the relevant quantities actual values (information age nodes have not been displayed for simplicity reasons), the next state is a function of both the current state and the selected action (factors external to the scheduler are modelled in the state transition model). Four different blocks can be identified:

- the resources capacities, the tasks requirements and the scheduling overheads model the agent's belief on the environment current state;
- the decision variable lists all the actions available to the scheduling agent;
- the next state variables model the belief on the outcome of each action;
- the utility variable computes the expected utility for each action.

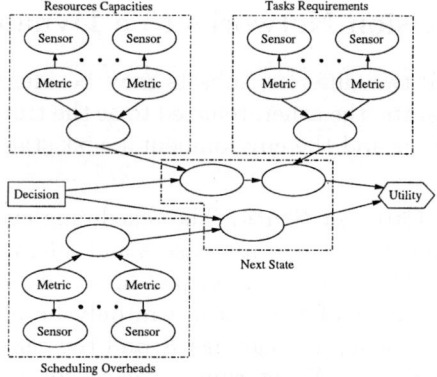

Fig. 1. A generic structure for a scheduling decision network.

The variety of available sensors determines the type of information that can be included in the execution model. These can include the node computing throughput, the network latency and bandwidth, the work completed at the sampling time, the communication volume, the available memory, etc. The actions available to the scheduler, represented by the decision variable, depend on the workload and on the distributed system being managed; these may include: assign task T_1 to node n_2, migrate task T_2 from node n_3 to node n_5, transfer 35% of node n_1 workload to node n_4, etc. The next state is composed by a set of variables that describe the most relevant aspects of the environment, after each action is executed. These should be quantities that are affected by the actions being selected, since the agent uses them to quantify the desirability of each action. These can include the tasks estimated new completion times, the resulting degree of load balancing, the actions expected overheads, etc. Finally, the utility variable represents the performance goals degree of achievement for each action.

4 Results and Concluding Remarks

Experimental Setup. To validate this approach experiments were conducted using a parallel ray tracer with image space decomposition. Four different scenes were used: balls3, balls3c, teapot9 and balls4pv[1]; these are listed by increasing order of computational and communication requirements. To simulate different levels of system sharing, four predefined synthetically generated stochastic background workload patterns were applied: dedicated (no sharing), light, medium and heavy background workloads. The evaluation results were obtained on a cluster of seven workstations, interconnected by a Myrinet network. The results obtained with the decision network based scheduler (DN) were compared with other decision making mechanisms: a static uniform work distribution, a demand driven work allocation (dd), and a deterministic sensor based strategy (det) with the same capabilities as the stochastic strategy, to enable direct comparisons between deterministic and stochastic approaches of identical complexity. The performance metric used to compare the various schedulers was the time required to render the scenes.

Results Analysis. Figure 2 shows the performance improvements for each dynamic scheduling strategy compared to the uniform work allocation for the two most complex scenes and different levels of system sharing. The performance improvement is computed as $(T_{unif} - T_{sched})/T_{unif}$, where $sched$ refers to each of the scheduling strategies and $unif$ refers to the uniform work distribution.

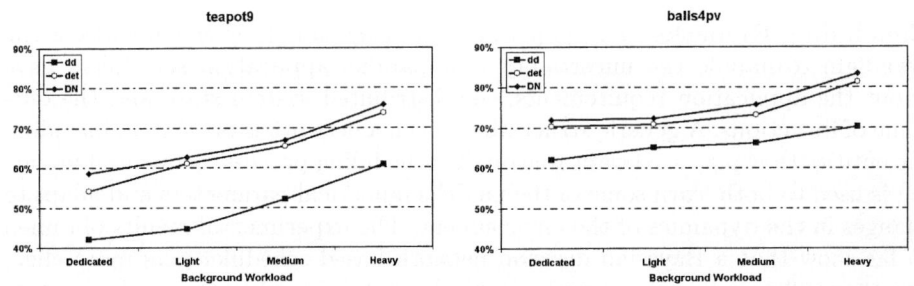

Fig. 2. Performance improvement.

---

[1] Taken from Eric Haines rendering benchmarks (ftp.princeton.edu).

Some remarks can be made from this figure:

- the improvements obtained with any of the dynamic scheduling strategies increase with the complexity of the data set and with the weigh of the background workload: the dynamic schedulers can redistribute the workload in runtime, counterbalancing changes on the nodes background workload;
- the sensor based dynamic strategies get more effective as the background workload increases when compared to the demand driven approach: the better quality of the generated schedules clearly overcomes the overheads associated with dynamically collecting data;
- the improvements obtained with DN may not seem significantly larger than those achieved with det; however, the difference between them tends to increase with the background workload weigh, the complexity of the data set and the number of nodes (Fig. 3). Additional results [12] also show that these improvements are achieved by reducing both idle times and task migrations.

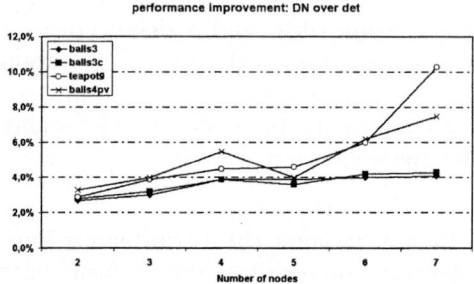

Fig. 3. Performance improvements relative to deterministic (dedicated mode).

Concluding Remarks. This paper proposes Bayesian decision networks as the paradigm to handle the uncertainty that parallel application schedulers have about the application requirements, the distributed system state and the outcome of its actions. A generic structure for such a network is presented, including the entities that play a relevant role on the scheduling process. Sequential updating is used to both learn some of the model's numerical parameters and adapt to changes in the dynamics of the environment. The experimental results obtained so far show that a Bayesian decision network based scheduler gets more effective than a deterministic one with identical capabilities when the level of system sharing and the application's workload irregularity increase.

These results corroborate our initial hypothesis: the explicit representation of uncertainty on the scheduler execution model and decision making mechanism, using decision networks, increases its effectiveness. Further research is required to confirm these results with different types of applications and environments.

Additional enhancements may also be considered, namely the assessment of the state transition model. Numerical imprecisions can be corrected by updating the conditional probabilities tables whenever new data is available: either Bayesian learning techniques or reinforcement learning algorithms can be applied to improve the model accuracy.

References

1. Dinda, P., O'Hallaron, D.: Host Load Prediction Using Linear Models. Cluster Computing **3** (2000)
2. Horvitz, E., Breese, J., Henrion, M.: Decision Theory in Expert Systems and Artificial Intelligence. Technical report, Palo Alto Laboratory (1988)
3. Pearl, J.: Probabilistic Reasoning in Intelligent Systems: Networks of Plausible Inference. Morgan Kaufmann Publishers (1988) ISBN 1-55860-479-0.
4. Russell, S., Norvig, P.: Artificial Intelligence: A Modern Approach. Prentice-Hall (1995) ISBN 0-13-103805-2.
5. Stankovic, J.: An Application of Bayesian Decision Theory to Decentralized Control of Job Scheduling. IEEE Transactions on Computers **C-34** (1985) 117–129
6. Schopf, J.: Structural Prediction Models for High–Perfomance Distributed Applications. In: Cluster Computing Conference. (1997)
7. Schopf, J., Berman, F.: Stochastic Scheduling. In: SuperComputing'99, Portland, OR, USA (1999)
8. Mitzenmacher, M.: How Useful is Old Information. IEEE Transactions on Parallel and Distributed Systems **11** (2000) 6–20
9. Dahlin, M.: Interpreting Stale Load Information. IEEE Transactions on Parallel and Distributed Systems **11** (2000) 1033–1047
10. Schaerf, A., Shoham, Y., Tennenholtz, M.: Adaptive Load Balancing: A Study in Multi-Agent Learning. Journal of Artificial Intelligence Research **2** (1995) 475–500
11. Zomaya, A., Clements, M., Olariu, S.: A Framework for Reinforcement–Based Scheduling in Parallel Processor Systems. IEEE Transactions on Parallel and Distributed Systems **9** (1998) 249–259
12. Santos, L., Proenca, A.: A Systematic Approach to Effective Scheduling in Distributed Systems. In: VECPAR'2002 – 5th Int. Meeeting on High Performance Computing for Computational Science, Porto, Portugal (2002) 813–825
13. Heckerman, D.: A Tutorial on Learning with Bayesian Networks. Technical report, Microsoft Research — Advanced Technology Division (1995) MSR–TR–95–06.
14. Cowell, R., Dawid, A., Lauritzen, S., Spiegelhalter, D.: Probabilistic Networks and Expert Systems. Springer-Verlag (1999) ISBN 0-387-98767-3.
15. Jensen, F.: An Introduction to Bayesian Networks. Springer-Verlag (1996) ISBN 0-387-91502-8.
16. Jensen, F.: Bayesian Graphical Models. Encyclopedia of Environmetrics (2000)
17. Jensen, F., Lauritzen, S.: Probabilistic Networks. Handbook of Defeasible and Uncertainty Management Systems: Algorithms for Uncertainty and Defeasible Reasoning **5** (2000) 289–320
18. Jordan, M.: Learning in Graphical Models. MIT Press (1998) ISBN 0-262-60032-3.
19. Friedman, N., Goldszmidt, M.: Sequential Update of Bayesian Network Structure. In: 13th Conference on Uncertainty in Artificial Intelligence, Morgan–Kaufmann (1997) 165–174
20. Spiegelhalter, D., Lauritzen, S.: Sequential Updating of Conditional Probabilities on Directed Graphical Structures. Networks **20** (1990) 579–605

From Heterogeneous Task Scheduling to Heterogeneous Mixed Parallel Scheduling*

Frédéric Suter[1], Frédéric Desprez[2], and Henri Casanova[1,3]

[1] Dept. of CSE, Univ. of California, San Diego, USA
[2] LIP ENS Lyon, UMR CNRS - ENS Lyon - UCB Lyon - INRIA 5668, France
[3] San Diego Supercomputer Center, Univ. of California, San Diego, USA

Abstract. Mixed-parallelism, the combination of data- and task-parallelism, is a powerful way of increasing the scalability of entire classes of parallel applications on platforms comprising multiple compute clusters. While multi-cluster platforms are predominantly heterogeneous, previous work on mixed-parallel application scheduling targets only homogeneous platforms. In this paper we develop a method for extending existing scheduling algorithms for task-parallel applications on heterogeneous platforms to the mixed-parallel case.

1 Introduction

Two kinds of parallelism can be exploited in most scientific applications: data- and task-parallelism. One way to maximize the degree of parallelism of a given application is to combine both kinds of parallelism. This approach is called *mixed data and task parallelism* or *mixed parallelism*. In mixed-parallel applications, several data-parallel computations can be executed concurrently in a task-parallel way. This increases scalability as more parallelism can be exploited when the maximal amount of either data- or task-parallelism has been achieved.

This capability is a key advantage for today's parallel computing platforms. Indeed, to face the increasing computation and memory demands of parallel scientific applications, a recent approach has been to aggregate multiple compute clusters either within or across institutions [4]. Typically, clusters of various sizes are used, and different clusters contain nodes with different capabilities depending on the technology available at the time each cluster was assembled. Therefore, the computing environment is at the same time attractive because of the large computing power, and challenging because it is heterogeneous.

A number of authors have explored mixed-parallel application *scheduling* in the context of homogeneous platforms [8–10]. However, heterogeneous platforms have become prevalent and are extremely attractive for deploying applications at unprecedented scales. In this paper we build on existing scheduling algorithms for heterogeneous platforms [6, 7, 11, 14] (*i.e.,* specifically designed for task-parallelism) to develop scheduling algorithms for mixed-parallelism on heterogeneous platforms.

---

* An extended version of this paper is given by [13].

This paper is organized as follows. Section 2 discusses related work. Section 3 shows how scheduling algorithms for task-parallel applications on heterogeneous platforms can be adapted to support mixed-parallelism, which is illustrated with a case study in Section 4. Section 5 presents our evaluation methodology and Section 6 presents our evaluation results. Section 7 concludes the paper with a summary of our contributions and a discussion of future work.

2 Background

Most existing mixed-parallel scheduling algorithms [8–10] proceed in two steps. The first step aims at finding an optimal *allocation* for each task, that is the number of processors on which the execution time of a task is minimal. The second step determines a *schedule* for the allocated tasks, that is the ordering of tasks that minimizes the total completion time of the application.

In [1], we proposed an algorithm that proceeds in only one step. The allocation process is substituted by the association of a list of configurations to tasks. This concept of configuration has been developed to be used in this work and will be detailed in Section 3. Furthermore, the algorithm we present in this paper, unlike that in [1], explicitly accounts for platform heterogeneity, and is thus applicable to real-world multi-cluster platforms.

The problem of scheduling a task graph onto a heterogeneous platform is as follows. Consider a Directed Acyclic Graph (DAG) that models a parallel application. Each node (or task) of the DAG has a computation cost which leads to different computation times on different processors. An edge in the DAG corresponds to a task dependency (communication or precedence constraint.) To each edge connecting two nodes is associated the amount of data in bytes to communicate. Each such transfer incurs a communication cost that depends on network capabilities. It is assumed that if two tasks are assigned to the same processor there is no communication cost and that several communications may be performed at the same time, possibly leading to contention on the network.

The target architecture is generally a set of heterogeneous processors connected via a fully connected network topology, *i.e.*, as soon as a task has completed, produced data are sent to all its successors simultaneously. We assume that computation can be overlapped with communication and the execution time to be known for each task-processor pair.

The objective is to assign tasks to processors so that the schedule length is minimized, accounting for all interprocessor communication overheads. This problem is NP-complete in the strong sense even when an infinite number of processors are available [2]. A broad class of heuristics for solving the scheduling problem is *list-scheduling*. In the context of heterogeneous platforms popular heuristics list-scheduling heuristics include *PCT* [6], *BIL* [7], *HEFT* [14], and *DLS* [11]. All these heuristics are based on the same two components: a *priority* function, which is used to order all nodes in the task graph at compile time; and an objective function, F_{obj}, which must be minimized.

3 Towards Heterogeneous Mixed Parallel Scheduling

While the list-scheduling algorithms described in Section 2 operate on a fully heterogeneous platform (compute resources and network links), in our first attempt at using these heuristics for mixed-parallelism we impose three restrictions on the platform and its usage: (i) the platform consists of a heterogeneous collection of homogeneous clusters; (ii) the network interconnecting these clusters is fully connected and homogeneous; (iii) data-parallel tasks are always mapped to resources within a single cluster. These restrictions are justified as follows. Restriction (i) clearly makes the problem more tractable but is in fact highly representative of currently available Grid platforms. Indeed, these platforms typically consist of clusters located at different institutions, and institutions typically build homogeneous clusters. Restriction (ii) is more questionable as end-to-end network paths on the wide-area are known to be highly heterogeneous. However, one key issue for scheduling mixed-parallel application is that of *data redistribution*: the process of transferring and remapping application data from one subset of the compute resources to another subset. Data redistribution on heterogeneous networks is a completely open problem, which is known to be NP-complete in the homogeneous case [3]. Assuming a homogeneous network among clusters allows us to more easily model redistribution costs. Finally, restriction (iii) is just a convenient way to ensure that we can reuse the parallel application models (*e.g.*, speed-up models) traditionally employed in the mixed-parallelism literature. This restriction can be removed in cases in which models for parallel application on heterogeneous platforms are available. In summary, our computing platform consists of several homogeneous clusters of different speeds and sizes interconnected via a homogeneous network.

To utilize list-scheduling algorithms for the purpose of scheduling mixed-parallel application, we have adapted the concept of *configuration* of [1]. A configuration is now defined as a subset of the set of the processors available within, and only within, a cluster. Moreover, we keep only information about the size and shape of the virtual grid represented by the configuration. While in task scheduling the smallest computational element is a processor, in our work the smallest element is a configuration. We limit configurations to contain numbers of processors that are powers of 2, to be rectangular, and to span contiguous processors, which is usual for the vast majority data-parallel applications. These limitations are for simplicity and to reduce the total number of configurations to consider, and can be removed easily if needed. For a given configuration size, we only consider a non-overlapping tiling of the cluster. This ensures that all configurations of the same size can be utilized concurrently. Removing this restriction would mandate sophisticated application performance models that are currently rarely available. For the purpose of scheduling, the target architecture is then abstracted as collections of configurations subject to the restrictions listed above.

Mixed-parallel programs can also be modeled by a DAG, in which nodes represent data-parallel tasks. Whereas in the purely task-parallel case each task was assigned a compute cost, here we assign a *cost vector* to each parallel task, that represents the compute cost of the task when mapped to each configuration.

4 Case Study: HEFT

To illustrate our approach we chose to extend the task-parallel scheduling heuristic proposed in [14], Heterogeneous Earliest Finish Time (HEFT), to the mixed-parallel case. It is important to note that our approach is general and applicable to other task-parallel scheduling heuristics. We chose HEFT because it is simple, popular, and was shown to be competitive.

The priority function used by HEFT is based on "upward ranking". Basically, it is the length of the critical path from a task to the exit task, including the computation cost of this task. The upward rank of a task is the sum of the average execution cost of this task over all available processors and a maximum computed over all its successors. The terms of this maximum are the average communication cost of an edge and the upward rank of the successor.

The Earliest Start Time (EST) is the moment when the execution of a task can actually begin on a processor. An execution can start either when a processor becomes available or when all needed data has arrived on the processor. Adding the execution cost, we obtain the Earliest Finish Time (EFT) of a task.

HEFT uses the EFT as the objective function for selecting the best processor for a node. The rationale is that the schedule length is the EFT of the exit node.

To derive a mixed-parallel version of the HEFT algorithm, called M-HEFT (Mixed-parallel HEFT), the priority and objective functions of HEFT have to be adapted the new compute and communication units. We consider two approaches to define the average mixed-parallel execution cost of a task and average mixed-parallel redistribution cost of an edge.

We first follow exactly the same approach as HEFT, *i.e.*, compute the average mixed-parallel execution cost as the sum of the compute times for the task over all 1-processor configurations, divided by the number of such configurations (*i.e.*, the number of processors in the platform). Similarly we assume serial communications and compute the average mixed-parallel communication cost as the sum of the average network latency and of the data size divided by the average network bandwidth. In this paper we only consider homogeneous networks, so the average latency and bandwidth are equal to the latency and bandwidth of any network link. We denote this heuristic by M-HEFT1.

The second approach is to adapt HEFT directly to the mixed-parallel case. One can compute average mixed-parallel execution cost of task as the sum of each element of the cost vector of the task divided by the number of configurations and the average redistribution cost of an edge as the sum of the redistribution cost for each couple of configurations divided by the number of such couples.

But in this case several identical configurations are accounted for in each cluster, which may result in a biasing of the average. For instance, in an 8-processor cluster, the execution cost formula will incorporate compute costs for 8 identical 1-processor configurations. To avoid the biasing of the average, we enforce that, for a given configuration size, only one configuration be taken into account for each cluster. For the same reason, we enforce that for a given source configuration size and a given destination configuration size, only one redistribution be taken into account. We denote this heuristic by M-HEFT2.

5 Evaluation Methodology

We use simulation for evaluating our approach as it allows us to perform a statistically significant number of experiments and makes is possible to explore a wide range of platform configurations. We use the SIMGRID toolkit [5, 12] as the foundation of our simulator.

We consider platforms that consist of 1 to 8 clusters. Each cluster contains a random number of processors between 4 and 64. Cluster processor speeds (in GFlop/sec) are sampled from a uniform probability distribution, for various means ([1–1000]) and ranges ([0–1.8]), with the range being proportional to the mean. For all experiments we assume that the network has a 5 ms latency and a 10 GBit/sec bandwidth. We keep the network characteristics fixed and vary the processor speeds to experiment with a range of platform communication/computation ratios. These parameters generate 280 different platform configurations. We generate several samples for each of them.

We consider two classes of application DAGs, depicted in Figure 1. The left part shows the DAG corresponding to the first level of decomposition of the Strassen algorithm. We instantiate 6 DAGs with different matrix sizes.

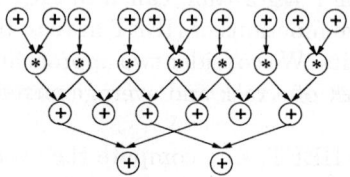 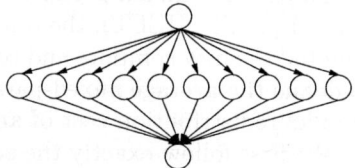

Fig. 1. Strassen (left) and Fork-Join (right) DAGs.

We also experiment with simple fork-join DAGs so that we can evaluate our algorithms for graphs with high degree of task-parallelism. In particular, the task-parallel HEFT heuristic should be effective for such graphs. We instantiate fork-join DAGs with 10, 50, and 100 tasks. Each task is either a matrix addition or matrix multiplication. We also assume that the entry task send two matrices to each inner task, and that each inner task send the result of the matrix operation, *i.e.*, one matrix, to the end task. We generate DAGs with 25%, 50%, and 75% of multiplication tasks. We have thus 9 possible fork-join graph configurations.

We compare the effectiveness of HEFT, M-HEFT and a third algorithm, HEFT*, which is a simple data-parallel extension of HEFT. For each cluster in the platform, HEFT* first determines the number of processors in the largest feasible configuration, and then computes p^*, the smallest such number over all clusters. The HEFT* algorithm then executes each task on configurations with p^* processors. The rationale behind HEFT* is that, unlike HEFT, it can take advantage of data-parallelism, and is thus a better comparator for evaluating M-HEFT. However, unlike M-HEFT, HEFT* cannot adapt the configuration size to achieve better utilization of compute and network resources.

6 Simulation Results

For the Strassen application we only present results for the first variant of M-HEFT, M-HEFT1. The Strassen DAG has a layered structure and the upward rank of two tasks at the same layer differ at most by the cost of an addition, which is generally insignificant. Therefore, the different methods used in M-HEFT1 and M-HEFT2 lead to virtually identical task orderings.

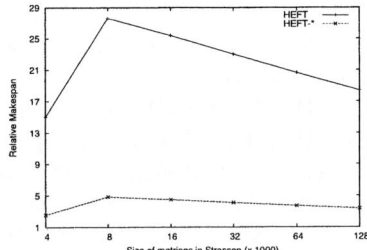 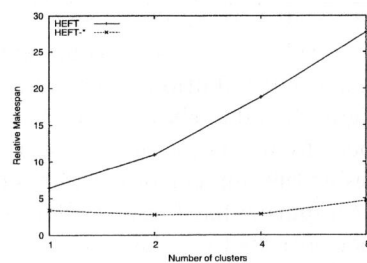

Fig. 2. Relative makespans for HEFT and HEFT* versus the matrix size used in the Strassen DAGs (left) and the number of clusters in the platform (right).

Over all our simulation results, the average makespans of HEFT and HEFT*, relative to that of M-HEFT, are 21.67 and 3.85 respectively. M-HEFT outperforms HEFT by more than one order of magnitude, which was expected as HEFT cannot take advantage of any data-parallelism and thus is highly limited in the number of resources it can use. More interestingly, the fact that M-HEFT clearly outperforms HEFT* demonstrates that our scheme for determining appropriate configuration sizes for data-parallel tasks is effective.

Figure 2(left) shows relative makespans for HEFT and HEFT* versus the matrix size used in the Strassen DAGs. The original decrease can be explained as follows. M-HEFT maps data-parallel matrix additions to very large configurations as they can be performed without communications. On the other hand, data-parallel matrix multiplications involve inter-processor communications and are thus mapped to smaller configurations. This results in many data redistributions that incur potentially large network latencies. For small matrices, these latencies are thus less amortized over the computation. For the same reason, as matrices get large, redistribution costs increase due to bandwidth consumption. We observe similar trends for HEFT*, but its performance is more on par with that of M-HEFT. Nevertheless, M-HEFT is still a factor 3.85 better on average. This is because HEFT*, unlike M-HEFT, is limited in the size of processor configuration that could be used for expensive tasks (*e.g.*, matrix multiplications).

In summary, while M-HEFT clearly outperforms its competitors, its performance is negatively impacted by redistribution costs, which indicates that these costs are not accounted for adequately. This points to a subtle limit of our approach. The mixed parallel *EFT* function includes only a computation cost, and no redistribution cost. Therefore, the mapping of the entry tasks of the application graphs are chosen without consideration of initial data-distribution costs. This in turn leads to these tasks being distributed on potentially very large configurations (typically for matrix additions in the case of the Strassen application), and thus to expensive data redistributions for subsequents tasks in the application task graph. This explains why M-HEFT leads to overly expensive data redistributions in some cases. It is important to note that our adaptation of HEFT into M-HEFT is somewhat naïve as our goal in this paper is to provide a generic method for adapting task-parallel list-scheduling heuristics to the mixed-parallel. It would be possible to further improve M-HEFT (and possibly other mixed-parallel list-scheduling heuristics obtained via our methodology) by including redistribution costs explicitly into the objective function.

Figure 2(right) shows relative makespans for HEFT and HEFT* versus the number of clusters in the platform. As expected HEFT's relative makespan increases as the number of clusters increases because, while M-HEFT can exploit large computing platforms, HEFT is limited to using 10 processors (the maximal number of tasks that can be executed concurrently). We see that HEFT*'s relative makespan also increases after 4 clusters. Recall that the p^* value is computed as a minimum over all clusters. Therefore, even when many clusters are available and some of these clusters are large, HEFT* is limited to using configurations containing at most the number of processors of the smallest cluster.

Unlike for the Strassen application, we compared the two variants M-HEFT1 and M-HEFT2 for the fork-join DAG. We found that overall the experiments M-HEFT1 led to better schedules in 33% of the cases, that M-HEFT2 led to better schedules in 26% of the cases, and that in the remaining 41% the two led to the same performance. Furthermore, the average performance of M-HEFT1 relative to M-HEFT2 is 1.15, with a maximum of 23.69. These results show that the two variants exhibit roughly similar efficacy. Since M-HEFT2 has the highest complexity we only present results for M-HEFT1 hereafter.

The average performance of HEFT relative to M-HEFT over all experiments is 4.70 and that of HEFT* is 12.11. This is in sharp contrast with the results we obtained for the Strassen application. As mentioned before some of our fork-joins graphs exhibit high levels of task-parallelism, which can be exploited by HEFT. On the other hand, HEFT* uses data-parallelism for all tasks with the same configuration size for all tasks. This both leads to small tasks being executed in parallel with high overhead, and to a limited ability to exploit high levels of task-parallelism. M-HEFT adapts configuration sizes to avoid this pitfall.

7 Conclusion and Future Work

In this paper we have proposed a generic methodology for the conversion of any heterogeneous list-scheduling algorithm for task-parallel application into an algorithm for the mixed-parallel case. We have presented a case study for the popular HEFT scheduling algorithm, which we have extended to obtain the M-HEFT (Mixed-parallel HEFT) algorithm. Our simulation results showed that

M-HEFT achieves good performance over competitors in the vast majority of the scenarios.

The first future work of this paper is to validate the approach with more significant benchmarks. The first step will be to define a large set of generated DAGs. Parameters such as shape, density, regularity, communication to computation ratio will have to be part of the DAG generation. The second step is to study the data redistribution with heterogeneous configurations (both heterogeneous processors and network). Our approach would be to consider initially only special but relevant heterogeneous configuration (*e.g.,* two different sets of homogeneous processors connected over single network link), rather than the fully heterogeneous case. Third, as explained in Section 6, we will improve our generic methodology by incorporating redistribution costs in the objective function to reduce redistribution overheads in the application schedule.

References

1. V. Boudet, F. Desprez, and F. Suter. One-Step Algorithm for Mixed Data and Task Parallel Scheduling Without Data Replication. In *Proc. of the 17th International Parallel and Distributed Processing Symposium (IPDPS'03)*, Apr 2003.
2. P. Chretienne. Task Scheduling Over Distributed Memory Machines. In *Parallel and Distributed Algorithms*, pages 165–176. North Holland, 1988.
3. F. Desprez, J. Dongarra, A. Petitet, C. Randriamaro, and Y. Robert. Scheduling Block-Cyclic Array Redistribution. *IEEE TPDS*, 9(2):192–205, 1998.
4. I. Foster and C. Kesselman, editors. *The Grid: Blueprint for a New Computing Infrastructure*. Morgan-Kaufmann, 1998. ISBN 1-55860-475-8.
5. A. Legrand, L. Marchal, and H. Casanova. Scheduling Distributed Applications: The SimGrid Simulation Framework. In *Proc. of the 3rd IEEE Symposium on Cluster Computing and the Grid (CCGrid'03)*, pages 138–145, Tokyo, May 2003.
6. M. Maheswaran and H. J. Siegel. A Dynamic Matching and Scheduling Algorithm for Heterogeneous Computing Systems. In *Proc. of the 7th Heterogeneous Computing Workshop (HCW'98)*, pages 57–69, 1998.
7. H. Oh and S. Ha. A Static Scheduling Heuristic for Heterogeneous Processors. In *Proceedings of Europar'96*, volume 1124 of *LNCS*, pages 573–577, 1996.
8. A. Radulescu, C. Nicolescu, A. van Gemund, and P. Jonker. Mixed Task and Data Parallel Scheduling for Distributed Systems. In *Proc. of the 15th International Parallel and Distributed Processing Symposium (IPDPS)*, San Francisco, Apr 2001.
9. S. Ramaswany. *Simultaneous Exploitation of Task and Data Parallelism in Regular Scientific Applications*. PhD thesis, Univ. of Illinois at Urbana-Champaign, 1996.
10. T. Rauber and G. Rünger. Compiler Support for Task Scheduling in Hierarchical Execution Models. *Journal of Systems Architecture*, 45:483–503, 1998.
11. G. Sih and E. Lee. A Compile-Time Scheduling Heuristic for Interconnection-Constrained Heterogeneous Processor Architectures. *IEEE TPDS*, 4(2):175–187.
12. SimGrid. http://gcl.ucsd.edu/simgrid/.
13. F. Suter, H. Casanova, F. Desprez, and V. Boudet. From Heterogeneous Task Scheduling to Heterogeneous Mixed Data and Task Parallel Scheduling. Technical Report RR2003-52, Laboratoire de l'Informatique du Parallélisme (LIP), Nov 2003.
14. H. Topcuoglu, S. Hariri, and M.-Y. Wu. Performance-Effective and Low-Complexity Task Scheduling for Heterogeneous Computing. *IEEE TPDS*, 13(3):260–274, 2002.

Scheduling of MPI Applications: Self-co-scheduling

Gladys Utrera, Julita Corbalán, and Jesús Labarta

Departament d'Arquitectura de Computadors (DAC)
Universitat Politècnica de Catalunya (UPC)
{gutrera,juli,jesus}@ac.upc.es

Abstract. Scheduling parallel jobs has been an active investigation area. The scheduler has to deal with heterogeneous workloads and try to obtain throughputs and response times such that ensures good performance.
We propose a Dynamic Space-Sharing Scheduling technique, the Self Co-Scheduling, based on the combination of the best benefits from Static Space Sharing and Co-Scheduling. A job is allocated a processors partition where its number of processes can be greater than the number of processors. As MPI jobs aren't malleable, we make the job contend with itself for the use of processors applying Co-Scheduling.
We demonstrate that our Self Co-Scheduling technique has better performance and stability than other Time Sharing Scheduling techniques, especially when working with high communication degree workloads, heavy loaded machines and high multiprogramming level.

1 Introduction

An operating system must give support to different kind of parallel applications. The scheduler has to take into account the particular characteristics of each architecture and jobs to exploit the maximum performance of the overall system.

In our work, we will focus on shared memory multiprocessors (SMMs) and on Message Passing Interface [6] (MPI) jobs. This library is worldwide used, even on SMMs, due to its performance portability on other platforms, compared to other programming models such as threads and OpenMP. This is especially important for future infrastructures such as information power grids, where resource availability dynamically changes for submitted jobs. Our objective will be to obtain the best performance in service time and throughput.

In this study, we try to demonstrate that it is possible to combine Co-Scheduling (CS) policies with Space Sharing policies to build a Dynamic Space-Sharing scheduling policy, which we call the Self Co-Scheduling policy (SCS), in a dynamic environment where the number of processors allocated to a job may be modified during job execution without loss of performance. We implement, evaluate and analyze several schemes that exploit the low-level process management with the goal of minimising the loss of performance generated when the number of total processes in the system is greater than the number of processors, as MPI jobs aren't malleable[1].

---

[1] Malleable jobs are the ones that can modify their number of processes at execution time.

After that we select the two policies that work best and re-evaluate them in a more realistic environment varying the machine load and communication degree. We show that jobs obtain better responses times and stability when share resources with themselves, as under SCS, than when share resources with other jobs, as under the Time Sharing techniques. At processor level, we have observed that is better to free the processor as soon as the process detects it has no useful work to do and select the next process to run in the local queue, in a Round Robin manner.

The rest of the paper is organized as follows. In Section 2 is the related work. Then in Section 3 follows scheduling strategies evaluated and in Section 4 the execution framework and the performance results. Finally in Section 5 the conclusions and future work.

2 Related Work

There are three main approaches that share processors among jobs: Time Sharing (TS), Space Sharing (SS), and Gang Scheduling or its relaxed version, Co-Scheduling (CS).

When having more processes than processors, TS algorithms multiplex the use of processors among jobs in time. For parallel jobs the performance is degraded because of lack of synchronization and the context switching effect.

SS reduces the context switching effect, by partitioning the set of available processors among the jobs. These approaches have demonstrated to perform well on malleable jobs such as the OpenMP by adjusting the number of processes of a job to the number of available processors, which is not the case for MPI jobs.

From the combination of TS and SS, comes the CS, classified in the literature as: Explicit Scheduling (ES), Local Scheduling (LS), and Dynamic or Implicit Co-Scheduling (DCS). In [8] there is an interesting classification of CS techniques based on the combinations of two components in the interaction between the scheduler and the communication mechanism: what to do when waiting for a message and what to do on message arrival.

The ES proposed by Feitelson and Jette [4] based on a static global list, execution order of the jobs and simultaneous context switching over processors, suffers from processor fragmentation, context switching overhead and poor scalability.

In the LS [5] the scheduling decisions are made independently at each local node. Here the performance of fine grain communication jobs is deteriorated because there isn't any synchronization at all.

DCS [1] is an intermediate approximation where decisions are made locally but based on global communication events such as message arriving and round trip message time. This information is combined in several ways, resulting in different schedulers. The Periodic Boost (PB) [8] is a kind of DCS where some monitor process manipulates the process priorities based on the unconsumed message queue. In most of these cases the MPI library was modified.

With respect to coordination between the queuing system and the processor scheduler there is an interesting work in [3]. In [7] they determine at execution time the grain of the job. They classify an individual process as 'frustrated' if the scheduling policy actually applied doesn't satisfy its communication requirements.

3 Scheduling Policies Evaluated

We have implemented our SCS, the FIFO as a Static Space Sharing and other existing TS techniques. In Table 1 are summarized all the scheduling policies evaluated.

Let's notice that under TS, there may be from all to none of the processes from a job executing at the same time. But under SCS there will be always at least a fixed minimum number of processes executing at the same time.

3.1 Self-co-scheduling Technique

In order to combine SS and CS techniques we concentrate on three decisions: how many processors assign to a job, the process allocation in each partition and the scheduling of each processor local queue. These decisions are taken at low-level process management.

The number of processors assigned to a job is closely related to its number of processes and the multiprogramming level (MPL)[2]. This number is calculated by the equation (1):

$$Initial\_allocation = \text{Number Processes} / MPL$$
$$Re\_allocation = Initial\_allocation + (\text{No. Free Processors} / \text{No. Running Appls}) \quad (1)$$

Once the Queuing System launches a job, it starts execution if there are enough free processors that satisfy the *Initial_allocation* number given by **(1)**. Otherwise it must wait until other job finishes execution and free processors were recalculated. During execution, if there aren't any queued jobs, then free processors are redistributed in an equipartition way, between the jobs in the system, using the Re_allocation number **(1)**. As soon as a new job arrives if there aren't enough free processors for it, all the jobs are shrinked to their Initial_allocation number of processors.

A user-level scheduler does the process mapping to the set of assigned processors in an ordered way using the internal MPI identification. As the number of processors assigned to a job could be less than its number of processes, there may be a process local queue at each processor, which we schedule by applying the CS techniques.

When executing a MPI blocking function, if the process cannot continue execution it may blocks immediately (BI) or do Spin Blocking (SB) [1]. After blocking a Context Switching routine is invoked. This decides which process of the local queue follows executing: the next in a Round Robin (RR) fashion or the process that has the greater number of unconsumed messages (Msg).

So from these variations arise the following combinations of SCS: (1) With BI and RR; (2) With BI and Msg; (3) With SB and RR; (4) With SB and Msg.

3.2 Time Sharing Techniques

The TS techniques evaluated were: the scheduler of the IRIX 6.5 operating system as a pure TS scheduler and because is the native scheduler of the SMM where we are

---

[2] We call MPL to the number of processes subscribed to a processor.

working on, an implementation of the Periodic Boost (PB) [8] for being the one that demonstrated the best performance in [8], and some variations to other CS techniques based on the actions taken on the message waiting and arrival.

The CS variations arise at processor level management. We have implemented the analogous combinations already described in Section 3.1 for SCS.

Table 1. Configuration for the scheduling policies evaluated.

Next to run	How to wait for a message	
	BI	SB
RR	SCS, CS	SCS, CS
Msg	SCS, CS	SCS, CS

4 Evaluations

Firstly we present the architectural characteristics of the platform used. After that follows a brief description of the execution environment, the workloads and finally the performance results.

The metrics used were the slowdowns of the response times with respect to FIFO execution, which means jobs executing in a first come first served order with a SSS scheduling policy. In the rest of the paper we will refer to this policy as FIFO.

Our implementation was done on a shared memory multiprocessor, the SGI Origin 2000 [9]. It has 64 processors, organized in 32 nodes with two 250 MHZ MIPS R10000 processors each. The machine has 16 Gb of main memory of nodes (512 Mb per node) with a page size of 16 Kbytes. Each pair of nodes is connected to a network router. The operating system is IRIX 6.5 with its native MPI library with no modifications

4.1 Execution Environment

The Launcher is the user-level queuing system used in our execution environment. It performs a first come first served policy from a list of jobs belonging to a predefined workload. Once the Launcher starts executing a queued job, it enters under the control of a user-level scheduler, the CPUM, which implements the scheduling policies. It decides how many processors allocate to a job, where it will reside, the processors local queue scheduling and controls the maximum MPL for all the applications in the system, which is given as a parameter. The communication between the CPUM and the jobs is done through shared memory by control structures.

The CPUM wakes up periodically and at each quantum time expiration examines if new jobs have arrived to the system or have finished execution, updates the control structures and if necessary depending on the scheduling policy, redistributes processors. For a more detailed description about the CPUM mechanism refer to [10].

4.2 Applications and Workloads

To drive the performance evaluations we consider jobs from the MPI NAS Benchmarks suite [2] and the Sweep3D.

We have observed in initial experiments, that as we increment the MPL, the performance degrades due to the execution time dedicated to perform MPI blocking functions, like global synchronizations, barriers or blocking receives. In Table 2 and 3 we show the workloads constructed with the applications classified as high (hi), medium (me) and low (lo) comm. degree according to their MPI percentage execution time. The workloads were sized to run for 5 minutes.

Table 2. Workloads for the first part of the evaluation.

Workload	w1	w2	w3	w4
Applications	lu, cg, mg	cg, mg, ft	mg, ft	mg, ft, ep
Comm. Degree	hi, hi, hi	hi, hi, me	hi, me	hi, me, lo

Table 3. Workloads for the second part of the evaluation.

Workload	High comm	Low comm
Applications	bt, cg, mg, sweep3D	sweep3D, ep
Comm. Degree	hi, hi, hi, hi	hi, lo

4.3 Performance Results for the Scheduling Policies Evaluation

The evaluations were run on 64 processors, using the workloads from Table 2, MPL=4, with all the jobs having 64 processes each to avoid fragmentation, and the same arrival time. In Fig. 1 we present the execution times of the workloads evaluated under the different policies.

As we can the IRIX scheduler and the PB show the worst performance especially in high comm. degree workloads with slowdowns of 10 and 7 respectively.

The Spin Times evaluated were 1200 microsecs. when the next to run was by Msg and 800 microsecs. where the next to run was by RR. They were chosen in such a way they minimize the responses times. A Spin Time equal to zero, that is BI seemed the best option. This can be deduced from [1], as we have latency null because we are working on a SMM.

About the best slowdowns, we have SCS with 0.76 for high comm. and 0.58 for low comm. degree workloads. CS achieves 0.86 for high comm. and 0.72 for low comm. degree.

An interesting observation is that under SCS, the jobs execute always in the "same" environment, no matter the workload they are in, for being a kind of SS policy. But, under CS, applications become more sensible with respect to the environment, because they must share processors between them. We measure the impact with the standard deviation coefficient (CV) of the execution times of the applications between workloads. For example the MG has under the SCS techniques a CV of 5.5, while under CS techniques it can be greater than 26.9. This means that SCS show a greater stability than CS. This is important because an application, under SCS, will have a more predictable behavior.

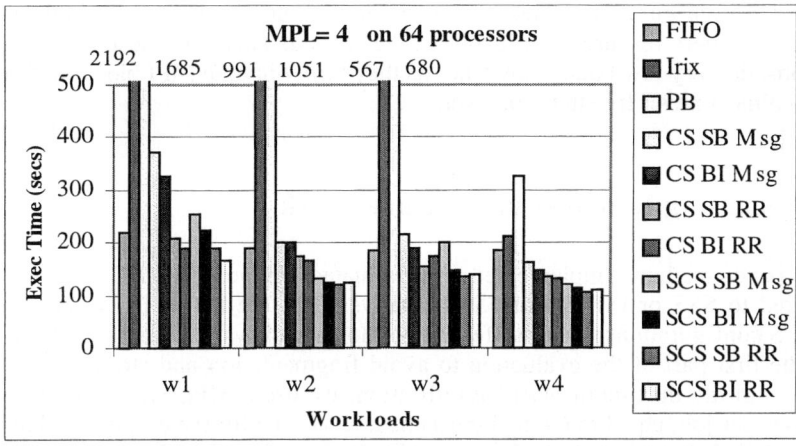

Fig. 1. Evaluation of different scheduling policies under the workloads from Table 2.

4.4 Comparing CS and SCS Varying Workload and System Characteristics

In the previous section we showed that SCS with BI and RR dominated the other strategies evaluated. In this section we show the results from re-evaluating it and comparing with CS. The workloads used are from Table 3 with different number of processes (between 32 and 60) and arrivals distributed in exponential form. In [7] Moreira et al showed that MPLs greater than 6 are equal to infinite MPL. So the greatest MPL evaluated was 6.

In Fig. 2 we can see graphically the avg. slowdowns of the response times with respect to FIFO execution, for jobs in high and low communication degree workloads, evaluated for MPL=2, 4 and 6 under CS and SCS policies with machine loads of 20% and 60%. In general SCS dominates CS especially for high comm. degree workload and MPL=4, in about 20%. For MPL=2, low comm. degree workloads and light loaded machines, SCS and CS are quite similar.

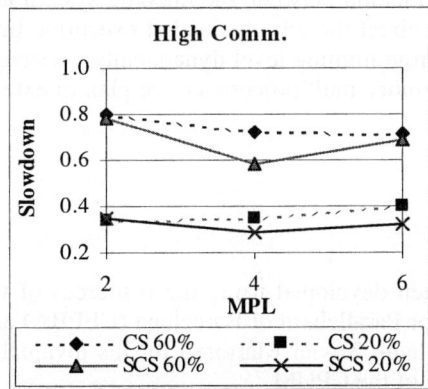 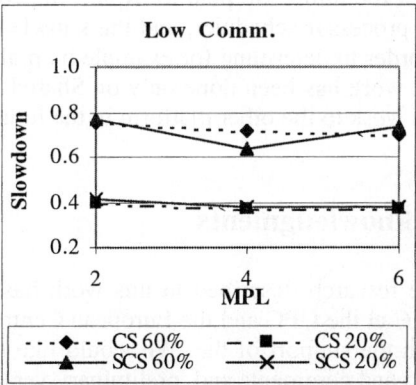

Fig. 2. Avg. Slowdowns for high and low comm. workloads under SCS and CS, varying MPL (2, 4 and 6) and machine load (20% and 60%).

Our CS configuration generates some kind of unfairness, when high and low comm. jobs share resources, in favour of the second. This is because low comm. degree jobs, do very few context switches as they rarely invoke MPI blocking functions, so they almost never free the processor.

5 Concluding Remarks and Future Work

In this paper we have implemented and evaluated our proposed SCS technique and compared to SSS or FIFO, and TS alternatives: PB, the native IRIX scheduler and some CS configurations, under different workloads and system characteristics.

In the first part of the evaluation to avoid fragmentation and strictly measure the impact with the maximum machine utilisation, we use a MPL=4, a fixed number of processes per job, equal to 64, and the same arrival time for the entire workload. Here we concluded that the SCS techniques with BI and choosing the next to run in the local queue in a Round Robin manner seemed the best option. They showed also a more predictable behaviour and stability than the TS techniques. SCS have an average improvement from 20% to 40% over FIFO executions depending on the communication degree. On the other hand for applications, which have global synchronization or do mostly calculations, CS showed better response times.

In the second part, we re-evaluate and compare the SCS and CS techniques in a more realistic environment varying machine utilisation, with different number of processes per job and arrivals time to the system. We observed that as in the first part, SCS dominate the CS techniques especially for high machine utilisation. For low machine utilisation, CS and SCS perform quite similar.

There is some fragmentation under the SCS generated due to the maximum MPL is fixed, which goes from 9% for MPL=6 to 24% for MPL=2. This can be avoided if MPI jobs were malleable. We are planning to work with moldable jobs, deciding the optimal number of processes to run the job, depending on its scalability and the current system fragmentation. As MPL and machine utilisation increment, the fragmentation in SCS decrements, improving its performance. On the contrary, with high MPL, CS cannot afford the synchronization problem as good as the SCS.

In the future we plan to exploit the coordination between the queuing system and the processor scheduler, and the knowledge about the job obtained at execution time in order to determine for example its multiprogramming level dynamically. Although this work has been done only on Shared Memory multiprocessors, we plan to extend this work to the other platform in the future.

Acknowledgments

The research described in this work has been developed using the resources of the DAC at the UPC and the European Centre for Parallelism of Barcelona (CEPBA) and with the support of the. We would like to thank Xavier Martorell for his invaluable help and comments and, preliminary version of the CPUM.

References

1. A.C. Arpaci-Dusseau, D. Culler. Implicit Co-Scheduling: Coordinated Scheduling with Implicit Information in Distributed Systems. ACM Trans. Compu. Sys. 19(3), pp.283-331, Aug. 2001.
2. D. Bailey, T. Harris, W. Saphir, R. Wijngaart, A. Woo and M. Yarrow, "The NAS Parallel Benchmarks 2.0", Technical Report NAS-95-020, NASA, December 1995.
3. J. Corbalan, X. Martorell, J. Labarta. Performance-Driven Processor Allocation. Proc. of the 4th Operating System Design and Implementation (OSDI 2000), San Diego, CA, October 2000.
4. D.G.Feitelson and M.A.Jette. Improved Utilization and Responsiveness with Gang Scheduling. Job Scheduling Strategies for Parallel Processing, volume 1291 of Lecture Notes in Computer Science. Springer-Verlag 1997.
5. A.Gupta, A.Tucker, and S. Urushibara. The Impact of Operating System Scheduling Policies and Synchronization Methods on the Performance of Parallel Jobs. In Proceedings of the 1991 ACM SIGMETRICS Conference, pages 120-132, May 1991.
6. Message Passing Interface Forum. MPI: A Message-Passing Interface standard. Int. Journal of SuperComputer Jobs, 8(3/4):165-414, 1994.
7. J.E.Moreira , W. Chan, L.L.Fong, H.Franke, M.A.Jette. An Infrastructure for Efficient Parallel Job Execution in Terascale Computing Environments. In Supecomputing'98, Nov. 1998.
8. S. Nagar, A.Banerjee, A.Sivasubramaniam, and C.R. Das. A Closer Look at Co-Scheduling Approaches for a Network of Workstations. In Eleventh ACM Symposium on Parallel Algorithms and Architectures, SPAA'99, Saint-Malo, France, June 1999.
9. Silicon Graphics, Inc. IRIX Admin: Resource Administration, Document number 007-3700-005, http://techpubs.sgi.com, 2000.
10. G. Utrera, J. Corbalán and J. Labarta. "Study of MPI applications when sharing resources", Tech. Report number UPC-DAC-2003-47, 2003.
http://www.ac.upc.es/recerca/reports/DAC/2003/index,en.html

Scheduling Tasks Sharing Files from Distributed Repositories

Arnaud Giersch[1], Yves Robert[2], and Frédéric Vivien[2]

[1] ICPS/LSIIT, UMR CNRS–ULP 7005, Strasbourg, France
[2] LIP, UMR CNRS–ENS Lyon–INRIA–UCBL 5668, Lyon, France

Abstract. This paper is devoted to scheduling a large collection of independent tasks onto a distributed heterogeneous platform, which is composed of a set of servers. Each server is a processor cluster equipped with a file repository. The tasks to be scheduled depend upon (input) files which initially reside on the server repositories. A given file may well be shared by several tasks. For each task, the problem is to decide which server will execute it, and to transfer the required files to that server repository. The objective is to find a task allocation, and to schedule the induced communications, so as to minimize the total execution time. The contribution of this paper is twofold. On the theoretical side, we establish a complexity result that assesses the difficulty of the problem. On the practical side, we design several new heuristics, including an extension of the `min-min` heuristic to such a decentralized framework, and several lower cost heuristics, which we compare through extensive simulations.

1 Introduction

In this paper, we are interested in scheduling independent tasks onto collections of heterogeneous clusters. These independent tasks depend upon files (corresponding to input data, for example), and difficulty arises from the fact that some files may well be shared by several tasks. Initially, the files are distributed among several server repositories. Because of the computations, some files must be replicated and sent to other servers: before a task can be executed by a server, a copy of each file that the task depends upon must be made available on that server. For each task, we have to decide which server will execute it, and to orchestrate the file transfers, so that the total execution time is kept minimum.

This paper is a follow-on of two series of work, by Casanova, Legrand, Zagorodnov, and Berman [3] on one hand, and by Giersch, Robert, and Vivien [5] on the other hand. In [3], Casanova et al. target the scheduling of the tasks typically submitted to APST, the AppLeS Parameter Sweep Template [1]. Casanova et al. have considered three heuristics designed for completely independent tasks (no input file sharing) that were proposed in [6]. They have modified these three heuristics (originally called `min-min`, `max-min`, and `sufferage` in [6]) to adapt them to the additional constraint that input files are shared between tasks. The number of tasks to schedule is expected to be very large, and special attention should be devoted to keeping the cost of the scheduling heuristics reasonably low.

In [5], Giersch et al. have introduced several new heuristics, which are shown to perform as efficiently as the best heuristics in [3] although their cost is an order of magnitude lower.

However, all the previous references restrict to a very special case of the scheduling problem: they assume the existence of a master processor, which serves as the repository for all files. The master distributes the files to the processors, so that they can execute the tasks. This master-slave paradigm has a fundamental limitation: communications from the master may well become the true bottleneck of the overall scheduling scheme.

In this paper, we deal with the most general instance of the scheduling problem: we assume a fully decentralized system where several servers, with different computing capabilities, are linked through an interconnection network. To each server is associated a (local) data repository. Initially, the files are stored in one or several of these repositories (some files may be replicated). After having decided that server S_i will execute task T_j, the input files for T_j that are not already available in the local repository of S_i will be sent through the network. Several file transfers may occur in parallel along disjoint routes.

The contribution of this paper is twofold. On the theoretical side, we establish in Section 3 a complexity result that assesses the difficulty of our scheduling problem. On the practical side, we design several heuristics. The first heuristic is the extension of the min-min heuristic to the decentralized framework (Section 4). The next heuristics aim at retaining the good performances of the min-min variants while reducing the computational cost by an order of magnitude (Section 5). We compare all these heuristics through extensive simulations (Section 6). We start by describing in Section 2 the specifications of our scheduling problem.

Due to space limitations, we refer to [4] for missing details and proofs.

2 Framework

Tasks and files. The problem is to schedule a set of n independent tasks $\mathcal{T} = \{T_1, T_2, \ldots, T_n\}$. The weight of task T_j is t_j, $1 \leq j \leq n$. The execution of each task depends upon one or several files, and a given file may be shared by several tasks. Altogether, there are m files in the set $\mathcal{F} = \{F_1, F_2, \ldots, F_m\}$. The size of file F_i is f_i, $1 \leq i \leq m$.

We use a bipartite graph $\mathcal{G} = (\mathcal{V}, \mathcal{E})$ to represent the relations between files and tasks. The set of nodes in the graph $\mathcal{G}$ is $\mathcal{V} = \mathcal{F} \cup \mathcal{T}$, and there is an edge $e_{i,j} : F_i \to T_j$ in $\mathcal{E}$ if and only if task T_j depends on file F_i. Each node in $\mathcal{V} = \mathcal{F} \cup \mathcal{T}$ is weighted by f_i or t_j.

Platform graph. The tasks are scheduled and executed on an heterogeneous platform composed of a set of *servers*, which are linked through a platform graph $\mathcal{P} = (\mathcal{S}, \mathcal{L})$. Each node in $\mathcal{S} = \{S_1, \ldots, S_s\}$ is a server, and each link $l_{i,j} \in \mathcal{L}$ represents a communication link from server S_i to server S_j. We assume that the graph $\mathcal{P}$ is connected, i.e., that there is a path linking any server pair.

Each server $S_i = (R_i, C_i)$ is composed of a local repository R_i, associated to a local computational cluster C_i. The files are stored in the repositories. We assume that a file may be duplicated, and thus simultaneously stored on several repositories. We make no restriction on the possibility of duplicating the files, which means that each repository is large enough to hold a copy of all the files.

For cluster C_i to be able to process task T_j, repository R_i must hold all files that T_j depends upon. Therefore, before C_i can start the execution of T_j, the server S_i must have received from the other server repositories all the files F_k such that $e_{k,j} \in \mathcal{E}$, and which were not already stored in R_i. For communications, we use the one-port model: at any given time-step, there are at most two communications involving a given server, one sent and the other received.

As for the cost of communications, first consider the case of adjacent servers in the platform graph. Suppose that server S_i sends the file F_j to another server S_k, through the network link $l_{i,k} = l$. We denote by b_l the bandwidth of the link l, so that f_j/b_l time-units are required to send the file. Next, for communications involving distant servers, we use a store-and-forward model: we route the file from one server to the next one, leaving a copy of the file in the repository of each intermediate server. The communication cost is the sum of the costs of the adjacent communications. Leaving copies of transferred files on intermediate servers multiplies the potential sources for each file and is likely to accelerate the processing of the next tasks, hence the store-and-forward model seems quite well-suited to our problem. Finally, we assume no communication time between a cluster and its associated repository: the cost of intra-cluster messages is expected to be an order of magnitude lower than that of inter-cluster ones.

As for computation costs, each cluster C_i is composed of c_i heterogeneous processors $C_{i,k}$, $1 \leq k \leq c_i$. The speed of processor $C_{i,k}$ is $s_{i,k}$, meaning that $t_j/s_{i,k}$ time-units are needed to execute task T_j on $C_{i,k}$. A coarser and less precise approach is to view cluster C_i as a single computational resource of cumulative speed $\sum_{k=1}^{c_i} s_{i,k}$. This is the approach used in all of our heuristics when the completion time of a task on a server needs to be evaluated.

Objective function. The objective is to minimize the total execution time. The execution is terminated when the last task has been completed. The schedule must decide which tasks will be executed by each processor of each cluster, and when. It must also decide the ordering in which the necessary files are sent from server repositories to server repositories. We stress two important points: (i) some files may well be sent several times, so that several clusters can independently process tasks that depend upon these files, and (ii) a file sent to some repository remains available for the rest of the schedule (if two tasks depending on the same file are scheduled on the same cluster, the file must only be sent once).

We let TSFDR($\mathcal{G}, \mathcal{P}$) (Tasks Sharing Files from Distributed Repositories) denote the optimization problem to be solved.

3 Complexity

Most scheduling problems are known to be difficult [7]. The TSFDR optimization problem is no exception. Heterogeneity may come from several sources: files or tasks can have different weights, while clusters or links can have different speeds. Simple versions of these weighted problems already are difficult. For instance the decision problem associated to the instance with no files and two single-processor clusters of equal speed already is NP-complete (it reduces to the 2-PARTITION problem as tasks have different weights). Conversely, mapping equal-size files and equal-size tasks on a single server platform with two heterogeneous processors and two links of different bandwidths is NP-hard too [5]. Even in the un-weighted version of our problem, where all files have same size and all communication links have same bandwidth, to decide where to move the files so as to execute the tasks is a difficult combinatorial problem due to file sharing. This is what formally state Definition 1 and Theorem 1.

Definition 1 (TSFDR-Move-Dec($\mathcal{G},\mathcal{P},K$)). *Given a bipartite application graph $\mathcal{G} = (\mathcal{F} \cup \mathcal{T}, \mathcal{E})$, a platform graph $\mathcal{P} = (\mathcal{S}, \mathcal{L})$, assuming: (i) uniform file sizes ($f_i = 1$), (ii) homogeneous interconnection network ($b_i = 1$), and (iii) zero processing time ($t_i = 0$ or $s_j = +\infty$); and given a time bound K, is it possible to schedule all tasks within K time-steps?*

Theorem 1. TSFDR-MOVE–DEC*($\mathcal{G},\mathcal{P},K$) is NP-complete.*

4 Adapting the `min-min` Scheme

Due to the presence of weights in tasks, in files, and in the platform graph, approximation algorithms are not likely to be feasible. Hence, we look for polynomial heuristics to solve TSFDR, and we will compare these heuristics through extensive simulations. Considering the work of Casanova et al. [3] for master-slave systems with a single server, we start by adapting the `min-min` scheme.

The principle of the `min-min` scheme is quite simple:

> While there remain tasks to be scheduled do
> 1. for each task T_k that remains to be scheduled and each processor $C_{i,j}$, evaluate the Minimum Completion Time (MCT) of T_k if mapped on $C_{i,j}$;
> 2. pick a couple $(C_{i,j}, T_k)$ with minimum MCT and schedule T_k on $C_{i,j}$.

The problem then is to evaluate a task MCT. When trying to schedule a task on a given processor, one has to take into account which required files already reside in the corresponding repository, and which should be brought from other servers. One can easily determine which communications should take place. But scheduling these communications is an NP-complete problem in the general case.

Scheduling the communications. We deal with the situation where all communications have the same destination (namely the server that will execute the candidate task T_k). In the 1-port model, if the routing in the platform graph is not fixed, we show that scheduling a set of communications is already NP-hard [4] in our context. As the routing is usually decided by table lookup, one can assume the routing to be fixed. But when trying to schedule the communications required for a task T_k, one must take into account that the communication links are already used at certain time slots due to previously scheduled communications. Even under a fixed routing, this also leads to an NP-complete problem [4].

Therefore, we rely on heuristics to schedule the necessary communications. A first idea would be to memorize for each link the date and length of all communications already scheduled, and to use an *insertion scheduling* scheme to schedule new communications. This scheme should be rather precise but may be very expensive. A second idea is to use a simple *greedy* algorithm which schedules new communications as soon as possible but always *after* communications using the same links. In both cases we assume a fixed routing. If a file is available on several servers, we heuristically decide to bring it from the closest server.

Complexity of the adapted min-min scheme. The complexity of the whole min-min heuristic is $O(n^2 s \Delta T(s + \Delta \mathcal{P} + log(\Delta T)) + n \max_{1 \leq i \leq s} c_i)$, when using the greedy communication scheduling heuristic, if we denote by ΔT the maximum number of files that a task depends upon, and by $\Delta \mathcal{P}$ the diameter of the platform graph. Of course, the complexity is larger with the insertion scheduling scheme. The last term in the expression comes from scheduling tasks on clusters. Indeed, once the communications are scheduled, on each cluster, we greedily schedule the tasks on the available processors. Among the tasks of lowest availability date, we take the largest one and schedule it on the processor on which it will reach its MCT.

The sufferage heuristic is a variant of min-min which sometimes delivers better schedules, but whose complexity is slightly greater [3, 5, 4].

5 Heuristics of Lower Complexity

As appealing as the min-min scheme could be because of the quality of the scheduling it produces [5], its computational cost is huge and may forbid its use. Therefore, we aim at designing heuristics which are an order of magnitude faster, while trying to preserve the quality of the scheduling produced.

The principle of our heuristics is quite simple. The min-min scheme is especially expensive as, each time it attempts to schedule a new task, it considers all the remaining tasks and compute their MCTs. On the opposite, we worked on solutions where we only consider a single task candidate per cluster. This leads to the following scheme:

While there remain tasks to be scheduled do
1. for each cluster C_i pick the "best" candidate task T_k that remains to be scheduled;
2. pick the "best" couple (C_i, T_k) and schedule T_k on C_i.

The whole heuristic then relies on the definition of the "best" candidate. For that, we design a *cost* function that we use as an estimate of the MCT of a task on a given server. Using the results of [5], we simply defined the *cost* of a task T_i on a server S_j as the sum of the time needed to send the required files to S_j (communication time), plus the time needed by the cluster C_j to process the task, when the cluster is seen as a single computational resource (computation time). The communication time can be evaluated using either of the two heuristics described in Section 4. In practice, we approximate it by using either the "sum" of all the communications required (over-approximation by sequentialization of all communications) or its "max" (under-approximation by considering all communications to take place in parallel).

Static heuristics. In our static heuristics, for each cluster we first build a list of all tasks sorted by increasing *cost* function. Then, each time we schedule a new task: 1) we define as local candidate for cluster C_i the task which has lowest *cost* on C_i and has not already been scheduled; 2) among all the local candidates we take the one of lowest *cost* and we assign it on the corresponding cluster; 3) we schedule the necessary communications and the computation as we did for the min-min scheme. The overall complexity of this scheme is: $O(n|\mathcal{E}| + s^3 + sn \log n + n\Delta T(\Delta \mathcal{P} + \log(\Delta T)) + ms^2 + n \max_{1 \leq i \leq s} c_i)$ (with greedy communication scheduling) which is an order of magnitude less than the complexity of the min-min scheme: we no longer have a n^2 term.

In the MCT variant of our static heuristic we make two modifications to our scheme. First, if on one cluster there is a task which only depends on files residing on the corresponding server, it becomes the local candidate (whatever its *cost*). Then, we compute the actual MCT of each local candidate (only on its candidate cluster) and we pick the task of lowest MCT. The two modifications only lead to an additional term of $O(ns\Delta T)$ in the complexity.

Dynamic heuristics may seem a better approach than static ones. Indeed, as scheduling decisions are taken, files are copied between servers and the original *cost* estimate become less pertinent. Hence the desire to use dynamically updated *costs* to drive the mapping decisions. But the updates should be kept as computationally cheap as possible. Hence, each time a file F_j is duplicated we update only the *costs*(C_i, T_k) where T_k depends on F_j. Using a clever data structure (see [4] for details) the overall complexity of maintaining dynamic *costs* is only $O(s^3 + sm(1 + u))$ where $u = 1$ when we over-approximate the communications and $u = \Delta T$ when we under-approximate them.

In the dynamic1 heuristics, for each cluster we maintain a heap of dynamic *costs*, and the selection of the local candidate for each cluster is cheap. Then, the overcost due to the dynamicity is only: $O(sm(1 + u) + sm \log n + ns\Delta T)$.

Another way of decreasing the complexity of the selection of the local candidate is to select, on each server, k tasks of lowest *costs*, instead of only one task. Such a selection can be realized in linear time in the worst case [2]. The dynamic2 heuristic uses this approach. Then, the overcost due to the dynamicity is: $O(sm(1+u) + \frac{n}{k}(ns + ks\log(ks)) + ns\Delta T)$. For the simulations described in the next section, we used $k = 10$.

6 Simulation Results

In order to compare our heuristics, we have simulated their executions on randomly built platforms and graphs. We have conducted a large number of experiments, which we summarize in this section.

Simulation framework. We generated three types of server graphs (7 servers): clique, random tree, or ring. Each server is associated a cluster of 8, 16, or 32 processors. The processors speeds and communication links bandwidths were chosen from a set of values recorded on real machines. These values were normalized so as to model three communication-to-computation cost ratios: computation intensive, communication intensive, and intermediate. For the tasks graphs, we generated four types of graphs, from very regular ones with lots of file sharing, to fully randomly built ones. The initial distribution of files to server is built randomly. A fully detailed description of the simulations can be found in [4].

Results. We report the performance of our heuristics together with their cost (i.e., their CPU time). The randommap heuristic randomly picks the local candidate (the same for all clusters) but uses the same optimizations and scheduling schemes than the other heuristics. Table 1 summarizes all the simulations, i.e., 36,000 random tests (1,000 tests over each combination of task graph, platform graph, and communication-to-computation cost ratio). For each test, we compute the ratio of the performance of all heuristics over the best heuristic for the test, which gives us a *relative performance*. The best heuristic differs from test to test, which explains why no heuristic in Table 1 can achieve an average relative performance of 1. The optimal relative performance of 1 would be achieved by picking, for any of the 36,000 tests, the best heuristic for this particular case. (The *relative cost* is computed along the same guidelines, using the fastest heuristic.)

The basic versions of our heuristics are far quicker than the min-min versions but at the cost of a great loss in the quality of the schedules produced (two times worse). The MCT variant greatly improves the quality of our heuristics (this is exemplified by randommap) while their costs remain very low. For example, the

Table 1. Relative performance and cost of the heuristics: basic versions, MCT variants, and MCT variants with communication scheduling with insertion scheduling. Standard deviations are in parentheses (for relative costs, all are between 128% and 211%).

Heuristic	Basic version		MCT variant		MCT variant + insert.	
	Perf.	Cost	Perf.	Cost	Perf.	Cost
min-min	1.14 (± 7.9%)	31,050	-	-	1.08 (± 7.5%)	61,771
sufferage	1.16 (± 14%)	33,985	-	-	1.07 (± 12%)	77,991
static	2.20 (± 33%)	16	1.46 (± 23%)	44	1.18 (± 11%)	56
dynamic1	2.32 (± 37%)	42	1.35 (± 17%)	67	1.11 (± 8.8%)	77
dynamic2	2.42 (± 39%)	310	1.92 (± 36%)	82	1.40 (± 31%)	90
randommap	141 (± 317%)	11	1.32 (± 18%)	41	1.08 (± 6.9%)	53

MCT variant of `dynamic1` produces schedules which are only 19% longer than those of `min-min`... but it produces them 460 times more quickly. We also ran the heuristics with the insertion scheduling heuristic for communication scheduling (rather than with the greedy scheduling as previously). As predicted, the quality of results significantly increased. The overhead is prohibitive for the `min-min` variants (these versions only differ from the original by the insertion scheduling scheme). Surprisingly, this overhead is reasonable for our heuristics. For example, this version of `dynamic1` produces schedules which are 3% *shorter* than those of the original `min-min`... and it produces them 400 times more quickly.

7 Conclusion

In this paper, we have dealt with the problem of scheduling a large collection of independent tasks, that may share input files, onto collections of distributed servers. On the theoretical side, we have shown a new complexity result, that shows the intrinsic difficulty of the combinatorial problem of deciding where to move files. On the practical side, our contribution is twofold: 1) we have shown how to extend the well-known `min-min` heuristic to this new framework, which turned out to be more difficult than expected; 2) we have succeeded in designing a collection of new heuristics which have reasonably good performance but whose computational costs are orders of magnitude lower than `min-min`: our best heuristic produces schedules whose makespan is only 1.1% longer than those of the best `min-min` variant, and produces them 585 times faster than the quickest `min-min` variant.

We plan to deploy the heuristics presented in this paper for a large medical application, with servers in different hospitals in the Lyon-Grenoble area, and we hope that the ideas introduced when designing our heuristics will prove useful in this real-life scheduling problem.

References

1. F. Berman. High-performance schedulers. In I. Foster and C. Kesselman, editors, *The Grid: Blueprint for a New Computing Infrastructure.* Morgan-Kaufmann, 1999.
2. M. Blum, R. W. Floyd, V. Pratt, R. R. Rivest, and R. E. Tarjan. Time bounds for selection. *Journal of Computer and System Sciences*, 7(4):448–461, 1973.
3. H. Casanova, A. Legrand, D. Zagorodnov, and F. Berman. Heuristics for Scheduling Parameter Sweep Applications in Grid Environments. In *HCW'2000*. IEEE Computer Society Press, 2000.
4. A. Giersch, Y. Robert, and F. Vivien. Scheduling tasks sharing files from distributed repositories (revised version). Research Report RR-2004-04, LIP, ENS Lyon, 2004.
5. A. Giersch, Y. Robert, and F. Vivien. Scheduling tasks sharing files on heterogeneous master-slave platforms. In *PDP'2004*. IEEE Computer Society Press, 2004.
6. M. Maheswaran, S. Ali, H. Siegel, D. Hensgen, and R. Freund. Dynamic Mapping of a Class of Independent Tasks onto Heterogeneous Computing Systems. *Journal of Parallel and Distributed Computing*, 59(2):107–131, Nov. 1999.
7. B. A. Shirazi, A. R. Hurson, and K. M. Kavi. *Scheduling and load balancing in parallel and distributed systems.* IEEE Computer Science Press, 1995.

A Scheduling Algorithm for Running Bag-of-Tasks Data Mining Applications on the Grid

Fabrício A.B. da Silva, Sílvia Carvalho, and Eduardo R. Hruschka

Universidade Católica de Santos (Unisantos)
R. Dr. Carvalho de Mendonça, 144 – 11070-100 – Santos (SP)
{fabricio,mygridgene,erh}@unisantos.edu.br

Abstract. Data mining applications are composed of computing-intensive processing tasks, which are natural candidates for execution on high performance, high throughput platforms such as PC clusters and computational grids. Besides, some data-mining algorithms can be implemented as Bag-of-Tasks (BoT) applications, which are composed of parallel, independent tasks. Due to its own nature, the adaptation of BoT applications for the grid is straightforward. In this sense, this work proposes a scheduling algorithm for running BoT data mining applications on grid platforms. The proposed algorithm is evaluated by means of several experiments, and the obtained results show that it improves both scalability and performance of such applications.

1 Introduction

Knowledge discovery in databases is the non-trivial process of identifying valid, novel, potentially useful, and ultimately understandable patterns in data [1]. Data Mining (DM) is a step in this process that centers on the automated discovery of new facts and relationships in data. DM consists of three basic steps: data preparation, information discovery and analysis of the mining algorithm output. All these steps exploit huge amounts of data and are computationally expensive. In this sense, several techniques have been proposed to improve the performance of DM applications, such as parallel processing [3] and implementations based on cluster of workstations [4].

A computational grid, or simply grid for short, provides access to heterogeneous resources in a geographically distributed area, allowing the integration of these heterogeneous and distributed resources into a unified computer resource. Computational grids are usually built on top of specially designed middleware platforms, the so-called grid platforms. Grid platforms enable the sharing, selection and aggregation of a variety of resources including supercomputers, servers, workstations, storage systems, data sources and specialized devices that are geographically distributed and owned by different organizations [5].

Among the most suitable applications for running on a grid are the Bag-of-Tasks (BoT) applications, which are parallel applications whose tasks are independent of each other. Examples of BoT applications include Monte Carlo simulations, massive searches (such as key breaking), parameter sweeps, image manipulation, and data mining. In this work, we analyze the use of computational grids for data mining.

BoT data mining applications can present scalability problems when executing on the grid. In this paper, we define scalability as the ability of a grid platform to reduce the total execution time of a BoT application as the number of machines added to the grid increases, given that the number of tasks of the BoT application is larger than the number of machines in the grid. This paper proposes a scheduling algorithm that improves scalability of BoT data mining applications on the grid, by grouping sets of independent tasks into larger ones, which are executed on grid machines. For the experimental results we used the MyGrid platform [6], which is specially designed to run BoT applications.

The remaining of this paper is organized as follows: Section 2 approaches the employed data mining algorithm (K-Means) in the context of BoT applications. This section also presents experimental results related with the scalability problem to be tackled by our scheduling algorithm. This algorithm is described and evaluated in Section 3. Finally, Section 4 presents the conclusions and points out some future work.

2 Running Data Mining Algorithms as BoTs Applications

Some recent works suggest that grids are natural platforms for developing high performing data mining services [7,8,9]. For instance, Orlando et al. [8] describe the application of two data mining algorithms in the Knowledge Grid [7]: DCP and K-means. The former algorithm enhances the popular Apriori Algorithm [10], which is an algorithm for frequent set counting, whereas the latter one is a clustering algorithm. In this work, we have also employed the K-Means algorithm (briefly described in Section 2.1) implemented as a BoT application on the MyGrid platform (depicted in Section 2.2). Section 2.3 reports experimental results that illustrate the need for a specific scheduling algorithm specially designed to run BoT applications.

2.1 K-Means Algorithm

Clustering algorithms involving the calculation of the mean (centroid) of each cluster are often referred to as K-means algorithms [2]. Basically, these algorithms partition a dataset of n objects into K clusters. The K value, which represents the number of clusters, is specified by the user. These algorithms try to minimize the error sum of distances (e.g. Euclidean) between objects of a cluster and its centroid. The implementation employed in our work can be summarized as follows:

1. Generate a random initial partition of objects into K nonempty clusters;
2. Compute the cluster centroids (mean vectors) of the current partition;
3. Assign each object to the cluster with the nearest centroid;
4. If the convergence criterion has been met, stop. Otherwise, go to step 2.

The convergence criterion can be defined either as the maximum number of iterations (t) of steps 2 and 3 or as the maximum absolute difference between centroids in two consecutive iterations. The K-Means Algorithm has two main drawbacks: (i) it may get stuck at local optimal partitions; (ii) the user has to provide the number of clusters (K). Besides, the K-Means can sometimes end up with less than K clusters if, in a set of K centroids, at least one is worse than all the other ones. This situation can happen, for instance, if all objects are nearer to the other (K-1) centroids. However, this is not a hard limitation, because it can indicate that there are less than K clusters. Indeed, the choice of K is very hard and a usual practical approach is to try several different solutions (one for each K-value) and then choose the most suitable one, by plotting the value of the clustering criterion against the number of clusters. In a grid environment, this approach can be performed in a parallel way, in which each machine runs the algorithm for a specific K-value. In addition, it is also useful to run the K-Means for different initial partitions, and it can also be performed in parallel (this approach is employed in our work). Due to those characteristics, the K-Means algorithm is suitable for execution as a BoTs application on the MyGrid platform. In addition, K-Means is suitable for our experiments, because of its efficiency - $O(tKn)$.

2.2 The Platform MyGrid

The MyGrid platform [6] was conceived to support the execution of BoT applications, which constitute a class of parallel applications that can be partitioned in several independent tasks. Usually, these tasks have an infrequent need for communication.

The main benefits of MyGrid are twofold: minimal installation effort and ease of use. Most grid platforms (for example see [11]) can only be installed and configured by system administrators. Moreover, installation procedures are usually manually repeated in a considerable number of machines. MyGrid enables regular users to create their own grid to run applications on whatever resources they have access to, without the need for these users to get involved into grid details and administrative procedures for heterogeneous platforms.

Since MyGrid focuses on BoT applications, its working environment consists of a small set of services to enable its users to manipulate their files on the grid. Consequently, no previous software installation nor shared file system are needed on machines. A user is required to install and configure MyGrid only on one machine, which is called home machine. Interactions with other machines are supported by the Grid Machine Interface (GMI), i.e., a minimal set of services that must be available in a machine so it can be used as a machine for grid computing, the so-called grid machine. These services consist of: (1) remote executing on a grid machine; (2) termination of a running task; and (3) file transfers between the home and grid machines.

Mygrid implements two task scheduling algorithms, *Work Queue*[6] and *Work Queue with Replication* (WQR) [12]. WQR starts the execution of each task on idle grid machines just like the Workqueue algorithm. Once the queue is empty, Mygrid starts execution of replicated instances of unfinished tasks on idle machines. Once a task completes, all other replicas are terminated. If a machine is overloaded or fails, the WQR will eventually re-execute the corresponding task on another machine.

2.3 Running K-Means as a Bag-of-Tasks Application

Our experiments were performed in a dataset formed by nine overlapping clusters. These clusters contain 100 objects, randomly generated using bi-dimensional Gaussian distributions with standard deviations equal to 0.5 and centers [x,y] given by [1,1],[3,1],[5,1],[1,3],[3,3],[5,3],[1,5],[3,5],[5,5]. We have employed K=150 clusters and t=500 iterations. These parameter values were chosen because they are suitable to characterize the scalability of the grid platform and to assess the performance of the proposed scheduling algorithm.

Initially, experiments to estimate the performance gains were executed on a grid of 25, 30 and 35 machines. Twenty machines are located at UniSantos and the others at the Federal University of Campina Grande, Brazil. A distance of about 3000 kilometers separates these two sites. The machines at UniSantos are 1.8 GHz Pentium IV machines with 128 MB of RAM. Machines at Campina Grande have 1.8GHz Pentium IV processors with 630 MB of RAM. The results shown in Figure 1 are the average of ten executions at different times of the day. A Pentium IV with 1GB of main memory was used as the *home machine*. This machine is dedicated to provide dataset distribution, task dispatching to grid machines and output file gathering. The scheduling strategy used was the standard Workqueue algorithm. Version 1.1 of MyGrid was used in experiments.

The execution time of DM applications can be significantly reduced on a Grid environment, but the scalability of a grid with more than 30 machines is poor, as illustrated in Figure 1. It is due to both the application behavior and the way MyGrid interacts with grid machines to manage input and output files. Thus, a more detailed view of the MyGrid working environment becomes necessary.

Each MyGrid task is divided in three subtasks (*init, remote* and *collect*), which are performed sequentially, in that order. Both *init* and *collect* subtasks are executed on the home machine, to send input files to grid machines and to collect the output files back to the home machine, respectively. The *remote* subtask runs on grid machines, and it is responsible to perform the computation itself. Depending on the size of input and output files and the number of grid machines, the execution of *init* and *collect* *subtasks* on the *home machine* may become a bottleneck. The overhead related to file transfers in the home machine depends on both the size of the input and output files and the number of machines added to the grid. In addition, the data transfer rate observed in the home machine can also be affected by the running time of *remote* subtasks. The shorter is the execution time of *remote* subtasks, the more frequent is the transfer of input and output files, which increases the data transfer rate in the home machine.

The main cause of the poor scalability depicted in Figure 1 is the bottleneck that appeared at the home machine. For each application task, a small number of processes are created at the home machine to manage the transferring of input and output files to and from grid machines. Such processes are eliminated as soon as their corre-

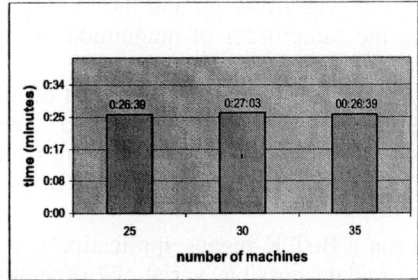

Fig. 1. Average execution times in a Grid environment (350 tasks).

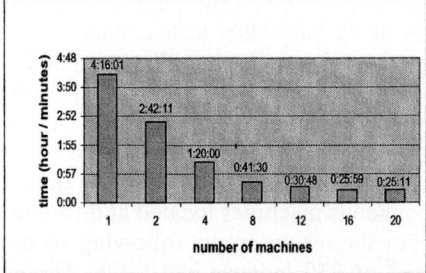

Fig. 2. Average execution times in a dedicated cluster environment.

sponding tasks have been completed. Thus, short-duration application tasks can increase the rate of creation/destruction of such processes to critical levels. Moreover, these processes execute concurrently, competing for memory, I/O subsystem, network bandwidth, CPU cycles, and other resources. Our tests have shown that the scalability of the application can be severely impacted when we execute a large number of tasks, due mainly to the bottleneck created at the home machine. This fact is evidenced by a new set of experiments generated on a dedicated cluster of Pentium IV machines. These results are shown in Figure 2.

For the results shown in Figure 2, we run the set of tasks sequentially on a single, stand alone machine with a Pentium IV (1.8 GHz) processor, and 1 GB of main memory. Then, the same set of tasks were run on 2, 4, 8, 12, 16 and 20 dedicated machines with similar hardware characteristics, interconnected by a 100 Mbps Ethernet LAN. The schedulling strategy was the standard Workqueue algorithm. By examining Figure 2, it is clear that scalability is poor for cluster platforms with more than 16 machines.

3 A Scheduling Algorithm for Data Mining Tasks

The bottleneck represented by the home machine can be characterized by examining the execution times of each independent task and isolating the times required for the corresponding *init*, *remote* and *collect* subtasks. For very short executions, like the ones here performed, the time waste transferring data sometimes is larger then the computational time. Beyond that, there is a reduction of performance when more machines are aggregated, because the home machine needs to manage more processes which compete for resources. In Mygrid 1.1, for each application task, a small number of processes are created to manage the transfer of input and output files to and from grid machines. Such processes are terminated as soon as their tasks have been accomplished. Thus, short application tasks can increase the rate of creation/destruction of such processes to critical levels. Moreover, these processes execute concurrently, competing for memory, I/O subsystem, network bandwidth, CPU cycles, and other resources in the home machine.

One way of reducing the impact of the bottleneck represented by the home machine on the scalability of the platform is to group sets of tasks in one larger task and then execute the "big" task in a grid machine. In the following we will refer to this big task as a job. The main idea is to group a set on tasks into one execution in a remote grid machine, when the number of machines of the grid grows up and the execution time of an individual task is small, i.e., having the same order of magnitude of the time need to transfer input and output files over the network.

In the following we propose scheduling algorithms for reducing the impact of the bottleneck at the home machine. As a consequence, scalability is improved. Two different platforms are considered: a dedicated cluster, in which all machines are located in the same administrative domain, and a grid environment composed of heterogeneous machines located at different sites.

For the results in the following sections we run a BoT K-means application composed of 350 independent tasks. These tasks simulate possible variations of initial partitions in the K-means algorithm (described in Section 2.1).

3.1 Dedicated Cluster

In a dedicated cluster we used the following algorithm to group tasks: (i) Tasks are grouped in a way that only one job (group of independent tasks) is created per machine. By considering just one job per machine we reduce the amount of work to be done by the home machine in managing jobs; (ii) Tasks are clustered into jobs in a way that the time needed for processing each job is about the same for every machine. In order to do so, static information about the configuration of each machine (such as processor speed and memory) should be available. Therefore, the amont of work to be executed by each processor is statically balanced.

We performed initial experiments using a cluster comprised of 20 machines at Unisantos in the same administrative domain. All machines have one Pentium IV (1.8 GHz) processor, and 128 MB of main memory. For those executions we used a home machine with 1 GB of main memory. Results for clusters composed of 8, 12, 16 and 20 machines are shown in Figure 3.

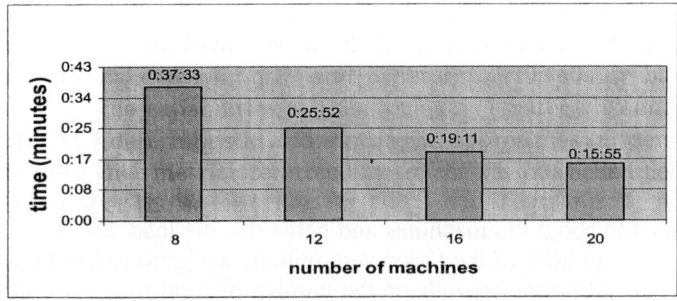

Fig. 3. Average execution times in a dedicated cluster - tasks grouped into jobs.

Since the platforms are composed of homogeneous machines, the number of independent tasks assigned to each machine is an integer in the interval $[\lfloor T/P \rfloor, \lceil T/P \rceil]$, where T is the total number of tasks (in this case 350) and P is number of machines. It is worth noting the improvement in scalability when tasks are grouped into jobs.

Since the files to be processed by each task are the same in this application, it is possible to send the files just once. As a consequence there is a reduction on both the amount of processing needed to manage remote tasks in home machine and network traffic. As a consequence there is significant reduction in the overall makespan, which can be verified by comparing the results in Figure 3 to the results in Figure 2.

3.2 Grid Platform

In grid platforms, faults are common, and this fact should be taken into account explicitly to define a scheduling algorithm. The degree of heterogeneity is also much larger in a grid, when comparing to a cluster. We used the algorithm proposed for the dedicated cluster as a starting point, and modified it as described below:

(i) Just like the dedicated cluster algorithm, tasks are grouped in a way that only one job is created per machine.
(ii) Cluster tasks into jobs in a way that the time needed for processing each job is about the same for every machine, *as if the machines are dedicated*, by using static information. In the case of a grid the static information should include the machine is local, i.e., if the machine is in the same administrative domain of the home machine. A larger amount of work should be assigned to local machines, as network delays are considerably smaller in this case.
(iii) For every x tasks completed, the grid machine should send the corresponding results back to the home machine. This mechanism is similar to a regular checkpointing. If the grid machine fails, the only tasks that have to be executed are those for which the results were not received. It is possible to obtain information about the current load of a grid machine by measuring the time needed for executing x tasks. Beyond that, the home machine can assert the failure of a grid machine by considering a dynamically adjustable timeout. If the home machine does not receive any results for a period of time equal or larger than the timeout, the grid machine is considered offline.
(iv) If a machine becomes idle, send a replica of the remaining tasks of the slowest machine to the idle machine. If the slowest machines' tasks have already been replicated, consider the second slowest machine, and so on (only one replica per task).

We also run experiments in a grid environment comprised of 25, 30 and 35 heterogeneous machines, 20 machines from Unisantos and the others from Federal University of Campina Grande. Initial results are shown in Figure 4. Given the static information available about the machines and networks, the load was distributed in a way that about 60 % to 80% of the tasks were initially assigned to local machines at Unisantos. The actual figure depends on the number of local machines that compose the grid. The other tasks were assigned to remote (Campina Grande) machines. Despite the short execution times and the relatively small number of machines, the results of Figure 4 show a very significant improvement when compared to results shown in Figure 1. Both the scalability and the performance are considerably improved. For instance, for the 35 machines plataform, the performance improvement in the overall makespan is about 56%.

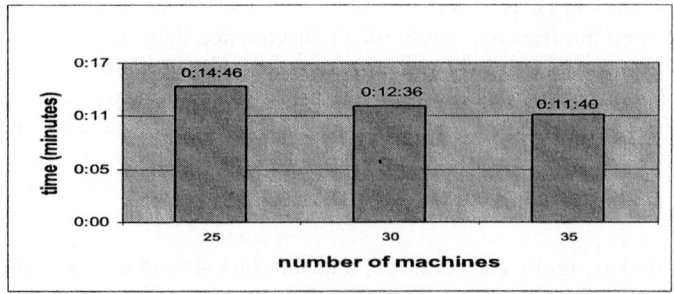

Fig. 4. Average execution times in a grid environment – tasks grouped into jobs.

4 Conclusion

This paper evaluated the scalability of a data mining method (K-Means Algorithm) implemented as a BoT application, using the middleware Mygrid. We proposed a scheduling strategy aimed to improve scalability, by grouping together sets of independent tasks to be executed in a grid machine. Our group is now investigating ways of improving even further the scalability of the platform by distributing the functions associated to the home machine among several machines.

Acknowledgements

We would like to acknowledge Walfredo Cirne and all the MyGrid/OurGrid team for the support provided during the execution of experiments. This work was developed in collaboration with HP Brazil R&D. Eduardo Raul Hruschka acknowledges CNPq (proc. 301.353/03-4) for its financial support.

References

1. Fayyad, U. M., Shapiro, G. P., Smyth, P. "From Data Mining to Knowledge Discovery : An Overview". In: Advances in Knowledge Discovery and Data Mining, Fayyad, U.M., Piatetsky-Shapiro, G., Smyth, P., Uthurusamy, R., Editors, MIT Press, pp. 1-37, 1996.
2. Witten, I. H., Frank, E., Data Mining – Practical Machine Learning Tools and Techniques with Java Implementations, Morgan Kaufmann Publishers, USA, 2000.
3. Freitas, A.A., Lavington, S.H., *Mining Very Large Databases with Parallel Processing*, Kluwer Academic Publishers, 1998.
4. Baraglia, R., Laforenza, D., Orlando, S., Palmerini, P., Perego, R., Implementation Issues in the Design of I/O Intensive Data Mining Applications on Clusters of Workstations, Proceedings of the 3rd Workshop on High Performance Data Mining, International Parallel and Distributed Processing Symposium 2000, Cancun, Mexico, May 2000.
5. Baker, M., Buyya, R., Laforenza, D. *Grids and Grid Technologies for Wide-area Distributed Computing*, Software, Pratice and Experience, v. 32, pp. 1437-1466, John Wiley and Sons, 2002.
6. Walfredo Cirne, Daniel Paranhos, Lauro Costa, Elizeu Santos-Neto, Francisco Brasileiro, Jacques Sauvé, Carla Oshtoff, Fabrício Silva, Cirano Silveira. *Running Bag-of_Tasks Applications on Computational Grids: The MyGrid Approach*. Proceedings of the 2003 International Conference on Parallel Processing, October 2003.
7. Canataro, M., Talia, D. *The Knowledge Grid*, Communications of ACM, v.46, n.1, 2003.
8. Orlando, S., Palmerini, P., Perego, R., Silvestri, F., *Scheduling High Performance Data Mining Tasks on a Data Grid Environment*, Proceedings of Int. Conf. Euro-Par 2002, 27-30 August 2002, Paderborn, Germany, LNCS 2400 - Springer-Verlag - Pag. 375-384.
9. Hinke, H., Novotny, J., *Data Mining on NASA´s Information Power Grid*, HPDC 2000, Pittsburgh, Pennsylvania, USA, pp.292-293, IEEE Computer Society.
10. Agrawal, R., Mannila, H., Srikant, R., Tiovonen, H., Verkamo, A.I., *Fast Discovery of Association Rules* In: Advances in Knowledge Discovery and Data Mining, Fayyad, U.M., Piatetsky-Shapiro, G., Smyth, P., Uthurusamy, R., Editors, MIT Press, pp. 307-328, 1996.

11. I. Foster, C. Kesselman. *Globus: A Metacomputing Infrastructure Toolkit*. Intl J. Supercomputer Applications, 11(2):115-128, 1997.
12. Daniel Paranhos, Walfredo Cirne, Francisco Brasileiro *Trading Cycles for Information: Using Replication to Schedule Bag-of-Tasks Applications on Computational Grids* Proceedings of International Conference on Parallel and Distributed Computting, 2003.

Lookahead Scheduling for Reconfigurable GRID Systems

Jesper Andersson[1], Morgan Ericsson[1], Welf Löwe[1], and Wolf Zimmermann[2]

[1] Software Technology Group, MSI, Växjö universitet
{jesan,mogge,wlo}@msi.vxu.se
[2] Martin-Luther-Universität Halle-Wittenberg, Institut für Informatik,
D-06099 Halle/Saale, Germany
zimmer@informatik.uni-halle.de

Abstract. This paper proposes an approach to continuously optimizing parallel scientific applications with dynamically changing architectures. We achieve this by combining a dynamic architecture and lookahead malleable task graph scheduling.

1 Introduction

There is a rich body of research on both off-line scheduling and on-line load balancing. The former produce good results for statically know programs, especially if the communication behavior is know statically. The latter is commonly used when programs or hardware configurations change during execution. The disadvantage of load balancing is its unawareness of the applications, leading to limited anticipation of future load situations. Re-scheduling of changed applications or hardware configurations is a non option since the algorithms are too slow.

This paper contributes by making off-line scheduling algorithms applicable for dynamic scenarios. We approach the problem by defining a Dynamic Service Architecture (DSA), responsible for managing event-driven changes to applications and continuously generating schedules for possible future situations. In order to make the continuous generation of schedules feasible we apply malleable task graph scheduling. This allows for faster scheduling by the reuse of partial schedules.

The paper is organized as follows. Section 2 gives the foundations for the execution and programming model, and introduces the DSA. Section 3 defines the mapping between the two architectural levels of the DSA. This section introduces the composition and the scheduling models. Finally, Section 4 concludes the paper.

2 Application Scenario

This section introduces the foundations of the programming and execution model of the intended system and defines the Dynamic Service Architecture.

2.1 Programming and Execution Model

The hardware infrastructure, *GRID System* in the following, consist of: Sensors called *input nodes*, generating a stream of *input values*, computation processors (*nodes*) and an interconnecting network for communication. On this GRID system *applications* are executed processing data from sensors. These applications are data parallel programs. Their inputs are sensor input values directly or the output of other applications. If the input of application a is the output of an application a', a is called *data-dependent* on a'. This is denoted by $a' \rightarrow a$. Applications are stateless, data driven, functions.

For individual applications we assume a High Performance Fortran (HPF)-like programming model, with data parallel synchronous programs without any data distribution. We can model such an application by a family of taskgraphs $G_x = (V_x, E_x, \tau_x)$. Scientific applications can automatically be compiled to such a family of task-graphs [5]. The tasks $v \in V_x$ model local computations without access to shared memory. $\tau(v)$ is the execution time of task v on the target machine, and there is a directed edge from v to w iff v writes a value to shared memory, that is read later by task w. Task-graphs are always acyclic. G_x does not always depend on the actual input x. In many cases of practical relevance, it only depends on the problem size n. We call these program *oblivious* and denote their task graphs by G_n. We write G instead of G_n if n is arbitrary but fixed. The *height* of a task v, denoted by $h(v)$, is the length of the longest path from a task with in-degree 0, to the task v.

The hardware is modeled by the LogP [2] model: in addition to the computation costs τ, it models communication costs with parameters *L*atency, *o*verhead, and *g*ap (which is actually the inverse of the bandwidth per processor). In addition to L, o, and g, parameter P describes the number of processors. Moreover, there is a capacity constraint: at most $\lceil L/g \rceil$ messages are in transmission in the network from any processor to any processor at any time. A send operation that exceeds this constraint stalls.

A LogP-schedule is a schedule that obeys the precedence constraints given by the task-graph and the constraints imposed by the LogP-machine, i.e., sending and receiving a message takes time o, there must be at least time g between two consequential send or receive operations, there must be at least time L between the end of a send task and the beginning of the corresponding receive task, and the capacity constraint must be obeyed. $TIME(s)$ denotes the execution time of schedule s, i.e., the time at which the last task finishes.

The *configuration* of a *system* is defined by: (i) its set of applications, (ii) the data dependencies, and (iii) Quality of Service (QoS) parameters for the different applications.

2.2 Dynamic Service Architecture

A configuration might change over time, i.e. it is dynamic. These changes are *user-*, *application-*, or *system-triggered*. The Dynamic Service Architecture (DSA) manages change the running system. The DSA can be seen as a conceptual control system listening to events from and reconfiguring a physical processing accordingly. The architecture of the processing system is defined by a composition of data-parallel applications. The focus of reconfiguration is, for this paper, the rescheduling.

The user in our scenario controls a certain set of applications. A typical user-triggered change is adding an application to the system or removing it after gaining the results. Some applications act as detectors, recognizing certain patterns in the processed data, that will require a reconfiguration. Application-triggered change is the detection of an interesting sensor activity requiring changed QoS parameters. The complexity of applications might be input dependent. Certain inputs may lead to load peaks in applications. In order to guarantee the required quality of service to applications, certain others applications are postponed to off-line computations. This is a typical system-triggered change.

The DSA use two architecture levels. A scheduler establishes a one-to-one mapping M between the *conceptual* processing architectures, A, and the actual *physical* running systems $M : A \rightarrow I$. User-triggered changes occur on the conceptual level. As a result a new conceptual system-architecture $a \in A$ is activated. This triggers computation of a corresponding implementation $i = M(a)$, including the computation of a new schedule that is distributed to the physical level. Inversely, application- and system-triggered changes occur on the physical level. The level should reflect the situation of the level in order to handle subsequent user-triggered changes in a correct manner. Hence, application- and system-triggered changes also affect the conceptual. However, we distinguish between conceptual events implementing user-triggered change requests and physical events implementing application- and system-triggered events. Both event-classes initiates the generation of new implementations using the translation and scheduling $M : A \rightarrow I$.

3 Mapping Conceptual to Physical Architecture

This section continues the discussion of the reconfigurations performed by the DSA, more specifically the mapping from a conceptual to a physical implementation. This mapping consists of the composition of applications to systems, discussed in Section 3.1 and scheduling. The scheduling is implemented as a continuous process, generating all possible schedules for a given state. This *lookahead* scheduling is defined in Section 3.2. The *lookahead*-scheduling must be as fast as possible, since it defines the rate of change events that can be dealt with. This means that we need to speed-up the classic static scheduling algorithms. This is accomplished using malleable task scheduling, which is described in Section 3.3.

3.1 Static Composition of Applications to Systems

So far, we defined the components: data-parallel applications, translated to task graphs and scheduled to the infrastructure. For each component, we can determine an upper time bound for its execution. Each component implements a function mapping an input array a_i to an output array a_o.

To this end, composition of components can be done by defining a program using these components and assigning the output array of one component to the input of another. Obviously, this system is a data-parallel program, too. It can be compiled and scheduled just like the individual components.

The computation of an optimum LogP-schedule is known to be NP-hard. However, good approximations and heuristics, c.f. our own contributions in task scheduling, e.g. [3], guarantee a small constant factor in delay. In practice, the results are even closer to the optimum. Moreover, an upper time bound for $TIME(s)$, the execution time of a schedule s, can be determined statically.

Adding/removing a component requires a complete re-translation of the system to a new task graph and a rescheduling of the new task-graph. This is, in principle, not a problem since adding/removing a component can be planned and prepared off-line. After the complete schedule is computed for the new system, the computation can replace the running system with the new system. However, for application- and system-triggered changes the delay to prepare for a change is not acceptable. Therefore, we compute new schedules before they are actually needed. We call this *lookahead scheduling*.

3.2 Lookahead Scheduling

As mentioned before, we distinguish between conceptual events E_c implementing user-triggered change requests and physical events E_p implementing application- and system-triggered events. If an $e_c \in E_c$ occurs, we compute the new architecture $a' \in A$ of the processing system, map it to a implementation architecture $i' = M(a')$, compute the delta between the current implementation architecture i and i and deploy that delta to the GRID system. If a $e_p \in E_p$ occurs, the reconfiguration must be finished in milliseconds. However, we exploit the assumption that the expectation on the rate of such events E_p is rather low. The fundamental principle for the optimization of our dynamic reconfiguration is to employ a continuous implementation mapping with a *lookahead schema*. For each system architecture A, we pre-compute possible changes Δ w.r.t. possible system events E_p. More specifically: given a current (baseline) architecture $a \in A$ and each possible system event $e_p \in E_p$, we compute the evolved architecture $a' = \Delta(e_p, a)$. After having determined possible deltas, we have a set of *lookahead(1)* architectures. These are mapped in the same way as the base-line-architecture is mapped to *lookahead(1)*- implementations: $i' = M(a')$. Together with the current baseline implementation $i = M(a)$, these possible *lookahead(1)*-implementations are deployed. If this process is not interrupted by other change events, we can react on events to come very swiftly. Details for both scenarios are defined in [1].

The frequency at which we can tolerate change events is the inverse of the delay for computing the lookahead schedules. This delay can be reduced by two means: (i) performing composition on task graph level instead of application level and (ii) using predefined schedules for the task graphs. Both will be discussed below.

Instead of composing data-parallel applications to a data-parallel system, which is then translated to task graphs and scheduled to the infrastructure, we bookkeep the task graphs of the individual applications and just compose these task graphs. Only a new application is translated into a new task graph. Inversely, removal leads to disconnecting the corresponding task graphs and deleting transitively depending tasks.

While reusing task graphs is straight forward, reusing schedules is not, since an optimum schedule (or its approximation) does not necessarily keep the schedules for the different task graphs distinct. Instead, it might merge tasks of different task graphs into one process. Moreover, optimum schedules of individual task graphs (or their approximations) are, in general, not part of the optimal schedule for a composed system (or its approximation).

This problem is approached by modeling task graphs as *malleable tasks* and systems with *malleable task graphs*. A malleable task is a task that can be executed on $p = 1\ldots P$ processors. Its execution time is described by a non-increasing function τ of the number of processors p actually used. For each task graph the schedules s_p can be pre-computed for $p = 1\ldots P$ and $\tau(p) = TIME(s_p)$. A malleable task graph is recursively defined as a task graph over malleable tasks, i.e. nodes are ordinary task graphs or malleable task graphs and edges are the data-dependencies between them.

3.3 Scheduling Malleable Task Graphs

We now show how malleable task-graphs stemming from oblivious programs can be scheduled. The basic idea is to schedule a malleable task-graph layer by layer. By definition, a layer contains only independent tasks. Hence, the results of [4] can be applied when scheduling a layer. After scheduling a layer one builds a communication phase. In order to determine the weight function of a malleable task v, a malleable task graph within v is scheduled for $p = 1, \ldots, P$ processors. If the task-graph only contains atomic tasks then traditional algorithms are applied. The following algorithm implements these ideas:

Algorithm $schedule(G, P)$
INPUT: Malleable Task Graph $G = (V, E)$ with Layers $\Lambda_0, \ldots, \Lambda_m$
    Number of available processors P
OUTPUT: A schedule s for G
    if each $v \in V$ is atomic **then**
        determine the weights τ_v;
        compute a schedule s of G; //any traditional scheduling algorithm suffices
        return s;
    end;
    for $i := 0, \ldots, m$ **do**
        for each $v \in \Lambda_i$ **do**
            if v is malleable **then**
                let G_v be the malleable task graph contained in v;
                for $j := 1, \ldots, p$ **do**
                    $s_v(j) := schedule(G_v, j)$;
                    $T_v(j) := \min(makespan(s_v(j)), T_v(j-1)$; //$T_v(0) = \infty$
                endfor;
            else determine τ_v;
                for $j := 1, \ldots, p$ **do** $T_v(j) := \tau_v$;
            endif;
        endfor;
        compute a schedule s_i for the tasks in Λ_i using [4];
        schedule the communication from the schedule s to s_i;
        extend s by this communication and s_i;
    endfor;
    return s;
end *schedule*

The computation of the schedules in the malleable tasks need only be done once.

In order to analyze the make-span of schedules computed by the algorithm, the time required for a communication phase, the make-span of schedules for task-graphs with atomic tasks, and the make-span of scheduling independent malleable tasks has to be determined. The worst case for a communication phase is an all-to-all communication. One could extend the underlying machine model with such communications or use a scheduling approach for implementing such a communication phase (e.g the approach of [3] for the LogP-model). In this article it is sufficient that a communication phase for p processors costs time $\tau_{comm}(n,p)$ where n is the total amount of data communicated. For the purpose of this article it is just sufficient to assume that the make-span of a schedule for a task-graph G containing only atomic tasks can be estimated by an approximation factor c, i.e. the make-span is at most $c \cdot T_{\mathrm{opt}}(G,p)$ where $T_{\mathrm{opt}}(G,p)$ is the make-span of an optimal schedule for G on p processors. For scheduling n independent tasks Λ onto p processors, we use the results of [4], i.e., any schedule has a make-span of at most $\sqrt{3} \cdot T_{\mathrm{opt}}(\Lambda(p))$. A better approach will reduce the approximation factor.

We define a degree of malleability for the analysis of the make-span. Since the number of hierarchy levels play a role, we inductively define the *hierarchy level* of a malleable task $v \in V$ and the hierarchy level of a malleable task-graph as follows:

- If v is not malleable the hierarchy level of v is 0
- If v is a malleable task-graph $G = (V, E, T)$ then the hierarchy level of v is that of G
- The hierarchy level of a malleable task graph $G = (V, E, T)$ containing only atomic tasks is 0
- The hierarchy level of a malleable task graph $G = (V, E, T)$ containing at least a malleable task is $k + 1$ where k is the maximal hierarchy level of a task $v \in V$

The *work* of a malleable task graph $G = (V, E, T)$ is defined as $W(G) \triangleq \sum_{v \in V} \tau_v(1)$.
The work of a set of tasks $V' \subseteq V$ is defined as $W(V') \triangleq \sum_{v \in V'} \tau_v(1)$.

The *degree of malleability*, $\mu(G,P)$ of a malleable task graph $G = (V, E, T)$ with layers $\Lambda_0, \ldots, \Lambda_n$ is inductively defined as follows:

i. $\mu(G,p) \triangleq 1$ iff each $v \in V$ is atomic
ii. $\mu(\Lambda_i, p) \triangleq \min\left(\min_{w \in \Lambda_i} \mu(w,p), \frac{W(\Lambda_i)/p}{W(\Lambda_i)/p + \tau_{comm}(p)}\right)$ where $\mu(w,p) = 1$ if w is atomic.
iii. $\mu(G,p) \triangleq \min_{i=0}^{n-1} \mu(\Lambda_i, p)$

Theorem 1. *Let $G = (V, E, T)$ be a malleable task-graph with hierarchy level k and s be a schedule for P processors computed by the above approach. Then the make-span for the schedule s is at most: $TIME(s) \leq c \cdot 3^{k/2} \frac{T_{\mathrm{opt}}(G)}{\mu(G,P)}$ where a scheduling algorithm with approximation factor c is used for each malleable task in any level of the hierarchy which has only atomic tasks.*

Proof. For $k = 0$, the claim states that $TIME(s) \leq c \cdot T_{\mathrm{opt}}(G)$ since $\mu(G,P) = 1$ in this case. The claim holds since G only contains atomic tasks and G is scheduled by an algorithm guaranteeing an approximation factor c. For $k \geq 1$, we prove the slightly stronger claim that $TIME(s) \leq c \cdot 3^{k/2} \cdot T_{opt}(G)/\mu(G,P) - \tau_{comm}(P)$ by induction on k.

CASE $k = 1$: We have to show that $TIME(s) \leq c \cdot \sqrt{3} \cdot T_{opt}(G)/\mu(G,P) - \tau_{comm}(P)$.

Let $TIME(s_i)$ the make-span of the schedule s_i for layer Λ_i. Observe that for the case $k = 1$, any $v \in V$ is malleable task that is a task-graph containing only atomic tasks. Thus, for any of these tasks x it holds that they have a schedule s_x with $TIME(s_x) \leq c \cdot T_{opt}(x)$. Hence, the layer Λ_i, using the result of [4], it holds $TIME(s_i) \leq c \cdot \sqrt{3} \cdot T_{opt}(\Lambda_i)$ where $T_{opt}(\Lambda_i)$ is the make-span of an optimal schedule for Λ_i. It holds:

$$TIME(s) \leq \sum_{i=0}^{n-1}(TIME(s_i) + \tau_{comm}(P)) + TIME(s_n)$$

$$= \sum_{i=0}^{n}(TIME(s_i) + \tau_{comm}(P)) - \tau_{comm}(P)$$

$$\leq \sum_{i=0}^{n}(c \cdot \sqrt{3} \cdot T_{opt}(\Lambda_i) + \tau_{comm}(P)) - \tau_{comm}(P) \text{ (see above)}$$

$$\leq c \cdot \sqrt{3} \cdot \sum_{i=0}^{n}(T_{opt}(\Lambda_i) + \tau_{comm}(P)) - \tau_{comm}(P)$$

$$\leq c \cdot \sqrt{3} \cdot \sum_{i=0}^{n}\left(T_{opt}(\Lambda_i) + \frac{W(\Lambda_i)}{P}(\frac{1}{\mu(G,P)} - 1)\right) - \tau_{comm}(P)$$

$$\leq c \cdot \sqrt{3} \cdot \sum_{i=0}^{n}\frac{T_{opt}(\Lambda_i)}{\mu(G,P)} - \tau_{comm}(P) \text{ since } W(\Lambda_i)/P \leq T_{opt}(\Lambda_i)$$

$$\leq c \cdot \sqrt{3} \cdot T_{opt}(G) - \tau_{comm}(P)$$

CASE $k > 1$: We argue similar as in the case $k = 1$. By induction hypothesis, we have for any $v \in V$ a schedule s_v such that $TIME(s_v) \leq c \cdot 3^{(k-1)/2} \cdot T_{opt}(s_v)/\mu(G,P) - \tau_{comm}(P)$. Using [4] we obtain $TIME(s_i) \leq c \cdot 3^{k/2} \cdot T_{opt}(\Lambda_i)/\mu(G,P) - \tau_{comm}(P)$ for the schedule s_i of layer Λ_i. With these observations, we calculate

$$TIME(s) \leq \sum_{i=0}^{n}(TIME(s_i) + \tau_{comm}(P)) - \tau_{comm}(P) \text{ cf. case } k=1$$

$$\leq \sum_{i=0}^{n} c \cdot 3^{k/2}\frac{T_{opt}(\Lambda_i)}{\mu(G,P)} - \tau_{comm}(P) \qquad \text{see above}$$

$$= c \cdot 3^{k/2} \cdot \sum_{i=0}^{n}\frac{T_{opt}(\Lambda_i)}{\mu(G,P)} - \tau_{comm}(P)$$

$$\leq c \cdot 3^{k/2}\frac{T_{opt}(G)}{\mu(G,P)} - \tau_{comm}(P)$$

Remark 1. The bounds can be improved if a better approximation algorithm for scheduling independent malleable tasks is used. If there is an algorithm guaranteeing an approximation factor δ then the factor $3^{k/2}$ can be replaced by δ^k.

4 Conclusions

This paper discussed systems of data-parallel applications requiring high performance. Additionally, applications could be added/removed dynamically. In our scenario, the system architecture could even change due to the results of applications.

We introduced a Dynamic Service Architecture for these systems, based on static compositions and optimizations, but also allows for high performance and flexibility, by use of a lookahead scheduling mechanism. In order to realize the lookahead scheduling, malleable task scheduling is required. The lookahead scheduling and the results in malleable task scheduling are the main contributions of this paper.

Future work include an implementation of the DSA for our test bed [1]. Here, we are are also concerned with practical questions like administrating and prioritizing applications.

On a theoretical level, we are interested in extending our cost model towards the compilation and scheduling processes of the applications. Together with a modeling of the expectations of different system events, we might then be able to prioritize the creation of specific evolved systems including even the creation of systems for more than one evolution step in the future.

References

1. J. Andersson, M. Ericsson, and W. Löwe. An adaptivehigh-performance service architecture. In *Software Composition Workshop (SC) at ETAPS'04*. Electronic Notes in Theoretical Computer Science (ENTCS), 2004.
2. D. Culler, R. Karp, D. Patterson, A. Sahay, K. E. Schauser, E. Santos, R. Subramonian, and T. von Eicken. LogP: Towards a realistic model of parallel computation. In *4th ACM SIGPLAN Symposium on Principles and Practice of Parallel Programming (PPOPP 93)*, pages 1–12, 1993. published in: SIGPLAN Notices (28) 7.
3. W. Löwe and W. Zimmermann. Scheduling balanced task-graphs to logp-machines. *Parallel Computing*, 26(9):1083–1108, 2000.
4. G. Mounie, C. Rapine, and D. Trystram. Efficient approximation algorithms for scheduling malleable tasks. In *11th ACM Symposium on Parallel Algorithms and Architectures SPAA'99*, pages 23–32. ACM Press, 1999.
5. Wolf Zimmermann and Welf Löwe. An approach to machine-independent parallel programming. In *CONPAR '94, Parallel Processing*, volume 854 of *LNCS*, pages 277–288. Springer, 1994.

---

[1] *LOFAR (LOw Frequency ARray), www.lofar.org* is a radio infrastructure aiming at multi-disciplinary research of astronomers, cosmologists, physicists, climatologists, cosmologists, radio scientists, and IT researchers. It consists of geographically distributed digital sensors connected to computing nodes with a high-speed network. Sensors are distributed over distances of 400 km and the system will produce data at a rate 25 Tbits/s. The Swedish initiative *LOIS (LOFAR Outrigger Scandinavia)* aims, among others, at extending and enhancing the IT infrastructure capabilities of LOFAR.

Topic 4
Compilers for High Performance

Hans P. Zima, Siegfried Benkner, Michael F.P. O'Boyle, and Beniamino Di Martino

Topic Chairs

This topic deals with a range of subjects concerning the compilation of programs for high performance architectures, from general-purpose platforms to specific hardware designs, including language aspects, program analysis, program transformation systems for optimization of resource utilization, the interaction between compiler and hardware/operating system, runtime systems, and feedback-oriented and adaptive compilation.

Out of the 19 papers submitted to this topic 5 were accepted for presentation at the conference. We provide here a short outline of the topics addressed in these contributions.

The paper "More Legal Transformations for Locality" by Cedric Bastoul and Paul Feautrier deals with an important problem in compilers applying program transformations to improve locality by proposing a method that makes such transformations applicable in situations with complex data dependences.

Bertrand Anckaert, Frederick Vandeputte, Bruno de Bus, Bjorn De Sutter, and Koen De Bosschere address the issue of exploiting link-time optimization of IA64 binaries on an EPIC platform.

The paper "A Polyhedral Approach to Ease the Composition of Program Transformations" by Albert Cohen, Sylvain Girbal, and Oliver Temam uses a formal framework based on a unified polyhedral representation of loops an statements for improving the robustness of loop transformations and supporting their composition.

The research note "Compiler-Guided Code Restructuring for Improving Instruction TLB Energy Behavior" by I.Kadayif, M.Kandemir, and I.Demirkiran describes a method for restructuring the code to increase the chances that virtual-to-physical address translation can be performed without going through the instruction TLB.

M. Kandemir, O.Ozturk, M.J. Irwin, and I.Kolcu explore the possibility of compressing infrequently used data for increasing the effectiveness of low-power operating modes.

More Legal Transformations for Locality

Cédric Bastoul[1] and Paul Feautrier[2]

[1] Laboratoire PRiSM, Université de Versailles Saint Quentin
45 avenue des États-Unis, 78035 Versailles Cedex, France
cedric.bastoul@prism.uvsq.fr
[2] École Normale Supérieure de Lyon
46 Allée d'Italie, 60364 Lyon, France
paul.feautrier@ens-lyon.fr

Abstract. Program transformations are one of the most valuable compiler techniques to improve data locality. However, restructuring compilers have a hard time coping with data dependences. A typical solution is to focus on program parts where the dependences are simple enough to enable any transformation. For more complex problems is only addressed the question of checking whether a transformation is legal or not. In this paper we propose to go further. Starting from a transformation with no guarantee on legality, we show how we can correct it for dependence satisfaction with no consequence on its locality properties. Generating code having the best locality is a direct application of this result.

1 Introduction

Exploiting data locality is one of the keys to achieve high performance level in most computer systems and hence one of the main challenges for optimizing compilers. The basic framework for increasing the cache hit rates aims at moving references to a given memory cell (or cache line) to neighboring iterations of some innermost loop. Let us consider for instance two accesses to the same memory cell. It seems probable that the longer the time interval between these accesses is, the higher the probability of the first reference to be evicted from the cache. Since such a transformation modifies the operation execution order, the existence of a good solution highly depends on data dependences.

To bypass the dependence problem, most of the existing methods apply only to perfect loop nests in which dependences are non-existent or have a special form (fully permutable loop nests) [18]. To enlarge their application domain some preprocessing, e.g. *loop skewing* or *code sinking*, may enable them [18, 1, 8]. More ambitious techniques do not lay down any requirement on dependences, but are limited to propose *solution candidates* having some locality properties then to *check* them for legality [10, 5]. If the candidate is proved to violate dependences, then another candidate having less interesting properties is studied. In this paper, we present a method that goes beyond checking by adjusting an optimizing transformation for dependence satisfaction, without modifying its locality properties. This technique can be used to correct a transformation candidate as well as to replace preprocessing.

This paper is organized as follows. In section 2 is outlined the background of this work. Section 3 deals with the transformations in the polyhedral model and focuses on both their dependences constraints and locality properties. Section 4 shows how it is possible to correct a transformation for legality. Lastly, section 6 concludes and discusses future work.

2 Background and Notations

A loop in an imperative language like C or FORTRAN can be represented using a n-entry column vector called its *iteration vector*:

$$x = \begin{pmatrix} i_1 \\ i_2 \\ \vdots \\ i_n \end{pmatrix},$$

where i_k is the kth loop index and n is the innermost loop. The surrounding loops and conditionals of a statement define its *iteration domain*. The statement is executed once for each element of the iteration domain. When loop bounds and conditionals only depend on surrounding loop counters, formal parameters and constants, the iteration domain can always be specified by a set of linear inequalities defining a polyhedron [11]. The term *polyhedron* will be used in a broad sense to denote a *convex set of points in a lattice* (also called $\mathbb{Z}$-polyhedron or lattice-polyhedron), i.e. a set of points in a $\mathbb{Z}$ vector space bounded by affine inequalities [15]. A maximal set of consecutive statements in a program with such polyhedral iteration domains is called a *static control part* (SCoP) [4]. Figure 1 illustrates the correspondence between static control and polyhedral domains. Each integral point of the polyhedron corresponds to an *operation*, i.e. an instance of the statement. The notation $S(x)$ refers to the operation instance of the statement S with the iteration vector x. The execution of the operations follows *lexicographic order*. This means in a n-dimensional polyhedron, the operation corresponding to the integral point defined by the coordinates $(a_1...a_n)$ is executed before those corresponding to the coordinates $(b_1...b_n)$ iff

$$\exists i, 1 \leq i < n, (a_1...a_i) = (b_1...b_i) \wedge a_{i+1} < b_{i+1}.$$

Each statement may include one or several *references* to arrays (or scalars, i.e. some particular cases of arrays). When the subscript function $f(x)$ of a reference is affine, we can write it $f(x) = Fx + a$ where F is called the *subscript matrix* and a is a constant vector. For instance, the reference to the array B in figure 1(a) is $B[f(x)]$ with $f\begin{pmatrix}i\\j\end{pmatrix} = \begin{bmatrix}1 & 1\\2 & 0\end{bmatrix}\begin{pmatrix}i\\j\end{pmatrix} + \begin{pmatrix}0\\1\end{pmatrix}$.

In this paper, matrices are always denoted by capital letters, vectors and functions in vector spaces are not. When an element is statement-specific, it is subscripted like A_S, excepted in general formulas where all elements are statement-specific in order to avoid too heavy notations.

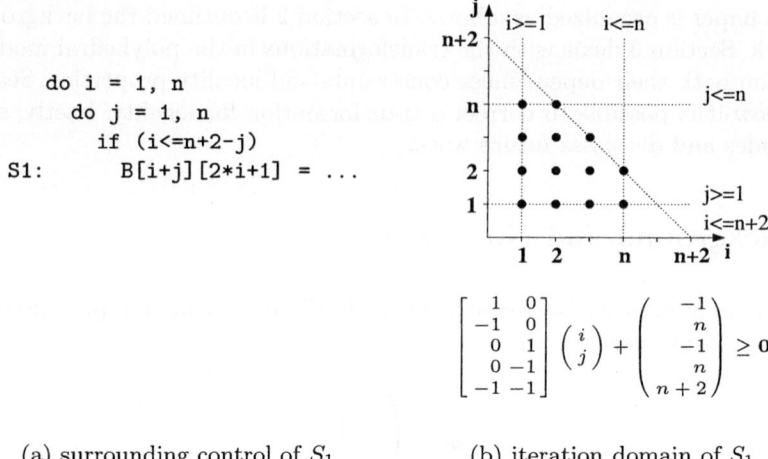

(a) surrounding control of S_1 (b) iteration domain of S_1

Fig. 1. Static control and corresponding iteration domain.

3 Affine Transformations for Locality

3.1 Formulation

The goal of a transformation is to modify the original execution order of the operations. A convenient way to express the new order is to give for each operation an execution date. However, defining all the execution dates separately would usually require very large scheduling systems. Thus optimizing compilers build schedules at the statement level by finding a function specifying an execution time for each instance of the corresponding statement. These functions are chosen affine for multiple reasons: this is the only case where we are able to decide exactly the transformation legality and where we know how to generate the target code. Thus, scheduling functions have the following shape:

$$\theta_S(x_S) = T_S x_S + t_S, \qquad (1)$$

where x_S is the iteration vector, T_S is a constant transformation matrix and t_S is a constant vector (possibly including affine parametric expressions using the structure parameters of the program i.e. the symbolic constants, mostly array sizes or iteration bounds).

It has been extensively shown that linear transformations can express most of the useful transformations. In particular, loop transformations (such as loop reversal, permutation or skewing) can be modeled as a simple particular case called unimodular transformations (the T_S matrix has to be square and has determinant ±1) [2, 16]. Complex transformations such as tiling [17] can be achieved using linear transformations as well [19]. These transformations modify the source polyhedra into target polyhedra containing the same points, thus with a new lexicographic order. Considering an original polyhedron defined by the system of affine constraints $Ax + c \geq 0$ and the transformation function θ leading to the target index $y = Tx$, we deduce that the transformed polyhedron can be defined by $(AT^{-1})y + c \geq 0$ (there exits more convenient way to describe the target polyhedron as discussed in [3]). For instance, let us consider the polyhedron in figure 2(a) and the transformation function $\theta\binom{i}{j} = \begin{bmatrix} 1 & 1 \\ 0 & 1 \end{bmatrix}\binom{i}{j}$. The corresponding transformation is a well known *iteration space skewing* and the resulting polyhedron is shown in figure 2(c).

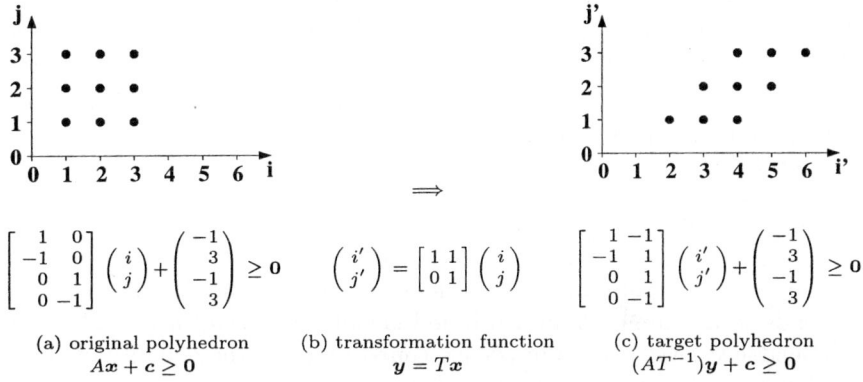

(a) original polyhedron $Ax + c \geq 0$

(b) transformation function $y = Tx$

(c) target polyhedron $(AT^{-1})y + c \geq 0$

Fig. 2. A skewing transformation.

3.2 Legality

It is not possible to apply any transformation to a program without changing its semantics. Statement instances reading and writing the same memory location should be kept in the same order. These operations are said to be *dependent* on each other. The dependence relations inside a program must direct the transformation construction. In this section, we recall how dependences can be expressed exactly using linear (in)equalities. Then we show how to build the legal transformation space where each program transformation has to be found.

Dependence Graph. A convenient way to represent the scheduling constraints between the program operations is the *dependence graph*. In this directed graph, each program statement is represented using a unique vertex, and the existing dependence relations between statement instances are represented using edges. Each vertex is labelled with the iteration domain of the corresponding statement and the edges with the dependence polyhedra describing the dependence relation between the source and destination statements. The dependence relation can be defined in the following way:

Definition 1. *A statement R **depends** on a statement S (written SδR) if there exits an operation $S(x_1)$, an operation $R(x_2)$ and a memory location m such that:*

1. *$S(x_1)$ and $R(x_2)$ refer the same memory location m, and at least one of them writes to that location;*
2. *x_1 and x_2 respectively belong to the iteration domain of S and R;*
3. *in the original sequential order, $S(x_1)$ is executed before $R(x_2)$.*

From this definition, we can easily describe the *dependence polyhedra* of each dependence relation between two statements with affine (in)equalities. In these polyhedra, every integral point represents a dependence between two instances of the corresponding statements. The systems have the following components:

1. *Same memory location*: assuming that m is an array location, this constraint is the equality of the subscript functions of a pair of references to the same array: $F_S x_S + a_S = F_R x_R + a_R$.
2. *Iteration domains*: both S and R iteration domains can be described using affine inequalities, respectively $A_S x_S + c_S \geq 0$ and $A_R x_R + c_R \geq 0$.
3. *Precedence order*: this constraint can be separated into a disjunction of as many parts as there are common loops to both S and R. Each case corresponds to a common loop depth and is called a *dependence level*. For each dependence level l, the precedence constraints are the equality of the loop index variables at depth lesser to l: $x_{R,i} = x_{S,i}$ for $i < l$ and $x_{R,l} > x_{S,l}$ if l is less than the common nesting level. Otherwise, there are no additional constraints and the dependence only exists if S is textually before R. Such constraints can be written using linear inequalities: $P_S x_S - P_R x_R + b \geq 0$.

Thus, the dependence polyhedron for $S\delta R$ at a given level l and for a given pair of references p can be described using the following system of (in)equalities:

$$\mathcal{D}_{S\delta R, l, p} : D \begin{pmatrix} x_S \\ x_R \end{pmatrix} + d = \begin{bmatrix} F_S & -F_R \\ A_S & 0 \\ 0 & A_R \\ P_S & -P_R \end{bmatrix} \begin{pmatrix} x_S \\ x_R \end{pmatrix} + \begin{pmatrix} a_S - a_R \\ c_S \\ c_R \\ b \end{pmatrix} \begin{matrix} = \\ \geq \\ \geq \\ \geq \end{matrix} 0 \qquad (2)$$

There is a dependence $S\delta R$ if there exists an integral point inside $\mathcal{D}_{S\delta R,l,p}$. This can be easily checked with some linear integer programming tool like PipLib[1] [6]. If this polyhedron is not empty, there is an edge in the dependence graph from the vertex corresponding to S up to the one corresponding to R and labelled with $\mathcal{D}_{S\delta R,l,p}$. For the sake of simplicity we will ignore subscripts l and p and refer in the following to $\mathcal{D}_{S\delta R}$ as the only dependence polyhedron describing $S\delta R$.

Legal Transformation Space. Considering the transformations as scheduling functions, the time interval in the target program between the executions of two operations $R(x_R)$ and $S(x_S)$ is

$$\Delta_{R,S}\begin{pmatrix} x_S \\ x_R \end{pmatrix} = \theta_R(x_R) - \theta_S(x_S). \qquad (3)$$

---
[1] PipLib is freely available at http://www.prism.uvsq.fr/~cedb

If there exists a dependence $S\delta R$, i.e. if $\mathcal{D}_{S\delta R}$ is not empty, then $\Delta_{R,S}\begin{pmatrix}\boldsymbol{x}_S\\\boldsymbol{x}_R\end{pmatrix}-\mathbf{1}$ must be a nonnegative form in $\mathcal{D}_{S\delta R}$ (intuitively, the time interval between two operations $R(\boldsymbol{x}_R)$ and $S(\boldsymbol{x}_S)$ such that $R(\boldsymbol{x}_R)$ depends on $S(\boldsymbol{x}_S)$ must be at least 1, the smallest time interval: this guarantees that the operation $R(\boldsymbol{x}_R)$ is executed after $S(\boldsymbol{x}_S)$ in the target program). This affine function can be expressed in terms of D and $\boldsymbol{d}$ by applying Farkas Lemma (see Lemma 1) [7].

Lemma 1. *(Affine form of Farkas Lemma [15]) Let $\mathcal{D}$ be a nonempty polyhedron defined by the inequalities $A\boldsymbol{x} + \boldsymbol{b} \geq \boldsymbol{0}$. Then any affine function $f(\boldsymbol{x})$ is nonnegative everywhere in $\mathcal{D}$ iff it is a positive affine combination:*

$$f(\boldsymbol{x}) = \lambda_0 + \Lambda^T(A\boldsymbol{x}+\boldsymbol{b}), \text{ with } \lambda_0 \geq \boldsymbol{0} \text{ and } \Lambda^T \geq 0,$$

where λ_0 and Λ^T are called Farkas multipliers.

According to this Lemma, for each edge in the dependence graph, we can find a positive vector λ_0 and matrix Λ^T (the Farkas multipliers) such that:

$$T_R\boldsymbol{x}_R + \boldsymbol{t}_R - (T_S\boldsymbol{x}_S + \boldsymbol{t}_S) - \mathbf{1} = \lambda_0 + \Lambda^T\left(D\begin{pmatrix}\boldsymbol{x}_S\\\boldsymbol{x}_R\end{pmatrix}+\boldsymbol{d}\right), \lambda_0 \geq \boldsymbol{0}, \Lambda^T \geq 0. \quad (4)$$

This formula can be split in as many equalities as there are independent variables ($\boldsymbol{x}_S$ and $\boldsymbol{x}_R$ components, parameters and scalar value) by equating their coefficients in both sides of the formula. The Farkas multipliers can be eliminated by using the Fourier-Motzkin projection algorithm [15] in order to find the constraints on the transformation functions. The constraint systems describe the legal transformation space, where each integral point corresponds to a legal solution.

3.3 Properties

Program transformations for locality aim at bringing the processing of some memory cells closer. The general framework using affine schedules is to find partial transformation functions (only the first few dimensions of the functions are defined) such that the partial execution dates of the operations referring to a given datum are the same. In this way, the operations have neighboring schedules and the datum may stay in the cache during the time intervals between the accesses. The framework ends by applying a completion procedure to achieve an invertible transformation function (see [18] for references).

For instance, let us consider self-temporal locality and a reference $B[f(\boldsymbol{x})]$ to an array B with the affine subscript function $f(\boldsymbol{x}) = F\boldsymbol{x} + \boldsymbol{a}$. Two instances of this reference, $B[f(\boldsymbol{x}_1)]$ and $B[f(\boldsymbol{x}_2)]$ refers the same memory location iff $f(\boldsymbol{x}_1) = f(\boldsymbol{x}_2)$, that is when $F\boldsymbol{x}_1 + \boldsymbol{a} = F\boldsymbol{x}_2 + \boldsymbol{a}$, then iff $F\boldsymbol{x}_r = \boldsymbol{0}$ with $\boldsymbol{x}_r = \boldsymbol{x}_1 - \boldsymbol{x}_2$. Thus there is self-temporal reuse when $\boldsymbol{x}_r \in \ker F$. The basis vectors of $\ker F$ give the reuse directions for the reference $B[f(\boldsymbol{x})]$, if $\ker F$ is empty, there is no self-temporal reuse for the corresponding reference. The reuse can be exploited if the transformed iteration order follows one of the reuse directions. Then we have to find an orthogonal vector to the chosen reuse direction to be the first part of the transformation matrix T. If this partial transformation do not violate dependences, we have many choices for the completion procedure in order for the transformation function to be instance-wise, by considering artificial dependences [13, 9] or not [3]. For instance, let us consider the following pseudo-code:

```
         do i = 1, n
            do j = 1, n
S1:               ... B[j] ...
```

the subscript function of the reference $B[j]$ is $f\begin{pmatrix}i\\j\end{pmatrix}=\begin{bmatrix}0&1\\0&0\end{bmatrix}\begin{pmatrix}i\\j\end{pmatrix}$, the kernel of the subscript matrix is then $\ker F = \mathrm{span}\ \{(1,0)\}$. Thus there is reuse generated by the reference $B[j]$, and we can exploit it thank to a transformation matrix built with an orthogonal vector to the reuse direction, e.g. [0 1] and its completion to a unimodular transformation matrix as described in [9]: $T=\begin{bmatrix}0&1\\1&0\end{bmatrix}$. The transformation function would be $\theta\begin{pmatrix}i\\j\end{pmatrix}=\begin{bmatrix}0&1\\1&0\end{bmatrix}\begin{pmatrix}i\\j\end{pmatrix}$, i.e. a loop interchange (the reader may care to verify that this solution do exploit the reuse of the reference $B[j]$). It is easy to generalize this method for several references by considering not only a reuse direction vector, but a reuse direction space (built with one basis vector per reference). It appears that there are a lot of liberty degrees when looking for a transformation improving self-temporal locality, since it is possible to choose the reuse direction space, the completion method and the constant vector of the transformation function.

Let us consider self-temporal locality and a transformation candidate before completion $\theta_{Sc}(\boldsymbol{x}_S) = T_{Sc}\boldsymbol{x}_S$. This function has the property that, modified in the following way:

$$\theta_S(\boldsymbol{x}_S) = C_S T_{Sc} \boldsymbol{x}_S + \boldsymbol{t}_S, \tag{5}$$

where C_S is an invertible matrix and $\boldsymbol{t}_S$ is a constant vector, the locality properties are left unmodified for each time step. Intuitively, if θ_{Sc} gives the same execution date for $\boldsymbol{x}_1$ and $\boldsymbol{x}_2$, then the transformed function θ_S does it as well. In the same way if the dates are different with θ_{Sc}, then the transformed function θ_S returns different dates. But while the values of C_S and $\boldsymbol{t}_S$ do not change the self-temporal locality properties[2], they can change the legality of the transformation.

Transformation expressions similar to (5) and having the same type of degrees of freedom can be used to achieve every type of locality (*self* or *group - temporal* or *spatial*) [16, 5]. The challenge is, considering the candidate transformation matrices T_{Sc}, to find the *corrected matrices* $C_S T_{Sc}$ and the constant vectors $\boldsymbol{t}_S$ in order for the transformation system to be legal for dependences.

---

[2] A more formal discussion on this property, showing that locality transformations have to respect rank constraints not modified by C_S and $\boldsymbol{t}_S$ can be found in [5].

4 Finding Legal Transformations

Optimizing compilers typically decouple the properties that the transformation functions have to satisfy to achieve optimization and legality. The basic framework is first to find the best transformations (e.g. for data locality improvement, which references carry the most reuse and necessitate new access patterns, which rank constraints should be respected by the corresponding transformation functions, etc.), then to *check* if a candidate transformation is legal or not[3]. If not, build and try another candidate, and so on. The major advantage of such a framework is to focus firstly on the most interesting properties, and the main drawback is to forsake these properties if a legal transformation is not directly found after a simple check of a candidate solution. We saw in section 3.3 that there exists an infinity of transformation functions having the same properties as a candidate transformation (see formula 5). Thus, it is not possible to check all these transformations to find a legal one. In this section we study another way: we show how to find, when possible, the unknown components $C_S T_{Sc}$ and t_S of formula 5 in order to correct the transformations for legality.

This problem can be solved in an iterative way, each dimension being considered as a stand-alone transformation. Each row of $C_S T_{Sc}$ is a linear combination of the rows of T_{Sc}. Thus, the unknown in the ith algorithm iteration are, for each statement, the linear combination coefficients building the ith row of $C_S T_{Sc}$ from T_{Sc} and the constant factor of the corresponding t_S entry. After each iteration, we have to update the dependence graph because there is no need to consider the dependences already satisfied. Thus, to find a solution is easier as the algorithm iterates. The algorithm is shown in figure 3.

Let us illustrate how the algorithm works using the example in figure 4. Suppose that an optimizing compiler would like to exploit the data reuse generated by the references to the array A of the program in figure 4(a) and that it suggests the transformation candidates in figure 4(b). As shown by the graph describing the resulting operation execution order, where each arrow represents a dependence relation and each backward arrow is a dependence violation, the transformation system is not legal. The correction algorithm modifies successively each transformation dimension. Each stand-alone transformation splits up the operations into sets such that there are no backward arrows between sets. The algorithm stops when there are no more backward arrows or when every dimension has been corrected. Then any polyhedral code generator, like CLooG[4] [3], can generate the target code. Choosing transformation coefficients as small as possible (step 1(c)i) is a heuristic helping code generators to avoid control overhead.

The correctness of the algorithm comes from two properties: (1) the target transformations are legal, (2) the C_S matrices are invertible. The legality is achieved because each transformation part is chosen in the legal transformation

---
[3] This can be done easily by instantiating the transformation functions in the legal transformation space as defined in section 3.2 then checking for the feasibility of the constraint system with any linear algebra tool.

[4] CLooG is freely available at http://www.prism.uvsq.fr/~cedb

Correction Algorithm: Adjust a transformation system to respect dependences.

Input: a dependence graph DG, the transformation candidates $\theta_{Sc}(x_S) = T_{Sc}x_S$.
Output: the legal transformations $\theta_S(x_S) = C_S T_{Sc} x_S + t_S$.

1. for dimension i = 1 to maximum dimension of T_{Sc}
   (a) build the legal transformation space with:
       - for each edge in DG, the constraints of (4) for the ith row of T_{Rc} and T_{Sc}
       - the constraints equating the ith row entries of each $C_S T_{Sc}$ with a linear combination of T_{Sc} entries whose coefficients are unknown
   (b) for each statement, remove from the solution space the trivial solution where $\forall j \geq i$ the linear combination coefficient of the jth row of T_{Sc} is null
   (c) if the solution space is empty, return ∅, else
       i. pick the solution giving for each statement the minimum values for the entries of the ith row of $C_S T_{Sc}$ and the ith element of t_S
       ii. update DG: for each edge in DG, add to the dependence polyhedron the constraint equating the ith dimension of $C_S T_{Sc} x_S + t_S$ of the statements labelling the source and destination vertices (this may empty the polyhedron for integral solutions)
       iii. if every dependence polyhedra in DG are empty, goto 2
       iv. for each statement, update the candidate transformation T_{Sc}:
           - replace a row such that the corresponding linear combination coefficient is not null with the ith row
           - replace the ith row with the ith row of $C_S T_{Sc}$
2. return the transformation functions $\theta_S(x_S) = C_S T_{Sc} x_S + t_S$.

Fig. 3. Algorithm to correct the transformation functions.

space (step 1a). The second property follows from the updating policy (step 1(c)iv): at start the C_S matrices are identities. During each iteration, we exchange their rows, multiply some rows by non null constants (as guaranteed by step 1b) and add to these rows a linear combination of the other rows. Each of these transformations does not modify the invertibility property.

5 Related Work

In compensation of the need for very simple dependences, first works on compiler techniques for improving data locality discuss *enabling transformations* to modify the program in such a way that the proposed method can apply. Wolf and Lam [16] proposed in their seminal *data locality optimizing algorithm* to use *skewing* and *reversal*[5] to enable *tiling* as in previous works on automatic parallelization. McKinley et al. [14] proposed a technique based on a detailed cost model that drives the use of *fusion* and *distribution* mainly to enable loop *permutation*. Such methods are limited by the set of directives they use (like *fuse*

---

[5] An exhaustive survey on loop transformations can be found in [18].

```
         do i = 1, n
            do j = 1, n
               do k = 1, n
S1:                 A(j,k) = A(j,k) + B(i,j,k) / A(j,k-1)
S2:      c = A(n,n) + 1
```

(a) Original program

$$\theta_{S1c}\begin{pmatrix}i\\j\\k\end{pmatrix} = \begin{bmatrix}1&0&0\\0&1&0\\0&0&1\end{bmatrix}\begin{bmatrix}0&1&0\\0&0&-1\\0&0&0\end{bmatrix}\begin{bmatrix}i\\j\\k\end{bmatrix} + \begin{pmatrix}0\\0\\0\end{pmatrix} \; ; \; \theta_{S2c} = \begin{pmatrix}0\\0\\0\end{pmatrix}$$

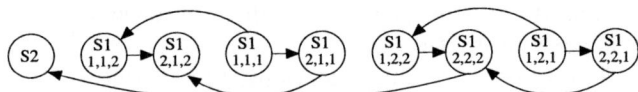

(b) Transformation function candidates

$$\theta_{S1c}\begin{pmatrix}i\\j\\k\end{pmatrix} = \begin{bmatrix}1&0&0\\\overline{0\;1\;0}\\0&0&1\end{bmatrix}\begin{bmatrix}0&1&0\\0&0&-1\\0&0&0\end{bmatrix}\begin{bmatrix}i\\j\\k\end{bmatrix} + \begin{pmatrix}0\\\overline{0}\\0\end{pmatrix} \; ; \; \theta_{S2c} = \begin{pmatrix}n\\0\\0\end{pmatrix}$$

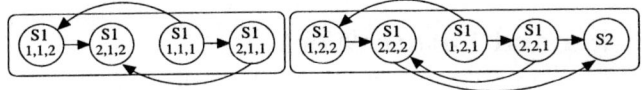

(c) First correction iteration

$$\theta_{S1c}\begin{pmatrix}i\\j\\k\end{pmatrix} = \begin{bmatrix}1&0&0\\0&-1&0\\0&0&1\end{bmatrix}\begin{bmatrix}0&1&0\\0&0&-1\\0&0&0\end{bmatrix}\begin{bmatrix}i\\j\\k\end{bmatrix} + \begin{pmatrix}0\\0\\\overline{0}\end{pmatrix} \; ; \; \theta_{S2c} = \begin{pmatrix}n\\n\\0\end{pmatrix}$$

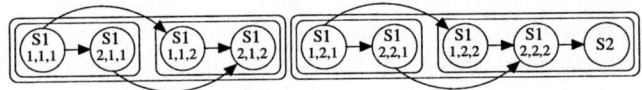

(d) Second and last correction iteration

```
         do j = 1, n
            do k = 1, n
               do i = 1, n
S1:                 A(j,k) = A(j,k) + B(i,j,k) / A(j,k-1)
S2:      c = A(n,n) + 1
```

(e) Target program

Fig. 4. Iterative transformation correction principle ($n = 2$ for graphs).

or *skew*) and because they have to apply them in a definite order. We claim that giving (and correcting) scheduling functions is more complete and has better compositionality properties.

A significant step on preprocessing techniques to produce fully permutable loop nests has been achieved by Ahmed et al. [1]. They use the Farkas Lemma to find a valid *code sinking*-like transformation if it exists. But this transformation is still independent from the optimization itself and it is limited to produce a fully permutable loop nest. The method proposed in this paper may find solutions while it is not possible to achieve such a loop nest. Recently, Griebl et al. [8] proposed to use well known space-time mapping [12] as a preprocessing technique for *tiling*. Our method can be included in their framework to find legal time mapping with locality properties.

Reasoning directly on scheduling functions, Li and Pingali proposed a completion algorithm to build a non-unimodular transformation function from a partial matrix, such that starting from a legal transformation, the completed transformation stay legal for dependences [13]. In the same spirit, Griebl et al. extended an arbitrary matrix describing a legal transformation to a square invertible matrix [9]. In contrast, we show in this paper how to find the valid functions before completion.

6 Conclusion and Future Work

In this paper is presented a general method correcting a program transformation for legality with no consequence on the data locality properties. It has been implemented in the Chunky prototype [5], advantageously replacing usual *enabling* preprocessing techniques and saving a significant amount of interesting transformations from being ignored. It could be used combined with a wide range of existing data locality improvement methods, for the single processor case as well as compiling techniques for parallel systems using space-time mappings [12].

Further implementation work is necessary to handle real-life benchmarks in our prototype and to provide full statistics on corrected transformations. Moreover, the question of scalability is left open since, for several tenth of deeply nested statements, the number of unknown in the constraint systems can become embarrassingly large. Splitting up the problem according to the dependence graph is a solution under investigation.

Acknowledgements

The authors would like to thank the CC'12 International Conference on Compiler Construction anonymous reviewers for having inspired this paper by pointing out their interest on this part of our work. We also wish to thank the Euro-Par anonymous reviewers for their help in improving the quality of the paper.

References

1. N. Ahmed, N. Mateev, and K. Pingali. Tiling imperfectly-nested loop nests. In *SC'2000 High Performance Networking and Computing*, Dallas, november 2000.
2. U. Banerjee. Unimodular transformations of double loops. In *Advances in Languages and Compilers for Parallel Processing*, pages 192–219, Irvine, august 1990.
3. C. Bastoul. Efficient code generation for automatic parallelization and optimization. In *ISPDC'03 IEEE International Symposium on Parallel and Distributed Computing*, pages 23–30, Ljubljana, october 2003.
4. C. Bastoul, A. Cohen, S. Girbal, S. Sharma, and O. Temam. Putting polyhedral transformations to work. In *LCPC'16 International Workshop on Languages and Compilers for Parallel Computers, LNCS 2958*, pages 209–225, College Station, october 2003.
5. C. Bastoul and P. Feautrier. Improving data locality by chunking. In *CC'12 International Conference on Compiler Construction, LNCS 2622*, pages 320–335, Warsaw, april 2003.
6. P. Feautrier. Parametric integer programming. *RAIRO Recherche Opérationnelle*, 22(3):243–268, 1988.
7. P. Feautrier. Some efficient solutions to the affine scheduling problem: one dimensional time. *International Journal of Parallel Programming*, 21(5):313–348, october 1992.
8. M. Griebl, P. Faber, and C. Lengauer. Space-time mapping and tiling – a helpful combination. *Concurrency and Computation: Practice and Experience*, 16(3):221–246, march 2004.
9. M. Griebl, C. Lengauer, and S. Wetzel. Code generation in the polytope model. In *PACT'98 International Conference on Parallel Architectures and Compilation Techniques*, pages 106–111, 1998.
10. I. Kodukula, N. Ahmed, and K. Pingali. Data-centric multi-level blocking. In *ACM SIGPLAN'97 Conference on Programming Language Design and Implementation*, pages 346–357, Las Vegas, june 1997.
11. D. Kuck. *The Structure of Computers and Computations*. John Wiley & Sons, Inc., 1978.
12. C. Lengauer. Loop parallelization in the polytope model. In *International Conference on Concurrency Theory, LNCS 715*, pages 398–416, Hildesheim, August 1993.
13. W. Li and K. Pingali. A singular loop transformation framework based on non-singular matrices. *International Journal of Parallel Programming*, 22(2):183–205, April 1994.
14. K. McKinley, S. Carr, and C. Tseng. Improving data locality with loop transformations. *ACM Transactions on Programming Languages and Systems*, 18(4):424–453, july 1996.
15. A. Schrijver. *Theory of linear and integer programming*. John Wiley & Sons, 1986.
16. M. Wolf and M. Lam. A data locality optimizing algorithm. In *ACM SIGPLAN'91 Conference on Programming Language Design and Implementation*, pages 30–44, New York, june 1991.
17. M. Wolfe. Iteration space tiling for memory hierarchies. In *3rd SIAM Conference on Parallel Processing for Scientific Computing*, pages 357–361, december 1987.
18. M. Wolfe. *High performance compilers for parallel computing*. Addison-Wesley Publishing Company, 1995.
19. J. Xue. On tiling as a loop transformation. *Parallel Processing Letters*, 7(4):409–424, 1997.

Link-Time Optimization of IA64 Binaries

Bertrand Anckaert, Frederik Vandeputte, Bruno De Bus,
Bjorn De Sutter, and Koen De Bosschere

Ghent University, Electronics and Information Systems Department
Sint-Pietersnieuwstraat 41 9000 Gent, Belgium
{banckaer,fgvdeput,bdebus,brdsutte,kdb}@elis.UGent.be

Abstract. The features of the IA64 architecture create new opportunities for link-time optimization. At the same time they complicate the design of a link-time optimizer. This paper examines how to exploit some of the opportunities for link-time optimization and how to deal with the complications. The prototype link-time optimizer that implements the discussed techniques is able to reduce the code size of statically linked programs with 19% and achieves a speedup of 5.4% on average.

1 Introduction

On the EPIC (Explicitly Parallel Instruction Computer) platform, the compiler determines which instructions should be executed in parallel. This responsibility corresponds to the belief that better performance can be achieved by shifting the parallelism extraction task from hardware (as in superscalar out-of-order processors) to the compiler: the hardware becomes less complex and the compiler can exploit its much wider view on the code [8].

Unfortunately compilers only have a fragmented program view: most compilers compile and optimize all source code files independently of each other. Even when all source code is compiled together, the libraries are still compiled separately, and hence not optimized for any specific program. The resulting lack of compile-time whole-program optimization is particularly bad for address computations: As the linker decides on the final program layout in memory, code and data addresses are not known at compile time. The compiler therefore has to generate *relocatable* code, which is most often far from optimal.

Optimizing linkers try to overcome these problems by adding a link-time optimization pass in the tool chain. Optimizing linkers take compiled object files and precompiled code libraries as input, and optimize them together to produce smaller or faster binaries. In this paper we present our link-time optimizer for the IA64 architecture. Our main contributions are:

- We extend the existing work on Global Offset Table optimizations by creating a second global pointer at link-time.
- We show how existing link-time liveness analysis can be adapted to deal with the rather peculiar register files of the IA64 architecture.
- We demonstrate how the set of branch registers can be exploited more effectively with whole-program optimization.

This paper is organized as follows. Section 2 presents a short overview of our link-time optimizer. IA64-specific whole-program analyses and optimizations are the topic of Section 3. Our results are summarized in Section 4, and Section 5 discusses related work. We conclude in Section 6.

2 Link-Time Optimizer Overview

Our link-time optimizer for the IA64 architecture is developed on top of Diablo [3] (http://www.elis.ugent.be/diablo), a portable and retargetable link-time program editor framework. Any application developed with Diablo first links the compiled program object files and the needed library code. The linked program is disassembled, and an interprocedural control flow graph (ICFG) is constructed via call-backs to object file format and architecture back-ends. Given the rather clean nature of the IA64 application binary interface, the ICFG construction is trivial. Nodes in the ICFG model basic blocks, while edges model execution paths. Basic blocks consist of an instruction sequence, in which each instruction has both an architecture-independent and architecture-dependent part.

On the ICFG all whole-program analyses and optimizations are performed iteratively, since applying one optimization may trigger other optimization opportunities. The core of Diablo provides a number of architecture-independent analyses and optimizations, such as interprocedural liveness analysis and unreachable code elimination that operate on the architecture-independent part of the instruction representation. Additional architecture-dependent analyses and optimizations, such as peephole optimization, and semantics-based analyses, such as constant propagation, rely on call-backs.

Once all optimizations are applied, the code layout is determined (optionally using profile information), and the code is scheduled into the parallel instruction bundles of the EPIC, and assembled into binary code again.

3 IA64 Whole-Program Optimizations

3.1 Global Offset Table Optimizations

Since the linker determines the final memory layout of the code and data in a program, the compiler does not know the final addresses of the global data. It must therefore assume that the data may be distributed throughout the 64-bit address space. Since 64-bit addresses cannot be encoded into a single instruction efficiently, the compiler generates code that indirectly accesses this data. Before each access, the data's address is loaded from a Global Offset Table (GOT) using a special purpose global pointer (GP) register that always points to the GOT.

Unfortunately the compiler has to assume that one GOT will not suffice for the final program. First, the compiler does not know how much data will end up in the final program. Moreover, the size of a GOT is limited: all addresses in a GOT need to be accessed through the same base GP value, and the GP-relative offsets used to access the elements in the table is limited by 22-bit width of immediate instruction operands. As a result, each compiler module (a single source code file or a group of files compiled together) is given a separate GOT. Every time control flow enters a module, the GP's value is reset to point to the corresponding GOT. The major drawbacks of this solution are that (1) global data accesses require additional loads because of the indirection through the GOT, and (2) the GP value needs to be reset again and again.

The latter drawback can be overcome at link-time by combining the small GOTs of different modules into fewer larger GOTs, eliminating all GP resets when control flow crosses the corresponding module boundaries. The former drawback can be avoided for global data that is allocated nearby the GOT itself, by computing the data's address with an addition to the GP instead of loading it from the GP. Both solutions are well known and were implemented on the Alpha 64-bit platform [9]. On the IA64 however, the read-only data section is not located nearby the GOT, as on the Alpha, and therefore much fewer address loads can be converted into additions.

Our link-time solution to this problem is to create a second GP to point to the read-only data. This second GP is created by eliminating all existing uses of the general-purpose register GR3, after which we use GR3 as a second GP: loads through the original GP are then converted to additions to the second GP.

To eliminate the existing uses of GR3, we rename them. At each program point where renaming is required, either a free register already is available, which is detected through liveness analysis, or we create a free register by adding spill code that spills a register to the stack. Fortunately this spilling is rarely needed. In single-threaded applications on the IA64/Linux platform, the special-purpose Thread Pointer register (GR13) can always be used. In almost all other applications we examined, only the special `setjmp()` and `longjmp()` C-library procedures use all registers, forcing us to insert register spills.

By merging smaller GOTs into larger GOTs – mostly one GOT suffices – 4% of all instructions can be eliminated on average. By converting loads from the GOT into additions, the static number of load instructions is reduced on average with 16.1%, while the number of executed loads decreases with 11.1%. Roughly one third of these improvements results from using a second GP.

3.2 Liveness Analysis

As a result of the optimizations of the previous section, most of the address computations involving the GOT become superfluous. Other optimizations, such as copy propagation, also render certain instructions useless, meaning that the values they produce are never used, which are hence dead. In order to actually eliminate those useless instructions, interprocedural liveness analysis is needed. This backward data flow analysis solves the following flow equations [6]:

$$\forall n \in N : live_{in}(n) = Consumed(n) \cup (live_{out}(n) \setminus Defined(n)), \quad (1)$$

$$\forall n \in N \setminus C : live_{out}(n) = \bigcup_{s \in succ(n)} live_{in}(s), \quad (2)$$

$$\forall c \in C : live_{out}(c) = (Saved(p) \cap live_{in}(r)) \cup Consumed(p). \quad (3)$$

N denotes the set of basic blocks and C denotes the set of basic blocks ending with a procedure call. Equation (1) states that all registers used (= consumed) in a block before being defined and all registers that are live at the end without being defined, are live upon entry to the block. Equation (2) implements the confluence of edges and equation (3) tells us which registers are live at procedure call-site c. Muth [6] describes in detail how to solve these equations.

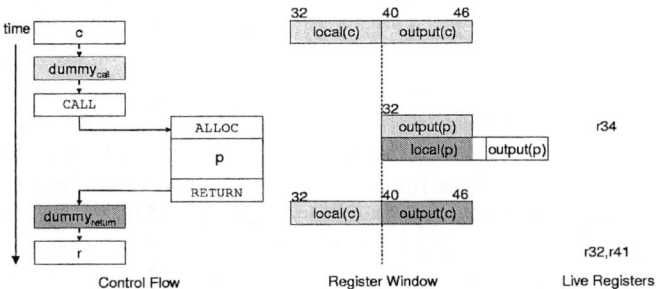

Fig. 1. Dummy blocks simulate the behavior of the register stack.

Straightforward application of these equations is incorrect on the IA64 architecture, since register windows are used to ease parameter passing. With register windows, every procedure call involves automatic register renaming. This process is depicted in Figure 1. Each procedure has a limited view on the register stack file, which always begins at register $r32$, and is divided in local (including the input registers) and output registers. When a call is executed, the register window moves and the local registers of the caller are no longer visible to the callee. Supposing the first output register of the caller was $r40$, this register is named $r32$ after the call. The caller can resize its window with the `alloc` instruction. When a `return` is executed, the register window of the caller is restored.

This register renaming needs to be modeled correctly in our liveness analysis. Our solution consists of adding dummy blocks to the ICFG prior to calls and at continuation points, as depicted in Figure 1. Note that they are added at the call site, as the number of local registers at the call site determines the renaming. Their flow equations are:

$$\forall d \in Dummy_{return} : live_{in}(d) = Ren_{callee}(live_{in}(r) \cap output(c)) \quad (4)$$
$$\forall d \in Dummy_{call} : live_{in}(d) = (live_{in}(r) \cap local(c)) \quad (5)$$
$$\cup (Ren_{caller}(live_{in}(p)) \cap output(c))$$

The functions $local()$ and $output()$ return the set of the local and output registers of a given procedure respectively. $Ren_{caller}()$ maps the name of a register in the callee to the name it has at the the call-site. $Ren_{callee}()$ does the opposite. In our example $Ren_{caller}(r32) = r40$ and $Ren_{callee}(r40) = r32$.

As equation (1) operates only within a basic block, no additional measures need to be taken to assure its correctness. Equation (4) corrects the liveness information that is propagated to the exit of a callee as a result of equation (2).

Without this equation the registers with names $r32$ and $r41$ would be considered to contain live values at the callee's exit, while the others might be considered to contain dead values (depending on the other successors). This is however incorrect due to the register stack mechanism: the value of the register named $r32$ in the register window of the caller is not visible to the callee, while the register named $r41$ is called $r33$ in the register window of the callee. The resulting liveness information should be that $r33$ contains a live value and that the other registers might contain dead values. As a result of equation (4) $live_{in}(dummy_{return}) = Ren_{callee}(\{r32, r41\} \cap \{r40, \ldots, r46\}) = Ren_{callee}(\{r41\}) = \{r33\}$, and the correct information is propagated into the callee's exit.

At first sight the correctness of equation (3) can be restored by adapting only the computation of the $Consumed(p)$ and $Saved(p)$ sets. The renaming cannot take place within these functions as these sets depend only on the callee p, while the renaming depends on the number of local registers of the caller, and this number is not necessarily constant for all callers of a procedure. Clearly a more complex solution is needed, resulting in equation (5). The first part states that local registers that are live at the return site are live at the call-site. Essentially these are live registers that are saved across the call. The second part assures that values in output registers, live at the called procedure p are live at the $dummy_{call}$ block. The Ren_{caller} is needed because the output registers will be renamed after the call. In our running example this results in $live_{in}(dummy_{call}) = (\{r32, r41\} \cap \{r32, \ldots, r39\}) \cup (Ren_{caller}(\{r34\}) \cap \{r40, \ldots, r46\}) = \{r32\} \cup (\{r42\} \cap \{r40, \ldots, r46\}) = \{r32, r42\}$.

3.3 Branch Register Optimization

The IA64 architecture has eight branch registers, which can be used to store return addresses of procedure calls or to store target addresses of indirect jumps and procedure calls. However, our measurements on the code produced by the GCC compiler have learned us that only 2 of these 8 registers are used frequently: register B0 to store return addresses, and register B6 to store target addresses.

This inefficient use of the branch registers is due to calling conventions. When a procedure calls another procedure and stores the return address in a branch register, the second procedure has to know where the return address has been stored. Since a procedure from a separately compiled module or library cannot know all its callers, a fixed branch register has to be used, in this case B0. When the second procedure in turn calls another procedure, it also uses register B0 to store the return address. Therefore the first return address has to be saved before the call and restored afterwards.

To avoid this spilling of return addresses as much as possible, we have implemented the following link-time solution: all procedures get assigned a value ranging from 0 to 6, each indicating one of the branch registers B0-B5 and B7 in which the return addresses of the the procedure's callers will be stored. In the call graph of the whole program, leaf procedures (i.e. procedures that do not call any other procedure) are assigned the value 6. Other procedures are then iteratively given the minimum value of their callees minus 1, or, if this would result in a negative number, the (standard) value 0. Once all procedures are assigned a value, all instructions using the branch registers are adjusted accordingly. Whenever a procedure still has a callee with the same value, a register spill remains, but otherwise they are eliminated.

Please note that exceptions to this simple solution have to be made in the presence of indirect procedure calls, i.e. calls through function pointers. For such calls, the possible targets can only be estimated at link-time. The set of procedures that can be called indirectly is limited however, and a link-time optimizer can derive this set from the relocation information available in the object files of the program. All procedures in this set are assigned the value 0.

Still our branch register optimization is applied to 10% of all procedures, reducing the number of saves and restores of branch registers by 5% on average.

3.4 Code Layout and Scheduling

When the analyses and optimizations are finished, the control flow graph is serialized and the instructions are scheduled. Code placement may have an important impact on the performance of an application as it may improve caching and reduce the number of page faults. We implemented a profile-based closest-is-best technique, based on Pettis and Hansen [7]. As a result the number of cycles lost waiting for instructions is reduced by 38% on average. The average performance impact is moderate however because, except for the vortex benchmark, instruction latency is not a major bottleneck. In vortex instruction latency accounts for 7% of the execution time, and there we achieve a speedup of almost 3%.

More important than code layout is code scheduling. This is particularly the case on the IA64 architecture, where the compiler needs to convey explicit information on the parallelism between instructions to the processor. Instructions that can be executed in parallel need to be clustered into instruction groups. These instruction groups are then mapped onto bundles. Each bundle has three 41-bit instruction slots and a 5-bit template that indicates the types of the instructions in the slots and the borders between instruction groups. The fact that the number of allowed combinations of instruction types and borders is limited to 32 complicates the scheduling process and as a result, it is often necessary to insert no-op instructions when no useful instruction can be found.

We implemented two local scheduling algorithms in our link-time optimizer: a list scheduler [1] and the so called *noptimizer*. The noptimizer is based on [4]. It can operate with different cost functions, to either minimize the number of no-op instructions, or to minimize the number of instruction groups and as a result the number of execution cycles. The noptimizer is a branch and bound version of the optimal scheduling algorithm, but its search depth is severely limited, to limit the execution time of the scheduler. Still, compared to the original binaries produced by the GCC compiler, the noptimizer is able to reduce 18% of all no-ops on average, with an overall compaction of 5%.

In order to increase the amount of parallelism, we also developed a global scheduling algorithm, the so-called *globtimizer*, which is roughly based on [10]. It is a branch and bound algorithm as well and it uses a cost function (number of no-op instructions, number of instruction groups, etc.) to optimize the instruction sequence. The idea here is to move instructions up and down between basic blocks using predication, schedule those basic blocks with a local algorithm and compare the quality of the solution with those of other configurations. The structure of the flow graph remains untouched however, so no basic blocks are merged or split like many other global scheduling algorithms do.

Currently some simple combinations and structures of basic blocks are considered to move instructions between blocks. Nevertheless, combined with the noptimizer, it is able to further reduce the number of no-op instructions by 23%.

4 Experimental Evaluation

To evaluate our link-time optimizer, we used 7 programs from the SPECint2000 benchmark suite. We compiled them with the GCC compiler (v3.2) and linked them with the glibc library (v2.3.1). The experiments were performed on a 4-way Intel Itanium multiprocessor system, running Linux 2.4.18.

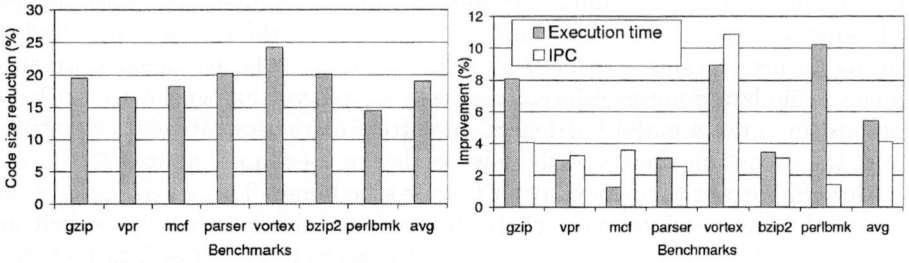

Fig. 2. Experimental results: code compaction, speedup and IPC improvement.

Our results are summarized in Figure 2. The code size is reduced on average with 19%. The major contributions come from unreachable code elimination (9.64%), instruction scheduling and bundling algorithms (5.22%) and reduction of load instructions (3.27%). We also achieve an average speedup of 5.4%. The IPC (instructions per cycle) improves up to 11%, and 4.1% on average. The speedup is mainly caused by the reduction of load instructions and, in the case of vortex, by profile-guided code layout.

5 Related Work

Srivastava and Wall describe the optimization of GOT accesses on the Alpha architecture [9], for which both Alto [6] and Spike [2] are link-time optimizers. By contrast, Diablo [3], the framework we used for link-time optimization, is portable and retargetable. We also extended the existing GOT optimizations by introducing a second GP that fully exploits the IA64 architecture.

Numerous EPIC-specific scheduling algorithms have been developed. In [5] and [10], Integer Linear Programming is used to obtain an optimal local and global schedule respectively. We adopted the filosophy in [10] and designed a branch and bound version, in which the structure of the flow graph is also preserved. A technique for minimizing the number of no-op instructions is presented in [4]. We extended this work by integrating branch instructions tighter into the bundling process and by filling up partially filled bundles more carefully. More details on our extensions, including source code, can be found at http://www.elis.ugent.be/diablo.

6 Conclusions

We have shown how link-time optimization is able to improve the code quality of IA64 code, which is crucial given the EPIC paradigm. By optimizing the Global Offset Table address computations, the use of the branch registers and by improving the code schedules, code size reductions of 19% on average were achieved, together with an average speedup of 5.4%.

Acknowledgements

This research was funded by Ghent University, by the Institute for the Promotion of Innovation by Science and Technology in Flanders (IWT) and by the Fund for Scientific Research-Flanders (FWO-Flanders).

References

1. E. Coffman: Computer and Job-Shop Scheduling Theory. Jon Wiley & Sons. (1976).
2. R. Cohn, D. Goodwin and G. Lowney: Optimizing Alpha Executables on Windows NT with Spike. Digital Technical Journal. **9** (1998) 3–20.
3. B. De Bus, D. Kästner, D. Chanet, L. Van Put and B. De Sutter: Post-pass compaction techniques. Communications of the ACM. **46** (2003) 41–46.
4. S. Haga and R. Barua: EPIC instruction scheduling based on optimal approaches. Annual workshop on Explicitly Parallel Instruction Computing Architectures and Compiler Techniques. **1** (2001) 22–31.
5. D. Kästner and S. Winkel: ILP-based instruction scheduling for IA64. Proc. of Languages, Compilers and Tools for Embedded Systems. (2001) 145–154.
6. R. Muth, S. Debray, S. Watterson and K. De Bosschere: alto: A Link-Time Optimizer for the Compaq Alpha. Software Practice and Experience. **31** (2001) 67–101.
7. K. Pettis and R. Hansen: Profile guided code positioning. Proc. of the ACM SIGPLAN Conf. on Programming Language Design & Implementation. (1990) 16–27.
8. M. Schlansker and B. Ramakrishna Rau: EPIC: Explicitly Parallel Instruction Computing. IEEE Computer. **33** (2000) 37–45.
9. A. Srivastava and D. Wall: Link-time optimization of address calculation on a 64-bit architecture. Programming Languages Design and Implementation. (1994) 49–60.
10. S. Winkel: Optimal global scheduling for Itanium processor family. Explicitly Parallel Instruction Computing Architectures and Compiler Techniques. (2002) 59–70.

A Polyhedral Approach to Ease the Composition of Program Transformations

Albert Cohen[1], Sylvain Girbal[1,2], and Olivier Temam[1]

[1] ALCHEMY Group, INRIA Futurs and LRI, Université Paris-Sud
[2] CEA LIST, Saclay

Abstract. We wish to extend the effectiveness of loop-restructuring compilers by improving the robustness of loop transformations and easing their composition in long sequences. We propose a formal and practical framework for program transformation. Our framework is well suited for iterative optimization techniques searching not only for the appropriate parameters of a given transformation, but for the program transformations themselves, and especially for compositions of program transformations. This framework is based on a unified polyhedral representation of loops and statements, enabling the application of generalized control and data transformations without reference to a syntactic program representation. The key to our framework is to clearly separate the impact of each program transformation on three independent components: the iteration domain, the iteration schedule and the memory access functions. The composition of generalized transformations builds on normalization rules specific to each component of the representation. Our techniques have been implemented on top of Open64/ORC.

1 Introduction

Todays compilers ability to apply and search for compositions of program transformations is limited. Compilers can embed a large array of optimizations, but they are often expressed as a collection of ad-hoc syntactic transformations based on pattern-matching. In addition, control structures are regenerated after each transformation, making it harder to apply the next transformations. Finally, compilers follow a rigid ordering of phases, so that only short and fixed sequences of program transformations can be applied [26]. Current approaches to iterative optimization [1, 11, 8] substitute empirical search strategies to the usual model-driven heuristics, but to not improve the transformation framework itself. Indeed, iterative/adaptive compilers usually choose a rather small set of transformations, e.g., cache tiling, unrolling and array padding, and focus on finding the best possible parameters, e.g., tile size, unroll factor and padding size. O'Boyle et al. [11] and Cooper et al. [8] outlined that the ability to perform long sequences of composed transformations is key to the emergence of practical iterative optimization frameworks. Another recent study [22] confirms that complex compositions of many distinct transformations can bring significant performance benefits.

This article introduces a framework for easily expressing compositions of program transformations, based on the polyhedral representation of programs [10] and on a robust code generation technique [23, 3]. We distinguish three different types of actions performed by program transformations: modification of the iteration domain (loop bounds and strides), modification of the iteration schedule of each statement, and modification of the memory access functions (array subscripts). Current program representations do not clearly reflect this separation, making the implementation and composition of program transformations more complicated. E.g., current implementations of loop fusion incur loop bounds and array subscript modifications that are only byproducts of a schedule transformation (the fused loops are often peeled, increasing code size and making further optimizations more complex). Within our representation, loop fusion is expressed as a schedule transformation with no explicit impact on the iteration domain and memory access. Similarly, a domain transformation like unrolling has no impact on the schedule or memory access functions representations; or a memory access transformation like privatization has no impact on the schedule or domain representations, thus not conflicting with the later application of skewing or unrolling. While our framework is geared toward iterative optimization techniques, it can also facilitate the implementation of statically driven program transformations.

To date, the most thorough application of the polyhedral representation is the Petit dependence analyzer and loop restructuring tool [15] within the Omega project [16]. These tools show that most single loop transformations (both unimodular and non-unimodular) can be modeled as geometric transformations of polyhedra. However, traditional polyhedral representations do not separate the three above-mentioned actions induced by program transformations. Indeed, space-time transformations in the polytope model [10, 25, 15, 18] were aimed at model-based optimizations through operation research algorithms (e.g., linear programming) with no real need for composition sequences. Some polyhedral approaches [15, 9, 19, 12, 24] reproduce or extend classical loop transformations, but ultimately rely on the program syntax for the identification of the loops to operate on. These works require the explicit generation of source code and reconstruction of polyhedra at each transformation step, whether our framework sticks to the polyhedral representation along the whole sequence of transformations. There is a large amount of related works and projects targeting loop-restructuring compilers, see e.g., [7, 13, 5, 14, 16, 17, 15, 21] for representative examples. The associated research report outlines the main comparison points with our approach [6].

This paper does not present performance numbers. The goal is to revisit some theoretical and engineering cornerstones of the design of loop-restructuring compilers. We refer to classical techniques for validation and optimization heuristics [26].

2 Separate Polyhedra for Unified Program Transformations

The polytope model is based on a *semantics*-based representation of loop nests. This representation clearly identifies three separate components: *array access functions* – affine functions describing the mapping of iterations to memory locations – from the *iteration domain* – a geometrical abstraction of loop bounds and strides shaping loop structures – and from the *affine schedule* – another geometrical abstraction of the ordering of iterations and statements. In addition to classical characterization of affine schedules, we also separate the description of iteration ordering of a single statement from inter-statements scheduling.

We assume constant strides and affine bounds for loops; affine array subscripts are hoped for but not mandatory. Within a function body, a *Static Control Part* (SCoP) is a maximal set of consecutive statements without while loops, where loop bounds and conditionals only depend on invariants (symbolic constants and surrounding counters) within this set of statements. These invariants are called the *global parameters* of the SCoP. We do not consider procedures, pointers, and inter-SCoP transformations.

The following definitions assume some familiarity with the polytope model: the unfamiliar reader may refer to an associated research report [6]; in addition, Section 3 ends with a short example. Formally, a SCoP is a pair $(\mathcal{S}, \mathbf{i}_{\text{gp}})$, where $\mathcal{S}$ is the set of consecutive *statements* and $\mathbf{i}_{\text{gp}}$ is the vector of *global parameters* (known at run-time). $d_{\text{gp}} = \dim(\mathbf{i}_{\text{gp}})$ denotes the number of global parameters and d^S the depth of statement S, i.e., the number of nested loops enclosing the statement in the SCoP. A statement $S \in \mathcal{S}$ is a quadruple $(\mathcal{D}^S, \mathcal{L}^S, \mathcal{R}^S, \theta^S)$, where $\mathcal{D}^S$ is the d^S-dimensional *iteration domain* of S, $\mathcal{L}^S$ and $\mathcal{R}^S$ denote array references written by S (left-hand side) and read by S (right-hand side) respectively, and θ^S is the *affine schedule* of S, defining the *sequential execution ordering* of iterations of S.[1]

Iteration domains. We denote matrices by capital letters. $\mathcal{D}^S$ is a *convex polyhedron* defined by matrix $\Lambda^S \in \mathcal{M}_{n, d^S + d^S_{\text{lv}} + d_{\text{gp}} + 1}(\mathbb{Z})$ such that

$$\mathbf{i} \in \mathcal{D}^S \iff \exists \mathbf{i}_{\text{lv}}, \Lambda^S \cdot (\mathbf{i}, \mathbf{i}_{\text{lv}}, \mathbf{i}_{\text{gp}}, 1)^t \geq 0$$

where Λ^S is the matrix defining the domain inequalities; n is the number of inequalities necessary to define the domain (the number of matrix rows, a priori not limited); 1 adds a matrix column to specify the affine component of each domain inequality; and d^S_{lv} is the number of *local variables* required to implement integer division and modulo operations via *affine projection*.

Statements guarded by non-convex conditionals – such as $1 \leq i \leq 3 \vee i \geq 8$ – are split into separate statements with convex domains in the polyhedral representation.

Memory access functions. $\mathcal{L}^S$ and $\mathcal{R}^S$ are *sets* of $(\mathtt{A}, f)$ pairs, where $\mathtt{A}$ is an array variable and f is the *access function* mapping iterations in $\mathcal{D}^S$ to locations in $\mathtt{A}$.

---

[1] The term polyhedron will be used in a broad sense to denote a linearly-bounded lattice, i.e., a set of points in a $\mathbb{Z}$ vector space bounded by affine inequalities.

The access function f is defined by a matrix $F \in \mathcal{M}_{\dim(\mathtt{A}),\, d^S + d^S_{\mathrm{lv}} + d_{\mathrm{gp}} + 1}(\mathbb{Z})$ such that
$$f(\mathbf{i}) = F \cdot (\mathbf{i}, \mathbf{i}_{\mathrm{lv}}, \mathbf{i}_{\mathrm{gp}}, 1)^t.$$
Access functions only describe *affine* references; other references are allowed if the dependence analysis framework can manage them [27, 2].

Affine schedules. θ^S is the *affine schedule* of S; it maps iterations in $\mathcal{D}^S$ to multidimensional *time stamps*, i.e., logical execution dates. Multidimensional time stamps are compared through the *lexicographic ordering* over vectors, denoted by $\ll$: iteration $\mathbf{i}$ of S is executed before iteration $\mathbf{i}'$ of S' if and only if $\theta^S(\mathbf{i}) \ll \theta^{S'}(\mathbf{i}')$.

θ^S is defined by a matrix $\Theta^S \in \mathcal{M}_{2d^S+1,\, d^S + d_{\mathrm{gp}} + 1}(\mathbb{Z})$ such that
$$\theta^S(\mathbf{i}) = \Theta^S \cdot (\mathbf{i}, \mathbf{i}_{\mathrm{gp}}, 1)^t.$$
Θ^S does not involve local variables, they would be redundant with the iterators they are related to. Notice the number of rows is $2d^S + 1$ and not d^S: to define the relative ordering of statements across iterations at depth k, we need d^S dimensions; to define the relative ordering of statements within each iteration, we need an additional dimension for each depth plus depth 0, hence the $2d^S + 1$ dimensions. This encoding was proposed before [10, 15], to model classical transformations into the polytope model.

The schedule matrix is decomposed in a form amenable to transformation composition and scalable code generation; it consists of three sub-matrices: a square *iteration ordering matrix* $A^S \in \mathcal{M}_{d^S, d^S}(\mathbb{Z})$ operating on iteration vectors, a *statement ordering vector* $\beta^S \in \mathbb{N}^{d^S+1}$, and $\Gamma^S \in \mathcal{M}_{d^S, d_{\mathrm{gp}}+1}(\mathbb{Z})$ called a *parameterization matrix*. The structure of the schedule matrix Θ^S is shown below. $A^S_{i,j}$ capture the iteration order of S with respect to surrounding loop counters. β^S_i specify the ordering of S among all other statements executed at the same iteration; the first row of Θ^S corresponds to depth 0, the outermost level[2]. $\Gamma^S_{i,j}$ extend the nature of possible transformations, allowing iteration advances and delays by constant or parametric amounts.

$$\Theta^S = \begin{bmatrix} 0 & \cdots & 0 & 0 & \cdots & 0 & \beta^S_0 \\ A^S_{1,1} & \cdots & A^S_{1,d^S} & \Gamma^S_{1,1} & \cdots & \Gamma^S_{1,d_{\mathrm{gp}}} & \Gamma^S_{1,d_{\mathrm{gp}}+1} \\ 0 & \cdots & 0 & 0 & \cdots & 0 & \beta^S_1 \\ A^S_{2,1} & \cdots & A^S_{2,d^S} & \Gamma^S_{2,1} & \cdots & \Gamma^S_{2,d_{\mathrm{gp}}} & \Gamma^S_{2,d_{\mathrm{gp}}+1} \\ \vdots & \ddots & \vdots & 0 & \ddots & 0 & \vdots \\ A^S_{d^S,1} & \cdots & A^S_{d^S,d^S} & \Gamma^S_{d^S,1} & \cdots & \Gamma^S_{d^S,d_{\mathrm{gp}}} & \Gamma^S_{d^S,d_{\mathrm{gp}}+1} \\ 0 & \cdots & 0 & 0 & \cdots & 0 & \beta^S_{d^S} \end{bmatrix}.$$

Towards a normalized representation. A given program can have multiple representations, and that, in turn, can limit the application of transformations. E.g., a condition for fusion is that statements must be consecutive; otherwise, if there is a statement in between, one must first decide where to move it (and check dependences). Normalization conditions avoiding these pitfalls are called

---

[2] Notice the first component of β is numbered β_0.

invariants. Besides avoiding useless composition prohibitions, these invariants also serve to avoid matrix parameters overflow.

The schedule density invariant is to ensure that all statements at an identical depth have a consecutive β ordering (no gap). As a side-effect, this invariant also avoids integer overflows on the β parameters, $\beta^S_k > 0 \Rightarrow \exists S' \in \mathcal{S}, \mathrm{pfx}(\beta^S, k) = \mathrm{pfx}(\beta^{S'}, k) \wedge \beta^{S'}_k = \beta^S_k - 1$, where $\mathrm{pfx}(\beta^S, k)$ denotes the k first dimensions of vector β^S. The condition states that, for any non-null β parameter at dimension k, there necessarily exists another statement S' with the same k-prefix and the preceding value at dimension k.

The domain parameter invariant avoids redundant inequalities and integer overflows in the domain matrix Λ^S parameters. For that purpose, we impose that the coefficients in a row of Λ^S are always relatively prime.

The sequentiality invariant states that two distinct statements, or two identical statements in distinct iterations, cannot have the same time stamp: $S \neq S' \vee \mathbf{i} \neq \mathbf{i}' \Rightarrow \theta^S(\mathbf{i}) \neq \theta^{S'}(\mathbf{i}')$. A sufficient (though not necessary) condition to enforce that property is the following: $|\det(\mathrm{A}^S)| = 1$ (unimodular) and $S \neq S' \Rightarrow \beta^S \neq \beta^{S'}$.

3 Polyhedral Program Transformations

In our framework, program transformations take the form of a set of elementary operations on matrices and vectors describing a SCoP.

We first define elementary operations called *constructors*. Given a vector v and matrix M with $\dim(v)$ columns and at least i rows, $\mathsf{AddRow}(\mathrm{M}, i, v)$ inserts a new row at position i in M and fills it with the value of vector v, $\mathsf{RemRow}(\mathrm{M}, i)$ does the opposite transformation. $\mathsf{AddCol}(\mathrm{M}, j, v)$ and $\mathsf{RemCol}(\mathrm{M}, j)$ play similar roles for columns. Moving a statement S forward or backward is a common operation: the constructor $\mathsf{Move}(P, Q, o)$ leaves all statements unchanged except those satisfying
$$\forall S \in \mathcal{S}, P \sqsubseteq \beta^S \wedge (Q \ll \beta^S \vee Q \sqsubseteq \beta^S) : \beta^S_{\dim(P)} \leftarrow \beta^S_{\dim(P)} + o,$$
where $u \sqsubseteq w$ denotes that u is a prefix of v, where P and Q are *statement ordering prefixes* s.t. $P \sqsubseteq Q$ defining respectively the context of the move and marking the initial time-stamp of statements to be moved, and where offset o is the value to be added/subtracted to the component at depth $\dim(P)$ of any statement ordering vector β^S prefixed by P and following Q. If o is positive, $\mathsf{Move}(P, Q, o)$ inserts o free slots before all statements S preceded by the statement ordering prefix Q at the depth of P; respectively, if o is negative, $\mathsf{Move}(P, Q, o)$ deletes $-o$ slots. These constructors make no assumption about the representation invariants and may violate them.

3.1 Composition of Primitives

From the earlier constructors, we define invariant-enforcing transformation *primitives* to serve as building blocks for transformation sequences. Figure 1 lists typical primitives affecting the polyhedral representation of a statement. $\mathbf{1}_k$ denotes the vector filled with zeros but element k set to 1, i.e., $(0, \ldots, 0, 1, 0, \ldots, 0)$; likewise, $\mathbf{1}_{i,j}$ denotes the matrix filled with zeros but element (i, j) set to 1.

Like the Move *constructor, primitives do not directly operate on loops or statements, but target a collection of statements and polyhedra whose statement-ordering vectors share a common prefix P. There are no prerequisites on the program representation to the application and composition of primitives.*

We also specified a number of optional *validity prerequisites* that conservatively check for the semantical soundness of the transformation, e.g., there are validity prerequisites to check that no dependence is violated by a unimodular or array contraction transformation. When exploring the space of possible transformation sequences, validity prerequisites avoid wasting time on corrupt transformations.

FUSION and FISSION best illustrate the benefit of designing loop transformations at the abstract semantical level of our unified polyhedral representation. First of all, loop bounds are not an issue since the code generator will handle any overlapping of iteration domains. For the fission primitive, vector (P, o) prefixes all statements concerned by the fission; and parameter b indicates the position where statement delaying should occur. For the fusion primitive, vector $(P, o+1)$ prefixes all statements that should be interleaved with statements prefixed by (P, o). Eventually, notice that fusion followed by fission (with the appropriate value of b) leaves the SCoP unchanged.

UNIMODULAR implements any unimodular transformation, extended to arbitrary iteration domains and loop nesting. U and V denote unimodular matrices.

SHIFT is a kind of source-level hierarchical software pipeline, extended with parametric forward/backward iteration shifts, e.g., to delay a statement by N iterations of one surrounding loop. Matrix M implements the parameterized shift of the affine schedule of a statement. M must have the same dimension as Γ.

RESTRICT constrains the domain with an additional inequality, given in the form of a vector c with the same dimension as a row of matrix Γ.

EXTEND inserts a new intermediate loop level at depth ℓ, initially restricted to a single iteration. This new iterator will be used in following code transformations.

ADDLVDOM and ADDLVACC insert a fresh local variable in the domain and and access functions, respectively. This local variable is typically used by RESTRICT.

PRIVATIZE and CONTRACT implement basic forms of array privatization and contraction, respectively, considering dimension ℓ of the array. Privatization needs an additional parameter s, the size of the additional dimension; s is required to update the array declaration (it cannot be inferred in general, some references may not be affine). These primitives are simple examples updating the data layout and array access functions.

Although this table is not complete, it demonstrates the expressiveness of the unified representation through classical *control and data* transformations.

Syntax	Effect
UNIMODULAR(P, U, V)	$\forall S \in \mathcal{S} \mid P \sqsubseteq \beta^S, A^S \leftarrow U.A^S.V$
SHIFT(P, M)	$\forall S \in \mathcal{S} \mid P \sqsubseteq \beta^S, \Gamma^S \leftarrow \Gamma^S + M$
RESTRICT(P, c)	$\forall S \in \mathcal{S} \mid P \sqsubseteq \beta^S, \Lambda^S \leftarrow \mathrm{AddRow}(\Lambda^S, 0, c/\gcd(c_1, \ldots, c_{d^S + d^S_{\mathrm{lv}} + d_{\mathrm{gp}} + 1}))$
EXTEND(P, ℓ, c)	$\forall S \in \mathcal{S} \mid P \sqsubseteq \beta^S, \begin{cases} d^S \leftarrow d^S + 1; \; \Lambda^S \leftarrow \mathrm{AddCol}(\Lambda^S, c, 0); \; \beta^S \leftarrow \mathrm{AddRow}(\beta^S, \ell, 0); \\ A^S \leftarrow \mathrm{AddRow}(\mathrm{AddCol}(A^S, c, 0), \ell, 1_\ell); \; \Gamma^S \leftarrow \mathrm{AddRow}(\Gamma^S, \ell, 0); \\ \forall (\mathtt{A}, \mathrm{F}) \in \mathcal{L}^S \cup \mathcal{R}^S, \mathrm{F} \leftarrow \mathrm{AddRow}(\mathrm{F}, \ell, 0) \end{cases}$
ADDLVDOM(P)	$\forall S \in \mathcal{S} \mid P \sqsubseteq \beta^S, d^S_{\mathrm{lv}} \leftarrow d^S_{\mathrm{lv}} + 1; \; \Lambda^S \leftarrow \mathrm{AddCol}(\Lambda^S, d^S + 1, 0)$
ADDLVACC$(P, \mathtt{A})$	$\forall S \in \mathcal{S} \mid P \sqsubseteq \beta^S, \forall (\mathtt{A}, \mathrm{F}) \in \mathcal{L}^S \cup \mathcal{R}^S, \mathrm{F} \leftarrow \mathrm{AddCol}(\mathrm{F}, d^S + 1, 0)$
PRIVATIZE$(\mathtt{A}, \ell)$	$\dim(\mathtt{A}) \leftarrow \dim(\mathtt{A}) + 1, \; \forall S \in \mathcal{S}, \forall (\mathtt{A}, \mathrm{F}) \in \mathcal{L}^S \cup \mathcal{R}^S, \mathrm{F} \leftarrow \mathrm{AddRow}(\mathrm{F}, \ell, 1_\ell)$
CONTRACT$(\mathtt{A}, \ell)$	$\dim(\mathtt{A}) \leftarrow \dim(\mathtt{A}) - 1; \; \forall S \in \mathcal{S}, \forall (\mathtt{A}, \mathrm{F}) \in \mathcal{L}^S \cup \mathcal{R}^S, \mathrm{F} \leftarrow \mathrm{RemRow}(\mathrm{F}, \ell)$
FUSION(P, o)	$b = \max\{\beta^S_{\dim(P)+1} \mid (P, o) \sqsubseteq \beta^S\} + 1$ $\mathrm{Move}((P, o+1), (P, o+1), b); \; \mathrm{Move}(P, (P, o+1), -1)$
FISSION(P, o, b)	$\mathrm{Move}(P, (P, o, b), 1); \; \mathrm{Move}((P, o+1), (P, o+1), -b)$

Fig. 1. Some classical transformation primitives

This table is not complete (e.g., it lacks index-set splitting and data-layout transformations), but it demonstrates the expressiveness of the unified representation.

Primitives operate on program representation while maintaining the structure of the polyhedral components (the invariants).

Despite their familiar names, the primitives' practical outcome on the program representation is widely extended compared to their syntactic counterparts. Indeed, transformation primitives like fusion or interchange apply to sets of statements that may be scattered and duplicated at many different locations in the generated code. In addition, these transformations are not proper *loop* transformations anymore, since they apply to sets of statement iterations that may have completely different domains and relative iteration schedules. For example, one may interchange the loops surrounding one statement in a loop body without modifying the schedule of other statements, and without distributing the loop first. Another example is the fusion of two loops with different domains without peeling any iteration.

Previous encodings of classical transformations in a polyhedral setting – most significantly [25] and [15] – use Presburger arithmetic as an expressive *operating* tool for implementing and validating transformations. In addition to operating on polytopes, our work *generalizes* loop transformations to more abstract *polyhedral domain* transformations, without explicitly relying on a nested loop structure with known bounds and array subscripts to define the transformation.

Instead of anchoring loop transformations on a syntactic program form, limiting ourselves to what can be expressed with an imperative semantics, we define higher level transformations on the polyhedral representation itself, abstracting away the overhead (versioning, duplication) and constraints of the code generation process (translation to an imperative semantics).

Naturally, this higher-level framework is beneficial for transformation composition. Figure 2 composes primitives into typical transformations. INTERCHANGE swaps the roles of $\mathbf{i}_o$ and $\mathbf{i}_{o+1}$ in the schedule of the matching statements; it is a fine-grain extension of the classical interchange making no assumption about the shape of the iteration domain. SKEW and REVERSE define two well known

Syntax	Effect		Comments
INTERCHANGE(P, o)	$\forall S \in \mathcal{S} \mid P \sqsubseteq \beta^S,$	$\begin{cases} V = I_{d^S} - 1_{o,o} - 1_{o+1,o+1} + 1_{o,o+1} + 1_{o+1,o}; \\ \text{UNIMODULAR}(\beta^S, I_{d^S}, V) \end{cases}$	swap rows o and $o+1$
SKEW(P, ℓ, c, s)	$\forall S \in \mathcal{S} \mid P \sqsubseteq \beta^S, V = I_{d^S} + s \cdot 1_{\ell,c};$ UNIMODULAR(β^S, I_{d^S}, V)		add the skew factor
REVERSE(P, o)	$\forall S \in \mathcal{S} \mid P \sqsubseteq \beta^S, V = I_{d^S} - 2 \cdot 1_{o,o};$ UNIMODULAR(β^S, I_{d^S}, V)		put a -1 in (o,o)
STRIPMINE(P, k)	$\forall S \in \mathcal{S} \mid P \sqsubseteq \beta^S,$	$\begin{cases} \ell = \dim(P); \ c = \dim(P); \\ \text{EXTEND}(\beta^S, \ell, c); \\ \text{ADDLVDOM}(\beta^S); \\ p = d^S + 1; \ u = d^S + d^S_{lv} + d_{gp} + 1; \\ \text{RESTRICT}(\beta^S, A^S_{\ell+1} - 1_c); \\ \text{RESTRICT}(\beta^S, -A^S_{\ell+1} + 1_c + (k-1)1_u); \\ \text{RESTRICT}(\beta^S, 1_c - k \cdot 1_p); \\ \text{RESTRICT}(\beta^S, k \cdot 1_p - 1_c) \end{cases}$	insert interm. loop insert local var. local var. and const. $i_c \leq i_{c+1}$ $i_{o+1} \leq i_o + k - 1$ $k \times p \leq ii$ $ii \leq k \times p$
TILE(P, o, k)	$\forall S \in \mathcal{S} \mid (P, o) \sqsubseteq \beta^S,$	$\begin{cases} \text{STRIPMINE}(P, k); \\ \text{STRIPMINE}((P, o), k); \\ \text{INTERCHANGE}((P, 0), \dim(P)) \end{cases}$	strip outer loop strip inner loop interchange

Fig. 2. Composition of transformation primitives

unimodular transformations, with respectively the skew factor s with it's coordinates (ℓ, c), and the depth o of the iterator to be reversed. STRIPMINE introduces a new iterator to unroll k times the schedule and iteration domain of all statements at the depth of P (where k is a *statically known integer*). This transformation is a complex sequence of primitives, see Figure 2. TILE extends the classical loop tiling at of the two nested loops at the depth of P, using $k \times k$ blocks, with arbitrary nesting and iteration domains. Tiling and strip-mining always operate on *time* dimensions; it is possible to tile the surrounding time dimensions of any collection of statements with unrelated iteration domains and schedules.

Other properties of our framework include confluence and commutativity, when operating on distinct components of the representation. Further exploration of these properties is under way, in an attempt to improve the structure of the transformation space for iterative optimization purposes.

3.2 Composition Example

Code complexity after loop transformations is mainly due to control optimizations (hoisting of conditionals, unrolling) which do not affect the complexity of our representation. The main asset of our framework is to hide the code complexity along the sequence. At intermediate steps, the complexity of the code representation within our framework remains fairly low, i.e., it only depends linearly on the number of original statements and statement-insertions. Using syntactic program transformations, the code complexity may increase at each intermediate step, sometimes even preventing further optimizations [6]. Notice that transformation with an intrinsic syntactical behavior like loop unrolling may also benefit from a polyhedral representation: one may strip-mine the loop instead, delaying the proper unrolling to the code generation step, and separate transformations may still be applied to each virtually unrolled iteration.

Let us compare our framework with syntactic transformations, studying the evolution of the representation along a sequence of transformations. The example in Figure 3 performs two matrix-vector multiplications, yielding $D =^t BEC$ (typical of quadratic form computations), where arrays B and E store $M \times N$ rectangular matrices.

We apply a sequence of three transformations to this program. In Figure 4, we interchange the loops in the first nest to optimize spatial locality on B. In Figure 5, we fuse the outer loops to improve temporal locality on A. Figure 6 shows part of the resulting code after advancing assignments to A by 4 iterations, in order to cover the latency of floating-point multiplications. This sequence corresponds to the following composition of primitives:

$$\text{INTERCHANGE}(\beta^{S_1}, 1); \text{ FUSION}((), 0); \text{ SHIFT}(\beta^{S_1}, [\begin{smallmatrix}0 & 0 & 0\\0 & 0 & -4\end{smallmatrix}]).$$

Based on the final polyhedral representation, the code generation phase will generate control-optimized code quite similar to the hand-optimized fragment in Figure 6, without redundant guards or dead iterations.

```
        do i = 1, M
          do j = 1, N
(S1)|     | A(i) += B(i,j)*C(j)
        do k = 1, M
          do l = 1, N
(S2)|     | D(k) += E(l,k)*A(l)
```

$\mathcal{D}^{S_1} = \{i, j \mid 1 \leq i \leq M \ ; \ 1 \leq j \leq N\}, A^{S_1} = [\begin{smallmatrix}1 & 0\\0 & 1\end{smallmatrix}], \varGamma^{S_1} = [\begin{smallmatrix}0 & 0 & 0\\0 & 0 & 0\end{smallmatrix}], \beta^{S_1} = [\begin{smallmatrix}0\\0\\0\end{smallmatrix}]$

$\mathcal{D}^{S_2} = \{k, l \mid 1 \leq k \leq M \ ; \ 1 \leq l \leq N\}, A^{S_2} = [\begin{smallmatrix}1 & 0\\0 & 1\end{smallmatrix}], \varGamma^{S_2} = [\begin{smallmatrix}0 & 0 & 0\\0 & 0 & 0\end{smallmatrix}], \beta^{S_2} = [\begin{smallmatrix}1\\0\\0\end{smallmatrix}]$

$\mathcal{L}^{S_1} = (\text{A}, [1\ 0\ 0\ 0\ 0]), \mathcal{L}^{S_2} = (\text{B}, [1\ 0\ 0\ 0\ 0])$

Fig. 3. Original code and its representation

```
        do j = 1, N
          do i = 1, M
          | A(i) += B(i,j)*C(j)
        do k = 1, M
          do l = 1, N
          | D(k) += E(l,k)*A(l)
```

Changes to the polyhedral representation:
$A^{S_1} = [\begin{smallmatrix}0 & 1\\1 & 0\end{smallmatrix}]$

Fig. 4. After interchange

```
MN = min(M,N)
do x = 1, MN
  do y = 1, MN
    A(y) += B(y,x)*C(x)
    D(x) += E(y,x)*A(y)
  do y = MN+1, M
    A(y) += B(y,x)*C(x)
    D(y) += E(y,x)*A(y)
do x = MN+1, N
  do y = 1, MN
    A(y) += B(y,x)*C(x)
    D(y) += E(y,x)*A(y)
```

Changes to the polyhedral representation:
$\beta^{S_2} = [\begin{smallmatrix}0\\0\\1\end{smallmatrix}]$

Fig. 5. After fusion

```
MN = min(M-4,N)
do x = 1, MN
  A(1) += B(1,x)*C(x)
  A(2) += B(2,x)*C(x)
  A(3) += B(3,x)*C(x)
  A(4) += B(4,x)*C(x)
  do y = 1, MN
    A(y+4) += B(y+4,x)*C(x)
    D(x) += E(y,x)*A(y)
  D(x) += E(MN-3,x)*A(MN-3)
  D(x) += E(MN-2,x)*A(MN-2)
  D(x) += E(MN-1,x)*A(MN-1)
  D(x) += E(MN,x)*A(MN)
do y = MN+1, M-4
  A(y+4) += B(y+4,x)*C(x)
  D(y) += E(y,x)*A(y)
D(M-3) += E(MN-3,x)*A(MN-3)
D(M-2) += E(MN-2,x)*A(MN-2)
D(M-1) += E(MN-1,x)*A(MN-1)
D(M)   += E(MN,x)*A(MN)
do x = MN+1, N
  do y = 1, M-4
  ...
```

Changes to the polyhedral representation:
$\varGamma^{S_1} = [\begin{smallmatrix}0 & 0 & 0\\0 & 0 & -4\end{smallmatrix}]$

Fig. 6. After four shifts (partial code fragment)

4 Implementation

This framework uses the PolyLib [20] and CLooG – a robust new code generator [3]. It is implemented as a plug-in for Open64/ORC[3]. The availability of CLooG is a major reason for making polyhedral approaches applicable to real codes. Our tool converts the WHIRL, the intermediate representation of Open64, to an augmented polyhedral representation with maps to the symbol table and syntax tree, and regenerates WHIRL from this representation [4]. 12 SpecFP 2000 and PerfectClub benchmarks were run through the whole source-to-polyhedra-to-source conversion cycle (without loop transformations); it takes from one second to two minutes on a typical Pentium III, see [4] and [3] for details. The software component in charge of the polyhedral transformations is driven through a script language, with a specific syntax to define primitives and to compose new transformations. The syntax is based on C++ with overloaded operators for vector and matrices. From the script, the tool generates the actual source code of the related transformations: it effectively generates a class for each transformation with its own methods for prerequisite checking and application.

5 Conclusion

We presented a polyhedral framework that enables the composition of long sequences of program transformations. Coupled with a robust code generator, this method avoids the typical code complexity explosion of long compositions of program transformations; these techniques have been implemented in the Open64/ORC compiler. While revisiting the design of a polyhedral program representation and the definition of transformation primitives, we have shown the advantages of using such advanced techniques in the software engineering of a loop-restructuring compiler. In particular, we have decoupled the polyhedral representation, transformation and code generation techniques from their historical applications to automatic affine scheduling and mapping (automatic parallelization and model-based optimization). More practically, the ability to perform numerous compositions of program transformations is key to the extension of iterative optimizations to finding the appropriate program transformations instead of just the appropriate program transformation parameters.

Acknowledgments

This work is supported by the "COP" RNTL grant from the French Ministry of Research. We are very thankful to Cedric Bastoul and Saurabh Sharma, whose algorithms, tools and support have been critical for the design and implementation of our framework.

---

[3] Publicly available at http://www-rocq.inria.fr/a3/wrap-it

References

1. M. Barreteau, F. Bodin, Z. Chamski, H.-P. Charles, C. Eisenbeis, J. R. Gurd, J. Hoogerbrugge, P. Hu, W. Jalby, T. Kisuki, P. M. W. Knijnenburg, P. van der Mark, A. Nisbet, M. F. P. O'Boyle, E. Rohou, A. Seznec, E. Stöhr, M. Treffers, and H. A. G. Wijshoff. Oceans - optimising compilers for embedded applications. In *Euro-Par'99*, pages 1171–1775, Aug. 1999.
2. D. Barthou, J.-F. Collard, and P. Feautrier. Fuzzy array dataflow analysis. *J. of Parallel and Distributed Computing*, 40:210–226, 1997.
3. C. Bastoul. Efficient code generation for automatic parallelization and optimization. In *ISPDC'2 IEEE International Symposium on Parallel and Distributed Computing*, Ljubjana, Slovenia, Oct. 2003.
4. C. Bastoul, A. Cohen, S. Girbal, S. Sharma, and O. Temam. Putting polyhedral loop transformations to work. In *Workshop on Languages and Compilers for Parallel Computing (LCPC'03)*, LNCS, College Station, Texas, Oct. 2003.
5. W. Blume, R. Eigenmann, K. Faigin, J. Grout, J. Hoeflinger, D. Padua, P. Petersen, W. Pottenger, L. Rauchwerger, P. Tu, and S. Weatherford. Parallel programming with Polaris. *IEEE Computer*, 29(12):78–82, Dec. 1996.
6. A. Cohen, S. Girbal, and O. Temam. Facilitating the exploration of compositions of program transformations. Research report 5114, INRIA Futurs, France, Feb. 2004.
7. K. D. Cooper, M. W. Hall, R. T. Hood, K. Kennedy, K. S. McKinley, J. M. Mellor-Crummey, L. Torczon, and S. K. Warren. The ParaScope parallel programming environment. *Proceedings of the IEEE*, 81(2):244–263, 1993.
8. K. D. Cooper, D. Subramanian, and L. Torczon. Adaptive optimizing compilers for the 21st century. *J. of Supercomputing*, 2002.
9. A. Darte, Y. Robert, and F. Vivien. *Scheduling and Automatic Parallelization*. Birkhaüser, Boston, 2000.
10. P. Feautrier. Some efficient solutions to the affine scheduling problem, part II, multidimensional time. *Int. J. of Parallel Programming*, 21(6):389–420, Dec. 1992. See also Part I, one dimensional time, 21(5):315–348.
11. G. Fursin, M. O'Boyle, and P. Knijnenburg. Evaluating iterative compilation. In *11th Workshop on Languages and Compilers for Parallel Computing*, LNCS, Washington DC, July 2002. Springer-Verlag.
12. A.-C. Guillou, F. Quilleré P. Quinton, S. Rajopadhye, and T. Risset. Hardware design methodology with the Alpha language. In *FDL'01*, Lyon, France, Sept. 2001.
13. M. Hall et al. Maximizing multiprocessor performance with the SUIF compiler. *IEEE Computer*, 29(12):84–89, Dec. 1996.
14. F. Irigoin, P. Jouvelot, and R. Triolet. Semantical interprocedural parallelization: An overview of the pips project. In *ACM Int. Conf. on Supercomputing (ICS'2)*, Cologne, Germany, June 1991.
15. W. Kelly. Optimization within a unified transformation framework. Technical Report CS-TR-3725, University of Maryland, 1996.
16. W. Kelly, W. Pugh, and E. Rosser. Code generation for multiple mappings. In *Frontiers'95 Symp. on the frontiers of massively parallel computation*, McLean, 1995.
17. I. Kodukula and K. Pingali. Transformations for imperfectly nested loops. In *Supercomputing (SC'96)*, Jan. 1996.

18. A. W. Lim and M. S. Lam. Communication-free parallelization via affine transformations. In 24th ACM Symp. on Principles of Programming Languages, pages 201–214, Paris, France, jan 1997.
19. A. W. Lim, S.-W. Liao, and M. S. Lam. Blocking and array contraction across arbitrarily nested loops using affine partitioning. In ACM Symp. on Principles and Practice of Parallel Programming (PPoPP'01), pages 102–112, 2001.
20. V. Loechner and D. Wilde. Parameterized polyhedra and their vertices. Int. J. of Parallel Programming, 25(6), Dec. 1997. http://icps.u-strasbg.fr/PolyLib.
21. M. O'Boyle. MARS: a distributed memory approach to shared memory compilation. In Proc. Language, Compilers and Runtime Systems for Scalable Computing, Pittsburgh, May 1998. Springer-Verlag.
22. D. Parello, O. Temam, and J.-M. Verdun. On increasing architecture awareness in program optimizations to bridge the gap between peak and sustained processor performance? matrix-multiply revisited. In SuperComputing'02, Baltimore, Maryland, Nov. 2002.
23. F. Quilleré, S. Rajopadhye, and D. Wilde. Generation of efficient nested loops from polyhedra. Intl. J. of Parallel Programming, 28(5):469–498, Oct. 2000.
24. R. Schreiber, S. Aditya, B. Rau, V. Kathail, S. Mahlke, S. Abraham, and G. Snider. High-level synthesis of nonprogrammable hardware accelerators. Technical report, Hewlett-Packard, May 2000.
25. M. E. Wolf. Improving Locality and Parallelism in Nested Loops. PhD thesis, Stanford University, Aug. 1992. Published as CSL-TR-92-538.
26. M. J. Wolfe. High Performance Compilers for Parallel Computing. Addison-Wesley, 1996.
27. D. Wonnacott and W. Pugh. Nonlinear array dependence analysis. In Proc. Third Workshop on Languages, Compilers and Run-Time Systems for Scalable Computers, 1995. Troy, New York.

Compiler-Guided Code Restructuring for Improving Instruction TLB Energy Behavior*

I. Kadayif[1], M. Kandemir[2], and I. Demirkiran[3]

[1] Canakkale Onsekiz Mart University, Canakkale, Turkey
kadayif@comu.edu.tr
[2] Pennsylvania State University, University Park, PA, USA
kandemir@cse.psu.edu
[3] Syracuse University, Syracuse, NY, USA
idemirki@eecs.syracuse.edu

Abstract. This paper presents a compiler-directed strategy for reducing energy consumption of instruction TLBs. The main idea is to restructure the code to increase the chances that one can generate virtual-to-physical address translation without going through the instruction TLB. The preliminary experimental results are promising.

1 Introduction and Motivation

TLB (translation lookaside buffer) is a crucial component that maintains recent virtual-to-physical address translations. Optimizing energy consumption of TLB is critical due to two main reasons. First, since this component is accessed at each memory reference, it can contribute to a significant fraction of on-chip energy budget. For example, instruction and data TLBs are known to contribute to over 15% of on-chip energy consumption in SH-3 and Intel StrongARM [14]. Second, since TLB is a very small component, its power density can be quite high as a result of frequent accesses. Therefore, reducing TLB energy consumption can be very important. TLB optimization has been focus of several circuit and architecture level studies. For example, Juan et al [7] proposed modifications to the basic cells and to the structure of TLBs that led to15% improvement in per access energy consumption. Choi et al [3] proposed a two-way banked filter TLB and a two-way banked main TLB. Balasubramonian et al [2] and Delaluz et al [5] proposed changing the TLB configuration dynamically, based on the need of the application at a given execution phase.

In a recent work, Kadayif et al [8] has demonstrated that using an optimizing compiler and a suitable help from the hardware, one can generate most of virtual-to-physical address translations automatically, thereby reducing the frequency of iTLB accesses. Their compiler-directed strategy operates as follows. Overall philosophy of this strategy is to perform the translation for a page once, and subsequently keep reusing it directly without going to the iTLB, as long as it does not change. This is achieved through the use of a register called Current Frame Register (CFR), whose format is of the form:

<Virtual Page Number, Physical Frame Number, Protection/Other Bits>.

---

* This research is partly supported by NSF Career Award #0093082.

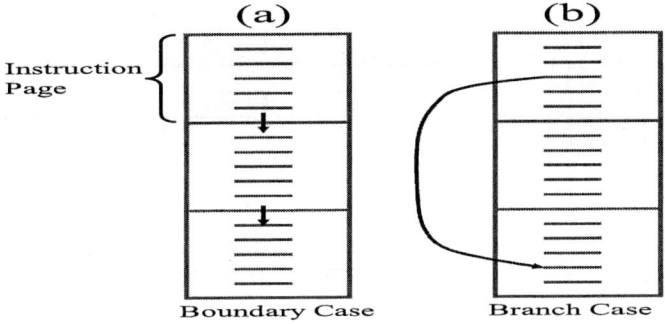

Fig. 1. Two possible instruction page transitions during execution. (a) Boundary case. (b) Branch case.

Basically, this register holds the current virtual-to-physical address translation. As long as we are sure that the current address translation does not change (i.e., we are within the same page), we take the translation from the CFR without going to the iTLB. Whenever there is a page change, we need to re-load (update) the CFR. This occurs in two scenarios in normal execution as shown in Figure 1: (a) two successive instructions, which are on page boundaries (we refer to this as the boundary case), i.e., one is the last instruction of a page, and the next is the first instruction of the next page (we assume that instructions are aligned so that a single instruction does not cross page boundaries), and (b) explicit branch instructions whose target is in a different page (we call this the branch case).

In the compiler-based scheme proposed in [8], the compiler is used to determine both the boundary and branch cases. The branch cases are handled as follows. The compiler assumes that the branch target is within the same page as the branch instruction if the static analysis of the code by the compiler can reveal (with 100% accuracy) that this indeed is the case (note that this typically occurs when branch targets are given as immediate operands or as PC relative operands). Otherwise, if the branch is unanalyzable or its target is proven to go outside the current instruction page, the compiler conservatively assumes a page change and updates the CFR contents via an iTLB access. The necessary compiler support for implementing this involves checking whether the target of a statically analyzable branch is on the same page of the branch itself. To handle the boundary case, the compiler inserts an explicit branch instruction at the end of each instruction page, with the target being the very next instruction (the first one on the next page). Note that this mechanism does not much affect iL1 and L2 hits or misses, and thus it does not affect the rest of the memory system energy consumption.

While the experimental results given in [8] indicate significant energy benefits without much performance overhead, note that *one can achieve even better energy savings by restructuring the code*. Specifically, what we need to do is to increase the number of branches whose targets can be proven to be within the same page (as the branch instruction itself). In the rest of this paper, we describe a compiler-directed strategy to achieve this.

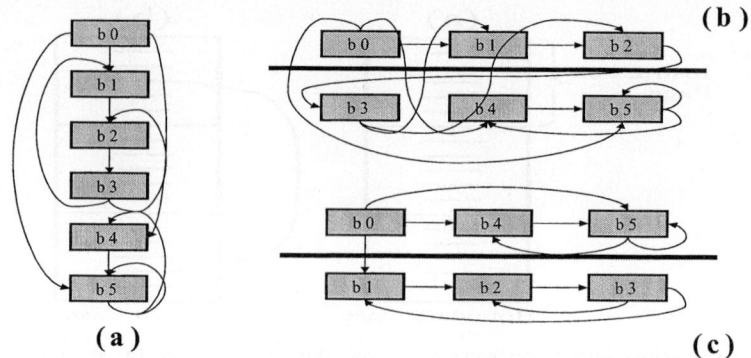

Fig. 2. (a) An example code fragment. (b-c) Two alternate page assignments.

2 Code Restructuring for Translation Reuse

We can define our problem as follows. Consider the code layout shown in Figure 2(a). Let b_i represent the size of basic block i, and P be the page size. We use B to denote the set of all basic blocks in the code. The connections between basic blocks (which correspond to the edges in the control flow graph representation of the program) are indicated using edges $e_{ij} = (b_i, b_j)$. Each edge e_{ij} also carries a weight, denoted w_{ij}, which indicates the importance of satisfying the edge. In this context, strictly speaking, "satisfying edge e_{ij}" means colocating b_i and b_j within the same instruction page. However, we will use a relaxed version of this definition, which says if b_i and b_j are stored in memory one after another, e_{ij} is satisfied. Given an assignment of basic blocks to memory (denoted using mapping M), it is likely that some edges will remain inside a page, whereas some other edges will cross the page boundaries; these two sets of edges are referred to as *in-page edges* and *out-page edges* in the rest of this paper. Then, the *cost* of such an assignment can be formulated as: $C(B,M) = \Sigma_i \Sigma_j w_{ij}$ such that e_{ij} is an out-page edge. That is, the total weights of all out-page edges give us the cost of mapping M. Obviously, our objective is to place basic blocks into pages such that $C(B,M)$ is minimized. In other words, we want to find an M that minimizes $C(B,M)$.

In this paper, we propose an algorithm for solving this problem of determining M (a memory assignment) for all blocks in B as shown below:

This strategy, which is actually a heuristic, operates in a page size (P) oblivious manner. Consequently, it adopts the relaxed definition of the concept of satisfying an edge, as described above. This heuristic is similar to the Kruskal's spanning tree algorithm [4], and is given below. In this algorithm, E is the ordered set of edges (according to their weights), and T is the (spanning) tree to be built.

```
1.   Order the edges according to non-increasing weights
2.   While (there is a node to be included in T)
    2.1.   Select the edge with the largest weight
    2.2.   If (adding the selected edge to T does not increase
           the degree of an node in T to 3 AND does not
           create a cycle in T) then
           add the selected edge to T
3.   Traverse T and store the nodes (basic blocks)
     that are directly connected consecutively in memory
```

Basically, this algorithm builds a spanning tree (which includes all nodes in B and a subset of edges in E), with the property that no node (basic block) is connected to more than 2 neighbors. This is done so to guarantee that each pair of nodes with a connected edge will be stored in memory consecutively (since in memory layout a basic block can have only two neighbors). This algorithm is similar to the one presented in [9] for assigning program variables to memory locations in DSPs. The main difference is that we work on basic blocks not variables since our objective is to restructure code layout for iTLB energy savings. Note also that this algorithm is completely different from prior work on instruction cache optimization [13, 6, 11, 10] as we use a different representation and solution method. It is to be noted, while this algorithm does not take into account the page size (P), we do not expect this to be a major problem in practice. This is because the algorithm will fail to optimize only in cases where the two basic blocks with high affinity fall into different pages. Given the large page sizes, we do not expect this to occur very frequently.

Figures 2(b) and (c) illustrate two possible page assignments, assuming, for simplicity, that each page can hold three basic blocks. The first alternative (shown in (b)) is the straightforward one, whereas the second (the one in (c)) is the one generated by our approach (under the assumption of same weights for all edges between the basic blocks). In both the cases, the thick line delineates the page boundary. Note that, in our case, there is only one out-page edge, while we have five out-page edges in the straightforward option.

3 Preliminary Experiments

We have implemented the proposed strategy and performed experiments with several Spec95 applications. The experiments have been performed using SimpleScalar [1], and the energy numbers have been obtained through CACTI [12]. In Figure 3(a), we present the *normalized* energy consumptions for two different strategies. The first strategy (marked opt-1) is the one presented in [8]. It uses the CFR to the fullest extent possible, but does not restructure code for the CFR reuse. The second strategy (marked opt-2) represents the results obtained through the compiler-based code restructuring strategy discussed in this paper. For each benchmark, both the bars are normalized with respect to the iTLB energy consumption of the original (default) case, where we do not make use of CFRs, and all instruction accesses go through iTLB. Note that in both opt-1 and opt-2, all extra energy consumptions due to CFR accesses have been included in the results. We observe from these results that, in five out of six benchmarks, our approach improves the iTLB energy consumption beyond the method presented in [8]. That is, restructuring application codes for CFR reuse is useful in practice. The performance (execution cycle) results are given in Figure 3(b). As before, all the bars are *normalized* with respect to the default version, where no iTLB energy-saving technique is used. We see that, while opt-1 generates (almost) the same results with the default version, opt-2 slightly improves performance as well. This is because our approach also enhances instruction cache locality by bringing the blocks with temporal affinity together. Therefore, we can conclude that the proposed scheme brings both energy and performance benefits. Our on-going work involves implementing an optimal scheme for basic block re-ordering (based on integer linear programming), and comparing it against the scheme proposed in this paper.

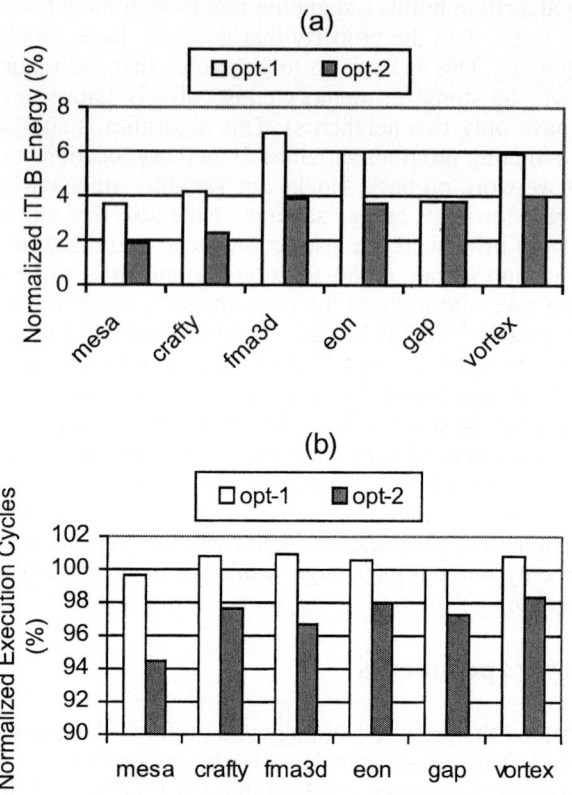

Fig. 3. (a) Normalized iTLB energy. (b) Normalized execution cycles.

References

1. T. Austin, E. Larson, and D. Ernst. SimpleScalar: an infrastructure for computer system modeling. *IEEE Computer Magazine,* pp. 59-67, Feb 2002.
2. R. Balasubramanian, D.H. Albonesi, A. Buyuktosunoglu, and S. Dwarkadas. Memory hierarchy reconfiguration for energy and performance in general-purpose processor architectures. In *Proc. 33rd International Symposium on Microarchitecture,* pp. 245--257, December 2000.
3. J-H. Choi, J-H. Lee , S-W. Jeong , S-D. Kim , and C. Weems. A low-power TLB structure for embedded systems. *IEEE Computer Architecture Letters,* Volume 1, January 2002.
4. T. H. Cormen, C. E. Leiserson, R. L. Rivest, and C. Stein. *Introduction to algorithms,* Second Edition, The MIT Press, 2001.
5. V. Delaluz, M. Kandemir, A. Sivasubramaniam, M. J. Irwin, and N. Vijaykrishnan. Reducing dTLB energy through dynamic resizing. In *Proc. the 21st International Conference on Computer Design,* San Jose, California, October, 2003.
6. N. Gloy, T. Blackwell , M. D. Smith, and B. Calder. Procedure placement using temporal ordering information. In *Proc.the 30th ACM/IEEE International Symposium on Microarchitecture,* p.303-313, December 01-03, 1997.

7. T. Juan, T. Lang, and J. J. Navarro. Reducing TLB power requirements. In *Proc. International Symposium on Low Power Electronics and Design,* 1997.
8. I. Kadayif, A. Sivasubramaniam, M. Kandemir, G. Kandiraju, and G. Chen. Generating physical addresses directly for saving instruction TLB energy. In *Proc. International Symposium on Microarchitecture,* Istanbul, Turkey, November 2002.
9. S. Liao, S. Devadas, K. Keutzer, S. W. K. Tjiang, and A. Wang. Storage assignment to decrease code size. In *Proc. International Symposium on Programming Language Design and Implementation,* pp. 186-195, 1995.
10. S. McFarling, Procedure merging with instruction caches. *ACM SIGPLAN Notices,* v.26 n.6, p.71-79, June 1991.
11. K. Pettis and R. C. Hansen. Profile guided code positioning, *ACM SIGPLAN Notices,* v.25 n.6, p.16-27, June 1990.
12. G. Reinman and N. P. Jouppi. CACTI 2.0: an integrated cache timing and power model. *Research Report 2000/7,* Compaq WRL, 2000.
13. A. D. Samples and P. N. Hilfinger. Code reorganization for instruction caches. *Technical Report UCB/CSD 88/447,* University of California, Berkeley, October 1988.
14. SH-3 RISC processor family.
http://www.hitachi-eu.com/hel/ecg/products/micro/32bit/sh_3.html.

Using Data Compression to Increase Energy Savings in Multi-bank Memories*

M. Kandemir[1], O. Ozturk[1], M.J. Irwin[1], and I. Kolcu[2]

[1] The Pennsylvania State University, University Park, PA, USA
{kandemir,ozturk,mji}@cse.psu.edu
[2] UMIST Manchester, M601QD, UK
ikolcu@umist.ac.uk

Abstract. New DRAM technologies such as SDRAMs, RDRAMs, EDRAMs, CDRAMs and others are vying to be the next standard in DRAMs and improve upon bandwidth limit of conventional DRAMs. With proliferation of power-aware systems, banked DRAM architecture has emerged as a promising candidate for reducing power. Prior work on optimizing applications in a banked memory environment has exclusively focused on uncompressed data. While this may be preferable from a performance viewpoint, it is not necessarily the best strategy as far as memory space utilization is considered. This is because compressing data in memory may reduce the number of memory banks it occupies and this, in turn, may enable a better use of low-power operating modes. In this paper, we explore the possibility of compressing infrequently used data for increasing effectiveness of low-power operating modes in banked DRAMs. Our experiments with five highly parallel array-based embedded applications indicate significant savings in memory energy over a technique that exploits low-power modes but does not use data compression/decompression.

1 Introduction

Low power dissipation in portable battery-operated platforms has drawn significant interest in the past decade. In many applications targeting at embedded platforms, a large fraction of energy consumption is due to main memory accesses. Recent work [1, 2, 3, 5] has suggested multi-banking as a way of reducing memory energy consumption. The main rationale behind this approach is that per access energy consumption is proportional to the size of the memory, and a small memory bank consumes much less (per access) energy than a large monolithic structure. In addition, unused memory banks in a multi-bank architecture can be placed into low-power operating modes to further energy savings.

Prior work on optimizing applications in a banked memory environment has exclusively focused on uncompressed data. In other words, the data manipulated by the application have been kept in memory in an uncompressed form throughout the execution. While this may be preferable from a performance viewpoint, it is not necessarily the best option as far as memory space utilization is considered. This is because compressing data in memory may reduce the number of memory banks it occupies

---

* This research is partly supported by NSF Career Award #0093082.

and this, in turn, may enable a better use of existing low-power operating modes (i.e., unused banks can be placed into low-power operating modes).

It is to be noted, however, that there are several important issues that need to be addressed before one has a reasonable data compression strategy.

- How should the available memory space be divided between compressed and uncompressed data?
- What should be the granularity of compression? A whole array, a data page, etc.?
- Which compression strategy should be employed when a given data block is to be compressed/decompressed?

The third question posed above is orthogonal to the ideas explored in this paper. That is, the proposed strategy can work in conjunction with any data compression/decompression algorithm. Therefore, without loss of generality, in this paper we use the algorithm proposed by Kjelso et al [4]. In this work, we concentrate on the remaining questions and propose a strategy for optimizing the effectiveness of low-power operating modes. While prior work employed data compression for energy savings and performance improvement [7, 9], to the best of our knowledge, this is the first study that considers data compression within the context of banked memory architectures.

Our strategy divides the available memory space (memory banks) into two disjoint parts (groups): one that holds compressed data and one that holds uncompressed data. The main objective here is to keep non-hot data (i.e., the data that are not very frequently used) in the compressed form to the extent possible. Another important issue here is to cluster the compressed data as much as possible for increasing the effectiveness of low-power modes. The timing for compressions/decompressions is also an important issue. Typically, one does not want to access data while it is in the compressed form (since it needs to be decompressed before it can be used; and this requires extra cycles in execution). In addition, when a new data is created or an old data is re-written, one needs to decide whether to store it in compressed or uncompressed form. In the rest of this paper, we focus on a banked memory architecture, and evaluate several compression-based memory bank management strategies for making best use of available low-power operating modes in a DRAM-based memory architecture.

This paper is structured as follows. The next section gives an overview of banked memory architecture and low-power operating modes. Section 3 presents details of our compression-based memory management strategy. Section 4 gives experimental data. Section 5 concludes the paper by summarizing our major results, and giving a brief discussion of ongoing work.

2 Banked Memory Architecture and Low-Power Operating Modes

We target a memory system that contains a number of banks, each of which can be energy-controlled independently. To save energy in this architecture, we put unused memory banks into a low-power operating mode. We assume the existence of three operating modes for a memory bank: active, napping, and power-down. Each mode is characterized by its power consumption per cycle and the time that it takes to transition back to the active mode (referred to as the resynchronization time or resynchronization cost). Typically, lower the energy consumption, higher the resynchronization time. These modes are characterized by varying degrees of the bank components being active. Table 1 shows our operating modes, their per cycle power consumptions, and resynchronization costs.

Table 1. Operating-modes for memory banks.

	Energy Consumption (nJ/cycle)	Resynchronization Cost (cycles)
active	3.570	0
napping	0.320	30
power-down	0.005	9,000

A memory controller that interfaces with the memory bus controls DRAM banks. The interface is not only for latching the data and addresses, but also to control the configuration and operation of the individual banks as well as their operating modes. For example, programming a specific control register in each memory bank could do the operating mode setting. Next is the issue of how the memory controller can be told to transition the operating modes of the individual banks. In this paper, we use a hardware-based approach to control mode transitions. In this approach, there is a watchdog hardware that monitors ongoing memory transactions. It contains some prediction mechanism to estimate the time until the next access to a memory bank and circuitry to ask the memory controller to initiate mode transitions. The specific hardware depends on the prediction mechanism that is implemented. In this work, we use an *adaptive next-mode prediction scheme*. In this scheme, if a memory bank has not been accessed for a while, then it is assumed that it will not be needed in the near future. A threshold is used to determine the idleness of a bank after which it is transitioned to a lower energy mode. The threshold is adaptive in the sense that it tries to adjust for any mispredictions it has made. Specifically, it starts with an initial threshold, and transitions to the lower energy mode if the bank is not accessed within this period. If the next access is to come soon after that (the resynchronization energy consumption is more dominant than the savings due to the lower energy mode), making the mode transition more energy consuming than if we had not transitioned at all, the threshold is doubled for the next interval. On the other hand, if we find that the next access comes fairly late, and we were overly conservative in the threshold value, then the threshold is reset to the initial value (one could potentially try more sophisticated techniques such as halving the threshold as well). Our objective in this work is to demonstrate that using data compression, one can increase the effectiveness of this low-power mode management scheme; i.e., we can either put more banks into low-power modes, or we can use more aggressive (i.e., more energy-saving) low-power mode for banks.

3 Compression-Based Memory Management

We explore design space in a systematic way, focusing on each design decision separately.

Memory Space Division. There are two different ways of dividing a given memory space between compressed and uncompressed data. In the "static" strategy, p out of a total of m memory banks are reserved for compressed data, whereas the remaining banks are allocated for uncompressed data. Obviously, if most of the data manipulated by the application are not hot (i.e, not frequently accessed), one may want to increase p. On the other hand, in the "dynamic" strategy, the number of memory banks allocated to compressed and uncompressed data changes dynamically during the course of execution. Note that the dynamic strategy is expected to perform better than the static one if the behavior of the application changes during the course of execution.

Compression Granularity. Data can be compressed/decompressed in different granularities. In the "fixed granularity" scheme, we divide each dataset (array) into blocks of fixed size (e.g., 4KB), and the block sizes for all datasets are the same. Alternately, in the "variable granularity" scheme, each dataset (e.g., an array) can have a different block size (though all the blocks belonging to a given dataset are of the same size). Note that, as a special case of this latter scheme, one can set the block size of each dataset to the dataset size (in which case we have only 1 block per array).

Data Creation/Re-writing Strategy. When the data is created or re-written, we have flexibility of storing it in compressed or uncompressed form. In the rest of this paper, these two strategies are termed as the "compressed store" and "uncompressed store." Ideally, this decision should be made considering the future use of the data in question. It should be noticed that it is also possible to make this decision by analyzing the data access pattern exhibited by the application. More specifically, if the compiler can determine that a data block will not be used for a long period of time, it can be stored in the compressed format. On the other hand, if it will be reused shortly, it can be stored in an uncompressed form. This *compiler-directed* strategy is called "dynamic" in the remainder of this paper, since it tunes the store strategy based on the reuse of each data block.

Decompression Strategy. Another important question is when to decompress data. One obvious choice is "on-demand" decompression, whereby data is decompressed (if it is in the compressed format) only when it is really needed. In contrast, in "pre-decompression," data is decompressed before it is actually accessed. This latter alternative is only possible when we have extra cycles to pre-decompress data. Also, this is a compiler-based strategy in that the compiler analyzes the code, identifies the potential reuses of each data block, and inserts explicit pre-decompress calls in the code to initiate pre-decompression of data blocks.

Based on these, Table 2 shows our four-dimensional design space, a subset of which is explored in this paper from both energy and performance (execution cycles) perspectives.

Let us now discuss the implementation details of these different memory management strategies. Static memory space division is easy to implement. Basically, we select a value for p, and use p banks for the compressed data. However, implementing dynamic strategy is more involved. This is because we need to decide when to change the value of p during the course of execution. To achieve this, we make use of the concept of the *miscompression rate* (or MCR for short), which is the fraction of the data accesses where the data is found in the compressed format in memory. Obviously, MCR should be reduced as much as possible. Based on MCR, our dynamic memory space division strategy operates as follows. If the current MCR is larger than a high-threshold (high-MCR) we reduce p, and if it is lower than a low-threshold (low-MRC), we increase p. We do not change p as long as the current MCR value is between low-MCR and high-MCR. In other words, as long as we are operating between

Table 2. Possible strategies for compression/decompression based memory bank management.

Method Id	Space Division	Comp. Gran.	Rewriting Strategy	Decomp. Strategy
1	static	fixed	dynamic	on-demand
2	static	fixed	dynamic	pre-decomp
3	static	fixed	uncompressed	on-demand
4	static	fixed	uncompressed	pre-decomp
5	static	variable	dynamic	on-demand
6	static	variable	dynamic	pre-decomp
7	static	variable	uncompressed	on-demand
8	static	variable	uncompressed	pre-decomp
9	dynamic	fixed	dynamic	on-demand
10	dynamic	fixed	dynamic	pre-decomp
11	dynamic	fixed	uncompressed	on-demand
12	dynamic	fixed	uncompressed	pre-decomp
13	dynamic	variable	dynamic	on-demand
14	dynamic	variable	dynamic	pre-decomp
15	dynamic	variable	uncompressed	on-demand
16	dynamic	variable	uncompressed	pre-decomp

low-MCR and high-MCR, we assume that the current division of the memory space is performing well. Also note that the current MCR value needs to be updated at regular intervals. In our current implementation, this is achieved as follows. Each time a block is accessed, we check whether it is compressed or not. If it is compressed, we increase a counter. At regular intervals (whose period is programmer-tunable), we compute the MCR, and change the memory space partitioning between compressed and uncompressed data if necessary.

As far as the compression granularity is concerned, implementing variable granularity is relatively easy if the block size for a given dataset is set to its total size (i.e., block size = array size). In this case, only bookkeeping necessary is a data structure that keeps track of which datasets are currently in the compressed form and which ones are not. However, implementing the fixed block strategy is more challenging, as we need to keep track of the status of each data block individually. In order to do this, our implementation employs a bit map, where each bit represents the status of a block. Consequently, each memory access goes through this bit map to determine the status of the data block to be accessed. In this paper, we experimented with only two re-writing strategies: uncompressed store and dynamic. We did not consider the compressed store strategy as our initial experiments showed that its performance is extremely poor (since, due to temporal locality of data, it is not a good idea to store the data always in the compressed format). Comparing the two strategies we experimented with, it is easy to see that the dynamic strategy requires extra compiler support (which will be explained shortly) to measure the reuse distance of the data (whereas the static strategy does not need such help).

Finally, while implementing on-demand decompression strategy is not very difficult, to implement the pre-decompression strategy, *the compiler needs to analyze the program and schedule data decompressions ahead of the time.* The necessary com-

Table 3. Default simulation parameters used for in experiments.

Simulation Parameter	Value
m	8
p	3
(High-MCF, Low-MCF)	(30%, 5%)
Block Size	2KB

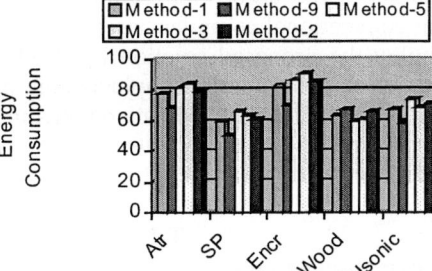

Fig. 1. Normalized energy consumption values with different methods.

piler support for this is similar to that of software-based data prefetching [11]. It is to be noted that this pre-decompression scheme can be successful only when we have extra cycles to pre-decompress the data. In our current implementation, we use an extra (pre-decompression) thread to perform pre-decompressions. This thread shares the same resources with the main execution thread. Whenever there are idle cycles, it kicks in and performs necessary (soon to be needed) decompressions. In the worst case, it may happen that the main execution thread wants to access a block and finds it compressed (which means that the pre-decompression thread could not find opportunity to decompress the block in question). If this happens, we proceed by allowing the pre-decompression thread to finish its job before the block is accessed.

In this work, we implement and study a subset of the strategies listed in Table 2. There are two main reasons for that. First, some of the strategies in Table 2 do not make much sense. Second, by being a bit careful in selecting the subset mentioned, one may have a good insight on the trends of interest as far as energy and performance behaviors are concerned.

4 Experimental Evaluation

Table 3 lists the default parameters used in our simulations. We use 4-issue processor core with 8K instruction and data caches. Later in our experiments we modify some of the parameters given in this table to conduct a sensitivity analysis. Recall that Table 1 gives the characteristics of the low-power operating modes used in this work.

All energy numbers presented in this paper have been obtained using a custom memory energy simulator (built upon SimpleScalar simulator). This simulator takes as input a C program and a banked memory description (i.e., the number and sizes of memory banks as well as available low-power operating modes with their energy saving factors and re-synchronization costs). As output, it gives the energy consumption in memory banks along with a detailed bank inter-access time profiles. By giving original and optimized programs to this simulator as input, we measure the impact of our compression-based strategy on memory system energy.

All necessary code restructurings for implementing "dynamic" and "pre-decompression" strategies have been implemented within the SUIF framework from Stanford University [8]. To implement the dynamic strategy, we keep track of data reuse information at the array block granularity (instead of array element granularity as done in conventional compiler-based locality optimization studies). For this purpose, we employ the proposed solution by Wolf and Lam [10], and feed the data block information to the compiler. Implementing pre-decompression is very similar to implementing compiler-directed data prefetching. Basically, the compiler predicts future use of data and inserts appropriate pre-compression calls (instead of prefetching calls) in the code.

Table 4. Our benchmark codes.

Bench	Brief Description	Numof Lines	Energy Consumption (mJ)
Atr	Network address ranslation	626	13.38mJ
SP	All-nodes shortest path algorithm	1028	29.31mJ
Encr	Digital signature for security	1411	36.06mJ
Wood	Color-based surface inspection	978	22.80mJ
Usonic	Feature-based estimation	1005	37.44mJ

We test the effectiveness of our optimization strategy using five benchmark codes. Table 4 presents descriptions and important characteristics of these benchmarks. The last column in this table gives memory energy consumption when no compression/decompression is employed but available low-power operating modes are fully exploited. While we also have access pointer-based implementations of these benchmark codes, in this study we used array-based implementations.

We focus mainly on five different methods (from Table 2), and Figure 1 gives their memory energy consumption values, normalized with respect to the case without compression (i.e., with respect to the last column of Table 4). The reason why we are focusing on these five methods can be explained as follows. We consider Method 1 as the base and obtain the other versions by changing one parameter at a time (see the column titles in Table 2). One can observe from Figure 1 that the average energy improvement brought about by Method 1 is 29.78%, indicating that our compression-based scheme can be very successful in practice. Now, if consider Method 9, we can see from Figure 1 that its average saving is significantly better than that of Method 1 (36.66%). To understand this better, we focus on one of our applications (Usonic) and present its bank usage behavior. The x-axis in the graph in Figure 2 represents the time, and the y-axis gives the total size of the compressed data (on a time quantum basis) in terms of the number of banks it occupies. For example, if the y-axis value is 4.3 at a given time quantum, it means that the total size of the compressed data can occupy 4.3 banks. One can conclude from this graph that the amount of bank space

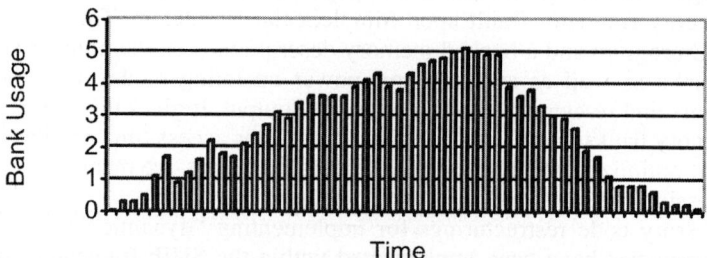

Fig. 2. Bank usage pattern off Usonic over the time (Method 9).

occupied for the compressed data varies over the time. Therefore, a strategy (such as Method 1), which allocates 3 banks for the compressed data (i.e., p=3 as shown in Table 3) throughout the execution, cannot generate the best energy behavior. In other words, for the best behavior, the amount of space allocated for the compressed data should be varied. In fact, we can see from Figure 1 that dynamic space division is beneficial for all applications except one.

5 Ongoing Work and Concluding Remarks

Banked-memory architectures enable power savings through low-power operating modes. While previous compiler, OS, and hardware based techniques showed significant savings in memory energy consumption by making use of these low-power modes, this paper demonstrated that further savings are possible if one employs data compression. We first described the potential search space, and then evaluated some select implementation methods. Our ongoing research includes experimenting with more sophisticated strategies (such as the one that allows different block sizes for different arrays), and evaluating the impact of our approach on multi-programmed workloads.

References

1. V. Delaluz, M. Kandemir, N. Vijaykrishnan, A. Sivasubramaniam, and M. J. Irwin. DRAM Energy Management Using Software and Hardware Directed Power Mode Control. In Proc. the 7th Int'l Symposium on High Performance Computer Architecture, 2001.
2. A. Farrahi, G. Tellez, and M. Sarrafzadeh. Exploiting Sleep Mode for Memory Partitions and Other Applications. *VLSI Design*, Vol. 7, No. 3, pp. 271-287.
3. M. Kandemir, I. Kolcu, and I. Kadayif. Influence of Loop Optimizations on Energy Consumption of Multi-Bank Memory Systems. In *Proc. International Conference on Compiler Construction,*, 2002.
4. M. Kjelso, M. Gooch, and S. Jones. Performance Evaluation of Computer Architectures with Main Memory Data Compression. Elsevier Science, *Journal of Systems Architecture*, 45 (1999), pp. 571-590.
5. A. Lebeck, X. Fan, H. Zeng, and C. S. Ellis. Power-Aware Page Allocation. In *Proc. 9th International Conference on Architectural Support for Programming Languages and Operating Systems*, Nov. 2000.
6. 128/144-MBit Direct RDRAM Data Sheet, Rambus Inc., May 1999.
7. B. Abali et al. Memory Expansion Technology (MXT): Software support and performance. *IBM Journal of Research and Development,* Vol 45, No 2, 2001.
8. S. P. Amarasinghe, J. M. Anderson, C. S. Wilson, S.-W. Liao, B. R. Murphy, R. S. French, M. S. Lam and M. W. Hall Multiprocessors from a Software Perspective, IEEE Micro, June 1996, pages 52-61.
9. L. Benini, D. Bruni, A. Macii, and E. Macii. Hardware-Assisted Data Compression for Energy Minimization in Systems with Embedded Processors. In *Proc. DATE*, 2002.
10. M. E. Wolf and M. S. Lam. A Data Locality Optimizing Algorithm. In *Proc. the ACM SIGPLAN'91 Conference on Programming Language Design and Implementation*, June, 1991.
11. T. Mowry. Tolerating Latency in Multiprocessors through Compiler-Inserted Prefetching. *ACM Transactions on Computer Systems,* 16(1):55-92, February 1998.

Automatic Customization of Embedded Applications for Enhanced Performance and Reduced Power Using Optimizing Compiler Techniques*

Emre Özer, Andy P. Nisbet, and David Gregg

Department of Computer Science
Trinity College, Dublin, Ireland
{emre.ozer,andy.nisbet,david.gregg}@cs.tcd.ie

Abstract. This paper introduces a compiler framework that optimizes embedded applications written in C, and produces high-level hardware descriptions of the applications for customization on Field-Programmable Gate Arrays (FPGAs). Our compiler performs machine-specific and machine-independent optimizations in order to increase the performance of an embedded application and reduce area/power requirements without explicit programmer intervention. Our experimental results show that our compiler framework can increase performance by 38% with loop and expression parallelism for eight embedded benchmarks. Also, area usage and power consumption are reduced by 69% and 55%, respectively through the efficient utilization of on-chip FPGA resources for **Xilinx** Virtex-II FPGA chip.

1 Introduction

The attraction of using FPGAs is the ability to generate application specific logic where the balance and mix of functional units can be altered. This has the potential to generate orders of magnitude speedup for computationally intensive algorithms. However, the time taken to develop and optimize an application for execution on an FPGA using a traditional hardware description language (HDL) design flow may be prohibitive.

Our goal is to develop a compilation framework (see **Fig. 1**) that takes an application written in *ANSI-C*, translates it into an intermediate representation (IR) for performing optimizations, and produces *Handel-C* code. *Handel-C* [12] is a high-level HDL language that can reduce both the time and the expertise required to develop embedded FPGA applications. However, the programmer must explicitly specify code parallelism, the storage type (memory or registers), operand bit-widths, resource sharing or private resource allocation using special language constructs. The detection and exploitation of parallelism in an application and explicit specification of operand bit-widths and decisions about how to allocate resources are tedious, error-prone tasks which can be largely automated.

---

* The research described in this paper is supported by an Enterprise Ireland Research Innovation Fund Grant IF/2002/035.

Our compilation framework is built upon the *Stanford SUIF1* compiler [8]. It performs several architecture-independent optimizations as well as architecture-specific optimizations on the *SUIF IR* such as loop parallelization, expression scheduling, shared hardware structuring, operand bit-width specifications, the efficient utilization of on-chip RAM and ROM, multi-ported RAM and dedicated multipliers in the target FPGA chip. The final step of our compiler is to generate *Handel-C* code from the optimized IR. The *Handel-C* compiler reads the generated *Handel-C* code and produces *EDIF netlists* for **Xilinx Virtex-II XC2V6000** FPGA chip [13].

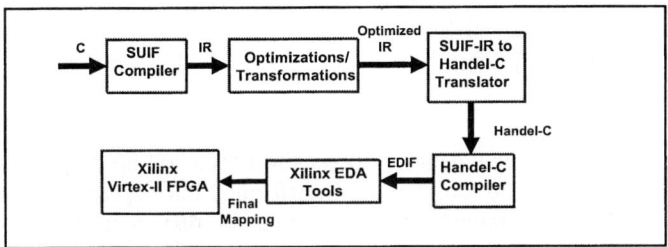

Fig. 1. Our compilation framework

The organization of the paper is as follows: *Section 2* explains compiler optimizations for improving performance and analyzes performance results. *Section 3* presents optimizations and experimental results of area and power consumption. *Section 4* discusses the related work in hardware compilation, and finally, *Section 5* concludes the paper with a discussion of future work.

2 Optimizations for Performance

This section introduces compiler optimizations to increase the performance of embedded applications by taking advantage of the *Handel-C* parallelism constructs. Each assignment statement in *Handel-C* takes exactly one clock-cycle to execute. The operand reads of variables are initiated at the start of a clock cycle and a write to the destination variable is finalized by the end of the clock cycle. This means that it is permissible to have multiple reads and a single write to a variable occurring in the same clock cycle.

2.1 Loop Iteration Parallelization

Handel-C has three different loop constructs: *for*, *par* and *seq*. Each construct has the same syntax:

for|par|seq(*loop_index*=LOWER_BOUND;*loop_index*<|>|=|!=UPPER_BOUND;*step*)
{Statements}

The way that the *for* loop in *Handel-C* works is the same as in C, i.e. the loop is executed by the amount of the loop trip count. In a *par* loop, every iteration and every statement in the iteration is executed in parallel. On the other hand, the *seq* loop fully unrolls or replicates the loop and executes the iterations sequentially. Loop-carried data dependency analysis can determine if loop iterations are independent and suitable for parallel execution. A *for* loop whose iterations are independent can be replaced by a *Handel-C par* construct. Note that the statements in the loop body are by default put into a *seq* construct in our system and will execute sequentially. For example, the C sample code on the left-hand side can be parallelized using *Handel-C* as shown on the right-hand side below.

```
/* C sample for loop */              /* Handel-C par loop */
for (i = 0; i<10; i++) {             par (i = 0; i<10; i++){
                                       seq {
    a[i] = b[i] + c[i];                  a[i] = b[i] + c[i];
    d[i] = e[i] * f[i];                  d[i] = e[i] * f[i];
                                       }
}                                    }
```

The sample loop has independent iterations since there are no data-carried dependences between iterations. The loop can be transformed into a *par* loop, and the loop body is conservatively contained in a *seq* statement that guarantees that the two statements execute sequentially. In this example, the two statements have no data dependency and can run simultaneously. However, our parallel expression scheduling algorithm described in the next section can detect this case and replace the *seq* with a *par* construct.

The loops that cannot be parallelized can be fully unrolled by using a *seq* loop although this may be of little benefit. The *Handel-C* compiler places each iteration sequentially, and separate hardware is synthesized for each iteration in the final FPGA. The advantage of fully unrolling is that there is no need for a loop increment and comparison operations. The disadvantage is the amount of FPGA area consumed by unrolled iterations. If area is of concern, then the *for* statement should not be replaced by a *seq* statement. The logic created by a *for* statement consists of the loop body and the loop test. It is important to note that the loop bounds must be known in order to translate a *for* into a *par* or a *seq*. Other loops such as *do..while* and *while* are translated into *par* and *seq* loops if they can be transformed into a canonical loop form.

2.2 Parallel Expression Scheduling

Expressions having no true data dependencies amongst them can be executed in the same cycle by enclosing them in a *par* block. A variable can be read multiple times and assigned to at most once in a *par* block. For instance, the three statements on the left-hand side of the following code are data-independent and can execute in parallel using a *Handel-C par* statement as shown on the right-hand side. All three statements in this *par* block will execute in a single cycle, whereas it takes three cycles to execute the code if no *par* is used since a statement in *Handel-C* takes a cycle to execute.

```
                                     par {
    A = B * C;                         A = B * C;
    D = E + F * G;                     D = E + F * G;
    H = M;                             H = M;
                                     }
```

```
foreach Procedure begin
   build_DAGs( );
   foreach DAG in Procedure begin
      topological_sort(DAG);
      foreach Node in DAG begin
         compute_depth(Node);
         foreach Predecessor of Node begin
            if (Is Edge(Node, Predecessor) anti-dependency?) then begin
               if (depth(Node) < depth(Predecessor)) then depth(Node) = depth(Predecessor);
            end
         end
      end
      sort_by_depth(DAG);
      schedule(DAG);
   end
end
```

Fig. 2. Pseudo-code for expression scheduling

Data dependency analysis is performed to find data-independent expressions and put them in *par* blocks. A directed acyclic graph (DAG) scheduling technique is used to schedule independent expressions into several parallel expression groups as shown in **Fig. 2**. Each node in the DAG represents an expression and an edge represents *true, output* or *anti-dependency* between two nodes. Anti and output dependencies are false data dependencies that can be eliminated by variable renaming. However, renaming uses temporary variables that demand more FPGA resources.

True and output dependencies restrict expressions to be grouped in the same *par* block. On the other hand, anti-dependencies may or may not restrict the scheduling of expressions. The expression that writes the anti-dependent operand (i.e. *the sink expression*) may be scheduled in an earlier cycle than the expression that reads the anti-dependent operand (i.e. *the source expression*). The depth of the sink expression must be made equal to the depth of the source one to guarantee that they will be at least in the same *par* block.

After the computation of all depths in the DAG, the DAG is converted into a flat list of nodes sorted in ascending order of their depths. The expressions with the same depth are put into the same *par* block. If there is only one expression associated with a depth, then a *par* block is not necessary.

Table 1. Benchmarks

Benchmark	Description
adpcm encoder	16-bit PCM to ADPCM speech encoder
matrix multiplication	8-by-8 matrix multiplication
shellsort	Shell Sort algorithm of 32 integer numbers
2D convolution	8-by-8 2D convolution algorithm with 3-by-3 kernel matrix
FIR	32-tap Finite Impulse Response filtering
IDFT	32-point Integer Inverse Discrete Fourier Transform
huffman encoder	32-character Huffman Encoding algorithm
g721decoder	CCITT G.721 ADPCM decoding

2.3 Experimental Results

The benchmark programs used in this study are shown in **Table 1**. They represent various embedded computing fields such as DSP, image processing, telecommunications and scientific computing. All benchmarks perform integer arithmetic because *Handel-C* supports only integer arithmetic. For each benchmark, our compiler carries out loop parallelism and expression scheduling and finally generates the *Handel-C* code. The *Handel-C* code is then passed to the *Handel-C* compiler, which in turn generates an EDIF netlist for the **Xilinx Virtex-II XC2V6000** FPGA device. **Xilinx Virtex-II XC2V6000** is a 6M-gate chip with 8448 Configurable Logic Blocks (CLBs). Each CLB is made of four slices, each of which has two 4-input look-up tables (LUTs), two 16-bit D-type registers, dedicated multiplexers and fast carry look-ahead chain. In addition to an array of CLBs, the **Xilinx Virtex-II XC2V6000** FPGA chip has also 324KByte block RAM/ROM and 144 18x18-bit multipliers. We used **Xilinx ISE 6.2** tool set with standard placement/route effort levels.

In order to form a baseline model for performance comparison, each benchmark is also transformed into *Handel-C* without performing any of the aforementioned optimizations. Our metric for performance comparison is *the total execution time* for each benchmark. The total execution time is determined by multiplying *the total cycles* required to execute the benchmark by *the minimum clock period*. The total cycles is computed as the sum of every individual or parallel statement. The minimum clock period is the clock period required to operate the FPGA after each benchmark is mapped and implemented onto the FPGA.

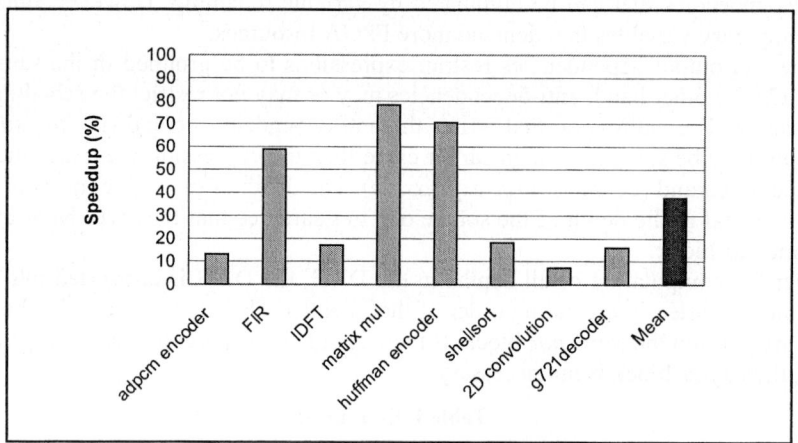

Fig. 3. Speedup results in the total execution time

Fig. 3 presents the speedups in terms of percentage reduction in the total execution time for each benchmark using two parallelization techniques. The highest speedups come from *matrix multiplication, huffman encoder* and *FIR* by *78%, 70%* and *59%*, respectively. *matrix multiplication* has three nested loops from which the outermost two loops can be translated into *par* loops because there are no loop-carried data dependencies across their iterations. For *huffman encoder* and *FIR*, the loops that are converted into *par* loops have a great amount of expression parallelism. The speedups of the other benchmarks vary between *7%* and *18%*. Smaller performance improvements in these benchmarks are caused by the fact that none of the loops can be parallelized. Only parallel expressions contribute to the speedups. In summary, an average of *38%* improvement in performance can be obtained over eight benchmarks.

3 Optimizations for Area and Power

This section discusses machine-specific optimizations to reduce area and power consumption in the **Xilinx Virtex-II XC2V6000** FPGA chip. Such optimizations typically attempt to utilize on-chip resources of block RAM/ROM and multipliers present on the FPGA chip so that precious FPGA area can be allocated to more compute-intensive operations or tasks.

3.1 On-chip RAM and ROM Utilization

Arrays are implemented using look-up tables and flip-flops in the FPGA slices. For FPGA devices that support on-chip block RAM and ROM such as in the **Xilinx Virtex-II** architecture, these arrays can be declared as *ram, rom* and *mpram* (multi-ported ram) in *Handel-C* to use on-chip RAM/ROM hardware resources so that the FPGA slices can be allocated for other uses. The difference between an array and a *ram* variable declaration is that any number of array elements can be read or written in a single clock cycle in an array, whereas only a single location in a single ported, single bank RAM array can be read or written in a single clock cycle. Similarly, ROM can be read only once in a clock cycle. If an array is to be used more than once in the same cycle, it can be declared as *mpram* to allow simultaneous read/writes through multiple ports in the same cycle. *mpram* supports only dual-port memory accesses.

A compiler algorithm is written to analyze the global and local array declarations and to find out whether they can be declared as *ram, rom* or *mpram*. An array variable can be declared as *rom* if it is not accessed more than **once** in a statement or a *par* block and all accesses are **reads**. The array variable can be declared as *ram* if it is not accessed more than **once** in a statement or a *par* block and **not** all accesses are **reads**. If it is accessed at most twice in a statement or a *par* block, then it can be declared as dual-port *mpram*. Array variable names can also be passed as arguments in procedure calls, and alias names can be used at the call sites. Thus, argument-parameter alias analysis is also performed for the whole program to determine the set of alias names for each array name passed as an argument to procedures. The array variable has to be kept as an array declaration if it is accessed more than **twice** in a statement or a *par* block.

3.2 On-chip Dedicated Multiplier Utilization

Customized multiplication units can take up a vast amount of slices in the FPGA. If the widths of multiplicands are wider, the multiplication operation takes longer and this can cause a drop in the overall clock frequency of the FPGA since every instruction in *Handel-C* takes exactly one cycle. Hence, it is requisite that the compiler must assign some of multiplication operations to the dedicated fast on-chip 18x18-bit multipliers in the FPGA chip. If the widths of the multiplicands are larger than 18 bits, several multipliers can be tied together to form wider multipliers. A multiplier can be assigned to more than one operation but only one of them can access it at any cycle. This will create a multiplexer logic in front of the multiplier to route only one set of multiplicands to the multiplier. As more sets of multiplicands share a multiplier, the multiplier becomes slower due to its wide multiplexer. The compiler must intelligently distribute multiplication operations among the dedicated multipliers so that none of the multipliers has prohibitively wide multiplexers. **Fig. 4** shows the on-chip 18 by 18 bit multiplier assignment algorithm.

```
foreach Procedure
   begin
      MULOP_LIST = find_multiplication_operations();
      sort_by_bitwidth(MULOP_LIST);
      initialize_multiplier_weights(MULTIPLIER_LIST);
      while (not end of MULOP_LIST)
         begin
            OP = pick_operation(MULOP_LIST);
            n  = bitwidth(OP_operand1);
            m  = bitwidth(OP_operand2);
            if (max(n,m) > 18) then number_of_required_multipliers = ⌈max(n,m)/18⌉;
            else number_of_required_multipliers = 1;
            MUL_SET = find_smallest_weighted_multipliers(MULTIPLIER_LIST,number_of_required_multipliers);
            assign(OP, MUL_SET);
            increment_multiplier_weights(MUL_SET,(n+m));
         end
   end
```

Fig. 4. The pseudo-code for the dedicated multiplier assignment algorithm

The algorithm puts each multiplication operation into a list by traversing each procedure. Then, the list is sorted by the sum of bit-widths of multiplicands of each operation in descending order. Each multiplier is assigned a weight that shows the estimated length of its multiplexer if more than one set of multiplicands share the multiplier. After all the weights are initialized to zero, an operation is popped off the list. If the bit-width of any of its multiplicands is less than or equal to 18, then only one multiplier is needed. Otherwise, more than one multiplier is needed to implement the current multiplication. The number of the required multipliers is computed by dividing the maximum bit-width by 18. After determining the exact number of multipliers, the multiplier list is searched to select the multipliers with the lowest weights. Then, the multiplication operation is assigned to these selected multiplier(s) and their weight(s) are incremented by the sum of the bit-widths of the two multiplicands. The same steps are applied to all operations in the multiplication operation list until all of them are assigned to the dedicated multipliers. The objective of the algorithm is to ensure that the multipliers to which long multiplications (i.e. the ones with wider operand bit-widths) are assigned are not multiplexed with many different sets of multiplicands of other multiplications.

3.3 Experimental Results

Area usage and power consumption are measured for each benchmark. Similar to *Section 2.3*, our base model for comparison are the same benchmarks with no on-chip RAM/ROM and multiplier utilization optimizations. For the area usage, we measure *the percentage reduction* in the number of FPGA slices on the chip. The *dynamic logic power consumption* consumed by the logic implemented on the FPGA chip is measured after a complete placement and routing of the design on the FPGA using **Xilinx's** *XPower* estimation tool. We set the FPGA's clock frequency and source voltage to 100MHz and 1.5V, respectively. Also, circuit switching activity rate is set to 100% or ½ the clock frequency, which is the rate at which a logic gate switches.

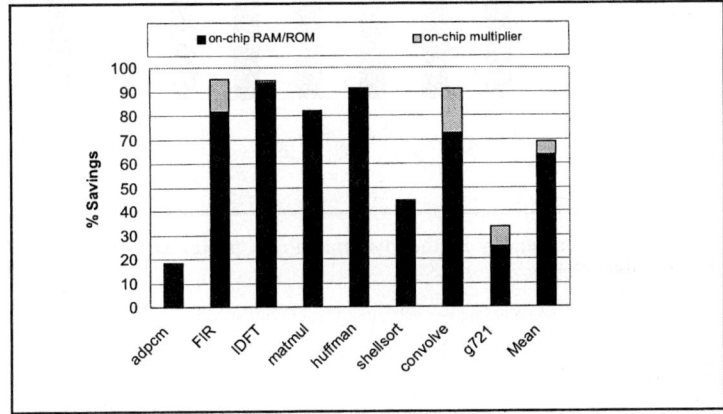

Fig. 5. The percentage reduction in the number of FPGA slices using on-chip RAM/ROMs and multipliers

Fig. 5 shows the percentage reduction in the number of FPGA slices on the chip. The black bar denotes the reduction contributed by only on-chip RAM/ROM and the grey bar represents the reduction contributed by only on-chip multiplier utilizations. It is possible to save an average of *69%* of FPGA slices for all benchmarks as shown in the last column. Of this average, *64%* comes from the on-chip RAM/ROM utilization and only *5%* savings from the slices are due to the on-chip multiplier utilization. The arrays in all benchmarks can fit into the on-chip RAM/ROM structures. Although the majority of FPGA slice savings is caused by efficient utilization of on-chip RAM/ROMs, a reasonable amount of reduction has been made using on-chip multiplier optimizations such as in *FIR, 2D convolution* and *g721decoder*. These three benchmarks have several multiplication operations that are allocated to on-chip multipliers. On the other hand, *adpcm, shellsort. huffman, matrix multiplication* and *IDFT* have either no multiplication or only one multiplication operation. Thus, the contribution of on-chip multiplier utilization to the area reduction for these benchmarks remains less than 1%.

Fig. 6 presents the percentage reduction in dynamic logic power consumption of on-chip RAM/ROM and multiplier utilizations for eight benchmarks. The large number of FPGA slice savings allows energy-efficient designs by reducing the logic power consumption proportional to the number of slices. Over eight benchmarks, an average of *55%* (i.e. 51% due to on-chip RAM/ROMs and 4% due to on-chip multipliers) reduction in dynamic logic power consumption is possible by utilizing on-chip FPGA resources.

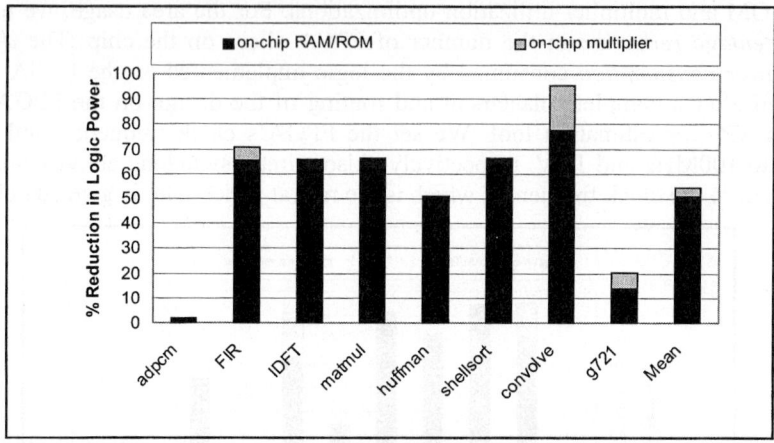

Fig. 6. The percentage reduction in the dynamic logic power consumption using on-chip RAM/ROMs and multipliers

4 Related Work

There have been several projects [1] [2] [3] [4] [5] [6] [7] [9] [10] [11] that attempt to transform modified C programs to custom hardware by way of producing low-level HDL languages such as Verilog or VHDL. In spite of their portable nature, these languages are not suitable to explore instruction-level and loop-level parallelism. Hence, we use *Handel-C* as our backend language whose high-level constructs allow us to exploit both instruction and loop-level parallelism with explicit control.

5 Conclusion

We have introduced a compilation framework that optimizes embedded applications written in C to improve performance, and reduce area/power requirements. Loop parallelism and expression scheduling are applied to decrease the total execution time of the programs. We have shown that an average of **38%** improvement in performance is possible with these parallelizations for all benchmarks. We have also presented the experimental results of machine-specific optimizations for utilizing on-chip block ROM/RAMs, and dedicated multipliers to reduce area and power consumption of the **Xilinx Virtex-II XC2V6000** FPGA chip. Our results have showed that an average of **69%** reduction in area and an average of **55%** reduction in logic power consumption over eight benchmarks can be attained through the efficient utilization of on-chip resources.

We are exploring various compiler optimizations such as shared expressions, storage reuse, expression splitting and combining, and software pipelining. Our ultimate aim is to develop an iterative hardware compilation environment that can apply compiler transformations using feedback-directed data from timing, resource mapping and power estimation tools to optimize for combinations of performance, power consumption and area.

References

1. M. Wazlowski, L. Agarwal, T. Lee, A. Smith, E. Lam, P. Athanas, H. Silverman, and S. Ghosh. "PRISM-II Compiler and Architecture", *Proceedings of IEEE Workshop on FPGAs for Custom Computing Machines*, Napa, California, April 1993.
2. B. A. Draper, A. P. W. Böhm, J. Hammes, W. Najjar, J. R. Beveridge, C. Ross, M. Chawathe, M. Desai, J. Bins, "Compiling SA-C Programs to FPGAs: Performance Results", *International Conference on Vision Systems*, Vancouver, July, 2001.
3. M. Hall, P. Diniz, K. Bondalapati, H. Ziegler, P. Duncan, R. Jain, and J. Granacki, "DEFACTO: A Design Environment for Adaptive Computing Technology", *Proceedings of the 6th Reconfigurable Architectures Workshop* (RAW'99), 1999.
4. J. Frigo, M. Gokhale, and D. Lavenier "Evaluation of the Streams-C C-to-FPGA Compiler: An Application Perspective", *9th ACM International Symposium on Field-Programmable Gate Arrays*, Monterey, CA, February, 2001.
5. T. J. Callahan, J. R. Hauser, and J. Wawrzynek, "The Garp Architecture and C Compiler", *IEEE Computer*, April 2000.
6. M. Budiu, and S. C. Goldstein, "Fast Compilation for Pipelined Reconfigurable Fabrics", *7th ACM International Symposium on Field-Programmable gate Arrays*, 1999.
7. D. C. Cronquist, P. Franklin, S. G. Berg, and C. Ebeling, "Specifying and Compiling Applications for RaPiD", *Field-Programmable Custom Computing Machines*, 1998.
8. R. P. Wilson, R. S. French, C. S. Wilson, S. Amarasinghe, J. M. Anderson, S. W. K. Tjiang, S. W. Liao, C. W. Tseng, M. W. Hall, M. S. Lam, and J. L. Hennessy, "SUIF: An Infrastructure for Research on Parallelizing and Optimizing Compilers", *Tech. Report*, Computer Systems Laboratory, Stanford University, CA, USA, 1994.
9. S. Gupta, N. Dutt, R. Gupta, and A. Nicolau, "SPARK: A High-Level Synthesis Framework For Applying Parallelizing Compiler Transformations", *the 16th International Conference on VLSI Design*, New Delhi, India, Jan. 2003.
10. Jonathan Babb, Martin Rinard, Andras Moritz, Walter Lee, Matthew Frank, Rajeev Barua, and Saman Amarasinghe, "Parallelizing Applications Into Silicon", *Proceedings of the IEEE Workshop on FPGAs for Custom Computing Machines '99* (FCCM '99), Napa Valley, CA, April 1999.
11. V. Kathail, S. Aditya, R. Schreiber, B. R. Rau, D. C. Cronquist and M. Sivaraman, "PICO: Automatically Designing Custom Computers", *IEEE Computer*, vol. 35, no. 9, pp. 39-47, September 2002.
12. Celoxica, *Handel-C Language Reference Manual*, Version 3.1, 2002.
13. Xilinx, *Xilinx Virtex-II Architecture Manual*, Sep. 2002.

Architecture-Independent Meta-optimization by Aggressive Tail Splitting

Michael Rock and Andreas Koch

Tech. Univ. Braunschweig (E.I.S.), Mühlenpfordtstr. 23, D-38106 Braunschweig, Germany
{rock,koch}@eis.cs.tu-bs.de

Abstract. Several optimization techniques are hindered by uncertainties about the control flow in a program, which can generally not be determined by static methods at compile time. We present a novel approach that aims to alleviate this limitation by explicitly enumerating all control paths through a loop nest, thus allowing wider and more efficient use of standard optimization passes later. Our analysis will address the possible explosion in code size, which despite high theoretical upper bounds appears to be acceptable in practice, as well as performance gains both for RISC and CISC target processors.

1 Introduction

The quality of many optimization algorithms is directly dependent on the amount and accuracy of the information retrieved by prior analysis steps. Despite continuous progress in the development of such tools, certain program structures invalidate the analysis efforts: One of these cases is the joining of two or more different flows of execution after a conditional at a so-called merge point in the control flow graph. At such a point, the analysis is uncertain which of the multiple paths was actually taken to arrive at the merge point. Thus, it cannot propagate beyond the merge point any assertions (e.g., on variable values and memory contents) that are not common to all of the execution paths arriving at the merge point.

2 Related Work

The criticality of analysis across merge points has been recognized previously. One technique that is already practically used in many compilers (e.g., using -ftrace for GCC [1]) is tail duplication [2]. This approach is used for the joining of blocks into single-entry point hyperblocks by copying a dedicated instance of all shared tail blocks into each individual hyperblock, thus reducing the number of merge points. The technique is also commonly used for improving the scheduling of the generated code [3] [4]. Some research has been done in combining loop unrolling and tail duplication [5], but the algorithm presented there does not re-splice the program as done in our approach.

3 Novel Approach

However, these existing approaches (both for hyperblock construction and scheduling) limit the scope of their operation to limited areas such as only within loop bodies. The algorithm presented here has all the advantages of conventional tail splitting and loop unrolling, but now the back edges are re-spliced between specialized versions of the loop body and do not lead back to the initial loop header. This unrolling, splitting and re-splicing operation was termed *Aggressive Tail Splitting* (ATS). Its effect is to keep disjunct paths of execution isolated for as long as feasible, thus avoiding the creation of analysis-impeding merge points. The approach was inspired by the basic use of the tail splitting transformation in an interpreter [6].

4 ATS Algorithm

We will introduce the algorithm using the trivial example of the C source code fragment shown in Figure 1.

```
TYPE b0(TYPE *a) {
    TYPE s=0;
    int i;
    for (i=1; i < SIZE; i++) {
        if (a[i] < 0) {
            a[i] = 0 ;
        }
        s += a[i] * a[i-1] ;
    }
    return s;
}
```

Fig. 1. Initial input source code

Since some of our later analyses of ATS effects will be performed with disabled compiler optimizations (to focus on ATS itself), we will perform some simple optimizations such as Common Subexpression Elimination (CSE) manually on b0(), leading to the version b1() shown in Figure 2.

```
TYPE b1(TYPE *a) {                              TYPE b2(TYPE *a) {
    TYPE s=0;                                       TYPE s=0;
    int i;                                          int i;
    for (i=1; i < SIZE; i++) {                      for (i=1; i < SIZE; i++) {
        // Block 1                                      TYPE t = a[i] ;
        TYPE t=a[i] ;   // perform CSE manually ...     if (t < 0) {
        if (t < 0) {    // ... here ...                     a[i] = 0 ;      // tail split here ...
            // Block 2                                  } else {
            a[i] = 0 ;                                      s += t*a[i-1] ; // ... and here
            t = 0 ;     // ... here ..                  }
        }                                           }
        // Block 3                                  return s;
        s += t * a[i-1] ;// ... and here        }
    }
    return s;
}
```

Fig. 2. Result of single intra-loop iteration tail splitting

4.1 Tail Splitting the Back Edge

We now begin by splitting all edges arriving at blocks with multiple predecessors, copying the original block backward along each of the split edges (see Figure 3). Initially, this occurs only within the loop body (as suggested in [6]), leading to the source code b2() shown in Figure 2.

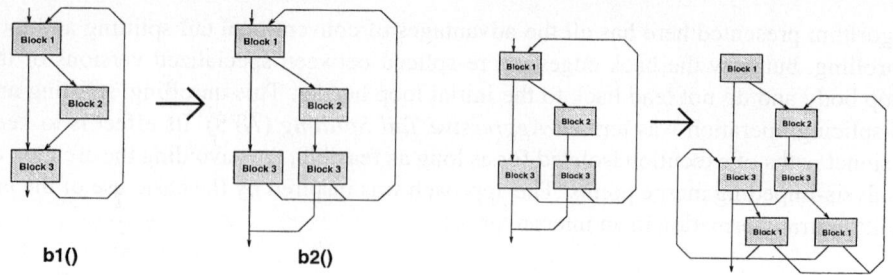

Fig. 3. First inter-iteration tail-splitting and beginning of head duplication

At this stage, we have not yet achieved our aim of reducing the number of merge points. On the contrary, in b1(), Block 1 only had two arriving execution paths, while in b2(), it now has three arriving paths. To avoid cluttering this figure, an *arriving* edge can carry more than one path of execution, the actual number is determined by the number of *originating* nodes.

Beyond conventional tail-splitting, ATS now *continues* this process along the back-edges, thus crossing loop iteration boundaries. This step is shown in Figure 3: The two back edges have been tail split and replaced by copies of the loop header Block 1. As before, however, the number of execution paths arriving at merge points has increased. Blocks 2 and 3 now both have three arriving paths. Due to the increasing convolution of the structure at this point, we will no longer show the source code of these intermediate steps.

Undeterred by the temporary decrease in solution quality, ATS continues in the same fashion. Later on, it generates the structure shown in Figure 4.a: Now, three distinct copies exist of the loop body. The entire first iteration has been peeled off and serves as a dispatcher to the specialized loops lower in the hierarchy. At this stage of the process, a specialized loop exists for each execution path arriving there. In the example, the loop B was created for the case that the condition in b2() was true during the previous (peeled-off) iteration, and the loop A for the inverse case.

4.2 Termination Conditions for the Tail Splitting Process

At this stage in the discussion, we have to address the question of when to stop splitting edges. Even with the structure shown in Figure 4.a, we could continue the process. It is thus useful to step back from the algorithm itself and re-examine the aim of our methods.

The main goal of ATS is to enable more aggressive optimizations inside of loop bodies. The manner in which this should be achieved is by providing other optimization passes with accurate information (assertions) about the execution path that lead to a given node (history). Based on these assertions, these optimizations can then perform their own transformations or scheduling decisions.

Thus, to arrive at the termination condition of the tail-splitting phase, we need to determine the minimal required length of the history. It turns out that for loops, this length depends on the distance vectors for loop-carried dependencies. In our example,

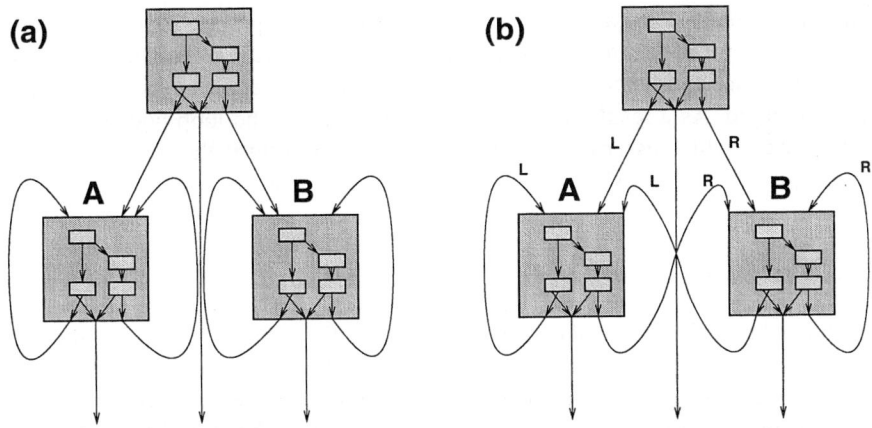

Fig. 4. (a) Loop unrolling and tail spliting, (b) ATS

b0() has a distance vector of $d : \{0, 1\}$, indicating that a calculation at a[i] depends on a[i-0] and a[i-1]. This implies that in this case, there is no need to keep track of data or control flow beyond the previous loop (the one that computed a[i-1]). Thus, the length of the history required is the maximum range of the distance vectors for loop carried dependencies.

By closely examining the digraph we use to represent the program structure (Figure 3 and 4.a), it can be determined that during the tail-splitting phase, this digraph will always have the same tree-like nature: The root and inner nodes serve as dispatcher to specialized loops represented by the leaf nodes (here A and B), with the only back-edges occurring at the leaf-level.

It is thus obvious that the depth of the digraph (the longest path from the root to a node) is limited by the length of the history: Each level of the graph represents a loop-carried dependency on a previous iteration.

4.3 Inter-leaf Node Re-splicing

Thus far, our approach relied on repeatedly applying a well-known technique to transform the program. But now that we have determined when to terminate this process, we can transform the leaf-level edges to accurately encode the histories without further expansion of the digraph. This is achieved by connecting each of the back-edges leaving a leaf-level node to the leaf-level node (same or different) that was specialized for that specific intra-loop execution path.

The method will become clearer when illustrated using the example in Figure 4.a. As stated previously, there are only two history-specific back-edges out of the loop bodies[1], depending on the condition $P \equiv $ if (t<0) ....

For the example, we will thus ensure that all paths R which had $P =$ true during the previous iteration will lead to the loop specialized for this case, namely B. Analogously,

---
[1] Since the center edge is not a back-edge (loop exit), it is disregarded here.

all paths L that had $P=$ false in the previous iteration will be processed in the specialized loop A. The result of this transformation is shown in Figure 4.b. The histories leading to a loop body are thus encoded in the selection of which specific specialized leaf-level loop is executed. As a result, a later optimizer processing the loop bodies A or B can now be certain which assertions can be exploited. For brevity, the actual source code for this structure, termed b3() in the following text, is omitted here.

5 Performance

To evaluate the performance of code subjected to ATS, we compiled the the processing stages b1() to b3() of our example onto the UltraSPARC III+ and Athlon XP targets. For the SPARC, we used the Sun Workshop C Compiler 5.3 with maximum optimization effort (option -fast). On the Athlon, we employed GNU GCC 3.2 with the -O3 option. The results are summarized in Table 1.

Table 1. Performance for the example code

Version		SPARC		Athlon	
		Time [s]	Speedup	Time [s]	Speedup
plain	b1	6.71	1.00	3.07	1.00
tail-split	b2	6.55	1.02	2.53	1.21
ATS	b3	5.90	1.13	1.95	1.57

So even for this trivial example, some modest performance gain (up to 57% on the Athlon) was realized. Before analyzing a larger, more meaningful application in Section 5.2, we will consider the effects of ATS on later optimization passes.

5.1 Increased Optimization Opportunities

As seen in Figure 4.b, even the post-ATS version b3() contains hyperblocks with multiple predecessors. However, now all paths even through different predecessors keep track of the same history (and thus assertion sets), increasing the opportunities for standard optimization techniques. Thus, constant propagation or CSE can now be applied across iteration boundaries. This is not possible when the loop is not tail split because of the uncertainty of the execution path leading to a block. Another technique that allows such inter-iteration optimization is loop unrolling. However, even here the path uncertainty problem remains and limits the applicability of the standard optimizers.

5.2 Effect of ATS on Control Flow

To evaluate the effect of ATS on a more meaningful program, we now consider a decoder for fixed-table Huffman codes. Since this algorithm actually contains a substantial control-content in its loops (in contrast to many DSP codes), it is a good candidate for

```
void huffDec(char* src, char* dst, int limit) {
    int i;
    int pos = 1<<7 ;
    for (i=0; i < limit; i++) {
        int index =0;
        int max ;
        /* Maximum bit length*/
        for (max=0; max < 6; max++) {
            if (*src & pos) {
                index++;
            }
            pos >>= 1 ;
            if (!pos) {
                pos = 1<<7 ;
                src++ ;
            }
            index = table[index] ;
            if (index & 1) {
                *dst++ = index>>1 ;
                break ; /* ready, early exit*/
            }
        }
    }
}
```

Fig. 5. Huffman decoder source

ATS. Due to the size of the intermediate and final stages, we will only list the original source program in Figure 5.

As before, the code is evaluated on the SPARC and Athlon targets. For each target, Table 2 shows three versions: The original, a fully-expanded ATS variant, and finally an optimized ATS version. The latter was subjected to agressive propagation, replacing constant accesses to constant memory locations with their contents. In both ATS cases, the inner loop has been completely unrolled to remove all internal merge points. The size given is the size of the compiler generated object code (.text segment).

Table 2. Results of the HuffDecode benchmark on Sparc and Athlon, time and growth

Compile	Stage	Size [B]	Time [s]	Growth	Speedup
SunCC	base code	164	6.08	1.00	1.00
-O1	complete ATS	480	9.65	2.93	0.63
SPARC	optimized ATS	1300	3.99	7.93	1.52
SunCC	base code	168	2.79	1.00	1.00
-fast	complete ATS	1600	6.1	9.52	0.45
SPARC	optimized ATS	1560	2.47	3.29	1.13
gcc	base code	103	1.78	1.00	1.00
-O1	complete ATS	193867	1.46	1882.20	1.23
Athlon	optimized ATS	5795	0.93	56.26	1.91
gcc	base code	109	1.59	1.00	1.00
-O3	complete ATS	30052	0.69	275.70	2.30
Athlon	optimized ATS	1745	0.45	16.01	3.53

The Huffman decoder has a theoretical upper bound of growth[2] in the number of instructions by a factor of 299593. However, the maximum observed growth factor of the optimized ATS digraph is only 56. More aggressive compiler optimization options

---

[2] Due to space limitations, the relevant analyses cannot be shown here.

(-fast and -O3) exploit the ATS'ed structure even further, leading to a maximum growth of only 16.01. This is well within the practical limits established by other techniques such as aggressive loop unrolling (e.g., by a factor of 96 in [7]).

In terms of performance gain, on the SPARC target, at best 52% were achieved (interestingly not at the highest compiler optimization level). On the Athlon however, the highest speedup was over 350%.

When examining the final assembly code, it becomes apparent that its post-ATS structure is completely different from the original source code: Now, the entire fixed Huffman table has been inlined into the program, resulting in an actual FSM for decoding the bit stream. This was possible due to the fact that the *index* variable now was subject to constant propagation, making the value constant for all reads. This reduces the number of memory accesses considerably, since the table walk is now encoded into the program's control flow. By lowering the pressure on the memory, more ILP can be exploited.

6 Interpreting the Benchmark Results

Since ATS itself is no optimization, the origin of the performance gains should be examined. In the initial example (b1()...b3()), the speed-up was realized by removing multiplications by zero after constant propagation.

The Huffman decoder of Figure 5 is more complex. A GCC-compiled version was analyzed in greater detail using processor-internal performance counters on both platforms. These measurements are shown in Table 3. Of interest are the actual number of instructions executed and the number of cycles stalled due to branch mispredictions.

Table 3. Results of the HuffDecode benchmark on Sparc and Athlon-XP, instructions and stalls

Stage	Instructions [10^9]	Mispredicts [10^6 Cycles]	Rel. Insts	Rel. Stalls
base code SPARC	3.85	12.11	1.00	1.00
complete ATS SPARC	3.58	72.25	0.93	5.96
optimized ATS SPARC	1.50	50.29	0.39	4.15
base code Athlon	5.04	73.04	1.00	1.00
complete ATS Athlon	2.30	63.95	0.46	1.03
optimized ATS Athlon	2.00	58.17	0.37	0.93

Since the ATS operation imposes a more complex control structure on the program (multiple value-specialized versions for each execution path), the quality of the processor's branch prediction logic becomes crucial. In the numbers shown above, it is obvious that the unit in the SPARC is not able to cope with the increase in complexity (more mispredict cycles) of the complete ATS form. The Athlon's predictor, however, can easily handle that version (only minimal increase in mispredict cycles).

The performance gains of ATS are mainly due to the reduced number of instructions actually executed (see Table 3). E.g., for the Athlon, only 30% of the original number of instructions is executed after ATS. This reduction is achieved by giving the later optimization stages (in this example: especially the constant propagation step) a larger set of assertions to work with. Since the constant propagation in the 'optimized' versions extends to removing references to the constant Huffman table, 71.9% of all memory accesses are eliminated in the post-ATS code (on both platforms).

7 Future Work

We are currently integrating an automatic ATS pass into a complete compile flow, thus allowing a more thorough experimental evaluation in the future. Furthermore, the ATS digraph itself is amenable to high-level minimization techniques which we expect to limit the practical code growth even further.

8 Conclusion

The ATS transformation can considerably increase the opportunities for applying standard optimization techniques to inner loops. In practice, the excessive theoretical growth in code size can often be avoided. By selectively applying the transformation, the growth penalty need only be accepted when economical (e.g., in the range of 100-200 and only for time-critical inner loops).

References

1. Cohn, R., Lowney P.G., "Design and Analysis of Profile-Based Optimization in Compaq's Compilation Tools for Alpha", *Journal of Instruction-Level Parallelism*, May 2000.
2. Gregg, D., "Comparing Tail Duplication with Compensation Code in Single Path Global Instruction Scheduling", in *Proceedings of the 9th International Conference on Compiler Construction (CC 2001)*, pp. 200-212, LNCS 2027, Genoa, April 2001.
3. Gao, G.R., Amaral, J.N., Dehnert, J. , Towle, R., "The SGI Pro64 Compiler Infrastructure", *Intl Conference on Parallel Architectures and Compilation Techniques (PACT)*, tutorial, 2000.
4. Huiyang, Z., Jennings, M.D., Conte, T.M. "Tree Traversal Scheduling: A Global Technique for VLIW/EPIC Processors", *Proc. 14th Workshop on Languages and Compilers for Parallel Computing (LCPC)*, LNCS, Springer Verlag, August 2001.
5. Hwu, W.W., Mahlke, S.A., Chen, W.Y., Chang, P.P., Warter, N.J., Bringmann, R.A., Ouellette, R.G, Hank, R.E., Kiyohara, T., Haab, G.E., Holm, J.G., and Lavery, D.M. "The superblock: an effective technique for VLIW and superscalar compilation." *The Journal of Supercomputing* 7, 1/2 (May 1993), 229-248.
6. Bala, V., Duesterwald, E., Banerjia, S., "Dynamo: A Transparent Dynamic Optimization System", *ACM SIG-PLAN Notices*, vol. 35, pp. 1-12, 2000.
7. Lowney, P.G., Freudenberger, S.M., Karzes, T.J., "The Multiflow Trace Scheduling Compiler", *The Journal of Supercomputing*, vol. 7, number 1-2, pp. 51-142, 1993.

If-Conversion in SSA Form

Arthur Stoutchinin[1] and Guang Gao[2]

[1] DMD Tools, STMicroelectronics
12, rue Jules Horowitz, Grenoble, France
Arthur.Stoutchinin@st.com
[2] ECE Department, University of Delaware
Newark, DE USA
ggao@capsl.udel.edu

Abstract. Predicated execution has become a standard feature in modern high-performance microprocessors, and is supported by compilers through if-conversion. The static single assignment (SSA) form has been demonstrated to be a very successful intermediate representation in modern compilers. Traditionally, SSA optimizations preceed the if-conversion because the classical SSA form is not adapted for if-converted programs. Recently, the Ψ-SSA extension of the classical SSA form has been proposed for optimizing predicated code. However, up until now, the if-conversion itself has not been applicable to programs in SSA form. As a result, optimizers that want to use SSA before and Ψ-SSA after the if-conversion would have to translate the program out of the SSA form before performing the if-conversion, and to rebuild the Ψ-SSA from scratch after the if-conversion. In this paper, we propose a simple model of if-conversion in SSA form such that if-conversion results in the Ψ-SSA. We implemented our if-conversion method in the Open64/ORC compiler code generator, and evaluated the size of the Ψ-SSA form resulting from if-conversion on a set of MediaBench programs.

1 Introduction

Predication has become a standard feature in modern high-performance microprocessor architecture. Compilers support predication through if-conversion [1], transforming programs into predicated code in such a way that execution of the instructions previously guarded by conditional branches is guarded by *predicates*. A predicate enables the instruction to execute if it is TRUE, or nullifies the instruction if it is FALSE.

The SSA form has been demonstrated to be a very successful intermediate representation for various compiler optimizations [2–9]. Traditionally, the SSA based optimizations are applied before the if-conversion since classical SSA form is not readily applicable to predicated code. Recently, it has been shown that SSA representation can benefit optimizations of predicated code as well [10]. Thus, the SSA representation may be desirable in the compiler back-end both, before and after the if-conversion. However, up until now, if-conversion itself could not be directly applied to programs in SSA form. Lacking such if-conversion algorithm, optimizers that want to use SSA before and after the if-conversion would have to translate the program out of the SSA form before performing the if-conversion, and rebuild the SSA representation from scratch in order to apply subsequent SSA based optimizations.

We have developed an approach that allows application of the if-conversion directly to programs in SSA form. Our approach builds upon the work on Ψ-SSA, an extension to the classical SSA representation suitable for predication [10]. In Ψ-SSA, Ψ-nodes are generated at program points where different predicated values merge. Similar to Φ-nodes, which indicate definitions reaching a given program point through different control flow path, Ψ-nodes indicate which predicated assignments to a variable may be active at a given program point. Our if-conversion method exploits the *duality* between the Φ- and the Ψ-nodes. In particular, the Φ-nodes in the original program are replaced by the Ψ-nodes in if-converted program. In contrast to [10], where the Ψ-SSA form is constructed taking predicated code as starting point, in our approach, the Ψ-SSA results naturally from if-converting a program in SSA form. As an important consequence, the number of Ψ-nodes in resulting Ψ-SSA is bound by the number of Φ-nodes existing before the if-conversion.

We have implemented our if-conversion method in the Open64/ORC code generator. We measured the size of the Ψ-SSA representation on a set of programs from the MediaBench testsuite [11]. Previous research focused on measuring the SSA size in the context of target-independent optimizations [2, 9]. We're not aware of any previous work that provides similar investigation into the size of the SSA representation at the assembly level. Our experience confirms that most functions display only a moderate increase in the intermediate representation size due to their translation to the Ψ-SSA form.

The rest of the paper is organized as follows. In the next section, we develop a code example that illustrates our if-conversion method. Section 3 reminds to the reader basics of the SSA and the Ψ-SSA representation. In section 4, we derive the sufficient condition for the correctness of our approach. Finally, section 5 presents experimental evidence that Ψ-SSA can be a practical alternative IR for compiler back-end optimizations.

2 Motivating Example

Consider well-known J.Fang's if-conversion algorithm [12]. This algorithm consists basically of two steps: the first step assigns predicates to basic blocks in the control flow graph; the second step predicates instructions in these basic blocks. In the process, instructions are concatenated into a straight-line code, eliminating merge points in the if-converted region. When program is in SSA form, the question arises: how can Φ-nodes be predicated ? Indeed, since if-conversion removes control flow merge points, Φ-nodes loose their meaning in the if-converted code.

The Ψ-SSA representation enables a simple solution to the problem: Φ-nodes in the if-converted region can be replaced by Ψ-nodes with the same operands. As

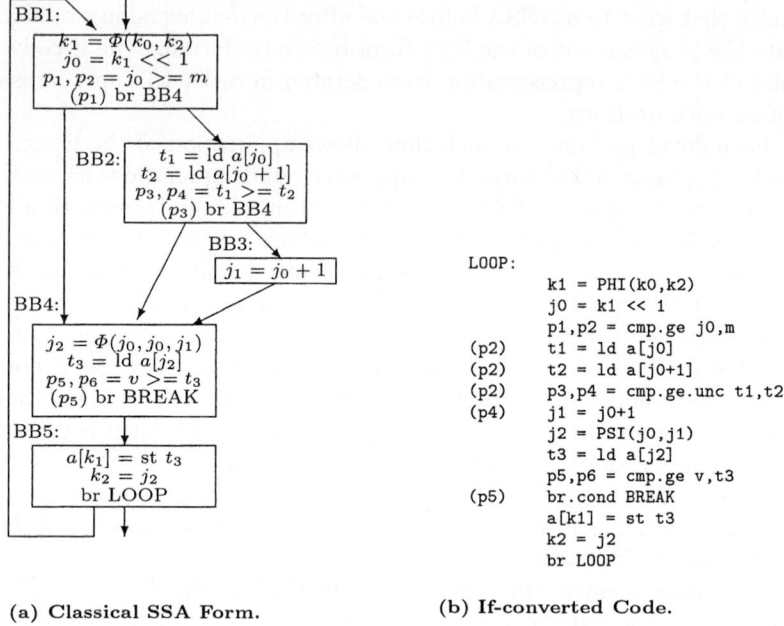

Fig. 1. Heapsort Inner Loop.

an example, consider if-converting the loop shown as psudo-assembly in Figure 1(a) (from the heapsort sorting algorithm). In the first step, predicate p_2 is associated with block BB2, predicate p_4 with block BB3, and predicate TRUE with blocks BB1, BB4, and BB5. In the second step, instructions in block BB2 are predicated with p_2, instructions in block BB3 are predicated with p_4, and conditional branches are removed from blocks BB1 and BB2. Figure 1(b) shows the if-converted code with the Φ-node in block BB4 replaced with a Ψ-node in the resulting Ψ-SSA form[1]. The new Ψ-node indicates that variable j_2 takes on value of either j_0 or j_1, depending on which of the assignments to the two variables is last executed at run-time. Thus, j_2 takes on value of $j1$ if the assignment $j_1 = j_0 + 1$ is executed (i.e. predicate p_4 is TRUE), otherwise it takes on value of j_0. Notice that variable j_2 has exactly the same value before and after the if-conversion: before the if-conversion, variable j_2 takes on value of j_1 if both branches, in blocks BB1 and BB2, are not taken; in if-converted program, j_2 takes on value of j_1 if p_4 is TRUE, which in turn is TRUE only if p_2 is TRUE, corresponding to both branches not being taken in the original program. Notice also that ordering of Ψ-node operands is significant and corresponds to the order in which j_0 and j_1 appear in if-converted sequence [10].

---

[1] Notice that resulting Ψ-node has two operands compared to three-operand Φ-node in the original control flow graph. This is because two or more identical operands in a Ψ-node can always be safely collapsed into a single operand [10].

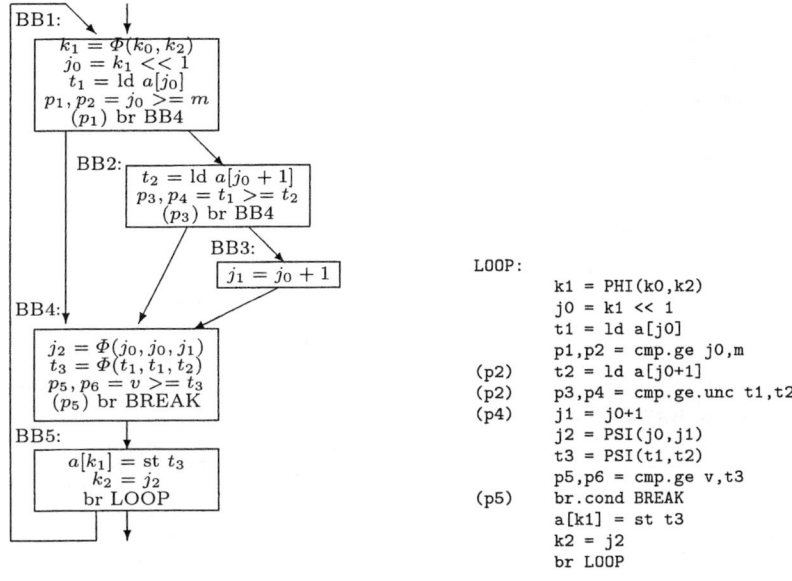

Fig. 2. Transformed Heapsort Inner Loop.

However, replacing Φ-nodes with Ψ-nodes during if-conversion is not as straightforward when transformations such as code motion or copy propagation have been applied to the program. Figure 2(a) shows the same loop after partially redundant load has been removed in block BB4. Replacing the second Φ-node with a Ψ-node in transformed loop results in incorrect program in Figure 2(b). In the if-converted code, variable t_3 takes on value of t_2 if p_2 is TRUE (corresponding to not taken branch in block BB1). This is different from its value before the if-conversion where it takes on the value of t_2 only if both conditions, p_2 and p_4, are TRUE.

In this paper, we prove that replacing Φ-nodes with Ψ-nodes during if-conversion results in correct if-converted program if the original program is in *conventional* SSA form [13]. When the SSA is first build it is conventional. However, the SSA program transformed through the code motion, copy propagation, etc. may not be conventional. The SSA form in the Figure 2(a) is no longer conventional as a result of the partial redundancy elimination. Conventional SSA form will be explained in details in the following section. For now, simply consider conventional SSA form of the transformed loop shown in Figure 3(a)[2]. If-converting loop in conventional SSA results in program in Figure 3(b), where variable t_3 takes on value of t'_2 only if both predicates, p_2 and p_4 are TRUE exactly like in the original program.

---

[2] In conventional form, a copy $t'_2 = t_2$ has been added to block BB3.

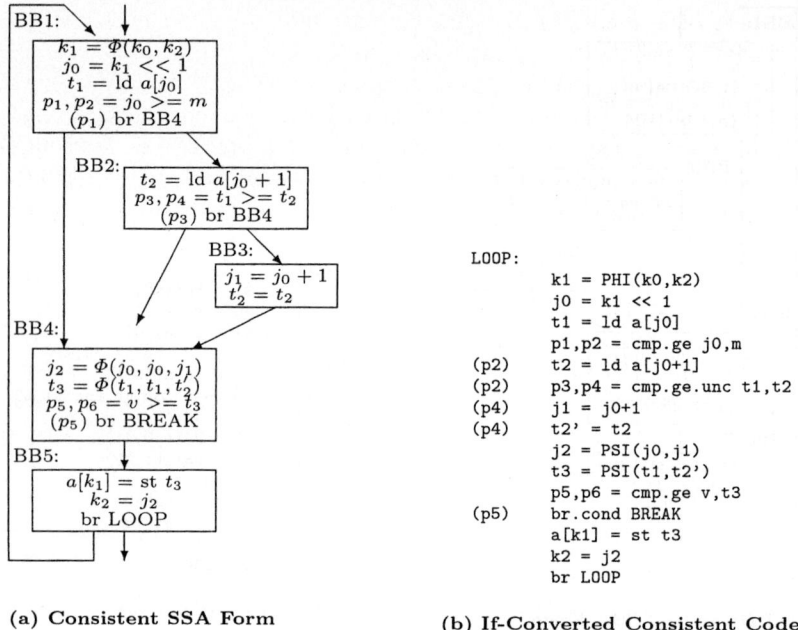

Fig. 3. Consistent Heapsort Inner Loop.

3 Preliminaries

We view an assembly-level program as a control flow graph of basic blocks. Each basic block contains a sequence of target processor instructions. Without loss of generality, we assume each instruction to be either a write to a register, a memory load or store, or a control flow transfer. Instructions can be predicated such that their associated guards determine whether the instruction is executed or not.

In the SSA form, each write to a register is given a unique name, a *SSA name*, and all of the uses reached by that write are renamed to match this new name. At the join basic blocks of the control flow graph, Φ-nodes indicate which SSA names reach the control flow join point. The SSA names in a SSA representation can be partitioned into Φ-*congruence classes* [13]:

Definition 1 (Φ-congruence). *Given a variable x, the Φ-congruence class of $x = \{y \mid x$ and y are referenced in the same Φ-node, or there exist a variable z such that y and z are referenced in the same Φ-node, and x and z are referenced in the same Φ-node $\}$.*

Intuitively, two names belong to a Φ-congruence class if they are connected via Φ-nodes.

When the Ψ-SSA form is first built, it satisfies the *conventional* SSA form property: *liveranges of SSA names sharing a Φ-congruence class do not interfere* [13]. The conventional SSA property allows sharing register assignment by the members of a Φ-congruence class in the final compiled program. As a result of various optimizations, the SSA form may loose this property. Sreedhar et al. developed an algorithm for restoring conventional SSA form by generating copy instructions and removing interferences within Φ-congruence classes. In the example in figure 2, the SSA form is not conventional because liveranges of t_1 and t_2 belonging to the same Φ-congruence class interfere. By adding a copy instruction $t'_2 = t_2$ in block BB3 in the figure 3(a), the SSA form consistency can be restored enabling our if-conversion method.

In Ψ-SSA, Ψ-nodes are placed at program points where register writes active under different predicates are merging. A Ψ-node has the form: $x_{k+1} = \Psi(x_1, x_2, \cdots, x_k)$, where the Ψ-node result, x_{k+1}, takes on value of the operand, x_i, whose assignment is last active assignment under its predicate at run-time. The order of Ψ-node operands is significant and corresponds to program order of assignments to these operands before any transformations. The original order of predicated assignments may be violated by the SSA optimizations, and may need to be restored during out-of-SSA translation based on the ordering information contained in Ψ-nodes [10].

4 If-Conversion in SSA Form

Following theorem establishes the sufficient condition for the correctness of our if-conversion method:

Theorem 1. *If a program is in conventional SSA form, replacing Φ-nodes by the Ψ-nodes with the same operands during if-conversion results in an equivalent new program.*

Proof. Let $\mathcal{S}$ be the block dominating the if-converted region. Consider a Φ-node $x_{n+1} = \Phi(x_1, x_2, \cdots, x_n)$ in block B_{n+1}. Let variable x_{n+1} take on value of x_k, such that at the run-time execution path $\mathcal{S} \to B_k \to B_{n+1}$ is taken, where assignment to x_k happens in basic block B_k.

In if-converted program, let this Φ-node be replaced with a Ψ-node $x_{n+1} = \Psi(x_1, x_2, \cdots, x_n)$. We need to show that if the original program is in conventional SSA form, x_{n+1} takes on value of x_k in if-converted code, i.e. following two conditions hold: (1) instruction defining x_k is active during execution (its predicate is TRUE), and (2) in predicated program, assignment to x_k is the last active assignment among assignments to $x_1, \cdots, x_k, \cdots, x_n$.

(1) A valid if-conversion algorithm guarantees that if instruction defining x_k is executed in original program, then this instruction must be active during execution of the if-converted program.

(2) Assume that the assignment to x_j, originally in basic block B_j is also active in the if-converted code. Thus, execution path $\mathcal{S} \to B_j \to B_{n+1}$ must have been taken at run-time in the program before the if-conversion. If the original program is in conventional SSA form, block $B_j \neq B_k$, otherwise liveranges of x_j and x_k would interfere. Furthermore, block B_j must preceed block B_k in the original program, i.e. execution path $\mathcal{S} \to B_j \to B_k \to B_{n+1}$ is taken,

Table 1. Summary Statistics of Media Bench Programs.

Program name	Functions in program	Instructions before SSA			Instructions expansion ratio		Variables before SSA			Variables expansion ratio	
		max	median	min	max	median	max	median	min	max	median
adpcm	5	144	92	57	1.23	1.04	89	83	19	1.61	1.14
epic	49	1471	68	12	1.24	1.06	1368	46	3	1.71	1.20
g721	24	556	41	17	1.13	1.06	414	34	14	1.38	1.17
ghostscript	3168	4104	38	4	1.52	1.03	2572	23	1	1.99	1.12
gsm	94	620	54	5	1.16	1.05	650	30	2	1.60	1.21
jpeg	465	723	74	4	1.24	1.03	608	49	1	1.71	1.11
mpeg2	209	1507	86	6	1.27	1.05	936	53	1	1.89	1.18
pegwit	97	1706	54	12	1.19	1.04	1334	28	2	1.65	1.14
pgp	371	2774	59	4	1.21	1.04	1469	34	1	1.83	1.18

otherwise the Φ-node in block B_{n+1} would assign value of x_j instead of x_k to x_{n+1}. Since if-conversion concatenates instructions in Breadth-First-Order of the basic blocks [12], the assignment to x_k is guaranteed to follow the assignment to x_j in the if-converted code sequence.

□

Because during if-conversion, Ψ-nodes are created as replacement to existing Φ-nodes, the number of Ψ-nodes resulting form the if-conversion is bound by the number of Φ-nodes before the if-conversion. Thus, the size of the Ψ-SSA representation is essentially the same as the size of the equivalent classical SSA representation before the if-conversion.

5 Measurements

5.1 Experimental Framework

We implemented our if-conversion method in the Open64/ORC compiler code generator (`ipf-orc.sourceforge.net`)[3]. Our goal was to investigate the size of the Ψ-SSA representation resulting from our if-conversion method for programs encountered in practice. Previous measurements of the size of SSA intermediate representation were obtained in the context of the target processor independent phase of the compiler optimization [2, 9]. We're not aware of any previous work that provides similar investigation into the size of the SSA representation at the assembly level.

Our benchsuite consists of the 9 MediaBench benchmarks [11] listed in the first column of table 1. These benchmarks include applications from multimedia and communications domain such as video and audio coding and compression, image processing, encription, etc. Benchmarks were cross compiled for the IA64 system under Linux using the NUE environment (`software.hp.com/ia64linux`).

---

[3] Details of our implementation are outside the scope of this paper.

5.2 Results and Observations

The summary statistics of our experiments are given in the table 1. The second column of the table shows the total number of routines in each benchmark program. The third column lists the maximal, minimal, and median size of routines in each of the benchmarks programs in the number of intermediate representation instructions. The fourth column shows the maximal and median increase in the number of intermediate representation instructions when each program is in Ψ-SSA form (the size of the Ψ-SSA representation is measured after the if-conversion). Next column in the table lists the maximal, minimal, and median size of routines in each program in the number of distinct variable names. Finally, the last column shows the maximal and median increase in the number of variable names (SSA names) in the Ψ-SSA form.

The maximal increase in the intermediate representation size of 1.52 times is smaller than 5.2 times reported by Cytron *et al.* [2]. We also observe smaller maximal increase in the number of distinct variable names in the program, 1.99 versus 3.8. These numbers are explained by the fact that we build *prunned* SSA representation instead of the *minimal* one discussed in Cytron's *et al.* work. This also confirms that prunned SSA representation can drastically reduce the number of Φ-nodes in program variables reported by Briggs *et al.* [9].

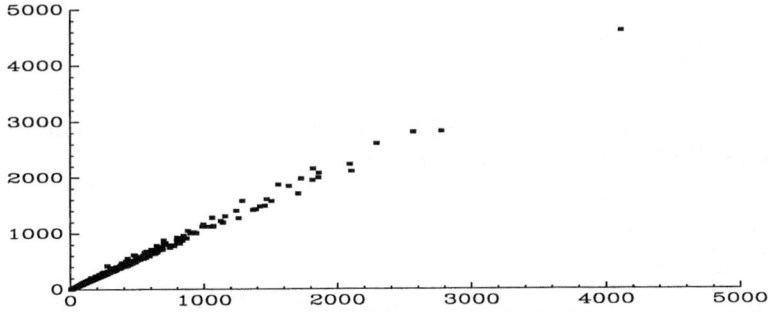

Fig. 4. Number of instructions in the Ψ-SSA versus number of instructions in the original program.

As the plot in figure 4 shows, increase in the number of intermediate representation instructions is linear in the size of the program. The increase in instruction counts reflects the number of Φ- and Ψ-nodes in the program. We didnot find any correlation between the original instruction count and the instruction count increase due to Ψ-SSA translation. More significant instruction count increases were related rather to the complexity of program's control flow. The median ratio varied from 1.03 to 1.06, and the largest increase of 1.52 occured in a function of 292 intermediate instructions.

The plot in figure 5 shows increase in the number of variable names in the Ψ-SSA. The median increase in the number of variables varied from 1.11 to 1.21.

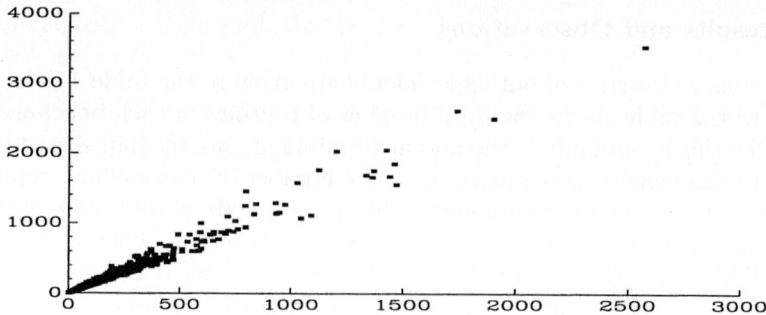

Fig. 5. Number of SSA names in the Ψ-SSA versus number of distinct variable names in the original program.

The largest increase of 1.99 occured in the same function with the largest instruction count increase, and this function had 193 intermediate variables. Alltogether, 90% of the functions had increase of the number of variables under 1.4 times.

6 Conclusions

SSA form benefits optimization of unpredicated code as well as predicated code after the if-conversion. We developed a method for if-converting programs in SSA form resulting in Ψ-SSA form. Our measurements of the size of Ψ-SSA present a strong evidence that Ψ-SSA is a practical alternative as intermediate representation for compiler code generators.

References

1. J.Allen, K.Kennedy, C.Porterfield, J.Warren: Conversion of control dependence to data dependence. In: SIGPLAN International Conference on Programming Languages Design and Implementation. (1983) 177 – 189
2. R.Cytron, J.Ferrante, B.Rosen, M.Wegman, K.Zadeck: Efficiently computing static single assignment form and the control dependence graph. ACM Transactions on Programming Languages and Systems **13** (1991) 451 – 490
3. M.Wegman, K.Zadeck: Constant propagation with conditional branches. ACM Transactions on Programming Languages and Systems **13** (1991) 181 – 210
4. B.Alpen, M.Wegman, K.Zadeck: Detecting equality of variables in programs. In: ACM Symposium on the Principles of Programming Languages. (1988) 1 – 11
5. K.Cooper, T.Simpson: SCC-based value numbering. Technical Report CRPC-TR95636-S, Center for Research on Parallel Computation, Rice University (1995)
6. K.Cooper, T.Simpson: Value-driven code motion. Technical Report CRPC-TR95637-S, Center for Research on Parallel Computation, Rice University (1995)
7. P.Briggs: Register Allocation via Graph Coloring. PhD thesis, Rice University, Houston, Texas (1992)

8. C.Click: Global code motion global value numbering. In: SIGPLAN International Conference on Programming Languages Design and Implementation. (1995) 246 – 257
9. P.Briggs, K.Cooper, T.Harvey, T.Simpson: Practical improvements to the construction and destruction of static single assignment form. Software Practice and Experience **28** (1998) 859 – 881
10. A.Stoutchinin, F.Ferriere: Efficient static single assignment form for predication. In: MICRO – International Symposium on Microarchitecture. (2001) 172 – 181
11. C.Lee, M.Potkonjak, H.W.Mangione-Smith: Mediabench: A tool for evaluating and synthesizing multimedia and communicatons systems. In: MICRO – International Symposium on Microarchitecture. (1997) 330–335
12. J.Fang: Compiler algorithms on if-conversion, speculative predicates assignment and predicated code optimizations. In: Ninth Workshop on Languages and Compilers for Parallel Computers. (1996) 135 – 153
13. V.Sreedhar, R.Ju, D.Gillies, V.Santhanam: Translating out of static single assignment form. In: Static Analysis Symposium, Italy. (1999) 194 – 204

Topic 5
Parallel and Distributed Databases, Data Mining and Knowledge Discovery

David Skillicorn, Abdelkader Hameurlain, Paul Watson, and Salvatore Orlando

Topic Chairs

This topic addresses the parallel storage and processing of data, ranging from traditional database systems, through OLTP and OLAP systems, to data mining. The primary motivation for parallelism in this setting is performance. Parallel database systems have been in use for several decades, but the use of parallelism for data mining is more recent. Interestingly, a new motivation for parallelism in handling data is becoming important – the need to compute results from multiple datasets without necessarily having full access to each of them. This idea of preserving some level of privacy for the contents of each part of a dataset generates some interesting problems and we look forward to papers in this area in years to come.

We received 15 submissions this year, and were able to accept 6, an acceptance rate of 40%.

The accepted papers cover a wide range of topics. Fagni, Perego and Silvestri address how to improve the performance of a concurrent web search cache, showing that a cache region dedicated to high-frequency queries improves overall performance. Lima, Mattoso and Valduriez use a new method for the virtual partitioning of OLAP queries on clusters to improve performance. Dash, Petrutiu and Scheuermann show how hierarchical clustering can exploit the property that clusters are not well-separated until the final stages of a clustering. Glimscher and Agrawal find a way to express both the E and M stages of the EM clustering technique as generalized reductions, which can then be implemented in parallel using a middleware system they have developed. Coppola, Pesciullesi, Ravazzolo and Zoccolo describe the design of a parallel customer relationship management system. Finally, di Giacomo, Martinez and Scott describe SearchPlus, a distributed system for bibliographic records.

We thank all of the referees whose work helped to select an interesting program.

A Highly Scalable Parallel Caching System for Web Search Engine Results

T. Fagni[1], R. Perego[1], and F. Silvestri[1,2]

[1] Istituto ISTI, Consiglio Nazionale delle Ricerche (CNR), Pisa, Italy
[2] Dipartimento di Informatica, Università di Pisa, Italy

Abstract. This paper discusses the design and implementation of SDC, a new caching strategy aimed to efficiently exploit the locality present in the stream of queries submitted to a Web Search Engine. SDC stores the results of the most frequently submitted queries in a *fixed-size read-only* portion of the cache, while the queries that cannot be satisfied by the static portion compete for the remaining entries of the cache according to a given cache replacement policy. We experimentally demonstrated the superiority of SDC over purely static and dynamic policies by measuring the hit-ratio achieved on two large query logs by varying cache parameters and the replacement policy used. Finally, we propose an implementation optimized for concurrent accesses, and we accurately evaluate its scalability.

1 Introduction

Due to the high locality present in the stream of queries processed by a Web Search Engine (WSE), caching the results of the queries submitted by users is a very effective technique to increase the throughput. WSE results caching, similarly to Web page caching, can occur at several places, e.g. on the client side, on a proxy, or on the server side. Caching on either the client or the proxy has the advantage of saving network bandwidth. Caching on the server side, on the other hand, has the effect of saving I/O and computational resources used by the WSE to compute the page of relevant results to be returned to a user. One of the issues related to server-side caching is the limited resources usually available on the server, in particular the RAM memory used to store the cache entries. However, the architecture of a scalable, large-scale WSE is very complex and includes several machines which take care of the various subtasks involved in the processing of user queries. The distributed architecture of a large-scale WSE [7,1] is composed by a farm of identical machines running multiple WSE *Core* modules, each of which is responsible for searching the index relative to one specific sub-collection of documents. In front of these searcher machines we have an additional machine hosting a *Mediator*. This module has the task of scheduling the queries to the various searchers, and of collecting the results returned back. Note that multi-threading is exploited extensively by all these modules in order to process concurrently distinct queries. Within this architecture the RAM memory is a very precious resource for the machines that host the WSE Core, which perform well only if the mostly accessed sections of their huge indexes can be buffered into the main memory. Conversely, the RAM memory is a less critical resource for the machine that hosts the Mediator. This machine can thus be considered as an ideal candidate to host a server-side cache. The performance improvement which may derive from the exploitation of query results caching at this level is remarkable. Queries resulting in cache-hits can be in fact promptly served thus enhancing WSE throughput, but also the queries whose results are not found in cache benefit substantially due to the lower load on the WSE and the consequent lower contention for the I/O, network and computational resources.

Beside the fact that several papers analyzed query logs to study the behavior of WSE users, only a few works propose effective techniques to exploit the locality present in the stream of user requests[5, 9, 4]. Markatos compared several caching policies on the basis of the hit-ratio obtained on a actual log of queries submitted to Excite [5]. The work by Saraiva *et al.* [9], discusses instead a two-level caching system which try to enhance the responsiveness of a hierarchically-structured search engine The first-level cache is similar to the one discussed in [5], while, the second-level cache is intended to store the posting lists of the keywords contained into the query strings. Finally, Lempel and Moran recently proposed *PDC* (Probabilistic Driven Caching), a new effective caching policy which associates a probability distribution to all the possible queries that can be submitted to a WSE. The distribution is built over the statistics computed on the previously submitted queries, and is used to compute a priority exploited to maintain an importance ordering among the entries of the cache.

In this paper, we are interested in studying the design and implementation of such a server-side cache of query results. Starting from the analysis of the content of two real query logs, we propose a novel replacement policy (called SDC- Static and Dynamic Cache) to adopt in the design of a fully associative cache of query results. According to SDC, the results of the most frequently accessed queries are maintained in a fixed size set of statically locked cache entries. This *Static Set* is rebuilt at fixed time intervals using statistical data coming from WSE usage data. When a query cannot be satisfied by the Static Set, it competes for the use of a *Dynamic Set* of cache entries. The management of the *Dynamic Set* can exploit, in principle, any replacement policy. In all the tests performed, the presence of a static portion of the cache resulted in large advantages both on the hit-ratio achieved and on the overall throughput of the caching system. We experimentally evaluated SDC by measuring the hit-ratio and the throughput achieved on actual query logs by varying the size of the cache, the percentage of cache entries of the Static Set, and the replacement policy used for managing the Dynamic Set. Differently from the other works, we also accurately assessed the scalability of our caching system with respect to the number of concurrent threads using it. The paper is organized as follows. Section 2 describes the query logs used, while in Section 3 we discuss our novel caching policy. Section 4 shows the results of the simulations performed on the different query logs. Finally, Section 5 presents some concluding remarks.

2 Analysis of the Query Logs

In order to evaluate the behavior of different caching strategies we used query logs from the Tiscali and Altavista search engines. In particular we used *Tiscali*, a trace of the queries submitted to the Tiscali WSE engine (www.janas.it) on April 2002, and *Altavista* a query log containing queries submitted to Altavista on the Summer of 2001[1]. Each record of a query log refers to a single query submitted to the WSE for requesting a *page* of results, where each page contains a fixed amount of URLs ordered according to a given rank. All query logs have been preliminarily cleaned by removing the useless fields. At the end of this preprocessing phase, each entry of a query log has the form *(keywords, page_no)*, where *keywords* corresponds to the list of words searched for, and *page_no* determines which page of results is requested. We further normalize the query log entries by removing those referring to requests of more than 10 results-per-page. After the cleaning, the *Altavista* query log contains 6, 175, 648 queries of which 2, 657, 410 are distinct. The *Tiscali* logs is instead composed of 3, 278, 211 queries of which 1, 538, 934 are distinct. Thus in both the logs about the 54% of the total number of queries are repeated more than once. The plots reported in Figure 1 assess the locality present in the query logs using a log-log scale. In particular Figure 1.(a) plots the number of occurrences within each log of the most popular queries, whose identifiers have been assigned in decreasing order of frequency. Note that, in all the two logs, more then 10, 000 different queries are repeated more than 10 times. Since the number of occurrences of a given query is a measure that might depend on the total number of records contained in the logs, to better highlight temporal locality present in the logs we also analyzed the time interval between successive submissions of the same query. The rationale is that if a query is repeatedly submitted within a small time interval, we can expect to be able to retrieve its results even from a cache of small size. Figure 1.(b) reports the results of this analysis. For each query log we plotted the cumulative number of resubmissions of the various queries as a function of the time interval (expressed as a distance measured in number of queries). Once more the results are encouraging: in the *Tiscali* log for more than 350, 000 times, the time interval between successive submissions of the same query is less than 100; in the *Altavista* log this temporal locality is slightly smaller than that in *Tiscali* but, again, for more than 150, 000 times the time interval is still less than 100.

3 The SDC Policy

SDC is a two-level policy which makes use of two different sets of cache entries. The first level contains the *Static Set* consisting in a set of statically locked entries filled with the most frequent queries appeared in the past. The Static Set is periodically refreshed. The second level contains the *Dynamic Set*. Basically, it is a set of entries managed by a classical replacement policy (i.e. *LRU, SLRU,*

---

[1] We are very grateful to Ideare S.p.A. and to Ronny Lempel for providing us with these query logs.

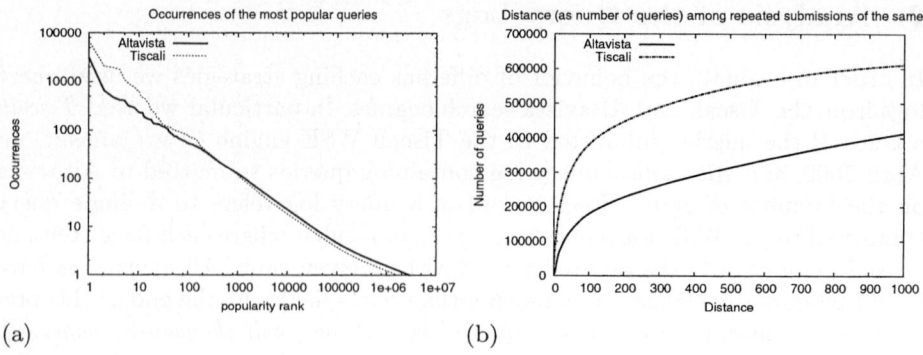

Fig. 1. Analysis of the locality present in the query logs.

etc.). The behavior of SDC in the presence of a query q is very simple. First it looks for q in the Static Set, if q is present it returns the associated page of results back to the user. If q is not contained within the Static Set, then it is looked for in the Dynamic Set. If q is not present, then SDC asks the WSE for the page of results and replaces an entry of the Dynamic Set according to the replacement policy adopted.

The rationale of adopting a static policy, where the entries to include in the cache are statically decided, relies on the observation that the most popular queries submitted to WSEs do not change very frequently. On the other hand, several queries are popular only within relatively short time intervals, or may become suddenly popular due to, for example, un-forecasted events (e.g. the 11th September 2001 attack). The advantages deriving from this novel caching strategy are two-fold. In fact the results of the most popular of the queries can always be usually retrieved from the Static Set even if some of these queries might be not requested for relatively long time intervals. On the other hand, the Dynamic Set of the cache can adequately cover sudden interests of users.

First Level – Static Set. The implementation of the first level of our caching system is very simple. It basically consists of a lookup data structure that allows to efficiently access a set of $f_{static} \cdot N$ entries, where N is the total number of entries of the whole cache, and f_{static} the factor of locked entries over the total. f_{static} is a parameter of our cache implementation whose admissible values ranges between 0 (a fully dynamic cache) and 1 (a fully static cache). The static cache has to be initialized off-line, i.e., with the results of most frequent queries computed on the basis of a previously collected query log.

Each time a query is received, SDC first tries to retrieve the corresponding results from the Static Set. On a cache hit, the requested page of results is promptly returned. On a cache miss, we also look for the query results in the Dynamic Set.

Second Level – Dynamic Set. The Dynamic Set relies on a replacement policy for choosing which pages of query results should be evicted from the cache as a consequence of a cache miss and the cache is full. Literature on caching proposes several replacement policies which, in order to maximize the hit-ratio, try to take the largest advantage from information about recency and frequency of references. SDC surely simplifies the choice of the replacement policy to adopt. The presence of a static read-only cache, which permanently stores the most frequently referred pages, makes in fact recency the most important parameter to consider. Currently, our caching system supports the following replacement policies: *LRU*, *LRU/2* [6] which applies a *LRU* policy to the penultimate reference, *FBR* [8], *SLRU* [5], *2Q* [3], and *PDC* [4] which consider both the recency and frequency of the accesses to cache blocks.

4 Experiments

All the experiments were conducted on a Linux PC equipped with a 2GHz Pentium Xeon processor and 1GB of RAM. Since SDC requires the blocks of the static section of the cache to be preventively filled, we partitioned each query log into two parts: a *training set* which contains 2/3 of the queries of the log, and a *test set* containing the remaining queries used in the experiments. The N most frequent queries of the training set were then used to fill the cache blocks: the first $f_{static} \cdot N$ most frequent queries (and corresponding results) were used to fill the static portion of the cache, while the following $(1 - f_{static}) \cdot N$ queries to fill the dynamic one. Note that, according to the scheme above, before starting the tests not only the static blocks but also the dynamic ones are filled, and this holds even when a pure dynamic cache ($f_{static} = 0$) is adopted. In this way we always starts from the same initial state to test and compare the different configurations of SDC, obtained by varying the factor f_{static}. (i.e. warm cache, using the terminology in [4]).

Figures 2 reports the cache hit-ratios obtained on the *Tiscali* (a), and *Altavista* (b) query logs by varying the ratio (f_{static}) between the sizes of the static and dynamic sets. Each curve corresponds to a different replacement policy used for the dynamic portion of the cache. In particular, f_{static} was varied between 0 (a fully dynamic cache) and 1 (a fully static cache), while the replacement policies exploited were *LRU*, *FBR* [8], *SLRU* [5], *2Q* [3], and *PDC* [4]. The total size of the cache was fixed to 256,000 blocks. Several considerations can be done looking at these plots. First, we can note that the hit-ratios achieved are in some cases impressive, although the curves corresponding to different query logs have different peak values and shapes, thus indicating different amounts and kinds of locality in the query logs analyzed. At a first glance, these differences surprised us. After a deeper analysis we realized that similar differences can also be found by comparing other query logs already studied in the literature [10, 2, 11], thus indicating that users' behaviors may vary remarkably from time to time. Another important consideration is that in all the tests performed SDC remarkably outperformed the other policies, whose performance are exactly those corresponding to a value of $f_{static} = 0$. The best choice of the value for f_{static} depends from the query log considered and, more importantly, the different replacement policies do not impact heavily on the overall hit-ratio for the optimal values of f_{static}.

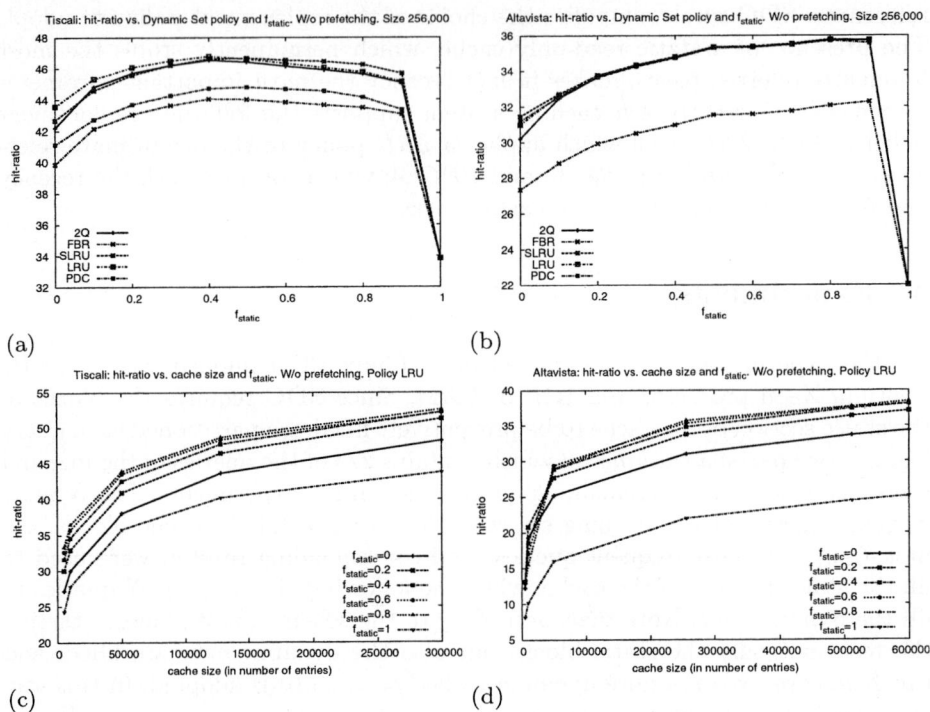

Fig. 2. Hit-ratios achieved on *Tiscali* and *Altavista* logs for different replacement policies and varying f_{static}.

To measure the sensitivity of SDC with respect to the size of the cache, Figures 2 plot the hit-ratios achieved on the *Tiscali* (c) and *Altavista* (d) query logs as a function of the number of blocks of the cache and the f_{static} parameter. As expected, when the size of the cache is increased, hit-ratios increase correspondingly. In the case of the *Tiscali* log, the hit-ratio achieved is about 37% with a cache of 10,000 blocks, and about 45% when the size of the cache is 50,000. Note that actual memory requirements are however limited: a cache storing the results as an *Html* page and composed of 50,000 blocks requires about 200MB of RAM.

We designed our caching system to allow efficient concurrent accesses to its blocks. This is motivated by the fact that a WSE has to process several user queries concurrently. This is usually achieved by making each query processed by a distinct thread. The methods exported by our caching system are thus thread-safe and also ensure the mutual exclusion. In this regard, the advantage of SDC over a pure dynamic cache is related to the presence of the *Static Table*, which is a read-only data structure. Multiple threads can thus concurrently lookup the *Static Table* to search for the results of the submitted query. In case of a hit, the threads can also retrieve the associated page of results without synchronization. For this reason our caching system may sustain linear speed-up even in configurations containing a very large number of threads. Conversely, the *Dynamic Table* must be accessed in the critical section controlled by a mutex. Note, in fact, that the *Dynamic Table* is a read-write data structure: while a cache miss obviously causes both the associative memory and relative list of pointers to be modified, also a cache hit entails the list pointers to be modified in order to sort the cache entries according to the replacement policy adopted.

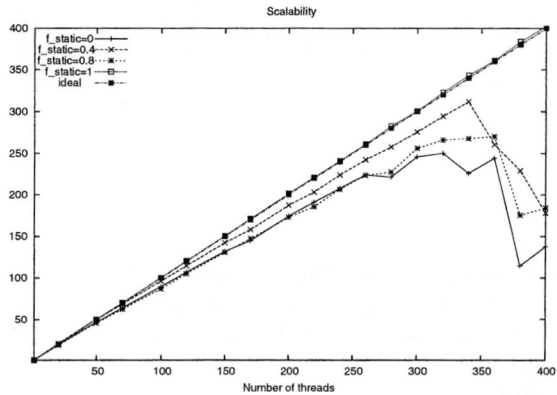

Fig. 3. Scalability of our caching system for different values of f_{static} as a function of the number of concurrent threads used.

Figure 3 shows the performance of our cache system in a multi-threading environment. In particular, the Figure plots, for different values of f_{static}, the scalability of the system as a function of the number of concurrent threads sharing a single cache. Scalability has been measured by considering the ratio between the wall-clock times spent by one and n threads to serve the same large bunch of queries. The replacement policy adopted in running the test was *LRU*. In the case of a cache hit, the thread serving the query returns immediately the requested page of results, and gets another query. Conversely, when a query causes a cache miss, the thread sleeps for 40 ms to simulate the latency of the WSE core in resolving the query. As it can be seen, the system scales very well even when a large number of concurrent threads is exploited. The scalability of a purely static cache is optimal since the cache is accessed read-only, but high scalability values are achieved also when SDC is adopted. Note that even when a purely dynamic cache is adopted ($f_{static} = 0$), our system scales linearly with up to 250 concurrent threads due to the very low cache management times (experimentally measured between 11 and $36\mu s$).

5 Conclusions and Future Works

In this paper we presented SDC, a new policy for caching the query results of a WSE which exploits the knowledge about the queries submitted in the past to make more effective the management of the cache. In particular, we maintain the most popular queries and associated results in a read-only static section of our cache. Only the queries that cannot be satisfied by the static cache section compete for the use of a dynamic cache. The benefits of adopting SDC were experimentally shown on the basis of tests conducted with two large query logs. In all the cases our strategy remarkably outperformed either purely static or dynamic caching policies. We evaluated the hit-ratio achieved by varying the percentage of static blocks over the total, the size of the cache, as well as the replacement policy adopted for the dynamic section of our cache. Moreover, we evaluated cost and scalability of our cache implementation when executed in a multi-threaded environment. The SDC implementation resulted very efficient due to an accurate software design that allowed to make cache hit and miss times negligible, and to the presence of the read-only static cache that reduces the synchronization between multiple threads concurrently accessing the cache.

References

1. Sergey Brin and Lawrence Page. The anatomy of a large-scale hypertextual Web search engine. *Computer Networks and ISDN Systems*, 30(1–7):107–117, 1998.
2. Bernard J. Jansen, Amanda Spink, and Tefko Saracevic. Real life, real users, and real needs: a study and analysis of user queries on the web. *Inf. Proc. and Manag.*, 36(2):207–227, 2000.
3. Theodore Johnson and Dennis Shasha. 2q: A low overhead high performance buffer management replacement algorithm. In *Proc. 1994 VLDB*, pages 439–450, 1994.
4. Ronny Lempel and Shlomo Moran. Predictive caching and prefetching of query results in search engines. In *Proc. of the twelfth international conference on World Wide Web*, pages 19–28. ACM Press, 2003.
5. Evangelos P. Markatos. On caching search engine results. In *Proc. of the 5th Int. Web Caching and Content Delivery Workshop*, 2000.
6. Elisabeth J. O'Neil, Patrick E. O'Neil, and Gerhard Weikum. The lru-k page replacement algorithm for database disk buffer. In *Proc. of the 1993 ACM SIGMOD International Conference On Management Of Data*, pages 297–306, 1993.
7. S. Orlando, R. Perego, and F. Silvestri. Design of a parallel and distributed web search engine. In *In proc. of ParCo 2001 int'l conf.*, 2001.
8. John T. Robinson and Murthy V. Devarakonda. Data cache management using frequency-based replacement. In *Proc. of the 1990 ACM SIGMETRICS Conference*, pages 134–142, 1990.
9. P.C. Saraiva, E. Silva de Moura, N. Ziviani, W. Meira, R. Fonseca, and B. Ribeiro-Neto. Rank-preserving two-level caching for scalable search engine. In *SIGIR'01*, 2001.
10. C. Silverstein, M. Henzinger, H. Marais, and M. Moricz. Analysis of a very large web search engine query log. In *ACM SIGIR Forum*, pages 6–12, 1999.
11. A. Spink, B.J. Jansen, D. Wolfram, and T. Saracevic. Searching the web: the public and their queries. *J. Am. Soc. Inf. Sc. & Tech.*, 53(2):226–234, 2001.

OLAP Query Processing in a Database Cluster*

Alexandre A.B. Lima[1], Marta Mattoso[1], and Patrick Valduriez[2]

[1]Computer Science Department, COPPE, Federal University of Rio de Janeiro, Brazil
{assis,marta}@cos.ufrj.br
[2]Atlas Group, INRIA and LINA, University of Nantes, France
Patrick.Valduriez@inria.fr

Abstract. The efficient execution of OLAP queries, which are typically read-only and heavy-weight, is a hard problem which has been traditionally solved using tightly-coupled multiprocessors. Considering a database cluster as a cost-effective alternative, we propose an efficient, yet simple, solution, called fined-grained virtual partitioning to OLAP parallel query processing. We designed this solution for a shared-nothing database cluster architecture that can scale up to very large configurations and support black-box DBMS using non intrusive, simple techniques. To validate our solution, we implemented a Java prototype on a 16 node cluster system and ran experiments with typical queries of the TPC-H benchmark. The results show that our solution yields linear, and sometimes super-linear, speedup. With 16 nodes, it outperforms traditional virtual partitioning by a factor of 6.

1 Introduction

Decision support applications require efficient support for On-Line Analytical Processing (OLAP) on larger and larger databases. OLAP queries are typically read-only and heavy-weight. In the TPC-H benchmark [12], specific to decision support systems, twenty-two database queries are complex, heavy-weight and read-only and only two have updates. Furthermore, OLAP queries have an ad-hoc nature. As users get more experienced about OLAP system features, they demand more efficient ad-hoc query support [5].

The efficient execution of OLAP queries, where "efficiency" means "as fast as possible", is still an open problem. High-performance of database management has been traditionally achieved with parallel database systems [13], implemented on tightly-coupled multiprocessors. Parallel data processing is then obtained by partitioning and replicating the data across the multiprocessor nodes in order to divide processing. Although quite effective, this solution requires the database system to have full control over the data and is expensive in terms of software and hardware. Clusters of PC servers provide a cost-effective alternative to tightly-coupled multiprocessors. Recently, the *database cluster* approach, i.e. clusters with off-the-shelf (black-box) DBMS nodes, has gained much interest for various database applications [1, 4, 9].

In this paper, we propose a solution to efficient OLAP query processing in a database cluster using simple parallel processing techniques. The basic technique we

---

* Work partially funded by CNPq, CAPES, INRIA and COFECUB.

employ is virtual partitioning [1] which gives more flexibility than physical (static) data partitioning [7] for parallel query processing. In its simplest form, it consists in fully replicating the database among the cluster nodes. To distribute the workload, predicates are added to queries to force each DBMS to process a different subset, called a *virtual partition*, of data items. Each DBMS processes exactly one sub-query. The problem is these sub-queries can take almost as long as the original query to be executed. Depending on the estimated amount of data to be processed, DBMS optimizers can opt for fully scanning the virtually partitioned table, reducing (or even eliminating) benefits obtained from virtual partitioning. Temporary disk-based structures demanded by sub-queries that deal with huge amounts of data can also limit virtual partitioning performance. In this paper, we propose a major improvement called *fine-grained virtual partitioning* (FGVP) which addresses these problems. As proposed in [1], virtual partitioning assigns each cluster node one sub-query, what can lead to problems aforementioned. Our approach is to work with a larger number of virtual partitions, much greater than the number of nodes. It is an attempt to keep sub-queries as simple as possible, avoid full table scans and expensive temporary disk-based structures. Our experimental results, based on our implementation on a 16-node cluster running PostgreSQL, show that linear, and sometimes super-linear, speedup is obtained for typical OLAP queries. In the worst cases, almost linear speedup is achieved. FGVP outperformed the traditional virtual partitioning for all queries when using more than two nodes. We think FGVP also provides a good basis for dynamic load balancing as it makes it possible to perform sub-query reallocation. This article is organized as follows. Section 2 presents our database cluster architecture. Section 3 describes our fine-grained virtual partitioning technique. Section 4 describes our prototype implementation as well as experimental results. Section 5 concludes.

2 Database Cluster Architecture

A database cluster [1] is a set of PC servers interconnected by a dedicated high-speed network, each one having its own processor(s) and hard disk(s), and running an off-the-shelf DBMS. Similar to multiprocessors, various cluster system architectures are possible: shared-disk, shared-cache and shared-nothing [13]. Shared-disk and shared-cache require a special interconnect that provides a shared space to all nodes with provision for cache coherence using either hardware or software. Shared-nothing (or distributed memory) is the only architecture that does not incur the additional cost of a special interconnect. Furthermore, shared-nothing can scale up to very large configurations. Thus, we strive to exploit a shared-nothing architecture as in PowerDB [11] and Leg@Net [4]. Each cluster node can simply run an inexpensive (non parallel) DBMS. In our case, we use the PostgreSQL [8] DBMS, which is freeware. Furthermore, the DBMS is used as a "black-box" component [4]. In other words, its source code is not available and cannot be changed or extended to be "cluster-aware". Therefore, extra functionality like parallel query processing capabilities must be implemented via middleware.

We use data replication to improve performance. As in [1, 4], we assume full database replication for simplicity: each database is replicated at each node. To maintain copy consistency, we can assume a preventive replication protocol [2] which scales up well in cluster systems. But since database updates are rare in OLAP applications, copy consistency is not an issue. The only potential problem with full database replication is database size, which can be huge. A good solution is to have a mix of data replication and data partitioning as in [9]. Our technique can be employed in such configurations.

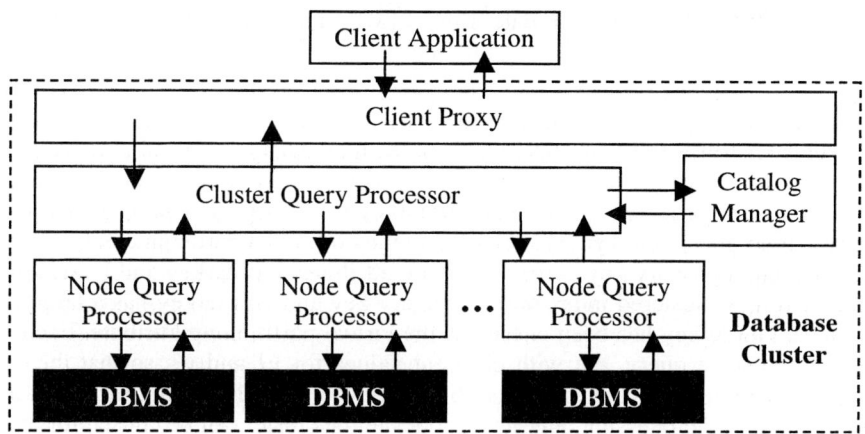

Fig. 1. Database Cluster Architecture

In our system architecture, query processing is done by independent DBMSs orchestrated by distributed middleware. Our middleware main components are Client Proxy, Cluster Query Processor, Catalog Manager and Node Query Processor (see Figure 1). The Client Proxy is the system entry point. It is used by client applications for query submission. Upon reception, the Client Proxy passes the query on to the Cluster Query Processor (CQP) which is responsible for elaboration and execution of Query Execution Plans (QEP). Elaborating a QEP means determining the virtual partitioning attributes, the number of virtual partitions and the participating nodes. For QEP elaboration, metadata information like schema definition and database statistics is obtained from the Catalog Manager. When the QEP is ready, the CQP starts the threads responsible for distributed query execution management and final result composition. This process involves interaction with the Node Query Processors. By definition, each node in a database cluster runs a DBMS instance [1]. In our architecture, each node has also a Node Query Processor (NQP) which is responsible for node-level query execution and result composition. After receiving its workload from the CQP, the NQP interacts with the local DBMS, submitting SQL queries and collecting their corresponding results. These results are locally computed and sent to the CQP which does global result composition. When all NQPs are finished, the CQP sends the final results to the Client Proxy for delivery to the client application.

3 Dynamic Virtual Partitioning

In this section, we introduce the principle of virtual partitioning and describe our proposal of fine-grained virtual partitioning.

Assuming a fully replicated database, virtual partitioning adds predicates to queries to force each DBMS to process a different subset, called a *virtual partition*, of data items. For example, let us consider query Q1 which accesses the Lineitem relation, the largest one from TPC-H:

Q1:   SELECT l_returnflag, l_linestatus, SUM(l_quantity), COUNT(*)
        FROM   lineitem
        WHERE l_shipdate <= date '1998-12-01' - interval '90 day'
        GROUP BY l_returnflag, l_linestatus;

Q1 is a typical OLAP query. It has an aggregate and accesses a huge table (in TPC-H's smallest configuration, Lineitem has 6,001,215 tuples). The select predicate is not very selective since there are 5,916,519 tuples that satisfy it. For simplicity, it contains no joins. Using virtual partitioning, this query would be rewritten by adding the predicate "and l_orderkey >= :v1 and l_orderkey < :v2" to the "where" clause of Q1. The rewritten query can then be submitted to the nodes that participate in Q1 processing. Lineitem's primary key is formed by the attributes l_orderkey and l_linenumber. Since there is a clustered index on the primary key and l_orderkey has a large range of values, l_orderkey has been chosen as the virtual partitioning attribute. Each node receives the same query, but with different values for *v1* and *v2*, so that the whole range of l_orderkey is scanned. This technique allows great flexibility for node allocation during query processing: any set of nodes in the cluster can be chosen for executing any query.

One basic goal of virtual partitioning is to reduce the amount of data read from disk by each DBMS. Clustered indexes play an important rule as, by using them, each DBMS can work on a different subset of disk pages. However, the existence of such indexes based on attributes used for virtual partitioning does not guarantee their use during query processing. DBMS query optimizers decide using them or not according to estimations on the amount of data to be retrieved, which depends on the attribute value range specified in each sub-query. Incidentally, the optimizer can even opt for performing a full scan on the virtually partitioned table. We experienced this when performing experiments with virtual partitioning and PostgreSQL.

To overcome this problem, we propose an optimization technique called "fine-grained virtual partitioning" (FGVP). Instead of assigning one sub-query per node (as in [1]), our approach is to produce an initial number of virtual partitions greater than the number of participating cluster nodes. For example, if four cluster nodes are chosen to participate in a query processing we could produce sixty-four virtual partitions, generating sixty-four sub-queries. Then, sixteen sub-queries would be submitted to each node. The number of initial partitions should be much greater than the number of participating nodes, each partition corresponding to a small range of data items. We believe small sub-queries can bring many improvements to traditional virtual partitioning. They make it possible to avoid full scans on the virtually partitioned table. Besides, some queries demand temporary structures to store data while being processed. According to the amount of data, disk resources have to be employed. Small queries can exclusively use main memory structures.

In [10], small physical database fragments are also used for OLAP query processing with good results. A data fragmentation technique to be applied on fact tables is proposed, called multi-dimensional hierarchical fragmentation (MDHF). Its success solely depends on a good fragmentation. FGVP has the advantage of not requiring physical data fragmentation, facilitating the migration of applications from sequential environments, as in [4].

FGVP can provide a good basis for the introduction of dynamic load balancing in virtual partitioning. As proposed in [1], traditional virtual partitioning assigns each node exactly one sub-query. Normally, when a DBMS starts executing a query, it is not possible to externally stop it, modify the query and resume execution. So, there is no room for workload redistribution. Working with more sub-queries than cluster nodes, as FGVP proposes, could make sub-query reallocation possible, easing load balancing. The more partitions we have, the more opportunities to perform dynamic optimization.

Obviously, if there are too many virtual partitions, performance can degrade as more inter-process communication would become necessary. The problem is thus to determine the number of virtual partitions. By now, we use a static approach based on database statistics and DBMS-specific information, like the threshold after which the DBMS starts performing full table scans instead of using clustered indexes. As our goal is just to investigate the general behavior of FGVP, we employ this simple approach by now in spite of not thinking it is appropriate for cluster databases with black-box DBMSs. Alternatively, as database clusters and multi-database systems [7] are similar in many aspects, one can reuse techniques of predicting query execution costs in such environments [3, 6, 14].

4 Experimental Validation

To validate our solution, we implemented a prototype on a cluster system and ran experiments with the TPC-H benchmark. The cluster system has 16 nodes, each with 2 Pentium III 1 GHz processors, 512 Mb main memory, and a disk capacity of 40 Gb. Cluster nodes are interconnected through a 1 Gb/s Myrinet network. We use the PostgreSQL 7.3.4 DBMS running on Linux Mandrake 8.0. We generated the TPC-H database as specified in [12] with a database size of 1.96 Gb. The fact tables (orders and lineitem) have clustered indexes on their primary keys. We also built indexes for all foreign keys. As our goal is to deal with *ad-hoc* queries, no other optimization was performed. The database is replicated at each cluster node.

Our prototype is implemented in Java and some components like the Cluster Query Processor (CQP) and parts of the Node Query Processor (NQP) are implemented as Java RMI objects. Our implementation exploits multi-threading. Each query is processed by a different thread of the CQP and NQP. Result composition is done in parallel at each NQP. Only the final global result composition is done by CQP in one of the participating nodes. To maximize system throughput and avoid bottlenecks, sub-query submission and result composition are processed by separate threads.

In our experiments, each query was run several times. To ease result presentation (Figures 2(b), 3(a) and 3(b)), we normalized their mean response time, by dividing each mean response time by the greatest response time of its associated query. Response time was measured by the client application from the moment it submitted the query till the moment it received the final result. We use TPC-H queries Q1, Q12 and Q14, each corresponding to a different OLAP query type according to the classification in [2]. Q1 accesses only the largest fact table and performs many aggregate operations. Q12 accesses both fact tables with a join between them. Q14 performs a join between the largest fact table and one dimension table. We concentrate on those three queries because they are most representative of OLAP applications.

The partitioning is as follows. For Q1, it is based on l_orderkey since it is the first primary key attribute of lineitem table and has few tuples for each value. For Q12, we use primary virtual fragmentation for the orders table based on primary key o_orderkey and derived virtual fragmentation for the lineitem table based on foreign key l_orderkey. For Q14 which accesses only one fact table (lineitem), we employ the same strategy as for Q1.

Figure 2 (a) shows the response time improvement of fine-grained virtual partitioning (FGVP) over traditional virtual partitioning (VP) varying the number of nodes. With one node, VP performs like sequential execution because there is only one partition to deal with. In this case, Q12 and Q14 yield better performance with VP than with FGVP. By analyzing the query plans generated by PostgreSQL, we observed that lineitem table is fully scanned in VP while an index is used in FGVP. Each sub-query generated by FGVP accesses only a small number of tuples. So, PostgreSQL decides to execute each one using an index as it is not aware that each sub-query is part of a larger query. For both strategies, all lineitem tuples need to be processed since there is no index on the predicate attributes of the initial query. With a single node, a full scan is more efficient in this case. Q1 performs better with FGVP than with VP even when only one node is used. This is because it consumes more CPU resources than Q12 and Q14 as it has many aggregate operations. In our prototype, we implemented modules for collecting intermediate results (from the intermediate queries) and performing the required aggregations. With two CPUs per node, query processing (performed by DBMS) and partial result aggregation (performed by our middleware) can be parallelized. By producing a large number of sub-queries, FGVP takes full advantage of this characteristic. On the other hand, as VP produces only one sub-query per node, it does not benefit too much from it. Then, FGVP performs better. With two nodes, only Q12 still yields better performance with VP. However, Q12 and Q14 start outperforming with FGVP. From 4 to 16 nodes, FGVP outperforms VP significantly, yielding an improvement factor of 6 for Q12 and Q14 with 16 nodes. This is due to different query plans generated for sub-queries produced by each technique. For instance, let us analyze the 16-node case. For Q1, VP produces a full scan of the fact table which is processed by all nodes, thus making difficult to benefit from the parallel execution. With FGVP, the table is accessed through an index and small intermediate results are generated. For Q12, VP produces a merge join algorithm [7] while FGVP produces a fast main-memory nested loop join (thus avoiding I/O operations) because the partitions are small. For Q14, VP produces full scans of the lineitem table for each (large size) virtual partition while FGVP uses an index to access the lineitem table, thus reducing response time.

Figure 2 (b) shows how FGVP scales up with the number of nodes. Q1 and Q12 have almost linear speedup. Q14 has slightly smaller speedup. In Q1 and Q12, all tables are accessed through clustered indexes while in Q14, the part table (200,000 tuples) is accessed through a non-clustered index which is less efficient.

Figure 3 shows the performance of FGVP with different numbers of partitions with 8 and 16 nodes. We observe that the best number (among those that are being showed) varies from query to query. Incidentally in these experiments, this number is the same for both configurations when we consider the same queries. Other experiments not described here show variations for the same query according to the number of nodes employed. This shows the importance of a good partition size estimation and dynamic adjustment.

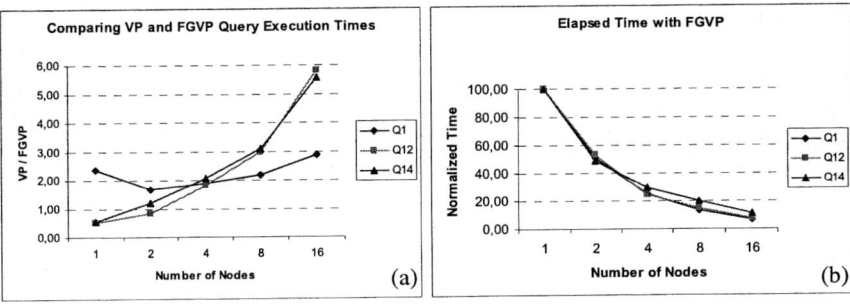

Fig. 2. Performance of VP versus FGVP

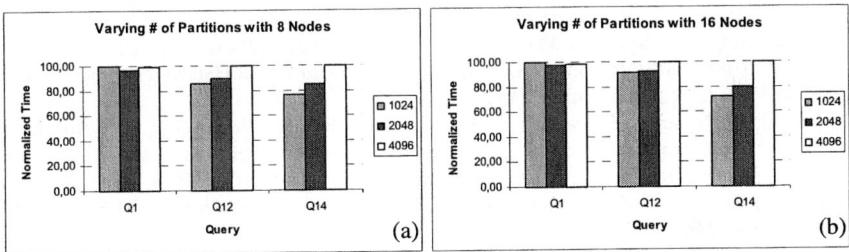

Fig. 3. FGVP with varying numbers of partitions

5 Conclusion and Future Work

In this paper, we proposed an efficient solution, called fined-grained virtual partitioning (FGVP), to OLAP parallel query processing in a database cluster. The idea behind FGVP is conceptually simple. Assuming replication of the database among the cluster nodes, queries are rewritten to deal with virtual partitions of the database. Unlike traditional virtual partitioning (VP), FGVP produces a number of sub-queries much larger than the number of nodes employed for query processing. Consequently, more light-weight sub-queries are generated, avoiding full table scans and expensive temporary disk-based structures. FGVP is a significant improvement over static virtual partitioning (VP). FGVP can also work with partially replicated databases.

Our database cluster has a shared-nothing architecture to provide for scale up to very large configurations. It supports black-box DBMS using non intrusive, simple techniques implemented by Java middleware. Thus, it can support any kind of relational DBMS. To validate our solution, we implemented a Java prototype on a 16-node cluster system and ran experiments with typical queries of the TPC-H benchmark. The results show that FGVP yields linear, and sometimes super-linear, speedup. With 16 nodes, FGVP outperforms VP by a factor of 6 which is excellent.

As a next step, we intend to introduce dynamic load balancing in FGVP. We also intend to investigate a DBMS-independent approach to calculate the number of sub-queries produced by FGVP.

Acknowledgements

Alexandre A.B. Lima would like to thank CAPES and Universidade do Grande Rio - Professor José de Souza Herdy for their support while visiting the University of Nantes during his doctoral studies.

References

1. Akal F., Böhm K., Schek H.-J., OLAP Query Evaluation in a Database Cluster: a Performance Study on Intra-Query Parallelism, *East-European Conf. on Advances in Databases and Information Systems (ADBIS)*, Bratislava, Slovakia, 2002.
2. Coulon C., Pacitti E., Valduriez P., Scaling up the Preventive Replication of Autonomous Databases in Cluster Systems, *VecPar Int. Conf.*, Valencia, 2004 (to appear).
3. Du W., Krishnamurthy R., Shan M.C., Query Optimisation in a Heterogeneous DBMS, *Int. Conf. on Very Large Data Bases*, 1992.
4. Gançarski, S., Naacke, H., Pacitti, E., Valduriez, P., Parallel Processing with Autonomous Databases in a Cluster System, *Int. Conf. on Cooperative Information Systems*, 2002.
5. Gorla N., Features to Consider in a Data Warehousing System, *CACM* 46(11), 2003.
6. Lu H., Ooi B., Goh C., On Global Multidatabase Query Optimization, *ACM SIGMOD Rec.* 21(4): 6-11, 1992.
7. Özsu T., Valduriez P., *Principles of Distributed Database Systems*, Prentice Hall, 1999.
8. PostgreSQL, http://www.postgres.org
9. Röhm U., Böhm K., Schek H.-J., OLAP Query Routing and Physical Design in a Database Cluster, *Int. Conf. on Extending Database Technology (EDBT)*, 2000.
10. Stöhr, T., Märtens, H., Rahm, E., Multi-Dimensional Database Allocation for Parallel Data Warehouses, *Proc. 26th Int. Conf. on Very Large Databases*, Cairo, Egypt, 2000.
11. The Project PowerDB, http://www.dbs.ethz.ch/~powerdb
12. TPC, TPC Benchmark™ H (Decision Support)), http://www.tpc.org/tpch
13. Valduriez P., Parallel Database Systems: open problems and new issues. *Int. Journal on Distributed and Parallel Databases*, 1(2), 1993.
14. Zhu Q., Larson P.A., A Query Sampling Method of Estimating Local Cost Parameters in a Multidatabase System, *Int. Conf. on Data Engineering (ICDE)*, Houston, Texas, 1994.

Efficient Parallel Hierarchical Clustering

Manoranjan Dash[1], Simona Petrutiu[2], and Peter Scheuermann[2]

[1] Department of Information Systems, School of Computer Engineering,
Nanyang Technological University, Singapore 639798
[2] Department of Electrical & Computer Engineering,
Northwestern University, Evanston, IL 60208*

Abstract. Hierarchical agglomerative clustering (HAC) is a common clustering method that outputs a dendrogram showing all N levels of agglomerations where N is the number of objects in the data set. High time and memory complexities are some of the major bottlenecks in its application to real-world problems. In the literature parallel algorithms are proposed to overcome these limitations. But, as this paper shows, existing parallel HAC algorithms are inefficient due to ineffective partitioning of the data. We first show how HAC follows a rule where most agglomerations have very small dissimilarity and only a small portion towards the end have large dissimilarity. Partially overlapping partitioning (POP) exploits this principle and obtains efficient yet accurate HAC algorithms. The total number of dissimilarities is reduced by a factor close to the number of cells in the partition. We present pPOP, the parallel version of POP, that is implemented on a shared memory multiprocessor architecture. Extensive theoretical analysis and experimental results are presented and show that pPOP gives close to linear speedup and outperforms the existing parallel algorithms significantly both in CPU time and memory requirements.

Keywords: hierarchical agglomerative clustering, partitioning, parallel algorithm, shared memory architecture.

1 Introduction

Hierarchical agglomerative clustering (HAC) is often used in various applications due to its capability to output a dendrogram showing all agglomerations. Unlike K-means and other types of clustering where objects are clustered into a given number of clusters, a dendrogram can be used to get any number of clusters. HAC algorithms are non-parametric, natural and simple in grouping objects, and capable of finding clusters of different shapes by using different similarity measures. However, they are limited in their application to real-world problems mainly due to high CPU time and memory complexities. Existing algorithms take $O(N^2 \log N)$ CPU time and require $O(N^2)$ memory. Parallel algorithms

---

* Research of the third author on this project was supported by NSF grant IIS-0325144.

are proposed to alleviate this limitation. Existing parallel algorithms either parallelize other clustering methods such as K-means (Dhillon and Modha [1]) and subspace clustering (Nagesh *et al.* [2]), or are not very efficient due to lack of performance enhancing partitioning [3].

In [4] we have shown that complexities of the existing sequential HAC algorithms can be reduced significantly by an efficient partitioning scheme without losing accuracy. The proposed methods are based on an observation that in HAC most iterations agglomerate very small clusters separated by very small dissimilarity. Only a small number of iterations towards the end agglomerate the large clusters. Using this observation a structure called *partially overlapping partitioning* (POP) divides the data into a number of overlapping cells. Analysis and experiments showed that POP-based sequential HAC algorithms reduce existing time and memory complexities by a factor close to the number of cells c.

In this paper we present parallel versions of POP, called pPOP. Due to the independent nature of each partitioned cell, parallelization is able to achieve similar reduction in time and memory complexities as POP, i.e., by a factor close to the number of cells c. We implement pPOP over a shared memory architecture. Experimental evaluations show that for large data sets pPOP obtains near linear speedup. In addition, for stored matrix implementations, pPOP results in a two order of magnitude improvement in computation time over the existing parallel HAC algorithms.

2   Background

Let us assume that there are N objects each with M attributes. We use real type data and Euclidean (L_2) distance to measure dissimilarity. Other distance measures, e.g. *Manhattan*, can be used (see [4]).

The 90-10 Rule: In an experiment we ran the centroid type HAC method over a 2-D data set with 100 clusters and some noise. In the centroid type, each cluster is represented by a centroid and the pair with the closest centroids is merged in each iteration. In Figure 1, we plot the closest pair distance for each iteration. Notice that most agglomerations except for a small portion towards the end have very small closest pair distance compared to the maximum closest pair distance. This maximum distance is taken over all agglomerations. If we plot the size of clusters merged in an iteration it also shows a similar plot. We experimented with many data sets having varying characteristics. For varying M, N (typically large – at least a few thousand objects), and K (number of clusters), the general trend is as follows: *if a majority of the objects are inside clusters then the shape of the distance plot is as shown in Figure 2*. We name this as '*90-10 rule*' to convey the idea that *in a dendrogram, most levels from the bottom merge pairs of very small clusters separated by a very small portion of the maximum closest pair distance*. The 90-10 rule extends to other HAC algorithms beyond the centroid method for both the geometric and the graph metrics. For space constraints, we restrict all discussions in this paper to centroid method.

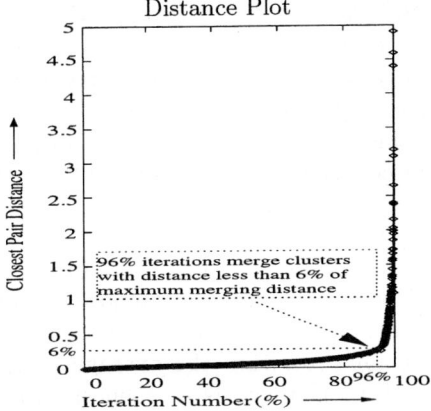

Fig. 1. An important property of HAC: the distance plot shows that the closest pair distance is very small even until last stage of agglomeration.

See [4] for detailed discussion on the 90-10 rule and other metrics. Next we show how to exploit this inherent characteristic of HAC.

2.1 Partially Overlapping Partitioning (POP)

An axis-parallel POP divides the data-space uniformly into c number of overlapping cells. The overlapping region is called δ-region where δ is the overlapping distance between two cells. Figure 2 depicts the axis-parallel POP. For the centroid metric (and other geometric metrics), if the representative point of a cluster falls in a δ-region then each affected cell that contains this δ-region holds it, otherwise only one cell holds it.

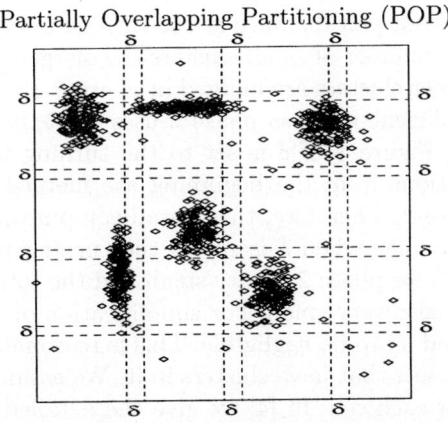

Fig. 2. The 90-10 rule is exploited by POP for efficient HAC.

Before discussing POP any further, we very briefly describe some existing HAC algorithms. HAC algorithms are mainly of two types: stored matrix (e.g., dissimilarity matrix and priority queues) and stored data (e.g., nearest neighbor). The *dissimilarity matrix method* stores dissimilarities between each pair of clusters. When a pair is merged dissimilarities are computed for the new cluster and the matrix is updated. The memory complexity of this method is $O(N^2)$ and the time complexity is $O(N^3)$. In the *priority queue* method a heap-based priority queue is maintained for each cluster. Because a priority queue requires $O(\log n)$ time for each insert and delete operation for n elements, the time complexity reduces to $O(N^2 \log N)$ although the memory complexity stays at $O(N^2)$. The *nearest neighbor array* method maintains nearest neighbors for each cluster in an array. If after each iteration the average number of clusters whose nearest neighbors need to be changed is α, then the time complexity reduces to $O(\alpha N^2)$ and the memory complexity reduces to $O(N)$. An upper bound for α is $2(3^M - 2)$. When memory is enough to store $O(N^2)$ dissimilarities, stored matrix algorithms are preferred as they do fewer computations. Otherwise, the stored data type is preferred.

2-Phase Algorithm: In [4] we proposed a new 2-phase algorithm for HAC based on the axis-parallel POP. In *phase 1* clusters are partitioned into c overlapping cells. The basic idea is that in each iteration the closest pair is found for each cell and from those the overall closest pair is found. If the overall closest pair distance is less than δ then the pair is merged and the priority queues (or the dissimilarity matrix or the nearest neighbor array) of only the container cell are updated. If the closest pair or the merged cluster is in a δ-region then the priority queues of the affected cells are also updated. *Phase 1* terminates when the closest pair distance exceeds δ. *Phase 2* merges the remaining clusters of *phase 1* using the existing algorithm, thus completing the dendrogram. **Accuracy**: POP in *phase 1* ensures that *any pair with distance less than δ must reside together in at least one cell*. Hence, as *phase 2* is the existing algorithm itself, the 2-phase algorithm guarantees the correct dendrogram. **Complexity Analysis**: By setting δ to the closest pair distance at the turning point of the distance plot (see Figure 1), a large number of small clusters are merged in *phase 1* while only a small number of larger clusters are merged in *phase 2*. Recall that phase 1 uses POP which is very efficient whereas phase 2 uses the existing algorithm which is not so efficient. In Figure 1, if δ is set to the turning point of the distance plot, 96% agglomerations from the beginning are merged in phase 1 and the remaining 4% in phase 2. Therefore, the overall computational time is reduced drastically. So, we see that when δ is set to the turning point, the number of clusters remaining (k') for phase 2 is very small and the total number of clusters in the δ-region ($|\delta|$) is also very small. For simplification of the complexity analysis, we consider k' and $|\delta|$ to be negligible. This is reasonable because the 90-10 rule holds for all data sets that have clusters in it. We assume equal cell size and equal δ-region size for each cell. In [4] we give the detailed complexity analysis comparison between the existing and the 2-phase algorithms. Following is a brief overview of that. Stored matrix type that requires $O(N^2)$ memory now requires

$O(\frac{N^2}{c})$ in the 2-phase algorithm. Hence memory is reduced by a factor close to c. Because of this reduction, the 2-phase dissimilarity matrix algorithm, whose time complexity is dominated by the time required to create the matrix, enjoys a reduction by a factor close to c. The time complexity of the priority queue algorithm is dominated by the update effort required to maintain the priority queues. After each agglomeration of the closest pair, the priority queues of all other clusters are updated. But in the 2-phase algorithm this effort is restricted only to the cell that holds the closest pair, and if it happens to be in a δ-region then it is restricted only to the affected cells. So after simplification the reduction factor is $\log_{\frac{N}{c}} N \times c$, i.e., the time complexity reduces from $O(N^2 \log N)$ to $O(\frac{N^2}{c} \log \frac{N}{c})$. In stored data type there is no reduction in the memory complexity of $O(N)$. The time complexity is dominated by the time required to update the nearest neighbors of the affected clusters. For the existing algorithm the time required to find the nearest neighbor of one affected cluster is $O(N)$ but for the 2-phase algorithm it is $O(\frac{N}{c})$. So, the overall reduction factor is close to c.

Setting δ and c – Nested Algorithm: The performance of the 2-phase algorithm depends on c and δ. As shown in the distance plot of Figure 1, there exists an ideal δ at the turning point at which the total time taken by the 2-phase algorithm is minimum. But it is not straightforward to compute. So, we adopted a nested approach where in the beginning POP partitioning starts with a very small δ and gradually increases it until a few or just one cluster remain. As δ increases, c which is set initially to a high value, is gradually reduced. *Accuracy* of this nested algorithm is assured from the accuracy of the 2-phase algorithm. Experiments show that the nested algorithm is more efficient than the 2-phase algorithm even when δ is set ideally for the 2-phase algorithm. For example, for the data set described in Section 2, the minimum time for the 2-phase algorithm is 125.4 cpu sec while the nested algorithm takes only 57.8 cpu sec.

Higher Dimensional Data: The above discussion focuses on 2-D data. For higher dimensions we proposed a very efficient data structure as a replacement for the axis-parallel partitioning. Due to space constraint we limit the scope of this paper to 2-D and refer the interested reader to [4].

3 pPOP Algorithms

Parallel HAC algorithms have been studied by Li [5], Li and Fang [6], Olson [3], and Wu et al. [7]. The common feature of these algorithms is: for 'stored matrix' type the task of computing and maintaining $O(N^2)$ dissimilarities is divided among the processors, whereas for 'stored data' type the task of computing and maintaining the $O(N)$ nearest neighbors is divided among the processors. For example, Olson used p processors to reduce the time complexity of the dissimilarity matrix method to $O(\frac{N^3}{p})$ and that of the priority queue method to $O(\frac{N^2 * \log N}{p})$ [3]. The time complexity for the nearest neighbor array method reduces to $O(\frac{\alpha*N^2}{p})$. These algorithms are not very efficient because they still require $O(N^2)$ total memory for 'stored matrix' type, and in each iteration they require to update all the priority queues or dissimilarity matrix. For 'stored data' type the existing methods need to check all the clusters after each agglomeration to determine whether the newly merged cluster is nearer than the previous nearest. So, the reduction in these parallel algorithms is mostly because of parallelization, but not due to efficient partitioning.

The advantage of pPOP is that each cell is sufficient by itself, and hence parallelization benefits by dividing the task of creating and maintaining the dissimilarities or priority queues or nearest neighbors of each *cell* among the processors. This reduces the total computation of searching for the closest pair and maintaining the data structure drastically. Below we give the complexities of sequential, existing parallel and pPOP algorithms. For complexity analysis we select the 2-phase algorithm of the stored matrix type since, as we shall show later, this algorithm achieves larger speedups compared to the existing algorithms. As before, we assume equal cell sizes, negligible $\|\delta\|$ size, and negligible phase 2 time. Among existing algorithms, those described by Olson [3] are selected. The number of processors is denoted by p.

Table 1. Comparison of time complexities of sequential, existing parallel, and pPOP algorithms. RF - Reduction Factor ($= \frac{ExistingParallel}{pPOP}$).

Priority Queues	Sequential	Existing Parallel	pPOP	RF
1. Create priority queues	$O(N^2)$	$O(\frac{N^2}{p})$	$O(\frac{N^2}{cp})$	
2. for $n = N$ to 2	$O(N)$	$O(N)$	$O(N)$	
3. find smallest distance	$O(n)$	$O(\frac{n}{p})$	$O(\frac{n}{p})$	
4. merge and update P	$O(n*\log n)$	$O(\frac{n*\log n}{p})$	$O(\frac{n*\log \frac{n}{c}}{p})$	
Overall	$O(N^2 * \log N)$	$O(\frac{N^2*\log N}{p})$	$O(\frac{N^2*\log \frac{N}{c}}{p})$	$\frac{\log N}{\log \frac{N}{c}}$
Overall (Dissimilarity Matrix)	$O(N^3)$	$O(\frac{N^3}{p})$	$O(\frac{N^3}{cp})$	c

In Table 1 (priority queues) step 1 of pPOP computes priority queues in $O(\frac{N^2}{cp})$ time. Recall that pPOP reduces the memory by a factor of c, i.e., $O(\frac{N^2}{c})$. pPOP divides the total computation for the c cells among p processors, and hence, assuming no synchronization delays the complexity becomes $O(\frac{N^2}{cp})$. Step 4 updates the priority queues of the affected clusters. In pPOP a priority queue holds $\frac{N}{c}$ elements in the beginning. Hence, due to parallelization the total time complexity of this step is $O(\frac{n*\log \frac{n}{c}}{p})$. So, the overall reduction factor is $\frac{\log N}{\log \frac{N}{c}}$. Table 1 shows the overall complexities for the dissimilarity matrix type as well. It has a reduction factor close to c. The memory requirement for priority queues and dissimilarity matrix types is reduced by a factor close to c. For the nearest neighbor type, the gain of pPOP over the existing parallel algorithms cannot be obtained directly from the complexity analysis. For the step where each cluster is checked to find whether it is affected by the agglomeration, pPOP needs to do it for one (or a few, if in $|\delta|$-region) cell whereas the existing algorithm needs to do it for all clusters. Similarly, the existing algorithm needs to check all the clusters to find the new nearest neighbor of each affected cluster. But pPOP requires only the container cell to be checked. Experimental results in the next section show that pPOP outperforms the existing algorithms substantially for all the above three types of HAC.

4 Experimental Results

We performed a number of experiments to study the performance and scalability of our proposed pPOP algorithms. Both stored matrix (priority queues) and stored data (nearest neighbors) types of pPOP were implemented using the 2-phase algorithm. For comparison purposes we implemented the corresponding existing parallel algorithms, hereby denoted as existing algorithms. These are described in [3]. The performance was measured in terms of CPU time, memory space and speedup. We experimented using several real, benchmark, and artificial data sets. Due to space constraint we show the results over an artificial data set that is used in [8]. Other results are available from www.ece.northwestern.edu/~manoranj/research.html. The experiments were run on the SGI Origin2000 multiprocessor system which is a shared memory machine consisting of 8 R10000 CPUs running at clock rate of 195MHs. The secondary cache size is 4MB. We used OpenMP which is an API for directed based parallel programming applications in a shared memory environment [9]. We decided to use it because it is designed for fine-grained parallelism, which was predominant in our algorithm.

The pPOP implementation in OpenMP uses guided self scheduling clause in the assignment of iterations to threads, i.e., processors. During each iteration of HAC each processor is assigned in turn a chunk of cells to work on, with the chunk size being reduced as we proceed with the iteration. After an iteration is finished, a critical region is established in order to find the overall closest pair of clusters and merge them. The priority queues of the cells affected by the agglomeration can be updated in parallel.

In Figure 3 we show the results over the synthetic data set whose size varies from 3K to 60K. The existing stored matrix algorithms require $O(N^2)$ memory, hence we could experiment only with a data size up to 5K; on the other hand for pPOP we report results for data sets up to 30K. The number of processors varies from 1 to 8. In Figure 3 (a-b) we report the speedups of pPOP. Although the speedup of pPOP is small for smaller data sets, we observe that for larger data sets (30K or higher) the speedup of pPOP improves substantially and approaches linear speedup for data sets of 60K. Figure 3 (c-d) gives the relative speedup of pPOP over the existing algorithm. pPOP is always superior over the existing algorithm because of its efficient partitioning, and independent nature of each cell. The relative speedup increases with data size. Among stored matrix and stored data types, pPOP's performance is much better for stored matrix. It

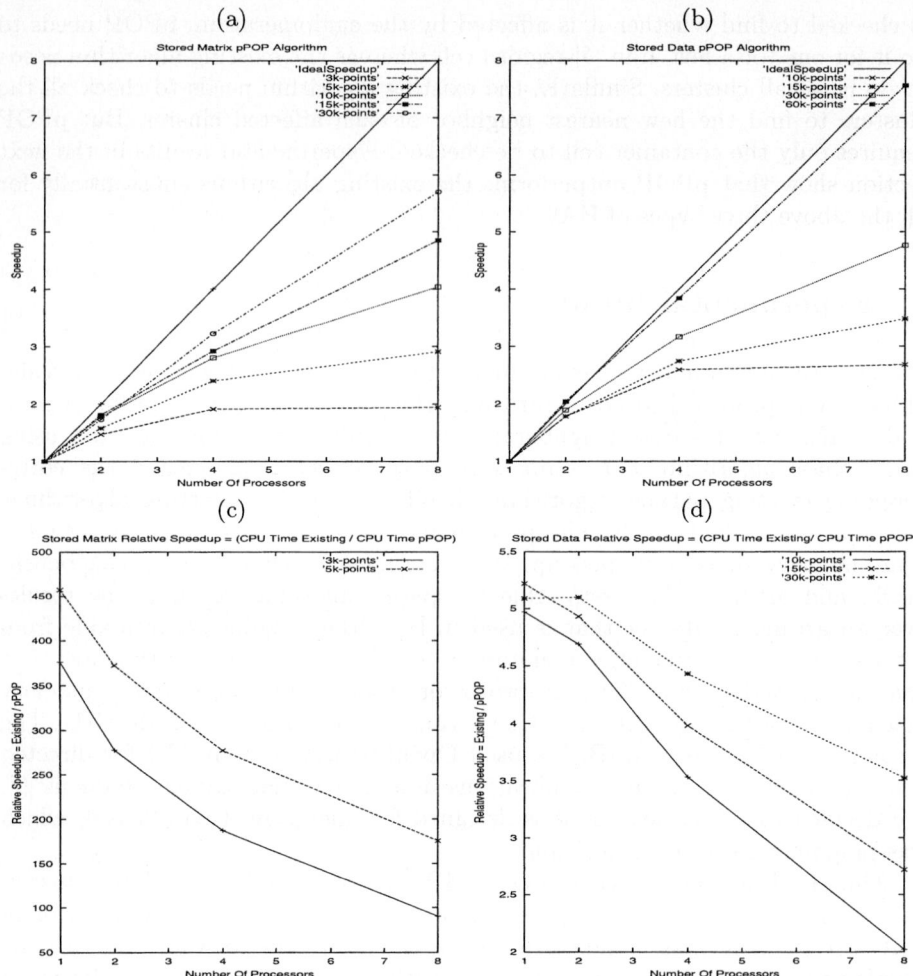

Fig. 3. Synthetic data results: For stored matrix and stored data types, and for varying #processors (1 to 8), (a-b) show performance of pPOP, and (c-d) show $RelativeSpeedUp = \frac{Existing}{pPOP}$.

achieves a two order of magnitude improvement in computation time over the existing algorithm.

As shown in Figures 3 (c-d) the relative speedup, $\frac{ExistingCPU}{pPOPCPU}$, decreases as the number of processors increases. This is due to the fact that for a small number of cells, when the number of processors is increased, some processors end up working on cells containing a very small number of clusters, and will therefore spend a lot of time being idle when they are done with the computation in a given iteration. However, as the data set size increases and/or the number of clusters increases, load balancing among the processors becomes better. This phenomena can be observed in our figures. Although for both 3K and 5K sizes for stored matrix type the relative speedup drops by approximately the same amount (285) when the number of processors increased from 1 to 8, the noticeable fact is that relative speedup for 1 processor for 3K size is 375, but that for 5K size is 460. That is to say as the number of processors increased, with increasing size of data the *rate of drop in speedup* decreased. Although due to the high memory requirement of the existing parallel algorithms we could not test for higher data sizes, we postulate that for larger data sets this trend of reduction in relative speedup for more processors will continue to slow down further.

We compared the memory for stored matrix type. For 3K and 5k pPOP reduced the memory requirement by a factor of 97 and 189 respectively. For stored data type both algorithms require similar amount of memory.

5 Conclusion and Future Directions

In this paper we proposed pPOP for efficient parallel HAC. Analysis and experiments showed that, for both stored matrix and stored data types, pPOP outperforms the existing algorithms significantly both in CPU time and memory requirements. This is achieved by exploiting a 90-10 rule of HAC which states that in a dendrogram, most levels from the bottom merge pairs of very small clusters separated by a very small portion of the maximum closest pair distance. The data space was partitioned by partially overlapping cells each of which could be processed independent of other such cells without affecting accuracy. Future work includes parallelizing the high-dimensional data structure.

References

1. Dhillon, I.S., Modha, D.M.: Large-scale parallel data mining. Lecture Notes in Artificial Intelligence **1759** (2000) 245–260
2. Nagesh, H., Goil, S., Choudhary, A.: PMAFIA: A scalable parallel subspace clustering algorithm for massive datasets. In: Proc. International Conference on Parallel Processing. (2000) 21–24
3. Olson, C.F.: Parallel algorithms for hierarchical clustering. Parallel Computing **21** (1995) 1313–1325
4. Dash, M., Liu, H., Scheuermann, P., Tan, K.L.: Fast hierarchical clustering and its validation. Data and Knowledge Engineering **44(1)** (2003) 109–138
5. Li, X.: Parallel algorithms for hierarchical clustering and cluster validity. IEEE Transactions on Pattern Analysis and Machine Intelligence **12** (1990) 1088–1092
6. Li, X., Fang, Z.: Parallel clustering algorithms. Parallel Computing **11** (1989) 275–290
7. Wu, C.H., Horng, S.J., Tsai, H.R.: Efficient parallel algorithms for hierarchical clustering on arrays with reconfigurable optical buses. Journal of Parallel and Distributed Computing **60** (2000) 1137–1153
8. Zhang, T., Ramakrishnan, R., Livny, M.: BIRCH: An efficient data clustering method for very large databases. In: Proceedings of ACM SIGMOD Conference on Management of Data, Montreal, Canada (1996) 103–114
9. Chandra, R., Dagum, L., Kohr, D., Maydan, D., McDonald, J., Menon, R., eds.: Parallel Programming in OpenMP. Morgan Kaufmann Publishers (2000)

Parallelizing EM Clustering Algorithm on a Cluster of SMPs

Leonid Glimcher and Gagan Agrawal

Department of Computer and Information Sciences
Ohio State University, Columbus, OH 43210
{glimcher,agrawal}@cis.ohio-state.edu

Abstract. In this paper, we report on parallelization of the EM clustering algorithm using the FREERIDE middleware developed in our prior work. FREERIDE is based upon the observation that the processing structure of a large number of data mining algorithms involves generalized reductions. FREERIDE offers a high-level interface and support both distributed memory and shared memory parallelization, besides efficient execution on disk-resident datasets. We show how the main processing loops in both the E and M steps of the EM algorithm essentially involve a generalized reduction, and therefore, the algorithm can be parallelized using FREERIDE.

1 Introduction

As the amount of data available for analysis has been increasing rapidly, scalability of data mining implementations has become an important issue. For dealing with large datasets, it's important to both parallelize the algorithms, and implement them to execute efficiently on disk-resident datasets. However, developing applications that can execute in parallel as well as efficiently process disk-resident data sets is a difficult task.

In our prior work, we have developed a middleware called FREERIDE (FRamework for Rapid Implementation of Datamining Engines) [4–6]. It has been used successfully for a number of common mining algorithms, including apriori and FP-tree based association-mining, k-means clustering, decision tree construction, k-nearest neighbor search, as well as a scientific feature extraction algorithm. FREERIDE is based upon an observation that the structure of parallel algorithms for the above problems essentially involves a generalized reduction. Besides allowing parallelization on both distributed memory and shared memory settings, FREERIDE allows efficient execution on disk-resident interface.

This paper focuses on the use of FREERIDE for creating a parallel and scalable implementation of the Expectation Maximization (EM) clustering algorithm. EM is a popular technique for clustering and has been widely used in machine learning and computer vision communities [2, 9, 8]. We show how each iteration of EM essentially involves two generalized reduction loops. We describe how FREERIDE can be used for parallelizing EM on cluster of SMPs and executing it efficiently on disk-resident datasets. Our implementation is experimentally evaluated on a cluster of SMPs, using several different datasets.

Input: k, # of clusters, $Y = \{y_1 \ldots y_n\}$ set of n $p-dimensional$ points,
ϵ, a tolerance for loglikelihood, $maxiterations$, maximum number of iterations.
Output: C, R, W, the matrices containing the updated mixture parameters.
X, a matrix with cluster membership probabilities.
Initialize: Set initial values for C, R, and W (random or approximate solutions)

WHILE: $\delta(llh) > \epsilon$ and $maxiterations$ has not been reached
  DO E and M steps
  E step
    $C' = R' = W' = llh = 0$
    for $i = 1$ to n
      $sump_i = 0$
      for $j = 1$ to k
        $\delta_{ij} = (y_i - C_j)^t R^{-1}(y_i - C_j)$
        $p_{ij} = \frac{w_j}{(2\pi)^{p/2}|R|^{1/2}} exp(-\frac{1}{2}\delta_{ij})$
        $sump_i = sump_i + p_{ij}$
      endfor
      $x_i = p_i/sump_i$, $llh = llh + ln(sump_i)$
      $C' = C' + y_i x_i^t$, $W' = W' + x_i$
    endfor
  M step
    for $j = 1$ to k
      $C_j = C'_j/W_j$
      for $i = 1$ to n $R' = R' + (y_i - C_j)x_{ij}(y_i - C_j)^t$ endfor
    endfor
  $R = R'/n$, $W = W'/n$

Fig. 1. Sequential Pseudo-code for the Expectation Maximization Algorithm

As compared to other efforts that reported parallelization of EM algorithm [3, 7, 1], this paper makes the following two important contributions. First, we combine both distributed memory and shared memory parallelization, and support execution on disk-resident datasets. Second, we show that EM can be parallelized using a middleware that has been used for a number of other data mining algorithms.

2 The EM Clustering Algorithm

This section gives a brief overview of the Expectation Maximization (EM) clustering algorithm. The EM algorithm was first introduced in the seminal paper [2]. EM is a distance-based based algorithm that assumes the data set can be modeled as a linear combination of multivariate normal distributions. There are several advantages to using EM for clustering data: it has a strong statistical basis, it is robust to noisy data, it can accept the desired number of clusters as input, it provides a cluster membership probability per point, it can handle high dimensionality and it converges fast given a good initialization [8].

The pseudo-code for EM algorithm is presented in Figure 1. Complete details of the statistical basis and the algorithm itself can be found in [2, 9, 8]. The goal of the EM algorithm is to estimate the means C, the covariances R and the mixture weights W of a Guassian probability function [8]. The algorithm works by successfully improving the solution found so far. The algorithm stops when the quality of the current solution becomes stable; this measured by a monotonically increasing statistical quantity called *loglikelihood*.

3 FREERIDE Middleware and Parallelization Techniques

This section gives a brief background on the functionality and interface of the FREERIDE system [4–6].

The FREERIDE (Framework for Rapid Implementation of Datamining Engines) is based on the observation that a number of popular data mining algorithms share a relatively similar structure. Their common processing structure is essentially that of *generalized reductions*. During each *phase* of the algorithm, the computation involves reading the data instances in an arbitrary order, processing each data instance, and updating elements of a *reduction object* using associative and commutative operators.

In a distributed memory setting, such algorithms can be parallelized by dividing the data items among the processors and replicating the reduction object. Each node can process the data items it owns to perform a local reduction. After local reduction on all processors, a global reduction can be performed. In a shared memory setting, parallelization can be done by assigning different data items to different threads. The main challenge in maintaining the correctness is avoiding race conditions when different threads may be trying to update the same element of the reduction object. We have developed a number of techniques for avoiding such race conditions, particularly focusing on the memory hierarchy impact of the use of locking. However, if the size of the reduction object is relatively small, race conditions can be avoided by simply replicating the reduction object.

Our middleware incorporates techniques for both distributed memory and shared memory parallelization and offers a high-level programming interface. For distributed memory parallelization, the interface requires the programmers to specify pairs of local and global reduction functions, and an iterator that invokes these and checks for termination conditions. For shared memory parallelization, the programmers are required to identify the reduction object, and also the updates to those, and also specify a way in which different copies of the reduction object could be merged together.

A particular feature of the system is the support for efficiently processing disk-resident datasets. This is done by aggressively using asynchronous operations for reading *chunks* or disk-blocks from the dataset.

In the past, FREERIDE has been used for a number of well-known data mining algorithms, including apriori and FP-tree based association mining, k-means clustering, decision tree construction, nearest neighbor search, and a scientific feature mining application. The details of the functionality and results from evaluation of the system are available in our earlier publications [4–6].

4 Parallelizing EM with FREERIDE

In this section, we describe the parallelization of EM using FREERIDE. For using FREERIDE, an application must comprise of one or more steps involving generalized reductions. Therefore, we initially discuss how the EM algorithm meets this requirement.

As stated previously, FREERIDE supports both distributed memory and shared memory parallelization, while also enabling scaling to disk-resident datasets. Our discussion will initially focus on distributed memory parallelization, and will then subsequently describe shared memory parallelization and scaling to disk resident datasets.

4.1 EM Algorithm and Generalized Reductions

Consider the sequential EM algorithm shown in Figure 1. Here n is the number of points or data instances, p is the number of dimensions, and k is the number of clusters that are desired. Let us consider the arrays that are involved. Y is a $n \times p$ array that is input to the algorithm. There are four output arrays, C, R, W and X, whose sizes are $p \times k$, $p \times p$, $k \times 1$, and $n \times k$, respectively. The algorithm also involves three temporary arrays, C', R', and W'. They have the same sizes as C, R, and W, respectively.

Let us consider the parallelization of this algorithm. The input array Y and the output array X can be partitioned between the nodes. Based upon these partitions, the loops iterating over the data instances can be parallelized. Both the E and M steps involve such loops. Based upon the computations performed in these loops, we can see that no communication is required for the arrays X and Y. However, the arrays C, R, W, C', R', W' cannot be partitioned, but instead should be replicated on all processors. However, now the question is, can we correctly update these arrays when the loops are executed in parallel.

In the E step, C' and W' are the two arrays that are updated. It can be easily seen that these updates are done using an associative and commutative operation (addition). Also, the scalar llh is also updated using an associative and commutative operation. Therefore, the E step is essentially a generalized reduction.

Now, consider the nested loop involved in the M step. The array C is computed from the arrays C' and W. If these two arrays are replicated on each node, then the computation for C can be performed on each node. Again, we can see that the array R' is updated using associative and commutative operations. Thus, this loop is also a generalized reduction.

To summarize, the parallel loops in the EM algorithm match the structure of generalized reductions and FREERIDE is well suited for parallelizing this algorithm.

4.2 Distributed Memory Parallelization

In view of the discussion above, it is easy to see how the EM algorithm can be parallelized on distributed memory machines.

The input data instances (the array Y) are distributed between the nodes. The arrays C, R, and W, whose initial values are provided by the user, are replicated on all nodes. The arrays C', R', W' are allocated and initialized on all nodes. The E step is carried out on all nodes, using the set of data instances they have. After such local processing, a *global combination* is done. In this step, all nodes communicate their values of C', W', and llh to one node, which aggregates these values and broadcasts them to all nodes.

Next, each node computes a new value of C, using the updated value of C' and W. Note that W has not been modified since the start of this iteration, so all nodes have the same and correct value. Then, the nested loop is again executed on each node's data instances. Another round of global combination is done to compute the final value of R'.

Finally, each node computes a new value of R and W. Thus, at the end of the M step, each node has an updated value of C, R, W, and llh. Using llh, each node independently decides if another iteration is needed. If another iteration is needed, the values of C, R, and W are used in E and M steps.

The above steps can be implemented easily using the FREERIDE interface. Two different local processing functions are created, corresponding to the processing in E and M steps, respectively. There is one global combination function associated with each of these. The global function associated with the E step is responsible for aggregating C', W', and llh. Similarly, the global function associated with the M step is responsible for aggregating R'. The iterator is responsible for checking the terminal conditions, and performing all computations that are replicated on all the nodes.

4.3 Shared Memory Parallelization

If multiple CPUs are available within a node, each of these can be used for local processing. A set of data instances can be assigned to each thread, which can perform the computation associated with these data instances. The main challenge in maintaining correctness arises because of race conditions when multiple threads may try to update the same array (C' and W' for E step, and R' for the M step).

In our prior work, we have reported on a number of techniques for avoiding race conditions that are implemented within the FREERIDE framework [5,6]. One of the techniques is *full replication*, where the updated arrays or datastructures are replicated. Another set of techniques is based upon the use of locking.

The size of the arrays that are updated by different threads is quite small in the case of EM algorithm, unless both the number of dimensions and the number of clusters is quite large. Therefore, full replication can work without a significant memory overhead and is used by our implementation. In using this technique, a separate copy of updated arrays is allocated and initialized for each thread. Each thread can update its copy without any possibility of race conditions. Before global combination, the different copies are aggregated.

4.4 Scaling to Disk-Resident Datasets

The existing support in FREERIDE can be used for scaling the EM algorithm to large datasets that do not fit into main memory. The data instances are divided into chunks. The middleware provides support for reading these chunks using asychronous operations, and applying the local processing function on each chunk.

5 Experimental Results

In this section, we evaluate the performance of our EM implementation. We conducted a number of experiments with the following two goals: 1) Studying of parallel scalability of our implementation, in both distributed and shared memory settings, and 2) Evaluating the scalability with increasing dataset size, i.e., does the execution time remain proportional to the dataset size as the dataset becomes disk resident.

Our experiments were conducted on a cluster of 700 MHz Pentium machines. The nodes in the cluster are connected through Myrinet LANai 7.0. The memory on each node is 1GB, but contention for main memory space from other processes was created, to make sure that the datasets were disk resident. We used 3 datasets each one containing a different number of 10-dimensional points. The datasets were: 1) A 253 MB dataset, containing $3.3*10^6$ points to be clustered, 2) A 498 MB dataset, containing $6.2*10^6$ points to be clustered, and 3) A 996 MB dataset, containing $12.5*10^6 points$ to be clustered.

Each one of there datasets was partitioned into 8192 chunks, thus leaving the number of chunks across the dataset the same, but growing the chunk size.

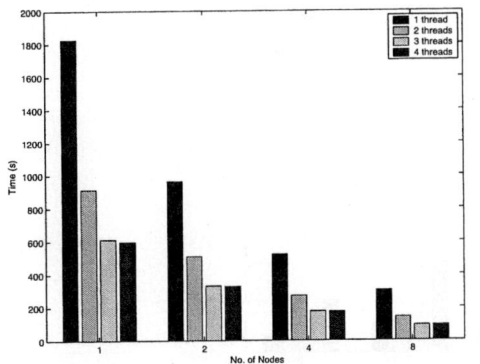
Fig. 2. Shared and distributed memory parallel performance on a small dataset (253 MB)

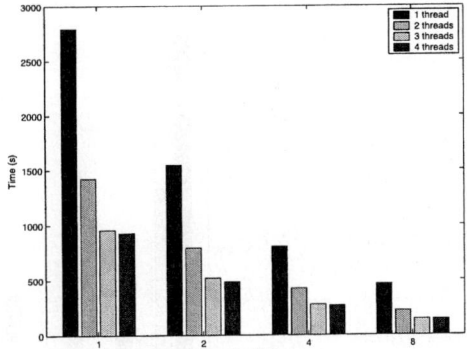
Fig. 3. Shared and distributed memory parallel performance on a medium dataset (498MB)

Let us consider speed-ups achieved through distributed memory parallelization on the smallest dataset first (Dataset size = 253MB, results summaraized in Figure 2). We report the speedup by increasing the number of nodes, but keeping the number of threads fixed. On 2 nodes the speed-ups were 1.76 for 1 thread, 1.79 for 2 threads, 1.84 for 3 threads, and 1.89 for 4 threads. On 4 nodes the speed-ups were 3.47 for 1 thread, 3.36 for 2 threads, 3.45 for 3 threads, and 3.54 for 4 threads. On 8 nodes the speed-ups were 6.00 for 1 thread, 6.44 for 2 threads, 6.65 for 3 threads and 6.53 for 4 threads. The results are not quite linear, but yet speed-ups are significant, achieving parallel efficiency between 75% and 85%. Now, we focus on speed-ups achieved through shared memory parallelization on the same dataset (253MB). With 2 threads running, the speed-ups were 1.96 for 1 node, 1.99 for 2 nodes, 1.90 for 4 nodes, and 2.10 on 8 nodes. With 3 threads running, the speed-ups were 2.88 on 1 node, 3.01 on 2 nodes, 2.86 on 4 nodes, and 3.19 on 8 nodes. The speedups are linear (and in some cases super-linear), because the reduction object is small enough to be cached in some of the instances. With 4 threads running the speed-ups were 2.94 for 1 node, 3.16 for 2 nodes, 2.99 for 4 nodes, and 3.20 for 8 nodes. Thus, adding the 4th thread doesn't result in additional speed-ups.

The results from the 498 MB dataset are summarized in Figure 3). On 2 nodes, the speed-ups were 1.80 for 1 thread, 1.79 for 2 threads, 1.84 for 3 threads, and 1.91 for 4 threads. On 4 nodes the speed-ups were 3.47 for 1 thread, 3.37 for 2 threads, 3.46 for 3 threads, and 3.50 for 4 threads. On 8 nodes the speed-ups were 6.02 for 1 thread, 6.46 for 2 threads, 6.65 for 3 threads and 6.46 for 4 threads. With 2 threads running, the speed-ups were 1.96 for 1 node, 1.95 for 2 nodes, 1.89 for 4 nodes, and 2.11 on 8 nodes. With 3 threads running, the speed-ups were 2.92 on 1 node, 2.98 on 2 nodes, 2.91 on 4 nodes, and 3.22 on 8 nodes. Once again, the 4th thread creates CPU contention and, therefore, doesn't result in any additional significant speed-up.

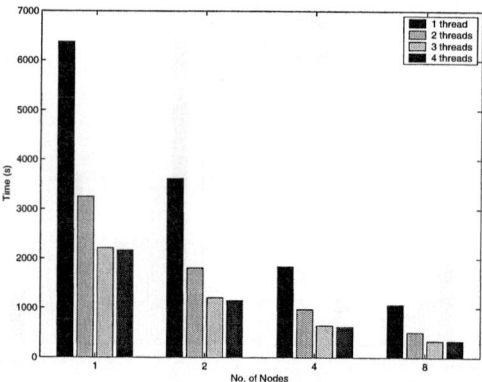

Fig. 4. Shared and distributed memory parallel performance on a large dataset (996MB)

The results from the largest dataset are summarized in Figure 4. On 2 nodes the speed-ups were 1.76 for 1 thread, 1.79 for 2 threads, 1.84 for 3 threads, and 1.89 for 4 threads. On 4 nodes the speed-ups were 3.47 for 1 thread, 3.37 for 2 threads, 3.45 for 3 threads, and 3.54 for 4 threads. On 8 nodes the speed-ups were 6.00 for 1 thread, 6.44 for 2 threads, 6.65 for 3 threads and 6.53 for 4 threads. With 2 threads running, the speed-ups were 1.96 for 1 node, 1.99 for 2 nodes, 1.90 for 4 nodes, and 2.10 on 8 nodes. With 3 threads running, the speed-ups were 2.87 on 1 node, 3.01 on 2 nodes, 2.86 on 4 nodes, and 3.18 on 8 nodes.

These experiments demonstrate that in the shared memory parallelization setting our implementation scales well up to 3 threads on 4 CPU machines. Also, as the size of a dataset has grown, the execution time has grown in a manner proportional to the dataset size, so the implementation scales well with increased dataset size, which is very important for a data intensive application. Lastly, the above experiments also demonstrate that a significant (although not strictly linear) speed-up is achieved by our implementation when parallelized in the distributed memory setting. In a cluster with 8 nodes, the parallel efficiency varied between 75 and 85 percent.

6 Conclusions

We have reported on parallelization of the EM clustering algorithm using the FREERIDE middleware developed in our prior work. We have shown how the main processing loops in both the E and M steps of the EM algorithm essentially involve a generalized reduction, and therefore, the algorithm can be parallelized using FREERIDE. As compared to other efforts that reported parallelization of EM algorithm, this paper has made two important contributions. First, we combine both distributed memory and shared memory parallelization, and support execution on disk-resident datasets. Second, we show that EM can be parallelized using a middleware that has been used for a number of other data mining algorithms.

References

1. P.E. Lopez de Teruel, , J. M. Garcia, and M. Acacio. A Parallel Algorithm and Its Application to Computer Vision. In *Proceedings of PDPTA*, 1999.
2. Arthur Dempster, Nan Laird, and Donald Rubin. Maximum Likelihood Estimation from Incomplete Data via the EM Algorithm. *Journal of the Royal Statistical Society*, 39(1):1–38, 1977.
3. G. Forman and B. Zhang. Distributed data clustering can be efficient and exact. *SIGKDD Explorations*, 2(2), December 2000.
4. Ruoming Jin and Gagan Agrawal. A middleware for developing parallel data mining implementations. In *Proceedings of the first SIAM conference on Data Mining*, April 2001.
5. Ruoming Jin and Gagan Agrawal. Shared Memory Parallelization of Data Mining Algorithms: Techniques, Programming Interface, and Performance. In *Proceedings of the second SIAM conference on Data Mining*, April 2002.
6. Ruoming Jin and Gagan Agrawal. Shared Memory Parallelization of Data Mining Algorithms: Techniques, Programming Interface, and Performance. *IEEE Transactions on Knowledge and Data Engineering (TKDE)*, 2003. to appear.

7. C. Kruengkrai and C. Jaruskulchai. A parallel learning algorithm for text classification. In *Proceedings of ACM SIGKDD 2002*, pages 201–206. ACM Press, August 2002.
8. C. Ordonez and P. Cereghini. SQLEM: Fast Clustering in SQL Using the EM Algorithm. In *Proceedings of the ACM SIGMOD Conference on Management of Data*, pages 559–570. ACM Press, June 2000.
9. S. Roweis and Z. Ghahramani. A Unifying Review of Linear Gaussian Models. *Journal of Neural Computation*, 1999.

A Parallel Knowledge Discovery System for Customer Profiling

M. Coppola[2], P. Pesciullesi[1], R. Ravazzolo[1], and C. Zoccolo[1]

[1] University of Pisa, Dip. di Informatica, Via Buonarroti 2, 56127 Pisa, Italy
[2] Ist. di Scienza e Tecnologie dell'Informazione, CNR,
Via Moruzzi 1, 56124 Pisa, Italy

Abstract. We describe a parallel KDD architecture we are developing as part of an open-source based customer relationship management system, in the framework of the SAIB industrial research project. The design of the prototype, leveraging on the features of the ASSIST programming environment, results in a high-performance parallel data mining core, tightly integrated with parallel data management and interfaced to business standard technologies and systems.

1 Introduction

In this paper we describe the system architecture and the implementation of a parallel Knowledge Discovery in Databases (KDD) system we are developing for Customer Profiling in the framework of the SAIB project (System for Internet Banking Applications).

Developing a parallel KDD system is an interesting challenge in itself, and also with respect to the main focus of our research on parallel programming environments. Beside developing efficient parallel mining algorithms, a key issue is the degree of integration the system can reach, both internally, as simplicity and performance of the interaction between mining algorithms and parallel data management, and externally, as ease of cooperation with different software technologies. Industry standard languages and technologies like XML, Java, or component programming have to be exploited to integrate advanced parallel modules within larger applications.

We describe the efforts made so far toward this goal, also focusing on the advantages that a high-level parallel programming environment like ASSIST [1] can bring in designing a parallel KDD architecture. In Sect. 2 we introduce the SAIB research project and present the overall architecture of the system. Sect. 3 summarizes our past research and the features of the ASSIST environment. We describe in Sect. 4 the design of the parallel KDD system, and in Sect. 5 the implementation of the parallel mining primitives and a simple case study we are using to test system integration. Sect. 6 summarizes results and future work directions.

---

* This work has been supported by the SAIB Project on High-performance infrastructures for financial applications, funded by MIUR and leaded by SchlumbergerSEMA, owned by ATOS Origin, and by the Italian MIUR Strategic Project L.449/97-2000 on High-performance distributed enabling platforms.

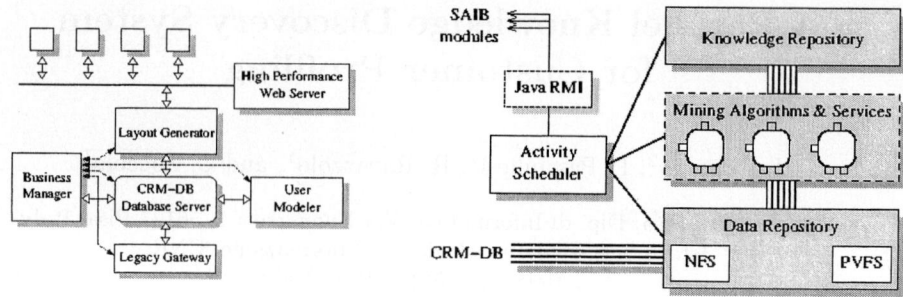

Fig. 1. (a) Overall Architecture of the SAIB system — (b) The UMS architecture.

2 A Data Mining Engine for the SAIB Project

SAIB is a large research project which brings together several Italian academic institutions and industrial partners in the effort of producing a flexible, open-source based Customer Relationship Management (CRM) solution for Internet Banking and Insurance.

The SAIB project has as essential goals to provide: (1) compatibility and cooperation with existing Hw/Sw infrastructures, in order to enhance their performance, (2) flexibility and programmability of the CRM solution, (3) privacy and security of end-user interaction, (4) multi-channel (e.g. kiosk, e-mail, mobile) and Internet-based interaction, (5) high performance customer profiling functionalities. The ultimate goal of the project goes beyond the Internet Banking solution, and it is to design a development environment for the broader class of CRM applications, including e.g. IT support systems for the public administration and call centers. In the SAIB system architecture (a partial overview is shown in Fig. 1a) a workflow interpreter (the Business Manager) executes a number of tasks, mostly triggered by local or remote user interaction. It controls a set of application modules which perform actual computation and interaction. The system can be easily tailored to different applications by adding new functional modules and new workflow programs.

Interchange of parameters and data among SAIB modules exploits industry standards like EJB interfaces, XML encoded Java RMI, and ODBC connection with a database server (the CRM-DB), which is a centralization point of the system. The Legacy Gateway module provides translation and integration of the Business Manager protocols to let SAIB cooperate with existing legacy systems.

Our KDD prototype is one of the main modules of the system, the User Modeler Server (UMS), providing static and dynamic customer profiling functions. We can roughly divide SAIB users into customers and administrative users. The UMS provides a restricted class of the administrative users (e.g. marketing analysts) with Knowledge Discovery and Data Mining services, in order to build customer profiles from the CRM-DB database. Knowledge models can then be deployed to the CRM main core, where they can be used by workflow programs to customize user interaction at different levels. Workflow programs can either access information that has been downloaded to the main database, or issue on-line, per-user queries to the data mining engine.

The UMS thus performs both *batch* operations (heavyweight, but that do not condition the main business flow) and *on-line* ones, which are subject to near-real-time constraints. To avoid interference with the main system, and to allow higher performance to the parallel mining algorithms, the UMS operates primarily on data cached in its internal, parallel data management module, tightly coupled with the mining engine. The UMS controlling interface allows the user to load data from the CRM-DB, and to put back new information. The example we describe at the end of Sect. 5 is a simple customer segmentation and classification process. It is just one of the feasible applications of KDD to CRM, that range from user interface personalization to potential fraud detection.

We wanted the Data Mining engine of SAIB to be able to deal with databases of several Gigabytes in size, to distribute computation and I/O in parallel and to scale with available computing resources to higher performance and throughput. In HPC these goals are often hard to meet with a portable, high-level software design. On the contrary, the programming approach of the ASSIST environment allowed us to efficiently develop parallel application modules and high-performance libraries, merging them into a complex application.

3   The ASSIST Parallel Programming Environment

The adoption of a high-level parallel programming environment to develop the high-performance modules is one of the central assumptions of the SAIB project.

Our research in the field stems from the skeletons model [2], one in the class of structured parallel programming (SPP) models. SPP models, by describing the parallel semantics of programs in a high-level way, provide increased portability, ease of code reuse and application evolution with respect to low-level parallel programming approaches based on communication libraries. The approach is characterized by the (hierarchical) composition of modules with completely defined interfaces, each composition mapping to a set of known implementation templates. However, too strong constraints can make it difficult to develop complex and dynamically behaving applications.

With ASSIST [1, 3] we close in to component-based parallel programming. A program is an unrestricted graph of sequential and parallel modules interacting through data streams. Parallel modules (parmods) can express mixed data-parallel and task-parallel computations, can explicitly manage load balancing and non-determinism if needed, have an internal state, and can interface to external resources. Each parmod (or combination of parmods) is then easily used as a component of larger applications.

The design of the ASSIST model explicitly targets the needs of large applications over massively parallel platforms and Computational Grids, taking into account issues like dynamic resource adaptiveness, dependability, heterogeneity. Some of the corresponding features have already been implemented in the ASSIST environment at present time (we refer the reader to more specific works like [4]), while the more general problem of devising a GRID programming model is being tackled in the framework of the ongoing national research project *"Grid.it"*.

In the KDD architecture we have implemented, we exploited several features of the **ASSIST** coordination language. Among them the support for *external objects* [1], that are object-like interfaces used from inside **ASSIST** code modules to access heterogeneous software resources. External objects can have their own run-time support, that must not interact with that of the application. In this case we have used the SMReference objects to manage large dynamic data structures in virtual shared memory, and we implemented the support for a parallel file system as an external object library for parallel and sequential modules.

4 Knowledge Discovery Architecture

The overall architecture of the KDD engine we have designed (Fig. 1b) is quite straightforward, based on four main modules providing

1. data management functionalities (the Data Repository, **DR**)
2. knowledge and meta-data management (the Knowledge Repository, **KR**),
3. a set of mining algorithms (**MA**s) and
4. a control interface, the **Activity Scheduler**.

The **Scheduler** interfaces to the rest of the SAIB system by means of Java RMI. Since it is not performance critical, the scheduler is actually implemented in Java, cooperating with the rest of the KDD engine by Java native methods, file system I/O and through the **ASSIST** program loader.

The scheduler accepts synchronous and asynchronous operation requests, enqueues them internally and manages the corresponding parallel programs and their results. Though the current SAIB architecture does not use it yet, concurrent operation execution with task priorities and dependencies can be dealt with in the scheduler. For testing purposes, a simple graphical Java front-end has been developed that connects to the scheduler and controls it.

The set of **MA modules** contains **ASSIST** parallel programs that perform actual mining tasks, and simpler ones used to manage the DR data (e.g. selection and sorting). We reused parallel mining applications designed in our previous research [5] and extended them as well with new functionalities. Taking advantage of the modular structure of **ASSIST** programs, we have evolved them so that they (1) interface to the DR module for most of the I/O, and (2) expose a common set of program interfaces (streams for data I/O, knowledge I/O and an XML-encoded file with running parameters).

This second requirement allows MAs to be viewed as a kind of software components within our system, decoupling their actual implementation from that of the mining engine as a whole. Adding new MA modules to the set known to the scheduler is simply a matter of defining their specific parameters as an XML schema. Moreover, standard conforming modules can be automatically composed into a larger **ASSIST** program to be compiled and run. This is still a work in progress to simplify and improve the performance of complex mining/validation processes that have to be executed routinely.

The **DR module** is implemented as an external object of ASSIST. It provides high performance I/O support for large files with simple record structure (the kind of regular data tables we have to host in a mining warehouse), a set of data types suitable to meet mining needs, and a mechanism of block-oriented views to allow parallel operation on the same file (a *dataset*, in the following) in a controlled way. It is a common choice to have data management functionalities integrated within the mining architecture. We aimed at a software layer that offered less overhead and more control on low-level issues w.r.t relational DBMS, while still providing a richer and more portable interface to data than working with flat files. For instance we use a block-oriented interface to allow explicit secondary memory management and concurrent operation within the MAs, since a large I/O grain is used anyway in the MA, and the waste of space is negligible.

Current DR implementation provides data types encoding floating-point numbers, nominal values from unordered list of labels, date values, booleans, unique keys and fixed-size uninterpreted raw data. We chose to have a fixed set of machine-dependent representations, to allow direct memory loading of data tables without translation, minimizing memory requirements and computational overhead in the algorithms. Special UNKNOWN values are provided for all types, and meta-data is kept linked with each dataset.

Most of these functions are provided as a library linked to the ASSIST generated code. A lower implementation layer moves data blocks in and out of each process memory, wrapping the actual file system layer. This design exploits parallel file system performance and bandwidth from within portable ASSIST programs, by allowing the data block engine to initiate concurrent data transfers across multiple I/O and processing nodes. The prototype is based on PVFS [6], and sequential UNIX file systems are supported too, including NFS, with lower I/O performance.

The **KR module** manages the knowledge produced by the mining algorithms, allowing to store, retrieve, refine knowledge models, and to track their history. Models are represented using the standard PMML 2.0 language [7], exploiting the PMML extension mechanism in a few cases where non-standard model semantics is needed. Each model is a PMML file containing the results of one or more mining algorithms, a link to the source dataset, and all the relevant meta-data and algorithmic parameter information. Models are interconnected so that a full KDD process can be designed and stored as a unit in the KR.

Knowledge models also have different states and attributes that condition their use inside the UMS, and their visibility in the whole SAIB system (e.g. a model has to be validated before its information is downloaded to the CRM-DB). A standard CVS server is used to store the actual PMML data, access to models being mediated by a custom server which performs additional controls and attribute caching. A client-side linked library provides XML parsing and serialization on models, as well as interface to the server processes.

5 Mining Primitives

The mining algorithms we have implemented are derived from our earlier experience in Data Mining with structured parallel programming environments [5].

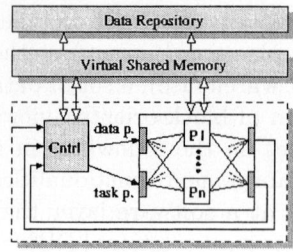 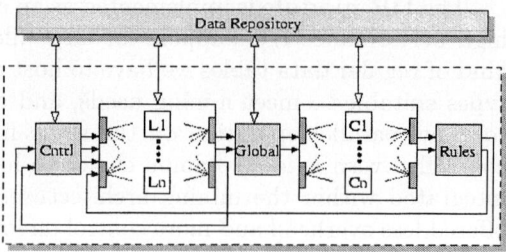

Fig. 2. (a) The parallel classifier. (b) Association-rule based clustering.

They have been extended, making use of the new features of ASSIST. We summarize here the basic UMS functions and the key points of their implementation.

We also show a few performance results measured on a cluster of 8 Pentium4/2GHz/512Mb RAM processing nodes with ATA UDMA disks, connected by a Gbit Ethernet switch and running Red Hat Linux 7.3.

Classification is performed with a decision tree induction algorithm. It uses the same score functions as the C4.5 classifier by Quinlan, but it currently works on nominal attributes only. Tree induction is a typical divide and conquer process, where a tree is built from the root (the whole input), at each node evaluating and partitioning the available data, each node expansion being independent from those happening on separate branches of the tree. Beside parallelism in the tree visit, large partitions call up for data-parallel decomposition of the splitting computation. Our prototype (Fig. 2a) keeps in the DR module data partitions associated with nodes, loading them on demand. The tree structure is local to the controller module, and statistic data are shared using SMReference external objects, thus the prototypes exploits a two-layer distributed memory hierarchy.

The expansion policy is given by a single process, controlling a parmod that performs all the computation. Expanding a node is a (possibly large) task, on which data-parallel, globally synchronized operations as well as task-parallel, concurrent ones are possible. Each task can result in more tasks to be produced (new nodes added to the tree) or in a fully sequential sub-computation (a complete subtree is generated down to the leaves). For the sake of conciseness, we disregard the fact that further decomposition happens of the parallel activities into evaluation and splitting steps.

After a first phase of the execution where the data parallel decomposition is used (current prototype actually employs this strategy only for the root node), we switch to the task-parallel behaviour, and below a certain node size to sequential computation. Dynamic load balancing in the task parallel case is guaranteed by the ASSIST support. The expansion policy, based on node size, determines the computation switch points and the relative priorities of different nodes.

ASSIST features allowed us to improve the program structure reported in [5] following the idea outlined in [1], i.e dynamically mixing data and task parallelism in the same high-level program. W.r.t. [5] we exploited the flexibility of the ASSIST parmod construct to express at the coordination level the different but correlated functionalities of the Conquer module (e.g. task parallel expansion, data parallel counting, and sorting). Stream guards controlled by shared variables make it clear and manageable the transition among the different behaviors. Data-parallel expansion allows to exceed the main-memory limits of a single processing node.

W.r.t. the solution in [1] we exploit a two-level distributed memory hierarchy (shared memory and parallel file system) using two kinds of external object. However, data-parallel expansion is still limited to the first node. We are currently developing the adaptive behaviour of the classifier to allow tuning of the transition from data to task parallelism, which is needed for large datasets, and to let the application deal with numeric attributes. We think that these improvements will also benefit performance on medium-size datasets. Figure 3a shows the speedup in such a case (sequential completion time is 43 seconds).

Association Rules are computed by an Apriori-like algorithm. Its parallel version is based on the partitioning method and requires two phases, each one performed in parallel on separate partitions of the input. As in [5], the input dataset is scanned two times in full, load-balancing being guaranteed by the on-demand distribution of ASSIST, but here we exploit the DR module to dispatch partitions from the hosting nodes to the requesting ones.

The partitioned method has a good parallel speed-up and it is scalable w.r.t. the number of transactions in the dataset. Fig. 3b shows almost linear speed-up for a medium-size, synthetic dataset (1.2M transactions of av. length 30).

Clustering is developed around the association rules module. We use a notion of clustering derived from [8], grouping together records that satisfy the longest, most popular association rule of a dataset. To produce more clusters, rules are mined again and again on unclustered records until a threshold on support or a prefixed number of clusters is reached.

The parallel implementation reuses the association rule main modules, the local tree build and the counting parmods, exploiting additional streams and guarded channels to control the iteration process, and to continuously reorganize a temporary copy of the input dataset. Figure 2b shows a simplified representation of the clustering application, including streams that carry on block indexes, frequent itemset information and cluster defining rules.

The rearranging strategy employed by the L_i modules during local frequent set search accumulates already clustered records into a "black list" of data blocks to be subsequently skipped, in order to enhance locality and progressively reduce the amount of I/O. Active blocks saved on disk are reassigned to any waiting process, ensuring proper load balancing. After the first phase, a global counting phase follows and an association rule is selected as a cluster definition. Clustered records are then discarded by the rearranging policy while searching for the next best rule.

Filters to perform basic manipulations on datasets can be implemented within the DR interface of an algorithm, or as stand-alone parallel programs that are run by the Scheduler like any MAs. In the first class there are field and record

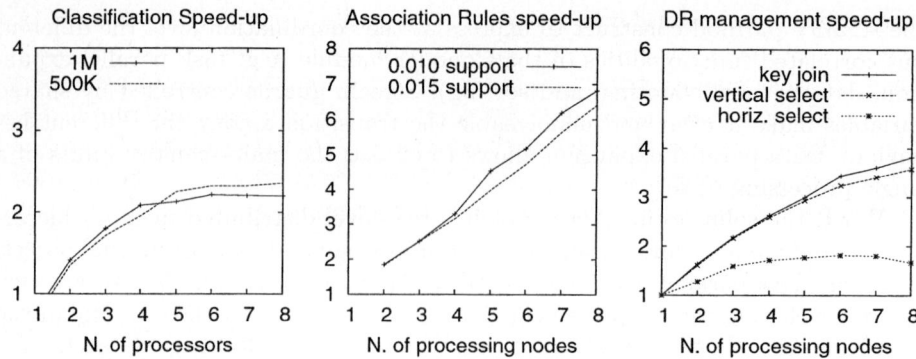

Fig. 3. Speed-up tests with 8 I/O nodes and N processing nodes. (a) Tree induction from 4M instances of the LED database, 20 attributes, 4% noise, 10 classes; switch to sequential computation at 500K, 1M node size. (b) Association Rule computation: 156Mb dataset, 1.2M trans., 661s and 1468s sequential time. (c) Simple key join, vertical and horizontal select DR routines: 2Gb dataset; peak data transfer bandwidth achieved is in the 86Mb/s – 141Mb/s range.

selections from a dataset, while the second class contains more complex operation like k-way merge-sorting, merging of datasets w.r.t. a common key attribute, summarization operators. In Fig. 3c we see the speedup obtained by three simple filter programs (key join, column and record selection) operating on datasets of 1 – 2Gbytes. Here communication bandwidth is the main limit to the speedup. Knowledge validation functions are also implemented as MAs. As an example, current UMS prototype provides functions to compute confusion matrices, in order to evaluate classification or clustering models.

Case Study. We have chosen a simplified case study to check the overall system design, showing correct interoperability of the UMS basic blocks. The target is to customize commercial advertising by developing a set of customer profiles, and a classifier that allows us to assign new customers to a base profile, on the ground of the limited amount of information that is initially available.

The initial data comes as a special purpose table stored within the CRM-DB, one record per customer, containing a large set of summarized attributes about customer behaviour. The input data is periodically imported to a DR dataset. New users may either be missing from this dataset, or correspond to records with many missing values.

The knowledge extraction process is based on the distinction of customer attributes into *factual* ones, that identify the user and do not change quickly over time, and *behavioral* ones, historical data about user interaction which are derived from transactional databases, and reflect evolving customer's commercial attitude. A process based on a similar data model is reported in [9].

We start the mining process extracting clusters from the dataset, by applying the association-rule-based clustering algorithm to the available behavioral data of customers. We try to identify classes of customers with similar habits, e.g. using the same bank services in the same period of time. After expert's validation, the set of cluster-defining rules is turned into a class label definition for all user with sufficient behavioral information. The second step is to infer a classification tree for customers, modeling the class label in terms of the factual attributes. After another validation step and possibly more iterations of the process, we can classify new users using the tree even if some of the factual attributes, usually available, are missing. Clearly, the distinction of attributes into factual and behavioral is not dogmatic, and trying different attribute sets for both process steps is part of the experimental part of the KDD process.

6 Conclusions and Future Work

We have implemented a parallel KDD architecture, written in Java, C++ and the ASSIST coordination language, with support for parallel I/O and mining algorithms, that is integrated with industry standard protocols and provides a Java RMI control interface. In the process we exploited a modular approach to integrate sequential and parallel code, reusing and modifying existing parallel kernels to integrate them into a single system. As a side effect of the project we have integrated in the ASSIST environment support for the PVFS parallel file system, using the abstraction of external objects.

The resulting design distinguishes from distributed mining frameworks like Papyrus [10]. Our KDD prototype fits in a general framework for high performance application development, not limited in scope to data mining. As a consequence, we designed a block-oriented, shared-memory like data distribution layer, instead of a sophisticated data transport layer like Papyrus' one.

Moreover, we aim at efficiently exploiting the processing power of Beowulf clusters. With respect to existing distributed mining systems relying on coordination of many independent sequential mining engines, like [11], our approach can be applied to a broader set of mining tasks, and it is much less influenced by the memory and computing power constraints of any single machine.

Work is still in progress to improve the different components of the system, especially to enrich the set of mining functionalities and to improve the management of meta-data within the KDD process. Tests are ongoing according to the simple KDD process outlined to verify the degree of integration with the SAIB CRM solution.

Our system is not constrained to a specific class of computing platforms. The current testbed is a small dedicated cluster of tightly coupled machines, but the ASSIST environment allows us to seamlessly run applications on clusters, LANs and WANs, as well as on grids and heterogeneous clusters (w.r.t. CPU architecture, O.S., and performance). While not every combination of different settings is already supported, the current implementation of ASSIST and of the KDD system makes it feasible to run mining algorithms on local networks with a common CPU architecture.

Moving to more distributed platforms opens up new research issues about the scheduling of parallel activities, and about the forecast of performance and scalability of the resulting KDD engine, as bandwidth and latency constraints affect the efficiency of many of the mining tasks. The problem in perspective merges with our current research on high-level Grid Programming environments.

Acknowledgments

We wish to thank SchlumbergerSEMA S.p.A., M. Vanneschi and M. Danelutto, and all the people of the Parallel Architectures Group in Pisa, who work on various aspects of the SAIB project or contribute to the development of ASSIST: M. Aldinucci, S. Campa, P. Ciullo, S. Magini, A. Paternesi, A. Petrocelli, E. Pistoletti, L. Potiti, M. Torquati and P. Vitale.

References

1. Vanneschi, M.: The programming model of ASSIST, an environment for parallel and distributed portable applications. Parallel Computing **28** (2002) 1709–1732
2. Cole, M.: Algorithmic Skeletons: Structured Management of Parallel Computations. Research Monographs in Parallel and Distributed Computing, Pitma (1989)
3. Aldinucci, M., Campa, S., Ciullo, P., Coppola, M., Magini, S., Pesciullesi, P., Potiti, L., Ravazzolo, R., Torquati, M., Vanneschi, M., Zoccolo, C.: The Implementation of ASSIST, an Environment for Parallel and Distributed Programming. In Kosch, H., László Böszörményi, Hellwagner, H., eds.: Euro-Par 2003: Parallel Processing. Number 2790 in LNCS (2003) 712–721
4. Aldinucci, M., Campa, S., Magini, S., Pesciullesi, P., Potiti, L., Ravazzolo, R., Torquati, M., Zoccolo, C.: Targeting Interoperability and Heterogeneous Architectures in ASSIST. To appear in Proc. of the Euro-Par 2004 Int.Conf., Pisa (2004)
5. Coppola, M., Vanneschi, M.: High-Performance Data Mining with Skeleton-based Structured Parallel Programming. Parallel Computing, special issue on Parallel Data Intensive Computing **28** (2002) 793–813
6. Carns, P.H., Ligon, W.B. III., Ross, R.B., Thakur, R.: PVFS: A Parallel File System For Linux Clusters. In: Proc. of the 4th Annual Linux Showcase and Conference. (2000) 317–327
7. The Data Mining Group: Pmml 2.0 specification. http://www.dmg.org/pmml-v2-0.html (2003)
8. Kosters, W.A., Marchiori, E., Oerlemans, A.A.J.: Mining clusters with association rules. In Hand, D., Kok, J., Berthold, M., eds.: Advances in Intelligent Data Analysis: 3rd Int.l Symp., IDA-99. Volume 1642 of LNCS. (1999) 39–50
9. Adomavicius, G., Tuzhilin, A.: Using data mining methods to build customer profiles. Computer **34** (2001) 74–82
10. Bailey, S., Creel, E., Grossman, R., Gutti, S., Sivakumar, H.: A High Performance Implementation of the Data Space transfer Protocol (DSTP). In Zaki, M.J., Ho, C.T., eds.: Large-Scale Parallel Data Mining. Volume 1759 of LNAI. Springer (1999) 55–64
11. Musicant, D., Celis, S.: Weka-parallel: Machine learning in parallel. Technical report, Carleton College of Computer Science (2000)

A Large-Scale Digital Library System to Integrate Heterogeneous Data of Distributed Databases

Mariella Di Giacomo, Mark Martinez, and Jeff Scott

Los Alamos National Laboratory, Los Alamos, NM 87545, USA
{mariella,mlbm,jscott}@lanl.gov

Abstract. The Web has become the primary means for information dissemination of all kinds; our interest is in dissemination of scientific information from on-line digital libraries. We have designed a Web application, called *SearchPlus*, based on a distributed, scalable, fault-tolerant, and secure architecture, to allow access to tens of millions of scientific bibliographic records and their citations, integrating information from multiple heterogeneous data sources, and making this information available for querying and analysis. A full-scale test-bed environment has been developed to assess hardware and software configuration and performance. This paper gives the motivations for building such a system, describes the architecture of our distributed database system, and highlights performance analyses and subsequent improvements.

1 Motivation

The Los Alamos National Laboratory (LANL) Research Library (RL) [2] focuses on digital information services. It provides commercially available scientific data, through Web applications, to LANL scientists as well as several external institutions. One such application is *SearchPlus*. The primary objective of *SearchPlus* is the construction of a comprehensive distributed database of scientific journal articles – now over 55 millions – and citation information in a common format – now over 500 million entries, and providing access to this information to RL customers. Scientists now rely on these resources to meet deadlines, write articles, and vie for funds in highly competitive fields. Such a critical research resource must exhibit as little service disruption as possible. To fulfill this requirement, a robust, fast, flexible, scalable, and secure system has been developed. Commercial products and those deriving from other research projects have been explored, but no complete solutions have been found. The underlying mass-storage systems and search engine are commercial products; the rest of the application is home-grown.

The the architecture design of *SearchPlus* was driven by the following functional requirements. 1) Transformation of bibliographic data for scientific publications in different formats into a common XML format, storage for indexing and retrieval. 2) Processing of data in a secure environment behind a firewall and making it available to users through a web application outside the firewall. 3) Providing an information retrieval system in the form of a web application with a flexible interface, allowing search and retrieval of bibliographic data, linking to cited and citing articles, linking to full-text articles, and providing weekly alerts. 4) Delivering a responsive, interactive service. 5)

Providing a reliable, fault-tolerant and highly-availability system[1], tolerating no loss of data. 6) Building a scalable and adaptable system capable of handling weekly updates, new data sources, formats, and content.

This paper makes two main contributions. A first is the description of the architecture of *SearchPlus*, which comprises a large collection of software and hardware components. We argue that the lessons learned may be of wide interest in the research community. The second contribution is the description of the methodology that has been used to optimize the performance, enhance the usability, and improve the robustness of *SearchPlus*. The rest of this paper is organized as follows: Section 2 outlines the software and hardware architecture of *SearchPlus*; section 3 provides insight into a large number of optimizations that have been performed on *SearchPlus*, together with their impact on the overall performance; section 4 provides concluding remarks.

2   Architecture

This section describes the architecture of the system, focusing on the main components and the reasons for choosing them. The overall architecture of the proposed environment is shown in Figure 1, the elements of the hardware configuration are listed in Table 1.

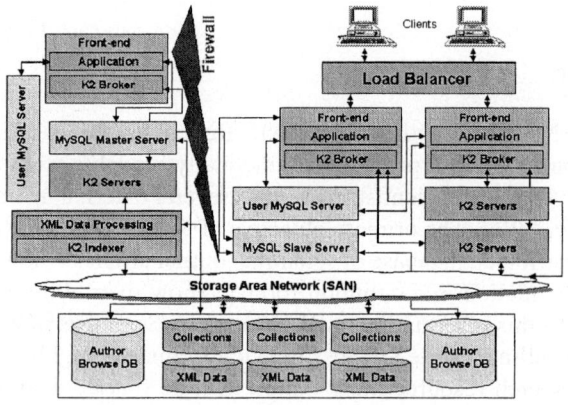

Fig. 1. The architecture of *SearchPlus*.

Table 1. Elements of the hardware architecture of *SearchPlus*.

Processing Nodes	Processors	Main Memory	Disk Storage
12	46	234 GB	7 TB

The physical architecture outside the firewall consists of a load balancer, two front-end systems running Verity K2 brokers and web applications, two systems running Verity K2 servers, a system running a MySQL server for an authentication/authorization database, and a system with a MySQL slave server for the author browse database. The architecture inside the firewall includes all the components residing outside the firewall with the exception of the load balancer and MySQL slave server. The MySQL master server and the XML data processing and K2 indexer system also reside inside the firewall. At a high level, there is a single user interface that provides a unified environment for both data retrieval and citation linkage. Users can access all the functionalities independent of the physical architecture of the system. At a low level, the load balancer accepts client connections and balances them between two front-end systems. Each front-end server runs the web application and a K2 broker. The application examines a MySQL database to verify the user authentication and authorization rights to the integrated XML data. If that is successful the application sends a query to the K2 broker running on the same machine. The K2 broker forwards the query to the appropriate K2 servers which search the collections. The K2 servers return results to the K2 broker which sorts and returns them to the application. At this point, the application may also query the author browse and citation database to build bibliography and citation counts. Java and Apache/Tomcat have been chosen as the platforms to provide web accessibility. Access to the MySQL databases is handled using servlets and connection pooling. The main components of the hardware and software architecture on which we will focus our attention are the following: storage architecture, XML data layout, Verity K2 Enterprise, and MySQL database.

Storage Architecture. Digital library data centers have demanding size, speed, reliability and flexibility requirements. In addition we also need to provide a secure environment for our data and systems. Our institution has a firewall with services inside and outside. All the data, systems, and services behind the firewall benefit from better security. However, we need to run our application and access data from outside the firewall for our external customers. Our choices for storage, file system, and distribution of processing have been predicated on protecting the data while keeping it easily accessible. Redundant Arrays of Inexpensive Disks (RAID) and Storage Area Network (SAN) technologies have been deployed to prevent data lost and provide storage capacity. The most complicated part of our design has been sharing data among servers located inside and outside the firewall. The combination of a SAN and a shared-access file system gives us the capability to achieve our objective. Sun's Quick File System (QFS), a Large Storage Configurations (LSC) file system, is designed to solve file system performance bottlenecks by maximizing the performance of the file system in conjunction with the underlying disk technology. QFS is implemented using a standard Solaris virtual file system interface and can be shared among Solaris environments. Several servers can read the data distributed in a file system, while another server can write and modify the same data. Using a shared-access file system, we can build, update, and modify data inside the firewall, while the web application outside the firewall has read-only access.

XML Data Layout. Scientific articles typically have associated metadata which provides such information as author, title, abstract, keywords, source, volume, issue, number of page, etc. An article may also have a bibliography of citations (cited references). There are two sets of records, stored in XML format: those containing the metadata for scientific articles and those containing citations for the same articles. A record containing the metadata for an article is stored in a single file on a file system reserved for metadata records. For those articles with a bibliography, citation data is stored in a single file on a file system reserved for citation records. For citations that have a corresponding metadata record in our XML repository, a link is established from the cited reference to the corresponding metadata record.

Verity K2 Enterprise. Native XML search engines have been evaluated and found to be underdeveloped, so we have investigated full-text search engines, settling on Verity K2 Enterprise (K2). Two of the compelling features of K2 are its distributed design and scalability. The design makes it easy to distribute indexing, search and retrieval, and administration. The K2 architecture consists of client, broker, server, admin server, and indexer components. The K2 client, in our case, refers to the Web application, developed in-house, which is integrated with K2 using Java. A K2 server is the core of the search and retrieval component. The K2 server contains the search engine for a specific set of indexed documents (collections) and the viewing service which renders documents returned by a search. The K2 broker manages communications between K2 clients and one or more K2 servers. When multiple collections are searched, each K2 Server performs a search against its collections, returning the results to the broker responsible for merging, sorting, and presenting results to the client. A broker can communicate with all its K2 servers simultaneously, whether or not they are on the same machine. Brokering enables scaling the system as the amount of information being searched for grows: brokers can be added according to demand. Similarly, as the number of documents to search grows, more K2 servers can be added, and collections can be mirrored to balance an increased load.

MySQL/Relational Database. Why has a relational database been used to store data related to XML bibliographic records when they are already stored on file systems and searchable with the Verity search engine? The XML data repository consists of millions of small files, and backup and recovery of such a file system can be problematic. Mirroring the data stored on disk in a relational database offers the additional benefit of faster backup time and useful data redundancy. In addition to searching and retrieval, we need to provide browsing capability on authors and cited and citing articles, and dynamic citation counts. A relational database provides flexibility to build browsing functionality. MySQL is an open source relational database. Four reasons for choosing MySQL are (1) Speed: MySQL has proven to be fast at handling links among 1,435,000,000 rows of data in several virtual tables; (2) Data storage capabilities: we currently manage over 400GB of data; to limit individual table size, we take advantage of merged tables; (3) Fault tolerance: As mentioned, we use MySQL in production and as disk-based backup for our data; (4) Security: MySQL replication is used to protect and update our data; the master MySQL server runs behind a firewall, while a slave server accesses the data from outside the firewall, in read-only mode.

3 Performance Analysis and Improvements

In order to determine why application performance was not as good as expected – the initial response time for a simple search was tens of seconds – we undertook a number of performance studies. To simplify the process we concerned ourselves with examination of individual problems. We analyzed each component of the overall architecture: layout of disk arrays and file systems, memory use, the benefits of running in a 32-bit versus a 64-bit environment, network infrastructure, tools used by the application (MySQL DB, Verity K2 Engine, Java Virtual Machine, Java compiler, XSL, JSP, Apache/Tomcat) and of course, the application code itself. We estimated the impact of every component on performance and scalability and focused our tuning efforts on those which would provide the most benefit.

3.1 Hardware Architecture

Hardware performance tuning efforts were focused on evaluating the number of database/collection servers, the number of CPUs per server, the amount of memory needed, and the number of files per file system. Initial assessments of performance typically involve processor speed or memory consumption, not transfer rates to and from disk storage. Since disks are several orders of magnitude slower than RAM, avoiding access to disk and making necessary access as fast as possible can have a huge impact on application performance. To solve this problem, we have used RAID technology. If configured properly (several experiments were performed using different disk striping strategies), the disk arrays are fast, cheap (considering the amount of data stored) and safer. We have stored over 7 TB of data, distributed in two categories: the first set consists of millions of small files laid out on file systems, the second of MySQL database tables. These require different choices for disk configuration. We looked at disk organization and layout, making sure that all parameters were appropriately tuned for the type of data stored. To sum up, important decisions impacting the I/O performance were choosing a correct page size, the unit of disk, the metadata, and data distribution of each file system. Looking at the network infrastructure, all the servers and storage devices that must communicate have been connected on the same network path and controlled by the same network switch to minimize latency and network delays.

3.2 Verity Tuning Optimizations

We knew before the project started that there would be a large number of users accessing the application, but we did not have a clear picture of the search distribution. Some time was spent monitoring how the Verity collections were queried and how many users access the system in a specific time frame. We also performed a number of optimizations, mostly on the allocation and caching policies of Verity K2, with improvements shown in Table 2. As can be seen, many of them had a significant performance impact.

Table 2. Impact of Verity K2 optimizations on the basic performance.

Broker Caching	Broker Thread Allocation	Sever Caching	Server Thread Allocation
20%	15%	22%	18%

3.3 MySQL Optimization

Over 400 GB of data are stored in the MySQL database. MySQL server configuration, table structure, table allocation, query handling, concurrency, and replication were examined for optimization.

Server Optimization. The first component analyzed was the MySQL server, its compilation, and linkage. By using a suitable compiler and appropriate compiler options, a 10-30% speed increase was realized. MySQL was compiled to take advantage of the 64-bit Solaris architecture and address up to 32 GB of memory, very helpful for query caching. We observed how MySQL server uses system memory, how the memory is shared, and how it is used by MySQL threads when performing queries. Several experiments were done tuning the size of the memory buffer that MySQL uses for storing the indexed data. Several tests were made using different block size for storing and retrieving the indexed data.

Table Structure Optimization. The first table optimization involved structuring the data and indexes to take as little space as possible on the disk and in memory. This results in significant improvements because disk reads are faster and less memory will be used. Indexing also takes fewer resources if done on smaller columns. The second optimization was to lay out the tables accessed at the same time on different file systems. The third and ongoing process of optimization involves running the MySQL table check mechanism to remove fragmentation and re-sort the indexes after updates. The fourth performance improvement that has been undertaken, which has not yet been completed, is restructuring the data inside the tables so that we can take advantage of some of the newer features of MySQL.

Replication. MySQL replication has been used in our architecture to protect and update our data. When using the MyISAM index mechanism, MySQL has extremely fast table locking (multiple readers/single writer). The biggest problem with this table type occurs when one has a mix of a steady stream of updates and slow selects on the same table. One way to address this problem is to use MySQL replication, where a master server updates the data in a secure environment and a slave server gets the data from the master and provides the data to the users. Using this mechanism, we have been able to reduce the MySQL query time while updates are made on the same data. Replication also provides a secure environment for the data to be updated. Table 3 summarizes the impact of the major MySQL optimizations.

Table 3. Impact of MySQL optimizations on the basic performance.

Compilation	Tables	Block Size	Table Structure	Buffer Cache	Query Cache
15%	from 24 to 2 hours	5%	20%	25%	30%

4 Conclusion

The main motivation of *SearchPlus* has been to develop a powerful, responsive, robust, and intelligent distributed database environment for knowledge discovery information. We have described the architecture of our relatively complex, multi-TB system and outlined our performance optimization methodology. All the optimizations performed have reduced the response time of the system to less than three seconds, a substantial performance improvement with respect to the original system. *SearchPlus* is therefore very responsive and fault-tolerant and is currently used by a sizable number of customers.

References

1. A. Fox and D. Patterson. Self-Repairing Computers. *Scientific American*, 288(6):54–61, 2003.
2. R. Luce. Evolution and Scientific Literature: Towards a Decentralized Adaptive Web. *Nature*, May 2001.

Topic 6
Grid and Cluster Computing

Thierry Priol, Craig Lee, Uwe Schwiegelshosh, and Diego Puppin

Topic Chairs

Grid and Cluster Computing are becoming the two main approaches for building large-scale and cost-effective computing infrastructures. Cluster computing relies on the gathering of just a few, or up to thousands, of inexpensive personal computers combined into a single machine, i.e., a cluster. However, despite that cluster computing is a more economically viable approach for building a supercomputing platform, the effective programming of such a computing infrastructure remains difficult. Grid computing is pushing this idea further by clustering a large variety of geographically distributed resources like computer, storage, data and other services, at a wider scale. Grids rely heavily on the Internet to connect all these resources together and offer them to the users as transparently as possible. Although much progress has been made in the deployment of grid infrastructures, many challenges still lie ahead of us before the ultimate goal of the Grid can be realized. This topic reports recent advances in Grid models, services, application development, data storage and management.

The topic received 48 submitted papers dealing with various aspects of Grid and Cluster computing. This is the highest number of submissions compared to the other topics, clearly indicating that Grid and Cluster Computing represent two very active research fields. Finally, after a careful review procedure, involving 115 international reviewers who did an awesome job, we only selected 8 regular papers and 2 short papers. These papers cover important, current research areas in Grid and Cluster computing.

The regular papers cover the main following issues: (1) an agreement-based infrastructure to enable quality of service in the Grids, (2) a double auction economic model to address Grid economy, (3) a distributed resource management application API to aid the development and distribution of application across the Grid, (4) a Grid service programming framework for the composition of mesh-based MPI applications, (5) the integration, into a Grid infrastructure, of a MPI-based system for the simulation of action potential propagation on a cardiac tissue, (6) the profiling of two Grid data transfer protocols and servers to enable full system optimization, (7) an approach to disk storage power management for a clustered server environment, and finally (8) a path selection-based algorithm for data dissemination and scheduling for distributed real-time application over the Grid.

The short papers deal with the following problems: (1) a component based grid-aware library service to integrate HPC legacy software modules, and (2) a monitoring framework which provides a uniform and general purpose interface to build worldwide information services.

Agreement-Based Interactions for Experimental Science

Katarzyna Keahey[1], Takuya Araki[1,2], and Peter Lane[1]

[1] Argonne National Laboratory, 9700 South Cass Avenue, Argonne, IL 60439 USA
{keahey,araki,lane}@mcs.anl.gov
[2] NEC Internet Systems Research Laboratories, Kanagawa 216-8555, Japan

Abstract. Enabling quality of service (QoS) in the Grids requires not only resource management strategies but also the development of protocols enabling structured negotiation for the use of resources. Such protocols will allow the creation of policies dynamically and automatically broadening the scope of Grid applications. In this paper, we describe design, implementation and application of an agreement-based infrastructure. We then discuss its use in the virtual control room developed for the National Fusion Collaboratory.

1 Introduction

Over the last decade computational Grids [1] became a very successful tool at providing distributed environments for the secure and coordinated execution of applications. More recently we have seen an increased demand for Grid technologies in areas with stringent quality of service (QoS) requirements such experimental science [2, 3]. This resulted in stronger emphasis on providing QoS in Grid technologies [4] and focus on technologies enabling it. The most recent work in Grid computing [5-8] indicates that the next-generation Grids will include policy-based resource management, a variety of authorization services, and support dynamic resource procurement as well as adaptation to changing system conditions.

If such dynamic, need and opportunity driven environment is to be achieved, it is fundamental to *establish mechanisms and protocols* enabling clients to negotiate and renegotiate policies *dynamically* rather than rely on static policy sets. Agreement services provide such mechanism. A client can negotiate with an agreement service to meet specific objectives in a Grid environment. An agreement service evaluates the client's request in the context of a potentially complex set of policies. As a result of the negotiation, an agreement is created representing a concretization of those policies to suit the client's requirements. An agreement can be subsequently renegotiated, amended, or otherwise dynamically and automatically updated.

Providing Grid services based on such agreements and managing them to automatically adjust to agreement changes allows many advantages to clients as well as providers in the Grid. An agreement-based services infrastructure combines information, negotiation and execution services that allow clients to query the availability of a particular service in the context of their priority needs, as well as to compare offers from different providers. Service providers can use agreements as a provisioning target driving resource management as well as to estimate future demand and analyze client needs. Combining multiple agreements allows for the creation of agreements of arbitrary complexity. Automatic resource management based on such agreements allows for adaptation that can both leverage and counteract the changing conditions in the Grid.

In this paper, we describe the implementation of an agreement-based infrastructure loosely based on the WS-Agreement specification [9] currently developed at the Global Grid Forum (GGF). We describe terms for specific applications including combined agreements, dependency based agreements, and agreement templates. To manage uncertainty, we associate agreements with confidence levels representing the strength of the agreement to the client. Finally, we demonstrate how our implementation satisfies a client's point of view, working under to constraints of a virtual control room developed by the National Fusion Collaboratory for use in fusion experiments.

2 Agreements: Architecture and Implementation

In this section, we first give an overview of the architecture of our system loosely based on WS-Agreement [9]. We then describe our implementation of this architecture and our definition of agreement terms used to capture agreements for services described in section 3.

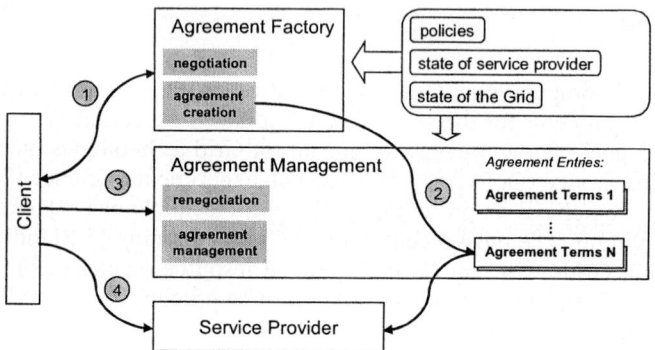

Fig. 1. Interactions in an agreement-based system

Figure 1 illustrates a basic interaction in an agreement based approach to resource management. The interaction starts with a negotiation process which can be viewed as a discovery phase in which clients advertise their needs to the agreement factory, and the factory represents what capabilities can be provided depending on policies, state of the Grid and other potential factors (1). This phase ends in the creation of an agreement when both sides commit (2). Agreements represent state that can be accessed and managed in terms of its lifetime and other properties. For example, in [9] they represented as Grid services [10] while we used an approach that is closer to Web Services Resource Framework (WS-RF) [11]. *Agreement terms* describe the objectives of a particular agreement. The client can manage (e.g., destroy or renegotiate) and monitor an agreement throughout its lifetime (3).

Once an agreement is created, it can be used to create or influence a service so that it meets specified objectives. Depending on the agreement terms, this may happen automatically (i.e., without requiring any further action from a client), or it may be triggered by an event from a client (4). Further, a client may be required to explicitly point a provider at an agreement, or the agreement may be available to the provider through other means.

Although in practice both agreement management and the service provider may be implemented in the same service, the agreement format and interface are distinct and the same across multiple services. By standardizing it, we enable service providers to integrate agreements into their implementation model.

2.1 Implementation

We implemented agreement-based interactions using the Globus Toolkit 3 (GT3). While our implementation was influenced by WS-Agreement [9] and Web Service Level Agreement (WSLA) [12], our use case did not require a full implementation of it. Instead, we focused on defining terms and functionality required by the application and practical experiences with the system.

Instead of representing each agreement as a Grid service [10], we implemented the factory to create and maintain a table of current "agreement entries" exposed as factory Service Data Elements and managed as factory state. The factory also implemented the agreement management interface. Although based on an implementation of Grid services, this approach makes our implementation much closer to WSRF [11]; in general we found this model to be simpler and lighterweight.

Our negotiation process is simplified and emphasizes discovery. An agreement factory allows a client to retrieve an "agreement template" (based on the `Agreement-TermType` in the section below) advertising some initial values of the agreements it supports: for example, a factory may support only services of a fixed description. The clients can then fill out some or all fields in this template and propose an agreement. By filling out more or fewer fields, the client can effectively ask a more or less concrete question about the availability of a specific service. The agreement may be rejected (if the terms specified by the client cannot be satisfied) by returning an exception. Alternatively, the factory can supply values for fields not filled out by the client and return it as provider's pre-committed offer together with an agreement handle identifying the agreement. Pre-commitment on the provider's side results in creating an "agreement entry" with a short expiration time that can be extended if the client commits (or expire if the client abandons negotiation). After receiving factory response, the client can either commit to the proposed agreement or try again. Our negotiation model is simpler than WS-Agreement as it does not implement multi-phase negotiations or support renegotiation once an agreement has been created. Further, it allows negotiation on the level of the whole agreement only. We also support a simpler commitment model: only the provider can pre-commit and client commit. Client's commitment extends the agreement time to the end of availability time.

Although for the purposes of our application domain an agreement should be claimed through an event (the requisite calculations are performed when the data becomes available) we decided to simplify this process in order to reduce the impact of service creation overhead on agreement claiming. Thus, agreement commitment causes the requisite application service to be instantiated as soon as its availability period starts. The client can then obtain the handle to the application service from the factory and claim the agreement from the application service which triggers the execution of desired actions.

2.2 Agreement Term Type

An agreement represents a commitment that services described by the service description will be provided during a specified time of service availability with a specified QoS (whenever applicable). At most one such service will be provided at a time, but it may be claimed multiple times as the availability period allows. Our agreement terms are described as follows:

```xml
<xsd:complexType name="AgreementTermType">
    <xsd:sequence>
        <xsd:element name="parties" type="tns:AgreementPartiesType"/>
        <xsd:element name="serviceInstanceHandle" type="xsd:anyURI"/>
        <xsd:element name="dependency" type="xsd:anyURI"
                minOccurs="0"
                maxOccurs="unbound"/>
        <xsd:element name="availability" type="tns:ScheduleType"/>
        <xsd:element name="expirationTime" type="xsd:dateTime"/>
        <xsd:element name="serviceLevel" type="tns:serviceLevelType"/>
        <xsd:element name="serviceDescription" type="xsd:anyType"/>
    </xsd:sequence>
</xsd:complexType>

<xsd:complexType name="AgreementPartiesType">
    <xsd:sequence>
        <xsd:element name="client" type="xsd:anyURI"/>
        <xsd:element name="provider" type="xsd:anyURI"/>
    </xsd:sequence>
</xsd:complexType>

<xsd:complexType name="ScheduleType">
    <xsd:sequence>
        <xsd:element name="startTime" type="xsd:dateTime"/>
        <xsd:element name="endTime" type="xsd:dateTime"/><
    </xsd:sequence>
</xsd:complexType>

<xsd:complexType name="serviceLevelType">
    <xsd:sequence>
        <xsd:element name="timeBound" type="xsd:duration"/>
        <xsd:element name="confidenceLevel" type="xsd:int"/>
    </xsd:sequence>
</xsd:complexType>
```

Listing 1. Generic term types used for any agreement in our system

The first three items of the schema correspond to the `wsa:ContextType` of the specification. They describe the parties of the agreement and include the Grid Service Handle (GSH) of the client and the provider. The `serviceInstanceHandle` element holds the GSH of the application service created as a result of the agreement. The `dependency` element contains the agreement handle(s) which the agreement is dependent on (see section 3.2 and section 3.4 for illustration).

The `availability` element defines the time period when the services specified in the agreement are available. The `expirationTime` element corresponds to the lifetime of the agreement (not the created service).

The `serviceLevel` element refers to QoS guaranteed by the agreement. The `timeBound` element describes guaranteed execution time. While some entities can be managed in a deterministic fashion (CPU reservation for example), others are not (for example, data transfer over the Internet). In order to account for this uncertainty, the service provider accompanies the agreement terms with a confidence level with which it can provide the terms.

The `serviceDescription` element is a domain-specific element; its content depends on the service; examples will be described in the next section.

3 Agreements and Services

In this section, we describe service description terms for a few simple services and corresponding service implementations that can be used within an agreement-based framework described in the previous section. We describe how service levels were implemented and how the corresponding levels of confidence were estimated.

3.1 CPU Reservation Service

Agreements for this service allow clients to reserve a CPU resource. The `serviceDescription` term is defined as follows:

```
<xsd:sequence>
    <xsd:element name="CPUUtilization" type="xsd:int"/>
    <xsd:element name="hostname" type="xsd:string"/>
</xsd:sequence>
```

This service description is sufficient to reserve one CPU per host and possibly time-sharing that CPU with other jobs. Thus, CPUUtilization describes the percentage of CPU to be used. When the agreement is claimed, the job ID of a job to which the reservation should be applied is passed as an argument to the claim. Note, that the services description here is not related to QoS: the agreement simply states that a certain service will be provided.

The implementation of the resource reservation service is similar to GARA [6]: to implement resource reservation, this service utilizes DSRT [13], which has functionality to allocate specified percentage of CPU cycles to a certain process. The service maintains a reservation table; when an agreement is proposed, the table is first consulted to make sure there is a slot available, and if a specified percentage of CPU cycles is free during the period of availability, the proposed agreement is accepted.

To meet the needs of more complex hardware configurations, these service description terms can be extended (see [14] for a more extensive definition of terms). We are currently working on generalizing this infrastructure to allow reservations in clusters using for example the Maui scheduler plug-in in conjunction with the PSB batch system [15].

3.2 Application Execution

The terms below allow a client to make agreements for the execution of an application. The service description represents the name of the program and concrete argument values. For the specific application used in our experiment (the magnetohydrodynamics equilibrium fitting code EFIT [16]) the `serviceDescription` is as follows:

```
<xsd:sequence>
    <xsd:element name="application" type="xsd:string"/>
    <xsd:element name="timeSteps" type="xsd:int"/>
    <xsd:element name="executionMode" type="xsd:string"/>
</xsd:sequence>
```

The agreement for this service guarantees execution of a specific service description with certain service level as described in listing 1 (in this case: the execution time is bounded by a certain value). It is important that the service description externalizes all the arguments that QoS may depend on; in this example both `executionMode` and `timeSteps` influence the execution time of our application.

To meet the QoS, this service reserves CPU resources using the CPU reservation service. In the current implementation, resource reservation is made by job execution agreement factory when the level of the execution service is negotiated, but we also envision scenarios where the client can use a preexisting reservation as input to negotiation. The GSH of CPU reservation service is stored as the `dependency` element of `AgreementTermType` (see listing 1). It should be noted that the agreement does not indicate in what way or to what extent the service depends on the dependency agreement; the knowledge of how to "consume" the dependency is application-specific.

The time bound on which the service will finish, is calculated by combining information about the resource reservation and prediction of execution time based on historical data and scaled to the number of timesteps and the CPU share. The confidence level of the time bound is modeled as prediction error. Thus, although the terms of the agreement are based on resource management, they are to some extent informational, that is, the estimate of execution time is based on prediction rather than adaptive management of the application.

When the agreement is claimed, the service starts executing the job using GT3's Grid Resource and Allocation Manager (GRAM) service. The CPU is claimed by associating the job ID of the job started in this way with the reservation and the execution time of the application is monitored by the provider and reported after the execution finishes.

3.3 Data Transfer Service

The Data Transfer Service is implemented based on the GT3 reliable file transfer (RFT) [17] service and uses RFT's `transferRequest` as part of its service description. Among other qualities, `transferRequest` contains information about the source and destination of the transfer, needed to calculate QoS. The exact parameters are as follows:

```
<xsd:sequence>
    <xsd:element name="transferRequest"
        type="rft-types:TransferRequestType"/>
    <xsd:element name="size" type="xsd:int"/>
</xsd:sequence>
```

Again, estimates of execution time (transfer time, in this case) are based on a simple prediction depending on historical data for this transfer, and confidence level on associated error. Although we have explored more sophisticated ways of QoS enforcement for data transfer [18], we have not yet integrated them into this system.

Since fusion codes produce multiple files as a result of a run, the data transfer service has been customized to operate on directories of data rather than individual files: the data is tarred before RFT is invoked and untarred at destination. As with application execution, the transfer time is monitored by the provider and reported after the service finishes.

3.4 End-to-End Application Execution

Providing an end-to-end application service based on remote execution requires coordinating several subsidiary services. In our case the workflow scenario is very simple and consists of application execution and data transfer of output data. The `serviceDescription` exposes interface similar to application execution:

```
<xsd:sequence>
    <xsd:element name="application" type="xsd:string"/>
    <xsd:element name="timeSteps" type="xsd:int"/>
    <xsd:element name="executionMode" type="xsd:string"/>
    <xsd:element name="outputDestination" type="xsd:string"/>
</xsd:sequence>
```

As before, the service description externalizes arguments on which the QoS depends; in this case we add the argument describing the destination of the data to those on job execution. In this way, the end-to-end timebound calculation can be adjusted to the location in which the execution is eventually scheduled.

The end-to-end application service directly depends on the job execution and data transfer services for its service level. We currently store those dependencies as the dependency element of `AgreementTermType`. Negotiating a composite agreement is more complex as it requires the factory to in turn to negotiate subsidiary agreements. Further, dependencies between multiple components may impose an order on negotiating subsidiary agreements. The end-to-end time is calculated by combining execution times of the services in appropriate ways (in our case by adding, but in general we could use min/max, etc.) and using the confidence level of subsidiary services to calculate a weighted error. As with the other services, the workflow and its subsidiary services are instantiated when the availability period starts. Claiming an agreement on an application service will trigger claims on subsidiary services.

In principle, by externalicing the application description as a workflow rather than an opaque service we could both express a stronger dependency and subject much of what is currently embedded implementation to automatic management. While full implementation of this concept would require incorporating a workflow language into our agreement structure, we made some modest steps in that direction by for example externalizing service monitoring.

4 Case Study: Interactions in the Virtual Control Room

Our prototype infrastructure and services were put to the test in the virtual control room experiment at SC03 illustrating how Grids can be used in fusion science experiments. Fusion experiments operate in a pulsed mode producing plasmas of up to 10 seconds duration every 15 to 20 minutes, with multiple pulses per experiment.

Decisions for changes to the next plasma pulse are made by analyzing measurements from the previous plasma pulse (hundreds of megabytes of data) within roughly 15 minutes between pulses. This mode of operation could be made more efficient by the ability to leverage Grid resources to do more analysis and simulation in the short time between pulses. Hence, the ability to do time-bounded execution in the Grids is of critical importance.

The virtual control room experiment followed the script of typical experiment preparation and interaction. Before an experiment, a scientist can negotiate an agreement for the execution of a remote fusion code and request for data to be delivered to a specific location. This process allows the scientist to experiment with, and fine-tune the parameters for the execution of the code. Thus the agreement-based system is used not only to perform management actions but also to structure and automate experimental process that has grown more complex with the use of Grids.

The agreement formed in this way promises to deliver an end-to-end QoS on execution time of the service as long as the execution is requested within a certain availability window. Delivering the QoS entails combining data transfers with application execution and CPU management. At the time of the experiment, the client can request service execution against a previously formed agreement and expect it to be satisfied with the agreed on QoS.

In the experiment our implementation and services discussed earlier were used to obtain agreements and claim execution of an end-to-end EFIT application service. The agreement based execution was triggered form the SC03 show floor in Phoenix Arizona when experimental data became ready. The execution comprised: (1) reservation-based remote execution of EFIT at Princeton Plasma Physics Lab (PPPL), and (2) data transfer to the control room team at General Atomics in California. Servers of each site executed on Pentium 4 1.5GHz CPU under Red Hat Linux 7.1.

	application service	data transfer service	end-to-end execution
Measured	95 (92-99) sec	54 (52-57) sec	173 (167-180) sec
Agreement	95 sec, 90%	53 sec, 93%	172 sec, 92%

The table above shows how our actual execution values compared to what was promised in the agreement. The "measured" row shows the mean of 10 values and their range for each quality measured. The "agreement" row shows promised value and the level of confidence with which it is promised. The results show good agreement with estimated values. The overhead (difference between total execution time and sum of application execution and data transfer time) is large mainly due to the fact that while the time spent on the respective services was measured locally, the end-to-end execution time was measured from the SC show floor accumulating the high latencies of acknowledgement messages from the services. Despite that, the overall execution time was satisfactory.

5 Conclusions and Future Work

Although our implementation provides only a simple negotiation model, we found that it fulfilled the needs of our use case very well. The negotiation phase worked well as a capability discovery customized to the needs of a client. In fact, some of our current agreements are used in "advisory" capacity and enable the scientist to do, in a structured way, what was previously done and an ad hoc manner: estimate times for codes that will be run during the experiment. Underpinning this interaction are the resource management actions ensuring the success of such preparations.

Given the dynamic and unreliable nature of a Grid environment, any guarantee must be qualified: resources may become unavailable or policies and priorities may change at any moment. Furthermore, while some qualities in the Grid can be managed (CPU reservations for example), others cannot: we cannot reserve bandwidth on the Internet or predict exactly the runtime of an application. For this reason, we have introduced *levels of confidence* used by the provider to represent the strength of a QoS guarantee. We modeled them as the probability that a certain QoS will be achieved. While this measure is correct from a provider's perspective, it is not very helpful for the client since it does not give it the means of verifying failure rate. However, with the addition of resource management it is possible to convert a provider's failure rate into a failure rate for a specific user. Such guarantee would be more appropriate from the perspective of our use case.

References

1. Foster, I., C. Kesselman, and S. Tuecke, *The Anatomy of the Grid: Enabling Scalable Virtual Organizations.* International Journal of High Performance Computing Applications, 2001. **15**(3): p. 200-222.
2. Keahey, K., M.E. Papka, Q. Peng, D. Schissel, G. Abla, T. Araki, J. Burruss, S. Feibush, P. Lane, S. Klasky, T. Leggett, D. McCune, and L. Randerson. *Grids for Experimental Science: the Virtual Control Room.* in *Challenges of Large Applications in Distributed Environments (CLADE).* 2004.
3. Pearlman, L., C. Kesselman, S. Gullapalli, B.F. Spencer, J. Futrelle, K. Ricker, I. Foster, P. Hubbard, and C. Severance, *Distributed Hybrid Earthquake EngineeringExperiments: Experiences with a Ground-Shaking Grid Application.* 13th International Symposium on High Performance Distributed Computing (HPDC-13), 2004.
4. Foster, I., *What is the Grid? A Three Point Checklist.* 2002: http://www-fp.mcs.anl.gov/~foster/Articles/WhatIsTheGrid.pdf.
5. Foster, I., C. Kesselman, J. Nick, and S. Tuecke, *The Physiology of the Grid: An Open Grid Services Architecture for Distributed Systems Integration.* 2002: Open Grid Service Infrastructure WG, Global Grid Forum
6. Foster, I., C. Kesselman, C. Lee, R. Lindell, K. Nahrstedt, and A. Roy. *A Distributed Resource Management Architecture that Supports Advance Reservations and Co-Allocation.* in *Proc. International Workshop on Quality of Service.* 1999.
7. K. Czajkowski, I. Foster, C. Kesselman, V. Sander, and S. Tuecke, *SNAP: A Protocol for Negotiating Service Level Agreements and Coordinating Resource Management in Distributed Systems.* 8th Workshop on Job Scheduling Strategies for Parallel Processing, July 2002.
8. Pearlman, L., V. Welch, I. Foster, C. Kesselman, and S. Tuecke. *A Community Authorization Service for Group Collaboration.* in *IEEE Workshop on Policies for Distributed Systems and Networks.* 2002.
9. Czajkowski, K., A. Dan, J. Rofrano, S. Tuecke, and M. Xu, *Agreement-based Grid Service Management (OGSI-Agreement) Version 0.* https://forge.gridforum.org/projects/graap-wg/document/Draft_OGSI-Agreement_Specification/en/1/Draft_OGSI-Agreement_Specification.doc, 2003.

10. Tuecke, S., K. Czajkowski, I. Foster, J. Frey, S. Graham, and C. Kesselman, *Grid Service Specification*. 2003: Open Grid Service Infrastructure WG, GGF
11. Foster, I., J. Frey, S. Graham, S. Tuecke, K. Czajkowski, D. Ferguson, F. Leymann, M. Nally, I. Sedukhin, D. Snelling, T. Storey, W. Vambenepe, and S. Weerawarana, *Modeling Stateful Resources with Web Services*. 2004: www.globus.org/wsrf.
12. Ludwig, H., A. Keller, A. Dan, and R.P. King, *A Service Level Agreement Language for Dynamic Electronic Services*. IBM Research Report RC22316 (W0201-112), January 24, 2002.
13. Nahrstedt, K., H. Chu, and S. Narayan. *QoS-aware Resource Management for Distributed Multimedia Applications*. in *Journal on High-Speed Networking, IOS Press*. December 1998.
14. Andrieux, A., K. Czajkowski, J. Lam, C. Smith, and M. Xu, *Standard Terms for Specifying Computational Jobs*. http://www.epcc.ed.ac.uk/%7Eali/WORK/GGF/JSDL-WG/DOCS/WS-Agreement_job_terms_for_JSDL_print.pdf, 2003.
15. Henderson, R. and D. Tweten, *Portable Batch System: External Reference Specification*. 1996.
16. Lao, L.L., H. St. John, R.D. Stambaugh, A.G. Kellman, and W. Pfeiffer, *Reconstruction of Current Profile Parameters and Plasma Shapes in Tokamaks*. Nucl. Fusion, 1985. **25**: p. 1611.
17. Madduri, R., C. Hood, and W. Allcock, *Reliable File Transfer in Grid Environments*. LCN, 2002: p. 737-738.18. Zhang, H., K. Keahey, and B. Allcock, *Providing Data Transfer with QoS as Agreement-Based Service*. submitted to IEEE International Conference on Services Computing (SCC 2004), 2004.

A Double Auction Economic Model for Grid Services

Liviu Joita[1], Omer F. Rana[1], W. Alex Gray[1], and John Miles[2]

[1] School of Computer Science and Welsh eScience Centre
l.joita@cs.cardiff.ac.uk
[2] School of Engineering, Cardiff University, UK

Abstract. The use of a service-based approach in Grid computing will lead to such services becoming valuable economic commodities. Current economics models on the Internet are concerned with the creation, sale and purchase of commodities (generally information resources), and ways in which buyers and sellers interact in markets for them. Although several initiatives are engaged in the development of Grid technologies, Grid economy issues are yet to be fully addressed. A "Grid market" architecture based on the double auction economic model is proposed. Components of the design, and a prototype of the double auction model framework, which makes use of the Globus and Java CoG toolkits, are subsequently presented.

Combining commodity technologies like Web Services with Grid technologies like Globus/UNICORE, has led to the development of Grid Services. This change is significant, as it enables the use and provision of such services on a variety of different platforms and operating systems. The price at which services can be sold depends on factors such as demand, supply and how the market operates. Participants in a Grid services market may be suppliers of computational or storage resources, or consumers as application or data users. We make the following assumptions within our market model: the demand for a particular Grid service is determined by the value users place upon the service, and the price they are willing to pay to obtain it. We also assume that the number of Grid services that are supplied in the market depends on how much suppliers can expect to charge for them, and costs of any existing similar services. The nature of competition amongst suppliers, how they interact with customers, and how the market is regulated play an important role on the pricing of Grid services. Based on the price charged and market rules, a provider has to recover cost and remain active on the market. The economic models and charging rules for Grid computing have to evolve according to market driven forces. A trading mechanism based on the auction economic model for Grid services should provide tools to deliver some overall value (or utility) to users, and it should allow them to express their requirements and overall goals.

The components of a Grid auction system, (referred to as the Grid Market Model Flow) based on a double auction model, are described. In a double auction system buyers and sellers are treated symmetrically, with buyers submitting *bids* and sellers submitting *asks*. The architecture combines workflow approaches with a resource trading model provided through the double auction system. We focus on the security and auction issues, which differs in focus from Grid Bank [6] and the procurement process (as investigated by others). The main contribution of this paper is the particular architecture for a Grid market, based on the notion of a Virtual Organisation (VO), and a partial implementation that demonstrates our ideas using Java CoG. The COllaborative VIrtual TEams (COVITE) project [8] develops a software tool in which such of VOs can form and evolve and offers also the possibility of developing and integrating instances of Grid markets. The Figure 1(a) shows the PSCD application integrated with the Grid market component. Resources can belong to multiple concurrent VOs, participants may buy or sell their resources (for example CPU time, CPU power and memory in order to host Slave Grid Services of their VOs) to Grid market instances, so the utilisation of their own surplus can be maximised. The paper is organised as follows: Section 1 illustrates related work of auction systems for Grid resources. Section 2 presents the Grid Market Model Flow architecture. Section 2.1 illustrates the double auction Grid economic market architecture scenarios and some implementation. Conclusion and further work follows in Section 3.

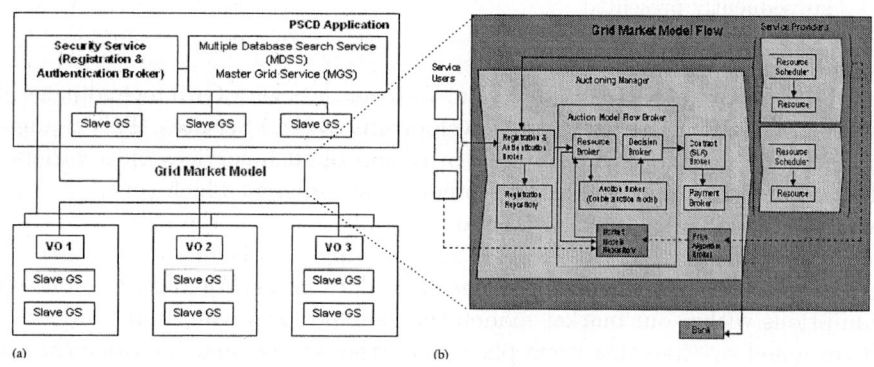

Fig. 1. Grid Market Module integration within the PSCD application.

1 Related Work

Various economic models including microeconomic and macroeconomic principles can be applied when sharing Grid resources in a market. Some of the commonly used economic models include: Commodity Market Model (service providers specify the service price and charge users according to the amount of service they utilise), Posted Price Model (similar to commodity market model except that there are special offers for services from time to time in order to attract more users), Bargain Model (resource providers offer their resources at a price as low as possible for use over longer time periods), Auctioning Model (such as those based on e-commerce, e-auctioning, e-marketplace, etc), Tendering/Contract-Net Model (based on the contracting mechanism – which governs the agreement between users and service providers for consuming the resources), Cooperative Bartering Model (resources are shared within a community – and members are allowed to barter for resources owned by the community), Monopoly (the case when a single service provider is the only one on the market offering a particular service and therefore dictates the resource price), Market-based control (whereby an 'artificial economy' is established for sharing resources) [2]. In addition to these economic models, different pricing schemes apply on the markets, such as: Flat price model (once the price is established it remains unchanged for a certain period of time), Competitive economic models (such as e-auctioning and e-marketplaces), Usage timing (peak/off-peak – like telephone services), Demand and supply based (prices change as supply and demand change), Foresight-based (based on the ability to model and predict responses by competitors), Loyalty based (special prices for loyal and regular users), Advance agreement/contract based (the price is agreed and a contract established a time before the use of resources) [2]. Some of these economic models and pricing schemes have already been used in different Grid projects to demonstrate their applicability in a Grid economy. In [3] the commodity market based approach to allocated resources is used, where resources are classified into different categories based on their hardware components, network connectivity and operating systems. An algorithm is presented to determine the price of the resources based on the demands of the resources (or bundled resources) and the supply for commodities. This price is based on the tâtonnement process – that is postulated in economic theory to explain price determination in competitive market. In such a scheme the allocation is not changed until the price reaches equilibrium. The initial price is pre-assigned. The resource is allocated on a first come-first serve base.

In [4] two economic models are used: commodities market and the Vickrey auction. First, the commodity market model treats disparate resources as interchangeable commodities. The supply and demand pricing scheme is applied to determine when transactions are completed. Second, the Vickrey (second-price sealed-bid) auction model is used for determining transaction prices for two commodities – the CPU (processor) and storage. In [5] a market based mechanism for trading CPU time is presented, in which the Vickrey (second-price sealed-bid) auction and a simple sealed-bid double auction (both buyers and sellers offer a low price and a high price, as well as a rate of change) are implemented.

Economic theory models have generally been used in the context of resource allocation, although the important question about which model is the *most* appropriate for supporting resource and allocation on the Grid is often not fully addressed. Often it is also difficult to pinpoint the exact model to use in a particular instance – requiring multiple models to co-exist. However, numerous applications of double auctions in electronic commerce, including stock exchange and financial markets, business-to-business commerce, market-trading of equities, commodities etc, have lead to a great deal of interest in fast and effective algorithms of the double auction model [8]. This economic model therefore provides a useful approach in the context of resource allocation on the Grid. We make use of this research, and feel that it best represents the exchange of resources within a Grid environment. The double auction model proposed aims to force the market to a theoretical equilibrium price, as there are posted "bids" (buy orders) or "asks" (sell orders) and the auction process rules determine prices. The Grid services have a price based on the demand, supply, and value in the Grid economy. The real-time process auction periodically closes the bidding process and chooses the winners. Members of the Global Grid Forum GESA working group [6] have started to define the protocols and service interfaces needed to support a variety of economic models for the charging of Grid Services using the Open Grid Service Architecture (OGSA). This effort is concentrated on developing an infrastructure to enable the trading of Grid services, as well as defining additional service data and ports needed to describe an economic Grid service. This has led to the definition of a "Chargeable" Grid Service and the Grid Banking Service, as a step to a Market for Computational Services. In [7] a computational economic framework for resource allocation and for regulating supply and demand in a Grid computing environment is presented. Various economic models from the real world market are described for setting the price for Grid resources such as: commodity market, posted price, tender and auction models, as described earlier in this section. The prices of the resources are based on supply and demand, and their perceived value to a user.

The architecture proposed in this paper is complementary to the GESA working group. It also describes an auction mechanism service, in part presented in [7], and shows the interaction with other components in a Grid market. The main focus is on an auction broker (making use of workflow operations) based on the double auction model. The advantage of the double auction model is that it provides a robust process of price discovery. Bidders can initially bid conservatively, and then raise their bids in response to market forces.

2 Grid Market Model Flow

A Grid market is a virtual place where Grid services are traded at specified time periods. The concept of a market may be related to that of a "Virtual Organisation" (VO). A typical Grid project often involves many individuals and institutions collaborating for the duration of the project – forming a VO. Each such VO may be an instance of a Grid market in which resources can be traded.

Grid Market Architecture: Components of the Grid Market Model Flow architecture are illustrated in Figure 1(b). A double auction model in which a 'bid' and 'ask' are submitted to a common Grid market any time during the trading period is used. An ask starts at the highest price and is decreased, and a bid starts at the lowest price and is increased. Customers issue a bid for the resources needed, and Grid resources issue an ask with the price and resources offered. If any bid and ask match their requirements and prices, the trade is executed. The Grid Market Model Flow architecture has two types of components: Brokers and Data Repositories – Brokers operate on the data in Repositories. The Brokers include a Registration & Authentication Broker, an Auction Broker, a Contract Broker, and a Payment Broker.

Registration & Authentication Broker is used for security management. The broker defines a security framework using the Globus Security Infrastructure (GSI) – requiring users to provide a private key and an X.509 certificate to authenticate themselves [1]. GSI provides a single sign-on capability, and supports delegation, via a proxy, to perform the authentication on users and Grid resources' behalf. Further details regarding the full implementation of the Registration & Authentication Broker can be find in [11]. *Auction Model Flow Broker* is used for auction management, and comprises of a Resource, Auction and Decision Broker. The Resources Broker is used to access the Market Models Repository for displaying the current active bids. The Resource Broker contains rules which apply to the market, such as: the definition of a bid string, the reserve price for the ask, time period over which an ask is alive on the market, etc. All participants have to agree with the rules defined in the Resource Broker in order to place their bids or asks in a market. The Auction Broker is used for the real-time auction process, while the Decision Broker decides when the auction process will close and a winner chosen. The Decision Broker also contains rules such as: how long a bid or an ask can stay alive, when a bid or an ask are fired, etc. *Contract Broker* is responsible for managing the Service Level Agreement (SLA) after the transaction is executed. The SLA is part of the contract that guarantees the quality and the performance of Grid services acquired during the auction process by users. An SLA must cover performance parameters, such as: networking parameters (e.g. bandwidth), computational parameters (e.g. number of processors, CPU time, CPU power, storage capacity, etc), the duration for which the Grid service is required, the cost of the transaction, etc. We explore the use of SLAs in other work [9]. *Payment Broker* is used to charge for the transaction via a Bank [6]. The Contract Broker initiates, and the Payment Broker closes the procurement process. The Bank is an established infrastructure used to transfer money from the account of a service user to a service provider accounts, to pay for Grid services purchased within the auction.

Data repositories include a *Market Models Repository*, which both service users and providers have access to. This repository contains a list of prices for different services available on the market. Service providers can define a fair price for their resources using the Price Algorithm Broker, and service users can issue a bid for resources with the price which is in demand on the market at that time. The second repository – *Registration Repository* is used to keep information about all participants on the market.

2.1 Scenarios

Two Grid economic market architecture case scenarios are presented from the point of view of how the market is defined: first, a common single market for trading resources, and second, a market that contains sub-markets for each type of resources traded.

Common Single Market: The following resources can be traded in a common single market: CPU time, CPU power, memory, and storage capacity. The auctions will run according to the following high-level protocol:
1. A Service User submits bids with preferences, threshold and expenditure limits (prices) to the Auction Broker.
2. A Service Provider submits their asks with a "reserved" price to the Auction Broker.
3. The Auction Broker updates its price quote every T minute, indicating the current going price(s). A value of 5 minutes for T is used as the default in our system.
4. The Auction Broker uses a schedule to decide for how long a bid and an ask are alive on the market.

The rules defined in the Decision Broker specify when or under what conditions a bid matches an ask, and record the transaction. A bid or an ask become inactive if they do not lead to a transaction within the scheduled time for which they are alive, as specified in the Auction Broker. Both a service user and provider may decide to once again issue a bid and an ask, or to retrieve them from the market.

The bid format is defined as a *bid string* – representing a willingness to buy or sell the service in an auction. A service user bid string contains a list of bid points in the following format: $(\gamma_j \tau_j, j=\{1,...,4\})$, where: τ_i is the bid substring in the following format: $\tau_i = (t_i, p_i)$. In this bid, parameter 1 represents CPU time, parameter 2 represents CPU power, parameter 3 represents memory, and parameter 4 represents storage capacity. t_i represents the type of the Grid service a participant is willing to buy. p_i represents the price of the Grid service a participant is willing to pay for. An Auction Broker uses the following rule: a service user is willing to buy services of type t_i for a price p_i, but not bigger than the expenditure limit prices defined when the bid is placed on the market. γ_i is the preference (threshold) parameter, and lies between 1 to 10. The threshold has to be chosen when a bid is placed on the market. A service provider bid string contains a list of bid points in the following format: $(\tau_k, k=\{1,...,4\})$, where $\tau_i = (t_i, p_i)$. Parameter definitions are the same as before. An Auction Broker uses the following rule: a service provider is willing to sell goods of type t_i for a price p_i, but not less than the reserved price defined when the ask is placed on the market. The bid is placed on the market with a threshold. The bid string contains preferences for each service traded on the market.

The following example illustrates the rules: consider a service user bid string of (10(1 £30), 2(2 £25), 4(3 £35), 7(4 £32)); the threshold is £5 and the expenditure limit £33 for each service; as soon as substrings 1 and 4 are fulfilled (because their preferences are bigger than the threshold) then the bid is fired and the transaction recorded; even if substrings 2 and 3 are not fulfilled, or only one of them is fulfilled. Alternatively, consider a service provider ask string of (-(1 £35), -(2 £0), -(3 £0), -(4 £0)) with a reserved price of £29. The requirement of the ask is fully fulfilled if the price offered for it is no less than the reserved price of the ask. If the price is less than the reserved price, the ask is fired but the transaction is not recorded.

Common Market with Multiple Sub-markets: We now consider a common market which contains multiple sub-markets – with each sub-market being used for trading specific services. As before, the common market contains four sub-markets: for CPU time, CPU power, memory, and storage capacity. The auction uses the same high-level protocols and assumptions as in the single market. A service user bid format is defined as follows: $(\gamma_i \tau_i)$, where $\tau_i = (p_i)$, in which p_i represents the price of the Grid service. The following rules will apply in the Auction Broker: (1) the service user is willing to buy the good at the auction, for a price p_i, but not bigger than the expenditure limit price. (2) γ_i is the preference (threshold), on a scale from 1 to 10 – chosen when a bid is placed on the market. The service provider bid string contains a list of bid points in the following format: (τ_i), where $\tau_i = (p_i)$, in which p_i is the selling price of the Grid service. As before, the service provider wants to sell the service at the auction for p_i, but not less its reserved price.

Consider the following scenario: if a user needs, for example, three out of four services, bids with their preferences may be placed on specific sub-markets, with a threshold for each bid. A transaction is fulfilled as soon as all the bids from sub-markets complete. If we consider the following service: 1=CPU time, 2=CPU power, 3=memory, and 4=storage capacity, and only 1, 2, and 4 are required, then the bids in specific sub-markets are as follows: for service 1, $10\tau_i$; for 2, $4\tau_i$; and for 4, $6\tau_i$. The threshold t chosen is 5. As soon as the bids with the $\gamma_i > t$ (service 1 and 4 here) have fulfilled their requirements, service user has traded on each sub-market, but for a price not bigger than the limit expenditure price. The transaction is recorded even if the bid with $\gamma_i < t$ (service 2 here) is not matching his requirements. Conversely, a service provider's ask is fully fulfilled if the price offered for it is no less than the reserved price of the ask. If the price is less than the reserved price, the ask is fired but the transaction is not recorded.

The prototype uses RedHat Linux 7.3, and Globus 2.2.4. The programming tools are: Java2 SDK Version 1.4.1_01, Java CoG version 1.1., Servlets, and JSP, with an Apache Tomcat Web server. MS-Access database is used as a Market Model Repository hosted by another machine running the WindowsXP operating system. Further development will upgrade the database to MS SQL Server, which is more robust and scalable than MS-Access. The Service Providers may make available their Grid resources and publish them in the Market Repository, as show in Figure 2. The interface to the Market Repository is an XML based Web Service. The Auction Broker can invoke and search the repository using Simple Object Address Protocol (SOAP) messages. The prototype currently supports security and auction brokers only – work is underway to add additional components.

3 Conclusion and Further Work

A Grid market architecture based the double auction theory is described. This auction model is the dominant market model for real-world trading of equities, commodities, derivatives, etc. This paper focuses on two Grid market model

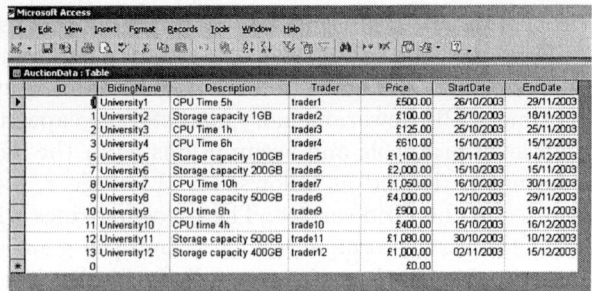

Fig. 2. The Market Repository.

architectures (the designing and development) of an auctioning system, based on the double auction model postulated in the economic literature, used for Grid resource acquisition based on the "supply and demand" principle.

References

1. Globus Security. See Web site at: www.globus.org/security
2. R. Buyya, D. Abramson, J. Giddy, and H. Stockinger, "Economic Models for Resource Management and Scheduling in Grid Computing". Available at: http://hst.home.cern.ch/hst/publications/emodelsgrid.pdf, 2002.
3. K. Subramoniam, M. Maheswaran, and M. Toulouse, "A Micro-Economic Model for Resource Allocation in Grid Computing Systems", IEEE Canadian Conference on Electrical and Computer Engineer (CCECE '02), May, 2002.
4. R. Wolski, J. S. Plank, T. Bryan, J. Brevik, "G-commerce: Market Formulations Controlling Resource Allocation on the Computational Grid", International Parallel and Distributed Processing Symposium (IPDPS), San Francisco, CA, April, 2001.
5. O. Regev, N. Nisan, "The POPCORN Market - an Online Market for Computational Resources", First International Conference on Information and Computation Economies ICE-98, October 25-28, 1998.
6. S. Newhouse, "GWD-R (draft-ggf-gesa-services-1)", Global Grid Forum, June 2003.
7. R. Buyya, "Grid Economy: A Market Paradigm for Distributed Resource Management and Scheduling for Service Oriented Grid Computing", PhD Thesis, Monash University, Australia, 2002.
8. D. Friedman, J. Rust, "Double Auction Market: Institutions, Theories, and Evidence", Perseus Publishing, 1993.
9. R. Al-Ali and O. Rana and D. Walker and S. Jha and S. Sohail, "G-QoSM: Grid Service Discovery Using QoS Properties", Computing and Informatics Journal, Special Issue on Grid Computing, Vol. 21, No.4, pp 363–382, 2002.
10. COVITE project Web site: http://www.wesc.ac.uk/projectsite/covite/
11. L. Joita, O. Rana, P. Burnap, J. S. Pahwa, A. Gray, J. Miles, "A Grid-Enabled Security Framework for Collaborative Virtual Organisations", 5th IFIP Working Conference on Virtual Enterprises (PRO-VE), Toulouse, France, 22-27 August 2004 (to appear).

Towards a Monitoring Framework for Worldwide Grid Information Services

Serafeim Zanikolas and Rizos Sakellariou

Department of Computer Science, The University of Manchester, Manchester, UK
{zanikolas,rizos}@cs.man.ac.uk

Abstract. Despite important advances in Grid computing, scalability and interoperability issues obstruct the integration of the existing isolated grids into *the* Grid. We propose a worldwide monitoring framework which forms the basis for building worldwide grid information services that can have a global notion of the Grid. This work motivates the monitoring framework in terms of applications and challenges, sets its architecture and design space, and offers indicative performance results in a variety of hypothetical Grid instances.

1 Introduction

Grid computing is emerging as a new paradigm for on-line resource sharing and collaboration; numerous yet small and isolated grids exist as of early 2004. The existing grids, however, are missing the properties that are required for scaling and integration into *the* Grid (i.e., the capitalisation denotes a worldwide Grid where millions of nodes can collaborate to form "virtual organisations" as needed), in the same way that the Internet bridges together dispersed networks around the globe. The technical issues of this integration include the restructuring or extension of basic grid services in order (i) to satisfy the *scalability* and *robustness* required for operation into the Grid, and (ii) to support *interoperability* between grids with potentially different middleware implementations. Among the basic grid services, information services (e.g., [3]) include the collection of data about resources (monitoring) and the use of those data to support resource discovery and enquiry. However, existing information services are not interoperable, and suffer in terms of scalability because they are intended for small to medium size deployments. In this respect, the execution of a large scope resource discovery query (e.g., search throughout the Grid for the cheapest resource with given characteristics) would need a significant amount of time.

This paper proposes a worldwide monitoring framework that operates on top and hides the diversity of existing monitoring and information services, and thus supports the visibility and eventually utilisation of resources across grids. The suggested framework can cope with the scalability requirement of a Grid because resource information is collected in advance of queries, as opposed to on demand. Consequently, the overhead is proportional to the total number of grid resources, in contrast to the total number of query requests, which is the case for existing information services. The framework provides a uniform and general-purpose interface for worldwide grid information services, and allows analysis of the Grid in a similar way Netcraft and Alexa (netcraft.com, alexa.com) analyse the World Wide Web. Besides motivating a worldwide monitoring framework and setting the design issues that need to be addressed for a scalable, flexible, and robust realisation, an additional contribution of the paper is a feasibility experiment with simulation results that indicate the performance of a minimal version of the proposed framework in a variety of hypothetical Grid instances. This paper considers architectural and performance issues of the framework but does not contribute towards interoperability.

The next section outlines potential applications of the suggested monitoring framework and identifies the involved challenges. Section 3 proposes a general architecture and highlights the relevant design space. Section 4 presents simulation results to illustrate the framework's scalability and Section 5 concludes.

2 Applications, Challenges and Related Work

The proposed monitoring framework can be used two-fold: (i) to build efficient, Grid-wide resource discovery end-user applications or services, such as search engines or directories of grid resources, or (ii) to provide a variety of "global view" services to enhance the understanding of the Grid.

Grid search engines, as envisaged in [4], may provide an efficient means of resolving complex, large-scope queries (i.e., involving the combination of information from many organisations.) However, end-users may not be able to specify exactly their needs or be unaware of the resources that are available throughout the Grid. In this respect, *resource directories* could support end-users to find the appropriate matches through a browsable list of resources, offered in a variety of classifications. The latter can be based on criteria such as discipline (in case of collaborations, data repositories, and on-line instruments), sharing policy, physical location (e.g., continent, country, city), administration hierarchy (e.g., virtual organisation, organisation, department), supported quality of service guarantees, and hardware and software specifications.

On the other hand, *global view services* will provide a comprehensive quantitative and qualitative characterisation of the Grid throughout time. This will help to understand the dynamics of the Grid, support long-term capacity planning and contribute to its realistic modelling and simulation by answering questions such as: how many are the grid resources in the public (commercial) Grid; what is their distribution among sites; which resource types are rare or popular; which grid sites are most reliable; how these numbers evolve throughout time; and how does the Grid's evolution relates to that of the Internet.

Challenges. A realisation of the proposed monitoring framework needs to address a set of contradicting requirements; namely scalability, extensibility, adaptivity, flexibility, completeness and freshness of the collected information. The system must be *highly scalable* with respect to the number of resources available throughout the Grid, and to the imposed network intrusiveness. The latter is a crucial concern given the high utilisation costs of wide area network links. The design must also allow for *extensibility* to readily accommodate the new types of resources that are introduced for sharing through the Grid. This concern has motivated the grid and web communities to adopt XML and RDF based representations (www.w3.org/{XML,RDF}); their applicability on a worldwide monitoring framework, however, should be investigated with respect to the stringent scalability requirement. The monitoring framework also needs to be *adaptive* to unforeseen conditions, such as a rapid growth of the number of grid resources or an Internet slowdown caused by a worm. In addition, it must be *flexible* with respect to (i) the overhead that resource providers are willing to accept; and (ii) the supported query language for resource-related information retrieval.

Moreover, the framework has to support a high degree of coverage of the Grid. Although this *completeness* may not always be needed, it is what differentiates the proposed framework from the existing, scope-limited grid information services. The completeness requirement is an engineering challenge (e.g., to achieve the required scalability and accommodate the diversity in monitoring and information systems) as well as a political one (e.g., why should resource providers expose detailed information concerning their resources.) Last, the *freshness* of the collected information refers to its correctness with respect to the dynamic nature of resources. Obviously optimal freshness cannot be achieved given the scale and dynamism of the Grid. Highly dynamic features of resources (e.g., CPU load) could be considered as statistically annotated aggregates.

Related Work. Many research activities relate to the integration of grids, including middleware interoperation [2] and semantically rich (e.g., domain dependent) resource matching [5]. In the context of large-scale information services, however, the only work we are aware of is that of Balaton *et al* [1]. In contrast to our work which extends the web crawler paradigm, their proposal resembles the architecture of the USENET Internet service. In particular, their system appears to have a more decentralised structure; it would be interesting, however, to see how a USENET-like system could support user-level flexibility of data collection, delivery and scheduling policies (discussed in the next section).

3 Overall Architecture and Design Space

This section describes the considered framework's architecture and the relevant design space that needs to be explored. The presented architecture is a generalised extension of the one described in [4]. For the needs of the following discussion, resource provider and resource information collector sites are referred as *grid* and *collector sites* respectively. Grid sites host resources that are exposed through the monitoring framework; collector sites systematically collect information about available resources. The exact granularity of grid sites (e.g., departments or organisations) is an issue that needs to be investigated. Generally, higher granularity suggests more potential for aggregating resource information, but also more network links affected by the traffic generated by the monitoring process. The proposed architecture consists of proxies, crawlers, and collectors (Fig. 1). *Proxies* are located on grid sites for continuously collecting data about local resources (i.e., configuration and status properties), either from the actual resources or from the local information services. Also, proxies aggregate and normalise these data according to a universal and compact data representation. On the collection side, multi-threaded *crawlers* traverse the network of previously discovered proxies for retrieving monitoring information, which is later used in application-specific ways. Additionally, proxies actively push a minor subset of the monitoring information to multi-threaded *collectors*, located at the collector site, to achieve timely delivery of previously designated important events (e.g., that an important resource has gone through a hardware upgrade.) Note also that proxies, in addition to the important functionality already described, are necessary to achieve the decoupling, in time and space, between resources on the one hand, and crawlers and collectors on the other[1].

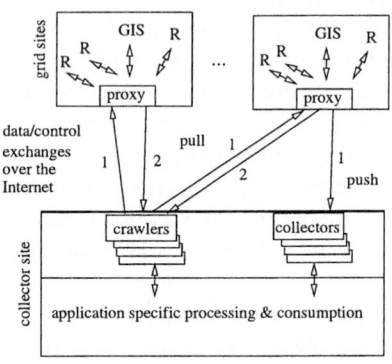

Fig. 1. Overall Architecture of the proposed monitoring framework (R:Resources, GIS:Grid Information Services).

Design Space. The described architecture sets the components and their interactions, but still many design variations need to be evaluated. These relate to four core themes: resource representation and addressing; data collection and reduction; data delivery and scheduling policies; and proxy membership management.

An extensible, expressive, yet highly compact *resource representation* is needed to capture, in addition to standard resource properties, the multi-dimensional relationships between resources (e.g., physical, logical, administrative.) Additional information regarding usage policies and disciplines needs to be supported. Another issue to be tackled is the provision of a universal *resource addressing scheme* that, among others, features high performance filtering; property-, structure-based schemes (e.g., feature vectors and uniform resource identifiers (URIs) respectively) and combinations of them have to be considered.

Exchanges between proxies and crawlers/collectors need to be relatively rare (given the sheer number of proxies.) To this end, crawlers have to set, and communicate to proxies, a *data collection policy* concerning the type, level of detail, and temporal granularity (e.g., fifteen-minute average values of, say, processor load) of the desired information. Proxies systematically collect and keep this information for the period between consecutive crawler "visits." The collection

---

[1] Time decoupling means that two parties need not be on-line at the same time in order for them to communicate; space decoupling means that the parties need not know the other end's address/location.

activity may vary significantly depending on the capabilities of local information services (e.g., lack of subscriptions requires polling.) In addition, *data reduction* techniques need to be evaluated for minimising the size of the transmitted information, and therefore the imposed network overhead.

A *data delivery policy* determines whether proxies passively provide (pulled from crawlers) or actively send (pushed to collectors) a given type of information. This should be decided on the basis of a given information's lifetime and significance, as well as the importance of the associated resource. Once such a policy is in place, a *scheduling strategy* is needed to determine the frequency and ordering of pull and push exchanges, in order to achieve optimum freshness of the delivered information, taking into account the resource limitations of the collector site (e.g., an inappropriate scheduling of pushes could easily overwhelm a collector site's network or processing capacity.)

An equally important design axis is the effective *management of proxy membership*. That is, the procedure that determines how new proxies join the monitoring framework (e.g., register with a collector site or passively discovered.) Also, collector sites have to set and identify the conditions in which a proxy's membership should be maintained or terminated (e.g., during a temporary network partition or a permanent shutdown of a grid site).

4 Simulation-Based Preliminary Results

A simulation-based evaluation (based on SimPy, `simpy.sourceforge.net`) of a minimal version of the framework was carried out to illustrate its scalability and estimate its performance in terms of the time needed for a (theoretically) complete Grid crawl. Although performance needs may vary among different applications of the framework, a Grid crawl per 24 hrs is assumed for resource discovery purposes. (Note that the framework *is* suitable for more frequently changing information through the use of push updates, which however are not considered in this simulation.) At this point, data reduction, failures, access control, and proxy management issues are not considered (i.e., proxies are assumed to be known to the collector site.) Proxies are collecting information about the latest configuration of resources and their status since the previous crawler visit; crawlers visit proxies sequentially and retrieve all information for all resources (i.e., no filtering is applied.) No collectors exist since proxies do not push any information.

A collector site has a 16 Mbps full-duplex Internet connection, and adequate crawler hosts to saturate the connection. Connections between grid sites and the collector site are assumed to have an average end-to-end bandwidth of 256 Kbps. The end-to-end latency is sampled from a normal distribution, with parameters $\mu = 0.2, \sigma = 0.1$ (seconds.) Given the network settings and an average resource description size, the total transmission time of crawler and proxy messages is calculated by taking into account packet size, end-to-end bandwidth and latency.

The simulation has been run against three Grid instances (see Tab. 1), each one consisting of 10^6 grid sites, with a different static distribution of resources among sites (10 resources each, uniform in the range [1,100] and upper-bounded Zipf(a=1.5), respectively.) The number of average crawls per second and the needed time for a complete Grid crawl may be misleading because the three Grid instances do not have the same total number of grid resources and the same number of resources per grid site. A more useful metric, the average crawled resources per sec (see Tab. 1), suggests that the performance is dependent on the average number of resources per site; this is because the cost of network latency dominates the cost of network connections to proxies with only a few associated resources. This further advocates higher granularity of grid sites (e.g., organisation vs department.) The distribution of resources among sites does not seem to affect performance significantly except in the Zipf instance probably due to proxies with numerous associated resources being visited at the end of the Grid crawl (leaving the collector site's bandwidth under-utilised since every connection can have only up to 256 Kbps.) Last, an estimation of the achievable number of crawls of a Grid consisting of 10^7 resources within 24 hrs, ranging from more than 2 or 3 crawls per 24 hrs, is quite encouraging with respect to scalability, especially considering that this is only a minimal version of the framework with no elaborate optimisations (such as not re-retrieving valid data.)

Table 1. Configuration & results of a simulation of the framework in three different Grid instances.

	balanced	uniform	Zipf
total resources	1e7	5e7	6.315e6
avg resources/site	10	50.0089	6.3831
sites with one resource	0	10173	607894
avg crawled resources/sec	362.5294	399.0035	312.1786
Grid crawls per 24 hrs assuming 10^7 resources	3.1323	3.4474	2.6972

5 Summary

This paper introduced and motivated a monitoring framework that forms the basis of (to be developed) worldwide information services, which are important for the bridging of grids into the Grid. The paper proposed a general-purpose architecture, identified the relevant design space, and presented preliminary performance results to illustrate the framework's scalability in a variety of hypothetical Grid instances. Further exploration of the design space is needed to fully define a working framework and improve overall performance.

References

1. Balaton, Z., Gombás, G., Németh, Z.: Information System Architecture for Brokering in Large Scale Grids. In: Proc. of DAPSYS 2002, Linz, Kluwer (2002) 57–65
2. Brooke, J., Fellows, D., Garwood, K., Goble, C.: Semantic Matching of Grid Resource Descriptions. In: Proc. of AxGrids04, (2004)
3. Czajkowski, K., Fitzgerald, S., Foster, I., Kesselman, C.: Grid Information Services for Distributed Resource Sharing. In: Proc. of HPDC-10 (2001)
4. Dikaiakos, M., Ioannidis, Y., Sakellariou, R.: Search Engines for the Grid: A Research Agenda. In: Proc. of AxGrids03, Santiago de Compostela, Spain (2003)
5. Miles, S., Papay, J., Decker, K., Moreau, L.: Towards a Protocol for the Attachment of Semantic Descriptions to Grid Services. In: Proc. of AxGrids04, (2004)

A Model for a Component Based Grid-Aware Scientific Library Service

S. Lombardo and A. Machì

ICAR/CNR Department of Palermo
{s.lombardo,machi}@pa.icar.cnr.it

Abstract. The paper presents a hierarchical model for integrating structured HPC legacy software modules into a grid-adaptive scientific library service able to deliver quality computing-service on a performance contract basis. The service proactively administers both computing resources and configurable library software modules on behalf of client applications, and seamlessly supports their re-configuration according to grid-aware strategies of active managers of client applications. The library administrator logic mimics functionalities of components containers of service-oriented architectures. The authors discuss the coordination of library-serviced components into grid-aware applications composed according to the programming model being developed by the Italian *Grid.it* Project. Sample architecture for the library service, based on *Grid.it* component technology, is sketched.

1 Introduction

A scientific library is a set of structured software modules developed to solve problems in a specific domain and it is used to build applications according to the Software Engineering reuse concept. In specific domains, where high performance computing is needed, the library is often implemented by encapsulating parallel and distributed computation. In particular, the structured parallel programming approach has embodied such knowledge into patterns for the management of set of processes described by notable Processes Graphs, called *skeletons* and *parmods* [1]. Skeletons are automatically coded by a compiler to keep high parallelism efficiency and software portability, while maintaining low user parallel programming effort.

In the past decade code has been developed with such a structured approach for computing environments mapped on networks of static resources, controlled by stable policies and providing services for exclusive resource allocation.

When deployed on grid environments managed as a Virtual Organization of resources dynamically discoverable and shareable, such a code is inefficient because of the unreliable behaviour of resources, intrinsic in the Virtual Organization model [2].

To interact properly with other grid systems parallel modules must be encapsulated in services, enabling the scientific library on the grid, according with the checklist definition of a Grid system that: "coordinates resources that are not subject to centralized control", "using standard, open, general-purpose protocols and interfaces", "to deliver non trivial qualities of service" [3].

The first requirement implies allowing serviced library modules be orchestrated by client application. The second feature requires adoption of OGSA compliant interfaces [4], protocols and life cycle support service. Last requirement implies to implement a complex software engine for contract submission, violation detection (monitoring) and decision about reconfiguration of allocated resources.

To deliver non-trivial Quality of Service the library service should be able to honour performance contracts tailored to support both application biased and system biased optimisation strategies. Sample templates for performance contract are:

1. **Real-Time:** (application optimisation) execution of the library module with respect of performance indexes as maximum execution time, or minimum number of instructions per second (template used in Project GRADS[5]).
2. **Resource Co-allocation:** (application optimization) execution on specific nodes and links, with management of grid-resources performed by client application manager or framework execution manager as in TRIANA [6]).
3. **Priority:** (resource optimisation) pre-emptive best-effort optimisation on a pool of grid resources administered.
4. **Low cost:** (resource optimisation) cheapest service cost, execution on free grid-resources (supported by grid middleware like CONDOR).

Several grid projects focused on reuse of executable legacy code describe application workflow as a directed a-cyclic graph including data as pre/post conditions, control tasks and library execution tasks. They commit graph enacting to a centralized process operating as a job manager. The manager performs optimal mapping of executable tasks at deploy time on available computing resources according to performance profiles maintained by network-wide performance monitor services.

In Netsolve [7], a moderate degree of fault tolerance is maintained by an agent, attempting to find among the pool of available grid resources an appropriate server to optimally service client requests, keeping track of failed servers.

In GRADS the application development environment recompiles library source code and generates a configurable object program instrumented for performance monitoring. The run-time support monitors execution and reconfigure the application workflow. Library cost models are required to be known to the programming environment at compile time for proper object code optimisation.

In this scenario we propose a component-based model for implementing an HPC library service, enabling code execution on grid resources on behalf of client workflows, subject to quality performance constraints. The required grid-awareness is distributed among various software elements of the environment playing different roles [8] developed with component technology and interacting through ports

Section 2 introduces the reference component architecture. Section 3 sketches the distributed adaptivity model with entity and actors involved. Section 4 describes the integration of the adaptivity model with application management strategy.

2 The *Grid.it* Component Model

Software component technology is a young programming-paradigm, even though its definition is quite old. Its aim is to enable the development of applications by composing existing software elements in an easy way. The technology used for implementing component-to-component binding depends on required performance, interoperability and may change in according the scope of the connection. The Globus Project proposed the OGSA architecture for grid services and a component architecture for adaptive grid programming OGSA compliant has been defined in [4].

A new component architecture focusing on HPC grid programming is presently being developed by the Italian national *Grid.it* project [9]. *Grid.it* components are intended to support design of HPC applications over a grid middleware. In the working architecture model [10] components expose their functionalities through a series of interfaces that differ for the interaction paradigm:

- **RPC:** (required) interface conforms to the Remote Procedure Call standard model;
- **Event:** ((optional) interface receives and to send asynchronous events;
- **Stream:** optional interface implements unidirectional data-flow channels;
- **Configuration (active and introspection):** (required) interface supports component configuration and status monitoring

An application is modelled by composing compatible components in a graph that:

- may change arcs connecting components during the execution of the application;
- may have connections to components implemented according to other open standard component architectures (CCM, Grid Services...);
- includes a logic unit named *application manager* able to instantiate other application components, to connect them and (re) configure the application graph.

The *application manager* may be implemented as a single component or as a coordination of components implementing the application coordination *strategy*.

3 A Hierarchical Component-Based Model for Grid Adaptivity

As mentioned above, we suppose that grid-adaptivity can be modelled using different actors playing hierarchically cooperative roles. These roles model may be mapped onto a component-based grid software infrastructure.

At top level stays the *active resource&execution manager*: its role involves (re) selection of proper resources (nodes and library). It maintains grid discovery ability, detailed grid-awareness, reservation privileges and an adequate policy to coordinate resource provision in order to ensure application performance (ability in performance contract negotiation).

At second level is the *proactive resource administrator:* this role requires definition of each application performance in terms of a performance contract, monitoring of performance and a policy for reconfiguration. It embodies library cost models It represents the front-end of the library-services and its goal is to monitor contracted performance and to adapt management of available grid resources for optimal execution of the library modules. This action takes advantage of self-optimisation capability embodied in parallel skeleton templates.

A *reactive quality-service coordinator* is already implemented in some parallel skeletons. His role is load (re) balancing of physical processes over a cluster of virtually privates inhomogeneous resources labelled with their effective quality indexes plus partial reconfiguration of the processor graph after in the event of their variations.

Monitoring of resource status, support for application deployment, detection and registration of events requiring attention and possible adaptation may be performed by a *passive resource coordinator* staying at lowest level.

The *resource administrator* and the *resource coordinator* roles may be assigned to grid middleware while the quality-service coordinator role is assigned to skeletons. The *resource administrator* mimics functionalities of components containers of service-oriented architectures. The *active resource&execution manager* instead are external actors for the proposed service architecture and they may be elements of a Problem Solving Environment (PSE) [12], of a Grid-Portal or of a generic environment for grid programming. For example, the Application Manager of the Grid.it component-programming model could play the execution manager role.

4 Integration of the Library Server with Grid.it Application Strategy

Our proposed library server is a grid-aware application, which exposes to several external grid applications a service for library modules orchestration.

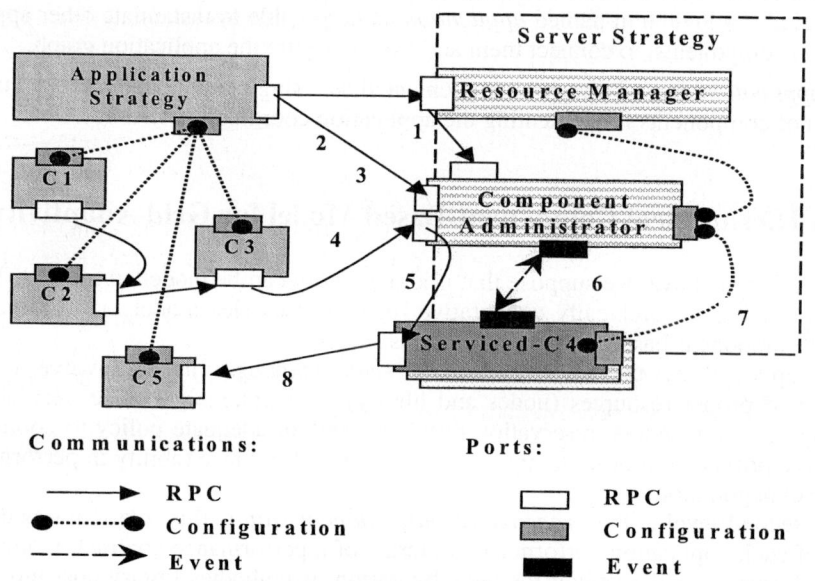

Fig. 1. Client Application interaction

Even if it uses the *Grid.it* component architectural model the library service provides an open standard OGSA-based interface to not limit client applications architecture. *Grid.it* supplies a rich and expressive model to design a hierarchical and distributed application management that is particularly useful for the implementation of our hierarchical roles.

Figure 1 shows a sample a scenario of interaction between the library server and a *Grid.it* client-application to implement a two-fold optimisation strategy.

In this scenario, RPC library module ports are not available to client applications because they are hidden by the *Administrator*, but library module uses RPC port could be connected with other components as established by the strategy of client application *Manager* during the factory operation. The *Administrator* schedules client service requests in a priority queue and processes them on the basis of their associated performance contract. According to the contract, an application biased or a global resource policy is selected when selecting the proper subset of pool resources to map the Virtual Process Graph of each request. Requests are served when the minimum required resource set is available.

Here is a description of the interaction sequence between client application and the library server:

- The *Resource Manager* selects a pool of appropriate resources (hosts of the VPG) and library modules (discovered from an external repository) and communicates them to the Component Administrator (1).
- The *Component Administrator* acquires the selected resources and updates its internal *register* to provide the available library modules.
- The *Resource Manager* after investigating about the available libraries (2), negotiates invocation of a library module via the Component Administrator *factory* port (3) with a contract specifying performance template and workflow information like bindings URI use ports. The administrator deploys (if not already done) the library module over each node in the pool independently from its effective configuration.
- When the library service is invoked (4) by a client *Application* component the *Administrator* schedules the request and, when the requests come to queue top, it dispatches the request to the *Library module* (5).
- Start the contract monitoring with adaptivity (re) configuration. Using the event bus provided by the run-time support, the *Library module* throws events (6) to communicate the progress status of its job (checkpointing). It can also register itself to receive events regarding workflow modifications. The *Manager* monitors the respect of performance contracts and in case of violation it issues a (re) configuration command. The *Administrator*, in turn, throws proper (re) configuration events to *Library module*, which performs self-reconfiguration.

5 Conclusion

A hierarchical model for coordinating a grid-aware library service into optimised application workflows has been presented.

The service configures parallel library modules for execution with different QoS modalities. For each module it exposes several performance contract templates according to its capabilities of grid resource management and job queue scheduling policy.

The client application manager performs application workflow optimisation by selecting the most convenient performance template and delegates to the service administrator local control (deployment, enactment, monitoring) of the service workflow sub-graph, constrained by the specifications of the performance contract.

A prototype of the library service is currently being developed as a task of the Scientific Libraries WorkPackage of the Italian Grid.it project.

References

1. M. Vanneschi: The programming model of ASSIST, an environment for parallel and distributed portable applications. Parallel Computing 28(12): 1709-1732 (2002).
2. F.Berman, G.C. Fox, A.J.G.Hey: Grid Computing. Making the Global Infrastructure a Reality. Wiley 2003.
3. What is the Grid? A Three Point Checklist. I. Foster, GRIDToday, July 20, 2002
4. R. Armstrom, D. Gannon, K. Keahey, S. Kohn, L. McInnes, S. Parker, and B. Smolinsk. "Toward a common component architecture for high-performance scientific computing". In Conference on High Performance Distributed Computing, 1999.
5. S. Vadhiyar and J. Dongarra, "GrADSolve - A Grid-based RPC system for Remote Invocation of Parallel Software", Journal of Parallel and Distributed Computing, February 2003.
6. I. Taylor, M. Shields and I. Wang. "Resource Management of Triana P2P Services". Book, Grid Resource Management, edited by Jan Weglarz, Jarek Nabrzyski, Jennifer Schopf and Maciej Stroinski, June 2003.
7. S. Agrawal, J Dongarra, K Seymour and S. Vadhiar "NetSolve: past,present and future – a look at a Grid enabled server" in Grid Computing: Making the grid infrastructure a reality.J. Wiley and Sons 2003.
8. A. Machì, S. Lombardo "A conceptual model for grid-adaptivity of HPC applications and its logical implementation with components technology" ICSSA04/AGCPA Krakow, Poland June 2004.
9. M. Vanneschi "Grid.it: a National Italian Project on Enabling Platforms for High-performance Computational Grids" GGF Intern. Grid Summer School on Grid Computing Vico Equense Italy July 2003 www.dma.unina.it/~murli/SummerSchool/session-14.htm
10. M. Aldinucci, M. Coppola, M. Danelutto, M. Vanneschi, C. Zoccolo "Grid.it component model" Project Grid.it WP8 Deliverable, Jan 2004.
11. A. Pugliese, D. Talia, "Application-Oriented Scheduling in the KNOWLEDGE GRID: A Model and Architecture" Proc. ICCSA 2004 Perugia Italy 2004, LNCS 3044, pp55-65.
12. F. Berman, A. Chien, K Cooper, J Dongarra, I. Foster, D. Gannon, L Johnson, K Kennedy., C Kasselman, J Mellor-Crummery, D Reed, L Torczon,R Wolski: "The GrADS Project: Software Support for High-Level Grid Application Development.

Developing Grid-Aware Applications with DRMAA on Globus-Based Grids*

J. Herrera[1], E. Huedo[2], R.S. Montero[1], and I.M. Llorente[1,2]

[1] Departamento de Arquitectura de Computadores y Automática, Universidad Complutense, 28040 Madrid, Spain
[2] Laboratorio de Computación Avanzada, Simulación y Aplicaciones Telemáticas, Centro de Astrobiología (CSIC-INTA), 28850 Torrejón de Ardoz, Spain

Abstract. In spite of the great research effort made in Grid technology in the last years, application development and execution in the Grid continue requiring a high level of expertise due to its complex and dynamic nature. The Distributed Resource Management Application API (DRMAA) has been proposed to aid the rapid development and distribution of applications across the Grid. In this paper we present the first implementation of the DRMAA standard on a Globus-based testbed, and show its suitability to express typical scientific applications. The DRMAA routines have been implemented using the functionality provided by the GridWay framework.

1 Introduction

In recent years a great research investment has been made in Grid computing technologies. However, deployment of existing applications across the Grid requires a high level of expertise and a significant amount of effort. The most important barriers arise from the nature of the Grid itself:

- Complexity, in order to achieve any functionality the user is generally responsible for manually performing all the scheduling steps.
- Heterogeneity of Grid resources, that potentially belongs to multiple administration domains.
- Dynamism of the availability, cost and load of Grid resources.
- High fault rate, resource or network failures are the rule rather than the exception.

In a previous work [1] we have developed a Globus submission framework, GridWay, that allows an easier and more efficient execution of jobs on a dynamic Grid environment in a "submit and forget" fashion. GridWay automatically performs all the job scheduling steps [2] (resource discovery and selection, and job preparation, submission, monitoring, migration and termination), provides fault recovery mechanisms, and adapts job execution to the changing Grid conditions [3].

---

* This research was supported by Ministerio de Ciencia y Tecnología through the research grant TIC 2003-01321 and Instituto Nacional de Técnica Aeroespacial.

On the other hand, the Grid lacks of a standard programming paradigm to port existing applications among different environments. Grid technologies are specified, standardized and implemented within the *Global Grid Forum* (GGF)[1] framework. GGF is structured in working and research groups focused on specific aspects of Grid Computing. In particular, the *Distributed Resource Management Application API Working Group* (DRMAA-WG)[2] has developed an API specification for job submission, monitoring and control that provides a high level interface with *Distributed Resource Management Systems* (DRMS). In this way, DRMAA could aid scientists and engineers to express their computational problems by providing a portable direct interface to DRMS.

There are several projects underway to implement the DRMAA specification on different DRMS, like Sun Grid Engine (SGE) or Condor. However, to the best of the authors' knowledge, DRMAA has never been implemented in a Globus-based DRMS. In this work we discuss several aspects of the implementation of DRMAA within the GridWay framework, and investigate the suitability of the DRMAA specification to distribute typical scientific workloads across the Grid.

In Section 2, we describe the DRMAA standard and its implementation. Section 3 analyzes several aspects involved in the efficient execution of distributed applications, and how they are addressed by GridWay. Then, in Section 4, we study the implementation of several applications using DRMAA. Finally, the paper ends in Section 5 with some conclusions.

2 Distributed Resource Management Application API

One of the most important aspects of Grid Computing is its potential ability to execute distributed communicating jobs. The DRMAA specification constitutes a homogenous interface to different DRMS to handle job submission, monitoring and control, and retrieval of finished job status. In this sense the DRMAA standard represents a suitable and portable framework to express this kind of distributed computations.

Although DRMAA could interface with DRMS at different levels, for example at the intranet level with SGE or Condor, in the present context we will only consider its application at Grid level. In this way, the DRMS (GridWay in our case) will interact with the local job manager (i.e PBS, SGE,...) through the Grid middleware (Globus Toolkit 2.2). This development and execution scheme with DRMAA is depicted in figure 1.

In the following list we describe the DRMAA interface routines implemented within the GridWay framework:

– Initialization and finalization routines: `drmaa_init` and `drmaa_exit`.
– Job template routines: `drmaa_set_attribute`, `drmaa_allocate_job_template` and `drmaa_delete_job_template`. These routines enable the manipulation of job definition entities (job templates) to set parameters such as the executable, its arguments or the standard output streams.

---
[1] http://www.gridforum.org (2004)
[2] http://www.drmaa.org (2004)

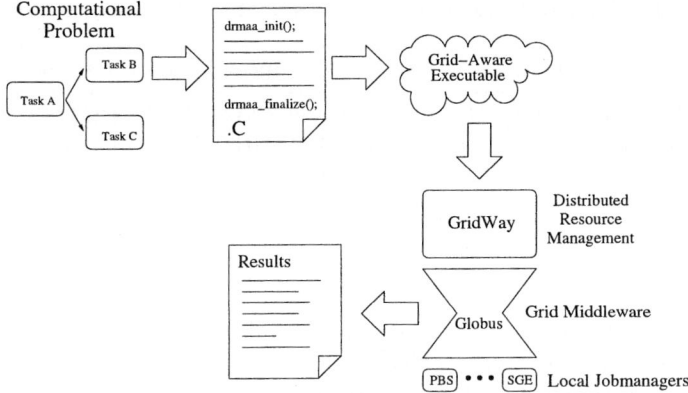

Fig. 1. Development and execution cycle using the DRMAA interface.

- Job submission routines: drmaa_run_job and drmaa_run_bulk_jobs. The GridWay has native support for *bulk* jobs, defined as a group of n similar jobs with a separate job id.
- Job control and monitoring routines: drmaa_control, drmaa_synchronize, drmaa_wait and drmaa_job_ps. These routines are used to control (killing, resuming, suspending, etc..) and synchronize jobs, and monitor their status.

The DRMAA interface (see [4] for a detailed description of the C API) includes more routines in some of the above categories as well as auxiliary routines that provides textual representation of errors, not implemented in the current version. All the functions implemented in the GridWay framework are thread-safe.

3 Efficient Execution of Grid Applications

In spite of the DRMAA standard can help in exploiting the intrinsic parallelism found in some application domains, the underlying DRMS is responsible for the efficient and robust execution of each job. In particular the following aspects are considered by the GridWay framework:

- Given the dynamic characteristics of the Grid, the GridWay framework periodically adapts the schedule to the available resources and their characteristics [3]. GridWay incorporates a *resource selector* that reflects the applications demands, in terms of requirements and preferences, and the dynamic characteristics of Grid resources, in terms of load, availability and proximity (bandwidth and latency) [5].
- The GridWay framework also provides adaptive job execution to migrate running applications to more suitable resources. So improving application performance by adapting it to the dynamic availability, capacity and cost of Grid resources. Moreover, an application can migrate to a new resource to satisfy its new requirements or preferences [3].

Table 1. Characteristics of the machines in the UCM research testbed.

Name	Model	OS	Speed	Memory	Job Mgr.
babieca	5×Alpha DS10	Linux 2.2	466MHz	256MB	PBS
hydrus	Intel P4	Linux 2.4	2.5GHz	512MB	fork
cygnus	Intel P4	Linux 2.4	2.5GHz	512MB	fork
cepheus	Intel PIII	Linux 2.4	662MHz	256MB	fork
aquila	Intel PIII	Linux 2.4	568MHz	128MB	fork

We expect that DRMAA will allow to explore several common execution techniques when distributing applications across the Grid [6], for example fault tolerance could be improved by replicating job executions (redundant execution), or several alternative task flow paths could be concurrently executed (speculative execution).

4 Experiences

In this section we describe the ability of the GridWay framework when executing different computational workloads distributed using DRMAA. The following examples resembles typical scientific problems whose structure is well suited to the Grid architecture. These experiments were conducted in the UCM research testbed, based on the Globus Toolkit 2.2 [7], briefly described in table 1.

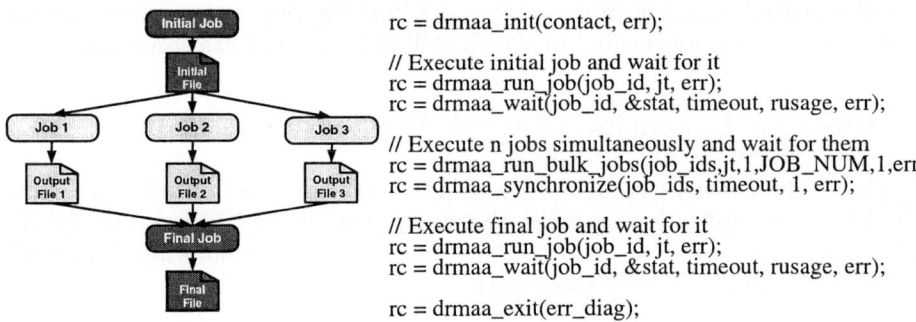

Fig. 2. High-throughput scheme and its codification using the DRMAA standard.

4.1 High-Throughput Computing Application

This example represents the important class of Grid applications called *Parameter Sweep Applications* (PSA), which constitutes multiple independent runs of the same program, but with different input parameters. This kind of computations appears in many scientific fields like Biology, Pharmacy, or Computational Fluid Dynamics. In spite of the relatively simple structure of this applications, its efficient execution on computational Grids involves challenging issues [8].

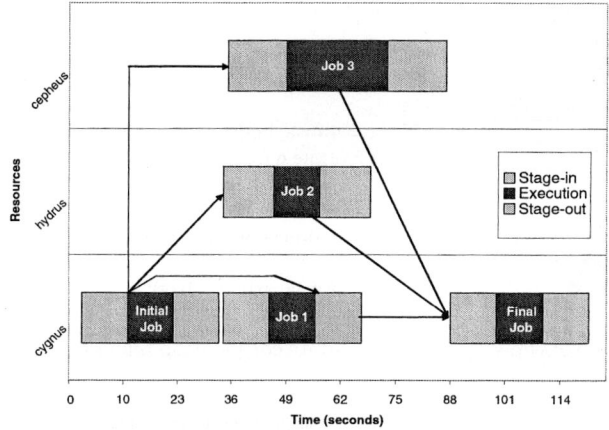

Fig. 3. Execution profile for the high-throughput application.

The structure of the PSA is shown in figure 2 (left-hand side). A initial job is submitted to perform some pre-processing tasks, and then several independent jobs are executed with different input parameters. Finally a post-processing job is executed.

Figure 3 shows the execution profile of the PSA, each computational task is executed in a different Grid resource. The total turnaround time of the experiment is 121 seconds, with an average execution and file transfer times for each computational task of 15 seconds and 22 seconds, respectively. In this case the average CPU utilization during the PSA execution was 20%. In this case the overhead induced by job scheduling (i.e. querying the MDS Grid service to obtain a preliminary list of potential hosts, and to assign a rank to them [1]) is 5% of the overall execution time.

4.2 Pipelined Workflow Application

The pipelined workflow comprises the execution of a long chain of n jobs. Each job in the sequence uses the computed solution of its predecessor to initialize. Considering this dependencies each job in the chain can be scheduled by Grid*W*ay once the previous job has finished. This computational scheme typically appears in long running simulations that can be broken up into a series of tasks (see figure 4).

In this case all the jobs in the pipeline sequence are submitted to the same host, cygnus, with an average turnaround time per job of 33 seconds. The total turnaround time for this application is 203 seconds, and an average resource CPU utilization of 81%.

4.3 Master-Worker Optimization Loop

We now consider a generalized Master-Worker paradigm, which is adopted by many scientific applications like genetic algorithms, N-body simulations or Monte

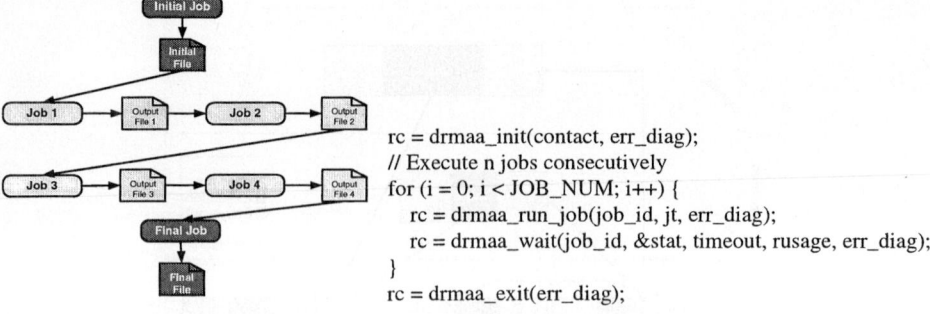

```
rc = drmaa_init(contact, err_diag);
// Execute n jobs consecutively
for (i = 0; i < JOB_NUM; i++) {
    rc = drmaa_run_job(job_id, jt, err_diag);
    rc = drmaa_wait(job_id, &stat, timeout, rusage, err_diag);
}
rc = drmaa_exit(err_diag);
```

Fig. 4. Pipelined workflow application and its codification using the DRMAA standard.

```
rc = drmaa_init(contact, err_diag);

// Execute initial job and wait for it
rc = drmaa_run_job(job_id, jt, err_diag);
rc = drmaa_wait(job_id, &stat, timeout, rusage, err_diag);

while (exitstatus != 0) {
    // Execute n Workers concurrently and wait for them
    rc = drmaa_run_bulk_jobs(job_ids, jt, 1, JOB_NUM, 1, err_diag);
    rc = drmaa_synchronize(job_ids, timeout, 1, err_diag);

    // Execute the Master job, wait for it and get exit code
    rc = drmaa_run_job(job_id, jt, err_diag);
    rc = drmaa_wait(job_id, &stat, timeout, rusage, err_diag);
    rc = drmaa_wexitstatus(&exitstatus, stat, err_diag);
}
rc = drmaa_exit(err_diag);
```

Fig. 5. Master-Worker application and its codification using the DRMAA standard.

Carlo simulations among others. A Master process assigns a description (input files) of the task to be performed by each Worker. Once all the Workers are completed, the Master process performs some computations in order to evaluate a stop criterion or to assign new tasks to more workers (see figure 5).

Figure 6 shows the execution profile of three generations of the above Master-Worker applications. The average execution time per iteration is 120 seconds, with an average computational and transfer times per worker of 15.7, and 23.3 seconds respectively. In this case the total turnaround time is 260 seconds with an average CPU utilization of 22%.

5 Conclusions

We have shown how DRMAA can aid the rapid development and distribution across the Grid of typical scientific applications, and we have demonstrated the robustness and efficiency of its implementation on top of the GridWay framework and Globus.

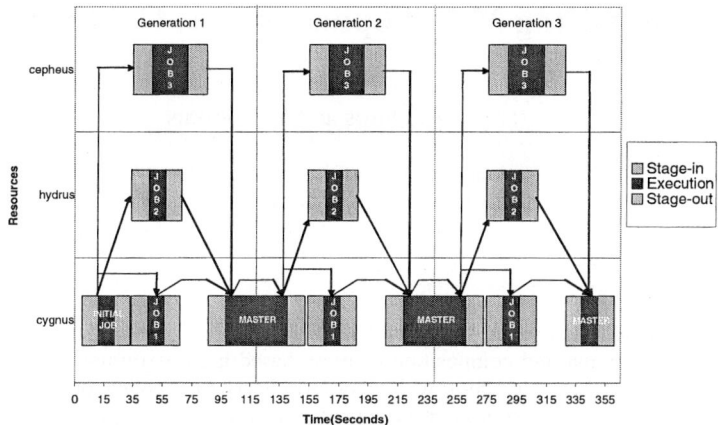

Fig. 6. Execution profile for three iterations of the Master-Worker application.

It is foreseeable, as it happened with other standards like MPI or OpenMP, that DRMAA will be progressively adopted by most DRMS, making them easier and worthier to learn, thus lowering its barrier to acceptance, and making Grid application portable across DRMS adhered to the standard.

References

1. Huedo, E., Montero, R.S., Llorente, I.M.: A Framework for Adaptive Execution on Grids. J. of Software – Practice and Experience (2004) (in press).
2. Schopf, J.M.: Ten Actions when Superscheduling. Technical Report GFD-I.4, Scheduling Working Group – The Global Grid Forum (2001).
3. Huedo, E., Montero, R.S., Llorente, I.M.: Adaptive Scheduling and Execution on Computational Grids. J. of Supercomputing (2004) (in press).
4. Rajic, H., et al.: Distributed Resource Management Application API Specification 1.0. Technical report, DRMAA Working Group – The Global Grid Forum (2003).
5. Montero, R.S., Huedo, E., Llorente, I.M.: Grid Resource Selection for Opportunistic Job Migration. In: Proc. of the 9th Intl. Conf. on Parallel and Distributed Computing (Euro-Par 2003). Volume 2790 of Lecture Notes in Computer Science., Springer–Verlag (2003) 366–373.
6. Badia, R.M., Labarta, J., Sirvent, R., Cela, J.M., Grima, R.: GridSuperscalar: A Programming Paradigm for Grid Applications. In: Workshop on Grid Applications and Programming Tools (GGF8). (2003).
7. Foster, I., Kesselman, C.: Globus: A Metacomputing Infrastructure Toolkit. International Journal of Supercomputer Applications **11** (1997) 115–128.
8. Huedo, E., Montero, R.S., Llorente, I.M.: Adaptive Grid Scheduling of a High-Throughput Bioinformatics Application. In: Proc. of the 5th Intl. Conf. on Parallel Processing and Applied Mathematics (PPAM 2003). Volume 3019 of Lecture Notes in Computer Science., Springer–Verlag (2004) (in press).

Exposing MPI Applications as Grid Services

E. Floros and Y. Cotronis

National and Kapodistrian University of Athens,
Department of Informatics and Telecommunications
{floros,cotronis}@di.uoa.gr

Abstract. This paper presents a Grid Services programming framework for the virtualization and composition of mesh-based high performance MPI applications and their interaction with other Grid Services. Applications are abstracted using a Uses / Provides port scheme where ports represent access points to data quantities. Quantities are modeled using Service Data Elements and Grid Service Handles. Clients can query services based on provides quantities and subscribe to related notification sources. Service clients execute MPI applications through a customized application management service, passing requirements regarding the mesh-topology and the execution environment. The framework defines and provides reference implementations of core portTypes used to instantiate and control the execution of a parallel application. Finally services can be composed using their Uses / Provides Quantities based on service workflow descriptions.

1 Introduction

The advent of grid computing has stimulated development of a new breed of applications targeted for deployment in highly distributed and heterogeneous computing platforms, exploiting disperse computational resources. Nevertheless practice has proved that Grids are hard to program and currently a universal Grid programming model remains a highly desired goal. While it may be possible to build grid applications using established programming tools, they are not particularly well-suited to effectively manage flexible composition or deal with heterogeneous hierarchies of machines, data and networks with heterogeneous performance [8].

Recently there is a shift of grid programming towards the Service-oriented paradigm for application development. The Open Grid Services Architecture (OGSA) [3] leverages Web Service technologies and introduces the notion of Grid Services. A Grid service is a Web service that conforms to a set of conventions relating to its interface definitions and behaviors [13].

An area of interest for applying Grid Services is the virtualization of legacy high-performance applications. Virtualization is a common approach for exposing and extending the functionality of software assets. Various approaches have been introduced applying primarily Object Based and Service Oriented technologies. Among them Common Component Architecture (CCA) defines a component model tailored towards high performance applications [1]. XCAT [10] extends CCA by introducing a Web Services based framework extensively utilizing XML and outlining Grid-oriented Application Factories. XCAT is a predecessor to OGSA providing different service functionality and semantics. Pardis [6] and GridCCM [12] propose CORBA-based frameworks and extensions to IDL and ORB. In these frameworks component stubs are used to implement parallel servers and clients, each instance corresponding to a single MPI processes.

One class of such legacy applications is the high performance parallel applications developed using MPI [11] and in particular mesh based simulation models like Meteorological, Hydrological, Pollution, Fire Propagation etc. Such models are usually developed as isolated MPI applications, typically applying the SPMD programming paradigm. Data processed form 2D or 3D meshes which map directly to real-world coordinates. Models most of the times can benefit from their interoperation (for instance a Hydrological model can interact with a Meteorological model) to produce more accurate results. The composition of these applications is not a straightforward process since source codes are developed by separate teams which are difficult or unwilling to cooperate. As a result ad-hoc approaches are followed.

In this paper, MPI applications are exposed as OGSA Grid Services. The virtualization of an MPI application with Grid Service semantics and tools introduces new potentials since there are obvious benefits both in terms of functionality and interoperability. Programmers can use the well defined value-added infrastructure of Grid Services to search for applications (e.g. in UDDI repositories), retrieve formal descriptions in a standard defined XML (WSDL) document and easily bind their functionality to diverse clients. These client applications can be developed in various languages relieving the programmer from the burden of understanding the inner engineering of the application or even of the MPI message-passing semantics per se. Diverse clients may leverage the capabilities of high performance applications and utilize high-end expert code which till now was isolated due to inherent complexity of MPI and lack of high-level composition semantics. Moreover multiple MPI services can be put together using workflow-based composition, in order to exchange data and interoperate.

The rest of the paper is organized as follows. Section 2 presents the proposed MPI application virtualization model and introduces the notion of Uses and Provides Quantities. Section 3 describes the inner details of the framework and provides examples of simple client-service interaction and more complex, workflow-like service composition. Finally section 4 provides conclusions and future directions of this research.

2 Virtualization of MPI Applications

To demonstrate the framework we use a real life application scenario based on Flood Forecasting [5]. In this scenario three different models cooperate, in a cascading pattern, to execute a flood crisis application (Fig. 1). A Meteorological Model (parallel) provides quantitative precipitation forecasts to a Hydrological simulation (sequential) which in turn feeds a Hydraulic (parallel) simulation with Hydrographs. The three models interoperate to provide weather forecasts, discharge forecasts and flood scenarios.

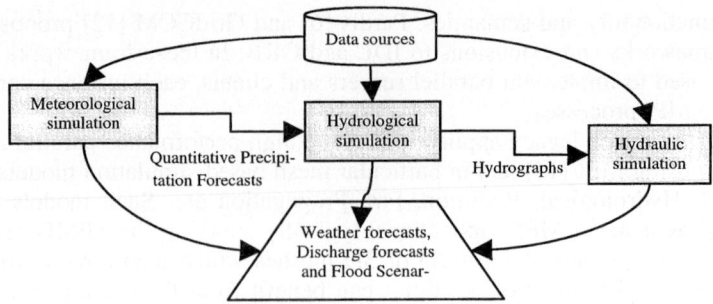

Fig. 1. Sample scenario with three cascading simulation models

2.1 Data-Centric Virtualization

In order to expose a parallel MPI application as a Grid Service one has to bridge two different distributed programming worlds: the service-oriented, operation-centric world of OGSA Grid Services and the two-sided communications world of MPI. The common denominator for these two approaches is the data that are being exchanged between two communication parties. In service-oriented and object-oriented approaches like OGSA, Web Services, CORBA etc., data are bound to operations that receive them as method parameters and return them as results of method execution. In MPI, messages do not trigger explicit method calls but carry data and also act as synchronization mechanism.

We can follow two approaches to wrap a Grid Service around the MPI application: wrapping of processes or wrapping of data. In the first approach the Grid Services runtime interacts directly with the processes that comprise the MPI application. This requires the Grid Service processes or at least a subset of them, to immerse inside the MPI context that the MPI processes define and be an MPI program itself.

A more general approach is to have the Grid Service directly access the output data that a given MPI application produces. In this approach the implementation of the service is independent of the implementation and execution details of the underlying MPI system. In this paper we have followed the later approach since not only is the more general one but also because the current implementations of OGSA support mainly Java based servers, which cannot participate in an MPI communication world.

We abstract an MPI application by defining the following: The application implements an algorithm (e.g. a Meteorological or Hydraulic model) that accepts as input a set of external data in the form of *uses quantities* executes and produces a set of result data in the form of *provides quantities*. For instance Precipitation is a *Provides Quantity* of the Meteorological model and a *Uses Quantity* of the Hydrological model.

The application may also utilize a set of meta-data that define execution requirements of the algorithm and other behavioral aspects and execution details. *Uses* and *Provides Quantities* may be available in various forms and mediums: a simple binary or text file in the local file system, a remote file replicated in various hosts and controlled by a replication service, a database in a network available RDBMS or a networked server application that provides or consumes data by exchanging them in a message-passing fashion.

Following the above approach a Grid service can describe public interfaces of the MPI application by exposing the uses and provides quantities of the application, together with the required meta-data for execution. The implicit operation that the MPI application implements, and is being exposed as a Grid Service operation is the execution of the algorithm.

2.2 Modeling of Provides and Uses Quantities

We model *Provides Quantities* using Service Data Elements (SDEs). Service Data is a structured collection of information that is associated to a Grid Service. This information is easy to query, so that Grid Services can be classified and indexed according to their Provides Quantities. Although originally intended to provide attributes and meta-information of a Grid Service, the SDE conception is in accordance with the Provides Quantity notion.

For example in the Flood crisis scenario the Meteorological model may provide a Precipitation Quantity which is comprised by precipitation related information on a specific 2D x-y axis. The *Precipitation* SDE will have the following simplified XML Schema definition (XSD):

```xml
<complexType name="Precipitation">
  <complexContent>
    <extension base="Quantity2D">
      <sequence>
        <element name="value" type="float" maxOccurs="unbounded"/>
        <element name="time" type="float" maxOccurs="unbounded"/>
      </sequence>
    </extension>
  </complexContent>
</complexType>
```

The above defines a Provides Quantity named "Precipitation" which extends a framework defined Quantity2D SDE with two additional fields: the *value* of precipitation and the relevant *time* step. Quantity2D defines an abstract quantity that is enclosed in the mesh rectangle (*xmax,ymax*) and (*xmin,ymin*):

```xml
<complexType name="Quantity2D">
  <sequence>
    <element name="xmax" type="float" maxOccurs="unbounded"/>
    <element name="ymax" type="float" maxOccurs="unbounded"/>
    <element name="xmin" type="float" maxOccurs="unbounded"/>
    <element name="ymin" type="float" maxOccurs="unbounded"/>
  </sequence>
</complexType>
```

A service may implement a variable number of Provides Quantities. Clients can search for services which provide specific quantities and acquire a reference to them (in the form of a GSH). An implementation code (e.g. a JavaBean if Java is the target language) is generated automatically by the above XML. The service programmer has to customize it and extend it to retrieve and prepare the Quantity information from the application output data. In most cases the MPI application source code should not require any modification to be used from the service. The programmer having knowledge of the result dataset format can derive with little effort the requested data and return them to the client.

A *Uses Quantity* is modelled using a {GSH, QuantityName} pair that uniquely identifies the name of the quantity (e.g. "Precipitation") and the grid service that provides it. For instance the *Precipitation Uses Quantity* of the Hydrological application may have the form: *{http://www.gsmpi.org/models/Meteo, "Precipitation"}*

An application (simple client or another grid service) may define zero or more {GSH,QuantityName} pairs whose values can be set during the preparation of the application or dynamically during the execution time. A null value for a given GSH means that this *Uses Quantity* is not available and the algorithm should perform all required computations without taking advantage of them.

3 Programming Framework

3.1 Standard PortTypes

The framework defines and implements a set of standard portTypes and respective operations, which can be extended by the application developer. These portTypes are:

MPIAppPrepare: Provides the required functionality that handles pre-execution details of the MPI application, such as the setup the execution environment, the definition of the application topology, the definition of special requirements for computational resources (memory, cpu, disk space) etc.

MPIRun: Provides the *MPIRun::mpiRun* operation that is invoked in order to actually start the MPI application. This operation consults the requirements defined previously, prepares the underline MPI system and executes the *mpirun* command. Multiple versions of the operation can be implemented in order to support different MPI environments (e.g. MPICH, LAM etc.).

MPIAppMonitor: Implements basic operations for job monitoring and management. Operations can be used to query the state (*starting*, *running*, *completed* and *aborted*) of the application and control its life-cycle (*kill* or *restart* the application).

3.2 Notifications

The framework extensively utilizes the Notification facilities provided by Grid Services. Clients can register as notification sinks requesting from the service to be informed when a *Provides Quantity* has changed; usually when a computation has finished and the *Provides Quantity* is ready to be retrieved, or when intermediate results are available. When a *Provides Quantity* is changed the event is propagated to all interested parties. Clients are able to pull available results by sending an appropriate request to the SDE. Programmers may also choose to implement a quantity push schema where the service application itself assumes the responsibility to communicate the event together with the relative data to the client application. This ability is especially useful for supporting flexible service composition based on uses / provides quantities.

3.3 Programming in the Framework

The framework is applied as follows: First the programmer writes the *GWSDL* description of the service and the *XSDs* of the Provides Quantities. The service extends one or more of the framework's *portTypes* to generate a custom MPI Application Management Factory. The programmer edits and extends the SDE stubs (which in Globus Toolkit 3 are materialized in the form of JavaBeans) generated from the XSD specifications of the *Provides Quantities*. This is the most important and laborious part of the framework. The programmer has to implement in the JavaBean the data access logic in order to extract and deliver from the complete output data set the quantity information that this SDE provides. The service is then deployed in a Grid Services container (e.g. Apache Tomcat). The stubs generated from the *GWSDL* file are used to implement service clients.

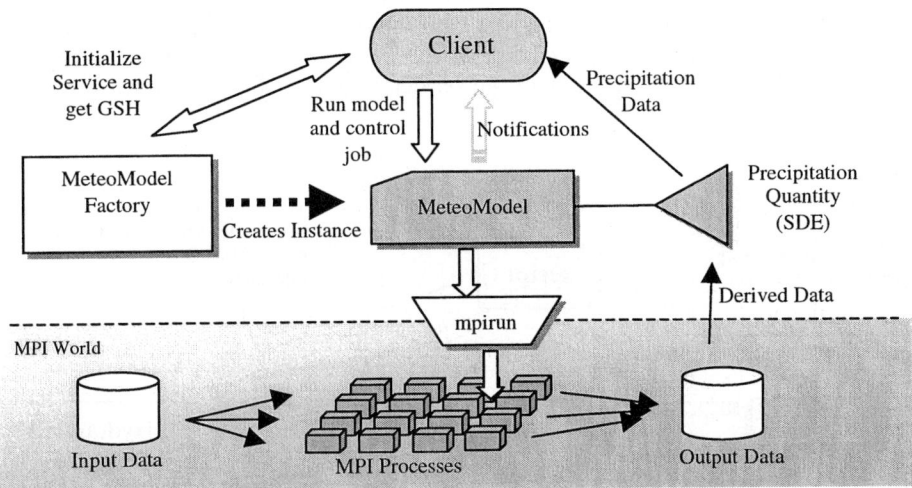

Fig. 2. Overview of Client / Service interaction within the framework

Fig. 2 depicts a sample application execution. A client application uses the *MeteoModelFactory* to instantiate a new *MeteoModel*. Then it passes the initialization parameters of the application. For instance a client may request a 2D dimensional mesh topology within a given range of X and Y coordinates. The client is not concerned with how many processes will be created and where. These details are derived from the requested mesh topology and are handled by the service. Additional requirements may be passed such as a time limit of execution or minimum model error.

The application is executed by calling *MPIRun::mpiRun* operation. Currently the service takes the initial requirements and produces an RSL file to be passed as a parameter to MPICH-G2 *mpirun* script. Issues of security and credentials delegation are handled using GT3 transport level and message level security.

The client may poll periodically the status of the application using *MPIAppMonito::mpiAppGetState* operation or can be notified using Notifications when results are available. To avoid having to extend from all these portTypes mpiRun can use default execution values thus an application can be started by extending only the *MPIRun* portType and issuing a single call to the *MPIRun::mpiRun*.

Moreover a client may request premature end of an application (*MPIAppMonitor::mpiAppKill* operation) either by keeping any intermediate results up to then or by flashing all output rendering them useless in order to start a new simulation.

Since Grid Services can be state-full many clients can dynamically acquire a reference of the running *MeteoModel* instance, connect and retrieve results from the executing model. For example there may be two service clients the first being the Hydrological model and a second visualization client that retrieves and displays graphical precipitation images.

3.4 Service Composition

The described framework facilitates the composition of virtualized MPI applications with other Grid Services (either MPI or non-MPI). Composition can be performed both in space and in time [4]. In the first case the two composed services either have prior knowledge of the *Uses Quantities* each other exposes or the {GSH,QtyName} pair is passed during the execution of the service as a parameter (using for instance a Perl script to instantiate them).

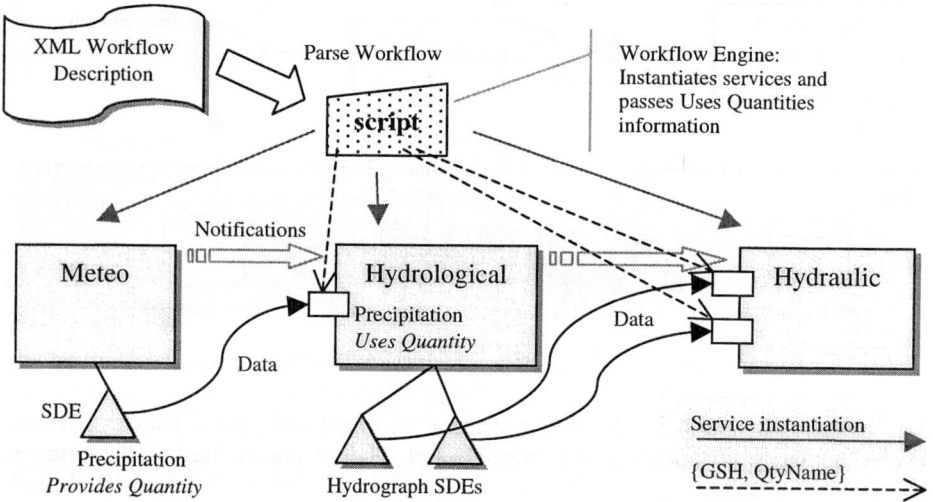

Fig. 3. Services composition through Uses / Provides Quantities

The framework can further be extended to support workflow based composition (composition in time). Currently, workflow systems for Grid and Web Services are evoking a high degree of interest, with initiatives such as WSFL [9], and Grid Services Flow Language (GSFL) [7] investigating the various aspects of workflow in their respective domains. Our approach of Quantities-based composition resembles the GSFL *notificationModel* which is the recommended solution for peer-to-peer, high-performance inter-service bulk data exchange.

In the example (Fig. 3) a simple workflow engine parses a description of the service composition in an XML format, extracts the Uses/Provides Quantities information and instantiates the services by passing the required Uses Quantities information.

4 Conclusions and Future Work

In this paper we have presented a programming framework for exposing high-performance parallel MPI applications as OGSA Grid Services. We have introduced the notion of *Uses* and *Provides Quantities* which are used to describe an abstract application interface, permit data exchange between applications and facilitate the composition of two or more applications in a service workflow.

The framework is currently work-in-progress and is being implemented on top of Globus Toolkit 3 as the Grid Services middleware and MPICH-G2 as the MPI execution environment. The recent refactoring of OGSA that has led to the introduction of the Web Services Resource Framework (WSRF) [2] is pushing for a similar refactoring of all research activities in this area. As a result our imminent steps will be to adapt our framework to WSRF semantics. Moreover, we are evolving the framework at various levels: Application execution semantics are extended to support interactivity and check-pointing. Service composition is enhanced with formal workflow syntax and extended to support semantic information and compatibility assertions. Furthermore, we investigate the capability of Web Services and MPI programs to co-operate at the process level. Finally, we plan to further evolve the dynamic capabilities of the framework especially in the context of dynamic workflow transformations.

References

1. Armstrong R., Gannon D., et al: Towards a Common Component Architecture for High-Performance Scientific Computing. 8th IEEE International Symposium on High Performance Distributed Computation, August 1999.
2. Czajkowski K. et al.: The WS-Resource Framework (WSRF) v1.0, GGF, March 2004.
3. Foster I., Kesselman C., Nick M. J., Tuecke S.: The Physiology of the Grid: An Open Grid Services Architecture for Distributed Systems Integration. Open Grid Service Infrastructure WG, Global Grid Forum, June 2002.
4. Gannon D., et al..: Grid Web Services and Application Factories. In: Grid Computing: Making the Global Infrastructure a Reality, p251-p264. Willey, April 2003.
5. Hluchy L. et al.: Problem Solving Environment for Flood Forecasting. 7th World Multiconference on Systemics, Cybernetics and Informatics (SCI 2003), July 2003.
6. Keahey K. Gannon D.: PARDIS: A Parallel Approach to CORBA. IEEE 6th International Symposium on High Performance Distributed Computing, August 1997.
7. Krishnan S. Wagstrom P. von Laszeswki G.: GSFL: A Workflow Framework for Grid Services. Argonne National Laboratory, Preprint ANL/MCS-P980-0802, August 2002.
8. Lee C., Talia D.: Grid Programming Models: Current Tools, Issues and Directions. In: Grid Computing: Making the Global Infrastructure a Reality, p555-p576, Wiley, April 2003.
9. Leymann F. Web Services Flow Language (WSFL 1.0), IBM Software Group. http://www-4.ibm.com/software/solutions/webservices/pdf/WSFL.pdf, May 2001.
10. Madhusudhan G., Krishnan S., Slominski A., Merging the CCA Component Model with the OGSI Framework. Proc. of CCGrid2003, May 2003.
11. Message Passing Interface Forum: MPI: A Message Passing Interface Standard, June 1995.
12. Perez C, Priol T., Ribes A.: A Parallel CORBA Component Model, INRIA/RR-4552. September 2002.
13. Tuecke S. et al.: Open Grid Services Infrastructure (OGSI). Version 1.0, Global Grid Forum, June 2003.

Globus-Based Grid Computing Simulations of Action Potential Propagation on Cardiac Tissues

José M. Alonso, Vicente Hernández, and Germán Moltó*

Departamento de Sistemas Informáticos y Computación,
Universidad Politécnica de Valencia. Camino de Vera s/n 46022 Valencia, Spain
Tel. +34963877356, Fax +34963877359
{jmalonso,vhernand,gmolto}@dsic.upv.es

Abstract. With the advent of Grid technologies, the study of the electrical activity of the heart, by means of concurrent parametric simulations of the action potential propagation on cardiac tissues, can be greatly benefited. Studies of the electrical behaviour, such as late ischemia require the execution of multiple computational and memory intensive parametric simulations. This paper describes the integration, into a Grid infrastructure, of a parallel MPI-based system for the simulation of action potential propagation on a three-dimensional parallelepiped-modelled cardiac tissue. Developed upon the Globus Toolkit, it features state-of-the-art capabilities such as data compression, simulation failure recovery, and the combination of parallel execution on distributed resources, what has enabled an outstanding increase in research productivity.

1 Introduction

The simulation of action potential propagation on cardiac tissues represents a major computational challenge. The fine spatial and time discretization steps required to solve the equation (1) that governs this phenomenon on a monodomain cardiac model makes this problem only affordable with High Performance Computing techniques. This is particularly important for three-dimensional executions, where a simulation of action potential propagation during few milliseconds on a medium-sized tissue may last for several days on a sequential platform.

$$\nabla \cdot \sigma \nabla Vm = C_m \cdot \frac{dVm}{dt} + I_{ion} + I_{st}. \qquad (1)$$

The previous equation relates the membrane potential of the cells, Vm, the ionic currents that traverse the membrane, I_{ion}, the membrane capacitance, C_m, the anisotropy tensor, σ, and the electrical stimulus, I_{st}. The comprehensive Luo-Rudy Phase II [1] cellular model has been employed to calculate the I_{ion} term.

---

* The authors wish to thank the financial support received from The Spanish Ministry of Science and Technology to develop the project GRID-IT (TIC2003-0131). This work has been partially supported by the Structural Funds of the European Regional Development Fund (ERDF).

In addition to the inherent computational cost of a single simulation, there are many research studies that require the execution of a huge amount of parametric simulations. Studies of vulnerable window in ischemia require to vary the time interval between two consecutive stimulus in order to detect the range of values which provokes a reentry, a phenomenon that can derive into heart fibrillation. Besides, to study the effects of late ischemia it is necessary to vary the coupling resistances in all the dimensions of the tissue and analyze the evolution of the electrical activity for different anisotropy ratios. Moreover, to evaluate the influence of certain medicines, it is crucial to alter the concentration of these drugs, over a determined range, to study how it affects to the action potential propagation.

Even though there have been several parallel approaches to this computational problem [2], the good efficiency results achieved on a beowulf cluster, together with appearing to be the first cardiac simulation system to combine both a parallel and a Grid Computing approach, represent a step forward in the study of the electrical activity of the heart.

Therefore, both parallel computing techniques, that speedup a single simulation, and Grid Computing technology, that enhances the efficiency of multiple simulations, will be combined in order to achieve a global simulation system that increases the productivity for these computational demanding cardiac case studies.

The article is structured as follows: Section 2 describes the main functionality of the simulation system. Then, section 3 details the characteristics of the Grid Computing system designed. Next, in section 4, a case study is presented to expose the functionality in a production testbed. Finally, section 5 concludes the paper, exposing the relevant achievements.

2 Characteristics of the Simulation System

A cardiac tissue simulation consists of an iterative process that allows to calculate the membrane potential of the cells along a time period.

First, a parallel simulation system was developed for two-dimensional tissues [3] in order to reduce the simulation time on beowulf architectures with outstanding efficiency results (94% of efficiency with 32 processors). Next, the simulation system has been extended for three-dimensional anisotropic tissues, achieving good scalability results. For example, Fig. 1 shows the speedup and efficiency when simulating, on a cluster of PCs, an action potential propagation during 250 ms in a 100x100x100 cell cardiac tissue, with a timestep of 10 μs. Such a simulation lasts 177.97 hours on a sequential platform, but only 6.45 hours when using 32 processors.

The simulation system periodically generates a set of checkpoint files, using the MPI-2 parallel I/O routines, what allows a simulation to be restarted from the point that was stopped, even with a different number of processors. The checkpoint data consist of a snapshot of the tissue, that is, a double precision binary dump of all its cells, along with other ones of the membrane potential and ionic vectors.

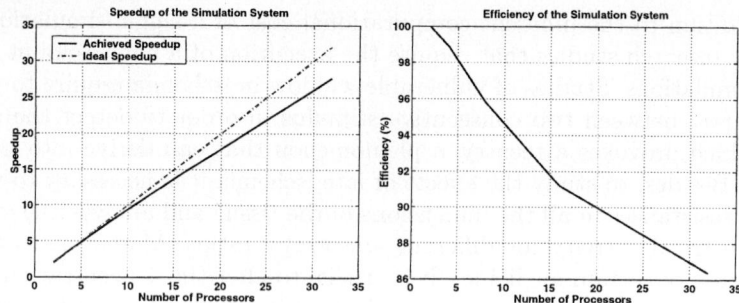

Fig. 1. Speedup and efficiency of the simulation system. Running on a 20 Pentium Xeon 2.0 Ghz biprocessor cluster, with 1 GByte of RAM and a interconnected by a SCI network.

Provided that the checkpoint data for this application can result in a very large data set, we have analyzed the effectivity of data compression on realistic checkpoint files, where a 100x100x100 cell cardiac tissue is stimulated to provoke an action potential that depolarizes all the tissue.

Table 1 shows the compression ratio that is achieved, for the best and the worst tissue state, using a Lempel-Ziv coding provided by the standard *gzip* Unix command. The best case corresponds to a tissue in rest state, i.e. at the beginning of the simulation, where similar values may be found for all the cells of the tissue. On the other hand, the worst case corresponds to a propagating wavefront (once applied the supra-threshold stimulus) on an anisotropic tissue, where changes between the cells are very frequent.

Table 1. Effectivity of checkpoint data compression for the best and the worst case of a 100x100x100 cell tissue state. Size is expressed in MBytes.

	Uncompressed Data	Compressed Data	Compression Ratio
Best Case	515.2	4.9	105.14
Worst Case	515.2	142.4	3.62

In both cases, data compression required an average 60 seconds, while decompression lasted for an average 14 seconds. Therefore, compression offers a significant reduction of the binary data generated by the simulator, as in the worst case the result files can be reduced to less than a third part.

3 Grid Computing System Developed

3.1 Portability and Interoperability

Enabling portability requires that all the platform-dependent optimizations, such as the compiler flags *-march* or *-mcpu*, are avoided. Besides, architecture-dependent optimized numerical libraries, such as the BLAS [4] and LAPACK [5] implementation by the Intel Math Kernel Library, should not be used, as they may result in executing illegal instructions on the remote host if both architectures do not match. Fortunately, traditional compiler optimization flags, i.e. *-O3*, can be used with no risk.

Our application has been statically linked to generate a self-contained simulation system. Even the MPI communication library has been introduced into the executable, using a MPICH [6] implementation, configured to disable shared-memory communication between processes on the same node of a cluster, which can potentially introduce memory-leak problems because of relying on the System V IPC facilities [7]. This procedure enables to perform a parallel execution without depending on the MPI implementation of the execution host.

This parallel self-contained simulation system can be executed on a wide range of Linux machines, thus isolating the application from the runtime environment, something that may, a priori, be unknown in a Grid. This has been ensured by executing parallel simulations in a variety of different architectures such as Pentium III, Pentium IV, Pentium Xeon and even Intel Itanium 2 running different Linux flavours such as Red Hat Linux Advanced Server, Red Hat 8.0, Fedora Core 1 and Debian GNU/Linux.

It should be pointed out that such a simulator, compiled on an Intel Pentium Xeon PC (32 bit), runs on compatibility mode on an Intel Itanium 2 (64 bit) platform, but it is up to 8 times slower than on the original architecture. Therefore, we have natively compiled on the Intel Itanium 2 platform in order to achieve comparable execution times on both architectures, and to be able to exploit Itanium Grid execution nodes. This results on two self-contained simulation systems, one for IA-32 and other one for IA-64, an strategy that could be refined to target more architectures.

3.2 Modules Developed

Figure 2 shows a conceptual view of the Grid Computing system developed based on the Globus Toolkit [8]. The *JobScheduler* is the module responsible for the allocation of simulations to computational resources. This module delegates, for each simulation, into a *JobSubmitter*, which is in charge of the proper execution of the task in the resource.

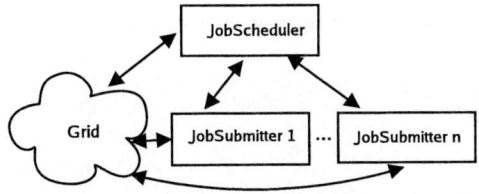

Fig. 2. Scheme of the Grid Computing system developed.

The JobScheduler Module. This module reads an input file with a parametric description of the multiple simulations that form the case study. For each simulation, it computes the best available resource, from a predefined list of machines, by consulting its number of available nodes, via the Monitoring and Discovery Service (MDS). Clusters with the Globus Resource Allocation Manager (GRAM) Reporter installed, report the number of free computing nodes, delegating in the local queue manager (LoadLeveler, PBS, etc). For workstations, an estimation of the CPU usage during the last minute serves as an indicator of the availability of the resource. This strategy allows to customize a parallel execution to the number of available nodes in the host. Then, this module selects an appropiate executable based on the architecture of the remote machine.

The JobScheduler is also responsible for submitting the unassigned simulations and restarting the failed executions, delegating, for each of them, on an instance of the JobSubmitter module. If no available resources exist, it periodically checks their availability to continue submitting pending tasks.

The JobSubmitter Module. This module is in charge of the proper execution of a single simulation. First of all, the input files that the simulation system needs are staged in, via the GridFTP service, to the execution host. Through the Globus native interface, the remote machine is queried about its availability to run MPI jobs, so the parallel or serial execution can be selected. The execution of the simulation is integrated, if configured, with the queue manager of the remote node (PBS, LoadLeveler, etc), thus respecting the execution policies of that organization.

While the simulation is running on the remote resource, a checkpoint job is periodically submitted by this module, which transfers, if not already done, a compressed image of the generated checkpoint data to the local machine. Thus, the latest checkpoint data always resides at the submission machine and a failed simulation can be automatically resumed on a new computational resource. A message digest mechanism ensures that no old checkpoint data is transferred twice, wasting bandwidth.

Once the execution has finished, all the result data are compressed, transferred back to the submission node and saved on the appropriate local folder created for this simulation. All the temporary created files in the execution node are deleted, and finally, the JobSubmitter module anotates whether the simulation has finished correctly or not. This information will be used by the JobScheduler module to be able to restart the failed simulations.

4 Case Study

4.1 Description

Myochardial ischemia is a condition caused by oxygen deprivation to the heart that can result in an angina. It is known that ischemia increases the extracellular potassium concentration in the affected area, what shortens the action potential duration. Therefore, it is possible to study the effects of several degrees of ischemia, by means of multiple parametric simulations, varying the extracellular potassium concentration for a group of cells in the tissue.

For a 50x50x50 cells cardiac tissue, a range from 5 to 12.9 milliMolar (mM.) potassium extracellular concentration will be studied, with an increment of 0.2 mM. between each simulation. Only 2 ms will be simulated with a timestep of 0.01 ms. This results in 40 independent parametric simulations that can be executed in the computational resources that a Grid testbed offers.

4.2 Testbed

The available testbed is composed of local resources, belonging to our research group, the High Performance Networking and Computing Group (GRyCAP-UPV), and remote resources from the Distributed Systems Architecture & Security group (ASDS-UCM), at Madrid Complutense University. Table 2 summarizes the main features of the machines.

The Globus Toolkit version 2.4 [9] has been installed on the testbed. Ramses cluster is the Certication Authority of GRyCAP-UPV and its credentials have been installed on ASDS-UCM machines to allow remote job submission.

Table 2. Detailed machine characteristics of the testbed.

Machine	Processors	Memory
Kefren (grycap)	20 (2 x Intel Xeon 2.0 Ghz)	1 GByte
Ramses (grycap)	12 (2 x Intel Pentium III 866 Mhz)	512 MBytes
Bastet (grycap)	2 x Itanium 2 900 Mhz	4 GBytes
Hydrus,Cygnus (asds)	1 x Pentium IV 2.53 Ghz	512 MBytes
Aquila (asds)	1 x Pentium III 666 Mhz	128 MBytes
Cepheus (asds)	1 x Pentium III 666 Mhz	256 MBytes

4.3 Execution Results

Table 3 summarizes the tasks distribution in the Grid. The maximum number of processors in a parallel execution has been limited to eight, a polite policy with the rest of the users that allows multiple concurrent simulations. In the table, an entry like 7 (8 p.) indicates that seven simulations were performed with eight processors each one. Machine Bastet does not appear on the table because it was heavily loaded and the scheduler never chose it for job submission. Each simulation generates 64 MBytes of data, that can be compressed to 1.7 MBytes.

The Grid execution of this short case study lasted for 1232 seconds (20.53 minutes). On the other hand, a traditional sequential execution in only one node of cluster Kefren required 154.05 minutes, what represents a speedup of 7.5 in the cardiac case study execution in the Grid. A parallel computing approach, performing 8-processors parallel executions sequentially in cluster Kefren, required 32.74 minutes.

Table 3. Distribution of the simulations in the testbed, for each machine. The number in parentheses indicates the number of processors involved in the execution.

Machine	Simulations	Machine	Simulations
Kefren	7 (8 p.), 5 (5 p.), 3 (1 p.), 2 (4 p.), 2 (7 p.)	Hydrus	3 (1 p.)
Ramses	7 (8 p.), 3 (1.p)	Cygnus	3 (1 p.)
Cepheus	2 (1 p.)	Aquila	3 (1 p.)

It can be seen that the scheduler has distributed the tasks proportional to the computational power of each machine, what represents a proper balance loading scheme. Had the machine Bastet been available, it would have received a task load adequate to its computational power. Besides, as the state of the Grid is investigated before each task submission, the job allocation is dynamically adjusted to the computational load of the resources during the scheduling process.

It is important to point out that a Grid execution is ideal for resource-starved cardiac case studies, as it broadens the computing resources available, no longer confined to those belonging to a single organization.

5 Conclusions

This paper has presented the integration of a system for the simulation of action potential propagation on three-dimensional monodomain modelled cardiac tissues, into a Globus-based Grid infrastructure.

The Grid Computing system developed features state-of-the-art capabilities such as data compression, self-contained executable and dependencies migration, cross-linux portability and parallel execution of simulations on cluster nodes of the Grid.

With the execution of cardiac case studies in a Grid environment, productivity has been largely enhanced compared to traditional sequential execution approaches. It is clear that the advent of new Grid technologies is getting possible to increase the research productivity by performing multiple concurrent, geographically distributed, parallel or sequential simulations of action potential propagation on cardiac tissues.

Therefore, having available a parallel simulation system that can be integrated with a Grid infrastructure enables to focus both on speedup, running on a cluster of PCs, and productivity, taking full advantage of the computational power of a Grid.

Acknowledgements

We would like to thank the Distributed Systems Architecture & Security group, belonging to Madrid Complutense University (Spain), for sharing both its computational resources, enlarging our testbed, and its knowledge through the results of the GridWay project [10].

References

1. Luo, C.H., Rudy, Y.: A Dynamic Model of the Cardiac Ventricular Action Potential. I Simulations of Ionic Currents and Concentration Changes. Circulation Research **74** (1994) 1071–1096
2. Vigmond, E.J., Aguel, F., Trayanova, N.A.: Computational Techniques for Solving the Bidomain Equations in Three Dimensions. IEEE Transactions on Biomedical Engineering **49** (2002) 1260–1269
3. Alonso, J.M., Ferrero (Jr.), J.M., Hernández, V., Moltó, G., Monserrat, M., Saiz, J.: High Performance Cardiac Tissue Electrical Activity Simulation on a Parallel Environment. Proceedings of the First European HealthGrid Conference (2003) 84–91
4. Lawson, C.L., Hanson, R.J., Kincaid, D., Krogh, F.T.: Basic Linear Algebra Subprograms for FORTRAN Usage. ACM Trans. Math. Soft. **5** (1979) 308–323
5. Anderson, E., Bai, Z., Bischof, C., Blackford, S., Demmel, J., Dongarra, J., Du Croz, J., Greenbaum, A., Hammarling, S., McKenney, A., Sorensen, D.: LAPACK Users' Guide. Third edn. Society for Industrial and Applied Mathematics, Philadelphia, PA (1999)
6. Gropp, W., Lusk, E., Doss, N., Skjellum, A.: A High-Performance, Portable, Implementation of the MPI Message Passing Interface Standard. Parallel Computing **22** (1996) 789–828
7. Gropp, W.D., Lusk, E.: User's Guide for MPICH, a Portable Implementation of MPI. Mathematics and Computer Science Division, Argonne National Laboratory. (1996)
8. Foster, I., Kesselman, C., Nick, J., Tuecke, S.: The Physiology of the Grid: An Open Grid Services Architecture for the Distributed Systems Integration. Infrastructure WG, Global Grid Forum (2002)
9. Foster, I., Kesselman, C.: Globus: A Metacomputing Infrastructure Toolkit. Intl. J. Supercomputer Applications **11** (1997) 115–128
10. Huedo, E., Montero, R.S., Llorente, I.M.: A Framework for Adaptive Execution on Grids. Software Practice and Experience (2004) To appear.

Profiling Grid Data Transfer Protocols and Servers

George Kola, Tevfik Kosar, and Miron Livny

Computer Sciences Department, University of Wisconsin-Madison
1210 West Dayton Street, Madison WI 53706
{kola,kosart,miron}@cs.wisc.edu

Abstract. The trend of data intensive grid applications has brought grid storage protocols and servers into focus. The objective of this study is to gain an understanding of how time is spent in the storage protocols and servers. The storage protocols have a variety of tuning parameters. Some parameters improve single client performance at the expense of increased server load, thereby limiting the number of served clients. What ultimately matters is the throughput of the whole system. Some parameters increase the flexibility or security of the system at some expense. The objective of this study is to make such trade-offs clear and enable easy full system optimization.

1 Introduction

The increasing trend towards data intensive grid applications [1] [2] [3] has brought grid data transfer protocols and storage servers into focus. We have done a full system profile of GridFTP [4] and NeST [5], two widely used data access and storage server and detailed how time is spent in each server.

Our profiling details server side CPU characteristics and shows the effects of concurrency level, number of parallel streams and protocol parameters like block-size on server load and transfer rate. This makes the trade-off between single client performance and server load clear. We also explain why certain parallelism level lowers the server load while increasing the transfer rate.

A good understanding of this helps computer architects to add processor features and operating system designers to optimize the operating system. It enables middleware and applications developers to optimize their software and helps grid deployers to choose appropriate machine configuration for their applications.

2 Methodology

We wanted to understand how time is spent in data access protocols and storage servers. We decided to study GridFTP and NeST, two widely used grid data access and storage servers. NeST server is interesting because it supports space reservation and a variety of interfaces: native chirp [5], NFS [6] and GridFTP.

Our desire to perform a full-system characterization including the path through the kernel while keeping the system perturbations minimal narrowed our choice of profiler to Oprofile [7], a Linux system-wide profiler based on Digital Continuous Profiling Infrastructure [8]. Oprofile uses the hardware performance counters on the Pentium family of processors.

For profiling, we setup two server machines: a moderate server, 1660 MHz Athlon XP CPU with 512 MB RAM, and a powerful server, dual Pentium 4 2.4 GHz CPU with 1 GB RAM. Both servers used Linux kernel 2.4.20. The moderate server had 100 Mbps connectivity while the powerful one had 1000 Mbps connectivity. We used three client machines, two of them were in local area and 1 was in wide area. The local area clients were dual 2.8 GHz Xeons and had 100 Mbps connectivity and were chosen randomly from a pool of 50 machines and the wide area client was quad 2 GHz Xeon with 100 Mbps connectivity. The powerful machines ensured that the clients were not the bottleneck and brought out the server characteristics.

We got full system profiles for both of GridFTP 2.4.3 and NeST servers using clients in the local area. For the extended study of GridFTP performance, we used clients both in local area and wide area.

Since we used real wide-area transfers we did not have any control over the loss rate. We did not trace it during the experiment because we felt such a packet trace collection at end hosts would interfere with our experiment. But we did periodic network traces and found that wide-area losses were negligible (less than 0.5%) at 100 Mbps. We have a 655 Mbps wide-area ATM connectivity and we found that the packet losses started showing up above 250 Mbps.

We tried out some commonly used options like parallel streams and concurrent number of file transfers in GridFTP and found the effect on server load.

3 Full System Characterization

We studied how the time is spent on the server side and present the results in this section. This characterization details the fraction of time spent in the different system parts including the kernel. This is significant for data servers because most of the time is spent in the kernel and plain user-level server profile is not sufficient.

3.1 GridFTP

Figure 1 shows the GridFTP server CPU characteristic when a single local area client reads/writes a set of 100 MB files. The read and write clients achieved a transfer rate of 6.45 MBPS and 7.83 MBPS respectively.

In terms of server CPU load, reads from the server are more expensive than writes to the server. The extra cost is spent in interrupt handling and in the ethernet driver. The machine has an Intel Ether Express Pro 100 network interface card(NIC). We found that interrupt coalescing lowered the interrupt cost during write to server. The NIC transfers the received packets via DMA to main

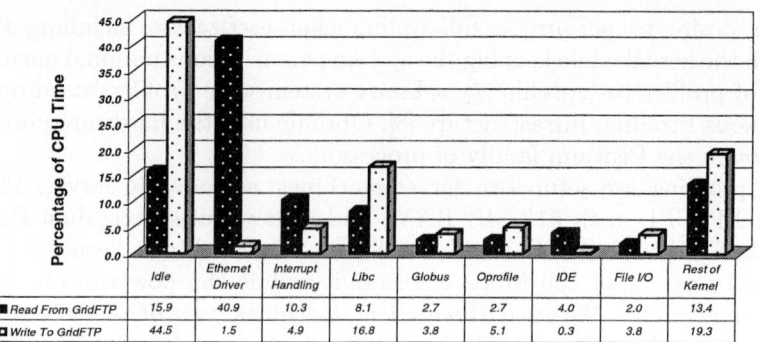

Fig. 1. GridFTP Server CPU Characteristics.

memory resulting in low CPU cost for writes to server. CPU is used to transfer output packets to the NIC resulting in high cost of read from the server. NIC with capability to DMA the output packets along with a driver capable of using that feature would reduce server read load considerably.

In the Libc, 65% of the time is spent in the getc function. The IDE disk has a greater overhead on reads compared to writes. Tuning the disk elevator algorithm may help here. The file I/O part includes the time spent in the filesystem. It is higher for writes because of the need for block allocation during writes. The rest of the kernel time is spent mostly for TCP/IP, packet scheduling, memory-copy, kmalloc and kfree.

3.2 NeST

Figure 2 shows the NeST server CPU profile when a single local area client reads/writes a set of 100 MB files using the chirp protocol. The read and write clients achieved a transfer rate of 7.49 MBPS and 5.5 MBPS respectively.

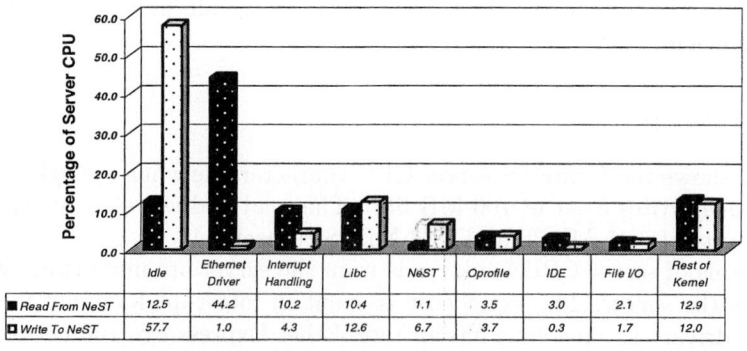

Fig. 2. NeST Server CPU Characteristics.

NeST server has a 16% higher read transfer rate and a 30% lower write transfer rate compared to GridFTP server. The lower performance of writes to NeST server is because of the space reservation feature called 'Lots'. Before each write, NeST server verifies that the client has not exceeded the storage allocation, and at the end of write, it updates this meta-data persistently. This causes the slowdown. NeST allows turning off 'Lots' and doing that makes the write performance close to that of GridFTP server. This shows that space reservation while being a useful feature comes with a certain overhead.

4 Effect of Protocol Parameters

GridFTP allows us to use different block-sizes and multiple parallel streams. Further, clients can concurrently transfer multiple files to/from the server. We studied the effect of the above parameters and concurrency on transfer rate and CPU utilization.

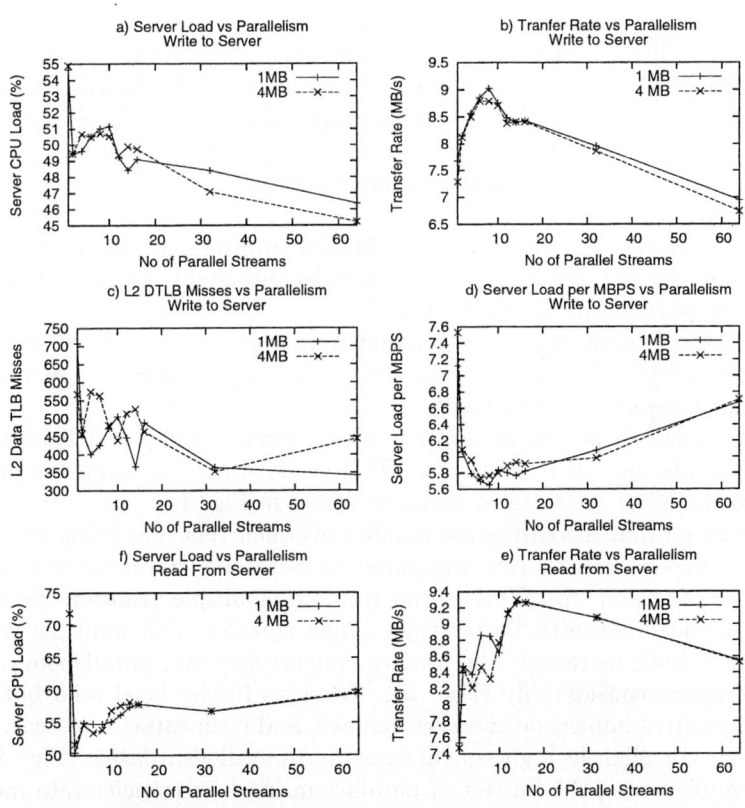

Fig. 3. The effect of block size and the number of parallel streams on GridFTP server load, transfer rate, and TLB misses.

The effect of using different block-sizes and parallelism while writing to the moderate GridFTP server is shown in Fig. 3. Interestingly, the server load drops and the transfer rate increases as we move from one stream to two streams. We repeated the experiment 20 times and got the same results. We analyzed further and decided to look at the Translation Look-Aside Buffer(TLB) misses. TLB is a cache that speeds up translating virtual addresses to physical addresses in the processor. As seen in Fig. 3c, the L2 Data TLB misses drops as the parallelism is increased from one to two. The drop in L2 DTLB misses explains the simultaneous decrease in server load and increase in transfer rate.

We went a step further and tried to find out what was causing the reduction in L2 DTLB misses and found that the Pentium processor family supports a large page size of 4 MB in addition to the normal page size of 4 KB. For data servers, using the large pages would be greatly beneficial. Unfortunately, the Linux kernel at present does not allow application to request such jumbo pages, but internally the kernel can use these large pages. We found that the internal kernel usage of jumbo 4 MB pages during use of parallel streams causes the drop in TLB misses. We also found that using a block size of 4 MB did not make the kernel use the jumbo page internally.

We tried the experiment with different machines and found that they had a different parallelism TLB miss graph. The variance in TLB misses was quite small till 10 parallel streams and starts rising after wards. Another interesting thing we found was that the TLB miss graph of a machine at different times was similar. At present, it appears that the Linux kernel usage of large pages internally depends mostly on the machine configuration. This requires a more thorough analysis.

Figure 3d shows the server load per MBPS transfer rate. Data movers may want to lower server CPU load per unit transfer rate and this graphs shows how they can use parallelism to achieve this.

The effect of block size when reading from the server is shown in Fig. 3e and 3f. We find that the optimal parallelism for reading from the server is different from that used to write to it.

We have studied the effects of different concurrency and parallelism levels on the transfer rate and CPU utilization. This study was done using the powerful server and the effect on write to server is shown in Fig. 4.

We observed that increasing the number of concurrent files being transferred results in a higher transfer rate compared to increasing the number of parallel streams (Fig. 4a and 4b). This is the result of multiple transfers being able to saturate the bandwidth better than single transfer with multiple streams. In wide area, both increasing the level of concurrency and parallelism improve the performance considerably (Fig. 4b). Whereas in the local area both have very little positive impact on the performance, and even cause a decrease in the transfer rate for slightly high concurrency and parallelism levels (Fig. 4a). As the file size increases, the impact of parallelism level on transfer rate increases as well (Fig. 4c). This study shows that increased parallelism or concurrency level does not necessarily result in better performance, but depends on many

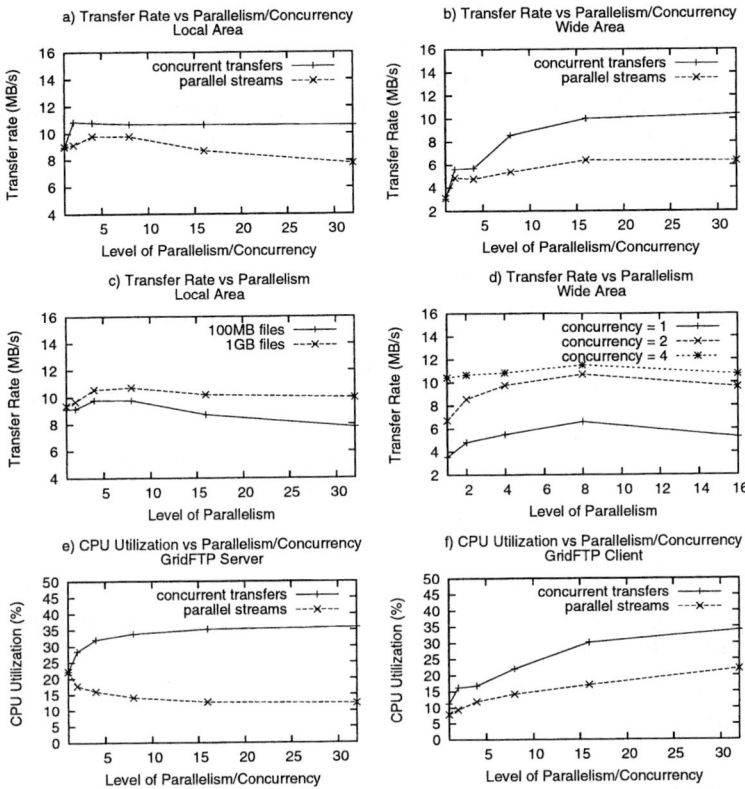

Fig. 4. The effect of concurrency and parallelism levels on the transfer rate and CPU utilization during write operations to the GridFTP server.

parameters. The users should select the correct parallelism or concurrency level specific to their settings.

Increasing the concurrency level also results in a higher load on the server (Fig. 4e), whereas increasing the number of parallel streams decreases server CPU utilization. On the client side, both increasing the concurrency and parallelism levels cause an increase in the CPU utilization of the client machine (Fig. 4f). We believe that using a combination of concurrency and parallelism can result in higher performance than using parallelism only or concurrency only (Fig. 4d). It can also help in achieving the optimum transfer rate by causing lower load to the server.

5 Related Work

CPU, memory and I/O characteristics of commercial and scientific workloads have been well studied [9] [10] [11] [12]. However, grid storage servers and data transfer protocols have not been profiled and characterized in detail.

Networked Application Logger (NetLogger) [13] toolkit enables distributed applications to precisely log critical events and thereby helps to identify system bottlenecks. It requires application instrumentation, which is difficult for complex and binary-only applications. It cannot be used to log frequent short events and kernel operations. The instrumentation may change the behavior of the program. Since, we wanted to perform a full system characterization that shows the time spent in kernel, we could not use NetLogger.

Vazhkudai et. al. [14] instrumented GridFTP [4] to log performance information for every file transfer and used it to predict the behavior of future transfers. They found that disk I/O takes up to 30% of the total transfer time and using disk I/O data improves end-to-end grid data transfer time prediction accuracy by up to 4% [15]. Our profiling gives a more complete picture of system performance and we believe that this information can be used to make more accurate predictions.

Silberstein et. al. [16] analyzed the effect of file sizes on the performance of local and wide-area GridFTP transfers and found that files sizes should be at least 10 MB for slow wide-area connections and 20 MB for fast local-area connection in order to achieve 90% of optimal performance, and small files do not benefit from multiple streams because of increased overhead of managing the streams. Our profiling work would help people find values for other parameters to achieve close to optimal performance.

6 Conclusion

In this work, we have done a full system profile of GridFTP and NeST, two widely used data access and storage servers and detailed how time is spent in these servers.

We have characterized the effect of concurrency and GridFTP protocol parameters block size and parallelism on data transfer rate and server CPU utilization. We have made clear the trade-off between single client performance and server load and shown how client performance can be increased and server load decreased at the same time and explained the reason behind this. This allows users to configure and optimize their systems for better end-to-end transfer performance and higher throughput.

We are planning to analyze our profiling data further in order to find better correlations between different parameters. This will help us to provide more useful information to users helping them to perform more sophisticated configuration and optimization.

Finally, we want to make our profiling dynamic and want to design and implement a feedback mechanism between our profiling model and higher level data movement schedulers like Stork [17]. This would allow systems to dynamically increase their knowledge, and enable schedulers to make intelligent scheduling decisions based on analysis of dynamically collected profiler data.

References

1. Sagal, B.: Grid Computing: The European DataGrid Project. In: IEEE Nuclear Science Symposium and Medical Imaging Conference, Lyon, France (2000)
2. CMS: The Compact Muon Solenoid Project. http://cmsinfo.cern.ch/ (2004)
3. LIGO: Laser Interferometer Gravitational Wave Observatory. http://www.ligo.caltech.edu/ (2003)
4. Allcock, B., Bester, J., Bresnahan, J., Chervenak, A., Foster, I., Kesselman, C., Meder, S., Nefedova, V., Quesnel, D., Tuecke, S.: Secure, efficient data transport and replica management for high-performance data-intensive computing. In: IEEE Mass Storage Conference, San Diego, CA (2001)
5. Bent, J., Venkataramani, V., LeRoy, N., Roy, A., Stanley, J., Arpaci-Dusseau, A.C., Arpaci-Dusseau, R.H., Livny, M.: Flexibility, manageability, and performance in a Grid storage appliance. In: Proceedings of the Eleventh IEEE Symposium on High Performance Distributed Computing (HPDC11), Edinburgh, Scotland (2002)
6. Sandberg, R., Goldberg, D., Kleiman, S., Walsh, D., Lyon, B.: Design and implementation of the Sun Network Filesystem. In: Proc. Summer 1985 USENIX Conf., Portland OR (USA) (1985) 119–130
7. Oprofile: A System Wide Profiler for Linux. http://oprofile.sourceforge.net (2003)
8. Anderson, J.M., Berc, L.M., Deanand, J., Ghemawat, S., Henzinger, M.R., Leung, S.A., Sites, R.L., Vandevoorde, M.T., Waldspurgerand, C.A., Weihl, W.E.: Continuous profiling: Where have all the cycles gone? In: Proceedings of the sixteenth ACM symposium on Operating systems principles. (1997)
9. Barroso, L., Gharachorloo, K., Bugnion, E.: Memory system characterization of commercial workloads. In: Proceedings of the International Symposium on Computer Architecture. (1998)
10. Lee, D., Crowley, P., Bear, J., Anderson, T., Bershad, B.: Execution characteristics of desktop applications on Windows NT. In: Proceedings of the 25th International Symposium on Computer Architecture. (1998)
11. Crandall, P., Aydt, R., Chien, A., Reed, D.: Input/output characteristics of scalable parallel applications. In: Proceedings of the IEEE/ACM Conference of Supercomputing, San Diego, CA (1995)
12. Kuo, S., Winslett, M., Cho, Y., Lee, J., Y.Chen: Efficient input and output for scientific simulations. In: Proceedings of I/O in PArallel and Distributed Systems. (1999)
13. Tierney, B., Gunter, D.: NetLogger: A toolkit for distributed system performance tuning and debugging. Technical Report LBNL-51276, LBNL (2002)
14. Vazhkudai, S., Schopf, J.M., Foster, I.: Predicting the performance of wide area data transfers. In: In Proceedings of the 16th Int. Parallel and Distributed Processing Symposium. (2002)
15. Vazhkudai, S., Schopf, J.M.: Using disk throughput data in predictions of end-to-end grid data transfers. In: Proceedings of the third International Workshop on Grid Computing, Baltimore, MD (2002)
16. Silberstein, M., Factor, M., Lorenz, D.: Dynamo - directory, net archiver and mover. In: Proceedings of the third International Workshop on Grid Computing, Baltimore, MD (2002)
17. Kosar, T., Livny, M.: Stork: Making Data Placement a First Class Citizen in the Grid. In: Proceedings of the 24th Int. Conference on Distributed Computing Systems, Tokyo, Japan (2004)

Storage Power Management for Cluster Servers Using Remote Disk Access

Jong Hyuk Choi and Hubertus Franke

IBM Thomas J. Watson Research Center
P.O. Box 218 Yorktown Heights, NY 10598, USA
{jongchoi,frankeh}@us.ibm.com
http://www.research.ibm.com

Abstract. This paper presents a novel approach to disk storage power management for the cluster server systems which accesses remote active storage devices instead of turning on local storage devices. It is more economical to access remote storage than to spin up a standby disk both in terms of latency and power consumption. The number of active storage in the system is controlled according to the disk I/O load and demand. Unlike previous approaches to storage power savings, the proposed scheme does not rely on the special types or combinations of disk drives. The proposed scheme can be used for various storage configurations including internal disks, NAS, and SAN.

1 Introduction

Recently, the power, energy, and thermal issues become essential issues in designing efficient data center infrastructure consisting of large number of clusters servers and storage subsystems. Although the storage systems are responsible for around 30% of data center energy consumption, it was not until recently that the disk storage power management began receiving research attentions, although the disk power management has been extensively studied in the context of portable computers. The portable computer disk power management puts the disk drive into one of the low power states such as standby and sleep when the inactivity period is expected to be large enough to offset the state change overhead both in power consumption and in latency [1].

For network servers, however, it has been known that the idle time between two disk accesses is not long enough to turn the disk drives into a low power states as in the portable computer disk power management scheme [2,3]. Although the memory caching for high performance extends the idle period, an extremely high degree of caching is required to reach the break-even point in power savings. Moreover, the miss penalty for the memory cache would be unacceptable if the disk drive is to be first powered up before an access.

Several solutions to the server disk power management have recently been proposed [4,5,3]. Gurumurthi et al. [4,5] proposed the use of the dynamic RPM (DRPM) disk drives [6] which can dynamically modulate its rotation speed in order to save spindle power which is responsible for over 80% of disk drive power consumption. Carrera et al. [3] proposed the use of dual disk drives - a high performance and a low power drives - in a single system as well as the use of the two speed DRPM disks according to load and demand.

While DRPM can reduce disk storage power consumption for a wide range of load and demand because it provides multiple power and performance levels in between the active and the standby modes, the DRPM disk drive is not yet readily available. The DRPM disk drive will have more complex and expensive disk spindle motor, servo, and head control mechanisms. A more limiting factor is that the savings would be limited by the power consumption at the minimum disk spindle speed which is much higher than the power consumption at the standby and the sleep modes.

This paper proposes a new approach to the server disk power management which does not rely on the special DRPM disk drives. The proposed approach utilizes remote disk drives in the storage hierarchy accessible through the file system or the device level network protocols, if the local disk drive has been put into a standby state. Because the latency of accessing a remote disk drive is far less than the disk spin-up latency, the miss penalty of the memory cache can be kept small enough to ensure seamless service.

The number of active disks in the system can be managed according to system load, demand in workload, and service quality levels. If disk I/O demand is expected to rise, more disk storage devices should be activated, while on the other hand, if it is expected to diminish, a selected number of disk drives should be put into the standby or into the sleep state to reduce power consumption.

Through a trace-driven simulation of a Web server cluster, we have shown that the proposed storage power management scheme outperforms the conventional cluster node concentration scheme [7, 8] in power and energy savings.

This paper first describes the underlying network service models and the architecture of the storage hierarchy required for the service in the next section. Then, the proposed power management mechanism for the disk storage system is described in Section 3. Section 4 presents the evaluation results in comparison with the existing cluster node power management mechanism. Section 5 summarizes the paper and presents future works of this research.

2 Server Storage System Models

The disk storage power management scheme proposed in this paper relates to typical cluster Web server configurations where a cluster consists of hundreds or thousands of Web server nodes to which client requests are directed by a network dispatcher for scalability and reliability. The content storage is cached and/or replicated for scalability and reliability as well. The consistency of data in the cache / replicated storage can be maintained either by lazy [9] or by eager [10] synchronization mechanism. Although the disk caching / replication can degrade write performance, it improves read performance significantly. Given the storage need for Web serving is by and large read-centric, it is common to rely on a high degree of replication in conjunction with partitioning.

In network servers, it is typical to utilize memory caching in order to reduce latency and improve throughput. Memory caching is also instrumental in reducing the disk drive power consumption because it can filter many disk accesses when its size is larger than the working set of the workload. With a sufficient amount of memory caching, the demand for the storage subsystem can be kept low while the demand for the processor and the main memory remains high.

The proposed storage power management scheme controls the number of active storages devices of the cached / replicated storage systems according to the predicted demand in workload. When a request dispatched to a server with a cached / replicated storage cannot be fulfilled in the memory cache, content should be retrieved from the disk storage. If the cached / replicated storage is not active, the proposed scheme forwards the storage access request to one of the active storage devices containing equivalent content.

The proposed power management scheme can also be applied to SAN environments. It can be used in the basic SAN configuration which has a shared nothing architecture in which each LUN of a disk array is assigned to a single host. It can also be used in the recent storage virtualization environments. With the storage virtualization, it is possible to make non-disruptive changes in the storage configuration and to create virtual disks over multiple storage devices.

3 Storage Power Management via Remote Access

3.1 Decoupling Processing and Storage I/O Demands

The existing power management approaches failed to decouple the disk I/O demand from the processing demand.

The load concentration approach for cluster nodes presented in [7, 8] performs dynamic provisioning of cluster nodes according to the system utilization and the request demand. Because it is the entire node including CPU and disk that is powered up and down, this approach cannot capture the difference in the processing and disk I/O demands. As a result, the disk drive can consume unnecessary power when many requests can be satisfied from the memory cache.

The DRPM [4, 5] does not completely decouple the I/O and the processing demands. The DRPM disk should operate at least at the minimum spindle speed to avoid high access cost. Even when the disk rotates at its lowest speed, the power consumption is much higher than that in the standby and sleep modes.

Our motivation is to decouple the I/O power management decision from the node power management decision, so that we can perform disk load concentration while leaving the rest of the node operational without the disk drive. When a disk-less cluster node could not satisfy client requests out of its memory, it can access an active remote storage device if internal storage is attached to each node. The number of active disks in a cluster is determined according to the disk storage load condition and is not totally dependent on the number of active cluster nodes. In an alternate configuration where NAS or SAN devices are used, the number of active cluster nodes and the number of active file servers / storage arrays can also be controlled in a decoupled way.

3.2 Operational Principles

In the proposed storage power management scheme, not all disk drives in the cluster should remain active. The number of active disk drives at one level of the storage hierarchy is controlled according to system load and request demand for the storage hierarchy subsystem. If the demand for disk I/O remains low or is expected to diminish, some number of disk storage devices can be put into the standby state. On the other hand, if the demand for disk I/O grows, additional disk storage devices should be spinned up to the active state.

If the local disk storage device is in the standby or sleep state when a client request cannot be fulfilled by the memory cache, the cluster node retrieves required contents from one of the remote active disk drives instead of spinning up the local disk. This strategy is based on a clear observation that the remote disk access through a remote file system or the storage area network has the latency which is orders of magnitude lower than that of the local disk spin-up.

When a cluster node decides to access an active remote disk rather than to activate local disk, it consults the states of the disk drives and the cluster nodes to select a cluster node which contains an equivalent, active disk drive. The disk drive selection decision should be based on the load and demand conditions of the disk storage devices and the cluster nodes.

Instead of having the storage device switch at the memory cache miss handler of each node, a request dispatcher could have utilize the storage power management status of each cluster node. However, this is not a viable option because it is not feasible to track the contents of the memory cache at the dispatcher. Although a form of cooperative caching can be utilized to increase the memory cache hit ratio, the dispatch mechanism at the memory cache miss handler is still required to forward storage requests to one of the active disk storage devices.

In the storage configurations where the cluster nodes access the networked storage such as SAN device, the storage switch can be located at various places, such as the cluster node, SAN switch, and the SAN storage array.

In order to benefit most from the proposed power management scheme, multiple disk drives of a storage array need to be spinned down independently without disrupting the operations of the other disk drives in the array. In SAN, multiple LUNs of a storage array that map to the same set of disk drives can be exported to different cluster nodes as separate storage units. Therefore, it is important to map multiple LUNs of a disk array to a replication group which can be turned on and off in tandem. In the storage virtualization environment, it would be important to map multiple LUNs of a disk array to the same MDG.

4 Evaluation Results

4.1 Simulation Model

We developed a simulation model of a Web server cluster in which each cluster node model consists of one CPU and two mirroring disk drives which have separate queues. Each cluster node maintains a main memory cache for the Web content objects with the LRU replacement policy. The Web request processing times are modeled with empirical parameters obtained by measuring the Apache 2.0 Web server on the RedHat 9 Linux OS on top of a 2.4GHz Intel Pentium 4 box (with hyper-threading enabled). The processor occupancy for a zero-byte content request is measured to be 300us and the content transmission throughput to be 80us per 1KB. The disk drive performance and power consumption parameters similar to the one described in [4] were used. The entire cluster system under test consists of 16 such cluster nodes. The disk seek and rotational latency are set to 4.2ms and the disk transfer time is set to 71us per 4KB. A seek and a rotational delay is introduced every 44KB which models the average transfer size between two seeks. The NFS block transfer time is set to double the local disk transfer time [11].

The processor power consumption was calculated as the proportions of its max / min power consumption (65W/10W) according to the measured processor utilization. The power management scheme is assumed not to use the DVS for processors nor to use DRPM for disk drives. The disk power consumption are modeled after a disk drive similar to the IBM Ultrastar 36ZX [2]. The active, idle, and standby power of the disk is set to 39W, 22.3W, and 4.15W, respectively. The rest of the cluster node, such as memory, video, and NIC, is assumed to consume 12W based on the data presented in [12].

In the experiment, we used a simple storage demand estimation mechanism which is similar to that of the UNIX process load average estimator. The demands for processing and disk I/O are monitored every 8 seconds while the load concentration decision is made every 80 seconds. Exponential moving average (EMA) for 8 minutes are used in making power management decision. Every 80 second, additional nodes can be activated if the EMA demand for processing is larger than the number of active nodes. If the EMA demand for disk I/O is larger than the number of active disks, additional disk drives can be activated. The node and the disks are deactivated when the EMA demand goes below the number of provisions. The increase and decrease amount is set proportional to the demand changes and the number of nodes / disks currently active.

4.2 Web Trace Data

The 1998 World Cup Web access trace [13] was used as the simulation input representing the real workload in the Internet. The one day trace of the 62nd day from the launch of the Web site was used. Because there was a soccer match in this day, a huge increase in demand is observed in the trace. In order to regenerate the workload to represent today's Internet environment, we scaled the workload by 8 along the time axis.

4.3 Evaluation Result

Fig. 1 shows the power consumption results for the cluster Web server system described above having a 20MB memory cache. The x-axis is the time of the day scaled by a factor of 8. The y-axis is the power consumption of the cluster in

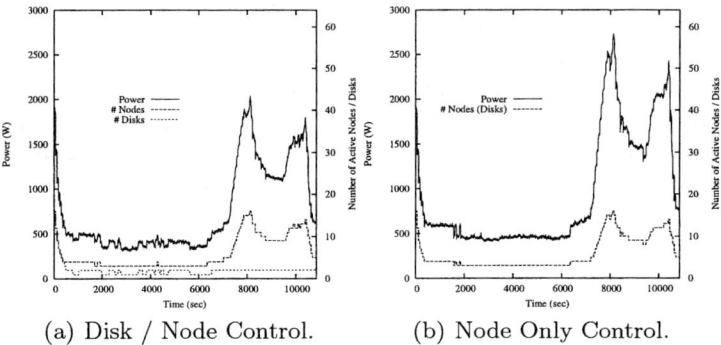

Fig. 1. Power Consumption with 20MB Memory Cache per Node.

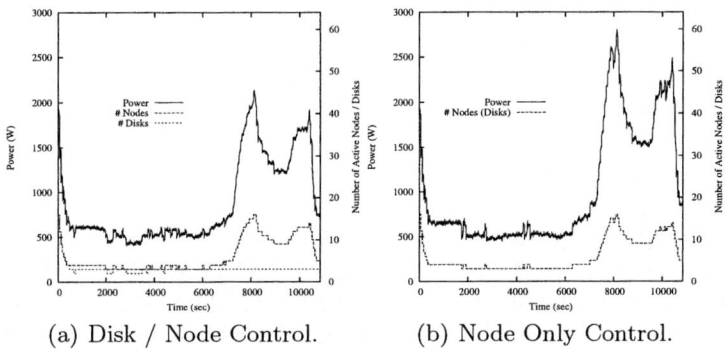

Fig. 2. Power Consumption with 8MB Memory Cache per Node.

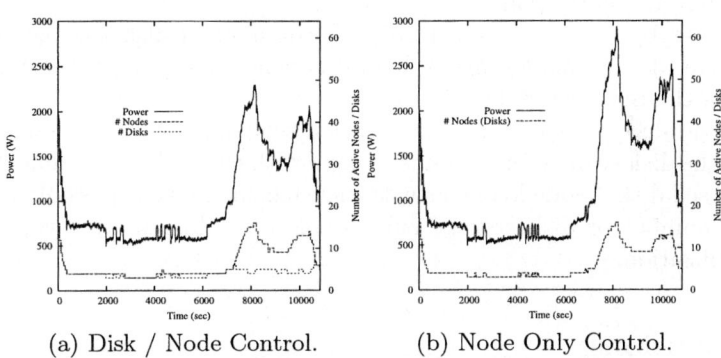

Fig. 3. Power Consumption with the Update Traffic (4MB Cache).

Watts. The number of active nodes and active disks are also shown along with the power consumption results (the second y-axis).

Fig. 1 (a) shows the power consumption of the proposed disk / node load concentration scheme which performs the disk and the node power management decision in a decoupled way. Fig. 1 (b) shows the power consumption of the conventional load concentration method where the number of active nodes are controlled by the processing demand. The power management decision for the disk is subordinate to that of the node.

The benefit of the proposed disk / node load concentration clearly manifests itself in Fig. 1. Due to the high locality of reference in the trace, the 20MB memory cache filters 98.5% of the disk accesses. The proposed scheme maintained 6.0 active nodes and 4.0 active disks on average, while the conventional load concentration scheme maintained 5.88 active nodes and disks. The increase in the number of active nodes is due to the increased processing loads for remote disk access. The proposed scheme improved 20.3% of energy savings over the conventional scheme. The amount of power savings also depends on the system parameters such as the types of the disk drives and the degree of memory caching.

Fig. 2 (a) and (b) show the power consumption of the two schemes with 8MB of memory cache per node. With 8MB memory cache, the hit ratio was measured to be 96.4%. The proposed scheme maintained 6.2 active nodes and 3.1 active disks on average while the conventional scheme maintained 6.0 active nodes and disks. The proposed scheme improved 13.3% of energy savings over the conventional scheme.

Fig. 3 (a) and (b) show the power consumption of the two schemes with 4MB of memory cache per node and under the presence of the synthetic update workload. The update workload was designed in consideration of content update such as news and images of the World Cup. It was assumed that the frequency and amount of update traffic increases along with the request demands. The hit ration was measured to be 93.9%. The proposed scheme maintained 6.2 active nodes and 4.1 active disks on average while the conventional scheme maintained 6.1 active nodes and disks. The proposed scheme improved 8.2% of energy savings over the conventional scheme.

As proved by the evaluation result, the number of disk storage that need to be active at any given time is highly dependent on the request rate, the hit ratio of the memory cache and degree of updates in the workload. Without the proposed disk / node load concentration scheme, it is unavoidable to over-provision the disk storage subsystems to cope with the highest possible demands. The proposed disk / node load concentration scheme make it possible to provide the right number of disk storage subsystems according to the workload and system utilization.

5 Conclusions

This paper proposed a novel disk storage power management scheme which decouples the power management decision for the disk storage from that of the cluster nodes, thereby opening a new design space for the storage power management. Through a trace-driven simulation of a 16 node cluster Web server for a real Web workload trace, the proposed disk storage power management scheme proved to reduce the system energy and power consumption up to 20% further over the savings from the conventional load concentration mechanism. We are currently investigating the use of the proposed disk power management scheme with the DRPM disks to enable more seamless provisioning of disk storage with reduced power consumption.

References

1. Li, K., Kumpf, R., Horton, P., Anderson, T.: A quantitative analysis of disk drive power management in portable computers. In: Proc. of Winter 1994 USENIX Conference. (1994) 279,292
2. Gurumurthi, S., Zhang, J., Sivasubramaniam, A., Kandemir, M., Franke, H., Vijaykrishnan, N., Irwin, M.J.: Interplay of energy and performance for disk arrays running transaction processing workloads. In: Proc. of the Int'l Symposium on Performance Analysis of Systems and Software (ISPASS). (2003) 123–132
3. Carrera, E.V., Pinheiro, E., Bianchini, R.: Conserving disk energy in network servers. In: Proc. of the 17th Int'l Conference on Supercomputing (ICS'03). (2003) 86–97
4. Gurumurthi, S., Sivasubramaniam, A., Kandemir, M., Franke, H.: DRPM: Dynamic speed control for power management in server class disks. In: Proc. of the 30th Int'l Symposium on Computer Architecture (ISCA'03). (2003) 169–179
5. Gurumurthi, S., Sivasubramaniam, A., Kandemir, M., Franke, H.: Reducing disk power consumption in servers with DRPM. IEEE Computer 36 (2003) 59–66
6. Okada, K., Kojima, N., Yamashita, K.: A novel drive architecture of HDD: Multimode hard disc drive. In: Proc. of Int'l Conference on Consumer Electronics,. (2000) 92–93
7. Pinheiro, E., Bianchini, R., Carrera, E.V., Heath, T.: Load balancing and unbalancing for power and performance in cluster-based systems. In: Proc. of the Int'l Workshop on Compilers and Operating Systems for Low Power. (2001)
8. Elnozahy, E.N., Kistler, M., Rajamony, R.: Energy-efficient server clusters. In: Proc. of the Workshop on Power-Aware Computing Systems. (2002)
9. Burns, R.C., Rees, R.M., Long, D.D.E.: Efficiently distributing data in a web server farm. IEEE Internet Computing 5 (2001) 56–65
10. Kemme, B., Alonso, G.: Don't be lazy, be consistent: Postgres-R, a new way to implement database replication. In: Proc. of the 26th VLDB Conference. (2000)
11. Ellard, D., Seltzer, M.: NFS tricks and benchmarknig traps. In: Proc. of the USENIX 2003 Annual Technical Conference FREENIX Track. (2003) 101–114
12. Bohrer, P., Elnozahy, E.N., Keller, T., Kistler, M., Lefurgy, C., McDowell, C., Rajamony, R.: The case for power management in web servers. In Graybill, R., Melhem, R., eds.: Power-Aware Computing. Kluwer Academic / Plenum Publishers (2002) 261–289
13. Arlitt, M., Jin, T.: Workload characterization of the 1998 World Cup web site. Technical Report HPL-1999-35(R.1), HP Laboratories Palo Alto (1999)

A Path Selection Based Algorithm for Maximizing Self-satisfiability of Requests in Real-Time Grid Applications

Mohammed Eltayeb[1], Atakan Doğan[2], and Füsun Özgüner[1]

[1] Department of Electrical Engineering, The Ohio State University,
2015 Neil Avenue, Columbus, Ohio 43210, USA
{eltayeb,ozguner}@ece.osu.edu

[2] Anadolu University, Department of Electrical and Electronics Engineering
26470 Eskişehir, Turkey
atdogan@anadolu.edu.tr

Abstract. Efficient data scheduling in Grid environments is becoming a seemingly important issue for distributed real-time applications that produce and process huge datasets. Thus, in this paper, we consider the data scheduling problem so as to provide reliable dissemination of large-scale datasets for the distributed real-time applications. We propose a new path selection-based algorithm for optimizing a criterion that reflects the general satisfiability of the system. The algorithm adopts a blocking-time analysis method combined with a simple heuristic (LCSP or SLCP). The simulation results show that our algorithm outperforms the algorithms existing in the literature.

1 Introduction

Research in real-time *Grid* computing is needed to enable Grid services for newly emerging class of *large-scale* real-time distributed applications. The amount of data produced and processed by these new large-scale applications poses a great challenge on the Grid infrastructure. Let us consider the following example. Assume a distributed industrial vision and inspection system that provides complex and sensitive inspection for industrial facility lines [1]. The vision equipments provide images for the product, which needs to be analyzed, matched, verified and stored in *real-time* fashion. A distributed computing system connected by a wide area network, then, provides an efficient computing environment for distributed inspection tasks on the datasets [2]. The *large-scale* datasets may include a combination of real-time still pictures, thermal images, video clips, etc. Such a system requires transferring huge datasets between distributed running tasks. Due to the fact that the datasets are large, an efficient mechanism must be devised to allow dataset transfer between tasks in remote locations. This mechanism must cater for cases by which a particular dataset is requested by more than one task in different locations. They also must account for future requests of a particular dataset and certainly the *deadline* by which the dataset is to be delivered to the final destination(s).

Examples of some other distributed real-time applications that share similar features of the industrial vision system include distributed medical information and imaging systems [3], computer vision [4], and distributed surveillance applications [5]. In [5], three different heuristics, referred to as PPH, FPH and FPA (Partial Path Heuristic, Full Path Heuristic, and Full Path All destinations heuristic, respectively), were proposed for data transfer scheduling with real-time constraints for a defense information system. In [5], deadlines for requests, each of which represents a data transfer, were assumed. The goal was to minimize the number of requests that miss deadlines. Dissemination of datasets was achieved by adopting a data *staging* technique by which a transferred data-item is cached in intermediate nodes along the path of the transfer from the source to the destination of the transfer.

The three aforementioned heuristics (PPH, FPH and FPA) schedule only one request for transfer along the shortest path from source to destination in each iteration. *Concurrent Scheduling* (CS), on the other hand, as proposed in [6], allows a communication step to include different request transfers simultaneously in an organized fashion. This is possible because some requests may be achievable through separate paths. The CS algorithm was built on top of the EPP (Extended Partial Path) heuristic proposed in [7].

Data replication problem in Grid is related to the dissemination problem and has been studied to minimize the latency of data transfer as well as to reduce bandwidth consumption, improve the system reliability and to load balance the requests [8], [9].

In this paper, we facilitate a discussion for real-time Grid computing and issues of data dissemination and scheduling requirements for large-scale data in distributed real-time applications. We propose an efficient algorithm for the data dissemination of these applications.

2 System Model

Our goal is to provide a solution to the data dissemination problem for applications with large-scale datasets. Thus, we focus on the scheduling the transfer of datasets rather than scheduling the application tasks. In our system model, we assume that applications arrive *aperiodically* online at a specific point (broker and scheduler) by which the tasks of the applications are mapped and scheduled on the distributed resources. The distributed tasks require large dataset transfers from remote locations that are determined upon the arrival of the applications. Each task may have one or more requests for data transfer.

- The network graph $G = (V, E)$ specifies the connectivity of a set of n vertices $V = \{V_1,...,V_n\}$ and m edges $E = \{E_{i,j}: V_i, V_j \in V$ and there is a communication channel between the vertices$\}$. Each vertex V_i is a node with limited storage capacity C_i. Each edge $E_{i,j}$ represents a time delay for the transfer between the end vertices V_i, V_j. This delay is assumed to be constant on $E_{i,j}$.
- A distributed application $A_j = \{T_{j,1},..., T_{j,kj}\}$ is composed of k_j tasks running on several predetermined processing nodes. Each task produces a set of *requests* to specific data-items of large, fixed sizes at different times during its execution.
- $R_{Tj,i} = \{r^j_{i,1},...,r^j_{i,li}\}$ is the set of l_i requests produced by the ith task $T_{j,i}$ of application A_j.

- Each request $r^j_{i,u}$ is associated with one of χ data-items I_t (t =1,...,χ) to be transferred to a *destination* node $N^j_{i,u}$ where the corresponding requesting task $T_{j,i}$ resides.
- Each request $r^j_{i,u}$ is assigned a deadline $Dl(r^j_{i,u})$ by which the data-item must be delivered to its destination.
- The request is also assigned a priority value $Py(r^j_{i,u})$ which is inherited from the application, which includes the task that produced the request. A request $r^j_{i,u}$ is hence summarized by the following tuple: $\langle I_t(r^j_{i,u}), N^j_{i,u}, Dl(r^j_{i,u}), Py(r^j_{i,u})\rangle$.

An *achieving* path $P_{r^j_{i,u}}$ of a request $r^j_{i,u}$ is defined here as a path that has a network latency less than or equal to the deadline of the request. An achieving path is also assumed to be *simple*. A simple path between a source V_S and a destination V_N ($V_S \neq V_N$) is given as P = {V_S,...,V_i,..., V_j,..., V_N} and $i \neq j$ for all V_i, $V_j \in$ P. The set of achieving paths of a request $\mathcal{P}_{r^j_{i,u}}$ is defined as the collection of all achieving paths from all sources of the data-item associated with the request $r^j_{i,u}$. We also assume that the arrival time of a request is its release time.

Our model also assumes a staging mechanism for data transfers which was presented in [5]. The data-item associated with a request will be stored in intermediate nodes for the duration of the deadline of the request associated with the transfer.

3 Problem Statement

Our goal is to satisfy all requests of all tasks present at any specific point of time. These requests form a batch of data-item transfers with specific deadlines. Due to the size of the individual data-items and the storage capacity of the intermediate nodes, it is not possible to accommodate all of the staging at the same time. We are also restricted by the need to allow multiple copies of the data-item exist during a specific time period for a request. An efficient heuristic should aim at maximizing the satisfiability at all times.

Let δ be a specific schedule of data transfers. A request $r^j_{i,u}$ is *satisfied* in δ if and only if the data-item associated with $r^j_{i,u}$ is delivered at the destination node $N^j_{i,u}$ on or before the deadline $Dl(r^j_{i,u})$. Let the set of satisfiable requests by the schedule δ be defined as $S(\delta) = \{r^j_{i,u}: r^j_{i,u}$ is *satisfied* in δ, $\forall j, i, u\}$. The optimization criterion of the staging heuristic is provided by the *effect* of the schedule δ which is defined as:

$$E(\delta) = \sum_{r^j_{i,u} \in S(\delta)} Py(r^j_{i,u}) \qquad (1)$$

4 Blocking Analysis Concurrent Scheduling Algorithm (BACS)

The heuristic proposed in this paper employs a data *concurrent scheduling* (CS) method and a data *blocking analysis* (BA) method (hence the name BACS) for data transfers to solve the data-scheduling problem. By concurrent scheduling, we mean that several data-items will be allowed to stage, allowing the service of multiple requests simultaneously. By blocking analysis, we mean that the delays encountered by the transferred data-items (due to blocking for intermediate storage) will be computed and used for assigning staging paths.

A blocking along the path occurs when a request competes with another request in one or more of the intermediate nodes due to limited capacity. Due to this situation, BACS enforces a special *blocking policy*. This policy compels the lower priory requests to *await* before the specific blocking point (a contention node on the path) until a space adequate for its data-item is available in the contention node. The lower priority request, in such a case, is called an *awaited* request or a *blocked* request. An awaited request will be blocked at a specific contention node for at least the amount of time needed to clear the node from the higher priority request.

4.1 BACS with Shortest Least Contending Path (SLCP) First Heuristic

The BACS algorithm attempts to generate an optimum set of paths for the individual requests. BACS iterates through three phases of execution. The algorithm starts execution by accepting a batch of requests, each defined by a tuple $\langle I_t(r^j_{i,u}), N^j_{i,u}, Dl(r^j_{i,u}), Py(r^j_{i,u}) \rangle$. BACS, then, finds a set of achieving paths for each request from a set of multiple sources. This is accomplished by running a version of Dijkstra's shortest path algorithm for each request which can find a shortest path to a specific node in the network from multiple sources [5], [10]. These paths are later sorted based on their lengths for each request.

BACS generates an initial set of paths composed of the shortest path for each request. It is obvious that this set neither guarantees the satisfiability of all requests nor maximizes this satisfiability. The reason is simply enforced blocking policy by which lower priority requests must be blocked for higher priority requests for an amount of time that is proportional to the individual path lengths.

From the initial set of paths, BACS uses a graphical method to compute the total delay incurred by each request, which is the *first phase* of the BACS:

- Determine the effective priorities for all requests. They are computed as the weight of the task multiplied by the weight of the application:

$$Py(r^j_{i,u}) = Py(A_j) \times Py(T_{j,i}) \qquad (2)$$

where $Py(r^j_{i,u})$ is the effective priority of $r^j_{i,u}$, $Py(T_{j,i})$ is the priority of task $T_{j,i}$, and $Py(A_j)$ is the priority of application A_j.
- Compute the direct blocking delays between all possible request pairs as explained in [11]. The amount of time a lower priority request will be blocked at a contention point is determined by the time needed for the higher priority request to clear its path (clear the intermediate nodes on the request's path).
- Finally, in this step, we develop the *Blocking Dependency Graph* (BDG) for the requests. Each node in this graph represents a request and each directed edge

represents the awaited time incurred on a request by a higher priority request sharing the same path as if the two are the only requests in the system. The developed BDG represents the dependencies between the requests.

Once we have a BDG, it is possible to compute the total end-to-end delays of all available requests in the batch, which is the *second phase*. This delay for each request is the length of the *critical path* of the request in the resultant DAG and must be equal to or less than the corresponding request's deadline in order for the request to be satisfied. The critical path of a request is the longest path to the request from all available nodes in the graph and its length represents the total blocking time.

Once the blocking delays are found, it is possible to compute the effect of the schedule from defined by (1). BACS then checks for the total satisfiability condition shown by the diamond in the second column. If the condition is not satisfied, the algorithm performs a path set modification phase.

In the *third phase*, BACS finds a subset of requests which is referred to as the set of *candidate requests*. This set is composed of all requests (represented by nodes) in the critical path of only unsatisfied requests (or some of the unsatisfied requests). Then, BACS attempts to modify the path set for the candidate requests. By changing the path of a higher (blocking) request, it is possible that the delay incurred by a lower (blocked) request is reduced. The following is performed in this phase:

- Starting from the highest requests in the chain, the algorithm searches for an alternative path.
- The alternative path is the *least contending* based on the *Contention Index* (CIX) function defined in [11] among all achieving paths of the request.
- If two requests with same contention amount exist the shorter is picked first (hence SLCP).

4.2 Least Contending Shortest Path (LCSP) First Heuristic

In the LCSP heuristic, the algorithm iteratively replaces the paths for each request in the candidate set produced in the third phase of the algorithm. The following steps are performed by this heuristic until reaching a feasible solution or exhausting the set:

1. Starting with the highest priority request in the set of candidates, LCSP replaces its current path with the next shortest path. If two paths are of the same length the heuristic selects the *least contending* of the two based on the CIX value defined in [11]. Note that computing the CIX value is performed only when arbitration is needed in this step.
2. The algorithm evaluates the total delays of the requests and computes the effect value by jumping to the first phase. If no better effect value is found, the next path for the current request is tested and so on. Once a better effect value is found for this request, the path of the particular request is fixed.
3. The heuristic moves to another request in the list of candidates and repeats Step 1 and Step 2 until a feasible solution is found or the candidate requests list is covered. The algorithm exits with the best solution found.

5 Simulation Results

The performance of the BACS algorithm is tested by simulation in a network of 30 machines with arbitrary topologies. These machines constitute the nodes which can be sources, destinations and/or intermediate storage locations. Each machine has a limited capacity equal to a data-item size and all data-items are of the same size. We tested the performance of the algorithm for the general situation in which the requests for a particular number of data-items are generated randomly by a subset of the 30 machines with random number of sources and destinations. Random requests are assumed to arrive at the centralized scheduling unit in batches. The parameters used to measure the algorithm performance are the number of requests in the batch and the deadlines of the requests (the urgency of the application). We allowed the load representing the number of requests to vary between 100 and 600 requests while the deadline is set to about 70% of the average path length (500 time units). The performance is measured as the percentage of satisfied requests as well as the effect value. The performance of algorithm was also tested by changing the level of urgency of the applications set be the deadline value. Here, we fixed the load at 500 requests in the batch (high load situation).

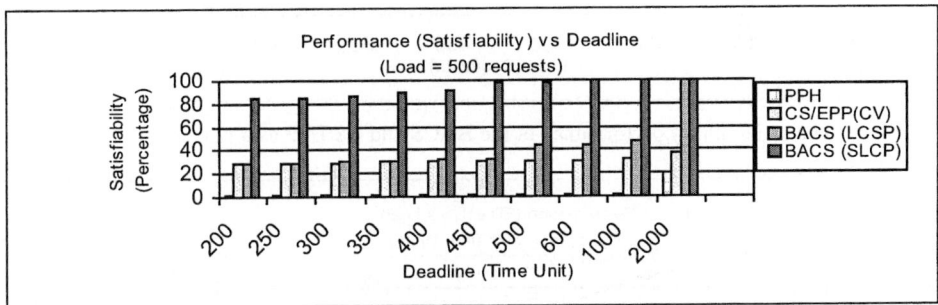

Fig. 1. The performance of the BACS, CS/EPP (CIX=CV) and PPH staging algorithms in terms of the percentage of satisfied requests at 500 request load.

Fig. 1 shows the performance of BACS, CS/EPP [6] and PPH [5] (FPH and FPA have been shown to have comparable performance with PPH even with different cost functions. See [5] for details). BACS shows better performance for the parameters set of the experiment. The PPH adheres to a method by which only one request is transferred at a time. This can result in high deadline miss rate for high load conditions. Although BACS was slightly better than CS/EPP in improving the number of satisfied requests, it showed considerable advantage over CS/EPP when the effect function was evaluated as shown in Fig. 2. This is mainly because BACS algorithm considers the entire batch and not only portions as in CS/EPP. BACS responds very well when deadlines are relaxed since many paths are considered for staging the requests. Fig. 3 and 4 show a comparison between the BACS and the CS/EPP when fixing the deadline and altering the load of the system. BACS shows superiority over CS/EPP especially at light load situations.

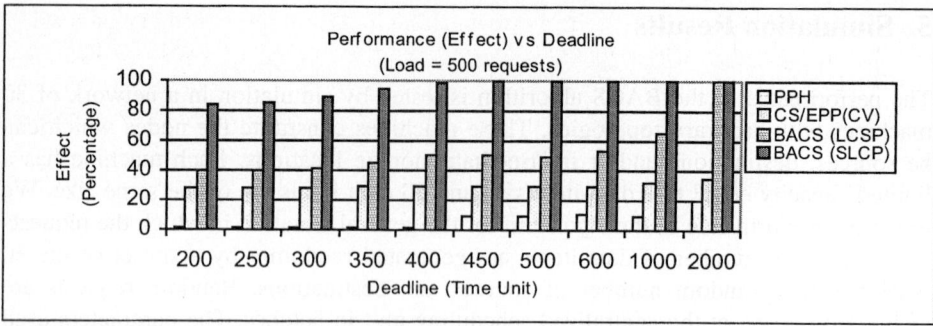

Fig. 2. The performance of the BACS, CS/EPP (CIX=CV) and PPH staging algorithms in terms of the percentage of the priorities of satisfied requests at 500 request load.

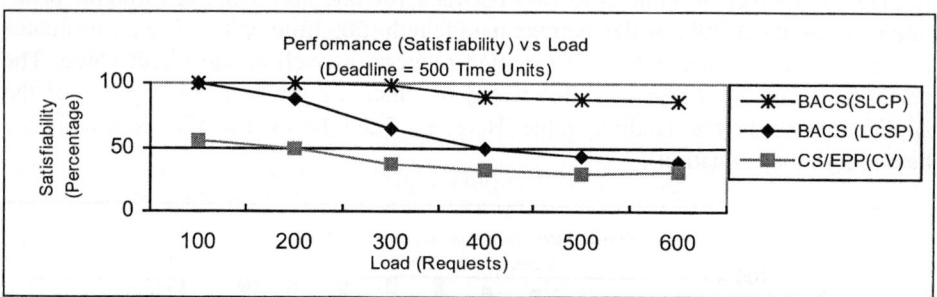

Fig. 3. The satisfiability performance of the BACS and CS/EPP as a function of the load.

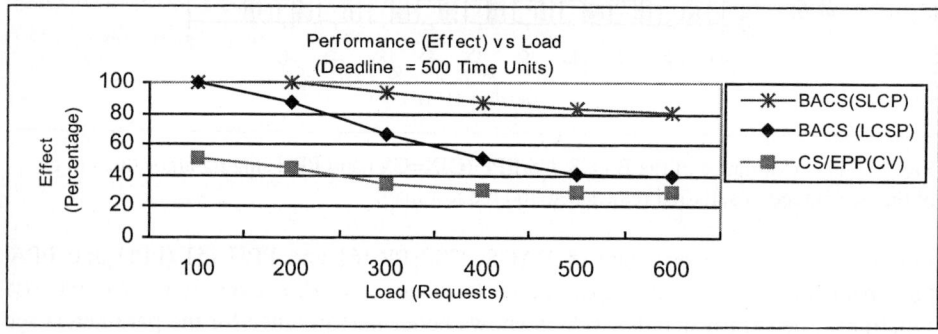

Fig. 4. The effect performance of the BACS and CS/EPP as a function of the load. The deadlines of all requests are fixed at 500 time units.

6 Conclusions

We have addressed the problem of data scheduling in distributed real-time systems with large-scale data communications and the need to consider real-time measures for Grid real-time applications in general. Our goal was to maximize the number of requests that meet their deadlines in a limited capacity environment by adopting the data staging scheme for the purpose of data dissemination. We proposed a path selection-based algorithm which maximized the objectives based on two new heuristics LCSP and SLCP. The performance of the BACS algorithm is shown by simulation to be superior to other static staging algorithms. BACS takes a batch of requests and generates a static schedule that is hopefully close to optimal. It is, however, clear that the complexity of BACS is higher than these other algorithms since multiple path search is performed for solving the problem.

References

1. Meliones, A., Baltas, D., Kammenos, P., Spinnler, K., Kuleschow, A., Vardangalos, G., Lambadaris, P.: A Distributed Vision Network for Industrial Packaging Inspection. High Performance Computing and Networking (1999) 1303-1307
2. Thomas, A., Rodd, M., Holt, J., Neill, C.: Real-Time Industrial Visual Inspection: A Review. Journal of Real-Time Imaging (1995) 139-158
3. Lee, J., Tierney, B., Johnston, W.: Data Intensive Distributed Computing: A Medical Application Example. High Performance Computing and Networking Conference (1999)
4. Shukla, S. B., Agrawal, D. P.: Scheduling Pipelined Communication in Distributed Memory Multiprocessors for Real-time Applications. In Annual International Symposium on Computer Architecture (1991) 222-231
5. Theys, M. D., Tan, M., Beck, N., Siegel, H. J., Jurczyk, M.: A Mathematical Model and Scheduling Heuristic for Satisfying Prioritized Data Requests in an Oversubscribed Communication Network. IEEE Transaction on Parallel and Distributed Systems, Vol. 11, No. 9. (2000) 969-988
6. Eltayeb, M., Do an, A., Özgüner, F.: Concurrent Scheduling for Real-time Staging in Oversubscribed Networks. The 16th International Conference on Parallel and Distributed Computing Systems (2003)
7. Eltayeb, M., Do an, A., Özgüner, F.: Extended Partial Path Heuristic for Real-time Staging in Oversubscribed Networks. The 18th International Symposium on Computer and Information Sciences (2003)
8. Lamehamedi, H., Shentu, Z., Szymanski, B. K., Deelman, E.: Simulation of Dynamic Data Replication Strategies in Data Grids. The Int'l Parallel and Distributed Processing Symposium (2003)
9. Ranganathan, K., Foster, I.: Decoupling Computation and Data Scheduling in Distributed Data Intensive Applications. The 11th Int'l Symposium for High Performance Distributed Computing (2002)
10. Cormen, T., Leiserson, C., Rivest, R.: Introduction to Algorithms. MIT Press, Cambridge, Massachusetts (1990)
11. Eltayeb, M., Do an, A., Özgüner, F.: A Data Scheduling Algorithm for Autonomous Distributed Real-Time Applications in Grid Computing. Int'l Conf. on Parallel Processing (2004)

Topic 7
Applications on High Performance Computers

Peter Arbenz, Rolf Hempel, and David Walker

Topic Chairs

Research in parallel and high-performance computing eventually has to be justified by the resulting improvements in real-world applications. The increase of availability, scalability and efficiency of PC clusters has led to increased interest in clusters for commercial and industrial applications. The recent development of middleware tools makes it possible to use computational or data grids to solve the largest classes of problems with requirements for highly heterogeneous computer resources.

This year nine papers were submitted for the applications topic. A major theme, both in terms of submission count and of paper quality, is visualization. Few papers were submitted in the "classical" area of numerical applications in science and engineering, only one of which was accepted. It seems that parallel computing in this area of research has matured to the point where corresponding papers are rather submitted to specialized application-oriented conferences. The following papers were selected for the applications session:

"*Using a Structured Programming Environment for Parallel Remote Visualization*". Remote visualization in a Grid environment requires flexibility in the placement of application components and techniques that reduce the requirements on network bandwidth. This paper presents a distributed visualization system based on the structured parallel programming environment ASSIST. The performance of the resulting ASSIST component for isosurface extraction is compared with the alternative approach in which MPI is used directly for inter-process communication.

"*FlowVR: a Middleware for Large Scale Virtual Reality Applications*". The paper introduces FlowVR, a middleware for virtual reality applications on clusters or Grids. While today most VR applications run on modestly parallel systems, preferably with shared memory, large-scale applications benefit from the computing power made available by large clusters or Grid environments. FlowVR offers a wide range of options for gathering the results in coherent views, depending on the required level of details, latency and refresh rates.

"*Parallel and Grid Computing in 3D Analysis of Large Dimension Structural Systems*". The "Grid Structural Analyser", a parallel application for the three-dimensional structural analysis of buildings, is presented. The performance of sparse linear system solvers from several publicly available libraries is compared for two test examples. Grid computing is used for the execution of many design variants in parallel, with the purpose of finding the one that best satisfies the given design requirements. The use of parallel and Grid computing leads to a speed-up of the design process.

Using a Structured Programming Environment for Parallel Remote Visualization

Pierluigi Ammirati[1], Andrea Clematis[1],
Daniele D'Agostino[1], and Vittoria Gianuzzi[2]

[1] IMATI-CNR, Via de Marini 6, 16149 Genova, Italy
{pierluigi,clematis,dago}@ge.imati.cnr.it
[2] DISI University of Genova, Via Dodecaneso 35, 16146 Genova, Italy
gianuzzi@disi.unige.it

Abstract. Remote visualization is a key issue in many Grid applications and a demanding task both from the design and the computational point of view. For these reasons adequate algorithms and programming tools are necessary in order to face problems that arise implementing a remote visualization system for Grid architectures. In this paper we report our experiences in the design of components for a remote visualization system using ASSIST [18] a high level environment for parallel programming. The use of ASSIST is considered at two levels: the high level design of the system, and the implementation of a component for isosurface extraction. Performances of the ASSIST component for isosurface extraction are assessed in comparison with a C-MPI based implementation.

1 Introduction

The evolution of the Grid, as a cooperative environment for virtual organizations, puts in evidence the need of new sets of tools and algorithms for remote interrogation and visualization of 3D volumetric data, produced by large and complex simulations or by data acquisition instruments.

A rapidly increasing attention is paid to this problem and new algorithms, methodologies and systems are being developed to satisfy the requirements of visualization in a Grid environment [1, 3, 6, 17], since it is a widely accepted opinion that current visualization tools do not satisfy emerging requirements of Grid architectures. Grid visualization software must provide remote visualization, and should permit reuse of existing software, provide a user friendly interface, be modular and extensible permitting to develop complex applications combining building blocks, and provide high performance. The NERSC Visualization Greennbook [8] indicates, among its recommendations, the improvement of remote visualization and the application of parallel computing to massive scientific data visualization. The use of "components" as the basic technology to develop this kind of system sounds like a natural possibility. Hereafter we adopt the term component in a broad sense without referring to a specific component model.

We indicate in remote visualization the key issue of our study and we would like to contribute to overcome the current limits of available systems. Our general objective is to contribute to the development of a more flexible and effective approach that permits to execute possibly heterogeneous visualization components on a distributed and parallel set of dynamic resources. We would like to explicitly deal with network bandwidth constraints, to interact with heterogeneous clients and to get high performance, permitting to achieve interactivity for large data sets. In this paper our main and specific goal is to study problems related with the use of parallel processing in this context. Other projects like Visapult [13] and Terascale Browser [12] pursue very similar goals. With respect to these projects our effort is much more limited in the availability of high-end computing resources, but in our opinion it provides some interesting experiences, especially for that it concerns the exploitation of parallel computing.

We are mostly interested in the design and implementation of efficient and portable parallel visualization software. Moreover we would like to assemble a visualization application properly combining suitable components. The possibility of obtaining a dynamic behaviour of the visualization system is another important issue. In order to achieve these objectives it is important to act on two aspects in a combined way:

- To develop effective and efficient 3D data interrogation, handling, and visualization parallel algorithms;
- To select an adequate programming environment that provides the support for the high level design of structured parallel algorithms, possibly using a modular or component based approach.

This is the effort we are currently pursuing in cooperation with other research groups within an Italian national research programme named Grid.it, whose general goals are to create a national Grid infrastructure, to develop an adequate programming environment for Grid applications, and to deploy advanced applications on this architecture. In this general context this paper discusses our experiences with the use of ASSIST [18], a high level structured parallel programming environment, for remote visualization.

The rest of the paper is organized in the following way. In Section 2 we shortly introduce remote visualization. Section 3 presents the ASSIST programming environment, while Section 4 is dedicated to visualization system design using ASSIST. Section 5 presents evaluations and experimental results and leads to future works and final considerations.

2 Remote Visualization for the Grid

We have concentrated our attention on isosurface extraction that is a basic operation for many tasks and interrogations of volumetric data. For details about isosurface extraction the reader may refer to [15]. The result of isosurface extraction is a Triangulated Irregular Network (TIN) that should be displayed using adequate volume rendering software and hardware. Rendering aspects are for the moment outside of our scope and will be not further considered, but the availability of adequate surfaces and volume rendering tools is assumed hereafter.

Meshes generated by isosurface extraction are generally large geometric models that may count hundreds of millions of triangles. Very often the model resolution is far beyond the necessity of the final user, who may be interested in getting a rough model of the entire surface, with the possibility of panning to select a zone of interest and zooming into it. The availability of a model at different Level of Details (LOD) [7] is thus an interesting possibility in many situations. It provides an aid in local visualization but it also makes visualization more *latency tolerant* for remote images display. For this reason we have developed a parallel module for TIN simplification [5]. This module may be coupled to the isosurface extraction component either tightly, by memory sharing, or loosely, by use of a file or networked interface.

Remote visualization has shown to be a communication bandwidth demanding application. Being aware of these communication requirements, we plan since the beginning of our study to use a parallel compression component, based on a TIN compression algorithm [16], which permits to reduce the size of the TIN. Also the parallel compression component can share memory or file interfaces with isosurface extraction and TIN simplification modules.

Figure 1 represents a complete pipeline at a high level of abstraction that may require different refinements and configurations, depending on the visualization task, and the available computing resources [17].

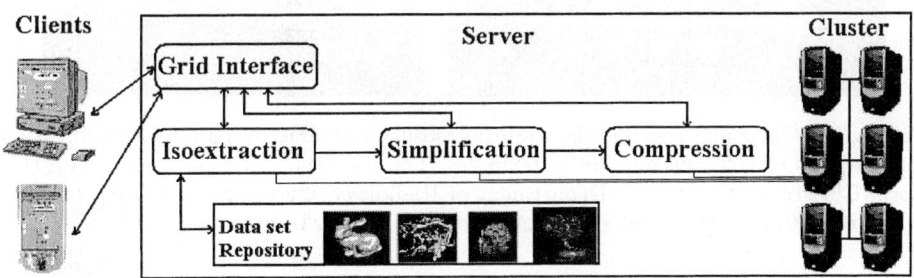

Fig. 1. A data processing pipeline for remote visualisation.

In Table 1 we have summarized I/O data flow and computing times for a sequential isosurface extraction process, based on an optimised version of the Marching Cubes algorithm [15]. Three different data sets have been considered, and for each data set different isovalues are used. The data sets are Computerized Tomography (CT) scans of a bonsai, a frog and a Christmas tree. Computing times have been collected using a Linux PC equipped with 2.66 GHz Pentium processor, 512 MB of Ram and two EIDE disks interfaced in RAID 0.

Figure 2 shows the original Christmas tree and the result of an isosurface extraction and a simplification step.

Table 1. This table summarizes I/O data volumes (in Mbyte for input and number of triangles for output) of the isosurface extraction component for different data sets and iso-values. An active cell is a basic volume element traversed by the isosurface.

Data set Id - Isovalue	Input file size (Mbyte)	% of Active cells	Output size (number of triangles)	Computing time (sec.)
Bonsai - 254	16	0.02	6,348	1.657
Bonsai - 180	16	0.87	286,954	1.790
Bonsai - 2	16	11.67	3,896,986	3.847
Frog - 150	30	0.49	306,892	3.221
Frog - 1	30	1.70	1,073,360	3.637
Frog - 75	30	4.10	2,655,552	4.539
Xmas tree -1 80	499.5	0.89	4,514,539	31.065
Xmas tree - 25	499.5	60.07	365,459,601	222.795

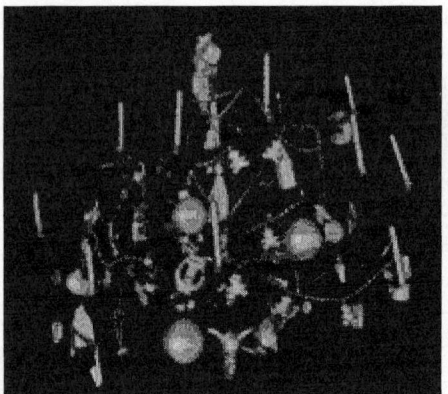 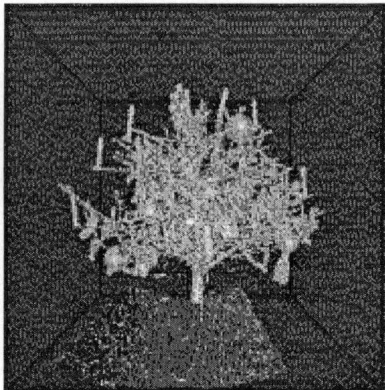

Fig. 2. The Christmas tree and the resulting isosurface (value 180) after a simplification that reduces of 60% the number of triangles. The data set was generated from a real world Christmas Tree by the Department of Radiology, University of Vienna and the Institute of Computer Graphics and Algorithms, Vienna University of Technology.

In Section 5 we will see that a parallel version of the isoextraction algorithm may greatly reduce the computing time, with some exception when considering small data sets. We have collected similar data for the simplification and compression stage. It is clear that depending on the data size, on the specific visualization task, on available computing resources, and client characteristics different pipeline configurations should be activated.

The very important aspect is that in order to develop effective and efficient software for Grid visualization, we must be able to use a high level design, programming, and integration approach, that permits to obtain the necessary flexibility in putting together components, and efficiency in exploiting parallel computing, dealing with changing architectures and application related parameters.

3 The ASSIST Programming Environment

ASSIST is a high level structured parallel programming system that integrates skeleton technology in a flexible and powerful environment in order to provide suitable support for the development of high performance portable applications in multidisciplinary environments. ASSIST is a research product developed at the Computer Science Department of University of Pisa - Italy. A complete description of ASSIST is far beyond space limits and goals of this paper and the interested reader may refer to [18].

An ASSIST program is a graph in which nodes represent modules or components, and arcs correspond to interfaces and are associated to a directional streaming of data. Streaming permits to compose modules in a complex program. Modules may also share information using external objects like a virtual shared memory. Each module can be a parallel module, *parmod* in ASSIST terminology and a rectangle in ASSIST graphs, or a sequential module, a circle in an ASSIST graph. It is possible to reuse a composition of modules as a component of a more complex program.

While the ASSIST graph permits to express interaction among program components, the parmod permits to express parallelism inside each component in a powerful and effective way. Parallel computation in the parmod is implemented by a set of *virtual processors* that interact using a *topology*, which provides a naming scheme for the virtual processors. The *internal state* of the parmod can be partitioned or replicated among virtual processors. The internal state can hold variables that permit to control communications with input and output streams. A parmod may have different *input streams* and through them interacts with the rest of the program selecting input with a nondeterministic behaviour similar to that of CSP [9] guarded commands. Moreover each input stream is associated to an independent distribution strategy like on demand, scatter, broadcast, multicast. Results of parmod computation are provided to other components through parmod *output streams*. From our point of view, one of the main relevant characteristics of the parmod construct is the fact that it is very similar to a component since it provides a single interface (input and output stream), and different implementation strategies can be associated to that interface in a semi-transparent way.

4 Visualization System Design and Implementation in ASSIST

It is quite natural to use ASSIST to design a modular and flexible pipeline that contains the three components depicted in Figure 1. Figure 3 represents the ASSIST graph for this pipeline. Here the producer is a source of 3D data, for example a simulator, while the consumer could be a rendering module, the two modules are no further detailed in this case. An interesting aspect is for the consumer the ability of non-deterministically receiving input from one of the three incoming streams.

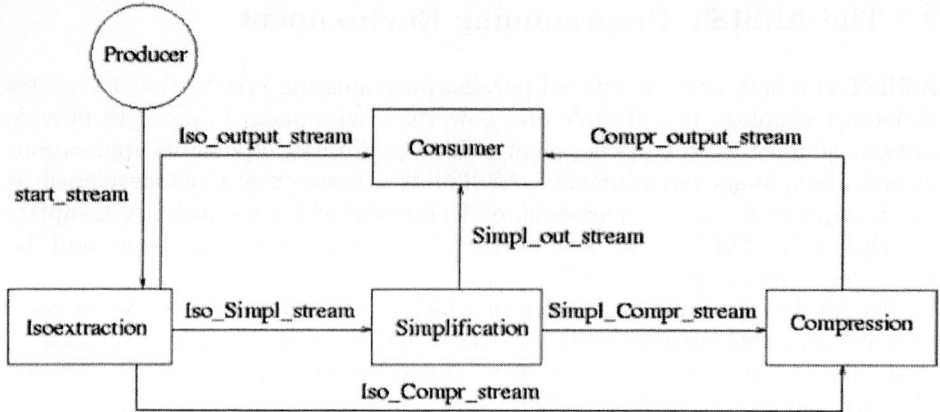

Fig. 3. The visualization pipeline expressed as an ASSIST graph.

Depending on the size of the produced TIN, and on other suitable parameters, the isoextraction parmod decides to activate the simplification and compression modules or to send data directly to the consumer. This decision is implemented by sending information and data on one of different output streams thus activating the proper component.

Different components could access a shared virtual memory that permits to optimise information flow especially when the access to large data structure is necessary. The use of shared memory is no further considered here.

At a further level of details we deal with components design and implementation. Hereafter we focus our attention on the isosurface extraction component. As already mentioned in our case this process is based on the Marching Cubes algorithm [15]. The selected parallelization strategy was originally described in [4]. The interested reader may refer to that paper in order to get more information about the Marching Cubes algorithm and our approach to its parallelization. However it is important to notice that with respect to that parallel version some improvements have been introduced.

First of all using the Parallel Virtual File System (PVFS) [14], instead of NFS, permits to reduce I/O times, when more processes access large data sets concurrently. Another point concerns load balancing for active cell processing. In a first processing phase the parallel algorithm determines active cells, i.e. volume elements that are intersected by the isosurface. For each active cell a further computation is necessary in order to properly define interpolation and construct the local triangulation that approximates the isosurface. Depending on the characteristic of the data set, and on the configuration of the pipeline a load balancing step may be dynamically executed under the supervision of the Coordinator parmod.

Figure 4 depicts the ASSIST graph for the isosurface component. It is designed around two parmods, the Coordinator and the Parallel_Marching_Cubes.

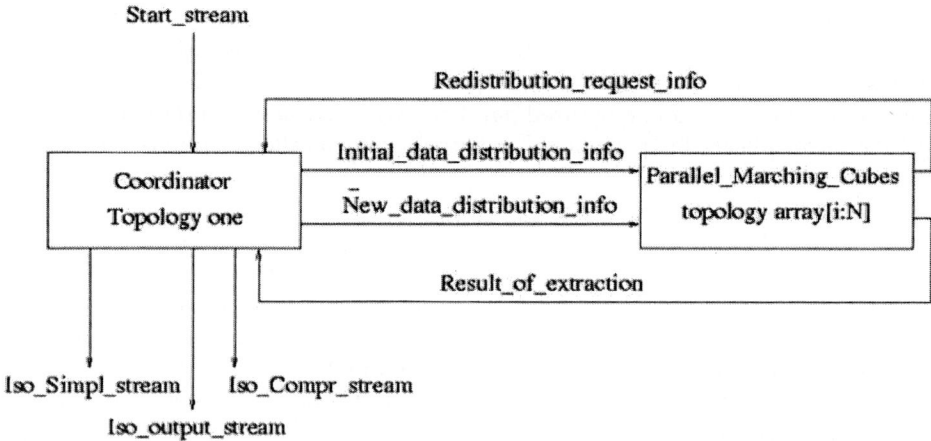

Fig. 4. The ASSIST isosurface extraction component of the visualization pipeline.

The Parallel_Marching_Cubes parmod is actually the parallel implementation of the Marching Cubes algorithm, while the Coordinator implements some control functions and provides support for dynamic load balancing.

5 Evaluation and Experimental Results

Our evaluation is mainly focused on the support provided by ASSIST for the design and implementation of parallel software for remote visualization for Grid applications. We started using ASSIST as a design and programming tool for our visualization system in summer 2003 and we are currently using version 1.1. Presently a new version (1.2), which includes implementation of external objects and a Corba interface, is available. We spent about one afternoon to install the system on a Linux cluster, but we have to spend some more time and we need some support from the developers in order to solve some compilation problems and to properly set up the system. In few days the ASSIST installation was ready and we were able to run our early program examples. The learning time of the system was quite short, at least for people having a previous knowledge of parallel programming. The system is in general well documented, despite the documentation of some specific features like the use of virtual shared memory could be improved. An important point is the easy reuse of sequential C/C++ code. We were able to plug the original sequential code in the parallel parmod with a minimal effort.

The most important aspect using ASSIST is the possibility of adopting a high level design and implementation approach for the kind of application we are interested in. The parmod construct, together with input and output streams specification, provide suitable abstractions to express parallel components with their used and provided ports. The availability of several parallel programming facilities, like parmod virtual processes topology and shared virtual memory, permit to design efficient solution. Finally the compilation and run time supports in our experience provide effective parallel program performances. Thus ASSIST seems to have an added value with respect to the use of lower level approaches, e.g. based on C-MPI, for structured parallel program design and implementation.

We have collected experimental results for our parallel version of the isosurface extraction component of the visualization system. We have experimented two versions of parallel isosurface extraction, one developed using C-MPI the other using ASSIST. The two versions are based on the same logical design, e.g. adopt the same load balancing policy, and perform the same amount of I/O operations. Also the two versions are implemented using the same sequential optimisations. The input sizes of each data set, output sizes and sequential computing times for different isovalues are collected in Table 1, presented in Section 2. Table 2 provides speed-up results for C-MPI and ASSIST versions of the isosurface parallel component on a Linux cluster for 2, 4 and 8 nodes. In the cluster each node has the same configuration of the node considered for sequential execution (see again Section 2), and nodes are interconnected through an Ethernet - Gigabit switch.

The results of Table 2 indicate that our parallel isosurface extraction algorithm provides acceptable performances and good scalability for demanding data sets and computation, and that ASSIST provides excellent performances. For small data sets, the overhead due to parallel system start up, and to contention on the file system overwhelm the benefits derived from the speed-up obtained for the computational kernel, that we have checked to scale up almost linearly.

Table 2. Measured speed-up of the C-MPI and ASSIST version of isosurface extraction algorithm for different data sets and different isovalue.

Data set -isovalue	ASSIST 2 procs	C - MPI 2 procs	ASSIST 4 procs	C - MPI 4 procs	ASSIST 8 procs	C - MPI 8 procs
Bonsai - 254	1,23	1,23	2,06	1,43	2,06	0,9
Bonsai - 180	1,31	1,26	2,19	1,38	2,95	0,95
Bonsai - 2	1,46	1,27	2,5	1,88	3,23	1,71
Frog - 150	1,47	1,40	2,42	1,84	3,38	1,49
Frog - 1	1,49	1,35	2,7	1,82	2,61	1,63
Frog - 75	1,50	1,32	2,56	1,85	3,56	1,87
Xmas tree - 180	1,54	1,62	2,49	2,68	4,21	4,01
Xmas tree - 25	1,74	1,76	3,51	3,46	6,53	6,41

We measured a start up time that is greater in the case of MPI especially for 8 processors. This leads to slow down speed-up for MPI in the cases Bonsai 254 and Bonsai 180 with 8 processors. The results obtained in most demanding cases (Xmas tree 180 and 25) indicate that with large data sets we are able to provide good efficiency figures (around 80% for 8 processors). We expect that scalability will go beyond the 8 nodes cluster used for this experiment, when larger data sets will be tested.

6 Conclusions and Future Works

In this paper we have discussed design and implementation issues for a parallel remote visualization system oriented towards analysis and visualization of volumetric data in Grid applications. We are particularly interested in using high level parallel programming languages and tools, in order to develop modular, flexible and dynamic visualization applications. Our experience of use of ASSIST is a positive one. The environment provides a very good support for the design and implementation of modular and adaptable parallel programming applications. While we expect to pay a performance fee passing from a lower level tool, MPI, to ASSIST, experimental data show exactly the opposite result. The reason of this result is in part related to the higher start-up time of MPICH-ch_p4.

We are currently experiencing also other systems like Ccaffeine [10] a Common Component Architecture (CCA) [11] compliant model. The early experiences using Ccaffeine show a higher order of complexity for installation, learning curve, and deployment of a first running example.

Future plans include the further development of simplification and compression components using ASSIST.

Finally a new Grid aware and truly component based version of ASSIST is currently being designed and developed [2]. The availability of this version will be very interesting for a better integration and awareness of our visualization application in Grid architectures.

Acknowledgements

This work has been supported by MIUR programme L.449/97-99 SP3 "Grid Computing: Enabling Technologies and Applications for eScience", and by FIRB strategic project on Enabling Technologies for Information Society, Grid.it. A particular thank is due to Pierpaolo Ciullo and all the ASSIST group at Computer Science Department of the University of Pisa - Italy.

References

1. P. Baker, W. Bethel, J. Clyne, S. Fulcomer, B. Hathaway, J. Kohl, P. Moran and S. Parker: DiVA: Distributed Visualization Framework Component Interface. In Workshop Findings Document July 8-9 2003, Eugene Oregon.
http://www-vis.lbl.gov/diva/documents/DiVA-July2003-Findings.pdf
2. R. Baraglia, M. Danelutto,D. Laforenza, S. Orlando,P. Palmeri, P. Pesciullesi, R. Perego and M. Vanneschi: AssistConf: a Grid configuration tool for the ASSIST parallel programming environment". In Proceedings of Euromicro PDP Conference 2003.
3. A. Clematis, D. D'Agostino, W. De Marco and V. Gianuzzi: A Web-Based Isosurface Extraction System for Heterogeneous Clients. In Proceedings of the 29th Euromicro Conference, 2003 IEEE Computer Society Press, pp. 148-156.

4. A. Clematis, D. D'Agostino and V. Gianuzzi: An Online Parallel Algorithm for Remote Visualization of Isosurfaces. In Proceedings of the 10th EuroPVM/MPI Conference, LNCS, No. 2840, pp. 160-169, 2003.
5. A. Clematis, D. D'Agostino, V. Gianuzzi and M. Mancini: Parallel Decimation of 3D Meshes for Efficent Web based Isosurface Extraction In Proceedings of the International Conference of Parallel Computation (PARCO 2003), North Holland/Elsevier, in press.
6. D. D'Agostino: Designing parallel programs for 3D data processing on GRID and Ubiquitous computing architectures. Ph.D Thesis Proposal, Dept. of Computer Science, University of Genova, Italy, Nov. 2002.
http://www.disi.unige.it/person/DagostinoD/dagoproposal.pdf
7. C. Gotsman, S. Gumhold and L. Kobbelt: Simplification and Compression of 3-D Meshes. In Tutorials on multiresolution in geometric modelling, A. Iske, E. Quak, M. Floater (eds.), Springer 2002.
8. B. Hamann, W. Bethel, H.D. Simon and J.C. Meza: NERSC Visualization Greenbook-Future Visualization Needs of the DoE Computational Science Community hosted at NERSC. In Int. Journal of High Performance Computing Applications, Vol. 17, No. 2 pp. 97-124, Summer 2003.
9. C.A.R. Hoare: Communicating Sequential Processes. In Comm. of the ACM Vol. 21, No. 8, pp. 666-677, Aug. 1978.
10. http://acts.nersc.gov/events/Workshop2003/slides/CCA/framework.pdf
11. http://www.cca-forum.org/
12. http://www.llnl.gov/icc/sdd/img/terascale.shtml#links
13. http://www-vis.lbl.gov/projects/visapult/
14. W.B. Ligon III, and R. B. Ross: PVFS: Parallel Virtual File System. In Beowulf Cluster Computing with Linux, Thomas Sterling, editor, pp. 391-430, MIT Press, Nov. 2001.
15. W. Lorensen and H. Cline: Marching cubes: A high resolution 3-D surface construction algorithm. In Computer Graphics, Vol. 21, pp.163-169, 1987.
16. J. Rossignac: Edgebreaker: Connectivity compression for triangle meshes. In IEEE Transactions on Visualization and Computer Graphics, Vol. 5, No. 1, pp.47-61, Jan-Mar. 1999.
17. J. Shalf and E.W. Bethel: The Grid and Future Visualization System Architectures. In IEEE Computer Graphics and Applications, Vol. 23, No. 2, pp. 6-9, Mar./Apr. 2003.
18. M. Vanneschi: The programming model of ASSIST, an environment for parallel and distributed portable applications. In Parallel Computing, Vol 28, No. 12, pp.1709-1732, Dec. 2002.

Parallel and Grid Computing in 3D Analysis of Large Dimmension Structural Systems

José M. Alonso, Carlos de Alfonso, Gabriel García, and Vicente Hernández*

Departamento de Sistemas Informáticos y Computación,
Universidad Politécnica de Valencia,
Camino de Vera s/n, 46022, Valencia, Spain
{jmalonso,calfonso,ggarcia,vhernand}@dsic.upv.es
http://www.grycap.upv.es

Abstract. In this article a parallel application for the 3D structural analysis of buildings is presented. Taking into account that solving the system of linear equations is the most time-consuming phase, several parallel public domain numerical libraries, that reflect the start-of-the-art, have been tested. A timing result comparison when analyzing two medium-sized buildings is included. Besides, a Grid-based demonstrator has been developed, integrating the parallel application and taking advantage of the different resources remotely distributed in the network. The Grid demonstrator enables the simulation of a larger number of different structural alternatives, with the purpose of finding the one which better accomplishes with all the economical limitations, safety requirements, aesthetical aspects and other criteria.

1 Introduction

Structural analysis of buildings is the process to determine the response of a structure to specified external loads or actions. This response is usually measured by obtaining tensions and displacements that take place at any point of each structural component.

This structural analysis plays a central role during the preliminary design stage of a building, when several alternatives must be considered. At this stage, construction standards obliges the designer to take into account a wide range of situations and load cases which, with a given probability, could occur during the lifetime of the structure. For each of these cases and model configurations, a structural analysis is required. A requirement of speed is imposed by the need of having comprehensive information in the preliminary design cycle. Usually, these initial designs have been based on simplified models to minimize the time and effort spent on this phase. Although, in some cases, all the designs are rejected,

---

* The authors wish to thank the financial support received from The Spanish Ministry of Science and Technology to develop the project GRID-IT (TIC2003-0131). This work has been partially supported by the Structural Funds of the European Regional Development Fund (ERDF).

returning to the initial design stage, a selection among these alternatives has to be made before proceeding a detailed design phase. Now, a large number of different structural configurations have to be analysed quickly and in a realistic way to achieve the most efficient solution (the cheapest and the safest one). An iterative trial-error process is developed by the structural engineer where, varying sections or load conditions, the whole structure is analysed and the results are interpreted. At this final stage, calculations must be performed accurately, complying criteria of safety, cost limitations or construction constrains.

3D structural models deal with 6N degrees of freedom (dof), where N indicates the number of nodes considered in the structure. Current trend of erecting more complex and larger buildings is providing structural problems that can reach dimensions of about several hundreds of thousands of equations.

Therefore, structural analysis is one of the most time consuming stage in the design cycle of a building. Memory and computing power requirements to manage efficiently large systems implies the utilization of high performance computing techniques, even when Grid strategies are applicable to enable a decrease in the simulation time, a realistic analysis of larger buildings, and to test a greater number of alternatives in the preliminary and detailed design stage.

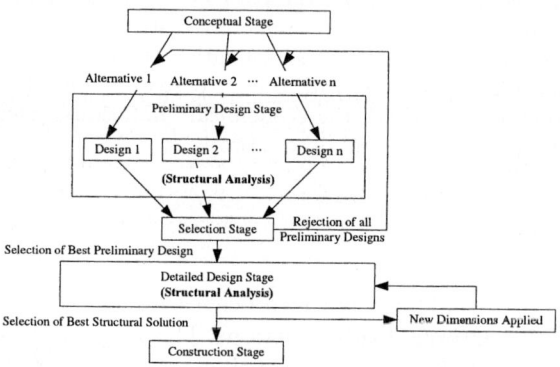

Fig. 1. Flow diagram describing the engineering design process.

The paper is structured as follows: Section 2 describes the parallelization of the structural analysis method applied. Advantages of using parallel numerical libraries with state-of-the-art capabilities, together with a brief summary of each of them, are detailed in Sect. 3. Performance of the parallel application developed is shown in Sect. 4. Section 5 presents a Grid Computing-based approach that allows performing multiple structural analyses. Finally, conclusions and further work are described in Sect. 6.

2 Parallelization of the Structural Analysis Method

The matrix methods [1] represent the most powerful design tool in structural engineering and they are appropriate to be implemented on computers. These methods are based on the idea of replacing the actual continuous structure by a mathematical model made up from structural elements of finite size with known elastic and inertial properties that can be expressed in matrix form.

This paper introduces a MPI-based parallelization of the *stiffness method*. It employs the stiffness properties of the structural elements to form the equations that represent the relationship between nodal forces acting on the structure and its displacements. The unknown joint movements are computed by solving the equilibrium equations at the joints. At each joint, the loading conditions are specified by six force components, and the response is described by six displacement components. This method is based on the following five phases:

1. Generation of the force matrix $F \in \Re^{6N \times M}$, where M is the number of combinations of actions applied to the structure, and N represents the total number of joints in the building.
2. Generation of the stiffness matrix $K \in \Re^{6N \times 6N}$ of the structure, by assembling the stiffness matrices of the different elements that compose it. The K and F matrices are generated in parallel, where each processor will assemble a local part of them. As a result, these matrices will be partitioned among the processors, following a row-wise block-striped distribution.
3. Calculation of the joint movement. For each joint, six displacements must be calculated. The relationship between external forces F and displacements $D \in \Re^{6N \times M}$ is given by

$$KD = F. \qquad (1)$$

Thus, a system of linear equations must be solved to compute the node movements. K matrix is sparse, symmetric and positive definite. Great precision is required at this point, as it determines the remaining structural phases.

4. Calculation of the member end forces. Once joint movements are known, the structural elements are divided into p groups. Each processor computes the 12 internal forces at both ends of its assigned structural elements, and the reactions at the points attached to the rigid foundation.
5. Computation of the deformations at any point of the structure. Finally, the bending moments and deformations at the predefined division points of the members are evaluated in parallel, to check that they do not exceed the established limits. Once the bending moments of an element have been worked out, its deformation is computed by means of the differential equation of an elastic curve, see (2). In this equation, $M(x)$ is the bending moment at the internal point x of a member, $y(x)$ represents its deflection function, E is its module of elasticity, and I the moment of inertia of the bending plane.

$$\frac{\partial^2 y(x)}{\partial x^2} = \frac{M(x)}{EI(x)}. \qquad (2)$$

3 Parallel Modern Numerical Libraries to Solve the System of Linear Equations

Solving the large sparse symmetric system of linear equations is the crucial problem in the 3D static structural analysis, where direct or iterative methods can be applied. Different efforts and initiatives have recently appeared encouraging the adoption and use of software tools that make it easier for programmers to write high-performance scientific applications and solving major scientific and technical problems [2]. The use of these mature tools provides important advantages, such as code portability, reusability, modularity and better performance.

In this way, multiple parallel software packages based on direct methods have emerged to solve large sparse symmetric systems: SPOOLES [3], MUMPS [4], PSPASES [5], WSMP [6], etc. WSMP and PSPASES employ the theoretically most scalable algorithms for Multifrontal Cholesky factorization and they also perform parallel ordering. Both the subtree-to-subcube mapping of the elimination tree among the processors and a two-dimensional distribution of frontal and update matrices among subgroups of processors are crucial to obtaining the highest scalability. SPOOLES employs subtree-to-subcube mapping but uses a one-dimensional distribution. MUMPS uses a two-dimensional distribution of data at only the topmost supernode of the elimination tree. However, PSPASES and WSMP apply both of them.

In addition, several parallel software packages based on iterative methods and preconditioners have been implemented: PETSc [7], Aztec [8], P-SPARSLIB [9], BlockSolve95 [10], HYPRE [11], etc. Even, some packages deal, exclusively, with parallel preconditioners for iterative methods: BPKIT [12], SPAI [13], etc. PETSc is nowadays the most popular, comprehensive and widely used.

3.1 Numerical Libraries Employed

In this work, WSMP, MUMPS, PETSc and BlockSolve numerical libraries have been applied to solve the linear system mentioned before.

Watson Sparse Matrix Package. WSMP is a high-performance, robust software package for solving large sparse symmetric and non-symmetric systems of linear equations. A node, composed of one or more processors of CPUs, communicates with other ones via message-passing (MPI). However, parallelism within multiprocessor nodes is exploited by threads or message-passing. The serial ordering heuristics used in WSMP is based on the Multilevel Nested Dissection (MND) algorithm implemented in METIS library [14]. Parallelization implemented in WSMP exploits the natural parallelism of MND ordering. The elimination tree is performed and adjusted in order to balance the numerical factorization work among the processors. The parallel symmetric numerical factorization is based on Cholesky Multifrontal algorithm. Both this phase and the parallel solution of triangular systems are guided by the supernodal elimination tree, following the same factor matrix distribution.

Portable, Extensible Toolkit for Scientific Computation. PETSc includes an expanding suite of parallel linear, nonlinear equation solvers and time integrators that may be used in application codes. It also provides many of the mechanisms needed in parallel application codes, such as parallel matrix and vector assembly routines. The combination of a Krylov subspace method (CG and variations, GMRES, etc.) and a preconditioner (Jacobi, Block Jacobi, Incomplete Cholesky/ILU, Additive Schwarz, etc.) is the heart of the parallel iterative methods in PETSc. In addition, it allows uniform and efficient access to external direct solvers (MUMPS, SPOOLES, etc.) and preconditioners (BlockSolve95, HYPRE, ParPre, SPAI, etc.).

MUltifrontal Massively Parallel Solver. MUMPS is a MPI-based parallel package for solving linear systems of equations where the coefficient matrix is sparse and can be either unsymmetric, symmetric positive definite, or general symmetric. MUMPS uses a Multifrontal technique which is a direct method based on either the LU or the LDLT factorization of the matrix. A range of orderings to preserve sparsity is available: approximate minimum degree ordering (AMD), approximate minimum degree ordering with automatic quasi dense row detection (QAMD), an approximate minimum fill-in ordering (AMF) and the possibility of using an ordering provided by the user. PORD [15] and METIS packages are also possible choices.

BlockSolve95. It is a scalable parallel software library for solving large, sparse systems of linear equations. It contains implementations of several well-known iterative methods (CG, GMRES, SYMMLQ) and different preconditioners (Incomplete LU/Cholesky, SSOR or Block Jacobi). It uses an efficient implementation of a parallel coloring algorithm that allows for the efficient computation of matrix orderings and scalable performance of the linear solver. Just ILU and ICC preconditioners can be invoked from PETSc.

4 Experimental Results

Two medium-sized buildings, presented in Table 1, have been chosen to compare the results when using different numerical libraries.

Table 1. Features of the buildings used in test.

	Degrees of Freedom	Actions Applied
Building No. 1	Θ (60,000)	7
Building No. 2	Θ (100,000)	1

Table 2 shows the time, and efficiencies, spent on the whole structural analysis of these buildings. The linear system has been solved by means of the public domain numerical libraries WSMP, MUMPS, PETSc and BlockSolve95. WSMP has demonstrated to be the fastest solver, even more scalable than other direct solvers such as MUMPS, as expected. MUMPS library has been employed via PETSc, together with the MND ordering provided by METIS. Other orderings have also been tested in MUMPS, providing worse parallel performance.

Since coefficient matrices are very ill-conditioned, iterative methods are much slower than direct methods, although they present better efficiencies. Moreover, the whole system need to be solved for every right hand sides (RHS) when using iterative methods, which is much less efficient than just computing the forward-backward substitution for each RHS as direct methods do. Regarding PETSc, the lowest solving times have been found for Conjugate Gradient (CG) solver, combined with Block Jacobi preconditioning, where Incomplete Cholesky (ICC) factorization has been applied as subblock preconditioner. The main disadvantage of Block Jacobi is that the number of iterations arises as the number of processors increases. Finally, the linear system has been solved applying the CG solver implemented in PETSc, and the ICC preconditioner of BlockSolve95, through the interface provided by PETSc. In order to achieve good performance, different coefficient matrix formats must be employed in PETSc, depending on the external numerical library used. Worse times have been obtained for any other combination of solver and preconditioner in PETSc.

Table 2. Structural analysis time (in seconds) and efficiencies (%) for the test battery buildings using up to 16 processors. Simulations have been run on a cluster of 20 Pentium Xeon@2GHz biprocessors with 1 Gbyte RAM, connected by a SCI network.

Building	Proc.	WSMP	MUMPS + METIS	PETSc (CG + Block Jacobi + ICC)	PETSc + Blocksolve95 (CG + ICC)
No. 1	1	10.5 100.0 %	10.4 100.0 %	308.9 100.0 %	620.9 100.0 %
No. 1	2	5.6 93.0 %	6.8 76.8 %	164.4 94.0 %	311.1 99.8 %
No. 1	4	3.8 69.2 %	5.3 48.9 %	99.3 77.8 %	155.4 99.2 %
No. 1	8	2.9 45.0 %	4.8 27.3 %	64.7 59.7 %	88.1 88.1 %
No. 1	16	3.5 18.6 %	10.4 6.3 %	88.7 21.8 %	56.5 68.7 %
No. 2	1	10.7 100.0 %	16.5 100.0 %	60.8 100.0 %	115.0 100.0 %
No. 2	2	6.4 83.5 %	10.8 76.3 %	34.9 87.0 %	60.3 95.3 %
No. 2	4	4.7 57.5 %	8.7 47.2 %	19.3 78.7 %	32.4 88.9 %
No. 2	8	4.1 32.3 %	7.6 26.9 %	10.8 70.1 %	16.9 84.8 %
No. 2	16	3.9 16.9 %	7.1 14.6 %	6.8 56.2 %	11.1 64.4 %

5 Grid Computing-Based Structural Analysis

Grid Technologies main objective is to ease and coordinate resource sharing for collaborative problem solving in a dynamic multi institutional Virtual Organization [16] [17]. This technology aims at sharing further elements than information, by allowing direct, coordinated and secure access to different kind of computing resources, applications, communication, data, special devices, etc. In order to create the Grid infrastructure, some kind of middleware is needed. There are some well-known projects in Grid middleware, such as the open source Globus Toolkit (GT) [18] or Unicore [19]. Some private companies have developed their own middlewares, such as InnerGRID [20] or Avaki Data Grid [21]. In our case, a Grid Computing-based system, employing GT 2.4, have been implemented to analyze concurrently a group of different structural solutions at the preliminary and final design stages, making use of the parallel application developed.

5.1 Grid Structural Analyser Architecture

Figure 2 shows a conceptual view of the Grid Structural Analyser developed. Starting from a set of different structural alternatives and a group of machines distributed throughout internet, the Grid Structural Analyser performs the necessary work to analyse all of them by running the parallel application on the available resources.

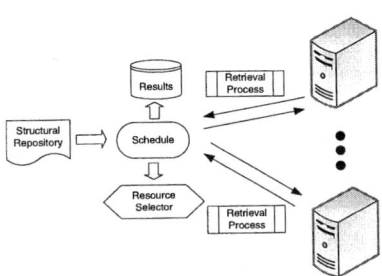

Fig. 2. Grid Structural Analyser architecture. The system consists in three main parts: the *scheduler*, which obtains designs from a structural repository; the *resource selector*, which selects an appropiate resource to run the parallel application; and the *retrieval processes*, which retrieve results files.

Scheduler. This module considers a pool of different structural configurations to be analysed and a warehouse where results will be stored. A structure is composed by a set of files which defines its geometry, properties and loads applied. All these input archives are needed by the executing host to be processed. Applying the well-known tar utility, all these files are compressed to lower the network usage time. The scheduler starts a Global Access to Secondary Storage (GASS) server which is used, by the executing hosts, to ask for the needed files.

Once the GASS server is launched, the scheduler gets a structure from the repository and asks the *resource selector* for a machine to run the parallel application. Then, the input files are staged in, via the GASS server, to the execution host. Next, the parallel application will be queued in the remote machine. Finally, the scheduler module starts a *retrieval process* which will recover results once the task has finished. If any part of the process fails, the scheduler will try to analyze it later. The scheduler finishes when all the structural alternatives have been analysed and the result files have been stored on the local host.

Resource Selector. This module uses the Monitoring and Discovery Service (MDS) utilities, provided by GT 2.4, to query the number of free processors of a computational resource in the Grid. This component handles with a list of ordered machines according to memory, computational power, workload, etc. These resources run the MDS daemon and they are enquired for their available processors. The system deals with a desirable minimum and a maximum number of processors to execute the parallel application. These numbers are thought both to increase productivity and to ensure executions with some minimum requirements. The maximum number provided should make it possible parallel application to run optimally, according to structure size and library used. The minimum number will allow maintaining more resources working concurrently. The first host with enough number of free processors is chosen. The system will try to launch parallel simulations using the optimal number of processors.

With the purpose of enabling portability, all the platform-dependent optimizations should not be used. In addition, versions of numerical libraries, such as BLAS and LAPACK, tuned for a determined architecture must be avoided. Numerical software packages employed, such as PETSc, MUMPS, etc., have to be linked statically. Moreover, the parallel application should not depend on the MPI implementation on the executing host and the standard MPICH implementation should be included in a self-contained executable. As a result, two executable archives have been created, one for IA32 and the other one for IA64. Thus, the resource selector is also responsible to provide a suitable executable for each machine.

Retrieval Processes. Since the parallel application is not aware of Grid, it uses neither GT 2.4 utilities nor API to send the result files to the scheduler. A retrieval process is started for each Grid task. This process will be in charge of waiting for the Grid job to be finished and recovering the result files. Once the task has concluded, the results are compressed and sent back to the local host, where they will be saved on the appropriate directories created. All the temporary files on the executing host will be deleted. The waiting time for a job termination query must be tuned according to the resources in the Grid and their average workload. In case of the job finishes and it produces no results or any of the files expected is missed, the retrieval process will inform the scheduler and the structure will be analysed later.

5.2 Case Study

To check the performance of the Grid demonstrator, a case study of a hotel is presented. Hotels are singular buildings with a high cost of construction. The designer must be extremely careful in the design stage because in countries, like Spain, where the tourism is one of the biggest industry, an adequate design can be the differential factor for economical profitability. The design of hotel facilities requires taking into account a great number of relevant factors leading to a high number of structural solutions for the same problem.

Four different layouts were presented: two of these designs presented a reticular (bi-directional) slab solution, with large spanning of the beams, leading to geometric models with more than 600.000 dof, and two more designs with one-directional slab with approximately 330.000 dof. Three alternatives were considered for the construction material: steel, reinforced concrete and steel-concrete composite frame. In each case (layout design plus material type) eight combinations of different structural member dimensions were provided. Therefore, in the preliminary design stage, a total of 96 possible combinations should be analysed before selecting the most suitable option.

Table 3 summarizes the task distribution in the Grid. As maximum, parallel executions were limited to four proccesors, a polite policy with rest of the remote users that allows multiple concurrent simulations. Due to the memory requirements, the minimum number of processors was set to two. In the table, an entry like 9 (2 p.) indicates that nine simulations were performed with two processors each one. As table shows, the scheduler assignes dinamically the number of tasks to the resources according to their free processors and computational power. The available testbed was composed of 3 machines belonging to our research group: one cluster of 8 Pentium Xeon@2Ghz biprocessors (Kefren); another cluster of 11 Pentium III@866Mhz biprocessors (Ramses); and an Itanium II@900Mhz biprocessor workstation (Bastet).

Table 3. Distribution of the task in the testbed. The number in parentheses indicates the number of processors involved in the execution.

Machine	Simulations
Kefren	40 (4 p.)
Ramses	27 (4 p.), 20 (3 p.)
Bastet	9 (2 p.)

The execution of the whole structural study lasted for 108.3 minutes by using just one node of cluster Kefren (traditional sequential alternative). Following a high performance computing approach, 20.26 minutes were needed. Each simulation were launched with four processors in Kefren cluster, allowing two concurrent executions. Finally, the Grid Structural Analyser required 16.31 minutes for the whole case study. MUMPS library was used in all these executions.

6 Conclusions and Further Work

Firstly, a parallel application for the 3D structural analysis of buildings has been described in this paper. In order to compute the joint displacements, several parallel numerical libraries, with state-of-the-art capabilities, have been tested to solve a large sparse symmetric linear system. Direct methods, and more concretely WSMP library, have demonstrated to be the fastest solver. Besides, this parallel application has been integrated into a Globus-based Grid infrastructure. The Grid Structural Analyser developed gives the possibility to analyse in detail, concurrently, a high number of different alternatives in the preliminary and final design stages, saving time and effort, providing the designer with a powerful tool to select the best option based on quantitative measures. Small and medium-sized enterprises can now easily increase its productivity and business volume by subcontracting resource usage or employing their own office computers. Since Grid Computing enables efficient resource usage when a high coordinated computational power is demanded, the system developed can be very useful for very large and singular buildings, and moreover in a time-consuming 3D dynamic structural analysis where a building must be simulated under the influence of multiple earthquakes.

References

1. Livesley, R.: Matrix Methods of Structural Analysis. 2 edn. Pergamon Press, Oxford, UK (1975)
2. Drummond, L., Marques, O.: The Advanced Computational Testing and Simulation Toolkit (ACTS): What Can ACTS Do for You? Technical Report LBNL-50414, Lawrence Berkeley National Laboratory (2002)
3. Ashcraft, C., Pierce, D., Wah, D.K., Wu, J.: The Reference Manual for SPOOLES, Release 2.2: An Object Oriented Software Library for Solving Sparse Linear Systems of Equations. Boeing Shared Services Group. (1999)
4. Amestoy, P., Duff, I., L'Excellent, J.Y., Koster, J.: MUltifrontal Massive Parallel Solver (MUMPS Version 4.3) Users' Guide. (2003)
5. Joshi, M., Karypis, G., Kumar, V., Gupta, A., Gustavson, F.: PSPASES: Scalable Parallel Direct Solver Library for Sparse Symmetric Positive Definite Linear Systems. Users's Manual (version 1.0.3). University of Minnesota. (1999)
6. Gupta, A.: WSMP: Watson Sparse Matrix Package Part I - Direct Solution of Symmetric Sparse System. Technical Report IBM Research Report RC 21886(98462), IBM (2000)
7. Balay, S., Buschelman, K., Smith, B.F., et al.: PETSc Users Manual. Technical Report ANL-95/11 - Revision 2.1.6, Argonne National Laboratory (2003)
8. Tuminaro, R.S., Heroux, M., Hutchinson, S.A., Shadid, J.N.: Official Aztec Users's Guide. Version 2.1. Sandia National Laboratories, SAND99-8801J. (1999)
9. Saad, Y., Lo, G.C., Kuznetsov, S.: PSPARSLIB Users Manual: A Portable Library of Parallel Sparse Iterative Solvers. University of Minnesota, Morgan-Stanley (New-York), Institute of Mathematics (Novosibirsk, Russia). (1998)
10. Jones, M., Plassmann, P.: BlockSolve95 Users Manual: Scalable Library Software for the Parallel Solution of Sparse Linear Systems. ANL-95/48. (1995)
11. Falgout, R.D., et al.: HYPRE: High Performance Preconditioners. User's Manual. Version 1.6.0. Lawrence Livermore National Laboratory. (2001)
12. Chow, E., Heroux, M.A.: BPKIT: Block Preconditioning Toolkit. Reference Manual. University of Minnesota. SGI/cray Research, Inc. (1996)
13. Barnard, S.T., Grote, M.J.: A Block Version of the SPAI Preconditioner, 9th SIAM Conf. on Parall. Proc. for Sci. Comp. (1999)
14. Karypis, G., Kumar, V.: METIS: A Software Package for Partitioning Unstructured Graphs, Partitioning Meshes, and Computing Fill-Reducing Orderings of Sparse Matrices. Version 4.0. University of Minnesota. (1998)
15. Schulze, J.: Towards a Tighter Coupling of Bottom-up and Top-down Sparse Matrix Ordering Methods. BIT **41** (2001) 800–841
16. Berstis, V.: Fundamentals of Grid Computing. Technical Report IBM Redbooks Paper, IBM (2002)
17. Foster, I., Kesselman, C., Tuecke, S.: The Anatomy of the Grid: Enabling Scalable Virtual Organizations. Intl. Journal Supercomputer Applications **15(3)** (2001)
18. Ferreira, L., Berstis, V.: Introduction to Grid Computing with Globus. Technical report, IBM (2002)
19. Unicore: Unicore Client User Guide. (2004)
20. GridSystems: Overview to InnerGrid. (2003)
21. Avaki: Avaki Data Grid 4.0 Software. (2003)

FlowVR: A Middleware for Large Scale Virtual Reality Applications

Jérémie Allard[1], Valérie Gouranton[2], Loïck Lecointre[1], Sébastien Limet[2], Emmanuel Melin[2], Bruno Raffin[1], and Sophie Robert[2]

[1] Laboratoire ID, CNRS/INPG/INRIA/UJF, Montbonnot, France
[2] LIFO, Université d'Orléans/CNRS, Orléans, France

Abstract. This paper introduces FlowVR, a middleware dedicated to virtual reality applications distributed on clusters or grid environments. FlowVR supports coupling of heterogeneous parallel codes and is component oriented to favor code reuse. While classical communication paradigms focus on either a synchronous approach (FIFO channels) or an asynchronous one (sampling), FlowVR enables a large range of intermediate policies to better balance the application performance between levels of details, latencies and refresh rates.

1 Introduction

Classically, a virtual reality (VR) application features a complex simulation using input and output devices to provide users with a sense of immersion in a synthetic world [7]. Most of today's VR applications only run on machines with a reduced number of processors, like visualization clusters or SGI Onyx. They do not take advantage of the computing power offered by large clusters and grid environments. One main limitation is the difficulty to assemble and distribute the different (potentially parallel) components and to maintain the overall application *coherent* while guaranteeing a good quality interaction with *low latency* and *high refresh rates*. We define the *coherency* as the fact that the information provided to the user senses at a given moment are related to the same simulated time.

To improve latency and refresh rates, VR applications can take advantage of a data exchange model based on sampling. The producer updates data in a shared buffer asynchronously read by the consumer. Some updates may be lost if the consumer is slower than the producer. While asynchronism leads to a performance improvement, the application coherency cannot be maintained. Depending on the context this may be acceptable. It is for example used when coupling haptic and visualization systems that run at very different frequencies (about 1000 Hz and 60 Hz respectively). Distributed virtual environments [9, 11] or VR middlewares like OpenMask [2] use such an approach, but parallel code coupling becomes difficult in this context as no coherency control is offered. The other approach classically used for parallel programming, parallel code coupling [8, 10], or distributed visualization environments [3–5], relies on a classical

FIFO synchronization semantics. It ensures proper application coherency, but it is difficult to efficiently implement a sampling approach.

In this paper, we propose a programming model that eases the implementation of a large range of synchronization policies, from FIFO to sampling. We present FlowVR [1], a middleware dedicated to VR and supporting coupling of heterogeneous parallel codes to build large scale applications. FlowVR reuses and extends the data flow paradigm commonly used for scientific visualization environments [3,4]. A VR application is seen as a set of possibly distributed modules exchanging data. Each module endlessly iterates, consuming and producing data. From the FlowVR point of view, modules are not aware of the existence of other modules, the FlowVR engine taking care of moving data between producers and consumers. This leads to a simple application programming interface (API) that eases turning an existing code into a FlowVR module (or several modules in case of a parallel code). For data exchange between modules, FlowVR defines an abstract network featuring from simple routing operations to complex message handling operations. Each message is associated with a *list of stamps*, a lightweight data used to route or filter messages. This list can also be routed separately from its message to special network nodes in charge of synchronization policies. Besides predefined FlowVR stamps, others, like a time or a 3D bounding box for instance, may be added to extend the network routing, filtering or synchronization abilities. The FlowVR network enables to build complex collective communications, a desirable feature for efficient parallel code coupling. It is also possible to go beyond the classical synchronization barrier, designing synchronizations waiting for the resolution of complex constraints based on stamps (a data semantically richer than a signal). Different FlowVR networks can be designed without modification of the module codes.

2 The FlowVR Application Model

In this section we introduce the FlowVR application model.

2.1 Running Example

All along this paper, we use a simple yet important example, an interactive VR application where the user can perturbate a fluid flow simulation with its hand. We distinguish three parts:

- A tracker that gives the user's hand position.
- A physical fluid simulation parallelized with MPI. The simulation is based on a 2D grid split in blocks amongst the different MPI processes. MPI communications take place at each iteration to exchange the values of the grid borders between neighbors. Each process should also receive the hand position, which acts as an obstacle for the fluid flow.
- A multi-projector visualization environment. Each projector, driven by its own PC, displays a tile of the entire scene. The distribution paradigm adopted

```
initialization
while not stop
    wait()
    get(position)
    computations
    put(grid)
```

Fig. 1. The algorithm of the simulation module.

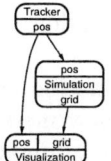

Fig. 2. Interactive fluid simulation with 3 modules.

is simple: all PCs run a copy of the visualization application, each one expecting the coordinates of the hand position and a density grid at each iteration. To ensure a strong coherency of the displayed images, these copies must receive the same input data at each iteration. Next, each copy computes its tile of the global image based on its own viewing frustum (viewing angle). All PCs must then display the new image synchronously, either using a hardware swaplock or a software barrier.

These codes can run independently at very different frequencies. The tracker is certainly the fastest one and the fluid simulation the slowest one. A sampling-based data exchange model will let the codes run independently at their highest frequency, but it may lead to incoherences. For instance, in a given image, the displayed hand position may not correspond to the one used to compute the displayed simulation state. On the opposite, a FIFO communication model will ensure the overall application coherency, but at the price of a lower performance. All codes will run at the same frequency, synchronized on the slowest one. The tracker will produce a new data as soon as room is available in the output channel buffer. The latency will increase by the time such data stay unused in this buffer, the time required by the fluid simulation to consume all data previously stored in this buffer. FlowVR has been designed to let the user specify these different policies and other *intermediate* solutions, without requiring any modification of the codes.

2.2 Modules

We first introduce the API used to program FlowVR *modules*. This API is kept as simple as possible to limit the effort required to convert an existing code into a FlowVR module. For that purpose we explicitly took advantage of the interactive nature of VR applications. A FlowVR module is a computation loop periodically reading input data and producing new results. To improve code reuse, a module cannot directly address another module. This way there is no explicit dependency between modules. Their only knowledge of the FlowVR environment is a list of input and output ports. The module API is based on three main methods:

- The *wait* defines the transition to a new iteration. It is a blocking call that ensures each connected input port holds a new message. Input ports not connected to any other port will never receive any message. They are *deactivated*.

- The *get* function enables a module to retrieve the message available on a port.
- The *put* function enables a module to write a message on an output port. Only one new message can be written per port and iteration. Each output message is automatically stamped by FlowVR with the current iteration number.

In our example, we would define:

- One module for the tracker with one output port (a position data).
- Each MPI process of the fluid simulation will define a module with one input (a position data) and one output (its block of the fluid density grid) (Fig. 1). To be able to distinguish the different blocks, each process stamps its output messages with the coordinates of its block.
- One module for each visualization process, with two input ports each, one to retrieve the tracker position and the other one to retrieve the whole density grid.

Each module has two additional predefined ports. The *input activation port* is used to lock the module to an external event (fixed frequency trigger for the tracker for instance). The *output activation port* is used to signal other components that the module has started a new iteration (see section 2.5).

2.3 Connections

Once modules are defined, they are assembled connecting their input and output ports. The simplest primitive used to build a FlowVR network is a *connection*. A connection is a typed FIFO channel with one source and one destination. Messages in a connection are numbered. Each message is stamped with this number and the source id.

Let us consider our example. We can build a simple first application with one tracker module, one fluid simulation module and one visualization module (Fig. 2). We add one connection from the tracker to the visualization, another one from the tracker to the simulation and a last one from the simulation to the visualization. This simple application implements a classical communication scheme using FIFO channels. The FIFO connections ensure a strong coherency. At each iteration the visualization module will always retrieve a tracker position and a density grid corresponding to the same simulated time. Therefore the resulting application will be synchronized on the slowest module, presumably the fluid simulation. If the tracker module is faster than the simulation module, there will be a significant lag between between user interactions and their effects on the virtual world. Also notice that adding the connections does not require to modify the code of the modules.

However, having only point to point FIFO connections, it is difficult to loosen the synchronizations imposed by the FIFO model or to express collective communications.

2.4 Filters

To extend the capabilities of the FlowVR network we introduce a new component, called *filter*.

A filter has typed input and output ports and can perform complex operations on messages. Filters have all the freedom to discard, combine or even generate messages. They are not restricted to receive only one message per port and per iteration like modules. They have free access to incoming buffers. Filters usually handle messages based on the associated lists of stamps. For instance, a filter can discard all incoming messages, which 3D bounding box falls outside of a given volume. Amongst filters, we distinguish the *routing nodes* as the filters that only forward all incoming messages on one or several outputs.

Let extend our example by now using four modules for the simulation and two modules for the visualization (Fig. 3(a)). The tracker messages must be broadcasted to these modules. For that purpose we introduce in our network several routing nodes. To broadcast the data to modules we choose to implement a binary-tree broadcast. The data exchange between simulation and visualization is more complex as we have to ensure that all visualization modules receive the whole density grid while each simulation module sends only one fourth of it. For that purpose we use a filter that combines two blocks of density grids into a larger one. This example implements a network with non trivial collective communications. A strong coherency is still ensured as the filter we use here does not suppress or generate new data (FIFO network).

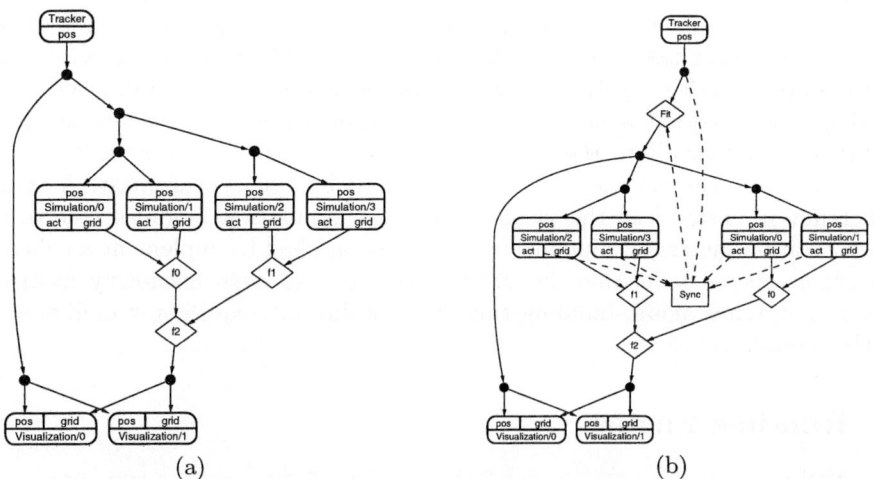

Fig. 3. (a) Fluid simulation with a FIFO network. Modules are represented as round-shaped squares, routing nodes as circles and filters as diamonds. (b) Fluid simulation with a coherent sampling network using one synchronizer (a square). Dashed lines correspond to connections carrying only stamps. The *act* port corresponds to the output activation port.

2.5 Synchronizers

We distinguish a special class of filters, called *synchronizers*, used to implement the resolution of non local constraints. A synchronizer works on stamps. Therefore all incoming and outgoing connections only carry message stamps. Generally a synchronizer activity is triggered by incoming stamps on some selected ports. As synchronizers do not receive the data part of the messages, their output ports are generally connected to filters. These filters typically have 2 input ports, one receiving full messages (the data and its list of stamps) from a module or a filter, and the other one receiving only stamps from a synchronizer. The filter processes the incoming full messages according to incoming stamps. For instance, such a filter can forward to its output only the full messages corresponding to the incoming stamps, discarding the other messages.

Classical synchronization schemes can often be expressed in term of signal handling. In this case the synchronizer only uses its inputs as signals. A sampling scheme is implemented by selecting the last received message each time an activation signal is received from the destination module (request for another input message). But synchronizers can implement more complex algorithms by taking advantage of the semantically rich information hold by stamps. For example, in VR environments some coherency constraints can be expressed in term of spatial relationships. A strong coherency is required for objects close to the user, while background or unseen parts of the scene require much less attention. A stamp holding a bounding box information can be used to implement such a coherency policy.

In our example, because the simulation will probably be slower than the tracker, we introduce a synchronizer to keep pace with the tracker (Fig. 3(b)). This synchronizer takes as input the stamps from the position messages, and the stamps from the activation output ports of the fluid modules. When all fluid modules request a new data, the synchronizer selects the newest stamp available and sends it to the filter Fit. This filter only forwards on its output port the messages having the stamps selected by the synchronizer. A strong coherency is ensured as the visualization and simulation modules receive the same position messages. Similar ideas could be applied to implement a coherent sampling scheme to enable the visualization to run asynchronously from the simulation. Once again, building this network did not require any modification of the module codes.

3 Runtime Engine

FlowVR is open source and currently ported on Linux for IA32, IA64 and Opteron.

The FlowVR runtime engine relies on daemons, one per participating node. Daemons are in charge of FlowVR networks. They act as brokers and relay messages between modules. Filters, including synchronizers, are implemented as dynamically loaded classes (*plugins*) within the daemon. Communications local to a node use a shared memory area. Care is taken to avoid unnecessary data copies and memory allocations by exchanging pointers and reusing allocated buffers. The current implementation of inter-node communications relies on TCP. Networks of heterogeneous nodes are easily exploited, as connections are dynamically created and each daemon can be launched independently. Several applications can safely run concurrently using the same daemons.

Each FlowVR application is managed by one special module called a *controller*, automatically loaded at starting time. The controller first starts the application's modules using their own launching command, ssh or mpirun for instance. Once the modules launched, they register themselves to their local daemon that sends an acknowledgment to the controller. Then, the controller sends to each daemon the list of plugins to load to implement the FlowVR network.

FlowVR integrates tools to generate the module launching commands and the list of plugins to load. It uses as input an XML description of the syntax of the launching command associated with each module code, as well as an XML description of the FlowVR network with an explicit placement of all components on target nodes. Ongoing work focuses on developing automatic and semi-automatic FlowVR network generation tools.

3.1 Experimental Results

We implemented the running example porting an existing fluid simulation code. The fluid simulation is parallelized with MPI, while the multi-projector visualization is handled by Net Juggler (also based on MPI) [6]. From the FlowVR point of view, each MPI fluid process and each Net Juggler process is seen as a module. Note that all fluid modules (respectively visualization modules) are synchronized through MPI communication calls FlowVR is not aware of. All results presented here run a fluid simulation based on a 2D 512×512 grid. The visualization modules integrate the fluid into a rich virtual environment (See Fig. 4(d)).

Two versions of the network were tested, a FIFO network (similar to Fig. 3(a)), and a coherent sampling network enabling the tracker, the fluid simulation and the visualization to run asynchronously. It extends the network presented in Fig. 3(b) by adding an extra synchronizer between the tracker and the visualization, and another one between the simulation and the visualization (Fig. 4(c)). Tests were performed on a PC cluster with dual Xeon PCs (2.66 GHz) connected through a Gigabit Ethernet network. Each machine was equipped with a GeForce FX 5600 graphics card.

The number of visualization and fluid modules vary from 1 to 4. Each module runs on its own PC. For instance when 8 nodes are used, 4 of them execute a fluid module, while each of the 4 other PCs run a visualization module. Each of these 4 PCs drives a video projector to display the result of its visualization module (1/4 of the global image).

We measured the refresh rate, i.e. number of iterations per second, for the visualization and the fluid simulation (see Fig. 4(a)). The FIFO networks impose the same refresh rate for the visualization and the fluid modules. For the coherent sampling network, the visualization and the fluid run asynchronously.

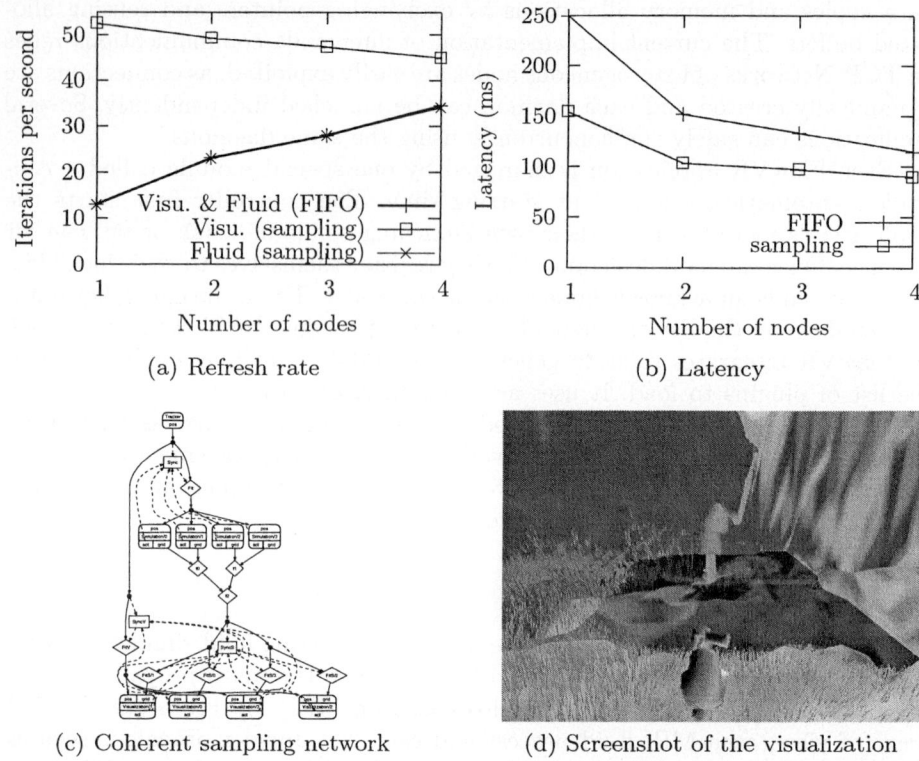

Fig. 4. Experimental results with a coherent sampling network and a FIFO network.

It enables the visualization to run significantly faster than the simulation. The fluid simulation keeps the same performance as in the FIFO case. It shows that the communications induced by synchronizers do not significantly affect the performance. As the number of nodes allocated to the fluid simulation increases, the fluid performance increases too. For the sampling approach this decreases the refresh rate of the visualization modules as they must upload to the graphics card new data from the fluid modules more frequently.

We also measured the overall latency, i.e. the time lag between the time a new tracker position is available and the end of the iteration of the visualization modules using this tracker position (see Fig. 4(b)). Allocating more nodes to the simulation also improves latency. Sampling leads to a better latency than FIFO, because sampling uses the more recent data available while FIFO uses the older one. Note that the FIFO was executed with intermediate buffers of size 2.

The synchronizers used for the sampling approach can be extended to enable a finer control over dependencies between modules. For instance, the synchronizer between the fluid modules and the visualization modules could take into account a user position data to know for each visualization module if the fluid is visible or not. If not, it could block the transmission of fluid grid to the visualization module, to let the visualization and network resources fully available for objects that are in the user field of view.

4 Conclusion

We introduced FlowVR, a middleware dedicated to distributed interactive applications. FlowVR distinguishes two main parts in an application, the modules and the network. Modules are endless loops reading and writing data on input and output ports. Modules are assembled in a network with advanced features for message handling. It enables parallel code coupling and the design of complex communication and synchronization schemes. First experiences show that FlowVR eases the development and deployment of interactive distributed applications, while leading to high performance executions.

Acknowledgment

This work is partly funded by the RNTL project Geobench.

References

1. FlowVR. http://flowvr.sf.net
2. OpenMASK. http://www.irisa.fr/siames/OpenMASK
3. Scirun: A scientific computing problem solving environment. http://software.sci.utah.edu/scirun.html
4. Covise Programming Guide, 2001. http://www.hlrs.de/organization/vis/covise
5. J. Ahrens, C. Law, W. Schroeder, K. Martin, and Michael Papka. A Parallel Approach for Efficiently Visualizing Extremely Large, Time-Varying Datasets. http://www.acl.lanl.gov/Viz/papers/pvtk/pvtkpreprint/
6. J. Allard, V. Gouranton, E. Melin, and B. Raffin. Parallelizing pre-rendering computations on a net juggler PC cluster. In *Immersive Projection Technology Symposium*, Orlando, USA, March 2002.
7. C. Cruz-Neira, D. J. Sandin, T. A. DeFanti, R. V. Kenyon, and J. C. Hart. The Cave Audio VIsual Experience Automatic Virtual Environement. *Communication of the ACM*, 35(6):64–72, 1992.
8. A. Denis, C. Pérez, and T. Priol. Padicotm: An open integration framework for communication middleware and runtimes. *Future Generation Computer Systems*, 2003.
9. E. Frécon and M. Stenius. Dive: A scalable network architecture for distributed virtual environments. *Distributed Systems Engineering Journa*, 5:91–100, 1998.
10. N. Karonis, B. Toonen, and I. Foster. Mpich-g2: A grid-enabled implementation of the message passing interface. *Journal of Parallel and Distributed Computing*, 63(5):551–563, 2003.
11. K. Watsen and M. Zyda. Bamboo - a protable system for dynamically extensible, real-time, networked, virtual environments. In *IEEE Virtual Reality Annual Internationnal Symposium*, Georgia, USA, 1998.

Topic 8
Parallel Computer Architecture
and Instruction-Level Parallelism

Kemal Ebcioğlu, Wolfgang Karl, André Seznec, and Marco Aldinucci

Topic Chairs

Parallel architecture design and ILP architectures are important topics at the core of every parallel system, affecting the total system performance in fundamental ways. Instruction-Level Parallelism has for decades represented a foremost performance booster of leading edge computing systems. The present topic, Euro-par Topic 8, includes parallel computer architectures, processor architecture (microarchitecture as well as compilation), the impact of emerging microprocessor architectures on parallel computer architectures, innovative memory designs to hide and reduce the access latency, multi-threading, and the impact of emerging applications on parallel computer architecture design.

A total of 24 papers were submitted to this topic. The overall high quality of the submissions rendered our task quite difficult. All papers were refereed by at least four experts in the field and some received five reports. At the end, we settled on 6 regular papers and 2 short papers spread across two sessions. Of course, all this was made possible by the referees who lent us their time and expertise with their high quality reviews.

Data memory improvement constitutes the common theme of five papers: V. De La Luz, M. Kandemir, A. Sivasubramaniam, and M. J. Irwin examine the possibility of data compression in caches, and avoiding decompression by getting programs to operate on data in the compressed domain. E. F. Torres, P. Ibañez, V. Viñals, and J. M. Llaberia discuss the influence of data replication and distribution polices on the performance of multibanked data caches. P.-F. Chuang, R. Sendag, and D. J. Lilja propose an address-correlation scheme that keeps track of multiple cache locations that store the same value. C. Kyriacou, P. Evripidou, and P. Trancoso study several schemes for exploiting cache locality in case of a data-driven multithreading execution. H. Zeffer, Z. Radovic, O. Grenholm, and E. Hagersten describe a technique to reduce the overhead due to software-based cache coherence for DSM.

Instruction fetch, control, and storage cover the rest of the papers in the topic: E. Fernandez, A. Ramirez, and M. Valero discuss how to target quality of service requirements by controlling instruction fetch in a SMT processor. E. Morancho, J. M. Llaberia, and A. Olive discuss a mechanisms for verifying data speculation in out-of-order issue processors. F. J. Cazorla, P. M. W. Knijnenburg, R. Sakellariou, G. Gaydadjiev and S. Vassiliadis compare and analyze the IA64 instruction and the SCISM tagging scheme.

Exploring the Possibility of Operating in the Compressed Domain*

V. De La Luz, M. Kandemir, A. Sivasubramaniam, and M.J. Irwin

Department of Computer Science and Engineering
The Pennsylvania State University
University Park, PA 16802, USA
{delaluzp,kandemir,anand,mji}@cse.psu.edu

Abstract. Data compression in caches has been studied from the performance and energy consumption points of view. In this paper, we study the possible benefits of operating with compressed operands. Operating in the compressed domain (i.e., executing instructions with compressed operands) as far as possible can offer several advantages. First, since it is less reliant on decompression, it may be possible to employ a fancier compression/decompression strategy, without the associated performance/power penalties, that can offer higher compression rates to further boost cache locality. Second, the transfers between cache and datapath can use fewer bits (to transmit codes instead of data values) and provide dynamic energy savings in the corresponding bus. In this paper, we demonstrate that there are cases where operations can be executed in the compressed domain, which can lead to performance improvements and energy savings. In particular, we show that by operating in the compressed domain, the effectiveness of prior techniques can be further improved.

1 Introduction

Prior studies have looked into compression of main memory [2, 1], program code [10, 8, 11, 6], and data structures [13]. Other studies [15, 14] have illustrated the benefits of compression in enhancing data cache locality. Compressing the data words accessed by a program can allow a lot more data to be stored in the caches to improve cache behavior. However, when the program needs to access this data, the values need to be decompressed and brought into the processor datapath for subsequent processing. In the process, a performance and possibly power penalty can be expended depending on how extensive a compression/decompression mechanism is employed. Two solution strategies to address this problem include: (i) using rather simple compression/decompression strategies that do not incur high performance or power costs, and (ii) recognizing that these costs may be high and avoiding to perform them as much as possible. While prior studies have used the first strategy, this paper explores the viability of the second strategy.

The benefits of compressing data cache values has been studied in [15, 14]. The goal is to use the cache space to hold more data values so that there would be fewer misses

---

* This research is partly supported by the NSF Career Award #0093082.

during execution. These studies have used a rather simple compression/decompression strategy wherein a certain number of frequently used values in the program are assigned a code (taking fewer bits than the data values themselves) to represent them in the cache. A cache of the codes themselves, maintaining their mapping to actual data values, is also maintained for returning these values back to the datapath once the cache lookup returns a code. This cache of frequent value codes needs to be fairly small for efficient lookup (from both performance and energy perspectives). However, this reverse lookup (decompression) is done at the cache and at every access, and does not reduce the number of decompression operations. Further, this mechanism does not reduce the number of bits transferred (which consumes dynamic energy) between the cache and the datapath either. Prior studies have also addressed these two deficiencies to a certain extent. For instance, Lee et al [9] proposed to selectively performing decompression, and Villa et al. [12] addressed the cache access (and data transfer) energy by optimizing on the data value "0", i.e., they use a single bit to represent this value.

In the interest of keeping decompression performance and power overheads low, earlier studies have used very simple techniques for compressing the data values in the cache. On the other hand, in this paper we examine this issue from a different perspective – *can we avoid decompressing the data values even when they are returned to the datapath by the cache?* In the earlier solutions, the presumption is that the datapath needs the values in uncompressed format for its execution. However, this may not always be the case. For instance, there could be situations when the execution simply reads some values from memory, and copies/stores them elsewhere without performing any arithmetic/logic operations on them. In such cases, the datapath can work with the encoded values, instead of decompressing and compressing them back. There could be other situations where operations could directly be performed in the compressed domain, e.g., comparing whether two compressed values are equal. There have also been previous proposals in the context of databases [7], wherein comparisons (greater/lesser than) can be performed in the compressed domain with certain order-preserving compression strategies. In all these cases, earlier proposals may be conservative by decompressing the values even if they may not be needed.

Operating in the compressed domain as far as possible can offer several advantages. First, since it is less reliant on decompression, it may be possible to employ a fancier compression/decompression strategy, without the associated performance/power penalties, which can offer higher compression rates to further boost cache locality. Second, the transfers between cache and datapath can use fewer bits (to transmit codes instead of data values) and provide dynamic energy savings in the corresponding bus. Third, many datapath operations can work with smaller width operands, providing further energy savings in the corresponding functional units and buses. It is also possible to employ the same functional unit to perform multiple functions (parallelism) with smaller width operands. In fact, a recent study [4] points out the energy and performance benefits of smaller width operands by exploiting the fact that most operands do not use the most significant bits of 64-bit words (however, they do not suggest or exploit the possibility of operands being compressed).

This paper examines the possibility of operating in the compressed domain as far as possible to achieve these benefits. A detailed study that quantifies all these benefits

Table 1. Our simulation parameters.

Parameter	Value
Processor Configuration	
RUU size	64 instructions
LSQ size	32 instructions
Fetch queue size	8 instructions
Fetch width	4 instructions/cycle
Decode width	4 instructions/cycle
Issue width	4 instructions/cycle
Commit width	4 instructions/cycle
Cache Configuration	
L1 cache (Split)	**Instruction** 16KB, 32-byte blocks direct mapped
	Data 16KB, 32-byte blocks direct mapped
L2 cache (Unified)	1 MB, 128-byte blocks 2-way associative

Table 2. Our benchmarks and their important characteristics.

Name	Brief Description	L1 Miss Rate
crafty	Game playing	0.1880
equake	Simulation of seismic wave propagation	0.0385
galgel	Computational Fluid Dynamics	0.2635
lucas	Primality testing	0.1321
mcf	Combinational optimization	0.0264
twolf	Placement and global routing	0.1551
vortex	Database	0.1198

Table 3. Improvement on L1 data cache miss rates due to data compression.

Name	Original Miss Rate	New Miss Rate	Improvement
crafty	0.1880	0.1454	22%
equake	0.0385	0.0231	40%
galgel	0.2635	0.2361	10%
lucas	0.1321	0.0743	43%
mcf	0.0264	0.0196	25%
twolf	0.1551	0.1163	25%
vortex	0.1198	0.1072	10%

by operating in the compressed domain and the hardware support for facilitating this mechanism, is beyond the scope of this paper. On the other hand, this paper explores the potential of such a mechanism, if the hardware were to support it, from the application perspective by presenting a characterization study. In particular, we identify different scenarios when data obtained from the caches can remain compressed (and until when), and profile the application execution characteristics for these scenarios.

This paper is organized as follows. In Section 2, we describe the experimental setup and the benchmarks used in our experiments. In Section 3, we discuss our results on executing instructions with compressed operands. In Finally, in Section 4, we present our conclusions.

2 Experimental Setup

In order to perform our experiments, we use SimpleScalar's [5] sim-outorder. SimpleScalar is a tool-set that provides a simulation environment for modern out-of-order processors. In particular, sim-outorder performs a detailed simulation of a out-of-order issue processor with speculative execution. The base parameters that were used in our simulations are given in Table 1.

We use a subset of randomly-selected applications from the Spec2000 benchmark suite. The benchmarks and some of their important characteristics are presented in Table 2. The last column gives the L1 data cache miss rates. In all our experiments, we fast-forwarded the first billion instructions (i.e., only functional simulation is performed), and executed a detailed simulation for the next 300 million instructions. Also, for all the experiments, we used the Reference input files that come with the benchmarks.

3 Operating in the Compressed Domain

It has been shown by prior studies that storing compressed data in the data cache (instead of the real values) can be useful in improving performance and energy behavior of applications. Performance is improved by reducing the miss rate of the data cache, requiring fewer number of accesses to higher levels of the memory hierarchy. Energy consumption can be reduced by accessing smaller portions of the cache. In order to be able to operate in compressed domain, we assume that all the values stored in the data cache are compressed. This brings an important advantage that the actual size of the data cache can be increased proportionally to the compression ratio for the values in the cache.

In Table 3, we show the improvements in L1 data cache miss rates when data in the cache is compressed to half its size (i.e., a compression ratio of 0.5 is employed). From this table, we see that we can obtain miss rate improvements ranging from 10% for vortex to 43% for lucas. These improvements can in turn help to reduce the overall execution time of the application. That is, storing compressed data in the cache can be very beneficial.

To the best of our knowledge, no prior studies have looked at the potential benefits of executing operations in the compressed domain, i.e., with compressed operands. Note that, normally, the compressed values might not be amenable to be used directly as operands to perform the operation, so these values would need to undergo a decompression process so that the original value can be retrieved. In order to ensure that an operation can be performed with compressed values, in this work, we consider that the operation can be performed in the compressed domain if at least one of the operands involved in the operation is equal to zero. As will be discussed later, in this case, the compressed operand does not need to go through a decompression process to obtain the original value.

Considering that an instruction takes two input operands (registers), there are different final states that a register can end up with after the instruction is executed. In Table 4, we list these different cases, and in the following we describe each case in detail. A C in an entry in this table means that the state of the register (either initial or final) is compressed. On the other hand, a U means that the state is uncompressed. In Cases 1 and 2, both input registers are compressed to begin with. In Case 1, one of the registers contains a value zero, and so both the registers remain compressed (after the operation) and the value will remain in the compressed domain. In Case 2, the values in both the registers are different than zero, so both the registers will be decompressed and the result of the operation is considered uncompressed. In Cases 3, 4, and 5, one of the registers is compressed and the final states of the registers (and the state of the output register) vary depending on whether a zero is present in at least one of the registers involved in the operation. In Case 3, the compressed register contain zero, so the output register will contain an uncompressed value. In Case 4, on the other hand, the uncompressed register is the one containing zero, so the final value can remain in compressed form. If none of the values in the input registers is zero (Case 5), then the compressed value is decompressed and the result of the operation will be uncompressed. Case 6 can occur only when a load operation is performed. Recall that we consider that the values are stored the cache in the compressed form. All other operations (where none of the

Table 4. Different possibilities when executing in the compressed domain (see the text for the detailed description of each case).

Case Number	Input Registers Initial States Reg. 1	Reg. 2	Input Registers Final States Reg. 1	Reg. 2	Output Register State
1	C	C	C	C	C
2	C	C	U	U	U
3	C	U	C	U	U
4	C	U	C	U	C
5	C	U	U	U	U
6	U	U	U	U	C
7	U	U	U	U	U

Table 5. Percentage of loaded values that are not touched by the execution. NA = Not available.

Load instruction	crafty %	equake %	galgel %	lucas %	mcf %	twolf %	vortex %
ldq	30.46	20.76	NA	NA	23.07	11.10	31.49
ldl	22.67	13.92	99.75	12.09	14.49	16.34	29.24
lda	26.10	32.22	1.05	34.46	30.19	34.07	55.16
ldah	7.96	0.0024	NA	NA	0.47	NA	NA
ldq_u	64.59	31.08	84.20	NA	32.77	57.35	70.65
ldbu	9.87	NA	NA	NA	NA	64.99	NA

registers contains data that are in compressed form to begin with) fall into Case 7. It should be noted that these seven cases together cover all possible scenarios. It should also be emphasized that this analysis differs from prior work significantly. In the first group of prior work, a "0" operand is detected and the corresponding instruction is handled in a special way. While we achieve the same thing, we also propagate the involved compressed operand to the other instructions (while the prior work has not considered operating on the compressed domain). In the second group of prior work, the idea is to use compression but always decompress the values before the instruction gets executed. We differ in that we propagate the compressed operands to the highest extent possible.

We are mainly interested in the cases where the final states of the registers are compressed. This means that:

• the input register values does not need to be decompressed, which can be translated into energy savings and performance improvement, and

• intermediate values can remain in the compressed domain, thereby extending the lifetime of compressed values, which can in turn lead to additional benefits in energy and execution time.

In Figure 1, we give the breakdown of the *final states* of the input registers and the output register. In this figure, notation $XY \to Z$ means that the final states of the input registers are X and Y, and Z is the state of the output register. From this figure, one can see that in up to 37% of the instructions executed (for equake and mcf), at least one of the registers involved can remain in the compressed form. Thus, these instructions present opportunities for energy and performance improvements. Therefore, we can conclude that there is potential for improving performance and saving energy by operating in the compressed domain. It should be mentioned that in obtaining our experimental data we always used the highest level of compiler optimization, in an attempt to eliminate the inefficiencies and extra operations that could come from insufficient compiler analysis.

While working with compressed values can generate benefits in performance and energy consumption, the decompression process itself may be harmful from a performance point of view due to the extra time needed to perform the decompression itself (though, in our proposal, this cost is incurred when it is absolutely necessary). In order to estimate the performance impact of the decompression process, we perform a series of experiments where we assume that the decompression of the registers is performed *on-demand*. In other words, a register is decompressed only at the time that its value

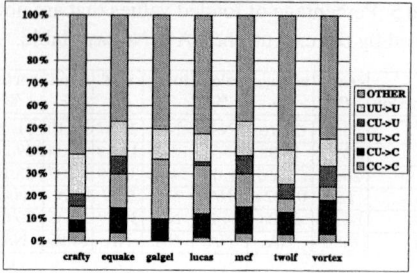
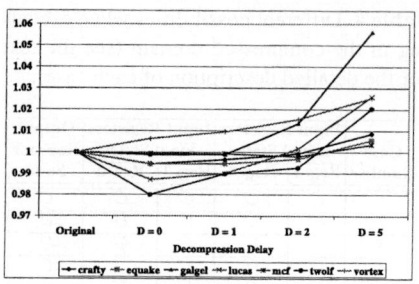
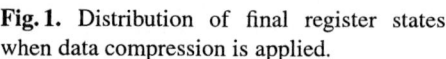

Fig. 1. Distribution of final register states when data compression is applied.

Fig. 2. Performance impact with different values of D (cost of decompression).

is actually required for the instruction to execute. The implementation of this policy results in the worst impact on execution time due to the extra time needed for the value to be decompressed; that is, these experiments represent the worst-case scenario for the proposed scheme. We ran our simulations with several values for the decompression delay D, and the results are presented in Figure 2. We believe that these D values cover a wide range of implementation alternatives. All the values in this graph are normalized with respect to the case labeled *Original*, where no compression is used. The normalized execution times for the cases where data compression is employed include the impact of the increased cache size due to compression.

We see that for the case where $D = 0$ (no decompression delay or ideal decompression), we can obtain performance benefits of up to 2%. We can also see from the figure that we can obtain performance benefits even if the decompression delay is 1 cycle, where the average improvement is 0.5%. When $D = 2$, the average performance improvement is reduced, though `twolf` still shows some benefits. When the delay is considerable bigger (5 cycles), all the benchmarks suffer from a performance impact. It is important to note that these results (for $D > 0$) are the worst case scenario and they can be reduced if a smart *pre-decompression* strategy can be implemented, by using either (both) hardware or (and) software techniques.

In order to give an idea of how important a pre-decompression strategy can be for improving execution time, we also implemented a policy in which all loaded values are decompressed beforehand such that the original values (decompressed) are ready by the time they are required by the executing program. Note that only the `load` operations are pre-decompressed. Intermediate values that might remain compressed are recovered using the *on-demand* policy. We show results for the case where the decompression delay $D = 5$ in Figure 3. The y-axis corresponds to the execution time obtained with the *on-demand* policy normalized to the execution time with the *pre-decompression* policy. Similar results where obtained with other values of D. We see that execution time can be reduced by using a smart decompression policy. In the case of our set of benchmarks, `galgel` is benefited the most, by improving the execution time by more that 5%. On the average (i.e., across all seven benchmarks), 2% improvement on performance is obtained, minimizing the negative side effects of performing data decompression *on-demand*. Of course, a better pre-decompression strategy could potentially achieve better results.

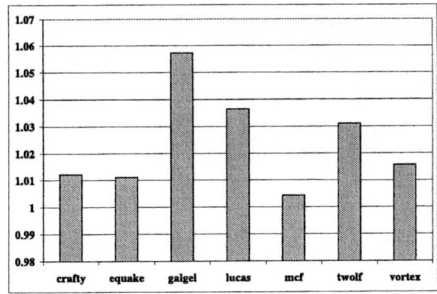

Fig. 3. Performance impact of different values of D. The y-axis corresponds to the execution time obtained with the *on-demand* policy normalized to the execution time with the *pre-decompression* policy.

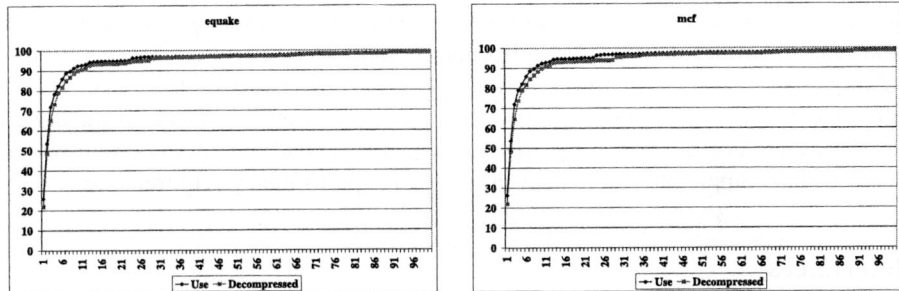

Fig. 4. CDF of the number of instructions before a value is decompressed/used.

Another important result obtained from our experiments is presented in Table 5. In this table, we give the percentage of different types of load instructions whose values are never used during execution. This table is important because these results can be translated into direct energy and performance benefits. Specially, given that these values are never touched, there is no need for them to be decompressed, thereby avoiding decompression delays and extra energy consumption. We observe from this table that for some types of load operations and some benchmarks, the percentage of non-used values are significant. For instance, 64% of ldq_u operations in crafty result in values that are not used during the execution of the application (therefore, these values do not need to be decompressed). It should also be mentioned that the results in this table are extracted from a single execution with a specific input file. It is conceivable that a different input file can generate different results.

We believe that simple decompression policies can be implemented, ensuring low impact on performance. In Figure 4, we show (for two of our benchmarks) the Cumulative Distribution Function (CDF) for the number of instructions executed before a loaded value is used (marked in the figure as "Use"). In the same figure, we also show the CDF for the number of instructions executed before a value is decompressed, marked as "Decompressed". The behavior of these curves indicates that values loaded from memory are not immediately decompressed in all the cases. This characteristic (and the results presented above) confirms the importance of developing techniques that initiate decompression ahead of time in order to minimize its potential negative impact on performance.

4 Conclusions

Prior work has demonstrated the benefits of compressing data at different levels of a given memory hierarchy. These benefits are obtained as a consequence of better utilization of the memory space. However, compressing data presents a potential drawback in the sense that the compressed values need to be decompressed before they can be used to execute the instructions that need them. In this work, we explore the possibility of operating in the compressed domain. We show that for the set of benchmarks used, up to 37% of the operations executed contain at least one register that remains compressed during the execution of the instruction. These compressed operands may contribute to further increase the performance and energy benefits obtained by other techniques such as narrow width operation. Our experimental results indicate that designing techniques that could take advantage of compressed values is a promising area for increasing performance and energy savings. This is particularly important with increasing chip complexity and market-driven power/performance requirements.

References

1. B. Abali and H. Franke. Operating system support for fast hardware compression of main memory contents. *Workshop on Solving the Memory Wall Problem*, 2000.
2. C. Benveniste, P. Franaszek, and J. Robinson. Cache-memory interfaces in compressed memory systems. *Workshop on Solving the Memory Wall Problem*, 2000.
3. D. Bhandarkar. *Alpha Implementations and Architecture. Complete Reference and Guide.* Digital Press, 1996.
4. D. Brooks and M. Martonosi. Dynamically exploiting narrow width operands to improve processor power and performance. *Proceedings of the Fifth Intl. Symposium on High-Performance Computer Architecture*, 1999.
5. D. Burger and T. M. Austin. The SimpleScalar tool set, version 2.0. *Computer Architecture News*, pages 13–25, June 1997.
6. S. Debray and W. Evans. Profile-guided code compression. *Proceeding of the Conference on Programming Language Design and Implementation (PLDI)*, 2002.
7. G. Graefe and L. Shapiro. Data compression and database performance. *Proceedings of ACM/IEEE-CS Symposium On Applied Computing*, 1991.
8. D. Kirovsky, J. Kin, and W.H. Mangione-Smith. Procedure based program compression. *Proceedings of the 30th International Symposium on Computer Architecture*, 1997.
9. J-S. Lee, W-K. Hong, and S-D. Kim. Design and evaluation of a selective compressed memory system. *Proceedings of the IEEE International Conference on Computer Design*, 1999.
10. C. Lefurgy, P. Bird, I-C. Chen, and T. Mudge. Evaluation of a high performance code compression method. *Proceedings of the 33th International Symposium on Computer Architecture*, 1997.
11. C. Lefurgy, E. Piccininni, and T. Mudge. Improving code density using compression techniques. *Proceedings of the 32th International Symposium on Microarchitecture*, 1999.

12. L. Villa, M. Zhang, and K. Asanović. Dynamic zero compression for cache energy reduction. *Proceedings of the 33rd International Symposium in Microarchitecture*, 2000.
13. J. Yang and R. Gupta. Data compression transformations for dynamically allocated data structures. *Proceedings of the International Conference on Compiler Construction*, 2002.
14. J. Yang and R. Gupta. Energy efficient frequent value data cache design. *Proceedings of the 35th Symposium on Microarchitecture*, 2002.
15. J. Yang, Y. Zhang, and R. Gupta. Frequent value compression in data caches. *Proceedings of the 33rd Symposium on Microarchitecture*, 2000.

Contents Management
in First-Level Multibanked Data Caches*

E.F. Torres[1], P. Ibañez[1], V. Viñals[1], and J.M. Llabería[2]

[1] DIIS, Universidad de Zaragoza, Spain
{enrique.torres,imarin,victor}@unizar.es
[2] DAC, Universidad Politécnica de Catalunya, Spain
llaberia@ac.upc.es

Abstract. High-performance processors will increasingly rely on multi-banked first-level caches to meet frequency requirements. In this paper we introduce *replication degree* and *data distribution* as the main multi-banking design axes. We sample this design space by selecting current data distribution policy proposals, measuring them on a detailed model of a deep pipelined processor and evaluating the trade-off introduced when the replication degree is taken into account. We find that the best design points use data address interleaving policies and several degrees of bank replication.

1 Introduction

Future superscalar processors will require a low-latency and high-bandwidth memory in order to attain high degrees of ILP. Among other things, a non-blocking, multiported, first-level cache able to feed data to ALUs in as few cycles as possible is needed[11]. This is specially true for integer codes without much data parallelism, where performance can be very sensitive to load-use delay.

To get a view of the involved trade-offs, Figure 1.a shows IPC (instructions committed per cycle) for an 8-issue out-of-order processor when varying three key parameters of a true-multiported first-level cache: size, latency (*lat*) and number of ports (*P*). Numbers come from executing SPEC2K integer codes in processors having three levels of on-chip cache (simulation details in Section 4). The uppermost line shows a 2-cycle latency, four-ported cache *2lat-4P*. The bottom line shows a 4-cycle latency, four-ported cache *4lat-4P*. Enclosed between the two lines there is a significant gap ranging from 1.98 to 2.37 IPC. As we can see size is an important parameter (roughly, a 2% IPC increase for each doubling), but it is not the most important one. The number of ports is increasingly important as we limit them: when moving from *2lat-4P* to *2lat-3P* and further to *2lat-2P* we see a 1.5% and 6% IPC decrease, respectively, no matter the size. On the other hand, increasing the cache latency by one cycle has a fixed 7% penalty, both from *2lat-4P* to *3lat-4P* and from *3lat-4P* to *4lat- 4P*.

---

* This work is supported by the Ministry of Education of Spain under grant TIC 2001-0995-C02-02 and the Diputación General de Aragón (BOA 58,14/05/03). We gratefully thank Elena Castrillo Soto for her contributions in editing this paper.

Contents Management in First-Level Multibanked Data Caches 517

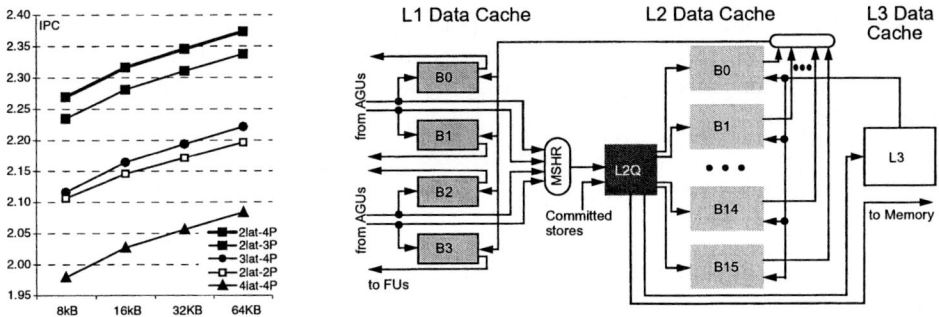

Fig. 1. a) Performance of a true multiported L1 cache. b) Data path of the memory data stream implemented in multibanking schemes. Load routing is made from IQ.

Clearly, we will want to include the 64kB design with four ports and a 2-cycle latency as a first-level cache. Unfortunately such a cache cannot be built with such low latency if we plan to use current or future technologies targeted to deliver fast clocking [1].

Multibanking has lately been proposed as a feasible alternative to a monolithic multiported cache [3, 4, 7–9, 11–13]. Multibanking distributes data (cache lines or words) into several disjoint banks which are physically located close to address generation units or even close to functional units, minimizing then the wiring delay.

Multibanking adds two main additional freedom degrees to the design, namely number of ports per bank and bank content management. There are no works exploring big areas of this new and broad design space, since research up to now has focused on the proposal and evaluation of specific multibanking schemes.

Our contribution is twofold, on the one hand we describe what we believe are the two main design axes for a multibanked scheme: *replication degree* and *distribution policy*, placing current proposals within our taxonomy. On the other hand, we simulate a representative sample of design options in detail taking as a design unit, exclusively, a simple single-ported bank. We do not consider banks having multiported bit cells intentionally, because we seek a fair comparison among alternatives totalling the same raw storage capacity and keeping constant access time. Instead, we replicate data into as many banks as needed in order to achieve the desired multiport effect which in turn will reduce effective capacity (mirror caching [11]).

Simulations assume a high-frequency 8-issue superscalar processor with three levels of cache within the chip. We carefully model the collateral effects of deep pipelining, such as speculative instruction issue by making latency-prediction and using a mechanism to recover from latency misprediction. As we will see, simple static distribution policies coupled with partial replication of data may outperform some dynamic distribution schemes [3, 9].

Next section gives details about the processor and memory we model. Section 3 introduces a bank content management taxonomy, mapping existing proposals into it and exposing the main performance trade-offs. Section 4 shows the experimental framework and discusses the performance of the alternatives selected in order to cover a broad design space. Section 5 ends the paper with our conclusions.

2 Processor and Memory

We model an 8-way processor with eight stages from Fetch to Issue Queue (IQ) and one stage between IQ and Execution. Other processor and memory parameters are listed in Figure 2. After issuing, instruction payload and operands are read and execution starts. Cache hit latency is assumed for a load instruction. Dependent instructions on the latency-predicted instructions can be issued speculatively while waiting for the resolution of all latency predictions. Because IQ is used as the Recovery stage, the predicted latency instruction itself and its dependent instructions remain in IQ until the last check arrives. We use a chained recovery mechanism which starts recovery as soon as a misprediction notification reaches IQ [12]. This mechanism is selective, meaning that only the instructions that depend on a mispredicted load will be re-executed.

fetch and decode width	8	L1 I-cache	64KB, 4-way	L2 Unified Cache	16 banks, 8-way,
branch predictor: hybrid (bimodal, gshare)	16 bits	L1 D-cache	2 cycles	256 kB	7 cycles
		banks	4	line size	128 B
reorder buffer entries	256	ports per bank	1 r/w	L2 MHSR	8 entries
in-flight load/store	128	bank size	2-16 KB, 4-way	L3 Unified Cache	4MB, 16-way, 19 cycles
integer/FP IQ entries	64 / 32	line size	32 B	line size	128 B
integer/FP units	8 / 4	L1 MHSR	16 entries	main memory lat.	200 cycles

Fig. 2. Microarchitectural parameters.

Figure 1.b shows the memory data path. The L1 cache is sliced into four logically independent banks. Each bank is tied only to one address computation unit (AGU) and placed as close to the functional units (FU) as possible. Each bank has only one read/write port shared among loads, committed stores and refills from the L2 cache. The banks are tied to the L2 cache through a single refill bus of 32 bytes.

The memory access takes one cycle to access the bank plus an extra cycle to reach the bypass network (line 1 in Figure 3). Load instructions that miss in L1 cache are reissued from IQ in time to catch the data when the refill is performed.

Requests to L2 cache are managed by the L2Q after accessing L2 tags (line 2 in Figure 3) in the same way Itanium II does [7]. The L2Q can send up to four independent and non-conflicting request to the sixteen address interleaved 2 cycles banks in each cycle. On an L2 cache miss a request is made to L3 cache and to memory (line 3 in Figure 3). A refill to L2 cache takes eight banks. The model can stand 16 primary L1 misses and 8 L2 misses.

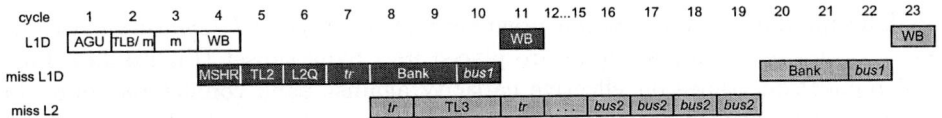

Fig. 3. Load pipeline. *tr* and *busX* mean transport cycles and bus use, respectively.

We assume memory disambiguation based on the Alpha 21264 approach, where loads place addresses into a load buffer, and stores place address and data into a store buffer at once. We model a multiported store buffer (not shown in figure) with the same latency as the first-level cache and assume an oracle predictor for store-load independency.

Store instructions use a memory port and are routed to the store buffer when issued. L1 is write-through and write-not-allocate. Store instructions are committed to L2 and, when they hit L1 cache (filtered by the L2 cache directory), placed in one or more local buffers (depending on the replication degree). The 8-entry coalescing local write buffers update the L1 cache banks in unused cycles.

3 Design Axes for Bank Content Management

3.1 Data Replication Degree

The replication degree sets the number of copies a data can have spread across the cache banks. In order to achieve a given replication degree, the item required in a bank miss is refilled into a single bank (no replication at all), a bank subset (partial replication) or into all banks (full replication).

Full replication refills all banks with each missed line (Figure 4.a). Effective capacity is the smallest possible (the single bank size), but the cache access is conflict-free because any four simultaneous requests are satisfied at once without issue memory ports contention (no *bank conflicts*).

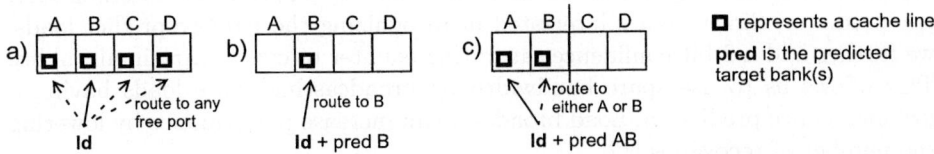

Fig. 4. Data replication options in a 4-bank (A,B,C,D) cache. a) Full Replication, b) No Replication line interleaving and c) Partial Replication line interleaving.

No replication allows only a single copy of the missed item (Figure 4.b). Here the effective capacity is the largest possible (the aggregated size of all banks) but cache access may frequently undergo issue port contention (*bank conflict*).

The middle ground between previous policies is *partial replication*: two or more copies of the missed item are placed in a bank subset (Figure 4.c). Partial replication trades off effective capacity against bank conflict reduction in an interesting way. A second-order effect of data replication is the accuracy of the bank predictor (if any) tied to the data distribution policy: the higher the replication the better the predictability, simply due to the fact that predicting one among n banks is usually harder than predicting one among $n/2$.

3.2 Data Distribution Policy

A distribution policy determines in which bank subset (one or more banks) a data has to be placed, trying to minimize bank conflicts and cache misses. A distribution policy has a mechanism that predicts or suggests the destination bank of every load. Static policies always place the same line in the same bank subset. Dynamic Policies allow lines to move or to replicate among subsets. Next, we briefly explain each policy.

Static Policies
Distribution by Data Type. It consists in distributing different data structures to different bank subsets. In the first-level cache context this idea has only been applied by Cho et al. to separate the stack region from the remaining ones [4]. Both this work and our simulation use a 4 KB Access Region predictor to guess the right region.
Distribution by Data Address. This is the most conventional approach and consists in distributing data according to a hash function applied to some bits of their addresses [8, 11–13]. The data can either be a word (word interleaving) or a line (line interleaving). This policy requires a bank or bank subset predictor because the load routing is made from IQ, prior to computing addresses. Bank mispredictions are corrected by re-routing loads to the correct place from IQ. As a bank predictor we use an *enhanced skewed binary predictor*, originally proposed by Michaud et al. for branch prediction [6, 13]. We choose to predict each address bit separately: up to 2 bits for 4 banks. Every bit predictor is identically sized: 8K entries for the three required tables and a history length of 13, totalling 9KB per predictor. Since we are interested in recognizing the hard-to-predict loads, we further add a 2-bit confidence saturating counter to every entry in all tables. This allows us to use spare bandwidth by broadcasting those loads having a low-confidence prediction. Load broadcast can increase performance by lowering the number of recoveries [12].

Dynamic Policies. They allow lines to move or replicate dynamically among bank subsets to avoid bank conflicts or maintain effective capacity. Two policies have been suggested up to now, both using load identities (program counter) to drive load routing and data placement.
Distributing the Working Set of Individual Loads (Instruction Working Set). This policy by Racunas and Patt tries to take the whole data working

set referenced by each load and place it in a single bank subset [9]. Lines referenced from several loads can migrate among subsets according to the relative confidence of predictions. Such predictions come from two tables (called iPAT and dPAT, respectively) requiring a total of 16KB for four banks. Bank mispredictions are corrected by accessing L2, and so they have the same cost that a first-level cache miss. The authors evaluate a first level made up of 8 two-ported banks in 16-way processor. We scale this proposal to two bank subsets with two replicated banks each and also evaluate a no replication version, which performs poorly due to frequent bank conflicts caused by lack of ports.

Conflict-Aware Distribution. This policy by Limaye et al. replicates a line as soon as a bank conflict appears [3]. A table remembers the last bank (a *cachelet*) assigned to each load and will repeatedly route it to such a bank unless a bank conflict appears, in which case a free bank will be referenced and refilled from level two. This proposal may allocate a line to one, several or all the banks. Our four single-ported banks are similar to the configuration they evaluated, requiring an 8 KB table to remember the last bank.

3.3 Studied Design Points

The policies we have described above have been evaluated by their authors using very different processor and memory hierarchy models. So, in Section 4 we evaluate them within the framework described in Section 2. We will use IPC as the main metric, correlating it with cache misses and bank conflicts. Data replication introduces a new trade-off into the data distribution policies because data replication reduces bank conflicts but at the same time the effective capacity reduction increases cache misses.

As design points with no replication we choose the following: distribution by data address (*Word* interleaving and *Line* interleaving) and distribution by instruction working set (*IWS*). Partial replication is evaluated by arranging the four banks into two subsets with two replicated banks. These design points are word interleaving (*Word-2R*), line interleaving (*Line-2R*), instruction working set (*IWS-2R*) and distribution by access region (*AR-2R*). Conflict-aware distribution (*CA*) is another design point with partial replication. Finally, we also evaluate a full replication organization (*4R*).

4 Results

We use SimpleScalar3.0c [2] to carefully model the processor and memory stated in Section 2. As workload we use all SPECint 2K but *mcf* because it has a very low IPC (0.23) no matter the L1 cache we consider. We use Alpha binaries simulating a contiguous run of 100M instructions from the SimPoints [10] after a warming-up of 200M instructions. All figures show the IPC harmonic mean.

Figure 5 shows the IPC achieved by the different distribution policies and replication degrees with banks ranging from 2KB to 16KB. Figure 5.a and Figure 5.b show policies with no replication and partial replication, respectively, while Figure 5.c compares the best ones to full replication, assuming a 2-cycle latency for L1.

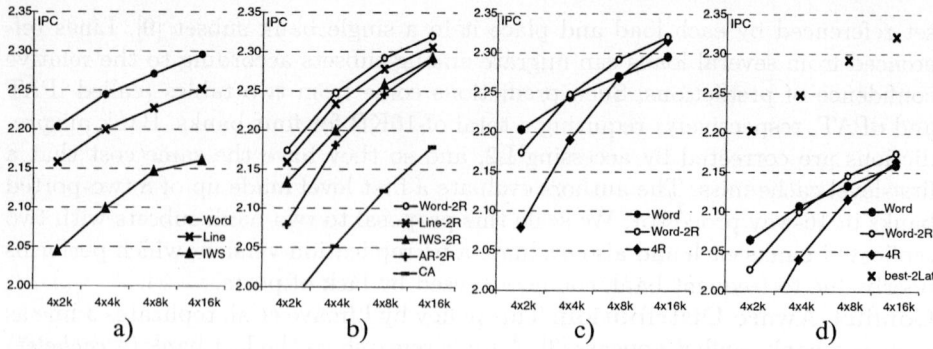

Fig. 5. IPC achieved with banks ranging from 2KB to 16KB with 2-cycle latency. a) no replication, b) partial replication, c) the best of the previous ones plus full replication, and d) merging the best points for latency 2 (x) with the best 3-cycle latency points.

In order to better understanding the IPC results we also show L1 cache miss ratio (Figure 6.a, grey), and bank misprediction ratio (Figure 6.a, black) on 8KB banks. The total length of each bar represents the load percentage undergoing latency misprediction. Besides, Figure 6.b exposes the average number of cycles a load, once ready, waits on IQ due to issue memory ports contention (*conflicts*).

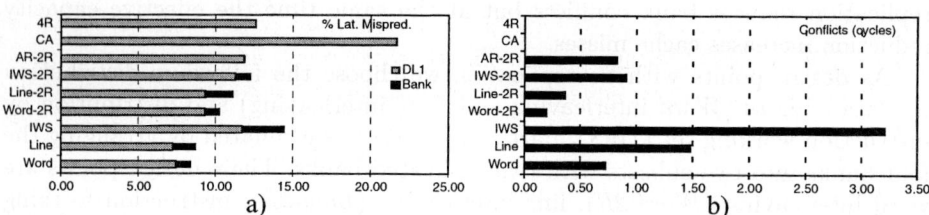

Fig. 6. a) Percentage of latency mispredictions, b) average number of lost cycles due to issue memory ports contention (*bank conflicts*).

Starting from Figure 5.a, notice that Instruction Working Set (*IWS*) behaves clearly worse than the interleaving policies (*Word* and *Line*) due to its higher number of bank conflicts (Figure 6.b) and of L1 cache misses (Figure 6.a). *Word* performs always better than *Line*. *Word* undergoes less bank conflicts, has a lower L1 cache miss rate and exhibits better bank predictability than *Line*.

In general, partial replication policies (Figure 5.b) reduce the number of bank conflicts but at the expense of increasing the L1 cache miss ratio. Conflict-Aware (*CA*) has an IPC lower than the other policies. *CA* has a very high L1 cache miss ratio (Figure 6.a) on account of the changes made in bank mapping to eliminate bank conflicts.

Access Region (*AR-2R*) and *IWS-2R* have a worse IPC than *Word-2R* and *Line-2R*. They experience a worse distribution across the bank subsets generating more L1 cache misses and bank conflicts (Figure 6). As in Figure 5.a, *Word-2R* is better than *Line-2R*.

Finally we analyze Figure 5.c, where we compare Full Replicated (*4R*) against *Word* and *Word-2R*, the best performing design points up to now. *Word-2R* provokes less bank conflicts and bank mispredictions than *Word*, but has more L1 cache misses because it is half the effective size than *Word* (Figure 6). As capacity is the main factor of performance losses (Figure 5), with 4 banks of only 2KB, *Word* is the best option. However, with banks greater than 2KB the best design point is *Word-2R*.

Increasing the bank size makes *4R* appealing. With 16KB the performance of *4R* is more or less the same as the best design point, but at much lower cost and complexity.

Summing up, the design point giving the best performance is *Word*. The optimal replication degree is subject to the bank size. For 2KB banks it is better not to replicate to keep effective capacity, whereas on 4KB and 8KB banks, partial replication removes enough bank conflicts to offset the effective capacity loss. Finally, full replication achieves the highest performance for 16KB banks with no predictor at all.

4.1 Effect of Rising L2 Latency

Doing over the same experiments with a 10 cycle latency to reach L2 shows up a lower IPC across all design points. The L1 cache miss penalty increase is more deeply suffered by those configurations with higher L1 miss rates. For example, with 2KB the IPC for *4R* decreases by 4.3% while *Word* does by 2.3%. The crossing between *Word-2R* and *Word* now goes to the right and *Word* becomes the best option for banks up to 4KB.

4.2 Effect of Rising L1 Latency

Up to this point we have assumed, regardless of the bank size, a 2-cycle L1 cache access latency. Bank latency, in cycles, depends on the particular technology used and on the design of the rest of the processor (cycle time, routing,...). It is out of the scope of this work to determine this latency. At the same time it is evident that banks of different sizes will have unequal latencies. This is why we present results over a range of values.

Figure 5.d shows performance for a 3-cycle L1. Moreover, for each bank size the best design points achieved with a 2-cycle L1 (extracted from Figure 5.c) are market with a cross. As we can see, even the smallest 2-cycle banks surpass any 3-cycle design points. Moreover a 2-cycle *Word* with 4 banks of 2KB outperforms most 4-port true-multiported schemes of Figure 1 with latencies of 3 or 4 cycles. Increasing the bank size is never a good option if it involves latency add-on.

5 Conclusions

There are two major decisions regarding contents management when facing the design of a first-level multibanked cache: *distribution policy* and *replication degree*. Choosing a given design point sets how many copies of a line are allowed and the identity of the banks able to hold them. In general, distribution policies that spread data flow by data address lead to the smallest number of cache misses and bank conflicts. Our results show that the most suitable distribution policy is *Word* interleaving.

The optimum replication degree is subject to the bank size. For a 2KB bank the optimal is no replication. Partial replication over two banks is the most effective method for sizes between 4KB and 8KB. On greater banks, Full replication performance is equivalent to the best bank content management at a lower cost. On the explored design space (banks between 2KB and 16KB with latencies of 2 and 3 cycles), increasing bank size is never a good option if it involves latency add-on.

References

1. V. Agarwal et al.: Clock Rate versus IPC: The End of the Road for Conventional Microarchitectures. Proc. of 27th ISCA (2000) 248–259.
2. D.C. Burger and T.M. Austin: The SimpleScalar Tool Set, Version 2.0. UW Madison Computer Science Technical Report #1342 (1997).
3. D. Limaye, R. Rakvic, and J.P. Shen: Parallel Cachelets. Proc. 19th ICCD 284–292. Sept. 2001.
4. S. Cho, P. Yew, and G. Lee: A High-Bandwidth Memory Pipeline for Wide Issue Processors. IEEE Trans. on Computers, vol. 50, no. 7 (2001) 709–723.
5. R.E. Kessler, E.J. MacLellan, and D.A. Webb: The Alpha 21264 Microprocessor Architecture. Proc. of ICCD'98 90–95. Oct. 1998.
6. P. Michaud, A. Seznec, and R. Uhlig: Trading Conflict and Capacity Aliasing in Conditional Branch Predictors. Proc. of 24th ISCA (1997) 292–303.
7. S. Naffziger et al.: The implementation of the Itanium 2 Microprocessor. IEEE J. Solid State Circuits, vol. 37, no. 11 (2002) 1448–1460.
8. H. Neefs, H. Vandierendonck, and K. De Bosschere: A Technique for High Bandwidth and Deterministic Low Latency Load/Store Accesses to Multiple Cache Banks. Proc. of 6th HPCA (2000) 313–324.
9. C. Racunas and Y.N. Patt: Partitioned First-Level Cache Design for Clustered Microarchitectures. Proc. of 17th ICS 22–31. June 2003.
10. T. Sherwood, E. Perelman, G. Hamerly, and B. Calder: Automatically Characterizing Large Scale Program Behaviour. Proc. of ASPLOS (2002).
11. G.S. Sohi and M. Franklin: High-Bandwidth Memory Systems for Superscalar Processors. Proc. 4th ASPLOS (1991) 53–62.
12. E. Torres, P.E. Ibañez, V. Viñals, and J.M. Llabería: Counteracting Bank Mispredictions in Sliced First-Level Caches. 9th EuroPar, LNCS 2790 586–596, Sept. 2003.
13. A. Yoaz, E. Mattan, R. Ronen, and S. Jourdan.: Speculation Techniques for Improving Load Related Instruction Scheduling. Proc. of 26th ISCA (1999) 42–53.

A Mechanism for Verifying Data Speculation*

Enric Morancho, José María Llabería, and Àngel Olivé

Computer Architecture Department, Universitat Politècnica de Catalunya, Spain
{enricm,llaberia,angel}@ac.upc.es

Abstract. High-performance processors use data-speculation to reduce the execution time of programs. Data-speculation depends on some kind of prediction, and allows the speculative execution of a chain of dependent instructions. On a misprediction, a recovery mechanism must re-issue the speculatively issued instructions. Some recovery mechanisms rely on keeping each instruction in the Issue Queue (IQ) until it is known that the instruction has used correct data. However, their authors either assume that the IQ and the Reorder Buffer (ROB) are unified, or do not detail how the instructions are removed from the IQ before reaching the ROB head entry.

We propose the Verification Issue Queue (VIQ), a mechanism fed with a verification flow graph; the VIQ decides if an instruction can either be removed from the IQ or must be re-issued; the VIQ also allows decoupling the IQ from the ROB. Our evaluations, in the context of load address prediction, show that the verification mechanism is crucial for exploiting the performance potential of data speculation, and that the kind of graph used by the VIQ has a performance impact similar to reducing first-level cache latency by one cycle.

1 Introduction

Data-speculative processors must use a recovery mechanism to guarantee that, for each misprediction, all the speculatively issued instructions dependent on the misprediction will be re-issued. The simplest recovery mechanism is similar to the one used in branch mispredictions: flushing the pipeline and re-fetching the instructions from the instruction cache. However, this mechanism is not feasible in current high-performance processors due to the large performance cost of re-fetching the instructions [1].

Other recovery mechanisms avoid *re-fetching* by keeping each issued instruction on a processor structure until it is known that the instruction has used correct operands. Several authors [2,3] propose the use of specific structures. On the other hand, some processors [4,5] keep the issued instructions in the Issue Queue until they are no longer speculative; on a misprediction, the instructions are re-issued when their operands become ready. In this work, we keep the speculatively issued instructions in the IQ.

---
* This work was supported by the spanish government (grant CICYT TIC2001-0995-C02-01), and the CEPBA (European Center for Parallelism of Barcelona).

An implicit approach [6,7] to verify an instruction is based on re-issuing the instruction after observing a new value for any of its operands, and checking when the instruction reaches the ROB head entry (i.e., it becomes the oldest instruction in flight). As the operands of the oldest instruction are known to be non speculative, then when the state of the oldest instruction is *completed* the instruction gets verified and it can be retired. However, this approach does not scale well because it requires the number of IQ entries to be equal to the number of ROB entries. Sazeides [7] presents a verification mechanism that uses the data flow graph also as a verification flow graph. However, he also unifies IQ and ROB and assumes that, on one cycle, verifications traverse all the verification flow graph, regardless of the number of graph levels.

This work presents the Verification Issue Queue (VIQ), a mechanism that decides if an instruction can be removed from IQ or must be re-issued; it also allows decoupling IQ from ROB. The VIQ is fed with a verification flow graph that can be built in several ways. We evaluate the VIQ in the context of load address prediction. We conclude that VIQ significantly outperforms the implicit mechanism, and that the verification flow graph used by VIQ has a performance impact similar to reducing first-level cache latency by one cycle. This paper is organized as follows: Section 2 describes the IQ used in this work, Section 3 presents the VIQ, Section 4 presents an evaluation of our proposal, and Section 5 concludes the work.

2   Issue Queue (IQ)

In dynamically-scheduled processors, instructions wait in the IQ for the availability of operands and functional units. To issue instructions out-of-order to the functional units, the IQ has two components: a) *wakeup logic* and b) *select logic*. The *wakeup logic* keeps monitoring the dependencies among the instructions in the IQ and, when the operands of a queued instruction become available, this logic will mark the instruction as ready. The *select logic* selects which ready instructions will be issued to the functional units on the next cycle. When latency of the selected instructions is elapsed, dependent instructions are woken up.

The IQ may be implented using CAM cells [8] or bit-matrices [9]. In this study, the IQ is similar to the one described in [10]; it uses a **dependence matrix** for tracking dependencies among instructions. The matrix has as many rows as the number of instructions analysed simultaneously for scheduling, and as many columns as the number of physical registers (registers for short). Each column is connected to a latency counter that starts counting-down when the related instruction is issued. On each crosspoint, a logical circuit determines if there is an unresolved dependence between the instruction and the register. For each row, the outputs of these logical circuits are used to compute a *ready bit* that indicates if the instruction is ready to be selected by the *select logic*. Ready bits are evaluated every cycle.

On processors that do not perform data speculation, instructions can be removed from IQ as soon as they are issued. However, on data-speculative processors, the recovery mechanism influences on the removal policy. In this work, we assume that the recovery mechanism keeps the speculatively issued instructions in the IQ until verifying them.

Re-issue Mechanism. While an issued instruction waits in the IQ for its verification, it is made non-visible to the *select logic*. This is handled by using a *no-request bit* in each matrix row; this bit it is set when the instruction is issued.

When an instruction must be reissued because it has used a misspeculated operand, the instruction is made visible again to the *select logic*. Moreover, its destination-register state is set to non-available; this change delays the issuing of its directly dependent instructions until the instruction is re-issued and a new data value is computed.

Two control circuits perform previous operations: the *removal circuit* and the *register-scoreboard circuit*. Fig. 1 shows the interface between the circuits and other IQ elements. Every cycle, the *removal circuit* is notified of the instructions that must be removed/ re-issued, and the *register-scoreboard circuit* is notified of the misspeculated instructions. The *removal circuit* removes verified instructions from the IQ, and makes visible again the instructions that must be re-issued. The *register-scoreboard circuit* activates latency counters (for each issued instruction, and when data of the predicted load instruction becomes available) or unsets columns (for each misspeculated instruction). As *ready bits* are re-evaluated every cycle, misspeculated instructions will re-evaluate their ready bits, waiting for operand availability. Furthermore, all instructions directly dependent on the misspeculated instructions will be slept.

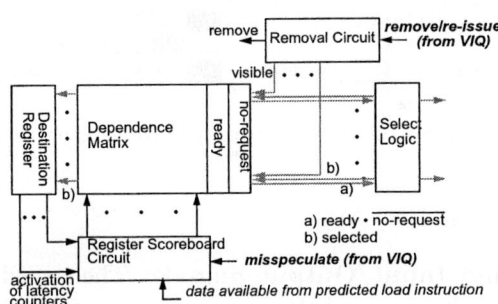

Fig. 1. Structure of the IQ.

3 Verification Issue Queue (VIQ)

In this section we explain how the VIQ decides whether an instruction can be removed from the IQ or must be re-issued. We will apply the VIQ in the context of address prediction. The effective addresses computed by load instructions will be predicted by an address predictor; this will allow the speculative execution of load instructions and their dependent instructions.

The VIQ works as follows. When the address-check result of a predicted load instruction is known, the VIQ initiates the verification/invalidation of the chain of speculatively issued instructions dependent on the load. The VIQ first verifies/invalidates issued instructions directly dependent on the load instruction. These verified/invalidated instructions are used, in next cycle, to verify/invalidate issued instructions that directly depend on them. Thus, each cycle, issued predicted load instructions and verified/invalidated instructions are used to verify/invalidate issued instructions.

In Fig. 2 we present an example of the propagation of verifications and invalidations (we only show significant pipeline stages). Assume that the effective address of load instruction a) has been predicted, memory has been accessed, and instructions b) and c) have been executed speculatively. On cycle 1, instruction a) issues (IQ), then reads registers (R) and computes its effective address (@). Concurrently, the address prediction is checked and turns out to be wrong; after that, VIQ propagates re-issue signals to the dependent instructions: on cycles 4 and 5, VIQ signals IQ that instructions b) and c) must re-issue, and they re-issue on cycles 5 and 6 respectively. Instruction a) access memory non speculatively (m), VIQ propagates its verification and notifies IQ that the instructions can be removed from IQ. In section 3.2, we will show that the VIQ can be also used to perform a faster propagation of the verifications/invalidations. Now, we describe the components of VIQ.

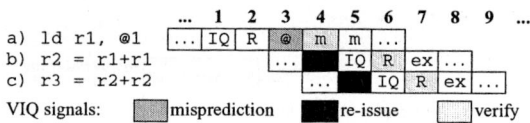

Fig. 2. Example of verification propagation using the VIQ.

3.1 Basic Structure

VIQ: Elements and Input/Output Signals. The issued instructions wait in the VIQ until it is kwnown whether its source operands have became valid or misspeculated. The VIQ has two components (Fig. 3): a) **Verify/Invalidate logic** and b) **Decision logic**. The **Verify/Invalidate logic** monitors if the source operands of the instructions are valid or misspeculated. When all the operands of an instruction become valid, or one of them is misspeculated, this logic marks the instruction as verified or invalidated respectively.

For each issued instruction, the **Decision logic** determines if: a) the instruction can be removed from the IQ, b) the instruction must be re-issued or c) for predicted load instructions, the data value has been misspeculated and the speculatively issued dependent instructions must be re-issued. The Decision logic considers: a) the instruction has been issued, b) verify/invalidate signals, c)

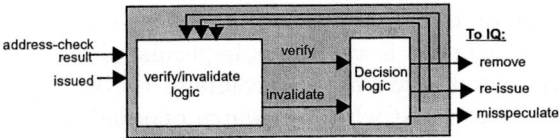

Fig. 3. Components of the VIQ.

instruction type (predicted load, other instruction), d) address-check result, and e) memorization elements related to the instruction (one bit per instruction). Decision-logic outputs are used in next cycle to propagate verify/invalidate information to directly dependent (in the verification-flow graph) issued instructions. In addition, these outputs are sent to the IQ in order to remove/re-issue instructions.

Structure of the Verify/Invalidate Logic. The core of this logic is a matrix structure (Fig. 4) with the same number of rows and columns as the dependence matrix of IQ. This matrix monitors if the source operands of the instructions are valid or misspeculated.

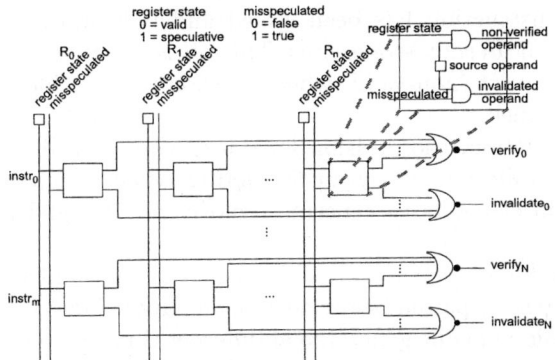

Fig. 4. Matrix structure of the VIQ.

The columns are wires that cross all rows and each row includes a bit for each column. Each column marks the register state (speculative, valid) and if the data value has been misspeculated. The state of a register is speculative until the Decision logic indicates that the producer instruction, which computes the data value, can be removed. The misspeculated signal is set only for one cycle, when Decision logic indicates that the instructions that have consumed misspeculated data value must be re-issued. Instructions are inserted in both the IQ and the VIQ at the same time. When an instruction is inserted in a row, the bits related to its source operands are set; this information is built at the Register Rename stage and inserted in VIQ after this stage.

Each crosspoint of the matrix contains a logical circuit that determines whether the source operand is valid or misspeculated (Fig. 4). For each row, the outputs of the circuits are used to compute the verify/invalidate signals. The verify signal is activated when all source operands of the instruction are valid. The invalidate signal is activated if any source operand of the instruction is misspeculated. Both signals are evaluated every cycle.

Predicted Load Instructions. A predicted load instruction establishes if its destination register is valid or misspeculated by using two pieces of information related to the same instance of the dynamic load instruction: a) address-check result and b) verify signal of the load instruction (its source operand is valid). Then, predicted load instructions must be tagged in order to take into account both pieces of information before the Decision Logic takes a decision. For this, we use the instruction type bit (predicted load instruction, other instruction). However, for incorrect address-check results, the chain of dependent instructions will be re-issued immediately since the computed address is used for initiating a new memory access after the address-check stage, and this chain of instructions have used misspeculated data values.

Decision Logic. Each entry of the matrix structure has a valid bit which indicates if the instruction has been issued and it is waiting for the verification/invalidation of its source operands. This bit is set each time the instruction is issued and it is unset (non-valid) when Decision logic decides that the instruction must be re-issued.

For each instruction, the decision logic uses verify/invalidate signals of the matrix structure, valid bits of the entries, address-check results, instruction type and memorization elements to generate remove, re-issue and misspeculate signals.

For a predicted load instruction, address-check result and verify signal of the source operand must be paired before deciding if data value is valid. Because both pieces of information may be generated at different time, a memorization element in Decision logic is needed. After pairing both pieces of information, on a correct address-check result, the remove signal is set. Otherwise, incorrect-address check result, the misspeculate signal is set only for one cycle, and in the next cycle the remove signal is set. For the other instruction type, verify/invalidate signals drive directly remove and re-issue&misspeculate signals respectively.

Fig. 5 shows the interface between VIQ components. At every cycle, the *removal circuit* and the *Register-Verify circuit* are notified of the removed/re-issued and the re-issued/misspeculated instructions respectively. The *removal circuit* removes an instruction when then remove signal is activated. It also sets the entry as non-valid when the re-issue signal is activated. The *register-verify circuit* sets the destination register state to valid on a remove signal, and activates the misspeculate column on a misspeculate signal.

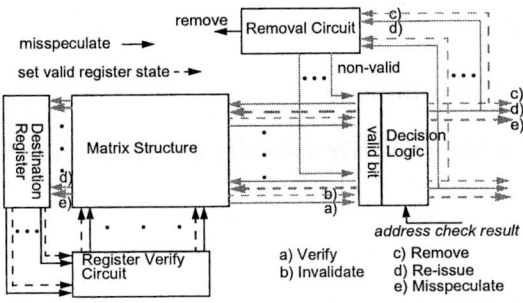

Fig. 5. Structure of the VIQ.

3.2 Verification Flow Graphs

The VIQ uses a verification flow graph to propagate the verifications/invalidations through the instructions. We can build several kinds of verification-flow graphs. Depending on the graph used, verifications/invalidations will be propagated faster to the instructions. However, the VIQ structure is independent of the verification flow graph used, the only difference is the dependence information stored in the matrix structure of the VIQ. We have considered two kinds of verification flow graphs:

Serial Verification Flow Graph. In this case, the verification flow graph is identical to the data flow graph, but the latency of all instructions is one cycle.

Enhanced Verification Flow Graph. In this case, we use a new verification-flow graph with fewer levels than the serial verification-flow graph. In this graph, every instruction is connected directly to the predicted load instructions it depends on (directly or indirectly). The goal of this design is to reduce the number of cycles needed to verify/invalidate instructions that depend on a predicted load instruction: only one step will be required to verify/invalidate source operands of instructions directly/indirectly dependent on a predicted load.

To build the rows of the matrix structure we use a table named Collapsed Graph Table (CGT) with as many entries as architectural registers, and each entry has a bit vector with as many bits as physical registers. The CGT is indexed by the architectural-register identifier. It is updated only with information known in the rename stage, and is also read in this stage to build row information. Predicted load instructions set the bit that identifies its destination register in the CGT entry of its destination architectural register. Any other instruction reads the CGT entries related to its source architectural registers, applies the OR function to the bit vectors, and stores the resultant bit vector in the CGT entry of its destination architectural register.

To insert an instruction in the VIQ, each instruction reads the CGT entries associated to its source architectural registers, applies the OR-function to the obtained bit vectors, and stores the result in its assigned entry of the matrix structure of the VIQ.

4 Evaluation

4.1 Evaluation Environment

To evaluate our proposal we derive a simulator from the SimpleScalar 3.0 (Alpha ISA) cycle-by-cycle simulator [11]. We simulate a 4-way processor (like 21264: same pipeline, branch prediction and functional units) with a 128-entry ROB, and an 8-way processor (scaling by 2 resources of the previous one) with a 256-entry ROB. Memory hierarchy has separated first-level caches (64Kb, direct mapped, variable latency), unified second-level cache (1Mb, dm, 12 cycles), and 80-cycle latency main memory.

To predict effective addresses, we use a hybrid address predictor composed of a stride and a context component. On average, this predictor predicts correctly about 83% of the effective addresses and its accuracy is about 96%. In our data-speculative processors, branch resolutions are resolved non-speculatively. Our proposals are evaluated on the SPEC95-INT benchmarks, and we present average results. Detailed information on simulation intervals and individual results can be found in [12].

4.2 Results

First, we evaluate the effectiveness of the implicit verification versus the use of the VIQ. We compare the performance of the baseline processor (without address prediction) and two data-speculative processors: one uses the implicit verification and the other uses the VIQ with the serial verification. In these evaluations, we assume that the number of IQ entries is equal to the number of ROB entries. We observe that the effectiveness of address prediction depends heavily on the verification mechanism. In fact, in most cases, the performance of the processor with implicit verification is worse than the performance of the baseline processor. Comparing both data-speculative processors, the use of the VIQ improves the performance about 5% (4-way processors) and from 13% to 17% (8-way processors). The instructions that take advantage of the faster verification process are the mispredicted branch instructions, because they can be resolved before reaching the ROB head-entry. We conclude that the verification mechanism is critical for dealing with mispredicted branch instructions.

Decoupling IQ from ROB. After this, we apply the VIQ both to remove the instructions from the IQ and to perform the verification process. Each instruction is removed from the IQ one cycle after the VIQ notifies the verification. Fig. 6.a shows the impact of the IQ size on the performance of a data-speculative processor with serial verification.

We can observe that the IQ size needed to saturate the performance is larger in data-speculative processors than in baseline processors. This is due to the fact that data-speculative processors use some IQ entries to maintain the speculatively issued instructions until they become non speculative. Then, these entries

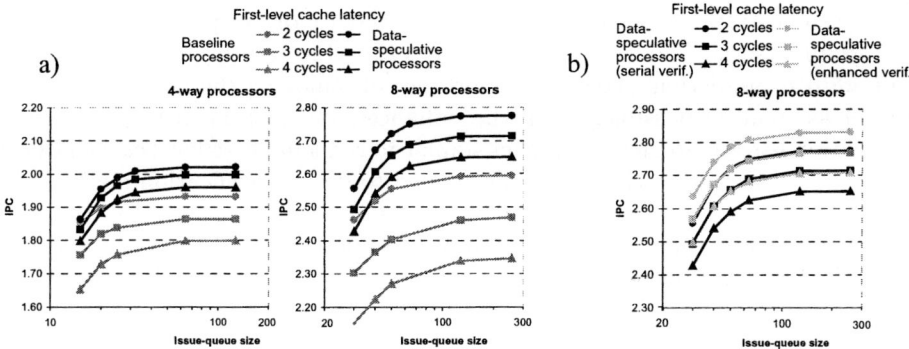

Fig. 6. a) IPC versus IQ size on baseline processors and data-speculative processors with serial verification, b) IPC versus IQ size on 8-way data speculative processors.

are not used by the scheduler to look ahead for independent non-issued instructions. For instance, in 4-way processors, using the 25-entry IQ, the performance is below saturation from 1.6% (2-cycle latency) to 1.9% (4-cycle latency). In 8-way processors, using the 50-entry IQ, the performance is below saturation from 2% to 2.3%.

Enhanced Verification Flow Graph. Finally, we use the enhanced verification flow graph to remove the instructions from the IQ. The use of the enhanced graph reduces the pressure on the IQ with respect to the use of the serial graph, since the enhanced mechanism frees some IQ entries earlier than the serial mechanism. Freeing IQ entries as soon as possible is useful because it may allow the insertion of younger instructions into the IQto be advanced. Fig. 6.b shows the impact of the IQ size on the performance of 8-way address-speculative processors, as well as both the serial and the enhanced verification mechanisms. Our results show the performance effect of the enhanced mechanism compared with the serial mechanism; it is similar to reducing the first-level cache latency by one cycle. For instance, the graph related to the 3-cycle latency, serial verification and 8-way processor, and the graph related to the 4-cycle latency, enhanced verification and 8-way processor, are almost overlapped. Furthermore, to achieve a target performance, processors with enhanced verification need fewer IQ entries than processors with serial verification. The behaviour in 4-way processors is similar to the behaviour in 8-way processors.

5 Conclusions

In this work we propose the detailed design of a verification mechanism: the Verification Issue Queue. This mechanism signals when an instruction can be removed from the Issue Queue or when it must be re-issued. Our evaluations show that the verification mechanism is crucial to exploit the performance potential of data-speculative processors. We also evaluate the use of two possible verification flow graphs (serial and enhanced) by the VIQ. Our results, in the context of load address prediction, show the importance of removing the instructions from the IQ as soon as possible: using the enhanced verification instead the serial verification has a performance impact similar to reducing first-level cache latency by one cycle. Future work must be developed to design verification mechanisms more scalable than the VIQ. Although the VIQ allows the IQ to be decoupled from the Reorder Buffer, some IQ and VIQ structures are proportional to the number of physical registers.

References

1. G. Reinman and B. Calder, "Predictive techniques for aggressive load speculation," in *International Symposium on Microarchitecture*, 1998, pp. 127–137.
2. T. Sato, "Quantitative evaluation of pipelining and decoupling a dynamic instruction scheduling mechanism," *Journal of System Architecture*, vol. 46, no. 13, pp. 1231–1252, 2000.
3. E. Morancho, J. M. Llabería, and À. Olivé, "Recovery mechanism for latency misprediction," in *Proceedings of the 2001 International Conference on Parallel Architectures and Compilation Techniques*, pp. 118–128.
4. R. E. Kessler, "The alpha 21264 microprocessor," *IEEE Micro*, vol. 19, no. 2, pp. 24–36, 1999.
5. K. Diefendorff, "Hal makes sparcs fly," *Microprocessor Report*, vol. 13, no. 5, Nov 1999.
6. T. Sato, "Data dependence speculation using data address prediction and its enhancement with instruction reissue," in *24 th. EUROMICRO Conference Volume 1 (EUROMICRO'98)*. IEEE Computer Society, Aug. 1998, pp. 285–292.
7. Y. Sazeides, "An analysis of value predictability and its application to a superscalar processor," Ph.D. Thesis, University of Wisconsin-Madison, 1999.
8. S. Palacharla, "Complexity-effective superscalar processors," Ph.D. Thesis, University of Wisconsin-Madison, Department Computer Science, 1998.
9. M. Goshima, K. Nishino, T. Kitamura, Y. Nakashima, S. Tomita, and S. ichiro Mori, "A high-speed dynamic instruction scheduling scheme for superscalar processors," in *Proceedings of the 34th annual ACM/IEEE international symposium on Microarchitecture*. IEEE Computer Society, 2001, pp. 225–236.
10. J. Farrell and T. C. Fischer, "Issue logic for a 600-mhz out-of-order execution microprocessor," *IEEE Journal of Solid-State Circuits*, vol. 33, no. 5, pp. 707–712, 1998.
11. D. Burger, T. Austin, and S. Bennet, "Evaluating future microprocessors: The simplescalar tool set," University of Wisconsin-Madison, Department Computer Science, Tech. Rep. CS-TR-96-1308, 1996.
12. E. Morancho, "Address prediction and recovery mechanisms," Ph.D. Thesis, Dept. of Computer Architecture, Universidad Politècnica de Catalunya, May 2002.

Feasibility of QoS for SMT

Francisco J. Cazorla[1], Peter M.W. Knijnenburg[2], Rizos Sakellariou[3],
Enrique Fernandez[4], Alex Ramirez[1], and Mateo Valero[1]

[1] Universidad Politécnica de Cataluña
{fcazorla,aramirez,mateo}@ac.upc.es
[2] LIACS, Leiden University, The Netherlands
peterk@liacs.nl
[3] University of Manchester, UK
rizos@cs.man.ac.uk
[4] Universidad de Las Palmas de Gran Canaria
efernandez@dis.ulpgc.es

Abstract. Since embedded systems require ever more compute power, SMT processors are viable candidates for future high performance embedded processors. However, SMTs exhibit unpredictable performance due to uncontrolled interaction of threads. Hence, the SMT hardware needs to be adapted in order to meet (soft) real time constraints. We show by a simple policy that the OS can exercise control over the execution of a thread which is required for real time constraints.

1 Introduction

To deal with real time constraints, current embedded processors are usually simple in-order processors with no speculation capabilities. However, embedded systems are required to host more and more complex applications and have higher and higher data throughput rates. Therefore, future embedded processors will resemble current high performance processors. Simultaneous Multithreaded (SMT) architectures [5][6] are viable candidates for future high performance embedded processors, because of their good cost/performance trade-off [2]. In an SMT, several threads are running together, sharing resources at the microarchitectural level, in order to increase throughput. A *fetch policy* decides from which threads instructions are fetched, thereby implicitly determining the way processor resources, like rename registers or IQ entries, are allocated to the threads. However, with current policies the performance of a thread in a workload is unpredictable. This poses problems for the suitability of SMT processors in the context of (soft) real-time systems.

The key issue is that in the traditional collaboration between OS and SMT, the OS only assembles the workload while it is the processor that decides how to execute this workload, implicitly by means of its fetch policy. Hence, part of the traditional responsibility of the OS has "disappeared" into the processor. One consequence is that the OS may not be able to guarantee time constraints even though the processor has sufficient resources to do so. To deal with this situation, the OS should be able to exercise more control over how threads are executed and how they share the processor's internal resources.

In this paper, we discuss our philosophy behind a novel collaboration between OS and SMT in which the SMT processor provides 'levers' through which the OS can fine tune the internal operation of the processor to achieve certain requirements. We want to reserve resources inside the SMT processor in order to guarantee certain requirements for executing a workload. We show the feasibility of this approach by a simple parameterized mechanism that assigns fetch slots and instruction and load queue entries to a High Priority Thread. This, in turn, is a first step toward enabling the OS to execute a thread at a given percentage of its full speed and thus enabling the use of out-of-order, high performance SMT processor in embedded environments.

This paper is structured as follows. In Section 2 we describe our novel approach to the collaboration between OS and SMT. In Section 3 we discuss a simple mechanism to enable such collaboration. We discuss related work in Section 4. Finally, Section 5 is devoted to conclusions and future directions.

2 QoS: A Novel Collaboration Between OS and SMT

In this paper, we approach OS/SMT collaboration as *Quality of Service (QoS)* management. This approach has been inspired by QoS in networks. In an SMT resources can be reserved for threads guaranteeing a required performance. We observe that on an SMT processor, each thread reaches a certain percentage of the speed it would achieve when running alone on the machine. Hence, for a given workload consisting of N applications and a given instruction fetch policy, these fractions give rise to a point in an N-dimensional space, called the *QoS space*. For example, Figure 1(a) shows the QoS space for two threads, eon and twolf, as could be obtained for the Pentium4 or the Power5. In this figure, both x- and y-axis span from 0 to 100%. We have used two fetch policies: *icount* [5] and *flush++* [1]. Theoretically it is possible to reach any point in the shaded area below these points by judiciously inserting empty fetch cycles. Hence, this shaded area is called the *reachable part* of the space for the given fetch policies. In Figure 1(b), the dashed curve indicates points that intuitively could be reached using some fetch and resource allocation policy. Obviously, by assigning all fetch slots and resources to one thread, we reach 100% of its full speed, that is, the speed it would reach when run alone. Conversely, it is impossible to reach 100% of the speed of each application at the same time since they have to share resources.

Each point or area (set of points) in the reachable subspace entails a number of properties of the execution of the applications: maximum throughput; fairness; real-time constraints; power requirements; a guarantee, say 70%, of the maximum IPC for a given thread; any combination of the above, etc. In other words, each point or area in the space represents a solution to a *QoS requirement*. It is the responsibility of the OS to select a workload and a QoS requirement and it is the responsibility of the processor to provide the levers to enable the OS to pose such requirements. To implement such levers, we add mechanisms to control how these

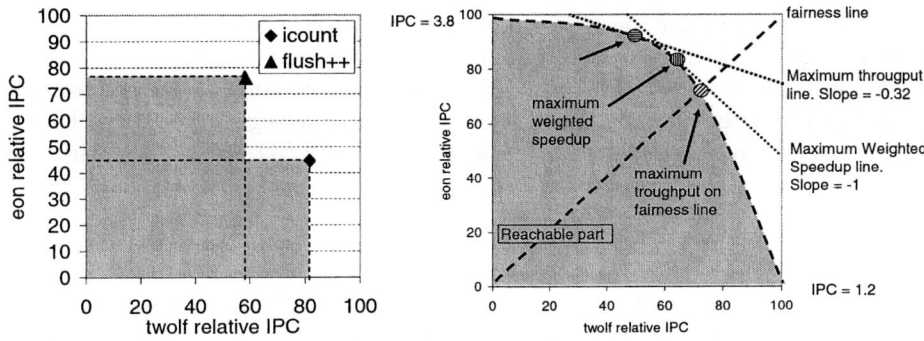

Fig. 1. (a) QoS space for three fetch policies; (b) important QoS points and areas.

resources are actually shared. These mechanisms include prioritizing instruction fetch for particular threads, reserving parts of the resources like instruction or load/store queue entries, prioritizing issue, etc. The OS, knowing the needs of applications, can exploit these levers to navigate through the QoS space.

In this paper, we present a first step toward this goal by studying the behavior of threads when a certain number of resources is reserved for a High Priority Thread (HPT). We show that such a simple mechanism is already capable of influencing the relative speed of threads considerably and hence can cover the QoS space to a great extent.

3 QoS by Resource Allocation

We use a standard 4-context SMT configuration. There are 6 integer, 3 FP, and 4 load/store functional units and 32-entry integer, load/store and FP IQs. There are 320 physical registers shared by all threads. Each thread has its own 256-entry reorder buffer. We use a separate 32K, 4-way data and instruction caches and a unified 512KB 8-way L2 cache. The latency from L2 to L1 is 10 cycles, and from memory to L2 100 cycles. We use an improved version of the SMTSIM simulator provided by Tullsen [5]. We run 300 million most representative instructions for each benchmark. We consider workloads of 2 threads that are of two different types: threads that exhibit a high number of L2 misses of over 1% of the dynamic load instructions, called *Memory Bounded* (MB) threads. These threads have a low full speed. Secondly, threads that exhibit good memory behavior and have a high full speed, called *ILP threads*. We consider 4 workloads in which the High Priority Thread (HPT) is ILP or MB, and the Low Priority Thread (LPT) is ILP or MB. The workloads are: gzip and bzip2 (ILP-ILP), gzip and twolf (ILP-MB), twolf and bzip2 (MB-ILP), and twolf and vpr (MB-MB).

3.1 Static Resource Allocation

We statically reserve 0, 4, 8, ..., 32 entries in the IQ and LSQ for the HPT. The remaining entries are devoted to the LPT. Moreover, we prioritize the instruction fetch and issue for the HPT: in each cycle, we first fetch/issue instructions from the HPT. If there are fetch opportunities left, then instructions from the LPT are fetched/issued. There are more resources in an SMT processor that are shared between threads, most notably the rename registers and the L1 and L2 caches. We have restricted attention to IQ and LSQ entries because these shared resources most directly determine which instructions from which thread are executed.

3.2 Results

We show the resulting QoS space for varying numbers of assigned resources in Figures 2(a) through 5(a). We also show the points obtained from the *round robin* (RR), *icount* (IC), and *flush* fetch policies for comparison. We immediately observe that our parameterized mechanism is capable of covering a large part of the reachable space by tuning its parameter. In contrast, the points reached by the three standard fetch policies show no coherent picture. For the ILP-ILP and MB-MB workloads, they reach points that are quite close together. For the other workloads, there is considerable difference and their relative position in the space changes. This shows that standard fetch policies provide little control over the execution of threads.

We show the resulting IPC values in Figure 2(b) through 5(b). We also show IPC obtained from the standard policies *icount* and *flush* for comparison.

ILP-ILP. Both threads have a high throughput and do not occupy IQs for a long time. As a result, reserving a number of these entries for the HPT and moreover prioritizing its fetch, quickly produces a situation in which the HPT dominates the processor and its speed comes close to its full speed. The total throughput is about the same as for *icount* (except the cases of 0 and 32) which means that what we take from one thread can successfully be used by the other.

ILP-MB. When the LPT thread misses in the L2, it tends to occupy resources for a long time which has an adverse effect on the speed of the other thread. Therefore, reserving resources for the HPT that is ILP causes its speed to sharply increase. As a result, the total throughput can be larger than for *icount*. *flush* needs to re-fetch and re-issue all flushed instructions, degrading its performance. The speed of the LPT does not degrade fast since it suffers many L2 misses and thus cannot use many resources.

MB-ILP. This case is the opposite to the previous one. Given that since the total throughput in comes largely from the LPT that is ILP, when it is denied many resources, its speed degrades fast and total throughput is degraded as well.

MB-MB. Throughput shows a flat curve that is about the same as for the *icount* and *flush* fetch policies as was the case for the ILP-ILP workload. Resources taken from one thread can effectively used by the other thread.

We conclude that by controlling resource allocation we can navigate through the QoS space and bias the execution of a workload to a prioritized thread. At the same time, we still reach considerable throughput for the LPT. Hence, our proposal to provide QoS in an SMT by means of resource allocation is a feasible approach.

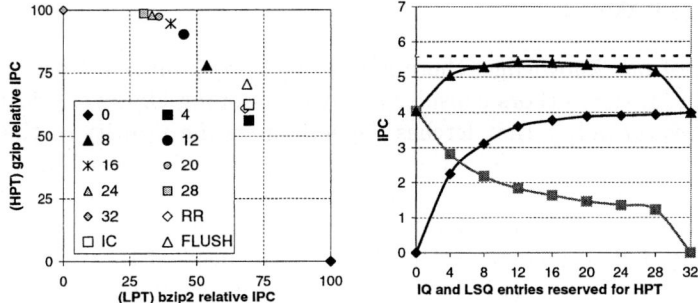

Fig. 2. (a) QoS space for ILP-ILP workload; (b) IPC values and overall throughput.

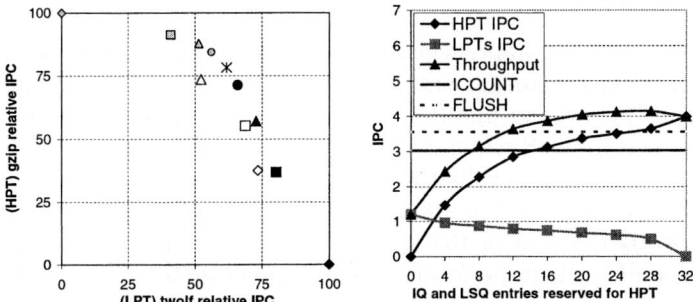

Fig. 3. (a) QoS space for ILP-MB workload; (b) IPC values and overall throughput.

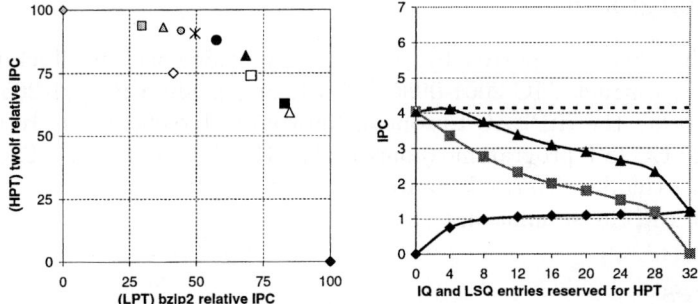

Fig. 4. (a) QoS space for MB-ILP workload; (b) IPC values and overall throughput.

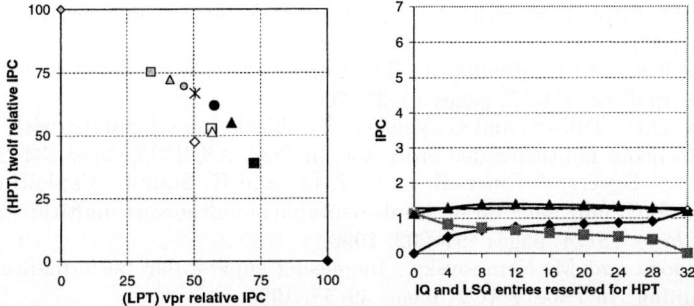

Fig. 5. (a) QoS space for MB-MB workload; (b) IPC values and overall throughput.

4 Related Work

To the best of our knowledge, there is not much work on real time constraints for SMTs. In [4] the authors consider a way of mapping OS-level priorities onto a modified *icount* policy that fetches depending on the priority of the threads. However, this approach exploits the fetch policy, obtaining very limited control, in contrast to our proposal. In [3] explicit static resource allocation is also studied and the authors conclude that resource allocation has little effect on throughput. However, they fail to recognize that the relative speed of threads does change significantly, which is precisely the property that we exploit to provide QoS.

5 Conclusions

In this paper, we have approached the collaboration between OS and SMT as Quality of Service management, where the SMT processor provides 'levers' through which the OS can fine tune the internal operation of the processor in order to meet certain QoS requirements, expressed as points or areas in the QoS space. We have shown, by evaluating a simple mechanism, that it is possible to influence to a great extend the execution of a thread in a workload, so that the OS can reach a large part of the QoS space. Hence, this mechanism is a first step toward enabling high-performance SMT processors to deal with real-time constraints and rendering them suitable for many types of embedded systems.

Acknowledgments

This work has been supported by the Ministry of Science and Technology of Spain under contract TIC-2001-0995-C02-01, and grant FP-2001-2653 (Francisco J. Cazorla), the HiPEAC European Network of Excellence, an Intel fellowship, and the EC IST programme (contract HPRI-CT-2001-00135). The authors would like to thank Oliverio J. Santana, Ayose Falcón, and Fernando Latorre for their work in the simulation tool.

References

1. F. J. Cazorla, E. Fernandez, A. Ramirez, and M. Valero. Improving memory latency aware fetch policies for SMT processors. In *Proc. ISHPC*, pages 70–85, 2003.
2. M. Levy. Multithreaded technologies disclosed at MPF. *Microprocessor Report*, November 2003.
3. S. E. Raasch and S. K. Reinhardt. The impact of resource partitioning on SMT processors. In *Proc. PACT*, pages 15–25, 2003.
4. A. Snavely, D.M. Tullsen, and G. Voelker. Symbiotic job scheduling with priorities for a simultaneous multithreaded processor. In *Proc. ASPLOS*, pages 234–244, 2000.
5. D. Tullsen, S. Eggers, J. Emer, H. Levy, J. Lo, and R. Stamm. Exploiting choice: Instruction fetch and issue on an implementable simultaneous multithreading processor. In *Proc. ISCA*, pages 191–202, 1996.
6. W. Yamamoto and M. Nemirovsky. Increasing superscalar performance through multistreaming. In *Proc. PACT*, pages 49–58, 1995.

Improving Data Cache Performance via Address Correlation: An Upper Bound Study

Peng-fei Chuang[1], Resit Sendag[2], and David J. Lilja[1]

[1] Department of Electrical and Computer Engineering
Minnesota Supercomputing Institute
University of Minnesota
200 Union St. S.E., Minneapolis, MN 55455, USA
{pengfei,lilja}@ece.umn.edu
[2] Department of Electrical and Computer Engineering
University of Rhode Island
4 E. Alumni Ave, Kingston, RI 02881, USA
sendag@ele.uri.edu

Abstract. Address correlation is a technique that links the addresses that reference the same data values. Using a detailed source-code level analysis, a recent study [1] revealed that different addresses containing the same data can often be correlated at run-time to eliminate on-chip data cache misses. In this paper, we study the upper-bound performance of an Address Correlation System (ACS), and discuss specific optimizations for a realistic hardware implementation. An ACS can effectively eliminate most of the L1 data cache misses by supplying the data from a correlated address already found in the cache to thereby improve the performance of the processor. For 10 of the SPEC CPU2000 benchmarks, 57 to 99% of all L1 data cache load misses can be eliminated, which produces an increase of 0 to 243% in the overall performance of a superscalar processor. We also show that an ACS with 1-2 correlations for a value can usually provide comparable performance results to that of the upper bound. Furthermore, a considerable number of correlations can be found within the same set in the L1 data cache, which suggests that a low-cost ACS implementation is possible.

1 Introduction

A recent study [1] has introduced address correlation, which is based on the dynamic linking of addresses that store the same value. This run-time correlation can be used to improve the performance of on-chip data caches by forwarding data to the processor on a miss or a partial hit[1] that is already resident in the cache at other correlated addresses. Source code-level analysis presented in [1] indicates that semantically equivalent information, duplicated references, and frequent values are the major

---

[1] A partial hit occurs when a request on an address is a hit in the cache, but the data at the address is not ready yet because it is in the process of being read by a previous miss in the same cache block. A partial hit can be as slow as a complete cache miss, depending on how close together the two accesses occur.

causes of the address correlations, and that taking advantage of duplicated references has excellent potential to benefit object-oriented programs due to their extensive usage of aggregation classes.

In this study, we focus on the upper bound potential of address correlation, and specific optimizations that will lead to better understanding of how it works. An *address correlation system* (ACS) is designed that stores the addresses evicted from the cache and their correlated addresses, which remain in the L1 data cache to supply the data for the evicted addresses. We show that there is substantial potential for hiding memory latency by providing the data from a correlated address instead of incurring a full miss penalty. When we have limited resources, where to find a useful correlation becomes important. We show that a useful correlation can usually be found in cache lines that are physically close to each other. Furthermore, the number of addresses that can be usefully correlated is usually bounded. Our results show that an ACS with 1-2 correlations for a value can usually provide comparable performance results to that of the upper bound results obtained in this study.

The remainder of this paper is organized as follows. Section 2 describes an address correlation system. Section 3 presents the experimental setup. The upper-bound performance results are given in Section 4. In Section 5, we present some results on how to reduce the cost of correlations. Section 6 discusses specific optimizations while related work is given in Section 7. Finally, we conclude in Section 8.

2 Address Correlation System

In this section, we start with a brief introduction of the concept of address correlation since it is new to the research community. We then describe the basic operation of an ACS. An exhaustive investigation of hardware design parameters for an address correlation mechanism and implementation details is beyond the scope of this paper.

2.1 What Is Address Correlation?

Address correlation is a new approach for exploiting value locality. A conventional cache system, which relies heavily on the temporal and spatial locality that programs exhibit, may exceed the limits of locality via address correlation. In [1], we investigate this program phenomenon, which links the addresses that reference the same data, and show the causes of address correlations with a detailed source-code level analysis of the programs. Address correlation is based on the observation that there is redundancy in data that was stored to memory (store value locality) [3], and that a few values appear very frequently in memory locations (frequent value locality) [4]. Extending these two observations, in [1], we proposed run-time correlation of the addresses that store the same data to improve the performance of on-chip data caches by forwarding data to the processor on a miss that is already resident in the cache at other correlated addresses. In this study, we focus on the upper bound potential of the ACS and specific optimizations that will lead to designs for the run-time correlation of addresses.

2.2 The Components of ACS

Address correlation requires tracking the contents of the cache, the relationships between different locations in the cache, and the memory access history. This section describes the basic operation of an ACS to study the upper bound potential of address correlation.

As shown in Fig. 1, ACS contains two main components: the Address Linking Table (ALT) and the Address Correlation Table (ACT). The ALT stores the correlations between addresses currently residing in the L1 data cache. The ACT keeps all the addresses evicted from the cache and their correlated addresses in the L1 data cache that can supply data for them. In other words, the ACT is the table providing the information about alternative data source on a cache miss.

When a cache line is brought into the L1 data cache, all the addresses in that cache line are linked with the addresses of other cache lines according to their values. An address will never be linked with the addresses in the same cache line since they all will be evicted from the cache at the same time. If a new value is stored to an address, all the links to the updated address are removed and new links are generated for the address based on its new value.

When a cache line is evicted from the cache, the linking information of each address in the line is moved from the ALT to the ACT. Lists of links for those addresses in the ACT that correlate with the evicted addresses are updated to properly reflect the availability of data (alternative data source). Removal of an entry from the ACT occurs when all the linked addresses of an ACT entry are evicted from the L1 data cache or stored with new values, or when the address of an ACT entry is put back to the L1 data cache again.

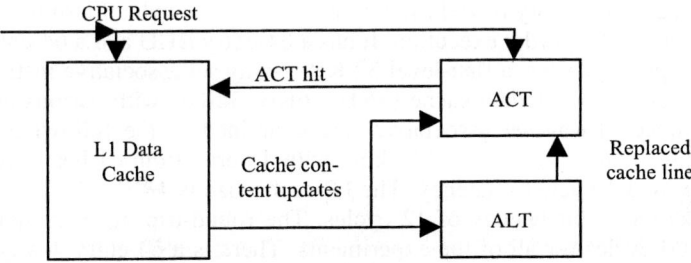

Fig. 1. *Address Correlation system.* The L1 data cache and the ACT are accessed at the same time. When there is a hit in the ACT (which is a miss in the L1 data cache), a correlated address is supplied from the ACT. The L1 data cache then uses this information to supply data in the next cycle.

2.3 Servicing Cache Misses and Partial Hits Faster

The ACT is accessed in parallel with the L1 data cache on a memory access. When the requested address misses in the cache but hits in the ACT, an address is then sent to the L1 data cache. The L1 data cache uses this information to find the value for a previous cache miss.

When a cache miss is supplied by an ACT hit, it is possible to cancel the data request to lower level memory such as L2 cache. In this study, we have explored two design alternatives on a cache miss: canceling the data request to lower level memory on an ACT hit (*cancel_L2*), and allowing the missed cache line to be brought into L1 from lower level memory (*with_L2*). The *cancel_L2* may change the cache behavior and its contents. However, it reduces the traffic between the L1 and L2 data caches substantially and the activities in the ACS system. The details of these designs are left as future work since it is not our intent in this study to design a specific mechanism for address correlation. The results given in the later sections are all for the *cancel_L2* configuration unless otherwise specified.

The ACS can also supply the data for partial hits. A partial hit occurs when a request for an address is a cache hit, but the data at the address is not available because it is still in transit from the lower level memory. Fast data delivery of partial hits is made possible by delaying the time for an entry to be removed from the ACT. An entry of addresses of the same cache line in the ACT is removed after the data in its cache line is in the cache instead of when the tag of its cache line is set in the cache. Since an address whose data exists in the cache will never be in the ACT, a hit in the ACT indicates that there was either a cache miss or a partial hit.

3 Experimental Setup

To determine the performance potential of an ACS, we used an execution driven simulation study based on the SimpleScalar simulator [5]. We modified the memory module to study the potential of address correlation and implemented an ACS for an upper-bound study.

The processor/memory model used in this study is capable of issuing 8 instructions per cycle with out-of-order execution. It has a 64-entry RUU and a 64-entry load/store queue. The processor has a first-level 32 KB, 2-way set associative instruction cache. Various sizes of the L1 data cache (8KB, 16KB, 32KB) with various associativities (direct-mapped, 4-way set associative) are examined in the following simulations. The first-level data cache is non-blocking with 4 ports. Both caches have block sizes of 32 bytes and 1-cycle hit latency. The L2 cache size is 1MB. The L2 cache has 64-byte blocks and a hit latency of 12 cycles. The round-trip main memory access latency is 200 cycles for all of the experiments. There is a 64-entry 4-way set associative instruction TLB and 128-entry 4-way set associative data TLB, each with a 30-cycle miss penalty.

The test suite used in this study consists of selected MinneSPEC [6] CPU2000 benchmark programs. All binaries are SimpleScalar PISA and compiled with SimpleScalar gcc at -O3 and each benchmark ran to completion.

4 Performance of an ACS

In this section, we examine the performance results of an ACS, which was described in Section 2. This analysis will give us the real potential that an address correlation mechanism can offer for exceeding the limits of locality. The results are given for a 32 KB L1 data cache with 4-way associativity, unless specified otherwise.

In Fig. 2, we show the reduction in L1 data cache misses with different cache sizes and the effect of address correlation on reducing the number of misses. The original superscalar processor with an 8KB, 4-way associative L1 data cache is used as the basis for these comparisons. We see that address correlation can eliminate most of the misses in the L1 data cache. That is, the data requested is usually already in the cache at other addresses.

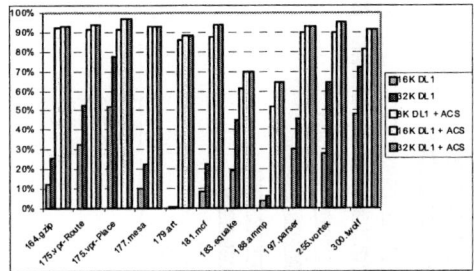 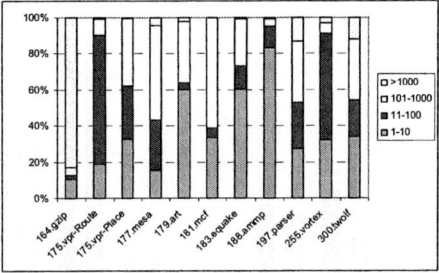

Fig. 2. The percentage cache miss reduction for varying L1 data cache sizes with and without the ACS. The original superscalar processor with an 8KB, 4-way associative L1 data cache is used as the basis for these comparisons.

Fig. 3. The fraction of memory references that are serviced by the L1 cache, the ACS, the L2 cache, and memory.

Fig. 3 shows how memory accesses are serviced. We can see that the ACS services 57 to 99% of the misses in the L1 cache. We categorized the memory accesses that are serviced by the ACS as the percentage of the misses that would have been serviced by the L2 cache and memory, in the absence of the ACS. For example, for *181.mcf*, 69% of the misses in the L1 data cache that would have been serviced by the L2 cache, and 37% of the misses that would have been serviced by the memory, in a system without address correlation are serviced by the ACS. Since memory access has a substantially higher latency than the L2 cache, the more misses that are serviced by the ACS instead of by memory, the more latency that can be hidden from the processor. These results show that the ACS is very effective in hiding the latency for memory references.

Finally, we give the potential overall processor speedup results that could be obtained in a system using address correlation. Here we assume 1 cycle is required for an ACT lookup and another 1 cycle is needed to supply data from the L1 data cache. While it is optimistic to choose only 2 cycles for the ACS to supply an alternative address for a miss, our aim is to show the potential rather than the actual speedups for this new mechanism.

In Fig. 4, the speedup results show a wide range from 0% to 243%. Our preliminary observation shows that supplying data from a correlated address may work especially well for memory intensive benchmarks. 175.vpr, 177.mesa and 300.twolf are the benchmarks that cannot benefit from address correlation even if there is a large reduction in the data misses and partial hits. These benchmarks do not exhibit memory-intensive behavior as is the case for 179.art and 181.mcf. In Fig. 3, we see that in the case of *179.art* and *181.mcf*, a large portion of the misses in the L1 data cache, which are eliminated by the ACS, would have been serviced by the memory in a system without an ACS. Therefore, it is easy to understand the performance results in Fig. 4 for the above benchmarks. Fig. 4 also shows that from the two variations of the ACS design, which is mentioned in section 2, *cancel_L2* performs better for *179.art, 181.mcf* and *188.ammp*, while for the other benchmarks *with_L2* performs better.

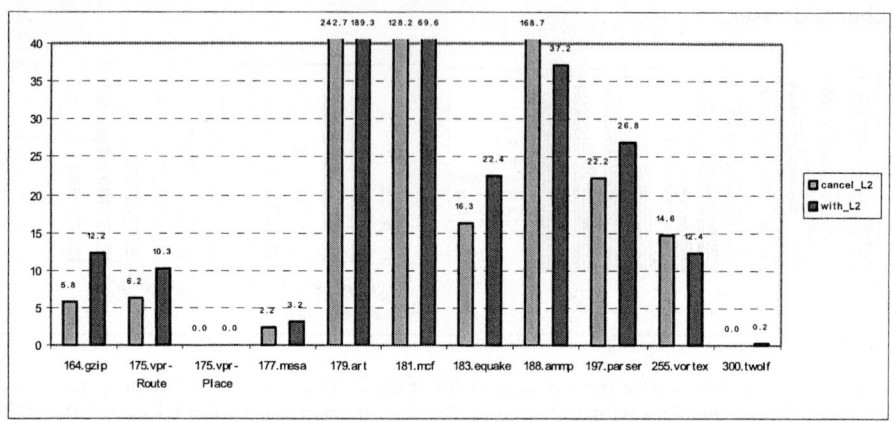

Fig. 4. The percentage speedup obtained by supplying data from a potential address residing in the L1 cache. In the case of *cancel_L2*, a hit in the ACT cancels the request for data from the lower level memory, while in *with_L2*, the request to lower level memory is allowed to continue and bring the data from lower level memory into the cache.

5 Reducing the Cost of Correlations

The performance results in Section 4 are only for an infinite table size ACS. They are nevertheless interesting since they gave us a sense of how much additional performance we can expect from an aggressive and accurate design of address correlation.

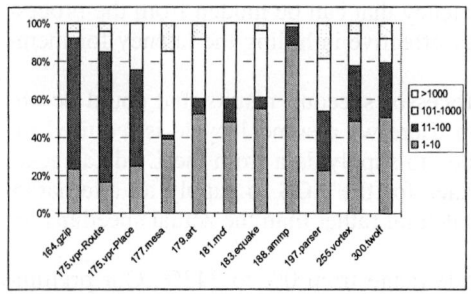

Fig. 5. The number of different addresses in the cache in which the missed data is found.

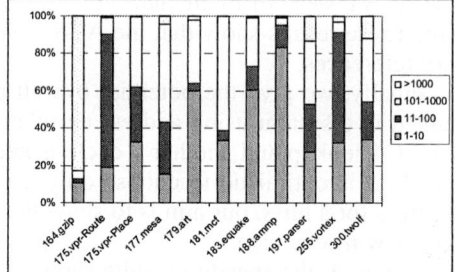

Fig. 6. The number of different addresses in the cache in which the partial hit data is found.

In Figs. 5-10, we show the results of some experiments that we have conducted to gain more insights into the performance potential of a realistic hardware implementation of an ACS system. Figs. 5 and 6 show the number of different addresses that can potentially supply the data on a miss and on a partial hit, respectively. We can see that for *181.mcf*, the same data can be found in 1 to 10 different addresses for 48% of the data misses, while for 12% of the misses it can be found in 11 to 100 different addresses, and so on. The interval ">1000" is more likely due to trivial values (0, 1, etc.), which can be a source of address correlation. Very similar results are obtained for the partial hits, as shown in Fig. 6.

When we have limited resources, where to find a useful correlation becomes an important issue. Memory allocation patterns of data, if there are any, may be one of the possible sources for finding data that can be correlated. In Fig. 7, we have plotted the distance of the addresses, in bytes, where the data can be found on a miss, for 3 out of the 10 benchmarks studied due to readability. We can see that, for 255.vortex, the data values for 35% of all misses can be found within a distance of 128 bytes. For 183.equake, the majority of the addresses that can be correlated fall within the range of 1000 bytes, and for 164.gzip, we can find useful data located within about 1500 bytes. This behavior implies there may be some certain storage allocation pattern that can be exploited for address correlation. While this information may be used for an individual benchmark, a common solution for all benchmarks is not straight-forward. We conclude that this figure does not show useful information for a hardware implementation since there is a random distribution of addresses that are possible candidates for supplying a miss. While we show a distance of up to 7KB in Fig. 7, we see the addresses scattered all over the address range when we plot the whole range.

Figs. 8 and 9 measure the distance of the potential correlated addresses, in sets, in which the data miss and partial hit are found, respectively. We can see that, usually 35% to 50% of the potential addresses for a data miss, and 30% to 45% of the potential addresses for partial hits, are in the same set in the cache. The distance of the data in sets gives us insights into efficient designs for an address correlation mechanism. Having found the data within the same set, the address correlation mechanism would require fewer searches for potential addresses with which to correlate.

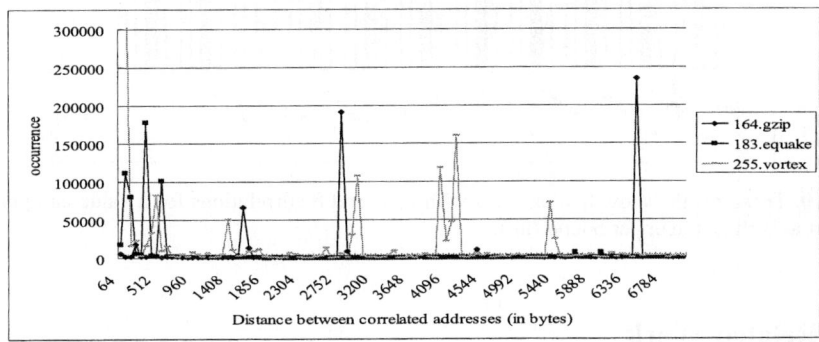

Fig. 7. The number of misses found within an address distance of 7KB, measured in bytes.

In Fig. 10, we show the effect of limiting the number of correlations. Figs. 5 and 6 have shown that, at a given time, thousands of addresses may contain the same data within a 32KB L1 data cache. It is impractical trying to correlate all these addresses. We have used a simple FIFO replacement policy to limit the number of correlations and compare the performance results with the upper bound where all addresses that can be correlated are taken into account. This figure shows that two addresses correlated to each other would be enough for comparable results to that of the upper bound. On average, correlating between two addresses can eliminate only 75% of the misses eliminated in the upper bound. Four and eight correlations can eliminate 85 and 88% of the misses eliminated in the upper bund, respectively. These results suggest that an ACS with 1-2 correlations for a value can usually provide comparable performance results to that obtained in the upper bound results study.

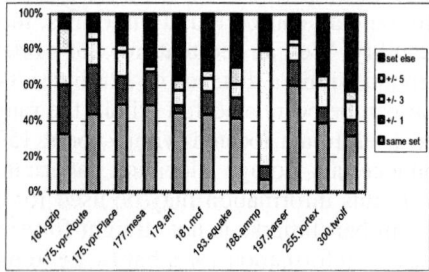 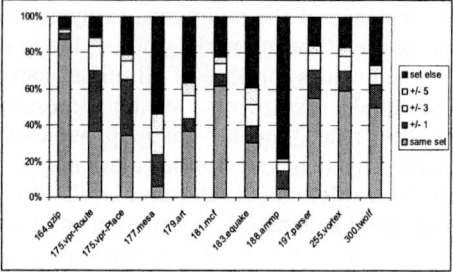

Fig. 8. The distance to the address in which data is found in the cache on a miss, measured by sets.

Fig. 9. The distance to the address in which data is found in the cache on a partial hit, measured by sets.

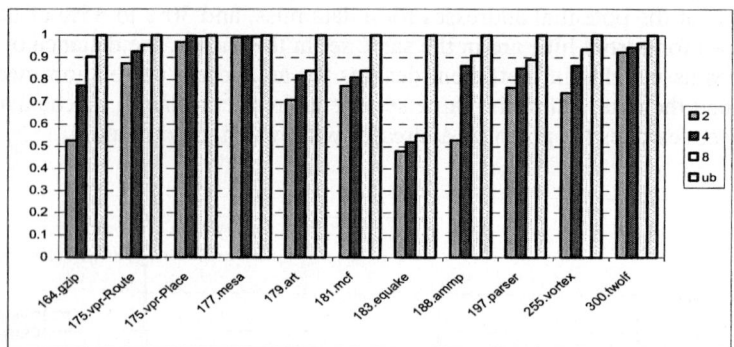

Fig. 10. These results show that an ACS with 2, 4, and 8 correlations for a value can perform almost as well as the upper bound (ub).

6 Related Work

Value locality [2] describes the recurrence of a previously seen value within a storage location. It allows the classical dataflow limit to be exceeded by executing instructions before their operand values are available.

There have been several studies on exploiting different kinds of value locality. Store value locality [3, 10, 11], a recently discovered program attribute that characterizes both memory-centric (based on message passing) and producer-centric (based on program structure) prediction mechanisms for stored data values, introduces the concept of redundancy in data words stored to memory by computer programs. In the study, many store instructions are shown to be silent; that is, they do not change the state of the system because they write a value that already exists at the write address. Consequently, they can safely be eliminated from the dynamic instruction stream.

Recent studies have also introduced Frequent Value Locality [2, 7-9], another type of locality in which a few values appear very frequently in memory locations and are, therefore, involved in a large fraction of memory accesses. Tracking the values involved in memory accesses has shown that, at any given point in the program's execution, a small number of distinct values occupy a very large fraction of these referenced locations. It has been shown [4] that, in the tested programs, ten distinct values occupy over 50% of all memory locations and on an average account for nearly 50% of all memory accesses during program execution.

Extending these two complementary studies, in our previous work, we propose a new technique, *Address Correlation* [1], which is based on the dynamic linking of addresses that store the same value. This previous study investigated the causes of address correlations using a detailed source code level analysis.

In this paper, we study the upper bound potential of address correlation and investigate specific optimizations to exploit it.

7 Conclusion

We have demonstrated the upper bound potential of an address correlation system (ACS). The ACS stores the addresses evicted from the cache and their correlated addresses, which remain in the L1 data cache to supply the data for the evicted addresses. We show that there is substantial potential for hiding memory latency by providing the data from a correlated address instead of incurring a full miss penalty. When we have limited resources, where to find a useful correlation becomes important. We show that a useful correlation can usually be found in cache lines that are physically close to each other. Our results also show that, usually, 40% of the misses and 50% of the partial hits in the L1 data cache can be found within the same set. Furthermore, the number of addresses that can be usefully correlated is usually bounded. Simulation results illustrate that an ACS with 1-2 correlations for a value can usually provide comparable performance to that of the upper bound results obtained in this study.

References

1. R. Sendag, P. Chuang, and D. J. Lilja, "Address Correlation: Exceeding the Limits of Locality", IEEE Computer Architecture Letters, Volume 2, May 2003.
2. M. H. Lipasti, C. B. Wilkerson, and J. P. Shen, "Value Locality and Load Value Prediction," in Proceedings of the second international conference on architectural support for programming languages and operating systems, pp. 138-147, 1996.
3. K. M. Lepak and M. H. Lipasti, "On the Value Locality of Store Instructions," in Proceedings of the 27th International Symposium on Computer Architecture, 2000.
4. Y. Zhang, J. Yang, and R. Gupta, "Frequent value locality and value-centric data cache design," ACM 9th International Conference on Architectural Support for Programming Languages and Operating Systems, pp. 150-159, Cambridge, MA, November 2000.

5. D.C. Burger, T.M. Austin, and S. Bennett, "Evaluating future Microprocessors: The SimpleScalar Tool Set," Technical Report CS-TR-96-1308, University of Wisconsin-Madison, July 1996.
6. AJ KleinOsowski and D. J. Lilja, "MinneSPEC: A New SPEC Benchmark Workload for Simulation-Based Computer Architecture Research," Computer Architecture Letters, Volume 1, May, 2002.
7. J. Yang and R. Gupta, "Energy Efficient Frequent Value Data Cache Design," ACM/IEEE 35th International Symposium on Microarchitecture, November 2002.
8. J. Yang and R. Gupta, "Frequent Value Locality and its Applications," Special Issue on Memory Systems, ACM Transactions on Embedded Computing Systems, 2002.
9. J. Yang, Y. Zhang, and R. Gupta, "Frequent Value Compression in Data Caches," ACM/IEEE 33rd International Symposium on Microarchitecture, pp. 258-265, December 2000.
10. G. B. Bell, K. M. Lepak, and M. H. Lipasti, "Characterization of Silent Stores," in Proceedings of the International Conference on Parallel Architectuers and Compilation Techniques, 2000.
11. K. M. Lepak, G. B. Bell, and M. H. Lipasti, "Silent Stores and Store Value Locality," IEEE Transactions on Computers, Vol. 50, No. 11, November 2001.

Exploiting Spatial Store Locality Through Permission Caching in Software DSMs

Håkan Zeffer, Zoran Radović, Oskar Grenholm, and Erik Hagersten

Uppsala University, Dept. of Information Technology,
P.O. Box 337, SE-751 05 Uppsala, Sweden
http://www.it.uu.se/research/group/uart

Abstract. Fine-grained software-based distributed shared memory (SW-DSM) systems typically maintain coherence with in-line checking code at load and store operations to shared memory. The instrumentation overhead of this added checking code can be severe. This paper (1) shows that most of the instrumentation overhead in the fine-grained SW-DSM system DSZOOM is store-related, (2) introduces a new write permission cache (WPC) technique that exploits spatial store locality and batches coherence actions at runtime, (3) evaluates WPC and (4) presents WPC results when implemented in a real SW-DSM system. On average, the WPC reduces the store instrumentation overhead in DSZOOM with 42 (67) percent for benchmarks compiled with maximum (minimum) compiler optimizations.

1 Introduction

The idea of implementing a shared address space in software across clusters of workstations, blades or servers was first proposed almost two decades ago [1]. The most common approach, often called shared virtual memory (SVM), uses the virtual memory system to maintain coherence. In this paper, we concentrate on fine-grain systems, which maintain coherence by instrumenting memory operations in the programs [2–4]. An advantage with these systems is that they avoid the high degree of false sharing, which is common in SVMs. Hence, they can implement stricter memory consistency models and run applications originally written for hardware-coherent multiprocessors. However, fine-grain systems suffer from relatively high instrumentation overhead [2–4]. Multiple schemes to reduce this overhead have been proposed. For example, Shasta [3] statically merges coherence actions at instrumentation time.

In this paper, we show that most of the instrumentation overhead in the sequentially consistent [5] fine-grained software DSM system, DSZOOM [4], comes from store instrumentation. We propose a dynamic *write permission cache* (WPC) technique that exploits spatial store locality. This technique dynamically merges coherence actions at runtime. We evaluate the proposed WPC technique in a parallel test bench. In addition, we present and evaluate two real WPC implementations as part of the DSZOOM system.

The WPC reduces DSZOOM's average store instrumentation overhead with 42 percent for SPLASH-2 benchmarks compiled with the highest optimization level and 67 percent for non-optimized applications. The parallel execution time for 16-processor runs for a 2-node configuration is reduced as much as 27 (32) percent and, on average, 7 (11) percent for benchmarks compiled with maximum (minimum) compiler optimization.

2 Write Permission Cache (WPC)

Blocking directory coherence protocols have been suggested to simplify the design and verification of hardware DSM (HW-DSM) systems [6]. DSZOOM's protocol is a distributed software-version of a blocking directory protocol. Only one thread at a time can have the exclusive right to produce global coherence activity at each piece of data. The blocking protocol avoids all corner cases of traditional coherence protocols. More protocol details can be found in [4, 7].

The applications studied have more loads than stores to shared memory. Yet, the store-related coherence checks stand for the largest part of the instrumentation overhead, as we will show in section 5.2. Most of this overhead comes from the fact that a locally cached directory entry (called "MTAG") must be locked before the write permission check is performed (blocking protocol).

The idea with the *write permission cache* (WPC) is to reduce locking and consulting/checking of MTAGs by exploiting spatial store locality. Instead of releasing the MTAG lock after a store is performed, a thread holds on to the write permission and the MTAG lock, hoping that the next store will be to the same coherence unit. If indeed the next store is to the same coherence unit, the store overhead is reduced to a few ALU operations and a conditional branch instruction. When a store to another coherence unit appears, a *WPC miss* occurs. Only then, a new lock release followed by a lock acquire must be performed.

The upper part of Figure 1 shows how an original store instruction expands into a *store snippet* when the DSZOOM system is used. Ry is a temporary register, Rx contains the value to be stored and addr is the effective address of this particular store operation. Lines 12 and 17 acquire and release the MTAG lock. Moreover, lines 13 and 14 load and check the MTAG value for permission. If the processor does not have write permission, the store protocol is called at line 15. Finally, at line 16, the original store is performed.

The lower part of Figure 1 shows how a WPC snippet is designed. The snippet consists of two parts: a fast- and a slow-path. Line 22 checks if the current coherence unit is the same as the one cached in the WPC (a WPC entry contains a coherence unit identifier, refered to as CU_id[addr] in Figure 1). If that is the case, then the processor has write permission and can continue its execution. The slow-path code is entered only if a WPC miss occurs. In that case, the processor actually checks for write permission in the MTAG structure. The slow path has much in common with the ordinary store snippet. However, one major difference is that the old lock, whose coherence unit identifier is cached in the WPC, has to be released (line 32). Moreover, at the end of the snippet,

```
01: original_store:                      11: original_store_snippet:
02:     ST Rx, addr;                     12:     LOCK(MTAG_lock[addr]);
03:                                      13:     LD Ry, MTAG_value[addr];
04:                                      14:     if (Ry != WRITE_PERMISSION)
05:                                      15:         call st_protocol;
06:                                      16:     ST Rx, addr;
07:                                      17:     UNLOCK(MTAG_lock[addr]);

21: wpc_fast_path_snippet:               31: wpc_slow_path_snippet:
22:     if (WPC != CU_id[addr])          32:     UNLOCK(MTAG_lock[WPC]);
23:         call slow_path               33:     WPC = CU_id[addr];
24:     ST Rx, addr                      34:     LOCK(MTAG_lock[addr]);
25:                                      35:     LD Ry, MTAG_value[addr];
26:                                      36:     if (Ry != WRITE_PERMISSION)
27:                                      37:         call st_protocol;
```

Fig. 1. Original and WPC-based store snippets.

the processor keeps the lock. At line 33, the processor inserts the new coherence unit identifier in its WPC. Memory mappings are created in such a way that the CU_id[addr] reference at lines 22 and 33 easily can be done with arithmetic instructions, i.e., a shift. Thus, the fast path contains no extra memory references since thread-private registers are used as WPC entries. In other words, an n-entry WPC system with t threads contains $n \times t$ WPC entries in total.

3 Experimental Setup

The benchmarks that are used in this paper are well-known workloads from the SPLASH-2 benchmark suite [8]. Data set sizes for the applications studied can be found in [7]. The reason why we cannot run volrend is that shared variables are not correctly allocated with the G_MALLOC macro. Moreover, all experiments in this paper use GCC 3.3.3 and a simple custom-made assembler instrumentation tool for UltraSPARC targets. To simplify instrumentation, we use GCC's -fno-delayed-branch flag that avoids loads and stores in delay slots. We also use -mno-app-regs flag that reserves UltraSPARC's thread private registers for our snippets. These two flags slow down SPLASH-2 applications with less than 3 percent (avg.). Compiler optimization levels are -O0 and -O3.

3.1 Hardware and DSZOOM Setup

All sequential and SMP experiments in this paper are measured on a Sun Enterprise E6000 server [9]. The server has 16 UltraSPARC II (250 MHz) processors and 4 Gbyte uniformly shared memory with an access time of 330 ns (*lmbench* latency [10]) and a total bandwidth of 2.7 Gbyte/s. Each processor has a 16 kbyte on-chip instruction cache, a 16 kbyte on-chip data cache, and a 4 Mbyte second-level off-chip data cache.

The HW-DSM numbers have been measured on a 2-node Sun WildFire built from two E6000 nodes connected through a hardware-coherent interface with a raw bandwidth of 800 Mbyte/s in each direction [6,11]. The WildFire system has been configured as a traditional cache-coherent, non-uniform memory access (CC-NUMA) architecture with its data migration capability activated while its coherent memory replication (CMR) has been kept inactive. The Sun WildFire access time to local memory is the same as above, 330 ns, while accessing data located in the other E6000 node takes about 1700 ns (lmbench latency). The E6000 and the WildFire DSM system are both running a slightly modified version of the Solaris 2.6 operating system.

All DSZOOM implementations presented in this paper run in user space on the Sun WildFire system. The WildFire interconnect is used as a cluster interconnect between the two DSZOOM nodes. Non-cachable block load, block store and ordinary SPARC atomic memory operations (`ldstub`) are used as remote put, get and atomic operations. Each node accesses a "private copy" of the shared memory. The DSZOOM system maintains coherence between these private segments, i.e., the hardware coherence is not used. Moreover, the data migration and the CMR data replication of the WildFire interconnect are inactive when DSZOOM runs.

3.2 Test Bench Setup

We have developed a parallel test bench environment to analyze new protocol optimizations such as the WPC. The test bench is called *protocol analyzer* (PA) and is designed for rapid prototyping and simulation of realistic workloads on parallel machines.

Our system has much in common with the Wisconsin Windtunnel II [12] simulator. It uses *direct execution* [13] and *parallel simulation* [12] to gain performance. The output from instrumented load and store operations of the studied benchmarks is used as input to PA. Moreover, an SMP (E6000) is used as host system during simulations. The SMP hardware guarantees coherence, memory consistency and correctness during the parallel execution of the program. PA implements a configuration system that makes it possible to model different memory systems. A PA model can simulate caches, cache coherence protocols and much more by using shared memory and simple counters. However, it is not possible to simulate target system's execution time.

Instrumentation overhead and calls to PA models can introduce timing errors or skewness in the simulation. It is important to consider these timing issues when analyzing data produced by the simulator.

4 Evaluating WPC

4.1 Simulated WPC Hit Rate

In this section, we investigate if it is possible to achieve high WPC hit rate with a few WPC entries. This is especially important for an efficient software WPC

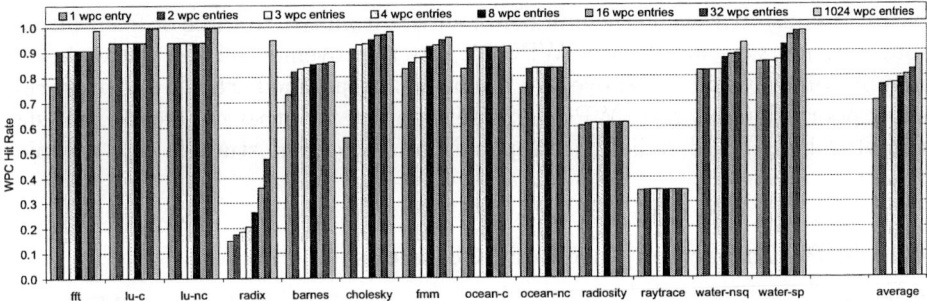

(a) WPC hit rate for benchmarks compiled with optimization level -O0.

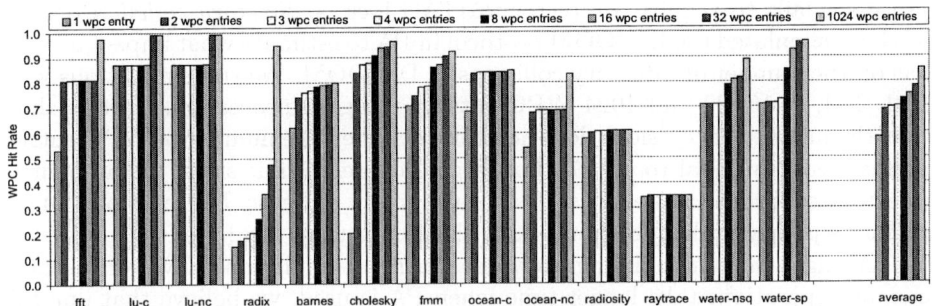

(b) WPC hit rate for benchmarks compiled with optimization level -O3.

Fig. 2. WPC hit rate for SPLASH-2 benchmarks compiled with -O0 (a) and -O3 (b) using a coherence unit size of 64 bytes.

implementation. We use a multiple-entry WPC model to simulate WPC hit rate in PA. WPC hit rate is measured as hits in the WPC divided by the number of stores to shared memory. All data are collected during the parallel execution phase of the applications when run with 16 processors. Because each processor has its own set of WPC entries, and each processor simulates its own WPC hit rate, a timing skewing introduced by PA is not a problem. This is especially true for the applications that only uses synchronization primitives visible to the runtime system. Moreover, our simulated 1-entry WPC hit rate numbers have been verified with a slightly modified DSZOOM implementation. The numbers are almost identical (maximum difference is less than 0.03 percent).

Figure 2 shows hit rate for thirteen applications when 1, 2, 3, 4, 8, 16, 32 and 1024 WPC entries and a coherence unit size of 64 bytes is used. Figure 2 contains WPC hit rate for applications compiled with (a) minimum and (b) maximum optimization levels. Applications compiled with a low optimization level seem to have higher WPC hit rates than fully optimized binaries. Still, almost all applications compiled with -O3 have a WPC hit rate above 0.7. In particular, this is true when two or more WPC entries are used. If the number of WPC entries is increased from one to two, applications such as barnes, cholesky and fft significantly improve their hit rate numbers. This is due to multiple simultaneous write streams. Increasing the number of entries from two to three or from three to four does not give such a large WPC hit rate improvement. Thus, increasing the number of WPC entries above two might not be justified. `radix` and `raytrace` show poor WPC hit rate. WPC hit rate numbers for other coherence unit sizes (32-8192 bytes) and individual applications have been studied in a technical report [7].

4.2 WPC Impact on Directory Collisions

Data sharing, such as multiple simultaneous requests to a directory/MTAG entry, might lead to processor stall time in a blocking protocol. If a requesting processor fails to acquire a directory/MTAG lock, a *directory collision* occurs. We have simulated the DSZOOM protocol in PA to estimate what impact a WPC has on the number of directory collisions in DSZOOM. We run simulations with 1, 2, 3, 4, 8, 16, 32 and 1024 WPC entries, using 16 processors. Results show that the number of collisions does not increase when the number of WPC entries is small (less or equal to 32) [7]. For larger coherence unit sizes, the amount of directory collisions (as well as false sharing) might increase. For example, `lu-nc` performs poorly with a coherence unit size larger than 128 bytes.

PA might introduce timing skewness during a simulation. However, because memory operations take longer time when PA is used, we believe that our collision numbers are unnecessarily negative, i.e., that the number of directory collisions will be even lower when run in DSZOOM than the simulation results currently show.

5 WPC Implementation and Performance

5.1 The DSZOOM-WPC Implementation

For an efficient DSZOOM-WPC implementation, it is necessary to reserve processor registers for WPC entries to avoid additional memory references in store snippets. With multiple WPC entries, the register pressure as well as the WPC checking code increases. As indicated in section 4.1, a 2-entry WPC may be a good design choice. Thus, in this paper, we implement and evaluate 1- and 2-entry WPC systems. We use SPARC's application registers as WPC entries.

The protocol of the base architecture maintains sequential consistency [5] by requiring all the acknowledges from the sharing nodes to be received before a global store request is granted. Introducing the WPC will not weaken the memory model. The WPC protocol still requires all the remotely shared copies to be destroyed before granting the write permission. WPC just extends the duration of the permission tenure before the write permission is given up. Of course, if the memory model of each node is weaker than sequential consistency, it will decide the memory model of the system. Our system implements TSO since E6000 nodes are used.

The WPC technique also raises dead- and livelock concerns. Most of the dead- and livelock issues are solved by the DSZOOM runtime system. A processor's

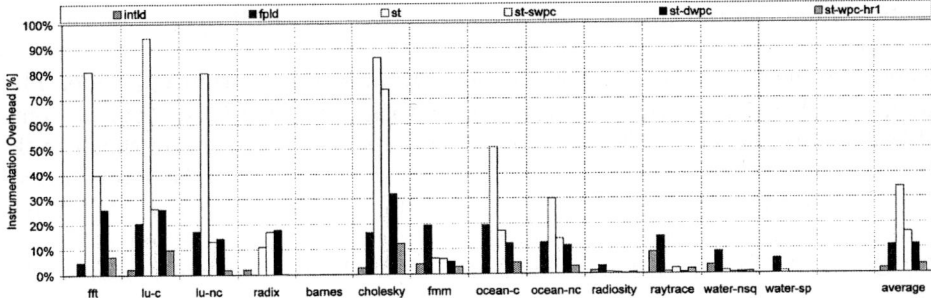

(a) Sequential instrumentation overhead breakdown for non-optimized binaries.

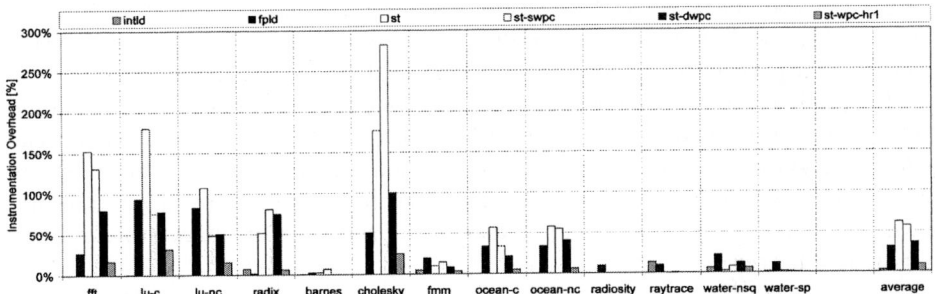

(b) Sequential instrumentation overhead breakdown for fully optimized binaries.

Fig. 3. Sequential instrumentation overhead breakdown for integer loads (*intld*), floating-point loads (*fpld*), the original store snippet (*st*), a 1-entry WPC store snippet (*st-swpc*), a 2-entry WPC store snippet (*st-dwpc*) and a store snippet with WPC hit rate 1.0 (*st-wpc-hr1*).

WPC entries have to be released at (1) synchronization points, at (2) failures to acquire directory/MTAG entries and at (3) thread termination. However, user-level flag synchronization can still introduce WPC related deadlocks. The WPC deadlock problem and three suggested solutions are discussed in [7]. In this study, applications that use flag synchronization (`barnes` and `fmm`) are manually modified with WPC release code.

5.2 Instrumentation Overhead

In this section, we characterize the overhead of inserted fine-grain access control checks for all of the studied SPLASH-2 programs. The write permission checking code (store snippets) is the focus of this section since the WPC technology is a store optimization technique. To obtain a sequential instrumentation breakdown for different snippets, we ran the applications with just one processor and with only one kind of memory instruction instrumented at a time. This way, the code will never need to perform any coherency work and will therefore never enter the protocol code (written in C).

Sequential instrumentation overhead breakdown for the benchmarks is shown in Figure 3. The store overhead is the single largest source of the total instrumentation overhead: 61 (34) percent for optimized (non-optimized) code. The single-WPC checking code (*st-swpc*) reduces this overhead to 57 (16) percent. Double-WPC checking code (*st-dwpc*) further reduces the store overhead to 36 (11) percent. As expected, the reduction is most significant for `lu-c` and `lu-nc` because they have the highest WPC hit rate, see Figure 2, and low shared load/store ratio [8]. `fft` and `cholesky` perform much better when a 2-entry WPC is used. For `radix`, the instrumentation overhead slightly increases for the *st-swpc* and *st-dwpc* implementations. The low WPC hit rate (see Figure 2) is directly reflected in this particular instrumentation breakdown.

Finally, the "perfect" WPC checking code (*st-wpc-hr1*) (a single-WPC snippet modified to always hit in the WPC) demonstrates very low instrumentation overheads: 9 percent for optimized and 3 percent for non-optimized code.

To summarize, we have seen that applications with low load/store ratio tend to have high store related instrumentation overhead. This store related overhead could be significantly reduced if the application has a high WPC hit rate and one or two WPC entries are used. On the other hand, if an application does not have a high WPC hit rate nor a low load/store ratio the ordinary store snippet might be a better alternative.

5.3 Parallel Performance

In this section, the parallel performance of two WPC-based DSZOOM systems is studied. Figure 4 shows normalized execution time for Sun Enterprise E6000 (SMP), 2-node Sun WildFire (HW-DSM) and three DSZOOM configurations:

1. DSZOOM-base: the original DSZOOM implementation.
2. DSZOOM-swpc: the DSZOOM implementation with a 1-entry WPC.
3. DSZOOM-dwpc: the DSZOOM implementation with a 2-entry WPC.

All DSZOOM configurations use a coherence unit size of 64 bytes. Both the HW-DSM configuration and the DSZOOM configurations run on two nodes and with eight processors per node (16 in total). The SMP configuration run on a single E6000 node with 16 processors and is used as an upper bound. The WPC technique improves the parallel DSZOOM performance with 7 (11) percent for benchmarks compiled with maximum (minimum) compiler optimization levels. The performance gap between the hardware-based DSM and the DSZOOM system is reduced with 14 (31) percent. Thus, the DSZOOM slowdown is in the range of 77 (40) percent compared to an expensive hardware implementation of shared memory, both running optimized (non-optimized) applications.

6 Related Work

The Check-In/Check-Out (CICO) cooperative shared memory implementation presented by Hill et. al. [14] uses similar ideas as the WPC technique. CICO

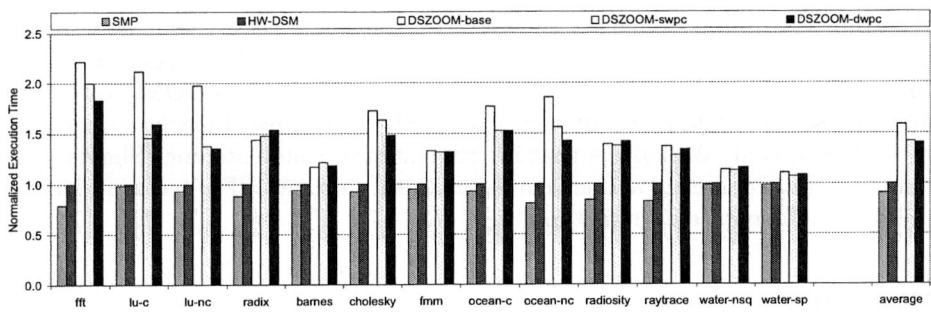

(a) 16-processor runs and non-optimized binaries.

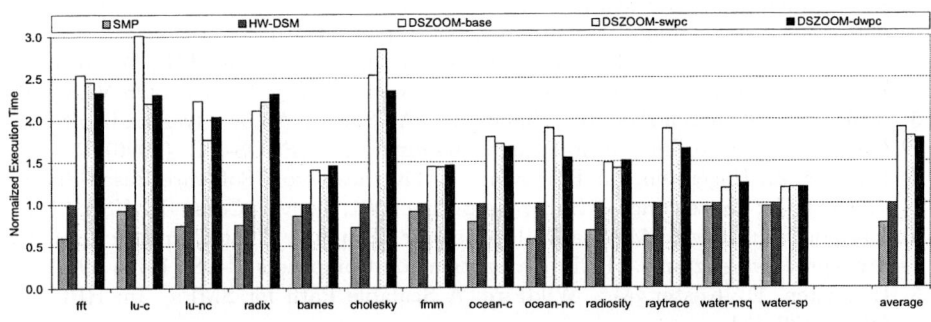

(b) 16-processor runs and fully optimized binaries.

Fig. 4. Parallel performance for 16-processor configurations.

is a programming model where a programmer can reason about access time to memory and give simple hardware coherence protocol performance hints. A *check out* annotation marks the expected first use and a *check in* annotation terminates the expected use of the data. Whereas CICO annotations are inserted as hints, a WPC entry actually "checks out" write permission since the directory/MTAG lock is not released until the next synchronization point or the next WPC miss.

Shasta [3] uses *batching* of miss checks, that is a "static merge" of coherence actions at instrumentation time. For a sequence of shared loads and stores, that touches the same coherence unit, the Shasta system combines/eliminates some of the access control checks (if possible). This way, all of the loads and stores in this sequence can proceed with only one check. The current WPC implementation works as a dynamic version of Shasta's batching technique.

7 Conclusions

In this paper, we introduce and evaluate a new *write permission cache* (WPC) technique that exploits spatial store locality. We demonstrate that the instrumentation overhead of the fine-grained software DSM system, DSZOOM [4], can be reduced with both 1- and 2-entry WPC implementations. On average, the original store instrumentation overhead, the single largest source of the total intrumentation cost, is reduced with 42 (67) percent for highly optimized (non-optimized) code. The parallel performance of the DSZOOM system for 16-processor runs (2-node configuration) of SPLASH-2 benchmarks is increased by 7 (11) percent. We believe that instrumentation-time batching (Shasta's approach [3]) of coherence actions combined with our new WPC technique might improve performance even further.

References

1. Li, K.: Shared Virtual Memory on Loosely Coupled Multiprocessors. PhD thesis, Department of Computer Science, Yale University (1986)
2. Schoinas, I., Falsafi, B., Lebeck, A.R., Reinhardt, S.K., Larus, J.R., Wood, D.A.: Fine-grain Access Control for Distributed Shared Memory. In: ASPLOS-VI. (1994)
3. Scales, D.J., Gharachorloo, K., Thekkath, C.A.: Shasta: A Low-Overhead Software-Only Approach to Fine-Grain Shared Memory. In: ASPLOS-VII. (1996) 174–185
4. Radović, Z., Hagersten, E.: Removing the Overhead from Software-Based Shared Memory. In: Proceedings of Supercomputing 2001, Denver, Colorado, USA (2001)
5. Lamport, L.: How to Make a Multiprocessor Computer That Correctly Executes Multiprocess Programs. IEEE Transactions on Computers **C-28** (1979) 690–691
6. Hagersten, E., Koster, M.: WildFire: A Scalable Path for SMPs. In: HPCA-5. (1999) 172–181
7. Zeffer, H., Radović, Z., Grenholm, O., Hagersten, E.: Evaluation, Implementation and Performance of Write Permission Caching in the DSZOOM System. Technical Report 2004-005, Dept. of Information Technology, Uppsala University (2004)
8. Woo, S.C., Ohara, M., Torrie, E., Singh, J.P., Gupta, A.: The SPLASH-2 Programs: Characterization and Methodological Considerations. In: ISCA-22. (1995) 24–36
9. Singhal, A., Broniarczyk, D., Cerauskis, F., Price, J., Yuan, L., Cheng, C., Doblar, D., Fosth, S., Agarwal, N., Harvey, K., Hagersten, E., Liencres, B.: Gigaplane: A High Performance Bus for Large SMPs. In: IEEE Hot Interconnects IV. (1996)
10. McVoy, L.W., Staelin, C.: lmbench: Portable Tools for Performance Analysis. In: Proceedings of the 1996 USENIX Annual Technical Conference. (1996) 279–294
11. Hennessy, J.L., Patterson, D.A.: Computer Architecture: A Quantitative Approach. 3rd edn. Morgan Kaufmann (2003)
12. Mukherjee, S.S., Reinhardt, S.K., Falsafi, B., Litzkow, M., Huss-Lederman, S., Hill, M.D., Larus, J.R., Wood, D.A.: Fast and Portable Parallel Architecture Simulators: Wisconsin Wind Tunnel II. IEEE Concurrency (2000)
13. Covington, R.C., Madala, S., Mehta, V., Jump, J.R., Sinclair, J.B.: The Rice Parallel Processing Testbed. In: ACM SIGMETRICS Conference on Measurement and Modeling of Computer Systems, ACM Press (1988)
14. Hill, M.D., Larus, J.R., Reinhardt, S.K., Wood, D.A.: Cooperative Shared Memory: Software and Hardware for Scalable Multiprocessor. In: ASPLOS-V. (1992)

CacheFlow: A Short-Term Optimal Cache Management Policy for Data Driven Multithreading

Costas Kyriacou[1], Paraskevas Evripidou[2], and Pedro Trancoso[2]

[1] Computer Engineering Depart., Frederick Institute of Technology, Cyprus
[2] Department of Computer Science, University of Cyprus, Cyprus
{cskyriac,skevos,pedro}@cs.ucy.ac.cy

Abstract. With Data Driven Multithreading a thread is scheduled for execution only if all of its inputs have been produced and placed in the processor's local memory. Scheduling based on data availability may be used to exploit short-term optimal cache management policies. Such policies include firing a thread for execution only if its code and data are already placed in the cache. Furthermore, blocks associated to threads scheduled for execution in the near future, are not replaced until the thread starts its execution. We call this short-term optimal cache management policy the CacheFlow policy.
Simulation results, on a 32-node system with CacheFlow, for eight scientific applications, have shown a significant reduction in the cache miss ratio. This results in an average speedup improvement of 18% when the basic prefetch CacheFlow policy is used, compared to the baseline data driven multithreading policy. This paper also presents two techniques to further improve the performance of CacheFlow: conflict avoidance and thread reordering. The results have shown an average speedup improvement of 26% and 31% for these two techniques, respectively.

1 Introduction

Multithreading is one of the main techniques employed for tolerating latency [1, 2]. In multithreading, a thread suspends its execution whenever a long latency event is encountered. In such a case, the processor switches to another thread ready for execution. This form of multithreading is usually referred as blocking multithreading [2]. Another form of multithreading is non-blocking multithreading. In this case, a thread is scheduled for execution only if all of its input values are available in the local memory, thus no synchronization nor communication latencies will be experienced.

Data driven multithreading (DDM) is a non-blocking multithreading model of execution evolved from the dataflow model of computation [3]. An implementation of a data driven multithreaded architecture is the Data Driven Network of Workstations (D^2NOW) [4]. D^2NOW utilizes conventional control-flow workstations, augmented with an add-on card called the Thread Synchronization Unit (TSU). The TSU supports data driven scheduling of threads.

A program in DDM is a collection of code blocks called the context blocks. A context block is equivalent to a function. Each context block comprises of several threads. A thread is a sequence of instructions equivalent to a basic block. A producer/consumer relationship exists among threads. In a typical program, a set of threads create data, the producers, which is used by other threads, the consumers. Scheduling of threads is done dynamically at run time by the TSU, based on data availability.

Data driven scheduling leads to irregular memory access patterns that affect negatively cache performance. This is due to the fact that threads are scheduled for execution based only on data availability without taking into account temporal or spatial locality. On the other hand, data driven scheduling allows for optimal cache management policies, by ensuring that the required data is prefetched into the cache, before a thread is fired for execution. Furthermore, we can ensure that data preloaded in the cache is not replaced before the corresponding thread is executed, thus reducing possible cache conflicts. We call this cache management policy the CacheFlow policy.

In this paper we examine three variations of the CacheFlow policy. In the first implementation we prefetch into the cache the data of the threads scheduled for execution in the near future. These threads are then placed in a firing queue and wait for their turn to be executed. We call this the *Basic Prefetch CacheFlow*. In the second implementation, called *CacheFlow with Conflict Avoidance*, we maintain a list of all addresses of the data prefetched for the threads in the firing queue, and make sure that this data is not evicted from the cache until the corresponding threads are executed. In the third implementation, called *CacheFlow with Thread Reordering*, we reorder the sequence of executable threads, before they enter the firing queue, in order to exploit spatial locality.

An execution driven simulator is used to evaluate the potential of the CacheFlow policy in reducing cache misses. The workload used on these experiments consisted of eight scientific applications, six of which belong to the Splash-2 suite [5]. Simulation results have shown a significant reduction in the cache miss ratio. This results in a speedup improvement ranging from 10% to 25% (average 18%) when the *Basic Prefetch CacheFlow* policy is used. A larger increase (14% to 34% with a 26% average) is observed when the *CacheFlow with Conflict Avoidance* is used. A further improvement (18% to 39% with a 31% average) is observed when the *CacheFlow with Thread Reordering* is employed.

2 Related Work

A variety of techniques such as data forwarding [6, 7], and prefetching [8, 9], have been proposed to tolerate the long memory access latency. With data forwarding a producer processor forwards data to the cache of consumer processors as soon as it generates it. The main drawback of data forwarding is that it may displace useful data from the consumer's cache. Our implementation of data driven multithreading model of execution employs the data forwarding concept in the sense that a producer node is responsible for forwarding remote data as soon as it is produced to the consumer node, and that it employs only remote write operations. The difference in our approach is that data is forwarded to the consumer's main memory, not to the cache, avoiding the possibility of displacing useful data from the cache.

Data prefetching reduces cache misses by preloading data into the cache before it is accessed by the processor. A review on prefetching is presented by Vanderwiel and Lilja [10]. Data prefetching can be classified as hardware prefetching [11, 12], software prefetching [13, 14], or thread based prefetching [15–17]. Thread based prefetching is employed in multithreaded processors [1, 18] to execute a thread in another context that prefetches the data into the cache before it is accessed by the computation thread. CacheFlow employs compiler-assisted hardware prefetching mechanisms. The difference between CacheFlow and other hardware prefetchers is that most of the other prefetchers attempt to predict possible cache misses based on earlier misses, while in CacheFlow the addresses of the data needed by a thread scheduled for execution is either specified at compile time or it is determined at run time when the thread becomes ready for execution. CacheFlow has the advantages of both software and hardware prefetching. In addition, it avoids unnecessary prefetching that would lead to extra bus traffic and cache pollution.

3 The CacheFlow Policy

One of the main goals of Data Driven Multithreading is to tolerate latency by allowing the computation processor do useful work while a long latency event is in progress. This is achieved by scheduling threads based on data availability. An argument against data driven multithreading is that it does not fully exploit locality, since threads are scheduled for execution based only on data availability. Scheduling based on data availability, on the other hand, allows the implementation of efficient short-term optimal cache management. This paper focuses on the implementation of these policies which we named CacheFlow.

The implementation of the CacheFlow policy is directly related to the Thread Issue Unit (TIU), a unit within the TSU [4] responsible to schedule threads which are ready for execution. The other two units of the TSU are the Post Processing Unit (PPU) and the Network Interface Unit (NIU). Each thread is associated with a synchronization parameter, called the *Ready Count*. The PPU updates the *Ready Count* of the consumers of the completed threads, and determines which are ready for execution. The NIU is responsible for the communication between the TSU and the interconnection network. In this paper we present four implementations of the TIU which are described in the following sections.

3.1 TIU with No CacheFlow

The TIU without CacheFlow support is a simplified version of the TIU depicted in Figure 1. It consists only of the Waiting Queue (WQ), the Firing Queue (FQ) and the IFP part of the Graph Cache. The Graph Cache serves as a look-up

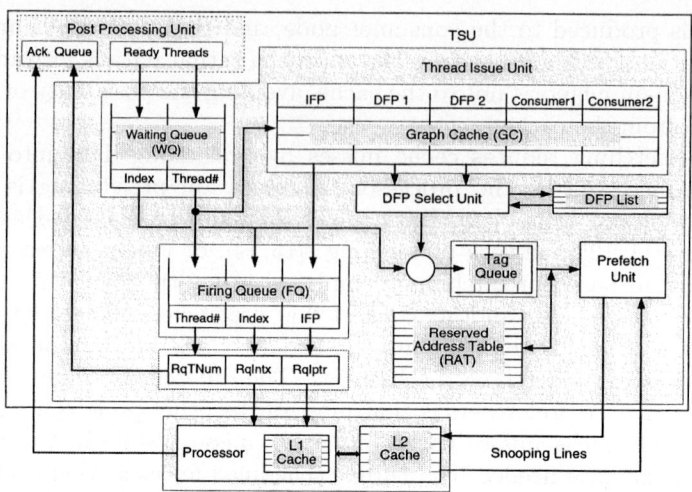

Fig. 1. The TIU that supports CacheFlow with Conflict Avoidance.

table. When a thread is deemed executable by the Post Processing Unit (PPU), its identification number (Thread#) and index are placed in the WQ. The TIU uses the thread identification number to determine the threads starting address by reading its IFP from the Graph Cache. After the thread's IFP is determined, the triplet Thread#, Index and IFP are shifted into the FQ and the thread waits for its turn to be executed. The last instructions in each thread read the starting address (IFP) of the next ready thread from the FQ and branch to that address.

3.2 TIU with Basic Prefetch CacheFlow Policy

To implement the CacheFlow policy with basic prefetch, two extra fields are added in the Graph Cache. These fields are determined at compile time and loaded in the Graph Cache at run time as Data Frame Pointer 1 (DFP1) and Data Frame Pointer 2 (DFP2). If a thread has only one input, then DFP1 contains the memory pointer to that value, while DFP2 is set to 0. If a thread has more than two inputs, then DFP1 is set to 0 while DFP2 is a pointer to the DFP list, a memory block within the TSU that contains a list of DFPs. Note that a thread input corresponds to an arrow in a dataflow graph. A thread input may be a single variable or a contiguous memory block that fits into one cache block.

When a thread becomes executable, i.e. all of its inputs have been produced, the Post Processing Unit (PPU) places the Thread# and index in the WQ. A thread is processed by first reading the Thread# and index from the WQ. The Thread# is used as a pointer to the Graph Cache that gives the IFP, DFP1 and DFP2 of the thread. The address of the data needed by the thread is obtained using the DFP with the index, thus the exact address of the data is determined dynamically, at run time. The Thread#, index and IFP are then shifted in the FQ and the thread waits for its turn to be executed. As soon as the addresses of the data needed by the thread are determined, they are sent to the Prefetch

Unit that snoops the processor to verify whether these addresses are already in the cache. If the required data is not in the cache, then a prefetch request is issued.

3.3 TIU with Conflict Avoidance (Optimization 1)

One disadvantage of the basic prefetch CacheFlow policy is that excessive traffic is placed on the processor's bus and snooping lines. Another disadvantage is that prefetching can cause cache conflicts, i.e. it is possible that a cache block required by a thread waiting in the FQ is replaced by another block, before the thread is executed. We call these conflicts *false cache conflicts* as they originate from the policy and not from the execution of the code. The possibility of false cache conflicts is reduced by keeping the size of the FQ as small as possible. A small FQ, on the other hand, increases the possibility that a thread is fired before its is prefetched. This becomes more critical for threads with a small number of instructions.

The TIU with *Conflict Avoidance* prevents the Prefetch Unit from replacing cache blocks required by the threads waiting in the FQ. This is achieved with the use of the Reserved Address Table (RAT) that contains the addresses of all cache blocks prefetched for the threads waiting in the FQ, as well as the thread currently running. All addresses required by a ready thread, removed from the WQ, are determined using the information from the Graph Cache, and placed in the Tag Queue. These addresses are then compared with the contents of the RAT to determine if prefetching would cause a cache conflict. A thread is shifted in the FQ if none of its addresses would result in a cache conflict. If it is detected that an address would result in a false cache conflict, then the tested thread is placed temporarily in a buffer, and the next thread from the WQ is tested. Threads waiting in the temporary buffer have precedence over the threads in the WQ. This is essential to avoid thread starvation, as a thread waiting in the temporary buffer is blocking its consumers from executing.

3.4 TIU with Thread Reordering (Optimization 2)

Both previous CacheFlow implementations address only the improvement of data locality. To exploit temporal code locality we have included a reordering mechanism that reorders the threads in the WQ. Threads with the same identification number are placed near each other in the WQ, increasing the probability that the code of a thread will be used many times before it is replaced from the cache. Furthermore, threads with the same identification number (Thread#), are ordered according to their index (iteration number), thus exploiting spatial data locality. Reordering reduces further snooping overheads on the processor and the bus. The thread reordering mechanism operates in parallel and asynchronously with the rest of the TIU and thus it does not add any extra delays in the datapath of the TIU.

The concept of thread reordering is depicted in Figure 2. Whenever a new thread becomes ready by the PPU, its Thread# is compared with the Thread#

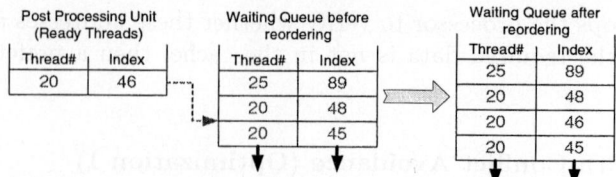

Fig. 2. Example of the WQ with the thread reorder Cache-Flow policy.

of the threads in the WQ. If a match is not found, the new thread is appended at the end of the WQ, otherwise the index of the new thread is compared with the index of the threads in the WQ with the same Thread#. The new thread is inserted in the WQ so that the index of the threads with the same Thread# appear in ascending order.

4 Evaluation Methodology

In order to evaluate the ability of the proposed cache management policy in reducing cache misses, we have built an execution driven simulator that uses native execution [19]. Both the host and the target processor is an Intel Pentium processor with a 256K L2 unified cache, and 16K L1 data and instruction caches. All caches are 4-way set associative with a 32-byte line size. Simulations were carried out for distributed shared virtual memory systems with 2, 4, 8, 16 and 32 processors. A clock cycle counter is maintained for each CPU, TSU unit and the interconnection network. The simulator uses the timings produced by the actual implementation of the TSU [4]. The time needed to execute each thread is obtained using the processor's time stamp performance counter [20]. Calls to functions that simulate the TSU and the interconnection network are interleaved with the execution of threads on the host processor, according to the clock cycle counter of each unit.

The cache miss rate is obtained using the processor's performance monitoring counters [20]. Since the machine used as the host is also the target, the state of the cache of the simulated application is affected by the simulation process. Therefore, the simulator performs extra operations to recover the system to the same state as it was before executing the simulation code.

Eight scientific applications are used to evaluate the three variations of the CacheFlow cache management policy. Six of these applications, *LU*, *FFT*, *Radix*, *Barnes*, *FFM* and *Cholesky* belong to the Splash-2 suite [5]. These applications have been modified to support data driven execution. The partition of the code into threads has been done manually. The creation of the data driven graph is done automatically by the simulator. The other two applications, *Mult* and *Trapez* represent standard algorithms used in large scientific applications such as the block matrix multiplication and the trapezoidal method of integration, respectively. To examine the effect of problem size on the effectiveness of the CacheFlow management we have used, for certain applications, two problem sizes: Data Size 1 (corresponding to 64K matrices) and 2 (corresponding to 1M matrices).

5 Results

5.1 Effect of Data Driven Sequencing on Miss Rate

Table 1 depicts the L2 cache miss rate for the sequential single threaded execution on a single processor and the different DDM configurations with CacheFlow on a 32-node system. Note that in order to avoid misleading results, for the measurement of the cache miss rate, we have scaled down the data size of the sequential single threaded execution to match the data size for each of the nodes in the 32-node DDM system, i.e. the data sizes used for the sequential single thread execution are the same as those used by each node in the data driven multithreaded execution. As expected, the baseline DDM configuration shows a higher miss rate than the sequential (increase from 7.1% to 9.8%), which corresponds to a 38% increase for the average of all applications. This reflects the loss of locality for both the code and data. The *Basic Prefetch CacheFlow* implementation reduces the miss rate from 9.8% to 3.2% (68% decrease compared to the baseline DDM). It is important to notice that the reduction achieved by the *Basic Prefetch CacheFlow* results in miss rate values lower than the original sequential execution. The use of the two CacheFlow optimizations results in further reductions on the miss rate, which becomes 1.9% and 1.4% respectively.

Table 1. Cache miss rate for Data Size 1.

Application	Miss Rate (Data Size 1 - DS1)				
	Sequential	DDM without Cache-Flow	Basic Prefetch (Buffer =16)	Optimization 1	Optimization 1 & 2
Mult	5.4%	7.5%	2.1%	1.4%	1.1%
Trapez	2.8%	3.6%	1.8%	1.3%	1.0%
LU	4.8%	6.1%	1.7%	0.9%	0.8%
FFT	7.5%	10.1%	3.0%	2.0%	1.2%
Barnes	7.4%	10.1%	2.6%	1.1%	0.8%
Radix	14.0%	18.1%	6.7%	4.1%	3.3%
FMM	8.3%	11.8%	3.2%	2.1%	1.6%
Cholesky	6.2%	11.3%	4.2%	2.4%	1.8%
Average Miss Rate	7.1%	9.8%	3.2%	1.9%	1.4%

5.2 Effect of Data Size on Miss Rate

The effect of problem size on the cache miss rate is presented in Table 2. By increasing the problem size by a factor of 16, the average cache miss rate for the sequential execution is increased from 7.9% to 9.7% (23% percentage increase). This increase is justified by the fact that the working set for the large problem size does not fit in the cache, resulting in more cache misses. The cache miss rate increase for the DDM execution without the CacheFlow management, is increased from 10.4% to 12.0% (16% percentage increase). The increase of the miss rate is significantly reduced when the different CacheFlow policies are used, (7%, 5% and 8% percentage increase respectively). This shows that CacheFlow is efficient in keeping the miss rate low independently of the problem size.

5.3 Effect of Firing Queue Size on Performance

Prefetching must be completed early enough to ensure that data is prefetched before the thread using that data is fired for execution. Nevertheless, prefetching must not be initiated too early, to avoid replacing cache blocks already prefetched by threads waiting in the Firing Queue (FQ). The effect, of the size of the FQ, for *Radix*, on the cache miss rate and the false conflicts when the *Basic Prefetch CacheFlow* is employed is depicted in Figure 3-(a). For these results a thread is shifted into the FQ as soon as the prefetching operation is initiated. The cache miss rate is higher when the FQ size is small. This is due to the fact that a thread might be fired before the prefetching is completed, resulting in cache misses. As the size of the FQ increases, there is more time for the prefetch unit to complete the prefetching operation, since the processor will execute other threads. Increasing the size of the FQ, increases also the number of false cache conflicts, resulting in more cache misses. The rest of the applications behave in a similar way. For all applications the minimum cache miss ratio is obtained when the FQ size is 16.

Table 2. Cache miss rate for Data Size 2.

Application	Miss Rate (Data Size 2 - DS2)				
	Sequential	DDM without Cache-Flow	Basic Prefetch (Buffer =16)	Optimization 1	Optimization 1 & 2
Mult	7.1%	9.2%	2.2%	1.5%	1.2%
LU	6.5%	7.4%	1.8%	1.0%	0.8%
FFT	9.1%	11.4%	3.1%	2.0%	1.3%
Radix	16.1%	19.8%	7.2%	4.3%	3.4%
Average Miss Rate	9.7%	12.0%	3.6%	2.2%	1.7%
Change (from DS1)	23%	16%	7%	5%	8%

To reduce the cache miss rate when the FQ is small, we have changed the FQ shifting policy. A thread is shifted only after prefetching is completed. This change affects also the CPU idle time. The effect, of the size of the Firing Queue (FQ), for *Radix*, on the CPU idle time when the *Basic Prefetch CacheFlow* is employed is depicted in Figure 3-(b). Measurements of the CPU idle time show that there is a significant increase in the CPU idle time when the FQ is too small. For all applications the CPU idle time when the FQ size is over 16 is the same as the idle time obtained without CacheFlow.

5.4 Effect of CacheFlow on Speedup

Figure 4 shows the effect of the three CacheFlow implementations on speedup, compared to sequential execution, for machine sizes ranging from 2 to 32 processors. A speedup improvement ranging form 10% to 25% (average 18%) is obtained when the it Basic Prefetch CacheFlow policy is used on a 32-processor system. A bigger increase (14% to 34% with a 26% average) is observed when the CacheFlow with *Conflict Avoidance* is used. A further improvement (18% to 39% with a 31% average) is observed when the *Thread Reordering* is employed.

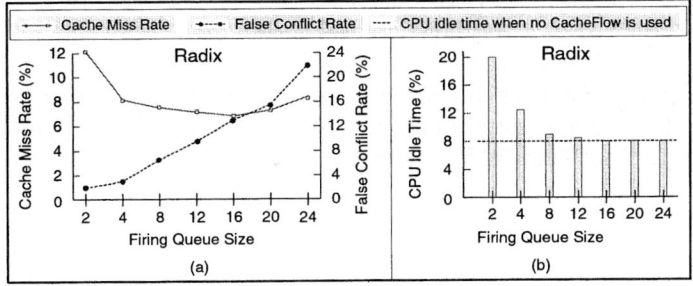

Fig. 3. FQ size effect on cache miss rate, false cache conflicts and CPU idle time.

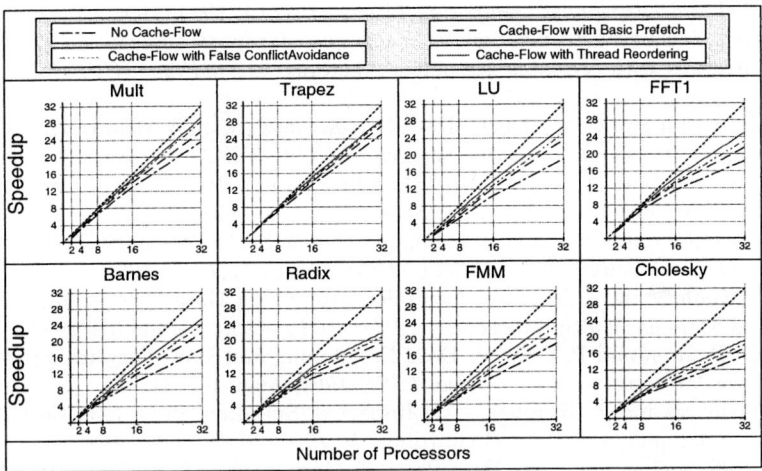

Fig. 4. Effect of CacheFlow and number of processors on speedup.

6 Conclusions and Future Work

Data Driven Multithreading is proposed as an execution model that can tolerate communication and synchronization latency. Nevertheless, data driven sequencing has a negative effect on performance due to loss of locality on the data access. In this paper we presented CacheFlow, a cache management policy that significantly reduces cache misses by employing prefetching. To avoid false cache conflicts and further exploit locality we proposed two optimizations: *Conflict Avoidance* and *Thread Reordering*.

Simulation results based on an execution driven simulator that runs directly on the host processor as well as measurements obtained from the developed hardware show that CacheFlow effectively reduces the miss rate. The basic prefetch implementation resulted in an average reduction in the cache miss rate of 67%, while the two optimizations resulted in further reductions: 81% and 86%, respectively. These reductions resulted in a speedup improvement on a 32-processor system of 18%, 26% and 31% respectively. An increase in the problem size by a factor of 16 resulted in a very low increase in the cache miss ratio (7%, 5% and 8% respectively). Overall the results show that CacheFlow is an effective technique in tolerating memory latency, and an important enhancement for the data driven multithreading system. In the future we plan on extending the CacheFlow policy to implement it on SMT systems.

References

1. S. Eggers et al. Simultaneous multithreading: A platform for next generation processors. *IEEE Micro*, pages 12–18, September/October 1997.
2. J. Sile, et al. Asynchrony in Parallel Computing: From Dataflow to Multithreading. *Parallel and Distributed Computing Practices*, Vol.1, No.1, 1998.
3. P. Evripidou and J-L. Gaudiot. A Decoupled Graph/Computation Data-Driven Architecture with Variable Resolution Actors. In *In Proc. ICPP-1990*, August 1990.
4. P. Evripidou and C. Kyriacou. Data Driven Network of Workstations (D^2NOW), *Journal of Universal Computer Science (J.UCS)*, Volume 6/ Issue 10, Oct. 2000.
5. S. Cameron Woo et al. The SPLASH-2 Programs: Characterization and Methodological Considerations, *In Proc. ISCA-1995*, Jun. 1995.
6. D. A. Koufaty et al. Data Forwarding in Scalable Shared Memory Multiprocessors, *In IEEE Transactions on Parallel and Distributed Systems*, 1996.
7. D. A. Koufaty and J Torrellas. Compiler Support for Data Forwarding in Scalable Shared Memory Multiprocessors, *In Proc. ICPP-1999*, September 1999.
8. D. K. Poulsen and P. C. Yew. Data Prefetching and Data Forwarding in Shared Memory Multiprocessors, *In Proc. ICPP-1994*, August 1994.
9. P. Trancoso and J. Torrellas, The Impact of Speeding up Critical Sections with Data Prefetching and Data Forwarding, *In Proc. ICPP-1996*, 1996.
10. S. Vanderwiel and D. Lilja Data Prefetch Mechanisms, *ACM Computing Surveys*, Vol. 32, No. 2, June 2000.
11. T. Sherwood et al. Predictor-directed stream buffers, *In 33rd International Symposium on Microarchitecture*, Dec. 2000.
12. J. Collins et al. Pointer cache assisted prefetching, *In Proc. MICRO-35*, Nov. 2002.
13. T.C. Mowry et al. Design and evaluation of a compiler algorithm for prefetching, *In Proc. ASPLOS-V*, Oct. 1992.
14. C.K. Luk and T.C. Mowry, Compiler based prefetching for recursive data structures, *In Proc. ASPLOS-VII*, Oct. 1996.
15. J. Collins et al. Dynamic Speculative Precomputation, *In Proc. MICRO-34*, Dec. 2001.
16. A. Roth and G. Sohi. Speculative data-driven multithreading, *In Proc. HPCA-7*, Jan. 2001.
17. Y. Solihin et al. Using User-Level Memory Thread for Correlation Prefetching, *In Proc. ISCA-2002*, May 2002.
18. C.K. Luk. Tolerating memory latency through software-controlled pre-execution in simultaneous multithreading processors, *In Proc. ISCA-2001*, June. 2001.
19. C. Kyriacou et al. DDM-SIM: An Execution Driven Simulator for Data Driven Multithreading. Technical report, University of Cyprus, 2004.
20. Intel Corp. IA-32 Intel Architecture: Software Developer's Manual. Volume 3: System Programming Guide, *Intel, 2003*.

SCISM vs IA-64 Tagging: Differences/Code Density Effects

Georgi Gaydadjiev and Stamatis Vassiliadis

Computer Engineering Lab, EEMCS, TU Delft, The Netherlands
{G.N.Gaydadjiev,S.Vassiliadis}@EWI.TUDelft.NL
http://ce.et.tudelft.nl/

Abstract. In this paper we first present two tagging mechanisms; the SCISM and IA-64; thereafter we describe the mapping of IA-64 ISA to a SCISM configuration without changing or reassigning the IA-64 instructions to preserve the original architectural properties. Under this limiting SCISM scenario, opcode reassignment will improve even more the SCISM performance, it is shown that SCISM tagging will significantly improve (between 21 and 29%) static code density. The results are based on analysis of various SPECINT2000 executables.

Keywords: Instruction Tagging, Instruction Level Parallelism, SCISM, IA-64.

1 Introduction

Tagging has been used extensively by microarchitects and designers as an efficient mechanism to facilitate implementation and potentially improve the performance of processors. Tagging for example has been used to enumerate and manage the hardware resources [1], to handle interrupts, e.g. [2], speculative execution (see for example [3], and to facilitate concurrent instructions routing see for example [4]. Instruction tagging for instruction level parallelism has been introduced for two main reasons namely: to reduce the complexity (and the cycle time), mostly the decode stage, of a pipelined machine implementation, and to potentially improve instruction level parallelism (ILP).

In this paper we analyze and compare the two ILP tagging mechanisms, SCISM [5] (the first known machine organization that employs tagging with the mentioned ILP characteristics) and IA-64 [6], [7] tagging mechanisms and provide evidence suggesting that the SCISM tagging provides some benefits when compared to IA-64 tagging. In particular we investigate and show the following: We provide evidence indicating that the SCISM tagging is a superset of the IA-64. We consider the side effects of tagging on code densities and show that the SCISM tagging of IA-64 instructions regarding code densities is clearly superior to the IA-64 for static code. In particular it is shown that the SCISM tagging reduces the IA-64 code size for SPECINT2000 benchmarks between 21% and 29%.

The paper is organized as follows. Section 2 gives a short description of IA-64 architecture[1] with the main focus on the *template* bit field role. In addition, the SCISM organization is described with emphasis on tags and their functionality. Section 3 maps

---

[1] We note that in this paper we use the original definition of the term of architecture as described by [8].

the IA-64 instruction set architecture to SCISM and shows how the original IA-64 aspects are preserved. In the same section the results concerning the static code density are discussed and the discussion is concluded.

2 The IA-64 and SCISM Tagging

IA-64 uses *bundles* as its compound instruction format. A bundle consists of three instruction slots and a template field. Each bundle in IA-64 is 128-bits long. Figure 1(a) shows the bundle's format. I_0, I_1 and I_2 represent the three instructions (41-bits each), while *template* (tag) is a 5-bit wide field. The template information is used for decoding, routing (dispersal) and ILP. Instruction groups can be seen as chained bundles in the absence of stops. The boundaries between instruction groups correspond directly to the *instruction level parallelism* (ILP) in a particular IA-64 implementation. IA-64 has five instruction slot types corresponding to the different execution unit types - Memory (M), Integer (I), Floating-point (F), Branch (B) and Long (extended) (L+X), or in shorthand (M, I, F, B, and L). IA-64 instructions are divided among six different instruction types - ALU (A), Memory (M), Integer (I), Floating-point (F), Branch (B) and Long (extended) (L+X), abbreviated as (A, M, I, F, B, and L). An interesting detail is that instruction of A-type, e.g. integer *add*, can be scheduled to either I or M execution unit. The L+X type uses two instruction slots and executes on I-unit or on B-unit. Due to the limited number of bits not all instruction triples are supported. There are 12 basic template types (each with two versions with stop on the bundle end or not): MII, MI_I, MLX, MMI, M_MI, MFI, MMF, MIB, MBB, BBB, MMB and MFB, where "_" (underscore) indicates a stop inside the bundle (not at the bundle boundaries).

Fig. 1. IA-64 and SCISM bundle formats.

In SCISM, instructions are *categorized* according to hardware utilization not op-code description. An obvious implication of this is that the number of rules needed to determine parallel execution depends on the number of categories, rather than on the number of individual instructions. Given that a category comprises of multiple instructions used by a single hardwired unit, the differences among category members are considered as "trivial" and are resolved by the hardware by means of some control signal or by minor hardware modifications. For example in an implementation [9] a set of fourteen IBM 370 ISA [10] operations is presented that belong to a single category and are executed

OPERATION	ALU Function	Operand Representation	Requires
Load Complement (LCR)	32-b signed addition	Two's complement	Adder
Load Positive (LPR)	32-b signed addition	Two's complement	Adder
Load Negative (LNR)	32-b signed addition	Two's complement	Adder
Load Register (LR)	32-b signed addition	Two's complement	Adder
Load and Test (LTR)	32-b signed addition	Two's complement	Adder
AND (NR)	bitwise logical AND	Binary	Logical
OR (OR)	bitwise logical OR	Binary	Logical
EXCLUSIVE-OR (XR)	bitwise logical EX-OR	Binary	Logical
Add (AR)	32-b signed addition	Two's complement	Adder
Subtract (SR)	32-b signed addition	Two's complement	Adder
Add Logical (ALR)	32-b unsigned addition	Unsigned Binary	Adder
Subtract Logical (SLR)	32-b unsigned addition	Unsigned Binary	Adder
Compare Logical (CLR)	32-b unsigned addition	Unsigned Binary	Adder
Compare (CR)	32-b signed addition	Two's complement	Adder

Fig. 2. RR-Format Loads, Logicals, Arithmetics and Compares operations [9].

by the same hardware (ALU) (see Figure 2). Another categorizations are obviously possible. Two different tagging mechanisms have been reported [11], [5], [12]. The first mechanism [5] requires $\lceil log_2(n) \rceil$ bits, with n being the number of instructions to be executed in parallel. The second (original) mechanism requires only one additional bit as depicted in Figure 1(b) for an example 4 instructions wide parallel machine, with I_1, I_2, I_3 and I_4 being the original instructions. In Figure 1(b), all instructions I_j are in their original form and $T \epsilon \{0,1\}$ represent the tags. The SCISM approach implies full binary compatibility, allowing straight-forward legacy code execution and parallelization. An interesting implicit property associated with the SCISM tagging is that it is allowed, contrary to the IA-64 tagging, to branch in the middle of a compound instruction allowing code compaction (complete elimination of *nops* and removing the need of branch alignment) [5]. Only tags are added to code thus if SCISM tagging is applied to existing code the original code remains unchanged. As a consequence there are no side effects such as branch target calculations.

3   Tagging Effects on Code Size

This section begins by showing how IA-64 instructions can be mapped onto SCISM without strict code mapping (no opcode space re-assignment). This straight-forward mapping is not an optimal approach for SCISM due to the shared major IA-64 opcodes, but is a quick way to demonstrate the SCISM potential and create a base for comparison. It should be taken into account that this is also the worst-case scenario with respect to SCISM when investigating binary code density. To transform IA-64 to SCISM code a three-way SCISM organization is assumed. This is to create an IA-64 bundle which corresponds to a SCISM compound instruction with a length of three. Please note that the discussion on code density differences is unrelated to any particular IA-64 implementation. To clarify this: the Itanium2 dispersal window (two bundles) corresponds to a six-way SCISM compound instruction leaving the code size differences between the two approaches unchanged. The SCISM organization by its definition is not restricted to certain number of instruction combinations (24 out of 32 possible when using 5 bits) while IA-64 is. The three-way SCISM compounding requires three tag bits for stop indication (see Section 2), leaving two out of five IA-64 template bits unused. On the other hand, the template removal will require additional information about the functional unit to be added to each individual instruction. In IA-64 all of the instructions are executed by one of the four execution units types: M, I, F or B. The two "remaining" IA-64 instruction types (A-type and X-type) are also executed by one of those execution units types (A-type by I or M and (X-type) by I or a B unit). This is why the additional bits are coupled to the designated functional units instead of the instruction types. This requires two additional bits for each basic instruction (or 6 for the total compound instruction). The SCISM instruction format for IA64 is depicted in Figure 3 (a). As stated earlier three single bit tags are needed to express the IPL (shown as T in the figure). In addition,

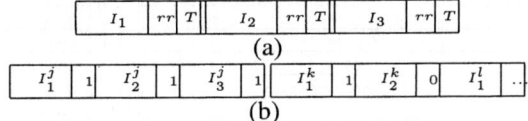

Fig. 3. Instruction format and bundle chaining.

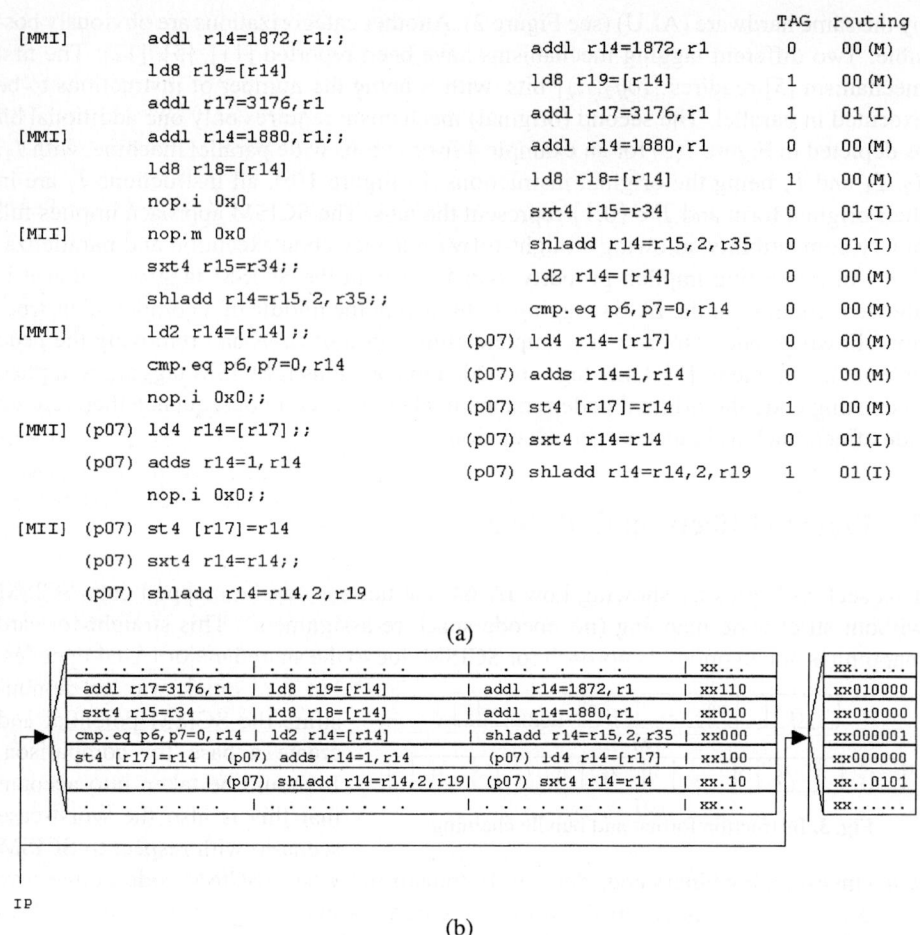

Fig. 4. IA-64 and SCISM code (*build_tree* function of gzip).

another two bits for routing (rr) per instruction are used to provide information about the targeted functional unit. Stated differently this is a 3-bit SCISM tagging [11]. The three 41-bit long IA-64 instructions are unmodified in their original form. Putting it together, SCISM instructions become 44-bit long (including tagging), hence the three way compound instruction will become 132 bits.

The template bits are not needed, since bits are added to indicate the position where a compounding is ending. When "wider" than the compound instruction implementation is used can be easily implemented in SCISM. Figure 3(b) shows an example of how 5 parallel instructions can be marked in the proposed 3-way SCISM organization, where I^j, I^k and I^l represent three subsequent SCISM compound instructions. To clarify the discussion above Figure 4(a) depicts a piece of the *build_tree* function code from the IA-64 binary of the *gzip* executable. In Figure 4(a) the left column shows the original IA-64 code, next column is the equivalent SCISM code, the TAG column represents the tagbits, and the routing information is shown in the last column. The encoding is for:

memory unit (M) = 00, and integer unit (I) = 01. Figure 4(b) shows a potential memory organization for IA-64 compounded instructions where the compound instructions are 136 bits long (the original 128 bit bundle plus an additional byte). Please note that this is one out of many implementations possible that can facilitate the proposed organization. The Template information of the original IA-64 bundle is replaced by the tags corresponding to the three instructions (filled with two don't care bits - 'x'), while the additional byte is addressed in parallel with the modified IA-64 bundle to access the routing information (along with the two additional spare bits). Furthermore, as in the case of S/370 example, except for the tagging, IA-64 instructions have not been modified preserving all the instruction properties. In order to find the improvement in code size we investigated the SPECINT2000 executables. Since the compiler optimizations may play significant role in the IA-64 approach, the effects of different compilers on code size where investigated in [13]. The differences on code sizes produced by different compilers, e.g. gcc and Intel where found marginal, so the results in this paper are independent on the compiler technology. The benchmarks where compiled using the CPU2000 default makefiles (optimization levels). The compilation was performed inside *Native User Environment* (NUE) developed at the Hewlett-Packard Labs [14]. This environment emulates a Linux/ia64 system, more precisely Itanium2, and was considered sufficient since the static binary code investigation was the primer concern. We considered only the code segments of the benchmark executables, leaving all other program segments out. The results are presented in Figure 5. The first column shows the total number of instructions involved, e.g. the complete code segment in instructions (not bundles). The second column in each pair shows the percentage of *nop* instructions found. It was found that for all of the benchmark executables approximately one third of the operations are *nop* operations. The 255.vortex benchmark was found as the one with the lowest *nop* count (26%), however this is an exception.

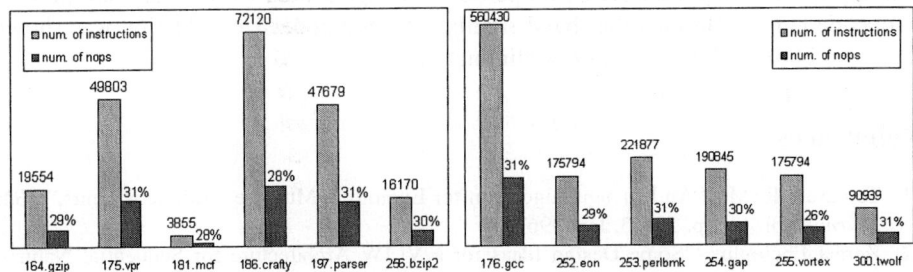

Fig. 5. IA-64 NOP utilization SPECINT2000.

The comparison on the code segment size between the IA-64 (Itanium2) and the SCISM is presented in Figure 6. The IA-64 results where estimated as follows: the total number of bundles (instead of instructions) was used, since the bundle size is predetermined to 128 bits it is a simple procedure to determine the code segment size in number of bits. In case of SCISM, the number of instructions was used and the worst-case instruction length of 44-bits was assumed. An even more restrictive scenario for SCISM

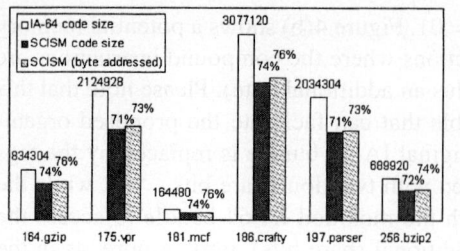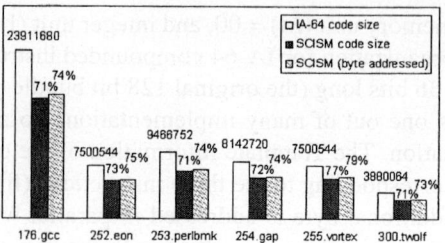

Fig. 6. IA-64 vs SCISM code size SPECINT2000.

assumes instructions to be byte rather than nibble addressed. This can be done with the addition of four more tag bits per bundle. Strictly speaking these four bits are not needed and can be used to improve performance and/or the hardware design. They will add however to the storage requirements. The results are presented in Figure 6, where the first column represents the number of bits (of the code segments) for the original benchmark executables. The second column shows how the code segment size when the SCISM approach is applied and the ratio is expressed in percent of the original size. The third column represents the byte addressed SCISM scenario when the instruction bundles are expanded to 136-bits by adding a second nibble. It can be seen that IA-64 code size for the SPECINT2000 executables will be compacted by 23% - 29% for the non byte addressed SCISM and 21% - 27% for the byte addressed SCISM.

4 Conclusions

This paper we have shown how the SCISM tagging can be applied to IA-64 instruction set. A straight-forward and hence very restricted scenario in respect to SCISM was applied. All important IA-64 properties where preserved with a marginal increase in opcode length. On the other hand significant static code size reduction was shown (between 21 and 29 %) due to *nop*s elimination.

References

1. Tomasulo R. M., "An Efficient Algorithm for Exploiting Multiple Arithmetic Units," *IBM Journal*, vol. 11, pp. 25–33, Jan. 1967.
2. Kemal Ebcioglu, "Some Design Ideas for a VLIW Architecture for Sequential Natured Software," in *Proceedings of IFIP WG 10.3 Working Conference on Parallel Processing*. Apr. 1988, pp. 3–21, Elsevier Science Publishers.
3. M. D. Smith, M. Lam, and M. A. Horowitz, "Boosting beyond static scheduling in a superscalar processor," in *Proceedings of the 17th Annual Symposium on Computer Architecture*, 1990, pp. 344–354.
4. Marketing Brochure, *The Series 10000 Personal Supercomputer*, Apollo Computer Inc., Chelmsford, MA, 1988.
5. S. Vassiliadis, B. Blaner, and R. J. Eickmeyer, "SCISM: A scalable compound instruction set machine," *IBM J. Res. Develop.*, vol. 38, no. 1, pp. 59–78, Jan 1994.
6. J. Huck, D. Morris, J. Ross, A. Knies, H. Mulder, and R. Zahir, "Introducing the IA-64 architecture," *IEEE Micro*, pp. 12–23, Sep-Oct 2000.

7. Intel Corporation, *Intel IA-64 Architecture Software Developer's Manual Vol.3, Rev. 1.0*, 2000.
8. Gerrit Blaauw and Frederick Brooks Jr., *Computer Architecture*, Addison-Wesley, One Jacob Way, 1997.
9. S. Vassiliadis, J.E. Philips, and B. Blaner, "Interlock collapsing ALUs," *IEEE Trans. Computers*, vol. 42, no. 7, pp. 825–839, July 1993.
10. IBM Corporation, *IBM enterprise System Architecture/370 Principles of operation*, 1989.
11. S. Vassiliadis and B. Blaner, "Concepts of the SCISM organization," Technical Report TR-01 C209, IBM Glendale Laboratory, Endicott, NY, Jan 1992.
12. R.J. Eickemeyer, S. Vassiliadis, and B. Blaner, "An in-memory preprocessor for SCISM instruction-level parallel processors," Technical Report TR-01 C407, IBM Glendale Laboratory, Endicott, NY, May 1992.
13. G. N. Gaydadjiev and S. Vassiliadis, "What SCISM tagging can do that IA64 can not," Tech. Rep. CE-TR-2004-02, TU Delft, 2004.
14. Stephane Eranian and David Mosberger, "The making of linux/ia64," *HPL*, Aug. 1999.

Topic 9
Distributed Systems and Algorithms

Henri E. Bal, Andrzej M. Goscinski, Eric Jul, and Giuseppe Prencipe

Topic Chairs

The wide acceptance of the internet standards and technologies as well as the emerging Grid structures make it hard to imagine a situation in which it would be easier to argue about the importance of distributed systems and algorithms than it is today. This topic intends to provide a forum for researchers from academia and industry interested in distributed systems and algorithms, including the areas of communications, Grids, distributed operating systems and databases.

This topic received 18 submissions, 5 of which have been accepted for the conference. The accepted papers cover a broad range of important issues in this area, including programming models, real-time games, object caching, fault tolerance, and load sharing. The accepted papers are:

- Exploiting Differentiated Tuple Distribution in Shared Data Spaces
- A Proxy Server-Network for Real-time Computer Games
- Accelerating Apache Farms Through ad-HOC Distributed Scalable Objects Repository
- Enhancing Efficiency of Byzantine-tolerant Protocols via Hash Functions
- An Effective Load-Sharing and Fault-Tolerance Algorithm in Heterogeneous Distributed Clustering Systems

Exploiting Differentiated Tuple Distribution in Shared Data Spaces

Giovanni Russello[1], Michel Chaudron[1], and Maarten van Steen[2]

[1] Eindhoven University of Technology
[2] Vrije Universiteit Amsterdam

Abstract. The shared data space model has proven to be an effective paradigm for building distributed applications. However, building an efficient distributed implementation remains a challenge. A plethora of different implementations exists. Each of them has a specific policy for distributing data across nodes. Often, these policies are tailored to a specific application domain. Thus, those systems may often perform poorly with applications extraneous to their domain. In this paper, we propose that implementations of a distributed shared data space system should provide mechanisms for tailoring data distribution policies. Through this flexibility the shared data space system can cope with a wide spectrum of application classes. The need for this flexibility is illustrated by experiments which show that there is no single distribution policy that works well in all cases.

1 Introduction

As distributed systems scale in the number of components and in their dispersion across large networks, the need for loose coupling between those components increases. This decoupling can take place in two dimensions: time and space [3]. Time decoupling means that communicating parties need not be active simultaneously. Space decoupling means that communicating parties need not have an explicit reference to each other.

Generative communication [10], also referred to as data-oriented coordination [12], provides both types of decoupling. In the literature several implementations of generative communication using shared data space systems have been proposed. To meet extra-functional system properties, such as scalability and timeliness, these distribution policies are often optimized for a specific application domain or technical infrastructure. This hard-wiring of a single policy limits the ability of these systems to suit different application characteristics.

Instead, we propose to cater for a wide variety of extra-functional requirements by a using flexible architecture. This architecture provides the possibility of adapting the distribution policies to application-level characteristics of access to the shared data space. In this way, the implementation provides a means to balance extra-functional properties of a system, such as performance, resource use and scalability, for a large class of applications.

In our design of a distributed shared data space, we apply the principle of separation of concerns. This means that we address functional requirements of an application separately from its extra-functional requirements. In particular, we propose to separate the policies for distributing data between nodes from the application functionality. Through this separation, tuning the distribution policy for extra-functional properties such as low latency or low bandwith use becomes transparent to the application. Also, through application of this principle, application logic and distribution logic are separate units of implementation. In this way, both the application code and the distribution code can be reused in different environments.

To substantiate this claim, we show in this paper, that matching the distribution policy with an application's needs, yields better performance than any single distribution policy. While, differentiation of policies has been applied to distributed shared memory systems [2, 5] and in shared data space [4, 15], this paper is the first to demonstrate with concrete results the need for differentiation in shared data spaces. Furthermore, we present experimental results that suggest that continuous adaptation of policies may also be needed.

The paper is organized as follows. In Section 2 we introduce the shared data space model and common distribution schemes. We also explain succintly our distributed shared data space implementation. In Section 3 we describe the setup for our experiments, followed by a discussion of the results in Section 4. We conclude in Section 5.

2 The Shared Data Space Model

A shared data space is capable of storing **tuples**. A tuple is an indivisible, ordered collection of named values. Tuples may be *typed*. Applications can interact with the data space via the three operations described in Figure 1. In this paper, we adopt the semantics of the corresponding operators as specified for JavaSpaces [11].

Operation	Description
put(tuple)	Stores a given tuple in the data space.
read(template)	Reads an arbitrary tuple that matches *template* from the data space. If no match can be found, the caller is blocked.
take(template)	Removes an arbitrary tuple that matches *template* from the data space. If no match could be found, the caller is blocked.

Fig. 1. The three data space operations.

Various approaches have been followed for constructing distributed shared data spaces. However, the most common approach is still to build a *centralized* data space, in which all tuples are stored at a single node. Examples of this approach include JavaSpaces [9] and TSpaces [18]. The obvious drawback is that the single node may become a bottleneck for performance, reliability and scalability.

For local-area systems, a popular solution is the *statically distributed* data space, in which tuples are assigned to nodes according to a systemwide hash function [14]. Static distribution is primarily done to balance the load between various servers, and assumes that access to tuples is more or less uniformly distributed. With the distributed hashing techniques as now being applied in peer-to-peer file sharing systems, hash-based solutions can also be applied to wide-area systems, although it would seem that there is a severe performance penalty due to high access latencies.

Fully replicated data spaces have also been developed, as in [8]. In these cases, which have been generally applied to high-performance computing, each tuple is replicated to every node. Since tuples can be found locally, search time can be short. However, sophisticated mechanisms are needed to efficiently manage the consistency amongst nodes. The overhead of these mechanisms limits the scalability to large-scale networks.

There are other examples of distributing shared data spaces, but in all cases the succes of these schemes has also been fairly limited. The main reason is that shared data spaces, like relational databases, essentially require content-based searching in order to read data. This type of searching is inherently expensive in large-scale settings, as has again recently been illustrated by the research on unstructured overlay networks [6, 7].

The approach we take, is that by dynamically differentiating how tuples should be distributed in a shared data space, we can achieve significant performance gains in comparison to any static, systemwide distribution scheme. The best scheme highly depends on the applications that access the shared data space. For this reason the supporting middleware should be able to support a myriad of schemes. Eilean [15], and in its successor OpenTS [4], are the first shared data space systems where tuples can be treated differently according to the use within the applications, similar to our approach. However, those systems present a static set of distribution strategies that can not easily extended. Moreover, the programmer has to provide explicit information about which distribution strategy has to be applied to each tuple, solely based on his/her knowledge of the application. Finally, the association tuples to distribution strategies is tangled within the application code, contrary to the principle of separation of concerns.

Our solution is called **GSpace**. A GSpace system consists of several *GSpace kernels* running on different nodes. Each kernel stores a part of the overall data space (called a *slice*), as shown in Figure 2. The kernels communicate with each other to present applications with a view of a logically unified data space, thus preserving its simple programming model.

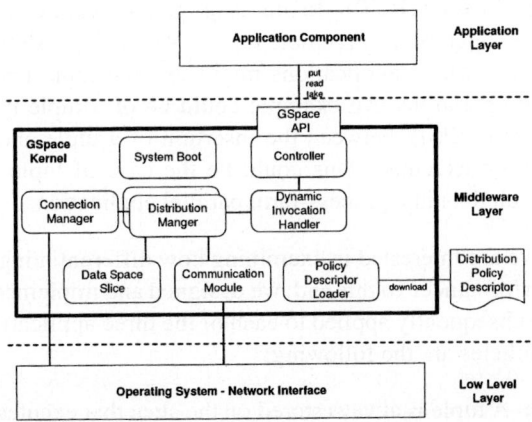

Fig. 2. The internal organization of a GSpace kernel.

Each kernel contains several *distribution managers* that are responsible for distribution of tuples. These modules each employ a different distribution policy for different tuple types, and are completely separated from application components. In other words: data distribution is carried out without specifing any details in the application code. Moreover, the set of policies is extensible such that new distribution policies can be defined. Distribution policies can be inserted in the middleware either at design or at run-time. Further details on GSpace internals can be found in [16].

3 Experiment Setup

To investigate the effect of using different distribution policies for different applications, we set up the following experiments.

We defined a number of patterns that characterize how distributed applications use the data space. Such a usage pattern consists of (1) the ratio of read, put and take operations, (2) the ordering in which these operations are executed, and (3) the distribution of the execution of these actions across different nodes. To avoid randomization anomalies, we generate a set of *runs* that comply with specific usage patterns. We execute the set of runs for different distribution policies. During execution of a run, we measure system parameters that are indicators of costs produced by a distribution policy.

We examined the following application usage patterns, which we considered to be representative for a wide range of applications.

Local Usage Pattern (LUP): In this case, tuples are retrieved from the slice on the same node where they have been inserted. This could be the case if components store some information for their own use or if producer and consumer of a tuple type are deployed on the same node.

Write-Many Usage Pattern (WUP): In this usage pattern applications on different nodes need to frequently and concurrently update the same tuple instance. This is problematic for the consistency of distributed shared-memory systems, since extra mechanisms are needed for mutual exclusion.

Read-Mostly Usage Pattern (RUP): In this usage pattern, application components execute mostly read operations on remote tuples. We distinguish two variants of this pattern: 1) RUP(i), where applications might execute tuple updates between sequences of read operations. An example could be of a tuple type representing a list-of-content. 2) RUP(ii), between the insertion of a tuple and its removal only read operations are executed. This could be the case of tuple type representing intermediate-result data in a process-farm parallel application.

As we mentioned, we are interested in examining how differentiating distribution policies can improve performance. To this end, we designed and implemented four different policies, which we subsequently applied to each of the three application usage patterns. The four different policies are the following:

Store Locally (SL): A tuple is always stored on the slice that excutes its put operation. Likewise, read or take operations are performed locally as well. If the tuple is not found locally then a request is forwarded to other nodes.

Full Replication (FR): Tuples are inserted at all nodes. The read and take operations are performed locally. However, a take has to be forwarded to all nodes by means of a totally-ordered broadcast, in order to remove all copies.

Cache with Invalidation (CI): A tuple is stored locally. When a remote location performs a read operation, a copy of the tuple is subsequently cached at the requester's location. When a cached tuple is removed through a take operation then an invalidation message is sent to invalidate all other cached copies of that tuple.

Cache with Verification (CV): This policy is similar to CI, except that invalidations are not sent when performing a take. On reading a cached tuple, the reader verifies whether the cached copy is still valid, that is the original has not been removed.

To compare the distribution policies we follow the approach described in [13]. We define a **cost function** (*CF*) as a linear combination of metrics that represent different aspects of the cost incurred by a policy. We used the following metrics in the cost function: *rl* and *tl* represent the average latency for the execution of read and take operations; *bu* represents the total network bandwidth usage; and *mu* represents the memory consumption for storing the tuples in each local data slice. For these parameters, the cost function for a policy *p* becomes:

$$CF_p = w_1 * rl_p + w_2 * tl_p + w_3 * bu_p + w_4 * mu_p \qquad (1)$$

Because put operations are non-blocking, application components do not perceive any difference in latency for different distribution policies. Therefore, the put latency is not used as a parameter for the cost function.

The w_i's control the relative contribution of an individual cost metric to the overall cost. An application engineer can set these parameters to match his preference of the relative importance of the different cost metrics. For the experiments in this paper, we take $w_i = 0.25$ for all *i*. Clearly, the settings of these weights determine the performance of the policies. The relevance of these results is not to identify the best setting of the weights, but to illustrate that different policies can be ranked according to a cost-criterium and that for different application characteristics different policies perform best (this holds for any setting of weights).

In our experiments, we simulated all application usage patterns with the policies described previously. The best policy for an application usage pattern is the one that produces the lowest cost value.

4 Results

All experiments were executed on 10 nodes of the DAS-2 [1]. Each usage pattern was simulated using runs of 500, 1000, 2000, 3000, and 5000 operations. For brevity reasons, the histograms in Figure 3 only illustrate the results obtained using runs of 5000 operations. In each histogram, the *X*-axis shows the distribution policies and the *Y*-axis represents the respective *CF* values. The results of shorter sequences of operations follow the same trend. The complete results of these other experiments can be found in the extended version of this paper [17].

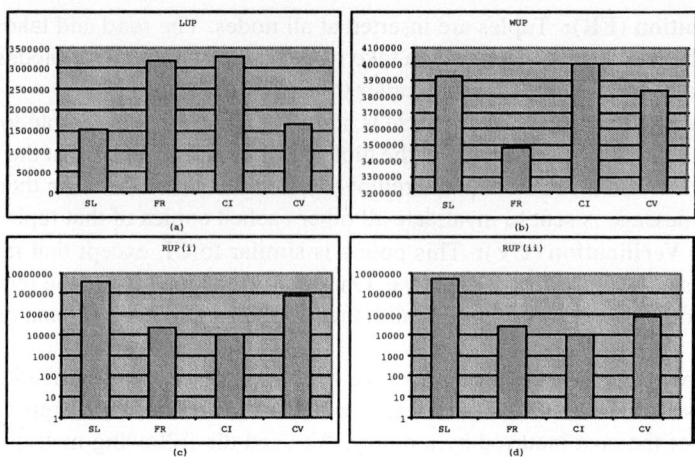

Fig. 3. Histograms showing the cost incurrend by each policy per application pattern.

Figure 3-(a) shows that **SL** is the best policy for the local usage pattern. Store-Locally guarantees low cost for the execution of space operations on local tuples. Figure 3-(b) shows that **FR** produces the lowest cost for the write-many usage pattern. This is because the extra resources spent on replicating tuples, reduce the time required for finding a matching tuple. Figure 3-(c) and (d) show the results for RUP(i) and RUP(ii), respectively. Note that a logarithmic scale is used. In both cases, the **CI** policy produces the lowest cost. This is because caching allows to execute most of the read operations locally. However, the CV policy performs considerably worse than CI policy because the former sends a validation message for each read executed on the local cache.

Figure 4 shows some unanticipated results collected for a set of experiments with the Read-mostly usage pattern RUP(i). Here, the ratio of *number of read operations* to *number of take operations* is decreased respect to the one used in the experiment in 3-(c), meaning that a greater number of take operations are executed. The X-axis shows the length of the run; i.e. number of operations. The Y-axis shows -on a logarithmic scale- the cost incurred by the distribution policies. The experiments described before suggest that the best policy for RUP(i) is CI. Instead, the graph shows that only for shorter runs, cost is minimized by the CI policy. As the number of the operations increases policy FR outperforms policy CI.

The reason for this changing of policy performances is due to the increased number of take operations executed for each run. This fact has two effects that jeopardize the performance of policy CI. Firstly, the execution of more take operations reduces the benefits introduced with caching since cached tuples are more often invalidated. Thus, read operations have to search for a matching tuple, increasing latency time and bandwidth use. On the other hand, policy FR replicates tuples at every insertion thus replicas are already available locally. Secondly, for each take operation policy CI uses point-to-point messages for cache invalidation. Instead, policy FR exploits the more effective atomic multicast technique for removing replicas, that reduces resource usage.

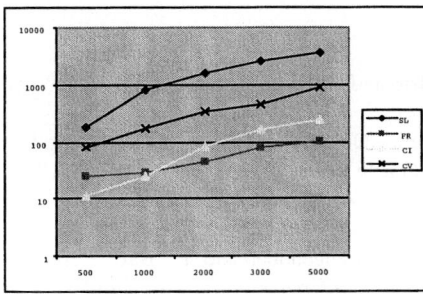

Fig. 4. Results of the simulation for the RUP(i) with different operation lengths.

What we see is that even given the behaviour of an application, it is difficult to predict which policy it fits best. One solution is to make more accurate models for predicting the cost of policies from behaviour. Building these models is quite intricate. For one thing, it is quite complex to determine all the parameters needed for such a model. An alternative approach is to let the system itself figure out which policy works best. In [13] an approach is reported in which a system automatically selects the best strategy for caching Web pages. This approach works by internally replaying and simulating the recent behaviour of the systems for a set of available strategies. Based on this these simulations, the system can decide which policy works best for the current behaviour of the system. We are extending GSpace to include such a mechanism that can dynamically select the best available distribution strategy.

5 Conclusion and Future Work

In this paper we discussed the use of a flexible architecture for distributed shared data space systems in which the strategy for distributing data amongst nodes can be configured without affecting application functionality. This flexibility enables the tailoring of distribution policies to balance the different extra-functional needs of applications. The separation of extra-functional concerns from application functionality enhances code reuse. Both application code and distribution policies are unit of reuse ready to be deployed in several enviroments.

The need for this flexibility is motivated by a series of experiments. These experiments show that there is no distribution policy that is best for different types of application behaviour.

Another important result of our experiments is the urge to have in the system a mechanism able to monitor at run-time the application behavior. In this way, the system is aware when the actual distribution policy is no more the most efficient one. When this happens, the system can adapt dynamically to the new needs of the application by switching distribution policy.

For future work we are currently optimizing migration strategies needed to dynamically change from one distribution policy to another, and are concentrating on developing accompanying mechanisms. At the same time, we are working on supporting real-time constraints in the same fashion as we are doing with distribution requirements.

References

1. H. Bal et al. "The Distributed ASCI Supercomputer Project." *Oper. Syst. Rev.*, 34(4):76–96, Oct. 2000.
2. H. Bal and M. Kaashoek. "Object Distribution in Orca using Compile-Time and Run-Time Techniques." In *Proc. Eighth OOPSLA*, pp. 162–177, Sept. 1993. Washington, DC.
3. G. Cabri, L. Leonardi, and F. Zambonelli. "Mobile-Agent Cooordination Models for Internet Applications." *IEEE Computer*, 33(2):82–89, Feb. 2000.
4. J. Carreira, J.G. Silva, K. Langendoen,and H. Bal. "Implementing Tuple Space with Threads". *In International Conference on Parallel and DistributedSystems (Euro-PDS97)*, 259–264, Barcelona, Spain, June 1997.
5. J. Carter, J. Bennett, and W. Zwaenepoel. "Techniques for Reducing Consistency-Related Communication in Distributed Shared Memory Systems." *ACM Trans. Comp. Syst.*, 13(3):205–244, Aug. 1995.
6. Y. Chawathe, S. Ratnasamy, L. Breslau, N. Lanham, and S. Shenker. "Making Gnutella-like P2P Systems Scalable." In *Proc. SIGCOMM*, Aug. 2003. ACM Press, New York, NY.
7. E. Cohen, A. Fiat, and H. Kaplan. "Associative Search in Peer-to-Peer Networks: Harnessing Latent Semantics." In *Proc. 22nd INFOCOM Conf.*, Apr. 2003. IEEE Computer Society Press, Los Alamitos, CA.
8. A. Corradi, L. Leonardi, and F. Zambonelli. "Strategies and Protocols for Highly Parallel Linda Servers." *Software – Practice & Experience*, 28(14):1493 – 1517, Dec. 1998.
9. E. Freeman, S. Hupfer, and K. Arnold. *JavaSpaces, Principles, Patterns and Practice*. Addison-Wesley, Reading, MA, 1999.
10. D. Gelernter. "Generative Communication in Linda." *ACM Trans. Prog. Lang. Syst.*, 7(1):80–112, 1985.
11. S. Microsystems. *JavaSpaces Service Specification*, Oct. 2000.
12. G. Papadopoulos and F. Arbab. "Coordination Models and Languages." In M. Zelkowitz, (ed.), *Advances in Computers*, volume 46, pp. 329–400. Academic Press, New York, NY, Sept. 1998.
13. G. Pierre, M. van Steen, and A. Tanenbaum. "Dynamically Selecting Optimal Distribution Strategies for Web Documents." *IEEE Trans. Comp.*, 51(6):637–651, June 2002.
14. A. Rowstron. "Run-time Systems for Coordination." In A. Omicini, F. Zambonelli, M. Klusch, and R. Tolksdorf, (eds.), *Coordination of Internet Agents: Models, Technologies and Applications*, pp. 78–96. Springer-Verlag, Berlin, 2001.
15. J. G. Silva, J. Carreira, and L. Silva. "On the design of Eilean: A Linda-like library for MPI." In *Proc. 2nd Scalable Parallel Libraries Conference*, IEEE, October 1994.
16. G. Russello, M. Chaudron, and M. van Steen. "Customizable Data Distribution for Shared Data Spaces." In *Proc. Int'l Conf. on Parallel and Distributed Processing Techniques and Applications*, June 2003.
17. G. Russello, M. Chaudron, and M. van Steen. "GSpace: Tailorable Data Distribution in Shared Data Space System." Technical Report 04/06, Technische Universiteit Eindhoven, Department of Mathematics and Computer Science, Jan. 2004.
18. P. Wyckoff et al. "T Spaces." *IBM Systems J.*, 37(3):454–474, Aug. 1998.

Enhancing Efficiency of Byzantine-Tolerant Coordination Protocols via Hash Functions

Daniela Tulone

Department of Computer Science
University of Pisa
Italy

Abstract. Distributed protocols resilient to Byzantine failures are notorious to be costly from the computational and communication point of view. In this paper we discuss the role that *collision-resistant* hash functions can have in enhancing the efficiency of Byzantine-tolerant *coordination protocols*. In particular, we show two settings in which their use leads to a remarkable improvement of the system performance in case of *large data* or *large populations*. More precisely, we show how they can be applied to the implementation of atomic shared objects, and propose a technique that combines randomization and hash functions. We discuss also the *earnings* of these approaches and compute their complexity.

1 Introduction

The widespread use of the Internet and the proliferation of on-line services involving sensitive data, has lead to increasing attacks against the infrastructure. This motivates the growing interest in the design of distributed protocols resilient to *arbitrary failures*. These failures, often referred as Byzantine failures, model well both malicious behavior and software bugs, since they make the process diverge arbitrarily from the protocol specification. For instance, a Byzantine process can send arbitrary messages, remain latent for a while and then mount a *coordination attack* with other malicious nodes. It is clear that asynchronous Byzantine-tolerant protocols play a crucial role in the implementation of systems that must stay always available and correct. However, commercial applications often implement protocols that are resilient only to crash failures rather than to Byzantine failures, because they are more efficient though less robust. Clearly, coping with Byzantine failures increases the complexity of the protocol. For example, the number of nodes required to guarantee liveness and safety is larger than in case of crash failures, and this affects the communication and computational complexity. The computational cost of masking faulty responses is costly in case of *large data* or *large populations*, since it involves comparisons.

This paper proposes two settings in which collision-resistant hash functions enhance the system performance in case of large objects and large populations. It is well-known that hash functions allow to represent any stream of data by means of a fixed-size string (i.e 128 bit-string) called *message digest*. This identification can be considered unique if the function is *strongly collision-resistant* that is, if it is *computationally infeasible* to find two different objects with the same mapping. Clearly, replacing an object by its message digest enhances the performance of the system, optimizes its resources, such as memory utilization and network bandwidth, and decreases the computational cost of masking faulty values. Hash functions have been applied to a number of fields to enhance efficiency, such as network protocols, peer to peer systems [12], web caching [13]. For instance, Broder and Mitzenmacher [12] proposed an approach for obtaining good hash tables on using multiple hashes of each input key to improve the efficiency of IP address look-ups. However, our paper diverges from these works because it focuses only on collision resistant hash functions (i.e. MD5, SHA, HMAC). These functions have been applied to *secure data dispersal* by Krawczyk [11] to verify the integrity of data stored at servers that might be malicious. Castro and Liskov in [5] apply hash functions to enhance the efficiency of their atomic protocol resilient to Byzantine failures. More precisely, they use Message Authentication Codes (keyed hash functions) to authenticate messages instead of public key signatures, and in the last phase of their protocol servers return a digest reply while a *designated server* returns the full object. Notice that this approach is *optimistic* since it improves the system performance only if the designated server is correct, in case of malicious failure the protocol turns to a slower mode. Our technique improves this idea because it does not only replace large replies with their message digests, but *embeds* hash functions in the design of coordination protocols, in particular atomic protocols, thus exploiting their inherent properties. This approach leads *always* to a significant improvement of the network bandwidth consumption and CPU overhead. Hence, our contribution consists in embedding hash functions in the design of Byzantine-tolerant atomic protocols to improve their efficiency in case of large data and large populations, and in analyzing their benefits over the previous approach. Notice that the techniques we propose are *general* and their applicability goes beyond the settings presented in this paper. We are not aware of similar use of hash functions in previous Byzantine-tolerant protocols. Our study was originally motivated by the performance evaluation of the Fleet system [4], a middleware implementing a distributed data repository for persistent shared objects resilient to malicious attacks against the infrastructure. Fleet is built on top of Byzantine quorum system techniques [3] that make the system highly scalable and available, and improve the system load balancing and the access cost per operation since each operation is performed across a *subset* of servers (quorums). Fleet is targeted to highly critical applications, for instance it was used to implement a prototype of an electronic voting application [2]. While evaluating the performance of Fleet we noticed a remarkable performance degrade as the number of processes or the size of the object increased. This motivated us to look into the underlying Byzantine-tolerant protocols, and investigate ways to boost their efficiency. In this paper we show first how hash functions can enhance the efficiency of protocols implementing shared atomic objects, a fundamental building block for distributed systems. In particular, we propose an optimized version based on message digests, of the protocol proposed by Chockler et al [1]. The choice of analyzing [1] was motivated by its quorum approach that improves the system performance for large populations and makes the use of hash functions particularly suitable. In addition, we propose an approach that combines randomization and hash functions and that for its generality can be embedded in most distributed coordination protocols to enhance efficiency in case of large objects and large universe. We evaluate the complexity in both settings and discuss their performance impact.

2 Hash Functions Embedded in Atomic Protocols

2.1 An Overview

In this section we show how hash functions can be embedded in the implementation of atomic shared objects with *linearizable semantics* [10] to enhance efficiency and guarantee safety. A replicated object with linearizable semantics behaves as one single object exists, and the sequence of the operations applied on it is consistent with the real-time order of the invocations and responses.

Our system model is asynchronous and consists of a static set of n servers, and a limited but unknown set of clients dispersed in a WAN. Clients can fail by crash and a fraction of b servers can be compromised. Since it is impossible in an asynchronous system to distinguish between a crash failure and a slow process, progress should rely on $n - b$ replies. As a consequence, servers might have different views of the system.

A protocol implementing a fault-tolerant atomic object can be decomposed in two building blocks [9]: a *leader election* protocol to choose a coordinator, and a *3-phase commit* protocol run by it. Notice that the 3-phase commit protocol is necessary to guarantee data correctness since a leader failure could leave the system in an inconsistent state. The leader collects in the first phase of the commit protocol data regarding pending operations, and the current state of the object. It orders the operations to be performed based on such data (it may apply them) and then, proposes and commits them. If a leader crashes during the execution of the commit protocol leaving the system in an inconsistent state, the next leader needs to re-establish consistency among the replicas. This can be done by completing the previous run based on data transmitted to servers prior to crash. For instance, if a leader crashes after the proposed phase and before contacting $b + 1$ servers, the next coordinator is unable to recover such data, unless it is self-verifying (i.e. by means of digital signatures). In fact, usually a process retrieves the most up-to-date value by computing the most recent value returned by at least $b + 1$ servers. To overcome this problem, the coordinator sends same information in the propose and in the commit phase. Clearly, this guarantees safety but increases the system overhead in case of large object states.

Such a symmetry can be broken by transmitting in the proposed phase only the message digest of the *full data* sent in the commit phase and eventual additional information. The idea is to reestablish consistency among the replicas by enabling the coordinator to retrieve the data sent by a previous faulty coordinator or to recompute it by means of previous state and additional data. This idea will be developed in the next subsection and become part of the optimizations of protocol [1]. Due to lack of space, we refer the reader to [1] for details on the protocol by Chockler et al, and to the technical report [2] for an analysis of [1] and details and correctness proofs of this optimized version.

2.2 A Specific Scenario: Ordering Operations

The main novelty of the protocol by Chockler et al [1] over previous atomic protocols resilient to Byzantine failures such as [5][8], lies in its quorum system approach that enhances its efficiency in case of large populations. The protocol uses Byzantine quorum systems [3], a variation of quorum systems introduced by Malkhi and Reiter. A Byzantine quorum system is a collection of subsets of servers (*quorums*) such that any two subsets intersects in at least $2b+1$ servers (*consistency*), and for any b faulty servers there is a quorum set containing only correct servers (*availability*). Communications are performed via quorums: only a subset of servers is accessed each time. As a result, some servers have out-of-date object state. This increases the complexity of processing data at the leader side, and makes the use of message digest particularly suitable in this case, since it saves a number of unnecessary comparisons. The choice of analyzing this protocol rather than others is given by its high scalability to large populations (our focus), its generality (i.e. it supports also *non-deterministic operations*), and its efficiency due to quorum systems. Notice that its design improves the client response time since concurrent clients do not have to wait to become a coordinator in order to receive a return result.

1) leader $\leftarrow$ contend()	1) $\langle leader, \{\sigma_i^c, \sigma_i^P, ops_i\}_{i\in Q}\rangle \leftarrow$ status()
2) if (*leader*)	2) if (*leader*)
3) $\quad \{\sigma_i^c, \sigma_i^P, ops_i\}_{i\in Q} \leftarrow$ getStatus()	3) $\quad (\sigma^c, \sigma^P) \leftarrow$ computeState()
4) $\quad (\sigma^c, \sigma^P) \leftarrow$ computeState()	4) $\quad$ if crash in commit phase
5) $\quad$ if crash in commit phase	5) $\quad\quad \sigma^c \leftarrow$ lastCommitted()
6) $\quad\quad \sigma \leftarrow \sigma^c$	6) $\quad$ else if crash in proposed phase
7) $\quad$ else if crash in proposed phase	7) $\quad\quad \sigma \leftarrow \sigma^P$
8) $\quad\quad \sigma \leftarrow \sigma^P$	8) $\quad$ else
9) $\quad$ else	9) $\quad\quad pendings \leftarrow$ computeOps()
10) $\quad\quad pendings \leftarrow$ computeOps()	10) $\quad\quad \sigma^P \leftarrow \langle$ hash(σ^c), $pendings\rangle$
11) $\quad\quad \sigma \leftarrow$ apply($pendings, \sigma^c$)	11) $\quad\quad \sigma^c \leftarrow$ apply($pendings, \sigma^c$)
12) $\quad$ propose(σ)	12) $\quad$ propose(σ^P)
13) $\quad$ commit(σ)	13) $\quad$ commit(σ^c)
14) fi	14) fi

Fig. 1. Client side original version.   **Fig. 2.** Client side optimized version.

An Overview of the Original Version. The protocol [1] works by applying operations to an object state to produce a *new state* and a return result. An object state σ, is an abstract data type containing the shared object, a sort of compressed history of the operations applied, and the return results for the last operations applied. The linearizability of the operations applied, is enforced by the order in which the object states are produced. The client side of the protocol consists of two concurrent threads: a non-blocking thread that simply submits an operation to quorum and waits for $b+1$ identical responses, and a thread that runs a leader election protocol and a *3-phase commit* protocol. An high level description of the client side of this thread is sketched in Figure 1; σ_i^c (or σ_i^p) denotes the object state that server S_i received in the last commit (or last propose), and ops_i the client requests received by S_i and still pending according to its local view. The client runs a leader election protocol to access replicas for some fixed time units, line 1:1. If the majority of the correct servers in the quorum grants such permission, the client executes the 3-phase protocol. It first collects information regarding the last proposed and commit phase, line 1:3 and, based on these data detects possible inconsistency among the replicas due to client crash. In case of leader failure, it completes the previous execution lines 1:5-6, 1:7-8, otherwise applies all the operations that have been submitted but not yet applied to the last object state, thus generating a *new object state*, lines 1:10-11. Then, it proposes this new state to quorum and commits it, line 1:12-13. Notice that in this protocol the proposed object state σ^p is equal to the commit state σ^c. An analysis of this protocol sketched in the next subsection, indicates the object state and the out-of-date data sent by servers that have not been recently contacted, as the major performance bottleneck. Indeed, because of the quorum system and the asynchronous system, the list of pending operations at each server can grow unbounded since a server can receive requests without being contacted in the commit phase. Since Fleet mechanisms rely on this protocol [1], this justifies the performance degrade observed by us in Fleet in case of large objects and large populations. We refer the reader to [2] for a detailed analysis of this protocol (communication and computational complexity). Partial results on the performance evaluation of Fleet can be found in [4].

Our Optimized Version. We propose an optimized version of the protocol whose efficiency is obtained by 1) introducing hash functions, 2) encapsulating the leader election protocol in the commit protocol and 3) reducing out-of-date data. An high level description of this optimization can be found in Figure 2; the variations consist in lines 2:1, 2:5, 2:10 and 2:12-13. Notice that *getStatus()* is piggybacked to the leader election protocol, line 2:1. The use of message digests lets us break the symmetry between the committed and the proposed state: in fact, the proposed state σ^p consists only of the message digest of the *previous committed* state and a signature of the operations applied in that run, line 2:10. In this version each time a coordinator applies operations, it generates a new σ^p and σ^c, lines 2:10-11. In case the coordinator crashes right after the proposed phase, the next leader is able to compute σ^p because the proposed phase was completed, and can compute the correspondent σ^c based on σ^p, line 2:5. Notice that for each σ^p there is one and only one committed state that immediately precedes it, and in case of inconsistencies the coordinator is able to retrieve a correct copy of it. It verifies the correctness of the previous committed state by means of its message digest contained in σ^p. Therefore, procedure *lastCommitted()* at line

2:5, looks for the most recent $b+1$ identical commit states, and if it cannot find them, computes the committed state correspondent to σ^p by: 1) retrieving the previous committed state (last state that was fully committed), and 2) applying to it the operations contained in σ^p. Details on the correctness proof can be found in [2].

Communication Complexity. The use of message digests reduces the communication complexity at least by a factor of $2(q-b)(1+|\sigma|-|ops|)$ with q size of the quorum and ops signature of operations applied by the coordinator. Notice that since malicious servers can transmit arbitrary data (though in this way they are easily detected), we consider only data transmitted by correct servers. The number of correct servers in a quorum is at least $q-b$. In addition, the number of operations applied depends on the degree of concurrency of the system; in absence of concurrent client requests it is equal to 1. To have a better feeling of the performance improvement, let us consider a Threshold quorum system [3] with $n = 101$, $b = 20$ and quorum size $q = 81$. If the shared object is a 100 Mbytes file and the compressed history of the operations up to that time is equal to 1Mbytes, and the size of the request is equal to 100 Kbytes, then our proposal reduces the data transmitted by at least 95 Gbytes. Notice that the communication complexity is also improved by saving one phase of the protocol and removing operations that have been already applied and that are stored at out-of-date servers. That is, it reduces the size of ops_i.

Computational Complexity. The workload of the protocol lies on the client side, in particular on *computeStatus()* at lines 1:4 and 2:3 and on *computeOps()*. Since their costs depend on the number of malicious and out-of-date servers contacted and on the data corruptness, we compute their computational complexity in the best and worst case for both versions of the protocol. Here σ represent the full object state. The computational cost in [1] is given by

$$\Omega\left(q|\sigma| + \sum_{i \in Q} ops_i + b \sum_{o \in \mathcal{O}} |o|\right) \quad O\left(b\,q|\sigma| + \sum_{i \in Q} ops_i + b \sum_{o \in \mathcal{O} \cup \mathcal{P}} |o|\right)$$

with $\mathcal{O}$ set of pending operations and $\mathcal{P}$ set of out-of-date operations. In the optimized version the complexity is

$$\Omega\left(b|\sigma| + \sum_{i \in Q} ops_i + b \sum_{o \in \mathcal{O}} |o|\right) \quad O\left(b^2|\sigma| + \sum_{i \in Q} ops_i + b \sum_{o \in \mathcal{O}} |o|\right)$$

Notice that since in the optimized version the proposed state is very small in most cases, the computational cost is reduced almost by half. Clearly, this approach is not convenient in case the parameters of the operations are greater than the object itself, but this case is uncommon. Evaluation data performed by us on Fleet, running this optimized version of the protocol, showed an improvement over the previous protocol [1] by a factor 10 for large objects (i.e ≥ 314 Kbytes) and large universe (i.e. 80 servers).

2.3 Performance Impact

It is worth noticing that the computation of hash functions such as SHA, HMAC, MD5, is very fast and that in our proposal the hash function is computed only in two cases: when an object is generated or updated, and when a coordinator crashes during the commit phase leaving the replicas in an inconsistent state. In addition, from a system point of view, each time that an object is transmitted it is serialized and then deserialized by the receiver to detect the most up-to-date object. An analysis of Fleet performance running protocol [1] with different object sizes and on a Java profiler [14] to find performance bottlenecks, identified the serialization and deserialization of the objects transmitted and received as main *hot spots* with respect to memory and CPU usage and time. Notice that the use of message digests let us bypass this performance bottleneck. This explains also the remarkable performance improvement obtained in Fleet.

3 A Randomized Approach

In this section we propose an *optimistic approach* that combines randomization and hash functions and that has broad applicability. Optimistic protocols run very fast if no corruptions occurs but may fall back to a slower mode if necessary. Our idea is based on the following intuition: it is sufficient for a process to receive *one correct full* return value, provided a guarantee from other servers of its correctness (i.e its message digest). Clearly, this approach does not guarantee a correct return value since it is hard to distinguish a correct process from a malicious one. Randomization overcomes this problem: in case of very large objects each correct server replies by sending the full data and its message digest with probability p, and its message digest with probability $1-p$. Therefore, if the client contacts n servers the expected number of full correct replies is equal to $(n-b)\,p$. For $p = \frac{1}{2}$ this approach reduces by half the amount of data transmitted and for $n = 3b + 1$ the expected number of correct full copies is greater than b. To retrieve the full correct data, the process performs these two steps:

1. it computes the message digest of the data, denoted by msg, by using a majority voting;
2. it verifies the correctness of one of the full data received (randomly chosen) by computing its hash function and comparing it with msg. It returns as soon as it finds a correct full response.

It might happen that no correct server has sent a full response. In this case, the client falls back to a slower mode and requests to servers the full response. Notice that the system could adopt a dynamic strategy and tune p dynamically depending on the object size and system resources.

If the process receives at least one full correct response (common case if p is tuned appropriately), the communication complexity is improved at least by a factor of $(1-p)(n-b)|data|$. This optimization is more evident for large populations and of course for large objects. The analysis of the computational complexity is less straightforward. In fact, it is very efficient to compute msg in step 1, but verifying the correctness of the full data in step 2 is a bit more expensive than a simple data comparison. Therefore, it is significant to evaluate the number of hash computations necessary to find a correct full reply, in order to assess the performance improvement over previous solutions. In the worst case the process needs to compute $b+1$ hash functions, this occurs when the coordinator picks first b corrupted data sent by malicious servers but with the correct message digest. Since the process chooses the data to verify uniformly at random, this bound is very pessimistic and is not representative for the actual computational cost of step 2. Therefore, we compute the *expected number* of hash computations performed by the process before finding a correct response. This gives us a measure of the computational cost in the average case. The following lemma computes an upper bound (not tight) of the expected number of hash computations in a more general case, when not all replicas are up-to-date. We refer to [2] for the proof.

Lemma 1. *If g is the number of correct up-to-date servers that are contacted, then the expected number of computations of the hash function in the worst case is less than $2 + \frac{b}{g}$.*

Notice that the expected number of g is equal to $p\,m$ with m number of the up-to-date correct copies returned by the servers contacted. Therefore, if $p = \frac{1}{2}$ then two computations of the hash function are enough on the average to detect a correct up-to-date reply.

4 Conclusions

In this paper we proposed two settings in which the use of hash functions leads to a remarkable improvement of system performance in a replicated system resilient to malicious attacks, and in case of large data and large populations. We have proposed two techniques, a deterministic and a randomized one, and applied them to Byzantine-tolerant atomic protocols. These techniques are general and can be applied to other settings requiring coordination among replicas, and for which both efficiency and safety are crucial.

Acknowledgments

The author would like to thank Mike Reiter for helpful discussions on protocol [1] and for his suggestion to explore a randomized approach.

References

1. G. Chockler, D. Malkhi, M. Reiter *Back-off protocols for distributed mutual exclusion and ordering*. In Proc. of the 21st International Conference on Distributed Systems, pp. 11-20, April 2001.

2. D. Tulone *On the Efficiency of Ordering Operations with Arbitrary Failures.* CNR ISTI Technical Report, April 2003.
3. D. Malkhi, M. Reiter, and A. Wool *The load and availability of Byzantine Quorum Systems.* SIAM Journal on Computing 29(6):1889-1906 (2000).
4. D. Malkhi, M.K. Reiter, D. Tulone, E. Ziskind. *Persistent objects in the Fleet system.* In Proc. of the 2nd Darpa Information Survivability Conference and Exposition (DISCEX II), June 2001.
5. M. Castro, B. Liskov. *Practical Byzantine Fault Tolerance and Proactive Recovery.* ACM Transactions on Computer Systems (TOCS), 20(4), pp. 398-461, November 2002.
6. M.K. Reiter. *Secure Agreement Protocols: Reliable and Atomic Group Multicast in Rampart.* In Proc. of the 2nd ACM Conference on Computer and Communication Security, pp. 68-80, Novemebr 1994.
7. K.P. Kihlstrom, L.E. Moser, P.M. Melliar-Smith. *The SecureRing Protocols for Securing Group Communication.* In Proc. of the IEEE 31st Hawaii International Conference on System Sciences, (3):317-326, January 1998.
8. G. Chockler, D. Malkhi. *Active disk paxos with infinitely many processes.* In Proc. of the 21st ACM symposium on Principles of Distributed Computing (PODC 2002), pp. 78–87, July 2002.
9. L. Lamport *The part-time parliament.* ACM Transactions on Computer Systems (TOCS), 16(2), pp. 133-169, May 1998.
10. M.P. Herley, J.M. Wing *Linearizability: A correctness condition for concurrent objects.* ACM Transactions on Programming Languages and Systems 12(3):463-492, July 1990.
11. H. Krawczyk *Distributed fingerprints and secure information dispersal.* In Proc. of the 12th annual ACM symposium on Principles of distributed computing (PODC 1993), September 1993.
12. A. Broder, M. Mitzenmacher *Using Multiple Hash Functions to Improve IP Lookups.* In Proc. of IEEE INFOCOM 2001, pp. 1454-1463, April 2001.
13. David Karger, Eric Lehman, Tom Leighton, Rina Panigrahy, Matthew Levine, Daniel Lewin *Consistent hashing and random trees: distributed caching protocols for relieving hot spots on the World Wide Web.* Proc. of the 29th annual ACM symposium on Theory of computing (STOC 1997), May 1997.
14. Jprofiler, http://www.codework.com/jprofiler/product.htm.

Accelerating Apache Farms Through Ad-HOC Distributed Scalable Object Repository*

M. Aldinucci[1] and M. Torquati[2]

[1] Inst. of Information Science and Technologies – CNR, Via Moruzzi 1, Pisa, Italy
[2] Dept. of Computer Science – University of Pisa – Viale Buonarroti 2, Pisa, Italy

Abstract. We present HOC: a fast, scalable object repository providing programmers with a general storage module. HOC may be used to implement DSMs as well as distributed cache subsystems. HOC is composed of a set of hot-pluggable cooperating processes that may sustain a close to optimal network traffic rate. We designed an HOC-based Web cache that extends the Apache Web server and remarkably improves Apache farms performances with no modification to the Apache core code.

Keywords: Web caching, cooperative caching, Apache, Web server, DSM.

1 Introduction

The demand for performance, propelled by both challenging scientific and industrial problems, has been steadily increasing in past decades. In addition, broadband networks growing availability has boosted Web traffic and therefore the demand for high-performance Web servers. Distributed memory Beowulf clusters are gaining more and more interest as low cost parallel architectures meeting such performance demand. This is especially true for industrial applications that require a very aggressive development and deployment time for both hardware solutions and applications, e.g. software reuse, integration and interoperability of parallel applications with the already developed standard tools.

We present HOC (Herd of Object Caches), a distributed *object* repository specifically thought for Beowulf class clusters. HOC provides applications with a distributed storage manager that virtualize processing elements (PEs) primary memories into an unique common memory. As we shall see, HOC is not yet another Distributed Shared Memory (DSM), it rather implements a more basic facility. It can be used as a DSM building block as well as for other purposes. Indeed, in this paper we present the design of a cooperative cache built on top of HOC for farms of the *Apache* Web server [1]. HOC provides Apache farms with a hot-pluggable, scalable cache that considerably improves Apache farms

---

* This work has been supported by the SAIB Project, funded by MIUR and leaded by SEMA S.p.A. by Athos Origin Group, on High-performance infrastructures for financial applications, and the Italian MIUR FIRB *Grid.it* project No. RBNE01KNFP.

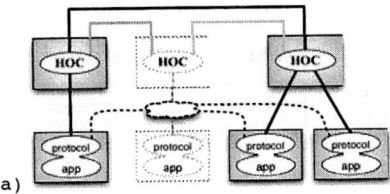 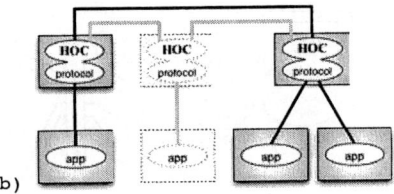

Fig. 1. Typical architectural schemes of applications based on HOC. a) With data access protocol within the application. b) Extending HOC external operations.

performance with no modification to the standard Apache 2.x core code, thus meeting key industrial requirements.

In Sec. 2 we present HOC design principles, HOC implementation and its raw performance. In Sec. 3 we discuss Web server test-bed peculiarities and HOC+Apache architecture. Experiments on such architecture are discussed in Sec. 4. The paper is completed by some conclusion.

2 HOC Design Principles

The HOC underlying design principle consists in clearly decoupling the management of computation and storage in distributed applications. The development of a parallel/distributed application is often legitimated by the need of processing large bunches of data. Therefore data storages are required to be fast, dynamically scalable and enough reliable to survive to some hardware/software failures. Decoupling helps in providing a broad class of parallel applications with these features while achieving very good performances.

HOC, like a DSM, virtualizes PEs primary memories in a common distributed repository. As a matter of fact, it realizes an additional layer of memory hierarchy that lays upon O.S. virtual memory, but that should be used instead of – or together with – disk-based virtual memory. The HOC implements an *external* storage facility, i.e. a repository for arbitrary length, contiguous segments of data (namely *objects*). Objects are identified by a configurable length *key* and have a *home node*, i.e. the reference node for the object (that is currently statically assigned to objects). It enables applications to fully utilize the underlying strength of a cluster, such as fast network communications and huge aggregated memory space. Current and near future network technologies clearly indicates that such a net-based virtual memory may perform much better than disk-based ones. Literature reports very good results both for general DSMs [2, 3] and for specific Web applications [4–6]. A large and fast data repository may be used as a cache facility to improve performance of I/O bound applications, and as a primary storage facility for CPU bound applications running out-of-core on a single PE memory.

HOC, unlike some DSMs, is quite robust and simple. It does not natively implement any consistency mechanism for data copies and does not force already developed applications to be rewritten. As shown in Fig. 1, the HOC-based

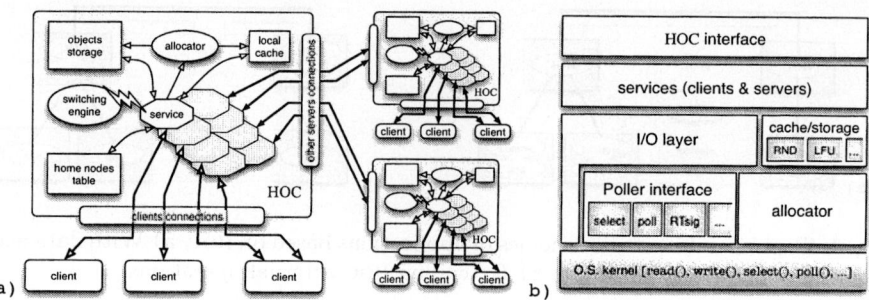

Fig. 2. HOC architecture. a) Functional view. b) Implementation layers.

architecture distinguishes the raw data access layer from the protocol each application requires to access data. In many cases, for example in the context of Web caching, a sophisticated data consistency mechanism is not really needed and may introduce unnecessary overheads. However, HOC may be enriched with locking and consistency mechanisms at the protocol level (as shown in Fig. 1 a).

2.1 HOC Implementation

A HOC server is implemented as a C++ single thread process; it relies on non-blocking I/O to manage concurrent TCP connections [7, 8]. The HOC core consists of an executor of a dynamic set of finite state machines, namely *services*, each of them realized as a C++ object. Each service reacts to socket-related events raised by O.S. kernel (i.e. connections become writable/readable, new connection arrivals, connection closures, etc.). In the case one service must wait on an event, it consolidates its state and yelds the control to another one. The HOC core never blocks on I/O network operations: neither on read()/write() system calls nor on HOC protocol primitives like remote node memory accesses. The event triggering layer is derived from the Poller interface [9], and may be configured to use several POSIX connections multiplexing mechanisms, such as: select, poll, and Real-Time signals.

HOC architecture is sketched in Fig. 2. A HOC may serve many clients, each of them exploiting many connections. A server may cooperate with other HOCs through a configurable number of connections. Connections both with clients and other servers may be established or detached at any time during HOC lifespan.

A connection is exclusively managed by a service. Services rely on an *allocator* in order to store objects into both a *object storage* and a write-back *cache* facilities. They may be both managed as a cache of a configurable but fixed amount of objects and used to store server home objects and to cache remote home objects respectively. As shown in Fig. 2 b) each of them may be managed by using a replacing policy chosen at configuration time. Currently HOC comes with RND (random) and LFU (Least Frequently Used) policies, however it is designed to be easily extended with other policies. HOC offers three basic object-related services (external operations): get, put, and remove. Assuming to have a set of nodes each of them running an HOC, and HOC_i receive:

- get(x). If $home(x) = i$, a local object $storage\_get(x)$ is issued. If otherwise $home(x) = j, j \neq i$ a local $cache\_get(x)$ and, in case of miss, a $remote\_get(x, j)$ is issued. In the latter case the $object(x)$ is stored in the i local cache. Anyway, depending on the get results, either the $object(x)$ or a miss message is sent to the client.
- put(x). If $home(x) = i$, $object(x)$ is put in the node i storage. Otherwise $object(x)$ is stored in the node i local cache. A capacity miss can occur in both cases. In the former case the victim object is simply deleted[1], in the latter the write-back reconciling protocol is started.
- remove(x). The $object(x)$ is purged from both object storages and caches of all nodes.

Overall, HOC implements a *Multiple Reader Single Writer* protocol. Observe that HOC does not natively implement any memory consistency and does not provide any locking facility. These features are supposed to be realized at the protocol level as shown in Fig. 1 a).

2.2 HOC Performances

HOC performances has been extensively tested on several homogeneous clusters. We report here tests performed on a 21 PEs RLX Blade; each PE runs Linux (2.4.18 kernel) and is equipped with an Intel P3@800MHz, 1GB RAM, a 4200rpm disk and three 100Mbit/s switched Eth devices. As shown in Fig. 3, HOC scales very well up to the maximum number of PEs available in our testing environment (10 HOCs and 10 clients). In the tests two different Ethernets are used for client-server and server-server connections. As shown in Fig. 3 b), HOC can deliver an aggregate throughput that is quite close to the 100Mbit/s per PE asymptotic network bandwidth. We experienced a little impact of local cache size on performances. Cache mainly acts as a network buffer.

Other tests performed on a Linux (2.4.22 kernel) cluster of 8 PEs P4@2GHz, 512MB RAM, connected through a 1Gbit/s Eth have confirmed for a single HOC and many clients an aggregate throughput of ∼91MB/s (∼96% of the measured 95MB/s maximum bandwidth) with 2048 concurrent stable connections for a test of 50,000 requests of 1MB objects. We also experienced more than 20,000 replies/s with 3,072 concurrent stable connections for a test of 200,000 requests of 512B objects. Actually, HOC has sustained a throughput close to the network bandwidth in any tested case.

3 Web Caching Test-Bed

Web caches have became the standard method to ensure high-quality throughput for Web access. Web caching can be adopted at different levels: 1) Web browser storing Web objects in its local memory or disk. 2) proxy server sitting somewhere between the clients and the Web server (typically at ISP level) serving a

---

[1] If configured as a cache, otherwise a *storage full* message is sent to the client.

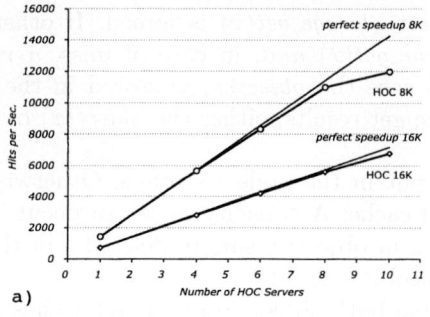 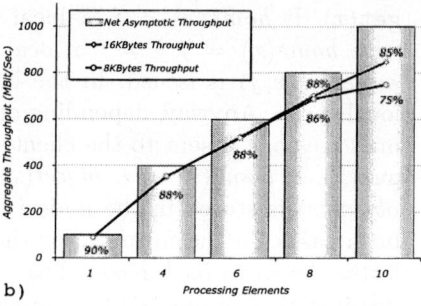

Fig. 3. a) HOC scalability of served requests. b) Aggregate throughput. Experiments are referred to 200,000 `get` operations w.r.t. a data set of 40,000 objects (8K and 16K average sizes) having cyclically distributed homes among HOCs. Each HOC serves a partition of requests referred to keys which are uniformly distributed and randomly chosen in the whole data set. Each HOC have a local cache of 1,000 objects.

large community of clients. 3) server accelerator (reverse proxy) sitting in front of one or more Web servers [10, 11].

Reverse proxies have been demonstrated to be the best solution among them. Differently from 1 and 2, they are under the site manager control and typically enable dynamic pages management through an API which allows application programs to explicitly add, delete, and update cached data [12, 13]. However, a reverse proxy may introduce unnecessary latencies due to the additional parsing and filtering of requests, including those leading to not cacheable replies. To mitigate these problems we adopt a different solution: we place a cache on back of the Web server in order to improve server performances. The cache cope with cacheable replies only and does not make any additional parsing on them. As we shall see in the next section, we implemented the solution by using a HOC-based architecture. Our approach is similar to others appeared in literature [14, 4, 6, 5]. Differently from other approaches, we did not designed another Web server, we are rather proposing a decoupled architecture which is composed of the standard Apache Web server and a HOC-based storage subsystem. Apache Web server is unmodified in core functionalities, we just modified the allocation policy of one optional Apache module. Therefore the architecture may benefit from Apache popularity, correctness and people expertise.

In a cache implemented on a single node (single or even multi-processor), the throughput is limited by the network interface. With cluster based Web cache the throughput can be increased simply adding more nodes. Since many HOCs may cooperate with each other, the cache throughput is not limited by the network interface, but by the aggregate cluster throughput.

3.1 HOC+Apache Architecture

The HOC+Apache architecture is compliant to Fig. 1 a). In this case the *app* is the Apache Web server, the *protocol* is a modified version of *mod_mem_cache* Apache

Table 1. Experimental environment summary. a) Raw data set characteristics. b) Access log characteristics. c) Apache Web server configuration.

a) Raw data set			b) Accesses log	
Total size	4 GB		Data transferred	~9 GB
N. of files	100,000		N. of distinct files requests	74226 (2.8 GB)
N. of requests	250,000		Avg. file size	36.8 KB
N. of files < 256 KB	96209 (2 GB)		N. of distinct files < 256 KB	71449 (1.5 GB)
Static pages	100%		Avg. size of distinct files < 256 KB	8 KB

c) Apache 2.0.47 MPM worker configuration (hybrid multi-threaded multi-process) [1]							
StartServers	4	MaxClients	512	ThreadPerChild	64	Log level	Notice
ServerLimit	8	MinSpareThreads	32	MaxRequestsPerChild	0	Access log	None

module [1] (compliant to RFC2616), and dashed lines are not present. In particular, *mod_mem_cache* has been modified by only substituting local memory allocation, read and write with HOC primitives. The work-flow of the *mod_mem_cache* modified to inter-operate with HOC is sketched in Fig. 4 a). No other functionalities have been modified. Both original and modified caches are able to cache static and dynamic pages. Note that the original version of *mod_mem_cache* implements a per process cache, thus different Apache processes never share the cache even if they are running on the same PE.

We used as objects key the MD5 digest of the *mod_mem_cache* native key. In order to enforce correctness and consistency we include in the object to be stored into HOC original HTTP request and reply headers. Protocol does not need any additional consistency mechanism but the ones ensured by *mod_mem_cache*.

Observe that HOC+Apache architecture is designed to improve Apache performance whether the performance bottleneck is memory size, typically in the case the data set does not fit the main memory. In all other cases (e.g. in-core data sets), the HOC+Apache architecture does not introduce performance penalties w.r.t. the stand-alone Apache equipped with the native cache.

4 Web Caching Experiments

We measured the performance of Apaches+HOC architecture on the RLX Blade described in Sec. 2.2. The main characteristics of the data set and accesses log are summarized in Table 1 a) and b). The data set is generated according to [15] by using a Zipf-like request distribution with $\alpha = 0.7$. 90% of files have small size (min. 2KB, max. 100KB, avg. ~13KB) and 10% of files have medium-large size (min. 101KB, max. 1MB, avg. ~280KB). In all tests we used the Apache 2.0.47 Web server in the *MPM Worker* configuration shown in Table 1 c). Files greater than 256KB are not cached. HTTP requests are issued by means of the *httperf* program [16] configured to count replies only within 1 second timeout. Each httperf is directly connected to an Apache with no switch/balancer in the middle. The whole site is replicated on all PEs running Apache.

In Fig. 4 b) we compare Apache against Apache+HOC performances. The test takes in to account three basic classes of configurations:

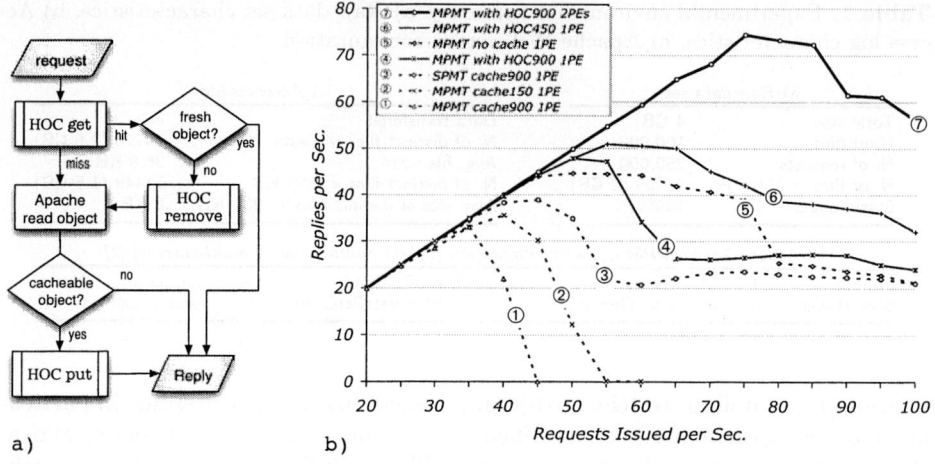

Fig. 4. a) High-level work-flow of *mod_mem_cache* changes needed to inter-operate with HOC. b) Replies rate for stand-alone Apache in the Multi-Process Multi-Threaded (MPMT) and Single-Process Multi-Threaded (SPMT) configurations and HOC+Apache architecture. Apache is tested with 900MB and 150MB native cache per process and without cache. HOC+Apache is tested with 450MB and 900MB devoted to HOC (1PE). HOC+Apache is also tested with Apache and HOC (900MB) on different PEs (2PEs).

i) a stand-alone multi-process multi-threaded Apache with no cache (⑤), with the *mod_mem_cache* (Apache native cache) exploiting both 900MB (①) and 150MB cache (②) per process. A stand-alone Apache configured as one server process exploiting 512 threads sharing the same 900MB cache (③);
ii) an Apache+HOC running on the same PE, HOC exploiting both 450MB (⑥) and 900MB (④) of total memory accessed by all Apache processes;
iii) an Apache+HOC running on different PEs, HOC exploiting 900MB of total memory accessed by all Apache processes (⑦).

The three cache sizes 150MB, 450MB, 900MB have an hit rate of 29%, 45%, 60% respectively when tested on a single Apache native cache. As clear from Fig. 4 b), in all cases the HOC+Apache architecture overwhelms the stand-alone Apache with or without the native cache, including the case the native cache is shared among all running threads.

i) We have observed that the Apache with the original cache lose its stability when the requests rate grows. This is mostly due to the lack of a common cache storage for all processes on the same PE, which leads to the replication of the same objects in several caches, and therefore to the harmful memory usage (①, ②). This rapidly leads the O.S. to the swap border resulting in a huge increase of reply latency. Indeed, Apache configured as SPMT thus exploiting a single shared native cache (③) perform better than MPMT configurations. Quite surprisingly the Apache with no cache performs even better (⑤). In reality this behavior is due to the File System buffer that acts as a shared cache for all Apache

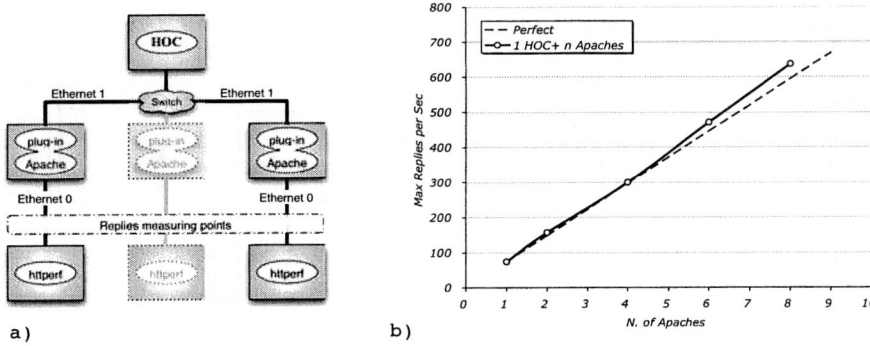

Fig. 5. One HOC supporting many Apaches. a) Experiment schema. b) Replies per second speedup. HOC is configured with 900MB of objects storage. Dataset and Apache configuration are summarized in Table 1.

processes coping with object replication problem. Actually, the FS buffer rise up to ∼700MB during test ⑤. In this case the performance also depends on site organization on disks. In general FS cache is unsuitable for Web objects since requests do not exploit spatial and temporal locality w.r.t. disk-blocks [15]. Moreover, FS cache is totally useless for dynamic Web pages, for which we experienced the effectiveness of the Apache native cache module [12].

ii) The Apache+HOC architecture performs better than stand alone Apache, even if Apache and HOC are mapped on the same PE (④, ⑥). Apache processes share a common memory through HOC. Since the accessed set of files does not fit in memory, performance may be influenced by replacing policies. Here, two different caching mechanism and three policies are active at the same time: 1) the FS buffer and HOC allocates memory in different bulks from the same physical memory; 2) the FS buffer replacing policy is active on the first bulk, the Apache GreadyDual-Size and the HOC LFU replacing policies are active in cascade on the second bulk. The knotty scenario prevents a fine analysis of system bottlenecks. Indeed, the decoupling approach we followed sought to simplify the design in order to make effective the system tuning.

iii) As a matter of fact, the 2PEs figures (⑦) confirm that mapping Apache and HOC on different PEs significantly improves performances. The same tests have been performed using as HOC replacing policy the random function instead of LFU. The performance degradation is just 8.7% (from 74.4 to 67 avg. replies). The replacing policy have a little impact on performances in this case, whereas the decoupling of cache management from the server activity significantly improves architecture stability and performance. In fact decoupling distributes the memory pressure due to the Apache server and its cache on different PEs.

Figure 5 highlights that HOC may support many Apaches with a very good, over linear scalability due to the partition of requests among several Apaches (which induces a lower disk load). Note that HOC+8Apaches sustain an aggregate of 637 hits/s with an average file size of 8KB on the Apache-HOC link (see Table 1 b). In this case HOC works at ∼45% of maximum reply rate since HOC reaches 1430 hit/s with N=1 and the same message payload, see Fig. 1.

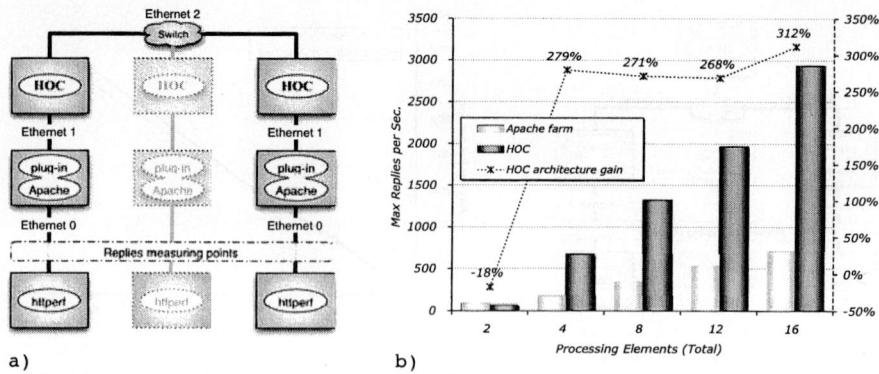

Fig. 6. Many cooperating HOCs experiment. a) Experiment schema. b) Dark bars shows the replies/s speedup w.r.t. the total amount of used PEs (n-HOCs and n-Apaches). Light bars shows the same measure for a reference architecture ($2n$-Apaches on the fastest tested configuration, see Fig. 4). Dashed line sketch the HOC-based architecture gain. The HOC local cache is fixed to 5,000 objects.

Figure 6 shows the scalability of nApache+nHOC. Also in this case we experienced a very good scalability up to our testing environment limit reaching a 312% gain w.r.t. an Apache farm (8HOCs+8Apaches vs 16Apaches). If compared with Apache with the original cache module the gain reaches the 406%. Observe that the big gap occurs from 1HOC+1Apache and 2HOC+2Apaches, where the whole set of all cacheable objects begins to fit in the aggregate memory.

5 Conclusion

We introduced HOC, a fast and scalable "storage component" for homogeneous cluster architectures. HOC implements a distributed storage service relying on state-of-the-art server technologies. As described in Sec. 2.2, these enable HOC to cope with a large number of concurrent connections and to sustain a very high throughput in both single and parallel configurations. We developed and tested a HOC-based Apache plug-in module which greatly improves Apache Web server farms performances. To the best of our knowledge, no other works target the problem with no modification to a preexisting centralized Web server.

We are currently improving HOC in two directions. First, by introducing multithreading to make it scalable also on SMP boxes. Second, by integrating HOC within the ASSIST parallel programming environment [17]. Overall, we envision a complex application made up of decoupled components, each delivering a very specific service. Actually, HOC provides the programmer with a data sharing service [18]. In this scenario, the application or the programming environment run-time support is supposed to provide the correct protocol to consistently use HOC external operations. At this end we are developing a protocol which offers standard hooks to implement several DSM consistency models [3, 2].

Acknowledgments

We wish to thank M. Danelutto and M. Vanneschi for many fruitful discussions, and A. Petrocelli who has contributed in experimenting HOC.

References

1. Apache Software Foundation. *Apache Web Server*, 2003. (http://httpd.apache.org).
2. M. Aldinucci. eskimo: experimenting with skeletons in the shared address model. *Parallel Processing Letters*, 13(3):449–460, September 2003.
3. G. Antoniu and L. Bougé. DSM-PM2: A portable implementation platform for multithreaded DSM consistency protocols. In *Proc. of the 6th Intl. HIPS Workshop*, number 2026 in LNCS. Springer, April 2001.
4. G. Chen, C. L. Wang, and F. C. M. Lau. A scalable cluster-based web server with cooperative caching support. *Computation and Currency: Practice and Experience*, 15(7–8), 2003.
5. F. M. Cuenca-Acuna and T. D. Nguyen. Cooperative caching middleware for cluster-based servers. In *Proc. of the 10th IEEE HPDC*, August 2001.
6. E. Cecchet. Whoops! : A clustered web cache for dsm systems using memory mapped networks. In *Proc. of the 22nd IEEE ICDCSW Conference*, July 2002.
7. D. Kegel. The C10K problem, 2003. (http://www.kegel.com/c10k.html).
8. A. Chandra and D. Mosberger. Scalability of Linux event-dispatch mechanisms. Technical Report HPL-2000-174, HP Labs., Palo Alto, USA, December 2000.
9. D. Kegel. Poller Interface, 2003. (http://www.kegel.com/poller/).
10. J. Wang. A survey of Web caching schemes for the Internet. *ACM Computer Communication Review*, 29(5), 1999.
11. M. Rabinovich and O. Spatsheck. *Web Caching and Replication*. A. Wesley, 2001.
12. A. Iyengar and J. Challenger. Improving Web server performance by caching dynamic data. In *Proc. of the USENIX Symposium on Internet Technologies and Systems Proceedings*, Berkeley, CA, USA, Dec. 1997.
13. B. Hines. Livin' on the Edge: Caching Dynamic Web Content with IBM WebSphere Application Server and WebSphere Edge Server. IBM, September 2001. (http://www-106.ibm.com/developerworks/websphere/techjournal/0109_hines/hines.html).
14. Whizz Technology. *EC-Cache Engine*, 2003. (http://www.whizztech.com/ec_cache.html).
15. L. Brelau, P. Cao, L. Fan, G. Phillips, and S. Shenker. Web caching and Zipf-like distributions: Evidence and implications. In *Proc. of the Infocom Conference*, 1999.
16. D. Mosberger and T. Jin. httperf – a tool for measuring Web server performances. In *Proc. on Internet Server Performance (WISP)*, Madison, USA, June 1998.
17. M. Vanneschi. The programming model of ASSIST, an environment for parallel and distributed portable applications. *Parallel Computing*, 28(12), Dec. 2002.
18. G. Antoniu, L. Bougé, and M. Jan. Juxmem: An adaptive supportive platform for data sharing on the Grid. In *Proc. of IEEE/ACM Workshop on Adaptive Grid Middleware (in conjunction with PACT 2003)*, New Orleans, USA, September 2003.

A Proxy Server-Network for Real-Time Computer Games

Jens Müller[1], Stefan Fischer[2], Sergei Gorlatch[1], and Martin Mauve[3]

[1] Westfälische Wilhelms-Universität Münster, Germany
[2] Technische Universität Braunschweig, Germany
[3] Heinrich-Heine-Universität, Düsseldorf, Germany

Abstract. Computer games played over the Internet have recently become an important class of distributed applications. In this paper we present a novel proxy server-network topology aiming at improved scalability of multiplayer games and low latency in client-server data transmission. We present a mechanism to efficiently synchronize the distributed state of a game based on the concept of eventual consistency. We analyse the benefits of our approach compared to commonly used client-server and peer-to-peer topologies, and present first experimental results.

1 Introduction

Computer games played over a network are categorized in the three main genres: *First Person Shooter Games (FPS)*, *Real-time Strategy Games (RTS)* and *Massive Multiplayer Online Roleplaying Games (MMORPG)*. For MMORPG, there exist several advanced networking topologies, which can support a high number of participating clients [1]. In contrast, the basic concepts of network communication for FPS and RTS games have not been altered for years. Commonly used architectures like client-server and peer-to-peer do not provide the required scalability for massive-multiplayer sessions. Although other communication architectures like peer-to-peer server-networks or hierarchical server-networks exist, their efficient usage for FPS and RTS games is still not well understood (an overview can be found in [2]).

In this paper, we present our work on a *Proxy Server-Network for Real-time Computer Games*. Our main goal is to improve network scalability for FPS and RTS games in comparison to traditional topologies, thus allowing games with higher numbers of players. A further objective is to enhance responsiveness of massive-multiplayer, Internet based games by providing low communication latency between clients and servers. The paper is organized as follows: In Sect. 2 we introduce our architecture and propose a concept for an efficient synchronisation of the distributed game state. We describe our implementation of the architecture as a C++ API and present experimental results on its scalability in Sect. 3. Section 4 compares our approach with related work and concludes the paper.

2 Architecture

Commonly used communication architectures like client-server or peer-to-peer topologies provide a variety of positive aspects and can be implemented quite easily, which leads to their high acceptance for computer games.

However, there are several disadvantages in both concepts, resulting in poor quality of service and inhibiting novel game designs with a higher number of participating users. Client-server and peer-to-peer topologies do not scale well when the number of users increases. Additionally, the server in a client-server setup forms a single point of failure for the whole session. While the peer-to-peer approach eliminates the problem of a single point of failure, it allows a hacked client to cheat, since submitted game updates are not filtered by a server instance and hidden information can be made visible to the player [3].

2.1 The Proxy Server-Network

The proposed proxy server architecture, which is derived from a peer-to-peer server-network [2], is shown in Fig. 1. The initial idea was sketched in [4].

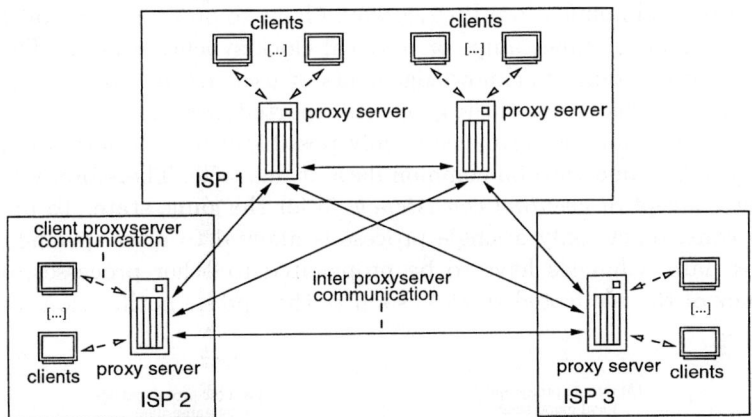

Fig. 1. Example session using the proxy server architecture.

The key idea of our architecture is the use of several interconnected proxy servers for a single game session. Each proxy server has a full view of the global game state, allowing it to serve arbitrary clients. Each client communicates with a single proxy to send user actions and receive updates of the game state. Proxy servers process user actions and forward them to other proxies using multicast (IP or application level) for synchronizing the distributed game state.

For the sake of low latency in Internet based sessions, proxy servers should be located at different Internet Service Providers (ISP), such that each client connects to a proxy at its ISP. In our current implementation, the setup of servers is done manually, but future work includes a dynamic setup of proxies respecting current user demand.

MMORPG commonly use an approach orthogonal to ours: Instead of replicating the game world, they partition it among servers, which forces clients to change the server when the user moves into another region. While this approach works well for the relatively slow game flow in MMORPG, it cannot be applied to fast paced games like FPS and RTS, in which recurrent pauses resulting from server changes are annoying for users. Since each proxy has full view of the game state, our proxy server-network avoids reconnects during a session.

As we show in Sect. 3.1, the described proxy network scales better with an increasing number of clients than traditional topologies.

2.2 Managing the Distributed Game State

Since the game state is replicated at the proxy servers, our architecture requires mechanisms to provide adequate consistency of the replicated data.

There is a classical trade off between scalability, responsiveness, and consistency in distributed simulations in general. Implementing strong consistency models like *causal* or even *sequential consistency* [5] would increase the amount of inter-proxy communication: Proxies would have to order changes of the game state using Lamport timestamps or physical clock synchronisation. This would delay the transmission of acknowledgements of user actions to clients, thereby reducing the responsiveness of the game. As reported in [6], a scalable distributed system with real-time performance is only possible if not all participating processes share the same view on common data at one time. Therefore, we propose to use the concept of *eventual consistency* [5] for the game state. To implement eventual consistency, only a single process is allowed to alter specific parts of replicated data. Changes have to be propagated to other processes, ensuring consistency of the replicated state as soon as the update notification arrives.

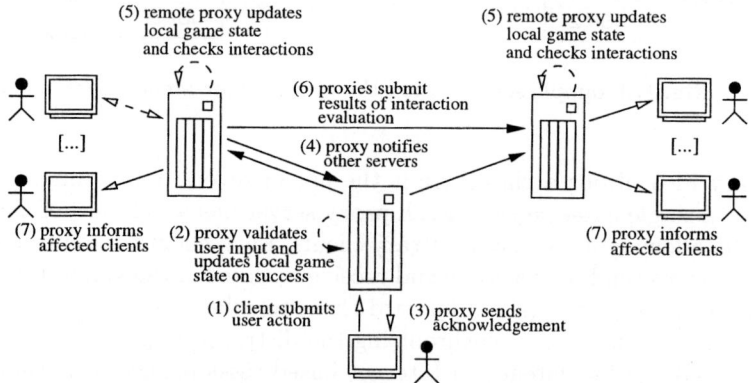

Fig. 2. Acknowledging and forwarding a single user action.

The main steps in our concept of transmitting and processing a single user action in the proxy server-network are depicted in Fig. 2. User actions transferred from clients (1) are categorized in *movements* and *interactions*. In case of movement commands, the server has to check if the user input is legal to prevent cheating before changing the state (2), and to inform other proxies about updated position values (4), which in turn update the local copies accordingly (5). In this way eventual consistency for the changed part of the game state is guaranteed and the clients receive acknowledgments for movement commands in a relatively short time (3). In case of interactions, notified proxies additionally have to check whether local clients are affected. If so, the responsible proxy updates the game state of its local client, and informs other servers in turn (6). For all state updates received from other proxy servers, each proxy evaluates which local clients are affected and informs them (7).

In our architecture, general parts of the game state that are not associated with a specific client are assigned to single proxy servers for maintenance, which ensures eventual consistency for these parts of the game state as well. Such data contain manipulable parts of the gameworld like collectible items or avatars not controlled by the users (*Non-Player Characters*). To avoid a single point of failure, applications have to implement a proper election method to reassign responsibility to another proxy in case of a server failure. For data representing states of clients, a reassignment of responsibility is not required: If a proxy process fails, clients connected to it will loose connection to the session anyway, while for all other processes the game will continue to be playable.

The presented concept for managing the distributed state allows an efficient synchronisation of the game state in combination with fast acknowledgment of user actions. However, due to the use of eventual consistency, execution of commands can occur in a different order than their initation by the users. In the following, we discuss the impact of this fact on the playing conditions.

2.3 Discussion of Game Correctness

A computer game provides equal game opportunities for all users, if the order of processed state updates is the same as the order of their initiation by the users. Additionally, in continuous simulations like real-time games, the point in time at which actions are processed has to be considered, too. Let us consider an example interaction between two clients, c_1 and c_2, in a real-time game: The user at c_1 issue an action affecting c_2 (e. g. shooting at the avatar managed by c_2 with an instant-hit weapon), while the user at c_2 issues a movement command some time later which would nevertheless prevent his avatar from being affected if processed before the action of the user at c_1.

Figure 3 depicts four situations for the communication between processes in our example, comparing a client-server setup (a) to our architecture (b), in which additional methods for improved correctness of the processing order are applied (c), (d). Let $l(c, p)$ denote communication latency between c and p.

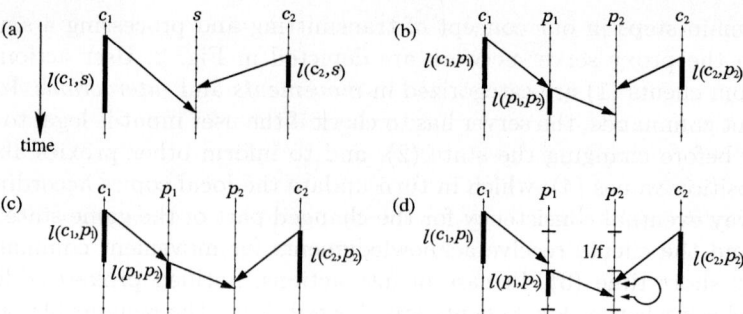

Fig. 3. Comparison of processing orders: Traditional client-server topology (a), proxy server-network (b), proxy server-network with client latency levelling (c), proxy server-network with client latency levelling and action reordering (d).

(a) An incorrect processing order is possible in the traditional client-server topology: client c_2 has an advantage in time over c_1 of $l(c_1, s) - l(c_2, s)$, due to which c_2 can dodge the hit although c_1 aimed correctly and shot first.

(b) The proxy server-network has lower latency $l(c_1, p_1) < l(c_1, s)$, but requires additional time $l(p_1, p_2)$ before the shot action reaches proxy p_2. Therefore, the processing order at p_2 is incorrect as well.

(c) Our implementation uses *client latency levelling*: communication between a proxy and clients with low latency is delayed; latencies are thus adjusted to an equal level $l(c_1, p_1) = l(c_2, p_2)$, which reduces the advantage in time for c_2 to $l(p_1, p_2)$. However, actions in the depicted example are still processed in an incorrect order. An upper limit for client latency levelling is configurable, e.g. by a session administrator, to avoid an unresponsive game for all clients due to a single slow connection of a specific client.

(d) In real-time games, updates of the state usually are done at fixed points in time, so called *ticks*, discretizing the actual continuous game simulation. With a *tickrate* f, proxy p_2 can reorder actions if they are received in the same tick of length $1/f$. This leads to the correct processing order.

This analysis shows that our architecture provides a degree of equal game opportunities comparable to the client-server setup, which is commonly used for FPS games. The provided correctness of the processing order is even better when actions are reordered by the game application. However, if proxies receive interactions at different ticks then reordering is not possible, since the first action received has already been processed. If the action processed first was a movement command of a local client, proxies can hold a backlog of old positions and apply remote actions with hindsight performing a *timewarp* [11]. In the relatively rare case that the acknowledgment sent for the first action is not reversible, we accept an incorrect processing order for the benefit of high responsiveness for clients.

3 Implementation and Analysis

The implementation of a game using our architecture consists of two parts: the functionality for the network communication, and the game specific data structures and algorithms for the simulation itself. We implemented the communication of the server-network as a C++ API using the TCP/IP protocol stack. Games using this API only have to implement the game specific part and to handle user commands according to the procedure depicted in Fig. 2.

Our implementation includes two different APIs, designated to clients and servers. Each of them provides a virtual callback-class, which has to be implemented by the game application to receive incoming data. For transmission of data, each API offers a sender-class providing methods to initiate the connection and send data to the network. The implementation provides functionality to send and receive byte arrays of arbitrary length. It offers different stages of reliability in data transfer, automatically fragmenting and acknowledging packets when using UDP and IP multicast for inter-proxy communication.

3.1 Scalability: Analysis and Experimental Results

In the computer games context, scalability is the ability of an architecture to maintain service when the number of participating players increases. The two kinds of resources utilized are computing cycles of the host machines and network bandwidth. With the common unavailability of IP Multicast for home computer Internet dial-up connections, utilized bandwidth of the peer-to-peer topology is quadratic to the number of participating clients, which disqualifies this architecture for large scale Internet-based sessions. Therefore we focus on a comparison of the proxy-network to a traditional client-server topology in our analysis.

The computation intensive tasks in simulating the game are the validation of user inputs and the filtering of client updates (steps (2) and (7) in Fig. 2). The proxy server-network scales better than the client-server topology, because these tasks are distributed among all participating proxies. Utilized bandwidth in client-server communication of the proxy architecture and client-server topology is equal: each client only receives state updates if it is affected. However, the proxy server-network utilizes additional bandwidth in inter-server communication, which will be taken into account in the analysis.

We implemented the following test game to measure the gain in scalability. Each client has control over a single avatar, other avatars are visible if they reside in the avatar's viewing radius. The environment, internally represented by a quadtree, is defined as a bordered plane, on which rectangular solid blocks define regions avatars cannot walk into. Avatars can catch each other if in range. After a successful catch, the winner gains a point and the caught avatar is taken out of the game for some time. Although this game is quite simple, its basic concept is comparable to sophisticated FPS games.

We measured the maximum number of supported clients by adding clients to the session until the servers become congested, i.e., when calculation of a tick takes longer than the time available. The tickrate of the evaluation sessions was 50 updates per second, which is comparable to tickrates in Internet based FPS and RTS games commonly ranging from 10 to 40. Each proxy server was run exclusively on a Linux host with 1,7 GHz Intel Pentium 4. Figure 4 depicts our experimental results. The left plot shows the maximum number of supported clients with different numbers of proxies. The right plot depicts the average total bandwidth at a single proxy server in different session setups.

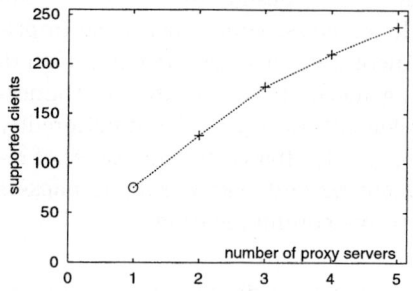
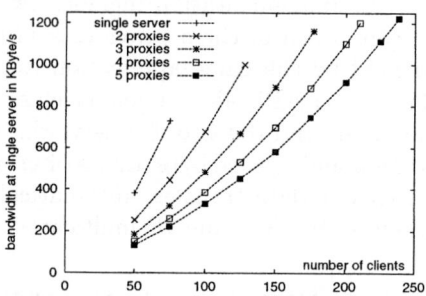

Fig. 4. Experimental results. Left: maximum supported clients with different numbers of proxy servers; Right: utilized bandwidth at a single proxy.

The results for computational scalability (left plot) show that our proxy architecture supports a much higher number of clients than a traditional client-server topology whose capacity is shown by ○. The absolute gain in the number of supported clients decreases slightly for each additionally attached proxy, due to increased calculation for processing remote clients' actions. Therefore we expect a game dependent, maximum number of clients for which an acceptable responsiveness is provided by our architecture.

The total utilized bandwidth at a single server (right plot) consists of the bandwidth of the client and proxy communication, each with incoming and outgoing bandwidth. While the bandwidth of inter-proxy communication grows only linearly with the the number of participating clients, the outgoing client bandwidth increases quadratically. In order to compensate the growing utilized bandwidth for increasing numbers of participating clients, additional proxy servers have to be added. Figure 4 (right) shows that in our test game, for approximately 25 additional clients joining the game one additional proxy server has to be added to the session in order to avoid an overload of servers due to high bandwidth.

The deviation of total average bandwidth was below 3% throughout multiple measurements for the same setup. The available number of host computers and network hardware capabilities of our testbed subnet limited us to a maximum of 240 clients with an overall bandwidth of about 6 MByte/s. With the bandwidth distributed to several subnets as in Internet based sessions, more proxies and therefore more clients can participate in a game.

4 Conclusion and Related Work

For multiuser simulations in general and real-time computer games in particular, there is a trade off between responsiveness, scalability, and consistency. Our proxy server-network weakens consistency for the benefit of higher responsiveness and improved scalability. At the same time, we do not sacrifice correctness of processing, which is still comparable to the commonly used client-server setup.

There has been active research in the domain of *Distributed and Collaborative Virtual Environments* (DVEs resp. CVEs). Existing DVE systems like MASSIVE [7] or DIVE [6], do not aim at maximized responsiveness for the benefit of stronger consistency. However, to achieve high responsiveness in games with direct, hand-eye coordinated movement like FPS, latency has to be below 150 ms [8]. Our architecture does not introduce additional delay for acknowledgment of actions, which only depend on the round trip times to the ISP.

OpenSkies, developed by Cybernet, is a commercial server-network developed for massive multiplayer online games [9]. In contrast to our proxy-network, servers in OpenSkies only filter communication to interested hosts and do not maintain the application state at all. In comparison to the spatial partitioning commonly used for MMORPG [1], our architecture does not require reconnects of clients in a running session. RING [10] proposes a network topology similar to ours, but updates are already filtered in inter-server communication. RING is specially designed for densely occluded virtual environments, while we aim at supporting arbitrary environments in computer games.

References

1. Cai et al.: A scalable architecture for supporting interactive games on the Internet. In: Proc. of the 16th Workshop on Par. and Dist. Simulation, IEEE (2002) 60–67
2. Smed, J., Kaukoranta, T., Hakonen, H.: Aspects of networking in multiplayer computer games. In: Proc. of Int. Conf. on Application and Development of Computer Games in the 21st Century, Hong Kong SAR, China (2001) 74–81
3. Pritchard, M.: How to hurt the hackers: The scoop on Internet cheating and how you can combat it. Game Developer (2000)
4. Fischer, S., Mauve, M., Widmer, J.: A generic proxy system for networked computer games. In: NetGames2002 Proceedings. (2002) 25–28
5. Tanenbaum, A., van Steen, M.: Distributed Systems. Prentice Hall (2002)
6. Frecon, E., Stenius, M.: DIVE: A scalable network architecture for distributed virtual environments. Distributed Systems Engineering Journal (1998)
7. Greenhalgh, C., Purbrick, J., D.Snowdon: Inside massive-3: Flexible support for data consistency and world structuring. In: Proceedings of the Third ACM Conference on Collaborative Virtual Environments (CVE 2000). (2000) 119–127
8. Armitage, G.: Sensitivity of Quake3 players To network latency, IMW2001 poster
9. Cybernet Systems: Openskies network architecture http://www.openskies.net/
10. Funkhouser, T.A.: RING: A client-server system for multi-user virtual environments. In: Symposium on Interactive 3D Graphics. (1995) 85–92
11. Mauve et al.: Local-lag and Timewarp: providing consistency for replicated continuous applications. IEEE Transactions on Multimedia, February 2004.

Topic 10
Parallel Programming:
Models, Methods and Programming Languages

Paul H.J. Kelly, Sergei Gorlatch, Christoph Kessler, and Dan Quinlan

Topic Chairs

Welcome to Euro-Par's Topic 10, which provides a forum for the presentation of the latest research results and practical experience in parallel programming. Topic 10 emphasises results which improve the process of developing high-performance programs. The general challenge is to target available hardware structures, while maintaining useful application structure. Of particular interest are novel techniques by which parallel software can be assembled from reusable parallel components – without compromising efficiency.

We thank the many helpful referees, who provided at least four reports on each of the 29 papers submitted. Nine of the papers submitted to Topic 10 were accepted. In addition, one paper, by Aldinucci et al, was transferred from Topic 19 (Demo Session). This year's submissions represent interesting new developments and concerns in the Parallel Programming research community:

MPI and OpenMP: Our distinguished paper, by Jarmo Rantakokko, on programming models for structured adaptive mesh refinement, concerns an extremely important and complex class of parallel numerical applications and studies OpenMP and MPI on a hardware platform which supports hybrid implementations. Very similar issues are explored by Fürlinger, Schenk and Hagemann who study OpenMP/MPI hybrid approaches to task-queue parallelism in their sparse linear solver.

Novel Languages: Meanwhile, novel languages are also represented. Rasmussen et al, present a short paper on the co-Array model in Python. Rocha, Silva and Santos Costa present an interesting study of lock alternatives in memoized logic programming. Ashby, Kennedy and O'Boyle present work on a powerful mixed functional-imperative language, showing the potential for optimization across language abstractions. An elegant approach to this important problem is offered by Kakehi et al.

Middleware and Operating Systems: Middleware issues form the third major theme. Ribes, Pérez and Priol discuss how to handle exceptions in parallel CORBA components. Magini et al of the ASSIST group at the University of Pisa contribute with interesting work on parallel composition in their CORBA-based system. Aldinucci et expand on the ASSIST architecture and discuss applications. Extending the envelope somewhat, Frachtenburg and colleagues from Los Alamos explore parallel programming models in scalable operating systems.

This year we rejected 20 papers, among them very promising work - we have worked hard to soften the blow through extensive and constructive refereeing - we hope to see the fruits of your work in 2005!

Comparison of Parallelization Models for Structured Adaptive Mesh Refinement

Jarmo Rantakokko

Scientific Computing, Department of Information Technology,
Uppsala University, Box 337, 751 05 Uppsala, Sweden

Abstract. We compare different parallelization approaches in different programming models using MPI and OpenMP on a SunFire 15K for structured adaptive mesh refinement. The results show that no static parallelization or programming model performs best in all cases and that different models are preferable in different settings of the application and grid hierarchy. We then suggest a dynamic approach using a mixed MPI-OpenMP model that adapts the number of threads during run time and gives good performance in all stages throughout the whole run as the solution state changes, i.e. the resolution in the computational grid changes.

1 Introduction

The dynamic features of adaptive numerical methods make them harder to parallelize than static grid computations. In the static case, domain decomposition methods are often used to split the grid among the processors. However, when the grid structure changes, as in the adaptive case, the grids have to be redistributed to avoid load imbalance. Also, it is more challenging to partition a domain with grids of different resolutions or with different work loads in different parts than a grid with a uniform resolution. A number of approaches and techniques have been developed for partitioning and parallelizing adaptive mesh refinement methods, see e.g. [1, 3, 4, 6, 7, 9–12, 15, 19]. Many of these efforts have shown to be very successful for a certain application and computer system. But, there is no approach that performs best on all applications and computer systems. The performance depends very much on the current state of the application and the computing environment as well as the programming model. This has been stated previously by many other researchers, see e.g. [14]. A difficulty in adaptive mesh refinement methods is that the state of the application is changing during runtime.

The contribution of this paper is twofold. First we carry out a comparison of different parallel programming models for structured adaptive mesh refinement. We compare shared memory programming using OpenMP, distributed memory programming using MPI, and mixed models of using both MPI and OpenMP. Secondly, we suggest a new dynamic approach using a mixed MPI-OpenMP model that adapts the number of threads during run time and gives a good performance in all stages through the whole run as the solution state and computational grid changes. It gives a high efficiency focusing the computational power in computationally intensive regions. The experiments were done a SunFire 15K and the adaptive numerical method we have studied is a structured block-based PDE solver, described in the next section.

2 The Adaptive Mesh Refinement Method

We solve hyperbolic partial differential equations, e.g. the Euler equations, on the form

$$\frac{\partial u}{\partial t} + \frac{\partial f}{\partial x} + \frac{\partial g}{\partial y} = 0 \qquad (1)$$

where $f = f(u)$ and $g = g(u)$ are the flux functions. This equation is discretized by a finite volume method in space (FVM). For the time discretization we use the classical fourth order Runge-Kutta method. To estimate the local error we use Richardson extrapolation. In space we have a structured cartesian grid and divide the domain into a fixed predefined number of blocks, controlled by the user. If the local error exceeds the error tolerance in any grid point within a block, the discretization is refined with a factor two in the entire block. On the contrary, if the error is much smaller than the error tolerance the discretization in the block is coarsened with a factor of two. This block-based approach with fixed number of blocks greatly simplifies the adaptive method compared to the Berger-Colella approach [2]. Figure 1 shows a pseudo code for the solver and Figure 2 shows a snapshot of the numerical solution and the corresponding adaptive grid. A thorough description with theoretical analysis of the numerical properties of the adaptive mesh refinement method can be found in [8, 17]. Moreover, a theoretical analysis of the performance of the solver can be found in [18].

```
Algorithm: adaptive mesh refinement
    (1)     while (t<Tstop)
    (2)         for (ib=1,grid.nblocks)
    (3)             EstimeError(grid.block(ib))
    (4)             Refine(grid.block(ib)) or
    (5)             Coarsen(grid.block(ib))
    (6)         end for
    (7)         UpdateGhostCells(grid)
    (8)         for (ib=1,grid.nblocks)
    (9)             AdvanceSol(grid.block(ib))
    (10)        end for
    (11)        UpdateGhostCells(grid)
    (12)    end while
```

Fig. 1. A sketch of the code structure for the block-based adaptive refinement algorithm.

3 Parallel Implementations

The code was originally written in Fortran 90 and parallelized with MPI. The parallelization is coarse grained over entire blocks, i.e. each processor is responsible for one or more blocks. The blocks have two layers of ghost cells. These ghost cells are updated

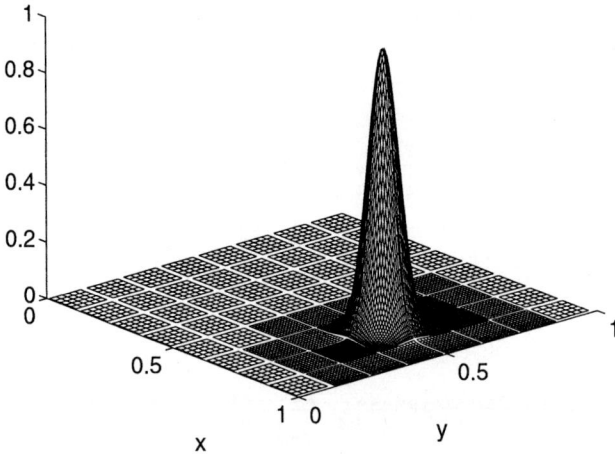

Fig. 2. Block-based adaptive refinement, entire blocks with large local errors are refined.

from neighboring blocks, requiring interprocessor communication. The number of messages is minimized by collecting data, the ghost cells, from several blocks that are going to the same processor into one large message. The communication is performed asynchronously with MPI_ISEND and MPI_PROBE, i.e. each processor first sends out all its messages with non-blocking calls and then probes for incoming messages receiving them in first-come first-serve order. This minimizes the synchronization overheads.

When the grid resolution changes in any of the blocks the entire grid structure is repartitioned and data redistributed, see Figure 3. A number of data partitioning methods are available, the recursive spectral bisection method (RSB) [13], the recursive coordinate bisection method (RCB) [13], a bin-packing method [5], and different variants of space filling curve techniques and diffusion schemes [12, 15, 16]. The performance of the different partitioning techniques applied to the adaptive method have been evaluated experimentally, see [16].

Three different parallelizations of the code with OpenMP exists. In a first implementation we have used a fine grained loop level parallelism within each block, i.e., all blocks are executed on all threads. The code inside the calls in lines 3, 4, 5, and 9 in Figure 1 are parallelized with OpenMP using the !$OMP PARALLEL DO directive for the grid points. This gives a one-dimensional parallelization of the blocks. (We will refer to this version as OMP1.) There are no problems with load imbalance in this version as long as the number of grid points in each dimension is large enough, but as some blocks may contain relatively few grid points the parallel tasks become small which can degrade the performance. Moreover, this version gives many synchronization points compared to a coarse grained parallelization.

In a second implementation we have a parallelization over the blocks using the !$OMP PARALLEL DO SCHEDULE directive for the lines 2 and 8 in Figure 1, i.e. entire blocks are executed on single threads. (We will refer to this version as OMP2.) Using STATIC scheduling gives a severe load imbalance as the blocks have different number of grid points and work load. Using chunk size equal to one improves load bal-

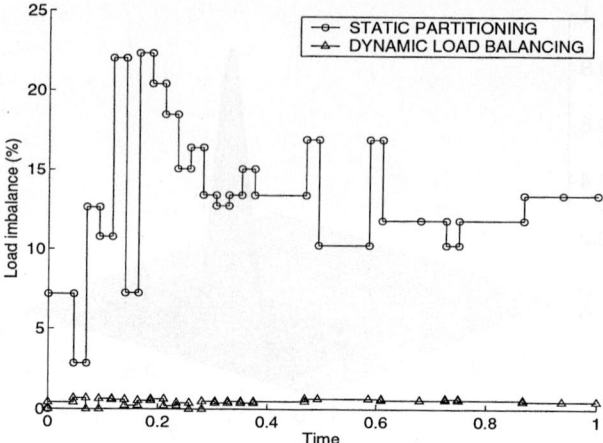

Fig. 3. The resolution of the grid changes frequently causing a load imbalance with static partitioning. Using re-partitioning the load imbalance can be kept down during the whole run.

ance as long as number of blocks is considerably larger than the number of threads. The DYNAMIC scheduling directive is another remedy for this but does not guarantee a good load balance as we cannot control which blocks are executed by the respective threads. As soon as one thread is idle it picks up the next block in line and starts processing it. This block may not be optimal for that thread in terms of load balance. Moreover, the threads can process different blocks from iteration to iteration causing unnecessary data movements (cache misses) even if the grid resolution does not change.

To avoid these data movements we have a third implementation (which we will call OMP3). This version is also over the blocks but now only using the PARALLEL region statement. Here we use a load balancing algorithm and then in the loop over the blocks check if this particular block belongs to this thread's partition before we execute the code. This version is similar to the coarse grained MPI version, ensuring that the threads will execute the same blocks as long as the partitioning stays constant.

4 Experiments

We run two test cases, one with many small blocks (tasks) and one with few large blocks. In the first case the grid is split into 8×8 blocks. We have maximally three grid levels ranging from 12×12 to 48×48 cells in the blocks. In the second test case we have only 8 blocks (2×4) but with resolutions ranging from 128×128 to 512×512 cells. We have run the code on a Sun Fire 15k server with 48 UltraSparcIII-cu processors (900 MHz). Each processor has a 8 MB L2 cache and the server has a total of 48 GB of shared primary RAM. The server is divided into two domains with the larger one containing 32 processors. Four CPUs are packaged into one CPU/memory board with 4 GB of RAM. The Sun Fire is a cache-coherent non-uniform memory access system. The NUMA ratio is close to 1:2 between accessing local (on board) and remote memory (from another board).

We compare fixed size speedup of the different strategies. All timings are compared to the one-processor-one-thread case, which is the same for all strategies. In our first test case with 64 blocks the strict MPI-version performs best, see Figure 4(a). Adding more threads to the processes only degrades the performance. The blocks are too small for OpenMP to be efficient. The synchronization and thread allocation overheads dominate the performance. From Figure 4(b), we can also see that the OMP1 version (loop-parallelism) does not scale above 2 in speedup. There is no gain in using more than 6 threads.

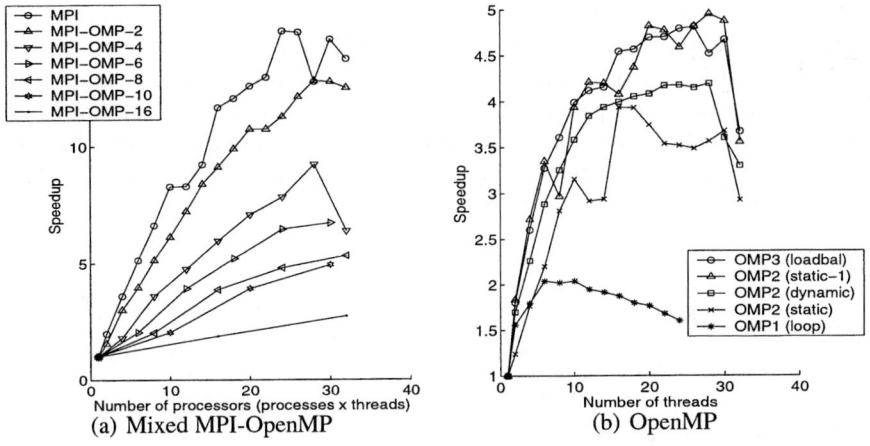

Fig. 4. Test case 1, speedup of different implementations.

Comparing the different OpenMP versions we can see that the coarse grain parallelism, OMP2/3, gives better performance than the loop-level parallelism, OMP1. Using only static scheduling over the blocks gives a poor load balance. The dynamic-clause improves the load balance but introduces non-locality of data increasing the number of remote accesses (and cache misses). Even if the grid does not change the individual blocks can be processed by different threads from one iteration to another. Using static scheduling with chunk-size one gives both good load balance and data locality, i.e., the blocks are always processed by the same threads even if the grid changes. As long as the number of blocks is significantly larger than the number of threads this strategy works well. The OMP3 version optimizes the load balance and gives data locality as long as the grid does not change. When the grid changes a partitioning algorithm "re-distributes" the blocks giving more remote data accesses. Compared to MPI the performance is significantly poorer. The computational work in the blocks is small compared to the serial (single thread) regions and the thread synchronization overheads.

Trying to improve the results (OMP3) by using different load balancing methods had little effect on the performance, the run times were not affected. The inherent overheads in OpenMP were much larger than the differences in load balance and communication requirements (edge-cut) produced by the different load balancing methods. Also, using SUN's migrate-on-next-touch, `madvise(3C)`, command to move pages close to threads to ensure correct data placement and avoid remote memory accesses, did only give 2-3% improvement in run time.

The first test case has too small blocks for efficient parallelization with OpenMP, still it shows some interesting characteristics. The second test case has much larger blocks, suitable for loop-level parallelism. Moreover, the number of blocks is small which makes it hard to load balance on block-level. These two features favor OpenMP as we can see from the results below. In Figure 5(a) we can see that the coarse level parallelism with MPI only gives a maximum speedup of 4.4 using 8 processors, one processor per block. The performance is completely dominated by the load imbalance. Adding more threads to the processes now improves the performance considerably giving a maximum speedup of 14 for 4 processes and 6 threads. The effects of the load imbalance can still be seen. Comparing the different OpenMP versions, Figure 5(b), we can see that the loop-level parallelism OMP1 performs best as the block-level variants OMP2/3 do not scale above 8 threads. In this case we can see that the load imbalance effects in OMP2/3 are of the same magnitude as the synchronization overheads in OMP1, the performance is similar up to 8 threads.

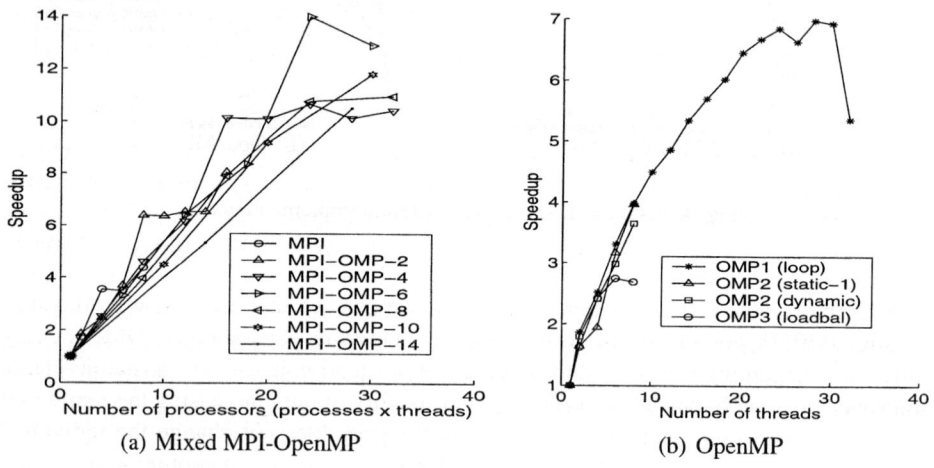

(a) Mixed MPI-OpenMP (b) OpenMP

Fig. 5. Test case 2, speedup of different implementations.

We have here compared different parallelization models for implementing structured adaptive mesh refinement. The results show that no particular model is the optimal. The results depend on the application and system state. These results confirm the conclusions from [14] where they state that no programming model, comparing different mixes of OpenMP and MPI in general, is the ideal model for all codes and applications. In an adaptive mesh refinement application the state can change during the simulation, hence giving benefit to different models or mixes of MPI and OpenMP during the same run. We therefore, in the next section, suggest a dynamic model that adapts the number of threads per process during the run and gives a good performance even if the state of the application changes.

5 Dynamic Threads in a Mixed MPI-OpenMP Model

All the previous approaches have some drawbacks making them non-optimal in different settings. A parallelization on the block-level has limited parallelism dictated by the number of blocks. A parallelization on the grid point level, i.e. within the blocks, has many synchronization points. Moreover, some blocks may be too small to parallelize giving too high parallel overheads. A static mixed coarse-fine level parallelization increases the parallelism compared to block level parallelization and decreases the parallel overheads compared to parallelization on the grid point level. Still, some blocks may be too small to parallelize with a fixed number of threads while some other blocks may be large enough to benefit from using additional threads than were given from start.

We suggest a dynamic MPI-OpenMP approach that sets the number of threads for each block individually depending on the size of the block. We can partition the grids assigning the blocks to the MPI-processes and then spawn an appropriate number of threads for each block in run time (using the function OMP_SET_NUM_THREADS before each parallel region). The most obvious strategy is to set the number of threads proportional to the number of grid points in the blocks. In our case the blocks have the same number of grid points at one level and are coarsened or refined with a factor of two in each space dimension between the levels. In two dimensions the workload differs with a factor of four between two levels. We can then assign 1 thread at level 0 grids, 4 threads at level 1 grids, 16 threads at level 2 grids, and so forth. With this approach we get the same work load per thread and each block can be processed in unit time, approximately. (We will refer to this approach as MPI-OMP-DYN1.) But, the parallel overheads grows significantly with the number of threads and the benefit of exploiting the low-level parallelism is lost if we use too many threads. Thus, a better strategy is to use less threads on the largest grids, e.g. we can use 1 thread on level 0 grids, 2 threads on level 1 grids and 4 threads on level 2 grids, and then give the blocks appropriate weights (1, 2 and 4) before load balancing between processes. (We will refer to this version as MPI-OMP-DYN2.) More generally, one can have a theoretical model for distribution of threads to blocks. What the optimal number of threads per block is depends on the absolute block sizes (not relative) and the computer system (synchronization overhead, cache performance, thread allocation time, etc).

In Figure 6(a), test case 1, we can see how the MPI-OMP-DYN1/2 versions perform compared to the strict MPI and OpenMP (OMP1) versions. The MPI-OMP-DYN1 version has perfect load balance but using up to 16 threads per block gives very high overheads, as discussed above. The MPI-OMP-DYN2 uses only 4 threads on the largest blocks and gives almost as good performance as strict MPI, which is optimal for this test case with many small blocks. For the second test case, the dynamic model (MPI-OMP-DYN2) gives the best scaling, see Figure 6(b). These two test cases demonstrate that the dynamic approach can adapt to different application states and give good performance in all stages. While the static approaches can be better in some settings they can fail to give good performance in others. In an adaptive mesh refinement code the state changes continuously.

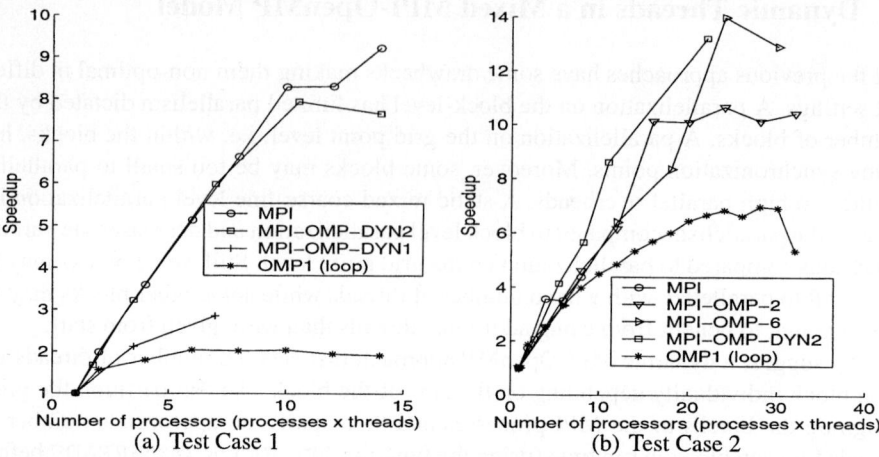

Fig. 6. Speedup of mixed MPI-OpenMP with dynamic thread allocation.

6 Conclusions

We have compared different parallelization models using MPI and OpenMP for structured adaptive mesh refinement. We have used coarse grained parallelization with MPI over the blocks and fine grained parallelization using OpenMP within the blocks. We have also tested different mixes of nested parallelization combining the two approaches above. Our results show that no static parallelization model is optimal in all cases. Depending on the system and application state different models gave the best performance. Many small blocks favor the coarse grain MPI parallelization while few large blocks were in favor for the fine grain OpenMP model. With this as background we suggest a new dynamic model that adapts to the solver. It is a mixed MPI-OpenMP model that allocates threads dynamically and independently within the different MPI processes. The number of threads varies during run time depending on which block is processed. Large blocks with high work load are assigned many threads and small blocks few threads. In an adaptive mesh refinement application the block sizes (and numbers) change during the run time shifting preference for different parallelization models. The dynamic MPI-OpenMP approach is an attempt to meet these needs and the results from the experiments are promising, even though more experiments are needed to fully explore its properties. Testing on other SAMR approaches, e.g., applied to the Berger-Collela method, and running on other computer system, e.g., clusters of SMP's, would be interesting. In connection, a theoretical model of how to choose the optimal number of threads per block depending on the grid size, numerical solver, and computer system would be very useful to explore. We consider this as continued work and research on the subject.

References

1. D.S. Balsara, C.D. Norton, *Highly parallel structured adaptive mesh refinement using parallel language-based approaches*, Parallel Computing, 27(2001): 37-70.
2. M. Berger, P. Colella, *Local adaptive mesh refinement for shock hydrodynamics*, Journal of Computational Physics, 82:64-84, 1989.
3. R. Blikberg and T. Sorevik, *Nested parallelism: Allocation of threads to tasks and OpenMP implementation*, Scientific Programming 9:185-194, 2001.
4. K. Devine et al., *Design of dynamic load-balancing tools for parallel applications*, technical report, Sandia National Laboratories, Albuquerque, NM, USA, 2000.
5. M.A. Iqbal, J.H. Saltz, S.H. Bokhari, *Performance tradeoffs in static and dynamic load balancing strategies*, Technical Report 86-13, ICASE, NASA Langley Research Center, Hampton, VA, 1986.
6. G. Karypis, K. Schloegel, and V. Kumar, *PARMETIS - parallel graph partitioning and sparse matrix ordering library, version 2.0*, University of Minnesota, Minneapolis, MN, 1998.
7. Z. Lan, V. Taylor, and G. Bryan, *Dynamic load balancing for structured adaptive mesh refinement applications*, in proceedings of ICPP 2001.
8. P. Lötstedt, S. Söderberg, A. Ramage, L. Hemmingsson-Fränden, *Implicit Solution of Hyperbolic Equations with Space-Time Adaptivity*, BIT, 42:1, p. 129-153, 2002.
9. P. MacNeice et al., *PARAMESH: A parallel adaptive mesh refinement community toolkit*, Computer physics communications, (126):330-354, 2000.
10. L. Oliker, R. Biswas, *PLUM: Parallel load balancing for adaptive unstructured meshes*, Journal of Parallel and Distributed Computing, 52(2):150-177, 1998.
11. M. Parashar, J. Browne, *System engineering for high performance computing software: The HDDA/DAGH infrastructure for implementation of parallel structured adaptive mesh refinement*, IMA Volume on Structured Adaptive Mesh Refinement (SAMR) Grid Methods, pages 1-18, 2000.
12. K. Schloegel, G. Karypis, and V. Kumar, *Multilevel diffusion schemes for repartitioning of adaptive meshes*, technical report 97-013, Department of Computer Science, University of Minnesota, Minneapolis, MN, 1998.
13. H.D. Simon, *Partitioning of unstructured problems for parallel processing*, Computing Systems in Engineering, 2(2/3):135-148, 1991.
14. L. Smith and M. Bull, *Development of mixed mode MPI/OpenMP applications*, Scientific Programming 9:83-89, 2001.
15. J. Steensland, *Efficient Partitioning of Dynamic Structured Grid Hierarchies*, Department of Information Technology, Uppsala University, Sweden, Ph.D. thesis, 2002.
16. J. Steensland, S. Söderberg, M. Thune', *A Comparison of Partitioning Schemes for a Blockwise Parallel SAMR Algorithm*, in Proceedings of PARA2000, T. Sorevik et.al (Eds), New Paradigms for HPC in Industry and Academia, 5th International Workshop, PARA2000, Lecture Notes in Computer Science, No. 1947, pp. 160-169, Springer-Verlag, Berlin, 2001.
17. S. Söderberg, *A Parallel Block-Based PDE Solver with Space-Time Adaptivity*, Department of Information Technology, Uppsala University, Sweden, Licentiate thesis, 2001.
18. M. Thune', S. Söderberg, *Optimal Parameter Values for a Parallel Structured Adaptive Mesh Refinement Algorithm*, in Proceedings of PARA2000, T. Sorevik et.al (Eds), New Paradigms for HPC in Industry and Academia, 5th International Workshop, PARA2000, Lecture Notes in Computer Science, No. 1947, pp. 177-186, Springer-Verlag, Berlin, 2001.
19. C. Walshaw, M. Cross, and M.G. Everett, *Parallel dynamic graph partitioning for adaptive unstructured meshes*, Journal of Parallel and Distributed Computing, 47(2):102-108, 1997.

Task-Queue Based Hybrid Parallelism: A Case Study

Karl Fürlinger[1], Olaf Schenk[2], and Michael Hagemann[2]

[1] Institut für Informatik,
Lehrstuhl für Rechnertechnik und Rechnerorganisation
Technische Universität München
karl.fuerlinger@in.tum.de
[2] Departement Informatik,
Universität Basel
{olaf.schenk,michael.hagemann}@unibas.ch

Abstract. In this paper we report on our experiences with hybrid parallelism in PARDISO, a high-performance sparse linear solver. We start with the OpenMP-parallel numerical factorization algorithm and reorganize it using a central dynamic task queue to be able to add message passing functionality. The hybrid version allows the solver to run on a larger number of processors in a cost effective way with very reasonable performance. A speed-up of more than nine running on a four-node quad Itanium 2 SMP cluster is achieved in spite of the fact that a large potential to minimize MPI communication is not yet exploited in the first version of the implementation.

1 Introduction

Hybrid parallelism (i.e., the combined usage of shared memory and message passing programming paradigms) seems to be a perfect fit for the hierarchical organization of today's popular SMP cluster systems. In most previous reports on the merits of hybrid parallelism (e.g., [12, 2, 1, 8]), existing MPI applications have been extended with OpenMP shared memory programming constructs. Often the authors arrive at the conclusion that the performance of pure MPI programs is generally somewhat better than those of their hybrid counterparts. Still, a hybrid approach might be advantageous in certain circumstances, such as when the MPI code scales poorly or when replication of the dataset limits the number of MPI processes per node.

In this paper we report on our experiences with the *other* path to hybrid parallelism: We start with PARDISO, a shared memory parallel direct solver for large sparse systems of linear equations [7] that has recently been included into the Intel Math Kernel Library (MKL) [5]. In order to support message passing, we introduce a modified version of the OpenMP-parallel central numerical factorization algorithm that accounts for most of the solver's runtime. The modified version uses a dynamic task-queue instead of OpenMP's `parallel for` construct for work sharing. The introduction of message passing functionality is then straightforward by adding new tasks related to message passing.

The rest of the paper is organized as follows: Section 2 provides some general information about PARDISO and describes the current OpenMP implementation of the central numerical factorization algorithm. Section 3 describes the new task-queue based implementation of this algorithm which is the basis for the message passing extensions described in Section 4. Then, in Section 5, we compare the performance of the different incarnations of our implementation (sequential, pure OpenMP, pure MPI, and hybrid OpenMP+MPI) on a number of test matrices. Finally, Section 6 concludes and discusses future work.

2 The Sparse Direct Solver PARDISO

PARDISO is a high performance direct solver for sparse systems of linear equations which arise in application areas such as semiconductor device simulation.

In this work we are only concerned with PARDISO's supernodal[1] left-right looking numerical factorization algorithm that accounts for most of the runtime of the solver. For a comprehensive discussion of PARDISO please consult [9–11].

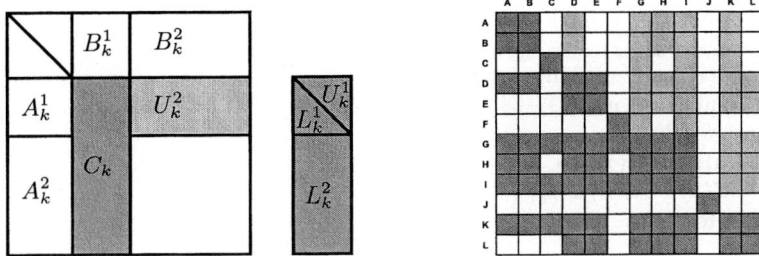

Fig. 1. Block-based LU factorization (left) and example sparse matrix with supernodes {A,B}, {C}, {D,E}, {F}, {G,H,I}, {J}, {K,L} (right).

To motivate the algorithm used by PARDISO consider the block-based LU factorization depicted in Fig. 1. In the k-th elimination step, the k-th block row and the k-th block column are computed by performing the following operations:

1. The external modifications ('*updates*') of the block columns of L and block rows of U:
$$C_k \leftarrow C_k - \begin{pmatrix} A_k^1 \\ A_k^2 \end{pmatrix} B_k^1, \quad U_k^2 \leftarrow U_k^2 - A_k^1 B_k^2. \tag{1}$$

2. The internal *factorization* of the diagonal block of C_k, to obtain the factors L_k^1 and U_k^1, followed by the internal factorization of the block columns of L and rows of U:
$$L_k^2 \leftarrow L_k^2 (U_k^1)^{-1}, \quad U_k^2 \leftarrow (L_k^1)^{-1} U_k^2. \tag{2}$$

---

[1] A supernode is defined as a group of consecutive rows/columns with similar non-zero structure.

Similar operations are performed in the supernodal approach used by PARDISO. However, instead of *one* external update involving all previously factored blocks (all blocks to the left), the update is split into several smaller supernode updates, of course only the non-zero supernodes need to be considered in this case. For example, in the sparse matrix shown in the right part of Fig. 1, supernode {D,E} is updated by {A,B} but not by {C}.

The original implementation of this algorithm is shown in Fig. 2 where $\text{nup}(j)$ denotes the number of outstanding external updates on supernode j, $\mathcal{U}(j)$ is the set of all supernodes that are updated *by* supernode j and $\mathcal{L}(j)$ holds all 'ready' supernodes that update supernode j.

The algorithm is called left-right looking, because for the factorization of supernode j, all updates from 'left' supernodes are considered (i.e, those with smaller index) and as soon as the factorization of j is finished, all 'right' supernodes that are updated by j are informed that j is ready.

```
 1: #pragma omp parallel for
 2: for j = 1 ... N_supernodes do
 3:     while nup(j) > 0 do
 4:         wait for i ∈ L(j)
 5:         L ← L\{i}
 6:         perform j ← i supernode update
 7:         nup(j) ← nup(j) −1
 8:     end while
 9:     for all k ∈ U(j) do
10:         L(k) ← L(k) ∪ {j}
11:     end for
12: end for
```

Fig. 2. PARDISO's original factorization algorithm.

It is not obvious how the algorithm shown in Fig. 2 can be extended with message passing functionality. The easiest way would be to use one OpenMP thread for MPI communication that could not participate in the main work-sharing `for` loop. Our approach is instead the re-organization of the algorithm using a central task-queue as described in the next section. Sending and receiving messages can then be handled quite easily by adding new task types related to message passing.

3 Task-Queue Based Re-organization

The central data structure of the new OpenMP version is a task-queue (a linked list of task descriptors). Fig. 3 shows a statechart of the new algorithm. The task-queue is initialized with all supernodes that are ready to be factored (i.e., those that do not require any external updates). Then, until all supernodes are

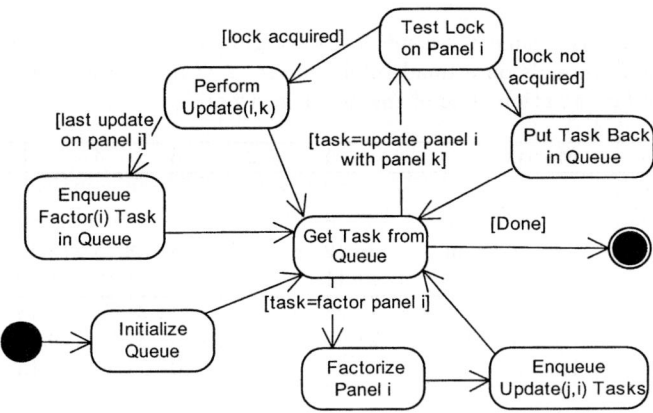

Fig. 3. Statechart of the new OpenMP factorization algorithm.

factored, each thread fetches a task descriptor from the head of the queue and performs the associated action.

There are two types of tasks: FACTOR(j) is the internal factorization of the supernode j. After the factorization of supernode j is finished, UPDATE(k,j) tasks are enqueued for all supernodes k that are updated by supernode j.

The UPDATE(k,j) tasks are the second type of tasks. A thread that performs an UPDATE(k,j) task checks if there are further outstanding UPDATE($k,\cdot$) tasks. If this is not the case, the supernode is ready to be factored and a FACTOR(k) task is enqueued.

In contrast to the original version, the new OpenMP version has the advantage of better load balancing since the fine grained tasks are dynamically distributed to the whole thread set. A potential disadvantage is the increased requirement for thread synchronization. Access to the task-queue is protected by using OpenMP's critical section construct. Furthermore, concurrent updates to the same supernode by different threads are avoided by using OpenMP locks.

Table 1 shows a performance comparison between the original and the new OpenMP version for the six test matrices listed in Tab. 2. Evidently, the new version is about ten percent slower for two threads and around 20 percent slower for four threads. The difference can be attributed to the increased synchronization requirements mentioned and a less advantageous order of the supernode updates with respect to cache misses.

4 Adding Message Passing Functionality

It is straightforward to add message passing functionality to the implementation described in section 3. The statechart of the hybrid version is shown in Figure 4, the new states related to MPI are shown in grey. Each MPI process consists of a number of OpenMP threads and "owns" a number of supernodes, implying its responsibility to perform all external updates on those supernodes as well

Table 1. Factorization time in seconds for the original and the new OpenMP algorithm for two and four threads on a quad Intel Itanium 2 SMP system. Sequential denotes the sequential factorization time of the new algorithm.

Matrix		1	2	3	4	5	6
Sequential		64.17	158.60	116.32	109.93	398.65	314.12
2 Threads	Original Version	32.58	76.41	56.02	53.05	192.88	153.30
	New Version	36.12	84.09	62.28	58.55	210.48	165.04
	Ratio	1.11	1.10	1.11	1.10	1.09	1.08
4 Threads	Original Version	17.91	41.24	30.44	28.78	103.23	82.91
	New Version	21.65	48.75	36.59	34.66	119.18	93.61
	Ratio	1.21	1.18	1.20	1.20	1.15	1.13

as the internal factorization. The assignment of supernodes to MPI processes is currently a simple static round-robin assignment.

Similar to the OpenMP version, the task-queue of each process in initialized with the supernodes that are ready to be factored. However, in the MPI version, each MPI process only adds the supernodes it owns. Furthermore, when process P_1 factors supernode k it enqueues only those UPDATE(j,k) tasks that affect the supernodes j owned by P_1. In addition P_1 sends the supernode k (i.e., the L and U factors and pivoting data) to all other MPI processes that own supernodes that are updated by k.

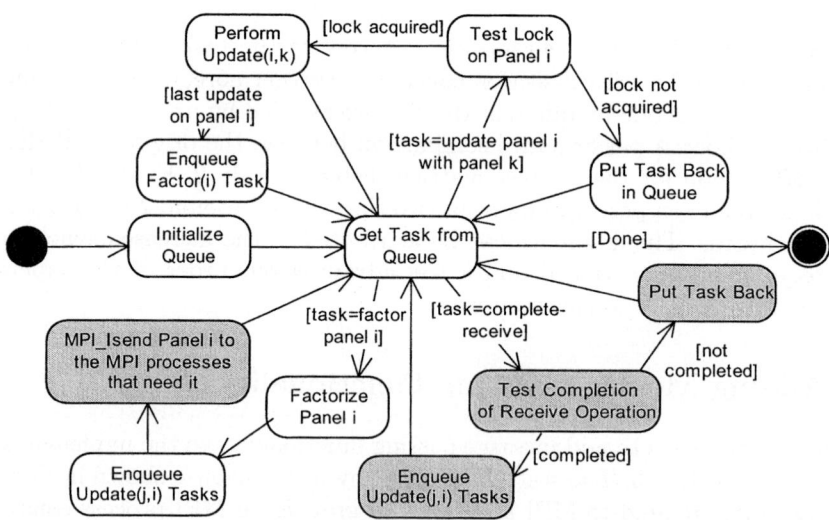

Fig. 4. Statechart of the hybrid OpenMP/MPI factorization algorithm.

To overlap communication with computation, asynchronous MPI operations are used. When an MPI process P_1 transfers a supernode to process P_2 it first sends an integer denoting the affected panel. Then it posts asynchronous send (MPI_Isend()) operations that transfer the actual panel data (the L and U factors and pivoting data).

On the receiving side, on each iteration of the main work loop, a thread first checks for incoming MPI messages. If an MPI_Iprobe() indicates an incoming supernode j, the thread posts corresponding asynchronous receive operations (MPI_Irecv()) for the L and U factors of the supernode and for the pivoting data. It then creates a COMPLETE_RECEIVE(j) task and places it on the task-queue. Then the thread proceeds as usual with its iteration of the work loop.

When a thread fetches a COMPLETE_RECEIVE(j) task from the queue, it uses MPI_Testall() to check whether all asynchronous operations related to panel j have finished. If this is not the case, the task is put back in the queue. Otherwise, all local UPDATE($\cdot,j$) tasks are enqueued just as if the panel would have been factored locally.

5 Results

The new OpenMP approach and the message passing extensions have been implemented in a way that allows us to derive four different versions of the application from the same source code. seq is the sequential version, omp.a is the pure OpenMP version with a threads, mpi.a (n) is a pure MPI version with a MPI processes on each of n nodes and hyb.a:b (n) is the hybrid version running on n nodes with a MPI processes on each node that consist of b OpenMP threads.

The programs were tested on an Itanium 2 cluster with four nodes. Each node ist built up of four Itanium 2 ('Madison') Processors with 1.3 GHz, 3 MB third level cache and 8 GB of main memory. The nodes are connected by Mellanox 4x Infiniband [3] HCAs. We used MVAPICH (based on MPICH 1.2.5, OSU patch 0.9.2) and the Intel C and Fortran compilers in version 7.1. The tests were performed with the matrices shown in Tab. 2.

Table 2. The test matrices. N is the order of the matrix, nnz is the total number of non-zeros and n_{super} is the number of supernodes identified by the pre-processing stages of PARDISO.

Number	Matrix	N	nnz	n_{super}
1	3D_75932_sef3D	75932	88,528,964	23,112
2	barrier2-4	113076	175,824,468	22,352
3	matrix-ibm-watson	125329	147,872,909	27,825
4	matrix_9	103430	129,396,404	23,447
5	ohne1	181343	417,744,155	24,099
6	para-2	153226	287,734,810	29,781

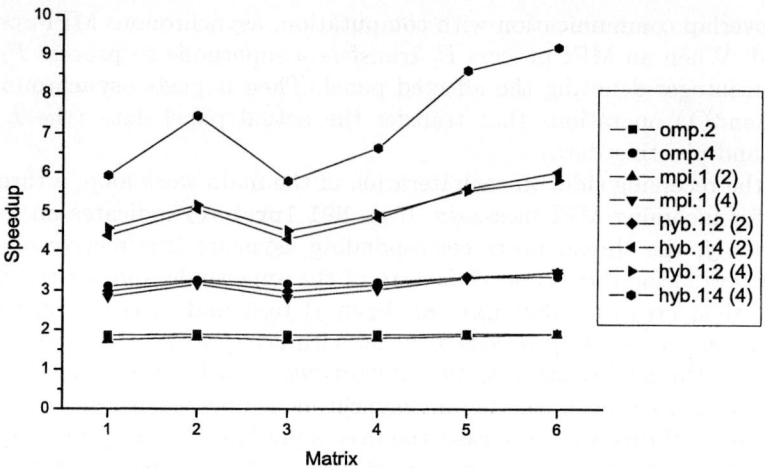

Fig. 5. Speedup relative to the sequential version.

Fig. 5 shows the speedup of various versions of our programs relative to the sequential version. The following conclusions can be derived:

- The MPI version is usually slightly slower than the OpenMP version on the same number of processors, the hybrid version is in between.
- The larger processor configurations perform better on the larger matrices.
- The hyb.1:2 (4) variant is usually slightly faster than the hyb.1.4 (2) version. This indicates that the decreased memory bus load (for two threads on one node) outweighs the additional message passing requirements.
- A speedup of more than nine can be achieved using a hybrid version running on all 16 processors.
- MPI variants with more than one MPI process on a node (not shown in Fig. 5) usually perform worse than variants with only one MPI process per node (e.g., mpi.1 (4) vs. mpi.2 (2)). This is due to the large memory requirement of the program (more than five GB for the largest matrices) and the fact that all data is replicated by the MPI processes.
- We additionally measured the communication time using mpiP [6]. Depending on the particular input matrix, the total aggregated time spent in MPI functions is in the range of 5–9%, 7–15% and 10–20% for 2, 3 and 4 MPI processes, respectively. The number is typically smaller for larger matrices and higher for the smaller exemplars. The most MPI time (around 30%) is spent in the MPI_Iprobe function (which is called very frequently). The receive operations for the panel (despite being asynchronous) and the MPI_Testall to check whether the panel has been transferred also contribute significantly.

6 Conclusion and Future Work

We have presented our experiences with the hybridization of an application using OpenMP work-sharing constructs. Our approach was to re-organize the application to use a central task-queue instead of the work-sharing constructs. This yields a very flexible solution and it is then straightforward to add message-passing extensions.

Adding message passing capabilities allows the solver to run on a much larger number of processors beyond the confines of a single SMP system. We have shown that reasonable performance can be achieved with a speedup of more than nine on 16 processors.

A number of optimizations have not yet been implemented in the current version of our hybrid code. Most notable, PARDISO already uses METIS [4] for graph-partitioning in its two-level scheduling approach [9] to optimize cache misses. This partitioning will be very effective to minimize the number of messages sent between processes once it is included in the hybrid version.

References

1. Edmond Chow and David Hysom. Assessing performance of hybrid MPI/OpenMP programs on SMP clusters. Technical Report UCRL-JC-143957, Lawrence Livermore National Laboratory, May 2001. submitted to J. Parallel and Distributed Computing.
2. D. S. Henty. Performance of hybrid message-passing and shared-memory parallelism for discrete element modeling. In *Proceedings of the 2000 ACM/IEEE conference on Supercomputing*. IEEE Computer Society, 2000.
3. Infiniband Trade Association. http://www.infinibandta.org/home.
4. G. Karypis and V. Kumar. A fast and high quality multilevel scheme for partitioning irregular graphs. *SIAM Journal on Scientific Computing*, 20(1):359–392, 1998.
5. Intel Math Kernel Library.
   http://www.intel.com/software/products/mkl/beta/features.htm.
6. mpip: Lightweight, scalable mpi profiling. http://www.llnl.gov/CASC/mpip/.
7. Pardiso website. http://www.computational.unibas.ch/computer_science/scicomp/software/pardiso/.
8. Rolf Rabenseifner. Hybrid parallel programming: Performance problems and chances. In *Proc. 45th Cray Users's Group (CUG) Meeting*, May 2003.
9. Olaf Schenk and Klaus Gärtner. Two-level scheduling in PARDISO: Improved scalability on shared memory multiprocessing systems. *Parallel Computing*, 28:187–197, 2002.
10. Olaf Schenk and Klaus Gärtner. Solving unsymmetric sparse systems of linear equations with PARDISO. *Future Generation Computer Systems*, 2003.
11. Olaf Schenk, Klaus Gärtner, and Wolfgang Fichtner. Efficient sparse LU factorization with left-right looking strategy on shared memory multiprocessors. *BIT*, 40(1):158–176, 2000.
12. Lorna Smith. Mixed mode MPI/OpenMP programming. Technical Report EH9 3JZ, Edinburgh Parallel Computing Centre, 2000.

Co-array Python: A Parallel Extension to the Python Language

Craig E. Rasmussen[1], Matthew J. Sottile[1], Jarek Nieplocha[2], Robert W. Numrich[3], and Eric Jones[4]

[1] Los Alamos National Laboratory, Los Alamos, NM 87545, USA
[2] Pacific Northwest National Laboratory
[3] Minnesota Supercomputing Institute, University of Minnesota, Minneapolis, MN 55455
[4] Enthought, Inc.

Abstract. A parallel extension to the Python language is introduced that is modeled after the Co-Array Fortran extensions to Fortran 95. A new Python module, CoArray, has been developed to provide co-array syntax that allows a Python programmer to address co-array data on a remote processor. An example of Jacobi iteration using the CoArray module is shown and corresponding performance results are presented.

1 Introduction

There have been several attempts at extending programming languages for use in a parallel processing environment. These language extensions include, but are not limited to, C* [1], Unified Parallel C (UPC) [2], Titanium (a Java extension) [3], F^{--} (a Fortran 77 extension) [4], High Performance Fortran (HPF) [5], and Co-Array Fortran (CAF) [6]. The existence of these parallel extensions indicate the popularity of the idea, but as yet, none have had the nearly universal acceptance of MPI [7] as a parallel programming model.

The Python language [8] provides a rapid prototyping environment and several extension modules (mostly serial) have been created for the scientific community [9]. A notable Python feature is that many elements of the language can be modified and extended, allowing tool developers to use existing syntactic elements to express new, more sophisticated operations. Furthermore, Python can be extended with "plugin" modules implementing these new linguistic features, rather than requiring a new or modified compiler. Portability is dictated by the portable implementation of these specific modules, unlike the significantly larger compiler suites for earlier parallel languages and extensions.

In this research note, we present a parallel Python extension module, CoArray, as a way of providing co-array notation and functionality to Python programmers. A co-array is a local data structure existing on all processors executing in a Single Program Multiple Data (SPMD) fashion. Data elements on non-local processors can be referenced via an extra-array dimension, called the co-dimension.

In this implementation, the Aggregate Remote Memory Copy (ARMCI) library [10] is used to facilitate memory transfer between co-array elements existing in memory on remote processors. ARMCI has also been used to implement the Global Array library, GPSHMEM – a portable version of Cray SHMEM library, and the portable Co-Array Fortran compiler from Rice University [11]. ARMCI provides simpler progress rules and a less synchronous model of RMA than MPI-2.

2   Co-array Python Implementation

The `CoArray` Python module implements co-arrays as an extension to Numeric Python [9], though other Python numerical array libraries could be used. The Numeric third-party module provides higher performance than do standard Python arrays. `CoArray` extends the Python array syntax by adding a set of parentheses (the co-dimension) to the standard array notation. For example, consider a two dimensional array A and a load operation from A into the scalar, $c = A[i, j]$ from row i and column j. If T is a co-array, then a similar load operation would be expressed as $c = T(k)[i, j]$, except in this case, the scalar is loaded with data from row i and column j on *processor k*, not necessarily the local processor. Note that the co-dimension k appears within parentheses to distinguish it from the normal Python array indices appearing within square brackets.

While this example illustrates load and store operations for a scalar, the truly powerful features are slicing operations. Using Python slices, one can transfer entire regions of a co-array in a single Python statement. For example, the statement $T(0)[-1,:] = T[0,:]$ puts the entire first row from T to the last row (index of -1) of T on processor 0. Assuming that the local processor index is 1, this could also have been written as $T(0)[-1,:] = T(1)[0,:]$.

The `CoArray` module uses the ARMCI library to transfer data (although nothing prohibits it from being implemented on top of other data-transport layers). ARMCI provides general-purpose, efficient, and widely portable remote memory access (RMA) operations (one-sided communication). ARMCI operations are optimized for contiguous and noncontiguous (strided, scatter/gather, I/O vector) data transfers. It also exploits native network communication interfaces and system resources such as shared memory [12]. Because very little processing occurs in Python, and because no extra copies of data are made (memory for the Numeric arrays are actually allocated by ARMCI), memory transfer operations using the `CoArray` module are roughly comparable to a C implementation, as will be shown in the example in next section.

It should also be noted that the local portion of a co-array can easily be shared with C as pointers to local co-array data are readily available. In addition, using the Chasm array descriptor library [13], one can assign local co-array data to an assumed shape, Fortran 90 array pointer.

2.1 Implementation Details

The key to the parallel `CoArray` module implementation is the Python `__call__`, `__setitem__` and `__getitem__` methods. The `__call__` method is invoked by the interpreter when the co-dimension is selected (using parenthesis notation) and returns the local Numeric array if the co-dimension index is local or a proxy to the remote co-array otherwise. The `__setitem__` method is invoked when the normal array dimensions are selected (using square bracket notation) on the left side of an assignment statement. When `__setitem__` is called on a remote proxy, the ARMCI library is used to put data to the remote co-array. Likewise, when `__getitem__` is called on a remote proxy, the ARMCI library is used to get data from the remote array. Otherwise, the `__getitem__` and `__setitem__` methods are forwarded to the local Numeric array.

3 Co-array Python Example

To illustrate the simplicity and expressive power of Co-Array Python syntax, we consider the two-dimensional Laplace equation on a square of size (M × M). We cut the square into horizontal strips by dividing the first dimension by the number of processors N=M/nProcs. Each processor allocates its own co-array,

```
T = coarray((N+2,M+2), Numeric.Float)
```

which contains a local strip with a halo of width one. The halo values on the boundary of the global square enforce Derichlet boundary conditions, but the halo values on interior boundaries, row boundaries in our case, must be updated with data from neighboring strips after each iteration.

The following variables are defined:

```
nProcs = mpi.size
me = mpi.rank

up = me - 1
dn = me + 1
if me == 0: up = None
if me == nProcs - 1: dn = None
```

where `me` is the local processor index and `up` and `dn` are the neighboring processor indices on which the logically up and down array strips are allocated. As can be seen by the use of the pyMPI [14] module elements, `mpi.size` and `mpi.rank`, the `CoArray` module has been designed to be used with MPI.

The inner, iterative Python loop, executed in SPMD fashion by each processor, is:

```
1 """ update interior values """
2 T[1:-1,1:-1] = ( T[0:-2,1:-1] + T[2:,1:-1] +
3                  T[1:-1,0:-2] + T[1:-1,2:] ) / 4.0
```

```
4 """ exchange boundary conditions """
5 mpi.barrier()
6 if up != None: T(up)[-1,:] = T[ 1,:]
7 if dn != None: T(dn)[ 0,:] = T[-2,:]
8 mpi.barrier()
```

Each processor replaces each value in its local strip by the average of the four surrounding values. Standard Python syntax allows us to represent this averaging in very compact notation (lines 2-3, above). The first two terms on the right side represent values up and down and the last two terms represent values left and right. This code corresponds to a true Jacobi iteration because the Numeric Python array module computes the entire result on the right side of the statement before storing the result to the left side. A Jacobi iteration written in C requires two arrays to avoid polluting the new solution with partially updated values. Two arrays are not needed in Python because temporary arrays are created as binary expressions on the right side of the equation are evaluated.

No communication between processors takes place during the averaging. It is all local computation. To update the halos after averaging, we need two barriers. The first barrier guarantees that all processors have finished computing their average using the old values before any processor updates the halos with new values. The second barrier guarantees that all processors have finished updating halos before any processor performs the next average using the new values.

We also coded the example in Python, using the pyMPI module, and in C with MPI to compare with Co-Array Python. We ran all three versions on a dual processor Macintosh G4 (1 GHz, 1.5 Gbyte) using Mac OS X version 10.3.2 and Python version 2.3. We used LAM/MPI version 7.0.4 for the C and Python code and ARMCI version 1.1 for the Co-Array Python code.

Table 1 shows timing results. Note that communication times for Co-Array Python are much shorter than those for pyMPI due to the need for Python to serialize (pickle) every message before sending it. This requires a heap allocation, a copy, and additional processing [14]. Co-Array Python is able to send data with no extra memory copies and communication times are roughly two times those of the C version.

Code complexity is an important metric for evaluating programming models. The Co-Array Python code requires less than half the number of statements,

Table 1. Timing data (seconds) for the Co-Array Python (CoP), MPI Python (PyMPI), and C (C) versions. Data are an average of 5 runs, each for 40 iterations. The Python MPI version failed to complete the 2048x2048 run.

Size	CoP_{comm}	CoP_{total}	$PyMPI_{comm}$	$PyMPI_{total}$	C_{comm}	C_{total}
128x128	0.017	0.33	0.07	0.38	0.013	0.05
256x256	0.023	1.28	0.13	1.41	0.015	0.14
512x512	0.041	6.28	0.28	6.47	0.020	0.55
1024x1024	0.068	28.4	0.52	28.78	0.032	2.49
2048x2048	0.089	113.5			0.047	10.13

six versus thirteen, to represent data transfer compared with the pyMPI code. The additional code in the MPI version was required to avoid synchronization deadlock.

Although not encountered in this simple example, the C MPI version would require an additional level of complexity, if data were partitioned across processors by blocks, rather than by strips. This would require the transfer of noncontiguous halo data along columns by the use of an MPI_Type_vector to specify strides. The Co-Array Python module transfers data stored in noncontiguous memory transparently, as the module provides the necessary stride information to the ARMCI library, rather than placing the burden on the user. This is also true of the pyMPI version, although it would suffer the same performance penalties discussed above.

4 Future Work

This research note describes an implementation of co-arrays in Python. The co-array syntax provides a concise mechanism for implementing parallel applications while hiding the underlying communication details, making it ideal for rapid prototyping. While computation was dominant in our simple example, other algorithms exist where communication is more of a concern. Therefore, we will pursue performance optimizations in the implementation of the CoArray module to take advantage of asynchronous transfers to provide overlap with computation.

Performance may be improved by changing when a data transfer occurs relative to when it is posted. An eager evaluation approach, which is currently implemented, forces immediate transfer. Varying degrees of laziness in this evaluation could defer the actual transfer until a later time, possibly eliminating them altogether if the data are not used. A lazy implementation of this extension could take advantage of data locality of several posted communication requests to aggregate multiple small communication requests into larger, coarser grained messages that perform well on modern communication hardware.

Finally, we would like to implement the CoArray module in C for performance and on top of other communications libraries to broaden the number of users who can take advantage of this library, with particular attention to compatibility with extensions and tools included as part of the Scientific Python (SciPy) distribution [9].

Acknowledgments

While completing this work, Craig Rasmussen and Matthew Sottile were funded by the Mathematical Information and Computer Sciences (MICS) program of the DOE Office of Science. This research was also supported in part by grant DE-FC02-01ER25505 from the U.S. Department of Energy as part of the Center for Programming Models for Scalable Parallel Computing sponsored by the Office of Science. We would like to thank Sung-Eun Choi for suggesting the idea of implementing co-arrays in Python.

References

1. Thinking Machines Corporation, Cambridge, Massachusetts: C* Language Reference Manual (1991).
2. Carlson, W.W., Draper, J.M., Culler, D.E., Yelick, K., Brooks, E., Warren, K.: Introduction to UPC and language specification. Technical Report CCS-TR-99-157, Center for Computing Sciences, 17100 Science Drive, Bowie, MD 20715 (1999) http://www.super.org/upc/
3. Yelick, K., Semenzato, L., Pike, G., on Miyamoto, C., Liblit, B., Krishnamurthy, A., Hilfinger, P., Graham, S., Gay, D., Colella, P., Aiken, A.: Titanium: A high-performance Java dialect. Concurrency: Practice and Experience **10** (1998) 825–836.
4. Numrich, R.W.: A parallel extension to Cray Fortran. Scientific Programming **6** (1997) 275–284.
5. Koebel, C.H., Loveman, D.B., Schrieber, R.S., Steele, G.L., Zosel, M.E.: The High Performance Fortran Handbook. The MIT Press, Cambridge, Massachusetts (1994).
6. Numrich, R.W., Reid, J.K.: Co-Array Fortran for parallel programming. ACM Fortran Forum **17** (1998) 1–31 http://www.co-array.org/
7. Gropp, W., Lusk, E., Skjellum, A.: Using MPI, portable parallel programming with the Message-Passing Interface. The MIT Press, Cambridge, Massachusetts (1994).
8. van Rossum, G., Drake, F. L., Jr. (ed): Python Reference Manual. (2003) http://www.python.org/
9. Scientific Python web site (2004) http://www.scipy.org/
10. Nieplocha, J., Carpenter, B.: ARMCI: A portable remote memory copy library for distributed array libraries and compiler run-time systems. In: Proceedings of RTSPP/IPPS99 (1999).
11. Coarfa, C., Dotsenko, Y., Eckhardt, J., Mellor-Crummey, J.: Co-array Fortran performance and potential: An NPB experimental study. In: Proceedings of LCPC 2003 (2003).
12. Nieplocha, J., Ju, J., Straatsma, T.P.: A multiprotocol communication support for the global address space programming model on the IBM SP. In: Proceedings of EuroPar 2000 (2000).
13. Chasm language interoperability web site (2004) http://chasm-interop.sf.net/
14. Miller, P.: pyMPI – An introduction to parallel Python using MPI (2002) http://pympi.sourceforge.net/

Targeting Heterogeneous Architectures in ASSIST: Experimental Results*

M. Aldinucci[2], S. Campa[1], M. Coppola[2], S. Magini[1], P. Pesciullesi[1], L. Potiti[1], R. Ravazzolo[1], M. Torquati[1], and C. Zoccolo[1]

[1] Dept. of Computer Science – University of Pisa – Viale Buonarroti 2, Pisa, Italy
[2] Inst. of Information Science and Technologies – CNR, Via Moruzzi 1, Pisa, Italy

Abstract. We describe how the ASSIST parallel programming environment can be used to run parallel programs on collections of heterogeneous workstations and evaluate the scalability of one task-farm real application and a data-parallel benchmark, comparing the actual performance figures measured when using homogeneous and heterogeneous workstation clusters. We describe also the ASSIST approach to heterogeneous distributed shared memory and provide preliminary performance figures of the current implementation.

Keywords: Structured parallel programming, heterogeneous workstation network, shared memory.

1 Introduction

A notable problem when dealing with parallel programming environments is the ability to produce code for heterogeneous networks/clusters of workstations. Although some versions of MPI (e.g. LAM-MPI) allow heterogeneous collections of PEs to be used, other versions do not support such feature, nor it is supported by most of the other parallel programming systems, including HPF.

ASSIST is a parallel programming environment based on the concepts of coordination languages and algorithmical skeletons, recently developed at the University of Pisa [1, 2]. The latest version (1.2) of the ASSIST programming environment supports both interoperability with other classical distributed-processing frameworks (e.g. CORBA) and the possibility to run portions of the same application on machines with different processors and operating systems.

The former is presented in [3], so this paper presents the latter. In Sect. 2 we discuss the current implementation of point-to-point communication and shared memory in an heterogeneous environment, highlighting strengths and weaknesses. In Sect. 3 the performance obtained running existing ASSIST applications and benchmarks on heterogeneous platforms are evaluated and compared with those obtained on homogeneous ones. Sect. 4 surveys related work.

---

* This work has been supported by the Italian MIUR FIRB Grid.it project, n. RBNE01KNFP, on High-performance Grid platforms and tools; the Italian MIUR Strategic Project L.449/97-2000, on High-performance distributed enabling platforms.

2 Heterogeneous Network/Cluster Targeting

The ASSIST compiler can produce code running onto an heterogeneous cluster/network, composed of processing elements with different processors (word size, endianness) and/or different operating systems. In the experiment section we will show that the incurred overhead is tolerable, therefore efficiency and performance figures are preserved w.r.t. the homogeneous case.

The compilation process has been designed [2, 4] to generate code for heterogeneous clusters/workstation networks[1]. When compiling for an heterogeneous platform, the ASSIST compiler produces a full set of object files for every configured architecture, and a global configuration file, that can be used to start the application on a set of heterogeneous computing nodes, by means of a simple invocation of the ASSIST loader. In an heterogeneous run, processes participating in the computation of the same ASSIST-CL parallel pattern (e.g. task farm, pipeline, parmod, ...) can be scheduled on processing elements having different CPU/operating system combinations, with minimal performance impact.

The ACE library [5] provides an hardware and operating system abstraction layer to the ASSIST runtime. It also provides standard routines, based on the CDR format (a flavor of XDR), to exchange data between processing elements with different architectures, preserving their original semantics independently of the endianness or of the machine word size used on the different machines. The provided routines handle all CORBA basic types (octets, integers, floating point values, ...) as well as unidimensional arrays of basic types. The ASSIST compiler inductively builds the conversion routines for more complex data structures (structured types, multidimensional arrays), using the ones for the basic types. This process (unlike in MPI), is done at compile time, and exploits the inlining facility of the native C++ compiler, so the produced conversion routines are quite fast. The CDR protocol prescribes that the sender doesn't encode data, but enriches the message with its byte-order. The receiver compares the sender byte-order with its own, and if they are different, applies the conversion.

The CDR based approach incurs only one significant overhead: since the memory layout of the data structures that must be communicated can be different on different architectures even if the byte-order is the same (due to word-size or alignment requirements), the protocol defines an architecture independent layout, that can be interpreted on every supported architecture. So data structures must be copied when sending and when receiving, even if byte-order conversion is not needed. This enlarges the memory footprint of the parallel program (a noticeable effect if the communicated structures are large), and can produce cache pollution effects.

We are currently working on an heterogeneity-enabled version of the shared memory library integrated in ASSIST [6]. The library provides the ASSIST programmer with a set of primitives that allow to allocate, read, write and deallocate

---

[1] Although at the moment only Linux and MacOS X machines has been configured and tested, the ASSIST compiler and runtime support already contains all the hooks needed to target other architectures/operating systems (namely the ones supported by the Hardware Abstraction Layer, based on the ACE library).

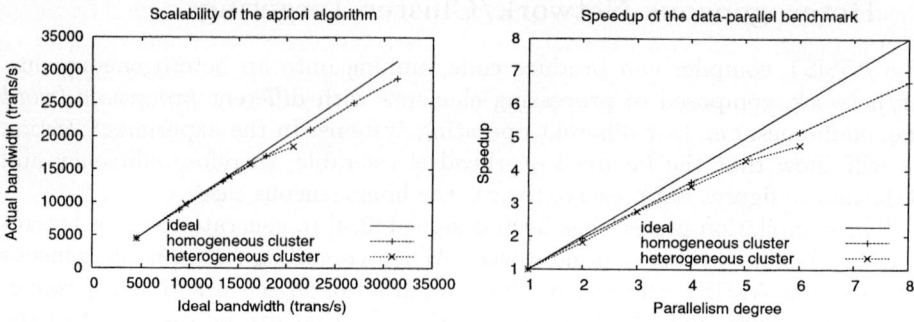

Fig. 1. Comparison of homogeneous and heterogeneous performance figures.

segments of distributed shared memory. Types are necessary in order to select the proper conversion routines, so we defined a typed API for the shared memory operations, exploiting the C++ template mechanism. In the implementation we reused the same conversion routines built for inter-process communication, but with a different philosophy: in the interaction through shared memory, the memory itself plays the role of a third party between the interacting entities, unaware of the data type that is being exchanged. Moreover, the receiver doesn't know the identity of the sender (it can happen that different entities write different entries in one array and a single entity read the array as a whole). To arrange for this, we decided to always encode data written to the shared space in an external data format, with a well defined layout and byte-order, and symmetrically decode data read from the memory. The chosen layout guarantees that all the elements of an array have the same layout (surprisingly, this is not true for plain CDR layout, when dealing with structured data types), in order to be able to read/write portions of arrays.

3 Performance Evaluation

We are going to show experimental results concerning application scalability, comparing the performance figures obtained with an homogeneous platform with those obtained with an heterogeneous one.

Environment. The homogeneous tests are run on a cluster of 8 Pentium IV 2GHz, equipped with 512MB of RAM and interconnected by 100/1000 Mbit/s Ethernet.

The heterogeneous tests are run on a cluster of 4 Pentium IV (same configuration), a fast PowerPC G4 (1.5GHz, 512MB RAM, 100/1000 Mbit/s Ethernet) and a slower one (800MHz, 256MB RAM, 100 Mbit/s Ethernet).

Apriori algorithm. The first algorithm we tested is a parallel implementation the Apriori data mining algorithm [7] for finding association rules in a transaction database. The transaction database is split in a stream of partitions that feeds a task-farm; the farm workers compute the frequent itemset independently in each partition, and then the partial results are accumulated to build a superset of the solution; in a second parallel scan of the database (another task-farm working on the stream of partitions) the partial solutions are ranked and a complete solution is found. In this implementation, the database partitions are sent over the ASSIST streams, to exploit the data conversion mechanisms between heterogeneous architectures.

The scalability for the homogeneous and the heterogeneous executions are compared in figure 1-left. The heterogeneous machines have different computational power: the Pentiums can process 4478 transactions per second, while the fast and the slow PowerPC can process respectively 5079 and 2174 transactions per second. The heterogeneous configurations tested are all the prefixes of the sequence [Pentium, fast PPC, Pentium, slow PPC, Pentium]. For different heterogeneous configurations, we computed the ideal bandwidth (x axis) as the sum of the bandwidths of the machines employed, and measured the delivered bandwidth (y axis). The curve obtained for the heterogeneous case is comparable with the homogeneous one. It departs from the ideal (reaching 0.9 of efficiency) only when the machine with slow NIC is employed: the application, in fact, has high network bandwidth requirements, and the introduction of a node with a slow network slightly decreases the efficiency of the task distribution process.

Data-parallel benchmark. The second algorithm tested is a synthetic data-parallel benchmark, with a variable communication stencil and featuring a good computation to communication ratio The benchmark implements an iterative computation over a square matrix M: at iteration h the hth row of M is broadcast to all the processing elements; the new value for the matrix M' is computed as $M'_{i,j} = N \cdot sin(\sum_k M_{h,k} \cdot M_{j,(j+k) \bmod N})$.

The ASSIST support currently implements only a naive partitioning strategy for data-parallel computations, in which all the partitions have the same size, even if the computation power of the employed machines differs. We can therefore compute the speedup of the program against the sequential time of the slowest machine. The sequential times for Pentiums, fast PPC and slow PPC were respectively 529, 113 and 197 seconds. The difference in the execution times between the Pentiums and the PowerPCs is considerable: the Pentiums are disadvantaged in number crunching codes because of the small number of general registers; register pressure, in fact, prevents the compiler from optimizing complex matrix access patterns.

The heterogeneous configurations tested are all the prefixes of the sequence [Pentium, fast PPC, Pentium, slow PPC, Pentium, fast PPC again]; in the maximal configuration we mapped two workers on the fastest machine, that otherwise would be idle most of the time. Figure 1-right displays the speedups obtained in homogeneous as well as heterogeneous runs. The overhead introduced by heterogeneity is negligible (less than 5%). The inability to adjust the partition sizes proportionally to the machine powers, instead, is limiting; moreover the solution of running more workers on faster machines introduces some inefficiencies (the heterogeneous speedup curve, in fact, loses linearity when the second worker is added). We are now considering to enhance the partitioning strategy to handle computational power heterogeneity, as well as dynamic changes in the available computational power.

Shared memory performance. Here we provide the first performance results regarding the described implementation of the heterogeneous shared memory library integrated in ASSIST. The benchmark employed allocates one segment of shared memory (10M integers); several processes in parallel read randomly chosen chunks (16k integers) of this memory and write them in other locations. The accesses are not synchronized. Figure 2 shows the aggregate bandwidth of the shared memory varying the parallelism degree (number of servers) on the 100Mbit/s network, both in homogeneous runs and in heterogeneous ones (the heterogeneous configuration has been enriched with two more Pentium IV machines, hosting two memory servers). The overhead introduced is always less than 6%, and decreases when the number of servers increase.

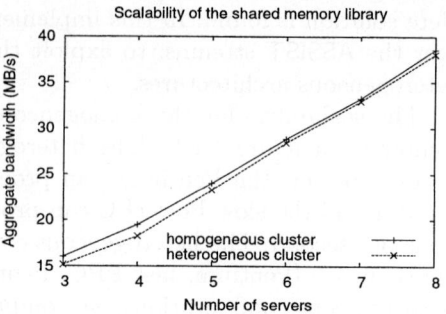

Fig. 2. Performance of the shared memory library ($\#clients = \#servers - 2$).

4 Related Work

PBIO [8] is a flexible communication library that handles heterogeneous communications using sender's native data representation: the receiver translates the message only if necessary, basing the decision on a message prefix describing the data format. Our approach, instead, adopts a consistent data layout, that is useful when we extend to shared memory. Mermaid [9] is the first example of an heterogeneous, transparent page-based DSM; in this scheme data is represented in the page holder's native form: this introduces several difficulties and limitations in the architectures that can be supported. Heterogeneous DSMs can benefit of type safe and reflective languages like Java. JavaDSM [10] is a prototypical implementation of a DSM that offers a single JVM image over a cluster of possibly heterogeneous workstations.

5 Conclusion

In this paper we described ASSIST support to heterogeneity, i.e. the ability to produce code for heterogeneous networks/clusters, and run portions of the same application (and even of the same parallel pattern) on machines with different processor architectures and operating systems.

We discussed the implementation of point-to-point communications, highlighting its strengths and weaknesses; we provide experimental evidence that this approach incurs in tolerable overhead, comparing the performance figures obtained by heterogeneous runs to those obtained by homogeneous ones of a real task-parallel program (the Apriori data mining algorithm) and a synthetic data-parallel benchmark.

Finally, we extended the approach to the shared memory library integrated in ASSIST and presented first experimental results showing that the impact of data translation on the achieved performance is small.

Acknowledgements

We wish to thank all the people that participated to the design and development of the ASSIST programming environment, in particular P. Ciullo, M. Danelutto, G. Giaccherini, A. Paternesi, A. Petrocelli, E. Pistoletti, P. Vitale, M. Vanneschi, G. Virdis.

References

1. Vanneschi, M.: The programming model of ASSIST, an environment for parallel and distributed portable applications. Parallel Computing **28** (2002) 1709–1732
2. Aldinucci, M., Campa, S., Ciullo, P., Coppola, M., Magini, S., Pesciullesi, P., Potiti, L., Ravazzolo, R., Torquati, M., Vanneschi, M., Zoccolo, C.: The Implementation of ASSIST, an Environment for Parallel and Distributed Programming. In Kosch, H., László Böszörményi, Hellwagner, H., eds.: Euro-Par 2003: Parallel Processing. Number 2790 in Lecture Notes in Computer Science (2003) 712–721
3. Magini, S., Pesciullesi, P., Zoccolo, C.: Parallel Software Interoperability by means of CORBA in the ASSIST Programming Environment. In: Euro-Par 2004 Parallel Processing. (2004) (to appear).
4. Aldinucci, M., Campa, S., Ciullo, P., Coppola, M., Danelutto, M., Pesciullesi, P., Ravazzolo, R., Torquati, M., Vanneschi, M., Zoccolo, C.: A Framework for Experimenting with Structured Parallel Programming Environment Design. In: ParCo 2003 Conference Proceedings, to appear, Dresden, Germany (2003)
5. Schmidt, D.C., Harrison, T., Al-Shaer, E.: Object-oriented components for high-speed network programming. In: Proceedings of the 1st Conference on Object-Oriented Technologies and Systems (COOTS), Monterey, CA, USENIX (1995)
6. Carletti, G., Coppola, M.: Structured Parallel Programming and Shared Objects: Experiences in Data Mining Classifiers. In Joubert, G.R., Murli, A., Peters, F.J., Vanneschi, M., eds.: Parallel Computing, Advances and current issues, Proceedings of the Int. Conf. ParCo 2001, Naples, Italy (4-7 September 2001), Imperial College Press (2002) ISBN 1-86094-315-2,.
7. Coppola, M., Vanneschi, M.: High-Performance Data Mining with Skeleton-based Structured Parallel Programming. Parallel Computing, special issue on Parallel Data Intensive Computing **28** (2002) 793–813
8. Eisenhauer, G., Bustamante, F.E., Schwan, K.: Native Data Representation: An Efficient Wire Format for High-Performance Distributed Computing. IEEE Transactions on Parallel and Distributed Systems **13** (2002) 1234–1246
9. Zhou, S., Stumm, M., Li, K., Wortman, D.: Heterogeneous Distributed Shared Memory. IEEE Trans. on Parallel and Distributed Systems **3** (1992) 540–554
10. W.Yu, A.Cox: Java/DSM: A Platform for Heterogeneous Computing. In: Proc. of ACM 1997 Workshop on Java for Science and Engineering Computation. (1997)

A Fusion-Embedded Skeleton Library

Kiminori Matsuzaki[1], Kazuhiko Kakehi[1], Hideya Iwasaki[2],
Zhenjiang Hu[1,3], and Yoshiki Akashi[2]

[1] Graduate School of Information Science and Technology,
University of Tokyo
{Kiminori_Matsuzaki,kaz,hu}@mist.i.u-tokyo.ac.jp
[2] Department of Computer Science,
The University of Electro-Communications
iwasaki@cs.uec.ac.jp,yoshiki@ipl.cs.uec.ac.jp
[3] PRESTO21, Japan Science and Technology Agency

Abstract. This paper addresses a new framework for designing and implementing skeleton libraries, in which each skeleton should not only be efficiently implemented as is usually done, but also be equipped with a structured interface to combine it efficiently with other skeletons. We illustrate our idea with a new skeleton library for parallel programming in C++. It is simple and efficient to use just like other C++ libraries. A distinctive feature of the library is its modularity: Our optimization framework treats newly defined skeletons equally to existing ones if the interface is given. Our current experiments are encouraging, indicating that this approach is promising both theoretically and in practice.

Keywords: Skeletal Parallel Programming, Optimization, Program Transformation, Fusion Transformation, Bird-Meertens Formalism.

1 Introduction

The increasing popularity of parallel programming environments, such as PC clusters, calls for a simple model and methodology that can assist programmers, including those who have little knowledge of parallel architecture and parallel algorithms, to develop efficient and correct parallel programs to solve various kinds of problems. Skeletal parallel programming, first proposed by Cole [1] and well-documented in [2], is such a methodology for building parallel programs in terms of a set of useful ready-made components (parallel primitives) called *skeletons*. Skeletons are generic and recurring patterns of parallel processing, whose parallel implementations are hidden from the programmers. These skeletons cannot only be efficiently implemented on various parallel architectures, but also be suitable as the target for systematic development by human [3–6].

The importance of equipping existing popular languages (like C or C++) with a well-structured skeleton library has been recognized [7–11], with which one can write parallel programs as sequential ones that call the library functions. However, the skeleton programs are slow in comparison with those directly coded in MPI, and this is an issue that prevents the skeletal parallel programming approach from being widely used. In fact, the simplicity of each skeleton often gives rise to a complicated combination of skeletons in a program, introducing a lot of data communication between them. Although individual skeletons can be efficiently implemented, their combination is inefficient unless the data communication between them can be eliminated.

There have been many attempts to apply transformation to optimizing combinations of skeletons [12, 13], where a set of transformation rules is defined, and an automatic or semi-automatic rewriting system is designed to apply these rules. There have been two problems with this approach. First, one would need a large set of rules to account for all possibilities of their combinations. Second, introducing a new skeleton would lead to large changes or extensions to the existing rule set. Lack of modularity causes these problems.

This paper proposes a new framework for designing of skeleton libraries that guarantees efficient combinations of skeletons, extending the theory developed in [14] in practice. Our idea was to associate each skeleton not only with an efficient parallel implementation but also with an interface for *efficient combination* with other skeletons. This interface contains information on how the skeleton consumes and produces its data. This idea is not new in the functional community, where we have seen the success of *shortcut deforestation (fusion)* [15] in optimizing sequential programs in Haskell compilers. However, as far as we know, we are the first to introduce this idea to the design of parallel skeleton libraries.

We designed and implemented a new skeleton library for skeletal parallel programming in C++. Our skeleton library has the following new features.

- *Single Optimization Rule*: Basically, we need just a single rule (Section 3) to optimize combinations of skeletons in the library, thanks to their structured interface. This is in sharp contrast to other transformation approaches [12, 13], where a large set of rules needs to be prepared. Furthermore, our rule can be applied to skeletal parallel programs in any way required, guaranteeing the same result and termination.
- *Modularity*: Our library allows new skeletons to be introduced without any change to the existing optimization framework, and ensures their efficient combination with existing skeletons in the library. This remedies the situation where transformation rules must take combinations of the skeletons with existing ones into account.
- *Simplicity*: From the programmers' point of view, as our library does not introduce any new syntax, a programmer who knows C++ should have no trouble in using it. We are able to construct a structured interface for the skeletons as well as apply a general optimization rule concisely and quickly (Section 4) with the help of the reflection mechanism provided with OpenC++ [16]. We found it very useful to use meta programming in implementing the transformation, which, we believe, is worth greater recognition in the skeleton community.

Our experiments in Section 4 demonstrate how promising our approach is.

In the rest of this paper, we will explain our idea based on the BMF data parallel programming model [17, 3], which provides us with a concise way of describing and manipulating parallel programs. After briefly reviewing the notations and basic concepts of BMF and skeletal parallel programming in Section 2, we show how to structure skeletons by standardizing their consumption and production of data. We then give our general rule for optimizing the combinations of skeletons in Section 3. We highlight our implementation of the library together with experiments in Section 4, and finally conclude the paper in Section 5.

2 BMF and Parallel Computation

We will now address our idea on the BMF data parallel programming model [17, 3]. Those familiar with the functional language Haskell [18] should have few problem in understanding the programs in this paper. From the notational viewpoint, the main difference is that we have used more symbols or special parentheses to shorten the expressions so that expressions can be manipulated more concisely.

2.1 Functions

Function application is denoted by a space and the argument which may be written without brackets. Thus $f\,a$ means $f(a)$. Functions are curried, and application associates to the left. Thus $f\,a\,b$ means $(f\,a)\,b$. A function application binds stronger than any other operator, so $f\,a \oplus b$ means $(f\,a) \oplus b$, not $f(a \oplus b)$. *Function composition* is denoted by a period. By definition, we have $(f\,.\,g)\,a = f(g\,a)$. Function composition is an associative operator, and the identity function is denoted by id.

Infix binary operators will often be denoted by $\oplus, \otimes$ and can be *sectioned*; an infix binary operator like $\oplus$ can be turned into unary or binary functions by $a \oplus b = (a \oplus)\,b = (\oplus b)\,a = (\oplus)\,a\,b$.

2.2 Parallel Data Structure: Join Lists

Join lists (or *append lists*) are finite sequences of values of the same type. A list is either empty, a singleton, or the concatenation of two other lists. We write $[\,]$ for the empty list, $[a]$ for the singleton list with element a (and $[\cdot]$ for the function taking a to $[a]$), and $x \mathbin{+\!\!+} y$ for the concatenation (join) of two lists x and y. Concatenation is associative, and $[\,]$ is its unit. For example, $[1] \mathbin{+\!\!+} [2] \mathbin{+\!\!+} [3]$ denotes a list with three elements, often abbreviated to $[1, 2, 3]$.

We also write $a : x$ for $[a] \mathbin{+\!\!+} x$. If a list is constructed with constructor $[\,]$ and $:$, we call it a *cons list*.

2.3 Parallel Skeletons: Map, Reduce, and Scan

It has been shown [3] that BMF [17] is a nice architecture-independent parallel computation model, consisting of a small fixed set of specific higher-order functions that can be regarded as parallel skeletons suitable for parallel implementation. Important higher-order functions are *map*, *reduce*, and *scan*.

Map is a skeleton that applies a function to every element in a list. It is written as infix $*$. Informally, we have

$$k * [x_1, x_2, \ldots, x_n] = [k\,x_1, k\,x_2, \ldots, k\,x_n].$$

Reduce is a skeleton that collapses a list into a single value by repeatedly applying a certain associative binary operator. It is written as infix $/$. Informally, for associative binary operator $\oplus$ and initial value e, we have

$$\oplus/_e \ [x_1, x_2, \ldots, x_n] = e \oplus x_1 \oplus x_2 \oplus \cdots \oplus x_n.$$

Scan is a skeleton that accumulates all intermediate results for computation of reduce. Informally, for associative binary operator $\oplus$ and initial value e, we have

$$\oplus\#_e \ [x_1, x_2, \ldots, x_n] = [e, e \oplus x_1, e \oplus x_1 \oplus x_2, \ldots, e \oplus x_1 \oplus x_2 \oplus \cdots \oplus x_n].$$

Note that this definition is slightly different from that in [17]; the e there is assumed to be the unit of $\oplus$. In fact, efficient implementation of the scan skeleton does not need this restriction.

2.4 An Example: Computing Variance

Consider, given the sequence $as = [a_1, a_2, \ldots, a_n]$, computing its variance var by $var = \sum_{i=1}^{n}(a_i - ave)^2/n$, where $ave = \sum_{i=1}^{n} a_i/n$, as a simple running example. This computation can be described using skeletons in BMF as follows.

$$\begin{aligned}
var \ as \ n &= sqSum/n \\
\textbf{where} \ ave &= (+/_0 \ as)/n \\
sqSum &= +/_0 \ (square * ((-ave) * as)) \\
square \ x &= x \times x
\end{aligned}$$

This BMF description is the best way of explaining our idea. In fact, to be processed by our system, it should be written in C++ using our skeleton library as follows. Note that the functions add, sub, and sq are defined by the user, and map1 is a variation of the map skeleton designed to treat sectioning notation; map1(sub, ave) corresponds to $(-ave)*$.

```
double variance( vector< double > *as, int size ) {
  double sum, ave, sq_sum;
  vector< double > *subs, *sqs;
  sum = as->reduce( add, 0.0 ); ave = sum / size;
  subs = as->map1( sub, ave ); sqs = subs->map( sq );
  sq_sum = sqs->reduce( add, 0.0 );
  return = sq_sum / size;
}
```

3 Shortcut Fusion on Skeleton Programs

To fuse the composition of skeletons into one to eliminate unnecessary intermediate data structures passed between skeletons, one may develop rules to do algebraic transformations on skeletal parallel programs like the authors in [12, 13]. Unfortunately, this would require a huge set of rules to take all possible combinations of skeletal functions into account. In this paper, we borrow the idea of shortcut deforestation [15], which optimizes sequential programs, and simplifies the entire set into just a single rule. The idea is to structure each skeleton with an interface that characterizes how it consumes and produces the parallel data structure, namely join lists.

3.1 Structuring Skeletons

To manipulate skeletal parallel programs, we structured skeletons in terms of the following three functions: acc, cataJ and buildJ.

Definition 1 (acc). Let g, p, q be functions, and $\oplus$ and $\otimes$ be associative operators. The skeleton acc, for which we write $[\![g, (p, \oplus), (q, \otimes)]\!]$, is defined by

$$[\![g, (p, \oplus), (q, \otimes)]\!] \ [\,] \ e = g \ e$$
$$[\![g, (p, \oplus), (q, \otimes)]\!] \ (a : x) \ e = p \ (a, e) \ \oplus \ [\![g, (p, \oplus), (q, \otimes)]\!] \ x \ (e \otimes q \ a) \ .$$

The function acc has an accumulation parameter, e. We define cataJ as a special case of acc, where the accumulation parameter is not used.

Definition 2 (cataJ). Given are function p and associative operator $\oplus$ with identity e. The skeleton cataJ, for which we write $(\!(\oplus, p, e)\!)$, is defined by

$$(\!(\oplus, p, e)\!) \ [\,] = e$$
$$(\!(\oplus, p, e)\!) \ (a : x) = p \ a \ \oplus \ (\!(\oplus, p, e)\!) \ x \ .$$

The last function, buildJ, is to standardize the production of join-lists with implicit parallelism.

$$\text{buildJ } gen = gen \ (+\!\!+) \ [\cdot] \ [\,]$$

We can now express our skeletons in terms of the above three functions.

Definition 3 (Skeletons in Structured Form).

$$f* = \text{buildJ } (\lambda c \ s \ e. \ (\!(c, f.s, e)\!))$$
$$\oplus/_e = (\!(\oplus, id, e)\!)$$
$$\oplus \#_e x = \text{buildJ } (\lambda c \ s \ e. \ [\![s, (\lambda(a, e). \ s \ e, c), (id, \oplus)]\!]) \ x \ e$$

3.2 Shortcut Fusion Rule

Following the thought in [15], we may define our shortcut fusion for join lists as follows.

Definition 4 (CataJ-BuildJ Rule).

$$(\!(c, s, e)\!) \ . \ \text{buildJ } gen = gen \ c \ s \ e$$

An example of applying this rule shows that reduce after map can be fused into a single cataJ.

$$\oplus/ \ . \ f* = \quad \{ \text{ map, reduce } \}$$
$$(\!(\oplus, id, \iota_\oplus)\!) \ . \ \text{buildJ } (\lambda \ c \ s \ e \ . \ (\!(c, s.f, e)\!))$$
$$= \quad \{ \text{ CataJ-BuildJ } \}$$
$$((\lambda \ c \ s \ e \ . \ (\!(c, s.f, e)\!)) \ \oplus \ id \ \iota_\oplus)$$
$$= \quad \{ \text{ lambda application } \}$$
$$(\!(\oplus, f, \iota_\oplus)\!) \ .$$

This CataJ-BuildJ rule, however, is not sufficient for some cases. Consider where we want to fuse $f * \ . \ g*$.

$$f * \ . \ g* = \text{buildJ } (\lambda c \ s \ e. \ (\!(c, s.f, e)\!)) \ . \ \text{buildJ } (\lambda c \ s \ e. \ (\!(c, s.g, e)\!))$$

We are blocked here, since the left cataJ is enclosed in buildJ. To proceed with the transformation, we may have to unfold the left buildJ, finally to have $([+\!\!+, [.].f.g, [\,]])$.

This result is unsatisfactory; this is not a form of the producer! Once transformed like this, further fusion by other consuming functions cannot take place. The trouble occurs due to uninvertible unfolding of buildJ, which we should avoid. The following rule does the trick in eliminating unnecessary unfolding of buildJ.

Definition 5 (BuildJ(CataJ-BuildJ) Rule).

$$\text{buildJ } (\lambda c\ s\ e.\ ([\phi_1, \phi_2, \phi_3])) \ .\ \text{buildJ } gen = \text{buildJ } (\lambda c\ s\ e.\ gen\ \phi_1\ \phi_2\ \phi_3)$$

This rule enables us to have

$$f * .\ g * $$
$$= \ \{ \text{ Map, BuildJ(CataJ-BuildJ), lambda application } \}$$
$$\text{buildJ } (\lambda c\ s\ e.\ ([c, s.f.g, e])),$$

which is exactly the structured form of $(f.g) *$.

Finally, we generalize the above rule to the following most generic fusion rule (for acc).

Definition 6 (BuildJ(Acc-BuildJ) Rule [14]).

$$\text{buildJ}(\lambda c\ s\ e.\ [\![g, (\oplus, p), (q, \otimes)]\!])\ (\text{buildJ } gen\ x)\ e$$
$$= fst\ (\text{buildJ } (\lambda c\ s\ e.\ gen\ (\odot)\ f\ d)\ x\ e)$$
where
$$(u \odot v)\ e = \text{let } (r_1, s_1, t_1) = u\ e$$
$$(r_2, s_2, t_2) = v\ (e \otimes t_1)$$
$$\text{in } (s_1 \oplus r_2,\ s_1 \oplus s_2,\ t_1 \otimes t_2)$$
$$f\ a\ e = (p\ (a, e) \oplus g\ (e \otimes q\ a), p\ (a, e), q\ a))$$
$$d\ e = (g\ e, \_, \_)$$

Here, *fst* returns the first element of a pair, and _ denotes a "don't care" value.

3.3 Modularity of Structured Forms

The shortcut fusion rule guarantees the fusibility of the composition of functions defined using acc (or cataJ in special) and buildJ. The implication of this is: User-defined skeletons are also the target of optimization once their structured form is known.

For example, we may want to introduce a new skeleton to capture a general polynomial computation pattern, which has many interesting applications including solving the maximum sum problems [17]. This new skeleton is parameterized by two associative operators $\oplus$ and $\otimes$ whose units are $\iota_\oplus$ and $\iota_\otimes$.

$$\text{poly } (\oplus)\ (\otimes)\ [x_1, x_2, \ldots, x_n] = x_1 \oplus (x_1 \otimes x_2) \oplus (x_1 \otimes x_2 \otimes \cdots \otimes x_n)$$

This new skeleton can be structured as follows, which can consume a join list produced by any other skeletons.

$$\text{poly } (\oplus)\ (\otimes) = fst\ .\ ([\odot, (\lambda\ a\ .\ (a, a)), (\iota_\oplus, \iota_\otimes)])$$
$$\text{where } (a_l, c_l) \odot (a_r, c_r) = (a_l \oplus c_l \otimes a_r,\ c_l \otimes c_r)$$

Note that the poly skeleton produces a single value, which could not be consumed by other skeletons. Therefore, we do not need to structure the output using buildJ.

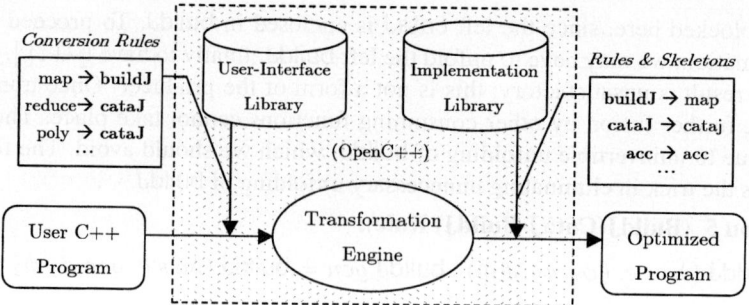

Fig. 1. Overview of our system.

4 Implementation of Transformation System

We implemented a prototype system with OpenC++, which transforms the skeletal parallel programs written in C++ with our skeleton system [10]. Fig. 1 overviews our system, which consists of three parts: (1) the user-interface library, (2) the generic transformation engine, and (3) the implementation library. Taking a C++ program, our transformation system first converts the skeletons in the program into structured form by applying rules given as meta-programs. The generic transformation engine manipulates and fuses the converted program with the shortcut fusion rules in Section 3. Finally, our system links the optimized program with efficiently implemented skeletons in our library.

4.1 Skeleton Interfaces

In OpenC++, the program text is accessible at the meta level in the form of a parse tree, represented as a nested list of logical tokens. A part of the C++ program to compute the variance in Section 2 is converted into the following parse tree.

```
[[sum = [[as -> reduce] ( [add , 0.0] )]] ;]
[[ave = [sum / size]] ;]
[[subs = [[as -> map1] ( [sub , ave] )]] ;]
[[sqs = [[subs -> map] ( [sq] )]] ;]
[[sq_sum = [[sqs -> reduce] ( [add , 0.0] )]] ;]
```

We define the rules to convert user skeletons to structured form and vice versa in OpenC++. For example, a meta program that converts a map skeleton into buildJ may be implemented as follows.

```
Ptree* map_to_buildJ( Ptree *sentence )
{
  Ptree *dst, *src, *function;
  if (Ptree::Match( sentence, "[[%? = [[%? -> map] ( %? )]] ;]",
                   &dst, &src, &function) ) {
    return make_buildJ( dst, src, Ptree::List( var_c ),
                       Ptree::List( var_s, function ), Ptree::List( var_e ));
  }
  ...
```

The reflection mechanism in OpenC++ enables pattern matching and function composition to be easily implemented. Thus, we can easily convert skeletons to their structured forms, e.g. for the poly skeleton, we can obtain the arguments $\oplus$, $\otimes$, $\iota_\oplus$, and $\iota_\otimes$ by pattern matching and derive structured forms after generating the new functions $\odot$ and $\lambda\, a\,.\,(a, a)$.

Using conversion with our user-interface library, the last three lines in the parse tree above are converted into the following structured forms.

```
['buildJ' subs as [[var_c] [var_s [sub ave]] [var_e]] ;]
['buildJ' sqs subs [[var_c] [var_s [sq]] [var_e]] ;]
['cataJ' sq_sum sqs [[add] [func_id] [0.0]] ;]
```

4.2 Generic Transformation Engine

Our system implements the fusion rule in Section 3, and it repeatedly applies the rule on structured forms. We restricted the elements in structured forms so that they were represented as a composition of functions. This simplified the application of the fusion rule so that just the occurrences of a bound variable to the corresponding argument had to be replaced. Note that such a restriction is insignificant since reflection can take care of it.

Our generic transformation engine applies the **CataJ-BuildJ** rule twice on the structured forms above in our running example, and optimizes it into a single cataJ form as follows.

```
['cataJ' sq_sum as [[add] [func_id [sq] [sub ave]] [0.0]] ;]
```

4.3 Experimental Results

The first program in Section 2 is inefficient because small functions are called one by one and unnecessary intermediate data are passed between skeletons. Our system automatically transforms the program so that it is efficient by composing skeletons.

To see how efficient the generated optimized program is, we compared it with the following two programs: (1) the original program using a map skeleton that produces new data, (2) another program using a map skeleton that overwrites the input data. In the second program, the individual skeletons are optimized so that they do not generate unnecessary data. We implemented these programs with our skeleton library in C++ and MPI, and did our experiments on a cluster of four Pentium 4 Xeon 2.0-GHz dual-processor PCs with 1 GB of memory, connected through a Gigabit Ethernet. The OS was FreeBSD 4.8 and we used gcc 2.95 for the compiler.

Fig. 2 plots the results of speedups to the original program with one processor for an array of 1,000,000 elements. The computation time for the original program with one processor is 1.67 (sec) and the computation times for (1), (2), and the optimized one with eight processors are 0.243, 0.197, and 0.138 (sec), respectively. As a natural consequence of using the skeletons, all programs demonstrated outstanding scalability. Comparison with (2) proves the success of our framework: The effect of fusion far exceeds individual refinements on each skeleton.

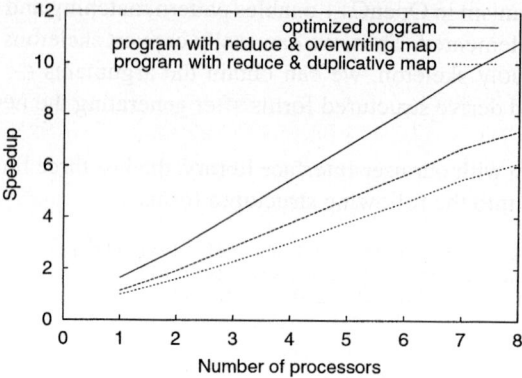

Fig. 2. Experimental results.

5 Conclusions

We proposed a new approach to the design and implementation of skeleton libraries. We implemented a parallel skeleton library in C++, which not only guaranteed each skeleton was efficiently implemented, but also efficiently combined the skeletons such that data communication between them could be eliminated. In contrast to popular approaches where the design of skeleton libraries has mainly focused on efficiently implementing single skeletons with little consideration to how combinations of skeletons are optimized, our approach unifies the two phases by structuring skeletons with an additional interface. This new approach is not only theoretically interesting, but also quite promising in practice. As we demonstrated, our library is easy to use, simple to implement, and suitable for extension.

We are still in the early stages of producing a really useful library supporting parallel programming in C++. In terms of theory, we have not yet taken functions like zip into account that traverse multiple data structures simultaneously; we are also interested in generalizing the approach from join lists to other parallel data structures such as matrices or trees. In terms of practice, our current implementation, whose main purpose is to test our idea, is expected to be improved through further analysis so that the generic optimization rule can be applied to more applications of skeletal parallel programming.

References

1. Cole, M.: Algorithmic skeletons : A structured approach to the management of parallel computation. Research Monographs in Parallel and Distributed Computing, Pitman, London (1989)
2. Rabhi, F., Gorlatch, S., eds.: Patterns and Skeletons for Parallel and Distributed Computing. Springer Verlag (2002)
3. Skillicorn, D.B.: The Bird-Meertens Formalism as a Parallel Model. In: NATO ARW "Software for Parallel Computation". (1992)

4. Gorlatch, S.: Systematic efficient parallelization of scan and other list homomorphisms. In: Annual European Conference on Parallel Processing, LNCS 1124, LIP, ENS Lyon, France, Springer-Verlag (1996) 401–408
5. Hu, Z., Iwasaki, H., Takeichi, M.: Formal derivation of efficient parallel programs by construction of list homomorphisms. ACM Transactions on Programming Languages and Systems **19** (1997) 444–461
6. Hu, Z., Takeichi, M., Chin, W.: Parallelization in calculational forms. In: 25th ACM Symposium on Principles of Programming Languages, San Diego, California, USA (1998) 316–328
7. Kuchen, H.: A skeleton library. In: EuroPar'02, LNCS, Springer-Verlag (2002)
8. Cole, M.: eSkel home page. http://homepages.inf.ed.ac.uk/mic/eSkel/ (2002)
9. Danelutto, M., Stigliani, M.: SKElib: parallel programming with skeletons in c. In: EuroPar'00, LNCS 1900, Springer-Verlag (2000) 1175–1184
10. Adachi, S., Iwasaki, H., Hu, Z.: Diff: A powerful parallel skeleton. In: The 2000 International Conference on Parallel and Distributed Processing Techniques and Application. Volume 4., Las Vegas, CSREA Press (2000) 525–527
11. Bacci, B., Danelutto, M., Orlando, S., Pelagatti, S., Vanneschi, M.: P3L: A structured high level programming language and its structured support. Concurrency Practice and Experience **7** (1995) 225–255
12. Gorlatch, S., Pelagatti, S.: A transformational framework for skeletal programs: Overview and case study. In Rohlim, J., et al., eds.: Parallel and Distributed Processing. IPPS/SPDP'99 Workshops Proceedings. Lecture Notes in Computer Science 1586 (1999) 123–137
13. Aldinucci, M., Gorlatch, S., Lengauer, C., Pelagatti, S.: Towards parallel programming by transformation: The FAN skeleton framework. Parallel Algorithms and Applications **16** (2001) 87–122
14. Hu, Z., Iwasaki, H., Takeichi, M.: An accumulative parallel skeleton for all. In: 11st European Symposium on Programming (ESOP 2002), Grenoble, France, Springer Verlag, LNCS 2305 (2002) 83–97
15. Gill, A., Launchbury, J., Peyton Jones, S.: A short cut to deforestation. In: Proc. Conference on Functional Programming Languages and Computer Architecture, Copenhagen (1993) 223–232
16. Chiba, S.: A metaobject protocol for C++. In: ACM Conference on Object-Oriented Programming Systems, Languages, and Applications (OOPSLA'95). SIGPLAN Notices 30(10), Austin, Texas, USA (1995) 285–299
17. Bird, R.S.: An introduction to the theory of lists. In Broy, M., ed.: Logic of Programming and Calculi of Discrete Design. Volume 36 of NATO ASI Series F., Springer-Verlag (1987) 5–42
18. Bird, R.: Introduction to Functional Programming using Haskell. Prentice Hall (1998)

Cross Component Optimisation in a High Level Category-Based Language

T.J. Ashby[1,2], A.D. Kennedy[1], and M.F.P. O'Boyle[2]

[1] School of Physics
University of Edinburgh
Scotland

[2] Institute for Computer Systems Architecture
University of Edinburgh
Scotland

Abstract. High level programming languages offer many benefits in terms of ease of use, encapsulation etc. However, they historically suffer from poor performance. In this paper we investigate improving the performance of a numerical code written in a high-level language by using cross-component optimisation. We compare the results with traditional approaches such as the use of high performance libraries or Fortran. We demonstrate that our cross-component optimisation is highly effective, with a speed-up of up to 1.43 over a program augmented with calls to the ATLAS BLAS library, and 1.5 over a pure Fortran equivalent.

1 Introduction

Superscalar architectures rely on instruction level parallelism (ILP) to get high performance. For codes with large data sets, bandwidth limitations to off-chip memory can mean that that the full parallelism of the CPU is not used during streaming operations, such as processing large arrays, because operands/results are not delivered/stored quickly enough. Optimising codes for cache locality ameliorates the situation by reducing the pressure on memory bandwidth, in effect increasing the available parallelism and removing the barrier to performance.

Modularity and encapsulation are fundamental tenets of software engineering, and we wish to adhere to them by writing programs in a high level way using layered composition of components. However, we do not wish to sacrifice performance. It is possible to capitalise on encapsulation and re-use by extensively optimising selected components for performance. Data locality is a program-wide property though, so optimisations must break the barriers introduced by modularity to maximise it. Hence we have two opposing strategies – encapsulate and heavily optimise individual components, or target global locality and optimise less extensively across components on a per program basis.

In this paper we present a case study of cross-component optimisation for a numerical benchmark that makes heavy use of abstraction and other advanced language features such as lexical scoping, dynamic allocation of objects etc, and consists mostly of vector operations. The language used is Aldor [3]. We show that the resulting program outperforms both a version that makes calls to highly tuned components instead of using our optimisations, and an equivalent program written purely in Fortran and compiled from source using Intel's Fortran compiler. The highly tuned components are the Level 1 binaries of the high-performance ATLAS BLAS library [12].

2 Language

Our approach is based on aggressive compiler optimisations for Aldor, a strict, statically typed (with type inference), mixed functional/imperative language. Aldor is similar to the ML family [9], with advanced module constructs including limited dependent types. The current Aldor compiler generates an intermediate representation (called FOAM), optimises it, and then translates it to C code that is linked against a small runtime library. This means that the language can be used anywhere that a C compiler is available as long as the run-time system is ported. The run-time system provides, amongst other things, initialisation of certain objects and garbage collection.

The core language of Aldor is very small, but offers great scope for abstraction and modularisation by building types. The language supports *domains* which are basic abstract data types, equivalent to modules in ML, and these are typed using *categories*, which are themselves roughly equivalent to (named) ML signatures. A category can inherit from and extend a simpler category, and the benchmark makes heavy use of user defined hierarchical categories to describe the rich mathematical structure of linear algebra (see sec. 3 for an example of category inheritance). This degree of abstraction is coupled with flexible syntax including overloading of both application syntax and the standard binary operators. The end result is very readable code that closely resembles the mathematical formulation.

Aldor incorporates a *whole program optimisation* strategy. This means that all the FOAM code used for a given executable must be available when it is linked, and thus can be analysed to perform extensive cross component optimisation at link-time. Aldor's language model is thus an important enabler for the exploitation of parallelism on single core chips.

3 Benchmark

In order to demonstrate the fundamentals of our approach, we discuss its application to a class of algorithms, being the family of iterative linear solvers (CG, BiCG etc.), with the specific example here being a variant of QMR [5]. The QMR algorithm consists of three nested parts – *basis vector generation*, which is called from the *search vector update*, which in turn is called from the *solution vector update*. The last of these is called repeatedly from a simple enclosing loop that iterates until some condition on the solution vector has been reached.

Each part of the algorithm maintains some state, and when called updates this state using the code extracts given in fig. 1. For more information, we

refer the reader to [10]. The extracts consist of expressions using and updating
lexically scoped variables from their environment, with tokens in upper case
letters denoting sparse matrices, (numbered) lower case letters denoting vectors,
and phonetic Greek letters denoting scalars. The expressions are essentially a
list of function calls attached to the overloaded binary operators (including the
overloading of juxtaposition), using the functions exported by the domains. So,
the components that we are optimising are the basic binary functions provided
by the abstract data types. Each of the functions contains, when unoptimised,
a fairly simple loop (or scalar manipulation), and allocates a new object to hold
its result.

Basis vector generation
1. w2 := AH w1 - conj(alpha) * w1
        - conj(beta) * w2;
2. deltaTemp := (v2 * w2);
3. deltaOld := delta;
4. delta := deltaTemp/gamma^2;
5. (v1, v2) := (v2/gamma, v1);
6. (w1, w2) := (w2/conj(gamma), w1);
7. beta := (gamma * delta)/deltaOld;
8. u := A v1;
9. alpha := (u * w1)/delta;
10. v2 := u - alpha * v1 - beta * v2;
11. gamma := norm(v2);

Search vector update
1. p2 := eta3 * (v2 - eta1 * p1
        - eta2 * p2);
2. (p1, p2) := (p2, p1);

Solution vector update
1. x := x + zeta * p1;

Fig. 1. Sections of Aldor code for the three parts of QMR.

Sparse matrix-vector multiplication for certain sparse matrices can be computed
by means of a "stencil" operation. We use these types of matrices in our example,
so the use of the matrices in lines 1 and 8 from the basis vector generation extract
are actually calls to stencil routines. The stencil we use is a 7 point Laplacian:

$$u_{i,j,k} := \kappa(v_{i+1,j,k} + v_{i-1,j,k} + v_{i,j+1,k} + v_{i,j-1,k} + v_{i,j,k+1} + v_{i,j,k-1} - 6v_{i,j,k})$$

The domain implementing this stencil belongs to the user defined category
LinearOperator (which is derived from LinearAlgebra, LinearSpace, Module,
AbelianGroup and AbelianMonoid, all of which are also user defined), to which
any other matrix representation would also belong. There is some re-use of the
vector in the application of a stencil (see [8]) that can be exploited for non-
stationary iterative methods, but the vectors have to be very large before tiling
has an effect, so we have not yet investigated this optimisation. Instead, we
look for locality across all the basic functions including the stencils. The code
consists mostly of simple operations on vectors, and so we maximise locality by
minimising the number of vector traversals in the program.

4 Transformations

The Aldor compiler has a number of standard optimisations that operate on FOAM code. These include aggressive inlining, which helps to remove the layering of abstraction. The ordinary compiler does not however yet include fusion and the subsequent removal of temporaries. We have developed and present here four different stages of transformation where each stage builds on the previous with more aggressive optimisation. The baseline presented here is not the most naive translation of the code – instead it uses BLAS routines written in Aldor. This allows a more meaningful comparison with the ATLAS augmented and Fortran versions.

- Baseline: Uses BLAS routines written in Aldor. That is, axpy, nrm2 etc. These already give destructive updates of vectors and some limited fusion.
- Level 1: Fully exploits locally available fusion in the vector operations without touching the stencil.
- Level 2: Fuses the action of the stencil with other vector operations within the basis vector generation.
- Level 3: Aggressively rearranges the code in the basis vector generation so that both stencil applications happen simultaneously, and the normalisation of the new v_1 vector is delayed until it is used. This latter optimisation is a direct trade of extra FLOPS for a saving in bandwidth.

Table 1. Breakdown of vector operations in basis vector generation.

	Baseline	Level 1	Level 2	Level 3
1.	stencil (line 1)	stencil (line 1)	lines 1, 2, 5, 6 (inc. stencil)	lines 1, 2, (5), 6, 8, 9 (inc. both stencils)
2.	1st axpy (line 1)	lines 2,5,6 and 1	lines 8, 9	lines 10, 11
3.	2nd axpy (line 1)	stencil (line 8)	lines 10, 11	
4.	inner prod. (line 2)	inner prod. (line 9)		
5.	scale (line 5)	lines 10, 11		
6.	scale (line 6)			
7.	stencil (line 8)			
8.	inner prod. (line 9)			
8.	1st axpy (line 10)			
9.	2nd axpy (line 10)			
10.	norm (line 10)			

The operations for the basis vector generation are broken down in more detail in table 1. The total number of vector traversals for this part of the algorithm along with count for the other parts of the algorithm is given in table 2. A full breakdown of the fusion in the other parts has been omitted due to space limitations.

Table 2. Total vector traversals count.

	Baseline	Level 1	Level 2	Level 3
basis vector generation	19	13	10	7
Search vector update	5	3	3	3
Solution vector update	2	2	2	2
Total	26	18	15	12

5 Results

5.1 Control Experiments

By way of comparison with our transformations we present results for two different approaches. The first substitutes the BLAS routines written in Aldor with calls to the ATLAS library. This amounts to extreme localised tuning of the basic vector operations in the algorithm to compare against cross-component optimisation. The second codes the entire problem in Fortran, based on a modification of the freely available QMRpack [4].

QMRpack was originally structured to use call-backs to invoke the user provided matrix-vector multiplication routines. This enables modularity for use with different operators, but introduces an artificial barrier to optimisation. To provide a fairer comparison with our approach, the stencil routine is now called directly from the main routine. The source files for the main routine and all the library subroutines (including the BLAS routines) are concatenated together before compilation. QMRpack does not try and separate the Lanczos process from the QR decomposition. Hence, the code just consists of a series of vector and scalar operations and does not attempt to mimic the extra structure in the Aldor code or any of the functional programming features. Indeed, it is written in Fortran 77.

5.2 Test Environment

The test environment comprised a 1 GHz Pentium III desktop machine (Coppermine with 256kB L2 cache) running Linux RedHat 7.1, the Intel C compiler icc and Intel Fortran compiler ifc, both version 8.0, the ATLAS BLAS binaries for Pentium III release 3.4.2, and the Aldor compiler version 1.0. All timing information was generated using gprof version 2.11 and the compiler profile generation switches provided by icc and ifc.

All compiler flags were chosen to try and maximise the performance of the code. The experiments are conducted using complex double floats (4 words = 16 bytes each) as the scalars, based on 3 dimensional regular grids with equal side lengths. Hence the size of an individual vector is $side^3 \times 16$ bytes. The problem sizes run from 15, which is just after the total data set no longer fits into cache, up until 40.

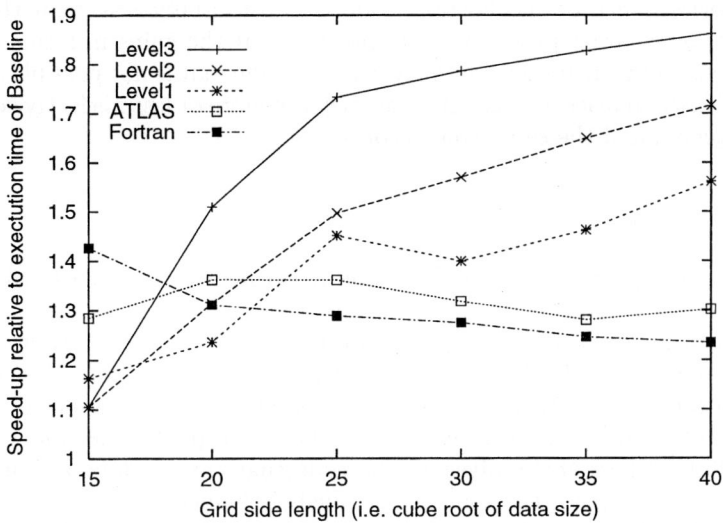

Fig. 2. Plot of problem size vs. speed-up over baseline.

5.3 Discussion

Fig. 2 shows a ratio plot comparing the total execution times of various programs against that of the baseline. For the ATLAS augmented version, the executable spends from ≈ 45% of its time in the ATLAS routines for the smallest datset size, rising steadily to ≈ 67% for the largest. As we can see, the Fortran code does well on the first data point. This seems reasonable, as the whole problem fits into the L2 cache until it reaches size 13, and superior overall scheduling (and prefetching) would make itself felt near the bottom end of the scale. Nonetheless, when the problem size reaches 20 the transformations of Level 3 outperform all the other approaches, and as the size of the dataset increases the relative gains improve.

The speed-up is up to 1.43 over the ATLAS version and up to 1.5 over Fortran, demonstrating significant gains for cross-component optimisation.

Latency, Prefetching and Code Generation. Due to the limits on out-of-order execution and the C compiler strategy for issuing prefetches, we may not be realising the full benefit of our optimisations. Both ifc and icc issue prefetches, but ifc issues more for the Fortran code, particularly in the vector expressions. Presumably the Fortran version is consequently closer to being simply bandwidth limited. As far as we can tell, ifc does no restructuring of the loops though. We wish to investigate improving our current approach by adding iterative optimisation to find transformations that reduce the suspected latency penalties.

The performance of the Fortran code is substantially less than the transformed Aldor at larger problem sizes, showing that the gains in terms of bandwidth outweigh the introduced latency costs. Although it is not possible to examine the ATLAS binaries, we assume that they are more or less perfectly scheduled and prefetched, and the same conclusion holds.

6 Related Work

Reducing the impact of modularisation on performance has been tackled many times in many different settings. To our knowledge though, the combination of language type, approach and target of optimisation presented here are unique. The most popular target for this sort of work is C++, with examples including expression templates [11], library annotations, and extensive analysis and preprocessing of source code. Expression templates are pragmatic (lots of people use C++) but ad hoc; the analysis of the code that can be done and hence the optimisations is limited, the semantic analysis is inadequate, and coupling the transformation system to an engine that searches for good transformations would be hard. Library annotations for domain specific languages work at a higher level [7]; we wish to invasively fuse operations on objects, not rewrite expressions at the ADT level. The approach taken in [1] is probably closest to our own; however, that project appears to be coming from the bottom up rather than the top down, in that they already support various optimisations and are edging toward effectively defining their own specialised language by adding parsing and semantic analysis of a language piggybacked onto C++. However, we do not require a user to specify the transformations as re-write rules. [2] is an automatic system for generating parsing and transformation tools similar to those used in [1]. Something like it might be used to generate a system for performing the transformations that we use, but we are focused on the transformations rather than developing a general tool. Also, although our transformations are quite specific, they are based purely on information that can be recovered from FOAM rather than requiring extra semantic input in terms of grammars and rules provided by a library programmer.

The work in [13] appears to be at a lower level than ours, and addresses a problem that is already dealt with by the existing Aldor optimisations. Future work on optimising Generic Java is likely to be of greater interest.

7 Conclusion

The results show that optimising whole program locality for ILP is important for this class of application. Hence, using an elegant language with lot of abstraction doesn't necessarily imply poor performance if the language infrastructure supports cross-component optimisation well. Indeed, focusing on cross-component optimisations can bring important speedups compared to traditional approaches, even if the result is almost certainly crudely scheduled and prefetched compared to individual extensively tuned assembly routines.

The most important immediate extension of this work is to fully investigate the interaction of the transformations with the functional features of Aldor. Following this we wish to investigate the impact of our optimisations on a wider set of codes with more complex operators, such as the red-black preconditioned QCD Wilson–Dirac operator, and more specialised solvers [6], as well as other benchmarks from the field of scientific computing.

References

1. Otto Skrove Bagge, Karl Trygve Kalleberg, Magne Haveraaen, and Eelco Visser. Design of the CodeBoost transformation system for domain-specific optimisation of C++ programs. In Dave Binkley and Paolo Tonella, editors, *Third International Workshop on Source Code Analysis and Manipulation (SCAM 2003)*, Amsterdam, The Netherlands, September 2003. IEEE Computer Society Press. (To appear).
2. Kei Davis and Dan Quinlan. ROSE II: An optimizing code transformer for C++ object-oriented array class libraries. In *Proceedings of the Third World Multiconference on Systemics, Cybernetics and Informatics (SCI'99) and the Fifth International Conference on Information Systems Analysis and Synthesis (ISAS'99)*.
3. Stephen Watt et al. Aldor. http://www.aldor.org.
4. R. W. Freund and N. M. Nachtigal. QMRpack. http://www.netlib.org/linalg/qmr/.
5. R. W. Freund and N. M. Nachtigal. QMR: a quasi-minimal residual method for non-Hermitian linear systems. In R. Beauwens and P. de Groen, editors, *Iterative Methods in Linear Algebra*, pages 151–154. Elsevier Science Publishers, 1992.
6. A. Frommer and B. Medeke. Exploiting structure in Krylov subspace methods for the Wilson fermion matrix. In A. Sydow, editor, *15th IMACS World Congress on Scientic Computation, Modelling and Applied Mathematics*, volume 3, pages 485–490. Wissenschaft und Technik Verlag, 1997.
7. Samuel Guyer and Calvin Lin. Broadway: A compiler for exploiting the domain-specific semantics of software libraries. In *Proceedings of the IEEE: Special Issue on Program Generation, Optimization, and Platform Adaptation*.
8. Gabriel Rivera and Chau-Wen Tseng. Tiling optimizations for 3D scientific computations. In *Proceedings of the 2000 ACM/IEEE conference on Supercomputing (CDROM)*, page 32. IEEE Computer Society, 2000.
9. Robert Harper Robin Milner, Mads Tofte and David MacQueen. *The Definition of Standard ML (Revised)*. The MIT Press, 1997.
10. T.J.Ashby, A.D.Kennedy, and M.F.P.O'Boyle. An iterative solver package in Aldor. Submitted to: Proceedings of the Third International Workshop on Numerical Analysis and Lattice QCD, November 2003.
11. Todd L. Veldhuizen. Expression templates. *C++ Report*, 7(5):26–31, June 1995. Reprinted in C++ Gems, ed. Stanley Lippman.
12. R. Clint Whaley, Antoine Petitet, and Jack J. Dongarra. Automated empirical optimizations of software and the ATLAS project. *Parallel Computing*, 27(1–2):3–35, 2001.
13. Peng Wu, Sam Midkiff, José Moreira, and Manish Gupta. Efficient support for complex numbers in Java. In *Proceedings of the ACM 1999 conference on Java Grande*, pages 109–118. ACM Press, 1999.

Concurrent Table Accesses in Parallel Tabled Logic Programs

Ricardo Rocha[1], Fernando Silva[1], and Vítor Santos Costa[2]

[1] DCC-FC & LIACC
University of Porto, Portugal
{ricroc,fds}@ncc.up.pt
[2] COPPE Systems & LIACC
Federal University of Rio de Janeiro, Brazil
vitor@cos.ufrj.br

Abstract. Tabling is an implementation technique that improves the declarativeness and expressiveness of Prolog by reusing answers to subgoals. The declarative nature of tabled logic programming suggests that it might be amenable to parallel execution. On the other hand, the complexity of the tabling mechanism, and the existence of a shared resource, the table, may suggest that parallelism might be limited and never scale for real applications. In this work, we propose three alternative locking schemes to deal with concurrent table accesses, and we study their impact on the OPTYap parallel tabling system using a set of tabled programs.

1 Introduction

Tabling (or *tabulation* or *memoing*) is an implementation technique where results for subcomputations are stored and reused. Tabling has proven to be particularly effective in logic programs: tabling can reduce the search space, avoid looping, and in general have better termination properties than traditional Prolog. The XSB Prolog system [1] is the most well known tabling Prolog system, with excellent results in application areas such as Natural Language Processing, Knowledge Based Systems, Model Checking, and Program Analysis.

One extra advantage of tabling is that tabled programs are most often pure logic programs, and are thus amenable to the implicit exploitation of parallelism. Because tabling has often been used to reduce search space, or-parallelism is most interesting. We thus proposed OPTYap [2], a design for combining implementation techniques for or-parallelism in shared-memory machines, namely environment copying [3], with the WAM extensions originally proposed in XSB [4]. Results have shown that OPTYap can achieve excellent speedups, while introducing low overheads for sequential execution [5].

The performance of tabling largely depends on the implementation of the table itself. The table will be called very often, therefore fast lookup and insertion is mandatory. Applications can make millions of different calls, hence compactness is also required. The XSB design used *tries* to implement this goal [6]. Tries are trees in which there is one node for every common prefix [7]. Tries have proven to be one of the main assets of XSB, because they are quite compact for most applications, while having fast lookup and insertion.

One critical issue in our parallel design was whether tries would be effective in the presence of concurrent accesses. One of the first implementations of tries in a parallel environment was the work by Chan and Lim [8], where tries were used to index words for alphabets with a finite number of symbols. In our work, we use tries to index Prolog terms, which can have an infinite number of symbols. We address concurrency by extending the trie structure originally proposed in XSB to support locking mechanisms. To achieve best performance, different implementations may be pursued. We can have one lock per table entry, one lock per path, or one lock per node. We can also have hybrid locking schemes combining the above. Our initial results did show that naive approaches could generate significant overheads or result in minimal concurrency. We thus studied alternative locking schemes that try to reduce overheads by only locking part of the tree when strictly necessary, and evaluated their performance. Our results show almost-linear speedups up to 32 CPUs on an Origin2000. Although our context is parallel tabling, we believe that these experiments will be of interest to application areas that rely on access to frequently updated trees.

The remainder of the paper is organized as follows. First, we introduce the trie data structure and the table space organization. Next, we describe the three alternative locking schemes implemented in OPTYap. We then end by presenting some initial results and outlining some conclusions.

2   Table Space

Tabling is about storing intermediate answers for subgoals so that they can be reused when a repeated subgoal appears. Execution proceeds as follows. Whenever a tabled subgoal S is first called, an entry for S is allocated in the *table space*. This entry will collect all the answers found for S. Repeated calls to *variants* of S are resolved by consuming the answers already stored in the table. Meanwhile, as new answers are generated, they are inserted into the table and returned to all variant subgoals.

The table space can be accessed in a number of ways: **(i)** to look up if a subgoal is in the table, and if not insert it; **(ii)** to verify whether a newly found answer is already in the table, and if not insert it; and, **(iii)** to load answers to variant subgoals. Hence, a correct design of the algorithms to access and manipulate the table data is critical to achieve an efficient implementation. Our implementation uses tries as proposed by Ramakrishnan *et al.* [6].

Tries were first proposed to index dictionaries [7] and have since been generalized to index recursive data structures such as terms (see [6, 9–11] for use of tries in tabled logic programs, automated theorem proving and term rewriting). An essential property of the trie structure is that common prefixes are represented only once. The efficiency and memory consumption of a particular trie largely depends on the percentage of terms that have common prefixes. For tabled logic programs, we often can take advantage of this property.

Each different path through the nodes in the trie, the *trie nodes*, corresponds to a term. The entry point is called the root node, internal nodes represent symbols in terms and leaf nodes specify completed terms. Terms with common prefixes branch off from each other at the first distinguishing symbol. Inserting a term requires in the worst case allocating as many nodes as necessary to represent its complete path. On the other hand, inserting repeated terms requires traversing the trie structure until reaching the corresponding leaf node, without allocating any new node. During traversal, each child node specifies the next symbol to be inspected in the input term. A transition is taken if the symbol in the input term at a given position matches a symbol on a child node. Otherwise, a new child node representing the current symbol is added and an outgoing transition from the current node is made to point to the new child. On reaching the last symbol in the input term, we reach a leaf node in the trie.

Figure 1 shows an example for a trie with three terms. Initially, the trie contains the root node only. Next, we insert $f(X,a)$. As a result, we create three nodes: one for the functor $f/2$, next for the variable X, and last for the constant a. The second step is to insert $g(X,b,Y)$. The two terms differ on the main functor, so tries bring no benefit here. In the last step, we insert $f(Y,1)$ and we save two common nodes with $f(X,a)$. Notice the way variables are represented. We follow the formalism proposed by Bachmair et al. [10], where each variable is represented as a distinct constant.

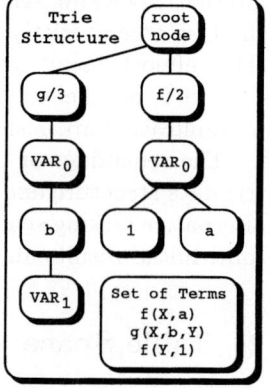

Fig. 1. Tries for terms.

The implementation of tries requires four fields per trie node. The first field (TrNode_symbol) stores the symbol for the node. The second (TrNode_child) and third (TrNode_parent) fields store pointers respectively to the first child node and to the parent node. The fourth field (TrNode_next) stores a pointer to the sibling node, in such a way that the outgoing transitions from a node can be collected by following its first child pointer and then the list of sibling pointers. We next present how tries are used to implement tabled predicates. Figure 2 shows an example for a predicate $f/2$ after the execution of the following operations:

```
tabled_subgoal_call:  f(X,a)
tabled_subgoal_call:  f(Y,1)
tabled_new_answer:    f(0,1)
tabled_new_answer:    f(n(0),1)
```

We use two levels of tries: one stores the subgoal calls, the other the answers. Each different call to a tabled predicate corresponds to a unique path through the *subgoal trie structure*. Such a path always starts from a *table entry* data structure, follows a sequence of the *subgoal trie nodes*, and terminates at a leaf data structure, the *subgoal frame*. Each subgoal frame stores information about the subgoal, namely an entry point to its *answer trie structure*. Each unique path through the *answer trie nodes* corresponds to a different answer to the entry subgoal. When inserting new answers, we only store the substitutions for the unbound variables in the subgoal call. This optimization is called *substitution factoring* [6]. Leaf answer nodes are chained in a linked list in insertion time order, so that we can recover answers in the same order they were inserted. The subgoal frame points to the first and last answer in this list. Thus, a variant subgoal only needs to point at the leaf node for its last loaded answer, and consumes more answers just by following the chain. To load an answer, the trie nodes are traversed in bottom-up order and the answer is reconstructed.

Traversing a trie to check/insert for new calls or for new answers is implemented by repeatedly invoking a `trie_check_insert()` procedure for each symbol that represents the term being checked. Given a symbol S and a parent node P, the procedure returns the child node of P that represents the given symbol S. Figure 3 shows the pseudo-code. Initially it traverses the chain of sibling nodes that represent alternative paths from the given parent node and checks for one representing the given symbol. If such a node is found then execution is stopped and the node returned. Otherwise, a new trie node is allocated and inserted in the beginning of the chain. To allocate new trie nodes, we use a `new_trie_node()` procedure with four arguments, each one corresponding to the initial values to be stored respectively in the `TrNode_symbol`, `TrNode_child`, `TrNode_parent` and `TrNode_next` fields of the new allocated node.

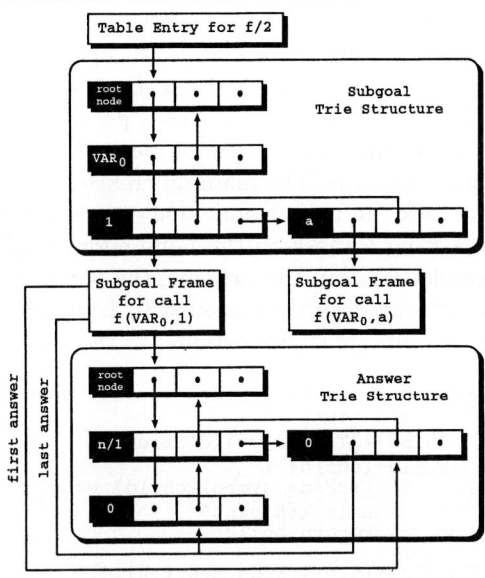

Fig. 2. Tries to organize the table space.

```
trie_check_insert(symbol s, trie node parent) {
    child = TrNode_child(parent)
    while (child) {            // check if a node for s was already inserted
        if (TrNode_symbol(child) == s)
            return child                                    // node found
        child = TrNode_next(child)
    }
    child = new_trie_node(s, NULL, parent, TrNode_child(parent))
    TrNode_child(parent) = child             // insert the new node for s
    return child
}
```

Fig. 3. Pseudo-code for `trie_check_insert()`.

3 Table Locking Schemes

There are two critical issues that determine the efficiency of a locking scheme for the table. One is *lock duration*, that is, the amount of time a data structure is held. The other is *lock grain*, that is, the number of data structures that are protected through a single lock request. It is the balance between lock duration and lock grain that compromises the efficiency of different table locking approaches. For instance, if the lock scheme is short duration or fine grained, then inserting many trie nodes in sequence, corresponding to a long trie path, may result in a large number of lock requests. On the other hand, if the lock scheme is long duration or coarse grain, then going through a trie path without extending or updating its trie structure, may unnecessarily lock data and prevent possible concurrent access by others. Unfortunately, it is impossible beforehand to know which locking scheme would be optimal.

The *Table Lock at Node Level (TLNL)* was our first implemented scheme. It only enables a single writer per chain of sibling nodes that represent alternative paths from a common parent node. Figure 4 shows the pseudo-code that implements it. The main difference from the original procedure is that in TLNL we lock the parent node while accessing its children nodes. Locking is done by applying a mask to the node address in order to index a global array of lock entries. Within this scheme, the period of time a node is locked is proportional to the average time needed to traverse its children nodes, and the number of lock requests is proportional to the length of the path.

```
trie_check_insert(symbol s, trie node parent) {
    lock(parent)                              // locking the parent node
    child = TrNode_child(parent)
    while (child) {
        if (TrNode_symbol(child) == s) {
            unlock(parent)                    // unlocking before return
            return child
        }
        child = TrNode_next(child)
    }
    child = new_trie_node(s, NULL, parent, TrNode_child(parent))
    TrNode_child(parent) = child
    unlock(parent)                            // unlocking before return
    return child
}
```

Fig. 4. Pseudo-code for the TLNL scheme.

The *Table Lock at Write Level (TLWL)* scheme improves TLNL by reducing lock duration. Like TLNL, TLWL only enables a single writer per chain of sibling nodes, but the common parent node is only locked when writing to the table is likely. Figure 5 shows the pseudo-code for TLWL. Initially, the chain of sibling nodes that follow the given parent node is traversed without locking. The parent node must be locked only when the given symbol is not found. This avoids locking when the symbol already exists in the chain. Moreover, it delays locking while insertion of a new node to represent the symbol is not likely. Note that we need to check if, during our attempt to lock, other worker expanded the chain to include the given symbol. Within this scheme, the number of lock requests is, on average, lower than TLNL. It ranges from zero to the number of nodes in path. Similarly, the amount of time a node is locked is also, on average, smaller. It includes the time needed to check the nodes that in the meantime were inserted by other workers, if any, plus the time to allocate and initialize the new node.

Last, the *Table Lock at Write Level - Allocate Before Check (TLWL-ABC)* scheme is a variant of TLWL. It also follows the *probable node insertion notion*,

```
trie_check_insert(symbol s, trie node parent) {
   child = TrNode_child(parent)
   initial_child = child                    // keep the initial child node
   while (child) {           // traverse the initial chain of sibling nodes
      if (TrNode_symbol(child) == s)
         return child
      child = TrNode_next(child)
   }
   lock(parent)
   child = TrNode_child(parent) // traverse the nodes inserted in the ...
   while (child != initial_child) {    // ... meantime by others, if any
      ...                                          // the same as TLNL
   }
   ...                                             // the same as TLNL
}
```

Fig. 5. Pseudo-code for the TLWL scheme.

but uses a different strategy to decide on when to allocate a node. In order to reduce to a minimum the lock duration, it anticipates the allocation and initialization of nodes that are likely to be inserted in the table before locking. However, if in the meantime a different worker introduces first an identical node, we pay the cost of having pre-allocated an unnecessary node that has to be additionally freed. Figure 6 presents the pseudo-code that implements this scheme.

```
trie_check_insert(symbol s, trie node parent) {
   ...                                             // the same as TLWL
   pre_alloc = new_trie_node(s, NULL, parent, NULL)  // pre-allocate ...
   lock(parent)                     // ... a node for s before locking
   child = TrNode_child(parent)
   while (child != initial_child) {
      if (TrNode_symbol(child) == s) {
         unlock(parent)
         free(pre_alloc)             // free the pre-allocated node
         return child
      }
      child = TrNode_next(child)
   }
   TrNode_next(pre_alloc) = TrNode_child(parent)
   TrNode_child(parent) = pre_alloc      // insert the pre-allocated node
   unlock(parent)
   return pre_alloc
}
```

Fig. 6. Pseudo-code for the TLWL-ABC scheme.

4 Preliminary Results

In order to evaluate the scalability of our locking schemes, we ran OPTYap for a set of selected programs in a Cray Origin2000 parallel computer with 96 MIPS 195 MHz R10000 processors. We selected the programs that showed significant speedups for parallel execution in previous work [2, 5]. The programs include the transition relation graphs for two model-checking specifications, a same generation problem for a 24x24x2 data cylinder, and a transitive closure of a 25x25 grid using left recursion. All programs find all the solutions for the problem.

In order to get a deeper insight on the behavior of each program, we first characterize the programs in Table 1. The columns have the following meaning: *time* is the execution time in seconds with a single worker; *sg* is the number of different tabled subgoal calls; *unique* is the number of answers stored in the table; *repeated* is the number of redundant answers found; *nodes* is the number of trie nodes allocated to represent the set of answers; and *depth* is the average number of trie nodes required to represent an answer. In parentheses, it shows the percentage of saving that the trie's design achieves on these data structures. Consider Figure 1 as an example, it requires 10 nodes to represent individually all answers but it uses only 8, thus achieving a saving of 20%. Smaller depth values or higher percentages of saving reflect higher probabilities of lock contention when concurrently accessing the table space.

Table 1 indicates that *mc-sieve* is the program least amenable to lock contention because it finds the least number of answers and has the deepest trie structures. In this regard, *lgrid* is the opposite case. It finds the largest number of answers and it has very shallow trie structures. Likewise, *samegen* also shows a very shallow trie structure, despite that it can benefit from its large number of different tabled subgoal calls. It is the case, because the answers found for different subgoals can be inserted without overlap. Finally, *mc-iproto* can also lead to higher ratios of lock contention. It shows the highest percentage of saving and it inserts a huge number of trie nodes in the table.

Table 1. Program characteristics.

Program	time	sg	unique	repeated	nodes	depth
mc-sieve	268	1	380	1386181	8624	53(57%)
mc-iproto	24	1	134361	385423	1554896	51(77%)
samegen	26	485	23152	65597	24190	1.5(33%)
lgrid	69	1	160000	449520	160401	2(49%)

Table 2 shows the speedups for the three locking schemes with varying number of workers. The speedups are relative to the single worker case. A main conclusion can be easily drawn from the results presented: TLWL and TLWL-ABC show identical speedup ratios and they are the only schemes showing scalability. In particular, for the *mc-sieve* program they show superb speedups up to 32 workers. Closer analysis allows us to observe other interesting aspects: the more refined strategy of TLWL-ABC does not show to perform better than TLWL; all schemes show identical speedups for *samegen*; and TLNL clearly slows down for more than 16 workers. The good behavior with *samegen* arises from the fact that this program calls 485 different tabled subgoals. This increases the number of entries where answers can be stored thus reducing contention points. The TLNL slowdown is related to the fact that this scheme locks the table even when writing is not likely. In particular, for repeated answers it pays the cost of performing locking operations without inserting any new node.

Table 2. Speedups.

Schemes	Workers			
	8	16	24	32
mc-sieve				
TLNL	7.2	11.8	3.9	4.7
TLWL	7.9	15.8	23.7	31.5
TLWL-ABC	7.9	15.8	23.7	31.4
mc-iproto				
TLNL	2.6	1.8	1.0	1.0
TLWL	5.0	9.0	8.8	7.2
TLWL-ABC	5.1	7.7	8.4	7.1
samegen				
TLNL	7.2	13.8	19.6	24.0
TLWL	7.2	13.9	19.7	24.1
TLWL-ABC	7.2	13.9	19.7	24.2
lgrid				
TLNL	6.7	12.1	6.2	5.3
TLWL	7.1	13.5	19.9	24.3
TLWL-ABC	6.9	13.4	18.9	24.2

During parallel execution, not only concurrency can be a major source of overhead. For some programs, the complexity of the tabling mechanism can induce some intricate dependencies that may always constraint parallel execution. Besides, as tabling, by nature, reduces the potential non-determinism available in logic programs, the source of parallelism may also be intrinsically limited.

To better understand the parallel execution behavior of our set of programs, we gathered in Table 3 a set of statistics regarding the number of contention points when using our best table locking scheme – TLWL. By contention points we mean the number of unsuccessful first attempts to lock a data structure. We distinguish three kind of locking attempts: **(i)** locking related with *trie nodes*, when inserting new subgoal calls or new answers; **(ii)** locking related with *subgoal frames*, when updating the subgoal frame pointers to point to the last found answer; and **(iii)** locking related with *variant subgoals*, when synchronizing access to check for available answers to variant subgoals. Note that TLWL only affects contention related with trie nodes.

The insignificant number of contention points obtained for *mc-sieve* supports the excellent speedups observed for parallel execution. In this regard, the contention also obtained for *samegen* indicates that locking is not a problem, and that the small overhead observed for *samegen* in Table 2 is thus mainly related with the complexity of the tabling mechanism. On the other hand, regarding *mc-iproto* and *lgrid*, lock contention is a major problem. For trie nodes they show identical numbers, but *mc-iproto* inserts about 10 times more answer trie nodes than *lgrid*. For subgoal frames and variant subgoals they show a similar pattern, but *mc-iproto* has higher contention ratios per time unit (remember from Table 1 that *mc-iproto* is about 3 times

Table 3. Contention points with TLWL.

| Contention | Workers | | | |
Points	8	16	24	32
mc-sieve				
trie nodes	188	415	677	1979
subgoal frames	0	0	0	2
variant subgoals	0	1	0	4
mc-iproto				
trie nodes	6579	10537	11816	11736
subgoal frames	9894	21271	33162	33307
variant subgoals	4685	25006	66334	81515
samegen				
trie nodes	119	201	364	417
subgoal frames	52	112	283	493
variant subgoals	0	1	0	0
lgrid				
trie nodes	5292	10341	12870	12925
subgoal frames	1124	7319	17440	27834
variant subgoals	1209	5987	23357	35991

faster than *lgrid*), hence justifying its worst behavior with the increase in the number of workers. For these programs, the sequential order by which leaf answer nodes are chained in the trie seems to be the key issue that reflects the high number of contention points. After inserting a new answer we need to update the subgoal frame to point to the last found answer. When checking for answers to variant subgoals we need to lock and possibly update the variant subgoal to point to the last loaded answer. For programs that find a large number of answers per time unit, this obviously increases contention when accessing such pointers, and thus obtaining good speedups in the presence of these conditions will always be a difficult task.

5 Concluding Remarks

We studied the impact of using alternative locking schemes to deal with concurrent table accesses in parallel tabling. We used OPTYap, that to the best of our knowledge, is the only available parallel tabling system for logic programming. Through experimentation, we observed that there are locking schemes that can obtain good speedup ratios and achieve scalability. Our results show that a main problem is not only how we lock trie nodes to insert items, but also how we use auxiliary data structures to synchronize access to the table space. In this regard, a key issue is the sequential order by which leaf answer nodes are chained in the trie structure. In the future, we plan to investigate whether alternative designs can obtain scalable speedups even when frequently updating/accessing tables.

Acknowledgments

This work has been partially supported by APRIL (POSI/SRI/40749/2001) and by funds granted to LIACC through the Programa de Financiamento Plurianual, Fundação para a Ciência e Tecnologia and Programa POSI.

References

1. Sagonas, K., Swift, T., Warren, D.S.: XSB as an Efficient Deductive Database Engine. In: ACM SIGMOD International Conference on the Management of Data, ACM Press (1994) 442–453
2. Rocha, R., Silva, F., Santos Costa, V.: On a Tabling Engine that Can Exploit Or-Parallelism. In: International Conference on Logic Programming. Number 2237 in LNCS, Springer-Verlag (2001) 43–58
3. Ali, K., Karlsson, R.: The Muse Approach to OR-Parallel Prolog. International Journal of Parallel Programming **19** (1990) 129–162
4. Sagonas, K., Swift, T.: An Abstract Machine for Tabled Execution of Fixed-Order Stratified Logic Programs. ACM Transactions on Programming Languages and Systems **20** (1998) 586–634
5. Rocha, R., Silva, F., Costa, V.S.: Achieving Scalability in Parallel Tabled Logic Programs. In: International Parallel and Distributed Processing Symposium, IEEE Computer Society (2002)
6. Ramakrishnan, I.V., Rao, P., Sagonas, K., Swift, T., Warren, D.S.: Efficient Access Mechanisms for Tabled Logic Programs. Journal of Logic Programming **38** (1999) 31–54
7. Fredkin, E.: Trie Memory. Communications of the ACM **3** (1962) 490–499
8. Chan, I.W., Lim, C.Y.: Parallel Implementation of the Trie Structure. In: International Conference on Parallel and Distributed Systems, IEEE Computer Society (1994) 538–543
9. McCune, W.W.: Experiments with Discrimination-Tree Indexing and Path Indexing for Term Retrieval. Journal of Automated Reasoning **9** (1992) 147–167
10. Bachmair, L., Chen, T., Ramakrishnan, I.V.: Associative Commutative Discrimination Nets. In: International Joint Conference on Theory and Practice of Software Development. Number 668 in LNCS, Springer-Verlag (1993) 61–74
11. Graf, P.: Term Indexing. Number 1053 in LNAI. Springer-Verlag (1996)

Handling Exceptions Between Parallel Objects*

Christian Pérez, André Ribes, and Thierry Priol

IRISA/INRIA, Campus de Beaulieu 35042 Rennes Cedex, France
Tel:+33 2 99 84 74 13, Fax:+33 2 99 84 71 71
{Christian.Perez,Andre.Ribes,Thierry.Priol}@irisa.fr

Abstract. Scientific computing is evolving from parallel processing to distributed computing with the availability of new computing infrastructures such as computational grids. We investigate the design of a component model for the Grid aiming at combining both parallel and distributed processing in a seamless way. Our approach is to extend component models with the concept of parallel components. At runtime, most component models rely on a distributed object model. In this paper, we study an exception handling mechanism suitable for dealing with parallel entities, such as parallel objects, that appear as collections of identical objects but acting as a single object from the programmer's viewpoint.

1 Introduction

Scientific computing has become a fundamental approach to model and to simulate complex phenomena. With the availability of parallel machines in the late 1980's, a strong emphasis was made to design algorithms and applications for scientific computing suitable for such machines. However, nowadays, science or engineering applications, such as multiphysics simulations, require computing resources that exceed by far what can be provided by a single parallel machine. Moreover, complex products (such as cars, aircrafts, etc.) are not designed by a single company, but by several of them, which are often reluctant to grant access to the source of their tools. From these constraints, it is clear that distributed *and* parallel processing cannot be avoided to manage such applications. More precisely parallel simulation codes will have to be interconnected through a specific middleware to allow a distributed execution, thus complying with localization constraints and/or availability of computing resources. Programming a computing resource, such as a Grid infrastructure, that has both dimensions, parallel and distributed processing, is challenging.

The objective of our research activities is to conciliate these two technologies in a seamless way achieving both performance and transparency. More specifically, we aim at providing a programming model based on the use of distributed software components. Since most component models are based on distributed objects, much of our work focuses on extending a distributed object model. This

---

* This work was supported by the Incentive Concerted Action "GRID" (ACI GRID) of the French Ministry of Research.

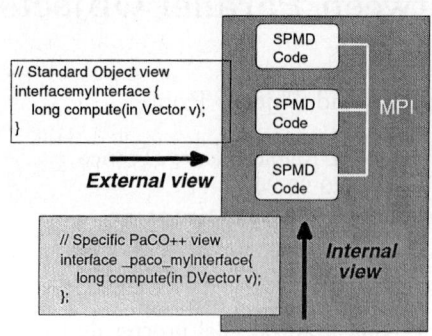

Fig. 1. A PaCO++ parallel object.

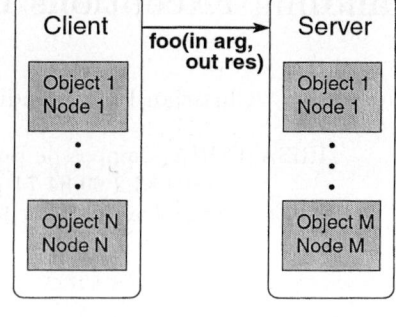

Fig. 2. Example of a distributed application.

work specifically aims at handling exceptions in the presence of a collection of distributed objects, called parallel objects in the remaining of the paper. A collection of distributed objects is the basic principle we introduced into CORBA for the encapsulation of parallel codes that rely on a SPMD execution model.

Section 2 introduces PaCO++ which implements the extension to the CORBA distributed object model for managing collections of identical objects. In Section 3, we describe our proposed model to manage exceptions in the presence of collections of objects. Related works aiming at handling exceptions in the context of concurrent or parallel systems are presented in Section 5. Section 4 gives some technical details about the implementation of the exception handling mechanism and Section 6 concludes the paper.

2   A Parallel Object Model: PaCO++

From the early 90's, several projects were set up to build distributed object middleware systems. However, most of these projects did not provide an efficient and adequate support for handling parallel codes. Research activities started in the late 1990's to support parallelism such as CORBA [5, 9, 6, 8]. All of them introduce the concept of parallel object. A parallel object is an entity that behaves like a regular object but whose implementation is parallel. Hence, a parallel object can be referenced and methods can be invoked on it as a single entity. Some of them also support data redistribution.

This section briefly describes PaCO++ [8], our research platform, so as to let Section 4 illustrates how parallel exceptions might be implemented. PaCO++ is the continuation of PaCO [9]. It is a *portable* extension to CORBA that is intended to be the foundation of GridCCM [7], our parallel extension of the CORBA Component Model. PaCO++ defines a parallel object as a collection of identical CORBA objects whose execution model is SPMD. A parallel object is implemented by a coordinated set of standard CORBA objects. As shown in Figure 1, a PaCO++ parallel object provides a standard external interface which is mapped to an internal interface. An invocation on a parallel object results in the simultaneous invocation of the corresponding operation of the internal interface on all the objects being part of the parallel object. Special attention has been devoted to the support of distributed arguments. The arguments can be distributed at both the client and the server sides. PACO++ provides mechanisms (static and dynamic) to specify the data distribution on both sides. As the two data distributions may be different, data redistribution may be required during the communication between the client and the server.

3 Managing Exceptions Within Parallel Objects

The problem addressed in this paper is the management of exceptions in the presence of parallel objects such as those provided by PACO++ (but it is not limited to them). If we consider a distributed application made of two parallel objects (a client and a server) like in Figure 2, it may happen that an operation invocation of the client on the server side raises an exception. However, since the server is a collection of objects, several scenarios may happen depending on which objects have raised an exception at the server side. One or several objects may raise exceptions with different values associated with them. It is thus necessary to define a model for exception management that gives a coherent view of the exceptions raised by the server to all the objects of a parallel client.

Scenarios. We have identified four scenarios that need to be handled. The simplest one is a *single exception* where only one object of the collection throws an exception. The *multiple exception* occurs when several objects of the server throw potentially different exceptions. A SPMD *exception* requires that all objects at the server side coherently throw the same type of exception. Last, a *chain of exceptions* occurs when the server calls a method on another remote object and this second object throws an exception.

Definition of a parallel exception. We define a parallel exception as the collection of exceptions thrown by one or several nodes of a method of a parallel object. We define an SPMD exception as an exception declared as SPMD by the parallel object. From an execution point of view, it is a parallel exception which is made of exceptions of identical type which have been coordinately raised by the all nodes of a parallel object.

3.1 Motivating Application

Let us introduce the EPSN project [1] as an application that motivates this work. Its goal is to analyze, design and develop a software environment for steering distributed numerical simulations from the visualization application. As shown in Figure 3, it adds a level of constraints which stem from the human interaction loop: data have to be extracted from the application and sent to a visualization machine but also the user actions have to be sent back to the application. As a user can connect to and disconnect from a running application from *a priori* any

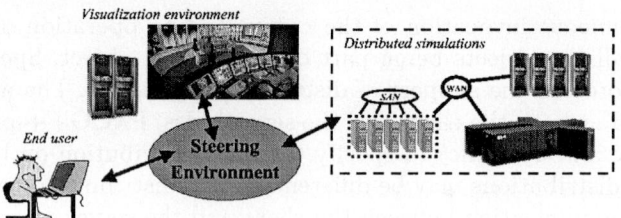

Fig. 3. Overview of the EPSN project.

machine, a distributed-oriented middleware appears well-suited to handle such interactions.

When the visualization application requests data, the simulation code might throw an exception because of memory limitation for example. Since the visualization code and the simulation code are parallel, the middleware has to manage the "parallel" exception. In particular, each node of the visualization application has to receive the exception. If the visualization tool can handle partial images, the exception may be able to carry a partial image. In the same time, it is possible to have several visualization clients for one simulation. Some of them may be able to manage partial data and some not. Hence, different exceptions need to be returned to the different clients.

3.2 The Three Types of Exceptions

Only three types of exception seem to be sufficient to handle all the previous scenarios. A *simple exception* type is just a plain exception. It can be obtained either when only an object of the server throws an exception or when an SPMD exception is raised. Section 3.4 will detail this. An *aggregated exception* type is composed of all the exceptions raised by the server objects. For example, it occurs when several objects of the server throw different exceptions. These exceptions have different meanings and cannot be grouped in a *simple* exception. A *complex* exception is composed of an *aggregated* exception with uncomplete data. Uncomplete data is the current value of the out arguments of the operation that generates the exception. Some restrictions may be applied to uncomplete data as described in Section 3.5.

3.3 Client Side

A client may not be aware of the proposed new types of exceptions or may not be able to catch them. Therefore, the model must allow a client to choose the kind of exceptions it wants to handle. The model defines three exception levels that correspond to the three types of exception defined in Section 3.2. By default, a client is assumed to only support simple exceptions (*simple level*). This level is the default level because of the legacy clients which are unaware of parallel objects. The second level which is the *aggregated level* is used by clients which are able to catch exceptions composed of several regular exceptions. The last level, *complex level*, indicates that the client may catch exceptions composed of an aggregated exception and uncomplete data. These levels form a hierarchy. A client that handles *aggregated exceptions* implicitly handles *simple exceptions*. Similarly, a client handling *complex exceptions* is assumed to also handle *simple* and *aggregated exceptions*.

3.4 Server Side

For each exception that a method of a parallel object may throw, additional information may be needed for a correct exception handling. First, SPMD exceptions have to be declared as such and data distribution information need to be specified for the distributed data the exception may contain (see Section 3.5). Second, priorities may be attached to exceptions. These priorities are fixed by the application designer (server side) and are not seen by a client. They are primarily used to decide which exception will be thrown to a client that only supports the *simple level* when several unrelated exceptions are raised. Third, the out arguments of a method that need to be returned as uncomplete data also have to be declared.

To determine the type of exceptions the server needs to send to a client, the server needs to know the level of exception the client supports. There are four cases depending on the type of exception raised by the parallel server and the level of exception supported by the client. First, if only one node of the server throws an exception, the server may safely send this exception as the client at least supports the *simple* exception type. Each node of a parallel client will receive this exception. Second, if every node of the server throws the same type of exception and the exception is defined as an SPMD exception, the server also applies the same algorithm as in the first case but also adds the management of distributed data. A sequential client is seen as a parallel client with one node. Third, if several nodes of the server throw exceptions and the client supports the *aggregated* level, the server sends an *aggregated* exception composed of all the exceptions raised. The implementation may send additional information like the object identifier where the exception has occurred. If the client only supports *simple* exceptions, the server throws the exception with the highest priority. Fourth, if the client supports the *complex* case and some out arguments are marked as valid for uncomplete data, a *complex* exception is returned.

3.5 Managing Data Within a Parallel Exception

Managing distributed data within a SPMD *exception.* An SPMD exception may contain data declared as distributed. When such an exception is raised, the runtime system may need to redistribute the data to the client (e.g., the client may be composed of a different number of nodes than the server). Since an exception can be seen as a structure with different data fields, there is no difference with the management of distributed arguments. As the model is not designed for a particular distributed object middleware, it just specifies that an implementation has to use the same mechanism as the out argument management in the parallel distributed middleware where it is implemented. Some implementation hints for PACO++ are given in Section 4.

Uncomplete data. Uncomplete data can only be encapsulated into complex exceptions. If the client does not support this exception type, the server cannot send uncomplete data. When the runtime system knows all the raised exceptions, it can determine which out arguments need to be sent to the client. Then, a complex exception made of several exceptions and the valid out arguments is created. Otherwise, a complex exception without out arguments are returned.

3.6 Managing Exception Chains

For example, an exception chain occurs when a client invokes a method A on a first server which invokes a method B on a second server that raised an exception. In the general case, an exception chain occurs when there are several – at least one – servers between the client which catches the exception, and the server which raises the exception. Ideally, we would like to send an aggregated exception combining the server exception and an information that this exception was thrown by an another server. This exception can be sent only if the client and the server support aggregated exceptions and that all intermediate servers have declared the exception type. Otherwise, either an unknown exception or a model specific exception like *chain of exception* should be raised.

4 Implementation Strategy

This section provides some hints on how to implement our model into PACO++.

PACO++ adds a layer between the user code and the CORBA code (e.g., stubs and the skeletons generated by an IDL compiler). This layer is generated thanks to the description of the distributed service (IDL) and the parallel description file specific to PACO++, which describes the parallel methods and the distributed arguments of these methods. The main implementation principle is to extend this file with exception related features. For each exception defined in the IDL file for a parallel method, the designer indicates the priority of the exception, whether this exception contains distributed data, etc. For each parallel method, two new exceptions are added into the generated IDL file: one for the aggregation case and one for the complex case.

The client and the server usually need to configure the PACO++ layer. For example, PACO++ layers need to obtain a reference to the distribution libraries. The client and server are configured using their contexts. These contexts need to be extended with information related to exceptions.

Each parallel method is mapped to two asynchronous methods, one for each direction. Thus, for the PACO++ layer, the exception is just an another kind of result. We only have to generate the code for these methods. The management

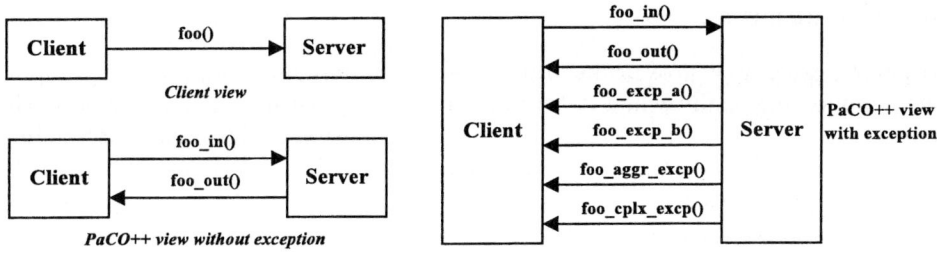

Fig. 4. Exception-related methods to transfer exceptions in a PaCO++ parallel object.

of data distribution is the same as for a regular invocation. Figure 4 shows these new methods for a parallel method foo that may throw exceptions of types A and B. The server code contains a barrier to synchronize all the objects before returning the results to the client. If an object throws an exception, the information is broadcast so that all objects have the same view of the exception state. After the barrier, the runtime system first checks if there is an exception; if not, the runtime system performs the normal code. So, a boolean test is added to the normal code. Thus, the overhead to supporting parallel exceptions in PaCO++ should be negligible for normal invocations.

5 Related Work

Several solutions have been proposed to support exceptions in parallel code. Some works [3, 10, 4] try to handle multiple exceptions within parallel loops. In [3], Modula-3 is extended so that if one or more exceptions of the same type are thrown within a parallel loop, the loop throws only one exception. But, if there are two or more different exception types, the runtime system sends an error. In [10], it is proposed to extend Java to have parallel loops with exception support. Each exception is stored into an aggregated exception that is thrown to the caller. The caller may retrieve some uncomplete data. In [4], when a node of the parallel loop throws an exception and another node communicates with it, the first node sends its exception to the other nodes. This exception is called a *global* exception. For the caller of the parallel loop, the runtime system generates a *concerted* exception from all the exceptions thrown.

In [2], it is proposed to manage exceptions between groups of distributed objects. Each group has one or more gates that enable to communicate with other gates. The invocations are asynchronous. The client has to call a method to get the result of the operation. In the server, if one object throws an exception, this exception is saved on the gate and the client will be aware of this exception when the synchronization method is called. Finally, Xu et al. [11] present a solution to manage exceptions between different processes that take part in the same *action*. They describe how to rollback this action when an exception occurs.

6 Conclusion

Handling exceptions is of prime importance when designing an object or component model, especially when it targets Grid infrastructures that combine both parallel and distributed aspects. The main contribution of this paper is to define the concept of parallel and SPMD exceptions for parallel objects implemented as a collection of identical sequential objects. The proposed model is able to manage all identified scenarios including concurrent uncoordinated exceptions raised by a parallel object. It also defines aggregated exceptions and complex exceptions with uncomplete data by integrating previous works on exceptions within parallel loops. Moreover, our proposed model gives a solution to hande distributed data for SPMD exceptions. We are currently implementing the proposed model in PaCO++. As outlined in this paper, the implementation appears straightforward and should exhibit a negligible overhead for normal operation. We are also working on defining a mathematical model of parallel exceptions.

References

1. The EPSN Project, http://www.labri.fr/Recherche/PARADIS/epsn/
2. Chris Exton and Ivan Rayner. Exception semantics in a parallel distributed object oriented environment. In *Proc. of the 21 International Conference TOOLS Pacific*, page 51, Melbourne, Australia, November 1996.
3. Ernst A. Heinz. Sequential and parallel exception handling in modula-3*. In P. Schulthess, editor, *Advances in Modular Languages: Proceedings of the Joint Modular Languages Conference*, pages 31–49, Ulm, Germany, September 1994.
4. Valérie Issarny. An exception handling model for parallel programming and its verification. In *Proc. of the ACM SIGSOFT'91 Conference on Software for Critical Systems*, pages 92–100, New Orleans, Louisiana, USA, December 1991.
5. K. Keahey and D. Gannon. PARDIS: A Parallel Approach to CORBA. In *Supercomputing'97*. ACM/IEEE, November 1997.
6. Object Management Group. Data parallel CORBA, November 2001. ptc/01-11-09.
7. Christian Pérez, Thierry Priol, and André Ribes. A parallel CORBA component model for numerical code coupling. *The International Journal of High Performance Computing Applications (IJHPCA)*, 17(4):417–429, 2003.
8. Christian Pérez, Thierry Priol, and André Ribes. PaCO++: A parallel object model for high performance distributed systems. In *Distributed Object and Component-based Software Systems Minitrack in the Software Technology Track of the 37th Hawaii International Conference on System Sciences (HICSS-37)*, Big Island, Hawaii, USA, January 2004. IEEE Computer Society Press.
9. C. René and T. Priol. MPI code encapsulating using parallel CORBA object. In *Proceedings of the Eighth IEEE International Symposium on High Performance Distributed Computing*, pages 3–10, Redondo Beach, California, USA, August 1999. IEEE.
10. Joel Winstead and David Evans. Structured exception semantics for concurrent loops. In *Fourth Workshop on Parallel/High-Performance Object-Oriented Scientific Computing*, Tampa Bay, October 2001.
11. J. Xu, A. Romanovsky, and B. Randell. Concurrent exception handling and resolution in distributed object systems. In *IEEE Trans. on Parallel and Distributed Systems. TPDS-11, N 10*, 2000.

Parallel Software Interoperability by Means of CORBA in the ASSIST Programming Environment*

S. Magini, P. Pesciullesi, and C. Zoccolo

University of Pisa, Dip. di Informatica - Via Buonarroti 2, 56127 Pisa, Italy
zoccolo@di.unipi.it

Abstract. Parallel software reuse and easy integration between parallel programs and other sequential/parallel applications and software layers can be obtained exploiting the software component paradigm. In this paper we describe the ASSIST approach to interoperability with CORBA objects and components, presenting two different strategies to export a parallel program in the CORBA world. We will discuss their implementations and provide some experimental results.

1 Introduction

The development of complex applications for the emerging high-performance large-scale computing platforms (ranging from huge Clusters up to computational GRIDS) poses productivity problems. High-level parallel programming tools give a partial solution, easing the development of complex parallel algorithms, while the exploitation of the software component paradigm can improve parallel software reuse and integration in larger applications.

ASSIST (A Software System based on Integrated Skeleton Technology) is a general-purpose high-level parallel programming environment, based on the skeleton and coordination language technology [1–3]. It combines the benefits of software reuse and integration with those of high-level programming, providing full interoperability with CORBA objects and components. It can easily import (or use) other pieces of software encapsulated in external CORBA objects and components, as in a traditional sequential language (i.e. C++); moreover, it can export parallel algorithms and applications as well: an ASSIST program can be encapsulated in a CORBA object and integrated into a larger application using standard CORBA invocation mechanisms.

CORBA, a distributed object- and component- based middleware, is a well established commercial standard, with thousands of users all over the world. It offers interoperability with several sequential languages and is supported by several

---

* This work has been supported by: the Italian MIUR FIRB Grid.it project, No. RBNE01KNFP, on High-performance Grid platforms and tools; the Italian MIUR Strategic Project L. 449/97-2000, on High-performance distributed enabling platforms; the Italian MIUR Strategic Project L. 449/97-1999 on Grid-computing for e-science.

big software companies. The interoperability with ASSIST provides the ability to develop computation- or data-intensive components of complex applications using an high-level parallel programming language, enabling the construction of scalable applications.

Recent studies recognized that CORBA technology could be leveraged to support the development of advanced Grid applications: the CORBA CoG kit [4], for example, provides access to the Grid services provided by the Globus Toolkit to CORBA applications. Several studies investigated how CORBA could be adapted to enable the construction of high-performance applications, enabling data-parallel programs to interact efficiently (see Sect. 5).

Our approach extends CORBA interoperability to stream-parallel programs, so we will focus on an efficient implementation of stream communication using standard CORBA mechanisms. This study is a preliminary step towards a complete CCM (CORBA Component Model, introduced in the CORBA 3 standard) integration and a possible generalization towards other component-based standards (e.g. Web/GRID services[5]). In previous works we showed that ASSIST programs could invoke external CORBA objects [6] and claimed that ASSIST programs could be automatically exported as CORBA objects [2]. In the following sections, we describe two different strategies to export an ASSIST program in the CORBA world (Sect. 2), we discuss their implementations (Sect. 3), provide experimental results (Sect. 4) and we survey related work (Sect. 5).

2 Exporting an ASSIST Program in the CORBA World

2.1 Structure of an ASSIST Program

ASSIST applications are developed by means of a coordination language [1] (ASSIST-CL), in which sequential as well as parallel computation units can be expressed and coordinated. Sequential code units, that are the bricks to build sequential and parallel modules, are described by a C-like procedure interface, in which input and output parameters are identified and associated with their type, and can be written in popular sequential languages (C, C++, Fortran 77), allowing maximal sequential code reuse. Parallel activities, designed following regular parallelism exploitation patterns, are marked, in the ASSIST-CL syntax, by means of the *parmod* construct: it can be considered a sort of "generic" skeleton, that can emulate the "classical" ones without performance degradation, but with more expressive power; for example, it allows both task and data parallelism to be exploited inside the same parallel module, and can be used to program lower (w.r.t. classical skeleton) level forms of parallelism. Parallel activities are decomposed in sequential indivisible units, which are assigned to abstract executors, called *virtual processors*. ASSIST allows them to communicate by reading and modifying in a controlled, consistence preserving way the shared computation state, which is partitioned among the virtual processors.

ASSIST modules interact by means of input and output interfaces. In module interfaces, ASSIST recognizes the standard CORBA types, which are mapped to native (C++) types following the standard C++ to CORBA type mapping[7].

The coordination model is based on the concept of *stream*. A stream is an ordered sequence, possibly infinite, of typed values. Streams connect modules in generic graphs, and drive the activation of their elaborations: either data-flow or nondeterministic behaviour can be selected, and the activation can be controlled by the internal module state. A composition of modules may be, in turn, reused as a component of a more complex application, provided that the interfaces match. The programming model includes also the concept of *external objects*, as a mean of inter-component interaction, as well as interaction with external world. Existing services (such as parallel file systems, databases, interactive user interfaces,etc.) can be encapsulated in an object (e.g. CORBA) and invoked inside an ASSIST application.

2.2 ASSIST Programs as Components of Larger CORBA Applications

A computation intensive component of a CORBA application can be conveniently expressed in ASSIST to exploit parallelism and achieve good performance, provided that it can interoperate with the application in a simple manner. We devised two possible ways of interconnecting an ASSIST subprogram to a distributed application, that address two different classes of problems:

RMI-like synchronous invocation of a subprogram (with arguments and return values), when the task to be computed in parallel is well defined and insulated;

stream-like asynchronous data passing (realized using the standard CORBA event channel mechanism), which is useful when the production of data to be processed takes longer and can be overlapped with the elaboration, or when the data structure to be processed is so big that it cannot be handled as a whole by the application, or as well when we want to receive partial results as soon as possible (for example when rendering images or videos).

An ASSIST subprogram, in order to be exported to the CORBA world, must be a composition of ASSIST modules (at the extreme, it can be a single module) in which one input stream and one output stream are left unconnected (i.e. no module produces/consumes them) and are elected to be the input and output of the entire exported component. In the RMI-like case, a further constraint to be satisfied (that cannot be statically checked) is that for every datum the program receives from the input source (by means of the input stream), it must produce one and only one datum on the output stream.

3 Implementation

The process of exporting an ASSIST program in the CORBA world has been automatized. We had two options: modify the original ASSIST compiler in order to insert specialized support code for interoperation with CORBA in the compiled modules or develop a separate transformation tool that produces an interoperable ASSIST program that can be compiled with the standard ASSIST compiler. The second options is more flexible: it allows experimenting with different solutions and different ORB implementations, and is easily extensible to other distributed objects middleware (CCM, GridServices). It can be slightly less efficient because the code interacting with CORBA runs as user code in the ASSIST program, and this imposes some restrictions on its interaction with the ASSIST runtime. We opted for the second solution: the ASSIST program undergoes an initial analysis phase (detailed in 3.1); then, if the conditions stated earlier are met, it is transformed, according to the option (RMI-synchronous vs. stream-asynchronous interaction, detailed in 3.2 and 3.3 respectively) chosen by the programmer, adding support code to interact with the CORBA runtime and services. Appropriate CORBA IDL interfaces are automatically generated for the transformed ASSIST program.

3.1 Analysis of the ASSIST Program

The analysis and transformation of the ASSIST program is carried out by a separate tool (the integration within the ASSIST compiler is ongoing), designed to parse and automatically manipulate an ASSIST program in source form.

The program is parsed and its representation is built as a set of C++ objects. Those objects provide methods to inspect and modify the structure of the program, that can finally be rewritten in the ASSIST syntax and compiled using the standard tools. The informations needed about the streams chosen to interface the program to CORBA are extracted from the parsed program. If no errors are encountered in this phase, the transformation proceeds according to the selected interaction method.

3.2 Exporting as a Synchronous Object Method

In the case of synchronous interaction, an ASSIST program (for example, see Fig. 1(a)) is exported as an object with a single `execute` method, with argument and return types equal respectively to the types of the input and output stream in the program interface. The transformation tool generates an IDL file (see Fig. 1(b)) describing the object interface, including any type definition (translated to IDL syntax) or preprocessor directive specified in the source program.

The original ASSIST program is enriched with a module that acts as a bridge towards CORBA (see Fig. 2). It instantiates an ORB in a separate execution thread, creates an instance of the CORBA object representing the exported algorithm and publishes it to a CORBA Naming Service, so that CORBA clients can easily search for and connect to it; finally it enters the ORB main loop, waiting for incoming requests.

Whenever a request is accepted by the ORB, the implementation of the execute method (running in the ORB thread) delivers the argument to the ASSIST program and then stops waiting for a response. The effective message delivery and reception is executed in the main thread of the server module, and controlled by the ASSIST runtime support. When the response is received, the execute method implementation returns it to its caller.

```
#define N 20
generic main() {
  stream long[N][N] Aaa;
  stream long[N][N] Bbb;

  spt(input_stream Aaa output_stream Bbb);
}
parmod spt (input_stream long A[N][N]
            output_stream long B[N][N])
{
// parallel code omitted
}
```
(a) ASSIST code.

```
// Autogenerated file - DO NOT EDIT
// CORBA interface for the spt ASSIST program

#define N 20

#define bool boolean

typedef long idl_ret_t[N][N];
typedef long idl_call_t[N][N];

interface spt {
  idl_ret_t execute(in idl_call_t _var);
};
```
(b) Generated IDL interface.

Fig. 1. Synchronous interaction example.

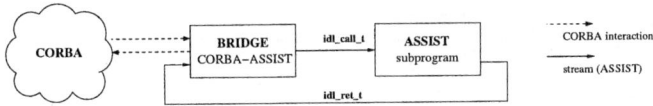

Fig. 2. Deployed ASSIST program (synchronous invocation).

3.3 Exporting as a Component Interconnected Through Event Channels

Using the former interaction method it is not possible to execute multiple requests simultaneously, therefore it is not effective for stream-parallel programs, like pipelines. To fully exploit stream parallelism, we developed a second mechanism based on CORBA event channels to implement asynchronous communications. The transformation tool generates an IDL file containing the definitions of event types for the input and output streams of the ASSIST subprogram, and a special termination event used to stop gracefully a running ASSIST server (see Fig. 3). In this case, no RMI interface is needed, because the ASSIST program communicates with CORBA only by means of events.

CORBA events can have any type and event reception is governed by the type, so we opted for the convention of encapsulating the datum in a named structure with a single field, to ease the discrimination of ingoing and outgoing messages.

The ASSIST program is transformed adding two modules (see Fig. 4), one (S1) that intercepts ingoing events and sends them to the input stream of the subprogram, and the other (S2) that receives message from the output stream and forwards them as outgoing events.

Each module, at initialization, connects to the CORBA naming service and obtains a reference to the COS Event Service, that is the CORBA service that manages event publishing, subscription and notification. Then the first one subscripts for the ingoing event and the special termination event and keeps forwarding messages until it receives a termination notification. At the same time,

```
// Autogenerated file - DO NOT EDIT        struct Termination {
// CORBA event definitions for               bool stop;
// the spt ASSIST program                  };

#define N 20                               struct source {
                                             idl_call_t data;
#define bool boolean                       };

typedef long idl_ret_t[N][N];              struct sink {
typedef long idl_call_t[N][N];               idl_ret_t data;
                                           };
```

Fig. 3. Generated IDL event definitions.

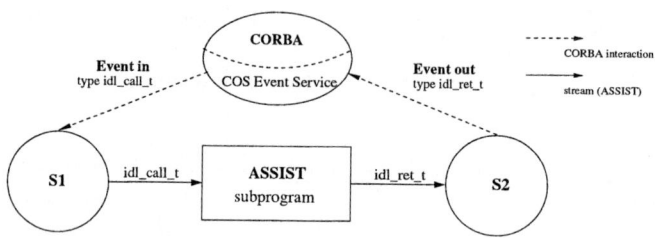

Fig. 4. Deployed ASSIST program (asynchronous message passing).

the other publishes the outgoing event type and produces an event whenever receives a datum from the ASSIST subprogram.

4 Performance Evaluation

The presented methodology is a viable solution to parallel software integration into larger applications.

In order to demonstrate this, we present some experiments, targeted to measure the performance of invoking parallel algorithms from sequential code.

Environment. The following experiments were performed on a Blade cluster consisting of 23 Intel Pentium III Mobile CPU 800MHz computing elements, equipped with 1GB of RAM and interconnected by a switched Fast Ethernet dedicated network. The CORBA implementation adopted is TAO, a free, portable and efficient ORB written in C++, based on the ACE library [8,9]. All the programs were compiled with gcc/g++ v.3 with full optimizations.

Basic performance metrics for the mechanisms employed. The first set of experiments are constructed to measure the maximum performance achievable by the current implementation of the two interaction methods, independently from the chosen algorithm. These performance metrics are compared with the ones obtained on the same platform using plain MPI and ASSIST equivalent programs. For the asynchronous interaction, we compared the bandwidth (varying the size of the message, see Fig. 5) of a pipeline of two processes implemented in

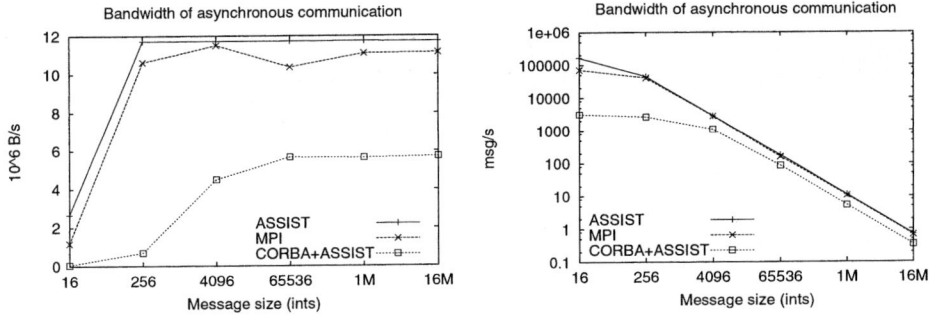

Fig. 5. Asynchronous communication bandwidth (left: 10^6 B/s, right: messages/s).

- MPI (plain MPI_Send/MPI_Receive),
- ASSIST (plain stream),
- C++ and ASSIST (CORBA event channels).

The first process produces the stream as fast as possible, while the second consumes it and computes the bandwidth. The communicating processes are mapped on different machines, in order to measure the bandwidth of a remote MPI/ASSIST/CORBA communication.

For the synchronous interaction, we measured the average round-trip time (see Table 1) of a remote service request, in which the server simply echoes the received message as a reply, varying the size of the request (and response). The server and the client are mapped onto different machines. We simulated this behaviour in MPI (by means of pairs of MPI_Send/MPI_Receives) and ASSIST (using two streams to send the request and the response). We compared the results to a CORBA+ASSIST version, in which the server is an ASSIST sequential module and the service, exported through RMI, is invoked by a C++ client.

Service time of an exported CORBA object. This experiment wants to show the impact of the invocation overhead introduced by CORBA on the performance of a data-parallel object. The service implements a synthetic data-parallel algorithm, operating on large matrices (the argument and result are 700×700 floats) and is written in ASSIST. The algorithm, given a matrix as input, transposes it in parallel (all to all communication) and then for every pair of rows (a_i, b_j) computes $c_{ij} = \sum_k a_{ik} \times b_{jk} \times \sin(k)$.

The performance of the CORBA object is compared to an equivalent solution expressed in the ASSIST-CL language. In this solution the sequential program

Table 1. Round trip times for MPI, ASSIST, and CORBA RMI encapsulating ASSIST.

Request/response size (ints)	MPI (s)	ASSIST (s)	CORBA+ASSIST (s)
4096	0.00533	0.0045	0.0048
65536	0.053	0.0452	0.0587
1048576	0.762	0.7178	0.923
16777216	12.09	11.47	14.79

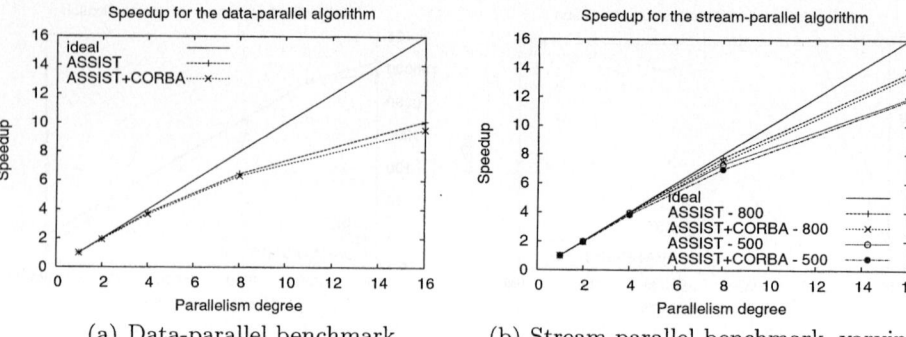

(a) Data-parallel benchmark.

(b) Stream-parallel benchmark, varying computational grain.

Fig. 6. Performance comparison: ASSIST vs. ASSIST+CORBA.

is encapsulated in an ASSIST sequential module and interacts with the parallel module implementing the algorithm by means of ASSIST streams[1].

The sequential version of the benchmark completes in 82.8s; a plain CORBA version in which the server is sequential (C++) completes in 83.1s, with an overhead of 0.3s. The overhead introduced by the program transformation is comparable (see table 2) to the one in the sequential case, and is in line with the raw overhead measured for the adopted mechanisms; this means that the implementation doesn't introduce inefficiencies, and that parallelism exploitation doesn't interfere with the transformation. So, if the overhead is negligible compared to the service time, as in this case, we obtain a good scalability, comparable with the one obtained with a pure ASSIST implementation (see Fig. 6(a)).

Table 2. Parallel service times for the data-parallel algorithm.

Par. degree	ASSIST-CL program	ASSIST + CORBA	overhead
2	42.2s	42.7s	0.5s (1.2%)
4	22.0s	22.6s	0.6s (2.7%)
8	12.8s	13.1s	0.3s (2.3%)
16	8.2s	8.7s	0.5s (5.7%)

Speedup of stream-parallel algorithms. The last experiment shows that stream-parallel programs with sufficient computation to communication ratio can gain high speedups. The stream-parallel algorithm chosen is the classical Mandelbrot set computation, parallelized as a task farm, in which each task is a set of contiguous points that must be evaluated; it produces a stream of computed lines (arrays of 500-800 values), with a rate that varies during the execution.

---

[1] This introduces overhead w.r.t. an ASSIST program in which the whole computation is carried in a parallel module, in fact this implies that the argument is scattered or gathered whenever a communication occurs between the sequential and the parallel module, but makes the performance comparable with the CORBA based solution.

Even in this case, the overhead introduced by the CORBA interaction vs. a pure ASSIST implementation is negligible (see Fig. 6(b)), allowing a good scalability up to 16 processors (the maximal configuration we tested).

5 Related Work

Several studies investigated how CORBA could be adapted to enable the construction of high-performance applications, focusing mainly on communication optimization with/between data-parallel programs.

PARDIS [10] project extended an existing CORBA ORB implementation, to allow both sequential and parallel clients to interact with sequential or parallel servers, where the processes composing a parallel unit can communicate via MPI message passing. They extended also the IDL language, in order to represent, by means of the new *distributed sequence type*, the possible data distribution policies often found in SPMD programs.

In PaCO [11] parallel (MPI) SPMD servers can be exported as a single parallel CORBA object and invoked by a sequential client; simple IDL extensions are introduced to support the classical BLOCK/CYCLIC data distributions.

GridCCM [12] extends the PaCO approach to CORBA components: SPMD MPI programs can be exported as CCM components, described by usual IDL3 interfaces and additional XML files describing the parallelism-oriented aspects.

Our approach can handle not only data-parallelism or SPMD programs, but also task-parallelism (e.g. pipeline / task farm) and mixed task- and data- parallelism, exploiting standard CORBA mechanisms such as object method invocation and event channels.

6 Conclusions

In this work we described the implementation of a compiling tool that automatizes the exportation of an ASSIST subprogram as a CORBA object. The ASSIST programming language eases this task, because it is based on a modular coordination model, that explicits all interactions between parallel components by means of interfaces. The translation of those interfaces to CORBA IDL notation and the construction of bridging ASSIST modules enables ASSIST subprograms to be integrated in larger applications using the CORBA standard as an interoperability layer. This simplifies the use of parallel algorithms within a sequential application and allows, as well, the composition of different parallel algorithms inside the same parallel application, configured at load- or run- time. The experiments showed that this is a viable solution to parallel software integration, in fact the two interaction methods allow the integration of low latency (data-parallel) or high-throughput (stream-parallel) parallel components with good performances, comparable to the ones that can be achieved in a parallel application written entirely in a high-level parallel programming language like ASSIST. These results makes ourselves confident that the integration between object- or component-based frameworks and high-performance computing is feasible, therefore we intend to extend the approach to generate bridges for CCM components (introduced in the CORBA 3 standard), and to generalize it to handle other component-based standards (e.g. Web/GRID services [5]) as well.

Acknowledgments

We wish to thank M. Vanneschi, M. Danelutto, M. Aldinucci, S. Campa, P. Ciullo, M. Coppola, G. Giaccherini, A. Paternesi, A. Petrocelli, E. Pistoletti, L. Potiti, R. Ravazzolo, M. Torquati, P. Vitale.

References

1. Vanneschi, M.: The programming model of ASSIST, an environment for parallel and distributed portable applications. Parallel Computing **28** (2002) 1709–1732
2. Aldinucci, M., Campa, S., Ciullo, P., Coppola, M., Magini, S., Pesciullesi, P., Potiti, L., Ravazzolo, R., Torquati, M., Vanneschi, M., Zoccolo, C.: The Implementation of ASSIST, an Environment for Parallel and Distributed Programming. In Kosch, H., László Böszörményi, Hellwagner, H., eds.: Euro-Par 2003: Parallel Processing. Number 2790 in LNCS (2003) 712–721
3. Aldinucci, M., Campa, S., Ciullo, P., Coppola, M., Danelutto, M., Pesciullesi, P., Ravazzolo, R., Torquati, M., Vanneschi, M., Zoccolo, C.: A Framework for Experimenting with Structured Parallel Programming Environment Design. In: ParCo 2003 Conference Proceedings, to appear, Dresden, Germany (2003)
4. Parashar, M., Laszewski, G., Verma, S., Gawor, J., Keahey, K., Rehn, N.: A CORBA Commodity Grid Kit. Concurrency Practice and Experience, special issue on GRID Computing Environments **14** (2002) 1057–1074
5. Foster, I., Kesselman, C., Nick, J.M., Tuecke, S.: Grid services for distributed system integration. Computer **35** (2002)
6. Aldinucci, M., Campa, S., Ciullo, P., Coppola, M., Danelutto, M., Pesciullesi, P., Ravazzolo, R., Torquati, M., Vanneschi, M., Zoccolo, C.: ASSIST demo: a High Level, High Performance, Portable, Structured Parallel Programming Environment at Work. In Kosch, H., László Böszörményi, Hellwagner, H., eds.: Euro-Par 2003: Parallel Processing. Number 2790 in LNCS (2003) 1295–1300
7. Object Management Group: The Common Object Request Broker: Architecture and Specification. (2000) Minor revision 2.4.1, http://www.omg.org
8. Schmidt, D.C., Harrison, T., Al-Shaer, E.: Object-oriented components for high-speed network programming. In: Proceedings of the 1st Conference on Object-Oriented Technologies and Systems (COOTS), Monterey, CA, USENIX (1995)
9. Schmidt, D.C., Levine, D.L., Mungee, S.: The design of the TAO real-time object request broker. Computer Communications **21** (1998)
10. Keahey, K., Gannon, D.: PARDIS: A parallel approach to CORBA. In: Proceedings of 6th High Performance Distributed Computing, IEEE (1997) 31–39
11. Pérez, C., Priol, T., Ribes, A.: PaCO++: A parallel object model for high performance distributed systems. In: Distributed Object and Component-based Software Systems Minitrack in the Software Technology Track of HICSS-37, Big Island, Hawaii, USA, IEEE Computer Society Press (2004) To appear
12. Denis, A., Pérez, C., Priol, T., Ribes, A.: Bringing high performance to the CORBA component model. In: SIAM Conference on Parallel Processing for Scientific Computing. (2004) To appear

Designing Parallel Operating Systems via Parallel Programming*

Eitan Frachtenberg, Kei Davis, Fabrizio Petrini,
Juan Fernandez, and José Carlos Sancho

Los Alamos National Laboratory, Los Alamos, NM 87545, USA
{eitanf,kei,fabrizio,juanf,jcsancho}@lanl.gov

Abstract. Ever-increasing demand for computing capability is driving the construction of ever-larger computer clusters, soon to be reaching tens of thousands of processors. Many functionalities of system software have failed to scale accordingly – systems are becoming more complex, less reliable, and less efficient. Our premise is that these deficiencies arise from a lack of global control and coordination of the processing nodes. In practice, current parallel machines are loosely-coupled systems that are used for solving inherently tightly-coupled problems. This paper demonstrates that existing and future systems can be made more scalable by using BSP-like parallel programming principles in the design and implementation of the system software, and by taking full advantage of the latest interconnection network hardware. Moreover, we show that this approach can also yield great improvements in efficiency, reliability, and simplicity.

1 Introduction

There is a demonstrable need for a new approach to the design of system software for large clusters. We claim that by using the principles of parallel programming and making effective use of collective communication, great gains may be made in scalability, efficiency, fault tolerance, and reduction in complexity.

Here *system software* refers to all software running on a machine other than user applications. For a workstation or SMP this is just a traditional microprocessor operating system (OS) (e.g. Linux kernel) but for a large high-performance cluster there are additional components. These include communication libraries (e.g. MPI, OpenMP), parallel file systems, the system monitor/manager, job scheduler, high-performance external network, and more.

Experience with large-scale machines such as Cplant, Virginia Tech's Terascale Cluster, and ASCI's Blue Mountain, White, Q, and Lightning, has shown that managing such machines is time comsuming and expensive. The components of the system software typically introduce redundancies in both functionality (communication and coordination protocols) and in hardware (multiple interconnection networks) and are typically 'bolted together,' each coming from a different developer or vendor, resulting

---

* This work is partially supported by the U.S. Department of Energy through Los Alamos National Laboratory contract W-7405-ENG-36 and the Spanish MCYT under grant TIC2003-08154-C06-03.

in a multiplication of complexity. Further, for the larger systems efficiency (delivery of theoretical capability), responsiveness, and reliability remain low, indicating that they have already outgrown current incoherent suites of system software.

We believe that the root of the problem is the use of largely independent, loosely-coupled compute nodes for the solution of problems that are inherently tightly coupled. A solution to this problem is to better integrate the compute nodes using modern interconnection network hardware. We propose a new methodology for the design of parallel system software based on two cornerstones: 1) BSP-like global control and coordination of *all* of the activities in the machine, and so, 2) treating the system software suite just like any other parallel application. In practice we are able to attain efficient and highly scalable coordination using a very small set of collective communication primitives. From a theoretical point of view this set of primitives should be amenable to formal semantic characterization and analysis (possibly following Lamport [12]), but this remains a potential direction for future research. More practically, this demonstrates new modalities for the use of collective communication.

Much of what we propose here has been implemented and shown to have achieved the desired goals – simplicity, efficiency, effectiveness, and very high scalability. Other components, based on the same primitives, are still in development. However, various subsets of these mechanisms are independent and so may be put into use in the absence of the others, so allowing incremental proof of concept.

2 Toward a Parallel Operating System

Distributed and parallel applications (including operating systems) are distinguished by their use of interprocessor communication. Distributed applications typically make much more use of local information and exchange a relatively small number of point-to-point messages. Parallel programs benefit from, and often require, mechanisms for global synchronization and exchange of information, such as barriers, reduction operations, etc. Many OS tasks are inherently collective operations, such as context switching and job launching; others benefit by being cast in terms of collective operations.

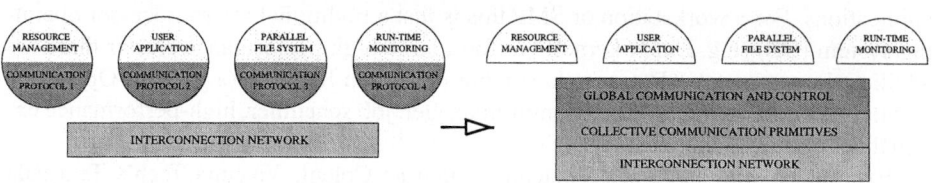

Fig. 1. Communication Protocol.

Our vision is that of a cohesive global operating system that is designed using parallel programming paradigms and techniques. Such a unified OS will not only be smaller and simpler than the sum of the parts currently used for system software, but also more efficient because of reduced overhead and redundancy (Fig. 1). All the roles of a cluster OS, such as job launching and scheduling, user-level communication, parallel I/O, transparent fault tolerance, and garbage collection can be implemented over a single communication infrastructure. This layer in turn is designed as a parallel program using the same collective communication primitives.

By using a carefully chosen set of low-latency, scalable communication primitives we can impose a global communication model where both system and user communication is tightly controlled at a fine granularity. The system as a whole behaves as a bulk-synchronous program, where computation and communication are divided into distinct, timed phases. In this model, called Buffered Coscheduling (BCS) [3], all the user and system-level communication is buffered and controlled. The entire cluster marches to the beat of a global strobe that is issued every few hundreds of microseconds. This is somewhat reminiscent of the SIMD model, except the granularity is in terms of time rather than instructions. In the periods between strobes, or *timeslices*, newly-issued communication calls are buffered until the next timeslice. At every strobe, nodes exchange information on pending communication, so that every node has complete knowledge of the required incoming and outgoing communication for the next timeslice. The nodes then proceed to globally schedule those communications that will actually be carried out during the timeslice, and proceed to execute them. The advantage of this model is that all the communication is controlled and can be maintained in a known state at every given timeslice, so that problems arising from congestion, out of order arrival, and hot spots can be avoided. As reported in Section 4, these constraints impose little or no overhead for scientific applications, while obtaining the advantages of a more deterministic, controllable machine.

3 Core Primitives and Operating System Functions

Suggested Primitives. Our suggested support layer consists of three network primitives. These may be implemented in hardware, software, or a combination of both; commodity hardware exists today that can implement them directly. While this set of primitives is complete (in terms of providing the needed functionality), other primitives might conceivably provide equivalent (or even more) functionality and efficiency.

XFER-AND-SIGNAL. Transfer (PUT) a block of data from local memory to a commonly registered memory area, or *global memory*, of a set of nodes, and optionally signal a local and/or a remote event upon completion.
TEST-EVENT. Check if a local event has been signaled. Optionally, block until it has.
COMPARE-AND-WRITE. Arithmetically compare a global variable on a set of nodes to a local value. If the condition is true on *all* nodes, then (optionally) assign a new value to a (possibly different) global variable.

Note that XFER-AND-SIGNAL and COMPARE-AND-WRITE are both atomic operations. That is, XFER-AND-SIGNAL either PUTs data to *all* nodes in the destination set or (in case of a network error) *none* of the nodes. The same condition holds for COMPARE-AND-WRITE when it writes a value to a global variable. Furthermore, if multiple nodes simultaneously initiate COMPARE-AND-WRITEs with identical parameters except for the value to write, then, when all the operations have completed, all nodes will see the same value in the global variable. In other words, both primitives are *sequentially consistent* operations [12]. TEST-EVENT and COMPARE-AND-WRITE are traditional blocking operations, while XFER-AND-SIGNAL is non-blocking. The only way to check for completion is to TEST-EVENT on a local event triggered by XFER-AND-SIGNAL. These semantics do not dictate whether the mechanisms are implemented by the host CPU or by a network co-processor, or where the global memory resides. Nor do they require that TEST-EVENT yield the CPU, though it may be advantageous to do so.

System Software Requirements and Solutions. In this section we discuss how the various responsibilities of a global OS can be realized with the aforementioned mechanisms and communication layer. These OS functions are all discussed and expressed in terms of the proposed abstract layer, showing its sufficiency and completeness for a global OS. Table 1 summarizes these arguments.

Table 1. Network mechanisms usage.

Characteristic	Function	Solution
Job Launching	Data dissemination	XFER-AND-SIGNAL
	Flow control	COMPARE-AND-WRITE
	Termination detection	COMPARE-AND-WRITE
Job Scheduling	Heartbeat	XFER-AND-SIGNAL
	Context switch responsiveness	Prioritized messages
Communication	PUT	XFER-AND-SIGNAL
	GET	XFER-AND-SIGNAL
	Barrier	COMPARE-AND-WRITE
	Broadcast	COMPARE-AND-WRITE + XFER-AND-SIGNAL
Storage	Metadata/file data transfer	XFER-AND-SIGNAL
Debugging	Debug data transfer	XFER-AND-SIGNAL
	Debug synchronization	COMPARE-AND-WRITE
Fault Tolerance	Fault detection	COMPARE-AND-WRITE
	Checkpointing synchronization	COMPARE-AND-WRITE
	Checkpointing data transfer	XFER-AND-SIGNAL
Garbage Collection	Live state synchronization	Determinism and COMPARE-AND-WRITE

Job Launching. The traditional approach to job launching, including the distribution of executable and data files to cluster nodes, is a simple extension of single-node job launching: data is transmitted using the network file system, and jobs are launched with scripts or simple utilities such as rsh or mpirun. These methods do not scale to large machines where the load on the network file system, and the time it would take to serially launch a binary on many nodes, make them inefficient and impractical. Several solutions have been proposed for this problem, all based on software tricks to reduce the distribution time. For example, Cplant and BProc use their own tree-based algorithms to distribute data with latencies that are logarithmic in the number of nodes [1, 9]. While more portable than relying on hardware support, these solutions are significantly slower and not always simple to implement [7].

Decomposing job launching into simpler sub-tasks shows that it only requires modest effort to make the process efficient and scalable:

- Executable and data distribution are no more than a multicast of packets from a file server to a set of nodes, and can be implemented using XFER-AND-SIGNAL.
- Launching of a job can be achieved simply and efficiently by multicasting a control message to the target nodes using XFER-AND-SIGNAL. The system software on each node then forks the process and waits for its termination.
- The reporting of job termination can incur much overhead if each node sends a single message for every process that terminates. This can be avoided by all processes of a job reaching a common synchronization point upon termination (using COMPARE-AND-WRITE) before delivering a single message to the resource manager.

Job Scheduling. Interactive responsiveness from a scheduler is required to make a large machine as usable as a workstation. This implies that the system must be able to perform preemptive context switching with latencies similar to uniprocessor systems, that is, on the order of a few milliseconds. Such latencies are almost impossible to achieve without scalable collective operations: the time required to coordinate a context switch over thousands of nodes can be prohibitively large when using point-to-point communication [10]. Even though the system is able to efficiently context switch between different jobs, concurrent (uncoordinated) application traffic and synchronization messages in the network can unacceptably delay response to the latter. If this occurs even on a single node for even just a few milliseconds it may have a severe detrimental effect on the responsiveness of the entire system [15].

Many contemporary networks offer some capabilities to the software scheduler to prevent these delays. The ability to maintain multiple communication contexts alive in the network securely and reliably, without kernel intervention, is already implemented in some state-of-the-art networks such as QsNet. Job context switching can be easily achieved by simply multicasting a control message to all the nodes in the network using XFER-AND-SIGNAL. Our communication layer can guarantee that system messages get priority over user communication to avoid synchronization problems.

Communication. Most of MPI's, TCP/IP's, and other communication protocols' services can be reduced to a rather basic set of communication primitives, e.g. point-to-point synchronous and asynchronous messages, multicasts, and reductions. If the underlying primitives and protocols are implemented efficiently, scalably, and reliably by the hardware and cluster OS, respectively, the higher level protocols can inherit the same properties. In many cases, this reduction is very simple and can eliminate the need for many of the implementation quirks of protocols that need to run on disparate network hardware. Issues such as flow control, congestion avoidance, quality of service, and prioritization of messages are handled transparently by a single communication layer for all the user and system needs.

Determinism and Fault Tolerance. When the system globally coordinates all the application processes, parallel jobs can be made to evolve in a controlled manner. Global coordination can be easily implemented with XFER-AND-SIGNAL, and can be used to perform global scheduling of all the system resources. Determinism can be enforced by taking the same scheduling decisions between different executions. At the same time, global coordination of all the system activities helps to identify the states along the program execution at which it is safe to checkpoint.

The tight control of global communication and the induced determinism that follows from this constraint allows for a seamless inclusion of various other important OS services and functionalities. One example is parallel I/O, which can benefit from the hot-spot and congestion avoidance of this model, since all the I/O operations can be scheduled as low-priority communications. The ability to synchronize an entire application to a known state at fine granularity (without having messages en route) is very important for performing global garbage collection, by keeping track of the live state of global objects [11]. Even more important is the ability to use these known states for automatic, fine-grained, coordinated checkpointing. Because of the frequency with which components can fail, one of the main challenges in using large-scale clusters is achieving fault tolerance. The difficulty of checkpointing for these clusters arises from the quantity of unknown system state at any given point in time, due largely to non-determinism in the communication layer. Eliminating and controlling most of these unknowns allows significant simplification of automatic checkpointing and restart at a granularity of a few seconds, far more responsively than current solutions. The checkpointing traffic is handled and scheduled by the communication layer together with all other traffic, again mitigating flow-control and non-determinism issues.

Implementation and Portability. The three primitives presented above assume that the network hardware enables efficient implementation of a commonly registered memory area. Such functionality is provided by several state-of-the-art networks such as QsNet and Infiniband and has been extensively studied [13, 14]. We note that some or all of the primitives have have already been implemented in several other interconnects; their expected performance is shown in Table 2. They were originally designed to improve the communication performance of user applications; to the best of our knowledge their usage as an infrastructure for system software has not been explored before this work.

Hardware support for multicast messages sent with XFER-AND-SIGNAL is needed to guarantee scalability for large-scale systems – software approaches do not scale well to thousands of nodes. In our case, QsNet provides PUT/GET operations, making the implementation of XFER-AND-SIGNAL straightforward.

COMPARE-AND-WRITE assumes that the network is able to return a single value to the calling process regardless of the number of queried nodes. Again, QsNet provides a global query operation that allows direct implementation of COMPARE-AND-WRITE.

4 Results

We have implemented a research prototype, called STORM [7], as a proof of concept of our approach. STORM is a full-fledged resource manager that provides job

Table 2. Measured/expected performance of the core mechanisms for n nodes.

Network	Comparison (μs)	Multicast (MB/s)
Gigabit Ethernet [17]	$46 \log n$	Not available
Myrinet [2]	$20 \log n$	$\sim 15n$
Infiniband [13]	$20 \log n$	Not available
QsNet ([14])	< 10	$> 150n$
BlueGene/L [8]	< 2	$700n$

launching, resource allocation and monitoring, load balancing, and various job scheduling algorithms including space-shared, time-shared, and backfilling variants. STORM's performance has been evaluated, modeled, and compared to several others from the literature [7]. STORM is an order of magnitude faster than the best reported results for job launching, and delivers two orders of magnitude better performance for context switching. STORM was later used to implement several new scheduling algorithms. In a comprehensive experimental evaluation [5, 6], our new algorithms improved the system's utilization and response times both for simple and dynamic workloads. By using our primitives for global resource coordination, the algorithms were better suited to avoiding and mitigating problems of internal and external fragmentation.

We have also implemented BCS-MPI, an MPI library based on the BCS model, on top of STORM. The novelty of BCS-MPI is its use of global coordination of a large number of communicating processes rather than an emphasis on the traditional optimization of the point-to-point performance. Several experimental results [4], using a set of real-world scientific applications, show that BCS-MPI is only marginally slower (less than 10%) than production-grade MPI implementations, but is much simpler to implement, debug, and reason about. Performance results for scientific applications provide strong evidence for the feasibility of our approach for transparent fault tolerance [16].

5 Conclusions

We have shown that a BSP-like approach to the design of system software is not only feasible but promises much-needed improvements in efficiency, simplicity, and scalability for all of the key functionalities of a cluster OS. Further, it provides a framework in which effective fault tolerance may be achieved. All of the functions may be implemented in terms of simple collective communication primitives directly supported by currently available interconnection networks.

Concretely, the resource manager STORM implements job launching, resource allocation and monitoring, job scheduling, and load balancing. BCS-MPI implements a high-level communication protocol. The full implementation of the fault tolerance and parallel I/O mechanisms is underway; experimental results provide ample evidence that the desired functionalities and behaviors are achievable.

References

1. R. Brightwell and L. A. Fisk. Scalable Parallel Application Launch on Cplant. In *Proceedings of IEEE/ACM Supercomputing 2001 (SC'01)*, Denver, CO, November 2001.
2. D. Buntinas, D. Panda, J. Duato, and P. Sadayappan. Broadcast/Multicast over Myrinet using NIC-Assisted Multidestination Messages. In *Workshop on Communication, Architecture, and Applications for Network-Based Parallel Computing (CANPC '00)*, Toulouse, France, January 2000.
3. Fabrizio Petrini and Wu-chun Feng. Buffered Coscheduling: A New Methodology for Multitasking Parallel Jobs on Distributed Systems. In *Proceedings of the International Parallel and Distributed Processing Symposium 2000*, volume 16, Cancun, MX, May 2000.
4. J. Fernandez, F. Petrini, and E. Frachtenberg. BCS MPI: A New Approach in the System Software Design for Large-Scale Parallel Computers. In *Proceedings of IEEE/ACM Supercomputing 2003 (SC'03)*, Phoenix, AZ, November 2003.
5. E. Frachtenberg, D. Feitelson, F. Petrini, and J. Fernandez. Flexible CoScheduling: Mitigating Load Imbalance and Improving Utilization of Heterogeneous Resources. In *Proceedings of the International Parallel and Distributed Processing Symposium 2003 (IPDPS03)*, Nice, France, April 2003.
6. E. Frachtenberg, D. G. Feitelson, J. Fernandez, and F. Petrini. Parallel Job Scheduling under Dynamic Workloads. In D. G. Feitelson, L. Rudolph, and U. Schwiegelshohn, editors, *Job Scheduling Strategies for Parallel Processing*, volume 2862 of *Lect. OPTnotes Comput. Sci.* Springer Verlag, 2003.
7. E. Frachtenberg, F. Petrini, J. Fernandez, S. Pakin, and S. Coll. STORM: Lightning-Fast Resource Management. In *Proceedings of IEEE/ACM Supercomputing 2002 (SC'02)*, Baltimore, MD, November 2002.
8. M. Gupta. Challenges in Developing Scalable Scalable Software for BlueGene/L. In *Scaling to New Heights Workshop*, Pittsburgh, PA, May 2002.
9. E. Hendriks. BProc: The Beowulf Distributed Process Space. In *Proceedings of the 16th Annual ACM International Conference on Supercomputing*, New York, NY, June 2002.
10. A. Hori, H. Tezuka, and Y. Ishikawa. Overhead Analysis of Preemptive Gang Scheduling. In D. G. Feitelson and L. Rudolph, editors, *Job Scheduling Strategies for Parallel Processing*. Springer Verlag, 1998.
11. T. Kamada, S. Matsuoka, and A. Yonezawa. Efficient Parallel Global Garbage Collection on Massively Parallel Computers. In G. M. Johnson, editor, *Proceedings of IEEE/ACM Supercomputing 1994 (SC'94)*, 1994.
12. L. Lamport. How to Make a Multiprocessor Computer That Correctly Executes Multiprocess Programs. *IEEE Transactions on Computers*, C-28(9), September 1979.
13. J. Liu, A. Mamidala, and D. K. Panda. Fast and Scalable MPI-Level Broadcast using Infiniband's Hardware Multicast Support. In *Proceedings of the 18th International Parallel & Distributed Processing Symposium*, Santa Fe, New Mexico, April 2004.
14. F. Petrini, W. Feng, A. Hoisie, S. Coll, and E. Frachtenberg. The Quadrics Network: High-Performance Clustering Technology. *IEEE Micro*, 22(1), January/February 2002.
15. F. Petrini, D. Kerbyson, and S. Pakin. The Case of the Missing Supercomputer Performance: Achieving Optimal Performance on the 8,192 Processors of ASCI Q. In *Proceedings of IEEE/ACM Supercomputing 2003 (SC'03)*, Phoenix, AZ, November 2003.
16. J. C. Sancho, F. Petrini, G. Johnson, J. Fernández, and E. Frachtenberg. On the Feasibility of Incremental Checkpointing for Scientific Computing. In *Proceedings of the 18th International Parallel & Distributed Processing Symposium*, Santa Fe, New Mexico, April 2004.
17. P. Shivam, P. Wyckoff, and D. Panda. EMP: Zero-copy OS-bypass NIC-driven Gigabit Ethernet Message Passing. In *Proceedings of IEEE/ACM Supercomputing 2001 (SC'01)*, Denver, CO, November 2001.

Topic 11
Numerical Algorithms

Emilio L. Zapata, Oscar Plata, David Keyes, and Pasqua D'Ambra

Topic Chairs

Fast and reliable parallel algorithms for the basic problems of numerical mathematics and their effective implementation in easy-to-use portable software components are crucial for computational solution of scientific and engineering problems. This Topic track is a forum for the presentation and discussion of new developments in the field of parallel numerical algorithms, covering all aspects from basic algorithms, efficient implementation on modern parallel, distributed and network-based architectures, to software design and prototyping in mathematical software components and performance analysis. Specific areas of interest include, among others, PDEs, large sparse or dense linear systems and eigensystems, non-linear systems, linear algebra and fast transforms.

Overall, fifteen papers were submitted to our Topic. Out of these fifteen submissions, six were accepted, all of them as regular papers. At the conference, the presentations were arranged into two sessions, containing three papers each of them. The first session starts out with a presentation investigating the old history effect in asynchronous iterations on a linear least squares problem. A partially asynchronous algorithm is developed. It is experimentally shown the effectiveness of a combined effort to decrease the effect of old history. The second paper presents a study of mixed parallel execution schemes for explicit and implicit variants of general linear methods. Experiments conducted in two different parallel platforms show interesting results. The final paper in this session proposes a data management and communication layer for an adaptive, hexahedral finite element method on distributed memory machines. The layer also provides an effective user interface.

The second session consists of three papers. The first concerns a splitting approach to the pressure-stabilized Petrov-Galerkin (PSPG) finite element method for incompressible flow. A parallel implementation of the method is described showing a good scalability for large systems. The next paper deals with a parallel direct solver for solving large sparse linear systems. A modification of a multilevel graph partitioning schema is proposed for a balanced domain decomposition of the problem. The last paper is about parallel hybrid algorithms for solving the triangular continuous-time Sylvester equation. It is shown that these algorithms outperform parallel implementations of the DTRSYL solver from LAPACK.

Finally, we would like to express our sincere appreciation to the members of the program committee for their invaluable help in the entire selection process. We would also like to thank the reviewers for their time and effort.

Reducing the Deteriorating Effect of Old History in Asynchronous Iterations[*]

Yasemin Yalçınkaya[1] and Trond Steihaug[2]

[1] Università degli Studi di Firenze, Dipartimento di Energetica "Sergio Stecco", Firenze, Italy
[2] University of Bergen, Department of Informatics, Bergen, Norway

Abstract. The deteriorating effect of old history in asynchronous iterations is investigated on an application based on the specialization of parallel variable distribution approach of Ferris and Mangasarian [4] to linear least squares problem. A partially asynchronous algorithm is developed which employs a combination of synchronization, a relaxation parameter and a certain form of overlap between subproblems. It is shown by numerical experiments that this combined effort to decrease the effect of old history is more effective than the single attempts considered in [9, 11].

The aim of the paper is to provide an efficient method to compensate for the deteriorating effect of old history on the convergence of asynchronous iterative methods. The issue of old history is investigated on an application based on the specialization of parallel variable distribution (PVD) approach of Ferris and Mangasarian [4] to linear least squares problem.

We have seen in [9, 11] that in the implementation of totally asynchronous iterations on linear least squares problems convergence is deteriorated due to the existence of *old history* in the system. Yet, some measures can be taken to reduce the effect of old history. It has been observed that increasing the number of processors increases the effect of old history, whereas introducing some form of synchronization, either in the form of local synchronization, or as barrier synchronization decreases the deterioration. The use of a relaxation parameter as a line search also improves the convergence rate.

Here we will formulate a partially asynchronous algorithm that combines most of the methods studied in [9, 11] plus a certain form of overlap to decrease/eliminate the effect of old history. This new algorithm is based on our observations in our former studies and the work of Dennis and Steihaug [2]. We will see that the new algorithm eliminates the deterioration in convergence rate observed due to the effect of old history.

In the following, we first give some notations and definitions that are used throughout the paper. In Sect. 2 we introduce the concepts of planar search and expanded blocks. The final algorithm is outlined and explained in Sect. 3. We give some experimental results in Sect. 4. Concluding remarks are included in Sect. 5.

---

[*] Work is a part of the first author's Dr. Scient. thesis completed at the University of Bergen, Department of Informatics, Bergen, Norway.

1 Preliminaries

The specialization of PVD to linear least squares problem was presented in [2, 8] and a more general approach was presented in [6]. The same domain decomposition approach to linear least squares problem has been employed in [1, 9–11] but under a different name.

Let A be an $m \times n$ real matrix, $b \in \mathbb{R}^m$, and M be an $m \times m$ positive definite matrix. The weighted linear least squares problem is

$$\min_{x \in \mathbb{R}^n} \|Ax - b\|_M, \tag{1}$$

where $\|y\|_M^2 = y^T M y$.

To apply the PVD approach, the space $\mathbb{R}^n$ is decomposed as a Cartesian product of lower dimensional spaces $\mathbb{R}^{n_i}$, $i = 1, \ldots, g$, where $\sum_{i=1}^{g} n_i = n$. Accordingly, any vector $x \in \mathbb{R}^n$ is decomposed as $x = [x_1 \cdots x_g]^T$, and the matrix A is partitioned as $A = [A_1 \cdots A_g]$, where $A_i \in \mathbb{R}^{m \times n_i}$, and each A_i is assumed to have full rank. Then, the least squares problem (1) is equivalent to

$$\min_{x \in \mathbb{R}^n} \left\{ \| \sum_{i=1}^{g} A_i x_i - b \|_M : x_i \in \mathbb{R}^{n_i}, \ i = 1, \ldots, g \right\}. \tag{2}$$

In a parallel processing environment, consistent with the application of the PVD approach we assume that there are g processors each of which can update one block component, x_i. These processors that solve the subproblems are called as *slave* processors and are coordinated by a *master* processor which computes the global solution vector.

Let $t \in \mathcal{T} = \{1, 2, \ldots\}$ be a normalized integer counter of actual times that one or more components x_i of x are updated, and let $x(t)$ be the approximation to x^*, the solution of (1), at a given time instant t. Further, let $\mathcal{T}^i \subset \mathcal{T}$ be the set of times component i is updated and $x_i(t)$ be the value of component x_i at time t. For a given t, each slave i for which $t \in \mathcal{T}^i$, is assigned the subproblem

$$\text{Solve for } d_i(t) \in \mathbb{R}^{n_i} : \min_{d_i \in \mathbb{R}^{n_i}} \|A_i d_i + r(t-1)\|, \tag{3}$$

where $d_i(t) = x_i(t) - x_i(t-1)$ is the *direction* vector and $r(t) = \sum_{i=1}^{g} A_i x_i(t) - b = r(t-1) + \sum_{i=1}^{g} A_i d_i(t)$ is the *residual*. The master computes the global solution

$$x(t) = x(t-1) + \sum_{i=1}^{g} \bar{d}_i(t), \tag{4}$$

where $\bar{d}_i \in \mathbb{R}^n$ is the vector d_i complemented with zeros in places of blocks $j \in \{1, \ldots, g\}$, $j \neq i$. Note that $\bar{d}_i(t)$ is nonzero only for the components i for which $t \in \mathcal{T}^i$.

If for all $t \in \mathcal{T}$ we have $t \in \mathcal{T}^i$, $i = 1, \ldots, g$, then we get a synchronous iteration in the form of Jacobi method on normal equations $A^T A x = A^T b$ [10], where the values of all the components $x_i(t)$ are computed using the values from the previous iteration, namely $x_j(t-1)$, $j = 1, \ldots, g$. However, in an asynchronous setting, the processor computing d_i may not have access to the most recent values of other components of x, and $d_i(t)$ is computed using a residual $r = \sum_{j=1}^{g} A_j x_j(\tau_j^i(t)) - b$, where $\tau_j^i(t)$ are *time-stamps* satisfying

$$0 \leq \tau_j^i(t) < t, \quad \forall\, t \in \mathcal{T}. \tag{5}$$

In (5) the difference $t - \tau_j^i(t)$ can be seen as a communication delay. This communication delay, or time-lag, causes that old data is used in the computation of a component x_i. Consequently, these updates computed using older values deteriorate the convergence property of the algorithm. We call these older values in the system as *old history* and their deteriorating effect as the *effect of old history*.

When the coefficient matrix A is partitioned, the blocks A_i are formed by non-overlapping consecutive columns. Nevertheless, in the implementation of (3) the effect of row dependence between the blocks comes into picture. The dependence between these subproblems can be represented by a directed graph called *dependency graph*, and for this particular problem the dependency graph is an adjacency graph since we have a symmetric system [10]. For the PVD formulation of the linear least squares problem as given above, the set of *essential neighbors* for a given node i in the dependency graph is defined as $E_i = \{j \mid$ block i and block j have nonzero elements on the same row positions$\}$.

2 From Jacobi to a Partially Asynchronous Algorithm

We first introduce *planar search* as an alternative to line search for finding a relaxation parameter. Then we show a way of making all the blocks essential neighbors of each other, which eliminates the need for waiting for an update from one's essential neighbors during the computations to avoid zero updates.

2.1 Planar Search

In (4), after each direction vector d_i is computed, a line search can be done to compute a relaxation parameter ω_i that will give the most decrease in the residual vector r for this direction. The relaxation parameter ω_i given in [11] is computed as the solution of the one dimensional least squares problem

$$\min_{\omega_i \in \mathbb{R}} \|(A_i d_i(t))\omega_i + r(t-1)\|_M, \tag{6}$$

and only the direction vectors that give ω_i values that are in the interval $(0, 2)$ are used to update the residual vector. However, instead of doing a line search to correct each of these d_i vectors one by one, we can wait for receiving all the direction vectors d_i, $i = 1, \ldots, g$, and introduce a planar search where the values of the relaxation parameters ω_i, $i = 1, \ldots, g$, are computed as the solution of a new least squares problem.

Let $\widehat{A} = [A\bar{d}_1 \cdots A\bar{d}_g]$, where $\widehat{A} \in \mathbb{R}^{m \times g}$. Then the planar search step is:

$$\text{Solve for } s(t) \in \mathbb{R}^g: \min_{s \in \mathbb{R}^g} \|\widehat{A}(t)s + r(t-1)\|_M \ . \tag{7}$$

The new iterate is

$$x(t) = x(t-1) + \sum_{i=1}^{g} s_i(t)\bar{d}_i(t) \ .$$

Planar search as defined in (7) is an extension of (6) where the restriction on the values of ω_i is eliminated. Obviously, the planar search step implemented in this fashion introduces a blocking synchronization point in the parallel implementation whereas line search is done without synchronization.

2.2 Expanded Blocks

We pointed out in [11] the need for waiting for an update from any one of the essential neighbors of a given block i between two consecutive updates from this block in order to avoid redundant zero computations. If the subproblems are formed such that each block is an essential neighbor of all the other blocks, the computations can go on continuously with no delay. One method to obtain blocks that are all essential neighbors of each other is to take the original blocks i, $i = 1, \ldots, g$, as they are and expand these blocks by $g - 1$ new columns, one for each block j, $j \neq i$, $j \in \{1, \ldots, g\}$. Each of these new columns is the one dimensional aggregate of subspace j, $j \neq i$, $j \in \{1, \ldots, g\}$.

Let $p \in \mathbb{R}^n$ be the *weight vector* partitioned as $p = [p_1 \cdots p_g]^T$, where $p_i \in \mathbb{R}^{n_i}$, $i = 1, \ldots, g$. Let $\bar{p}_i \in \mathbb{R}^n$ be the vector obtained by starting with a zero vector and placing the non-zero entries of p_i in the positions corresponding to the column indices in A of A_i. Define the $n \times (g - 1 + n_i)$ matrix P_i:

$$P_i = [\bar{p}_1 \cdots \bar{p}_{i-1} \ \bar{I}_i \ \bar{p}_{i+1} \cdots \bar{p}_g] \ , \tag{8}$$

where $I_i \in \mathbb{R}^{n_i \times n_i}$ is the identity matrix and $\bar{I}_i$ is the $n \times n_i$ matrix formed from columns of the $n \times n$ identity matrix so that $A\bar{I}_i = A_i$.

For $\tilde{n}_i = n_i + g - 1$, define the $m \times \tilde{n}_i$ matrix

$$\widetilde{A}_i = AP_i = [A\bar{p}_1 \cdots A\bar{p}_{i-1} \ A_i \ A\bar{p}_{i+1} \cdots A\bar{p}_g] \ . \tag{9}$$

The expanded subspace is now the range space of $\widetilde{A}_i$ and $A\bar{p}_j = A_j p_j$ is the one dimensional aggregate of A_j.

Unless the weight vector p is very sparse, all the newly formed expanded blocks $\widetilde{A}_i$, $i = 1, \ldots, g$, will be essential neighbors of each other.

For a given $p(t) \in \mathbb{R}^n$ at time t the subproblem (3) is replaced by

$$\text{Solve for } \tilde{d}_i(t) \in \mathbb{R}^{\tilde{n}_i}: \min_{\tilde{d}_i \in \mathbb{R}^{\tilde{n}_i}} \|\widetilde{A}_i(t)\tilde{d}_i + r(t-1)\|_M \ , \tag{10}$$

where $\widetilde{A}_i(t)$ is defined in (9) for the given vector $p(t)$.

The step $c \in \mathbb{R}^n$ is

$$c = \sum_{i=1}^{g} P_i \widetilde{d}_i \ . \tag{11}$$

For a given $c \in \mathbb{R}^n$, define the $n \times g$ matrix

$$C = [\bar{c}_1 \ \cdots \ \bar{c}_g] \ . \tag{12}$$

Consider the $m \times g$ matrix

$$\widehat{A} = AC = [A\bar{c}_1 \ \cdots \ A\bar{c}_g] \ . \tag{13}$$

Then the planar search step (7) is executed replacing the matrix $\widehat{A}(t)$ with the $\widehat{A}$ of (13). We use the vector $c(t)$ defined in (11), and the new iterate is

$$x(t) = x(t-1) + C(t)s(t),$$

where $C(t)$ is defined in (12).

One question is how to choose the weight vector p. When the only aim is to form new blocks that are all essential neighbors of each other any choice of a nonzero p that is not very sparse is acceptable. Yet, while forming the aggregations of each block we can also try to increase the convergence rate of the new algorithm. If we could choose $p(t) = x^* - x(t) \equiv e(t)$, then each $P_i(t)\widetilde{d}_i(t)$ would be $e(t)$ and we would get convergence in one step [2]. Since $e(t)$ is not known, a good approximate is taking $p(t) = x(t) - x(t-1) = C(t)s(t)$. This choice leads to a significant reduction in the number of iterations in sequential and synchronous implementations of this approach [2]. Other choices for p are constant values, e.g. $p = [1 \ \cdots 1]^T$, or a *predictor/corrector* scheme defined by keeping p fixed for several predictor iterations without having to redo any factorizations [2]. Note that when p is a zero vector, we get the block Jacobi iteration on normal equations.

3 A Partially Asynchronous Algorithm

The two issues introduced in the former section, if applied directly on the block Jacobi iteration on normal equations, will give a synchronous parallel algorithm since planar search brings out the need for a blocking synchronization step. We know that too frequent blocking synchronization points in parallel implementations should be avoided due to the performance penalties imposed on the system [10]. Hence, we decide to do a planar search after every l sweeps, i.e., after l updates received from each block i, $i = 1, \ldots, g$. Between two planar search steps the weight vector p and consequently $\widehat{A}_i$ matrices are kept constant. The l updates received between two synchronization points from each slave are accumulated and matrix $\widehat{A}$ is formed using these accumulated values. The accumulation of updates in this manner is an application of *reliable updating technique* which leads to very accurate approximations provided that the accumulations do not become too large [3].

Between two synchronization points a running copy of the residual is kept, which is updated on receiving a new result from any one of the slaves and its updated value is sent back to the slave who has communicated the last result. Hence, asynchronism is introduced in the system, which adds the flavor of Gauss–Seidel in the iterations. The original residual vector is used in the planar search step and is updated thereafter.

Observe that in (10), $\tilde{d}_i = [\delta_1^i \cdots \delta_{i-1}^i \, d_i \, \delta_{i+1}^i \cdots \delta_g^i]^T$, where $d_i \in \mathbb{R}^{n_i}$ is the variable representing the space spanned by the ith block of variables in x_i, and the scalar values δ_j^i are the aggregate variables expanding this space. Expanding the subproblems with these aggregate variables not only makes all the new blocks essential neighbors of each other, but also creates a sort of "overlap", though the overlapping part of the newly formed blocks is in aggregated form. In asynchronous implementations overlap to some degree is found to accelerate the overall iteration [7]. However, the benefit of overlap is shown to be in the inclusion of extra variables in the minimization for the local variables only. The updated value on the overlapped portion of the domain should not be utilized [5]. In our case, the inclusion of the scalar values δ_j^i in the update of the residual vector will increase the effect of old history in the system. Therefore, we form and solve the expanded subproblems, but throw away the calculated δ_j^i values. The c vector in (11) is replaced by $c = \sum_{i=1}^g \tilde{d}_i$, where d_i are extracted from $\tilde{d}_i$ of (10). Since we use accumulated updates in $\hat{A}$, the d_i used in forming c are, in fact, $d_i = \sum_{j=1}^l d_i^j$.

Algorithm 1 (PALSQ).

if *master*
    Partition A into g blocks and **Send** one to each *slave*.
    Initialize $x(0)$, $r(0)$, v^0, $c(1)$, $\hat{A}(1)$, $\bar{A}(1)$.
    Broadcast v^0, $\bar{A}(1)$.
else   {*slave i, $i = 1, \ldots, g$*}
    Receive A_i.
    Receive v^0, $\bar{A}(1)$.
    Form and factorize $\tilde{A}_i(1)$.
$t = 1$.
while not *converged* **do**
    if *slave i, $i = 1, \ldots, g$*
        for $j = 0, \ldots, l - 1$
            Solve for $\tilde{d}_i^j$: min $\|\tilde{A}_i(t)\tilde{d}_i + v^j\|_M$.
            Compute $A_i d_i$.
            Send $A_i d_i$, d_i.
            if $j \neq l - 1$ **Receive** v^{j+1}.
        Receive v^0, $\bar{A}(t+1)$.
        $t = t + 1$.
        Form and factorize $\tilde{A}_i(t)$.
    else   {*master*}

> **for** $j = 0, \ldots, l-1$
> **for** $i = 1, \ldots, g$
> **Receive** $A_i d_i, \bar{d}_i$.
> $v^{j+i/g} = v^{j+(i-1)/g} + A_i d_i$.
> **Send** $v^{j+i/g}$.
> $\widehat{A}_i(t) = \widehat{A}_i(t) + A_i d_i$.
> $c(t) = c(t) + \bar{d}_i$.
> Solve for $s(t) : \min \|\widehat{A}(t)s + r(t-1)\|_M$.
> $x(t) = x(t-1) + C(t)s(t)$.
> $r(t) = r(t-1) + \widehat{A}(t)s(t)$.
> Compute convergence criteria.
> **if not** *converged*
> $v^0 = r(t)$.
> $\bar{A}(t+1) = \widehat{A}(t)s(t)$.
> **Broadcast** v^0, $\bar{A}(t+1)$.
> $t = t+1$.
> Initialize $\widehat{A}(t), c(t)$.
> **else**
> **break**.

In Algorithm 1, initially the weight vector is $p = [1 \cdots 1]^T$. The matrix $\bar{A}$ is defined as $\bar{A}(t+1) = [A_1 p_1(t) \cdots A_g p_g(t)]$, and is utilized in forming the new $\widetilde{A}_i$ matrices after each planar search step. Notice that in the algorithm $\bar{A}(t+1)$ is set to $\widehat{A}(t)s(t)$. We know from (13) that $\widehat{A}(t) = AC(t)$. Then

$$\widehat{A}(t)s(t) = AC(t)s(t) = A(C(t)s(t)) = Ap(t) = \bar{A}(t+1) .$$

Thus, there is no explicit update of the weight vector $p(t)$, since its new value is packed in the matrix $\bar{A}(t+1)$. Obviously, if $p(t) = p$ is constant throughout the algorithm, the assignment and the broadcast of $\bar{A}$ is done only once at the start of the implementation, so as the forming and factorization of $\widetilde{A}_i$, $i = 1, \ldots, g$.

4 Numerical Experiments

The experiments are done using the "time-lagged" analytical model of [9]. The example test problem used in the graphs is problem ASH958 from the Harwell-Boeing sparse matrix test collection. The number of blocks and slaves g is 15. In the graphs the markers on the continuous lines give the value of the residual vector after each planar search. The dotted lines around the continuous lines illustrate the temporary residual vector v. The curve of continuous line with no markers depicts the behavior of a totally asynchronous implementation with no relaxation parameter and synchronization of any form, where g slaves operate on the original blocks A_i, $i = 1, \ldots, g$, receiving the latest available residual vector on the master whenever they send a new update.

We first check the effect of planar search on old history. In Fig. 1 totally asynchronous implementation is depicted against a partially asynchronous implementation where the blocks are the same as in the totally asynchronous one, but a planar search is done before updating the residual after l sweeps. The residual decreases monotonically for all choices of l although we observe a difference in the number of planar searches and total updates on the temporary residual vector v. Going from a synchronous implementation ($l = 1$) with planar search to a partially asynchronous implementation with a synchronization step after $l = 3$ sweeps, the number of synchronization points before convergence decreases from 21 to 9 (-57.1%), while the number of total updates computed increases from 315 to 405 (+28.6%). Considering that the updates are computed in parallel whereas during a synchronization step all the slaves remain idle, the case of $l = 3$ makes more efficient use of the available resources.

The results depicted in Fig. 1 compared to Fig. 5 of [9], where the effect of synchronization on old history is studied, point out that a synchronization step combined with planar search is more effective than a simple barrier synchronization in decreasing the magnitude of time-lag in the system.

In the second experiment we introduce expanded blocks formed using a constant weight vector $p = [1 \cdots 1]^T$. Here, $l = 1$ gives a Jacobi implementation with overlap and planar search. Figure 2 demonstrates that for all values of l the residual is monotonically decreasing. The number of planar searches before convergence decreases from 17 for $l = 1$ to 8 for $l = 3$ (-52.9%), whereas the total number of updates on the temporary residual vector increases from 255 to 360 (+41.2%). When we compare the results of this experiment with the former one, we see the positive effect of overlap on the convergence rate.

Lastly, we implement Algorithm 1 as it is given in the previous section, i.e., partially asynchronous implementation with expanded blocks and $p(t) = C(t)s(t)$. Here, different than the former two experiments, we observe that the cases of $l = 2$ and $l = 3$ have a better converge rate than the case of $l = 1$. Again, for all values of l we get a monotonically decreasing residual vector which is not contaminated by old history. Comparing the case of $l = 1$ with Dennis and Steihaug's Jacobi-Ferris-Mangasarian algorithm [2] with the "forget-me-not" variables of Ferris and Mangasarian [4] contributing to the planar search, we observe that inclusion of the δ^i_j values is necessary in a synchronous implementation, but in an asynchronous setting it is crucial and beneficial to discard them, since their contribution increases the effect of old history in the system if they are not discarded.

In this last experiment, the number of planar searches necessary before convergence decreases from 21 for $l = 1$ to 7 for $l = 3$, whereas the total number of updates on the temporary residual vector v remains constant at 315 for both $l = 1$ and $l = 3$. This means that the number of synchronization points is decreased by 66.7% while the same number of updates are computed asynchronously ($l = 3$) instead of following a synchronous pattern ($l = 1$).

All these experiments indicate that some degree of synchronization introduced in an asynchronous system combined with a form of overlap and a relax-

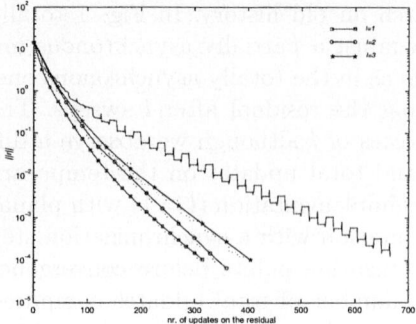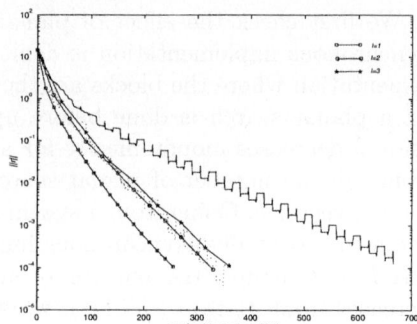

Fig. 1. Partially asynchronous implementation with planar search versus totally asynchronous implementation.

Fig. 2. Partially asynchronous implementation with planar search and expanded blocks in which $p = [1 \cdots 1]^T$ versus totally asynchronous implementation.

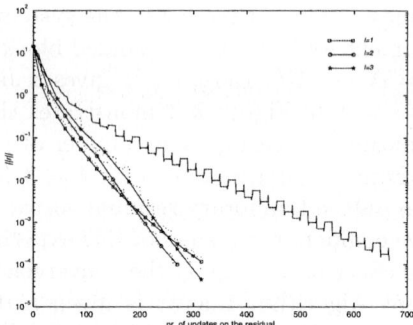

Fig. 3. Partially asynchronous implementation with planar search and expanded blocks in which $p(t) = C(t)s(t)$ versus totally asynchronous implementation.

ation parameter eliminate the deteriorating effect of old history. This approach of combining different issues that affect old history in the system is more effective than the single attempts carried out in [9, 11].

5 Concluding Remarks

Algorithm PALSQ is, in fact, a *predictor-corrector algorithm*, since in the asynchronous phase we accumulate updates (prediction) and only in the synchronization step the residual is updated (correction) with a combination of these accumulated updates and a relaxation parameter. During the asynchronous phase only the temporary residual vector v is updated. Therefore, the residual vector is concealed from the effect of old history.

We eliminate the effect of old history in a given asynchronous system introducing blocking synchronization, a relaxation parameter computed in a planar search, and overlap between subproblems in form of aggregated columns expanding the subproblem spaces, all applied together in Algorithm PALSQ. Slightly increasing the size of the subproblems for the sake of increasing dependency between slave processors and benefiting from the synchronization point for computing a relaxation factor turns out to be more effective than introducing only simple synchronization points or decreasing the number of processors in the system as means of decreasing the effect of old history.

We have considered only two values for the weight vector p, which is used in aggregating the blocks and forming overlap between the subproblems. Both the experiment with p having a constant value and the experiment with a variable $p(t)$ vector indicate that overlap is an effective factor in decreasing the deterioration caused by old history. Therefore, it is important to investigate further what other values can be assigned to the weight vector.

References

1. Dennis, Jr., J. E., Steihaug, T.: On the successive projections approach to least squares problems. SIAM J. Numer. Anal. **23** (1986) 717–733
2. Dennis, Jr., J. E., Steihaug, T.: A Ferris-Mangasarian technique applied to linear least squares problems. Tech. Rep. No. 150, Department of Informatics, University of Bergen, Norway (1998)
3. Duff, I. S., Van der Vorst, H. A.: Developments and trends in the parallel solution of linear systems. Parallel Comput. **25** (1999) 1931–1970
4. Ferris, M. C., Mangasarian, O. L.: Parallel variable distribution. SIAM J. Optim. **4** (1994) 815–832
5. Frommer, A., Pohl, B.: A comparison result for multisplittings and waveform relaxation methods. Numer. Linear Algebra Appl. **2** (1995) 335–346
6. Frommer, A., Renaut, R. A.: A unified approach to parallel space decomposition methods. J. Comput. Appl. Math. **110** (1999) 205–223
7. Frommer, A., Szyld, D. B.: On asynchronous iterations. J. Comput. Appl. Math. **123** (2000) 201–216
8. Renaut, R. A.: A parallel multisplitting solution of the least squares problem. Numer. Linear Algebra Appl. **5** (1998) 11–31
9. Steihaug, T., Yalçınkaya, Y.: Deteriorating convergence for asynchronous methods on linear least squares problem. In C. Lengauer, M. Griebl, S. Gorlatch (Eds.): Euro-Par'97 Parallel Processing, LNCS 1300, Springer Verlag (1997) 750–759
10. Yalçınkaya, Y.: Reducing the Effect of Old History in Asynchronous Iterations: An Empirical Study. Dr. Scient. Thesis, Department of Informatics, University of Bergen, Norway (2003)
11. Yalçınkaya, Y., Steihaug, T.: Asynchronous methods and least squares: An example of deteriorating convergence. In A. Sydow (Ed.): Proc. of the 15th IMACS World Congress on Scientific Computation, Modeling and Applied Mathematics, Vol. 1 Computational Mathematics, Wissenschaft & Technik Verlag (1997) 535–540

Execution Schemes for Parallel Adams Methods

Thomas Rauber[1] and Gudula Rünger[2]

[1] University Bayreuth
rauber@uni-bayreuth.de
[2] Chemnitz University of Technology
ruenger@informatik.tu-chemnitz.de

Abstract. Many recent solvers for ordinary differential equations (ODEs) have been designed with an additional potential of method parallelism, but the actual effectiveness of exploiting method parallelism depends on the specific communication and computation requirements induced by the equation to be solved. In this paper we study mixed parallel execution schemes for specific (explicit and implicit) variants of general linear methods, the Parallel Adams-Bashforth methods and the Parallel Adams-Moulton methods, which are new methods providing additional method parallelism. The implementations are realized with a library for multiprocessor task programming. Experiments on a Cray T3E and a dual Xeon cluster show good efficiency results, also for sparse application problems.

1 Introduction

For the integration of initial value problems for first-order ordinary differential equations (ODEs), several new parallel solution methods have been proposed which provide a potential of method parallelism in each time step. Many of those ODE solvers are based on multistage methods which offer a potential of method parallelism for the stage vector computation. Method parallelism allows a parallel execution of the solution method with a number of processors identical to the number of stages in the methods, which is usually less than ten. In the case of systems of ODEs there is an additional source of system parallelism in each stage vector computation corresponding to the system size. A mixture of method and system parallelism can be exploited to employ a larger number of processors and to improve scalability. Examples of those new ODE solvers are iterated Runge-Kutta (RK) methods [13,7] and diagonal-implicitly iterated Runge-Kutta methods [14,8]. For the parallel efficiency of iterated RK solvers the communication plays an important role. A global exchange is necessary after each iteration step of the stage vector computation and the efficiency strongly depends on the characteristics of the ODE application to be solved: for sparse ODE systems, the communication bandwidth of the target platform may limit the resulting performance. Experiments on different target platforms also show that a pure data parallel execution often leads to faster execution times than a parallel execution that exploits method parallelism [9]. On the other hand, parallel DIIRK methods for solving stiff ODEs show good efficiency for sparse and dense ODE systems on a wide variety of target platforms. Moreover, on most platforms, exploiting method parallelism leads to a significantly faster execution than pure data parallel executions.

In this paper, we explore another class of solution methods with different communication requirement. In particular, we consider a variant of general linear methods, the parallel Adams methods proposed by van der Houwen and Messina in [12]. The parallel Adams methods have the advantage that the computations of the parallel stages within each time step are completely independent from each other. Strong data dependencies occur only at the end of each time step. We have realized this method using multiprocessor tasks (M-tasks) for stage vector computations exploiting system parallelism internally within each M-task. We investigate the trade-off between reduced communication between different M-tasks and increased communication after each time step. The program is designed for arbitrary right hand sides of the ODE so that no application specific characteristics are exploited and the pure effect of M-Task programming can be seen. As application programs we consider dense and sparse ODE systems. Experiments have been performed on a Cray T3E and a dual Xeon cluster.

The M-Task programs are realized with the Tlib library, a library to program hierarchically organized M-task programs on top of SPMD modules [10]. The advantage is that different parallel program versions can be coded easily on top of given stage vector SPMD codes without causing an additional overhead. Several other models have been proposed for mixed task and data parallel executions, see [1, 11] for an overview of systems and approaches. Many environments for mixed parallelism in scientific computing are extensions to the HPF data parallel language, see [3] for an overview. Examples for parallel programming systems are HPJava [15], LPARX [6], and NestStep [5].

The rest of the paper introduces the parallel Adams method in Section 2. Section 3 presents the implementation scheme of the parallel Adams methods with mixed parallelism. Section 4 discusses runtime results and Section 5 concludes.

2 Parallel Adams Method

We consider the solution of initial-value problems (IVPs) of first order differential equations (ODEs) of the form

$$\mathbf{y}'(t) = \mathbf{f}(t, \mathbf{y}(t)) \quad \text{with } \mathbf{y}(t_0) = \mathbf{y}_0 \tag{1}$$

with initial vector $\mathbf{y}_0 \in \mathbb{R}^d$ at start time t_0, system size $d \geq 1$, and right hand side function $\mathbf{f} : \mathbb{R} \times \mathbb{R}^d \to \mathbb{R}^d$. Starting with $\mathbf{y}_0$ one-step methods for solving ODE systems of the form (1) determine a sequence of approximations $(\mathbf{y}_n)_{n=1,2,...}$ for the solution $\mathbf{y}(t_n)$ at time t_n, $n = 1, 2, ...$. Parallelism can only be exploited within the computations of each time step, since the time steps depend on each other. This can be realized by a distribution of the computation of components of $\mathbf{y}_n \in \mathbb{R}^d$ or by exploiting specific characteristics of potential method parallelism if available.

In recent years solvers have been designed that additionally provide a higher degree of method parallelism while providing good numerical properties. Many of them are based on classical implicit Runge-Kutta methods in which the implicit equation is treated with a predictor-corrector approach or fixed point iterations. Method parallelism can be exploited for the computation of different stage vectors, but there are still interactions between the parallel computations. Another class of parallel solvers is based on implicit multistage methods which are already parallel in the stage vector computation and which can be described by the class of general linear methods [2]. General linear methods compute several stage values $\mathbf{y}_{ni}$ (which are vectors of size d) in each time step which correspond to numerical approximation of $\mathbf{y}(t_n + a_i h)$ with abscissa vector (a_i), $i = 1, ..., k$ and stepsize $h = t_n - t_{n+1}$. The stage values of one time step are combined in the vector $\mathbf{Y}_n = (\mathbf{y}_{n1}, ..., \mathbf{y}_{nk})$ of size $d \cdot k$ and the computation in each step is given by:

$$\mathbf{Y}_{n+1} = (\mathbf{R} \otimes \mathbf{I})\mathbf{Y}_n + h(\mathbf{S} \otimes \mathbf{I})\mathbf{F}(\mathbf{Y}_n) + h(\mathbf{T} \otimes \mathbf{I})\mathbf{F}(\mathbf{Y}_{n+1}), n = 1, 2, ... \quad (2)$$

The matrices $\mathbf{R}$, $\mathbf{S}$ and $\mathbf{T}$ have dimension $k \times k$ and $\mathbf{R} \otimes \mathbf{I}$ denotes the Kronecker tensor product, i.e. the $d \cdot k \times d \cdot k$ dimensional block matrix with blocks $r_{ij} \cdot \mathbf{I}$ of size $d \times d$ for $i, j = 1, \ldots, k$. $\mathbf{I}$ denotes the $d \times d$ unit matrix and $\mathbf{F}(\mathbf{Y}_n) = (\mathbf{f}(\mathbf{y}_{n1}), \ldots, \mathbf{f}(\mathbf{y}_{nk}))$. Typical values for k lie between 2 and 8.

In this paper, we consider a variant of this method, the parallel Adams method proposed in [12]. The name was chosen due to a similarity of the stage equations with classical Adams formulas. The goal of this paper is to realize different computation schemes with mixed system and method parallelism exploiting several disjoint groups of processors and to investigate the effect of such an execution on the parallel efficiency.

The parallel Adams methods result from (2) by using specific matrices $\mathbf{R}$, $\mathbf{S}$ and $\mathbf{T}$. A diagonal matrix $\mathbf{T}$ yields an implicit equation system to be solved by fixed point iteration which results in the Parallel Adams-Moulton (PAM) method. A zero matrix $\mathbf{T} = 0$ results in the Parallel Adams-Bashforth (PAB) method.

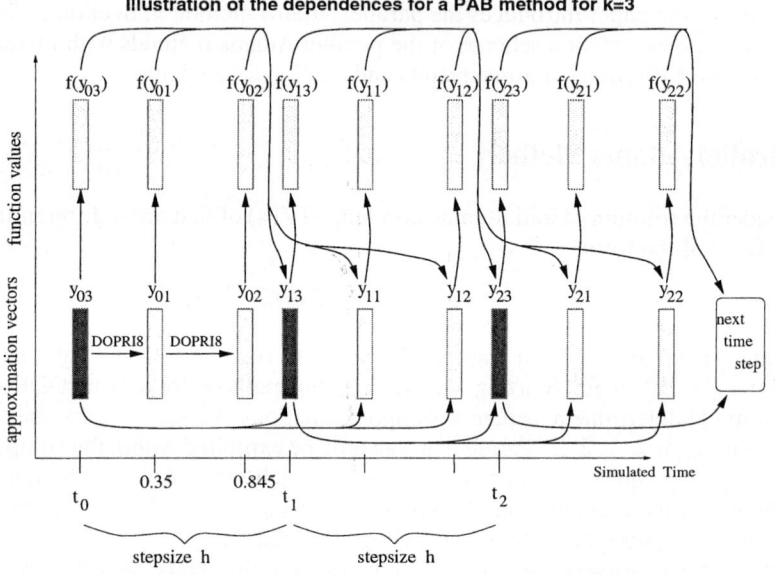

Fig. 1. Illustration of the dependence structure of a PAB method with $k = 3$ stage values. For abscissa values $\mathbf{a} = (a_1, \ldots, a_k)$ with $a_k = 1$ the stage values y_{nk} correspond to the approximations of $y(t_n), n = 0, 1, 2, \ldots$.

First, we consider PAB methods which result by setting $\mathbf{T} = 0$ and $\mathbf{R} = \mathbf{e} \cdot \mathbf{e_k}^T$ where $\mathbf{e} = (1, \ldots, 1)$ and $\mathbf{e_k} = (0, \ldots, 0, 1)$. With $V_\mathbf{x} = (\mathbf{x}, \mathbf{x}^2, \ldots, \mathbf{x}^k)$, $W_\mathbf{x} = (\mathbf{e}, 2\mathbf{x}, 3\mathbf{x}^2, \ldots, k\mathbf{x}^{k-1})$ and $b_i = a_i - 1$, we define $\mathbf{S} = V_\mathbf{a} W_\mathbf{b}^{-1}$ according to [12]. The abscissa values a_i, $i = 1, \ldots, k$, are determined such that a convergence order of $k + 1$ results. This can be obtained by using the Lobatto points.

To compute the stage vector $\mathbf{Y}_0$ of the initial time step from the given initial value y_0, the stage values y_{0i}, $i = 1, \ldots, k$ are computed by using another explicit method like DOPRI5. Starting from $\mathbf{Y}_n$, time step $n+1$ computes $\mathbf{Y}_{n+1} = (y_{n+1,0}, \ldots, y_{n+1,k})$ by first applying the right hand side function $\mathbf{f}$ to $y_{n0}, \ldots, y_{nk}$, i.e., by computing $\mathbf{F}(\mathbf{Y}_n)$. Then $\mathbf{Y}_{n+1}$ is obtained by adding $(\mathbf{R} \otimes \mathbf{I})\mathbf{Y}_n$ to $h(\mathbf{S} \otimes \mathbf{I})\mathbf{F}(\mathbf{Y}_n)$. Figure 1 shows the dependence structure of the resulting computation scheme for the case $k = 3$.

PAM methods result from (2) by using a diagonal matrix $\mathbf{T}$. The diagonal entries δ_i, $i = 1, \ldots, k$, are determined such that specific consistency conditions are satisfied. Moreover, $\mathbf{R} = \mathbf{e} \cdot \mathbf{e_k}^T$ and $\mathbf{S} = (V_{\mathrm{a}} - \mathbf{R} \cdot V_{\mathrm{b}} - \mathbf{T} \cdot W_{\mathrm{a}})W_{\mathrm{b}}^{-1}$ is used for the PAM method. Again, the Lobatto points can be used as abscissa values. Since $\mathbf{T}$ is diagonal, the resulting implicit relation is uncoupled and has the form

$$\mathbf{y}_{n+1,i} - h * \delta_i * \mathbf{f}(\mathbf{y}_{n+1,i}) = \mathbf{v}_{ni} \tag{3}$$

for $i = 1, \ldots, k$. The vectors $\mathbf{v}_{ni}$ are the d-dimensional vector components of $\mathbf{V} := (\mathbf{R} \otimes \mathbf{I})\mathbf{Y}_n + h(\mathbf{S} \otimes \mathbf{I})\mathbf{F}(\mathbf{Y}_n)$. Using a fixed point iteration results in the following computation scheme:

$$\begin{aligned} \mathbf{y}_{n+1,1}^{(j)} - h * \delta_1 * \mathbf{f}(\mathbf{y}_{n+1,1}^{(j-1)}) &= \mathbf{v}_{n1} \\ &\vdots \\ \mathbf{y}_{n+1,k}^{(j)} - h * \delta_k * \mathbf{f}(\mathbf{y}_{n+1,k}^{(j-1)}) &= \mathbf{v}_{nk} \end{aligned} \tag{4}$$

with $j = 0, 1, \ldots$. Equation (4) provides method parallelism and defines k independent tasks that can be executed in parallel on disjoint groups of processors. The right hand side of (4) uses the stage vectors from the previous time step:

$$\mathbf{v}_{ni} = \sum_{j=1}^{k} r_{ij} \mathbf{y}_{nj} + h * \sum_{j=1}^{k} s_{ij} \mathbf{f}(\mathbf{y}_{n,j}) \tag{5}$$

with matrices $R = (r_{ij})_{i,j=1,\ldots,k}$ and $S = (s_{ij})_{i,j=1,\ldots,k}$. The convergence order of the resulting implicit method is $k + 2$. The starting vectors for the iteration can be provided by a PAB method used as predictor, since both methods use the Lobatto points as abscissa values. The resulting predictor-corrector method is denoted as PABM in the following. The data dependencies of the PABM methods are illustrated in Figure 2.

3 Parallel Implementation

The PAB and PABM methods offer several possibilities for a parallel execution that differ in the order in which the computations in each time step are performed and in the way in which the required data exchange is obtained. The data dependencies require that the time steps are executed one after another.

Data parallel implementation of the PAB method. In each time step a pure data parallel realization computes the stage values one after another with all processors available. To compute $\mathbf{f}(y_{ni})$, $i = 1, \ldots, k$, each of the p processors evaluates about d/p

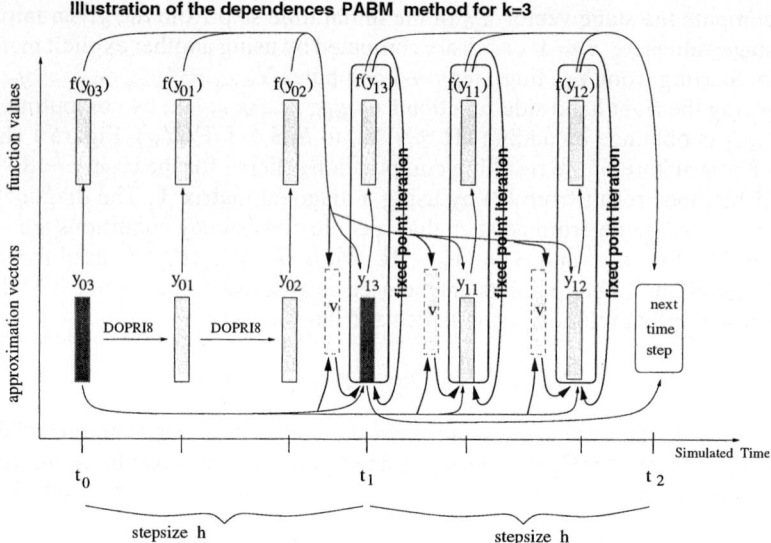

Fig. 2. Illustration of the dependence structure of a PABM method with $k = 3$ stage values. The stage value y_{ni}, $i = 1, \ldots, k$, computed by the PAB method is used as starting vector for the fixed point iteration of the PABM method. For each stage value, an additional vector v_{ni} is needed to hold the constant part of the right hand side of the fixed point iteration. The computation of v_{ni} is performed similarly as the computation of y_{ni}, but a different matrix S is used.

components of **f**. After this evaluation, each processor uses its local components of $\mathbf{f}(y_{ni})$ to compute the corresponding components of $y_{n+1,i}$ according to Equation (2). Since **f** is considered to be an unknown black-box function, we have to assume that all components of $y_{n+1,i}$ must be available for the computation of each component of $\mathbf{f}(y_{n+1,i})$ in the next time step. Thus after the computation of $y_{n+1,i}$, a global multi-broadcast operation is performed to make the entire stage value $y_{n+1,i}$ available to all processors for the next time step. Altogether, k global multi-broadcast operations are performed in each time step where each processor contributes about d/p components to each of these communication operations, thus realizing a globally replicated distribution of the entire stage vector.

Data parallel implementation of the PABM method. The data parallel implementation of the PABM method uses the data parallel implementation of the PAB method as predictor yielding the starting vector $y_{n+1,k}^{(0)}$ for each stage value of the PABM corrector in a replicated data distribution. A fixed number of corrector iterations is performed where in every iteration each processor first computes d/p components of $f(y_{n+1,k}^{(0)})$ and then uses its local components to compute the corresponding d/p components of $y_{n+1,k}^{(it+1)} = y_{n+1,k}^{(0)} + h \cdot \delta_k \cdot f(y_{n,k}^{(it)})$ with $it = 0, 1, \ldots$. Since only the local components are needed for this computation no data exchange has to be performed in between. For the next iteration, the vector $y_{n+1,k}^{(it+1)}$ is needed as argument vector for the function evaluation and so a global multi-broadcast operation is required to make all components of $y_{n+1,k}^{(it+1)}$ available to all processors.

M-task with internal data parallelism for PAB. An M-task realization exploiting method parallelism employs k disjoint processor groups $G_1, \ldots, G_k$ of about equal size $g = p/k$. In each time step n, processor group G_i is responsible for computing stage value $y_{n+1,i}$. The computation of $y_{n+1,i}$ requires to access $\mathbf{f}(y_{n1}), \ldots, \mathbf{f}(y_{nk})$ for $i = 1, \ldots, k$, which are computed in a distributed way. By an appropriate data exchange each processor gets exactly those components that it needs for its local computation. This can, e.g., be realized by first collecting the components of $\mathbf{f}(y_{ni})$ at a specific processor q_{ij} of G_i by a group-local gather operation and by then broadcasting $\mathbf{f}(y_{ni})$ from q_{ij} to the processors of all other groups, $1 \leq i \leq k$. The data exchange has to be performed for each stage value separately. Additionally, since $\mathbf{f}$ is a unknown function, $y_{n+1,i}$ has to be distributed among all processors of group G_i to establish a group-replicated distribution of the stage values.

Compared to the pure data parallel execution scheme, more communication operations are necessary, but most of the operations are group-local within a subgroup G_i, $1 \leq i \leq k$.

M-task with internal data parallelism for PABM. Using the PAB method as predictor, the PABM method performs a total number of It corrector steps, where each processor group G_k performs its iterations to compute $y_{n+1,k}^{(It)}$ independently from other processor groups. After the computation of the iteration vector $y_{n+1,k}^{(it+1)}$, $0 \leq it \leq It-2$, this vector is broadcasted to all other processor groups where it is required for the computation of $\mu_{n+1,k}^{(it+2)}$ in the next iteration. The final vector of the iteration $y_{n+1,k} = y_{n+1,k}^{(It)}$ is available on all processors of G_k in a replicated way and so the next time step can use this value for the predictor step without further communication.

Exploiting orthogonal structures of communication. The implementation with M-tasks uses two communication phases, one to make each processor those components of $\mathbf{F}(\mathbf{Y}_n)$ available that it needs for the computation of its local components of $\mathbf{Y}_{n+1}$, and one to establish a group-replicated distribution of $y_{n+1,i}$ in group G_i. The second communication phase is already based on group-local communication, but the first communication phase uses a global broadcast operation. We introduce an approach to re-organize the first communication phase such that it also uses group-local communication only.

In the following, we assume that the number k of stage vectors is a multiple of the number p of processors, so that all groups G_i contain the same number of processors $g = p/k$. Moreover, we assume that the dimension d of the stage values is a multiple of g. Based on the processor groups $G_1, \ldots, G_k$, we define orthogonal groups $Q_1, \ldots, Q_g$ with $|Q_k| = k$ and $Q_j = \{q_{lj} \in G_l \mid l = 1, \ldots, k\}$. For the computation of $\mathbf{F}(\mathbf{Y}_n)$, each processor q_{ij} of G_i computes d/g components of $\mathbf{f}(y_{ni})$. The corresponding components of $\mathbf{f}(y_{nl})$ for $l \neq i$ are computed by the processors q_{lj} of G_l. Exactly those components are needed by q_{ij} for the following computation of $y_{n+1,i}$. Therefore, the necessary components can be made available to each processor by a group-oriented multi-broadcast operation on the orthogonal processor group Q_j. These communication operations can be executed concurrently on all orthogonal groups $Q_1, \ldots, Q_g$. Thus, all group-oriented communication operations can be executed concurrently.

Exploiting orthogonal structures of communication for the PABM method. The PAM corrector is implemented identical to the M-task parallel version since disjoint processor groups are already exploited by the method. The PAB predictor also exploits the orthogonal communication structure.

4 Experiments

For the experiments, we have implemented PAB and PABM methods for $k = 2, 4, 8$. According to [12], we use a fixed stepsize strategy to clearly see the algorithmic effects. The implicit relations are solved by a fixed point iteration with a fixed number of iterations for all stages. As test problems we consider two classes of ODE systems: *Sparse* ODE systems are characterized by the fact that each component of the right-hand side function **f** of the ODE system has a fixed evaluation time that is independent of the size of the ODE system, i.e., the evaluation time of the entire function **f** increases linearly with the size of the ODE system. Such systems arise, e.g., from a spatial discretization of a time-dependent partial differential equations (PDEs). In particular, we consider an ODE system resulting from the spatial discretization of a 2D PDE describing the reaction of two chemical substances with a diffusion (Brusselator equation) [4, 7]. *Dense* ODE systems are characterized by the fact that each component of **f** has an evaluation time that increases linearly with the system size, i.e., the evaluation time of the entire function **f** increases quadratically with the size of the ODE system. In particular, we consider a spectral decomposition of a time-dependent Schrödinger equation resulting in a dense ODE system.

The parallel programs have been tested on two machines, a Cray T3E-1200 and a Beowulf cluster with 16 nodes where each node consists of two Xeon processors with 2.0 GHz. The nodes are connected by an SCI interconnection network.

Figures 3 and 4 show the runtimes in seconds of one time step of the PAB and PABM methods with $k = 8$ stage vectors for the Brusselator equation using system sizes 180000, 320000 and 500000. The figures show that for both methods, the orthogonal parallel version with M-tasks is considerably faster than the data parallel version and the standard M-task version, since the orthogonal version uses group-based communication only. For the PAB method, the data parallel version is faster than the standard M-task version, since the M-task version uses more communication operations and also uses a global gather operation in each time step. The orthogonal version especially shows a much better scalability than the other two versions. For the PABM method, the standard M-task version is faster than the data parallel version, since the data parallel version uses a global multi-broadcast operation in each iteration step whereas the M-task version uses a global gather operation only once in each time step. Orthogonal communication structures further improve the result. Similar results are obtained for $k = 4$. Figure 5 shows the results for the PAB method applied to the Brusselator equation on the Xeon cluster. The results are similar as for the T3E, but the machine contains only 32 processors. Figure 6 shows the execution times of the PABM method applied to the Schrödinger equation on the Xeon cluster. Here all versions show good scalability, since the execution time of **f** uses a large portion of the parallel runtime. But still, the

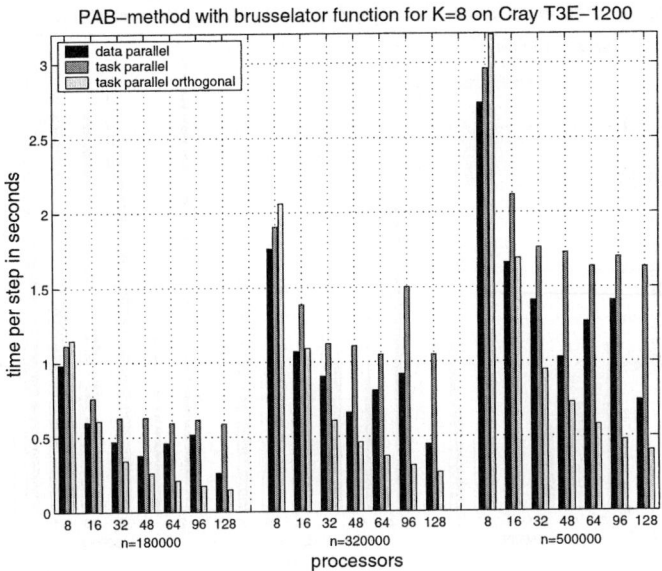

Fig. 3. Runtimes of the PAB method for Brusselator on Cray T3E with $k = 8$.

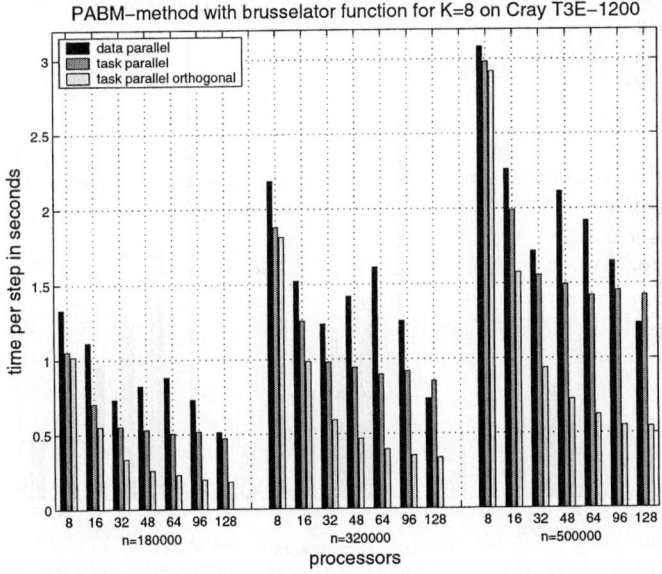

Fig. 4. Runtimes of the PABM method for Brusselator on Cray T3E with $k = 8$.

orthogonal task parallel version leads to the smallest execution time in most cases. The difference between the M-task version and the data parallel version is most significant for smaller ODE systems. Similar results are obtained for the PAB methods on both hardware platforms for $k = 4$ and $k = 8$.

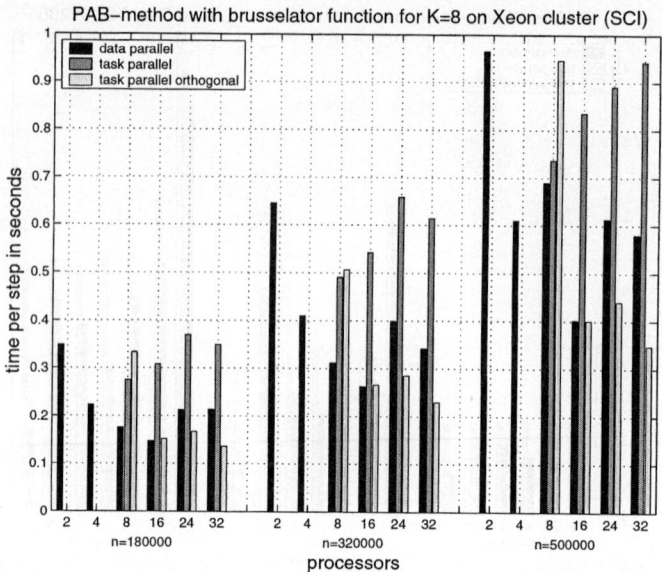

Fig. 5. Runtimes of the PAB method for Brusselator on SCI Xeon cluster with $k = 8$.

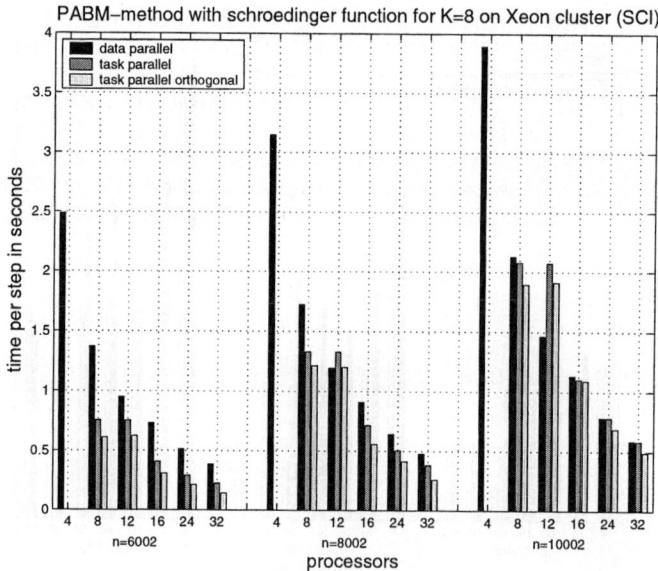

Fig. 6. Runtimes of the PABM method for Schrödinger on SCI Xeon cluster with $k = 8$.

5 Conclusions

The parallel Adams methods discussed in this paper provide potential method parallelism in the form of independent stage vector computation in each iteration step. But when exploiting method parallelism, collective communication is still required since information has to be exchanged between processor groups. By reordering the communication using an orthogonal organization of disjoint processor groups for the data exchange, the communication overhead can additionally be reduced. For the parallel Adams methods the combination of an implementation as M-task program with orthogonal processor groups for communication phases leads to the most efficient implementation on parallel platforms with different interconnection networks and yields a good scalability. A dual Xeon cluster and a Cray T3E have been used for the experiments. The experiments show that the implementation as M-task program is especially useful for parallel platforms with small bandwidth.

References

1. H. Bal and M. Haines. Approaches for Integrating Task and Data Parallelism. *IEEE Concurrency*, 6(3):74–84, July-August 1998.
2. J.C. Butcher. *The Numerical Analysis of Ordinary Differential Equations, Runge-Kutta and General Linear Methods*. Wiley, New York, 1987.
3. S.J. Fink. *A Programming Model for Block-Structured Scientific Calculations on SMP Clusters*. PhD thesis, University of California, San Diego, 1998.
4. E. Hairer and G. Wanner. *Solving Ordinary Differential Equations II*. Springer, 1991.
5. C.W. Keßler. NestStep: Nested Parallelism and Virtual Shared Memory for the BSP model. *The Journal of Supercomputing*, 17:245–262, 2001.
6. S.R. Kohn and S.B. Baden. Irregular Coarse-Grain Data Parallelism under LPARX. *Scientific Programming*, 5:185–201, 1995.
7. T. Rauber and G. Rünger. Parallel Iterated Runge–Kutta Methods and Applications. *International Journal of Supercomputer Applications*, 10(1):62–90, 1996.
8. T. Rauber and G. Rünger. Diagonal-Implicitly Iterated Runge-Kutta Methods on Distributed Memory Machines. *Int. Journal of High Speed Computing*, 10(2):185–207, 1999.
9. T. Rauber and G. Rünger. A Transformation Approach to Derive Efficient Parallel Implementations. *IEEE Transactions on Software Engineering*, 26(4):315–339, 2000.
10. T. Rauber and G. Rünger. Library Support for Hierarchical Multi-Processor Tasks. In *Proc. of the Supercomputing 2002*, Baltimore, USA, 2002. ACM/IEEE.
11. D. Skillicorn and D. Talia. Models and languages for parallel computation. *ACM Computing Surveys*, 30(2):123–169, 1998.
12. P.J. van der Houwen and E. Messina. Parallel Adams Methods. *J. of Comp. and App. Mathematics*, 101:153–165, 1999.
13. P.J. van der Houwen and B.P. Sommeijer. Iterated Runge–Kutta Methods on Parallel Computers. *SIAM Journal on Scientific and Statistical Computing*, 12(5):1000–1028, 1991.
14. P.J. van der Houwen, B.P. Sommeijer, and W. Couzy. Embedded Diagonally Implicit Runge–Kutta Algorithms on Parallel Computers. *Mathematics of Computation*, 58(197):135–159, January 1992.
15. G. Zhang, B. Carpenter, G.Fox, X. Li, and Y. Wen. A high level SPMD programming model: HPspmd and its Java language binding. Technical report, NPAC at Syracuse University, 1998.

A Data Management and Communication Layer for Adaptive, Hexahedral FEM

Judith Hippold* and Gudula Rünger

Chemnitz University of Technology, Department of Computer Science
09107 Chemnitz, Germany
{juh,ruenger}@informatik.tu-chemnitz.de

Abstract. The parallel realization of adaptive finite element methods (FEM) has to deal with several irregular and dynamic algorithmic properties caused by adaptive mesh refinement (AMR). For an implementation on distributed memory machines irregular communication behavior results from dynamically growing data structures and statically unknown communication partners. An efficient parallel implementation has to provide appropriate mechanisms to cope with the flexibility of the adaptive finite element approach at runtime. We have implemented a data management and communication layer for an adaptive, 3-dimensional, hexahedral FEM on distributed memory machines and use it to parallelize an existing sequential code. The data management and communication layer realizes duplicated data structures for boundaries of distributed data, additional hierarchical data structures to deal with uneven refinement, and coherence protocols to guarantee correctness of communication partners and messages. An easy to use interface provides access to the functionality of the layer.

1 Introduction

The discretization of the physical domain into a mesh of finite elements and the approximation of the unknown solution function by a set of shape functions on those elements makes the finite element method flexible and applicable for a wide range of different applications. For the simulation of large problems adaptive mesh refinement and parallel execution can help to reduce runtime. Especially for 3-dimensional problems, an efficient parallel realization is difficult to achieve and mostly interferes with the algorithmic structure. Parallelization modules accessible by interface functions can provide easy to use parallel functionality for application programmers, maintain the algorithmic structure, and offer reusability for a wide range of different finite element implementations.

Adaptive FEM belongs to the class of irregular algorithms. Irregularity is caused by AMR and hanging nodes: The adaptive refinement process with computations on different refinement levels creates hierarchies of data structures and requires the explicit storage of these structures including their relations. Hanging nodes can result from different refinement levels and are mid-nodes of faces

---
* Supported by DFG, SFB393 *Numerical Simulation on Massively Parallel Computers.*

or edges which are vertices of finite elements at the same time. Such nodes require projections on nodes of other refinement levels during the solution process. Both characteristics lead to irregular communication behavior. Thus the efficient parallel implementation is difficult, especially for distributed memory.

The contribution of this paper is the design and realization of a data management and communication layer for parallel, adaptive FEM with emphasis on hexahedral finite elements. Our layer encapsulates communication and the management of distributed data which allows easy extensibility and does not require modifications in the algorithmic structure of the existing sequential FEM code. The data management concept provides duplicated data storage with fast access and modification functions and also supports hanging nodes. A special communication mechanism reduces the number of messages considerably, decreases the overhead for managing asynchronous communication behavior, and allows SPMD programming style despite of irregular characteristics. Furthermore it offers wide facilities for optimization like additional overlapping of communication and computation phases, software caching, and monitoring of program behavior. We have incorporated the data management and communication layer into an existing sequential, adaptive, 3-dimensional, hexahedral FEM program package which was developed and implemented at the *SFB 393: Numerical Simulation on Massively Parallel Computers* at the Chemnitz University of Technology. However, the module is designed to be integrated into further FEM packages.

The paper is organized as follows: Section 2 summarizes the given adaptive FEM implementation. The concepts of our data management and communication layer are introduced in Section 3. Section 4 presents experimental results and Section 5 concludes.

2 The FEM Implementation

SPC-PM3AdH [1] is a 3-dimensional, adaptive finite element software package suitable to solve 2nd order elliptic partial differential problems like the Poisson equation or the Lamé system of linear elasticity. [2–4] present FE-software which differ from SPC-PM3AdH, e.g. in finite elements or solvers implemented or parallel realization. SPC-PM3AdH implements hexahedral elements with linear and quadratic shape functions as shown in Figure 1.

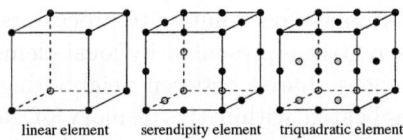

linear element   serendipity element   triquadratic element

Fig. 1. Finite elements implemented in SPC-PM3AdH.

Each hexahedron has the maximum number of 27 nodes. If only linear or serendipity elements with 20 nodes are used, the remaining nodes actually exist but have no associated shape functions. A finite element is implemented through a hierarchy of data structures: *Volumes* are the most coarse-grained structure and represent the finite elements. Volumes are composed of faces and each face consists of four *edges*. The smallest unit is the *node*. Nodes have three coordinates and, when they are associated with a shape function, a solution vector. Nodes can be vertices or mid-nodes of volumes, faces, and edges.

The program structure of SPC-PM3AdH consists of five phases as illustrated on the right:

I First the initial mesh is created from an input file.
II Volumes are subdivided into 8 children according to the estimated error and geometrical conditions like hanging nodes. The difference of refinement levels between two neighboring volumes is restricted to one.
III Each volume is associated with an element stiffness matrix. The matrices for new volumes, the right hand side vectors, and the global main diagonal are assembled in the third phase.
IV The system of equations is solved with the preconditioned conjugate gradient method. A Jacobi, Yserentant [5], or BPX [6] preconditioner can be selected for preconditioning.
V The error is estimated with a residual based error estimator [7]. Volumes are labeled for refinement if their estimated error is close to the maximum error.

3 The Data Management and Communication Layer

The basis of the data management and communication layer is a specific approach for distributed data storage and administrational data structures containing distribution information. The entire information concerning the distribution of data is hidden to the user but can be accessed and modified by functions supplied by the layer. The layer encapsulates the actual data exchange via MPI and communication optimizations.

3.1 Distribution of Data Structures

The parallel implementation assigns volumes to processors which leads to a partitioning of the stiffness matrix represented by local element stiffness matrices. The faces, edges, and nodes shared between neighboring volumes in different address spaces exist duplicated within the memory of each owning processor. This allows fast computations for each volume with minimal communication. Processors sharing duplicates are called *owners* of duplicates. The approach induces that the solution and auxiliary vectors are spread over the address spaces of the different processors. The entries for duplicated nodes exist redundantly and contain only subtotals which have to be accumulated to yield the total result. For duplicated hanging nodes *worker* processors can be selected which are allowed to perform the necessary projections exclusively. This avoids multiple accumulations.

Refinement of distributed data. The adaptive refinement process creates a hierarchy of faces and edges. Due to hanging nodes processors may perform computations on different refinement levels of the same edge or face. To gain fast access on the different levels faces are organized in a quadtree data structure and edges as binary trees. Refinement on duplicated faces and edges subdivides data structures only in local memory. Thus the remote hierarchy trees have to be kept consistently by a two-phase iterative refinement cycle:

The *first phase* iteratively subdivides local volumes according to the estimated error and geometrical demands, like unacceptable hanging nodes. In the *second phase* the remote refinement of subdivided duplicated faces and edges is done. A synchronization step ensures that the entire process is performed until no further subdivision of volumes is done on any processor.

Coherence lists. To keep data consistent and to support remote refinement the fast identification and localization of duplicates is necessary. For that reason the tuple *Tup(identifier, processor)* is introduced. *Identifier* denotes a data structure of type face, edge, or node and serves for local identification. The parameter *processor* allows global identification of a duplicate. This parameter is assigned after distributing a data structure and is the number of the processor where the duplicate is situated. The combination of both parameters allows unique identification.

The tuple *Tup* is used to implement *coherence lists*. Each duplicated data structure is tagged with a coherence list which contains information about the remote identifiers and the location of *all* existing duplicates. Figure 2 illustrates such lists for the edges *e1* and *e2* in the address space of processor *P2*. To access remote duplicates the owners and the corresponding remote identifiers can be extracted and used to send a message. A receiving processor is able to

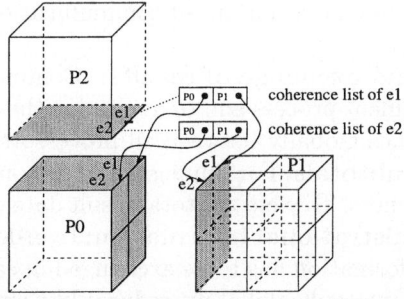

Fig. 2. Illustration of the coherence lists for the edges *e1* and *e2* in the address space of processor *P2* and illustration of remote refinement necessary for processors *P0* and *P2* after local refinement on *P1*.

determine a local structure by the received identifier and to perform actions like remote refinement. Coherence lists are built up and updated within the data management layer. The local identifier of a data structure serves as input parameter to find the corresponding list. Data structures without duplicates in remote address spaces have no lists associated.

Data consistency of refined volumes. Adaptive refinement on volumes distributed over different address spaces requires to maintain consistency for correct program behavior. This includes the remote refinement of duplicated faces and edges and the creation or update of coherence lists for new data structures. Figure 2 also illustrates remote refinement: The subdivision of the volume in the address space of processor *P1* requires the refinement of one face and of four edges located at *P0* and of one edge located at *P2*. The entire process can be generalized and structured into four steps. We refer to the data structures on higher levels of the hierarchy trees as *parents* and to the corresponding new structures arising from subdivision as *children*.
(1) To identify data structures which have to be remotely subdivided each parent volume is investigated for duplicated faces and edges. For each found duplicate, a data package has to be sent. It contains the remote identifier of the duplicate for identification and the local identifiers of the newly created data structures (e. g. nodes, children) for updating the coherence lists.
(2) The receiving owners refine the parent data structure identified by the received package and create and update the coherence lists of the new structures with the information of the message package. If the parent data structure is already subdivided, only the update of coherence lists is necessary.
(3) To update the lists of the remotely created, new structures in the address spaces of the remaining owners the identifiers of the new duplicates have to be sent back according to the package structure described in item (1).
(4) The other owners receive the messages and update their coherence lists.

3.2 Communication Approach

During a program run different situations of communication occur. They can be classified into three types:

Synchronization and exchange of results: Communication is necessary to synchronize the refinement process and to determine the maximum error. This communication takes place globally between all processors.

Accumulation of subtotals: For duplicated nodes only subtotals are computed by the different owners. To yield the total result data exchange is necessary.

Exchange of administrational information: In order to keep data consistent administrational information has to be exchanged in each program iteration step. This comprises for example global procedures like remote subdivision, the generation and update of coherence lists, and the identification of hanging nodes.

The last two communication situations have irregular characteristics. That means the communication partners, the number of messages, and the message sizes vary and depend on the specific program run. Furthermore the exact communication time is not known in advance. [8, 9] present a communication concept for irregular algorithms. However, the communication requirements for adaptive FEM are slightly different and suggest the following realization:

Communication protocol. Due to the special duplicated data storage of the data management layer and the program structure of SPC-PM3AdH the exchange of data within a computational phase can be delayed. This allows to displace the communication at the end of a computational phase and to separate computation from communication. The resulting communication mechanism can be described as follows:
(A) During computation each processor collects information about necessary data exchanges with different collect functions. These functions can be chosen adapted to the algorithmic needs and detect duplicates for a given local data structure. If there are duplicates, the remote identifiers and additional information are stored in send buffers for later exchange. During the computational phase a collect function can be called several times.
(B) After the local computations the gathered values are sent to the corresponding processors extracted from the coherence lists. This process is initialized by the first call of a get function. The application programmer uses get functions to obtain data from the receive buffer. Each function call returns an identifier of a local data structure and the received additional information for this structure.
(C) Afterwards specific actions can be performed. The return of an invalid identifier denotes an empty receive buffer.

The usage of the described collect&get communication mechanism is illustrated by the following pseudo-code for the parallel labeling of edges with hanging nodes.

```
for each local volume V { for each edge E of V {
            if(E is subdivided) {
                label children S1 and S2 of E as hanging;
                collect(S1, S2);    /* (A) */
            }
} } }
while(get(&id)) /* (B) */ { label the edge identified by id; /* (C) */ }
```

4 Experiments

This section shows some measurement results demonstrating the usability of the layer within the program SPC-PM3AdH. To gain experimental results two platforms are used: XEON, a 16x2 SMP cluster of 16 PCs with 2.0 GHz Intel Xeon processors running Linux and SB1000, a 4x2 SMP cluster of 4 SunBlade 1000 with 750 MHz UltraSPARC3 processors running Solaris. The MPI implementation is ScaMPI using an SCI network. We consider three examples: *layer3* a boundary layer for the convection-diffusion equation, *ct01* solving the Lamé equation, and *torte4d* a layer near a non-convex edge [1]. For parallel and sequential measurements linear finite elements and the Jacobi preconditioner are used.

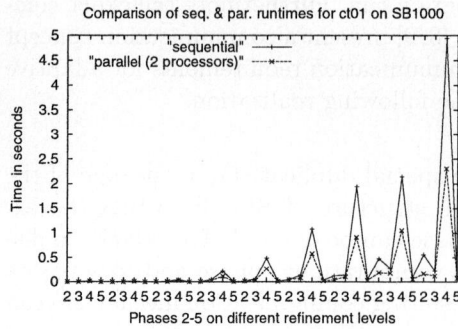
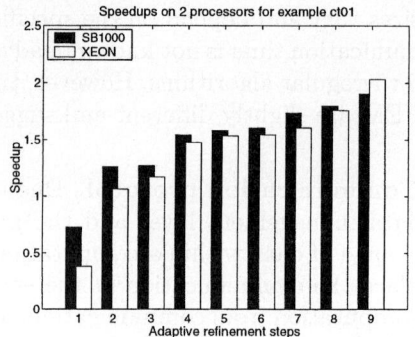

Fig. 3. Left: Execution times for the different algorithmic phases on different refinement levels for example *ct01* on SB1000 (2: AMR, 3: assembling element stiffness matrices, 4: solving the system of equations, 5: error estimation). Right: Speedups for example *ct01* on 2 processors.

Small numbers of processors are sufficient to achieve efficiency for the available examples.

The algorithmic phase consuming most of the sequential and parallel calculation time is to solve the system of equations (see Figure 3, left, phase 4). During this phase also the main portion of communication takes place in the parallel version. Figure 3 (right) shows speedup values for the example *ct01* on SB1000 (black) and XEON (white) after different adaptive refinement steps. Speedups improve with increasing refinement level because the computational effort compared to the communication overhead as well as the vector size and the number of volumes grow, e. g. for the specific example the ratio of the number of volumes to the number of sent messages is 8/100, 742/260, and 1884/1544 after the 1st, 5th, and 9th refinement step. For examples *torte4d* and *layer3* (Figure 4) we get superlinear speedup after 7 refinement steps, especially on XEON. This is mainly caused by the growing length of data structure lists and the operations on these lists which are distributed among multiple address spaces in the parallel version. Cache effects might have further influence.

5 Conclusion and Future Work

We have presented a data management and communication layer for adaptive, 3-dimensional, hexahedral FEM on distributed memory including a management for duplicated data storage and a special communication mechanism. Both are completely hidden to the user and accessible by functions of the layer. The advantages of the modular structure are the easy extensibility which does not require modifications in the algorithmic structure of the FEM code and the possibility for internal communication optimizations. First tests of the layer demonstrate usability and deliver good speedup results.

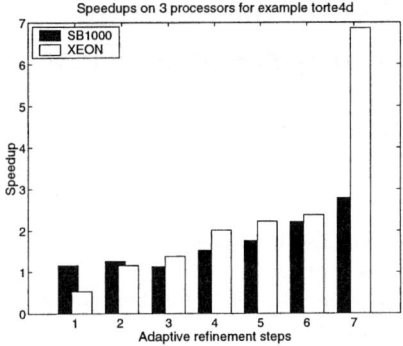 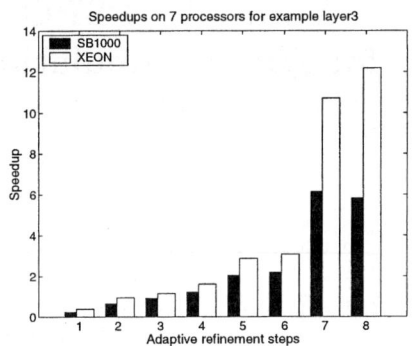

Fig. 4. Speedups for varying numbers of adaptive refinement steps on SB1000 and XEON. Left: example *torte4d* on 3 processors. Right: example *layer3* on 7 processors.

Acknowledgement

This work is supported by DFG, *SFB393: Numerical Simulation on Massively Parallel Computers*. Special thanks to S. Beuchler, A. Meyer, and M. Pester, Chemnitz University of Technology, Department of Mathematics, for their support and many interesting discussions.

References

1. Beuchler, S., Meyer, A.: SPC-PM3AdH v1.0, Programmer's Manual, Technical Report SFB393/01-08, Chemnitz University of Technology (2001)
2. Bangerth, W., Kanschat, G.: Concepts for Object-Oriented Finite Element Software - the deal.II Library, IWR Heidelberg, SFB359 Preprint 99-43 (1999)
3. Blazy, S., Kao, O., Marquardt, O.: padfem2 - An Efficient, Comfortable Framework for Massively Parallel FEM-Applications. In Dongarra, J., Laforenza, D., Orlando, S., eds.: Proc. of EuroPVM/MPI, LNCS 2840. Springer (2003) 681–685
4. Diekmann, R., Dralle, U., Neugebauer, F., Roemke, T.: PadFEM: A Portable Parallel FEM-Tool. In Liddell, H., Colbrook, A., Hertzberger, B., Sloot, P., eds.: Proc. of HPCN-Europe, LNCS 1067. Springer (1996) 580–585
5. Yserentant, H.: On the Multi-level-splitting of the Finite Element Spaces. Numer. Math. **49** (1986) 379–412
6. Bramble, J., Pasciak, J., J.Xu: Parallel Multilevel Preconditioners. Math. Comp. **55** (1991) 1–22
7. Kunert, G.: A Posteriori Error Estimation for Anisotropic Tetrahedral and Triangular Finite Element Meshes, PhD Thesis, TU-Chemnitz (1999)
8. Hippold, J., Rünger, G.: A Communication API for Implementing Irregular Algorithms on Clusters of SMPs. In Dongarra, J., Laforenza, D., Orlando, S., eds.: Proc. of EuroPVM/MPI, LNCS 2840. Springer (2003) 455–463
9. Hippold, J., Rünger, G.: Task Pool Teams for Implementing Irregular Algorithms on Clusters of SMPs. In: Proc. of the 17th IPDPS, CD-ROM. (2003)

A Parallel PSPG Finite Element Method for Direct Simulation of Incompressible Flow

Jörg Stiller[1], Karel Fraňa[1], Roger Grundmann[1],
Uwe Fladrich[2], and Wolfgang E. Nagel[2]

[1] Institute for Aerospace Engineering, TU Dresden, D-01062 Dresden, Germany
{stiller,frana,grundmann}@tfd.mw.tu-dresden.de
[2] Center for High Performance Computing, TU Dresden, D-01062 Dresden, Germany
{fladrich,nagel}@zhr.tu-dresden.de

Abstract. We describe a consistent splitting approach to the pressure-stabilized Petrov-Galerkin finite element method for incompressible flow. The splitting leads to (almost) explicit predictor and corrector steps linked by an implicit pressure equation which can be solved very efficiently. The overall second-order convergence is proved in numerical experiments. Furthermore, the parallel implementation of the method is discussed and its scalability for up to 120 processors of a SGI Origin 3800 system is demonstrated. A significant superlinear speedup is observed and can be attributed to cache effects. First applications to large-scale fluid dynamic problems are reported.

1 Introduction

We are interested in direct numerical simulations (DNS) of transitional and turbulent flows. Traditionally, specialized finite difference or spectral methods are used for this purpose. Though very efficient, these methods are often restricted to simple configurations. Unstructured finite volume methods and finite element methods are more flexible and offer the potential benefit of easier incorporating adaptive techniques. On the other hand, they are computationally less efficient and more difficult to parallelize. Also, the discretization scheme has to be carefully designed to meet the accuracy requirements for DNS.

In this paper, we consider a pressure-stabilized Petrov/Galerkin finite element method (PSPG-FEM) based on linear shape functions [1]. In Section 2, we describe a splitting approach that is similar to common projection and fractional step methods (see, e.g. [2]) but novel in the context of PSPG-FEM. The splitting yields an implicit Poisson-type equation for the pressure and an almost explicit predictor-corrector scheme for the velocity. In Section 3, we discuss the implementation on top of our in-house MG grid library [3]. Numerical accuracy and scalability of the method are examined in Section 4. In Section 5, we briefly discuss the application to DNS of electromagnetic stirring with rotating magnetic fields.

2 Finite Element Method

The flows under consideration are governed by the incompressible Navier-Stokes equations

$$\partial_t u + \nabla \cdot uu = -\nabla p + \nu \nabla^2 u + f$$

$$\nabla \cdot u = 0$$

where u is the velocity, p is the pressure divided by density, ν is the kinematic viscosity and f is the body force per unit mass. After triangulating the computational domain Ω into elements $\{\Omega_e\}$, the pressure-stabilized Petrov/Galerkin formulation of the problem can be stated as

$$\int_\Omega \left[w \cdot (\partial_t u + \nabla \cdot uu + \nabla p - f) + (\nabla w)^T : \nu \nabla u \right] d\Omega = \int_\Gamma w \cdot \nu \partial_n u \, d\Gamma$$

$$\int_\Omega q \nabla \cdot u \, d\Omega + \sum_e \int_{\Omega_e} \tau \nabla q \cdot r(u,p) \, d\Omega = 0$$

where w and q are the momentum and continuity weight functions, Γ is the boundary with normal n, τ is the elementwise defined stabilization coefficient, and

$$r(u,p) = \partial_t u + \nabla \cdot uu + \nabla p - \nu \nabla^2 u - f = r^*(u) + \nabla p$$

is the momentum residual. Dropping the stabilization term ($\tau = 0$) recovers the standard Galerkin formulation which, however, is unstable for equal-order interpolations. Appropriate choices for τ are given in [4]. In the following we use $\tau = $ constant, for simplicity.

The PSPG formulation can be formally integrated in time to give

$$\int_\Omega \left[w \cdot \left(\frac{u^{n+1} - u^n}{\Delta t} + \nabla \cdot \overline{uu} + \nabla \bar{p} - \bar{f} \right) + (\nabla w)^T : \nu \nabla \bar{u} \right] d\Omega = \int_\Gamma w \cdot \nu \partial_n \bar{u} \, d\Gamma$$

$$\int_\Omega q \nabla \cdot u^{n+1} \, d\Omega + \sum_e \int_{\Omega_e} \tau \nabla q \cdot \bar{r}(u,p) \, d\Omega = 0$$

Here, the overbar denotes the average over the time interval (t_n, t_{n+1}). Introducing the elementwise polynomial approximation yields the discrete equations which, symbolically, can be stated as

$$M \frac{U^{n+1} - U^n}{\Delta t} + \bar{N} + D\bar{P} + \nu L \bar{U} - \bar{F} = -\nu L^\Gamma \bar{U}$$

$$D \cdot U^{n+1} + \tau (D^T \cdot \bar{R}^* + L\bar{P}) = 0$$

where U, P, F and R represent the expansion coefficients, M is the mass matrix, $\bar{N}$ the contribution of the nonlinear term, D is the gradient matrix, L is the Laplace matrix and L^Γ the related boundary contribution. The momentum equation can be split into a velocity (predictor) and a pressure (corrector) step:

$$U^* = U^n - \Delta t M^{-1}(\bar{N} + \nu(L + L^\Gamma)\bar{U} - \bar{F})$$

$$U^{n+1} = U^* - \Delta t M^{-1} D \bar{P}$$

Employing $\bar{R}^* = (U^* - U^n)/\Delta t$, the following pressure equation can be obtained:

$$[(1-\delta)D \cdot M^{-1}D + \delta L]\bar{P} = -D \cdot U^*/\Delta t$$

where $\delta = \tau/\Delta t$. It is worth noting that the last three equations are still consistent with the exact solution of the continuous problem, if $\bar{N}$ is properly defined. We further remark that $\delta = 1$ is related to projection methods using the continuous pressure equation, and $\delta = 0$ corresponds to the standard Galerkin method.

In our piecewise linear approximation, the nonlinear term is evaluated as

$$\bar{N} = D \cdot \bar{U}\bar{U}$$

where $\bar{U}\bar{U}$ represents the nodal values of the momentum flux tensor $\bar{u}\bar{u}$. The time averages are evaluated using the second-order Adams-Bashforth method. Furthermore, the pressure gradient is modified using the Gresho-Chan trick [5]. The resulting, final splitting scheme is

$$U^* = U^n - \Delta t M^{-1}(D \cdot \bar{U}\bar{U} + \nu(L + L^\Gamma)\bar{U} - \bar{F})$$
$$[(1-\delta)D \cdot M_L^{-1}D + \delta L]\bar{P} = -D \cdot U^*/\Delta t$$
$$U^{n+1} = U^* - \Delta t M_L^{-1} D\bar{P}$$

where M_L denotes the lumped mass matrix.

3 Parallel Implementation

The numerical model was implemented on top of the MG grid library. MG provides data structures and procedures for handling unstructured grids as well as stable and fast methods for grid adaptation [3,6]. Both, the MG library and the flow solver are coded in Fortran 95 and use MPI for communication.

Parallelization with MG is based on grid partitioning. In the simple case considered here, the grid generated by an external grid generator is decomposed into a specified number of partitions using the MeTiS package [7]. For adaptive simulations, a recursive multilevel-partitioning strategy is available [3].

In each time step, the numerical model requires the solution of two linear algebraic systems. In the predictor step, the consistent finite element mass matrix must be resolved. Since this matrix is well conditioned, a few (1–3) damped Jacobi iterations are sufficient. The pressure equation is solved using the CG method. As a good initial approximation is available from the preceding time step, usually 10–100 CG iterations provide the new pressure with the required accuracy. Alternatively, a multigrid method using a damped Jacobi smoother and a CG coarse grid solver is under consideration. It may be expected that the multigrid solver is more effective in large-scale simulations, but this has no been verified yet.

Basically, the execution of one time step essentially involves two types of global operations which have to be performed in parallel:

- the computation of matrix-vector products
- the scalar product of coefficient vectors

The implementation of the latter is trivial, as only the contribution of multiple copies (in shared grid nodes) has to be canceled. The matrix-vector products can be expressed as, e.g.,

$$L_{ij}\bar{P}_j = \sum_p L_{ij}^p \bar{P}_j$$

where L_{ij}^p is the contribution from the elements of partition p. These local matrices are precomputed and stored in an edge-based data basis. Once the local products are evaluated, the final result is obtained by adding the individual contributions in all nodes shared between one or more partitions. In the actual implementation, this operation is realized just by calling the appropriate procedure of the MG library.

4   Accuracy and Performance

Two laminar flow configurations with known analytic solution were used to examine the accuracy of the method: The transient channel flow between two parallel plates, and the stationary flow of an electro-conducting fluid in an infinite cylinder that is exposed to a rotating magnetic field (see [8]). Unstructured tetrahedral grids with different mesh spacing h were used in both cases. The results shown in Fig. 1 clearly reveal a convergence rate of order 2.

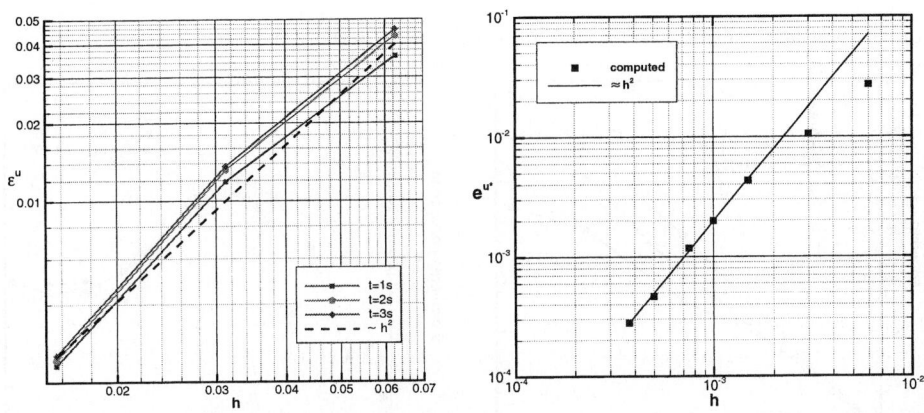

Fig. 1. L_∞ velocity error in transient channel flow (left) and in RMF-driven flow (right).

Performance tests have been conducted on a SGI Origin 3800 system at TU Dresden. Three cases with grids ranging from 110,000 up to 7.1 million elements were considered. In the tests, one time step with a fixed number of iterations was performed using one to 120 processors. The measured speedup and the computational rate are depicted in Fig. 2. Especially for the medium grid a significant superlinear speedup is observed. Since the parallel efficiency is unlikely to increase, this behavior can only be attributed to gains in local efficiency. A possible explanation is that our code heavily depends on indirect addressing, which on its part results in a suboptimal cache efficiency. With decreasing size the problem fits better into the cache hierarchy, leading to a gradual improvement of local computational efficiency. We remark that similar effects were observed by other authors [9].

For the other two cases the same arguments apply. The larger problem does not fit in cache in 64 or less processors but shows a superlinear performance gain for higher numbers. Finally, the smallest problem achieves optimal cache performance earlier, and thus shows an increasing loss of parallel efficiency. Therefore, the overhead introduced for such a small problem is obvious. However, when taking into account that the local grids contain only about one thousand elements (or 200 nodes) on 120 processors, the measured efficiency of 75% is still remarkable.

While the parallel speedup is dependent on the total problem size, the computational rate (average local size divided by wall clock time) shows a consistent behavior: Unless the local problem does not fit in cache a sustained rate of 2,500 elements per second is obtained in all three cases. The highest rate and, hence, optimal cache performance is achieved with 10 thousand (or less) local elements. From this point, decreasing parallel efficiency starts to reduce the computational rate.

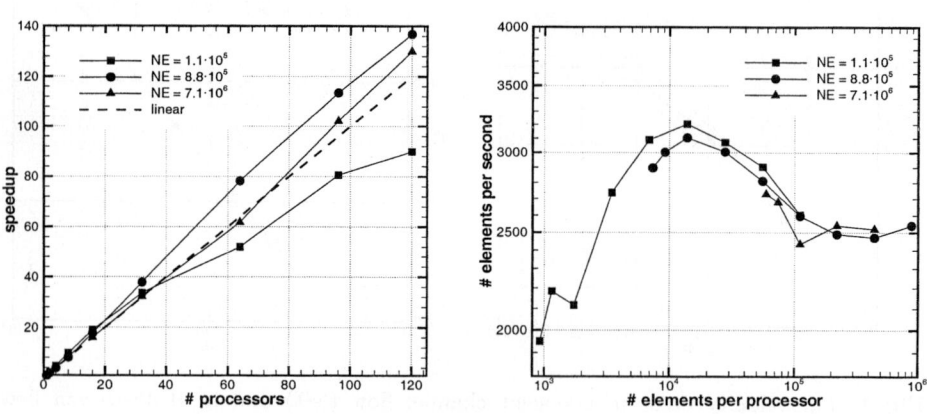

Fig. 2. Speedup (left) and computational rate (right).

5 Application

The solver is currently being used to perform direct numerical simulations of the flow in a cylindrical cavity that is exposed to a rotating magnetic field (RMF). This problem is related to electromagnetic stirring of melts in metallurgy and crystal growth in semiconductor production. Despite of its significance, the transitional and early turbulent regimes of flow are largely unexplored.

Figure 3 depicts the configuration. B is the flux density vector of the magnetic field which rotates with angular speed ω. Comparable to an induction motor, the field induces a primary rotating motion of the enclosed electro-conducting fluid. Additionally, a secondary recirculating flow develops due to the local action of friction forces. Under the so-called low-frequency/low-induction assumption, which is valid for most practical applications, the flow is described by the incompressible Navier-Stokes equations with an a priori known Lorentz force (see [10] for more details).

In our simulations computational grids ranging from 10^5 up to 20 million elements (3.5 million nodes) have been applied. Figure 3 shows a typical coarse grid divided into 32 partitions. An important result of the DNS study is the insight in the formation, evolution and finally dissipation of Taylor-Görtler-like vortices that obviously dominate the turbulence physics and provide an efficient mixing mechanism in this type of flow (Fig. 4). A detailed description of this and further results will be subject to a forthcoming paper.

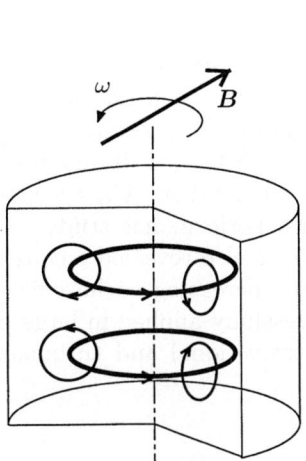
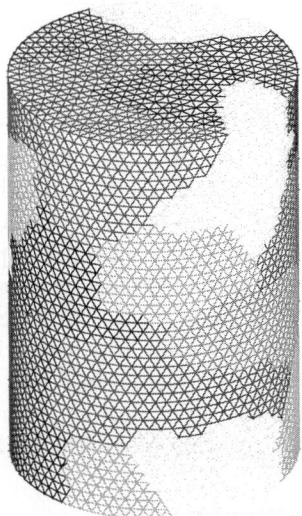

Fig. 3. Stirring with a rotating magnetic field: Sketch of configuration (left) and coarse grid divided into 32 partitions (right).

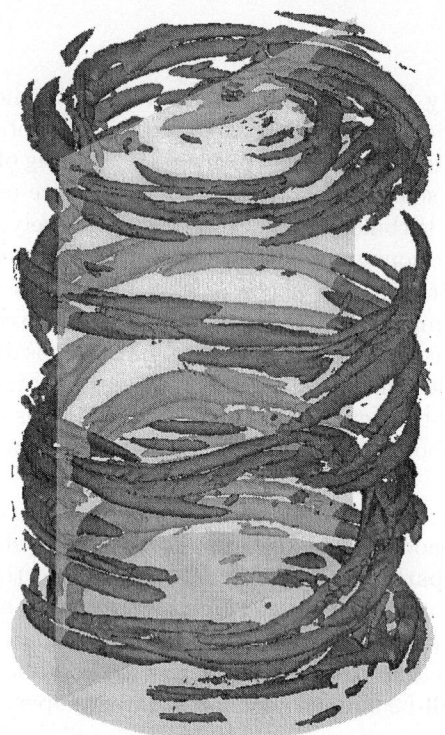

Fig. 4. Instantaneous vortex structures in weakly turbulent regime.

6 Conclusion

We have described a splitting approach to the pressure-stabilized Petrov-Galerkin finite element method for incompressible flow. Similar to conventional fractional step or projection methods it allows for segregating the pressure computation but retains a consistent formulation. The existing MG grid library provided a suitable basis for the parallel implementation of the method. The excellent scalability of the flow solver was demonstrated in a performance study on a SGI Origin 3800 using up to 120 processors. These tests also revealed a considerable superlinear speedup which can be explained by increasing cache efficiency at smaller local problem sizes. The solver was successfully applied in large production runs for direct numerical simulations of transitional and turbulent flows driven by rotating magnetic fields.

Acknowledgments

Financial support from German "Deutsche Forschungsgemeinschaft" in frame of the Collobarative Research Center SFB 609 is gratefully acknowledged. The measurements were done on a SGI Origin 3800 based on a grant from ZHR at TU Dresden.

References

1. Hughes, T.J.R., Franca, L.P., Balestra, M.: A new finite element formulation for computational fluid dynamics: V. Circumventing the Babuška-Brezzi condition: A stable Petrov-Galerkin formulation of the Stokes problem accommodating equal-order interpolations. Comput. Meth. Appl. Mech. Eng. **59** (1986) 85–99.
2. Codina, R.: Pressure Stability in Fractional Step Finite Element Methods for Incompressible Flows. J. Comp. Phys. **170** (2001), 112–140.
3. Stiller, J., Nagel, W.E.: MG – A Toolbox for Parallel Grid Adaption and Implementing Unstructured Multigrid Solvers. In: E.H. D'Hollander et al. (Eds.): Parallel Computing. Fundamentals & Applications. Imperial College Press 2000, 391–399.
4. Tezduyar, T., Osawa, Y.: Finite element stabilization parameters computed from element matrices and vectors. Comput. Meth. Appl. Mech. Eng. **190** (2001), 411–430.
5. Gresho, P.M.: On the theory of semi-implicit projection methods for viscous incompressible flow and its implementation via a nite element method that also introduces a nearly consistent mass matrix I: Theory. Int. J. Num. Meth. Fluids. **11** (1990), 587–620.
6. Stiller, J., Wienken, W., Fladrich, U., Grundmann, R., Nagel, W.E.: Parallel and Adaptive Finite Element Techniques for Flow Simulation. In C. Breitsamter et al. (Eds.): New Results in Numerical and Experimental Fluid Mechanics IV (Notes on Numerical Fluid Mechanics and Multidisciplinary Design, Vol. 87). Springer Verlag 2004, 366–373.
7. Karypis, G., Kumar, V.: Multilevel Algorithms for Multi-Constraint Graph Partitioning. Univ. Minnesota, Dep. Computer Science, TR 98-019, 1998.
8. Davidson, P.A.: An Introduction to Magnetohydrodynamics. Cambridge University Press 2001.
9. Baggag, A., Atkins, H., Keyes, D.E.: Parallel implementation of the Discontinous Galerkin Method. In: Proceedings of Parallel CFD'99, 1999, 233–240.
10. Gelfgat, Yu.M., Priede, J.: MHD flows on a rotating magnetic field (A review). Magnetohydrodynamics **31** (1995) 1–2, 188–200.

Load and Memory Balanced Mesh Partitioning for a Parallel Envelope Method[*]

Ondřej Medek, Pavel Tvrdík, and Jaroslav Kruis

Czech Technical University, Prague, Czech Republic,
{xmedeko,tvrdik}@fel.cvut.cz, kruis@fsv.cvut.cz

Abstract. We use a parallel direct solver based on the Schur complement method for solving large sparse linear systems arising from the finite element method. A domain decomposition of a problem is performed using a graph partitioning. It results in sparse submatrices with balanced sizes. An envelope method is used to factorize these submatrices. However, the memory requirements to store them and the computational cost to factorize them depends heavily on their structure. We propose a technique that modifies the multilevel graph partitioning schema to balance real computational load or memory requirements of the solver.

1 Introduction

Many engineering and scientific problems are solved using the finite element method (FEM). We consider a mesh of *finite elements*. One element consists of several *nodes*. Each node has its *degrees of freedom* (DOFs). In other words, the nodes contain *variables*. Globally, these variables form a system of *equations* $Ax = b$, where A is an $n \times n$ sparse matrix of coefficients. A mesh is usually represented by a *dual graph* G^{D} and a *nodal graph* G^{N}. The vertices in the dual graph represent the finite elements and 2 vertices are adjacent if and only if the corresponding elements share a common surface in 3D or a common edge in 2D. The vertices in the nodal graph represent the mesh nodes. All vertices that represent mesh nodes belonging to one element form a clique. An example of a FE mesh and its dual and nodal graph is on Fig. 1.

We use solver SIFEL, developed at the Czech Technical University, for solving problems by the FEM. Among others, it can solve problems in parallel by the popular method of the Schur complements, which are computed by an envelope method [1]. A solution of a problem consists of 6 following phases:

1. Domain decomposition.
2. Ordering of nodes.
3. Assembling of submatrices.
4. Factorization of submatrices (computation of the Schur complements).
5. Solution of the reduced problem.
6. Back substitution on subdomains.

---

[*] This work was supported by IBS3086102 grant of Czech Academy of Science and by MŠMT under research program #J04/98:212300014.

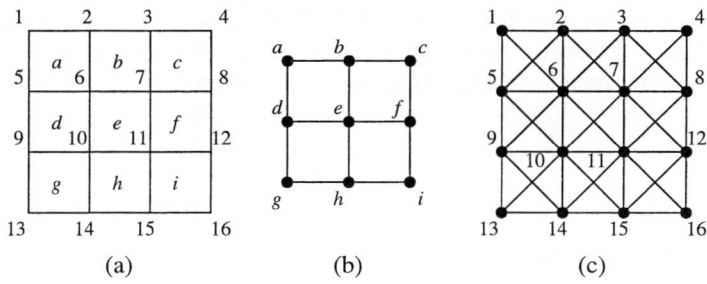

Fig. 1. A quadrilateral mesh (a) with 9 elements $a - i$ and 16 nodes 1 – 16. The dual (b) and the nodal (c) graph derived from the mesh.

The domain decomposition, and optionally ordering of nodes, are done as preprocessing steps. The solver does the rest. Phase 3 is the most memory consuming and Phase 4 is the most computationally intensive part of the whole solution.

Multilevel tools are widely used to solve the problem of domain decomposition. The dual graph is partitioned into k parts, inducing subdomains and corresponding submatrices, so that the sizes of submatrices are roughly equal.

Definition 1. *Consider a graph $G = (V, E)$ and an integer $k \geq 2$. An edge cut (node cut) is a set of edges (vertices, respectively) whose removal divides the graph into at least k partitions. The k-way graph partitioning problem is to partition V into k pairwise disjoint subsets $V_1, V_2, \ldots, V_k$ such that $|V_i| \doteq |V|/k$ and the size of the edge cut is minimized. A partitioning of a graph by a* node cut *is similar.*

Even though the sizes of submatrices are roughly equal, their memory requirements or their factorization time in Phase 4 are not equal. To define this formally, we use the term *quality* to denote the memory or computational complexity.

Definition 2. *Given an* unbalancing threshold $\delta \geq 1$*, we say that a partitioning $V_1, V_2, \ldots, V_k$ with a set of qualities $\{q_1, q_2, \ldots, q_k\}$ is* balanced *if*

$$\delta \geq (\max_{i=1}^{k} q_i) k / \sum_{i=1}^{k} q_i \ . \tag{1}$$

A partition V_i is overbalanced if $\delta < q_i k / \sum_i q_i$. The partitioning is disbalanced if at least one partition is overbalanced.

In general, the qualities of submatrices are influenced by the ordering of variables, as was already mentioned in [2, 3].

In this paper, we describe a novel approach to the domain decomposition that results into partitioning with balanced memory requirements or balanced factorization time estimations. This in fact leads to shorter execution time of the parallel solver. The idea is to integrate an ordering algorithm into a multilevel graph partitioning schema.

Section 2 describes the previous work. In Section 3, the new refinement heuristics that allows to balance better memory or computing complexity is explained. In Section 4, the results of experiments are presented and Section 5 concludes the paper.

2 Previous Work

2.1 Domain Decomposition (Phase 1)

Common methods for domain decomposition are based on graph partitioning, typically of dual graphs. A subdomain is made of elements from the same partition. The nodes belonging to more than one subdomain are called *boundary* and the remaining nodes are *internal*. Variables of the internal (boundary) nodes are called correspondingly. An example on Fig. 2 (a) shows a 2-way partitioning of G^D of the quadrilateral mesh from Fig. 1. The corresponding domain decomposition is shown on Fig. 2 (b). The boundary nodes are 3, 7, 10, 11, and 14. Note that this way of domain decomposition produces partitioning of the nodal graph G^N by a node cut, as shown on Fig. 2 (c).

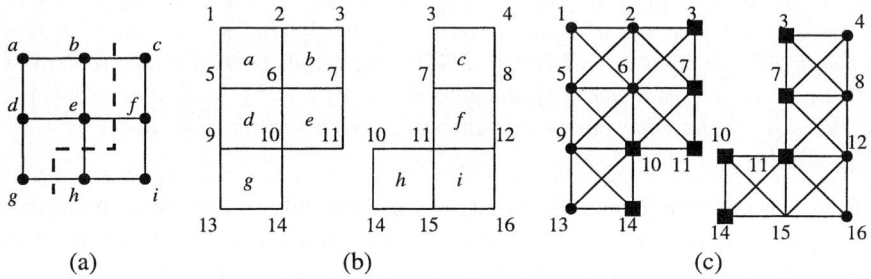

Fig. 2. Partitioning of the mesh from Fig. 1 using its dual graph (a) into 2 partitions (b) and corresponding partitioning of the nodal graph (c). The edge cut (a) is indicated by a dashed line and nodes in the node cut (c) as filled squares.

Multilevel Graph Partitioning. Popular multilevel graph partitioning software is METIS [4,5], CHACO [6], JOSTLE [7], and SCOTCH [8]. Our work is based on the multilevel k-way graph partitioning implemented in METIS [5]. This schema consists of the following 3 phases, shown on Fig. 3.

Coarsening. A sequence of smaller graphs $G_l = (V_l, E_l)$ is constructed from the original graph $G = G_0 = (V_0, E_0)$ so that $|V_l| > |V_{l+1}|$. The sequence of coarser graphs creates *levels*. A matching is often used to collapse 2 vertices into a *multivertex*. The SHEM heuristics for matching, introduced in [5], works well.

Initial partitioning. When the coarsest graph is sufficiently small, it can be partitioned by any graph partitioning technique.

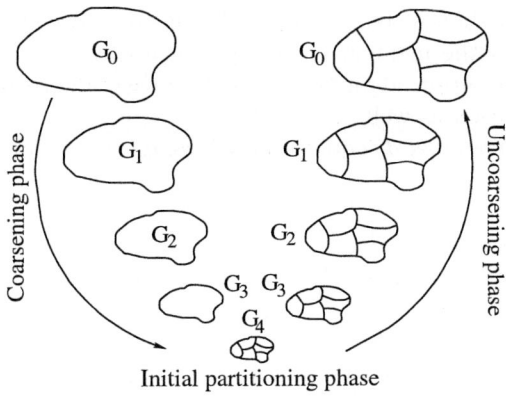

Fig. 3. The schema of multilevel *k-way* partitioning.

Uncoarsening. A coarser graph G_l is uncoarsed to G_{l-1} and the partitioning of G_l is projected to G_{l-1} and then refined. The Fiduccia-Mattheyses (FM) heuristics is a simple, fast, and sufficiently good option for the refinement. It searches candidate vertices in the set of *boundary vertices*, i.e., vertices adjacent to a vertex in another partition. Then it tries to move a selected vertex into other partitions. The move is *accepted* if one of the following conditions is fulfilled:
1. The size of the edge cut is decreased and the partitioning remains balanced.
2. The size of the edge cut is not increased, but the balancing is improved.

To improve the balancing of a strongly disbalanced partitioning, a balancing step may by added. It works like the FM heuristics, but the conditions for accepting a move are different:
1. The balancing is improved.
2. The size of the edge cut is decreased, but the balancing is not worsened.

2.2 Reordering and Assembling of Submatrices (Phases 2 + 3)

After the domain decomposition, the internal nodes in all partitions are reordered to minimize the size of the *envelope*.

Definition 3. *In the i-th step of the factorization of a symmetric matrix $\boldsymbol{A}$, the* wavefront *is set of pairs*

$$w_i(\boldsymbol{A}) = \{\{r, i\} : \exists a_{rs} \neq 0, r \geqq i, s \leqq i\} \ . \tag{2}$$

The envelope *of $\boldsymbol{A}$ is then (symbol $\setminus$ is the set difference)*

$$Env(\boldsymbol{A}) = \bigcup_{i=1}^{n} w_i(\boldsymbol{A}) \setminus \{i, i\} \ . \tag{3}$$

The envelope method stores the parts of $\boldsymbol{A}$ inside envelopes, which represents the main portion of the memory consumption of the solver. The SIFEL solver uses the Sloan reordering algorithm [9], but other algorithms (RCM, hybrid algorithm) can be used as well.

Every node i generates d_i equations, where d_i is the number of DOFs. Let n_i and n_b denote the number of internal and boundary, respectively, variables. Clearly, $n_i + n_b = n$. The SIFEL solver reads the subdomains and assembles submatrices with nodes in a given order. The boundary nodes come last. We estimate memory requirements $W(\boldsymbol{A})$ of the envelope solver and its computational complexity $OP(\boldsymbol{A})$ by

$$W(\boldsymbol{A}) = \sum_{i=1}^{n} |w_i(\boldsymbol{A})| = |Env(\boldsymbol{A})| + n \qquad (4)$$

$$OP(\boldsymbol{A}) = \sum_{i=1}^{n_i} |w_i(\boldsymbol{A})|^2 \; . \qquad (5)$$

3 The New Refinement Heuristics with Quality Balancing

In this section, we explain the main ideas of load and memory balancing for the parallel envelope method. At the beginning, it must be decided which quality will be balanced: the computational load or the memory complexity. Then the partitioning is started. To estimate the quality, the partitioner must have the information about the structure of the matrix $\boldsymbol{A}$. Since the dual graph G^D does not suffice, the partitioner must have at disposal also the weighted nodal graph G^N, in which each vertex is labelled with the number of DOFs of the node. We partition G^D and project the partitioning of G^D to G^N to estimate the quality. So, the partitioner needs also the information about the relation between G^D and G^N.

We use a multilevel k-way graph partitioning as described in Sect. 2.1. The first two phases are performed unchanged as in METIS. The SHEM heuristics is used for the coarsening and the multilevel graph bisection is used for the initial partitioning. We have modified the conditions of move acceptance of the refinement heuristics in the uncoarsening phase and extended it with a quality estimation process. This new heuristics is called QB. Assume that the current level is l. Similar to FM, QB chooses a vertex as a candidate for moving from a *source* partition $G^D_{l,s}$ to a *target* partition $G^D_{l,t}$. To decide whether the move will be accepted, QB starts for both partitions $G^D_{l,s}$ and $G^D_{l,t}$ the estimation process, sketched on Fig. 4, to obtain information about balancing of the qualities.

First, partition $G^D_{l,p}, p \in \{s,t\}$, is projected to $G^D_{0,p}$, the original partition. (This is skipped in the final level, where $l = 0$). After that, all nodes belonging to elements from $G^D_{0,p}$ correspond to a partition G^N_p of G^N. Then the internal vertices of G^N_p are reordered by the Sloan algorithm. Finally, the quality of G^N_p is estimated and returned to the refinement heuristics.

In the current implementation of the QB heuristics, nodes have either constant number of DOFs $d > 0$ or are *constrained*, i.e., the number of DOFs is 0.

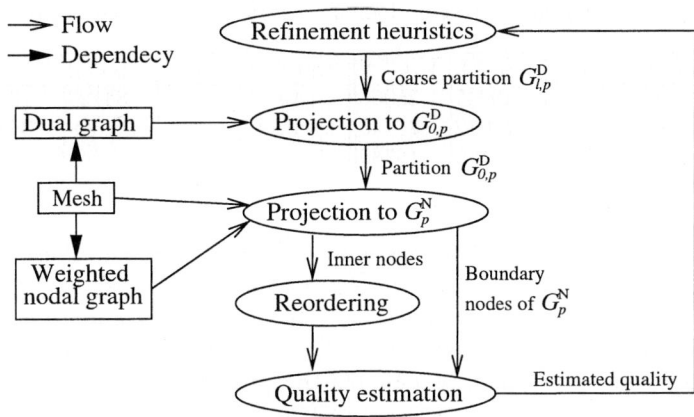

Fig. 4. Data flow of the QB heuristics.

All constrained nodes are omitted in the step of projection of $G_{0,p}^D$ to G_p^N, i.e., the reordering is performed only with nodes with the number of DOFs $d > 0$. After that, node i generates equations numbered $di, di+1, \ldots di+d-1$ and wavefronts $w_{di}(A), w_{di+1}(A), \ldots, w_{di+d-1}(A)$.

The original FM heuristics computes sums of weights of vertices in the source and target partitions for every candidate move. In fact, the weight of the candidate vertex is subtracted from the weight of the source partition and added to the weight of the target partition. However, in the QB heuristics, this would imply the reordering and estimation computing for every candidate move and this would extremely slow down the refinement. Thus, we had to modify the conditions of move acceptance as follows:

1. The size of the edge cut is decreased and the target partition is not overbalanced.
2. The quality q_s of the source partition is greater than the quality q_t of the target partition, but the size of the edge cut is not increased.

The conditions of move acceptance of the balancing step are also modified:

1. $q_s > q_t$.
2. The size of the edge cut is decreased and $q_s \geqq q_t$.

Only if a move is accepted, the qualities q_s and q_t are recomputed. Note that the new conditions may lead to overbalancing of the target, or even the source, partitions. Therefore, if the new value q_s is greater than its previous value, the vertex move is nullified.

4 Experimental Results

The SIFEL solver was modified to perform just the assembling and factorization of submatrices (Phases 3 and 4 in Sect. 1). All experiments were performed on

Table 1. Description of test problems. # stands for number of problem.

#	problem name	$|V(G^D)|$	$|V(G^N)|$	#	problem name	$|V(G^D)|$	$|V(G^N)|$
1	floor	49245	25065	4	jete	84123	22860
2	sieger	41012	21140	5	wheel	157529	34369
3	block	89196	16789				

a PC with Intel Pentium III, 1GHz, under GNU/Linux 2.4. The benchmarks are models of real problems of structural mechanics: "floor" and "sieger" are 2D problems discretized by triangles and "block", "jete", and "wheel" are 3D problems discretized by tetrahedrons. Their description is in Table 1.

Table 2 shows the results. # denotes the problem number as is quoted in Table 1. All problems were partitioned to $k = 4, 8, 16, 32$ partitions by METIS, called with its default parameters, and by the QB heuristics called with unbalancing threshold $\delta = 1.1$. First, QB was used to balance memory requirements of the solver and second, to balance computational complexity of the factorization phase. The size of the edge cut of G^D is denoted by $|E_c|$. Let $A_1, \ldots, A_k$ be the corresponding submatrices of A. Then $W_{max} = \max_{i=1}^{k} W(A_i)$, $\overline{W} = \frac{1}{k} \sum_{i=1}^{k} W(A_i)$, and $\Delta W = W_{max}/\overline{W}$. Similarly, let t_i be the factorization time of submatrix A_i. Then $t_{max} = \max_{i=1}^{k} t_i$, $\bar{t} = \frac{1}{k} \sum_{i=1}^{k} t_i$, and $\Delta t = t_{max}/\bar{t}$. The time for the domain decomposition of the QB heuristics is denoted by t_p. The values of t_{max} and t_p in Table 2 are given in seconds.

Table 2. Comparison of the FM and QB heuristics.

		METIS				Memory Balancing					Load Balancing			
#	k	$|E_c|$	W_{max} ΔW	t_{max}	Δt	$|E_c|$	W_{max} ΔW	t_{max}	Δt	t_p	$|E_c|$	t_{max}	Δt	t_p
1	4	317	9751224 1.34	52.3	1.68	343	7312581 1.09	31.0	1.27	49	358	30.3	1.03	34
	8	524	3889227 1.35	15.5	1.72	547	2798742 1.04	10.2	1.33	38	629	8.9	1.09	66
	16	894	1634397 1.40	5.6	1.99	944	1175811 1.07	3.2	1.33	49	1011	2.9	1.18	69
	32	1405	623400 1.36	1.5	1.79	1424	456873 1.08	0.9	1.36	75	1447	0.8	1.11	125
2	4	160	3540777 1.08	7.4	1.14	159	3294747 1.03	6.5	1.06	8	157	6.5	1.10	10
	8	298	2094126 1.30	5.3	1.72	288	1523460 1.06	3.2	1.33	16	301	2.7	1.11	34
	16	533	971229 1.35	2.3	1.91	576	700158 1.08	1.3	1.35	23	663	1.6	1.61	102
	32	1048	428196 1.40	0.8	1.75	1072	308319 1.09	0.5	1.35	46	1068	0.4	1.21	53
3	4	2561	18490254 1.05	236.4	1.07	2753	17972640 1.03	233.7	1.07	219	2741	218.3	1.01	257
	8	4255	7336176 1.12	65.1	1.20	4231	7005948 1.10	55.8	1.09	198	4303	51.9	1.03	222
	16	5896	3388941 1.34	25.3	1.72	6045	2747472 1.22	15.3	1.39	282	6026	12.5	1.08	244
	32	8744	1288023 1.33	5.1	1.51	8749	1008285 1.14	3.7	1.38	287	8986	3.0	1.12	350
4	4	915	17489307 1.24	167.9	1.39	922	14000742 1.08	113.4	1.14	64	930	97.6	1.03	76
	8	1560	6281361 1.18	40.7	1.26	1628	5662314 1.09	34.2	1.13	59	1674	34.4	1.08	85
	16	2398	2987406 1.48	14.6	1.67	2537	2148342 1.09	10.1	1.23	76	2460	9.0	1.13	92
	32	3591	1224027 1.65	4.7	2.16	3471	749844 1.09	2.2	1.21	104	3583	2.0	1.09	134
5	4	914	21716046 1.15	196.2	1.27	999	20269455 1.07	178.0	1.17	117	993	172.6	1.09	146
	8	1747	11810322 1.31	106.1	1.58	1823	9354273 1.08	77.9	1.26	120	1802	67.9	1.08	156
	16	3253	4625520 1.24	31.8	1.42	3173	3557904 1.06	21.3	1.15	157	3438	21.3	1.08	298
	32	5917	1838763 1.29	10.0	1.70	5826	1550547 1.18	6.4	1.32	504	6420	6.5	1.33	577

5 Evaluations of Results and Conclusions

The results demonstrate that the QB heuristics always produces partitionings with better ΔW and Δt than the FM heuristics. Also $\Delta W \leq \delta$ in nearly all cases of memory balancing, whereas $\Delta t > \delta$ in about 30% of cases of load balancing, for the sake of minimizing the edge cut size $|E_c|$. The balancing of the memory requirements always improves Δt as a side effect. On the other hand, the QB heuristics sometimes produces partitionings with slightly greater $|E_c|$.

The memory balancing is beneficial for the distributed systems with limited amount of main memory per processor. The balancing of the computational load leads to the shorter time of the Phase 4 of the solver. Of course, the QB heuristics slows down the domain decomposition phase. Whereas the standard METIS takes times of order of seconds, the time t_p of the QB heuristics is of order of tens or hundreds of seconds. Therefore, it should be used if the same decomposition can be reused several times, e.g., in nonlinear systems or when the same problem is solved with different materials, etc.

References

1. George, A., Liu, J.: Computer Solution of Large Sparse Positive Definite Systems. Prentice Hall, Englewood Cliffs, NJ (1981)
2. Hendricson, B.: Graph partitioning and parallel solvers: Has emperor no clothes? Irregular'98, Lecture Notes in Computer Science **1457** (1998) 218–225
3. Hendricson, B.: Load balancing fictions, falsehoods and fallacies. Applied Mathematical Modelling **25** (2000) 99–108
4. Karypis, G., Kumar, V.: A fast and high quality multilevel scheme for partitioning irregular graphs. SIAM J. Sci. Comput. **20** (1998) 359–392
5. Karypis, G., Kumar, V.: Multilevel k-way partitioning scheme for irregular graphs. J. of Parallel and Distrib. Comput. **48** (1998) 96–129
6. Hendrickson, B., Leland, R.: A multilevel algorithm for partitioning graphs. In: Proceedings of the 1995 ACM/IEEE conference on Supercomputing (CDROM), ACM Press (1995) 28
7. Walshaw, C., Cross, M.: Mesh Partitioning: a Multilevel Balancing and Refinement Algorithm. SIAM J. Sci. Comput. **22** (2000) 63–80 (originally published as Univ. Greenwich Tech. Rep. 98/IM/35)
8. Pellegrini, F.: Static mapping by dual recursive bipartitioning of process and architecture graphs. SHPCC'94 (1994) 486–493
9. Kumfert, G., Pothen, A.: Two improved algorithms for envelope and wavefront reduction. BIT **37** (1997) 559–590

Combining Explicit and Recursive Blocking for Solving Triangular Sylvester-Type Matrix Equations on Distributed Memory Platforms

Robert Granat, Isak Jonsson, and Bo Kågström

Department of Computing Science and HPC2N, Umeå University,
SE-901 87 Umeå, Sweden
{granat,isak,bokg}@cs.umu.se

Abstract. Parallel ScaLAPACK-style hybrid algorithms for solving the triangular continuous-time Sylvester (SYCT) equation $AX - XB = C$ using recursive blocked node solvers from the novel high-performance library RECSY are presented. We compare our new hybrid algorithms with parallel implementations based on the SYCT solver DTRSYL from LAPACK. Experiments show that the RECSY solvers can significantly improve on the serial as well as on the parallel performance if the problem data is partitioned and distributed in an appropriate way. Examples include cutting down the execution time by 47% and 34% when solving large-scale problems using two different communication schemes in the parallel algorithm and distributing the matrices with blocking factors four times larger than normally. The recursive blocking is automatic for solving subsystems of the global explicit blocked algorithm on the nodes.

Keywords: Sylvester matrix equation, continuous-time, Bartels–Stewart method, blocking, GEMM-based, level 3 BLAS, LAPACK, ScaLAPACK-style algorithms, RECSY, recursive algorithms, automatic blocking.

1 Introduction

This contribution deals with parallel algorithms and software for the numerical solution of the triangular continuous-time Sylvester equation (SYCT)

$$AX - XB = C, \qquad (1)$$

on distributed memory (DM) environments, where A of size $m \times m$, B of size $n \times n$ and C of size $m \times n$ are arbitrary matrices with real entries. The matrices A and B are in upper (quasi-)triangular Schur form. A quasi-triangular matrix is upper triangular with some 2×2 blocks on the diagonal that correspond to complex conjugate pairs of eigenvalues. SYCT has a unique solution X of size $m \times n$ if and only if A and B have disjoint spectra, or equivalently the separation $\text{sep}(A, B) \neq 0$. The Sylvester equation appears naturally in several applications. Examples include block-diagonalization of a matrix in Schur form and condition estimation of eigenvalue problems (e.g., see [17, 10, 19]).

Using the Kronecker product notation, $\otimes$, we can rewrite the Sylvester equation as a linear system of equations

$$Z_{\text{SYCT}} x = c, \qquad (2)$$

where $Z_{\text{SYCT}} = I_n \otimes A - B^T \otimes I_m$ is a matrix of size $mn \times mn$, $x = \text{vec}(X)$ and $c = \text{vec}(C)$. As usual, $\text{vec}(X)$ denotes an ordered stack of the columns of the matrix X from left to right starting with the first column. Since A and B are (quasi-)triangular, the triangular Sylvester equation can be solved to the cost $O(m^2 n + mn^2)$ using a combined backward/forward substitution process [1]. In blocked algorithms, the explicit Kronecker matrix representation $Zx = c$ is used in kernels for solving small-sized matrix equations (e.g., see [11, 12, 17]).

Our objective is to investigate the performance of our ScaLAPACK-style algorithms for solving SYCT [7, 6] when combined with recursive blocked matrix equation solvers from the recently developed high-performance library RECSY [11–13]. The recursive approach works very well on single processor architectures and shared memory machines utilizing just a few nodes. In a distributed memory environment, recursion can hardly be applied efficiently on the top-level of our parallel algorithms for solving SYCT. Still, we can gain performance by applying recursion when solving medium-sized instances of SYCT on the nodes, and this motivates our investigation of ScaLAPACK-style hybrid algorithms.

The rest of the paper is organized as follows; In Section 2, we review our ScaLAPACK-style algorithms for solving SYCT. Then the RECSY library, which is used for building the hybrid algorithms, is briefly presented in Section 3. In Section 4, we display and compare some experimental results of the standard and hybrid ScaLAPACK-style algorithms, respectively. Finally, in Section 5, we summarize our findings and outline ongoing and future work.

2 Parallel ScaLAPACK-Style Algorithms for Solving SYCT Using Explicit Blocking

To solve SYCT we transform it to triangular form, following the Bartels–Stewart method [1], before applying a direct solver. This is done by means of a Hessenberg reduction, followed by the QR-algorithm applied to both A and B. The right hand side C must also be transformed with respect to the Schur decompositions of A and B. Reliable and efficient algorithms for the reduction step can be found in LAPACK [2], for the serial case, and in ScaLAPACK [9, 8, 3] for distributed memory environments. Assuming that this reduction step has already been performed, we partition the matrices A and B in SYCT using the blocking factors mb and nb, respectively. This implies that mb is the row-block size and nb is the column-block size of the matrices C and X (which overwrites C). By defining $D_a = \lceil m/mb \rceil$ and $D_b = \lceil n/nb \rceil$, SYCT can be rewritten in blocked form as

$$A_{ii} X_{ij} - X_{ij} B_{jj} = C_{ij} - \left(\sum_{k=i+1}^{D_a} A_{ik} X_{kj} - \sum_{k=1}^{j-1} X_{ik} B_{kj} \right), \qquad (3)$$

```
for j=1, D_b
    for i=D_a, 1, -1
        {Solve the (i, j)th subsystem using a kernel solver}
        A_ii X_ij − X_ij B_jj = C_ij
        for k=1, i − 1
            {Update block column j of C}
            C_kj = C_kj − A_ki X_ij
        end
        for k=j + 1, D_b
            {Update block row i of C}
            C_ik = C_ik + X_ij B_jk
        end
    end
end
```

Fig. 1. Block algorithm for solving $AX - XB = C$, A and B in real Schur form.

where $i = 1, 2, \ldots, D_a$ and $j = 1, 2, \ldots, D_b$. The resulting serial blocked algorithm is outlined in Figure 1 [17, 19].

We now assume that the matrices A, B and C are distributed using 2D block-cyclic mapping across a $P_r \times P_c$ processor grid. We then traverse the matrix C/X along its block diagonals from South-West to North-East, starting in the South-West corner. To be able to compute X_{ij} for different values of i and j, we need A_{ii} and B_{jj} to be held by the same process that holds C_{ij}. We also need to have the blocks used in the general matrix-multiply and add (GEMM) updates of C_{ij} in the right place at the right time. In general, this means we have to communicate for some blocks during the solves and updates. This can be done "on demand": whenever a processor misses any block that it needs for solving a node subsystem or doing a GEMM update, it is received from the owner in a single point-to-point communication [7]. Because of the global view of data in the ScaLAPACK environment all processors know exactly which blocks to send in each step of the algorithm. Moreover, the subsolutions X_{ij} are broadcasted in block row i and block column j for use in updates of right hand sides. A brief outline of a parallel algorithm PTRSYCTD that uses this approach is presented in Figure 2. The matrices can also be shifted one step across the process mesh for every block diagonal that we solve for [19, 6]. This brings all the blocks needed for the solves and updates associated with the current block diagonal into the right place in one single global communication operation. A brief outline of such a parallel algorithm is presented in Figure 3. The "matrix-shifting" approach puts restrictions on the dimensions of the processor grid and the data distribution: P_r must be an integer multiple of P_c or vice versa, and the last rows/columns of A and B must be mapped onto the last process row/column [19]. Both communication schemes have been implemented in the same routine PGESYCTD [7, 6], which can solve four variants of SYCT with one or both of A and B replaced by their transposes.

The parallel algorithms presented in Figures 2 and 3 both tend to give speedup of $O(\sqrt{p})$, where p is the number of processors used in the parallel execution [19, 6, 7].

```
for k=1, the number of block diagonals in C
    {Solve subsystems on current block diagonal in parallel}
    if(mynode holds C_ij)
        if(mynode does not hold A_ii and/or B_jj)
            Communicate for A_ii and/or B_jj
        Solve for X_ij in A_ii X_ij − X_ij B_jj = C_ij
        Broadcast X_ij to processors that need X_ij for updates
    elseif(mynode needs X_ij)
        Receive X_ij
    if(mynode does not hold block in A needed for updating block column j)
        Communicate for requested block in A
    Update block column j of C in parallel
    if(mynode does not hold block in B needed for updating block row i)
        Communicate for requested block in B
    Update block row i of C in parallel
    endif
end
```

Fig. 2. Parallel "communicate-on-demand" block algorithm for $AX - XB = C$, A and B in real Schur form.

Notice that we are free to choose any kernel solver for the subsystems $A_{ii}X_{ij} - X_{ij}B_{jj} = C_{ij}$ in the algorithms presented in Figures 1, 2 and 3. Here A_{ii} and B_{jj} are of size $mb \times mb$ and $nb \times nb$, respectively, and C/X is of size $mb \times nb$. The original implementation of the parallel algorithms used LAPACK's DTRSYL as node solver, which is essentially a level-2 BLAS algorithm. For more information about the ScaLAPACK-style algorithms we refer to [19, 7, 6].

3 RECSY – Using Recursive Blocked Algorithms for Solving Sylvester-Type Subsystems

RECSY [13] is a high-performance library for solving triangular Sylvester-type matrix equations, based on recursive blocked algorithms, which are rich in GEMM-operations [4, 15, 16]. The recursive blocking is automatic and has the potential of matching the memory hierarchies of today's high-performance computing systems. RECSY comprises a set of Fortran 90 routines, all equipped with Fortran 77 interfaces and LAPACK/SLICOT wrappers, which solve 42 transpose and sign variants of eight common Sylvester-type matrix equations. Table 1 lists the standard variants of these matrix equations.

Table 1. The Sylvester-type matrix equations considered in the RECSY library. CT and DT denote the continuous-time and discrete-time variants, respectively.

Name	Matrix Equation
Standard Sylvester (CT)	$AX - XB = C$
Standard Lyapunov (CT)	$AX + XA^T = C$
Standard Sylvester (DT)	$AXB^T - X = C$
Standard Lyapunov (DT)	$AXA^T - X = C$
Generalized Coupled Sylvester	$(AX - YB, DX - YE) = (C, F)$
Generalized Sylvester	$AXB^T - CXD^T = E$
Generalized Lyapunov (CT)	$AXE^T + EXA^T = C$
Generalized Lyapunov (DT)	$AXA^T - EXE^T = C$

```
for k=1, number of block diagonals in C
    if(m = n) then
            if(P_c ≠ 1) Shift A East
            if(P_r ≠ 1) Shift B North
    elseif(m < n) then
            Shift A South-East
            if(P_r ≠ 1) Shift C South
    else
            Shift B North-West
            if(P_c ≠ 1) Shift C West
    endif
    {Solve subsystems on current block diagonal in parallel}
    if(mynode holds C_ij)
            Solve for X_ij in A_ii X_ij − X_ij B_ij = C_ij
            Broadcast X_ij to processors that need X_ij for updates
    elseif(mynode needs X_ij)
            Receive X_ij
            Update block column j of C in parallel
            Update block row i of C in parallel
    endif
end
```

Fig. 3. Parallel "matrix-shifting" block algorithm for $AX - XB = C$, A and B in real Schur form.

Depending on the sizes of m and n, three alternatives for doing a *recursive splitting* are considered [11, 13]. In Case 1 ($1 \leq n \leq m/2$), A is split by rows and columns, and C by rows only. Similarly, in Case 2 ($1 \leq m \leq n/2$), B is split by rows and columns, and C by columns only. Finally, in Case 3 ($n/2 < m < 2n$) both rows and columns of the matrices A, B and C are split:

$$\begin{bmatrix} A_{11} & A_{12} \\ & A_{22} \end{bmatrix} \begin{bmatrix} X_{11} & X_{12} \\ X_{21} & X_{22} \end{bmatrix} - \begin{bmatrix} X_{11} & X_{12} \\ X_{21} & X_{22} \end{bmatrix} \begin{bmatrix} B_{11} & B_{12} \\ & B_{22} \end{bmatrix} = \begin{bmatrix} C_{11} & C_{12} \\ C_{21} & C_{22} \end{bmatrix}.$$

This recursive splitting results in the following four triangular SYCT equations:

$$A_{11}X_{11} - X_{11}B_{11} = C_{11} - A_{12}X_{21},$$
$$A_{11}X_{12} - X_{12}B_{22} = C_{12} - A_{12}X_{22} + X_{11}B_{12},$$
$$A_{22}X_{21} - X_{21}B_{11} = C_{21},$$
$$A_{22}X_{22} - X_{22}B_{22} = C_{22} + X_{21}B_{12}.$$

First, X_{21} is solved for in the third equation. After updating C_{11} and C_{22} with respect to X_{21}, one can solve for X_{11} and X_{22}. Both updates and the triangular Sylvester solves are independent operations and can be executed concurrently. Finally, one updates C_{12} with respect to X_{11} and X_{22}, and solves for X_{12}. In practice, all four subsystems are solved using the recursive blocked algorithm. If a splitting point ($m/2$ or $n/2$) appears at a 2×2 diagonal block of A or B, the matrices are split just below this diagonal block.

The recursive approach is natural to SMP-parallelize, which is implemented in RECSY using OpenMP. The performance gain compared to standard algorithms is remarkable, including 10-fold speedups, partly due to new superscalar kernels. The software and documentation concerning RECSY is available for download [14]. For details we also refer to the papers by Jonsson and Kågström [11, 12].

4 Computational Experiments

In this section, we compare measured performance results for the parallel algorithms in Figures 2 and 3 solving SYCT using two different node solvers DTRSYL (from LAPACK) and RECSYCT (from RECSY) in PGESYCTD. The test results are for different values of $m = n$ and different process configurations $P_r \times P_c$ on the HPC2N Linux Super Cluster seth. The cluster consists of 120 dual Athlon MP2000+ nodes (1.667 GHz), where each node has 1–4 GB memory. The cluster is connected through the Wolfkit3 SCI high-speed interconnect with a bandwidth of 667 Mbytes/second. The network connects the nodes in a 3-dimensional torus organized as a $4 \times 5 \times 6$ grid, where each link is "one-way" directed. The theoretical peak system performance of seth is 800 Gflops/sec. The fraction t_a/t_w, where t_a and t_w denote the time for one flop and the per-word transfer time, respectively, is approximately 0.025. Compared to other more well-balanced systems, e.g., the HPC2N IBM SP system which has $t_a/t_w = 0.11$, communication is almost a factor 10 more expensive on seth.

The results are displayed in Tables 2 and 3. The variables q_s and q_d are the ratios between the execution times of PGESYCTD using the two different communication schemes. These ratios are presented for both node solvers (LAPACK, RECSY). If a ratio is larger than 1.0, the RECSY variant is the fastest, and represents the speedup gain compared to the LAPACK variant.

5 Discussion and Conclusions

The results in Table 2 show that the RECSYCT solver decreases the execution time up to 24% for moderate-sized block sizes $mb = nb = 128$ when "on-demand" communication is used, while the gain is only up to 8% for the "matrix-shifting" scheme, and even negative for a few cases.

From the results in Table 3, we conclude that the execution times for PGESYCTD using RECSYCT decrease for larger block sizes ($mb = nb = 512$), while the execution times for PGESYCTD using DTRSYL increase drastically compared to the results in Table 2.

In Table 4, we display the ratios of the shortest execution times of PGESYCTD using DTRSYL and RECSYCT, respectively, and one of the two communication schemes for a given processor grid and problem size. Overall the RECSYCT solver decreases the execution times between 15% and 43% compared to DTRSYL. The best results for RECSYCT are obtained when "on-demand" communication is used, while the best results for DTRSYL are obtained for the "matrix-shifting" scheme.

In conclusion, PGESYCTD with the RECSYCT solver has a large impact on the performance when mb and nb are several hundreds, mainly because the perfor-

Table 2. Timing results (in seconds) of PGESYCTD using different kernel solvers DTR-SYL (LAPACK) and RECSYCT (RECSY) and different communication schemes "matrix-shifting" (S) and "on-demand"(D). Here we use moderate-sized blocking factors $mb = nb = 128$.

$m=n$	$P_r \times P_c$	LAPACK S	D	RECSY S	D	Ratios q_s	q_d	$m=n$	$P_r \times P_c$	LAPACK S	D	RECSY S	D	Ratios q_s	q_d
2048	1×1	18.0	18.1	16.0	15.9	1.12	1.12	6144	2×2	573	215	528	200	1.08	1.08
2048	2×1	25.3	15.1	26.5	13.6	0.95	1.11	6144	4×2	277	156	276	148	1.00	1.05
2048	2×2	20.9	9.8	20.2	8.4	1.04	1.16	6144	4×4	160	112	160	103	1.00	1.09
2048	4×2	11.8	8.2	11.5	6.8	1.03	1.21	6144	8×4	74.2	73.0	73.4	62.3	1.01	1.17
2048	4×4	7.6	6.8	7.5	5.5	1.01	1.24	6144	8×8	68.9	65.2	68.4	59.5	1.01	1.09
2048	8×4	4.6	5.4	5.0	4.1	0.91	1.32	8192	4×2	662	359	651	347	1.02	1.03
2048	8×8	4.4	4.6	4.0	3.8	1.10	1.21	8192	4×4	369	247	367	231	1.01	1.07
4096	1×1	134	134	125	126	1.07	1.07	8192	8×4	172	152	169	133	1.02	1.14
4096	2×1	198	111	196	106	1.01	1.05	8192	8×8	153	136	152	127	1.00	1.08
4096	2×2	159	66.1	156	62.7	1.02	1.05	10240	4×4	742	462	714	442	1.04	1.04
4096	4×2	84.9	50.1	84.8	45.9	1.00	1.09	10240	8×4	362	272	336	245	1.08	1.11
4096	4×4	50.1	38.1	49.1	33.8	1.02	1.13	10240	8×8	302	247	301	234	1.00	1.06
4096	8×4	23.8	26.7	22.3	21.8	1.07	1.23	12288	8×4	559	441	556	406	1.01	1.08
4096	8×8	23.3	23.6	22.8	20.8	1.02	1.14	12288	8×8	490	405	488	385	1.00	1.05

Table 3. Timing results (in seconds) of PGESYCTD using different kernel solvers DTR-SYL (LAPACK) and RECSYCT (RECSY) and different communication schemes "matrix-shifting" (S) and "on-demand" (D). Here we use large blocking factors $mb = nb = 512$. The sign '–' means that the restriction on the data distribution imposed by the "matrix-shifting" scheme was not fulfilled (see Section 2).

$m=n$	$P_r \times P_c$	LAPACK S	D	RECSY S	D	Ratios q_s	q_d	$m=n$	$P_r \times P_c$	LAPACK S	D	RECSY S	D	Ratios q_s	q_d
2048	1×1	57.5	54.7	13.0	10.8	4.43	5.06	6144	2×2	425	410	187	123	2.28	3.34
2048	2×1	63.6	53.5	14.9	9.7	4.27	5.52	6144	4×2	329	381	109	93.7	3.02	4.07
2048	2×2	38.3	40.0	10.9	7.2	3.51	5.55	6144	4×4	198	335	75.3	72.5	2.63	4.63
2048	4×2	35.6	38.8	7.9	6.2	4.51	6.25	6144	8×4	–	297	–	59.2	–	5.02
2048	4×4	28.1	32.7	6.4	5.6	4.35	5.83	6144	8×8	–	245	–	50.9	–	4.81
2048	8×4	–	–	–	–	–	–	8192	4×2	580	707	247	202	2.35	3.51
2048	8×8	–	–	–	–	–	–	8192	4×4	350	614	158	152	2.21	4.04
4096	1×1	258	255	80.9	80.1	3.19	3.19	8192	8×4	288	521	107	113	2.68	4.60
4096	2×1	267	234	87.1	63.3	3.05	3.69	8192	8×8	183	413	90.7	91.7	2.02	4.50
4096	2×2	167	170	57.7	40.5	2.89	4.20	10240	4×4	542	989	296	275	1.83	3.60
4096	4×2	138	162	36.6	32.7	3.78	4.97	10240	8×4	–	848	–	200	–	4.24
4096	4×4	89.2	143	31.1	26.5	2.86	5.40	10240	8×8	–	688	–	170	–	4.06
4096	8×4	80.3	122	20.2	22.1	3.98	5.53	12288	8×4	657	1220	311	314	2.11	3.89
4096	8×8	55.7	95.4	17.1	17.1	3.26	5.57	12288	8×8	406	971	257	256	1.58	3.80

Table 4. Ratios q_{best} and gain $g = 1 - q_{\text{best}}^{-1}$ in percent between the best timing results from Tables 2 and 3 for PGESYCTD using different kernel solvers DTRSYL (LAPACK) and RECSYCT (RECSY) and different communication schemes "matrix-shifting" (S) and "on-demand" (D).

$P_r \times P_c$	$m=n$	q_{best}	$g(\%)$	$m=n$	q_{best}	$g(\%)$	$P_r \times P_c$	$m=n$	q_{best}	$g(\%)$	$P_r \times P_c$	$m=n$	q_{best}	$g(\%)$
1×1	2048	1.67	40	4096	1.67	40	2×2	6144	1.75	43	8×4	8192	1.42	30
2×1	2048	1.56	36	4096	1.75	43	4×2	6144	1.66	40	8×8	8192	1.50	33
2×2	2048	1.36	26	4096	1.63	39	4×4	6144	1.54	35	4×4	10240	1.68	40
4×2	2048	1.32	24	4096	1.53	35	8×4	6144	1.23	19	8×4	10240	1.36	26
4×4	2048	1.24	19	4096	1.44	31	8×8	6144	1.28	22	8×8	10240	1.45	31
8×4	2048	–	–	4096	1.18	15	4×2	8192	1.22	18	8×4	12288	1.42	30
8×8	2048	–	–	4096	1.35	26	4×4	8192	1.63	39	8×8	12288	1.58	37

mance gain provided by RECSY in solving the SYCT subsystems on the nodes makes the waiting time for the broadcasts much smaller. Typically, PGESYCTD with the DTRSYL solver is optimal for smaller block sizes. We also expect PGESYCTD with RECSYCT to give less speedup compared to using DTRSYL, since a much faster node solver makes overlapping of communication and computation harder. On the other hand, by the use of larger block sizes, i.e., larger SYCT subsystems are solved on the nodes, we also get less but larger messages to communicate, which may well compensate for the worse communication-computation overlap.

Future work includes extending the comparisons to other parallel platforms, e.g., the HPC2N IBM SP system which has much less compute power but provides a better "compute/communicate ratio". Our objective is to develop a software package *SCASY* of ScaLAPACK-style algorithms for solving all transpose and sign variants of the matrix equations listed in Table 1. The implementations will build on standard node solvers from LAPACK and SLICOT [18, 20, 5], and recursive blocked solvers from RECSY. By using the LAPACK/SLICOT wrappers provided in the RECSY library, the ScaLAPACK-style hybrid algorithms come for free.

Acknowledgements

This research was conducted using the resources of the High Performance Computing Center North (HPC2N).

Financial support has been provided by the *Swedish Research Council* under grant VR 621-2001-3284 and by the *Swedish Foundation for Strategic Research* under grant A3 02:128.

References

1. R.H. Bartels and G.W. Stewart Algorithm 432: Solution of the Equation $AX + XB = C$, *Comm. ACM*, 15(9):820–826, 1972.
2. E. Anderson, Z. Bai, C. Bischof. J. Demmel, J. Dongarra, J. DuCroz, A. Greenbaum, S. Hammarling, A. McKenny, S. Ostrouchov and D. Sorensen. *LAPACK User's Guide*. Third Edition. SIAM Publications, 1999.
3. S. Blackford, J. Choi, A. Clearly, E. D'Azevedo, J. Demmel, I. Dhillon, J. Dongarra, S. Hammarling, G. Henry, A. Petitet, K. Stanley, D. Walker, and R.C. Whaley. *ScaLAPACK Users' Guide*. SIAM Publications, Philadelphia, 1997.
4. J. J. Dongarra, J. Du Croz, I. S. Duff, and S. Hammarling, A set of Level 3 Basic Linear Algebra Subprograms, *ACM Trans. Math. Soft.*, 16(1):1–17, 1990.
5. E. Elmroth, P. Johansson, B. Kågström, and D. Kressner, A Web Computing Environment for the SLICOT Library, In P. Van Dooren and S. Van Huffel, *The Third NICONET Workshop on Numerical Control Software*, pp 53–61, 2001.
6. R. Granat, A Parallel ScaLAPACK-style Sylvester Solver, *Master Thesis*, UMNAD 435/03, Dept. Computing Science, Umeå University, Sweden, January, 2003.
7. R. Granat, B. Kågström, P. Poromaa. Parallel ScaLAPACK-style Algorithms for Solving Continous-Time Sylvester Matrix Equations, In H. Kosch et.al. (editors), *Euro-Par 2003 Parallel Processing*, Lecture Notes in Computer Science, Springer-Verlag, Vol. 2790, pages 800–809, 2003.

8. G. Henry and R. Van de Geijn. Parallelizing the QR Algorithm for the Unsymmetric Algebraic Eigenvalue Problem: Myths and Reality. *SIAM J. Sci. Comput.* 17:870–883, 1997.
9. G. Henry, D. Watkins, and J. Dongarra. A Parallel Implementation of the Nonsymmetric QR Algorithm for Distributed Memory Architectures. Technical Report CS-97-352 and Lapack Working Note 121, University of Tennessee, 1997.
10. N.J. Higham. Perturbation Theory and Backward Error for $AX - XB = C$, *BIT*, 33:124–136, 1993.
11. I. Jonsson and B. Kågström. Recursive Blocked Algorithms for Solving Triangular Matrix Equations – Part I: One-Sided and Coupled Sylvester-Type Equations, *ACM Trans. Math. Software*, Vol. 28, No. 4, pp 393–415, 2002.
12. I. Jonsson and B. Kågström. Recursive Blocked Algorithms for Solving Triangular Matrix Equations – Part II: Two-Sided and Generalized Sylvester and Lyapunov Equations, *ACM Trans. Math. Software*, Vol. 28, No. 4, pp 416–435, 2002.
13. I. Jonsson and B. Kågström. RECSY - A High Performance Library for Solving Sylvester-Type Matrix Equations, In H. Kosch et.al. (editors), *Euro-Par 2003 Parallel Processing*, Lecture Notes in Computer Science, Springer-Verlag, Vol. 2790, pages 810–819, 2003.
14. I. Jonsson and B. Kågström. RECSY – A High Performance Library for Sylvester-Type Matrix Equations. www.cs.umu.se/research/parallel/recsy, 2003.
15. B. Kågström, P. Ling, and C. Van Loan. GEMM-based level 3 BLAS: High-performance model implementations and performance evaluation benchmark. *ACM Trans. Math. Software*, 24(3):268–302, 1998.
16. B. Kågström, P. Ling, and C. Van Loan. GEMM-based level 3 BLAS: Portability and optimization issues. *ACM Trans. Math. Software*, 24(3):303–316, 1998.
17. B. Kågström and P. Poromaa. Distributed and shared memory block algorithms for the triangular Sylvester equation with Sep^{-1} estimators, *SIAM J. Matrix Anal. Appl.*, 13 (1992), pp. 99–101.
18. NICONET Task II: Model Reduction, website: www.win.tue.nl/niconet/NIC2/NICtask2.html
19. P. Poromaa. Parallel Algorithms for Triangular Sylvester Equations: Design, Scheduling and Scalability Issues. In Kågström et al. (eds), *Applied Parallel Computing. Large Scale Scientific and Industrial Problems*, Lecture Notes in Computer Science, Vol. 1541, pp 438–446, Springer-Verlag, 1998.
20. SLICOT library in the Numerics in Control Network (NICONET), website: www.win.tue.nl/niconet/index.html

Topic 12
High Performance Multimedia

Odej Kao, Harald Kosch, Subramania Sudharsanan, and Andreas Uhl

Topic Chairs

In the recent years multimedia technology has emerged as a key technology, allowing different media – from text, image to video and sound – to be stored, processed, indexed, retrieved, and delivered in digital form. A significant part of the current research community effort has emphasized the delivery of the data as an important issue of multimedia technology. However, the creation, processing, and management of multimedia forms are the issues most likely to dominate the scientific interest in the long run.

The aim to deal with information coming from multimedia sources already results in a data explosion. This requirement to store, process, and manage large data sets naturally leads to the consideration of programmable parallel processing systems as strong candidates in supporting and enabling multimedia technology. Therefore, this fact taken together with the inherent data parallelism in these data types makes multimedia computing a natural application area for parallel processing. Furthermore, the indexing and retrieval of multimedia data includes time-consuming algorithms, thus high-performance architectures and algorithms are necessary in order to allow the application of multimedia databases and archives in real-world scenarios. Traditional parallel processing concepts have to be adapted to the large volume of multimedia data, so a number of interesting topics regarding scheduling, workload balancing, Quality-of-Service (QoS) and other resource management issues arise. In addition, parallel and distributed algorithms/architectures offer a starting point for development of grid-enabled multimedia applications and for mobile multimedia. Finally, the adaptation of parallel and distributed algorithms in multimedia systems and the performance evaluation of implemented systems are interesting topics to be studied.

The broad research spectrum of high-performance multimedia was expressed by 13 submitted papers. The dominating topics this year were related to different aspects of Video-on-Demand (VoD) systems as well as QoS issues for multimedia delivery. All papers were reviewed by four persons. Based on these reviews 4 regular and 2 short papers were selected. The regular papers deal with memory organization supporting visual data, node placement for distributed multimedia servers, and queuing algorithms for guaranteed disk bandwidth. One regular paper reports on the Horus project for parallel image and video processing. Finally, a very interesting regular paper about parallel block-matching motion compensation – submitted and reviewed in topic 7 – completed the selection. The short papers present caching and mirroring strategies for VoD systems and extensive practical evaluation of stream reuse techniques in peer-to-peer VoD.

We would like to thank all authors for their submissions, EuroPar organizing committee and all reviewers for the great collaboration.

Towards User Transparent Data and Task Parallel Image and Video Processing: An Overview of the Parallel-Horus Project

F.J. Seinstra, D. Koelma, and A.D. Bagdanov

Intelligent Sensory Information Systems,
Faculty of Science, University of Amsterdam,
Kruislaan 403, 1098 SJ Amsterdam, The Netherlands
`{fjseins,koelma,andrew}@science.uva.nl`

Abstract. In the research area of image and video processing, the *Horus library* [5] has become a valuable system for software driven rapid prototyping, and an essential vehicle for knowledge transfer at the level of universities and expertise centers. Due to its strictly sequential implementation, however, Horus can not always satisfy all demands for high performance. As computational requirements for the processing of large image sets and video streams continue to increase, it is essential to provide a Horus implementation that applies to high performance computers.
As researchers in image and video processing can not be expected to also become experts in parallel computing, the *Parallel-Horus* project aims to shield Horus users from *all* intrinsic complexities of parallelization. This paper presents an overview of the Parallel-Horus project; it discusses the project's goals, as well as current and future research directions. Also, the efficiency of the current Parallel-Horus implementation is demonstrated by evaluating a strictly sequential state-of-the-art imaging application.

1 Introduction

The complexity and scale of scientific problems in many different research areas has been a driving force for the development of ever faster computer architectures. To effectively exploit the available processing power of such systems, it is essential to have a thorough understanding of their architectural complexity. Despite this requirement, large numbers of non-expert users are still tempted by the potential processing speeds – generally to emerge with nothing but a disappointing result. In [3] this problem is stated somewhat more dramatically as follows:

> Anecdotal reports abound about researchers with scientific and engineering problems who have tried to make use of parallel processing systems, and who have been almost fatally frustrated in the attempt.

Clearly, there is a major discrepancy between the desire to easily obtain increased performance, and the potential of high performance systems to satisfy this desire.

The specific research area of image and video processing – which is the field of focus of this paper – also demonstrates a persistent desire to access the speed potential of high performance computers. The desire partially stems from the fact that it has been recognized for years that the application of parallelism in imaging can be highly beneficial [21]. Yet, in image processing research the discrepancy between desire and reality is no less severe.

We ascribe this problem primarily to the fact that no programming tool is available that can effectively help non-expert parallel programmers in the development of efficient high performance imaging applications. Existing programming tools generally require the user to identify the available parallelism at a level of detail that is beyond the skills of non-expert parallel programmers [11, 14, 16]. Thus, the key to fast algorithm design and evaluation on large data sets, as well as to rapid high performance image and video application development, lies in the availability of a familiar programming tool that shields the image processing researcher from all intrinsic difficulties of parallel implementation.

Our work (referred to as the *Parallel-Horus* project, see also [13–16]) is an attempt to effectively bridge the gap between the expertise of the image processing community, and the additional expertise required for efficient employment of high performance computers. More specifically, the project aims at designing and implementing a software architecture that allows non-expert parallel programmers to develop image and video processing applications for efficient execution on Beowulf-type commodity clusters [19]. Hence, the Parallel-Horus project addresses the following fundamental research issue: *how to design a sustainable, yet efficient software architecture for data and task parallel image and video processing, that provides the user with a fully sequential programming model, and hides all parallelization and optimization issues from the user completely.*

This paper presents an overview of the Parallel-Horus project. Apart from presenting the project's aims, it discusses directions for current and future research. In addition, the efficiency of the current Parallel-Horus implementation is demonstrated by evaluating a strictly sequential state-of-the-art imaging application.

2 Parallel-Horus: Design for Sustainability and Efficiency

In the literature, a trend in the design of parallelization aids for high performance image and video processing is to provide a *software library* consisting of a set of pre-parallelized routines, and hide as much as possible the complexities of parallelization behind the library's API. Due to the relative ease of implementation, a multitude of such libraries exists [3, 4, 6–8, 18, 20]. An important design goal in much of this research is to provide operations that have optimal efficiency on a range of parallel machines. In general, this is achieved by hard-coding a number of different parallel implementations for each operation, one for each platform. Unfortunately, the creation of a parallel library in this manner has several major drawbacks. First, manually creating multiple parallel versions of the many operations commonly applied in image processing research is a laborious task. Second, obeying to requests for library extensions becomes even more troublesome than in the sequential case. Third, as new target platforms are made available at regular intervals, code maintenance becomes hard – if not impossible – on the long term. Finally, with each library expansion it becomes ever more difficult to incorporate a single elegant performance optimization strategy for full applications. For these reasons we take a different approach.

In the development of the Parallel-Horus architecture we strive to enhance sustainability by minimizing the implementation effort, without compromising on the efficiency of execution. For one, this is achieved by recognizing that there is a limited number of ways in which the pixels in images and video streams can be processed to produce meaningful results. Important in this respect is the classification of imaging operations made in Image Algebra [12]. Originating from this classification, a sequential image and video processing library is currently being implemented at the University of Amsterdam: *Horus* [5]. The Horus library is highly innovative in its design and implementation, most significantly because the library's core comprises of a very small set of so-called *algorithmic patterns*. The primary importance of these patterns is that each serves as a template operation for a large set of image operations with similar behavior, which greatly reduces the implementation effort. Also, the algorithmic patterns abstract from the actual datatype each operation is applied upon, to avoid a combinatorial explosion of code that deals with all possible kinds of image datatypes.

The next important step is to recognize that, for parallel implementation of each algorithmic pattern, much of the related sequential code can be reused. To that end, for each sequential algorithmic Horus pattern we have defined a so-called *parallelizable pattern* [16]. Such pattern constitutes the maximum amount of code of an algorithmic pattern that can be performed both sequentially and in parallel – in the latter case without having to communicate to obtain non-local data. Also, further reduction of the software engineering effort is achieved by implementing the parallelizable patterns such that they are capable of adapting to the performance characteristics of the parallel machine at hand [13,16].

The final step in reaching our goal is to recognize that for high performance of full applications it is not sufficient to consider parallelization of library operations *in isolation*. As shown in [15], it is possible to extend any library implemented on the basis of parallelizable patterns with a very efficient finite state machine based run-time system for communication minimization of applications running in data parallel fashion. Moreover, the run-time system that was integrated in the Horus library according to this approach proved to deliver close-to-optimal parallel performance for many strictly sequential imaging applications [14].

3 Parallel-Horus: Large-Scale Image and Video Processing

In the early days of image processing research, algorithms and applications had modest computational requirements. In general not more than a few small-sized low-dimensional images needed to be processed. In contrast, today's research in image and video processing is often confronted with large collections of large-sized and high-dimensional images, and hours of high resolution video. In addition, thorough evaluation (thus: calculation) of large algorithmic parameter spaces on image data sets is now commonplace.

Partially, this shift in computational requirements is driven by advances in techniques for high resolution image acquisition, e.g. in biomedical imaging [23] and light microscopy [10]. Also, applications requiring interactive access to large image collections are becoming ever more abundant [17, 22]. Finally, in the presentation of results obtained from imaging experiments there is a trend to incorporate evaluations on larger data sets. This is due to the fact that many imaging algorithms perform well on a limited set of (artificial) images, but are much less applicable to other (often real-world) images. Thus, to investigate the limitations and potential of algorithms, large-scale evaluations are essential. To this end, *the main focus of the Parallel-Horus project is on high performance processing of large image data sets, video streams, and large algorithmic parameter spaces.*

On the one hand, support for this type of processing is delivered in the form of pre-parallelized library operations, which are integrated in the existing sequential Horus library (component ① in Figure 1). Also, the run-time system for communication minimization (as referred to above) is integrated in the Horus library and to be expanded to cover all functionality for the processing of large parameter spaces, data sets, and video streams (component ② in Figure 1).

On the other hand, the usage of a single library operation to initiate the processing on a set of images, or a complete parameter space, may not fit the image processing researcher's programming and specification preferences. Instead, it is expected that many developers of imaging applications will keep to the more familiar specification of a chain of events by way of loop constructs provided by the host programming language (i.e. C or C++) [11]. Such constructs often incorporate inherent potential for (independent) parallel processing of each loop iteration. As this type of parallelism can not be included in the image library

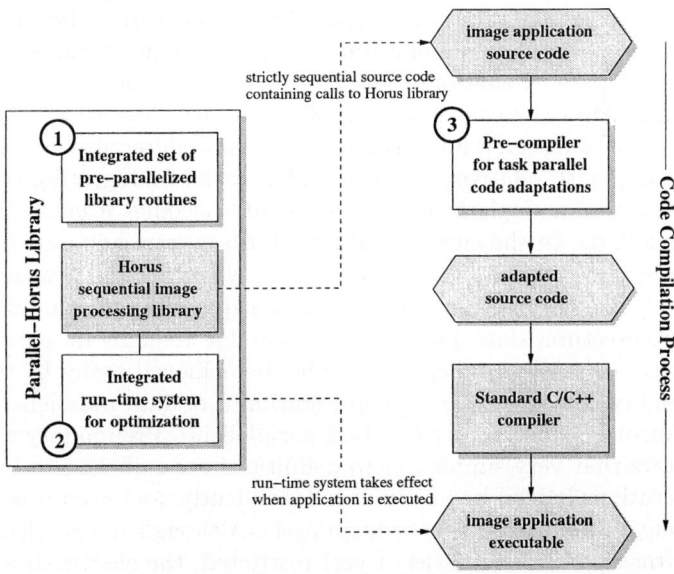

Fig. 1. Simplified overview of the Parallel-Horus architecture.

itself, initial solutions are to be incorporated in the form of macros or compiler directives for parallelization, similar to those described by Oliveira et. al. [9]. However, as the imaging researcher should not personally have to include such parallelization hints in a sequential application, at a later stage in the project code adaptations are to be generated automatically by means of a pre-compiler (see component ③ in Figure 1). The pre-compiler is to analyze the strictly sequential image processing code, and insert task parallelization macros wherever needed, to alleviate the programmer from personally having to perform this task.

4 Parallel-Horus: Towards Mixed Data/Task Parallelism

Parallel-Horus extends our earlier research, which was aimed at the design of a software architecture for user transparent data parallel processing of single (or just a few) images on Beowulf clusters. *A major aim of the Parallel-Horus project is to integrate task parallelism within the same software architecture to facilitate high performance processing of a much larger class of applications.*

Our earlier restriction to data parallelism was sensible for many reasons. First, the application of data parallelism is a natural approach for image processing, as many operations require the same function to be applied to each individual element of an image data structure. Second, in applications that need to process just few images, the number of independent tasks is generally small, especially when compared to the fairly abundant number of elements present in the image structures to be processed. Another important reason is that load balancing is generally much more difficult in the task parallel model. The decisive factor for not having integrated task parallelism in our work so far, however, was the expected difficulty of combining the two programming paradigms in a fully user transparent parallel environment.

The situation has changed, however. Because our previous work has shown that it is possible to design a fully user transparent architecture for data parallel image processing with relative ease [13–16], the task of integrating the two parallel paradigms into a single framework suddenly becomes feasible. The reason for this is as follows. In the current Parallel-Horus system any part of a strictly sequential program runs in data parallel mode, whenever the available number of processors is known. In other words: for any sequential program implemented using our architecture, data parallelism comes for free. In its simplest form, combined data and task parallelism can now be achieved easily by designating a certain number of nodes to each loop iteration (i.e. data parallelism), and let these iterations run concurrently (i.e. task parallelism). Preliminary work in this direction shows that very simple macro definitions are sufficient to let independent loop iterations indeed be executed independently, and even provide partial load balancing without extensive code analysis. Although the resulting mixture of data and task parallelism is (as of yet) restricted, the ease with which these results are obtained is remarkable. Therefore we strongly believe that further investigations in this direction are not only important, but also will be rewarding.

5 An Example Application

To evaluate the current Parallel-Horus implementation, this section describes the (data) parallel execution of an example imaging application: i.e. the extraction of rectangular size distributions from document images. Prototypical code of the Parallel-Horus architecture, as wel as for this and many other example applications, is available at http://www.science.uva.nl/~fjseins/ParHorusCode/.

Measurements were performed on the 72-node Distributed ASCI Supercomputer 2 located at the Vrije Universiteit in Amsterdam [2]. All nodes consist of two 1-Ghz Pentium-III CPUs, with 2 GByte of RAM, and are connected by a Myrinet-2000 network; each node ran the RedHat Linux 7.2 operating system. Our software architecture was compiled using gcc 2.96 and linked with MPICH-GM, which uses Myricom's GM as its message passing layer on Myrinet.

Rectangular Size Distributions

As discussed in [1], *rectangular size distributions* are an effective way to characterize visual similarities between document images. A particularly efficient sequential implementation for this problem, using recursive filters, is presented in Listing 1. It should be noted that such recursive filter operations make for a sequential implementation which is notoriously hard to parallelize efficiently.

Due to the Parallel-Horus architecture, the sequential code of Listing 1 directly constitutes a parallel program as well. Measurement results for the gener-

```
calculateRectangularSizeDistribution(IMAGE inIm, INT w, INT h) {
  vertIm = verDist(inIm, 0);
  area = reduceOp(inIm, "sum");
  FOR (y=0; y≤h; y++) DO
    oy = (y/2h)*(inIm.height+1);
    vThreshIm = horDist(binPixOpC(vertIm, oy, "grtrthan"), 0);
    filtered = -1;
    FOR (x=0; x≤w; x++) DO
      IF (filtered != 1.0) THEN
        ox = (x/2w)*(inIm.width+1);
        hThreshIm = binPixOpC(vThreshIm, ox, "lessequal");
        hThreshIm = binPixOpC(verDist(hThreshIm, MAXVAL), oy, "grtrthan");
        hThreshIm = binPixOpC(horDist(hThreshIm, MAXVAL), ox, "lessequal");
        filtered = (area - reduceOp(hThreshIm, "sum")) / area;
      FI
      ... and save 'filtered' for current x,y combination ...
    OD
  OD
}
```

Listing 1: *Pseudo code for fast calculation of rectangular size distributions; maximum size of calculated filters denoted by 'w' and 'h'. Functions 'horDist' and 'verDist' perform horizontal and vertical distance transforms, using recursive filter-pairs [1].*

# CPUs	'39x59' (s)	'79x119' (s)
1	1263.277	4589.818
2	779.486	2821.391
4	497.145	1789.319
8	216.679	781.175
16	109.284	393.763
24	68.652	253.647
32	53.967	197.962
48	37.048	134.180
64	27.926	100.792

(a)   (b)

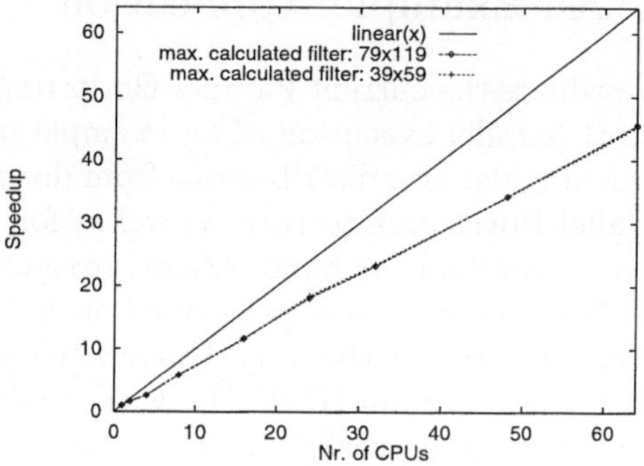

Fig. 2. (a) Performance and (b) speedup for computing rectangular size distributions for document image of size 2325×3075 (2-byte) pixels. Maximum size of calculated filters either 39×59 or 79×119. Speedup lines for either measurement essentially coincide.

ated parallel version of the presented program are given in Figure 2. The results show that, for realistic images of size 2325×3075 pixels, our architecture provides very good speedup characteristics: 45.5 on 64 processors – an efficiency of 71.2%. Therefore, we conclude that the Parallel-Horus architecture provides high parallel efficiency for the presented image processing problem. Remember, for the application programmer these results come without any parallelization effort whatsoever. More results for other imaging applications are found in [14, 16].

6 Conclusions

The Parallel-Horus project aims at the development of an effective programming tool that provides sustainable support in the implementation of data and task parallel image and video processing software by non-experts in high performance computing. With the implementation of the current Parallel-Horus architecture, which supports fully user transparent data parallel execution, we have largely succeeded in that mission. However, strictly data parallel execution does not always deliver sufficiently high performance. Also, due to the need for independent processing of large amounts of data, many state-of-the-art multimedia applications have a much higher degree of inherent task parallelism than before. Therefore, we are currently working on extensions that allow for integration of efficient task parallelism within the same Parallel-Horus framework. Solutions are to be provided in the form of library operations for the processing of large data sets, as well as through a pre-compiler that is to make straightforward task parallel adaptations to strictly sequential image processing source codes. In the end, the availability of a combined data and task parallel environment for user transparent high performance image and video processing will certainly have a stimulating effect on the investigation of the many computationally demanding problems the image processing community is so often confronted with.

References

1. A.D. Bagdanov et al. Multi-scale Document Description using Rectangular Granulometries. In *Document Analysis Systems V*, pages 445–456, August 2002.
2. H.E. Bal et al. The Distributed ASCI Supercomputer Project. *Operating Systems Review*, 34(4):76–96, October 2000.
3. L.H. Jamieson et al. A Software Environment for Parallel Computer Vision. *IEEE Computer*, 25(2):73–75, February 1992.
4. Z. Juhasz et al. A PVM Implementation of a Portable Parallel Image Processing Library. In *Proc. EuroPVM'96*, pages 188–196, Munich, Germany, October 1996.
5. D. Koelma et al. Horus C++ Reference, Version 1.1. Technical report, ISIS, Faculty of Science, University of Amsterdam, The Netherlands, January 2002.
6. C. Lee and M. Hamdi. Parallel Image Processing Applications on a Network of Workstations. *Parallel Computing*, 21(1):137–160, January 1995.
7. P.J. Morrow et al. Efficient Implementation of a Portable Parallel Programming Model for Image Processing. *Concurrency: Pract. & Exp.*, 11:671–685, 1999.
8. C. Nicolescu and P. Jonker. A Data and Task Parallel Image Processing Environment. *Parallel Computing*, 28(7–8):945–965, August 2002.
9. P. Oliveira and H. du Buf. SPMD Image Processing on Beowulf Clusters: Directives and Libraries. In *Proc. IPDPS 2003*, Nice, France, April 2003.
10. P. van Osta et al. The Principles of Scale Space Applied to Structure and Colour in Light Microscopy. *Proc. Royal Microscop. Soc.*, 37(3):161–166, September 2002.
11. C.M. Pancake and D. Bergmark. Do Parallel Languages Respond to the Needs of Scientific Programmers? *IEEE Computer*, 23(12):13–23, December 1990.
12. G.X. Ritter and J.N. Wilson. *Handbook of Computer Vision Algorithms in Image Algebra*. CRC Press, Inc, 1996.
13. F.J. Seinstra and D. Koelma. P-3PC: A Point-to-Point Communication Model for Automatic and Optimal Decomposition of Regular Domain Problems. *IEEE Transactions on Parallel and Distributed Systems*, 13(7):758–768, July 2002.
14. F.J. Seinstra and D. Koelma. User Transparency: A Fully Sequential Programming Model for Efficient Data Parallel Image Processing. *Concurrency and Computation: Practice and Experience*, 16(6):611–644, May 2004.
15. F.J. Seinstra, D. Koelma, and A.D. Bagdanov. Finite State Machine Based Optimization of Data Parallel Regular Domain Problems Applied in Low Level Image Processing. *IEEE Transactions on Parallel and Distributed Systems (in press)*, 2004.
16. F.J. Seinstra, D. Koelma, and J.M. Geusebroek. A Software Architecture for User Transparent Parallel Image Processing. *Parallel Computing*, 28(7–8):967–993, August 2002.
17. C.G.M. Snoek and M. Worring. Multimodal Video Indexing: A Review of the State-of-the-art. *Multimedia Tools and Applications*, 2004. (in press).
18. J.M. Squyres et al. Cluster-Based Parallel Image Processing. Technical report, Lab. for Scientific Computing, Univ. Notre Dame, Indiana, USA (TR 96-9), 1996.
19. T. Sterling et al. BEOWULF: A Parallel Workstation for Scientific Computation. In *Proc. ICPP'95*, pages I:11–14, Oconomowoc, USA, August 1995.
20. R. Taniguchi et al. Software Platform for Parallel Image Processing and Computer Vision. In *Proc. SPIE*, volume 3166, pages 2–10, San Diego, USA, July 1997.
21. S.H. Unger. A Computer Oriented towards Spatial Problems. *Proceedings of the Institute of Radio Engineers*, 46:1744–1750, 1958.
22. J. Vendrig. *Interactive Exploration of Visual Content*. PhD thesis, ISIS, Faculty of Science, University of Amsterdam, The Netherlands, October 2002.
23. A.G. Webb. *Introduction to Biomedical Imaging*. Wiley-IEEE Press, 2002.

Visual Data Rectangular Memory

Georgi Kuzmanov, Georgi Gaydadjiev, and Stamatis Vassiliadis

Computer Engineering Lab, Microelectronics and Computer Engineering Dept.,
EEMCS, TU Delft, Mekelweg 4, 2628 CD Delft, The Netherlands
{G.Kuzmanov,G.N.Gaydadjiev,S.Vassiliadis}@EWI.TUDelft.NL
http://ce.et.tudelft.nl/

Abstract. We focus on the parallel access of randomly aligned rectangular blocks of visual data. As an alternative of traditional linearly addressable memories, we suggest a memory organization based on an array of memory modules. A highly scalable data alignment scheme incorporating module assignment functions and a new generic addressing function are proposed. To enable short critical paths and to save hardware resources, the addressing function implicitly embeds the module assignment functions and it is separable. A corresponding design is evaluated and compared to existing schemes and is found to be cost-effective[1].

1 Introduction

Vector processor designers have been interested in memory systems that are capable of delivering data at the demanding bandwidths of the increasing number of pipelines, see for example [1,6,9,12]. Different approaches have been proposed for optimal alignment of data in multiple memory modules [1, 3, 9–12]. Module assignment and addressing functions have been utilized in various interleaved memory organizations to improve the performance. In graphical display systems, researchers have been investigating efficient accesses of different data patterns: blocks (rectangles), horizontal and vertical lines, forward and backward diagonals [11]. While all these patterns are of interest in general purpose vector machines and graphical display systems, rectangular blocks are the basic data structures in visual data compression (e.g., MPEG standards). Therefore, to utilize the available bandwidth of a particular machine efficiently, new scalable memory organizations, capable of accessing rectangular pixel patterns are needed.

In this paper, we propose an addressing function for rectangularly addressable systems, with the following characteristics: 1.) Highly scalable accesses of rectangular subarrays out of a two-dimensional data storage. 2.) Separable addressing of the memory modules per rows and columns, which potentially saves hardware. We also introduce implicit module assignment functions to further improve the designs. In addition, we propose a memory organization and its interface, which employs conflict free addressing and data routing circuitry with minimal critical path penalties.

---

[1] This research is supported by PROGRESS, the embedded systems research program of the Dutch organization for Scientific Research NWO, the Dutch Ministry of Economic Affairs, and the Technology Foundation STW (project AES.5021). The authors of this material express special acknowledgements to Jens Peter Wittenburg for his valuable opinions and expertise.

The remainder of the paper is organized as follows. Section 2 motivates the presented research and introduces the particular addressing problem. In Section 3, the addressing scheme and the corresponding memory organization are described. Related work is compared to ours in Section 4. Finally, the paper is concluded with Section 5.

2 Motivation

Most of the data processing in MPEG is not performed over separate pixels, but over certain regions (blocks of pixels) in a frame. Many computationally and data intensive algorithms access such blocks from an arbitrary position in a virtual two-dimensional storage where frames are stored. This generates problems with data alignment and access in system memory, see [7,8], described formally in the remainder of the section.

Formal Problem Introduction and Proposed Solution. Consider linearly addressable memories (LAM). Pixel blocks with their *upper-left pixel aligned as a byte at a first (word addressing) position of a LAM word will be referred to as aligned*. All other pixel blocks will be referred to as *non-aligned*. Assume a LAM with word length of w bits ($w = 8, 16, 32, 64, 128$) and the time for linear memory access to be T_{LAM}. The time to access a single $a \times b$ sub-array of 8-bit pixels, depending on its alignment is:

1.) Aligned sub-array: $\frac{8 \cdot a \cdot b}{w} \cdot T_{LAM}$; 2.) Not aligned sub-array: $(\frac{8 \cdot a}{w} + 1) \cdot b \cdot T_{LAM}$.
The time, required to access N $a \times b$ blocks with respect to their alignment will be:
1.) All N blocks aligned: $N \cdot \frac{8 \cdot a \cdot b}{w} \cdot T_{LAM}$;
2.) None of the blocks aligned: $N \cdot (\frac{8 \cdot a}{w} + 1) \cdot b \cdot T_{LAM}$;
3.) Mixed: $N \cdot [\frac{1}{a} \cdot \frac{8 \cdot a}{w} + \frac{a-1}{a}(\frac{8 \cdot a}{w} + 1)] \cdot b \cdot T_{LAM} = N \cdot (\frac{8 \cdot a}{w} + 1 - \frac{1}{a}) \cdot b \cdot T_{LAM}$.

By *mixed* access scenario we mean accessing both aligned and non-aligned blocks. We assume that the probability to access an aligned block is $\frac{1}{a}$, while for a non-aligned block it is $\frac{a-1}{a}$. For simplicity, but without losing generality, assume square blocks of $n \times n$, (i.e., $a=b=n$). We can estimate the total number of LAM cycles to access N square blocks, again with respect to their alignment:

1.) All N blocks aligned: $\boxed{\frac{8 \cdot n^2}{w} \cdot N}$; 2.) None of the blocks aligned: $\boxed{(\frac{8 \cdot n^2}{w} + n) \cdot N}$;

3.) Mixed: $\boxed{(\frac{8 \cdot n^2}{w} + n - 1) \cdot N}$. Obviously, the number of cycles to access an $n \times n$ block in a LAM, regardless of its alignment, is a square function of n, i.e., $O(n^2)$.

An appropriate memory organization may speed-up the data accesses. Consider the memory hierarchy in Figure 1 and time to access an entire $n \times n$ block from the 2-dimensionally accessible memory (2DAM) to be T_{2DA}. In such a case, the time to access N $n \times n$ sub-blocks in the mixed access scenario will be:

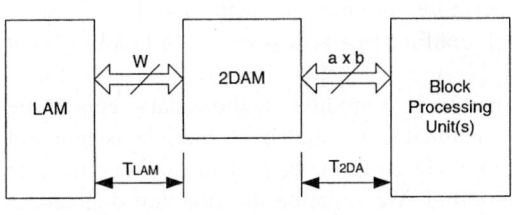

Fig. 1. Memory hierarchy with 2DAM.

$$\frac{N}{n} \cdot \frac{8 \cdot n^2}{w} \cdot T_{LAM} + N \cdot T_{2DA}, [sec] \Leftrightarrow (\frac{8 \cdot n}{w} + \frac{T_{2DA}}{T_{LAM}}) \cdot N, [LAM cycles].$$

That is the sum of the time to access the appropriate number of aligned blocks ($\frac{N}{n}$) from LAM plus the time to access all N blocks from the 2DAM. It is evident that in a mixed access scenario, the number of cycles to access an $n \times n$ block in the hierarchy from Figure 1 is a linear function of n, i.e., $O(n)$ and depends on the implementation of the 2D memory array. Table 1 presents access times per single $n \times n$ block. Time is reported in LAM cycles for some typical values of n and w. There are three cases: 1.) neither of the N blocks is aligned - worst case (WC); 2.) mixed block alignment (Mix.); and 3.) all blocks are aligned - best case (BC). The last two columns contain cycle estimations for the organization from Figure 1. In this case, both mixed and best case scenarios assume that aligned blocks are loaded from the LAM to the 2DAM first and then non-aligned blocks are accessed from the 2DAM. The 2DAM worst case (contrary to LAM) assumes that all blocks to be accessed are aligned. Even in this worst case, the 2DAM-enabled hierarchy may perform better than LAM best case if the same aligned block should be accessed more than once. For example, assume accessing k times the same aligned block. In LAM, this would take $k \cdot \frac{8 \cdot n^2}{w} = [\frac{8 \cdot n^2}{w} + (k-1) \cdot \frac{8 \cdot n^2}{w}]$, while with 2DAM, it would cost $[\frac{8 \cdot n^2}{w} + (k-1) \cdot \frac{T_{2DA}}{T_{LAM}}]$ LAM cycles per block. Obviously, to have a 2DAM enabled memory hierarchy, faster than pure LAM, it would be enough if $\frac{8 \cdot n^2}{w} > \frac{T_{2DA}}{T_{LAM}}$. All estimations above strongly suggest that *a 2DAM with certain organization may dramatically reduce the number of accesses to the (main) LAM, thus considerably speeding-up related applications.*

Table 1. Access time per $n \times n$ block in LAM cycles. $t = \frac{T_{2DA}}{T_{LAM}}$.

n	w	LAM			2DAM	
		WC	Mix.	BC	Mix./BC	WC
8	8	72	71	64	8+t	64+t
	16	40	39	32	4+t	32+t
	32	24	23	16	2+t	16+t
16	8	272	271	256	32+t	256+t
	16	144	143	128	16+t	128+t
	32	80	79	64	8+t	64+t

3 Block Addressable Memory

In this Section, we present the proposed mechanism by describing its addressing scheme, the corresponding memory organization and a potential implementation.

Addressing Scheme. Assume $M \times N$ image data stored in $k = a \times b$ memory modules ($1 \leq a \leq M; 1 \leq b \leq N$). Furthermore, assume that each module is linearly addressable. We are interested in parallel, conflict-free access of $a \times b$ blocks (B) at any (i,j) location, defined as: $B(i,j) = \{(i+p, j+q) | 0 \leq p < a, 0 \leq q < b\}, 0 \leq i \leq M-a, 0 \leq j \leq N-b$. To align data in k modules without data replication, we organize these modules in a two-dimensional $a \times b$ matrix. A module assignment function, which maps a piece of data with 2D coordinates (i,j) in memory module $(p,q) : 0 \leq p < a, 0 \leq q < b$, is required. We separate the function denoted as $m_{p,q}(i,j)$, into two mutually orthogonal assignment functions $m_p(i)$ and $m_q(j)$. We define the following module assignment functions for each module at position (p,q):

$$m_p(i) = (i-p) \bmod a, \quad m_q(j) = (j-q) \bmod b. \tag{1}$$

The addressing function for module (p,q) with respect to coordinates (i,j) is defined as:

$$A_{p,q}(i,j) = (i \ div \ a + c_i) \cdot \frac{N}{b} + j \ div \ b + c_j, \qquad (2)$$

$$c_i = \begin{cases} 1, i \ mod \ a > p \\ 0, otherwise \end{cases} ; \quad c_j = \begin{cases} 1, j \ mod \ b > q \\ 0, otherwise \end{cases} .$$

Obviously, if $p = a - 1 \Rightarrow c_i = 0$ for $\forall i$; if $q = b - 1 \Rightarrow c_j = 0$ for $\forall j$, respectively. In essence, c_i and c_j are the module assignment functions, implicitly embedded into the linear address $A_{p,q}(i,j)$. The proof of all properties of the proposed addressing scheme can be found in [7].

Memory Organization. The key purpose of the proposed addressing scheme is to enable performance-effective memory implementations optimized for algorithms requiring the access of rectangular blocks. Designs with shortest critical paths are to be considered with the highest priority, as they dictate machine performance. Equations (1)-(2) are generally valid for any natural values of parameters a, b and N (i.e., for $\forall \ a, b, N \in \mathbb{N}$). To implement the proposed addressing and module assignment functions, however, we will consider practical values of these parameters. Since pixel blocks processed in MPEG algorithms have dimensions up to 16×16, values of practical significance for parameters a and b are the powers of two up to 16 (i.e., 1, 2, 4, 8, 16). Figure 2 illustrates an example for a block size of $a \times b = 2 \times 4$.

Module Addressing. An important property of the proposed module addressing function is its *separability*. It means that the function can be represented as a sum of two functions of a *single* and *unique* variable each (i.e., variables i and j). The separability of $A_{p,q}(i,j) = Ai_p(i) + Aj_q(j)$ allows the address generators to be implemented per column and per row (see Figure 2) instead of implemented as individual addressing circuits for each of the memory modules.

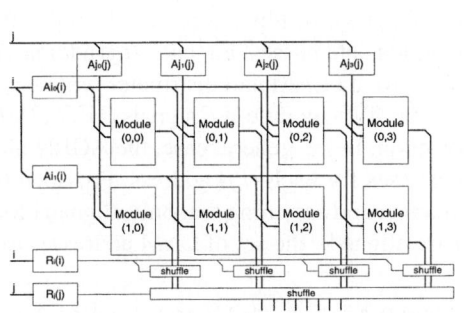

Fig. 2. 2DAM for $a=2$, $b=4$ and $N = 2^n \geq 16$.

(a) Generation Circuit of q-addresses for $1 \leq q < b$

(b) LUTs contents for $a=2$, $b=4$

Fig. 3. Module address generation.

The requirements for the frame sizes of all MPEG standards and for Video Object Planes (VOPs) [2] in MPEG-4 are constituted to be multiples of 16, thus, N is a multiple of 2^4. For the assumed values of N and b, further analysis of Equation (2) suggests that $j\ div\ b + c_j < \frac{N}{b}$ and $(j\ div\ b + c_j)_{max} = \frac{N}{b} - 1$, i.e., no carry can be ever generated between $Ai_p(i)$ and $Aj_q(j)$. Therefore, we can implement $A_{p,q}(i,j)$ for every module (p,q) by simply routing signals to the corresponding address generation blocks without actually summing $Ai_p(i) + Aj_q(j)$. Figure 3(a) illustrates address generation circuitry of q-addresses ($Aj_q(j)$) for all modules except the first ($1 \leq q < b$). With respect to (2), if c_j is 1 the quotient $j\ div\ b$ should be incremented by one, otherwise it should not be changed. To determine the value of c_j, a Look-Up-Table (LUT) with $j\ mod\ b$ inputs can be used. For the assumed practical values of a and b (≤ 16), such a LUT would have at most 4 inputs, i.e., c_j *is a binary function of at most 4 binary digits*. Row p-addresses are generated identically. For *p=1* or *q=3*, $c_i = 0$, $c_j = 0$ respectively. Therefore, address generation in these cases does not require a LUT and an incrementor. Instead, it is just routing $i\ div\ a$ and $j\ div\ b$ to the corresponding memory ports, i.e., blocks $Ai_1(i)$ and $Aj_3(j)$ in Figure 2 are empty. Figure 3(b) depicts all 4 LUTs for the case $a \times b = 2 \times 4$. The usage of LUTs to determine c_i and c_j is not mandatory, fast pure logic can be utilized instead.

Data Routing Circuitry. In Figure 2, the shuffle blocks, together with blocks $R_p(i)$ and $R_q(j)$, illustrate the data routing circuitry. The shuffle blocks are in essence circular barrel shifters, i.e. having the complexity of a network of multiplexors. An $n \times n$ shuffle is actually an $n \rightarrow 1$ n-way multiplexor. In the example from Figure 2, the i-level shuffle blocks are four ($2 \rightarrow 1$) 16-bit multiplexors and the j-level one is ($4 \rightarrow 1$) 64-bit. To control the shuffle blocks, we can use the module assignment functions for $p = q = 0$, i.e., $R_i(i) = i\ mod\ a$ and $R_j(j) = j\ mod\ b$. These functions calculate the (p,q)-coordinates of the "upper-left" pixel of the desired block, i.e., pixel *(i,j)*. For the assumed practical values of a and b being powers of two, the implementation of $R_i(i)$ and $R_j(j)$ is simple routing of the least-significant $log_2(a)$ -bits (resp. $log_2(b)$) to the corresponding shuffle level.

LAM Interface. Figure 4 depicts the organization of the interface between LAM and 2DAM (recall Figure 1) for the modules considered in Figure 2. The data bus width of the LAM is denoted by W (in number of bytes). In this particular example, W is assumed to be 2, therefore modules have coupled data busses. For each *(i,j)* address, the AGEN block sequentially generates addresses to the LAM and distributes write enable (WE) signals to a corresponding module couple. Two module WE signals (WE_i, WE_j) are assumed for easier row and column selection. In the general case, the AGEN block should sequentially generate $\frac{a \cdot b}{W}$ LAM addresses for each *(i,j)* address. Provided that pixel data is stored into LAM in scan-line manner and assuming that only aligned blocks will be accessed from the LAM (i.e., *(i,j)* are aligned), the set of LAM addresses to be generated is defined as follows:

$$A_{LAM}(i,j) = (i+k) \cdot N + j + l \cdot W, \quad k = 0, 1, ..., a-1; l = 0, 1, ..., \tfrac{b}{W} - 1 \ .$$

In the 2DAM, the data words should be simultaneously written in modules:

$(p,q) = (k, l \cdot W), (k, l \cdot W + 1), ..., (k, l \cdot W + W - 1)$ at local module address:
$A_{p,q}^{LAM}(i,j) = (i\ div\ a) \cdot \frac{N}{b} + j\ div\ b \ .$

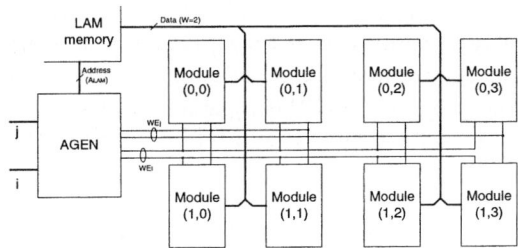

Fig. 4. LAM interface for W=2, a=2, b=4.

Note, that accessing only aligned blocks from the LAM enables thorough bandwidth utilization. When only aligned blocks are addressed, all address generators issue the same address, due to (2). Therefore, during write operations into 2DAM, the same addressing circuitry can be used as for reading. If the modules are true dual port, the write port addressing can be simplified to just proper wiring of both i and j address lines because the incrementor and the LUTs from Figure 3(a) are not required. Therefore, module addressing circuitry is not depicted in Figure 4.

Critical Paths. Regarding the performance of the proposed design, we should consider the created critical path penalty. Assuming generic synchronous memories where addresses are generated in one cycle and data are available in another, we separate the critical paths into two: address generation and data routing. For the proposed circuit implementation, the address generation critical paths are the critical path of either a $log_2(\frac{M}{a})$-bit or a $log_2(\frac{N}{b})$-bit adder, whichever is longer, and the critical path of one (max. 4-input) LUT. The data routing critical path is the sum of the critical paths of one $a \rightarrow 1$ multiplexor and one $b \rightarrow 1$ multiplexor. More details regarding the implementation of the memory organization and a case study design can be found in [7].

4 Related Work and Comparisons

Two major groups of memory organizations for parallel data access have been reported in literature - organizations with and without data replication (redundancy). We are interested only in those without data replication. Another division is made with respect to the number of memory modules - equal to the number of accessed data points and exceeding this number. Organizations with a prime number of memory modules can be considered as a subset of the latter. An essential implementation drawback of such organizations is that their addressing functions are non-separable and complex, thus slower and costly to implement. We have organized our comparison with respect to block accesses, discarding other data patterns, due to the specific requirements of visual data compression. To compare designs, two basic criteria have been established: scalability and implementation drawbacks in terms of speed and/or complexity. Comparison results are reported in Table 2. Budnik and Kuck [1] described a scheme for conflict free access of $\sqrt{N} \times \sqrt{N}$ square blocks out of $N \times N$ arrays, utilizing $m > N = 2^n$ memory modules, where m is a prime number. Their scheme allows the complicated full crossbar switch as the only possibility for data alignment circuitry and many costly $modulo(m)$ operations with m not a power of two. In a publication, related to the development of the Burroughs Scientific Processor, Lawrie [9] proposes an alignment scheme with data switching, simpler than a crossbar switch, but still capable to handle only $\sqrt{N} \times \sqrt{N}$ square blocks out of $m=2N$ modules, where $N = 2^{2n+1}$. Both schemes in [1] and [9] require a larger number of modules than the number of simultaneously accessed (image) points (N). Voorhis and Morin [12] suggest various addressing functions considering $p \times q$ subarray accesses and different number of memory modules m: both $m = p \times q$ and $m > p \times q$. Neither of the functions proposed in [12] is separable, which leads to an extensive number of address generation and module assignment logic blocks. In [3] the authors propose a module assignment scheme based on Latin squares, which is capable of accessing $\sqrt{N} \times \sqrt{N}$ square blocks out of $N \times N$ arrays, but not from random positions. Similar drawbacks has the scheme proposed in [10]. A display system memory, capable of simultaneous access of $p \times q$ rectangular subarrays is described in [11]. The design, proposed there, utilizes a prime number of memory modules, which enables accesses to numerous data patterns, but disallows separable addressing functions. Therefore, regarding block accesses, it is slower and requires more memory modules than our proposal. Large LUTs (in size and number) and a yet longer critical path with consecutive additions can be considered as other drawbacks of [11]. A memory organization, capable of accessing $N \times N$ square blocks, aligned into $(1 + N)^2$ memory modules was described in [5]. The same scheme was used for the implementation of the matrix memory of the first version of HiPAR-DSP [13]. Besides the restriction to square accesses only, that memory system uses a redundant number of modules, due to additional DSP-specific access patterns considered. A definition of a rectangular $p \times q$ block random addressing scheme from the architectural point of view dedicated for multimedia systems was introduced in [8], but no particular organization was presented there. In the latest version of HiPAR16 [4], the matrix memory was improved so that a restricted number of rectangular patterns could also be accessed. This design, however, still utilizes an excessive number of memory modules as p and M respectively q and N should not have common divisors. E.g., to access a 2×4 pattern, the HiPAR16 memory requires $3 \times 5 = 15$ memory modules, instead of only 8 for ours. The memory of [4] requires a complicated circuitry. Both [4] and [13] assume separability, however, the number of utilized modules is even higher than the closest prime number to $p \times q$. Compared to [1, 3–5, 9–11, 13], our scheme enables a higher scalability and a lower number of memory modules. This reflects to the design complexity, which has been proven to be very low in our case. Address function separability reduces the number of address generation logic and critical path penalties, thus enables faster implementations. Regarding address separability, we differentiate from [1, 3, 9–12], where address separability is not supported. As a result, *our memory organization is envisioned to have the shortest critical path penalties among all referenced works.*

Table 2. Comparison to other proposed schemes.

Related Work	scalability	# modules (m)	implementation drawbacks or limitations
Budnik, Kuck [1]	$\sqrt{N} \times \sqrt{N}$ from $N \times N$	prime $m > N = 2^n$	$mod(m)$, crossbar, no addressing
Lawrie [9]	$\sqrt{N} \times \sqrt{N}$	$m = 2.N; N = 2^{2n+1}$	$mod(m)$, no addressing
Voorhis, Morin [12]	$p \times q$ from $M \times N$	$m \geq p \times q$	not separable, $mod(pq), mod(pq+1)$,
Kim, Prasanna [3]	$\sqrt{N} \times \sqrt{N}$ from $N \times N$	$m = N$	certain blocks are inaccessible
De-lei Lee [10]	$\sqrt{N} \times \sqrt{N}$ from $N \times N$	$m = N$	many modules for higher N
Park [11]	$p \times q$ from $M \times N$	prime $m > p \times q$	not separable, many adders, big LUTs
HiPAR-DSP [5,13]	$N \times N$	$m = (1 + N)^2$	$2 \times N + 1$ additional modules, $mod(m)$
HiPAR-DSP16 [4]	$p \times q$ from $M \times N$	$m >> p \times q$	big number of modules, $mod(m)$
This proposal	$p \times q$ from $M \times N$	$m = p \times q$	none of the above, rectangular patterns only

5 Conclusions

We presented a scalable memory organization capable of addressing randomly aligned rectangular data patterns in a 2D data storage. High performance is achieved by a reduced number of data transfers between memory hierarchy levels, efficient bandwidth utilization, and short hardware critical paths. In the proposed design, data are located in an array of byte addressable memory modules by an addressing function, implicitly containing module assignment functions. An interface to a linearly addressable memory has been provided to load the array of modules. Theoretical analysis proving the efficiency of the linear and the two-dimensional addressing schemes was also presented. The design is envisioned to be more cost-effective compared to related works reported in the literature. The proposed organization is intended for specific data intensive algorithms in visual data processing, but can also be adopted by other general purpose applications with high data throughput requirements including vector processing.

References

1. P. Budnik and D. J. Kuck. The organization and use of parallel memories. *IEEE Transactions on Computers*, 20(12):1566–1569, December 1971.
2. ISO/IEC JTC11/SC29/WG11, N3312. MPEG-4 video verification model version 16.0.
3. K. Kim and V. K. Prasanna. Latin squares for parallel array access. *IEEE Transactions on Parallel and Distributed Systems*, 4(4):361–370, 1993.
4. H. Kloos, J. Wittenburg, W. Hinrichs, H. Lieske, L. Friebe, C. Klar, and P. Pirsch. HiPAR-DSP 16, a scalable highly parallel DSP core for system on a chip: video and image processing applications. In *IEEE International Conference on Acoustics, Speech, and Signal Processing*, volume 3, pages 3112–3115, Orlando, Florida, USA, May 2002. IEEE.
5. J. Kneip, K. Ronner, and P. Pirsch. A data path array with shared memory as core of a high performance DSP. In *Proceedings of the International Conference on Application Specific Array Processors*, pages 271–282, San Francisco, CA, USA, August 1994.
6. P. M. Kogge. *The Architecture of Pipelined Computers*. McGraw-Hill, 1981.
7. G. Kuzmanov, G. N. Gaydadjiev, and S. Vassiliadis. Multimedia rectangularly and separably addressable memory. Technical Report CE-TR-2004-01, TU Delft, Delft, January 2004. http://ce.et.tudelft.nl/publications.php.
8. G. Kuzmanov, S. Vassiliadis, and J. van Eijndhoven. A 2D Addressing Mode for Multimedia Applications. In *Workshop on System Architecture Modeling and Simulation (SAMOS 2001)*, volume 2268 of *Lecture Notes in Computer Science*, pages 291–306. Springer-Verlag, 2001.
9. D. H. Lawrie. Access and alignment of data in an array processor. *IEEE Transactions on Computers*, C-24(12):1145–1155, December 1975.
10. D. Lee. Scrambled Storage for Parallel Memory Systems. In *Proc.IEEE International Symposium on Computer Architecture*, pages 232–239, Honolulu, HI, USA, May 1988.
11. J. W. Park. An efficient buffer memory system for subarray access. *IEEE Transactions on Parallel and Distributed Systems*, 12(3):316–335, March 2001.
12. D. C. van Voorhis and T. H. Morrin. Memory systems for image processing. *IEEE Transactions on Computers*, C-27(2):113–125, February 1978.
13. J. P. Wittenburg, M. Ohmacht, J. Kneip, W. Hinrichs, and P. Pirsh. HiPAR-DSP: a parallel VLIW RISC processor for real time image processing applications. In *3rd International Conference on Algorithms and Architectures for Parallel Processing, 1997. ICAPP 97.*, pages 155–162, Melbourne, Vic. , Australia, December 1997.

Dynamic Granularity Switching in Parallel Block-Matching Motion Compensation

Florian Tischler and Andreas Uhl*

Salzburg University, Department of Scientific Computing
Jakob Haringer-Str. 2, A-5020 Salzburg, Austria
uhl@cosy.sbg.ac.at

Abstract. Different granularity levels for parallel block-based motion compensation as used in video compression are discussed. We introduce the concept of switching between granularities to resolve load balancing problems which limit scalability. The performance of the corresponding MPI implementations on a SGI Power Challenge and a Cray T3-E is compared.

1 Introduction

The widespread use of digital video in various environments has caused a high demand for efficient compression techniques. Unfortunately, many compression algorithms have prohibitive execution times using a single serial microprocessor [12], which leads to the use of high performance computing systems for such tasks. Software based approaches are becoming more popular in this area because of the rapid evolution of multimedia techniques which has dramatically shortened the time available to come up with a new hardware design for each improved standard. In this context, several papers have been published describing real-time video coding on general purpose parallel architectures (see [2] for an overview): MPEG encoding [1, 3], H.26X [9], wavelet-based video compression [6, 8], and fractal video compression [11].

Block-matching motion compensation is the most demanding part of current video coding standards and requires 60 – 80 % of the total computations involved. A significant amount of work discusses dedicated hardware for block-matching [4, 10], software based block-matching approaches have also been investigated [13]. However, in most cases, only one approach (in particular one parallelization granularity), is discussed in detail and it is left to the reader to compare the results to other research in this area. In recent work [14], we have systematically compared different levels of parallelization granularity from the scalability point of view and have also considered requirements and restrictions coming from the applications' side. We have as well compared these different granularity levels in the context of a complete wavelet packet based parallel video codec [5].

---

* The authors have been partially supported by the Austrian Science Fund FWF, project no. P13903.

In this work we further improve the so far best performing granularity (in terms of scalabiblity) by using a different granularity level at the end of the computation to resolve intrinsic load balancing problems. In section 2 we shortly review block-matching motion compensation in the video coding context. Section 3 discusses three granularity levels for parallel block-matching which are experimentally evaluated on two architectures using MPI based message passing. In section 4 we introduce the concept of switching granularities and provide experimental results to show the improved performance and also the limitations of this approach.

2 Block-Matching Motion Compensation in Video Coding

The main idea of motion compensated video coding is to use the temporal and spatial correlation between frames in a video sequence for predicting the current frame from previously (de)coded ones. Since this prediction fails in some regions (e.g., due to occlusion), the residual between this prediction and the current frame being processed is computed and additionally stored after lossy compression.

Because of its simplicity and effectiveness block-matching algorithms are widely used to remove temporal correlation [7]. In block-matching motion compensation, the scene (i.e. video frame) is classically divided into non-overlapping "block" regions. For estimating the motion, each block in the current frame is compared against the blocks in the search area in the reference frame (i.e. previously encoded and decoded frame) and the motion vector (d_1, d_2) corresponding to the best match is returned (see Fig. 1).

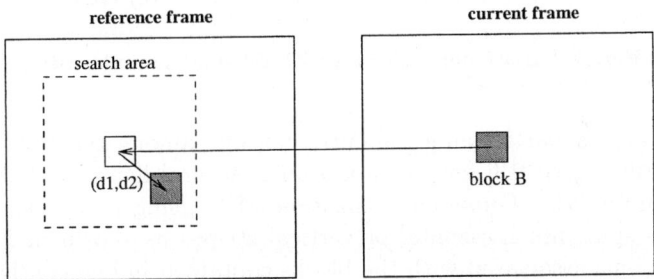

Fig. 1. Block-Matching motion estimation.

The "best" match of the blocks is identified to be that match giving the minimum mean square error (MSE) of all blocks in search area defined as

$$MSE(d_1, d_2) = \frac{1}{N_1 N_2} \sum_{(n_1, n_2) \in \mathcal{B}} [s(n_1, n_2) - \hat{s}(n_1 + d_1, n_2 + d_2)]^2$$

where $\mathcal{B}$ denotes a $N_1 * N_2$ block (the "search area") for a set of candidate motion vectors (d_1, d_2), s is the current frame and $\hat{s}$ the reference frame. The algorithm which visits all blocks in the search area to compute the minimum is called "full search" and delivers the optimal result in terms of visual quality.

3 Granularity of Parallel Block-Matching Algorithms

In general, granularity determines important characteristics of a parallelization approach. A coarse grained parallelization usually requires little communication, on the other hand, balanced load may be hard to achieve. Independent of the results corresponding to parallelization efficiency, granularity has also major impact with respect to hardware requirements and coding delay of a video compression system.

As described in the previous section, the current frame and the reference frame are segmented into equal sized blocks. Based on this partitioning, two intra-frame granularities may be used (fine grained parallelization). In the *Block-based* parallelization (BB) approach, the blocks of the current frame are distributed among the processing elements (PE), the computation of the motion vector for each block is done locally. To avoid interprocess communication, the entire search area surrounding the corresponding block in the reference frame (see Fig. 2.a) is sent to the PE in charge of the current block resulting in an overlapping data partition.

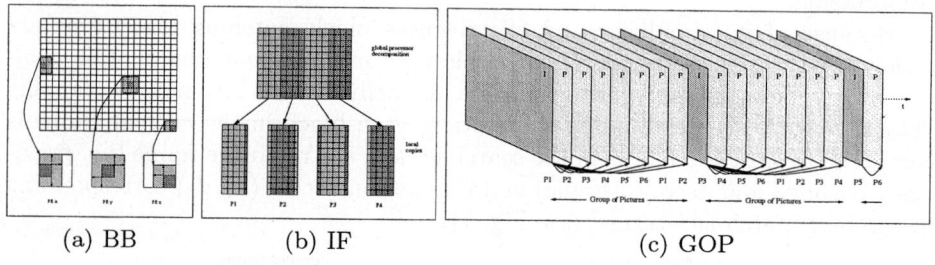

(a) BB   (b) IF   (c) GOP

Fig. 2. Partitioning schemes with different granularities.

Another way of partitioning the current frame among the PE is the stripe subimage method (see [13] for a comparison of several flavours of this method) which we denote *Intra Frame* parallelization (IF). Using this method, the current frame is split into horizontal or vertical stripes as shown in Fig. 2.b and the computations associated with the blocks contained in one contiguous stripe are assigned to one PE. Again, a search-overlap is used to avoid interprocess communication among the PEs.

To preserve the quality reached by the sequential version, the smallest allowed unit for splitting the frame is limited by the blocksize. These units are combined according to the number of PEs to form stripes as uniformly sized as possible. Therefore, this method should perform well especially in the case where the units can be distributed evenly among the compute nodes.

Additionally, two types of inter-frame granularity (coarse grained parallelization) may be considered. A group of pictures (GOP) is a collection of P-frames (predicted frames) which are all processed in relation to one reference frame, the I-frame. An obvious way to distribute the block-matching computations is to assign entire frames to single PEs. Fig. 2.c illustrates this technique which is denoted GOP parallelization. A second possibility is to assign complete GOPs to single PEs (denoted eGOP).

It is obvious that the lower communication demand (in terms of the number of messsages to be exchanged) of inter-frame granularity has to be paid with higher memory consumption and a possibly higher coding delay [14]. Consequently, coarse grained parallelization (GOP and eGOP) will not be suited for real-time and on-line applications like video conferencing, since especially the high coding delay is not acceptable for these types of applications. Additionally, the high memory requirements lead to high costs (especially for hardware solutions) and poor cache performance. Note that the eGOP approach is not investigated further since coding delay and memory requirements are too high for practical systems. Moreover, load imbalance is significant for short videos using this technique.

3.1 Experiments

All subsequent simulations use the standard test sequence "Football" in QCIF (quarter common intermediate format), having a size of 176×144 pixels. The blocksize is set to 8x8, motion vectors may point 7 pixels in each direction, the computations are performed on 40 successive frames of the sequence. Two different architectures using native MPI versions are employed: a SGI Power Challenge (20 MIPS R10000 processors and 2.5 GB memory) and a Cray T3E (DEC Alpha EV5 processors, 128 MB memory each, interconnected by a 3D torus with 3 GB/s bandwidth in each network node). Note that the two systems have opposite properties. Whereas the SGI has the slower compute nodes, the communication system is faster on the SGI (shared memory) as compared to the Cray (networked). Therefore, we may expect better results with respect to scalability on the SGI. SPMD-style programming with non-blocking MPI communication commands is used, we measure wall clock time from program start-up to program termination.

Block-based parallelization shows satisfactory speedup on the SGI only (see Fig. 3.a – but even here it shows the poorest performance of all three granularities). The communication demand is too high to deliver reasonable results, even though the load may be balanced perfectly. Intra Frame parallelization performs better but exhibits several plateaus in the speedup plots. See [14] for detailed explanations of this phenomenon.

The overall performance of GOP parallelization is the best compared to the other two approaches. Also, GOP parallelization leads to plateaus in the speedup plots. The reason for these plateaus is that a fixed number of frames can not be distributed evenly among an arbitrary number of PEs, therefore several PEs may remain idle when the last frames are processed (such a situation is visualized in Fig. 5.a). Note, that these plateaus disappear for long video sequences since unbalanced load in the last scheduling round is not important for a large number of scheduling rounds.

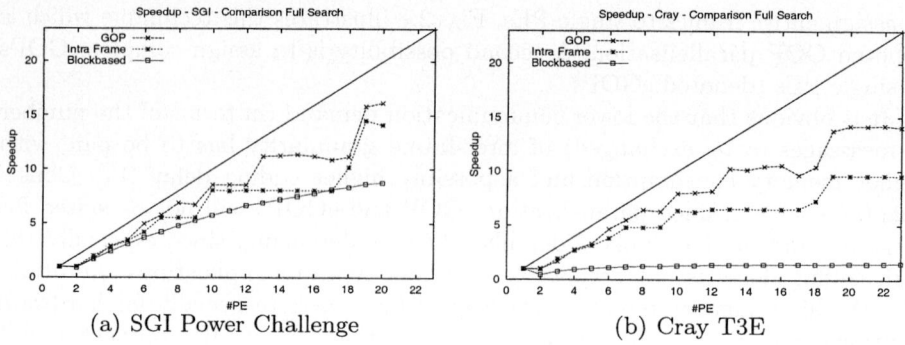

Fig. 3. Speedup Results.

4 Hybrid Granularity in Parallel Block-Matching

Due to its good efficiency, the GOP parallelization approach is an interesting candidate for video compression applications and other applications where large quantities of still image data need to be compressed. As we have noted, due to its high coding delay and memory demand it is not suited for on-line applications. However, for off-line applications where large amounts of data need to be processed in reasonable time, coding delay and memory demand are not critical issues. For example, we mention large surveillance systems and all types of storage applications in general. In an environment where long videos are processed (e.g. for insertion into a video-on-demand server) the load-inbalance phenomenon as discussed in the previous section does not pose any problem. However, in case of many short video sequences to be compressed this behaviour degrades execution efficiency significantly. Coming back to a surveillance application as an example, this would be the case if the video cameras only record if triggered by a sensor measurement, e.g. an infrared sensor. Here, many short videos are delivered to the storage system and need to be compressed as fast as possible. The aim of this section is to introduce a methodology to improve the efficiency of the GOP parallelization approach in such an environment.

Since the load-balancing problems are intrinsic to the GOP approach and both other approaches show limited scalability, we propose to extend the GOP approach in a way where we can keep its good efficiency and incorporate the good load-balancing facilities of the other two schemes. In particular we suggest to start the computations according to the GOP parallelization and switch to a finer granularity as soon as the efficiency of the first scheme degrades. Note that switching granularity implies a data redistribution procedure which is costly in terms of communication and synchronization demand. Of course, the location of the granularity switch needs to be determined in advance (i.e. to avoid a performance degradation) and should not react to an already poor behaviour. The GOP approach is restricted to a number of frames which is an integer multiple of the number of PEs. Frames exceeding this set are processed according

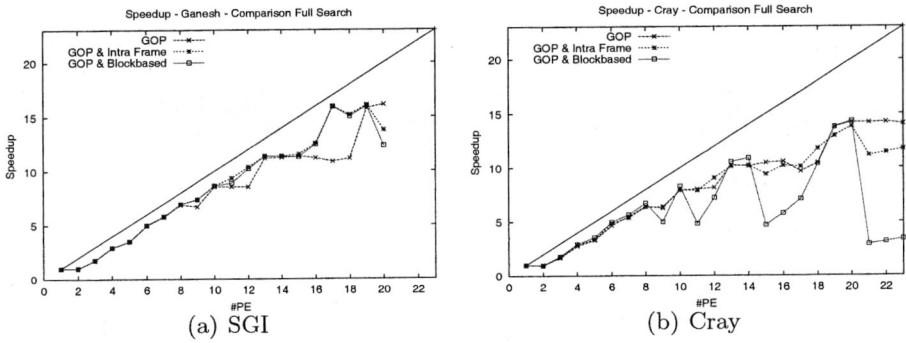

Fig. 4. Speedup of Hybrid Modes.

to an approach with finer granularity (BB or IF) and the data are distributed accordingly. Fig. 4 shows the speedup results of this approach on the SGI and the Cray, respectively.

Fig. 4.a shows that both hybrid schemes outperform the original GOP approach on the SGI, the plateaus disappear. However, on the Cray the results are not that simple (Fig. 4.b). The results are slightly improved for a selected number of PEs but for others we notice a significant degradation of speedup, especially for the hybrid GOP + BB mode. Fig. 5 shows a vizualization of the execution behaviour on the SGI with 9 PEs where horizontal gray bars denote the computation times on the different PEs and thin black lines depict communication events.

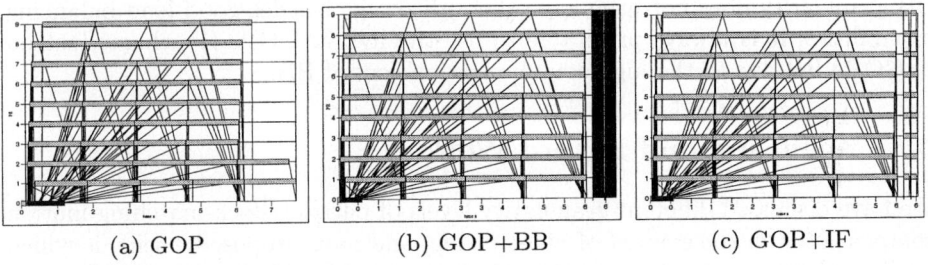

Fig. 5. Visualization of the execution behaviour.

In Fig. 5.b we clearly notice the intensive communication during the last phase of the computation which is caused by the Blockbased parallelization approach. On the SGI, the performance is still better as compared to the pure GOP approach due to the efficient communication subsystem, on the Cray the excessive communication causes the performance to go down. The question remains under which circumstances it makes sense to switch granularity and in which cases it is better to stick with the GOP approach only. Fig. 6 shows the amount of P frames left to be processed when switching the granularity, depending on the size of the GOPs, the number of PEs involved, and the length of the processed video.

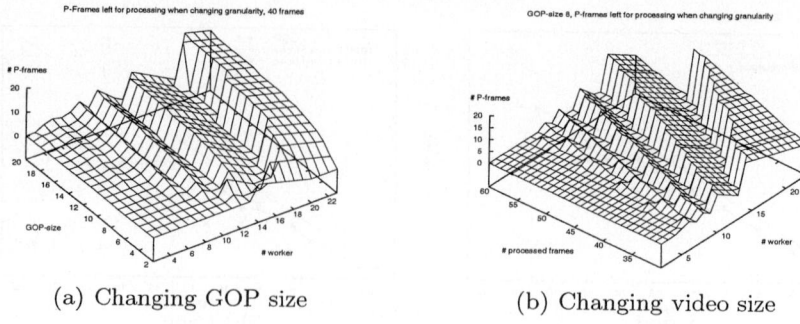

(a) Changing GOP size   (b) Changing video size

Fig. 6. Work to be done after changing granularity.

It turns out that the higher the amount of "remaining" P frames (i.e. frames processed not in GOP mode) is, the less efficient will it be to switch to finer granularity. The obvious reason is that the finer granular approaches simply do not scale that well, therefore, as their share of the overall computation time increases, the overall performance decreases. The second reason is that in case of a large amount of such frames the number of idle PEs when sticking to the GOP approach is relatively small which makes the drop in efficiency small as well. The actual value from which number of remaining P frames on it is better to use a pure GOP approach, of course depends on the parameters of the hardware (relation communication/computation cost) and on video size, GOP structure, and other values related to the coding process. Since the load balancing problems of the IF granularity are not significant for a low number of frames and the BB granularity does not have any other advantage besides good load balancing capabilities, it is always preferable to use the hybrid GOP+IF scheme instead of GOP+BB due to the high communication cost of BB.

5 Conclusion and Future Work

We have discussed different granularity levels of parallel block-matching motion compensation in the context of video coding and have proposed a scheme which involves switching to finer granularity in case of load-balancing problems. It turns out that the performance of inter-frame parallelization granularity may be improved under certain circumstances. In future work we will develop an analytical framework which will allow to determine the best strategy based on parameters of the hardware and of the video.

References

1. I. Ahmad, S. M. Akramullah, M. L. Liou, and M. Kafil. A scalable off-line MPEG-2 video encoding scheme using a multiprocessor system. *Parallel Computing*, 27(6):823–846, 2001.
2. I. Ahmad, Y. He, and M. L. Liou. Video compression with parallel processing. *Parallel Computing*, 28(7–8):1039–1078, 2002.

3. S.M. Akramullah, I. Ahmad, and M.L. Liou. Performance of software-based MPEG-2 video encoder on parallel and distributed systems. *IEEE Transactions on Circuits and Systems for Video Technology*, 7(4):687–695, 1997.
4. S.-C. Cheng and H.-M. Hang. A comparison of block-matching algorithms mapped to systolic-array implementation. *IEEE Transactions on Circuits and Systems for Video Technology*, 7(5):741–757, October 1997.
5. M. Feil and A. Uhl. ParWave: Granularity in parallel wavelet packet video coding. In E. Krause and W. Jäger, editors, *High Performance Computing in Science and Engineering 2002*, pages 479–490, Stuttgart, Germany, 2002. Springer-Verlag.
6. M. Feil and A. Uhl. Motion-compensated wavelet packet zerotree video coding on multicomputers. *Journal of Systems Architecture*, 49:75–87, 2003.
7. B. Furht, J. Greenberg, and R. Westwater. *Motion estimation algorithms for video compression*. Kluwer Academic Publishers Group, Norwell, MA, USA, and Dordrecht, The Netherlands, 1997.
8. R. Kutil. Approaches to zerotree image and video coding on MIMD architectures. *Parallel Computing*, 28(7–8):1095–1109, August 2002.
9. K.K. Leung, N.H.C. Yung, and P.Y.S. Cheung. Parallelization methodology for video coding – an implementation on the TMS320C80. *IEEE Transactions on Circuits and Systems for Video Technology*, 8(10):1413–1423, 2000.
10. S.B. Pan, S.S. Chae, and R.H. Park. VLSI architectures for block matching algorithms using systolic arrays. *IEEE Transactions on Circuits and Systems for Video Technology*, 6(1):67–73, February 1996.
11. A. Pommer. Fractal video compression on shared memory systems. In P. Zinterhof, M. Vajtersic, and A. Uhl, editors, *Parallel Computation. Proceedings of ACPC'99*, volume 1557 of *Lecture Notes on Computer Science*, pages 317–326. Springer-Verlag, February 1999.
12. K. Shen, G.W. Cook, L.H. Jamieson, and E.J. Delp. An overview of parallel processing approaches to image and video compression. In M. Rabbani, editor, *Image and Video Compression*, volume 2186 of *SPIE Proceedings*, pages 197–208, 1994.
13. M. Tan, J. M. Siegel, and H. J. Siegel. Parallel implementation of block-based motion vector estimation for video compression on four parallel processing systems. *Internaltional Journal of Parallel Programming*, 27(3):195–225, 1999.
14. F. Tischler and A. Uhl. Granularity levels in parallel block-matching motion compensation. In D. Kranzlmüller, P. Kacsuk, J. Dongarra, and J. Volkert, editors, *Recent advances in Parallel Virtual Machine and Message Passing Interface (EuroPVM/MPI) - 9th European PVM/MPI Users Group Meeting*, volume 2474 of *Lecture Notes on Computer Science*, pages 183 – 190. Springer-Verlag, September 2002.

Placement of Nodes
in an Adaptive Distributed Multimedia Server

Balázs Goldschmidt[1], Tibor Szkaliczki[2,*], and László Böszörményi[3]

[1] Budapest University of Technology and Economics
balage@inf.bme.hu
[2] Computer and Automation Research Institute
of the Hungarian Academy of Sciences
sztibor@sztaki.hu
[3] University Klagenfurt, Department of Information Technology
laszlo@itec.uni-klu.ac.at

Abstract. Multimedia services typically need not only huge resources but also a fairly stable level of Quality of Services. This requires server architectures that enable continuous adaptation. The Adaptive Distributed Multimedia Server (ADMS) of the University Klagenfurt is able to dynamically add and remove nodes to the actual configuration, thus realizing the *offensive adaptation* approach.
This paper focuses on the optimal placement of nodes for hosting certain ADMS components (the so-called *data collectors*, collecting and streaming stripe units of a video) in the network. We propose four different algorithms for *host recommendation* and compare the results gained by running their implementations on different test networks. The *greedy* algorithm seems to be a clear looser. Among the three other algorithms (*particle swarm*, *linear programming* and *incremental*) there is no single winner of the comparison, they can be applied in a smart combination.

1 Introduction

Even highly sophisticated multimedia servers with a distributed architecture, such as the Darwin server of Apple [1] or the Helix architecture of RealNetworks Inc. [2] are *static* in the sense that actual configurations of the the distributed server must be defined manually. The Adaptive Distributed Multimedia Server (ADMS) of the University Klagenfurt [3] is able to dynamically add and remove nodes to the actual configuration. Thus, ADMS realizes the *offensive adaptation* approach [4]. In case of shortage of resources, instead of reducing the quality of the audio-visual streams by usual, defensive, stream-level adaptation, it tries to migrate and/or replicate functionality (i.e. code) and/or audio-visual data on demand.

---

* Partial support of the EC Centre of Excellence programme (No. ICA1-CT-2000-70025) and the Hungarian Scientific Research Fund (Grant No. OTKA 42559) is gratefully acknowledged.

It is crucial for the performance of the system, to find optimal placement for the server nodes. The *host recommender* component of the ADMS system determines the host computers in the network where the server components (also called applications) should be loaded. The optimal location depends on the actual load and capacity of the available nodes and links, and on the actual set of client requests.

The distributed multimedia server architecture [3] has different components that can be located on different hosts of the network. *Data managers* or *servers* store and retrieve the media data. *Data collectors* or *proxies* collect the data from the servers and stream them to the clients.

This paper deals with the configuration recommendation algorithms. Unfortunately, the related problems are NP-complete. However, there are many approximation algorithms that can result in nearly optimal solutions within short time. We examine, what kind of mathematical approaches can be applied to the adaptive host recommendation.

The formal and detailed description of the model and the algorithms can be found in a technical report[1] [5].

2 Related Work

Finding the optimal deployment of proxies in a network is a well known problem in the literature. Most of the works, however, deal only with (1) static configurations, (2) web-proxies, and (3) caching-problems [6–8]. Static configuration means, that proxies are deployed once, and their placement can not change later, or with high cost only. On the other hand, web proxies have to serve the clients with relatively short documents and images that have to arrive unmodified. Online multimedia data delivery serves huge data-streams, where some modifications are still acceptable, having the emphasis on the timing constraints. Finally, we are currently not interested in caching problems, because they are orthogonal to the offensive adaptation problem. This issue was discussed elsewhere [4]. Therefore, we cannot use the former results directly.

We looked at the mathematical background and found the facility location problem(FLP), which is an intensively studied problem in operations research. The problem is to select some facility candidates and to assign each client to one of the selected facilities while minimizing the cost of the solution. A detailed description of the problem can be found in [9]. Despite the similarity, some significant differences prohibit the direct application of the approximation algorithms for the FLP to the host recommendation. First, while the cost of a facility is usually a constant value in case of the FLP, in the current problem it depends on the maximum bandwidth required by the clients assigned to the proxy. Furthermore, the limited bandwidth of the subnets must be taken into account as well.

In [9] it is shown that FLP is NP-hard for general graphs. Many constant approximation algorithms have been published for the facility location problem

---

[1] http://143.205.180.128/Publications/pubfiles/pdffiles/2004-0005-BGAT.pdf

with polynomial running time that are usually combined with each other. The linear programming techniques play key role in many algorithms with constant approximation ratio [10, 11]. The best approximation ratio was achieved combining the primal-dual method with the greedy augmentation technique [12].

As a different approach, evolutionary algorithms (EA) also provide an efficient way of finding good solutions for NP-hard problems. In [13] an EA is proposed to solve the P-median problem, which problem is in close relation to FLP. A kind of evolutionary algorithms, particle swarm optimisation proved to be effective in a wide range of combinatorial optimisation problems too [14].

3 The Problem Model

In order to implement and compare different algorithms that solve the problem, we have defined the following model and metrics.

From the actual point of view the Adaptive Distributed Multimedia Server consists of Data Managers and Data Collectors. Multimedia data (videos) are stored on the Data Managers. The videos are sliced, and the resulting stripe units are distributed to the Data Managers. When needed, the video is recollected and streamed to the clients by the Data Collectors. This technique helps both network and node resource load balancing.

According to [15], Data Managers that contain stripe units of the same video should be kept as close to each other as possible, practically on the same subnet, because that configuration gives the best performance. Based on this result our model considers such a group of Data Managers a single *server*.

The media is collected from the Data Managers by Data Collectors (proxies). They can be loaded on any node that hosts a Vagabond2 Harbour [16]. The nodes that may play the role of a Data Collector are considered *candidates*.

The *client* is the third kind of component in the model. The clients connect to Data Collectors and get the desired media via streaming from them. The clients define their requests as lists that contain QoS requirements in decreasing order of preference. We assume that the demand list has a last, default element, that represents the situation when the client's request is rejected. In the current model we also assume that all clients want to see the same video at the same time.

In our current model only the proxies can be deployed dynamically. The locations of the servers and that of the clients are not to be modified.

The network model is basically a graph where the nodes are called *areas*. An area is either a subnet (including backbone links) or a router that connects subnets. We handle them similarly because from our point of view they have the same attributes and provide the same functionality. The edges of the graph are the connections between the routers and the subnets they are part of.

We assume that we know the subnet location of the clients and servers, the candidate nodes, and the attributes of the network, the latter presented as the attributes of each area: bandwidth, delay jitter, etc. The route between any two nodes is described as a list of neighbouring areas.

The solutions of a problem are described as possible *configurations* that provide the following information:

- for each client, the index of the QoS demand that has been chosen to be satisfied, and the candidate that hosts the proxy for the client
- for each candidate, the server, where the video should be collected from

In order to compare different solutions, we have defined a cost function that gets the initial specifications and a possible configuration as input parameters. Using this information the function calculates the network resource needs (total allocation and over-allocation), the number of rejected clients (those whose chosen demand is the last, default one), the sum of the chosen demand indexes of the clients (*linear badness*), and the so called *exponential badness*, that is defined as $\sum_{c \in C} 2^{i_c}$ where C is the set of clients, and i_c is the index of the chosen demand for client c. This last metric is useful if we want to prefer 'fair' to 'elitist' configurations.

Given two costs, that cost is less, that has (in decreasing order of preference):

- less over-allocation (we can't afford over-allocation),
- less rejection (we want to increase the number of accepted clients),
- less exponential badness (we prefer fairness),
- less total allocation (we want to minimize the network load).

Using this cost metric we were able to compare several algorithms.

4 Solution Algorithms

In this chapter we provide a short introduction to the algorithms we have implemented and tested. More details can be found in [5].

4.1 Greedy

The greedy algorithm we use here is almost identical to that published in [17]. The difference is that the cost function is changed to that described in the previous chapter, and that the clients' demand list is also taken into account.

4.2 Particle Swarm

The particle swarm algorithm is based on the algorithm of Kennedy and Eberhardt [14]. The original algorithm uses a set of particles, each of them representing a possible configuration. Every particle is connected to several other particles thus forming a topology.

The particles are initialized with random values. Each particle knows its last configuration ($\underline{x}(t)$), its best past configuration ($\underline{b}_p$), and the configuration of that neighbour (including itself) that has the least cost ($\underline{b}_n$). In each turn, if it has less cost than any of its neighbours, it counts a new configuration by creating a linear combination of $\underline{b}_p$ and $\underline{b}_n$, using probabilistic weights, then it adopts this new configuration. The whole process runs until some condition is met. The combination is defined as:

$$\underline{x}(t+1) = \underline{x}(t) + \varphi_1(\underline{b}_p - \underline{x}(t-1)) + \varphi_2(\underline{b}_n - \underline{x}(t-1))$$

where φ_i is a random number from $[0, 1)$.

This original algorithm, however, can solve those problems only, where the dimensions of the configurations represent binary or real values. This is the consequence of the linear combination technique. In our case the configurations' dimensions represent unordered elements of sets. Therefore, instead of the linear combination that can not be applied to them, we use the following. For each dimension we first take the value with a certain probability from either $\underline{x}(t)$, or $\underline{b}_n$ (crossover). Then, with another probability, we change it to a random value (e) from its value set (mutation). Finally, we assign the value to the given dimension.

Crossover and mutation are also applied by genetic algorithms[18]. The difference between genetic algorithms and our algorithm is that in our case the connections of the partners (neighbours) are static. Particles do crossovers only with their neighbours, and the crossovers transfer information from the better configuration to the worse only, thus the best results are always preserved.

The algorithm runs until every particle has the same cost. But this condition combined with the mutation leads from the initial fast evolution to a final fluctuation. We apply the simulated annealing of the mutation rate in order to avoid this phenomenon [19]. The control variable of the annealing is the number of particles that have their configuration changed. The less particle changed, the less the mutation rate is.

4.3 Linear Programming Rounding

LP Model. We chose an algorithm based on linear programming rounding for the solution of the configuration problem. For simplicity, we minimize only the number of refused clients, the exponential goodness and sum of the reserved bandwidth for each subnet. Weights are assigned to the different optimization criteria to express their priority.

Inequations express three types of constraints as follows. The reserved bandwidth of a subnet is less than or equal than the available. A proxy-server connection needs at least as much bandwidth as the maximum among the accepted requests served through it. Each client is either assigned to a client-proxy-server route or rejected.

We introduce variables $X_{i,j,k}$ to indicate whether the request of client i is served by server k through proxy j. Their possible values are 0 and 1. Since the time complexity to find the exact solution for an integer linear programming problem is large, we consider the LP-relaxation of the problem, where the possible values of the variables can be any real number.

Rounding. First we solve the linear program and obtain an optimal solution. If $X_{i,j,k} = 1$ then let the request of client i be served by server k through proxy j. Unfortunately, the possible fractional values of X-type variables do not represent legal solutions. We round the solution in a greedy manner. We take each X variables with fractional value one after the other. If the client is still not served, we try to select the current client-proxy-server route denoted by variable $X_{i,j,k}$, and check the load conditions in the network. The client i is served by server k through proxy j in the solution if and only if these conditions are fulfilled after the selection of the route.

4.4 Incremental Algorithm

In order to find solutions quickly, we implemented a very simple but efficient algorithm. It operates on the so-called *FLP graph* which is a bipartite graph, where one set of nodes denotes the clients, the *facility* nodes represent the proxy-server routes and the edges between them correspond to the client-proxy-server routes that are able to satisfy client requests. The main steps of the algorithm are as follows.

After generating the FLP graph, the facilities are sorted in decreasing order of the bandwidth of the represented proxy-server routes. We take the facilities one after the other. The algorithm selects facility f_i if it can serve new clients or there is at least one client c_j already assigned to a facility denoted by f_0 where the QoS parameters of edge (c_j, f_i) is better than that of edge (c_j, f_0). If facility f_i is selected, we take the clients adjacent to it one after the other and client c_j is assigned to it if it fulfills the above condition and its request can be satisfied through facility f_i without overloading the network. A facility is deselected if no clients are assigned to it.

5 Results

We implemented the algorithms and tested them on simulated network environments. Each test network consists of 50 subnets. Six test series were generated; each of them consisted of ten cases. The number of servers is always ten, while the number of clients and proxies varies in different series; there are 5, 10, 15, 20, 25, 30 clients and 10, 20, 30, 40, 40, 40 proxies in the different series. We examined the cost of the solutions and the running time as a function of the number of clients. Figure 1 shows the costs (linear and exponential badnesses, numbers of rejected clients) and the runtime. The figures compare the results of the algorithms described above, namely linear programming rounding, swarm algorithm, greedy algorithm and incremental algorithm. The linear programming does not produce legal solutions, without rounding, but can be used as a lower bound for the cost measures.

According to the figure, the swarm algorithm produces the best results, and the linear programming rounding achieves results with slightly higher cost and the greedy algorithm fails to find nearly optimal algorithms for a high number of clients.

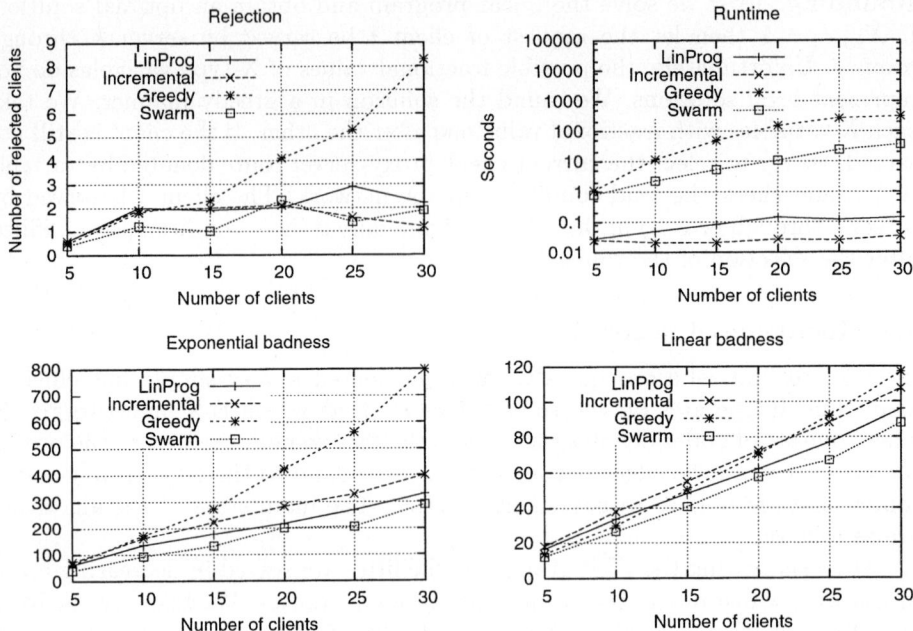

Fig. 1. The results of the measurements for linear programming rounding, incremental, greedy, and swarm algorithms.

On the other side, the running time of the incremental algorithm is clearly the best, the running time of linear programming with rounding is the second, while the swarm and the greedy algorithms run substantially slower.

6 Conclusions and Further Work

We introduced a number of algorithms for host recommendation in an Adaptive Distributed Multimedia Server. We did not find an algorithm that is the best in every aspect. A good idea is producing an initial solution quickly using the fast deterministic algorithms. Later, if time allows, a more sophisticated solution (admitting more clients) might be rendered, using the stochastic swarm algorithm.

In the future, we intend to improve the model of the network by incorporating node information also. Thus not only the predicted future values of the network parameters and client requests might be taken into account at the recommendation, but the performance of the host nodes also. Later the implementations of the algorithms will be integrated into the Adaptive Distributed Multimedia Server and tested also in a real network environment.

References

1. Apple Computer, Inc. QuickTime Streaming Server: Darwin Streaming Server: Administrator's Guide. (2002) http://developer.apple.com/darwin/projects/streaming/
2. Helix Community: Helix Universal Server Administration Guide. (2002) https://www.helixcommunity.org/2002/intro/platform
3. Tusch, R.: Towards an adaptive distributed multimedia streaming server architecture based on service-oriented components. In Böszörményi, L., Schojer, P., eds.: Modular Programming Languages, JMLC 2003. LNCS 2789, Springer (2003) 78–87
4. Tusch, R., Böszörményi, L., Goldschmidt, B., Hellwagner, H., Schojer, P.: Offensive and Defensive Adaptation in Distributed Multimedia Systems. Computer Science and Information Systems (ComSIS) **1** (2004) 49–77
5. Goldschmidt, B., Szkaliczki, T., Böszörményi, L.: Placement of Nodes in an Adaptive Distributed Multimedia Server. Technical Report TR/ITEC/04/2.06, Institute of Information Technology, Klagenfurt University, Klagenfurt, Austria (2004)
6. Steen, M., Homburg, P., Tannenbaum, A.S.: Globe: A wide-area distributed system. IEEE Concurrency (1999)
7. Li, B., Golin, M., Italiano, G., Deng, X., Sohraby, K.: On the optimal placement of web proxies in the internet. In: Proceedings of the Conference on Computer Communications (IEEE Infocom). (1999)
8. Qiu, L., Padmanabhan, V.N., Voelker, G.M.: On the placement of web server replicas. In: INFOCOM. (2001) 1587–1596
9. Cornuejols, G., Nemhauser, G.L., Wolsey, L.A.: The uncapacitated facility location problem. In Mirchandani, P., Francis, R., eds.: Discrete Location Theory. John Wiley and Sons, New York (1990) 119–171
10. Shmoys, D., Tardos, E., Aardal, K.: Approximation algorithms for facility location problems. In: Proceedings of the 29th ACM Symposium on Theory of Computing. (1997) 265–274
11. Charikar, M., Guha, S.: Improved combinatorial algorithms for the facility location and k-median problems. In: IEEE Symposium on Foundations of Computer Science. (1999) 378–388
12. Mahdian, M., Ye, Y., Zhang, J.: Improved approximation algorithms for metric facility location problems. In: Proceedings of 5th International Workshop on Approximation Algorithms for Combinatorial Optimization. (2002)
13. Dvorett, J.: Compatibility-based genetic algorithm: A new approach to the p-median problem. In: Informs Fall 1999 Meeting. (1999)
14. Kennedy, J., Eberhardt, R.C.: Swarm Intelligence. Morgan Kaufmann (2001)
15. Goldschmidt, B., Tusch, R., Böszörményi, L.: A corba-based middleware for an adaptive streaming server. Parallel and Distributed Computing Practices, Special issue on Dapsys 2002 (2003)
16. Goldschmidt, B., Tusch, R., Böszörményi, L.: A mobile agent-based infrastructure for an adaptive multimedia server. In: 4th DAPSYS (Austrian-Hungarian Workshop on Distributed and Parallel Systems), Kluwer Academic Publishers (2002) 141–148
17. Goldschmidt, B., László, Z.: A proxy placement algorithm for the adaptive multimedia server. In: 9th International Euro-Par Conference. (2003) 1199–1206
18. Davis, L., ed.: Handbook of Genetic Algorithms. Van Nostrand Reinhold (1991)
19. Kirkpatrick, S., Gelatt, C.D., Vecchi, M.P.: Optimization by simulated annealing. Science, Number 4598, 13 May 1983 **220, 4598** (1983) 671–680

A Practical Performance Analysis of Stream Reuse Techniques in Peer-to-Peer VoD Systems[*]

Leonardo Bidese de Pinho and Claudio Luis de Amorim

Parallel Computing Laboratory, COPPE Systems Engineering Program,
Federal University of Rio de Janeiro, RJ, Brazil
{leopinho,amorim}@cos.ufrj.br

Abstract. Although many works have reported simulated performance benefits of stream reuse techniques to the scalability of VoD systems, these techniques have been rarely evaluated in practical implementations of scalable VoD servers. In this work we investigate the behavior of representative stream reuse techniques in the GloVE system, a low-cost VoD platform whose scalable performance depends on the combination of the stream techniques it uses. More specifically, we show experimental results focusing on the requirements of the amount of server's channels and aggregate bandwidth that GloVE demands for several combinations of stream reuse techniques. Overall, our results reveal that stream reuse techniques in isolation offer limited performance scalability to VoD systems and only balanced combinations of batching, chaining, and patching techniques explains the scalable performance of GloVE on delivering popular videos with low startup latency while using the smallest number of server's channels.

1 Introduction

In recent years, considerable research efforts have been concentrated on the design of scalable Video on Demand (VoD) systems since they represent a key-enabling technology for several classes of continuous media applications, such as distance learning and home entertainment. In particular, a central problem VoD designers face is that in a typical VoD system with many videos, even a transmission of a single high-resolution video stream in a compressed format consumes a substantial amount of resources of both the video server and the content distribution network. Given that users can choose any video and start playback at anytime, a significant investment on server's hardware and network's infrastructure will be required to support large audiences. Therefore, it is fundamental that a VoD server supports stream distribution strategies capable of using efficiently the available network resources in order to reduce its cost per audience ratio.

Typically, a VoD server supports a finite pool of stream channels, where the amount of channels is often determined by the server's bandwidth divided by

---

[*] This work was partially sponsored by the Brazilian agencies CAPES and CNPQ.

the video playback rate. The reason that conventional VoD systems cannot scale is because they dedicate one different channel to each active client, in an one-to-one approach. As a result, the total number of clients a conventional VoD system supports is equal to the limited number of server's channels. Due to the scalability limitation of the one-to-one approach, several scalable streaming techniques have been proposed in the literature. Basically, such techniques allow multiple clients to share the stream contents that are delivered through each of the server's channels, in an one-to-many approach.

Three of the most well-known scalable streaming techniques are *Batching* [1] - where near requests to the same video are first enqueued and served afterwards by a single multicast stream from the server, *Chaining* [2] - in which clients on behalf of the server can send video streams to subsequent requests, provided the requests arrive within a certain time interval during which the prefixes are still in the playout buffers of the clients, and *Patching* [3] - where a new client that requests a video is first inserted into an available active multicast stream for the video and upon receiving the video stream the client will temporarily store it into a local buffer. In addition, the server will send to the new client an extra video stream, namely the patch, which contains the video segment that the client missed.

Recently, we introduced a novel scalable streaming technique, called the Cooperative Video Cache (CVC) [4] that combines chaining and patching under a P2P model. CVC implements a cooperative stream cache over the distributed collection of client's playout buffers[1] that store dynamically video streams the server sends to the clients. The key idea behind CVC is to use such a large dynamic stream cache as the primary source of video contents so that most client requests become cache hits with the server attending only the remainder cache misses. Experimental results of the CVC-based VoD prototype we developed, namely the Global Video Environment (GloVE) [5], demonstrated that CVC can reduce drastically VoD server's occupancy thus enabling scalable VoD systems to be built.

In contrast with existing works that evaluated performance of scalable streaming techniques through simulations, in this work we assess the efficiency of representative streaming techniques in practical situations. More specifically, since the GloVE prototype allows the evaluation of streaming techniques either separately or in any combination of them, we examined their relative performance contribution to the scalability of VoD systems that GloVE encompasses. In addition, we addressed the impact of the distribution of popularity of videos on the effective use of VoD system's resources.

The remainder of this paper is organized as follows. In Section 2 we describe the main characteristics of the GloVE platform and operation modes. In Section 3 we analyze GloVE's performance results for several scenarios of streaming techniques and workloads. In Section 4 we compare our work to related ones. Finally, we conclude in Section 5.

---

[1] Reserved memory space used by the client equipment to hide the network jitter and inherent variations of VBR videos.

2 The GloVE System

The GloVE (Global Video Environment) is a scalable VoD system that implements the cooperative video cache while supporting a P2P system with a centralized metadata component, namely the CVC Manager (CVCM). The CVCM monitors globally the video contents stored in the playout buffers of the clients. The CVC Client (CVCC) software allows the content of any client buffer to be shared with other connected clients, reducing the demand on the CVC Server (CVCS). GloVE assumes that the communication network has IP multicast support and that the VoD server manages the video as a sequence of blocks that can be randomly accessed. Further details of GloVE can be found in [6].

The design of CVCM allows to select different operation modes according to several combinations of stream reuse techniques (e.g., Batching, Chaining, and Patching)[5]. In this article, however, we restrict our analysis to the following modes: *Batching*, in which late requests for a video will form a group that will be serviced from a previous client that is in the prefetching phase; *Chaining*, CVCM implements a slightly different form of chaining in that every stream has only one receiver; *Patching+Batching*, which implements a combination of patching and batching; *CVC+Batching*, which adds batching to the CVC technique.

3 Experimental Analysis

3.1 Evaluation Methodology

The experiments we describe below were performed on a 6-node cluster of Intel Pentium IV 2.4 GHz, 1 GB RAM, running Linux kernel 2.4.18 using a 3Com Fast Ethernet switch with IP Multicast support. One node executed both CVCS and CVCM whereas each of the other nodes executed multiple instances of clients. We used a MPEG-1 video (352x240, 29.97 frames/s) of Star Wars IV movie, with average video playback rate near to 1.45 Mbps. To allow multiple clients per node, we developed an emulator of MPEG-1 decoder, which uses a trace file containing the playback duration of each segment of the video. Our workload ranges from medium to high client arrival rates according to a Poisson process with values of 6, 10, 20, 30, 60, 90, and 120 clients/min. We assumed that the CVCS can deliver at most 56 MPEG-1 streams simultaneously, and supports two types of client access: *smooth access* when clients request video blocks according to the video's playback rate, and *burst access* when clients request blocks as fast as possible and the playout buffers are filled at the transmitting rate determined by the availability of CVCS' bandwidth. We used 64 KB blocks as the system access unit and the prefetch limit[2] of 32 blocks, which can hold a video segments up to 11s. We evaluate performance of GloVE under practical network conditions, using playout buffer sizes of 64, 128, and 256 blocks, which can store video sequences that last near to 22, 44, and 88s, respectively. Given

---

[2] Minimal amount of video blocks that must be stored in the playout buffer in order to initialize video's playback.

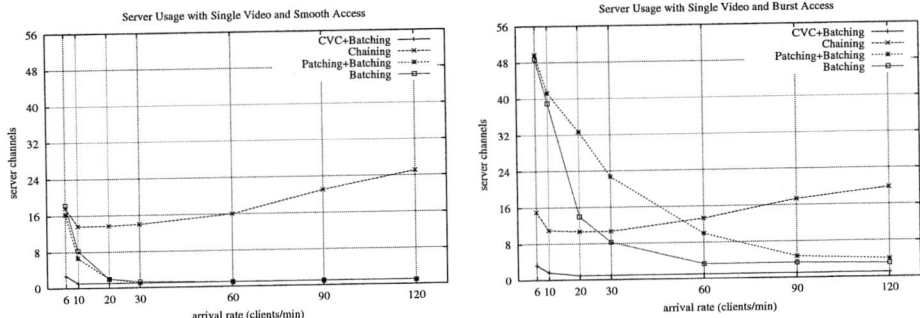

Fig. 1. Channels usage for each GloVE mode and access type. (a) Smooth. (b) Burst.

to the restricted space, we only show the results for 128 blocks, which hold video segments up to 44s. More detailed results are reported in [6]. In the experiments with multiple videos, we emulated a collection of eight videos, where the video popularity follows a general Zipf distribution with $\alpha = 0.7$ [1]. In addition, we also investigated the sensitivity of the VoD system to $\alpha = 0$ which represents the uniform distribution, and $\alpha = 1$, which represents Zipf without skew.

3.2 Experimental Results

We present the results of our experiments in three parts: single video, multiple videos, and a sensitivity analysis of server's performance to videos' popularity.

Figure 1 shows the occupancy of server's channels, or simply server occupancy, using block request rate with smooth and burst accesses for a single video and different GloVE modes. Server occupancy indicates the relative degree of scalability of a particular technique. Specifically, the lower server occupancy a technique produces for a given amount of active clients the higher is its scalability. CVC+Batching with smooth access occupies only one channel for almost all arrival rates we measured. This mode allows stream reuse for incoming requests up to 30s apart[3]. So, only for very low arrival rates with intervals between requests greater than 30s that misses to the playout buffers become significant. Intuitively, stream reuse is higher for smooth accesses because they generate long prefetches whereas burst accesses often issue short prefetches. Notice that, this difference in access behavior is not shown in the figure due to the buffer size we used. The Chaining mode is influenced negatively by prefetching for arrival rates higher than 10 clients/min. In this case, the higher is the arrival rate, the higher is the probability of a client to arrive while all previous ones either issued prefetching or are transmitting video streams, so that they can not be providers of video segments and a new server channel will be required. Also, as the prefetching phase is shorter with burst accesses, server's occupancy presented slightly smaller values in this mode. The Patching+Batching mode is influenced strongly by the client access type. The fact is that this mode initially

---

[3] Due mostly to playout buffer of 44s minus the prefetch limit of 11s.

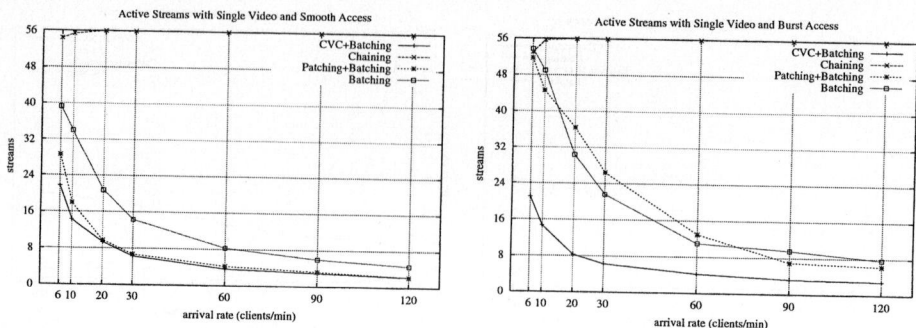

Fig. 2. Active streams for each GloVE mode and access type. (a) Smooth. (b) Burst.

can only use batching because patching requires multicast streams to be available. When using burst accesses the server occupancy tends to be very high for arrival rates less than 60 clients/min. In contrast, with smooth access the chances for Batching increase significantly, so that the minimum server occupancy is reached at 30 clients/min. The client access type also determines the efficiency of the Batching mode. Specifically, when using the server with smooth access the channel's occupation is very similar to that achieved by Patching+Batching. The main difference between the two modes is due to the higher average prefetching that Batching generates for burst accesses and arrival rates between 20 and 90 clients/min. When Patching is used, the average prefetching is substantially reduced, decreasing the opportunity of stream reuse for the next arriving clients.

Figure 2 shows the amount of active streams for each server mode when using a single video system. The values in the figure indicate the aggregated bandwidth that each mode requires. As shown in the figure, the minimum amount of active streams is two, one stream from each of the server and the first client. For the CVC+Batching mode the number of active streams is practically the same whether using smooth or burst accesses to the server. Another factor that affects the number of active streams is the type of stream reuse that is used most frequently. Whenever possible Patching is preferable since it generates fewer streams. Otherwise, the server will attempt to employ Chaining which generates a new stream if the server succeeds. The latter occurs often when the time interval between requests is not greater than than the capacity of the playout buffer in seconds. Intuitively, with Chaining the number of active streams is equal to the amount of clients, since the streams that Chaining generates have only one receiver and is independent of the client access type. Patching+Batching is highly influenced by client access type. For smooth access, the large amount of multicast streams that Batching generates increases the use of Patching as well. Burst accesses, particularly for arrival rates less than 60 clients/min, produce shorter prefetches, decrease the use of Batching, which in turn reduce Patching, and increase the amount of streams with a single receiver. In the Batching mode, the number of active streams depends on the client access type only for arrival rates less than 30 clients/min. In this case, when clients use burst accesses, the

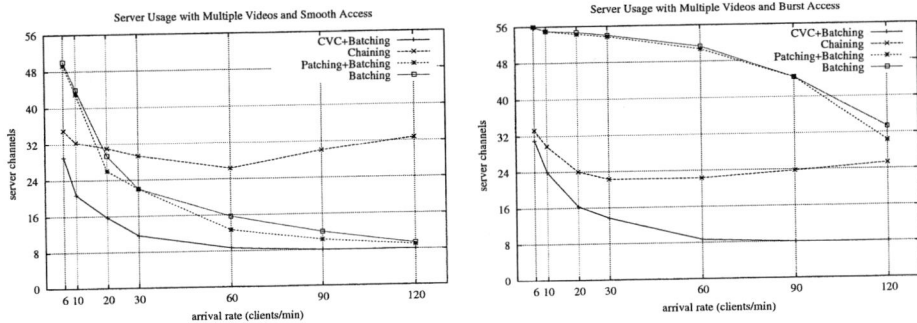

Fig. 3. Channels usage for each GloVE mode. (a) Smooth access. (b) Burst access.

majority of streams comes from the server itself, because there is no opportunity for Batching. For smooth accesses, Batching dominates stream generation.

Figure 3 presents server occupancy according to the different GloVE modes when the VoD server offers eight videos. The videos' popularity follows a Zipf distribution with $\alpha = 0.7$. In the best case, at least one channel will be used for each different video the clients request. In the worst case, one channel will be used for each client request, which is ultimately the behavior of a conventional VoD system. The curves related to CVC+Batching indicate that the minimum amount of busy channels occurred at 60 clients/min. This is not surprisingly since for a single video the referred minimum was near to 10 clients/min. The negative impact of the client access type on server occupancy was restricted to small arrival rates. Similarly to the single video case, Chaining still suffers from the negative impact of prefetching, especially for arrival rates higher than 30 clients/min. At this rate, we noticed the minimum server occupancy of 22 channels. Chaining achieved better performance when clients used burst accesses to the server. For Patching+Batching, most of the observations we made for single videos hold also for multiple videos. The main difference is that there exist considerably fewer opportunities to apply either Batching or Patching. For arrival rates ranging from 90 to 120 clients/min, the minimum server occupancy is reached with smooth accesses. Note that with burst accesses, 30 channels is the minimum value of server occupancy, which was achieved with the highest arrival rate. The behavior of Batching is very similar to that of Patching+Batching we analyzed above.

In Fig. 4 we show the amount of active streams for a server with multiple videos. The minimum number of active streams tends to be 16 provided that all the videos are requested twice at least. In CVC+Batching, the curves remained similar for both burst and smooth access types. As explained for a single video, the emphasis on using preferably Patching led to the least amount of active streams. Naturally, Chaining tends to be used more for requests within shorter interarrival rates. In Chaining the number of active streams is equal to the number of active clients, as in the case of single video. Comparing Patching+Batching with Batching, it becomes clear that both behave similarly. The main difference is that Patching+Batching generates fewer streams due to Patching. Also, while

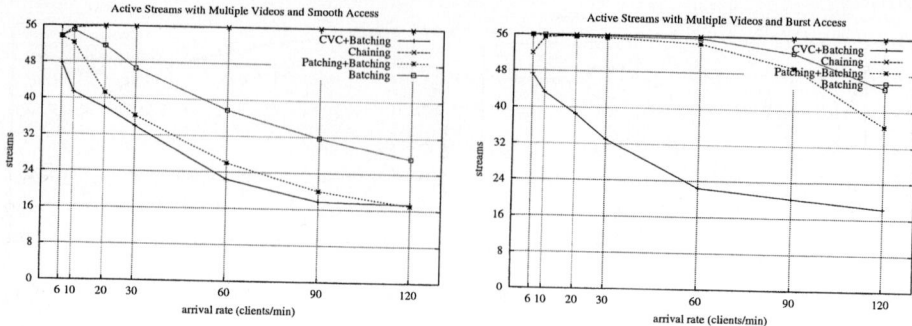

Fig. 4. Active streams for each GloVE mode. (a) Smooth access. (b) Burst access.

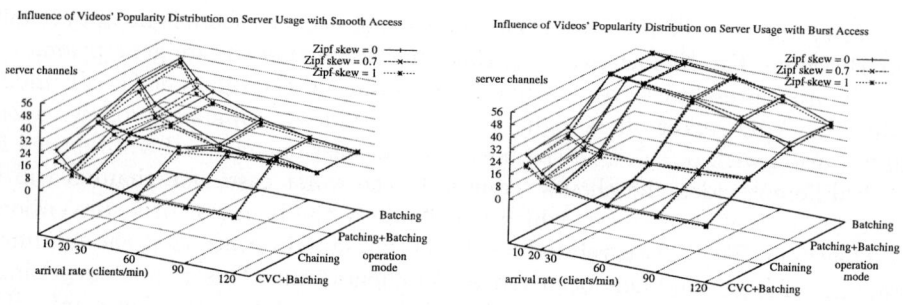

Fig. 5. Channels usage for different Zipf skews. (a) Smooth access. (b) Burst access.

the former achieved the minimum of 17 streams, the latter created 27 streams at least. Due to space limitation we do not present results for start-up latency. However, as reported in [6], the average start-up latency we measured was low ranging from 10 to 13s and 1.2 to 9.8s for smooth and burst access, respectively.

Figure 5 illustrates the impact of the distribution of video's popularity on the occupancy of server's channels. The variations of the popularity distribution of videos did not impact significantly the amount of channels the VoD system required to service the client requests. Indeed, none of the operation modes presented variations higher than 20% on the occupancy of server's channels.

4 Related Work

The original works that introduced the above techniques are Batching [1], Chaining [2], Patching [3], and CVC [4]. GloVE uses Chaining and Patching as originally proposed, and exploits Batching but in a different way from the original work. Specifically, in GloVE clients instead of the server can apply the Batching technique. The work in [7] presents a comparison of stream merging algorithms, but does not report results of any practical implementation. An additional analysis of related works focusing on P2P systems can be found in [6].

5 Conclusions

In this work we compared the performance of stream reuse techniques implemented in a practical P2P VoD system, namely the Global Video Environment (GloVE). In particular, we described different operation modes of GloVE according to the combination of several stream reuse techniques it uses, namely Batching, Patching, Chaining, and CVC. Also, we measured the influence of client access type (either smooth or burst) on server's performance. Finally, we analyzed the impact of video popularity distribution on system behavior.

Overall, the CVC+Batching mode outperformed the other modes for VoD servers with either single or multiple videos. Also, the client access type does not significantly affect CVC+Batching performance, suggesting that CVC+Batching will work efficiently for different VoD server designs. Furthermore, the results revealed that some skews on videos' popularity distribution will not impact substantially the resulting VoD system's performance. Thus, we speculate that a VoD system coupled with CVC+Batching will attain scalable performance for large audiences.

Currently, we are working on mechanisms that will support scalable VoD systems for mobile environments with heterogeneous devices. Also, we plan to extend GloVE to dynamically self-adapt to variations on network and peer conditions.

References

1. Dan, A., Sitaram, D., Shahabuddin, P.: Dynamic Batching Policies for an On-Demand Video Server. Multimedia Systems **4** (1996) 112–121
2. Sheu, S., Hua, K.A., Tavanapong, W.: Chaining: A Generalized Batching Technique for Video-On-Demand. In: Proceedings of the International Conference on Multimedia Computing and Systems. (1997) 110–117
3. Hua, K.A., Cai, Y., Sheu, S.: Patching: A Multicast Technique for True Video-on-Demand Services. In: Proceedings of the ACM Multimedia. (1998) 191–200
4. Ishikawa, E., Amorim, C.: Cooperative Video Caching for Interactive and Scalable VoD Systems. In: Proceedings of the First International Conference on Networking, Part 2. Lecture Notes in Computer Science 2094 (2001) 776–785
5. Pinho, L.B., Ishikawa, E., Amorim, C.L.: GloVE: A Distributed Environment for Scalable Video-on-Demand Systems. International Journal of High Performance Computing Applications (IJHPCA) **17** (2003) 147–161
6. Pinho, L.B., Amorim, C.L.: Assessing the Efficiency of Stream Reuse Techniques in Peer-to-Peer Video-on-Demand Systems. Technical Report ES-626/04, COPPE/UFRJ Systems Engineering Program (2004)
7. Bar-Noy, A., Goshi, J., Ladner, R.E., Tam, K.: Comparison of stream merging algorithms for media-on-demand. In: Proceedings of the SPIE Multimedia Computing and Networking (MMCN), San Jose, CA (2002)

Supporting Caching and Mirroring in Distributed Video-on-Demand Architectures[*]

X.Y. Yang[1], F. Cores[2], A. Ripoll[1], P. Hernández[1],
B. Qazzaz[1], R. Suppi[1], and E. Luque[1]

[1] Computer Science Department, ETSE, Universitat Autònoma de Barcelona,
08193-Bellaterra, Barcelona, Spain
[2] Computer Science & Industrial Engineering Departament, EPS,
Universitat de Lleida, 25001, Lleida, Spain

Abstract. Distributed Video-on-Demand (DVoD) systems are proposed as a solution to the limited streaming capacity and null scalability of centralized systems. In such full decentralized architectures with storage constraints, server content management is crucial to achieving an efficient resource utilization. We have proposed and evaluated, through an analytical model, a combined scheme with caching and mirroring to manage distributed servers storage. The results show that our approach achieves the best mean service distance, independently of underlying architecture. In order to support this scheme, we have modified the storage manager of our VoD prototype so as to guarantee that the new workload of this scheme does not affect disk throughput.

1 Introduction

Providing service for a large number of clients geographically dispersed on a metropolitan or country-wide, Distributed Video Demand Systems (DVoD) are desirable. With a naive design, employing only one large centralized server to support these distributed clients, results in inefficient resource allocations, rendering this design virtually impractical. To address this problem, some researchers have focused on various techniques to reduce either individual or overall bandwidth requirements by multicasting.

As an orthogonal solution, other researches have proposed service distribution to manage the dispersion of clients employing a number of servers with a limited storage capacity, each of which attends to clients located in a given locality. These are interconnected via a high-speed network infrastructure in order to be able to share / exchange video streams [1][4]. Such systems have shown themselves capable of providing an optimal solution, i.e. minimum communication-storage cost for distributed continuous media streaming applications.

In such distributed architectures, servers' storage management is crucial in guaranteeing the accessibility of all catalog contents, load balancing among

---

[*] This work was supported by the MCyT-Spain under contract TIC 2001-2592 and partially supported by the Generalitat de Catalunya- Grup de Recerca Consolidat 2001SGR-00218.

servers, fault tolerance and system efficiency (number of concurrent clients supported). In the literature [3], there are basically two main approaches to solving content management in a distributed environment: caching and mirroring.

With a caching scheme, server storage is used to keep the most popular contents. This scheme introduces a high volume of replication, resulting in some catalog contents not being available, and requiring one or several servers with a full copy of system videos for attending to cache-misses; these nodes can turn into a bottleneck for the system growth.

Mirroring consists of the replication of system catalog contents in different locations (servers). However, in a distributed architecture with storage constraints, it is not possible for a server to keep a full system contents mirror. Therefore, we propose the utilization of a distributed mirroring scheme, where each local server maintains only a portion of the full-content catalog. In this way, the mirror would be distributed among several partial mirrors handled by a set of adjacent servers. Depending on the number of servers, the system could have several distributed mirrors, thus replicating video information. In contrast to caching, mirroring replication is performed without taking video popularity into account.

The main drawback of mirroring is that it reduces server-storage efficiency, causing an increase in the volume of remote requests, which have to be managed by remote servers. Therefore, we propose a new scheme which combines both techniques: caching and mirroring. Caching can be used for increasing the number of locally served requests, while distributed mirroring can be used to distribute remote-request service, to avoid bottlenecks and to limit the required resources, increase fault-tolerance, system scalability and allow load balancing policies.

The remainder of this paper is organized as follows: in section 2, we evaluate the goodness of this approach via an analytical model. In section 3, we analyze the storage system design implications due to caching incorporation. Viability evaluation is shown in section 4 and, in the final section, we indicate the main conclusions to be drawn from our results.

2 Caching and Distributed Mirroring-Based Storage Management

To estimate system efficiency using a combined scheme, we have defined the analytical model shown in Table 1. As far as performance is considered, we use the system mean service distance (Dms) defined as the average number of networks over which the streams have to cross from servers to clients. For this analysis, we assume a distributed architecture (as shown in Figure 1) compounded by several independent networks, called local networks. A proxy-server is connected to each local network attending a subset of system clients.

In this model, we assume that all multimedia contents have the same size and that every server can hold V_S (V_C assigned to caching scheme and V_M assigned to mirroring) of N videos, where N is the number of contents in the

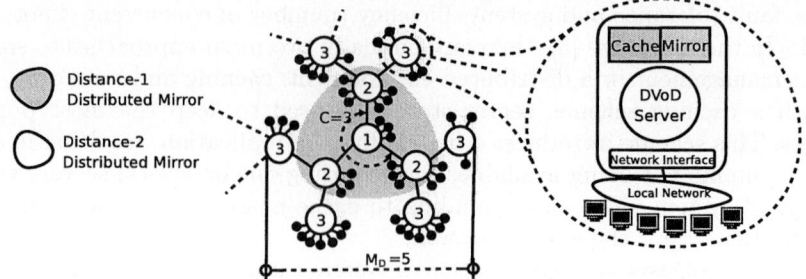

Fig. 1. Distributed VoD architecture.

system catalog. With respect to the DVoD architecture, we define C as the mean topology connectivity and M_D as the maximum service distance; this is defined as the longest path between the farthest nodes in the topology (see Figure 1).

The mean service distance is calculated as the sum of system-request percentages attended at different distances d, and is given by (1), where R_d can be evaluated as the total number of requests (R_{total}) minus the percentage of requests attended at a distance less than d, (R_{AT}), given by (2).

R_{AT} depends on the request percentage captured by cached videos and distance-d distributed mirror videos. However, the cached videos only influence local requests, due to the fact that all cache contents would be the same in all local servers. Therefore, the model assumes that remote requests can only be served from mirrored videos.

To evaluate the percentage of requests served from cache (3), we have to estimate cache-hit probability (P_{hc}), calculating the total access frequency of the most popular contents V_c. We assume that content popularity is modeled by the Zipf distribution shown in (3), where Z indicates degree of popularity [2]. On the other hand, the probability of a request being served by the distributed mirror at distance d basically depends on mirror size. This size is evaluated in (4) taking individual server-mirror capacity into account, as well as the number of partial-mirrors at distance d and the mean probability of accessing a mirror video, P_{hm} (evaluated as $\frac{1}{N}$).

Table 1. Analytical model to evaluate the mean service distance.

Measure	Expression	
Mean Service Distance	$D_{ms} = \sum_{d=1}^{M_D} R_d$	(1)
% Requests attended at distance **d**	$R_d = R_{Total} - R_{AT}^{d-1}$ $= MAX\left\{0, 1 - \left(R_{Cache} + R_{MirrorD}^{d-1}\right)\right\}$ $= MAX\left\{0, 1 - \left(P_{hc} + V_M \times P_{hm} \times \frac{1-C^d}{1-C}\right)\right\}$	(2)
% Requests attended by local cache	$R_{Cache} = P_{hc} = \sum_{m=1}^{V_c} \frac{1}{m^z \times \sum_{i=1}^{N} \frac{1}{i^z}}$	(3)
% Requests attended by mirrors at distance **d**	$R_{MirrorD}^d = V_M \times P_{hm} \times \sum_{i=0}^{d-1} C^i$ $= V_M \times P_{hm} \times \frac{1-C^d}{1-C}$	(4)

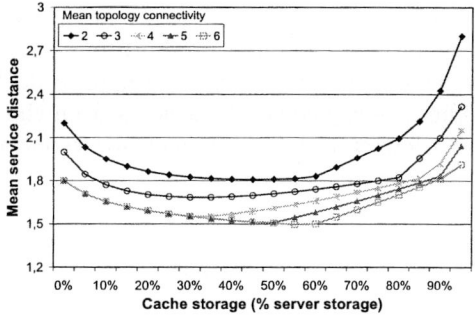

Fig. 2. Mean service distance according to the storage distribution.

In order to examine the influence of combined scheme (caching and mirroring), we use the Dms (1) for different storage distributions between both schemes and for different topology connectivities. Figure 2 shows the results with the assumption that the DVoD architecture has a maximum service distance of 10 (MD), a catalog with 100 videos (N), and a server storage capacity (V_S) of 20 videos. The x-axis represents the percentage of server-storage dedicated to caching, while the rest is assigned for mirroring.

As can be observed, the best mean service distance is never achieved by using only one scheme. When all server storage is dedicated to mirroring (left side) the Dms is too high, while, when only caching is used (right-hand side) the results are even worse. The best Dms are always obtained as result of a combination of both schemes. This behavior is independent of architecture connectivity. With these results, we can confirm that combining both schemes (caching and mirroring) improves the effectiveness of DVoD architecture.

3 Storage System Design Implication due to Cache

In this session, we will analyze the design issues in the storage system in order to incorporate cache and mirror in video servers. In these DVoD systems, the cache must capture a high degree of local client demands in order to reduce the interaction between distributed mirrors. This interaction is produced when a video is not in the local server and must be attended by an adjacent server at distance-1 or distance-2 (as indicated in Figure 1). The service of a client request by adjacent servers consumes the local network resource as well as interconnection links, producing performance deterioration. In this scene, the cache content must constantly (partially or totally) be replaced, following locality criteria, in order to capture clients' dynamic behavior. This replacement inevitably introduces writing operations in the storage system.

The incorporation of writing operations has design implications throughout the storage system. The main points that must be taken into account are: 1) buffer cache: writing operations could affect the cache performance of reading operations, 2) Admission control: the disk resource required for writing is usually higher and 3) allocation policy: this must guarantee disk performance after successive videos replacements.

The buffer cache consists of using the main memory to hold video blocks, which are supposed to be used again in a short period of time, in order to avoid disk access. We have added frequency information in the traditional LRU replacement policy. The new replace policy selects from the least recently used blocks, which have the lowest access frequency for replacement. The writing blocks are also controlled by frequency information, and the replacement policy assigns an initial frequency value for the writing blocks, in accordance with the video position which the block belongs to. In this way, writing blocks from the initial part of the video have more probability of being cached than the those from the tail part of the video.

The Admission Control has to manage all available resources. This module has to provide a mechanism for disk-block allocation and oversees disk-space assignment between the cache and mirror. Rather than an elaborated disk-allocating policy, our allocation policy is based on large disk blocks. Writing operations are also overseen by Admission control, in which resource requirements are differentiated according to disk-operation type.

4 Writing Operation Cost Evaluation

In order to evaluate the impact of cache and mirror on the storage system, we are interested in those parameters that could be affected by writing operations. 1)We will evaluate how the writing operations could affect the hit ratio of the buffer cache; 2) how they could affect the disk-transfer bit ratio and 3) how the successive writing operations can affect global storage system throughput. We carried out the experiment on our VoD protype system equipped with an Adaptec ATA RAID with four hard disks. Zipf and Poission distributions were used to generate the workload.

Figure 3 shows the cache-hit ratio in two caching replacement policies when 10% of disk operations are writing. Dedicating a cache with a size of 3% of the total data, LRU achieves a hit ratio of 15%, while our policy is able to achieve 27.5%. In general terms, our policy is able to achieve an improvement of between 5 and 11%. The writing block frequency assignment policy is seen to have a special impact on hit ratio when buffer-cache size is small (3%). This is due to the fact that LRU suffer from writing operations.

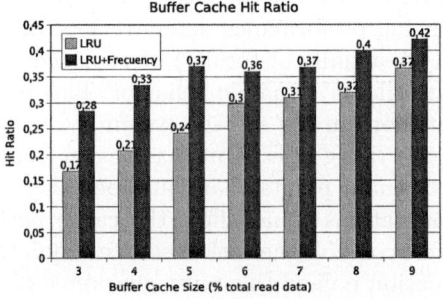

Fig. 3. Cache hit ratio with LRU and LRU+Frecuency.

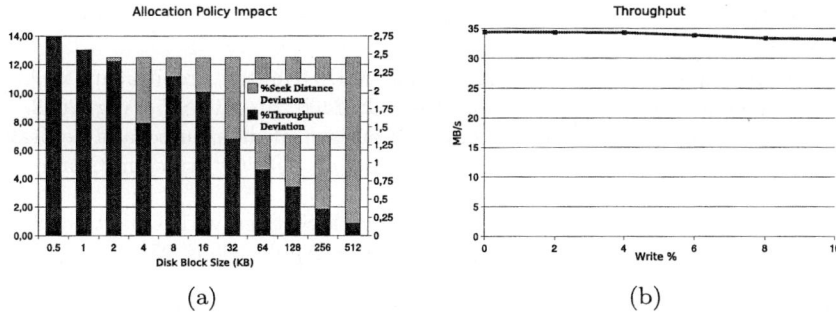

Fig. 4. (a) Allocation Policy Impact vs Block Size. (b) Disk Throughput with writing.

Figure 4 (a) shows disk throughput and seek-distance deviation in different allocation policies, depending on disk-block size. Regardless of disk block size, it shows a seek distance deviation of 12.5%. On the other hand, the impact of the allocation policy on disk throughput is reduced as the disk-block size increases. An optimal policy could only improve up to 0.91% when disk blocks are 64KB. This is because seek time is neglected when the disk scheduler reads 64KB for each disk request. These experimental results justify our allocation policy design. Furthermore, with a large disk block size such as 64KB, the cache replacement present in our storage system would affect storage performance by less than 1.00%.

In order to evaluate the impact of the writing operation on the storage system, we have generated different writing percentages. Figure 4 (b) shows disk throughput with a range of 2% to 10% of writing operations. A degradation of only 3.47% is observed when 10% of disk operations are cache replacements. This degradation is taken in account in the admission control policy in order to evaluate resource availability.

5 Conclusions

In this paper, we have presented a new combined storage management scheme that uses both caching and distributed mirroring. To evaluate the goodness of this approach, we have developed an analytical model based on mean service distance as a measurement of global system-performance. In accordance with this analysis, we have shown that a combined scheme achieves the best result, independently of the underlying distributed architecture. This model also provides a method by which to obtain optimal storage distribution between the cache and mirror.

The viability of this combined scheme has been validated in a prototype. Experimental results have shown that up to 10% of writing operations can be tolerated without affecting disk throughput. Furthermore, for writing operations, the modified replacement policy in the buffer cache can achieve up to 11% improvement with respect to a traditional design.

References

1. F. Cores, A. Ripoll, E. Luque, Double P-Tree: A Distributed Architecture for Large-Scale Video-on-Demand, Euro-Par 2002, LNCS 2400, pp. 816-825, Aug. 2002.
2. S.A. Barnett and G. J. Anido, "A cost comparison of distributed and centralized approaches to video-on-demand," IEEE Journal on Selected Areas in Communications, vol. 14, pp. 1173-1183, August 1996.
3. J.C.I. Chuang, M.A. Sirbu, Distributed Network Storage with Quality-of-Service Guarantees, Journal of Network and Computer Applications 23(3), pp 163-185, July 2000.
4. C. Shahabi, F. Banaei-Kashani, Decentralized Resource Management for a Distributed Continuous Media Server, IEEE Transactions on Parallel and Distributed Systems, Vol. 13, No. 11, pp 1183-1200, November 2002.

An Efficient Fair Queuing Algorithm for Guaranteed Disk Bandwidth*

Young Jin Nam and Chanik Park

Department of Computer Science and Engineering
Pohang University of Science and Technology/PIRL
Kyungbuk, Republic of Korea
{yjnam,cipark}@postech.ac.kr

Abstract. This paper proposes an efficient fair queuing algorithm for a disk resource shared by multiple I/O applications each of which demands guaranteed bandwidth. The proposed algorithm determines a set of eligible I/O requests that can be processed before any backlogged I/O request reaches its virtual finish time. Next, it selects the most eligible I/O request that minimizes disk head movement overhead.

1 Introduction

Disk QoS requirement is commonly represented with the amount of disk bandwidth because partitioning disk bandwidth is known as a solution to meet different QoS requirements altogether. Since processing I/O requests in disk resources typically involves high overhead of mechanical disk head movements, existing packet-based fair queuing algorithms for network QoS cannot be directly applied for disk QoS. A few research efforts have been made to devise an I/O traffic control scheme for disk resources. Bruno *et. al.* in [1] proposed a YFQ algorithm that is a derivative of a packet-based fair queuing algorithm. However, YFQ attempts to reduce disk head movement overhead only when more than one I/O requests have the same virtual finish time. Lee *et. al.* in [2] introduced a rate-based QoS control based on a leaky bucket [3]. Lumb *et. al.* proposed the Facade architecture [4] that assigns a deadline time to each incoming I/O request only if the current IOPS is not greater than its target IOPS and schedules the pending I/O requests according to the earliest deadline first scheme. Note that only YFQ is based on fair queuing among the previous schemes. This paper proposes an efficient fair queuing algorithm for the disk resource that fully takes into account the overhead caused by disk head movements when enforcing a given disk QoS requirement.

---

* The authors would like to thank the Ministry of Education of Korea for its financial support toward the Electrical and Computer Engineering Division at POSTECH through its BK21 program. This research was also supported in part by the Korea Science and Engineering Foundation (KOSEF) under grant number R01-2003-000-10739-0 and by HY-SDR IT Research Center.

2 The Proposed Algorithm

Each I/O application has its own I/O request queue denoted with Q_i. Each queue (Q_i) maintains its own virtual start time (S_i) and virtual finish time (F_i). We denote with ϕ_i a resource weight allotted to Q_i. The notation of r_i^k denotes the k-th pending I/O request in Q_i. The notation of l_i^k represents the request size of r_i^k. Given an arrival of an I/O request r_i, if Q_i is not empty, r_i is simply backlogged at the end of Q_i. Otherwise, S_i, F_i, and r_i^1 are updated as follows: $S_i = \max\{v(t), F_i\}$, $F_i = S_i + \frac{l_i^1}{\phi_i}$, and $r_i^1 = r_i$. We introduce the concept of *Virtual Finish Time (VFT) miss*, which occurs when any F_i exists such that $v(t) > F_i$ for each backlogged Q_i, $1 \le i \le N$, where N is the number of I/O applications. Algorithm 1 describes how to schedule an I/O request from a set of queues for disk I/O service. First, it determines a set of eligible I/O requests, denoted $R_{eligible}$, that entails no VFT miss from the requests at the head of I/O queues. Note that scheduling any I/O request in $R_{eligible}$ does not diminish a guaranteed QoS level. Second, it selects an I/O request denoted as r_c^1 that incurs the minimum disk head movement overhead. The disk head movement overhead for an I/O request represents the overhead time to move from the current head position to the position to serve the I/O request. The proposed algorithm simply defines the overhead time as the absolute distance between the start block addresses of the given I/O request and its previous I/O request. More accurate computation of the overhead time needs to estimate a seek time and a rotational delay between two I/O requests. Third, it removes r_c^1 from Q_c. If Q_c is backlogged, its subsequent I/O request becomes a new r_c^1. Fourth, it assigns the global virtual time of $v(t)$ as the virtual start time S_c of the scheduled I/O request r_c^1, as with SFQ algorithm. Finally, it updates the virtual start time (S_c) with the previous virtual finish time and then increases the virtual finish time (F_c) in proportion to the request size l_c^1 if Q_c is backlogged.

Algorithm 1: The proposed I/O scheduling scheme
---
  determine $R_{eligible} = \{r_i^1 | r_i^1$ does not cause VFT miss$\}$ when r_i^1 is scheduled;
  choose r_c^1 of the minimum disk overhead among $R_{eligible}$;
  remove r_c^1 from Q_c; update $v(t)$; /* using $v(t) = S_c$ */
  if $Q_c \ne \emptyset$ **then** $S_c = F_c$; $F_c = S_c + \frac{l_c^1}{\phi_s}$;
---

3 Performance Analysis

Worst-Case Fairness Analysis: The worst-case fairness can be represented by $\left|\frac{S_i(t_1,t_2)}{\phi_i} - \frac{S_j(t_1,t_2)}{\phi_j}\right|$, where $S_i(t_1,t_2)$ represents an amount of service received by C_i for a time interval $[t_1, t_2)$. We consider a condition where the worst-case fairness occurs with two competing I/O applications, C_i and C_j. Two I/O requests of size L_i^{max} are backlogged in Q_i. An I/O request of size L_j^{max} is queued at the head of Q_j, where L_i^{max} and L_j^{max} represent the maximum request size of C_i and C_j, respectively. We assume that $S_i = F_i^k$ for C_i and $F_j = F_i^k$ for C_j at that point. The proposed algorithm schedules the first I/O request of size L_i^{max} from C_i. Next, we have $S_i = F_i^{k+1} = F_i^k + \frac{L_i^{max}}{\phi_i}$ and $F_i = F_i^{k+2} = F_i^k + 2\frac{L_i^{max}}{\phi_i}$. Assume that t_1 represents the time when the processing of the first I/O request from C_i is completed. At time t_1, the proposed algorithm begins to schedule the I/O requests in C_j, because the scheduling of the I/O request in C_i will cause the VFT miss in C_j. Obviously, the I/O requests in C_j are scheduled unless $F_j > F_i = F_i^{k+2}$. When the proposed algorithm schedules the I/O request of C_j whose $F_j = F_i$, we obtain $S_j = F_i^{k+2}$. At this point, both I/O requests at the head of Q_i and Q_j belong to $R_{eligible}$. Assuming that the disk head movement overhead of the I/O request of size L_j^{max} in Q_j is smaller than the I/O request in Q_i, the proposed algorithm schedules the I/O request at the head of Q_j. Mark with t_2 the time when the processing of this I/O request is completed. Given this worst-case scenario, the maximum amount of scheduled I/O from C_j without serving I/O from C_i corresponds to the I/O amount processed in the time period between t_1 and t_2. Thus, the worst fairness between the two classes is written as $(L_j^{max} + 2\frac{L_i^{max}}{\phi_i}\phi_j + L_j^{max})/\phi_j$.

Simulation Results: We implement the proposed algorithm as a driver-specific disk scheduling algorithm within the DiskSim [5]. We use a IBM DNES 309170W SCSI disk. We employ four different sets of I/O applications: IOS_1, IOS_2, IOS_3, and IOS_4. Each of IOS_1 and IOS_2 is featured by two competing I/O workloads of C_1 and C_2 that have the same distribution of I/O request sizes ($U(4, 32)$ for IOS_1, $U(4,4)$ for IOS_2, where $U(\alpha, \beta)$ means a uniform distribution between αKB and βKB) and the same resource weights ($\{\phi_1, \phi_2\} = \{10, 10\}$ for IOS_1 and IOS_2). Each of IOS_3 and IOS_4 is characterized by two competing I/O workloads that have different distribution of I/O request sizes ($U(4, 32)$ and $U(4, 64)$ for C_1 and C_2 of IOS_1, $U(4,4)$ and $U(8,8)$ for C_1 and C_2 of IOS_2) and different resource weights ($\{\phi_1, \phi_2\} = \{20, 10\}$ for IOS_3 and IOS_4). The inter-arrival rate of each I/O application is configured fast enough to ensure that its queue is always backlogged. We devise three metrics to quantitatively evaluate fairness and I/O throughput: Φ_1, Φ_2, and Φ_3. Denote with T_I a monitoring interval for the performance measurement ($T_I = 1$ second in our simulations). The Φ_1 and Φ_2 represent the average value and the maximum value of $\mathcal{F}_{i,j}(k)$ over all the monitoring intervals, where $\mathcal{F}_{i,j}(k) = \left| \frac{S_i(kT_I,(k+1)T_I)}{\phi_i} - \frac{S_j(kT_I,(k+1)T_I)}{\phi_j} \right|$. Finally, Φ_3 represents the average value of the aggregate I/O throughput (MB/s) for each C_i over all the monitoring intervals. Table 1 shows that the average and worst-case fairness and I/O throughput of the proposed algorithm and YFQ for the four I/O application sets.

The Φ_1 of the proposed algorithm is about 1.6–2.0 times worse than that of YFQ. The Φ_2 of the proposed algorithm is about 1.4–3.2 times worse than that of YFQ. The Φ_2 of the proposed algorithm are bounded by the associated worst-case fairness values obtained from its theoretical analysis, except for IOS_3. We think that the exceptional case is mainly due to the fact that measuring

Table 1. Average and worst-case fairness(Φ_1, Φ_2) and I/O throughput(MB/s)(Φ_3) of the proposed algorithm and YFQ for the four I/O application sets.

Set	YFQ			Proposed			Set	YFQ			Proposed		
	Φ_1	Φ_2	Φ_3	Φ_1	Φ_2	Φ_3		Φ_1	Φ_2	Φ_3	Φ_1	Φ_2	Φ_3
IOS_1	3.41	10.40	1.22	5.56	15.20	1.28	IOS_3	4.48	16.80	1.39	9.03	55.20	1.43
IOS_2	0.63	1.60	1.12	1.19	3.20	1.79	IOS_4	0.61	1.60	1.57	1.07	3.20	2.22

the serviced I/O amount is simply on the basis of the completed I/O requests. Finally, the Φ_3 of the proposed algorithm is observed to be about 1.02–1.58 times higher than that of YFQ. A higher performance gain was obtained with IOS_2 and IOS_4 having sequential I/O access patterns, because a sequential I/O access pattern contains larger opportunities to optimize disk head movements than a random I/O access pattern.

4 Conclusion and Future Work

We proposed an efficient fair queuing algorithm for disk resources and conducted its performance analysis. The key of the proposed algorithm is to select a set of eligible I/O requests not causing a VFT miss and then schedule the most eligible I/O request that minimizes disk head movement overhead. Simulation results showed that the proposed algorithm outperformed YFQ by 1.02–1.59 times in terms of I/O throughput with tightly bounded fairness. In future work, we plan to implement the proposed algorithm in the Linux kernel and evaluate its performance and overhead under real-world I/O workload traffic featured by self-similarity. In addition, we need to examine the problem of bounding the maximum delay for a given I/O rate by adding an I/O traffic shaping model.

References

1. Bruno, J., Brustoloni, J., Gabber, E., Ozden, B., Silberschatz, A.: Disk scheduling with quality of service guarantees. In: Proceedings of the IEEE International Conference on Multimedia Computing and Systems. (1999)
2. Lee, H., Nam, Y., Park, C.: Regulating I/O performance of shared storage with a control theoretical approach. In: Proceedigns of the 21st IEEE Mass Storage Systems Symposium/12th NASA Goddard Conference on Mass Storage Systems and Technologies. (2004)
3. Turner, J.: New directions in communications. IEEE Communications **24** (1986)
4. Lumb, C., Merchant, A., Alvarez, G.: Facade: Virtual storage devices with performance guarantees. In: Proceedings of Conference on File and Storage Technologies. (2003)
5. Ganger, G., Worthington, B., Patt, Y.: The DiskSim Simulation Environment Version 2.0 Reference Manual. CMU. (1999)

Topic 13
Theory and Algorithms for Parallel Computation

Kaklamanis Christos, Nancy Amato, Danny Krizanc, and Andrea Pietracaprina

Topic Chairs

As theory of modelling and algorithms form the basis for parallel computing, the theory workshop is still a key part of Europar despite of the diversity of the topics of the conference. This year, the submissions mainly concentrate on algorithms to improve computation and communication issues.

In more detail, Hansen, Kubale, Kuszner, and Nadolski present a distributed probabilistic algorithm for coloring the nodes of a graph. They show that their algorithm is optimal on several graph classes and give bounds on its running time for arbitrary graphs.

Spielman, Teng, and Üngor propose off-centers as an alternative type of Steiner points for refining Delaunay triangulations. They give a parallel algorithm for Delaunay refinement with off-centers with improved parallel running time compared to previous approaches.

In the paper entitled "Near-Optimal Hot-Potatoe Routing on Trees", Busch, Magdon-Ismail, Mavronicolas, and Wattenhofer present packet routing algorithms in tree-shaped networks without buffers; these algorithms are optimal within logarithmic or polylogarithmic factors.

Cáceres, Dehne, Mongelli, Song, and Szwarcfiter present a new algorithm in the BSP/CGM model for computing a spanning tree and the connected components of a graph. This algorithm is significantly simpler than previous ones.

Summarizing, we found the collection of papers satisfying and diverse enough to enlight several theoretical aspects of parallel computation. Finally, we wish to thank all the authors and the referees for their efforts.

Distributed Largest-First Algorithm for Graph Coloring

Jennie Hansen[1], Marek Kubale[2], Lukasz Kuszner[2], and Adam Nadolski[2]

[1] Department of Actuarial Mathematics and Statistics, Heriot-Watt University, Edinburgh EH14 4AS, UK
[2] Department of Algorithms and System Modeling, Gdańsk University of Technology, Narutowicza 11/12, 80-952 Gdańsk, Poland*

Abstract. In the paper we present a distributed probabilistic algorithm for coloring the vertices of a graph. Since this algorithm resembles a largest-first strategy, we call it the distributed LF (DLF) algorithm. The coloring obtained by DLF is optimal or near optimal for numerous classes of graphs e.g. complete k-partite, caterpillars, crowns, bipartite wheels. We also show that DLF runs in $O(\Delta^2 \log n)$ rounds for an arbitrary graph, where n is the number of vertices and Δ denotes the largest vertex degree.

1 Introduction

We discuss the vertex coloring problem in a *distributed network*. Such a network consists of processors and bidirectional communication links between pairs of them. It can be modeled by a graph $G = (V, E)$. The set of vertices V corresponds to processors and the set E of edges models links in the network. To color the vertices of G means to give each vertex a color in such a way that no two adjacent vertices get the same color. If at most k colors are used, the result is called a *k-coloring*.

We assume that there is no shared memory. Each processor knows its own links and its unique identifier. We want these units to compute a coloring of the associated graph without any other information about the structure of G. We assume that the system is synchronized in *rounds*. The number of rounds will be our measure of the time complexity. Such a model of coloring can be used in a distributed multihop wireless network to eliminate packet collisions by assigning orthogonal codes to radio stations [1].

In evaluating the performance of a random distributed coloring algorithm $\mathcal{A}$ on a graph G there are at least two random variables of interest: $C_\mathcal{A}(G)$, the number of colors used by the algorithm to color graph G and $T_\mathcal{A}(G)$, the number of rounds used to color G.

A good distributed algorithm is one where $C_\mathcal{A}(G)$ is close to $\chi(G)$, the chromatic number of G, and where $T_\mathcal{A}(G)$ is small relative to the number of vertices

---

* Supported in part by KBN grants: 4T11C04725 and 4T11C00324.

in G. The difference $C_\mathcal{A}(G) - \chi(G)$ can be viewed as a measure of the effectiveness of the algorithm.

We note that it is not always easy to achieve both speed and effectiveness. In [3] a distributed algorithm for $(\Delta+1)$-coloring of graphs was given, where Δ is the largest vertex degree in a graph. Also, analysis of its time complexity was presented. It was proved to run in $O(\log n)$ time. We shall refer to this as the *trivial* algorithm (same as in [2]). The trivial algorithm is extremely simple and fast, however not optimal. In fact, the number of colors used by the algorithm is close to Δ even if the graph is bipartite. This is not surprising as the trivial algorithm has no mechanism of economizing on colors. Further improvements to the trivial algorithm were proposed in [2]. In that paper a new algorithm which is able to compute a coloring using $O(\Delta/\log\Delta)$ colors was given, but it works on triangle-free graphs only and fails on some instances of the problem.

One way to improve the performance of a distributed algorithm is to introduce a strategy into the algorithm which is known to be effective in non-distributive coloring algorithms. For example, it is well known that it is much better to color the vertices in a largest degree first order as in some sequential heuristics like largest-first (LF), smallest-last (SL) and saturation largest-first (SLF) (see e.g. [5]). That is why this strategy is incorporated into the distributed largest-first (DLF) algorithm which was introduced in [6] and which is analyzed thoroughly in this paper.

The paper is organized as follows. In the next section the DLF algorithm is carefully defined. In Section 3 we investigate the time complexity of the DLF algorithm in a special case of regular graphs with fixed degree and in the general case, where no assumptions are made about the structure of G. In the final section we consider the effectiveness of DLF and we show that for some classes of graphs we have $C_{\text{DLF}}(G) - \chi(G) \leq 1$. This holds even when $\chi(G) << \Delta$. In particular, on these classes the DLF algorithm is better than the trivial algorithm in the worst case.

2 DLF Algorithm

In the DLF algorithm each vertex has three parameters:

- its degree in graph: $\deg(v)$,
- random value, which is generated locally and independently: rndvalue(v),
- palette of forbidden colors, which have already been used by its neighbors: usedcolor(v) (initially empty).

Within each round the vertices are trying to obtain their colors. If the same color is chosen by neighboring vertices then parameters: $\deg(v)$ and rndvalue(v) determine the precedence. Specifically, for neighboring vertices $v_1, v_2 \in V$, we say that the priority of v_1 is higher than that of v_2 if:

$$\deg(v_1) > \deg(v_2)$$

or

$$(\deg(v_1) = \deg(v_2)) \text{ and } (\text{rndvalue}(v_1) > \text{rndvalue}(v_2))$$

Within each round every uncolored vertex v executes the following five steps:

1. Choose parameter rndvalue(v) uniformly distributed on [0..1].
2. Send to all neighbors the following parameters: deg(v), rndvalue(v), and the first legal color (not on vertex v's list of forbidden colors).
3. Compare its own parameters with those received from its neighbors and check which vertex has the highest priority.
4. If vertex v's proposed color does not clash with proposals from its neighbors or if v has the highest priority amongst its neighbors, keep the proposed color, send message to neighbors and stop.
5. If not, update list usedcolor(v).

3 Time Complexity

In this section we investigate the complexity of the DLF algorithm. Let $\mathcal{A}$ be a randomized algorithm working on a graph with n vertices. We say that the complexity of $\mathcal{A}$ is $O(F(n))$ if there exist a constant $c > 0$ and $q(n) \in (0..1)$ such that $1/(1 - q(n))$ is $O(F(n))$ and the probability that the processing time of $\mathcal{A}$ exceeds $cF(n) + t$ rounds is at most $q(n)^t$, i.e.

$$\Pr[T_{\mathcal{A}}(G) > cF(n) + t] \leq q(n)^t. \tag{1}$$

It is easy to see that if the complexity of $\mathcal{A}$ is $O(F(n))$, then the mean of $T_{\mathcal{A}}(G)$ is $O(F(n))$ as well.

Proposition 1. [6] *For any r−regular graph G with n vertices the algorithm DLF runs in $O(\log n)$ rounds.*

Proof. Let G be an r−regular graph and let v be an uncolored vertex at the beginning of some round k. Let $\Pr(v)$ denote the probability that v gets a color during this round and let $UN(v)$ denote the number of uncolored neighbors of vertex v at the start of round k. Suppose that rndvalue(v) = x, then the conditional probability that v gets a color during the k-th round, $\Pr[v \mid \text{rndvalue}(v) = x]$, is given by $\Pr[v \mid \text{rndvalue}(v) = x] = x^{UN(v)} \geq x^r$. It follows that

$$\Pr(v) \geq \int_0^1 x^r \, dx = \frac{1}{r+1}.$$

Using this bound and applying established techniques for analyzing probabilistic recurrence relations (see [4], Theorem 1.1), we obtain

$$Pr[T_{\text{DLF}}(G) \geq \lfloor \log_{(r+1)/r}(n) \rfloor + t + 1] \leq \left(\frac{r}{r+1}\right)^{t-1}. \quad \square$$

3.1 Time Complexity in a General Case

Now we establish a general upper bound on the complexity of the DLF algorithm.

Theorem 1. *For any n-vertex graph G the DLF algorithm runs in $O(\Delta^2 \log n)$ rounds.*

Proof. Let us consider a graph G with n vertices. The sketch of our reasoning is as follows: we divide the coloring process of G into several phases and estimate the number of rounds in each phase of the process. The first phase starts at the beginning of coloring process and terminates in the first round in which all vertices of degree Δ are colored. The second phase starts in the next round and is terminated in the round in which all vertices of degree $\Delta - 1$ are colored, and so on.

For $i = 1, \ldots, \Delta$, let V_i denote the set of vertices in G of degree i and let $n_i = |V_i|$. For each $i = 1, \ldots, \Delta$, define the random variable T_i to be equal to the first round in which all the vertices in G of degree at least i are colored, i.e. T_i is the completion time of the $(\Delta + 1 - i)$-th phase of the coloring process. For notational convenience we put $T_{\Delta+1} = 0$.

Now, let k be a positive integer such that $1 \leq k \leq \Delta$. We estimate the probability $\Pr[T_k > s_2 \mid T_{k+1} \leq s_1]$ for some positive integers $s_1 < s_2$, i.e. the conditional probability that the $(\Delta + 1 - k)$-th phase finishes after the s_2-th round given that the previous phase has been completed by the s_1-th round. Consider any uncolored vertex v of degree k in some round l, $s_1 < l \leq s_2$. Since $T_{k+1} \leq s_1$, v has no uncolored neighbors of degree greater than k. Moreover, it has at most k uncolored neighbors of degree k demanding the same color as v, so the probability that v will be colored in round l is at least $1/(k+1)$ (see the proof of Fact 1). Therefore, since the number of uncolored vertices of degree k is at most n_k after the start of the $(s_1 + 1)$-st round, we can apply Theorem 1.1 from [4] again to obtain the following bound

$$\Pr[T_k > s_2 \mid T_{k+1} \leq s_1] \leq \left(\frac{k}{k+1}\right)^{t-1}, \qquad (2)$$

where $s_2 = s_1 + \lfloor \log_{(k+1)/k} n_k \rfloor + 1 + t$ and integer $t \geq 1$.

Now, let $t_1 > \ldots > t_\Delta > t_{\Delta+1} = 0$ be a sequence of integers. We use the Bayes identity to obtain

$$\Pr[T_k \leq t_k] \geq \Pr[T_k \leq t_k \mid T_{k+1} \leq t_{k+1}] \cdot \Pr[T_{k+1} \leq t_{k+1}], \qquad (3)$$

for $k = 1, \ldots, \Delta$. Iterating (3) and noting that $T_{\text{DLF}}(G) = T_1$ we have

$$\Pr[T_{\text{DLF}}(G) \leq t_1] \geq \prod_{k=1}^{\Delta} \Pr[T_k \leq t_k \mid T_{k+1} \leq t_{k+1}]. \qquad (4)$$

Now fix $t \geq 1$ and for $k = 1, \ldots, \Delta$, set

$$t_k = \sum_{i=k}^{\Delta} \left(\lfloor \log_{1+1/i} n_i \rfloor + t + 1 \right).$$

From (2) and (4) it follows that

$$\Pr[T_{\text{DLF}}(G) \leq t_1] \geq \prod_{k=1}^{\Delta}\left(1-\left(\frac{1}{1+1/k}\right)^{t-1}\right) \geq$$

$$\left(1-\left(\frac{1}{1+1/\Delta}\right)^{t-1}\right)^{\Delta} \geq 1-\Delta\left(\frac{1}{1+1/\Delta}\right)^{t-1}. \quad (5)$$

The last estimate follows from the Bernoulli inequality. The obvious bound $n_i \leq n$ implies that $t_1 \leq \Delta \log_{1+1/\Delta}(n) + (t+1)\Delta$. Therefore, (5) leads us to the following bound for all real $t \geq 1$,

$$\Pr[T_{\text{DLF}}(G) > \Delta\log_{1+1/\Delta}(n) + (t+1)\Delta] \leq \Delta\left(\frac{1}{1+1/\Delta}\right)^{t-2}. \quad (6)$$

After putting $u = t\Delta$ into (6) we obtain

$$\Pr[T_{\text{DLF}}(G) > \Delta(\log(n)/\log(1+1/\Delta)+1)+u] \leq 2(\Delta+1)\left(\frac{\Delta}{1+\Delta}\right)^{u/\Delta}. \quad (7)$$

Then for $s = u + \log_q(2(\Delta+1))$, where $q = (\Delta/(\Delta+1))^{1/\Delta}$, we obtain the following inequality

$$\Pr[T_{\text{DLF}}(G) > \Delta(\log(2(\Delta+1)n)/\log(1+1/\Delta)+1)+s] \leq q^s, \quad (8)$$

From (8) it follows that the algorithm runs in $O(\Delta^2 \log n)$ time. □

Notice that the estimates we have used are far from tight, and the bound for the complexity is not effective if $\Delta = \Theta(n)$. However, graphs which occur in applications of the model of distributed graph coloring are typically sparse. For example, if we consider graphs for which $\Delta = O(\log n)$ then the complexity of DLF is $O(\log^3 n)$. Further, if we cannot bound Δ, but we have some information about the structure of graph, so that a better bound on t_1 can be derived, then using inequality (5) we can obtain a tighter bound for the complexity of the algorithm.

In the next section we show that the DLF algorithm works particularly well on certain graph classes and for some of these classes it can even be shown that the number of rounds is bounded by a fixed constant independent of the number of vertices in the graph.

4 Effectiveness

In this section we consider the accuracy of the DLF algorithm and compare its effectiveness to the effectiveness of the trivial algorithm.

Much work has been done on characterizing the performance of the sequential LF coloring algorithm (complexity, accuracy, hard to color graphs, etc.). So it

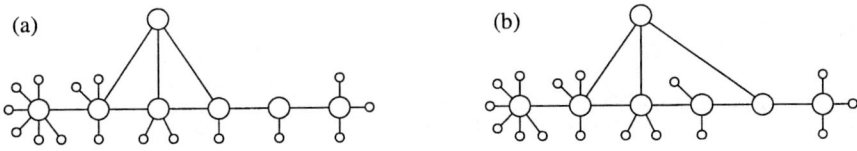

Fig. 1. Example graphs.

would be useful if known performance results for the LF algorithm could provide bounds on the effectiveness of the DLF algorithm since the latter incorporates the largest-first heuristics.

Unfortunately, this is not the case. For example, consider the three-colorable graphs in Figure 1. It is easy to check that graph (a) is colored optimally by the LF algorithm, whereas it is colored with 4 colors by the DLF algorithm. On the other hand, graph (b) is colored optimally by the DLF algorithm and with one additional color by the LF algorithm.

So what can be said about the effectiveness of the DLF algorithm? Recall that $C_{\text{DLF}}(G)$ denotes the number of colors used by the DLF algorithm to color G. It is clear that for any graph G we have the following trivial bounds

$$\chi(G) \leq C_{\text{DLF}}(G) \leq \Delta + 1,$$

but these bounds are useless in assessing the effectiveness of the DLF algorithm when $\chi(G) \ll \Delta$. On the other hand, in [6] experimental results are obtained which indicate that the expected number of colors used by DLF is much smaller than Δ on random graphs. For some classes of graphs it is even possible to show that the DLF algorithm is always "nearly optimal". In fact, the examples given below show that there are infinite classes of graphs $\mathcal{G}$ such that for all $G \in \mathcal{G}$

$$C_{\text{DLF}}(G) - \chi(G) \leq 1$$

Proposition 2. *The DLF algorithm produces an optimal coloring for complete k-partite graphs.*

Proof. When $k = 1$ the fact is trivial. Suppose that Fact 2 is true for $i = 1, 2, \ldots k - 1$. We will show it to be true for $i = k$. Let $V = P_1 \cup P_2 \cup \ldots \cup P_k$, where P_i are independent sets in G. Let $v \in P_i$ be a vertex colored during the first round. Observe that the first color can be assigned to vertices from P_i only. So after the first round only a number of vertices from P_i are colored. In the second round, vertices from $\overline{P_i} = V \backslash P_i$ have color number 1 added to their lists of forbidden colors. So, vertices in P_i which were not colored in the first round obtain color 1 without conflicts during round 2. Now, we can consider vertices from $\overline{P_i}$ separately and $\overline{P_i}$ forms a complete $(k-1)$-partite graph which will be colored with $k - 1$ colors. □

Corollary 1. [6] *For all $p, q \geq 1$, $C_{\text{DLF}}(K_{p,q}) = 2$, where $K_{p,q}$ is a complete bipartite graph.*

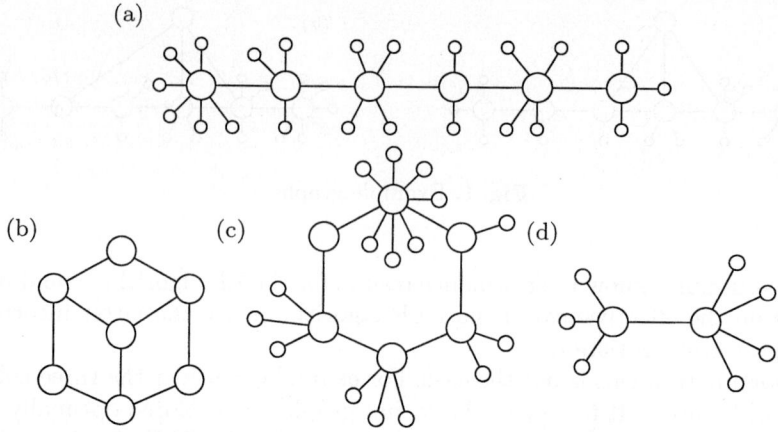

Fig. 2. An example of: (a) caterpillar, (b) BW_3 - bipartite wheel, and (c) crown, (d) double-star.

Complete graph K_n is colored with $n = \Delta + 1$ colors by the DLF algorithm. In addition, when we remove one or two edges from K_n the DLF algorithm still produces an exact coloring.

Proposition 3. *For $n \geq 2$, $C_{\mathrm{DLF}}(K_n - e) = \chi(K_n - e) = n - 1$.*

Now consider some special classes of sparse graphs: caterpillars, double-stars, bipartite wheels and crowns.

Proposition 4. *Caterpillars are colored with at most three colors by the DLF algorithm.*

Proof. A caterpillar is a tree whose leaves are attached to a path (see Fig. 2(a)). Vertices from the path will be colored with at most three colors and each leaf gets a color 1 or 2. □

Proposition 5. *For $k \geq 4$ the bipartite wheel BW_k is colored optimally within two rounds and BW_3 is colored optimally within three rounds.*

Proof. A bipartite wheel is a graph obtained from an even cycle of length $2k$ by adding a central vertex connected to every second vertex of the cycle (see Fig. 2(b)). It is clear that for $k \geq 4$ in the first round the central vertex is colored with color 1. In the second round all vertices obtain colors without conflict. Those connected to the central vertex get color 2 and the others receive color 1. The result for $k = 3$ follows by a similar argument. □

Proposition 6. *Crowns are colored with at most three colors.*

Proof. A crown is a graph obtained from a cycle by attaching some pendant edges (see Fig. 2(c)). This case is similar to Proposition 4. A cycle will be colored with at most three colors and each leaf will be colored by either color 1 or 2. □

Proposition 7. *Double-stars are colored optimally by the DLF algorithm.*

Proof. A double-star is a tree with $n-2$ leaves (see Fig. 2(d)). In the first two rounds the vertices of degree greater than 1 are colored with colors 1 and 2. Next, all leaves are colored in the following round. Each of them has exactly one colored neighbor, so it is colored with 1 or 2. □

Table 1. Comparison of the number of colors used by the trivial and DLF algorithms in the worst case.

G	$\chi(G)$	$C_T(G)$	$C_{\text{DLF}}(G)$
complete k-partite	k	$\Delta+1$	k
$K_{p,q}$	2	$\Delta+1$	2
$K_n - e$	$n-1$	$\Delta+1$	$n-1$
BW_k	2	5	2
caterpillar	2	$\Delta+1$	3
double-star	2	4	2
crown	2 or 3	$\Delta+1$	3

In Table 1 we summarize the results given above and compare these results to the worst-case performance of the trivial algorithm on the same classes of graphs. In the first column we list the graph classes and in the second we give the corresponding chromatic number of any graph in the class. For each class the third column gives the worst-case value of $C_T(G)$, the number of colors used by the trivial algorithm, and the fourth column gives the corresponding worst-case value of $C_{\text{DLF}}(G)$. Table 1 shows that for these classes of graphs the DLF algorithm is more effective than the trivial algorithm and, in fact, it always produces an optimal or nearly optimal coloring.

References

1. Battiti, R., Bertossi, A. A., and Bonuccelli, M. A.: Assigning codes in wireless networks. Wireless Networks **5** (1999) 195–209.
2. Grable, D. A., and Panconesi, A.: Fast distributed algorithms for Brooks-Vizing colorings. J. Algorithms **37** (2000) 85–120.
3. Johansson, Ö.: Simple distributed $\Delta+1$ - coloring of graphs. Inf. Process. Lett. **70** (1999) 229–232.
4. Karp, R. M.: Probabilistic recurrence relations. J. ACM **41** (1994) 1136–1150.
5. Kubale, M.: Introduction to Computational Complexity and Algorithmic Graph Coloring. GTN, Gdańsk, 1998.
6. Kubale, M. and Kuszner, L.: A better practical algorithm for distributed graph coloring. Proc. of IEEE International Conference on Parallel Computing in Electrical Engineering (2002) 72–75.

Parallel Delaunay Refinement with Off-Centers

Daniel A. Spielman[1], Shang-hua Teng[2], and Alper Üngör[3]

[1] Dept. of Mathematics, M.I.T., Cambridge, MA 02139, USA
spielman@math.mit.edu
[2] Dept. of Computer Science, Boston Univ., Boston, MA 02215, USA
steng@cs.bu.edu
[3] Dept. of Computer Science, Duke Univ., Durham, NC 27708, USA
ungor@cs.duke.edu

Abstract. Off-centers were recently introduced as an alternative type of Steiner points to circum-centers for computing size-optimal quality guaranteed Delaunay triangulations. In this paper, we study the depth of the off-center insertion hierarchy. We prove that Delaunay refinement with off-centers takes only $O(\log(L/h))$ parallel iterations, where L is the diameter of the domain, and h is the smallest edge in the initial triangulation. This is an improvement over the previously best known algorithm that runs in $O(\log^2(L/h))$ iterations.

Keywords: Delaunay refinement, parallel algorithms, triangulations.

1 Introduction

Mesh generation problems ask for the discretization of an input domain into small and simple elements. These discretizations are essential in many applications including physical simulations, geographic information systems, computer graphics, and scientific visualization [9]. In addition to having the mesh conforming the input domain, most applications require that the mesh elements are of good quality and that the size of the mesh is small. A mesh element is considered good if its smallest angle is bounded from below. A bad element is likely to cause interpolation errors in the applications. Hence, mesh quality is critical for the accuracy and the convergence speed of the simulations. Mesh size, naturally, is also a factor in the running time of the applications algorithms.

Two main approaches for solving the mesh generation problem are the *quadtree methods* [2, 10] and the *Delaunay refinement methods* [5, 13]. Both methods compute quality-guaranteed size-optimal triangular meshes. However, Delaunay refinement methods are more popular than the quadtree methods mostly due to their superior performance in practice in generating smaller meshes. Over the years, many versions of the Delaunay refinement have been suggested in the literature [5, 7, 11, 13, 14, 16].

Delaunay triangulation of a given input is likely to have bad elements. Delaunay refinement iteratively adds new points, called *Steiner points*, into the domain to improve the quality of the mesh. A sequential Delaunay refinement

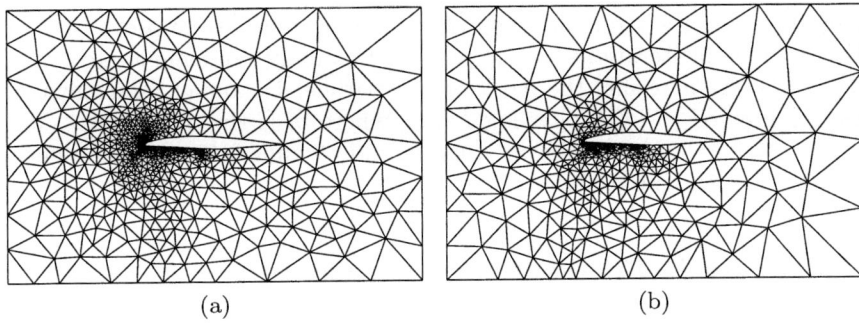

Fig. 1. Circumcenter vs. off-center insertion on an airfoil model. The smallest angle in both meshes is 32°. Delaunay refinement with circumcenters inserts 731 Steiner points and results in a mesh with 1430 triangles (a). On the other hand, Delaunay refinement with off-centers inserts 441 points and generates a mesh with 854 triangles (b).

algorithm typically adds one new vertex in each iteration. Each new vertex is chosen from a set of candidates – the circumcenters of bad triangles (to improve mesh quality) and the mid-points of input segments (to conform to the domain boundary). Ruppert [13] was the first to show that proper application of Delaunay refinement produces well-shaped meshes in two dimensions whose size is within a small constant factor of the best possible. Recently, we introduced a new type of Steiner points, called *off-centers* and proposed a new Delaunay refinement algorithm [16]. We proved that this new Delaunay refinement algorithm has the same theoretical guarantees as the Ruppert's refinement, and hence, generates quality-guaranteed size-optimal meshes. Moreover, experimental study indicates that our Delaunay refinement algorithm with off-centers inserts 40% fewer Steiner points than the circumcenter insertion algorithms and results in meshes 30% smaller in the number of elements. This implies substantial reduction not only in mesh generation time, but also in the running time of the application algorithms. Fig. 1 illustrates the performance difference between off-center and circumcenter insertion in meshing a region around an airplane wing. See [16] for further analysis.

Parallelization of Delaunay refinement methods is important for large scale applications. Recently, we gave the first parallel complexity analysis of the Delaunay refinement with circumcenters [15]. A main ingredient of this parallel algorithm is a notion of independence among candidate Steiner points for insertion at each iteration. The parallel algorithm consists of two main steps at each iteration. First, we generate an independent set of points for parallel insertion and then update the Delaunay triangulation in parallel. The independent sets have some nice properties. Insertion can be realized sequentially by Ruppert's Delaunay refinement method. Hence, an algorithm that inserts all the independent points in parallel will inherit the size and quality guarantees of Ruppert's method. The independent sets can be generated efficiently in parallel. In addition, they are "large enough" so that the number of parallel iterations needed is shown to be $O(\log^2(L/h))$, where L is the diameter of the domain and h is the smallest edge in the input triangulation [15]. In this paper, we show that by replacing the circumcenters with the off-centers, we improve the bound on the number of iterations of the parallel Delaunay refinement algorithm to $O(\log(L/h))$. As a result, the work of our parallel Delaunay refinement algorithm is improved to $O(m \log m \log(L/h))$, where m is the output size. This is close to $O((n \log(L/h) + m) \log m)$ time bound of Miller's sequential algorithm [11], where n is the input size.

2 Delaunay Refinement with Off-Centers

In two dimensions, the input domain Ω is represented as a *planar straight line graph* (PSLG) – a proper planar drawing in which each edge is mapped to a straight line segment between its two endpoints [13]. Due to space limitation we present our parallelization results only on periodic point sets, a special type of PSLG. If P is a finite set of points in the half open unit square $[0,1)^2$ and $\mathbb{Z}^2$ is the two dimensional integer grid, then $S = P + \mathbb{Z}^2$ is a *periodic point set* [6]. The periodic set S contains all points $p + v$, where $p \in P$ and v is an integer vector. As P is contained in the unit square, the diameter of P is $L \leq \sqrt{2}$. It is worth to note that some of the pioneering theoretical mesh generation work, such as sliver removal algorithms, are first studied on periodic point sets [4].

Let P be a point set in $\mathbb{R}^d$. A simplex τ formed by a subset of P points is a *Delaunay simplex* if there exists a circumsphere of τ whose interior does not contain any points in P. This empty sphere property is often referred to as the *Delaunay property*. The Delaunay triangulation of P, denoted $Del(P)$, is a collection of all Delaunay simplices. If the points are in general position, that is, if no $d+2$ points in P are co-spherical, then $Del(P)$ is a simplicial complex. The Delaunay triangulation of a periodic point set is also periodic. The Delaunay triangulation of a point set can be constructed in $O(n \log n)$ time in two dimensions [6].

Radius-edge ratio of a triangle is the ratio of its circumradius to the length of its shortest side. A triangle is considered *bad* if its radius-edge ratio is larger than a pre-specified constant $\beta \geq \sqrt{2}$. This quality measure is equivalent to other well-known quality measures, such as smallest angle and aspect ratio, in two dimensions [13].

The line that goes through the midpoint of an edge of a triangle and its circumcenter is called the *bisector* of the edge. Given a bad triangle pqr, suppose that its shortest edge is pq. Let c denote the circumcenter of pqr. We define the *off-center* to be the circumcenter of pqr if the radius-edge-ratio of pqc is smaller than or equal to β. Otherwise, the *off-center* is the point on the bisector, which makes the radius-edge ratio of the triangle based on p, q and the off-center itself β (Figure 2). The circle that is centered at the off-center and goes through the endpoints of the shortest edge is called the *off-circle*. In the first case, off-circle is same as the circumcircle of the triangle.

The description of the Delaunay refinement algorithm with off-centers is very simple for the periodic point set input. We maintain the Delaunay triangulation of the point set. As long as there exists a bad triangle in the triangulation, we

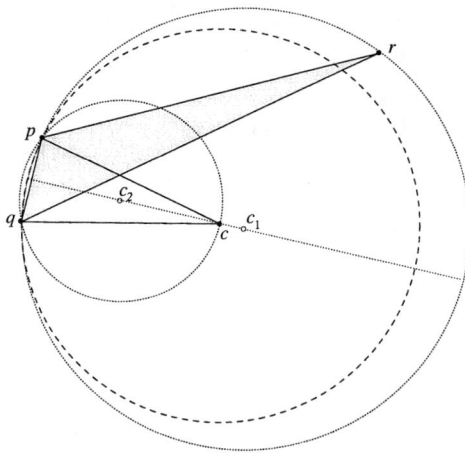

Fig. 2. The off-center and the circumcenter of triangle pqr is labeled c and c_1 respectively. The circumcenter of pqc is labeled as c_2. The off-circle of pqr is shown dashed.

insert its off-center as a Steiner point and update the Delaunay triangulation. We refer to [16] for a detailed description of the algorithm for PSLGs and also for the termination and size-optimality proofs.

3 Parallel Delaunay Refinement with Off-Centers

When parallelizing a Delaunay refinement algorithm, at each parallel iteration we would like to insert as many Steiner points as possible. However, some off-centers can be arbitrarily close to each other, hence neither a sequential nor a parallel refinement algorithm insert them all and can still provide termination guarantee. Furthermore, we would like to insert a set of points that has a sequential realization, i.e. there exist a provably good sequential algorithm that inserts the same set of points. We select the set of insertion points in a parallel iteration based on the following definition of *independence* among candidate off-centers.

Definition 1. *Two off-centers $\dot{c}_a$ and $\dot{c}_b$ (and also the corresponding off-circles c_a and c_b) are said to* conflict *if both c_a and c_b contain each other's off-center. Otherwise, $\dot{c}_a$ and $\dot{c}_b$ (respectively c_a and c_b) are said to be* independent.

Our parallelization of the off-center insertion is based on the same edge classification that we used in parallelizing the Ruppert's refinement [15]. Let h be the shortest edge length in the initial Delaunay triangulation. The class $\mathcal{E}_i$ contains the edges whose length are in the half open interval $\left[\sqrt{2}^{i-1}h, \sqrt{2}^i h\right)$. Therefore, there are at most $\lceil \log_{\sqrt{2}}(L/h) \rceil$ edge classes. A triangle is said to be associated with edge class $\mathcal{E}_i$ if its shortest edge is in $\mathcal{E}_i$. The ith iteration of the outer loop in the following algorithm removes all the bad triangles associated with $\mathcal{E}_i$.

Algorithm 1 PARALLEL DELAUNAY REFINEMENT WITH OFF-CENTERS.
───
Input: A periodic point set P in $\mathbb{R}^2$
  Let T be the Delaunay triangulation of P
  for i=1 to $\lceil \log_{\sqrt{2}}(L/h) \rceil$ **do**
    Let $\dot{\mathcal{C}} = \{\dot{c}_1, \ldots, \dot{c}_n\}$ be the set of off-centers of bad triangles associated with $\mathcal{E}_i$
    while $\dot{\mathcal{C}}$ is not empty **do**
      Let $\mathcal{I}$ be a maximal independent subset of $\dot{\mathcal{C}}$
      Update Delaunay triangulation inserting all points in $\mathcal{I}$ in parallel
      Update $\dot{\mathcal{C}}$
    end while
  end for
───

Lemma 1. *Suppose all the triangles associated with edge classes $\mathcal{E}_j$, $\forall j < i$ are good. Then, off-circle of every bad triangle associated with class $\mathcal{E}_i$ is empty of other points.*

Proof. Suppose the off-circle is not empty. Then, there exists a vertex inside off-circle but outside the circumcircle (Delaunay property). Consider a morph of the circumcircle into off-circle, where the morphing circle passes through the endpoints of the shortest edge pq and hence its center moves from the circumcenter to the off-center along the bisector. Let w be the vertex that the morphing circle hits first. The triangle pqw is bad because its circumradius is larger than the radius of the off-circle. Moreover, its shortest edge is less than $|pq|/\sqrt{2}$. This implies that pqw is associated with an edge class $\mathcal{E}_j$ where $j < i$. This contradicts the assumption. □

Lemma 2. *Suppose c_a and c_b are two conflicting off-circles of two triangles associated with class $\mathcal{E}_i$ at a parallel iteration, and let r_a and r_b be their radii. Then, $r_b/2 < r_a < 2r_b$.*

Proof. Off-circle c_a contains $\dot{c}_b$ and (by Lemma 1) no vertices of the triangulation. As some vertex on c_a lies outside c_b, the diameter of c_a is greater than the radius of c_b. Thus, $r_a > r_b/2$. A symmetric argument implies $2r_b > r_a$. □

We next show that the mesh generated by Algorithm 1 is realizable by the sequential algorithm introduced in [16] and described in Section 2.

Theorem 1. *Suppose M is a mesh produced by an execution of the* PARALLEL DELAUNAY REFINEMENT WITH OFF-CENTERS. *Then, M can be obtained by some execution of the sequential* DELAUNAY REFINEMENT WITH OFF-CENTERS.

Proof. Let $\mathcal{I}_1, \mathcal{I}_2, \ldots, \mathcal{I}_k$ be the sets of vertices inserted by the PARALLEL DELAUNAY REFINEMENT WITH OFF-CENTER at iterations $1, \ldots, k$, respectively. We describe a sequential execution that inserts all the points in $\mathcal{I}_i$ before any point of $\mathcal{I}_j$ for $i < j$. In other words, first all the points in set $\mathcal{I}_1$ are inserted sequentially, then all the points in set $\mathcal{I}_2$, and so on. To determine the order within each maximal independent set, we use a dependency graph. For any two off-centers $\dot{a}, \dot{b} \in \mathcal{I}_i$, $\dot{a}$ has to wait for the insertion of $\dot{b}$ if $\dot{a}$ is inside the corresponding off-circle b of $\dot{b}$. In the dependency graph we put an edge from $\dot{a}$ to $\dot{b}$. Because each edge is directed from a smaller circle to a larger circle, this dependency graph is acyclic. Topological sorting of the vertices of the dependency graph gives us a valid sequential insertion within each $\mathcal{I}_i$. Notice that in a sequential realization an off-center $\dot{a} \in \mathcal{I}_i$ remains a candidate until it is inserted. This is because the shortest edge of the corresponding triangle remains to be the shortest edge of a bad triangle (not necessarily the same one).

Furthermore, in any sequential execution, a point $p \in \mathcal{I}_i$ cannot eliminate a point $q \in \mathcal{I}_j$ for any $i < j$. Otherwise, q would not be inserted in the jth iteration of the parallel execution. Finally, the parallel and sequential executions terminate after inserting exactly the same set of points. An extra element in the insertion set of one would indicate that the execution of the other one is not terminated, as this implies existence of a bad element. Since the same points are inserted effectively in the same order, the output mesh is the same. □

We recall that β is the threshold of the ratio of the radius to shortest edge-length defining a bad triangle. Thus, for $\beta \geq \sqrt{2}$, inserting the off-center of a bad triangle whose shortest edge length is l, introduces new Delaunay edges of length at least $\sqrt{2}l$.

Lemma 3. *Suppose the shortest edge associated with any bad triangle is in $\mathcal{E}_i$. Let $e \in \mathcal{E}_i$ be the shortest edge of a bad triangle that exists before the first iteration of the inner loop. Then, after $O(1)$ iterations, either e does not exist anymore or all the triangles associated with e are good.*

Proof. Suppose e exists and still is the shortest edge of a bad triangle pqr after iteration 51 of the inner loop. This implies that the off-center $\dot{c}$ of pqr is not inserted because it was in conflict with another vertex at each of the iterations 1 through 51 of the inner loop during the ith iteration of the outer loop. So, for each iteration $k = 1, \ldots, 51$, an off-center $\dot{c}'_k$ in conflict with $\dot{c}$ is inserted. Moreover, by Lemma 2 the radius r'_k of the corresponding off-circle c'_k is at least half the size of c, i.e. $r'_k > r/2$, where r is the radius of the off-circle associated with e. Let $\dot{C}' = \{\dot{c}'_1, \dot{c}'_2, \ldots, \dot{c}'_{51}\}$ be the set of off-centers that were inserted in iterations $k = 1, \ldots, 51$. Let C' be the corresponding set of off-circles. The radius of each one of these circles in C' is at least $r/2$. By Lemma 1, each circle $c'_k \in C'$ is empty of all the centers in $\dot{C}'$ inserted prior to $\dot{c}'_k$. So the centers in $\dot{C}'$ are pairwise at least $r/2$ apart from each other. This in turn implies that the circles of radius $r/4$ on the centers in $\dot{C}'$ do not overlap. These circles can spread into an area of size at most $2\pi(r + r/4)^2$ because each of the corresponding off-centers is in conflict with an off-center on either side of the edge e. One can fit at most

$$\left\lfloor \frac{2\pi(r + r/4)^2}{\pi(r/4)^2} \right\rfloor = 50$$

circles of radius $r/4$ in that region. Therefore, the number of centers in $\dot{C}'$ is at most 50, which is a contradiction. □

Our algorithm, one at a time, handles the bad triangles associated with each edge class. In the next Lemma we justify the double loop structure of our parallel refinement algorithm. We prove that during and after the ith iteration of the outer loop no bad triangle associated with an edge in $\mathcal{E}_i$ is introduced.

Lemma 4. *Suppose the shortest edge associated with any triangle is in $\mathcal{E}_i$. During and after iteration i of the outer loop of Algorithm 1, the following are true:*

I *if an edge $e \in \mathcal{E}_i$ disappears it never appears again;*
II *no new edges are introduced to edge class $\mathcal{E}_j$, $\forall j \leq i$;*
III *the radius-edge ratio of a triangle associated with $\mathcal{E}_i$ does not increase.*

Proof. (I) If there is no empty circle containing e before a Steiner point insertion, clearly there is no such circle after the insertion. (II) During and after the ith iteration of the outer loop the smallest edge that can be introduced has length at least $\sqrt{2}^i h$. (III) Consider an edge $pq \in \mathcal{E}_i$ and a triangle pqr associated with pq. We claim that the quality of a triangle pqu replacing pqr is better than that of pqr. The new vertex u must be inside the circumcircle of pqr. Otherwise, pqr is intact. On the other hand, u can not be too close to edge pq, i.e., $\min\{|pu|, |qu|\} \geq |pq|$. Otherwise, a shorter edge than pq would be introduced contradicting (II). The radius of the circumcircle of pqu is smaller than that of pqr when u is inside the circumcircle of pqr but outside the diametral circle of pq. This in turn, implies that the radius-edge ratio of a triangle associated to an edge in $\mathcal{E}_i$ can only be improved through refinement. □

Theorem 2. PARALLEL DELAUNAY REFINEMENT WITH OFF-CENTERS *algorithm takes $O(\log(L/h))$ iterations to generate a size-optimal well-shaped uniform mesh, where L is the diameter of the input and h is the length of the shortest edge in the initial triangulation.*

Proof. By Lemma 3 and 4, the ith outer loop of the algorithm takes $O(1)$ parallel iterations to fix all the triangles associated with $\mathcal{E}_i$. Overall the algorithm takes at most $51 \log_{\sqrt{2}}(L/h)$ parallel iterations. The size optimality and quality guarantee of the parallel algorithm follows from Theorem 1. □

4 Discussions

We should note that in our design and analysis of the parallel Delaunay refinement, we focused and gave a bound on the number of parallel iterations. To complete the parallel time complexity analysis at each iteration we employ the parallel maximal independent set algorithm presented in [15] and one of the parallel Delaunay triangulation algorithms given in [1, 12]. As a future research we plan to extend the off-center algorithm to three dimensions and explore its benefits both in theoretical and practical fronts.

Acknowledgements

Research of the first author is partially supported by NSF under grant CCR-0112487. Research of the second author is partially supported by NSF under grants CCR-0311430 and CCR-0325630. Research of the third author is partially supported by NSF under grant CCR-00-86013.

References

1. N. M. Amato, M. T. Goodrich, and E. A. Ramos. Parallel algorithms for higher-dimensional convex hulls. *Proc. 35th IEEE Symp. Foundations of Computer Science*, 683–694, 1994.
2. M. Bern, D. Eppstein, and J. Gilbert. Provably good mesh generation. *J. Comp. System Sciences* 48:384–409, 1994.
3. M. Bern, D. Eppstein, and S.-H. Teng. Parallel construction of quadtrees and quality triangulations. *Int. J. Comp. Geometry & Applications* 9(6):517–532, 1999.
4. S.-W. Cheng, T. K. Dey, H. Edelsbrunner,M. A. Facello, and S.-H. Teng. Sliver exudation. *Proc. 15th ACM Symp. Comp. Geometry*, 1–13, 1999.
5. P. Chew. Guaranteed-quality triangular meshes. TR-89-983, Cornell Univ., 1989.
6. H. Edelsbrunner. *Geometry and Topology for Mesh Generation*. Cambridge Univ. Press, 2001.
7. H. Edelsbrunner and D. Guoy. Sink insertion for mesh improvement. *Proc. 17th ACM Symp. Comp. Geometry*, 115–123, 2001.
8. W.H. Frey. Selective refinement: a new strategy for automatic node placement in graded triangular meshes. *Int. J. Numerical Meth. Eng.*, 24, 2183–2200, 1987.
9. P. Frey, and P.-L. George. *Mesh Generation*. Oxford, Paris, 2000.
10. S. Mitchell and S. Vavasis. Quality mesh generation in three dimensions. *Proc. 8th ACM Symp. Comp. Geometry*, 212–221, 1992.
11. G. L. Miller. A time efficient Delaunay refinement algorithm. *Proc. ACM-SIAM Symp. on Disc. Algorithms*, 400–409, 2004.
12. J. H. Reif and S. Sen. Parallel computational geometry: An approach using randomization. *Handbook of Comp. Geometry*, 765–828, Elsevier, Amsterdam, 1999.
13. J. Ruppert. A new and simple algorithm for quality 2-dimensional mesh generation. *Proc. 4th ACM-SIAM Symp. on Disc. Algorithms*, 83–92, 1993.
14. J. R. Shewchuk. *Delaunay Refinement Mesh Generation*. Ph.D. thesis, Carnegie Mellon University, 1997.
15. D. A. Spielman, S.-H. Teng, and A. Üngör. Parallel Delaunay refinement: Algorithms and analyses. *Proc. 11th Int. Meshing Roundtable*, 205–217, 2002.
16. A. Üngör. Off-centers: A new type of Steiner points for computing size-optimal quality-guaranteed Delaunay triangulations. *Proc. of Latin American Theoretical Informatics (LATIN)*, 151–162, 2004,

Near-Optimal Hot-Potato Routing on Trees

Costas Busch[1], Malik Magdon-Ismail[1],
Marios Mavronicolas[2], and Roger Wattenhofer[3]

[1] Rensselaer Polytechnic Institute, Troy, NY 12180, USA
[2] University of Cyprus, Nicosia CY-1678, Cyprus
[3] ETH Zurich, 8092 Zurich, Switzerland

Abstract. In *hot-potato (deflection) routing,* nodes in the network have no buffers for packets in transit, which causes conflicting packets to be deflected away from their destinations. We study *one-to-many* batch routing problems on arbitrary tree topologies. We present two hot-potato routing algorithms, one deterministic and one randomized, whose routing times are asymptotically near-optimal (within poly-logarithmic factors from optimal). Both algorithms are *distributed* and *greedy*, so, routing decisions are made locally, and packets are advanced towards their destinations whenever possible.

1 Introduction

Packet routing is the general task of delivering a set of packets from their sources to their destinations. *Hot-potato* (or *deflection*) routing is relevant in networks whose nodes cannot buffer packets in transit – any packet that arrives at a node must immediately be forwarded to another node at the next time step, as if it were a "hot potato." A *routing algorithm* (or *protocol*) specifies at every time step the actions that each node takes while routing the packets. Hot-potato routing was introduced by Baran [4], and since then, hot-potato routing algorithms have been extensively studied formally and observed to work well in practice for various network architectures [1, 5, 7, 8, 10, 12, 16, 17].

Here, we consider *synchronous tree* networks in which a global clock defines a discrete time. At each time step t, a node may receive packets, which it forwards to adjacent nodes according to the routing algorithm. These packets reach the adjacent nodes at the next time step $t + 1$. At each time step, a node is allowed to send at most one packet per link[1]. Bufferless routing in tree networks has generated considerable interest in the literature: most existing work is for the matching routing model, [2, 14, 18] or the direct routing model, [3, 9]; hot-potato routing of permutations on trees has been studied in [15].

We consider *one-to-many* batch routing problems on trees with n nodes, where each node is the source of at most one packet; however, each node may be the destination of multiple packets. The *routing time* of a routing-algorithm is

---

[1] At any time step, at most two packets can traverse an edge in the tree, one packet along each direction of the edge.

the time for the last packet to reach its destination. Denote by rt^* the minimum possible routing time for a given routing problem. For a given routing problem, and a set of paths to be followed from the sources to destinations, the *dilation* D is the maximum length of a path, and the *congestion* C is the maximum number of paths that use any link (in either direction). Since at most one packet can traverse an edge in a given direction at each time step, the routing time for the specified paths is $\Omega(C + D)$. On a tree, any set of paths must contain the shortest paths, from the sources to the destinations. Let C^* and D^* be the congestion and dilation for the shortest paths respectively. We immediately get that $rt^* = \Omega(C^* + D^*)$. For *store-and-forward* routing, in which nodes have buffers for storing packets in transit, there are routing algorithms whose performance on trees is close to rt^* [11, 13]. However, such algorithms are not applicable when buffers are not available.

We consider *greedy* hot potato routing. A routing algorithm is greedy if a packet always follows a link toward its destination whenever this is possible. In hot-potato routing, a problem occurs if two or more packets appear at the same node at the same time, and all these packets wish to follow the same link at the next time step. This constitutes a *conflict* between the packets because only one of them can follow that particular link. Since nodes have no buffers, the other packets will have to follow different links that lead them further from their destination. We say that these packets are *deflected*. In a greedy algorithm, a packet π can be deflected only when another packet makes progress along the link that π wished to follow.

Our Contributions. We present two hot-potato routing algorithms on trees with near optimal routing time. Our algorithms are *local,* and thus *distributed*: at every time step, each node makes routing decisions locally based only on the packets it receives at that particular time step. In our algorithms, every source node determines the time at which its packet will be injected. From then on, the packet is routed greedily to its destination. We assume that each source node knows the tree topology, as well as C^* and D^* for the batch routing problem; we emphasize, however, that it need not know the specific sources and destinations of the other packets. The assumption that C^* and D^* are known is common to many distributed routing algorithms [7, 13]. Our two algorithms are sumarized below:

i. Algorithm Deterministic has routing time $O((\delta \cdot C^* + D^*) \lg n) = O(\delta \cdot rt^* \cdot \lg n)$, where δ is the maximum node degree in the tree. All routing choices are deterministic.
ii. Agorithm Randomized has routing time less than $O((C^* + D^*) \lg^2 n) = O(rt^* \cdot \lg^2 n)$ with probability at least $1 - \frac{1}{n}$. Randomization is used when packets select priorities. These priorities are then used to resolve conflicts.

For bounded-degree trees, algorithm Deterministic is within a logarithmic factor of optimal. Algorithm Randomized is only an additional logarithmic factor away from optimal for arbitrary tree topologies.

The general idea of our algorithms is to divide packets into levels based on the position of their source in the tree. Packets at different levels are routed in different phases. We show that there are at most $O(\lg n)$ such phases. In algorithm Deterministic, each phase has a duration $O(\delta \cdot C^* + D^*)$, while in algorithm Randomized, each phase has duration $O((C^* + D^*) \lg n))$. Combining these bounds with the bound on the number of phases leads to our routing time bounds. The heart of both of our algorithms lies in the use of *canonical deflections*, in which packets are only deflected onto edges used by other packets that moved forward in the previous time step.

Paper Outline. We first introduce trees in Section 2, and packet behavior in hot-potato routing in Section 3. We then present our deterministic and randomized routing algorithms in respective Sections 4 and 5.

2 Trees

A tree $T = (V, E)$ is a connected acyclic graph with $|V| = n$ and $|E| = n - 1$. The *degree* of node v is the number of nodes adjacent to v. Let $v \in V$; then, T induces a subgraph on $V - \{v\}$ which consists of a number (possibly zero) of connected components. Each such connected component is a *subtree of v in T*[2]. If v is adjacent to K nodes in T, then there are k disjoint subtrees $T_1, \ldots, T_k$ of v, one for each node $v_i \in T_i$ that is adjacent to v. The *distance* from v to u, is the number of edges in the (unique) shortest path from v to u.

The main idea behind our algorithms is to look at the tree from the point of view of a short node. A node v in the tree is *short* if every subtree of v contains at most $n/2$ nodes. At least one short node is guaranteed to exist, and by starting at an arbitrary node and moving along a path of largest subtrees (if the size of the subtree is greater than $n/2$), a short node can be found in in $O(n)$ time. A tree T may have many short-nodes, however, a deterministic algorithm always starting at a particular node will always return a particular short-node. So, from now on, we will assume that the short-node of a tree is uniquely determined.

We now define (inductively) the *level* ℓ of a node, and the *inner-trees* of T as follows. The tree T is the only inner-tree at level $\ell = 0$. The only node at level $\ell = 0$ is the short node of T. Assume we have defined inner-trees up to level $\ell \geq 0$. Every connected component obtained from the inner-trees of level ℓ by removing the short nodes of these inner-trees at level ℓ is an inner-tree at level $\ell + 1$. The level $\ell + 1$ nodes are precisely the short nodes of the inner-trees at level $\ell + 1$.

It is clear that the above definition inductively defines the inner-trees at all levels; it correspondingly assigns a level to every node. We can easily construct an $O(n^2)$ procedure to determine the node levels and inner-trees of T at every level. Further, the following properties (which we state here without proof) hold:

---

[2] Note that for unrooted trees which we consider here, a subtree of a node v originates from every adjacent node of v; in contrast, the convention for rooted trees is that a subtree of v is any tree rooted at a child of v.

(i) every inner-tree is a tree, (ii) the maximum level of any node and inner-tree is no more than $\lg n$, (iii) an inner-tree T' at level ℓ contains a unique node x at level ℓ, which is the short node of the inner-tree (we say that x is the inducing node of T'), (iv) any two inner-trees at the same level are disconnected, and (v) all nodes in a level ℓ inner-tree other than the inducing node have a level that is smaller than ℓ.

3 Packets

Packet Paths. A *path* is any sequence of nodes $(v_1, v_2, \ldots, v_k)$. The length of the path is the number of edges in the path. After a packet has been routed from its source to its destination, it has followed some path. We define the *original path* of a packet π as the shortest path from the source node of the packet to its destination node. This will be the path that would be greedily followed if the packet experiences no deflections.

Let ℓ be the smallest level of any node in the original path of π. Then, there is a unique node v with level ℓ in the path of π (since otherwise inner-trees of the same level would not be disconnected). Let T' be the inner-tree that v is inducing. The whole original path of π must be a subgraph of T' (from the definition of inner-trees). We say that the level of packet π is ℓ, and that the inner-tree of π is T'.

After injection, the *current path* of packet π, at any time step t, is the shortest path from its current node to its destination node. At the moment when the packet is injected, its current path is its original path. While packet π is being routed to its destination, it may deviate from its original path due to deflections. However, the packet traverses each edge of its original path at least once before reaching its destination.

A packet moves *forward* if it follows the next link of its current path; otherwise, the packet is deflected. When a packet moves forward, its current path gets shorter by removing the edge that the packet follows. If a packet is deflected, its current path grows by the edge on which the packet was deflected, and its new current path is the shortest path from its current node to its destination node.

Packet Routing and Deflections. In our algorithms, a packet remains in its source node until a particular time step at which the packet becomes *active*. When the packet becomes active, it is injected at the first available time step on which the first link of its original path is not used by any other packets that reside at its source node. We call such an injection a *canonical injection*.

At each time step, each node in the network does the following: (i) the node receives packets from adjacent nodes, (ii) the node makes routing decisions, and (iii) according to these decisions, the node sends packets to adjacent nodes. We say that two or more packets *meet* if they appear in the same node at the same time step; they *conflict* if they also wish to follow the same link forward. In a conflict, one of the packets will successfully follow the link, while the other packets must be deflected. Our algorithms are greedy: a packet always attempts move forward unless it is deflected by another packet with which it conflicts.

In our algorithms, packets are deflected in a particular fashion so as to ensure that the congestion of any edge (with respect to the set of current paths) never increases. Consider a node v at time t. Let S_f denote the set of packets which moved forward during the previous time step and now appear in v. Let E_f be the set of edges that the packets in S_f followed during the previous time step. Let π be a packet in node v that is deflected at time t. Node v first attempts to deflect π along an edge in E_f. Only if all the edges in E_f are being used by packets moving forward, or by other deflected packets, then some other edge adjacent to v is used to deflect π. We call this process of deflecting packets *canonical deflection*. The deflection is *safe* if it was along an edge in E_f. One can show that if injections are canonical, then all deflections are safe. Safe deflections simply "recycle" edges from one path to another path, and thus cannot increase an edge's congestion.

4 A Deterministic Algorithm

Here we present the algorithm Deterministic in pseudo-code format.

Algorithm: Deterministic

Input: Tree T with max. node degree δ; A set of packets Π with shortest path congestion C^* and dilation D^*; each node knows T, C^*, D^*;

for *every packet π at level ℓ* **do**
1   π gets active at time $\tau \cdot \ell$, where $\tau = 2(\delta \cdot C^* - 1) + D^*$;
2   The injection and deflections of π are canonical;
3   π moves greedily to its destination;
end

Theorem 1. *The routing time of algorithm* Deterministic *is $O((\delta \cdot C^* + D^*) \lg n)$.*

We proceed by sketching the proof of Theorem 1. Let m be the maximum level in T. Since a level $l+1$ inner-tree has fewer than half the nodes of the level l inner-tree that gave rise to it, it is easy to see that $m \leq \lg n$. We divide time into consecutive phases $\phi_0, \phi_1, \ldots, \phi_m$, such that each phase consists of τ time steps. Denote the level i packets by Π_i, $0 \leq i \leq m$. The packets in Π_i become active at the first time step of phase ϕ_i. We will show that all packets of level i are absorbed during phase ϕ_i. In particular, we will show that the following invariants hold, where $i \geq 0$:

P_i: all packets of $\Pi_0 \cup \Pi_1 \cup \cdots \cup \Pi_i$ are absorbed by the end of phase ϕ_i.

Set P_{-1} to true. It suffices to show that the following statement holds, for $i \geq 0$:

Q_i: if P_{i-1} holds, then all packets in Π_i are absorbed during phase ϕ_i.

Consider a particular level $\ell \geq 0$ and phase ϕ_ℓ. Suppose that $P_{\ell-1}$ holds. In phase ϕ_ℓ the only packets in the network are those of Π_ℓ. During phase ϕ_ℓ, let π be the first packet to leave its inner-tree going from node u to node v. Since deflections are safe, this means that some packet π' moved from v to u in the previous time step. Since the entire original path of π' was contained in some level ℓ inner-tree, this means that π' left its inner-tree before π did, a contradiction. Thus, we have:

Lemma 1. *During phase ϕ_ℓ, each packet of Π_ℓ remains inside its inner-tree.*

Since only packets of the same inner-tree may conflict with each other, we only need to show that every level ℓ packet with a particular inner-tree T' is absorbed during phase ϕ_ℓ.

We adapt a technique developed by Borodin et al. [6, Section 2], called a "general charging scheme", based upon deflection sequences. Their result implies that for greedy routing on trees, whenever a packet is deflected, some other packet makes it to its destination. Thus if the number of packets is k, then a packet can be deflected at most $k-1$ times, giving:

Corollary 1 ([6]). *Each packet is absorbed in at most $D^* + 2(k-1)$ time steps after injection, where k is the number of packets with inner-tree T'.*

All packets with inner-tree T' use the inducing short node r. Since the degree of r is at most δ, and the congestion on an edge never increases, $k \leq \delta \cdot C^*$. Subsequently, all packets of inner-tree T' are absorbed in at most $2(\delta \cdot C^* - 1) + D^* \leq \tau$ time steps, i.e., by the end of phase ϕ_ℓ. Since there are at most $O(\lg n)$ phases, the theorem is proved.

5 A Randomized Algorithm

Algorithm Randomized is similar to algorithm Deterministic, except that packets now have priorities: low or high. High priority packets have precedence (to move forward) over low priority packets in a conflict. Conflicts between equal priority packets are arbitrarily resolved canonically.

Algorithm: Randomized
Input: A tree T; A set of packets Π with shortest path congestion C^* and dilation D^*; each node knows T, C^*, D^*;
    for *every packet π at level ℓ do*
1   π gets active at time step $\tau \cdot \ell$, where $\tau = 16 \cdot (C^* + D^*) \cdot (2 \lg n + \lg \lg 2n) + 3D^* + 1$;
2   The injection and deflections of π are canonical, and initially π has low priority;
3   π moves greedily to its destination;
4   When packet π becomes active it has low priority;
5   Let $p = \frac{1}{4(C^* + D^*)}$; if π is deflected, on the next time step, its priority becomes high with probability p, and low with probability $1-p$, independent of its previous priority; π keeps its new priority until the next deflection;
    end

Theorem 2. *With probability at least $1 - 1/n$, the routing time of algorithm Randomized is at most $\kappa(C^* + D^*) \lg^2 n$, for some constant $\kappa \approx 33$.*

We very briefly sketch the proof of Theorem 2. We define P_i and Q_i as in Section 4, and we show that P_i holds with high probability by showing that Q_i holds with high probability. Once again we focus on the level ℓ packets of a particular inner-tree T', and suppose that all packets of the previous phases are absorbed. We will show that every packet with inner-tree T' will be absorbed in phase ϕ_ℓ, with high probability. Let $T_1, T_2, \ldots, T_w$ denote the subtrees of inducing node r in T'. We summarize some useful properties of these subtrees (without proof).

Lemma 2. *At any time step during phase ϕ_ℓ, at most C^* packets are in any subtree T_j.*

Using Lemma 2, since at most one new packet can enter a subtree in one time step, we obtain:

Lemma 3. *During any time period of length x time steps, at most $C^* + x$ different packets have appeared in any subtree T_j.*

Corollary 2. *If packet π is not deflected for a period of x time steps, then it may have conflicted at most once with at most $2C^* + x$ different packets.*

The *depth* of node v (or packet currently at node v), is its distance from r.

Lemma 4. *At any time during phase ϕ_ℓ, the depth of a packet is at most D^*.*

Consider some packet π. By Lemma 4, the current path length of a packet is always at most $2D^*$, so if it is not deflected for $2D^*$ time steps, then it reaches its destination. Suppose that π has high priority. It can only be deflected by another high priority packet σ. Each such packet σ has at most one chance to deflect it, with probability at most p, since σ is in high priority. If π is not deflected for $2D^*$ time steps, then at most $2C^* + 2D^*$ packets had a chance to deflect it (Corollary 2). Thus the probability of deflection is at most $2(C^* + D^*)p = \frac{1}{2}$, giving the main lemma:

Lemma 5. *A high priority packet reaches its destination without deflections with probability at least $1/2$.*

Thus, each time a packet is deflected, it has probability at least $\frac{1}{2}p = \frac{1}{8(C^*+D^*)}$ to reach its destination. If a packet has not reached its destination after τ steps, then it has been deflected at least $\frac{1}{2}(x - D^*)$ times, over the interval of times $[t_1, t_x]$ where $x = \tau - 2D^* - 1$. The probability of not reaching its destination after so many deflections is at most $(1 - \frac{1}{2}p)^{\frac{1}{2}(x-D^*)}$, which after substituting the expression for τ yields:

Lemma 6. *Packet π is absorbed in phase ϕ_ℓ with probability at least $1 - \frac{1}{n^2 \lg 2n}$.*

Since there are at most n packets in a phase, and at most $\lg n + 1$ phases, we use a union bound to obtain a lower bound on the probability that every phase is succesful, giving Theorem 2.

References

1. A. S. Acampora and S. I. A. Shah. Multihop lightwave networks: a comparison of store-and-forward and hot-potato routing. In *Proc. IEEE INFOCOM*, pages 10–19, 1991.
2. N. Alon, F.R.K. Chung, and R.L.Graham. Routing permutations on graphs via matching. *SIAM Journal on Discrete Mathematics*, 7(3):513–530, 1994.
3. Stephen Alstrup, Jacob Holm, Kristian de Lichtenberg, and Mikkel Thorup. Direct routing on trees. In *Proceedings of the Ninth Annual ACM-SIAM Symposium on Discrete Algorithms (SODA 98)*, pages 342–349, 1998.
4. P. Baran. On distributed communications networks. *IEEE Transactions on Communications*, pages 1–9, 1964.
5. Constantinos Bartzis, Ioannis Caragiannis, Christos Kaklamanis, and Ioannis Vergados. Experimental evaluation of hot-potato routing algorithms on 2-dimensional processor arrays. In *EUROPAR: Parallel Processing, 6th International EUROPAR Conference*, pages 877–881. LNCS, 2000.
6. A. Borodin, Y. Rabani, and B. Schieber. Deterministic many-to-many hot potato routing. *IEEE Transactions on Parallel and Distributed Systems*, 8(6):587–596, June 1997.
7. C. Busch. $\tilde{O}$(Congestion + Dilation) hot-potato routing on leveled networks. In *Proceedings of the Fourteenth ACM Symposium on Parallel Algorithms and Architectures*, pages 20–29, August 2002.
8. C. Busch, M. Herlihy, and R. Wattenhofer. Randomized greedy hot-potato routing. In *Proceedings of the Eleventh Annual ACM-SIAM Symposium on Discrete Algorithms*, pages 458–466, January 2000.
9. Costas Busch, Malik Magdon-Ismail, Marios Mavranicolas, and Paul Spirakis. Direct routing: Algorithms and Complexity. In *Proceedings of the 12th Annual European Symposium on Algorithms (ESA)*, September 2004.
10. W. D. Hillis. *The Connection Machine*. MIT press, 1985.
11. T. Leighton, B. Maggs, and A. W. Richa. Fast algorithms for finding O(congestion + dilation) packet routing schedules. *Combinatorica*, 19:375–401, 1999.
12. N. F. Maxemchuk. Comparison of deflection and store and forward techniuques in the Manhattan street and shuffle exchange networks. In *Proc. IEEE INFOCOM*, pages 800–809, 1989.
13. Friedhelm Meyer auf der Heide and Berthold Vöcking. Shortest-path routing in arbitrary networks. *Journal of Algorithms*, 31(1):105–131, April 1999.
14. Grammati E. Pantziou, Alan Roberts, and Antonios Symvonis. Many-to-many routing on trees via matchings. *Theoretical Comp. Science*, 185(2):347–377, 1997.
15. Alan Roberts, Antonios Symvonis, and David R. Wood. Lower bounds for hot-potato permutation routing on trees. In *Proc. 7th Int. Coll. Structural Information and Communication Complexity, SIROCCO*, pages 281–295, June 2000.
16. C. L. Seitz. The Caltech Mosaic C: An experimental, fine-grain multicomputer. In *Proc. 4th Symp. on Parallel Algorithms and Architectures*, June 1992. Keynote Speech.
17. B. Smith. Architecture and applications of the HEP multiprocessor computer system. In *Proceedings of the 4th Symp. Real Time Signal Processing IV*, pages 241–248. SPIE, 1981.
18. L. Zhang. Optimal bounds for matching routing on trees. In *Proceedings of the 8th Annual ACM-SIAM Symposium on Discrete Algorithms*, pages 445–453, 1997.

A Coarse-Grained Parallel Algorithm for Spanning Tree and Connected Components*

E.N. Cáceres[1], F. Dehne[2], H. Mongelli[1], S.W. Song[3], and J.L. Szwarcfiter[4]

[1] Universidade Federal de Mato Grosso do Sul, Campo Grande, Brazil
{edson,mongelli}@dct.ufms.br
http://www.dct.ufms.br/~{edson,mongelli}
[2] Carleton University, Ottawa, Canada K1S 5B6
frank@dehne.net
http://www.dehne.net
[3] Universidade de São Paulo, São Paulo, Brazil
song@ime.usp.br
http://www.ime.usp.br/~song
[4] Universidade Federal do Rio de Janeiro, Rio de Janeiro, Brazil
jayme@nce.ufrj.br
http://www.cos.ufrj.br/docentes/jayme.html

Abstract. Dehne et al. present a BSP/CGM algorithm for computing a spanning tree and the connected components of a graph, that requires $O(\log p)$ communication rounds, where p is the number of processors. It requires the solution of the Euler tour problem which in turn is based on the solution of the list ranking problem. We present a new approach that does not need to solve the Euler tour or the list ranking problem. It is based on the integer sorting algorithm which can be implemented efficiently on the BSP/CGM model [1].

1 Introduction

Computing a spanning tree and the connected components of a graph are basic problems and arise as subproblems in many applications. Parallel algorithms for these problems have been proposed by Hirschberg et al. [2]. An efficient CRCW PRAM algorithm takes $O(\log n)$ time with $O((m+n)\alpha(m,n))/\log n$ processors, where $\alpha(m,n)$ is the inverse of the Ackermann's function [3]. Dehne et al. [4] present a coarse-grained parallel algorithm that requires $O(\log p)$ communication rounds, where p is the number of processors. It is based on the Euler tour problem which in turn is based on the list ranking problem.

We present a new approach that does not need to solve the Euler tour or the list ranking problem. It still requires $O(\log p)$ communication rounds and has

---

* Partially supported by FINEP-PRONEX-SAI Proc. No. 76.97.1022.00, CNPq Proc. No. 30.0317/02-6, 30.5218/03-4, 47.0163/03-8, 55.2028/02-9, FUNDECT-MS Proc. No. 41/100117/03, and the Natural Sciences and Engineering Research Council of Canada. We would also like to thank the anonymous referees for their review and helpful comments.

the advantage of avoiding the list ranking computation which has been shown to require large constants in practical implementations. The proposed algorithm is based on the integer sorting algorithm which can be implemented efficiently on the BSP/CGM model.

2 Preliminaries and Main Ideas

We use the *Coarse-Grained Multicomputer* (CGM) model [5], with p processors each with an $O(N/p)$ local memory, where N is the input size. A CGM algorithm consists of alternating local computation and global communication rounds. The communication cost is modeled by the number of communication rounds.

Consider a bipartite graph $H = (V_1, V_2, E)$ with vertex sets V_1 and V_2 and edge set E where each edge joins one vertex of V_1 and one vertex in V_2. If v is a vertex of a subgraph H' of H, then $d_{H'}(v)$ denotes the degree of v in H'. Let the vertices of V_1 be $u_1, u_2, \cdots, u_{n_1}$ and the vertices of V_2 be $v_1, v_2, \cdots, v_{n_2}$.

We define a *strut ST in V_1* as a spanning forest of H such that each $v_i \in V_2$ is incident in ST with exactly one edge of E, and (u_j, v_i) is an edge of ST implies (u_k, v_i) is not an edge of H, for any $u_k \in V_1$, $k < j$. To define a *strut in V_2*, the roles for the sets V_1 and V_2 in the above definition are exchanged.

A vertex $u \in V_1$ is called *zero-difference* in ST if $d_H(u) - d_{ST}(u) = 0$. Otherwise, the vertex is referred to as *non-zero-difference*.

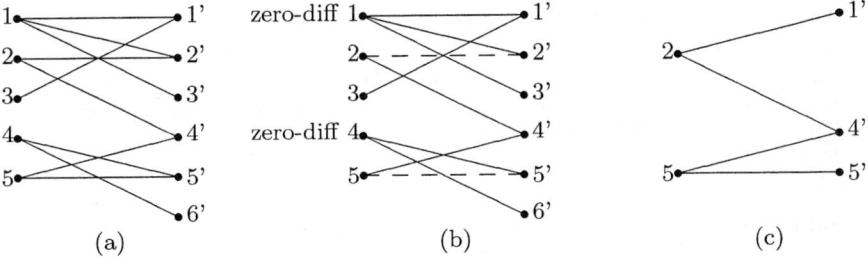

Fig. 1. (a) A bipartite graph (b) a strut (solid lines) (c) the compacted graph.

Fig. 1(a) shows a graph $H = (V_1, V_2, E)$ with $V_1 = \{1, 2, 3, 4, 5\}$, $V_2 = \{1', 2', 3', 4', 5', 6'\}$ and $E = \{(1, 1')\ (1, 2')\ (1, 3')\ (2, 2')\ (2, 4')\ (3, 1')\ (4, 5')\ (4, 6')\ (5, 4')\ (5, 5')\}$. We first compute a spanning forest for H by determining a strut ST in H (see Fig. 1(b)). Now compute the zero-difference vertices in V_1. Consider vertex 1. All the (three) edges in H incident with this vertex is also in ST. Thus $d_H(1) - d_{ST}(1) = 0$ and vertex 1 is zero-difference. Likewise vertex 4 is also zero-difference. Notice that vertex 2 is not zero-difference.

In the example we have two zero-difference vertices. If we have only *one* zero-difference vertex, then the problem is easily solved by adding to ST one arbitrary edge of $H - ST$ incident to each non-zero-difference vertex of ST. In case there are two or more zero-difference vertices we can do the following. For each zero-difference vertex $u \in V_1$ compact all the vertices $v_i \in V_2$ incident with u by compressing all the vertices v_i onto the smallest of the v_i. Repeat this until only one zero-difference vertex remains.

3 The CGM Algorithm for Bipartite Graphs

Let $H(V_1, V_2, E)$ be a bipartite graph with $|V_1| = n_1$, $|V_2| = n_2$ and $|E| = m$. Each of the p processors has $O(m/p)$ or $O((n_1+n_2)/p)$ local memory. Algorithm 1 computes a spanning tree of H, in $O(\log p)$ communication rounds.

Algorithm 1 - CGM Algorithm for Spanning Tree
Input: A bipartite graph $H(V_1, V_2, E)$ where $V_1 = \{u_1, \ldots, u_{n_1}\}$, $V_2 = \{v_1, \ldots, v_{n_2}\}$ and $|E| = m$. An edge (u_i, v_i) of E has a vertex u_i in V_1 and a vertex v_i in V_2. The m edges are equally distributed among the p processors at random.
Output: A spanning tree of G.
Phase I:
1: Initialize $\bar{V}_1 := V_1$ and $\bar{V}_2 := V_2$ and $\bar{E} := E$.
2: **for** $\log p$ times **do**
3:   Sort the edges (u, v) of $\bar{E}$ by v and then by u.
4:   **for** each v_i of V_2 **do**
5:    Choose the smallest vertex u_j among all edges (u, v_i) and mark the edge (u_j, v_i). Let ST be the set of the marked edges.
6:   **end for**
7:   Compute the degree of each vertex $u \in \bar{V}_1$ in $H(\bar{V}_1, \bar{V}_2, \bar{E})$.
8:   Compute the degree of each vertex $u \in \bar{V}_1$ in $H_{ST}(\bar{V}_1, \bar{V}_2, ST)$.
9:   Using the degrees computed in the previous steps compute the number of zero-difference vertices.
10: **if** number of zero-difference vertices = 1 **then**
11:   the algorithm finishes
12: **end if**
13: Compact the graph to produce the compacted graph $H(\bar{V}_1, \bar{V}_2, \bar{E})$.
14: **end for**
Phase II:
1: Compute a spanning forest with the edges of graph $H(\bar{V}_1, \bar{V}_2, \bar{E})$ that do not belong to ST and removing those with degree($\bar{u}$)=1 where $\bar{u} \in \bar{V}_1$.
2: Set all processors to *active* mode.
3: **for** k:=1 to $\log p$ **do**
4:   Partition the active processors into groups of size two.
5:   **for** each group P_i, P_j of active processors, $i < j$, in parallel **do**
6:    Processor P_j sends its edge set $\bar{E}_j$ to processor P_i.
7:    Processor P_j is set to *passive* mode.
8:    Processor P_i computes the spanning forest $(\bar{V}_1, \bar{V}_2, \bar{E}_s)$ of the graph $SF = (\bar{V}_1, \bar{V}_2, \bar{E}_i \cup \bar{E}_j)$ and sets $\bar{E}_i := \bar{E}_s$.
9:   **end for**
10: **end for**

Consider the graph of Fig. 1(a). We use array $EDGE$ to store edges of E: $(1,1')(1,2')(1,3')(2,2')(2,4')(3,1')(4,5')(4,6')(5,4')(5,5')$.

Make a copy of $EDGE$ in $EDGE'$. Lines 3 to 6 of Algorithm 1 obtain a strut ST. Line 3 sorts the edges in $EDGE'$ lexicographically in the following way. Given two edges (i,j) and (k,l) then $(i,j) < (k,l)$ when $j < l$ or $((j = l)$ and $(i <$

k)). Array $EDGE'$ contains the sorted edges: $(1,1')(3,1')(1,2')(2,2')(1,3')(2,4')$ $(5,4')(4,5')(5,5')(4,6')$.

Lines 4 to 6 find a strut ST in V_1. It is represented by solid lines of Fig. 1(b). Array $EDGE'$ represents ST: $(1,1')(1,2')(1,3')(2,4')(4,5')(4,6')$.

A strut ST in V_1 determines a spanning forest of H. Lines 7 to 9 find the zero-difference and non-zero-difference vertices of the strut ST. Determine the degrees of each of the vertices in V_1 and store in D_H. In our example $D_H = (3,2,1,2,2)$. Determine now which vertices of V_1 are zero-difference. For this, determine the degree of each of the vertices of V_1 in $EDGE'$ and store in D_{ST}. Again for our example, $D_{ST} = (3,1,0,2,0)$. Thus the zero-difference vertices are vertices $\{1,4\}$ and the non-zero-difference vertices are vertices $\{2,3,5\}$.

Line 13 produces a compacted graph. For each zero-difference vertex $u \in V_1$ compact all the vertices $v_i \in V_2$ incident with u by merging all the vertices v_i onto the smallest of the v_i. The new compacted graph $H(\bar{V_1}, \bar{V_2}, \bar{E})$ is shown in Fig. 1(c). Note that vertices $2'$ and $3'$ are compressed onto vertex $1'$ and therefore the original edge $(2,2')$ now becomes $(2,1')$.

Algorithm 1 computes the spanning tree of $H = (V_1, V_2, E)$ in $O(\log p)$ communication rounds. The proof can be found in [6].

4 Generalization and Main Results

To transform any graph into a bipartite graph, subdivide each edge by adding a new vertex on each edge. Consider the vertices of the original graph as belonging to V_1 and the new added vertices as V_2, then we have a resulting bipartite graph.

To determine the connected components of a graph, in each iteration of Algorithm 1, determine each of the sublists of $EDGE'$ formed by edges $(u,v), u = u_i$ that forms a tree, labeled by $EDGE'_{u_i}$. At the end of the algorithm, we can represent each tree with the smallest vertex. Each of the different vertices represent a connected component of the graph.

References

1. Chan, A., Dehne, F.: A note on coarse grained parallel integer sorting. Parallel Processing Letters **9** (1999) 533–538
2. Hirschberg, D.S., Chandra, A.K., Sarwate, D.V.: Computing connected components on parallel computers. Comm. ACM **22** (1979) 461–464
3. Karp, R.M., Ramachandran, V.: 17. In: Handbook of Theoretical Computer Science - J. van Leeuwen (ed.). Volume A. Elsevier/MIT Press (1990) 869–941
4. Dehne, F., Ferreira, A., Cáceres, E., Song, S.W., Roncato, A.: Efficient parallel graph algorithms for coarse grained multicomputers and BSP. Algorithmica **33** (2002) 183–200
5. Dehne, F., Fabri, A., Rau-Chaplin, A.: Scalable parallel geometric algorithms for coarse grained multicomputers. In: Proc. ACM 9th Annual Computational Geometry. (1993) 298–307
6. Cáceres, E.N., Dehne, F., Mongelli, H., Song, S.W., Szwarcfiter, J.L.: A coarse-grained parallel algorithm for spanning tree and connected components. Technical report, USP http://www.ime.usp.br/~song/papers/span.pdf (2003)

Topic 14
Routing and Communication in Interconnection Networks

Jose Duato, Fabrizio Petrini, Olav Lysne, and Angelos Bilas

Topic Chairs

Parallel and distributed computing has expanded its scope and application areas during the last decade. Besides relying on the traditional parallel computers, the availability of an ever increasing set of services through Internet together with continuous improvements in communication bandwidth has led to new programming paradigms, deployment of high-speed communication infrastructure, and installation of very large servers. In turn, these servers are usually based on clusters of processors interconnected by means of some interconnection network. Therefore, communication networks, protocols, and application programming interfaces (APIs) are crucial factors for the performance of parallel and distributed computations.

This topic of Euro-Par 2004 is devoted to all aspects of communication in on-chip interconnects, parallel computers, networks of workstations, and more widely distributed systems such as grids. Contributed papers were sought to examine the design and implementation of interconnection networks and communication protocols, advances in system area and storage area networks, routing and communication algorithms, and the communication costs of parallel and distributed algorithms. On-chip and power-efficient interconnects, I/O architectures and storage area networks, switch architectures as well as multimedia and QoS-aware communication were topics addressed in this year's Call for Papers.

The CfP attracted 24 submissions to this topic, of which eight papers (33%) were selected for publication and presentation at the conference. Seven of them were accepted as regular papers, and one as distinguised. The selected papers cover a wide scope, ranging from MPI implementations on supercomputers to on-chip multiprocessors for routing applications, also covering topics like buffer management, load balancing, QoS support and forwarding table distribution in InfiniBand, and novel aspects like reduction of power consumption.

We would like to thank all the authors for their submissions to this Euro-Par topic. We owe special thanks to the 40 external referees who provided competent and timely review reports. Their effort has ensured the high quality of this part of Euro-Par 2004.

Implementing MPI on the BlueGene/L Supercomputer

George Almási[1], Charles Archer[2], José G. Castaños[1], C. Chris Erway[1], Philip Heidelberger[1], Xavier Martorell[1], José E. Moreira[1], Kurt Pinnow[2], Joe Ratterman[2], Nils Smeds[1], Burkhard Steinmacher-burow[1], William Gropp[3], and Brian Toonen[3]

[1] IBM Thomas J. Watson Research Center
Yorktown Heights, NY 10598-0218
{gheorghe,castanos,cerway,philip,xavim,jmoreira,nsmeds,steinmac}
@us.ibm.com
[2] IBM Systems Group
Rochester, MN 55901
{archerc,kwp,jratt}@us.ibm.com
[3] Mathematics and Computer Science Division, Argonne National Laboratory
Argonne, IL 60439
{gropp,toonen}@mcs.anl.gov

Abstract. The BlueGene/L supercomputer will consist of 65,536 dual-processor compute nodes interconnected by two high-speed networks: a three-dimensional torus network and a tree topology network. Each compute node can only address its own local memory, making message passing the natural programming model for BlueGene/L. In this paper we present our implementation of MPI for BlueGene/L. In particular, we discuss how we leveraged the architectural features of BlueGene/L to arrive at an efficient implementation of MPI in this machine. We validate our approach by comparing MPI performance against the hardware limits and also the relative performance of the different modes of operation of BlueGene/L. We show that dedicating one of the processors of a node to communication functions greatly improves the bandwidth achieved by MPI operation, whereas running two MPI tasks per compute node can have a positive impact on application performance.

1 Introduction

The BlueGene/L supercomputer is a new massively parallel system being developed by IBM in partnership with Lawrence Livermore National Laboratory (LLNL). BlueGene/L uses system-on-a-chip integration [5] and a highly scalable architecture [2] to assemble a machine with 65,536 dual-processor compute nodes. When operating at its target frequency of 700 MHz, BlueGene/L will deliver 180 or 360 Teraflops of peak computing power, depending on its mode of operation. BlueGene/L is targeted to become operational in early 2005.

Each BlueGene/L compute node can address only its local memory, making message passing the natural programming model for the machine. This paper describes how we implemented MPI [10] on BlueGene/L.

Our starting point for MPI on BlueGene/L [3] is the MPICH2 library [1], from Argonne National Laboratory. MPICH2 is architected with a portability layer called the Abstract Device Interface, version 3 (ADI3), which simplifies the job of porting it to different architectures. With this design, we could focus on optimizing the constructs that were of importance to BlueGene/L.

BlueGene/L is a feature-rich machine and a good implementation of MPI needs to leverage those features to deliver high-performance communication services to applications. The BlueGene/L compute nodes are interconnected by two high-speed networks: a three-dimensional torus network that supports direct point-to-point communication and a tree network with support for broadcast and reduction operations. Those networks are mapped to the address space of user processes and can directly be used by a message passing library. We will show how we architected our MPI implementation to take advantage of both memory mapped networks.

Another important architectural feature of BlueGene/L is its dual-processor compute nodes. A compute node can operate in one of two modes. In *coprocessor* mode, a single process, spanning the entire memory of the node, can use both processors by running one thread on each processor. In *virtual node mode*, two single-threaded processes, each using half of the memory of the node, run on one compute node, with each process bound to one processor. This creates the need for two modes in our MPI library, with different performance impacts.

We validate our MPI implementation on BlueGene/L by analyzing the performance of various benchmarks on our 512-node prototype. This prototype was built using first-generation BlueGene/L chips and operates at 500 MHz. We use microbenchmarks to assess how well MPI performs compared to the limits of the hardware and how different modes of operation within MPI compare to each other. We use the NAS Parallel Benchmarks to demonstrate the benefits of virtual node mode when executing computation-intensive benchmarks.

The rest of this paper is organized as follows. Section 2 presents an overview of the hardware and software architectures of BlueGene/L. Section 3 discusses those details of BlueGene/L hardware and software that were particularly influential to our MPI implementation. Section 4 presents the architecture of our MPI implementation. Section 5 describes and discusses the experimental results on our 512 node prototype that validate our approach. Finally, Section 6 contains our conclusions.

2 An Overview of the the BlueGene/L Supercomputer

The BlueGene/L hardware [2] and system software [4] have been extensively described elsewhere. In this section we present a short summary of the BlueGene/L architecture to serve as background to the following sections.

The 65,536 compute nodes of BlueGene/L are based on a custom system-on-a-chip design that integrates embedded low power processors, high performance network interfaces, and embedded memory. The low power characteristics of this architecture permit a very dense packaging. One air-cooled BlueGene/L rack contains 1024 compute nodes (2048 processors) with a peak performance of 5.7 Teraflops.

The BlueGene/L chip incorporates two standard 32-bit embedded PowerPC 440 processors with private L1 instruction and data caches, a small 2 kB L2 cache/prefetch buffer and 4 MB of embedded DRAM, which can be partitioned between shared L3 cache and directly addressable memory. A compute node also incorporates 512MB of DDR memory.

The standard PowerPC 440 cores are not designed to support multiprocessor architectures: the L1 caches are not coherent and the processor does not implement atomic memory operations. To overcome these limitations BlueGene/L provides a variety of custom synchronization devices in the chip such as the lockbox (a limited number of memory locations for fast atomic test-and-sets and barriers) and 16 KB of shared SRAM.

Each processor is augmented with a dual floating-point unit consisting of two 64-bit floating-point units operating in parallel. The dual floating-point unit contains two 32 × 64-bit register files, and is capable of dispatching two fused multiply-adds in every cycle, i.e. 2.8 GFlops/s per node at the 700 MHz target frequency. When both processors are used, the peak performance is doubled to 5.6 GFlops/s.

In addition to the 65,536 compute nodes, BlueGene/L contains a variable number of I/O nodes (1 I/O node to 64 compute nodes in the current configuration) that connect the computational core with the external world. We call the collection formed by one I/O node and its associated compute nodes a processing set. Compute and I/O nodes are built using the same BlueGene/L chip, but I/O nodes have the Ethernet network enabled.

The main network used for point-to-point messages is the *torus*. Each compute node is connected to its 6 neighbors through bi-directional links. The 64 racks in the full BlueGene/L system form a $64 \times 32 \times 32$ three-dimensional torus. The network hardware guarantees reliable, deadlock free delivery of variable length packets.

The *tree* is a configurable network for high performance broadcast and reduction operations, with a latency of 2.5 microseconds for a 65,536-node system. It also provides point-to-point capabilities. The *global interrupt* network provides configurable OR wires to perform full-system hardware barriers in 1.5 microseconds

All the torus, tree and global interrupt links between midplanes (a 512-compute node unit of allocation) are wired through a custom link chip that performs redirection of signals. The link chips provide isolation between independent partitions while maintaining fully connected networks within a partition.

BlueGene/L system software architecture: User application processes run exclusively on compute nodes under the supervision of a custom Compute Node Kernel (CNK). The CNK is a simple, minimalist runtime system written in approximately 5000 lines of C++ that supports a single application running by a single user in each BG/L node. It provides exactly two threads running one on each PPC440 processor. The CNK does not require or provide scheduling and context switching. Physical memory is statically mapped, protecting a few kernel regions from user applications. Porting scientific applications to run into this new kernel has been a straightforward process because we provide a standard Glibc runtime system with most of the Posix system calls.

Many of the CNK system calls are not directly executed in the compute node, but are function shipped through the tree to the I/O node. For example, when a user application performs a write system call, the CNK sends tree packets to the I/O node managing the processing set. The packets are received on the I/O node by a daemon called ciod. This daemon buffers the incoming packets, performs a Linux write system call against a mounted filesystem, and returns the status information to the CNK through the tree. The daemon also handles job start and termination on the compute nodes.

I/O nodes run the standard PPC Linux operating system and implement I/O and process control services for the user processes running on the compute nodes. We mount a small ramdisk with system utilities to provide a root filesystem.

The system is complemented by a control system implemented as a collection of processes running in an external computer. All the visible state of the BlueGene/L machine is maintained in a commercial database. We have modified the BlueGene/L middleware (such as LoadLeveler and mpirun) to operate through the ciod system rather than launching individual daemons on all the nodes.

3 Hardware and System Software Impact on MPI Implementation

In this section we present a detailed discussion of the BlueGene/L features that have a significant impact on the MPI implementation.

The torus network. guarantees deadlock-free delivery of packets. Packets are routed on an individual basis, using one of two routing strategies: a *deterministic* routing algorithm, in which all packets follow the same path along the x, y, z dimensions (in this order); and a minimal *adaptive* routing algorithm, which permits better link utilization but allows consecutive packets to arrive at the destination out of order.

Efficiency: The torus packet length is between 32 and 256 bytes in multiples of 32. The first 16 bytes of every packet contain destination, routing and software header information. Therefore, only 240 bytes of each packet can be used as payload. For every 256 bytes injected into the torus, 14 additional bytes traverse the wire with CRCs etc. Thus the efficiency of the torus network is $\eta = \frac{240}{270} = 89\%$.

Link bandwidth: Each link delivers two bits of raw data per CPU cycle (0.25 Bytes/cycle), or $\eta \times 0.25 = 0.22$ Bytes/cycle of payload data. This translates into 154 MBytes/s/link at the target 700 MHz frequency.

Per-node bandwidth: Adding up the raw bandwidth of the 6 incoming + 6 outgoing links on each node, we obtain $12 \times 0.25 = 3$ bytes/cycle per node. The corresponding bidirectional payload bandwidth is 2.64 bytes/cycle/node.

Reliability: The network guarantees reliable packet delivery. In any given link, it resends packets with errors, as detected by the CRC. Irreversible packet losses are considered catastrophic and stop the machine. The communication library considers the machine to be completely reliable.

Network ordering semantics: MPI ordering semantics enforce the order in which incoming messages are matched against the queue of posted messages. Adaptively routed packets may arrive out of order, forcing the MPI library to reorder them before delivery. Packet re-ordering is expensive because it involves memory copies and requires packets to carry additional sequence and offset information. On the other hand, deterministic routing leads to more network congestion and increased message latency even on lightly used networks.

The tree network. serves a dual purpose. It is designed to perform MPI collective operations efficiently, but it is also the main mechanism for communication between I/O and compute nodes. The tree supports point-to-point messages of fixed length (256 bytes), delivering 4 bits of raw data per CPU cycle (350 Mbytes/s). It has reliability guarantees identical to the torus.

Efficiency: The tree packet length is fixed at 256 bytes, all which can be used for payload. 10 additional bytes are used with each packet for operation control and link reliability. Thus, the efficiency of the tree network is $\eta = \frac{256}{266} = 96\%$.

Collective operations: An ALU in the tree network hardware can combine incoming and local packets using bitwise and integer operations, and forward the resulting packet along the tree. Floating-point reductions can be performed in two phases, one to calculate the maximum exponent and another to add the normalized mantissas.

Packet routing on the tree network is based on packet classes. Tree network configuration is a global operation that requires the configuration of all nodes in a job partition. For that reason we only support operations on MPI_COMM_WORLD.

CPU/network interface: The torus, tree and barrier networks are partially mapped into user-space memory. Torus and tree packets are read and written with special 16-byte SIMD load and store instructions of the custom FPUs.

Alignment: The SIMD load and store instructions used to read and write network packets require that memory accesses be aligned to a 16 byte boundary. The MPI library does not have control over the alignment of user buffers. In addition, the sending and receiving buffer areas can be aligned at different boundaries, forcing packet re-alignment through memory-to-memory copies.

Network access overhead: Torus/tree packet reads into aligned memory take about 204 CPU cycles. Packet writes can take between 50 and 100 cycles, depending on the whether the packet is being sent from cache or main memory.

CPU streaming memory bandwidth. is another constraint of the machine. For MPI purposes we are interested mostly in the bandwidth for accessing large contiguous memory buffers. These accesses are typically handled by prefetch buffers in the L2 cache, resulting in a bandwidth of about 4.3 bytes/cycle.

We note that the available bandwidth of main memory and the torus and tree network are in the same order of magnitude. Performing memory copies on this machine to get data into/from the torus results in reduced performance. It is imperative that network communication be zero-copy wherever possible.

Inter-core cache coherency: The two processors in a node are not cache coherent. Software must take great care to insure that coherency is correctly handled in software. Coherency handled at the granularity of the CPUs' L1 cache lines: 32 bytes. Therefore, data structures shared by the CPUs should be aligned at 32-byte boundaries to avoid coherency problems.

4 Architecture of BlueGene/L MPI

The BlueGene/L MPI is an optimized port of the MPICH2 [1] library, an MPI library designed with scalability and portability in mind. Figure 1 shows two components of the MPICH2 architecture: message passing and process management. MPI process management in BlueGene/L is implemented using system software services. We do not discuss this aspect of MPICH2 further in this paper.

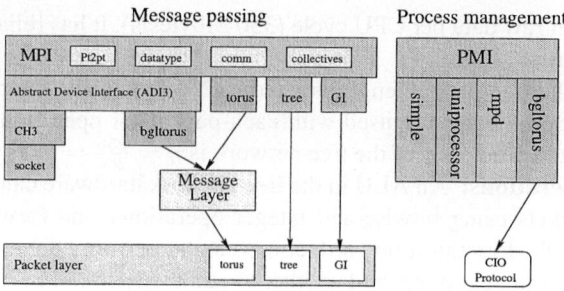

Fig. 1. BlueGene/L MPI software architecture.

The upper layers of the message passing functionality are implemented by MPICH2 code. MPICH2 provides the implementation of point-to-point messages, intrinsic and user defined datatypes, communicators, and collective operations, and interfaces with the lower layers of the implementation through the Abstract Device Interface version 3 (ADI3) layer [8]. The ADI3 layer consists of a set of data structures and functions that need to be provided by the implementation. In BlueGene/L, the ADI3 layer is implemented using the BlueGene/L Message Layer, which in turn uses the BlueGene/L Packet Layer.

The ADI layer is described in terms of MPI requests (messages) and functions to send, receive, and manipulate these requests. The BlueGene/L implementation of ADI3 is called bgltorus. It implements MPI requests in terms of Message Layer messages, assigning one message to every MPI request. Message Layer messages operate through callbacks. Messages corresponding to send requests are posted in a send queue. When a message transmission is finished, a callback is used to inform the sender. Correspondingly, there are callbacks on the receive side to signal the arrival of new messages. Those callbacks perform matching of incoming Message Layer messages to the list of MPI posted and unexpected requests.

The BlueGene/L Message Layer is an active message system [7, 9, 12, 13] that implements the transport of arbitrary-sized messages between compute nodes using the torus network. It can also broadcast data, using special torus packets that are deposited on every node along the route they take. The Message Layer breaks messages into fixed-size packets and uses the Packet Layer to send and receive the individual packets. At the destination, the Message Layer is responsible for reassembling the packets, which may arrive out of order, back into a message.

The Message Layer addresses nodes using the equivalent of MPI_COMM_WORLD ranks. Internally, it translates these ranks into physical torus x, y, z coordinates, that are used by the Packet Layer. The mapping of ranks to torus coordinates is programmable by the user, and can be used to optimize application performance by choosing a mapping that support the logical communication topology of the application.

Message transmission in the Message Layer is implemented using one of multiple available communication protocols, roughly corresponding to the protocols present in more conventional MPI implementations, such as the eager and rendezvous protocols.

The Message Layer is able to handle arbitrary collections of data, including non-contiguous data descriptors described by MPICH2 *data loops*. The Message Layer incorporates a number of complex data packetizers and unpacketizers that satisfy the multiple requirements of 16-byte aligned access to the torus, arbitrary data layouts, and zero-copy operations.

The Packet Layer is a very thin stateless layer of software that simplifies access to the BlueGene/L network hardware. It provides functions to read and write the torus/tree hardware, as well as to poll the state of the network. Torus packets typically consist of 240 bytes of payload and 16 bytes of header information. Tree packets consist of 256 bytes of data and a separate 32-bit header. To help the Message Layer implement zero-copy messaging protocols, the packet layer provides convenience functions that allow software to "peek" at the header of an incoming packet without incurring the expense of unloading the whole packet from the network.

4.1 Operating Modes

Coprocessor mode: To support the concurrent operation of the two non-cache-coherent processors in a compute node, we have developed a *dual core library* that allows the use of the second processor both as a communication coprocessor and as a computation coprocessor. The library uses a non-L1-cached, and hence coherent, area of the memory to coordinate the two processors. The main processor supplies a pool of work units to be executed by the coprocessor. Work units can be *permanent*, executed whenever the coprocessor is idle, or *transient* functions, executed once and then removed from the pool. An example of a permanent function would be the one that uses the coprocessor to help with the *rendezvous* protocol (Section 4.2). To start a transient function, one invokes the `co_start` function. The main processor waits for the completion of the work unit by invoking the `co_join` function.

Virtual node mode: The CNK in the compute nodes also supports a virtual node mode of operation for the machine. In that mode, the kernel runs two separate processes in each compute node. Node resources (primarily the memory and the torus network) are evenly split between both processes. In virtual node mode, an application can use both processors in a node simply by doubling its number of MPI tasks, without having to explicitly handle cache coherence issues. The now distinct MPI tasks running in the two CPUs of a compute node have to communicate to each other. We have solved this problem by implementing a virtual torus device, serviced by a virtual packet layer, in the scratchpad memory.

4.2 MPI Messaging Protocols

The one-packet protocol in the Message Layer handles short messages that fit into a single packet. That is, messages with length less than 240 bytes. Short message packets are always sent with deterministic routing, in order to avoid the issue of out-of-order arrival.

The eager protocol is designed to deliver messages between 200 bytes and 10 kbytes in size at maximum net bandwidth. The receiver of an eager message has to accept and process each incoming packet. Since the network is reliable, no provisions for packet retransmission exist in the Message Layer.

The processing of eager protocol packets is much simpler when the network guarantees in-order delivery. When packets arrive out of order, software on the receive side spends processor cycles finding the destination message buffer and the offset in that buffer based on information in the packet.

The rendezvous protocol optimizes processor usage and multi-link bandwidth. Whereas the eager protocol is able to maximize single link bandwidth, the per-packet processor overhead is too large to support the full bandwidth of the network. Reading a packet from the network requires 204 CPU cycles. Sending a packet takes between 50 and 100 cycles. When the network is at maximum capacity, data can flow at the rate of 3 Bytes/cycle on every node. At 270 raw bytes per packet, a processor has 90 cycles to handle each packet. Clearly, that is not possible with a single processor, and only marginally possible with two.

The rendezvous protocol minimizes the amount of CPU processing for most packets, by having packets carry the destination buffer address with them. This technique requires an initial dialog between the sender and the receiver to establish the destination address. The initial handshake costs 1500 cycles of processor time *each* on the sender and the receiver; but since handling rendezvous packets takes about 150 cycles less than handling eager packets, the handshake cost is amortized for larger messages, making the rendezvous protocol viable beyond message lengths of 20 packets, or 5 kbytes.

Self-contained packets are also more suitable to be handled by the co-processor. All packets carry their destination addresses and processing them is a local operation suitable for the non-cache-coherent coprocessor. The software coherency protocol necessary for the hand-over of received data from the coprocessor to the main processor costs about 4000 CPU cycles.

Another advantage of self-contained packets is that they are insensitive to arrival order. Thus the bulk of rendezvous messages can be transmitted with adaptively routed packets, allowing for better network utilization.

5 Experimental Results

In this section we present preliminary performance results using the first BlueGene/L 512-node prototype, which became operational in October 2003. We first present microbenchmark results that analyze different aspects of our current MPI implementation. We then compare different message passing protocols. Next, we analyze BlueGene/L-specific implementations of common collectives. Finally, we present results for the NAS parallel benchmarks in both coprocessor and virtual node modes. None of the results presented here use link chips – we restrict our studies to three-dimensional meshes rather than tori.

Roundtrip latency analysis: We measured the latency of very short messages between two neighbor nodes on BlueGene/L using Dave Turner's mpipong program [11]. Current $\frac{1}{2}$-roundtrip latency stands at approximately 3000 cycles (6 microseconds at the 500 MHz CPU frequency of the prototype), consisting of multiple components.

26% of the total latency, or 800 cycles, is incurred by MPICH2 code. This overhead can be ameliorated by deploying higher compiler optimization levels and/or using better compilers.

The ADI3 glue layer (`bgltorus`) and the Message Layer together contribute about 400 cycles (13%) of overhead, mostly testing data types, translating ranks and creating Message Layer objects to handle the data.

The packet layer costs 1000 cycles (29%) of the total, mostly because handling single packets is more expensive per packet than handling multiple packets belonging to the same message.

Finally, actual hardware latency is currently at 32%. We estimate that this time could be reduced to 5-6% for very short messages by using shorter network packets (down to 32 bytes instead of 256 bytes long when the message we are transmitting fits into such a packet), resulting in 20-25% overall savings in latency. We expect latency numbers to improve as the MPI implementation matures.

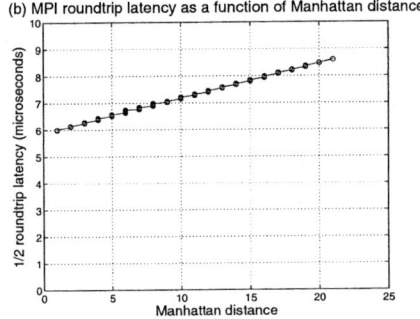

Fig. 2. Roundtrip latency as a function of Manhattan distance.

Latency as a function of Manhattan distance: Figure 2 shows $\frac{1}{2}$-roundtrip latency as a function of the Manhattan distance between the sender and the receiver in the torus. The figure shows a clear linear dependency, with 120 ns of additional latency added for every hop. We expect the per-hop latency to diminish as CPU frequency increases.

Single-link bandwidth: Figure 3(a) shows the available bandwidth measured with MPI on a single bidirectional link of the machine (both sending and receiving). The figure shows both the raw bandwidth limit of the machine running at 500 MHz ($2 \times 125 = 250$ MBytes/s) and the net bandwidth limit ($\eta \times 2 \times 125 = 220$ MBytes/s), as well as the measured bandwidth as a function of message size. With the relatively low message processing overhead of the MPI eager protocol, high bandwidth is reached even for relatively short messages: $\frac{1}{2}$ bandwidth is reached for messages of about 1 KByte.

A comparison of point-to-point messaging protocols: Figure 3 (b), (c) and (d) compare the multi-link performance of the eager and rendezvous protocols, the latter with and without the help of the coprocessor. We can observe the number of simultaneous active connections that a node can keep up with. This is determined by the amount of time spent by the processor handling each individual packet belonging to a message; when the processor cannot handle the incoming/outgoing traffic the network backs up.

In the case of the eager and rendezvous protocols, without the coprocessor's help, the main processor is able to handle about two bidirectional links. When network traffic increases past two links, the processor becomes a bottleneck, as shown by Figures Figure 3 (b) and Figure 3 (c). Figure 3 (d) shows the effect of the coprocessor helping out in the rendezvous protocol: MPI is able to handle the traffic of more than three bidirectional links.

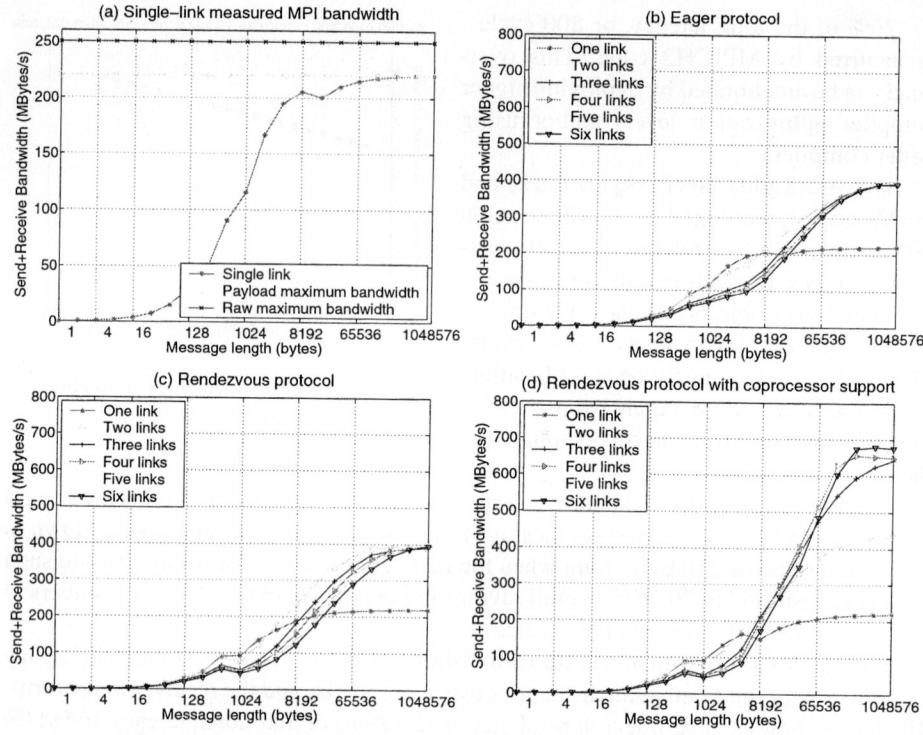

Fig. 3. Comparing multi-link bandwidth performance of MPI protocols.

Optimized broadcast on the torus network: One of the challenges of implementing an efficient MPI library on the BlueGene/L machine is to find the efficient algorithms for the MPI collectives that are well suited to the machine's network. The MPICH2 implementation of the `MPI_Bcast` primitive has a limited efficiency on the BlueGene/L machine because it is designed for a machine with a crossbar type network.

The BlueGene/L MPI implementation includes an optimized broadcast algorithm based on MPI point-to-point communication that can be used with any communicator that maps onto a regular mesh in the physical torus.

For an n-dimensional mesh or torus, the algorithm consists of n concurrently executing stages (illustrated in Figure 4 (a) for the two-dimensional mesh case). The basic operation of the algorithm is a broadcast of a part of the message along a one-dimensional line in the n-dimensional topology. On an n-dimensional mesh the algorithm has the property that each process receives $\frac{1}{n}$ of the complete message from each of the n directions. On a torus topology each process receives $\frac{1}{2n}$ of the full message from each of the incoming links. Each block of the message is further subdivided to pipeline the broadcast process. The optimal subdivision size is a function of total message length, communicator topology and Manhattan diameter of the network.

The BlueGene/L algorithm has proved superior to the standard MPICH2 broadcast algorithms, because those are oblivious to underlying network topology. The current

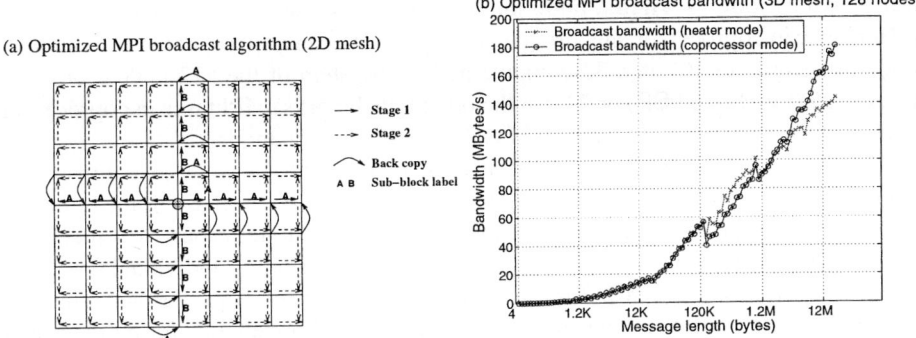

Fig. 4. Optimized mesh broadcast algorithm.

implementation (Figure 4) is limited by the CPU processing capability of the node processors. For a mesh mapped communicator of size 8x4x4 an overall performance of 140MB/s is seen in single processor mode and 170MB/s in co-processor mode.

Tree bandwidth: As mentioned in Section 3, the tree supports collective operations, including broadcast and reduction operations. The MPI library currently uses the tree network to implement broadcast and integer reduce and allreduce operations on the `MPI_COMM_WORLD` communicator. Tree-based reduction of floating-point numbers is under development.

Figure 5 shows the measured bandwidth of tree-based MPI broadcast and allreduce measured on the 512-node prototype. Broadcast bandwidth is essentially independent of message size, and hits the theoretical maximum of $0.96 \times 250 = 240$ Mbytes/s. Allreduce bandwidth is somewhat lower, encumbered by the software overhead of re-broadcasting the result.

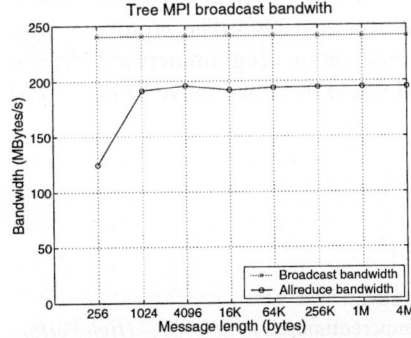

Fig. 5. Tree-based MPI broadcast and allreduce: measured bandwidth.

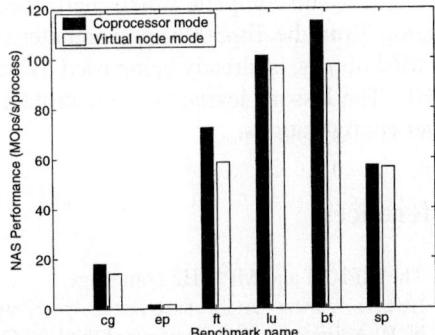

Fig. 6. Comparison of per-node performance in coprocessor and virtual node mode.

Coprocessor mode vs. virtual node mode: Figure 6 shows a comparison of per-task performance in coprocessor and virtual node modes. We ran a subset of the class B NAS parallel benchmarks [6] on a 32-compute node subsystem of the 512-node prototype. We used 25 (for BT and SP) or 32 (for the other benchmarks) MPI tasks in coprocessor mode, and 64 (for all benchmarks) MPI tasks in virtual node mode.

Ideally, per-task performance in virtual node mode would be equal to that in coprocessor mode, resulting in a net doubling of total performance (because of the doubling of tasks executing). However, because of the sharing of node resources − including the L3 cache, memory bandwidth, and communication networks − individual processor efficiency degrades between 2-20%, resulting in less than ideal performance results. Nevertheless, the improvement warrants the use of virtual node mode for these classes of computation-intensive codes.

6 Conclusions

With its 65,536 compute nodes, the BlueGene/L supercomputer represents a new level of scalability in massively parallel computers. Given the large number of nodes, each with its own private memory, we need an efficient implementation of MPI to support application programmers effectively. The BlueGene/L architecture provides a variety of features that can be exploited in an MPI implementation, including the torus and tree networks and the two processors in a compute node.

This paper reports on the architecture of our MPI implementation and also presents initial performance results. Starting with MPICH2 as a basis, we provided an implementation that uses the tree and the torus networks efficiently and that has two modes of operation for leveraging the two processors in a node. The results also show that different message protocols exhibit different performance behaviors, with each protocol being better for a different class of messages. Finally, we show that the coprocessor mode of operation provides the best communication bandwidth, whereas the virtual node mode can be very effective for computation intensive codes represented by the NAS Parallel Benchmarks.

BlueGene/L MPI has been deployed on our 512-node prototype, a small system compared to the complete BlueGene/L supercomputer, but powerful enough to rank at position 73 in the Top500 supercomputer list of November 2003. The prototype, with our MPI library, is already being used by various application programmers at IBM and LLNL. The lessons learned on this prototype will guide us as we move to larger and larger configurations.

References

1. The MPICH and MPICH2 homepage.
   http://www-unix.mcs.anl.gov/mpi/mpich.
2. N. R. Adiga et al. An overview of the BlueGene/L supercomputer. In *SC2002 − High Performance Networking and Computing*, Baltimore, MD, November 2002.

3. G. Almási, C. Archer, J. G. C. nos, M. Gupta, X. Martorell, J. E. Moreira, W. Gropp, S. Rus, and B. Toonen. MPI on BlueGene/L: Designing an Efficient General Purpose Messaging Solution for a Large Cellular System. Lecture Notes in Computer Science. Springer-Verlag, September 2003.
4. G. Almási, R. Bellofatto, J. Brunheroto, C. Caşcaval, J. G. C. nos, L. Ceze, P. Crumley, C. Erway, J. Gagliano, D. Lieber, X. Martorell, J. E. Moreira, A. Sanomiya, and K. Strauss. An overview of the BlueGene/L system software organization. In *Proceedings of Euro-Par 2003 Conference*, Lecture Notes in Computer Science, Klagenfurt, Austria, August 2003. Springer-Verlag.
5. G. Almasi et al. Cellular supercomputing with system-on-a-chip. In *IEEE International Solid-state Circuits Conference ISSCC*, 2001.
6. D. Bailey, T. Harris, W. Saphir, R. van der Wijngaart, A. Woo, and M. Yarrow. **The NAS Parallel Benchmarks 2.0**. Technical Report NAS-95-020, NASA Ames Research Center, December 1995.
7. G. Chiola and G. Ciaccio. Gamma: a low cost network of workstations based on active messages. In *Proc. Euromicro PDP'97, London, UK, January 1997, IEEE Computer Society.*, 1997.
8. W. Gropp, E. Lusk, D. Ashton, R. Ross, R. Thakur, and B. Toonen. MPICH Abstract Device Interface Version 3.4 Reference Manual: Draft of May 20, 2003. http://www-unix.mcs.anl.gov/mpi/mpich/adi3/adi3man.pdf.
9. S. Pakin, M. Lauria, and A. Chien. High performance messaging on workstations: Illinois Fast Messages (FM) for Myrinet. In *Supercomputing '95, San Diego, CA, December 1995*, 1995.
10. M. Snir, S. Otto, S. Huss-Lederman, D. Walker, and J. Dongarra. *MPI - The Complete Reference, second edition*. The MIT Press, 2000.
11. D. Turner, A. Oline, X. Chen, and T. Benjegerdes. Integrating new capabilities into NetPIPE. Lecture Notes in Computer Science. Springer-Verlag, September 2003.
12. T. von Eicken, A. Basu, V. Buch, and W. Vogels. U-net: A user-level network interface for parallel and distributed computing. In *Proceedings of the 15th ACM Symposium on Operating Systems Principles, Copper Mountain, Colorado*, December 1995.
13. T. von Eicken, D. E. Culler, S. C. Goldstein, and K. E. Schauser. Active Messages: a mechanism for integrated communication and computation. In *Proceedings of the 19th International Symposium on Computer Architecture*, May 1992.

Modular On-Chip Multiprocessor for Routing Applications

Saifeddine Berrayana, Etienne Faure, Daniela Genius, and Frédéric Pétrot

Laboratoire LIP6, Département ASIM, Université Paris-VI,
4 place Jussieu, Paris, France
{saifeddine.berrayana,etienne.faure,daniela.genius,frederic.petrot}
@lip6.fr

Abstract. Simulation platforms for network processing still have difficulties in finding a good compromise between speed and accuracy. This makes it difficult to identify the causes of performance bottlenecks: Are they caused by application, hardware architecture, or by a specificity of the operating system? We propose a simulation methodology for a multiprocessor network processing platform which contains sufficient detail to permit very precise simulation and performance evaluation while staying within reasonable limits of both specification and simulation time. As a case study, we show how a model can be developed for a IPv4 packet routing application, exhibiting the performance and scalability bottlenecks and can thus be used to reason about architectural alternatives.

1 Introduction

This paper proposes an efficient methodology for the performance tuning of networking applications on network processors by means of describing a flexible platform composed of hardware, software and operating system. Network processors are programmable routers used in specialized telecommunication equipment [1]. Among systems on chip they are distinguished by highly parallel hardware: one network processor can contain dozens or even hundreds of microprocessor cores, each executing a massively task parallel application.

In order to arrive at a reasonable performance, the application must be fine-tuned, and hardware and operating system must be adapted: Our approach is thus typically System-on-Chip. The remainder is structured as follows: First, we present our design methodology. We then show IPv4 as a case study and argue that our platform (hardware and operating system) can be seen as a shared-memory multiprocessor. Our validation methodology, as well as some performance results are shown before we conclude by outlining current limitations and perspectives.

2 Description Methodology

The network processor modeling platform we took as a reference is ST Microelectronics' STepNP [2], which allows to study network processor architectures on transaction level [3]. Transaction level modeling reduces simulation time considerably, at the cost of a loss of precision. Our principal aim is to propose a methodology that enables to expose the multiple causes for multiprocessor-related bottlenecks, hence our need for a fine-grained simulation at register and signal level.

SystemC [4] is a hardware modeling language and simulation kernel that consists of a C++ class library. It permits the instantiation and assignment of models of hardware components and their signals in a hierarchic, building-block manner. Different levels of detail are possible, from coarse grain transaction level to fine grain register transfer level (RTL) [5].

We opted for a cycle-accurate simulation for which the values present at the connectors of a hardware component are known at every clock cycle. The SOCLIB component library [6] provides cycle-accurate, bit-accurate (CABA) simulation models, written in SystemC. Though equivalent to synthesizable models concerning external architecture, these models cannot be directly synthesized –for reasons of protection of intellectual property– and moreover, require much less processing power for their simulation.

The building blocks are easily connected by a unique VCI (*Virtual Component Interconnect*) interface [7], which permits to encapsulate arbitrary interconnection networks and to refer to them by a standardized protocol which fixes the number and type of signals required for a communication between components.

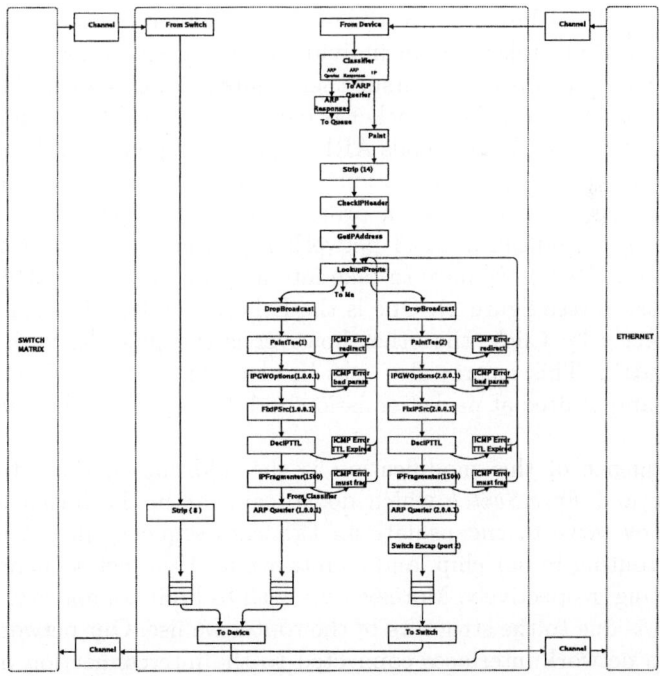

Fig. 1. Router Software Configuration.

3 Porting an IPv4 Routing Application

A functioning system on chip consists of three parts: The software (the application itself), the hardware and the operating system. Together they build the platform; the challenge is to have them cooperate efficiently. In the next section, we will explain the choices we made to achieve this aim.

3.1 Software

IPv4 routing [8], well-known in the networking community, serves as our reference application. Essentially it consists in taking two or more input streams analyze their headers, and redirect the packets to the appropriate outputs. This application serves as an example in publications on Click [9, 10]. Click is a modular router configuration language consisting of two description levels: A simple high-level language to describe the structure of networking applications by composing so-called *elements* and a library of C++ components containing elementary functionality (such as IP packet header identification, buffering, discarding).

In section 3.3 we will restrict to discussing the application already modified to run on our platform: As can be seen in Figure 1, predefined elements are for example *FromDevice* and *ToDevice*, representing entering and exiting packet streams, respectively. Other important elements are *Paint* and *Strip* to mark packets for loops and delete a packet header. Within compound elements, several other elements' functionalities are summed up hierarchically. The central functionality of IPv4 lies in such a compound element, *LookupIProute*, which determines the route taken by an individual packet according to the information in its header. The packets pursue their course along two possible routes: In the direction of *ToDevice* if the packet is damaged, in the direction of *ToSwitch* otherwise. A third exit, *ToMe*, sends ARP requests/responses to the router itself.

Let us now explain the modifications we made for SoC implementation. The first modification concerns I/O. In order to have the application running on one element of a (multiprocessor) network processing platform (one Network Processing Unit, NPU), we have to take into account the fact that its interfaces are asymmetrical (see figure 2). One is the Ethernet link and as such is already treated correctly by Click; the other however is the junction with an on-chip switching matrix. This matrix also uses headers for routing, which have to be constructed and grafted at packet emission and deleted when the packet arrives, respectively.

A consequence of this modification is the addition of these two modules *SwitchEncap* and *FromSwitch* which do not appear in the initial list of Click modules. They serve to encapsulate an Ethernet sequence in the header used for internal routing in our chip, and to retrieve an Ethernet sequence after this internal routing, respectively. Moreover we wish to limit complexity. The second modification is due to the structure of the router we use. Our network processor only has two network interfaces connected to the interconnection matrix; likewise IPv4 only has four communication points: Two each for ingress and egress packets.

Thus, it is known that the packets arriving from this interface have already been treated by another block, and are moreover destined for the Ethernet link to the outer world. We can conclude that the verification and routing have already been accomplished by another unit. The only work still to be done is to direct the block to the egress interface. For this reason, Figure 1 shows an additional straight line between *fromSwitch* and *toEth*. The only work remaining is to strip the header (delete the first 8 bytes) for the interconnection matrix routing which is expressed in element *Strip(8)*. The modified application is shown in Figure 1.

3.2 Hardware Architecture Model

The application has been made as independent as possible from the context in which it is used, retaining the essentials of its structure. We look for a modular system architecture where we can easily add, exchange, and regroup hardware models. Such a system is shown in Figure 2. The first step in a modular approach is to define an elementary (SystemC) building block which can be instantiated to the required quantity. This building block is itself a router, and elementary in the sense that it has only two interfaces. This choice enables us to treat platforms with only two interfaces as well as to dimension our platform according to the number of network interfaces required. We have thus a two-level hierarchy of one router composed of four NPUs, interconnected by a switch matrix, each with a certain number of processors inside. From the hardware architecture point of view, two separate parts are to be realized on-chip (we will concentrate on the latter):

- an interconnection matrix that links network processors; possible simple but realistic topologies are a full crossbar for a small number of processors, a fat tree for a larger number [11]
- the network processor itself, i e. the unit which does the packet treatment

The hardware architecture we implement was originally inspired by the architecture developed by ST Microelectronics in the context of the STepNP platform. It regroups multiple processors, two network interfaces around a on-chip interconnection network. The platform contains on the one hand calculation units, represented by MIPS R3000 processors with their caches, as well as SRAM memory containing the application code and data; all of these already exist as SOCLIB CABA models. On the other hand, we designed dedicated I/O co-processors

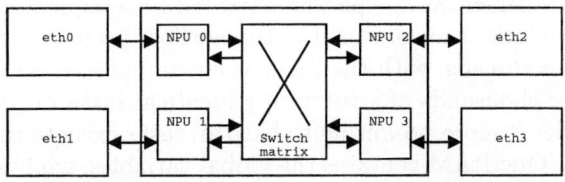

Fig. 2. Router with four network interfaces.

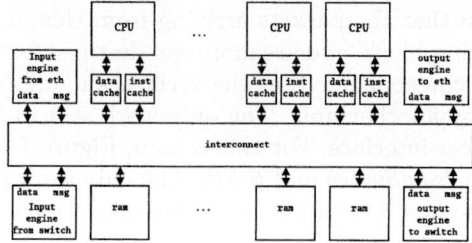

Fig. 3. Network processing unit hardware architecture.

to insert Ethernet packets into the system and to extract them, once they are treated. There are two such co-processors for each network interface, one each for ingress and egress, named respectively *input-engine* and *ouput-engine*. The architecture is shown in Figure 3. All these elements communicate via an on-chip interconnect carrying a VCI interface. Note that the interconnection network used in the simulations is a virtual model called VGMN (for VCI Generic Micro Network) whose parameters are the number of ports and latency. We are well aware that some precision is lost by using this abstract model instead of a real interconnect model.

A first noteworthy advantage of modularity is the fact that the number of co-processors is fixed for each NPU, whatever the number of interfaces that have to be connected to the system. In particular, this avoids to take into account problems due to saturation of interconnect bandwidth in the presence of multiple packet injectors. It will consequently be much easier to extrapolate our performance results to a larger model composed of several instances of the basic model and functioning independently of each other.

3.3 The Embedded Software

Once the functionality of our application being fixed, the next step is to port it to the parallel target architecture. Basically, there are two options available: On the one hand, exploiting coarse grain parallelism, which means looking for independent treatments and transform them into separate communicating tasks. On the other hand, duplicating the application in order to obtain identical clones, each treating an IP packet throughout its passage.

The simple kind of routing application we used as benchmark has relatively weak intrinsic parallelism, as one packet is treated by a sequence of functionality corresponding to Click elements. In [12], the authors propose to decompose one instance IPv4 into threads, with the Fifos as cutting points. As existing routers treat hundreds or thousands of structurally identical tasks in parallel from end to end, we consider it more promising to have all tasks execute the same code on different packets. One RAM contains the global variables, while all other RAMs are allocated to one processor each. However, the accesses to the shared memory are critical, as will show the experimental results.

3.4 Operating System Mutek

Once application and hardware fixed, it remains to determine the operating system. Our choice was the Mutek[13] micro kernel. This micro kernel is able to handle multiple tasks running on multiple processors. It provides a C standard library as well as support for POSIX threads.

An important point is that we wish to assign tasks statically to the processors, in order to avoid the cost of task migration between processors, and to avoid migration-related coherence problems. Mutek gives us that possibility.

Once the decisions on kernel and scheduling are made, it remains to determine the number of tasks assigned to on single processing unit. We made performance measures concluding that the time required for context switching outweights time gained by data latency. In the experiments presented here we show one task statically assigned to one processing unit.

4 Validation and Performance

The hardware and software part of our network processor have to be tested together. The entire platform simulated under SystemC will serve as proof of concept. The validation is made using several Ethernet packets benchmarks injected using the *input engine* and analyzed by the *output engine*.

The most relevant performance measure is the throughput in bits per cycle our processor can achieve. Our system is globally synchronous, i e. all of its components share the same clock domain. Thus, we can choose a clock cycle as basic time unit. On the other hand, we have left open the choice of the frequency; clearly, the higher the frequency at which our system works, the better its performance. Our first measure will serve as baseline throughout our experiments: The throughput the system guarantees when one single processor is instantiated, which is the maximal throughput of a single NPU.

When adding processors to our NPU, as can be expected, the performance is not linearly improved. The more adequate question is, how many processors can be added in order to still obtain a profit? Running the simulation during 2 000 000 cycles, and counting the number of 32 bit words arriving at each egress interface allows to neglect the impact of boot time (around 16 000 cycles); also packet(s) that have been treated but not yet been emitted are ignored. All IP packets have 56 bytes, which constitutes the worst case in our application context. This simulation yields in total 689 words of 32 bits each that have been read at egress, a throughput of 0.011 bit/cycle. This rather bad result is due to the non-optimized application. The weak throughput actually corresponds to an average time for treatment of 40 000 cycles for an IP packet. However, keep in mind that our goal is to exhibit performance bottlenecks on a detailed level, for a variety of architectures. Details on the execution times of the individual functions are summed up in table 7 for the case of a packet entering via the Ethernet link and exiting via the interconnection matrix. This corresponds to the longest possible path which a packet can take between two interfaces (around 110 000 cycles).

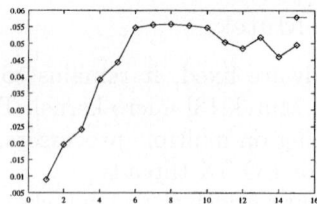

Fig. 4. Number of processors and resulting throughput in bits/cycle.

This is nevertheless our reference for multiprocessor experiments. The first consists in simply varying the number of on-chip processing units. Figure 4 illustrates our results, presenting on the x-axis the number of MIPS R3000 processors, on the y-axis the throughput achieved, in bit/cycle. The measurements show that the best throughput is reached for 8 processors, but it is only slightly higher than that for six processors. All units added after the sixth will not yield a significant improvement, they rather lead to performance degradation. The reason for this is that all processors share the same resources, particularly the memory banks where the IP packets are copied to before or after being treated. This contention is only due to the fact that there is only one memory bank for all tasks, not because the data are shared. Thus, we distribute the memory around the interconnection network by replacing the four big memory segments used by the ingress/egress co-processors by as many segments as processors. Such a segment has to fulfill three functions: 1) contain the local data of the thread, 2) allow the ingress processors to write the entering packets there, 3) supply a space for the processor to copy the outgoing packets.

More precisely, each memory bank will be divided into three non overlapping regions. The second region is subdivided into two sections, one for each *input engine*. As none of the two co-processors knows the behavior of the other, we avoid to make them share variables or address space to not further complicate matters.

This system architecture modification permits a significant performance gain, also for larger numbers of processors. The graph for this second experiment is shown in Figure 5. Here, the performance peak is reached at 27 processors, with a throughput of 0.3 bit/cycle. Here again, a raise in performance to a certain point is followed by an abrupt degradation, caused by contention for the access to shared resources, more precisely, the access to the two segments of shared memory which have remained unchanged and are used by all threads: The code segment and the segment containing the global variables of the system. The presence of contention means that several initiators wish to establish a communication with the target at one time: In consequence, at least one initiator stays blocked waiting for access to the the resource.

The graph in Figure 6 shows the number of cycles due to bottlenecks stemming from accesses to global data in the RAM during an interval of time of 2 000 000 cycles. A closer look reveals that for 28 and more processors, the number of conflict cycles exceeds 1.8 million, which means that the system is paralyzed. Taking a closer look, most of the conflicts happen when processors try to access the RAM containing the operating system data. At the same time, they stay in the scheduler function, which wakes up another thread ready to run when the current thread falls asleep. In our case, there is only one thread per processor, thus we decided to prevent this function from being executed. To do this, the thread should never stop, even if it cannot access the ressource. We changed the lock associated with the shared ressources (input and output engines) from a mutex to a spinlock. From the thread point of view, both have the same behavior: A function call will return only when the resource is available. From the OS point of view, a mutex will suspend the calling thread if the resource is unavailable, whereas a spin lock will continue to demand the resource. More information about mutex and spin locks can be found in [14].

New performance results after this software modification are shown in figure 8: The throughput is still growing for more than 27 CPUs, the maximal throughput of 0.56 bit/cycle is reached with 45 processors. Figure 9 compares the three different sets of results we obtained with the different implementations. Obviously, the last change, replacing mutex with spinlocks, does not improve greatly unless we use more than 28 CPUs. This change does not or only marginally improve the throughput per processor, but it allows far more processors to share the same resources without any loss of performances.

5 Conclusion and Perspectives

By specifying and implementing a simulation model for a network processing platform, we have fulfilled two goals. Firstly, we have determined a detailed and efficient simulation system. The cycle accurate abstraction level is a good compromise between precision and simulation speed. The simulation speed allows us to consider real applications, while the level of detail allows to observe in detail the behavior of our system and helps remedy its insufficiencies. The second goal concerns architecture itself: We have successfully described a network processor

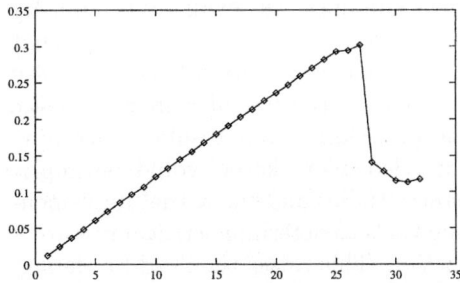

Fig. 5. Throughput in bits/cycle for the distributed memory implementation as a function of the number of CPUs.

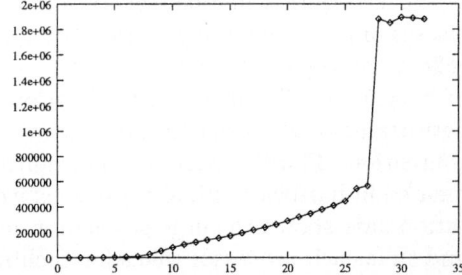

Fig. 6. Number of conflict cycles for the data RAM as a function of the number of CPUs, for a period of 2 000 000 cycles.

Element	execution time (in cycles)	ratio	Element	execution time (in cycles)	ratio
Acquisition	30534	27.14	PaintTee	346	0.31
Classifier	608	0.54	IPGWOptions	634	0.56
Paint	25332	22.52	FixIPsrc	314	0.28
Strip	356	0.316	DecIPTTL	998	0.89
Checkipheader	8506	7.56	IPFragmenter	422	0.38
Getipadress	2934	2.61	EtherEncap	15308	13.61
Lookupiprouter	598	0.53	Extraction	25218	22.42
Dropbroadcast	394	0.35	Total	112502	100%

Fig. 7. Execution time in cycles required for each element (single-thread, MTU = 56).

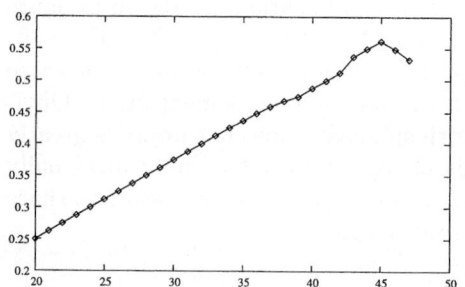
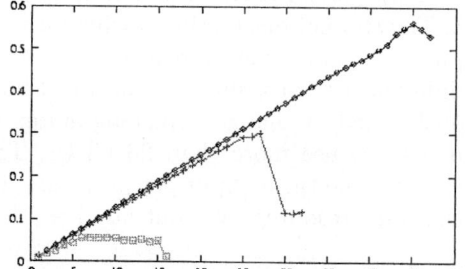

Fig. 8. Spin lock memory implementation: Throughput in bits/cycle, relative to the number of processors.

Fig. 9. Three sets of results obtained for the three different architectures, for the same simulation time.

core as well as run a multi task application. If performances remain modest, this is mainly due to the software part which will have to undergo profound optimization. Results are on the other hand only slightly impaired by the multiple instantiation of our processing unit.

Future work will take several directions: Our next step will be to sum up classes of networking applications by a few "generic" application templates that reflect the typical thread structure - examples for such templates are Quality of Service or Classification applications. Secondly, the use of a more realistic network-on-chip model would improve the precision of our results concerning contention. Thirdly, parts of the multi-threaded micro kernel could be implemented in hardware. Finally, our performance studies and the actual implementation have shown the urgent need for debug tools to determine critical resources and time each processor takes for each function. This raises the need for debugging tools for SystemC.

Our platform clearly constitutes a starting point for further experimentation; it is very open in the sense that it allows for a large range of applications from IPv4 routing to traffic analysis and encryption protocols.

References

1. N. Shah: Understanding network processors. Master's thesis, Dept. of Electrical Eng. and Computer Science, Univ. of California, Berkeley (2001)
2. Paulin, P., Pilkington, C., Bensoudane, E.: STepNP Platform. Ottawa, Canada (2002)
3. L. Cai and D. Gajski: Transaction level modelling in system level design. Tr, Univ. of California, Irvine (2003)
4. Open SystemC Initiative: SystemC.
   Technical report, OSCI (2003) http://www.systemc.org
5. Groetker, T., Liao, S., Martin, G., Swain, S.: System Design in SystemC. Kluwer (2002)
6. SOCLIB Consortium: Projet SOCLIB: Plate-forme de modélisation et de simulation de systèmes integrés sur puce (the SOCLIB project: An integrated system-on-chip modelling and simulation platform). Technical report, CNRS (2003) http://soclib.lip6.fr
7. VSI Alliance: Virtual Component Interface Standard (OCB 2 2.0). Technical report, VSI Alliance (2000)
   http://www.vsi.org/library/specs/summary.htm#ocb
8. Baker, F.: Requirements for ip version 4 router. Internet Eng. Task Force, ftp://ftp.ietf.org/rfc/rfc1812.txt (1995)
9. E. Kohler: Click system free software. URL=www.pdos.lcs.mit.edu/click (1995)
10. Kohler, E.: The Click modular router. PhD thesis, Massachusetts Institute of Technology, Dept. of Electrical Engineering and Computer Science (2000)
11. Andriahantenaina, A., Charléry, H., Greiner, A., Mortiez, L., Zeferino, C.: SPIN: a scalable, packet switched, on-chip micro-network. In: Design Automation and Test in Europe Conference (DATE'2003) Embedded Software Forum, Muenchen, Germany (2003) pp. 70–73
12. Chen, B., Morris, R.: Flexible control of parallelism in a multiprocessor PC router. In: Proceedings of the 2001 USENIX Annual Technical Conference (USENIX-01), Berkeley, CA, The USENIX Association (2001) 333–346
13. Pétrot, F., Gomez, P., Hommais, D.: Lightweight implementation of the POSIX threads API for an on-chip mips multiprocessor with VCI interconnect. In Jerraya, A.A., Yoo, S., Verkest, D., Wehn, N., eds.: Embedded Software for SoC. Kluwer Academic Publisher (2003) 25–38
14. Tanenbaum, A. In: Distributed Operating Systems. Prentice Hall (1995) 169–185

OMULT: An Optical Interconnection System for Parallel Computing*

Bhabani P. Sinha[1] and Subir Bandyopadhyay[2]

[1] Indian Statistical Institute
203 Barrackpore Trunk Road, Kolkata 700 000, India
bhabani@isical.ac.in
[2] School of Computer Science, University of Windsor
401 Sunset Avenue, Windsor, Ontario, N9B 3P4, Canada
subir@uwindsor.ca

Abstract. In this paper, we introduce a new interconnection system OMULT (Optical Multi-Trees) using both electronic and optical links among processors. The processors are organized in the form of an $n \times n$ array of trees, each containing $2n - 1$ nodes with n leaf nodes. The total number of nodes is $2n^3 - n^2$. Nodes in a single tree are connected by usual electronic links, while the leaf nodes of different trees are suitably interconnected by bi-directional optical links in both horizontal and vertical directions. The resulting network topology is almost 3-regular with an $O(\log n)$ diameter and the maximum node degree of only three. We show that the common computations on multiprocessors can be carried out faster on this architecture compared to OTIS - a recent proposal also using optical communication.

1 Introduction

In a parallel/distributed computing environment, the optical interconnection system has become popular over the recent years, particularly because of the improved speed, power and crosstalk properties of optical links compared to the electronic links. When a given network topology is embedded on a 2-dimensional plane, very often some of the nodes, which are quite far apart, need to be connected by a long link, causing a large signal propagation delay and limiting the overall speed of computing. Use of optical links for connecting distant processor nodes and electronic links for connecting the near-by ones may offer a solution to such a problem with almost uniform delay among all links. Recently, Marsden et al. [3], Hendrick et al. [2] and Zane et al. [11] have proposed an architecture in which the processors are partitioned into groups. Processors within each group are interconnected by electronic links, while those in different groups are connected using optical interconnections.

---

* This work was supported by a grant from the Natural Sciences and Engineering Research Council of Canada to Subir Bandyopadhyay and was carried out when Bhabani Sinha was visiting the University of Windsor.

The Optical Transpose Interconnect System (OTIS) proposed by Marsden et al [3] is an example of such a hybrid architecture in which the processors are partitioned into groups of the same size, and processor i of group j is connected to processor j of group i via an optical link. Krishnamoorthy et al. [4] have shown that when the number of processors in each group is equal to the total number of groups, then the bandwidth and power efficiency are maximized, and system area and volume are minimized. The OTIS mesh optoelectronic computer is a class of OTIS computers where the processors in each group are interconnected by electronic links following the two-dimensional mesh paradigm. An N-processor OTIS-Mesh [9], [11] has a diameter of $4N^{1/4} - 3$. Mapping algorithms for various fundamental on the OTIS-Mesh has been studied by several authors, e.g., Wang and Sahni [8], [9], [10], Rajasekaran and Sahni [6], Osterloh [5].

In this paper, we propose a new optical interconnect system called Optical Multi-Trees (OMULT) which uses multiple binary trees as the basic building blocks instead of the 2D meshes. The proposed topology has a total of N ($2n^3 - n^2$) nodes, partitioned into n^2 complete binary trees. All links within a tree are electronic, while the inter-tree links are all optical. The optical links are attached to the leaf nodes following a somewhat similar approach in providing interblock connections in a Multi-Mesh topology [1], so that ($n^3 - n$) leaf nodes have degree 3 (including electronic and optical links), and the remaining n leaf nodes have degree 2. A number of fundamental algorithms occurring in real-life applications can be very efficiently solved on the OMULT topology with lesser order of time complexities than those on the OTIS mesh. For example, summation/average/maximum/minimum of n^3 elements and prefix computation of n^2 elements can be computed on this network in O(log n) time, two n x n matrices can be multiplied in O(log n) time, and n^2 elements can be sorted in O($\log^2 n$) time. These time complexities may be compared to O(n) time for finding summation/maximum/minimum and prefix computation of n^4 elements, O(n^4) time for multiplying two n4 x n4 matrices, and O(n) time for sorting n^4 elements on the OTIS mesh with n4 processors.

2 OMULT Topology

The Optical Multi-Trees (OMULT) interconnection system consists of n^2 complete binary trees T_{ij}'s, $1 \le i, j \le n$, arranged in the form of an n x n array, each tree having n leaf nodes and $n-1$ internal nodes. Each node is a processor. The nodes within each tree are interconnected by usual electronic links, while the leaf nodes of different trees are interconnected by optical links according to rules given below. To fix our idea, we label the nodes in each tree T_{ij}, $1 \le i, j \le n$, by distinct integers from 1 to $2n-1$ in reverse level order, i.e., the leaf nodes in each tree are numbered from 1 to n, in order from left to right, and the internal nodes are also numbered from left to right in successive lower levels (the root node being at the lowest level - level 0). Thus, the root node in each tree is given the node number $2n-1$, and the node k in a tree T_{ij} will be referred to by the processor node $P(i, j, k)$, $1 \leq i, j \leq n$, $1 \leq k \leq 2n-1$. The total number of nodes in the system is $N = n^2 (2n-1) = 2n^3 - n^2$. The optical links interconnect only the leaf nodes in different trees as follows:

1) Processor $P(i, j, k)$, $1 \leq i, j \leq n$, $1 \leq k \leq n$, $j \neq k$, is connected to processor $P(i, k, j)$ by a bi-directional (full-duplex) optical link *(horizontal inter-tree link)*,
2) Processor $P(i, j, k)$, $1 \leq i, j \leq n$, $1 \leq k \leq n$, $i \neq k$, is connected to processor $P(k, j, i)$ by a bi-directional optical link *(vertical inter-tree link)*,
3) For $i = k$ and/or $j = k$, processor $P(i, j, k)$, $1 \leq i, j \leq n$, $1 \leq k \leq n$, is connected to processor $P(i, j, 2n-1)$ by a bi-directional optical link.

An example for the OMULT topology for $n = 4$ is shown in Fig. 1. These rules for interconnecting the leaf nodes in different trees have some similarities with those for interconnecting boundary/corner nodes in different meshes of the Multi-Mesh topology [1]. It follows from the above interconnection scheme that each of the leaf nodes $P(i, j, k)$, $1 \leq i, j \leq n$, $1 \leq k \leq n$, excepting those of the form $P(i, i, i)$, has 2 optical links and one electronic link connected to it, while each of the leaf nodes $P(i, i, i)$, $1 \leq i \leq n$, has only one optical link and one electronic link connected to it. All non-root internal nodes in a tree have 3 electronic links each, while each root node of the form $P(i, i, 2n-1)$, has one optical link and two electronic links All remaining root nodes has two optical links and two electronic links. The total number of optical links in the network is equal to $2(n^3 - n) + n + n + 2(n^2 - n) = 2(n^3 + n^2 - n)$.

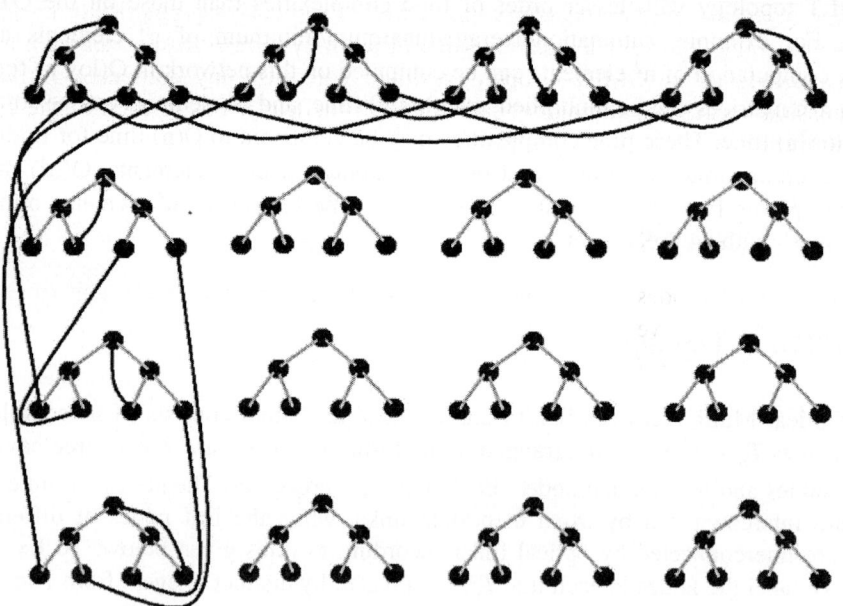

Fig. 1. Optical Multi-Trees (OMULT) Interconnection System (All optical interconnection links are not shown).

2.1 Diameter

To find the diameter of the OMULT interconnection system, we note that, starting from a leaf node $P(i, j, k)$, $1 \leq i, j \leq n$, $1 \leq k \leq n$, we can always reach another leaf node $P(i, j', k')$, $1 \leq k' \leq n$, in the same tree, traversing a maximum of $2 \log n$ electronic links. From this node $P(i, j, k')$, $1 \leq k' \leq n$, we can reach the leaf node $P(i, k', j)$ using only one optical link. From this node $P(i, k', j)$, we can reach any other node, say $P(i, k', j')$, in the tree $T_{ik'}$ using a maximum of $2 \log n$ electronic links. Thus, starting from any node in a tree T_{ij}, we can reach any other node in a tree $T_{ik'}$ in the same row of trees using at most one optical link and $4 \log n$ electronic links. This property also holds if we want to reach any node in another tree in the same column of trees. Hence, to reach a node $P(i', j', k')$, $1 \leq i', j' \leq n$, from $P(i, j, k)$, $1 \leq i, i', j, j' \leq n$, $1 \leq k, k' \leq 2n-1$, we traverse the path $P(i, j, k) \rightarrow P(i, j, j') \rightarrow P(i, j', j) \rightarrow P(i, j', i') \rightarrow P(i', j', i) \rightarrow P(i', j', k')$.

The resulting path consists of at most $6 \log n$ electronic links and 2 optical links. Further, it is always possible to find a pair of nodes which are exactly $6 \log n + 2$ distance apart (summing both the electronic and optical path lengths). Hence, we have the following result.

Theorem 1: The diameter of the OMULT topology is $6 \log n + 2 = O(\log N)$, with $6 \log n$ electronic links and 2 optical links.

Since $N = 2n^3 - n^2$, the diameter of the OMULT network asymptotically tends to assume the value $2 \log N$ - the same as that of a binary tree. However, unlike a tree, the root nodes in the OMULT topology would not cause any bottleneck for data transmission because of alternative paths through the leaf nodes of each of the constituent trees.

2.2 Node Connectivity

Consider two leaf nodes $P(i, j, k)$ and $P(i', j', k')$. We can find two node-disjoint paths Path 1 and Path 2 between $P(i, j, k)$ and $P(i', j', k')$ for different cases in the following way, where the node $P(i, j, k_s)$ denotes the sibling of the node $P(i, j, k)$:

Path 1 : $P(i, j, k) \rightarrow P(i, j, j') \rightarrow P(i, j', j) \rightarrow P(i, j', i') \rightarrow P(i', j', i) \rightarrow P(i', j', k')$
Path 2 : a) for $k \neq i$ and $k' \neq j'$: $P(i, j, k) \rightarrow P(k, j, i) \rightarrow P(k, j, k') \rightarrow P(k, k', j) \rightarrow P(k, k', i) \rightarrow P(i', k', k) \rightarrow P(i', k', j') \rightarrow P(i', j', k')$

b) for $k = i$ and $k' = j'$: $P(i, j, k) = P(i, j, i) \rightarrow P(i, j, i_s) \rightarrow P(i_s, j, i) \rightarrow P(i_s, j, i') \rightarrow P(i', j, i_s) \rightarrow P(i', j, k'_s) \rightarrow P(i', k'_s, j) \rightarrow P(i', k'_s, j') \rightarrow P(i', j', k'_s) \rightarrow P(i', j', k')$

c) for $k = i$ and $k' \neq j'$: $P(i, j, k) \rightarrow P(i, j, i_s) \rightarrow P(i_s, j, i) \rightarrow P(i_s, j, i') \rightarrow P(i', j, i_s) \rightarrow P(i', j, k') \rightarrow P(i', k', j) \rightarrow P(i', k', j') \rightarrow P(i', j', k')$

d) for $k \neq i$ and $k' = j'$: $P(i, j, k) \rightarrow P(k, j, i) \rightarrow P(k, j, k'_s) \rightarrow P(k, k'_s, j) \rightarrow P(k, k'_s, i) \rightarrow P(i', k'_s, k) \rightarrow P(i', k'_s, j') \rightarrow P(i', j', k'_s) \rightarrow P(i', j', k')$

If any one of the nodes $P(i, j, k)$ and $P(i', j', k')$, say $P(i, j, k)$, is not a leaf node, then starting from $P(i, j, k)$ we can reach the two leaf nodes which are the leftmost and the rightmost leaf nodes of the subtree rooted at $P(i, j, k)$. From these two leaf nodes, we can find two node disjoint paths to the node $P(i', j', k')$ in the same way as above. Hence, we get the following result.

Theorem 2: The node connectivity of the OMULT topology is two.

Thus, the OMULT topology can tolerate single node/link failure.

Theorem 3: The diameter of the OMULT topology under single node/link failure is equal to *8 log n + 6*.

Proof: Considering Path 1 and Path 2 above, the path lengths for different possible cases can be as follows. Path 1): *6 log n + 2*, Path 2a): *6 log n + 4*, Path 2b): *8 log n + 6*, Path 2c) : *6 log n + 6*, Path 2d): *6 log n + 6*. Hence, the maximum path length under single node/link failure is *8 log n + 6* corresponding to the path 2b) above.

3 Mapping Algorithms for Some Basic Operations

In this section, we briefly mention the mapping of several fundamental operations on the OMULT topology. We have found that the performance of the proposed OMULT topology in executing these basic algorithms is better in terms of the order of complexities than that on the OTIS-Mesh. We have considered the following operations on the OTIS-Mesh: a) Data Broadcast, b) Row/Column Group-Broadcast, c) Complete Group-Broadcast, d) Summation/Average/Maximum/Minimum, e) Prefix Computation, f) Matrix Transpose, g) Matrix Multiplication, h) DFT Computation, i) Sorting j) Computational Geometry algorithms. Due to lack of space we will only present the algorithms for prefix computation and Row/Column Group-Broadcast as examples and summarize the performance of the others for comparison with OTIS.

3.1 Row/Column Group-Broadcast

We may use only the leaf nodes in a tree for performing input/output operations. Assuming that we have n data elements $d_1, d_2, ..., d_n$ in the n leaf nodes in a tree T_{ij}, for different applications, we may need to broadcast all these n data elements to the respective leaf nodes in all trees in the same column (row). We can perform this operation in two phases. In phase 1, the group of data elements is distributed over all trees in the same row i (column j). In phase 2, we broadcast them to all trees in column j (row i). The whole process can be performed in $O(log\ n)$ time as follows.

Without loss of generality, we assume that initially the n data elements $d_1, d_2, ..., d_n$ are stored in the leaf nodes of the tree T_{11}; data d_k, $1 \leq k \leq n$, being stored in the processor node $P(1, 1, k)$, and we want to broadcast these to all trees in the first column. First, using the horizontal optical links, data element d_k is moved to $P(1, k, 1)$ for all k, $1 \leq k \leq n$. Data d_k is then broadcast to all nodes in the tree T_{1k}, $1 \leq k \leq n$, which needs $2 \log n$ steps along electronic links. This completes phase 1 of distributing the n data elements to all trees in the same row (row 1).

In phase 2, for broadcasting the data elements to all trees in the same column (column 1), the data d_j, $1 \leq j \leq n$, now stored in the processors $P(1, j, k)$, $1 \leq k \leq n$, is sent to the processors $P(k, j, 1)$ using the vertical optical links. Using the horizontal optical links once again, data d_j is then moved to the processors $P(1, j, k)$. If the data elements were initially stored in any other tree T_{ij}, the same method can be applied to broadcast them to all trees in i^{th} row and j^{th} column. It follows from above that the total number of communication steps needed for the whole process is $2 \log n$ (electronic links) + 3 (optical links).

3.2 Prefix Computation

Suppose we have $n^2 (= p)$ data elements $d_1, d_2, ..., d_p$ stored in the n^2 leaf nodes of all the diagonal trees in OMULT. We assume that each processor $P(i, j, k)$ has two registers $A(i, j, k)$ and $B(i, j, k)$. Let $\otimes$ be a binary operation defined over the given data elements. Prefix computation involves computing $d_1 \otimes d_2 \otimes ... \otimes d_i$ for all i, $1 \leq i \leq n^2$. We assume that the A-registers of the leaf nodes $P(i, i, k)$ of the diagonal tree T_{ii}, $1 \leq i, k \leq n$, initially store the data values $d_{(i-1)n+k}$. We want to store the final prefix values in the leaf nodes of the trees T_{ni} in the nth row. The algorithm PC for prefix computation on the OMULT topology is given below.

Algorithm PC :

Step 1: Using the A-registers, broadcast the initial data values in each tree T_{jj}, $1 \leq j \leq n$, to all trees in the same column (using the row/column group-broadcast algorithm above).

Step 2: Compute the prefix $d_{(j-1)n+1} \otimes d_{(j-1)n+2} \otimes ... \otimes d_{(j-1)n+k}$ using the A-registers in the tree T_{kj}, $1 \leq j, k \leq n$, and store it in the A-register $A(k, j, 2n-1)$. Broadcast the content of $A(k, j, 2n-1)$ to the A-registers of all leaf nodes $A(k, j, 1), A(k, j, 2), ..., A(k, j, n)$ in T_{kj}.

Step 3: For $1 \leq j \leq n$ do in parallel
begin
       if $k < n$, then $A(n, j, k) \leftarrow A(k, j, n)$;
       if $j < l$ then $B(n, j, l) \leftarrow A(n, l, j)$;
       end;

Step 4: For all j, $1 \leq j \leq n$, compute the prefix values $B(n, j, 1) \otimes B(n, j, 2) \otimes \ldots \otimes B(n, j, j-1)$, store these values in $B(n, j, 2n-1)$ the root node of the tree T_{nj}), and then broadcast the content of $B(n, j, 2n-1)$ to the B-registers of all leaf nodes in T_{nj}.

Step 5: For all j, k, $1 \leq j, k \leq n$, do in parallel
$A(n, j, k) \leftarrow A(n, j, k) + B(n, j, k)$;

Step 1 needs $2 \log n + 3$ time units. Each of steps 2 and 4 needs $2 \log n$ time units. Each of steps 3 and 5 needs one time unit. Hence, the prefix values of n^2 data elements are computed in $6 \log n + 5$ time units.

3.3 Performance of Algorithms on the OMULT

A summary of the performances of some algorithms, along with the topological properties of the OTIS mesh and the OMULT network has been given in table 1 for comparison. In the table Electronic links are identified by E and optical links by O.

Table 1. Comparisons between OTIS-Mesh and OMULT Network

	OTIS-Mesh	OMULT
Number of Nodes (N)	$N = n^4$	$N = 2n^3 - n^2$
Diameter	$4n - 3$	$6 \log n + 2$
Broadcast time	$4n - 1$ (E) + 1 (O)	$6 \log n$ (E) + 2 (O)
Data Sum time	$8n - 1$ (E) + 1 (O)	$6 \log n$ (E) + 2 (O)
	(for n^4 elements)	(for n^3 elements)
Prefix Sum time	$7n - 1$ (E) + 2 (O)	$6 \log n$ (E) + 5 (O)
(# of elements)	(for n^4 elements)	(for n^3 elements)
Matrix multiplication	$O(n^4)$	$6 \log n$ (E) + 8 (O)
Time (size of matix)	($n^4 \times n^4$ matrices)	($n \times n$ matrices)
Sorting time	$22n$ (E) +	$5 \log n$ (E) + 5 (O)
(# of elements)	$O(n^{3/4})$ (O)	(n elements)
	(n^4 elements)	

4 Conclusions

Algorithm mapping for the common basic operations involved in real-life applications for numerical and scientific processing, image and signal processing can be more efficiently done using the OMULT network than on the OTIS-Mesh, with comparable investments on establishing optical links among the processor nodes, and lesser cost for the electronic links. Also, the topology of the proposed OMULT network is very simple, making it particularly attractive for parallel computing.

Acknowledgement

Mr Rabiul Islam's help in evaluating some of the algorithms is acknowledged.

References

1. D. Das, M. De and B. P. Sinha, "A new network topology with multiple meshes", *IEEE Transactions on Computers,* Vol. 48, No. 5, pp. 536-551, May 1999.
2. W. Hendrick, O Kibar, P. Marchand, C. Fan, D. V. Blerkom, F. McCormick, I. Cokgor, M. Hansen and S. Esener, "Modeling and optimization of the optical transpose interconnection system" *Optoelectronic Technology Center, Program Review,* Cornell Univ. , Sept. 1995.
3. G. C. Marsden, P. J. Marchand, P. Harvey and S. C. Esener, "Optical transpose interconnection system architectures, *Optical Letters,* Vol. 18, No. 13, pp. 1083-1085, July 1993.
4. A. Krishnamoorthy, P. Marchand, F. Kiamilev and S. Esener, "Grain-size consierations for optoelectronic multistage interconnection networks", Applied Optics, Vol. 31, No. 26, pp. 5480-5507, September 1992.
5. A. Osterloh, "Sorting on the OTIS-Mesh", *Proc. 14th International Parallel and Distributed Processing Symposium (IPDPS 2000),* pp. 269-274, 2000.
6. S. Rajasekaran and S. Sahni, "Randomized routing, selection and sorting on the OTIS-Mesh optoelectronic computer", *IEEE Transactions on Parallel and Distributed Systems,* Vol. 9, No. 9, pp. 833-840, 1998.
7. I. D. Scherson and S. Sen, "Parallel sorting in two-dimensional VLSI models of computation", *IEEE Transactions on Computers,* Vol. 38, No. 2, pp. 238-249, February 1989.
8. C.-F. Wang and S. Sahni, "Basic operations on the OTIS-Mesh optoelectronic computer", *IEEE Trans. on Parallel and Distributed Systems,* Vol. 9, No. 12, pp. 1226-1236, 1998.
9. C.-F. Wang and S. Sahni, "Image processing on the OTIS-Mesh optoelectronic computer", *IEEE Transactions on Parallel and Distributed Systems,* Vol. 11, No. 2, pp. 97-109, 2000.
10. C.-F. Wang and S. Sahni, "Matrix multiplication on the OTIS-Mesh optoelectronic computer", *IEEE Transactions on Computers,* Vol. 50, No. 7, pp. 635-646, 2001.
11. F. Zane, P. Marchand, R. Paturi and S. Esener, "Scalable network architectures using the optical transpose interconnection system (OTIS)", *Journal of Parallel and Distributed Computing,* Vol. 60, No. 5, pp. 521-538, 2000.

Distributing InfiniBand Forwarding Tables*

Aurelio Bermúdez, Rafael Casado, and Francisco J. Quiles

Department of Computer Science, University of Castilla-La Mancha
02071 Albacete, Spain
{aurelio.bermudez,rafael.casado,francisco.quiles}@uclm.es

Abstract. InfiniBand is an emerging technology both for communication between processing nodes and I/O devices, and for interprocessor communication. After the occurrence of a topology change, InfiniBand management entities collect the current topology, compute new forwarding tables, and upload them to routing devices. Traditional distribution techniques prevent deadlock but, at the same time, they affect negatively user traffic. In this paper, we propose two alternative deadlock-free mechanisms to distribute forwarding tables. These proposals adhere InfiniBand specification, can be easily implemented, and reduce significantly the impact of the distribution process.

1 Introduction

The InfiniBand Architecture (IBA) [7] defines a technology for interconnecting processor nodes (hosts) and I/O nodes to form a system area network. Hosts and I/O nodes are interconnected using an arbitrary (possibly irregular) switched point-to-point network, instead of using a shared bus. End nodes use channel adapters (CAs) to connect to the fabric. The network is composed of one or more subnets interconnected by routers. Each port within a subnet has a 16-bit local identifier (LID). Switches perform intra-subnet routing using the packet's destination LID included in the header of the packet. A forwarding table (FT) specifies which port forwards the packet.

IBA subnets are managed in an autonomous way. There is a subnet management mechanism capable of assimilating any topology change without external intervention, guaranteeing service availability. The specification defines various subnet management entities, describing their functions and the structure of the control packets used to exchange information among them. An entity called the subnet manager (SM) is in charge of discovering, configuring, activating, and maintaining the subnet. This entity exchanges subnet management packets (SMPs) with subnet management agents (SMAs) present in every device. Fig. 1(a) shows an example of irregular subnet including these management entities. In [2], we presented a completely functional prototype of a subnet management protocol which adheres to IBA specifications. This initial approach covers the detection of topology changes, device discovery, and

---

* This work was partly supported by the following projects: CICYT TIC2003-08154-C6-02, and JCCM PBC-02-008.

computation and distribution of subnet routes. The discovery process was optimized in [3], and the computation of subnet routes was improved in [4].

In this work, we focus on the last step in assimilating a topology change, i.e., the distribution of switch forwarding tables. The SMPs for updating these tables are completely defined in the IBA specification. However, the update order is not detailed. Updating FTs in an uncontrolled way could generate deadlock situations [10]. The reason is that although the new and the previous sets of subnet routes are deadlock-free, the coexistence of both routing schemes during the distribution process is not necessarily deadlock-free.

In order to prevent deadlock situations during this process, static reconfiguration [5, 12] was assumed in our initial implementation. This means that user traffic is stopped while forwarding tables are being updated. As we showed in [2], this technique has negative effects over user traffic; in particular, a temporary lack of service and, consequently, a massive packet discarding during the period of time in which subnet ports are inactive.

The impact of static reconfiguration could be reduced by using dynamic reconfiguration techniques, such as *Partial Progressive Reconfiguration* [6], *Skyline* [8], and *Double Scheme* [11]. All these techniques allow the distribution of forwarding tables without stopping network traffic, guaranteeing deadlock-freedom during the process. However, the adaptation to IBA of these dynamic reconfiguration mechanisms is not trivial, due to it implies the modification of the current specification in some way (addition of new elements, the use of provided elements for different purposes, the assumption that SMAs perform special functions, etc.).

In this paper we present and analyze several alternatives that do not involve any change in the IBA specification. In particular, the idea is to relax the traditional static reconfiguration technique, either by reducing the amount of subnet ports that must be actually deactivated, or by preventing the use of certain transitions during the distribution of switch forwarding tables.

The remainder of this paper is organized as follows. First of all, Section 2 describes the way in which static reconfiguration is performed in IBA, and informally presents two optimized distribution processes. In Section 3 we comparatively analyze the different distribution techniques through several simulation results. Finally, Section 4 gives some conclusions and future work.

2 Improving the Distribution Process

2.1 Static Distribution

The static distribution of forwarding tables is performed in three sequential steps. First, all subnet ports are deactivated by the SM. In particular, the SM sends a *SubnSet(PortInfo)* SMP [7] to change the state of each port to *INITIALIZE*. In this state, the port can only receive and transmit SMPs, discarding all other packets received or presented to it for transmission. The next step is the sending of the FTs itself. This phase is performed using either *SubnSet(LinearForwardingTable)* SMPs or *SubnSet(RandomForwardingTable)* SMPs. Finally, after the SM verifies that all subnet switches have received their tables, user traffic must be allowed again, by activating subnet ports. By means of *SubnSet(PortInfo)* SMPs, the SM sets the state of each port to *ACTIVE*. In this state, the port can transmit and receive all packet types.

SMPs used to perform the two first steps must employ directed (source) routing [7]. The reason is that new FTs have not been configured yet. Instead, SMPs for the activation phase can use either directed or LID (destination) routing.

2.2 Deactivation of Break Node Ports

This distribution technique assumes that the management mechanism uses the *up*/down** routing algorithm [12] to compute the set of subnet FTs. Up*/down* is a popular deadlock-free algorithm valid for any topology. It is based on a cycle-free assignment of direction to the operational links in the network. For each link, a direction is named *up* and the opposite one is named *down*. In this way, the network is configured as an acyclic directed graph with a single sink node. As an example, Fig. 1(b) shows a possible assignment of directions for the topology in Fig. 1(a). To avoid deadlocks, legal routes never use a link in the *up* direction after having used one in the *down* direction.

In a directed graph, a *break node* [6] is a node that is the source of two or more arcs. Break nodes prevent certain transitions (input port – output port) from being used by the packets crossing them. These restrictions are necessary to guarantee deadlock-freedom. In Fig. 1(b), transitions $6_{1 \to 2}$ and $6_{2 \to 1}$ (from port *1* to *2* and vice versa in node *6*) are forbidden. Similarly, transitions $10_{2 \to 3}$ and $10_{3 \to 2}$ are not allowed. In other words, the corresponding dependencies between channels are deactivated [6].

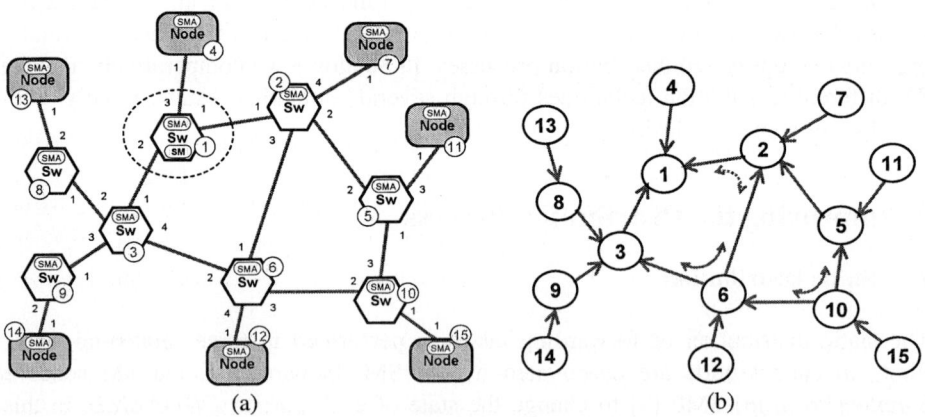

Fig. 1. (a) Example of irregular subnet topology composed of 8 switches and 7 end nodes. Circled numbers represent the LID assigned to each subnet device by the SM (located in node *1*) during the discovery process. Small numbers at the ends of the links represent switch and channel adapter port numbers. (b) A possible directed graph for this topology.

Deadlocks can appear during the distribution process only if the break node in a cycle changes its position. Let us suppose that the previous break node in the left cycle of Fig. 1(b) was the node labeled as *2*. That means that the direction assigned to the link connecting nodes *6* and *2* in the previous configuration was the opposite one of the currently assigned. Therefore, transitions $2_{1\to3}$ and $2_{3\to1}$ were not allowed. If the distribution process activates the dependencies for node *2* before the deactivation of the dependencies in node *6*, there is a potential deadlock in the cycle. The reason is that node *2* could route packets from *1* to *6* and, simultaneously, node *6* could route packets from *2* to *3*, closing the cycle. Deadlock could also appear in the opposite direction.

Obviously, a static distribution process prevents these situations, because of subnet ports are not activated until new FTs have been completely distributed. An optimized distribution mechanism could be based on the deactivation of only break node dependencies, instead of deactivating subnet ports, before the sending of tables. That would imply not to allow input port – output port transitions in break nodes. Unfortunately, we cannot program switch FTs to prevent these transitions. The reason is that IBA switches do not take into account the input port used by the packet to route it. This information is not stored in the tables.

As an intermediate step, we could derive a deadlock-free distribution mechanism that only deactivates the ports of the switches that will act as break nodes in the new configuration. Moreover, note that it is not necessary to deactivate all ports in those nodes. Instead, it is enough to select those ports associated with *up* links, and deactivate all of them, except one. Thus, we ensure that forbidden transitions will not be used by any packet. For the directed graph in Fig. 1(b), the distribution process only must deactivate port 6_1 or 6_2. Similarly, it is only necessary to deactivate port 10_2 or 10_3. In this way, we are allowing the use of many subnet routes during the distribution of tables. As in the static distribution process, there is a third step, after the distribution of tables, in which break node ports are activated.

2.3 Deactivation of Break Node Dependencies

The previous mechanism is easy to implement and, as we will see, it reduces the negative effects of the "pure" static distribution process. However, there are still a lot of subnet routes that could be used during the sending of tables, without introducing potential deadlock situations. For example, let us suppose that, in Fig. 1(a), port 10_2 has been selected for deactivation. In this situation, we are not allowing those packets generated or destined to node *10* that must use the link connecting *10* and *6*.

We can improve the distribution mechanism by only preventing a few input port – output port combinations, allowing the rest of routing options. To do that, we can conveniently program the set of subnet SL to VL mapping tables. The IBA specification itself suggests that these tables can be used to avoid routing deadlocks [7].

IBA uses a virtual lane (VL) based mechanism to create multiple virtual links within a single physical link. Each port could implement 1, 2, 4, 8, or 15 data VLs. As a packet traverses the subnet, each link along the path can support a different number of VLs. Therefore, it is not possible to transmit a packet only using a VL specified in its header. Instead, IBA uses a more abstract criterion, based on the concept of service level (SL). Each switch has a SL to VL mapping table to establish, for each pair input port – output port, a correspondence between the service level of the packet (a number from 0 to 15) and the VLs supported by the output port assigned to the packet.

Packets are discarded if the SL to VL mapping table returns the value VL15. This is the value that we are going to use to prevent forbidden transitions. The resulting distribution mechanism does not require deactivating any subnet port. Instead, it replaces the initial deactivation step with the sending of SL to VL mapping tables to new break nodes. In particular, it is necessary to send a SL to VL mapping table for each forbidden input port – output port combination in each break node. In this table, all entries contain the value VL15. Therefore, packets trying these transitions will be automatically discarded.

For the example in Fig. 1(b), it is necessary to configure four SL to VL mapping tables, in order to deactivate forbidden dependencies in *6* and *10*. The sending of SL to VL mapping tables is performed by the SM by using directed routed *SubnSet(SLtoVLMappingTable)* SMPs.

3 Performance Evaluation

All the results presented in this work have been obtained through simulation. Before showing and analyzing them, we briefly describe the simulation methodology.

3.1 Simulation Methodology

Our model embodies key physical and link layer features of IBA, allowing the simulation of various IBA-compliant network designs. Also, it incorporates the subnet management entities and packets defined in the specification. To develop it, we have used the OPNET Modeler [9] simulation software. The current IBA model is composed of copper links (supporting different data rates), 4-port fully demultiplexed switches, and end nodes containing a channel adapter (hosts). Several physical and link layer details are described in [1, 2, 3, 4].

We have evaluated randomly generated irregular subnets with 8, 16, 24, and 32 switches, assuming that there is at least a host connected to each switch, if a port is available. Also, not all switch ports are connected. Each subnet switch supports a linear forwarding table (LFT) with 1,024 entries.

We have considered a packet maximum transfer unit (MTU) of 256 bytes (the minimum MTU value allowed by the IBA specification). The packet generation rate is Poisson, and it is expressed in packets/sec/node. Traffic sources also use a uniform distribution to obtain the packet destination (among all the active hosts) and the SL value (from 0 to 15). The traffic load applied is different for each subnet topology. We have selected low load values (25% of saturation rate), in order to prevent network saturation during the analysis.

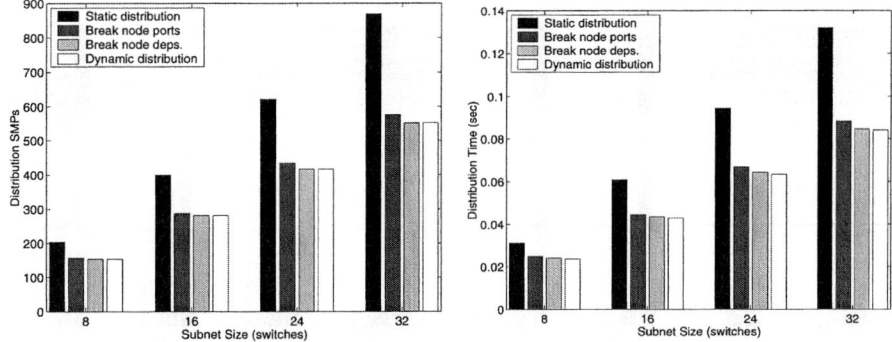

Fig. 2. Control packets and time required to distribute switch forwarding tables as a function of subnet size. The change consists of the addition of an individual switch. Results for switch removal (not shown here) are very similar.

For each simulation run, after a transient period we have programmed a topology change, consisting of the addition or removal of a switch. The experiment is repeated for each switch in the subnet, and average values are shown in the plots. Traps support is disabled, and the period of time between consecutive sweepings has been tuned according to the subnet topology. The simulation is stopped once the topology change has been completely assimilated.

3.2 Simulation Results

In this section we analyze the behavior of the distribution techniques presented and their impact over application traffic. For the sake of comparison, we have added to some plots a series showing the results for a basic dynamic distribution process. This process consists of only one step, in which FTs are directly distributed to subnet switches, without deactivating ports or dependencies. Therefore, it is deadlock-prone.

Fig. 2 shows the amount of SMPs and the time required by the subnet management mechanism to update switch FTs after the occurrence of a topology change, as a function of the distribution mechanism used and the subnet size. Results also consider both the deactivation and activation phases (when applicable).

We can see that the technique based on the deactivation of break node ports achieves an important reduction of both parameters. Also, we can observe an additional reduction for the distribution mechanism based on the deactivation of break node dependencies. In fact, results for dynamic distribution are almost identical.

Fig. 3 shows the amount of packets discarded as a function of the distribution mechanism used, the type of topology change, and the subnet size. Note that, in case of switch removal, there is a big amount of packets that are discarded due to many subnet routes disappear, at least until new tables have been distributed and alternative routes are provided. Independently of the distribution technique applied, this massive discarding is inevitable, because of IBA routes are deterministic.

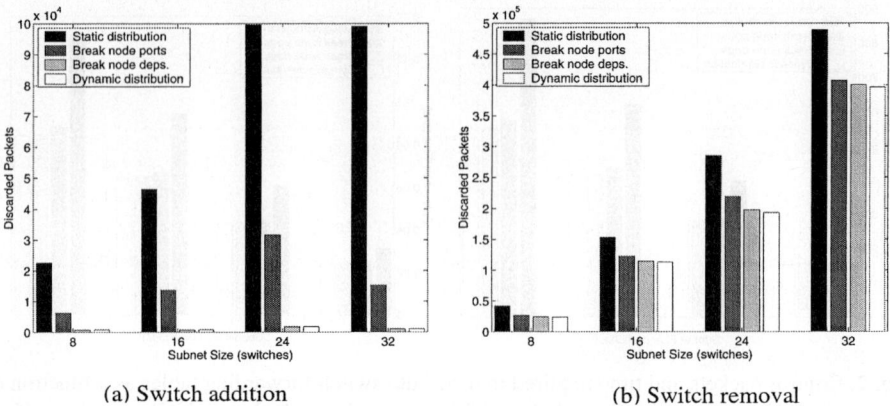

(a) Switch addition   (b) Switch removal

Fig. 3. Number of packets discarded during the change assimilation as a function of subnet size. Note that vertical scales are different.

In Fig. 3(a) we can appreciate a considerable improvement if we only deactivate break node ports, due mainly to the number of packets dropped by deactivated ports decreases significantly. The number of discarded packets decreases even more for the deactivation of break node dependencies. Packets discarded by deactivated ports have been replaced by packets dropped by the SL to VL mapping process. As happens before, results provided by this technique are very similar to dynamic distribution.

Note that there are other minor causes for packet discarding. For example, during the transient period, there are packets that can not be routed because of the corresponding DLID is not found in a FT, or because of the output port returned by the table coincides temporary with the packet's input port.

Finally, Fig. 4 shows the effects of the distribution mechanisms over user traffic. Note that the figure only shows a detail of the distribution phase, instead of the complete change assimilation process. For all plots, the X-axis represents the simulation time. The first plot shows the aggregate amount of SMPs exchanged by the management entities. The big step in this plot corresponds exactly with the distribution of subnet ta-

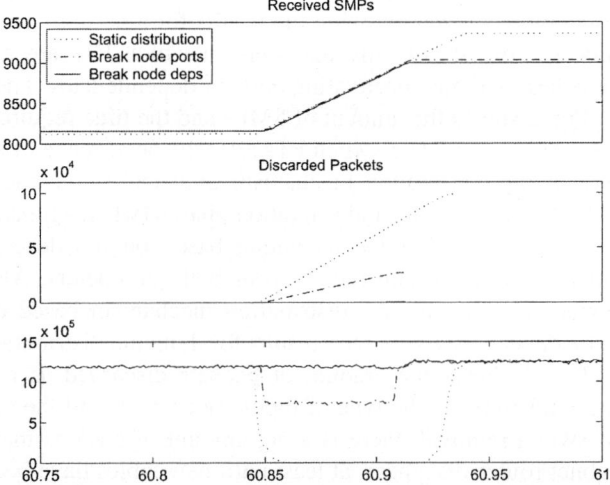

Fig. 4. Detail of the distribution process for an irregular subnet composed of 24 switches and 22 hosts and a change consisting of the addition of a switch (at time 60.1 sec.).

bles. Second plot represents the aggregate amount of discarded packets during the simulation. As the change considered is a switch addition, the period of time during packets are discarded coincides with the distribution process. The last plot shows instantaneous network throughput.

In the top plots we can observe (again) a reduction in the amount of distribution SMPs and time, and in the number of packets discarded, when the proposed distribution mechanisms are used. Bottom plot shows the way in which the optimized techniques improve network throughput during the process. Static reconfiguration has a very negative effect over instantaneous throughput. We can appreciate that the proposed distribution techniques improve considerably this behavior. In fact, when the mechanism based on the deactivation of break node dependencies is used, the gap in this plot completely disappears.

4 Conclusions

We have presented two deadlock-free mechanisms to distribute InfiniBand subnet forwarding tables. These mechanisms have been directly derived from the traditional static reconfiguration process. However, they have a much better behavior. We have seen that the impact of the distribution process over application traffic is practically avoided. The reason is that the proposals allow the use of many subnet routes during the period of time in which forwarding tables are being distributed. The main advantage of these proposals is that they respect the InfiniBand specification and, therefore, they can be easily implemented over real systems.

References

1. Bermúdez, A., Casado, R., Quiles, F.J., Pinkston, T.M., Duato, J.: Modeling InfiniBand with OPNET. Workshop on Novel Uses of System Area Networks, February 2003
2. Bermúdez, A., Casado, R., Quiles, F.J., Pinkston, T.M., Duato, J.: Evaluation of a subnet management mechanism for InfiniBand networks. In Proc. IEEE Int. Conference on Parallel Processing, October 2003
3. Bermúdez, A., Casado, R., Quiles, F.J., Pinkston, T.M., Duato, J.: On the InfiniBand subnet discovery process". In Proc. IEEE Int. Conference on Cluster Computing, December 2003
4. Bermúdez, A., Casado, R., Quiles, F.J., Duato, J.: Use of provisional routes to speed-up change assimilation in InfiniBand networks. In Proc. Workshop on Communication Architecture for Clusters, April 2004
5. Boden, N.J., et al.: Myrinet: A gigabit per second LAN. IEEE Micro, February 1995
6 Casado, R, Bermúdez, A., Quiles, F.J., Sánchez, J.L., Duato, J.: A protocol for deadlock-free dynamic reconfiguration in high-speed local area networks. IEEE Transactions on Parallel and Distributed Systems, vol. 12, no. 2, February 2001
7. InfiniBand Architecture Specification (1.1), November 2002. http://www.infinibandta.com/
8. Lysne, O, Duato, J.: Fast dynamic reconfiguration in Irregular networks. In Proc. Int. Conference on Parallel Processing, August 2000

9. OPNET Technologies, Inc. http://www.opnet.com/
10. Pinkston, T.M., Pang, R., Duato, J.: Deadlock-free dynamic reconfiguration schemes for increased network dependability. IEEE Transactions on Parallel and Distributed Systems, vol. 14, no. 6, June 2003
11. Pinkston, T.M., Zafar, B., Duato, J.: A method for applying Double Scheme dynamic reconfiguration over InfiniBand. In Proc. Int. Conference on Parallel and Distributed Processing Techniques and Applications, June 2003
12. Schroeder, M.D., et al.: Autonet: a high-speed, self-configuring local area network using point-to-point links. IEEE Journal on Selected Areas in Communications, vol. 9, no. 8, October 1991

Tuning Buffer Size in InfiniBand to Guarantee QoS*

Francisco J. Alfaro and José L. Sánchez

Universidad de Castilla-La Mancha, 02071-Albacete, Spain
{falfaro,jsanchez}@info-ab.uclm.es

Abstract. InfiniBand (IBA) is a new industry-standard architecture both for I/O server and interprocessor communication. IBA employs a switched point-to-point network, instead of using a shared bus. IBA is being developed by the InfiniBand^{SM} Trade Association to provide present and future server systems with the required levels of reliability, availability, performance, scalability, and quality of service (QoS).
In previous papers we have proposed an effective strategy for configuring the IBA networks to provide users with the required levels of QoS. This strategy is based on the proper configuration of the mechanisms IBA carries to support QoS. In this paper, we determine the minimum virtual lines' buffer size required to guarantee QoS to the applications.

1 Introduction

The InfiniBand Trade Association (IBTA) [1] was formed in 1999 to develop a new standard for high-speed I/O and interprocessor communication. InfiniBand defines a technology for interconnecting processor nodes (hosts) and I/O devices to form a system area network [2]. In a first stage, instead of directly replacing the PCI bus with a switch-based interconnection to access I/O devices, these devices are attached to a Host Channel Adapter (HCA), which is connected to the PCI bus. In this way, the communication is switched from HCA, affording the desired reliability, concurrency and security. Moreover, it is foreseen that the PCI bus could be replaced in a near future by other advanced technologies like PCI Express Advanced Switching.

InfiniBand implements some mechanisms to provide each kind of application with the required QoS. In previous works, [3] and [4], we have developed a methodology to configure such mechanisms. The proposed methodology successfully provides applications with both bandwidth and latency guarantees. In this paper, we determine the minimum buffer size in the virtual lanes of the switch ports and host interfaces required to achieve this goal.

The structure of the paper is: Section 2 presents a summary of the most important mechanisms included in IBA to support QoS; in Section 3, we review our proposal to give QoS guarantees; Section 4 presents the evaluation methodology used to determine the buffer size required in the VLs to provide applications with QoS, as well as the obtained results; finally, some conclusions are given.

---

* This work was partly supported by the Spanish CICYT under Grant TIC2003-08154-C06 and Junta de Comunidades de Castilla-La Mancha under Grant PBC-02-008.

2 InfiniBand

InfiniBand hardware provides highly reliable, fault-tolerant communication, improving the bandwidth, latency, and reliability of the system. The InfiniBand architecture offers a new approach to I/O. It simplifies and speeds server-to-server connections and links to other server-related systems, such as remote storage and networking devices, through a message-based fabric network.

Specifically, IBA has three mechanisms to support QoS: Service levels (SLs), virtual lanes (VLs), and a virtual lane arbitration for transmission over links. IBA defines a maximum of 16 SLs, although it does not specify which characteristics the traffic of each service level should have. Therefore, the distribution of the different existing traffic types among the SLs may be stated by the manufacturer or the network administrator. By allowing the traffic to be segregated by categories, we will be able to distinguish between packets from different SLs and to give them a different treatment according to their needs.

IBA ports support VLs as a mechanism for creating multiple virtual links within a single physical link. Each VL must be an independent resource for flow control purposes. A VL represents a set of transmission and reception buffers in a port. IBA ports can support a minimum of two and a maximum of 16 VLs. Since systems can be constructed with switches supporting a different number of VLs, the number of VLs used by a port is configured by the subnet manager. Moreover, packets are marked with a Service Level (SL), and a relation between SL and VL is established at the input of each link by means of a *SLtoVLMappingTable*.

When more than two VLs are implemented, the priorities of the data lanes are defined by the *VLArbitrationTable*. This arbitration affects only to data VLs, because control traffic uses its own VL, which has greater priority than any other VL. The VLArbitrationTable consists of two tables, one for scheduling packets from high-priority VLs and another for low-priority VLs. However, IBA does not specify what is high and low-priority. The arbitration tables implement weighted round-robin arbitration within each priority level. Up to 64 table entries are cycled through, each one specifying a VL and a weight, which is the number of units of 64 bytes to be transmitted from that VL. This weight must be in the range from 0 to 255, and is always rounded up as a whole packet.

Moreover, a *LimitOfHighPriority* value specifies the maximum number of high-priority packets that can be sent before a low-priority packet is transmitted. Specifically, the VLs of the high-priority table can send $LimitOfHighPriority \times 4096$ bytes before a packet from the low-priority table can be transmitted. If at a given time, no high-priority packets are ready for transmission, low-priority packets can also be transmitted.

3 Our Proposal to Give QoS Guarantees

In previous works we have proposed a simple strategy to treat the requests of latency guarantee. Specifically, when an application requests latency guarantee, the maximum distance allowed between two consecutive entries in the high-priority table must be computed in order to allocate entries on that table to that application. Moreover, the application could also request a mean bandwidth that would result in a weight to put in the entries of the arbitration table. Therefore, for a certain connection that requests a maximum delay that results in a distance d, and a mean bandwidth that results in a weight w, the number of entries needed is $max\{\frac{64}{d}, \frac{w}{255}\}$.

Obviously, the maximum distance between two consecutive entries in the high-priority table requested by a connection ranges from 1 to 64. However, in order to optimize the filling up of the table, we only consider the following distances: 2, 4, 8, 16, 32, and 64 [5]. Therefore, the applications' requests of a maximum distance between two consecutive entries in the high-priority table are turned into the closest lower power of 2 [4].

Traffic is grouped in SLs according to its maximum latency. Specifically, all connections using the same SL need the same maximum distance between two consecutive entries in the high-priority table, regardless of their mean bandwidth. For the most requested distances, we could distinguish between two or four different SLs considering the mean bandwidth. In this way, if we have enough available VLs, each kind of traffic could use a different VL.

Moreover, in [4] and [5] we proposed an algorithm to select a free sequence of entries in the high-priority table to meet a new application's request. This algorithm successfully allocates a new sequence in the table if there are enough available entries. This is achieved because the available entries are always in the best situation to treat the most restrictive request. For a connection requesting $\frac{64}{d}$ entries with a maximum distance d between them, the algorithm looks for a previously established sequence, for the corresponding VL, with enough available weight. If there is no available sequence, a new free sequence with those characteristics is looked for.

In a more formal way, in a table T, let the sequence $t_0, t_1, \ldots, t_{62}, t_{63}$ represent the entries of the table. Each t_i has an associated weight w_i whose value can vary between 0 and 255. Thus, we say an entry t_i is *free* if and only if $w_i = 0$. For a table T and a request of distance $d = 2^i$, we define the sets $E_{i,j}$ with $i = \log_2 d$ and $0 \leq j < d$, as

$$E_{i,j} = \left\{ t_{j+n \times 2^i} \quad n = 0, \ldots, \frac{64}{2^i} - 1 \right\}$$

Each $E_{i,j}$ contains the entries of the table T spaced by an equal distance d which are able to meet a request of distance $d = 2^i$ starting with the entry t_j. We say a set $E_{i,j}$ is free if $\forall t_k \in E_{i,j}$, t_k is free. Other properties derived from this definition are available in [5].

In [4] we also presented a simple algorithm to maximize the number of requests allocated in the arbitration table. For a new request of distance $d = 2^i$, the algorithm studies all possible sets $E_{i,j}$ for this kind of request, in a certain order, and selects the first set that is free (so, all its entries are free). The order the sets are inspected is based on the application of the bit-reversal permutation to the distance values in the interval $[0, d-1]$. Specifically, for a new request of maximum distance $d = 2^i$, the algorithm selects the first free $E_{i,j}$ in the sequence $E_{i,iR_0}, E_{i,iR_1}, \ldots, E_{i,iR_{d-1}}$ where $_iR_j$ is the bit-reversal function applied to j codified with i bits. Note that this algorithm is only applied if there is no previously allocated sequence for the same requested distance with available room in its entries.

For example, the order the sets are inspected for a request of distance $d = 2^3$ is $E_{3,0}$, $E_{3,4}$, $E_{3,2}$, $E_{3,6}$, $E_{3,1}$, $E_{3,5}$, $E_{3,3}$, and $E_{3,7}$. Note that this algorithm first fills in the even entries, and later the odd entries. In this way, if we have available entries, we can always meet a request of distance 2, which is the most restrictive. The same consideration can be made for longer distances.

In [5], we have also proved several theorems showing that the algorithm can always allocate a new request if there are enough available entries. This is achieved because the algorithm always selects the sequences in the optimal way for satisfying later the most restrictive possible request.

When a connection finishes, its bandwidth is deducted from the accumulated bandwidth in the entries it was occupying. When this accumulated bandwidth is zero those entries must be released. When some entries are released, a disfragmentation algorithm must be applied to leave the table in a correct state, such that the proposed filling in algorithm can be used. This disfragmentation algorithm and its properties are also described in [5]. Basically, it puts together small free sets to form a larger free set, moving the content of some entries.

Both algorithms together permit the allocation and release of sequences of entries in the arbitration table in a optimal and dynamical way [5]. This allows us to provide applications with QoS using in an optimal way the IBA mechanisms.

4 Performance Evaluation

In [4] and [5] we have evaluated the behavior of our proposals using a large buffer size. We have shown that our proposals are able to provide applications with QoS guarantee. In this section, we are going to determine the minimum switch port and host interface VLs buffer size our proposals require to satisfy the QoS requirements. In the following points, we explain the network and the traffic models we have used.

4.1 Network Model

We have used irregular networks randomly generated. All switches have 8 ports, 4 of them having a host attached, the other 4 being used for interconnection between switches. We have evaluated networks with sizes ranging from 8 to 64 switches (so, with 32 to 256 hosts, respectively). We have also tested several packet sizes ranging from 256 to 4096 bytes, and the three link rates specified in IBA. Taking into account that all these variations present similar results and the space limitation, we have only included here results for the 16 switches-network using a packet size of 256 bytes and a link rate of 2.5 Gbps.

In the switches both at input and output ports, there are 16 VLs in order to assign a different VL to each SL. Each switch has a multiplexed crossbar. We will test several buffer sizes. As IBA uses virtual cut-through we have only considered buffer sizes that allow to store completely whole packets. Specifically, we have considered buffer sizes of 1, 2, 3 or 4 whole packets of capacity.

Table 1. Features of the SLs used.

SL	Maximum Distance	Bandwidth Range (Mbps)
0	2	0.064 - 1.55
1	4	0.064 - 1.55
2	8	0.064 - 1.55
3	16	0.064 - 1.55
4	32	0.064 - 1.55
5	32	1.55 - 64
6	64	0.008 - 0.064
7	64	0.064 - 1.55
8	64	1.55 - 64
9	64	64 - 255

4.2 Traffic Model

We have used 10 SLs for traffic needing QoS. Each SL presents a different maximum distance and bandwidth requirements. We have used CBR traffic, randomly generated among the bandwidth range of each SL. For the most requested distances several SLs have been considered using the mean bandwidth of the connections. Specifically, the SLs used and their features are shown in Table 1.

The connections of each SL request a maximum distance between two consecutive entries in the high-priority table and a mean bandwidth in the range shown in Table 1. Note that this is similar to requesting a maximum deadline and computing the maximum distance between two consecutive entries in the virtual lane arbitration tables.

Each request is considered by each node along its path and is accepted only if there are available resources. Connections of the same SL are grouped into the same sequence of entries. The total weight of the sequence is computed according to the accumulated bandwidth of the connections sharing that sequence. When the connection cannot be settled in a previously established sequence (or there is no previous one), our algorithm looks in the high-priority arbitration table for a new free sequence of entries with the correct distance between its entries.

When no more connections can be established, we start a transient period in order for the network to reach a stationary state. Once the transient period finishes, the steady state period begins, where we gather results to be shown. The steady state period continues until the connection with a smaller mean bandwidth has received 100 packets.

Although the results for BE and CH traffic are not the main focus of this paper, we have reserved 20% of available bandwidth for these types of traffic, which would be served by the low-priority table. So, connections would only be established up to 80% of the available bandwidth.

4.3 Simulation Results

We can see in Table 2 several metrics measured for the different buffer sizes considered. Specifically, we have computed the injected and delivered traffic (in

Table 2. Traffic and utilization for different buffer sizes.

	Buffer size (in packets)			
	1	2	3	4
Injected traffic (Bytes/Cycle/Node)	0,7262	0,7262	0,7258	0,7258
Delivered traffic (Bytes/Cycle/Node)	0,6788	0,7262	0,7258	0,7258
Av. utilization for host interfaces (%)	67,88	72,63	72,58	72,58
Av. utilization for switch ports (%)	68,06	73,49	73,48	73,48
Av. reservation for host interfaces (Mbps)	1849,81	1849,81	1848,67	1848,67
Av. reservation for switch ports (Mbps)	1871,84	1877,84	1871,75	1871,75

bytes/cycle/node), the average utilization (in %) and the average bandwidth reserved (in Mbps) in host interfaces and switch ports. Note that the maximum utilization reachable is 80%, because the other 20% is reserved for BE and CH traffic. So, we are close to the maximum utilization achievable. Obviously, we could achieve a higher utilization establishing more connections, but we have already made many attempts for each SL. We could establish other connections, but these connections would be of SLs of a small mean bandwidth because the network is already heavily loaded, and it is unlikely that these new connections would provide us with more information. So, it seems reasonable to assume that with this load we can study the network behavior in a quasi-fully loaded scenario.

Note also that the behavior is quite similar for all the buffer sizes considered. Regardless of the buffer size used, the network reaches a similar throughput and the reservation performed in the host interfaces and switch ports is quite similar. However, there is an important difference for the buffer size of one packet. In this case, the network cannot transmit all the packets it receives, and delivered traffic is lower than injected traffic. As we will see, this causes long waitings for the packets. However, for other buffer sizes the injected traffic is equal to the delivered traffic, thereby showing that the network is able to transmit all the packets it receives.

We have also computed the percentages of packets that meet a certain deadline threshold. These thresholds are different for each connection and are related

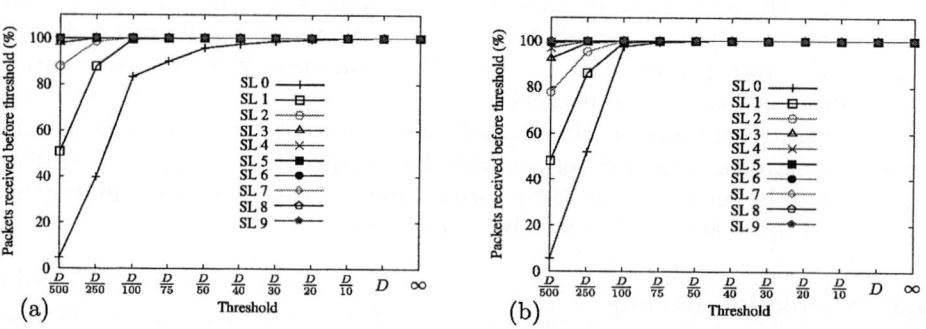

Fig. 1. Packet delay for buffer sizes of (a) 1 packet and (b) 2 packets.

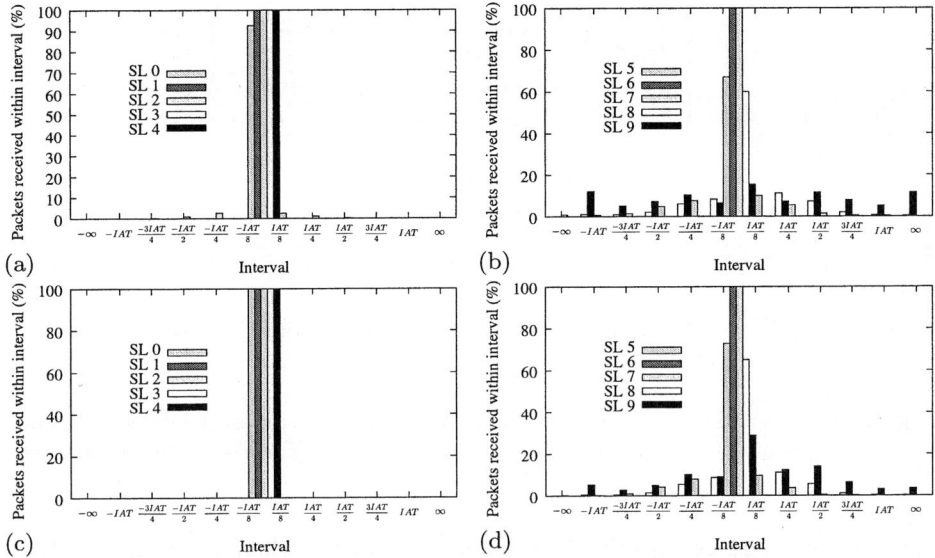

Fig. 2. Packet jitter for buffer sizes of (a) and (b) one packet, (c) and (d) two packets.

to their requested maximum deadline. This maximum deadline is the maximum delay that has been guaranteed to each connection. In the figures, this deadline is referred to as D. The results for each SL are presented in Fig. 1 for buffer size of one and two packets. Results for larger buffer sizes are quite similar to those obtained for two packets. In these figures, we can see that all packets of all SLs arrive at their destinations before their deadlines. However, packets of SLs with stricter deadlines arrive at their destination closer to their deadline, although in time to meet their requirements.

We have also measured the average packet jitter. We have computed the percentage of packets received in several intervals related to their interarrival time. Obviously, these intervals are different for each connection. The results for each SL are shown in Fig. 2. We have only shown results for buffer sizes of one and two packets. We can see that results for buffer size of one packet are worse than for two packets. Results for a larger buffer size are quite similar to those of buffer size of two packets. Moreover, results for the buffer size of two packets are much better than those obtained for just one packet, although almost all packets arrive at their targets in the central interval $[\frac{-IAT}{8}, \frac{IAT}{8}]$. For other SLs with bigger mean bandwidth the jitter has a Gaussian distribution never exceeding $\pm IAT$. Note that packets of SLs with less priority (SLs 9 and 8) present a worse behavior for buffer size of one packet than for buffer size of two packets.

Finally, for a given deadline threshold, we have selected the connections that deliver the lowest and the highest percentage of packets before this threshold. In the figures, these connections will be referred to as the worst and the best connections, respectively. We have selected a very tight threshold so that the

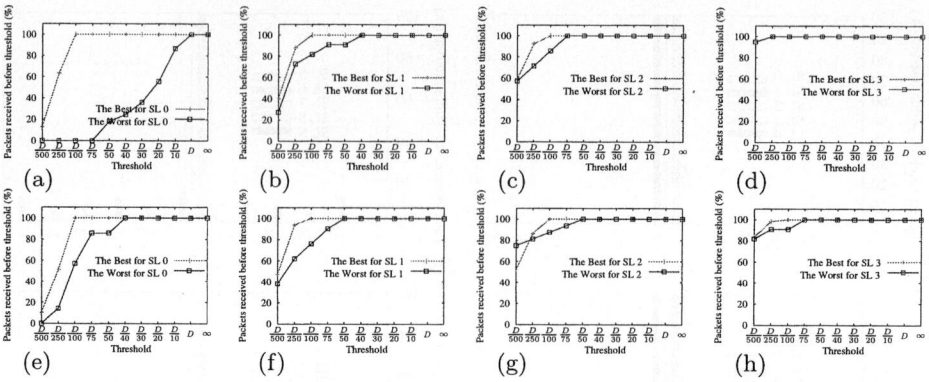

Fig. 3. The best and the worst connection for SLs with the strictest delay requirements, for buffer sizes of (a), (b), (c) and (d) one packet, and (e), (f), (g) and (h) two packets.

percentage of packets meeting the deadline was lower than 100% in Fig. 1. In particular, we have selected the threshold equal to $\frac{Deadline}{100}$. Note again that this threshold is different for each connection and depends on its own maximum deadline. Fig. 3 shows the results for buffer sizes of one and two packets and for the SLs 0, 1, 2, and 3, which are the SLs with the highest deadline requirements. The results for other SLs are even better than these shown in the figures. It is noteworthy that, in all cases, even the packets of the worst connection arrive at their destination before their deadline. We can also see that for buffer size of two packets, results are very similar in the best and the worst case. However, for buffer size of one packet, SL 0 (which has the highest deadline requirements), has a very different behavior. This is due to the long waitings of the packets in the buffers caused by the lack of available room in the next buffer of their paths.

5 Conclusions

In [4] and [5] we proposed a new methodology to provide each kind of applications with the previously required QoS level. We also proposed an algorithm to select a free sequence of entries in the arbitration table. This algorithm successfully allocates a request in the arbitration table if there are enough available entries. It manages the requests in an optimal way, being able later to satisfy the most restrictive possible request. Some formal properties and theorems derived from this algorithm are shown in [5].

In this paper, we have determined the minimum virtual lanes' buffer size required to provide applications with QoS guarantee. We have tested several buffer sizes ranging from 1 to 4 packets. The most important result may well be that for all buffer sizes our proposals meet the QoS requirements. However, important waitings for buffer size of one packet have been observed. These waitings disappear when we have a buffer size of two or more packets.

These are clearly important results, which show that our methodology does not require a great deal of space resources to achieve QoS in InfiniBand environments. We configure in an accurate way the mechanisms that InfiniBand provides to support QoS, with a minimum of buffer room in switch ports and host interfaces.

References

1. InfiniBand^{TM} Trade Association: http://infinibandta.com. (1999)
2. InfiniBand Trade Association: InfiniBand Architecture Specification Volume 1. Release 1.0. (2000)
3. Alfaro, F.J., Sánchez, J.L., Duato, J., Das, C.R.: A Strategy to Compute the InfiniBand Arbitration Tables. In: Proceedings of International Parallel and Distributed Processing Symposium (IPDPS'02). (2002)
4. Alfaro, F.J., Sánchez, J.L., Duato, J.: A New Proposal to Fill in the InfiniBand Arbitration Tables. In: Proceedings of IEEE International Conference on Parallel Computing (ICPP'03). (2003) 133 – 140
5. Alfaro, F., Sánchez, J., Menduiña, M., Duato, J.: Formalizing the Fill-In of the Infiniband Arbitration Table. Technical Report DIAB-03-02-35, Dep. de Informática Universidad de Castilla-La Mancha (2003)

Reducing Power Consumption in Interconnection Networks by Dynamically Adjusting Link Width*

M. Alonso, J.M. Martínez, V. Santonja, and P. López

DISCA, Universidad Politécnica de Valencia, Valencia, Spain
malonso@disca.upv.es

Abstract. The huge increase both in size and complexity of high-end multiprocessor systems has triggered their power consumption. Air or liquid cooling systems are needed, which, in turn, increases power consumption. Another important percentage of the consumption is due to the interconnection network.
In this paper, we propose a mechanism that dynamically reduces the available network bandwidth when traffic becomes low. Unlike other approaches that completely switch links off when they are not fully utilized, our mechanism is based on reducing their bandwidth by narrowing their width. As the topology of the network is not modified, the same routing algorithm can be used regardless of the power consumption level, which simplifies the router design.
By using this strategy, the consumption may be strongly reduced. In fact, the lower bound of this reduction is a design parameter of the mechanism. The price to pay is an increase in the message latency with low network loads.

1 Introduction and Motivation

Many compute-intensive applications require a huge amount of processing power, which can only be achieved with massively parallel computers. The interconnection of these systems has been considered as one of the key factors to achieve a high performance. Hence, a lot of research effort has been made in order to increase interconnection network performance. However, as power consumption has arisen as an important problem in current commercial multiprocessors, it is also a significant aspect in the interconnection network design of these machines. For example, the integrated router and links of the Alpha 21364 microprocessor consume about 20% of the total power (23 W of a total of 125 W), where 58% of this power is consumed in the link circuitry [10]. Designers of an IBM InfiniBand 8-port 12X switch estimate it consumes 31 W, with links taking up to 65% (20W) [10].

A greater power consumption requires sophisticated high temperature dissipation mechanisms. Furthermore, the probability of failures increases with temperature. All these aspects make us to be aware of the importance that the control of the power consumption has in the system and, in particular, in the interconnection network. Several works have developed power models for networks, characterizing the power profile of network routers and links [9, 12, 13], showing that power consumption is a function of

---

* This work was supported by the Spanish CICYT under Grant TIC2003-08154-C06-01.

the square of voltage and the transmission frequency. In order to increase the data transmission rates, high frequencies are used. Indeed, for a successful communication, it is necessary to maintain the frequency tuning of the link even when no data is transmitted.

An alternative is the Dynamic Voltage Scaling (DVS) approach. In [5], links track and adjust their voltage levels to the minimum supply voltage that guarantees correct operation as link frequency is also changed, providing a mechanism that can potentially lower link power dissipation. In [10], a history-based DVS policy is proposed that uses past network utilization to predict future traffic and tune link frequency and voltage dynamically to minimize network power consumption. The drawback is that a sophisticated hardware mechanism is required for ensuring correct link operation during scaling, thus incurring significant delay overhead and additional CMOS area. Also, DVS links continue to consume power even while idle.

A more interesting option is based on selectively turning off links when they have a low utilization index. A Dynamic Power Management (DPM) policy [11] is proposed which turns links on/off in response to communication traffic variations. The problem here is the resynchronization time when they are connected again. Current links have a startup time of approximately 800 ns [6], and link designers are optimistic that can be improved down to 100 clock cycles [11]. But in order to overcome turned off links, a proper routing algorithm is required. A deadlock-free fault-tolerant routing algorithm serves for this purpose [1,4]. This proposal is easier to implement as compared with DVS and incurs less power overhead. A drawback for this proposal is that it is necessary the use of a fault-tolerant routing algorithm or a specifically designed algorithm that increases the complexity, the hardware needed, and that may introduce some penalty on network performance. On the other hand, the lower bound for the reduction in the power consumption is 37.5% for an 8-ary 2-mesh topology, which may not be a great deal compared to the added network complexity.

In this paper, we propose a novel strategy to reduce power consumption in the interconnection network that eliminates the aforementioned drawbacks. Our approach is based on dynamically adjusting link width. As the topology remains the same, the same routing algorithm can be used regardless of the power consumption level, which simplifies the router design. On the other hand, the lower bound of power consumption may be significantly lowered. In fact, this lower bound is a design parameter of the mechanism. The price to pay is an increase in the message latency with low network loads.

The rest of the paper is organized as follows. Section 2 describes the proposed power saving mechanism, evaluating it in Section 3. Finally, some conclusions are drawn.

2 Description of the DALW Power Saving Mechanism

The proposed mechanism is based on measuring network traffic and dynamically reducing the available network bandwidth when traffic becomes low. Instead of completely switching off links when they are not fully utilized, their bandwidth is reduced by narrowing their width (phit size). For this reason, the mechanism is referred to as DALW: Dynamic Adjustment of Link Width.

As links are never completely switched off, the same routing algorithm can be used both when all the links have full width and when some power saving features are in use.

By adjusting the width of the links, different power saving degrees can be achieved. However, by completely turning off links, power saving levels are coarse. Moreover, the lower bound of the power consumption can be as low as the minimum width of the link that guarantees network connectivity (i.e. one-bit link). For instance, in a 2-D torus network with 32-bit links, by completely switching off links, the lower bound of power consumption is 50%, because no more than two links per node can be switched off or the network becomes disconnected or the routing becomes complex. In this case, there are only three power consumption states: 100% (4 links on), 75% (3 links on) and 50% (2 links on). However, by reducing the width of links, power could be dynamically reduced to many different consumption states. Although, it could be possible to narrow a link to any width, in order to make the implementation simple, only some predefined widths should be used. In particular, the reduced widths should be an exact division of the original width. As the width is usually a power of two, each attainable reduced width results from halving the previous width. For example, the width of a 32-bit link could be reduced to 16,8,4,2 or 1 bit. In this case, one 32-bit flit is transmitted by means of 2,4,8,16 or 32 phits, respectively, and the design of the serializing/deserializing logic is relatively simple.

This mechanism is especially suitable for those switches that can be configured to support either a high number of narrow ports or a low number of wider ports. For instance, the Mellanox InfiniScale III InfiniBand switch can be configured either as a 24 4X-ports or as eight 12X-ports [8]. When a port is working in "wide" mode, it is actually composed of several (three, in this case) independent narrow ones (serial ports, in this case) working in parallel. Therefore, our mechanism can be easily implemented by selectively turning off these individual ports and using the serializing/deserializing logic that already exists at both sides of the link.

The price to pay with the DAWL mechanism is that message latency will be increased at low loads, when messages cross narrow links, as one of the latency components depends on link bandwidth. While this can be a problem for latency sensitive applications, we expect that the benefits on power saving will compensate latency increase. We try to mitigate this problem by using a selection function [2] that gives a higher priority to wider links. Indeed, both the current link width and the degree of virtual channel multiplexing are considered. Finally, it is always possible to disable the power saving mechanism when a latency sensitive application is run.

The mechanism should reduce link width when network traffic is not intense. Hence, each node must locally estimate network traffic. There are several ways proposed in the literature: the number of busy virtual channels [7, 11], the length of the buffers associated to links [11] or the link utilization [11]. Link utilization has the advantage of faithfully representing the traffic level of the link [11]. However, it has the disadvantage of being affected by the performance degradation of the network at saturation [7]: accepted traffic (and link utilization) drastically decreases and message latency grows exponentially. As a consequence, by using link utilization alone, a link may be incorrectly disconnected when the network is congested, thus making a bad situation worse. In [11] a litmus test based on link buffer occupancy is used. However, our experiments showed that when the network is congested, actual traffic pattern becomes somewhat atypical due to routing constraints (i.e. availability of escape channels, atomic channel allocation requirements,...).

Assuming that all nodes are injecting messages into the network, when the network is congested newly generated messages can not be routed because all the required channels are busy for a long time. As a consequence, these messages are locally queued. Therefore, the power saving mechanism should reduce link width only if network utilization is low and the message injection queue of the node is empty.

On the other hand, when network utilization becomes high, link width should be progressively increased, up to its nominal value. Again, network congestion may complicate things. Assume that, as traffic was low in the past, most of the links are working at a fraction of their full width. Then, a traffic burst arises into the network. Even if the traffic value is not enough to saturate the network with all its links working at 100%, it may congest the actual (width-reduced) network, thus decreasing utilization (as stated above) and never switching links to 100%. Therefore, again, the local pending message queue should be used as a litmus test to enable link width recovery. In particular, link width should be completely enabled (at their nominal width) if this queue grows.

Link utilization can be easily measured by increasing a counter every time a phit is transferred. This counter is periodically checked in order to apply power saving actions. The link utilization level is obtained by dividing the counter value by the number of elapsed link clock cycles. The counter is reset after every check period. Taking into account that the mechanism will increase or decrease the width of individual links, it operates on a per-link basis rather than on a per-node basis [11]. In the latter case, a node would measure the utilization of all its links and then it would decide which of them are completely disabled or enabled. Obviously, measuring link utilization on a per-link basis makes no sense in this case, as a disabled link is not utilized at all.

Once link utilization is measured, link width will be halved when its utilization is lower than a threshold value (u_{off}). On the contrary, link width is doubled when its utilization reaches another threshold (u_{on}). On the other hand, every time the width of a link is modified, the destination node of the link must be notified in order to properly manage the deserializing logic. This can be done by means of dedicated control signals or sending in-band control packets.

The number of allowed nested reductions is a design parameter that imposes a lower bound on link power consumption. We propose up to two nested link width reductions, thus imposing a lower bound of 25% of nominal power consumption. Hence, a link may work at 100%, 50% or 25% of its nominal width. Further link width reductions are possible, but they impose a severe penalty in latency when the workload is low. In any case, the lowest possible bound is achieved when link bandwidth narrows down to one bit. Concerning threshold values, when u_{off} is high, the mechanism will reduce link bandwidth at heavier traffic rates, thus obtaining higher power saving benefits. On the other hand, if u_{on} is also high, higher network traffic will be required in the network in order to recover link with, thus again improving power saving. Indeed, both values are closely related. Notice that when traffic is lower than u_{off}, link width will be halved, thus transmitting twice as many phits for the same network traffic value. As a consequence, link utilization will also suddenly grows to twice its value. Hence, in order to avoid widening the link that has just been reduced, the value of u_{on} has to be set, at least, to a

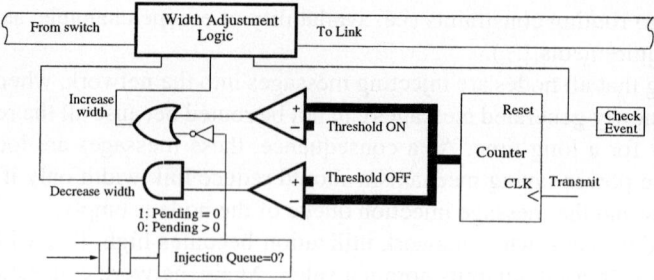

Fig. 1. Hardware required to implement the proposed power saving scheme.

value $2 * u_{\text{off}}$. In practice, some hysteresis value h should be added to this lower bound. Figure 1 shows the hardware required to implement the mechanism.

After the first link width reduction, traffic may continue decreasing. Our mechanism allows to reduce link width even more without requiring additional comparators. Notice that in this case, the measured link utilization will also decrease, dropping again below u_{off} and thus triggering another link width reduction by a half.

Notice that if link utilization was measured as the ratio of the number of transmitted flits (instead of phits) and the number of elapsed link clock cycles, utilization does not longer grow every time the link width is reduced. In this case, further link width reductions (i.e., nested width reductions) would require the definition of new thresholds, thus making implementation more complex. In other words, the number of required thresholds and comparators would be equal to twice the number of nested reductions.

3 Performance Evaluation

In this section, we evaluate the behavior of the power saving strategy proposed in this paper. Using simulation we assess the power reduction achieved by the DALW mechanism and quantify its impact on network performance. We compare the results obtained by this technique with the performance of the default system (without applying the power saving mechanism). The metrics are the average latency of a message (measured from generation to delivery time) and the relative power consumption of the links as compared with the default system.

Our simulator models a wormhole switching network at the flit level [2]. Each node of the network consists of a processor, its local memory and a router. The router contains a routing control unit, a switch, and several physical links. There are four independent memory channels between the router and the local memory. A deadlock-avoidance-based fully adaptive routing algorithm [3] is used. Physical channels are split into three virtual channels (one adaptive plus two escape). Each virtual channel has an associated buffer with capacity for four flits. Nodes also include the power saving mechanism presented in this paper. For network size, a 32-ary 2-cube (1K-node 2D torus) and a 8-ary 3-cube (512-node 3D torus) will be evaluated.

Message traffic and message length depend on the applications. Two kind of experiments were run. First, we have evaluated network behavior with a constant message

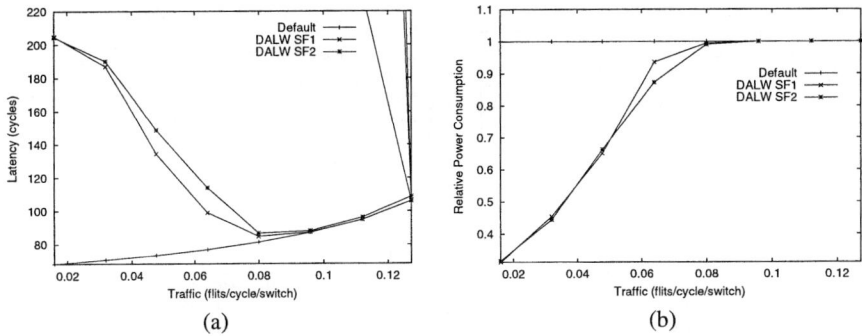

Fig. 2. (a) Latency and (b) relative power consumption comparison for a 32-ary 2-cube (1K node 2-D torus) when the DALW is active and when it is not.

generation rate, evaluating the full range of traffic, from low load to saturation. We have also run other experiments, aimed at studying the dynamic behavior of the network, using variable generation rates during the simulation run. In both kinds of experiments, all the nodes inject messages at the same rate. We use a synthetic workload based on the *uniform* distribution. The destination node of each message is chosen among all the nodes in the network (except the source node) with the same probability. The message length is 16 flits and the interarrival time is based on a uniform distribution.

As explained in section 2, we use three possible states of the links: 100% (full width), 50% and 25%. Network links are initially set to 100% of their nominal width. We assume that the power consumption of each link is proportional to its actual width.

The time needed to increase the width of a link, T_{on}, mainly depends on the delay required to re-enable some of its lines. Based on the best value reported in [6, 11], we have used $T_{on} = 100$ cycles. Reducing the width of a link also needs some time to decrease the circuit voltage level to zero [11]. We assume $T_{off} = T_{on} = 100$ cycles. When narrowing a link, we assume that the available width is instantly reduced but it continues consuming power until T_{off} has elapsed. On the contrary, when widening a link, the new link width is available to messages after T_{on} cycles, but power consumption increases at once. Unlike other works [11], link width adjustment decisions are made with a period greater than T_{on} or T_{off} in order to allow stabilizing links after changes. In particular, we use a check period of 1000 clock cycles.

As stated in Section 2, the value of the utilization thresholds is based on the maximum utilization achievable by the network. We have analyzed the average utilization of the network links in the 2-D and 3-D torus as a function of delivered traffic using *uniform* workload. We conclude that the utilization increases linearly with the traffic up to a value slightly greater than 50%. With higher injection rates, the utilization of the network decreases due to congestion. Based on this maximum we choose a threshold to narrow a link of $u_{off} = 0.2$. When the utilization of a link decreases below this value and there are not any pending messages, the width of the link is halved (except in state 25%). The threshold used to widen a link is $u_{on} = 0.5$. When the utilization of a link surpasses this value, the width of the link is doubled (except in state 100%). If messages begin to queue at the pending messages queue, the state of the link is set to 100%.

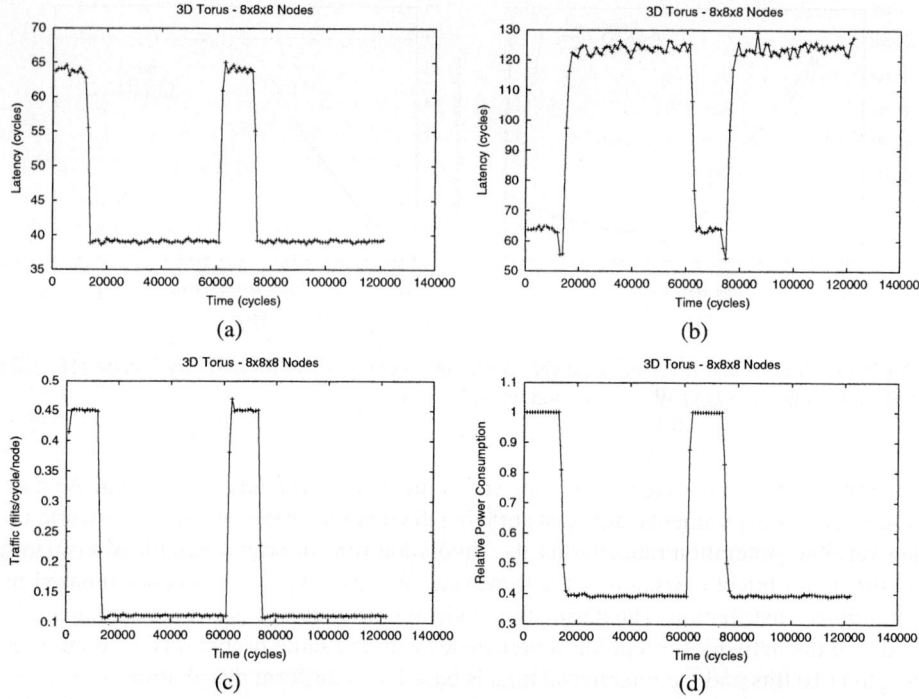

Fig. 3. Results with variable injection rates. Graph (a) shows the latency for the default system. Graphs (b,c,d) depict latency, traffic and power consumption with DALW.

Figure 2 compares results obtained for a 2-D torus when DALW mechanism is applied with the default system. Figure 2a compares the average message latency and 2b shows the relative power consumption. Two selection functions have been evaluated. The SF2 selects the output link for a message following a static priority, whereas the SF1 selects it taking into account the current width of the feasible links and their degree of virtual channel multiplexing. SF2 offers slightly better latency values than SF1. For very low load, most of the links are in the 25% state. Consequently, the relative power consumption of the network is approximately a 26% of the nominal value. For this traffic point, we have a 3-fold increase in the latency of the messages. For higher traffic the relative power consumption increases, indicating that some links are changing to states with more avaliable bandwidth (to 50% or 100%). At the same time, latency decreases. For example, when relative power consumption is 50%, we obtain a 100% increase in latency. For heavy network traffic, the power saving approaches to zero and latency matches the default network values, as almost all the links have moved to state 100%. On the other hand, Figure 2a shows that the throughput is not reduced when the DALW technique is applied.

In order to analyze the dynamic behavior of the DALW mechanism, we have run simulations using a two-level load. The simulation finishes when a given number of messages (700,000) have arrived at their destinations. We initiate the simulation with heavy traffic (just before entering saturation). When 25% of messages reach their destinations, the load changes abruptly to a low value (a quarter of the saturation load). After 50% of messages reach their destination, the input traffic is set again to the high value, and after 75% of messages have been received, it returns to the low value.

The results for a 3-D torus and a uniform traffic pattern are shown in Figure 3. In the default system, all the links are always using its maximum width (100%) and the relative power consumption is always one. When DALW is applied and the traffic is heavy, most of the links are in the 100% state, thus latency and delivered traffic equal the default values. In the transitions from heavy to light traffic there is an instantaneous reduction of the latency (see Figure3b). In this transient, the load is low but the links provide full bandwidth. This happens while nodes are monitoring their output links, detecting the decrease in their utilization and activating the power saving mechanism. When the width of some links is set to 50% or 25%, power consumption is reduced (Figure 3d) and latency increases (Figure 3b). In the transition from light to heavy traffic most of the links transit to state 100%. This process last t_{on} cycles. During this period the traffic is high but a large number of links provide low bandwidth. Thus, some new messages must wait in the injection queues of the nodes. When finally the link operates at its full width, these messages take profit of the increased capacity of the links to enter the network and reach their destination. This is shown in Figure 3c as a traffic peak in the low to high load transition.

4 Conclusions

In this paper we have proposed a novel power saving mechanism (DALW) for interconnection networks, based on dynamically adjusting the width of network links as a function of the network traffic. As links are never completely disconnected, the network topology remains the same regardless of the applied power saving measures. Therefore, routing is not changed, which simplifies router design. Moreover, the lower bound of power consumption can be as low as the one required to keep network connectivity. In fact, it is a design parameter of DALW.

We have evaluated the DALW mechanism on a k-ary n-cube network. The results show that there is a trade-off between power consumption and latency increase. At low traffic loads, the relative power consumption of the network is approximately a 26% of the nominal value, with a 3-fold increase in latency. For higher traffic, the relative power consumption increases and latency decreases. For heavy network traffic, the power saving approaches zero and latency matches the default network values.

As future work, we plan to analyze in detail the impact of the DALW design parameters (power saving levels, thresholds, startup and shutdown times and check period) on network behavior, especially with dynamic workloads.

References

1. B.V. Dao, J. Duato, and S. Yalamanchili, Dynamically configurable message flow control for fault-tolerant routing, *IEEE Trans. on Parallel and Distributed Systems*, vol. 10, no. 1, pp. 7–22, Jan. 1999.

2. J. Duato, A new theory of deadlock-free adaptive routing in wormhole networks, *IEEE Trans. on Parallel and Distributed Systems*, vol. 4, no. 12, pp. 1320–1331, Dec. 1993.
3. J. Duato and P. López, Performance evaluation of adaptive routing algorithms for k-ary n-cubes, *Parallel Computer Routing and Communication Workshop*, May 1994.
4. J. Duato, A theory of fault-tolerant routing in wormhole networks, in *IEEE Trans. on Parallel and Distributed Systems, Vol 10, No. 1, pp. 7-22*, 1999.
5. J. Kim and M. Horowitz, Adaptive supply serial links with sub-1V operation and per-pin clock recovery, *Int. Solid-State Circuits Conf.*, Feb. 2002.
6. E.J. Kim, et al. Energy Optimization Techniques in Cluster Interconnects. *Int. Symp. on Low Power Electronics and Design*, Aug. 2003.
7. P. López and J. Duato, Deadlock-free adaptive routing algorithms for the 3D-torus: limitations and solutions, *Parallel Architectures Languages Europe 93*, June 1993.
8. Mellanox Technologies home page. http://www.mellanox.com.
9. C. Patel, S. Chai, S. Yalamanchili and D. Schimmel, Power constrained design of multiprocessor interconnection networks, *Int. Conf. on Computer Design*, Oct. 1997.
10. L. Shang, L-S. Peh and N.K. Jha, Dynamic Voltage Scaling with Links for Power Optimization of Interconnection Networks, *9th Int. Symp. on High-Performance Computer Architecture*, Jan. 2003.
11. V. Soteriou, and L-S. Peh, Dynamic Voltage Power Management for Power Optimization of Interconnection Networks Using On/Off Links, *11th Hot Interconnects*, Aug. 2003.
12. H-S. Wang, L-S. Peh and S. Malik, A power Model for Routers: Modeling Alpha 21364 and InfiniBand Routers, *10th Hot Interconnects*, Aug. 2002.
13. H-S. Wang, X. Zhu, L-S. Peh and S. Malik, Orion: A Power-Performance Simulator for Interconnection Networks, *35th Int. Symposium on Microarchitecture*, Nov. 2002.

A Methodology to Evaluate the Effectiveness of Traffic Balancing Algorithms*

J.E. Villalobos[1], J.L. Sánchez[1], J.A. Gámez[1], J.C. Sancho[2], and A. Robles[2]

[1] Dept. de Informática
Universidad de Castilla-La Mancha
02071- Albacete, Spain
{jvillalobos,jsanchez,jgamez}@info-ab.uclm.es
[2] Dept. de Informática de Sistemas y Computadores
Universidad Politécnica de Valencia
46071- Valencia, Spain
{jcsancho,arobles}@gap.upv.es

Abstract. Traffic balancing algorithms represent a cost-effective alternative to balance traffic in high performance interconnection networks. The importance of these algorithms is increasing since most of the current network technologies for clusters are either based on source routing or use deterministic routing. In source-routed networks, the host is responsible for selecting the suitable path among the set of paths provided by the routing algorithm. The selection of an optimal path that maximizes the channel utilization is not trivial because of the huge amount of combinations. Traffic balancing algorithms are based on heuristics in order to find an optimal solution. In this paper, we propose a new methodology based on the use of metaheuristic algorithms to evaluate the effectiveness of traffic balancing algorithms. Preliminary results show that the set of paths provided by current traffic balancing algorithms are still far from an optimized solution. Thus, it is worth continuing to design more efficient traffic balancing algorithms.

1 Introduction

Clusters of PCs represent a cost-effective alternative to parallel computers. The use of high performance switch-based interconnects, such as Myrinet [1], Gigabit Ethernet [2], and InfiniBand [3], provides the flexibility, low latency and high bandwidth required in these environments. Often, the interconnection pattern between switches is defined by the user, which may be irregular. To manage such an irregularity, generic routing algorithms can be used, such as *up*/down** [4], *smart* [5], *adaptive-trail* [6], and *minimal adaptive* [7]. Up*/down* is the most popular routing algorithm used in clusters. However, this algorithm imposes a large number of routing restrictions in order to remove cyclic channel dependencies in the network graph, which prevent most of the messages from being

---

* This work was partly supported by the Spanish CICYT under Grant TIC2003-08154-C06 and JCC de Castilla La-Mancha under Grants PBC-02-008 and PBC-02-002.

routed through minimal paths. Unfortunately, this fact has a negative impact on the traffic balance since it tends to concentrate the traffic in a few channels, which dramatically limits its performance.

In order to balance the traffic in the network, dynamic or static strategies can be applied. Dynamic strategies select the routing paths depending on the network status. This status can be known either locally (adaptive routing strategies [8],[9],[10]) or globally (source-based routing schemes [11]). However, current commercial interconnects do not support adaptive routing mainly due to the possible increase in switch complexity and the difficulties in guaranteeing in order packet delivery. In addition, acquiring global information about the network status can introduce high overhead in the network due to the generated control traffic. Therefore, dynamic strategies to balance traffic are not suitable for application to clusters of PCs.

On the other hand, unlike dynamic strategies, static strategies do not require any hardware support. Thus, they can be applied to any commercial interconnect. These strategies try to achieve an even routing path distribution among the network channels to maximize the network utilization. To this end, traffic balancing algorithms can be used. These algorithms focus on selecting only one path between every source-destination pair, without taking into account the network status. Traffic balancing algorithms select each routing path among the set of paths provided by the applied routing algorithm. Usually, these algorithms assume a uniform packet destination distribution due to the fact that the traffic pattern cannot be known in advance. The use of these algorithms is suitable in networks that either apply source routing [1] or where the routing tables at switches provide deterministic routing [3]. Moreover, unlike adaptive routing strategies, traffic balancing algorithms guarantee the message delivery in order, which is required by many parallel applications.

Traffic balancing algorithms are based on an iterative procedure by which the routing paths provided by the applied routing algorithm are progressively discarded until only one routing path remains between each source-destination pair. Different heuristics can be applied to discard routing paths. The simplest criterion is the random selection. Its main drawback is that it cannot guarantee an even distribution of routing paths over the network channels. More efficient algorithms have been proposed in the literature, such as Summatory of Crossing paths [1], Deviation of Crossing paths [12], and Maximum Crossing path [13].

Given that these algorithms are based on an iterative procedure, it is clear that their effectiveness will strongly depend on the order in which routing paths are discarded. In this sense, we wonder whether these algorithms are able to obtain an optimal solution that achieves the highest performance within the bounds imposed by the applied routing scheme. To answer this question, we should firstly try to obtain the optimal solution. To this end, it would be necessary to evaluate all the possible combinations among the routing paths provided by the applied routing algorithm. However, this approach is non-viable in terms of computational time due to the huge amount of possible combinations.

An alternative to the enumeration of all the possible combinations, is the use of a heuristic algorithm. In this way, we will be sure to obtain a *good* solution (possibly a *local optima*) to the problem in a reasonable time. Concretely, we propose to use *metaheuristic* algorithms [14] because they can be applied to a wide set of different optimization problems, requiring less effort to adapt them to a specific problem. Nowadays, metaheuristics are widely used to solve important practical combinatorial optimization problems, and so we think that this can be an attractive approach to tackle with the problem posed in this work.

In this paper, we propose a new methodology to evaluate the effectiveness of traffic balancing algorithms, which is based on the use of metaheuristic techniques. This evaluation methodology allows us to know at what extent the performance provided by current traffic balancing algorithms differs from that provided by the optimal solution.

The rest of this paper is organized as follows. In Section 2 we describe the traffic balancing algorithm selected in the evaluation of the new methodology. Section 3 briefly describes the metaheuristics used in this paper. Section 4 describes the methodology followed in the evaluation and Section 5 shows some preliminary results achieved. And finally, Section 6 provides some concluding remarks and indications on future research.

2 Computing Routing Tables

We can identify three main stages in the routing table computation process. First, all the shortest paths between every source-destination pair are computed. To this end, the routing algorithm must be able to provide several routing paths between every pair of nodes. Secondly, a traffic balancing algorithm will be applied to select a unique routing path for every source-destination pair. Finally, routing tables must be filled.

To illustrate the application of the proposed evaluation methodology, we have selected the up*/down* routing algorithm, and the Maximum Crossing Path (MaxCp) [13] traffic balancing algorithm. Up*/down* routing is able to provide several paths between every pair of nodes, especially when the DFS methodology [15] is applied to compute its routing paths. Moreover, MaxCp achieves better performance than other traffic balancing algorithms at a lower computation cost when using only a virtual channel.

The MaxCp algorithm is based on minimizing the crossing-path metric, which represents the number of routing paths crossing each channel (i.e., the channel utilization). This algorithm tries to achieve a uniform channel utilization, preventing a few channels from becoming a bottleneck in the network. To this end, the algorithm associates a counter to every channel in the network. Each counter is initialized to the value of the crossing-path metric. Next, a procedure is repetitively applied to the channel with the highest counter value. In each step, a path crossing the channel is selected to be removed if there is more than one path between the source and the destination nodes of that routing path, thus preventing the network from becoming disconnected. When a routing path is removed, the counters associated with every channel crossed by the path are updated. Among the routing paths belonging to a channel, the algorithm will firstly choose the path whose source and destination hosts have the highest number of routing paths between them. The algorithm finishes when the number of routing paths between every pair of nodes is reduced down to the unit.

3 Metaheuristics

As defined in the Metaheuristics Network[1] home page, *"a metaheuristic is a set of concepts that can be used to define heuristic methods that can be applied to a wide set of different problems"*. That is, a metaheuristic can be viewed as a general algorithmic framework which allows us to apply a combinatorial optimization technique to a great variety of problems, with relatively few modifications when instancing the method to a particular problem. In this work, we have selected two metaheuristics which belongs to two different approaches: *genetic algorithms* (from evolutionary computation) and *simulated annealing* (from the neighborhood search approach).

3.1 Genetic Algorithms

Genetic algorithms (GAs) ([16], [14] chapter 3) are evolutionary algorithms based on the principles of natural evolution. GAs try to evolve a population of individuals (potential solutions to the problem) during the search process. Evolution is based on the application of *genetic*-based operators as *selection, crossover* and *mutation* to the individuals contained in the population. A GA works in three main steps:

1. Selection: some individuals are selected from the current population by considering their fitness (the goodness of the solution they codify). Different strategies can be used: proportional to fitness, rank-based, tournament, etc.
2. Reproduction: the selected individuals are grouped (married) in pairs and some *offsprings* are obtained by crossover, that is, by interchanging genetic material between the parents. For example, from `aaaaaa` and `bbbbbb` we obtain `aabbbb` and `bbaaaa` when position two is selected as the crossover point. After crossover, some individuals are mutated (slightly changed) in order to introduce some diversity in the search. For example, following our example, `aabbab` corresponds to mutate position five of the first offspring.
3. Replacement: The new population/generation is obtained by selecting the best individuals from the old and new ones.

This process (steps 1 to 3) is repeated until a certain termination condition is achieved. In this work we use *Steady State* GAs, a model of GA in which only a pair of individuals is selected for reproduction in step 1. The good behavior of this model has been largely demonstrated in the specific literature.

---

[1] http://www.metaheuristics.org

3.2 Simulated Annealing

Simulated Annealing (SA) ([17], [14] chapter 10) is probably the oldest metaheuristic known and has its origin in statistical mechanics. In SA instead of maintaining a population of solutions, only a solution is considered at each state, being its neighborhood explored at each iteration.

SA behaves as a hill climbing algorithm, that is, given the current solution s, the algorithm moves to the best solution selected among the neighborhood ($N(s)$) of s (a neighbor s' is a solution obtained from s by making a small change or movement). The key idea in SA is to allow moves resulting in worse solutions than the current one in order to avoid being trapped in local optima (as usually happens in hill climbing). To implement this behavior, a parameter T called *temperature* is used. Concretely, the probability of accepting a neighbor s' of s is computed following the Boltzmann distribution: $exp(-\frac{f(s')-f(s)}{T})$. Therefore, when T is high, the probability of accepting a worse configuration is also high, but when the value of T decreases, the probability of accepting a worse solution also decreases. The value of T is decreased/cooled during the search process by following a cooling schedule.

4 Methodology of Evaluation

The methodology proposed to evaluate the effectiveness of the traffic balancing algorithms is based on the two metaheuristics previously described. As has been mentioned, the goal of these algorithms is to provide a high quality (the optimal one ideally) solution from a set of possibles configurations. In the problem of evaluating the effectiveness of traffic balancing algorithms, the optimal solution corresponds to the set of paths which achieve the highest performance in the network when considering only a unique route for every source-destination node pair. We will use the solution found by the metaheuristics to evaluate the effectiveness of the traffic balancing algorithm described in Section 2.

Below we will specify the parameter choices for both metaheuristic algorithms, but first let us to introduce two crucial points which are common to them: individual *representation* and *evaluation*.

- *Individual* **representation**. An individual will be a vector of integers, where each position corresponds to a pair of (source, destination) nodes, and its value is in the range $2..k$, being k the number of different paths from source to destination. Pairs having only a possible route between them are not included in the solution. So, the representation length is actually considerably smaller than the upper bound $n(n-1)$, with n the number of nodes in the network.
- *Individual* **evaluation**. Due to the large number of individuals to be evaluated during the search process, the use of a simulator, although very precise, is prohibitive in terms of CPU time. As an alternative we propose to use a performance metric, which can be easily computed from the routing paths codified by the individual. In our initial experiments we have tried with several routing metrics, as minimizing the maximum use of a channel, the

deviation in channels use, and the sum of channels use. Among them, we have selected the minimization of the deviation in channel use, because it is the one which achieves the results closer to the simulator use.

4.1 Genetic Algorithm Design

In order to completely specify the Steady State GA we have to consider the following parameters:
− Initial Population. Individuals in the initial population are generated at random.
− Selection. We use rank-based selection. That is, individuals are sorted according to its fitness and each individual receives a probability of selection proportional to its position in the ranking. This mechanism of selection helps to avoid premature convergence.
− Crossover operator. Two offsprings are generated by using the classical one-point crossover.
− Mutation operator. Each position of each individual in the population is selected for mutation with a small probability (0.01 or 0.02 in our case).
− Stopping Criterion. The algorithm is allowed to complete a fixed number of generations which depends on the network size.

4.2 Simulated Annealing Design

The main point here is to define the neighborhood of a given solution s. In our case, s' it is said a neighbor of s if they differ in exactly one position. That is, given s=aaaaaa, then abaaaa or aaaaba are in $N(s)$, but abaaba is not a neighbor of s. On the other hand, as in our GA design, the initial solution is randomly generated and a maximum number of iterations is considered as stopping criterion. The remaining parameters/decisions are:
− Iterations per Temperature. The number of iterations to be carried out before to modify the temperature is the minimum between 1000 and $\prod_{i=1}^{m} k_i$, being m the individual length and k_i the number of different paths between the pair (source-destination) associated with position i. In this way we allow the algorithm to carry out a high number (in function of the network complexity) of iterations per temperature, in order to achieve a stable behavior.
− Cooling schedule. The temperature decreases geometrically by using the expression: $T_{i+1} = \alpha T_i$, with $\alpha = 0.95$. This is without any doubt the most popular cooling scheme in simulated annealing, because it allows a fast decreasing during the exploration phase and a slow cooling during the exploitation phase.

5 Performance Evaluation

In this section, we evaluate by simulation the performance of the optimal routing path configuration resulting from applying the metaheuristic algorithm proposed in Section 4 and compare it with that of the MaxCp traffic balancing algorithm.

We use the up*/down* routing algorithm based on the DFS methodology to compute the routing tables. To obtain realistic simulation results, we have used timing parameters for the switches taken from a commercial network. We have selected Myrinet because it is becoming increasingly popular due to its good performance/cost ratio. Myrinet uses source routing. Packet sizes of 32 and 1024 bytes are used. Also, we assume a uniform distribution of packet destinations. Traffic injection rate is the same for all the hosts. As shown by many works, uniform distribution can be considered the most suitable synthetic traffic pattern for analyzing interconnection networks under general purpose applications exhibiting undetermined and variable traffic patterns with different locality degrees.

5.1 Network Model

The network is composed of a set of switches and hosts, all of them interconnected by links. Irregular network topologies are considered in the evaluation. The topologies have been generated randomly. We have generated ten different topologies for each network size analyzed. Results in this paper correspond to the topologies that achieve the average behavior for each network size.

Each switch has 8 ports wherein 4 ports connect to hosts and the remainder connect to other switches. Network sizes of 8, 16, 32, and 64 switches have been considered in order to evaluate the influence of the network size on performance. These network sizes can be considered representative of clusters of PCs used nowadays. Note that connecting 4 hosts per switch, a 64-switch network would support up to 256 hosts.

We assume short LAN cables to interconnect switches and hosts. These cables are 10 meters long, offer a bandwidth of 16 MB/s, and have a delay of 4.92 ns/m. Flits are one byte wide. Physical channel is also one flit wide. Transmission of data across channels is pipelined. Hence, a new flit can be injected into the physical channel every 6.25 ns and there can be a maximum of 8 flits on the link at a given time. A hardware *stop and go* flow control protocol [1] is used to prevent packet loss. The first flit latency through the switch is 150 ns.

5.2 Simulation Results

For the sake of brevity, the graphs showed in this paper correspond to a 32-switch network that exhibits an average behavior. The results for the remainder are qualitatively similar. Figure 1 shows the average latency for the *metaheuristic algorithm* (MetaHe) and *MaxCp traffic balancing algorithm* (MaxCp) when short (a) and long (b) packets are used. As can be seen, the routing path configuration obtained by applying the metaheuristic algorithm achieves better performance than the one selected by applying the MaxCp traffic balancing algorithm. The difference in throughput is roughly a 20 percent, regardless of the network size and the packet length used. We have also analyzed bimodal traffic under different rates of short and long messages, obtaining a behavior very similar, in relative terms, to that exhibited with only short or long messages.

As the performance of the MaxCp algorithm is higher than that of other existing algorithms, the results obtained in this evaluation can be generalized to other low-performance algorithms. These results show that there is still enough margin to improve the performance of traffic balancing algorithms. Notice that the performance improvements resulting from applying traffic balancing algorithms are achieved without requiring any hardware support that would increase the overall system cost. Therefore, these algorithms represent a cost-effective approach to improve the performance of current commercial interconnects.

On the other hand, despite the fact that metaheuristic algorithms often allow us to obtain a set of routing paths able to provide an optimal traffic balance, they would not be advisable to use them in their own right, rather than for benchmarking other algorithms. The reason is that metaheuristic algorithms exhibit a high computation cost which may not be acceptable in most cluster environments. The execution time of the applied metaheuristic algorithm depends on the network size. It ranges between one hour and one hour and a half for 8-switch and 64-switch networks, respectively, when running on a Myrinet cluster of 16 PCs (350 MHz Pentium III with 128 Mbytes RAM).

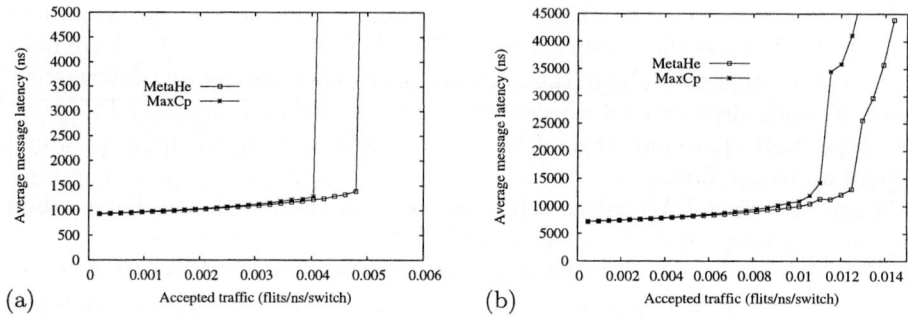

Fig. 1. Average message latency vs accepted traffic for a 32-switch network. Message length is (a) 32 bytes and (b) 1024 bytes.

6 Conclusions

In this paper, we proposed a new methodology to evaluate the effectiveness of traffic balancing algorithms. This methodology is based on comparing the performance achieved by the set of paths selected by applying a traffic balancing algorithm against that achieved by the set of routing paths selected by using a metaheuristic algorithm.

Preliminary results obtained for a representative traffic balancing algorithm have shown that its performance is significantly lower than that obtained by using a metaheuristic algorithm. In particular, a difference in throughput as high as 20 percent is obtained. However, the metaheuristic algorithms cannot guarantee that in all the configurations they will always provide an optimized set of routing paths that achieve the best performance, and can even be detrimental to performance. These results encourage us to continue to searching for more effective traffic balancing algorithms suitable to improve the performance of current commercial interconnects.

For future work, we plan to analyze the causes that limit the performance of current traffic balancing algorithms in order to find new criterions to select the final set of routing paths. The resulting algorithms must be viable in terms of computational time. Also, we plan to study the effectiveness of traffic balancing algorithms in the context of regular networks, such as fat trees.

References

1. Boden, N., Cohen, D., Felderman, R., Kulawik, A., Seitz, C., Seizovic, J., Su, W.: Myrinet - a gigabit per second local area network. IEEE Micro (1995) 29–36
2. Sheifert, R.: Gigabit Ethernet. Addison-Wesley (1998)
3. Association, I.T.: InfiniBand Architecture Specification Vol. 1, Release 1.0. (2000)
4. Schroeder, M., Birrell, A., Burrows, M., Murray, H., Needham, R., Rodeheffer, T., Satterthwate, E., Thacker, C.: Autonet: a high-speed, self-configuring local area network using point-to-point links. IEEE Journal on Selected Areas in Communications **9** (1991) 1318–1334
5. Cherkasova, L., Kotov, V., Rockiki, T.: Fibre channel fabrics: Evaluation and design. In: Procs. of the 29th Hawaii Int. Conference on System Science. (1995)
6. Qiao, W., Ni, L.: Adaptive routing in irregular networks using cut-through switches. In: Procs. of the 1996 Int. Conference on Parallel Processing. (1996)
7. Silla, F., Duato, J.: High-performance routing in networks of workstations with irregular topology. IEEE Trans. on Parallel and Distributed Systems **11** (2000)
8. Dally, W., Aoki, H.: Deadlock-free adaptive routing in multicomputer networks using virtual channels. **4** (1993) 466–475
9. Duato, J.: A new theory of deadlock-free adaptive routing in wormhole networks. **4** (1993) 1320–1331
10. Konstantinidou, S., Snyder, L.: The Chaos router. IEEE Transactions on Computers **43** (1994) 1386–1397
11. Franco, D., Garcés, I., Luque, E.: A new method to make communication latency uniform. In: Procs. of ACM Int. Conference on Supercomputing. (1999) 210–219
12. Flich, J., Malumbres, M., López, P., Duato, J.: Improving routing performance in myrinet networks. In: Procs. of the Int. Parallel and Distributed Processing Symposium. (2000)
13. Sancho, J., Robles, A.: Improving the up*/down* routing scheme for network of workstation. In: Procs. of the European Conference on Parallel Computing. (2000)
14. Glover, F., Kochenberger, G.: Handbook of Metaheuristics. Kluwer Academic Publishers (2003)
15. Sancho, J., Robles, A., Duato, J.: A new methodology to compute deadlock-free routing tables for irregular topologies. In: Procs. of the 4th Workshop on Communication, Architecture and Applications Network-based Parallel Computing. (2000)
16. Michalewicz, Z.: Genetic Algorithms + Data Structures = Evolution Programs. Springer-Verlag (1996)
17. Laarhoven, P.V., Aarts, E.: Simulated annealing. Reidel Publishing Company (1988)

Load Unbalance in k-ary n-Cube Networks[*]

J. Miguel-Alonso[1], J.A. Gregorio[2], V. Puente[2], F. Vallejo[2], and R. Beivide[2]

[1] The University of the Basque Country,
Department of Computer Architecture and Technology,
P.O. Box 649, 20080 San Sebastián, Spain
miguel@si.ehu.es
[2] University of Cantabria, Computer Architecture Group, ETSIIT,
Av. de Los Castros s/n, 39005 Santander, Spain
{jagm,vpuente,fernando,mon}@atc.unican.es

Abstract. This paper studies the effect that HOL (Head-of-Line) blocking in the packet injection queue has on the performance of bidirectional k-ary n-cubes, for values of k over a certain threshold (around 20). The HOL blocking causes an unbalanced use of the channels corresponding to the two directions of bidirectional links, which is responsible for a drop in the network throughput and a rise in the network delay. Simulation results show that this anomaly only appears in those rings where most injections are performed (normally, those in the X axis), and that the elimination of the HOL blocking in the injection queue enables the network to sustain peak throughput after saturation.

1 Introduction

The performance of the interconnection network of a parallel computer has a great impact in the system's performance as a whole. *K*-ary *n*-cubes are the most common direct interconnection network topologies, encompassing rings, meshes, and tori. A central element of this kind of networks is the packet router that injects packets from (and delivers packets to) the compute node to which it is connected, and also routes packets coming from other routers which have to be delivered to other nodes.

The architecture of the router has a fundamental impact on the execution time of applications running in the parallel machine [4],[7]. Typical design objectives are to keep it simple (to reduce cycle time) while getting as much functionality as possible. Simplicity leads to the use of input transit queues and injection queues with FIFO policy. It is well known, however, that this policy has negative effects, HOL (Head-of-Line) blocking among the most harmful of them. Several works have dealt with ways of reducing this effect on transit queues [5]. However, HOL blocking in the injection, to the best of our knowledge, has not been reported in the literature, and as this paper will prove it also has a negative impact on performance that severely affects the scalability of this kind of networks. In fact, HOL at the network interface is one of the key reasons why performance suddenly drops when the network surpasses its saturation point. As we increment the applied load, before reaching the saturation

---

[*] This paper has been done with the support of the Ministerio de Ciencia y Tecnología, Spain, under grants TIC2001-0591-C02-01 and TIC2001-0591-C02-02, and also by the Diputación Foral de Gipuzkoa under grant OF-758/2003.

point, the network accepts the entire offered load. However, sometimes when this maximum point is reached, the accepted load falls *below* the levels reached before saturation – instead of staying at that maximum level.

This anomaly shows up in bidirectional links as an uneven usage (unbalance) between channels going in opposite directions, but only on those rings of the k-ary n-cubes were most injections are performed – typically, those in the X axis – and only for ring sizes over a certain threshold (around 20 nodes per ring). This threshold depends on many characteristics of the network (such as deadlock-avoidance policy, number of virtual channels, etc.) so it must be taken as a reference point, not as an absolute value. For different values of k we have observed variations in the way this unbalance materializes. In some cases, channels X+ are full during most of the simulation time while X- are almost empty (it may well happen the other way around). In some others there are oscillations. In none of the cases studied we have observed this phenomenon disappearing spontaneously. To avoid this anomaly, it is necessary to eliminate the HOL blocking from the injection queue. In the simulation (with uniform traffic) we achieve this by dropping the packets that are blocking the queue. A realistic implementation would require the use of non-FIFO queues, or the implementation of separate injectors: one per direction.

The rest of this paper is organized as follows. In Section 2 we describe the anomaly and the context where it arises. Section 3 presents its effect in rings (1-cubes). Section 4 is devoted to describing the causes of the unbalance and possible measures to avoid it. Section 5 studies the load unbalance in 2D and 3D tori. Section 6 analyzes the appearance of unbalance when running actual applications. Finally, in Section 7, the main conclusions of the work are summarized.

2 General Description of the Anomaly and Its Context

We have observed that, in k-ary n-cube networks with bidirectional links, when the applied load exceeds the saturation limit of the network for the corresponding traffic pattern (we have limited our studies to uniform traffic), the occupation of network resources is not balanced between both directions. This anomaly shows up in any of the mentioned networks, and it originates mainly in the structure of the packet router. Figure 1 describes our basic router organization, showing the usual hardware modules: crossbar, buffers, arbitration logic, synchronization, etc.

The design of a router has to maximize the use of the network resources avoiding communication anomalies such as packet deadlock, livelock and starvation. In our experiments, we use a router that has two or three virtual channels (FIFO queues) per input link to support fully Adaptive Bubble Routing (ABR) [7]. When using ABR, a subset of the total virtual channels is configured as a safe (or escape) virtual network [3] in which packet deadlock never occurs. The remaining virtual channels are configured as a fully adaptive virtual network. Packets move under two different policies. In the adaptive virtual network the injection and transit of packets are regulated by both Virtual Cut-Through flow control (VCT) and minimal adaptive routing. In the safe virtual network the injection of packets is regulated by Bubble Flow Control (BFC), a mechanism for avoiding packet deadlock in topologies based either on a single ring or on a set of rings. In the case of topologies composed of a set of rings, these rings must be visited under Dimension Order Routing (DOR) and packets traveling from one ring to another inside the safe network are considered as new injections. Packets in transit inside a safe ring are regulated by VCT. Packets can move freely from the safe to the adaptive network, but the change of packets from the adaptive to the safe network is regulated by BFC.

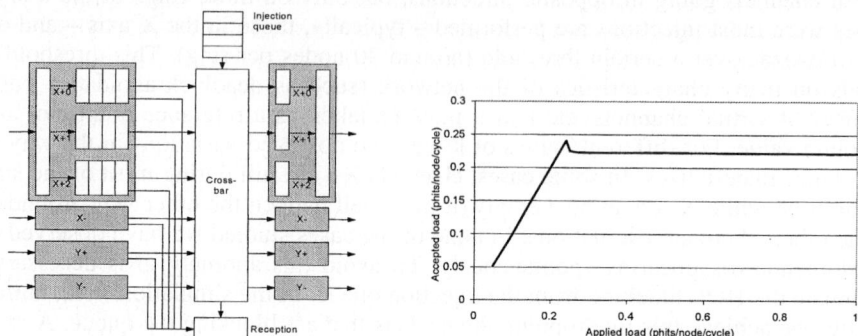

Fig. 1. Left: router model for a k-ary 2-cube. In this case, each physical channel is shared by three virtual channels (0 or Escape, 1 and 2). Each virtual channel has its own input queue to store packets in transit. Right: Accepted vs. applied load in a 30-ary 1-cube. Uniform traffic.

This interconnection network is very similar to the one being used in IBM's BlueGene/L supercomputer [1], in topology (a 3D torus), flow-control, deadlock-avoidance mechanism, etc. We have observed, however, that the mentioned unbalanced use of resources also appears when utilizing different deadlock-avoidance policies, such as the classic use of virtual channels proposed by Dally [2].

3 Load Unbalance in Rings

Let us consider a ring (k-ary 1-cube) with nodes such as those represented in Fig. 1. Each router is connected to its neighbors via two physical links: X+ (for packets moving from left to right) and X- (for packets moving from right to left). The bandwidth of each physical channel is one phit per cycle. Additionally, the router is connected (via an interface) with a computing element. For the rest of this section, we will consider these additional characteristics of the network: a) Each physical channel is shared between two virtual channels. We will name the VCs as follows: X+0, X+1, X-0 and X-1. Channels 0 are Escape channels. Channels 1 are adaptive; although there is no possible adaptation in a ring (there is only one way to reach the destination), the bubble restriction to inject does not apply in these channels; b) Packets are of 16 phits; c) Each VC has a transit queue of 8 packets (128 phits). The injection queue capacity is also of 8 packets; d) Traffic pattern is uniform.

We have studied the performance of this network by injecting a variable load (measured in phits/node/cycle) of traffic and determining the actual load delivered by the network. Data for a 30-node ring is plotted in Fig.1 (right). All applied load is delivered until the saturation point is reached. After that point, a sharp drop in the accepted load is observed (the magnitude of this drop also depends on many design parameters). One cause of this drop is that, when reaching the saturation point, network resources (queues) are not used in a balanced way. Channels moving packets to the right (X+) may be saturated while queues in the X- direction are almost empty. Of course, it may happen the other way around (X+ empty while X- full).

Fig. 2 plots the occupation of queues for different ring sizes (10, 20, 30 and 100), all with the characteristics stated before. We have performed simulations under a heavy applied load (always beyond saturation point) of uniform traffic. The occupation of the queues of one node in the ring has been sampled every 2Kcycles. In the case of a 10-node ring, occupation is always very low throughout the simulated time. The average distance traversed by packets when going from source to destination is very short and therefore saturation is reached before queues are fully used.

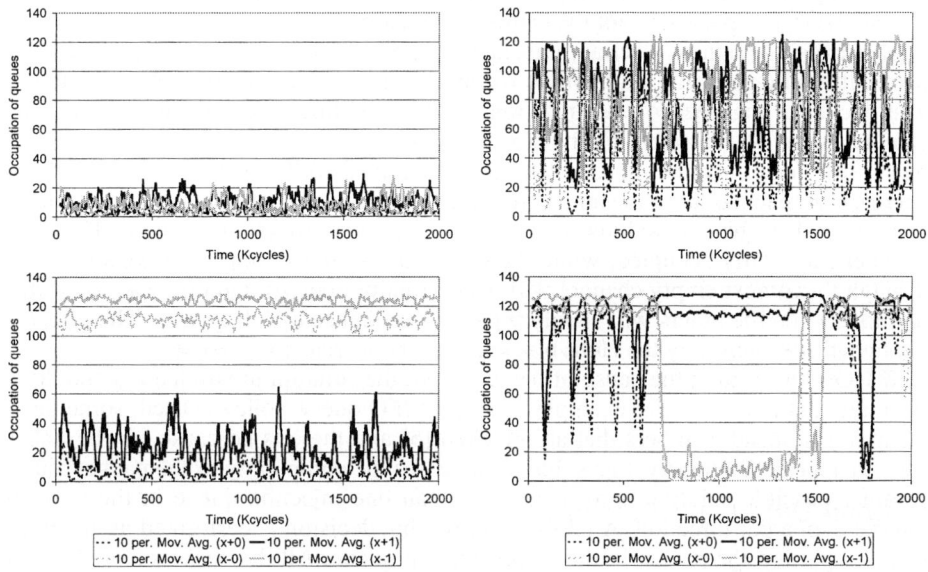

Fig. 2. Queue occupation in rings of 10 (top-left), 20 (top-right), 30 (bottom-left) and 100 (bottom-right) nodes; uniform load at maximum level. Measurements were taken every 2000 simulated cycles; graphs represent the moving average of them. Possible values are between 0 (totally empty queue) and 128 (queue full with 8 packets, 128 phits).

For larger rings, the behavior of the network drastically changes. For a 20-node ring, we observe a fast alternation of the unbalance: for a short time X+ is full while X- is empty but then the situation reverses and X+ is empty while X- is full, for a few cycles before returning to the previous situation. For a 30-node ring this alternation does not happen: almost immediately the unbalance appears and stays for all the simulation time: channels X- are saturated, while channels X+ have an occupation of about 30 phits, 1/4 of their capacity. The graph for a 100-node ring shows yet another scenario, which is actually a combination of the previous two: for 500 Kcycles channels X+ utilization oscillate, but below saturation, while X- are saturated; then, for about 1000 Kcycles, X+ are saturated while X- are less occupied. The last (rightmost) part of the graph shows how the unbalance then reverses. At any rate, in none of the experiments performed with rings over 20 nodes have we seen the anomaly disappearing spontaneously.

4 Causes of the Unbalance and Measures to Avoid It

We hypothesize that the cause of this unbalance is the Head-of-Line blocking of packets in the injection queue of the router. Packets injected in the network from a computing element (the only source/sink of traffic) have to go through the injection queue associated to the network interface, as shown in Fig. 1. Let us imagine that the packet in the head of the injection queue has to be routed through X+ (because the routing algorithm orders this) and this channel is saturated (actually, both X+0 and X+1 are saturated). The packet must wait. However, if the next packet in the queue has to be routed through X-, it must (unnecessarily) wait. Under uniform traffic, the probability of this situation arising is the same for X+ blocking X- as for X- blocking X+, so the traffic pattern is not the cause of the anomaly.

However, as soon as a certain unbalance happens (which is something we can expect with uniform traffic), a positive feedback effect exists that provokes instability. The packet at the head of the injection queue is willing to enter into the saturated channel and is waiting (forcing the next packets in queue to wait too). Meanwhile, the other (less busy) channel does not receive additional traffic, so part of its load will decrease as packets are consumed. As soon as the saturated channel has room for another packet, the one waiting uses it – so the channel is saturated again. Thus, one channel stays fully occupied, while the other one is almost empty. Packets traveling through the almost empty channel will, when finally injected, rapidly reach their destination, while the others will slowly traverse a sequence of full queues. The effect of the unbalance caused by the HOL blocking will be increasingly worse.

If HOL blocking is the cause of the anomaly, the solution to eliminate it should be avoiding this blocking. For example, if a non-FIFO queue policy is used, some packets may get ahead of others that are blocked [5] and there is no cause for the unbalance to appear. Obviously, this solution increases the complexity of the packet router. Another possible solution is to have more than one injection queue in the network interface: one per direction. A preliminary routing decision is performed at this interface, by putting the packet in the corresponding queue. All the packets stored in one of these queues will follow the same direction, so there is no HOL blocking. This solution, in addition to increasing the complexity of the interface, requires sufficiently long injection queues; otherwise, the HOL blocking would reappear in higher levels when queues are full.

A trivial solution would be to simply drop the packet causing the HOL blocking (the one at the head of the injection queue that cannot be injected). From a practical point of view this is not a valid solution, because it will force higher levels in the communication protocols (or even applications) taking care of recovering lost packets, and this will drastically reduce performance. However, it is perfectly applicable when dealing with a simulation and with synthetic workloads.

5 Studies with 2D and 3D Tori

A 2D torus (k-ary 2-cube) consists of a set of rings in the X dimension (X-rings) and another set in the Y dimension (Y-rings). In a simulation study of this network, we have observed that queue usage in the X-rings is almost identical to that of rings (section 3). However, unbalance does not appear in Y-rings. This is due to the policy followed to make packets advance: for adaptive channels, it is not compulsory to inject in an X-ring, but there is a certain preference to do so (X-rings are tested for availability before Y-rings). In the Escape channels packets advance following dimensional order routing (DOR), so under saturation (when Escape channels are more heavily used) most packets will necessarily be injected into an X-ring. Furthermore, Escape channels in Y-rings have more injectors, because they accept traffic from X-rings; thus, the probability of the HOL blocking happening is lower than in X-rings. Fig. 3 plots the queue occupation in an X-ring and in a Y-ring of a 30x30 torus. As we stated before, the graph for the X-ring looks like the one for the 30-node ring. However, curves for the Y-ring show a different, but very reasonable behavior: adaptive channels Y+1 and Y-1 are heavily (and equally) used. Escape channels Y+0 and Y-0 are less heavily (but equally) used.

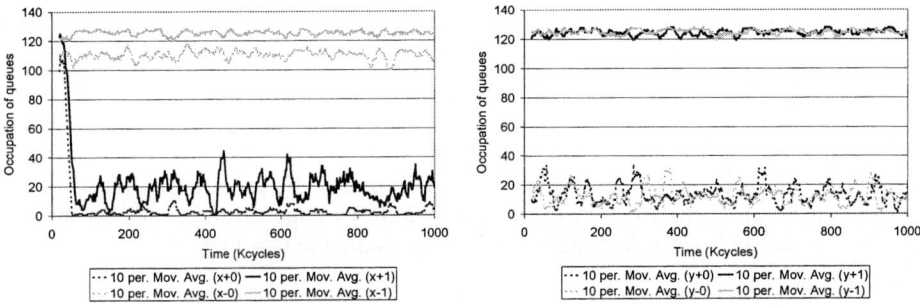

Fig. 3. Queue occupation in a 30x30 torus for the X-rings (left) and Y-rings (right) under maximum uniform load. Measurements were taken every 2000 simulation cycles; graphs represent the moving average of them. Possible values are between 0 (totally empty queue) and 128 (queue full with 8 messages, 128 phits).

What happens now with network performance? Fig. 4 (left) shows accepted vs. offered load for four variants of 30x30 torus with and without using any mechanism to avoid HOL blocking. We observe that using two adaptive VCs (2CVA) improves performance (compared with just one (1CVA)). However, the unbalance in the X-rings does cause drops in performance in both cases – although in the case "2CVA" this happens when applying a slightly higher load. We also observe that the elimination of HOL blocking allows a high level of accepted load to be reached (close to the theoretical maximum) and, what most importantly, maintained under applied loads above the saturation point. Even more noteworthy is the impact that this unbalance has on the average delay experienced by packets when traversing the network, as well as on the standard deviation of this delay. Fig. 4 (right) shows that, avoiding HOL blocking, std. dev. of packet delay drops to about 1/3 of its original value. It is easy to understand that the existence of one saturated path while the other is almost empty causes a great dispersion of the number of cycles that packets require to reach their respective destinations: while some packets experience an empty network, others have to compete to pass through a collection of saturated channels.

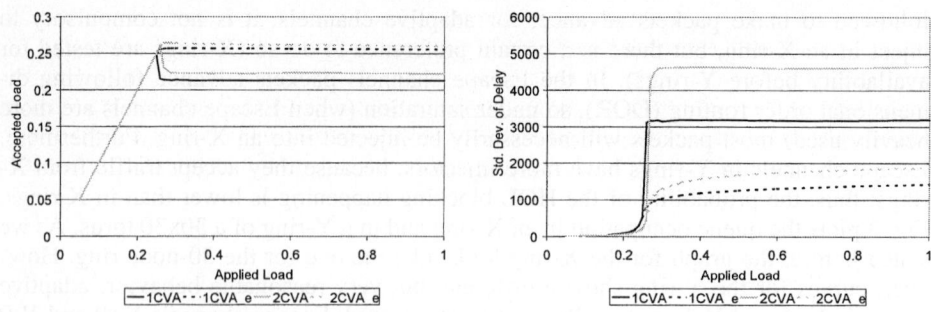

Fig. 4. Left: Accepted vs. applied (uniform) load in a 30x30 torus for 1 and 2 VCs (in addition to the Escape VC) and with/without avoiding HOL blocking. Right: standard deviation of the average packet delay in this network.

For 3D tori the unbalance also appears in the X-rings when their sizes surpass the threshold already stated (around 20). An obvious consequence is that this phenomenon can only be observed in very large networks (around 8000 nodes). As an example, the previous 30x30 torus (900 nodes) suffers from this effect, while a similar 3D torus (10x10x10, or 1000 nodes) does not. In addition to the better topological characteristics (in terms of network bisection bandwidth, average distance, and so on) this may be another reason to make 3D torus the most suitable topology.

6 Unbalance Using Actual Applications

This section proves that the unbalance is not a consequence of using a synthetic uniform load. This unbalance is also present under real applications traffic load. Using a simulator of multiprocessor systems [6] integrated with the SICOSYS simulator of interconnection networks [8], we have performed an execution-driven simulation of the Radix application (part of the SPLASH-2 benchmark suite). Fig. 5 shows the results of measuring each 5000 cycles queue occupation of each virtual channel's transit queues. Packet length is 20 phits; queue capacity is 4 packets.

Fig. 5 shows the results for an application characterized by its uniform traffic pattern such us radix. Between cycles 1M and 3M channel X-1 (adaptive) is more heavily used than X+1. This effect is particularly visible around cycle 2.5M. Additionally,

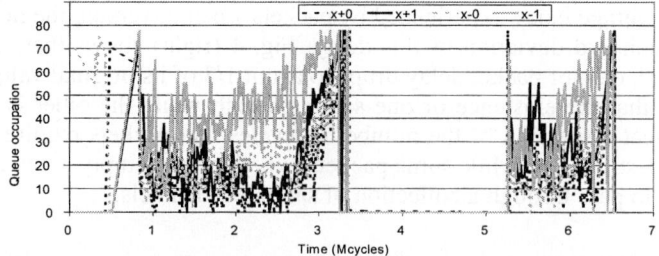

Fig. 5. Occupation of queues when running the Radix application on a 30-node ring with 512K integer keys. Data taken from an execution-driven simulation combining RSIM and SICOSYS.

around cycle 6M we can observe the alternation phenomenon previously described for rings. The unbalance cannot be attributed to the characteristics of the Radix application (it is not true that nodes in Radix send more data towards one direction that towards the other), because for the time periods represented in the figure the application interchanges keys in a highly random, uniform way. Thus, this behavior confirms our hypothesis about the occurrence of the anomaly not only with synthetic traffic but also with actual applications.

7 Conclusions

We have identified and analyzed the load unbalance that appears in bidirectional rings of k-ary n-cubes when uniform traffic is applied, which is caused by the HOL blocking in the injection queues. We show that the elimination of this blocking allows throughput to be sustained at its peak level and, additionally, may reduce the standard deviation of network latency by up to 30%. This unbalance appears under different deadlock-avoidance techniques, in 1, 2 and 3-D networks, but only for sizes over 20 nodes per dimension, and only in the rings where most packet injections are performed (usually, the rings in the X axis). Finally, we have shown that the anomaly can also be present with traffic generated by real applications.

References

1. NR Adiga, GS Almasi, Y Aridor, M Bae, Rajkishore Barik, et al., "An Overview of the BlueGene/L Supercomputer", Proc. of SuperComputing 2002, Baltimore, Nov. 16-22, 2002
2. W. J. Dally and C. L. Seitz, "Deadlock-free message routing in multiprocessor interconnection networks", IEEE Transactions on Computers, vol. 36, no.5, pp. 547-553, May 1987
3. J. Duato. "A Necessary and Sufficient Condition for Deadlock-Free Routing in Cut-Through and Store-and-Forward Networks". IEEE Trans. on Parallel and Distributed Systems, vol. 7, no. 8, pp. 841-854, 1996
4. J. Duato, S. Yalamanchili, and L. Ni, Interconnection networks. "An engineering approach", IEEE Computer Society, 2003
5. G. L. Frazier and Y. Tamir, "Dynamically-Allocated Multi-Queue Buffers for VLSI Communication Switches". IEEE Trans. on Computers, vol. 41, no.6, pp. 725-737, June 1992
6. V.S.Pai, P. Ranganathan, and S.V.Adve, "RSIM: An Execution-Driven Simulator for ILP-Based Shared-Memory Multiprocessors and Uniprocessors". IEEE TCCA New., Oct. 1997
7. V. Puente, C. Izu, R. Beivide, J.A. Gregorio, F. Vallejo and J.M. Prellezo, "The Adaptive Bubble Router", J. of Parallel and Distributed Computing. Vol 61, no. 9, September 2001
8. V. Puente, J.A. Gregorio, R. Beivide, "SICOSYS: An Integrated Framework for studying Interconnection Network in Multiprocessor Systems", Euromicro Workshop on Parallel and Distributed Processing, 2002

Understanding Buffer Management for Cut-Through 1D Rings

Cruz Izu[1] and Ramon Beivide[2]

[1] School of Computer Science, The University of Adelaide,
Adelaide SA 5001, Australia
cruz@cs.adelaide.edu.au

[2] University of Cantabria, Avda Los Castros s/n,
39005 Cantabria, Spain
mon@atc.unican.es

Abstract. This paper describes the impact that buffer management has on network performance for a cut-through 1D ring. Such network provides only one routing alternative between each pair of nodes. The network is kept simple to illustrate the significance of buffer management for all virtual cut-through networks. Besides, 2D and 3D torus, among other networks, can be seen as a collection of interlinked 1D rings. The simulation results will show that the key to maximum sustained throughput is to reserve half of our buffer capacity for transit packets. This increases link utilization by maximising packets at the sending node and empty buffers at the receiving node.

1 Introduction

Ring networks have been widely used to implement local, campus and metropolitan area networks. Extensions of 1-D rings such as Chordal Rings of degree three and four have been considered in the literature [2]. Popular extensions of the ring are the 2-D and 3-D Torus, frequently used to interconnect highly-coupled parallel systems [11].

An ideal router has infinite buffer space but will only buffer a packet when the output ports to which the packet is routed are busy. In real life, buffers are finite and have finite bandwidth. Buffer capacity in the order of a few packets per input link is sufficient to cope with the traffic variations of a non-saturated network [5]. Further buffer capacity has little impact on network response.

Most routers use input FIFO queues because of their simple implementation that translates into low access time and larger buffer capacity [6]. On the other hand, buffers may be better used in centralized or output queues, with the additional advantage of eliminating the head-of-line blocking (HOB) exhibited by input FIFO queues.

Full link utilization is achieved *if* we always have a packet requesting the link and *if* the flow control mechanism allows that packet to use the link. Most of the network research has focused on issuing the maximum number of requests from their incoming packets, by finding organizations that eliminate HOB within a reasonable cost in terms of router delay [13][9][8]. We should emphasize, though, the dual role played by the router buffer space. Maximum link utilization relies also on keeping the buffers non-full as to provide the guaranteed space required for packets under cut-through to progress. Hence, the goal of this paper is to understand how to best use of the network buffer space.

We have considered a simple 1D network with oblivious routing as there are only two possible outputs: the next router or the destination node. Consequently, the router architecture is very simple. We will see that in such simple scenario, buffer management can dramatically impact on network throughput. Blocked packets will release one buffer as they move to another one in the adjacent node. Thus, the variations on buffer utilization are mainly due to the allocation of buffers to new messages as they are injected from each source node. Injection policies that maintain buffer occupancy around 50% will maximize throughput for any ring and buffer size. This result provides an invaluable insight into the design of VCT interconnection networks able to reach and sustain high throughput under heavy loads with practically no additional hardware cost.

The rest of the paper is organized as follows. Section 2 presents the network under study, describes its router node and its theoretical performance limit. Section 3 presents a performance evaluation of the ring network obtained by a cycle-driven simulator. Finally, Section 4 collates the contributions of this paper.

2 Ring Networks

A ring is a collection of point-to-point links that connect N nodes in a loop as shown in figure 1. The links can be unidirectional or bidirectional. Routing in a ring is very simple: a packet is forwarded from one node to the next until it reaches its destination. In the case of bidirectional rings, at injection time the packet is sent in the direction of its minimal path.

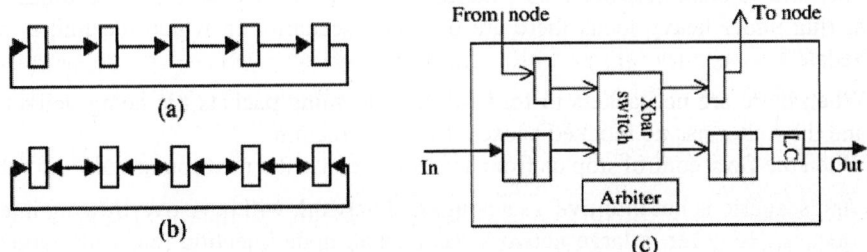

Fig. 1. Ring topology for N = 5 with (a) unidirectional and (b) bidirectional links. (c) ring node for unidirectional ring.

We can view a bidirectional ring as a set of two unidirectional rings. Thus, we will first study the simplest unidirectional ring. Figure 1(c) shows the router node for such a network. A ring router has a very simple architecture: it accepts messages from both its neighbour node (or nodes if bidirectional) and its local injection interface and according to their destination the routing logic selects whether to forward or deliver it. The crossbar switch connects the input and injection buffers to the output and delivery buffers. Finally, the link controller implements VCT flow control.

As the ring topology contains a cycle, it is possible for this network to reach deadlock when all its nodes have full input and output buffers and none of their messages have reached their destinations yet. No message can be forwarded; all of them are waiting for a packet buffer to become empty.

Network deadlock can be prevented by using virtual channels to break this cyclic dependency. This method, initially proposed for wormhole torus networks[4], has been widely used in multicomputer systems such as the Cray T3D [11].

Deadlock can also be avoided if each node is prevented from filling its local buffers as described in [10]. This means that packets at the injection queue must satisfy more restrictive flow control conditions than VCT (the existence of two free packet buffers instead of one). A switching technique based on this mechanism has recently been implemented in the BlueGene/L supercomputer [1]. In this paper we present results using the latter although similar responses were observed with the former method.

2.1 Performance: Theoretical Limits

The goal of a good design is to achieve high throughput, close to 100% link utilization at high loads, while providing fast switching at low loads. Maximum throughput under random uniform traffic is limited by the network bisection bandwidth - the minimum number of channels that must be removed to partition the network into two equal sub-networks [5]. In a unidirectional ring of N nodes, the maximum injection rate is $4/N$ flits/cycle/node.

Real networks, though, exhibit peak throughputs in the range 70 to 90% of this theoretical limit. If we know the causes behind the 10 to 30% performance loss, we may be able to push network performance closer to its limit. In fact, in a unidirectional ring under heavy loads there are only two scenarios in which the link is not utilized:

1) When there are no packets to forward: the incoming packets are being delivered and the router has no blocked packets for that direction.
2) When the flow control stop us from sending because the next router buffer is full

The first scenario is not intuitive, as a congested network will have overflowing injection queues. However, a large network has a small node injection rate with variable arrival rates, so it is possible that while the router is delivering one packet it has no new packets to keep the output link busy. This effect was identified and quantified by Pertel [14] and amounts for a small percentage, which is a function of the network size N; for a small N, such as 8, the maximum link utilization is 85% but for a large N such as 128, the value obtained (98.4%) is close to full utilization. In any case, this effect does not depend on the router design.

The second scenario can be prevented by managing the buffer space so that it not likely to be full. In other words, buffer resources should be allocated in such a way that they provide enough packets to keep the link busy most of the time as well as reducing the likelyhood of becoming full. Transit packets release one buffer at the leaving node, while occupying another at the receiving node (in our router, a packet buffer is available to receive the next packet as soon as the packet header leaves). Therefore, our buffer management will focus on restricting access to new messages, so that buffers are not overflowed.

There is a small body of research on restrictive injection in low degree networks. Stamatopoulos[12] proposes a congestion control mechanism which uses control packets to implement barriers in a cut-through 2D mesh. Lopez [7] restricts message injection in a wormhole torus network in order to reduce the probability of deadlock. Dally and Aoki [3] throttle injection in a wormhole mesh by limiting the number of virtual channels a new packet can request. All of them show that holding back packets at congested loads is a successful strategy regardless of flow control. Therefore, we want to understand why and how limited injection works under VCT.

3 Evaluation of Ring Injection Policies

In order to evaluate buffer management policies in a ring network, we have used a cycle driven simulator which models the basic router node described in section 2.1.

To simplify the model and eliminate trailing effects the packet size is set to 1 *phit*. Simulations with larger message size exhibit similar trends so they won't be shown. Our injection policy will be based on local buffer occupation. We have set the buffer size to 8 packets per physical link We will vary the free local buffer space required to inject a packet from the minimum, 2, to 7. The ring size varies from 16 to 128 nodes. The traffic pattern is random or uniform traffic: each node sends messages to any other node with equal probability. We should note that in a 1D ring, head-of-line blocking only occurs when a packet at the input FIFO, which is to be delivered, prevents another packet to move to the output queue. As the processor node is model as an ideal sink, this will never occur for the unidirectional ring.

As the goal is to increase link utilization, we will examine network performance versus offered load. We will measure the average network population and its impact on peak throughput at and beyond saturation. We will not show values for network latency as it does not provide any further insight.

3.1 Applying Restrictive Injection to Unidirectional Rings

Figure 2 shows the throughput as a function of the offered load achieved with and without additional restrictive injection. The acronym R64-f4 means a simulated ring of 64 nodes allowing packet injection into an output channel when it has space for four packets. Note the minimum under bubble routing is two packets (f=2).

Looking at network response for that base case, its peak throughput ranges from 80% for the 16-node ring to 90% for the 128-node ring.

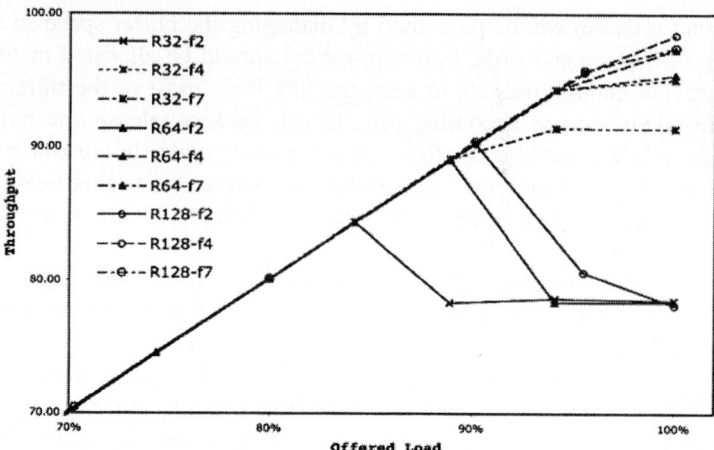

Fig. 2. Throughput versus offered load for unidirectional rings of variable size under increasingly restrictive injection policies.

Table 1. Ring population and buffers occupation at peak and full network loads.

Ring size	16	32	64	128
Peak Throughput	80 %	85 %	89 %	90 %
Net pop	48	91	198	367
% occupied	0.38	0.35	0.39	0.36
Throughput (100% Load)	79 %	79 %	78 %	78 %
Net pop	109	223	455	918
% occupied	0.85	0.87	0.89	0.90

However, with 100% offered load throughput degrades down to 78%. To explain this loss we refer to buffer occupation at and beyond saturation as shown in Table 1. We can see that rings saturate with loads that use approximately 35% of buffers regardless of network size. In all cases, ring population soars beyond saturation. Note that the deadlock avoidance method prevents new packets from filling the last free buffer, thus the packets queuing at the node interface will access 7 of the 8 local buffers (87% buffer capacity). This reduces the mobility of transit packets – compare buffer occupations at peak and full loads - and results in 10% throughput loss. These results indicate the need for an injection policy that reserves additional buffer space for packets already in transit.

Figure 3 shows the impact of restrictive injection policies on maximum sustained throughput. For any node size, reserving an additional free buffer result in a significant performance gain. On the other hand, reserving 7 buffers reduces peak throughput down by 5% by preventing too many packets from entering the ring so that the link is slightly under-utilized. Note that in this network all packets in transit request the same output link, so a low population of one to two packets per link is enough to keep the link fully utilized.

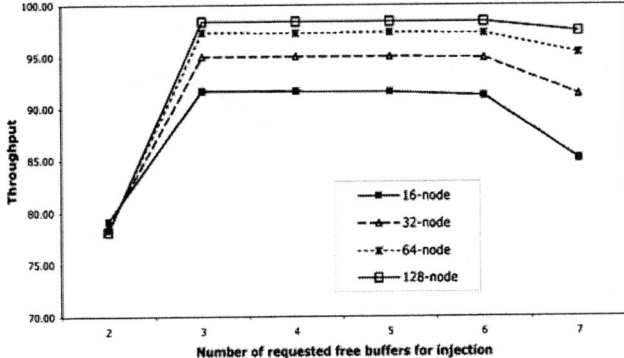

Fig. 3. Maximum sustained throughput as a function of the injection policy for rings of size 16 to 128 nodes.

We should note that holding packets back does not only eliminates the 10% performance drop but it results in higher peak throughput as well. For example, a 128-node ring is saturated at 95% load, delivering only 80% as shown in figure 2. However, if we hold some packets back at the node interface, the network does not saturate. Besides, at full load is able to deliver not only 95% but up 98.4% of packets.

The network behaviour as seen by the network interface when reserving 3 to 6 free packets for transit traffic is remarkable similar in terms of throughput and latency. For example at 90% load, average latency is either 206 or 207 cycles for any of the above policies. With R128-f3, the packet will spent on average 4 cycles waiting at the injection frame and 202 traveling towards it destination. With R128-f6 it will wait longer, 27 cycles, but it will travel faster as well, 180 cycles. There is also a clear difference in buffer occupation, ranging from 70% for R128-f3 to 35% for R128-f6. Note the latter figure matches the occupation at peak loads. So, we are not reducing the number of packet required to keep the links busy in spite of the restrictive conditions. Furthermore, the network delivers more packets if is kept nearly empty than if is allowed to flood all but one packet.

3.2 Applying Restrictive Injection to Bidirectional Rings

This section evaluates the performance of bidirectional rings. We have provided two injection queues, one per direction, so the main difference with the previous network is that messages travel shorter paths. Besides, they share access to the delivery channel. Thus, is not surprising that the network behaviour is similar to that observed for its unidirectional counterpart (refer to fig 2).

Figure 4 shows the network performance as a function of offered load for three bidirectional rings of size 32, 64 and 128 nodes. As the ring size grows, so it is does its peak throughput. However, for loads above saturation, throughput performance degrades back to 70%, a drop of up to 20%.

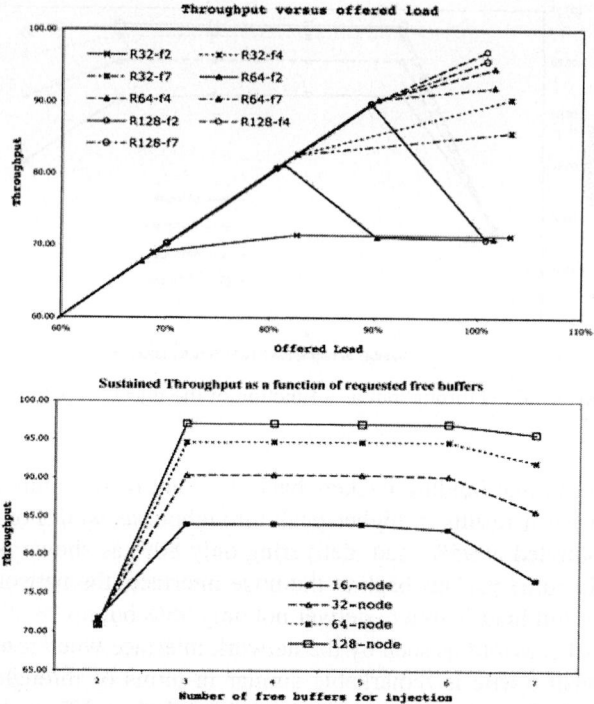

Fig. 4. Throughput for bidirectional rings under increasingly restrictive injection policies.

Again, requesting more free buffers before entering the network not only eliminates the drop but it also achieves slightly higher throughput. As Figure 4 shows, the networks works better if is prevented from filling up its buffers. Note that in all cases is better to be too restrictive (f=7) than too generous (f=2). Thus, a safe bet for any network size will be to reserve half of the buffer capacity to transit packets.

We can conclude that any given ring has an optimal population that allows the network to deliver maximum throughput. The population should be large enough to use all the network links, cycle per cycle, but small enough so that it maintains free space at each input buffer. For the network under study, a population of 2 to 3 packets per node is sufficient to keep the links busy. The remainder buffer is used more effectively when reserved for transit traffic.

4 Conclusions

In this paper we have explored injection policies that limit packet population in the network as a strategy to increase and sustain network peak throughput. We have seen that buffers are a critical resource in a virtual cut-through network. To utilize a network link we require a packet in the sending node and a free buffer in the receiving node. At heavy loads, buffer allocation should restrict new packets from accessing the scarce free buffers, which are compulsory to drain the network.

This work presents the first study of the impact that packet population has on peak throughput. Our simulations show that a ring work best when reserving at least 3 packet buffers to hold transit packets while the remaining buffers are used sporadically to cope with temporary congestions. The understanding gained from simple 1D rings will allows us to develop better buffer management policies for cut-through networks such as 2D and 3D torus currently used to built parallel systems.

Acknowledgments

This work has been supported by the Spanish CICYT TIC2001-0591-C02-01.

References

1. NR Adiga, GS Almasi, et al., "An Overview of the BlueGene/L Supercomputer", Proc. of SuperComputing 2002, Baltimore, Nov. 16-22, 2002.
2. B.W. Arden and H. Lee, "Analysis of Chordal Ring Networks". *IEEE Transactions on Computers*, vol 30 no. 4 pp. 291-295. April 1981.
3. W.J. Dally, H. Aoki, "Deadlock-Free Adaptive Routing in Multicomputer Networks Using Virtual Channels". IEEE Transactions on Parallel and Distributed Systems vol 4 no. 4 pp. 466-475. April 1993.
4. W.J. Dally, C.L. Seitz, "The Torus Routing Chip" *Distributed Computing* vol **1** pp. 187-196. 1987.
5. J. Duato, S. Yalamanchili and L. Ni, "Interconnection Networks: an engineering Approach", IEEE Computer Society Press, 1997.
6. M.J. Karol, M. G. Hluchyj, S. P. Morgan, Input Versus Output Queuing on Space Division Packet Switch, IEEE Transactions On Communications, vol **35** no. 12, pp.1347-1356.
7. P. López, J. M. Martínez, J. Duato, and F. Petrini, "On the Reduction of Deadlock Frequency by Limiting Message Injection in Wormhole Networks. *PCRCW Lecture Notes in Computer Science* 1417: pp. 295-307, 1997.
8. R. Sivaram, C.B. Stunkel, and D.K. Panda, "HIPQS: a High-Performance Switch Architecture Using Input Queuing", Proc. IPPS/SPDP'98, March 1998.
9. M. Katevenis, P. Vatsolaki, and A. Efthymiou, "Pipelined Memory Shared Buffer for VLSI Switches", ACM SIGCOMM, August 1995.
10. V. Puente, C. Izu, J.A. Gregorio, R. Beivide, and F. Vallejo, "The Adaptive Bubble Router", J. of Parallel and Distributed Computing. vol 61, no. 9, pp. 1180-1208.September 2001.
11. S.L. Scott and G. Thorson, "The Cray T3E networks: adaptive routing in a high performance 3D torus", Proc. of Hot Interconnects IV. 1996.
12. J. Stamatopoulos and J. A. Solworth, "Universal Congestion Control for Meshes". 7th Annual ACM Symposium on Parallel Algorithms and Architectures. SPAA '95 (165-174) ACM Press 1995.
13. Y. Tamir, and G.L. Frazier (1992), Dynamically-allocated Multiqueue buffers for VLSI Communication Switches, IEEE Trans. on Computers, vol 41 no. 2 pp. 725-737. 1992.
14. M.J. Pertel, "A Critique of Adaptive Routing" Technical Report Caltech-CS-TR-92-06. 199.

Topic 15
Mobile Computing

Sajal K. Das, Jiannong Cao, Jie Wu, and Gianluigi Ferrari

Topic Chairs

The rapid growth of small, portable and powerful computing devices along with wireless, mobile and sensor communication technologies offers tremendous opportunities for wireless networking and mobile/pervasive computing applications and services. This topic of the EUROPAR 2004 conference aims at bringing together computer scientists and engineers from academics and industry working in these exciting areas, and share their ideas and results with their peers.

The response to the call for papers was very good, attracting 22 papers from nine countries. Each paper was reviewed by three independent referees. After careful discussions on the reviews and judging the originality and the suitability of the topics, the topic chairs selected only 6 high quality papers for presentation at the conference. We sincerely thank all the reviewers for their help and timely reviews, as well as all the authors for submitting their fine work for consideration.

The accepted papers deals with cutting-edge research in various aspects of mobile computing systems including algorithmic issues, end-to-end protocols, multimedia applications and service-based architectures in mobile wireless systems, ad hoc and sensor networks, and wireless LANs.

Coverage and Connectivity in Networks with Directional Sensors
(Extended Abstract)

Evangelos Kranakis[1], Danny Krizanc[2], and Jorge Urrutia[3]

[1] School of Computer Science, Carleton University,
Ottawa, Ontario, K1S 5B6, Canada*
[2] Department of Math and Comp. Sci., Wesleyan University,
Middletown CT 06459, USA
[3] Cubiculo 222, Instituto de Matematicas,
Universidad Nacional Autonoma de Mexico, D.F., Mexico City, Mexico

Abstract. We consider the problem of providing full coverage of a planar region with sensors. Likewise, we consider the connectivity of the sensor network of directional antennas formed by sensors in this region. Suppose that n sensors with coverage angle (also known as beam width) $\alpha(n)$, and reachability radius $r(n)$ are thrown randomly and independently with the uniform distribution in the interior of the unit square. Let $p(n)$ be the probability that a given sensor is active. We prove that if $p(n) = \Omega\left(\frac{\log(n/r(n)^2 \sin^2(\alpha(n)/4))}{nr(n)^2 \alpha(n) \sin(\alpha(n)/4)}\right)$ then the probability the sensors provide full coverage of the unit square is at least $1 - n^{-O(1)}$. Likewise, we consider the connectivity of the resulting sensor network. We show that if $p(n) = \Omega\left(\frac{\log(n/r(n)^2 \sin^2(\alpha(n)/4))}{nr(n)^2 \alpha(n) \sin^2(\alpha(n)/4)}\right)$ then the probability that a connected subnetwork of sensors provides full coverage is at least $1 - n^{-O(1)}$.

1 Introduction

Sensors are low-power communication and sensing devices that can be embedded in the physical world and spread throughout our environment (see Kahn et al [7], Estrin et al [5], Hollar [6]). Large scale sensor networks are formed by sensors that can be automatically configured after being dropped over a given region. Such networks are deployed in order to handle various services over remote and/or sensitive regions, including monitoring, structural stability of buildings, quality of air, etc, and form the basis of numerous military and civilian applications. It is expected that the cost of such devices will drop significantly in the near future (see the study in [11], Agre et al [1], and Warneke et al [13]). Sensor nodes enable autonomy, self-configurability, and self-awareness. Sensor networks are built from such sensor nodes that create spontaneously an adhoc network. They

---

* Research supported in part by NSERC (Natural Sciences and Engineering Research Council of Canada) and MITACS (Mathematics of Information Technology and Complex Systems) grants.

can assemble themselves automatically, adapt dynamically to failures, manage movement, react to changes in network requirements, etc.

In this paper we are concerned with sensor networks formed in an adhoc manner from unreliable sensors dropped, e.g., from an airplane, on a given planar region, e.g., a unit square region. A sensor is a device that can sense events within a certain reachability radius if they are within a given coverage angle. The purpose of such a sensor network is to report events, e.g., to a certain base station located within the unit square region. To ensure that all events within this region are reported, it is important that coverage of the entire region is guaranteed as well as that the resulting sensor network is connected. There have been several studies on routing and location identification in adhoc networks in which sensors use only constant memory and local information. Several such interesting studies include Bose et al [2], Braginsky et al [3], Doherty et al [4], Kranakis et al [8], Kuhn et al [9], Meguerdichian et al [10], Ye et al [14]).

The first study on the coverage and connectivity problem we consider in this paper appeared in Shakkottai et al [12] for a restricted model in which the sensors are assumed to occupy the vertices of a square grid the size of a unit square, the reachability radius of the sensors is r, and the angular coverage of the sensors is 360 degrees. We follow closely the model of Shakkottai et al [12]. To accommodate faults we assume there is a probability p that a given sensor is active. However our sensor model is more general. First, unlike Shakkottai et al [12] where sensors may occupy only the vertices of a square unit grid, our sensors may occupy any position inside the unit square randomly and independently with uniform distribution. In addition to the position of the sensors, their orientation is also random and may occupy any angular position in the range from 0 to 360 degrees. Second, our sensors have limited angular visibility. Thus our sensors can be thought of as flood-lights with a given coverage angle of size α that can see every point in the part of the region subtended by their visibility angle, and reachability radius r. Equivalently, we can think of the sensors as antennas that can communicate provided they are within range and angle of each other.

Consider n sensors in the interior of a unit square. In order to study the limits of the coverage and connectivity problem we find it convenient to parametrize the characteristics of the sensors as a function of n. More specifically, let $\alpha := \alpha(n)$ be the coverage angle, and $r := r(n)$ the reachability radius of the sensor, respectively. Let $p := p(n)$ be the probability that a given sensor is active, in which case $1 - p(n)$ is the probability that the sensor is inactive.

We are interested in the problem of using sensors in order to cover every point of the given region, which for simplicity is assumed to be a unit square in the plane over which sensors are dropped, say from an airplane. More specifically we can formulate the following problem.

Problem 1. n sensors (i.e., directional antennas) are dropped randomly and independently with the uniform distribution (i.e., in position and rotation) over the interior of a unit square. What relation between p, α, n must exist in order to ensure that surviving sensors provide full coverage of the unit square? interior of the unit square?

We want to ensure that the network of directional sensors is connected, i.e., there is a communication path between any pair of sensors in the network. Note that since the placement and orientation of the sensors is random there is a reasonable probability that some sensors will be unable to communicate with any other sensors. Rather than achieve full connectivity we require that a connected subnetwork exists that achieves full coverage. In particular, we can formulate the following problem.

Problem 2. n sensors (i.e., directional antennas) are dropped randomly and independently with the uniform distribution (i.e., in position and rotation) over the interior of a unit square. What relation between p, α, n must exist in order to ensure the existence of a connected subnetwork of sensors that provides full coverage?

Our approach is sufficiently general to solve the coverage problem for a unit square region when the sensors can be placed only on its perimeter. In this case we assume that the reachability radius of the sensors is equal to 1. More specifically we consider the following problem.

Problem 3. n sensors (i.e., directional antennas) of *reachability radius* 1 are dropped randomly and independently with the uniform distribution (i.e., in position and rotation) over the perimeter of a unit square. What relation between p, α, n must exist in order to ensure the surviving sensors provide full coverage of the interior of the unit square?

1.1 Results of the Paper

We show that if $p(n) = \Omega\left(\frac{\log(n/r(n)^2 \sin^2(\alpha(n)/4))}{nr(n)^2 \alpha(n) \sin(\alpha(n)/4)}\right)$ then the probability that the sensors provide full coverage of the unit square is at least $1 - n^{-O(1)}$. Conversely, if the sensors provide full coverage of the unit square with probability at least $1 - 1/n$, then $p(n) = \Omega\left(\frac{\log(n/r(n)^2)}{nr(n)^2 \alpha(n)}\right)$. Likewise, we consider the probability that the resulting sensor network is connected. We show that if $p(n) = \Omega\left(\frac{\log(n/r(n)^2 \sin^2(\alpha(n)/4))}{nr(n)^2 \alpha(n) \sin^2(\alpha(n)/4)}\right)$ then the probability that a connected subnetwork of sensors provides full coverage is at least $1 - n^{-O(1)}$.

2 The Network of Sensors

Consider two sensors at A, B. If the coverage angle is 2π and A is reachable from B then also B is reachable from A. Clearly, this makes the sensor network an undirected graph. The situation is different when the coverage angle α of the sensors is less than 2π. Two sensors at A and B may well be within reachability range of each other but either A or B or both may not be within the coverage angle of the other sensor. Clearly, this makes such a sensor network a directed graph.

2.1 Covering a Circle with a Sensor

Consider a sensor at A. We would like that a circle of radius R fits inside the coverage range of A and is visible from A with an angle of size $\alpha/4$. Let d be the distance of the sensor from the center K of this circle. Since the reachability radius of the sensor is r, the circle at K is entirely visible from the sensor at A if $d + R \leq r$. Since $d = R/\sin(\alpha/4)$ the above inequality is equivalent to $\frac{R}{\sin(\alpha/4)} + R \leq r$. It is now easy to see that the following result is true.

Lemma 1. *A circle of radius R may lie within the coverage range of sensor if and only if*

$$R \leq \frac{\sin(\alpha/4)}{1 + \sin(\alpha/4)} \cdot r. \tag{1}$$

Moreover, the probability that a given sensor at distance d from a given circle of radius R such that $d + R \leq r$ is active and covers the circle, is $\Theta(p\alpha)$. ∎

3 Achieving Coverage

We are interested in specifying conditions on the three main parameters $\alpha := \alpha(n), r := r(n), p := p(n)$ so that the active sensors in the unit square provide full coverage. Using Lemma 1 we are interested in finding sensors that with high probability cover a given set of circles of radius $R = \frac{\sin(\alpha/4)}{1+\sin(\alpha/4)} \cdot \Theta(r)$. Any sensor within a strip of thickness $R/2$ at distance $\Omega(r)$ can potentially cover a circle of radius R. The probability that a given sensor within this strip will cover fully such a circle is at least $\Omega(p\alpha)$. The expected number of sensors within the strip at distance $\Omega(r)$ from A is at least $\Theta(nrR)$. It follows that up to a constant

$$\Pr[\text{An active sensor in the strip covers the circle at } A]$$
$$= 1 - \Pr[\text{No active sensor in the strip covers the circle at } A]$$
$$\geq 1 - \prod_{i=1}^{\Theta(nrR)} (1 - \Omega(p\alpha)) \geq 1 - e^{-\Theta(np\alpha rR)}.$$

It is now easy to see that in order to ensure that $1 - e^{-\Theta(np\alpha rR)} \geq 1 - n^{-O(1)}$ it is enough to assume that $p = \Omega\left(\frac{\log n}{n\alpha rR}\right)$.

To obtain coverage of the whole unit square we decompose the given region into circular overlapping subregions $C_1, C_2, \ldots, C_N$ each of radius R, where $N = \frac{1}{R^2} = \frac{1}{\Theta(r^2 \sin^2(\alpha/4)/(1+\sin(\alpha/4))^2)} \approx \Theta\left(\frac{1}{r^2 \sin^2(\alpha/4)}\right)$. This overlap is necessary in order to guarantee that if each such circle is fully covered by a sensor the whole unit square will then be also covered by the sensor network. Each circular subregion can be covered by a sensor with high probability. Let E_i be the event that C_i can be covered by a sensor. It follows that

$$\Pr[\text{Sensors provide full coverage}] \geq \Pr[\text{Each } C_i \text{ is covered by a sensor}]$$

$$= \Pr\left[\bigcap_{i=1}^{N} E_i\right] = 1 - \Pr\left[\bigcup_{i=1}^{N} \overline{E_i}\right] \geq 1 - \sum_{i=1}^{N} \Pr[\overline{E_i}] \geq 1 - \sum_{i=1}^{N} e^{-\Theta(np\alpha r R)}$$

$$\geq 1 - Ne^{-\Theta(np\alpha rR)} \geq 1 - \frac{1}{\Theta(r^2 \sin^2(\alpha/4))} e^{-\Theta(np\alpha rR)}.$$

It is now easy to see that in order to ensure that $1 - \frac{1}{\Theta(r^2 \sin^2(\alpha/4))} e^{-\Theta(np\alpha rR)} \geq 1 - n^{-O(1)}$ it is enough to assume that $p(n) = \Omega\left(\frac{\log(n/r^2 \sin^2(\alpha/4))}{nr^2\alpha \sin(\alpha/4)}\right)$. We have proven the following theorem.

Theorem 1. *Suppose that n sensors with coverage angle $\alpha(n)$ and reachability radius $r(n)$ are thrown randomly and independently in the interior of a unit square. If the probability $p(n)$ that a given sensor is active satisfies $p(n) = \Omega\left(\frac{\log(n/r(n)^2 \sin^2(\alpha(n)/4))}{nr(n)^2\alpha(n) \sin(\alpha(n)/4)}\right)$ then the probability the sensors provide full coverage of the unit square is at least $1 - n^{-O(1)}$.* ∎

As a special case we also obtain the result of Shakkottai et al [12] for $\alpha = 2\pi$. If $p(n) \geq \Omega\left(\frac{\log n}{nr(n)^2}\right)$ then $\Pr[$Sensors provide full coverage of the unit square$] \geq 1 - n^{-O(1)}$. It is also interesting to look at the reverse problem: if coverage is assured with high probability asymptotically in n, what is a lower bound on the probability $p(n)$ that a sensor is active? Partition the square into N pairwise disjoint circles each of radius r. If any given of these circles has no active sensor (say the circle centered at K) then K cannot be covered by any sensor lying in a neighboring circle. Consider the event E: the center of a circle is covered by an active sensor. It is clear that $p\frac{\alpha}{2} \leq \Pr[E] \leq p\alpha$. There are approximately $N = \Theta(1/r^2)$ such circles. It follows that

$$\Pr[\text{Unit square is covered}] \leq \Pr[\text{Centers of all the circles are covered}]$$

$$= \prod_{i=1}^{N} \Pr[E] = (1 - \Pr[\overline{E}])^N \leq (1 - (1 - \Theta(p\alpha))^{\Theta(nr^2)})^N$$

$$\leq \exp\left(-(1-\Theta(p\alpha))^{\Theta(nr^2)} N\right) = \exp\left(-(1-\Theta(p\alpha))^{\Theta(nr^2)}/\Theta(r^2)\right)$$

It follows from the inequalities above that if $\Pr[\text{Unit square is covered}] \geq 1 - 1/n$ then also $\exp\left(-(1-\Theta(p\alpha))^{\Theta(nr^2)}/\Theta(r^2)\right) \geq 1 - 1/n$, which in turn implies that $\frac{(1-\Theta(p\alpha))^{nr^2}}{r^2} \leq \log\left(\frac{n}{n-1}\right) \leq \frac{1}{n-1}$. It follows that $p = \Omega\left(\frac{\log(n/r^2)}{nr^2\alpha}\right)$. We have proven the following theorem.

Theorem 2. *Suppose that n sensors with coverage angle $\alpha(n)$ and reachability radius $r(n)$ are thrown randomly and independently in the interior of a unit square. If $\Pr[\text{Sensors provide full coverage of the unit square}] \geq 1 - 1/n$ as $n \to \infty$, then the probability $p(n)$ that a given sensor is active satisfies $p(n) = \Omega\left(\frac{\log(n/r(n)^2)}{nr(n)^2\alpha(n)}\right)$.* ∎

3.1 Achieving Coverage Only from the Perimeter

Our approach in this section is sufficiently general to solve the coverage problem for a unit square region when the sensors can be placed only on its perimeter. In this case we assume that the reachability radius of the sensors is equal to 1. We mention without proof the following result whose proof is similar to the proofs of Theorems 1 and 2 above.

Theorem 3. *Suppose that n sensors with coverage angle $\alpha(n)$ and reachability radius 1 are thrown randomly and independently on the perimeter of a unit square. If the probability $p(n)$ that a given sensor is active satisfies $p(n) = \Omega\left(\frac{\log(n/\sin^2(\alpha(n)/4))}{n\alpha(n)\sin(\alpha(n)/4)}\right)$ then*

$$\Pr[\text{Sensors provide full coverage of the unit square}] \geq 1 - n^{-O(1)}. \quad (2)$$

Furthermore, if Inequality 2 is valid then $p(n) = \Omega\left(\frac{\log n}{n\alpha(n)}\right)$. ■

4 Achieving Coverage and Connectivity

Connectivity is not always assured in a network of directional antennas. For example, it can happen that a sensor lying in the convex hull formed by the sensors is such that its angle of coverage points outside of the convex hull. Thus it cannot connect to any other sensor. We resolve this problem by showing in the sequel that there exists a connected subgraph of the graph of sensors that provides complete coverage of the unit square. We also provide an efficient routing algorithm in this setting.

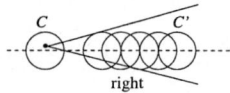

Fig. 1. Finding a sensor in C in order to cover the circle C'.

Consider the configuration decomposing the unit square into circles of radius $R = \frac{\sin(\alpha/4)}{1+\sin(\alpha/4)} \cdot \Theta(r)$. Recall that the coverage range of each sensor is r and consider the radius R. For each circle in this configuration we can find with high probability four active sensors within the circle whose directional antennas fully cover a circle in the *up, down, left, right* directions, respectively. For example, for the two circles at distance $\leq r - 2R$ depicted in Figure 1 we can find with high probability a sensor in C to cover the circle C'.

$$\Pr[\text{An active sensor in } C \text{ covers the circle } C']$$
$$= 1 - \Pr[\text{no active sensor in } C \text{ covers the circle } C']$$

$$\geq 1 - \prod_{i=1}^{\Theta(nR^2)} (1 - \Theta(p\alpha)) \geq 1 - e^{-\Theta(np\alpha R^2)}.$$

It is now easy to see that in order to ensure that $1 - e^{-\Theta(np\alpha r R^2)} \geq 1 - n^{-O(1)}$ it is enough to assume that $p = \Omega\left(\frac{\log n}{n\alpha R^2}\right) = \Omega\left(\frac{\log n}{nr^2\alpha \sin^2(\alpha/4)}\right)$.

The same idea as before can now be used to achieve full coverage of the unit square by a connected subgraph of the resulting network of sensors. The circles must be such that there is sufficient overlap in order to guarantee connectivity. To obtain coverage of the whole unit square we partition the given region into circular subregions $C_1, C_2, \ldots, C_N$ each of radius R, where $N = \frac{1}{R^2} = \frac{1}{\Theta(r^2 \sin^2(\alpha/4)/(1+\sin(\alpha/4))^2)} = \Theta\left(\frac{1}{r^2 \sin^2(\alpha/4)}\right)$. Each circular subregion can be covered by a sensor with high probability. Let E_i be the event that C_i can be covered by a sensor. It follows that

$$\Pr[\text{Sensors provide full coverage}] \geq \Pr[\text{Each } C_i \text{ is covered by a sensor}]$$

$$= \Pr\left[\bigcap_{i=1}^{N} E_i\right] = 1 - \Pr\left[\bigcup_{i=1}^{N} \overline{E_i}\right] \geq 1 - \sum_{i=1}^{N} \Pr[\overline{E_i}] \geq 1 - \sum_{i=1}^{N} e^{-\Theta(np\alpha R^2)}$$

$$\geq 1 - Ne^{-\Theta(np\alpha R^2)} \geq 1 - \frac{1}{\Theta(r^2 \sin^2(\alpha/4))} e^{-\Theta(np\alpha R^2)}.$$

It is now easy to see that in order to ensure that $1 - \frac{1}{\Theta(r^2 \sin^2(\alpha/4))} e^{-\Theta(np\alpha R^2)} \geq 1 - n^{-O(1)}$ it is enough to assume that $p = \Omega\left(\frac{\log(n/r^2 \sin^2(\alpha/4))}{nr^2\alpha \sin^2(\alpha/4)}\right)$.

It remains to prove that a connected subgraph of the sensor network provides full coverage of the unit square. Consider the set S of active sensors that fully cover at least one of the circles of radius R. The previous argument shows that this set of sensors indeed provides full coverage of the unit square. It remains to prove that it is also connected. Let s, t be two arbitrary active sensors in S. It is enough to show that there is a path from s to t with high probability asymptotically in n. By definition of S there is a circle, say C, of radius R which is fully covered by the sensor s. Let C' be a circle of radius again R in whose interior the sensor t belongs. It was shown before that we can find a path from circle C to circle C' like in the Manhattan routing on a two dimensional mesh in order to reach every node at C' (and hence also the target node t). Summing up we have proven the following theorem.

Theorem 4. *Suppose that n sensors with coverage angle $\alpha(n)$ and coverage radius $r(n)$ are thrown randomly and independently in the interior of the unit circle. If $p(n) = \Omega\left(\frac{\log(n/r(n)^2 \sin^2(\alpha(n)/4))}{nr(n)^2\alpha(n) \sin^2(\alpha(n)/4)}\right)$ then the probability there exists a connected subnetwork of sensors which provides full coverage is at least $1 - n^{-O(1)}$.* ∎

As a special case we also obtain the result that for $\alpha = 2\pi$, if $p(n) = \Omega\left(\frac{\log n}{nr(n)^2}\right)$ then $\Pr[\text{A connected subnetwork of sensors provides full coverage}] \geq 1 - n^{-O(1)}$.

5 Conclusion

We have considered the problem of coverage and connectivity of a sensor network over a unit square. We established conditions on the probability that a given sensor is active in order to guarantee that a subnetwork of the resulting sensor network is connected and provides full coverage of the unit square. We also note that a similar approach will work for coverage and connectivity of sensor networks in more general planar regions.

References

1. J. Agre and L. Clare, An integrated architecture for cooperative sensing networks, IEEE Computer, vol. 33, no. 5, May 2000, 106-108.
2. P. Bose, P. Morin, I. Stojmenovic, and J. Urrutia, Routing with guaranteed delivery in ad hoc wireless networks. Wireless Networks, 7(6):609-616, 2001.
3. D. Braginsky and D. Estrin, Rumor routing algorithm for sensor networks, 2001. Available at http://lecs.cs.ucla.edu/ estrin/.
4. L. Doherty, L. E. Ghaoui, and K. S. J. Pister, Convex position esti- mation in wireless sensor networks, in Proceedings of IEEE Infocom, (Anchorage, AK), April 2001.
5. D. Estrin, R. Govindan, J. Heidemann and S. Kumar: Next Century Challenges: Scalable Coordination in Sensor Networks. In Proc. 5th ACM/IEEE International Conference on Mobile Computing – MO-BICOM'1999.
6. S.E.A. Hollar: COTS Dust. Msc. Thesis in Engineering-Mechanical Engineering, University of California, Berkeley, USA, 2000.
7. J. M. Kahn, R. H. Katz, and K. S. J. Pister, Mobile networking for smart dust, in Proceedings of MobiCom 99, (Seattle, WA), August 1999.
8. E. Kranakis, H. Singh, and J. Urrutia, Compass Routing in Geometric Graphs, in proceedings of 11th Canadian Conference on Computational Geometry, CCCG-99, pages 51-54, Vancouver Aug. 15-18, 1999.
9. F. Kuhn, R. Wattenhofer, Y. Zhang, A. Zollinger. Geometric Ad-Hoc Routing: Of Theory and Practice, Proceedings of the 22nd ACM Symposium on the Principles of Distributed Computing (PODC), Boston, Massachusetts, USA, July 2003.
10. S. Meguerdichian, F. Koushanfar, M. Potkonjak, and M. Srivastava, Coverage problems in wireless ad-hoc sensor networks, in Proceedings of IEEE Infocom, (Anchorage, AK), 2001.
11. National Research Council, Embedded, Everywhere: A Research Agenda for Systems of Embedded Computers, Committee on Networked Systems of Embedded Computers, for the Computer Science and Telecommunications Board, Division on Engineering and Physical Sciences, Washington, DC, 2001.
12. S. Shakkottai, R. Srikant, N. Shroff, Unreliable Sensor Grids: Coverage, Connectivity and Diameter, In proceedings of IEEE INFOCOM, 2003, held in San Francisco, March 30 to April 2, 2003.
13. B. Warneke, M. Last, B. Leibowitz, and K. Pister, SmartDust: communicating with a cubic- millimeter computer, IEEE Computer, vol. 34, no. 1, January 2001, 44-51.
14. W. Ye, J. Heidemann, and D. Estrin, An energy-efficient MAC protocol for wireless sensor networks, in Proceedings of IEEE Infocom, (New York, NY), June 2002.

Local Route Recovery Algorithms for Improving Multihop TCP Performance in Ad Hoc Wireless Networks

Zhi Li and Yu-Kwong Kwok*

Department of Electrical and Electronic Engineering
The University of Hong Kong, Pokfulam Road, Hong Kong
ykwok@hku.hk

Abstract. TCP (transmission control protocol) will for sure continue to be the major transport protocol in wireless environments, due to its large install-base in existing applications. Indeed, in future ad hoc wireless networks where devices communicate among each other over multihop routes, TCP is expected to be the prominent protocol for data exchange (e.g., multihop wireless FTP). However, there is a severe performance degradation in using TCP over a wireless link. Despite that there are a large number of wireless TCP adaptation schemes proposed in the literature, improving TCP performance over a multihop wireless route is still very much unexplored. In this paper, we propose local route recovery approaches for improving the performance of multihop wireless TCP. Our simulation results generated by NS-2 indicate that the local recovery approaches outperform complete replacement approaches (i.e., using full-blown ad hoc routing protocols such as AODV and DSR) in terms of end-to-end delay, throughput, packet delivery rate, and control overhead.

Keywords: wireless TCP, multihop communications, ad hoc networks, routing, local recovery.

1 Introduction

In the past few years, we have witnessed that the Internet has extended its reach to the wireless networking environments. Thus, traditional Internet protocols [4, 7, 8] are also heavily used over wireless links. However, a straightforward migration of such protocols to wireless networks will result in poor performance [6]. In particular, the TCP (transmission control protocol), which has many salient features that are useful in a wired network, needs significant modifications in order that it can deliver packets over the wireless links efficiently. Indeed, the congestion control mechanisms are particularly problematic in wireless environments.

---

* This research was supported by a grant from the Research Grants Council of the HKSAR Government under project number HKU 7162/03E.
  Corresponding Author.

Congestion is assumed to be the primary reason for packet losses in wired networks. When packet losses are detected, the sender's TCP exponentially throttles down the congestion window size before retransmitting lost packets. It then backs off its retransmission timer, and enters the congestion avoidance phase. All these mechanisms are aimed at reducing the load injected into the network. However, wireless networks have different characteristics compared with wired networks, such as high bit error rate and occasional blackout. Thus, packet losses can be due to channel errors rather than congestion. Unnecessary reduction in network load over a long period of time (TCP's timers are on the order of tens of seconds) leads to very inefficient use of the precious channel bandwidth and high delays.

Recently, many adaptive TCP approaches for various wireless environments have been suggested. The major objective of these schemes is to make TCP respond more intelligently to the lossy wireless links. According to [1,6], there are three major classes of wireless TCP approaches: end-to-end, link layer, and split-connection approaches. Unfortunately, all these previous approaches are only suitable for use in a single wireless link. For ad hoc networks where devices communicate in a multihop manner, these protocols are inapplicable because we cannot afford to have each pair of intermediate devices on a multihop route to execute these wireless TCP protocols [3,11]. Indeed, if a multihop ad hoc route is broken (e.g., due to deep fading in one of its links), the performance of a TCP session over such a route can be severely affected. The most obvious result is that the TCP sender will eventually discover such breakage after several unsuccessful retransmissions (i.e., after a long delay due to the large TCP timers) and then initiate a new session after setting up a new route. This can lead to unacceptably long delay at the receiver side.

In this paper, we study the performance of two local recovery approaches, which work by swiftly repairing the broken link using a new partial route. In Section 2, we describe our proposed local recovery approaches. Section 3 contains our simulation results generated by the NS-2 [12] platform. We give some concluding remarks in Section 4.

2 The Proposed Approach

2.1 Overview

When the original route is down, we do not simply inform the source that the route cannot be used. Instead, we suppress the notification which is transmitted to the source by TCP, and then find a new partial route between the separated nodes to replace the broken part of the old route. Our approach, remedial in nature, is a *local recovery* (LR) technique [5,13]. The essence of LR is to shield the route error from the source in the hope that we can avoid incurring the excessive delay induced by TCP. Indeed, since the problem is found locally, the remedial work should be done locally.

For example, suppose that due to channel fading and nodes' mobility, the link between node N and $N+1$ is broken. Firstly, we suppress the upstream notification generated by TCP. Afterward, we find if the route table of node N has another route to node $N+1$. If there is a new route to the $N+1$ (i.e., the next node of such a route is not $N+1$), then the broken route is immediately repaired by using this route. If no such route exits, local recovery packets will be sent to repair the route.

A local recovery timer is set to make sure the local recovery process will not consume more time than to re-establish a new route by the source. Thus, if the local recovery timer is expired, we give up local recovery and make use of the full blown ad hoc routing protocol. Figure 1 shows the flow chart of the local recovery algorithm. We explain the whole process in detail below.

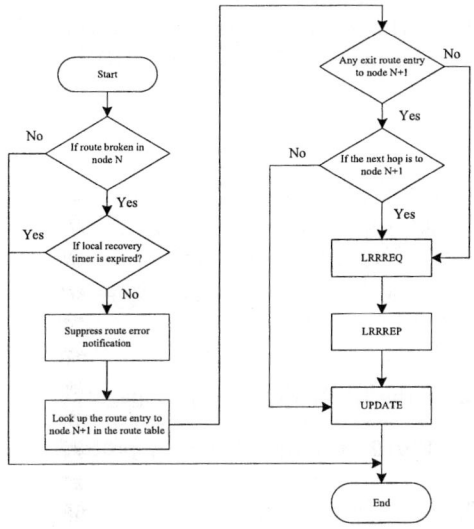

Fig. 1. The proposed local recovery algorithm.

2.2 Discovering Local Recovery Route

In the remedial process, a node N generates the local recovery route request (LRRREQ) packet, which includes the following information: type of the packet, local recovery source address, local recovery destination address, original destination address, local recovery broadcast identifier (ID), and hop count. Whenever node N generates a LRRREQ, the local recovery broadcast ID is increased by one. Thus, the local recovery source and destination addresses, and the local recovery broadcast ID uniquely identify a LRRREQ. Node N broadcasts the LRRREQ to all nodes within the transmission range. These neighboring nodes then relay the LRRREQ to other neighboring nodes in the same fashion. An intermediate node, upon receiving the LRRREQ, first checks whether it has seen this packet before by searching its LRRREQ cache. If the LRRREQ is in the cache, the newly received copy is discarded; otherwise, the LRRREQ is stored in the cache and is forwarded to the neighbors after the following major modifications are done: incrementing the hop count, updating the previous hop node, and updating the time-to-live (TTL) field. Figure 2(a) illustrates how the LRRREQ propagates and how the reverse paths are set up.

When node $N+1$ or some other intermediate node, which has a fresh route to the node $N+1$, receives the LRRREQ, it then generates a local recovery route reply (LRRREP) packet, which includes the following information: type of the packet, local recovery source address, local recovery destination address, original destination address, hop count, and TTL. The LRRREQ is then unicast to the local recovery source along the reverse path until it reaches the local recovery source. During this process, each intermediate node on the reverse path updates its routing table entry to the local recovery destination and original destination. Figure 2(b) illustrates the process where the LRRREP is unicast through the reverse path.

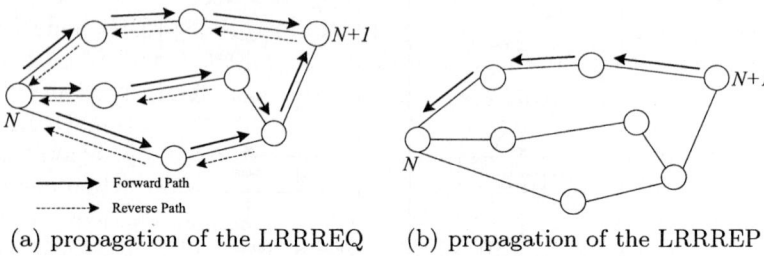

(a) propagation of the LRRREQ   (b) propagation of the LRRREP

Fig. 2. The route recovery process.

2.3 Route Updating

Although the new partial route is found from node N to node $N+1$, updating is needed for the original route. As described above, there are two cases where updating of the original route must be done. The first case is the event that the local recovery destination receives the LRRREQ. The second case is the event that an intermediate node gets the LRRREQ and it has a fresh route to the local recovery destination in its routing table.

According to the different directions, forward and backward updatings are carried out. The forward updating process is triggered by receiving the update packet, which contains the following information: type of packet, update destination address, original destination address, hop count, and TTL. The backward updating process is triggered by receiving the LRRREP packet. In any updating, the original route should be re-established. In the first case, only backward updating is done, while in the second case, both forward and backward updatings are needed.

Figure 3(a) illustrates how the updating process is carried out in the two cases. In the former case, node $N+1$ receives the LRRREQ, and thus, backward

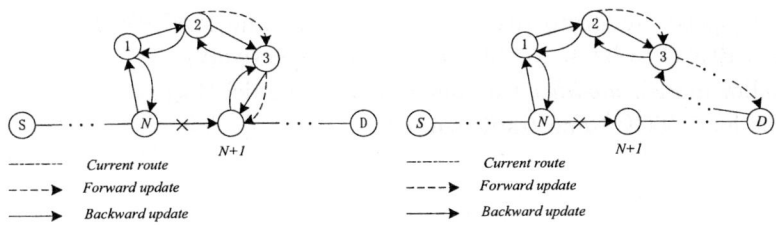

(a) forward and backward updating (b) local recovery algorithm for finding a new partial route directly to the destination

Fig. 3. Illustration of the two route recovery approaches.

updating is done through the route of nodes $N+1, 3, 2, 1, N$. In the latter case, forward updating is done through the route of nodes $2, 3, N+1$, while backward updating is done through the route of nodes $2, 1, N$. The detailed updating process is as follows: when node 2 receives the LRRREQ and it has a route entry to node $N+1$, node 2 sends the update packet to node 3 according to the route entry to node $N+1$. Upon receiving the update packet, node 3 should update the route entry to the original destination node D and then check if it is the local recovery destination. The same forward updating process continues until the update packet is received by the local recovery destination. On the other hand, LRRREP is sent to node 1 following the reverse route. Upon receiving the LRRREP, node 1 should update the route entry to the original destination node D and then check if it is the local recovery source. The same backward updating process continues until the LRRREP is received by the local recovery source.

2.4 Local Recovery of a Route to Destination

This variant of our approach is similar to the mechanism we described above. The only difference is that the goal of route reconstruction is to find a new partial route from node N directly to the destination. Figure 3(b) illustrates this algorithm.

3 Performance Results

3.1 Simulation Environment

In our study, we use packet level simulations to evaluate the performance of TCP in ad hoc networks. The simulations are implemented in Network Simulator (NS-2) [12] from Lawrence Berkeley National Laboratory (LBNL) with extensions for wireless links from the Monarch project at Carnegie Mellon University [2]. The simulation parameters are as follows:

- number of nodes: 50;
- testing field: $1500m \times 300m$;

- mobile speed: uniformly distributed between 0 and MAXSPEED (we choose MAXSPEED to be 4, 10, 20, 40, 60m/s, respectively);
- mobility model: *modified* random way point model [14];
- traffic load: TCP Reno traffic source;
- radio transmission range: 250m;
- MAC layer: IEEE 802.11b.

Each simulation is run for 200 seconds and repeated for ten times. We compared four protocols in our simulations. They are DSR (Dynamic Source Routing) [9], AODV (Ad Hoc On-Demand Distance Vector) [10], LR1 and LR2. LR1 is the local recovery protocol in finding the new route between node N to the destination. LR2 is the local recovery protocol in finding the new route between node N and node $N+1$.

3.2 Performance Metrics

To evaluate TCP performance in different routing protocols, we compare them using four metrics:

1. Average End-to-End Delay: the average elapsed time between sending by the source and receiving by the destination, including the processing time and queuing time.
2. Average Throughput: the average effective bit-rate of the received TCP packets at the destination.
3. Delivery Rate: the percentage of packets reaching the destination (note that some packets are lost during the route breakage and the route reconstruction time).
4. Control Overhead: the data rate required by the transportation of the routing packets.

3.3 Simulation Results

Due to space limitations, our simulation results are summarized in Figure 4. Compared with the AODV protocol, the two LR protocols have lower end-to-end delays, higher throughput, and lower control overhead. The reason is that when the route is broken, the local recovery process can efficiently reconstruct the route, without incurring excessive TCP time-out delays. LR2 exhibits a better performance than LR1. The reason is that on average, the time consumed by node N to find node $N+1$ is less than that to find the destination.

DSR has the worst performance among all compared protocols. Since it drops much more packets than other protocols. In particular, when the mobility is increased, the delivery rate decreases rapidly. However, the other three protocols can keep a stable delivery rate with increasing speed.

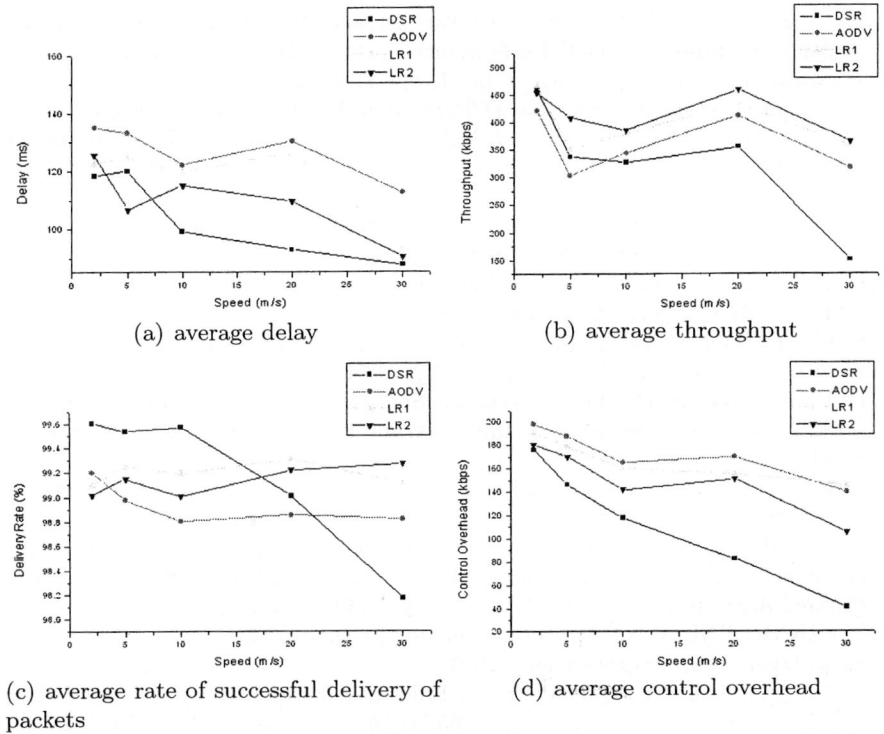

Fig. 4. Simulation results.

4 Conclusions

In this paper, we study the problem of improving multihop wireless TCP performance in an ad hoc network. TCP performance over a single wireless link is notoriously poor due to its long retransmission delays in error recovery. The long delay problem is even more acute in a multihop wireless environment. We compare four approaches in route error recovery: two of them are based on complete replacement of the old route, while the other two are based on local reconstruction of the broken route. Our simulation results generated by the NS-2 platform using TCP Reno traffic sources show that the local recovery approaches significantly outperform the complete replacement approaches.

References

1. H. Balakrishnan, V. N. Padmanabhan, S. Seshan, and R. H. Katz, "A Comparison of Mechanisms for Improving TCP Performance over Wireless Links," *IEEE/ACM Transactions on Networking*, vol. 5, no. 6, pp. 756–769, Dec. 1997.
2. J. Broch, D. A. Maltz, D. B. Johnson, Y.-C. Hu, and J. Jetcheva, "A Performance Comparison of Multi-Hop Wireless Ad Hoc Network Routing Protocols," *Proc. MOBICOM*, pp. 85–97, Oct. 1998.

3. K. Chandran, S. Raghunathan, S. Venkatesan, and R. Prakash, "A Feedback-Based Scheme for Improving TCP Performance in Ad Hoc Wireless Networks," *IEEE Personal Communications*, vol. 8, no. 1, pp. 34–39, Feb. 2001.
4. D. Comer, *Internetworking with TCP/IP*, vols. 1-3, Prentice Hall, 1991.
5. R. Duggirala et al., "Performance Enhancements of Ad Hoc Networks with Localized Route Repair," *IEEE Trans. Computers*, vol. 52, no. 7, pp. 854–861, July 2003.
6. H. Elaarag, "Improving TCP Performance over Mobile Networks," *ACM Computing Surveys*, vol. 34, no. 3, Sept. 2002.
7. K. Fall and S. Floyd, "Simulation Based Comparisons of Tahoe, Reno, and SACK TCP," *ACM Computer Communications Review*, vol. 26, no. 3, pp. 5–21, 1996.
8. S. Floyd, "TCP and Explicit Congestion Notification," *ACM Computer Communications Review*, vol. 24, no. 5, pp. 10–23, 1994.
9. D. B. Johnson and D. Maltz, "Dynamic Source Routing in Ad Hoc Wireless Networks," in *Mobile Computing*, T. Imielinski and H. Korth (eds.), Chapter 5, Kluwer Academic Publishers, 1996.
10. C. E. Perkins, E. M. Royer, and S. R. Das, "Ad Hoc On-Demand Distance Vector(AODV) Routing," *IETF Internet Draft*, draft-ietf-manet-aodv-10.txt, 2002.
11. D. A. Maltz, J. Broch, J. Jetcheva, and D. B. Johnson, "The Effects of On-Demand Behavior in Routing Protocols for Multihop Wireless Ad Hoc Networks," *IEEE J. Selected Areas in Comm.*, vol. 17, no. 8, pp. 1439–1453, Aug. 1999.
12. The UCB/LBNL/VINT Network Simulator (NS), http://www.isi.edu/nsnam/ns/, 2003.
13. D. Tian and N. D. Georganas, "Energy Efficient Routing with Guaranteed Delivery in Wireless Sensor Networks," *Proc. WCNC 2003*, vol. 3, no. 1923–1929, 2003.
14. J. Yoon, M. Liu, and B. Noble, "Random Waypoint Considered Harmful," *Proc. INFOCOM 2003*.

Cluster Based Distributed Mutual Exclusion Algorithms for Mobile Networks

Kayhan Erciyes

California State University San Marcos,
Comp. Sci. Dept., 333 S.Twin Oaks Valley Rd.,
San Marcos CA 92096, USA
kerciyes@csusm.edu

Abstract. We propose an architecture that consists of a ring of clusters for distributed mutual exclusion algorithms in mobile networks. The mobile network is partitioned into a number of clusters periodically using a graph partitioning algorithm called *Fixed Centered Partitioning* first. Each cluster is represented by a coordinator node on the ring which implements various distributed mutual exclusion algorithms on behalf of any member in the cluster it represents. We show the implementation of Ricart-Agrawala and Token-based algorithms on this architecture. The message complexities for both algorithms are reduced substantially using the proposed architecture . . .

1 Introduction

Mobile ad hoc networks do not have fixed infrastructure and consist of mobile wireless nodes that have temporary interconnections to communicate over packet radios. Clustering, that is, partitioning of the mobile network graph into smaller subgraphs, can be used to solve various problems such as routing and mutual exclusion in such networks. In general, distributed mutual exclusion algorithms may be classified as permission based or token based. In the first case, a node would enter a critical section after receiving permission from all of the nodes in its set for the critical section. For token-based algorithms however, processes are on a logical ring and posession of a system-wide unique token would provide the right to enter a critical section. Susuki-Kasami's algorithm [10] (N messages) and Raymond's tree based algorithm [7] (log(N) messages) are examples of token based mutual exclusion algorithms. Examples of nontoken-based distributed mutual exclusion algorithms are Lamport's algorithm [5] (3(N-1) messages), Ricart-Agrawala (RA) algorithm (2(N-1) messages) [8] and Maekawa's algorithm [6]. *Safety, liveness* and *fairness* are the main requirements for any mutual exclusion algorithm. Lamport's algorithm [5] and RA algorithm [8] are considered as one of the only fair distributed mutual exclusion algorithms in literature.

Distributed mutual exclusion in mobile networks is a relatively new research area. A fault tolerant distributed mutual exclusion algorithm using tokens is discussed in [12] and a *k-way* mutual exclusion algorithm for ad hoc wireless networks where there may be k processes executing a critical section at any time is presented in [13]. In this study, we propose a model to perform distributed mutual exclusion algorithms in mobile networks. We first partition the mobile network into a number of clusters using the multilevel graph partitioning heuristic we developed called *Fixed Centered Partitioning* that is executed by a special node called the *Central Coordinator*. Upon any changes of configuration or periodically gathering of the changes, the central coordinator starts a new configuration process by partitioning the network graph into new clusters. The nodes in the cluster including the nodes that have connections to other clusters are called *neighbor nodes*. The coordinator chooses one of the neighbor nodes in each cluster as the cluster *coordinator* and sends the cluster connectivity information and neighbor connectivity information to it. These coordinators perform the required critical section entry and exit procedures for the nodes they represent. Using this architecture, we improve and extend the RA and Token Passing algorithms described in [2] to mobile networks and show that these algorithms may achieve an order of magnitude reduction in the number of messages required to execute a critical section at the expense of increased response times and synchronization delays. The rest of the paper is organized as follows. Section 2 provides the backround by describing FCP along with a review of performance metrics of mutual exclusion. The extended RA algorithm on the proposed model called *Mobile_RA* is described in Section 3. The second algorithm implemented on the model uses Token Passing and is called *Mobile_TP* as briefly described in Section 4. Finally, discussions and conclusions are outlined in Section 5.

2 Background

2.1 Partitioning of the Mobile Network

Graph partitioning algorithms aim at providing subgraphs such that the number of vertices in each partition is averaged and the number of edges cut between the partitions is minimum with a total minimum cost. An arbitrary network can be constructed as an undirected connected graph $G = (V, E, w)$ where V is the set of routing nodes, E is the set of edges giving the cost of communication between the nodes and $w: E \to \Re$ is the set of weights associated with edges. *Multilevel partitioning* is performed by coarsening, partitioning and uncoarsening phases [3]. During the coarsening phase, a set of smaller graphs are obtained from the initial graph. In the maximal matching, vertices which are not neighbors are searched. In Heaviest Edge Matching (HEM), the vertices are visited in random order, but the collapsing is performed with the vertex that has the heaviest weight edge with the chosen vertex. In Random Matching (RM) however, vertices are visited in random order and an adjacent vertex is chosen in random too. The coarsest graph can then be partitioned and further refinements can be achieved by suitable algorithms like Kernighen and Lin [4]. Finally, the partition of the coarsest graph is iteratively reformed back to the original graph.

We provide a partitioning method called *Fixed Centered Partitioning* (FCP) [1] where several fixed centers are chosen and the graph is then coarsened around these fixed centers by collapsing the heaviest or random edges around them iteratively. Different than [3], FCP does not have a matching phase, therefore iterations are much faster. FCP requires the initial marking of the fixed centers. One possible solution is to choose the fixed centers randomly so that they are all at least some bounded distance from each other. The heuristic we used is $h = 2d / k$ where d is the diameter of the network and k is the number of partitions (clusters) to be formed. The time complexity of the total collapsing of FCP is $O(n)$ and FCP provides much favorable partitions than CM and RM in terms of the average edge cost, time to partition a graph and the quality of the partitions experimentally [1]. Fig. 1 shows a mobile network of 20 nodes which is partitioned into clusters A, B, C and D using FCP. Nodes 16, 20, 2 and 14 are the initial centers, 17 is the central cordinator and also the cluster coordinator for cluster D, and the coordinators for clusters A, B, C and D are 10, 15 and 19 and they form the ring together with 4 and 7.

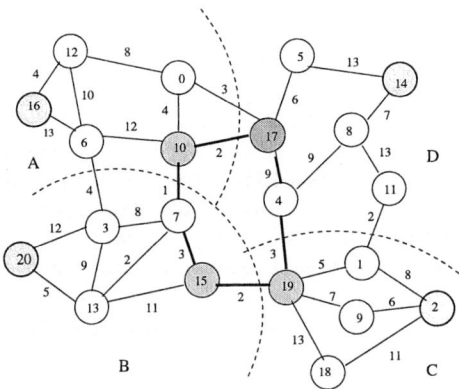

Fig. 1. Partitioning of the Mobile Network.

2.2 Performance Metrics

Performance of a distributed mutual exclusion algorithm depends on whether the system is *lightly* or *heavily* loaded. If no other process is in the critical section when a process makes a request to enter it, the system is lightly loaded. Otherwise, when there is a high demand for the critical section which results in queueing up of the requests, the system is said to be heavily loaded. The important metrics to evaluate the performance of a mutual exclusion algorithm are the Number of Messages per request (M), Response Time (R) and the Synchronization Delay (S). M can be specified for high load or light load in the system. The Response Time R is measured as the interval between the request of a node to enter critical section and the time it finishes executing the critical section. The synchronization delay S is the time required for a node to enter a critical section after anoother node finishes executing it. The minimum value of S is one message transfer time T since one message suffices to transfer the access rights to another node [9].

3 Extended Ricart-Agrawala Algorithm

For distributed mutual exclusion in mobile networks, we propose a hierarchical architecture where nodes form clusters and each cluster is represented by a *coordinator* in the ring. The relation between the cluster coordinator and an ordinary node is similar to a central coordinator based mutual exclusion algorithm. The types of messages exchanged are *Request*, *Reply* and *Release* where a node first requests a critical section and upon the reply from the coordinator, it enters its critical section and then releases the critical section. The algorithm for the Mobile_RA coordinator is shown in Fig. 2.

```
Process Mobile_RA_Coord;
begin
   repeat
      msg=receive_msg();
      switch (msg.type):
         case Node_Req : send(Coord_Req, next_coord);
                         insert_msg(Node_Req, wait_queue);
         case Coord_Req : If Idle OR all local requests > timestamps
                                send(Coord_Rep, next_coord)
                          Else Block the Coord_Req message;
                                insert_msg(Coord_Req, wait_queue);
         case Coord_Rep : send(Node_Rep, node);
         case Node_Rel  : If wait_queue <> empty and top_queue <> local
                          For every external request with < timestamp
                                send_all(Coord_Replies, next_coord);
   until forever
end.
```

Fig. 2. Mobile_RA Coordinator Algorithm.

The coordinator sends a critical section request (*Coord_Req*) to the ring for each node request (*Node_Req*) it receives. When it receives an external request (*Coord_Req*), it performs the operation of a normal RA node by checking the timestamps of the incoming request by the pending requests in its cluster and sends a reply (*Coord_Reply*) only if all of the pending requests have greater timestamps than the incoming request. When a node sends a *Node_Rel* message, the coordinator sends *Coord_Rel* messages to all of the requests in the *wait_queue* that have smaller timestamps than the local pending ones. Fig. 3 shows an example scenario for the Mobile_RA Algorithm in the network of Fig. 1. The following describes the events that occur:

1. Nodes 12, 2 and 5 in clusters A, C and D make critical section requests with messages $Node\_Req\_12(1)$, $Node\_Req\_2(3)$ and $Node\_Req\_5(2)$ with the shown timestamps.
2. The coordinator for clusters A, C and D (C_A, C_C, C_D) form request messages R_{12} and R_2 and R_5 and send these to the next coordinators on the ring, Steps 1 and 2 are shown in Fig. 3.a.

3. C_B passes R_{12} to its successor C_C as it has no pending requests in its cluster. C_C also passes R_{12} as R_{12} has a lower timestamp than the pending request in its cluster ($Node\_req\_2(3)$). C_D passes R_{12} back to C_A as this request has a lower time stamp than its own too, but it blocks the request R_2 by C_C as this incoming request has a higher timestamp.

4. C_A now has received its original request back meaning all of the coordinators have confirmed that either they have no pending requests or their pending requests all have higher timestamps. C_A now sends a $Coord\_Rep$ message to node 12 which can enter its critical section. Steps 3 and 4 are depicted in Fig. 3.b.

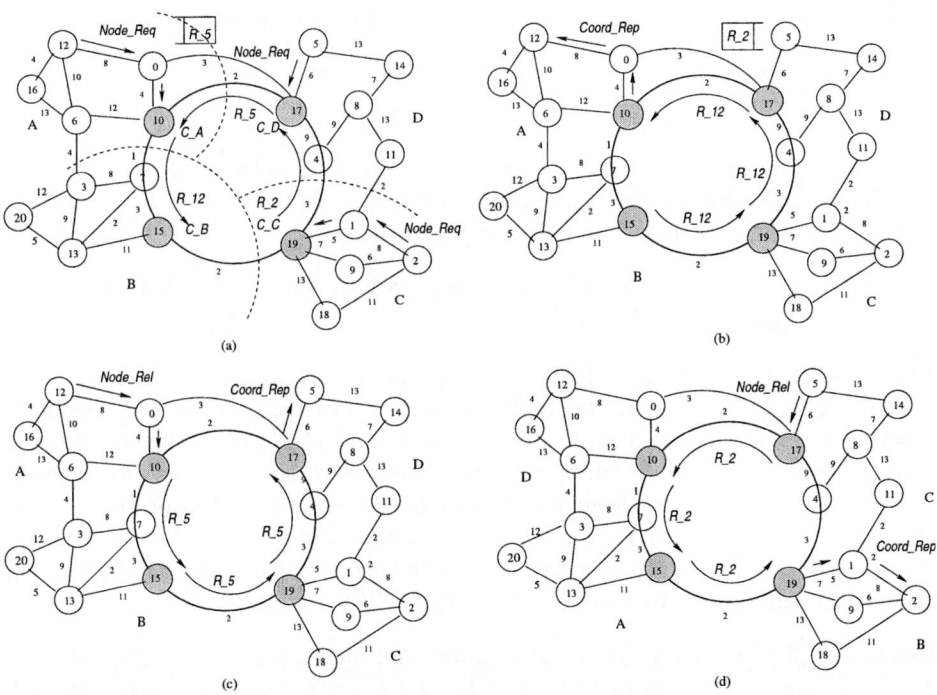

Fig. 3. Operation of the Mobile_RA Algorithm.

5. Node 12 in cluster A sends $Node\_Rel$ message to its coordinator C_A to inform that it has finished executing its critical section. C_A now sends a reply to the request it was blocking (R_5). C_C passes R_5 to C_D as it has a lower timestamp than its request R_2.

6. C_D now has its reply now for request $Node\_Req\_5(2)$ and sends a reply message ($Coord\_Rep$) to node 5 which can then enter its critical section. Steps 5 and 6 are shown in Fig. 3.c.

7. When node 5 finishes executing its critical section, it sends a release (*Node_Rel*) message to its coordinator C_D which now sends the reply message R_2 it was blocking to the ring.
8. C_A and C_B pass this message and C_C now has the reply from the ring and can send the reply to node 2 (*Coord_Rep*) which can enter its critical section. Steps 7 and 8 are shown in Fig. 3.d.

If there are multiple requests within the same cluster, time stamps are checked similarly for local request. The order of execution in this example is nodes 12 → 5 → 2 in the order of the timestamps of the requests.

Theorem 1. *The total number of messages per critical section using the Mobile_RA Algorithm is $k+3d$ where k is an upper bound on the number of neighbor nodes in the ring including the cluster coordinators and d is an upperbound on the diameter of a cluster.*

Proof. An ordinary node in a cluster requires three messages (*Request, Reply and Release*) per critical section to communicate with the coordinator. Each of these messages would require maximum d transfers between a node and the coordinator. The full circulation of the coordinator request (*Coord_Req*) requires k messages resulting in $k + 3d$ messages in total.

Corollary 1. *The Synchronization Delay (S) in the Mobile_RA Algorithm varies from $2dT$ to $(k + 2d - 1)T$.*

Proof. When the waiting and the executing nodes are in the same cluster, the required messages between the node leaving its critical section and the node entering are the *Release* from the leaving node and *Reply* from the coordinator resulting in $(2dT)$ message times for S_{min}. However, if the nodes are in different clusters, the *Release* message has to reach the local coordinator in d steps, circulate the ring through $k-1$ nodes to reach the originating cluster coordinator in the worst case and a *Reply* message from the coordinator is sent to the waiting nodes in d steps resulting in $S_{max}=(k-1)T + 2dT=(k+2d-1)T$.

Corollary 2. *In the Mobile_RA Algorithm, the response times are $R_{light}=(k+3d)T + E$ and R_{heavy} varies from $w(2dT + E)$ to $w((k+2d-1)T + E)$ where k is the number of clusters and w is the number of pending requests.*

Proof. According to Theorem 3, the total number of messages required to enter a critical section is $k + 3d$. If there are no other requests, the response time for a node will be $R_{light}=(k+3d)T+E$ including the execution time (E) of the critical section. If there are w pending requests at the time of the request, the minimum value R_{heavy_min} is $w(2dT + E)$. In the case of S_{max} described in Corollary 1, R_{heavy_max} becomes $w((k + 2d - 1)T + E)$ since in general $R_{heavy}=w(S + E)$.

Since the sending and receiving ends of the algorithm are the same as of RA algorithm, the safety, liveness and fairness attributes are the same. The performance metrics for the Mobile_RA Algorithm is summarised in Tab. 1.

Table 1. Performance of Mobile_RA Algorithm.

M_{light}	M_{heavy}	R_{light}	$R_{heavy-min}$	S_{min}	S_{max}
$k+3d$	$k+3d$	$(k+3)dT+E$	$w(2dT+E)$	$2dT$	$(k+2d-1)T$

4 Extended Token Passing Algorithm

We propose an extended implementation of the Token Passing (TP) Algorithm for mobile networks. Token is circulated in the ring only and the coordinator for each cluster consumes the token if it has a pending request for the token from any of its nodes, otherwise it passes the token to the next coordinator in the ring. A node that finishes execution of its critical section returns the token to its coordinator which may forward it to any other node waiting for the token in its cluster or pass it to the next coordinator in the ring. Since there is only one token, mutual exclusion is guaranteed. The message complexity for Mobile_TP is $O(k+3d)$ since $(k+3d)$ is an upperbound on the number of messages depending on the current location of the token when a request for a critical section is made. The Synchronization Delay (S) and the Response Time values of Mobile_TP are the same as the Mobile_RA Algorithm.

5 Conclusions

We proposed a framework to implement distributed mutual exclusion algorithms in mobile networks where the network is represented by a dynamically changing graph and this graph is partitioned into clusters at regular intervals. We showed that this model provides improvement over message complexity of Ricart and Agrawala and Token-based algorithms and also the time required to execute a critical section. A comparison of the two algorithms with their regular counterparts in terms of their message complexities is shown in Tab. 2. If we assume $k=m=d$ for simplicity, the message complexities of the mobile algorithms are in the order of $sqrt(N)$ where N is the total numner of nodes in the network. The order of magnitude of improvement over the classical RA and the Token-Based algorithms using our model is achieved at the expense of increased response times and synchronization delays. The coordinators have an important role and they may fail. New coordinators may be elected and any failed node member can be excluded from the cluster which is an improvement over both classical algorithms as they do not provide recovery for a crashed node in general. The

Table 2. Comparison of the Mobile Mutual Exclusion Algorithms with others.

	Regular	Mobile Algs.	Mobile (k=m=d)
Ricart-Agrawala Alg.	$2(N-1)$	$k+3d$	$\Theta(4sqrt(N))$
Token Passing Alg.	N	$O(k+3d)$	$O(4sqrt(N))$

recovery procedures can be implemented using algorithms as in [11] which is not discussed here. Our work is ongoing and we are looking into implementing k-way distributed mutual exclusion algorithms in mobile networks. Partitioning of the network graph using FCP can also be performed in parallel by the existing coordinators which would improve performance.

References

1. Erciyes, K, Marshall, G.: A Cluster-based Hierarchical Routing Protocol for Mobile Networks, ICCSA 2004, SV-Lecture Notes in Computer Science, to appear, (2004)
2. Erciyes, K,: Distributed Mutual Exclusion Algorithms on a Ring of Clusters, ICCSA 2004, SV-Lecture Notes in Computer Science, to appear, (2004)
3. Karypis, G., Kumar, V.: Multilevel k-way Partitioning scheme for irregular graphs. Journal of Parallel and Distributed Computing, Vol. 48, (1998), 96-129
4. Kernighan, B., Lin, S.: An Effective Heuristic Procedure for Partitioning graphs, The Bell System Technical Journal, (1970), 291-308
5. Lamport, L.: Time, Clocks and the Ordering of Events in a Distributed System, CACM, Vol. 21, (1978), 558-565
6. Maekawa, M.: A sqrt(n) Algorithm for Mutual exclusion in Decentralized Systems, ACM Transactions on Computer Systems, Vol. 3(2), (1985), 145-159
7. Raymond, K.: A Tree-based Algorithm for Distributed Mutual Exclusion. ACM Trans. Comput. Systems, 7(1), (1989), 61-77
8. Ricart, G., Agrawala, A.: An Optimal Algorithm for Mutual Exclusion in Computer Networks, CACM, Vol. 24(1), (1981), 9-17
9. Shu, Wu: An Efficient Distributed Token-based Mutual Exclusion Algorithm with a Central Coordinator, Journal of Parallel and Distributed Processing, Vol.62(10), (2002), 1602-1613
10. Susuki, I., Kasami, T.: A Distributed Mutual Exclusion Algorithm, ACM Trans. Computer Systems, Vol. 3(4), (1985), 344-349
11. Tunali, T, Erciyes, K., Soysert, Z.: A Hierarchical Fault-Tolerant Ring Protocol For A Distributed Real-Time System, Parallel and Distributed Computing Practices, Vol. 2(1), (2000), 33-44
12. Walter,J., E., Welch, J., L., Vaidya, N., H.: A Mutual Exclusion Algorithm for Ad Hoc Mobile Networks, Wireless Networks, Vol. 7(6), (2001), 585-600
13. Walter, J., E., Cao, G., Mohanty, M.: A K-way Mutual Exclusion Algorithm for Ad Hoc Wireless Networks, Proc. of the First Annual workshop on Principles of Mobile Computing POMC 2001, (2001)

An Improved Handoff Technique for a Seamless Multimedia Services

Seung-Won Lee[1], Hyeon-Ok Hong[2], Seong-Ho Park[3], Yong-Woon Park[2], and Ki-Dong Chung[1]

[1] Department of Computer Science, Pusan National University, Rep. of Korea
{swlee,kdchung}@pusan.ac.kr
[2] Department of Computer Science, Dongeui Institute of Technology, Rep. of Korea
{hohong,ywpark}@dit.ac.kr
[3] Information Technology Center, Pusan National University, Rep. of Korea
shpark@pusan.ac.kr

Abstract. Servicing mobile hosts over the Internet has some problems including nodes' location, independent addressing, address conversion and packet forwarding as well as mobile host's location management. However, packet transmission and independent addressing in the mobile environments are very similar to the ones in the multicasting environment. So, in this paper, we propose a new scheme about how to manage the multicasting group and to set the communication path in the mobile environment by applying two-level addressing mode, which is similar to mobile IP, for lcation-independent address setting. More specifically, we propose the smooth handoff scheme that minimizes the handoff delay. To check the performance of our proposed scheme, we modified the NS-2 network simulator and as a result, we showed that our proposed scheme is better than other techniques in terms of transmitted packets' throughput.

1 Introduction

With the rapid proliferation of mobile communication devices over the Internet, people can access anywhere they want at anytime. For instance, people do not need wired devices to read their mails or query to databases any more; they just access their data with their mobile devices such as PDAs or smart phones.

In IP[1] network environment, when a user moves to the current network segment to the other segment, his IP address has to be changed accordingly. This is because the IP address uniquely distinguishes the network segment to which the mobile host is connected. By using TCP[2], for example, a session needs its own address to sustain connection that the connection is broken down if the address is changed. So, to support mobility in IP network, the unique address mechanism such as Mobile IP[3] is required.

Mobile IP has some delay factors such as foreign agent's registration and triangle routing but they are not so serious as to hinder email delivery or web surfing. However, video objects, which require high transmissionbandwidth and real time delivery, have serious impact on the network. To service video objects in mobile envelopments properly, the delay that occurs between cells or network segment has to be minimized. So, for the efficient network resource management, multicasting facilities can be helpful if it is applied to support the mobile hosts.

Our research focus in this paper is to support multimedia service on the mobile environments that for this, we apply multicasting facility to the mobile hosts considering mobility. Multicasting is very similar to mobile communication in that both of them permit their members' dynamic jon/leave and independent addressing. In addition, we anticipate the mobile host's future movement by monitoring the overlapped area of base stations and by applying pre-join technique to this, we can minimize the delay factors. So, we propose the dynamic group management and path setting scheme exploiting multicasting facilities on the mobile environment.

The rest of paper is organized as follows. In section 2, we explain how to provide mobility to hosts by using the expanded mobile IP equipped with multicasting mechanism. In section 3, we propose seamless communication method for multimedia objects by applying the smooth handoff. In section 4, we test our proposed scheme by comparing it with other schemes. Finally we draw the conclusion and refer to the further research.

2 Expanded Mobile IP Based on Multicasting

Although multicasting facilities have been developed with the purposes than mobile IP, they are similar each other in some points. The packets transmitted to a multicasting group do not have information about the subscribers' location. This can be resolved by exploiting mobile IP's location-independent addressing and packets routing scheme for the multicasting group management. Based on this, we supports the mobile hosts' mobility by exploiting the multicasting group management and path setting scheme.

2.1 Network Architecture

In this paper, we assume that the overall network is composed of a number of LANs and the mobility in the network is classified into inter-region mobility and intra-region mobility. For the intra-region mobility, we apply the existing multicasting techniques and for the inter-region mobility, we apply the existing mobile IP. Fig.1 shows the overall network architecture.

2.2 Addressing

We apply two level addressing that is similar to Mobile IP but the addresses, which are set differently from mobile IP, is composed of multicasting group addresses that are used as edge router address and mobile hosts' address. Fig.2 shows what packet is composed of.

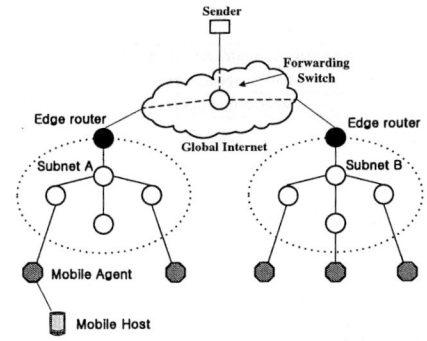

Fig. 1. The Overall Network Architecture.

Edge Router IP Address	Multicast Group Address	Data Packet

IP Encapsulation — IP Header — IP Data packet

Fig. 2. The Packet Structure.

2.3 Advertisement Message

As told earlier, the host mobility is classified into two categories; intra-region mobility and inter-region mobility. The intra-region mobility is identified by the address change of the mobile hosts and inter-region mobility is identified by the address change of the subnet's edge router. So, the mobile agent advertises routinely both its address and its' subnet edge router. To deliver the advertisement message, we modified the existing mobile IP as shown in Fig.3.

type	length	sequence number	
registration lifetime		RBHFMGV	reserved
mobile agent address			
edge router address			

Fig. 3. The Advertisement Message.

For mobile hosts' new path setting through join, a mobile agent has to know the mobile host's address or multicasting address. So, when a mobile host moves to the other segment, it sends such information as the addresses of the multicasting group, the edge router and current subnet's mobile agent to the next subnet's mobile agent through the registration message.

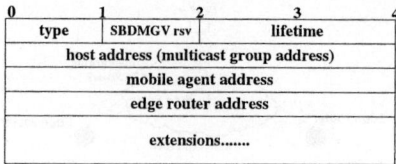

Fig. 4. The Registration Message.

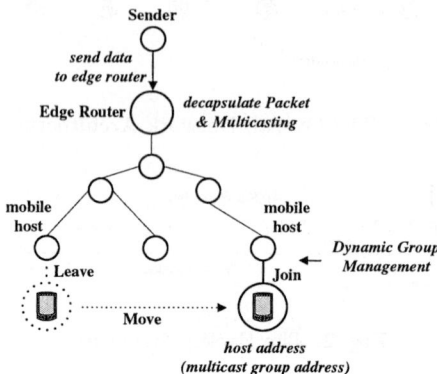

Fig. 5. An Intra-region routing in subnet.

2.4 An Intra-region Routing in Subnet

For intra-region routing in a subnet, we exploit the multicasting group management policy to support mobile host's addressing and packet routing. Fig.5 shows how the path is set as the mobile node moves. The movement of the mobile host is detected by comparing the current and next mobile agents' address after receiving the message.

Generally, the delay in mobile IP occurs between HA(Home Agent) and FA(Foreign Agent). On the other hand, the delay in the proposed intra-region mobility is the delay that occurs between mobile agents and edge routers for the multicasting join. As shown in Fig.5, the number of hops in the proposed policy is smaller.

2.5 An Inter-region Routing Between Subnets

In the inter-region routing, it is not trivial to support mobility by applying multicasting group management policy due to the multicasting facilities scalability. For this, in this paper, we exploit the characteristics of mobile IP based on the triangle routing and the route optimization. At this time, the edge router of each subnet plays a role as both the home agent and the foreign agent. By doing this, the delay can be minimized by reducing the intra-region delay.

An Improved Handoff Technique for a Seamless Multimedia Services   945

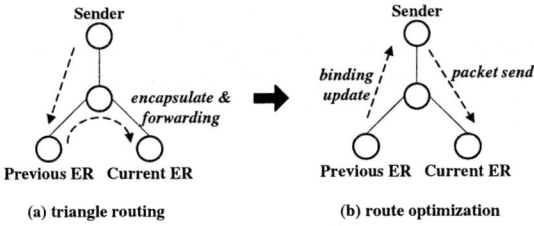

Fig. 6. An inter-region routing between subnet.

3 Seamless Communication for Multimedia Services

In servicing video objects, which usually require high transfer bandwidth and real time delivery, the delay factors including path setting have major impact on the efficient delivery of those objects. To minimize the impact, the handoff delay should be reduced. In this section, we expand the proposed schemes in the previous section for the continuous media delivery. For this, we anticipate the host's movement by monitoring the overlapped area.

Although it is smaller than the case of Mobile IP, the delay between subnets still exists in our proposed policy. Small delay makes major impact on the data delivery in case of continuous media objects. The intra-region handoff is the one that occurs during multicasting join. So, by reducing the multicasting join delay, the overall delay can be reduced. To do this, we use pre-join after anticipating the hosts' movement.

3.1 The Anticipation of Hosts' Movement

To anticipate to which direction the mobile host moves, we monitor the overlapped area among base stations. If a host is in the overlapped area between two mobile agents, both of those agents will receive the advertisement message. The host, then, will realize whether it is in the overlapped area or not. Fig.7 shows the overlapped area that the number means each host's location and the alphabet is the ID for each mobile agent(for example, MA-A means mobile agent A).

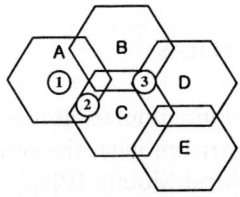

Fig. 7. The overlapped area on mobile environment.

In case a mobile host is in ①, it receives the advertisement message from MA-A. When the mobile host is in ②, it receives the message from both MA-A and MA-C whereas it receives the message from all of three agents in case the mobile host is in ③.

3.2 A Smooth Handoff in Intra-subnet

In case the handoff occurs in the intra-region, it is processed based on the pre-join. To do pre-join, we first decide where the pre-join has to be done. A channel is pre-allocated provided that a host is in the overlapped area, it is highly possible for the mobile host to move to the area where the newly contacted agent exists. This is easily implemented because it is based on the existing multicasting facilities. The mobile agent, which is requested to allocate a new channel for the mobile host in the overlapped area, does the multicast join process for the new channel. However, this channel is not used immediately; it is used only after the mobile host moves to the next mobile agent's area.

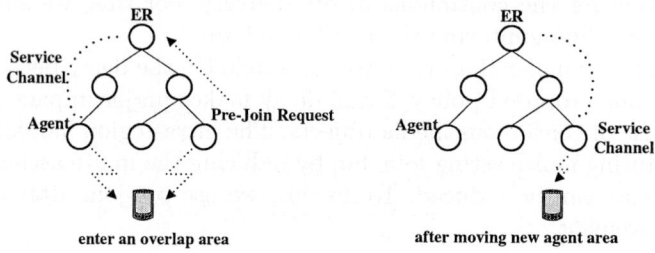

Fig. 8. The pre-join works in the intra-region.

3.3 A Smooth Handoff in Inter-subnet

In case the handoff occurs in the inter-region, the edge router in the newly contacted area is requested to set the packet delivery path. Moreover, the newly contacted edge router asks the existing or current router to copy the packets and as a result, it allocates one channel in advance.

4 Performance Evaluation

For simulation, we use the network simulation tool : NS-2. We modified the NS-2 to compare our proposed algorithm with the other policies including Cellular IP[10], Hawaii[11] and Hierarchical Mobile IP[8].

We assume that each wired connection is a 10Mb/s duplex link and each mobile host is connected to an access point that has NS-2's CSMS/CA wireless

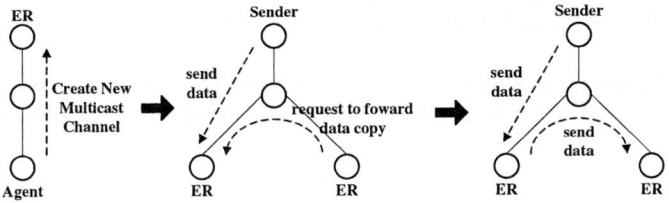

Fig. 9. The pre-join works in the inter-region.

link model, each of which has its unique frequency. We got the test result by monitoring one mobile host that moves constantly among access points at the variable speed. For the test's simplicity, we did not take the switching and message processing capacity of network nodes into consideration. The packets are assumed to be sent through UDP at every 10ms and TCP session is assumed as greedy download service like FTP.

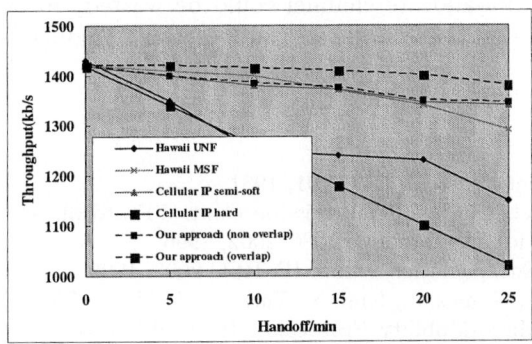

Fig. 10. The overall throughput of TCP connection.

As shown in Fig.10, as the handoff ratio increases, the throughput decreases due to packet loss. Fig.10 shows the overall throughput of TCP connection provided that the mobile host handoffs routinely with mobile agents. As shown in the figure, our propose policy works no less than the Hawaii's MSF and Cellular IP's semi-soft. This is because the subnet changing delay in our proposed policy is limited to the point between the edge router and mobile agent; this is similar to the Hawaii's MSF and Cellular IP's semi-soft. Moreover, in case the smooth handoff based on the overlapped area, the channel is allocated in advance that the connection delay can be hided and as a result, the our proposed policy shows better performance. On the other hand, the movement speed is fast, the pre-join can not applied properly that the pre-allocate channel could be wasted.

5 Conclusion and Further Research

In this paper, we proposed two policy for the efficient continuous media object delivery on the mobile environment. First, we provide the mobility to mobile hosts that other than the existing research, our policy exploits various aspects of the existing IP multicasting scheme. So, our proposed policy runs only with small modification of the existing network components and as a result, can be equipped at the reasonable cost. Moreover, because it is based on the multicasting facilities, it is easily expanded to support continuous media objects.

The second policy we proposed is the smooth handoff policy to minimize the delay factors. This policy is based on the fact that the mobile hosts on the edge of the overlapped area could receive the advertisement message from more than one agents; by using this phenomenon, we made the algorithm with which the moving host's location is detected. With the location information, we did pre-join so that mobile hosts could receive their packet without connection reestablishment.

For performance evaluation, we used the expanded NS-2 and as a result, our proposed algorithm works better than the existing policies such as Hawaii and Cellular IP. But we found out that in some cases, the pre-join can not applied properly that the pre-allocate channel could be wasted. So, we will do further research to improve the correct and fast anticipation of mobile hostss' movement.

References

1. J.Postel. Internet Protocol. RFC 791, 1981
2. DARPA Internet Program. Transmission Control Protocol. RFC 793, 1981
3. C.Perkins. IP Mobility Support. RFC 2002, 1996
4. C.Perkins. IP Encapsulation within IP. RFC 2003, 1996
5. A. T. Campbell, Gomez, J., Kim, S., Turanyi, Z., Wan, C-Y. and A, Valko "Comparison of IP Micro-Mobility Protocols", IEEE Wireless Communications Magazine, Vol.9, No.1, February 2002.
6. A.Aziz. "A Scalable and Efficient Intra-Domain Tunneling Mobile-IP Scheme", ACM Computer Communication Review, 24(1), Janurary 1994
7. L. H. Sahasrabuddhe and B. Mukherjee, "Multicast Routing Algorithms and Protocols: A Tutorial," IEEE Network, 14(1), pp. 90-102, January/February 2000
8. C.E.Perkins, "Mobile IP Local Registration with Hierarchical Foreign Agents." Internet Engineering Task Force, Internet Draft work in progress, February 1996
9. C.E.Perkins and D.B.Hohnson, "Route Optimization in Mobile IP" Mobile IP Working Group, Internet Draft work in progress, Nevember 1997.
10. A. Campbell et al., "Cellular IP," Internet draft, draftietf-mobileip-cellularip-00, work in progress, Dec. 1999.
11. R. Ramjee et al., IP micro-mobility Support using HAWAII, Internet draft, draft-ietf-mobileip-hawaii-00, work in progress, June 1999.

An Improved Approximation Algorithm for the Minimum Energy Consumption Broadcast Subgraph

Vittorio Bilò and Giovanna Melideo

Dipartimento di Informatica, Università di L'Aquila,
Via Vetoio loc. Coppito, I-67100 L'Aquila, Italy
{bilo,melideo}@di.univaq.it

Abstract. In an ad-hoc wireless network each station has the capacity of modifying the area of coverage with its transmission power. Controlling the emitted transmission power allows to significantly reduce the energy consumption and so to increase the lifetime of the network. In this paper we focus on the Minimum Energy Consumption Broadcast Subgraph (MECBS) problem [1,2,6], whose objective is that of assigning a transmission power to each station in such a way that a message from a source station can be forwarded to all the other stations in the network with a minimum overall energy consumption. The MECBS problem has been proved to be inapproximable within $(1-\epsilon)\ln n$ unless $NP \subseteq DTIME(n^{O(\log \log n)})$ [2,6], where n is the number of stations. In this work we propose a $2H_{n-1}$-approximation greedy algorithm which, despite its simplicity, improves upon the only previously known ratio of $10.8 \ln n$ [1] and considerably approaches the best-known lower bound on the approximation ratio.

1 Introduction

Ad hoc wireless networks have received significant attention during the recent years. In particular, they emerged due to their potential applications in emergency disaster relief, battlefield, etc [5,7]. Unlike traditional wired networks or cellular networks, they do not require the installation of any wired backbone infrastructure. The network is a collection of transmitter/receiver stations each equipped with an omnidirectional antenna which is responsible for sending and receiving radio signals. A communication is established by assigning to each station a transmitting power.

A fundamental problem in ad hoc wireless networks is to support *broadcasting*, that is to allow a source station to transmit a message to all stations in the network. A communication from a station s to another t occurs either through a single-hop transmission if the transmitting power of s is adequate, or through relaying by intermediate stations, otherwise. One of the main advantages of ad-hoc networks is the ability of the stations to vary the power used in a transmission in order to reduce the power consumption and so to increase the lifetime of the network.

In this paper we focus on the "energy-efficient" broadcasting, where the objective is to designate the transmission powers at which each station i has to transmit in such a way that a communication from a source station s to all the other stations can be established and the overall energy consumption is minimized. This problem is referred to as *Minimum Energy Consumption Broadcast Subgraph* (MECBS, for short) [1, 2].

An ad hoc wireless network is usually modelled by (i) a complete graph $G(S)$ whose n vertices $S = \{1, \ldots, n\}$ represent radio stations and (ii) a symmetric *cost function* $c : S \times S \mapsto I\!\!R^+$ which associates each pair of stations i and j with its *transmission cost*, that is the power necessary for exchanging messages between i and j. Clearly, $c(i, i) = 0$ for every station $i \in S$. A power assignment $\omega : S \mapsto I\!\!R^+$ to the stations induces a directed weighted graph $G_\omega = (S, E)$, called *the transmission graph*, such that an edge $\langle i, j \rangle$ belongs to E if and only if the transmission power of i is at least equal to the transmission cost from i to j, i.e., $\omega(i) \geq c(i, j)$. The *cost* of a power assignment ω is the overall power consumption yielded by ω, i.e., $cost(\omega) = \sum_{i=1}^{n} \omega(i)$.

The *Minimum Energy Consumption Broadcast Subgraph* (MECBS) problem described above is then defined as follows:
- *Input.* A set S of n sender/receiver stations and a source station $s \in S$.
- *Output.* A power assignment ω such that the overall energy consumption $cost(\omega)$ is minimized and the induced transmission graph G_ω contains a directed spanning tree rooted at s.

The MECBS problem has been proved to be inapproximable within $(1-\epsilon) \ln n$ unless $NP \subseteq DTIME(n^{O(\log \log n)})$ [2, 6] where n is the number of stations. The only known logarithmic approximation algorithm for the problem is due to Caragiannis et al. who presented a $10.8 \ln n$-approximation algorithm which uses a reduction to the Node-Weighted Connected Dominating Set problem [1].

This paper takes an important step toward the reduction of the gap between these two bounds. Indeed, we propose a $2H_{n-1}$-approximation greedy algorithm for the problem that is returning assignments of cost at most twice the cost of the best possible approximated solution. Besides the reduction of the existing gap, our algorithm confirms the strict relationship occurring between the MECBS problem and the Set Cover problem. In fact, as also suggested by the negative results in [2], our result reinforces the intuition that MECBS can be seen as a set covering problem with additional connectivity requirements.

The paper is organized as follows. In section 2 we first introduce the necessary notations and definitions, then we describe our algorithm. Section 3 is dedicated to the correctness proof and to the performance analysis of the proposed algorithm, and, finally, Section 4 presents future directions.

2 An Improved Logarithmic Approximation Algorithm

In this section we present our logarithmic approximation algorithm which, given (i) the complete weighted graph $G(S)$ with vertices $S = \{1, \ldots, n\}$ representing stations and with a cost function $c : S \times S \mapsto I\!R^+$ which associates each pair of stations i and j with its transmission cost, and (ii) a vertex s, returns a power assignment $\omega : S \mapsto I\!R^+$ which induces a transmission graph G_ω containing a directed spanning tree rooted at s.

Before presenting the algorithm, we introduce a few preliminary notations.

Notations. For any station i, let $C_i = \bigcup_{j \in S} \{c(i, j)\}$ denote the set of all the possible level powers allowing i to transmit to any other station j in the network through a single-hop transmission. It is important to notice that even if, by definition, for any station i, the transmission power $\omega(i)$ can range in the interval $[0, \infty]$, the set of possible power assignments is actually the discrete set C_i of at most $n - 1$ values.

For any transmission power $\varpi \in C_i$ assigned to a station i, $S_\varpi(i) = \{j \in S \mid j \neq i \text{ and } \varpi \geq c(i, j)\}$ is the set of all the stations j distinct from i into the area of coverage of i, and $\mathcal{P}_\varpi(i) = \{X \in \mathcal{P} \mid X \cap S_\varpi(i) \neq \emptyset\}$ is the family of sets belonging to a given partition $\mathcal{P}$ and containing stations covered by i.

Finally, given a power assignment ω associating a power transmission $\omega(i)$ to each station i, $\Pi(s, G_\omega)$ is the set of vertices i connected to a vertex s through a directed path from s to i, while $G_\omega[X]$ is the subgraph of G_ω induced by a set $X \subseteq S$.

2.1 The Greedy-Assignment Algorithm

We propose the "GREEDY-ASSIGNMENT" algorithm, which uses a greedy approach for solving the MECBS problem. Indeed, in order to compute a power assignment ω, it initially assigns a transmission power zero to every station in S; then it iteratively uses a specific greedy rule to select a station and increases its transmission power until ω induces on S a weakly connected transmission graph G_ω. Finally, this assignment ω is re-adjusted in order to grow the directed spanning tree rooted at s, one station at a time, without destroying any connection previously established.

The algorithm works as follows. The loop 3 iteratively manages a partition $\mathcal{P}$ of stations initially composed of n singleton sets and a power assignment ω identically zero at the beginning. At each stage, $\mathcal{P}$ represents all the weakly connected components of G_ω. This invariant is guaranteed by the "MERGE" procedure which, given the pair $\langle i, \varpi \rangle$ chosen at line 3.(a), merges all the sets in $\mathcal{P}_\varpi(i)$ together with $S_\varpi(i) \cup \{i\}$ in such a way that, after line 3.(b), their union induces on G_ω a new weakly connected subgraph. At the end of loop 3, all the sets contained in the partition $\mathcal{P}$ are merged together, denoting that the transmission graph G_ω is weakly connected.

Line 3.(a) is the greedy decision-making step: a station i and a transmission power $\varpi \in C_i$ are chosen such that $|\mathcal{P}_\varpi(i)|$ components of G_ω are connected with i with a minimum average power consumption.

Definition 1. *Given a partition $\mathcal{P}$ of S, a station $i \in S$ and a transmission power $\varpi \in C_i$, we define the* cost-effectiveness $\varepsilon_\varpi(i, \mathcal{P})$ *of ϖ at i, with respect to $\mathcal{P}$, as the average transmission power at which i covers sets in $\mathcal{P}_\varpi(i)$, i.e.,*

$$\varepsilon_\varpi(i,\mathcal{P}) = \frac{\varpi}{|\mathcal{P}_\varpi(i)|}.$$

If $\mathcal{P}_\varpi(i) = \emptyset$ we set $\varepsilon_\varpi(i,\mathcal{P}) = \infty$.

According to Def. 1, at line 3.(a) a pair $\langle i, \varpi \rangle$ is chosen such that, after setting $\omega(i) = \varpi$ (see line 3.(b)), the cost-effectiveness $\varepsilon_\varpi(i,\mathcal{P})$ is minimum. Then all the components in $\mathcal{P}_\varpi(i)$ are merged together with $S_\varpi(i) \cup \{i\}$ at line 3.(c).

Finally, The loop 5 re-adjusts the transmission powers of stations i not connected to s through a directed path, without destroying any connection previously established, until the transmission graph contains a directed spanning tree rooted at s.

GREEDY-ASSIGNMENT $(G(S), s)$

1. $\mathcal{P} \leftarrow \{\{1\}, \{2\}, ..., \{n\}\}$
2. $\omega(i) \leftarrow 0$ for every station $i \in S$.
3. **while** $|\mathcal{P}| > 1$ **do**
   (a) Choose the pair $\langle i, \varpi \rangle$ with the lowest cost-effectiveness $\varepsilon_\varpi(i,\mathcal{P})$, for $i \in S$ and $\varpi \in C_i$
   (b) $\omega(i) \leftarrow \varpi$
   (c) $\mathcal{P} \leftarrow$ MERGE $(G(S), \mathcal{P}, i, \varpi)$
4. $\nu = \omega$
5. **while** $\Pi(s, G_\omega) \neq S$ **do**
   (a) choose a vertex $i \in \Pi(s, G_\omega)$ and a vertex $j \notin \Pi(s, G_\omega)$ such that there exists a directed edge $\langle j, i \rangle$ in G_ν
   (b) $\omega(i) \leftarrow \nu(j)$
6. **return** ω

MERGE $(G(S), \mathcal{P}, i, \varpi)$

- $S_\varpi^+(i) \leftarrow (\bigcup_{X \in \mathcal{P}_\varpi(i)} X) \cup S_\varpi(i) \cup \{i\}$
- $\mathcal{P}_\varpi^-(i) \leftarrow \mathcal{P}_\varpi(i) \cup \{S_\varpi(i) \cup \{i\}\}$
- **return** $\mathcal{P} \setminus \mathcal{P}_\varpi^-(i) \cup \{S_\varpi^+(i)\}$

The correctness proof of the algorithm is based on the two following invariant properties.

2.2 Invariant Properties

Let us denote by $\mathcal{P}^{(k)}$ the partition of stations at the end of the k-th iteration of loop 3 and by $\omega_1^{(k)}$ and $\omega_2^{(k)}$ the power assignments returned at then end of the k-th iteration of loop 3 and loop 5, respectively.

The first important property states that, although different transmission powers $\varpi_1, \ldots, \varpi_m$ may be assigned to a same station i, such a sequence is increasing. This guarantees that re-adjustments (see lines 3.(b) and 5.(b)) never destroy connections previously established.

Property 1. For any station $i \in S$ and for any constant k,l,m,q such that $k<l$ and $m<q$ it holds: $\omega_1^{(k)}(i) \leq \omega_1^{(l)}(i) \leq \omega_2^{(m)}(i) \leq \omega_2^{(q)}(i)$.

Proof. Initially, $\omega(i) = \omega_1^{(0)}(i) = 0$ (line 2). Let $\omega(i)$ be set to ϖ at the k-th iteration of loop 3, i.e., $\omega_1^{(k)}(i) = \varpi$. The "MERGE" procedure (line 3.(c)) makes the union of all the sets in $\mathcal{P}_\varpi^{(k)}(i)$ and $S_\varpi(i) \cup \{i\}$. Hence, since for every $\varpi' \in \mathcal{C}_i$ such that $\varpi' \leq \varpi$ it holds $S_{\varpi'}(i) \subset S_\varpi(i)$, when $h > k$ it must be $\mathcal{P}_{\varpi'}^{(h)}(i) = \emptyset$ and, as a consequence, $\varepsilon_{\varpi'}(i, \mathcal{P}^{(h)}) = \infty$. This implies that it cannot be $\omega_1^{(k)}(i) > \omega_1^{(h)}(i)$ for every $h > k$. Let $\omega_2^{(0)}$ denote the assignment $\omega \ (= \nu)$ at line 4. Now, consider the m-th iteration of loop 5, for any $m \geq 1$. At line 5.(b) the transmission power $\nu(j)$ is assigned to i only if the edge $\langle j, i \rangle \in G_\nu$ but $\langle i, j \rangle \notin G_{\omega_2^{(m)}}$. Then it must be $\nu(j) > \omega_2^{(m)}(i)$. □

Now we are able to prove that, for any k, at the end of the k-th iteration of loop 3, the partition $\mathcal{P}^{(k)}$ always contains sets of stations that induce on $G_{\omega_1^{(k)}}$ weakly connected subgraphs. In other words, $\mathcal{P}^{(k)}$ represents all the weakly connected components of $G_{\omega_1^{(k)}}$.

Property 2. For each set $X \in \mathcal{P}^{(k)}$, $G_{\omega^{(k)}}[X]$ is weakly connected.

Proof. This is obviously true for $k = 0$. For any $k \geq 1$, at the k-th iteration of loop 3, a pair $\langle i, \varpi \rangle$ is chosen in such a way to increase to ϖ the transmission power associated with i. By construction, each set in $\mathcal{P}_\varpi(i)$ contains at least one station belonging to $S_\varpi(i)$, so $S_\varpi^+(i)$ necessarily induces on the transmission graph $G_{\omega_1^{(k)}}$ a weakly connected subgraph. □

3 Correctness and Performance Analysis

In the following section we provide a correctness proof of our algorithm, by showing that the power assignment ω returned by it actually induces a transmission graph G_ω which contains a directed spanning tree rooted at a source station s. Due to space limitations, a few details are omitted. Finally, we show that it has an approximation ratio of $2H_{n-1}$.

Theorem 1 (Correctness). *The output ω returned by the* GREEDY-ASSIGN-MENT *algorithm is feasible.*

Proof. The proof is made of two parts: first we prove that ω at the end of loop 3 (i.e., ν) induces on S a weakly connected transmission graph; then we show that the graph G_ω induced by the output ω admits a directed spanning tree rooted at s. The former is a trivial consequence of Prop. 2. By Prop. 1, no readjustment at line 5.(b) can destroy connections previously established, then the latter is clearly true if the algorithm terminates, as it is guaranteed by the while condition of loop 5. Thus, we only need to prove that loop 5 always terminates. At a generic iteration of loop 5, if $\Pi(s, G_\omega) \neq S$, since G_ν is weakly connected, there must exist two vertices i and j satisfying the conditions in line 5.(a). The claim follows by noting that adding the power assignment in 5.(b), by Prop. 1, we have that the set $\Pi(s, G_\omega)$ expands. □

Given a sequence of pairs $\mathcal{F} = \langle i_1, \varpi_1 \rangle, \ldots, \langle i_r, \varpi_r \rangle$, let $cost(\mathcal{F}) = \sum_{j=1}^{r} \varpi_j$. Each transmission power ϖ_j can be spread among the sets in $\mathcal{P}_{\varpi_j}(i_j)$ by assigning each set a cost equal to the cost-effectiveness of ϖ_j at i_j, with respect to $\mathcal{P}^{(j)}$. That is, for every $X \in \mathcal{P}_{\varpi_j}(i_j)$, $cost(X) = \varepsilon_{\varpi_j}(i_j, \mathcal{P}^{(j)})$. If $\mathcal{T}_{\mathcal{F}}$ denote the family of sets $\mathcal{T}_{\mathcal{F}} = \bigcup_{j=1}^{r} \mathcal{P}_{\varpi_j}(i_j)$, one has: $cost(\mathcal{F}) = \sum_{j=1}^{r} \varpi_j = \sum_{X \in \mathcal{T}_{\mathcal{F}}} cost(X)$. Observe that if we consider the sequence $\mathcal{A}$ of the pairs chosen by our algorithm, clearly $|\mathcal{T}_{\mathcal{A}}| = n - 1$. Number these sets in the order in which they are merged by the algorithm breaking ties arbitrarily.

Let OPT be the cost of an optimal solution ω^*. At any iteration k_i when T_i is merged the following lemma holds.

Lemma 1. *For any $i \leq n-1$ there exists a pair $\langle j, \varpi \rangle$ having a cost-effectiveness $\varepsilon_\varpi(j, \mathcal{P}^{(k_i)}) \leq \frac{OPT}{n-i}$.*

Proof. Let $\mathcal{F}^{(i)}$ be the sequence of the pairs chosen by the optimal algorithm such that $\omega^*(j) > \omega_1^{(k_i)}(j)$. Clearly, $OPT \geq cost(\mathcal{F}^{(i)}) = \sum_{j | \langle j, \varpi \rangle \in \mathcal{F}^{(i)}} \omega^*(j)$. Now, suppose by contradiction that for every pair $\langle j, \varpi \rangle \in \mathcal{F}^{(i)}$, it holds $\varepsilon_\varpi(j, \mathcal{P}^{(k_i)}) > \frac{OPT}{n-i}$. Since by definition, the cost-effectiveness of a pair $\langle j, \varpi \rangle$ cannot decrease as i increases and $|\mathcal{T}_{\mathcal{F}^{(i)}}| = n - i$, we have that $OPT \geq cost(\mathcal{F}^{(i)}) > (n-i)\frac{OPT}{n-i} = OPT$. □

Theorem 2 (Performance Guarantee). *The* GREEDY-ASSIGNMENT *algorithm has an approximation ratio of $2H_{n-1}$.*

Proof. We first show that $cost(\nu) \leq H_{n-1} \cdot OPT$, then that $cost(\omega) \leq 2 \cdot cost(\nu)$.

Since the algorithm chooses the power assignment with the lowest cost-effectiveness, from the above lemma we have that $cost(T_i) \leq \frac{OPT}{n-i}$. Thus, the solution at line 4 is such that $cost(\nu) = \sum_{i=1}^{|T|} cost(T_i) = \sum_{i=1}^{n-1} cost(T_i) \leq \sum_{i=1}^{n-1} \frac{OPT}{n-i} = OPT \sum_{i=1}^{n-1} \frac{1}{i} = H_{n-1} \cdot OPT$.

At any iteration of loop 5 a transmission power $\nu(j)$ is assigned to i only if the conditions of line 5.(a) are enjoyed by the pair of stations i and j. Such an assignment establishes a directed path from s to j, then j can be chosen at step 5.(a) at most once. Consequently, $cost(\omega) \leq 2 \cdot cost(\nu)$. □

Notice that the approximation ratio $2 \cdot H_{n-1}$ is essentially tight, as shown by the following example.

Example. When the GREEDY-ASSIGNMENT algorithm runs on the input graph in Fig. 1.a) it computes the partial solution ν depicted in Fig. 1.c). Thus, in the case in which n is an odd number:

$$cost(\nu) = \sum_{j=0}^{\lceil \frac{n-1}{2} \rceil - 1} \frac{1}{\lceil \frac{n-1}{2} \rceil - j} = \sum_{j=1}^{\lceil \frac{n-1}{2} \rceil} \frac{1}{j} = H_{\lceil \frac{n-1}{2} \rceil} = H_{n-1} - 1.$$

Next, loop 5 exactly doubles all the transmission powers, hence $cost(\omega) = 2H_{n-1} - 2$. Since the cost of the optimal solution in Fig. 1.b) is $1 + \epsilon$, we have that the approximation factor of our algorithm is tight up to low order terms.

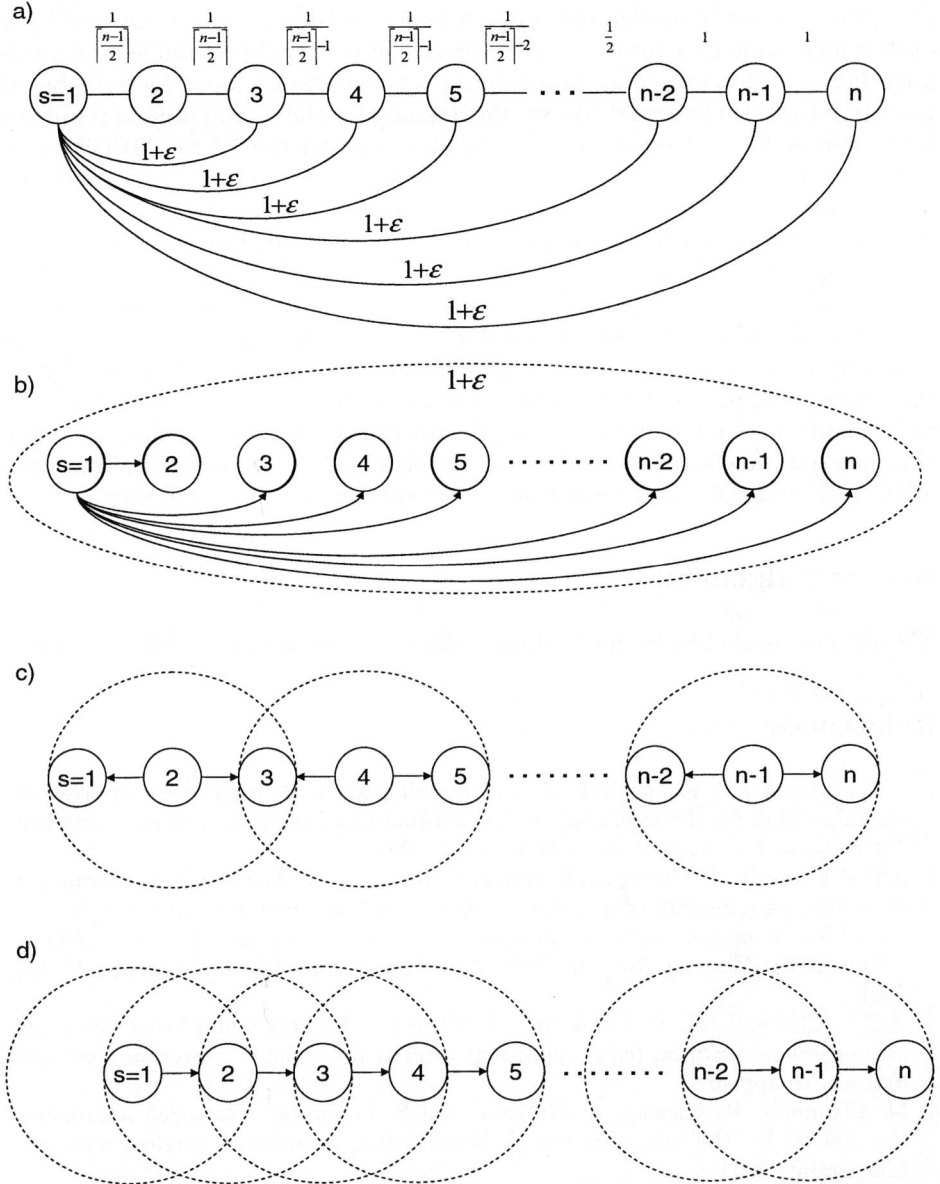

Fig. 1. a) The input graph. Missing edges are assumed to have an infinite cost. b) The optimal solution. c) The partial solution ν. d) The solution ω.

4 Future Directions

We presented a $2H_{n-1}$-approximation algorithm for the MECBS problem [1, 2, 6], which, despite its simplicity, improves the only known algorithm with a ratio of $10.8 \log n$ [1] and considerably approaches the best-known lower bound of $\log n$ on the approximation ratio [2, 6]. We leave the natural open problem of bridging this gap. Moreover, once broadcasting models are developed for node-based models, future studies could address the impact of mobility and limited resources (both bandwidth and equipment). A significant restriction of the MECBS problem, denoted MECBS[N_d^α], consists in considering the stations located in the d-dimensional Euclidean Space, and a cost function s.t. $c(i,j) = dist(i,j)^\alpha$, where α is the *distance-power gradient*. It has been proved that the MECBS[N_d^α] problem is NP-Hard for $\alpha > 1$ and $d > 1$, while it is in P if $\alpha = 1$ or $d = 1$ [1–3, 6]. The best known approximation algorithm, called *MST*, has been presented and compared with other heuristics (SPT, BIP) through simulations on random instances in the case $d = \alpha = 2$ [7]. Its performance has been investigated by several authors [2, 6, 4] and the evaluation of its approximation ratio progressively reduced till $3^d - 1$ for every $\alpha \geq d$ [4]. Starting from these results, a further left open question is that of comparing through simulations our algorithm with MST, SPT and BIP when restricting to the special case of MECBS[N_d^α].

Acknowledgements

The authors would like to thank Michele Flammini for several useful discussions.

References

1. I. Caragiannis, E. Kaklamanis, and P. Kanellopoulos. A logarithmic approximation algorithm for the minimum energy consumption broadcast subgraph problem. *Information Processing Letters*, 86:149–154, 2003.
2. A.E.F. Clementi, P. Crescenzi, P. Penna, G. Rossi, and P. Vocca. On the complexity of computing minimum energy consumption broadcast subgraph. In *Proceedings of the 18th Annual Symposium on Theoretical Aspects of Computer Science (STACS)*, volume 2010 of *Lecture Notes in Computer Science*, pages 121–131. Springer-Verlag, 2001.
3. A.E.F. Clementi, M. Di Ianni, and R. Silvestri. The minimum broadcast range assignment problem on linear multi-hop wireless networks. *Theoretical Computer Science.* to appear.
4. M. Flammini, R. Klasing, A. Navarra, and S. Perennes. Improved approximation results for the minimum energy broadcasting problem in wireless networks. Manuscript, 2003.
5. T.S. Rappaport. *Wireless communications: principles and practice.* Prentice-Hall, Englewood Cliffs, NY, 1996.
6. P. J. Wan, G. Calinescu, X. Li, and O. Frieder. Minimum energy broadcast routing in static ad hoc wireless networks. *Wireless Networks*, 8(6):607–617, 2002.
7. J. E. Wieselthier, G. D. Nguyen, and A. Ephremides. On the construction of energy-efficient broadcast and multicast trees in wireless networks. In *Proceedings of the 19th Annual Joint Conference of the IEEE Computer and Communications Societies (INFOCOM)*, pages 585–594. IEEE Computer Society, 2000.

UVOD:
A Jini-Based Ubiquitous VOD Service Architecture[*]

Tae Uk Choi and Ki-Dong Chung

Dept. of Computer Science, Pusan National University,
San-30, Jangjeon-dong, Keumjeong-gu, Pusan 609-735, South Korea
{tuchoi,kdchung}@pusan.ac.kr

Abstract. In this paper, we present a ubiquitous VOD service architecture (UVOD) that provides mobile users with a seamless VOD service.

1 Introduction

The conventional VOD system is a machine-oriented application, in which a user goes to a computer to request a movie. A VOD session is set up between the client and the server, and the content is delivered to the client. If the user wants to move to a different location, he/she needs to stop the video play at the current location and makes a new request at the new location. As well, the user needs to manually search for the last frame to be seen at the previous location. This is because there is no user mobility support in the VOD system. Current computing environments are changing from mobile computing to ubiquitous computing, where users compute at anytime and anywhere. In ubiquitous environments, a user is free to move one space to another. Thus, a ubiquitous VOD system should be able to support *user mobility*[1-4].

Fig. 1 shows a scenario of a VOD service in ubiquitous environments. A user starts watching a movie from a desktop in Space A. When the user moves out of Space A, the user can receive the VOD service continuously with a PDA. Upon arrival to Space B, the VOD session is shifted from the PDA to the desktop in Space B, and then the user watches the same video via the desktop. This scenario gives a new meaning to VOD applications: ubiquitous VOD applications should be able to trace the current location of a specific user and migrate the VOD session from one host to another. In this paper, we present a ubiquitous VOD service architecture (UVOD) that perceives user movement and provides mobile users with seamless VOD service. We design UVOD based on the Java and Jini[5], which is recently released by Sun Microsystems. Jini provides a distributed software platform for dynamically creating networked components, applications and services. The distinctive features of Jini are simplicity and platform independence, good security and robustness, and ability to be combined with other Java technologies such as JMF(Java Media Framework)[6].

---

[*] This work was supported by grant No. R05-2002-000-00345-0 from the Basic Research Program of the Korea Science & Engineering Foundation.

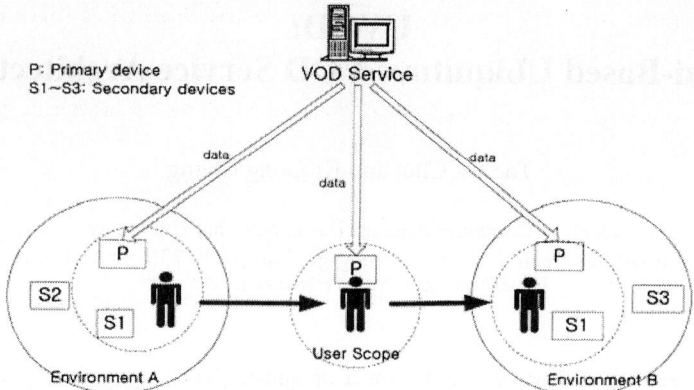

Fig. 1. VOD service in ubiquitous environments

2 Design Issues

As shown in Fig.1, we assume that a user has a *user scope*: the service area in which client hosts are able to serve the user. The user scope may not be the shape of a circle. There may be several client hosts in a user scope. We call the client host that currently serves a user as the *primary client* and the client hosts that are not the primary host as *secondary clients*. Note that the client hosts in the user scope change as the user moves from one space to another. One of the secondary hosts may become the primary client in the near future. This causes the selection problem of a new *primary client*. We believe that the client which is able to provide the user with maximum QoS should be the primary client. Thus, UVOD should select the primary client depending on the capacity and resource of client hosts.

After selecting the primary client host, the VOD session needs to migrate from the previous primary client to the new primary client, called *session handoff*. That is, as the user moves one space to another, the old session closes and a new session starts. This means that handoffs divide an entire session into *sub-sessions*. We represent a sub-session as follows.

SubSession (Sender Address, Receiver Address, Movie, Position)

Sender Address and *Receiver Address* denote the IP address of the client host transmitting the video and the IP address of the client receiving the video respectively. *Movie* is the movie title requested by the user. *Position* denotes the start position(frame number) in the movie file. The transmission of a sub-session starts from that position. For example, if a user moves according to the scenario of Fig.1, the subsesisons can be represented as follows.

SubSession (Sender, PC1 , "hyori.mpg", 0)
SubSessoin (Sender, PDA, "hyori.mpg", last_stop_position)
SubSession (Sender, PC2, "hyori.mpg", last_stop_position)

Thus, UVOD should support session handoff by closing the old sub-session and then starting a new sub-session. However, it is possible that the resource capacity of the new primary client is very different from that of the previous primary client host. In this case, the QoS of the session needs to be adjusted. For example, when the session shifts from PC(Personal Computer) to PDA(Personal Digital Assistant), the QoS such as frame size, frame rate and data format should be changed in order that the new sub-session can be played out on the PDA. Thus, UVOD should negotiate the QoS of the new sub-session according to the capacity and resource of the hosts and user preference.

3 UVOD Architecture

Fig. 2 shows the UVOD architecture, which consists of the UVOD Service, the UVOD Client and the UVOD Sender. UVOD has two communication interfaces: Jini for control data and JMF for media data. Jini is used to discover the UVOD Service and communicate internal control messages. JMF is exploited to transmit, receive and play out the video between a UVOD sender and a UVOD client.

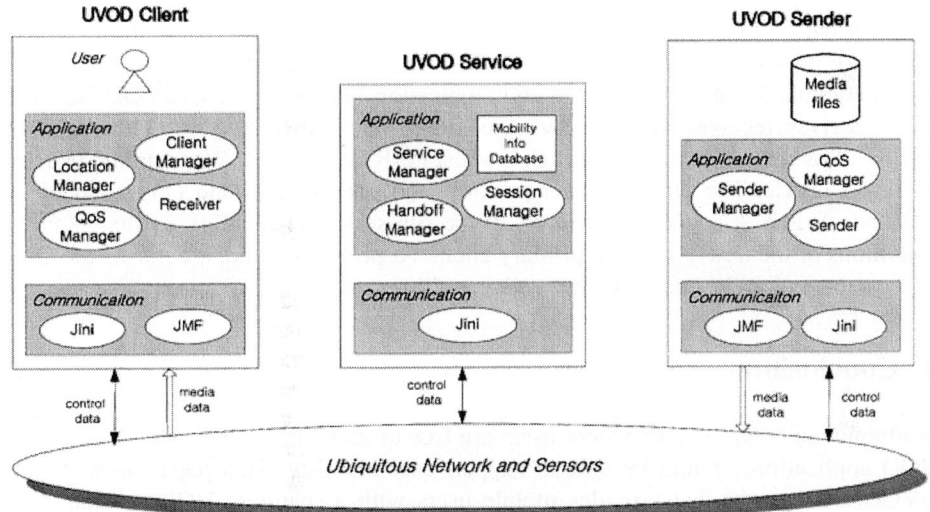

Fig. 2. UVOD architecture

The UVOD Service is a Jini service that controls the UVOD client and the UVOD sender. It manages the video sessions and the information of the user's mobility. The Service Manager creates a Jini service, which takes charge of the communications with the UVOD Client and the VOD Sender, and registers the proxy to a Jini Lookup Service. It receives the information associated with users and client hosts from UVOD Clients, and stores the information into the Mobile Info Database. The database includes the information of user mobility such as user locations, client locations, user scopes, client host lists in a user scope, movie lists, session lists, and the QoS information of sessions. The Session Manager allows the session to stop and resume when a user moves from one client host to another. To do this, it needs to hold the stopped position in the media file that is currently being displayed. The Handoff Manager selects a primary client when the user's location changes, considering the location of clients within the user scope. Once a new primary client is selected, it informs the Session Manager as well as the UVOD Clients and the UVOD Sender of the occurrence of handoff.

The UVOD Sender takes charge of transmitting video data to UVOD clients. The Sender Manager downloads the proxy from the Jini Lookup Service and communicates with the UVOD Service through the proxy. According to messages from the Service Manager, the Sender starts or stops transmitting the movie requested by the user. Note that it can transmit a part of the movie file, not a whole file during the sub-session. That is, the data is retrieved from the last stopped position to the frame position that a handoff occurs in the media file. It employs RTP to send the video data. The QoS Manager controls the frame rate depending on network conditions and resource conditions. To do this, it need to cooperate with the QoS Manager of the UVOD Client.

The UVOD Client takes charge of providing a mobile user with a seamless video play. The Location Manager collects the location information of users, and transmits the information to the UVOD Service. This manager utilizes the function of location sensors or proximity hardware such as AIR ID[7]. The Client Manager processes the user's commands and provides the communication interface of the UVOD Service. The Receiver receives and displays the video data transmitted from the UVOD Sender. The data is received through a separate RTP channel. We employ JMF to display a movie. The QoS Manager performs QoS negotiation and QoS adaptation. At handoff time, it negotiates the QoS of the sub-session depending on the resource conditions of the newly selected primary client. At provision time, it controls the QoS of the sub-session depending on network conditions.

4 Conclusion

In ubiquitous environments where users are free to move from one space to another, VOD applications should be able to support user mobility. This paper presents the UVOD architecture that provides mobile users with a seamless VOD service. The significant features of UVOD are the selection of the primary client out of the secondary hosts within the user scope, the support of sub-session handoff, and the QoS adaptation of the sub-session during the handoff time. In the future, we will implement the UVOD system.

References

1. Yi Cui, K. Nahrstedt, D. Xu, "Seamless User-level Handoff in Ubiquitous Multimedia Service Delivery", Multimedia Tools and Applications Journal, 2003.

2. J. P. Sousa and D. Garlan, "Aura: an Architectural Framework for User Mobility in Ubiquitous Computing Environments", IEEE/IFIP Conference on Software Architecture, Montreal, 2002.
3. H.J. Wang et al. "ICEBERG: An Internet-core Network Architecture for Integrated Communications", IEEE Personal Communications, Special Issue on IP-based Mobile Telecommunication Networks, 2000.
4. G. Appenzeller et al. "The Mobile People Architecture", ACM Mobile Computing and Communication Review, Vol. 1, No. 2, 1999.
5. Sun Microsystems, http://developer.java.sun.com/developer/products/jini
6. Sun Microsystems, http://java.sun.com/products/java-media/jmf
7. "Air ID System", in http://www.pcprox.com

Topic 16
Integrated Problem Solving Environments

Daniela di Serafino, Elias Houstis, Peter Sloot, and Domenico Talia

Topic Chairs

Problem Solving Environments (PSEs) can be defined as integrated computing environments for developing, executing and analysing applications in a specific domain. They provide a set of user-friendly mechanisms and tools that allow to "compose" an application, by gluing together, using some kind of problem-oriented language, different building blocks. Such building blocks range from libraries and application codes, to tools for I/O, data visualization and analysis, and interactive steering. PSEs may also incorporate some form of knowledge, in order to assist the users in formulating, solving and analysing their problems. The main motivation for developing PSEs is that they enable to build applications without dealing with most of the details related to hardware and software architectures, to solution algorithms and their implementations, and to analysis and monitoring tools, thus allowing end-users to concentrate on the application problems to be solved. PSEs can be used for different purposes, such as modelling and simulation, design optimisation, rapid prototyping, and decision support.

Developing fully integrated PSEs requires different expertise and a huge amount of programming effort. A significant evolution in the design, development and application of PSEs took place in the last decade, pushed by the rapid changes in hardware and software and by the requirements of applications. Nevertheless, more research is needed to realise fully integrated PSEs, enabling more complex simulations, higher levels of abstraction and more effective cooperation among multiple users in distributed collaborative environments. The exploitation of technologies such as parallel and distributed computing, component-based software engineering, advanced interactive visualization, and Grid computing plays a fundamental role in pursuing this goal.

Six papers were submitted to the topic Integrated Problem Solving Environments and each one received three reviews. Three regular papers were selected, dealing with different aspects of the PSE research. The paper by C. Gomes et al., "Pattern/Operator based Problem Solving Environments", presents an approach for extending PSEs and Grid Computing Environments with Design Patterns and Operators, that allow to manage the composition and the execution of the collection of components available within such environments. A prototype implementation of this approach is also described. The paper by Z. Jiao at al., "Databases, Workflows and the Grid in a Service Oriented Environment", discusses a toolkit which combines a Grid-enabled database-driven repository with a workflow system, in order to assist users in Engineering Design Search and Optimisation processes. Finally, the paper by M. Lettere at al., "A Parallel Programming Tool for SAR Processors", describes a parallel programming model for developing image processing algorithms, which is based on structured parallelism and object-oriented abstractions. A sample implementation of this model is also presented.

We wish to thank the people who submitted their contributions, the reviewers who helped us in selecting the papers and the Euro-Par 2004 Organizing Committee, for making this topic possible.

Pattern/Operator Based Problem Solving Environments

Cecilia Gomes[1], Omer F. Rana[2], and Jose C. Cunha[1]

[1] CITI Center, University Nova de Lisboa, Portugal
[2] School of Computer Science, Cardiff University, UK

Abstract. Problem Solving Environments (PSEs) provide a collection of tools for composition of scientific applications. Such environments are often based on graphical interfaces that enable components to be combined, and in some cases, subsequently scheduled on computational resources. A novel approach for extending such environments with Design Patterns and Operators is described – as a way to better manipulate the available components – and subsequently manage their execution. Users make use of these additional abstractions by first deploying 'Structural Patterns' and by refining these through 'Structural Operators'. 'Behavioural Patterns' may then be used to define the control and data flows between components – subsequent use of 'Behavioural Operators' manage the final configuration for execution control and dynamic reconfiguration purposes. We demonstrate the implementation of these Patterns and Operators using Triana [14] and the Distributed Resource Management Application (DRMAA) API [10].

1 Introduction and Motivation

A Problem Solving Environment (PSE) is a complete, integrated computing environment for composing, compiling, and running applications in a specific area [1]. In many ways a PSE is seen as a mechanism to integrate different software construction and management tools, and application specific libraries, within a particular problem domain. One can therefore have a PSE for financial markets [4], for Gas Turbine engines [5], etc. Focus on implementing PSEs is based on the observation that previously scientists using computational methods wrote and managed all of their own computer programs – however now computational scientists must use libraries and packages from a variety of sources, and those packages might be written in many different programming languages. Engineers and scientists now have a wide choice of computational modules and systems available, enough so that navigating this large design space has become its own challenge. A survey of 28 different PSEs by Fox, Gannon and Thomas (as part of the Grid Computing Environments WG) can be found in [6], and practical considerations in implementing PSEs can be found in Li et al. [2]. Both of these indicate that such environments provide "some backend computational resources, and convenient access to their capabilities". Furthermore, workflow features significantly in both of these descriptions. In many cases, access to data resources is also provided in a similar way to computational ones. Often PSE and Grid Computing Environment is used interchangeably – as PSE research predates the existence of Grid infrastructure. The aim of our work is to: (1) extend the capabilities of existing PSEs with support for Patterns and Operators, and (2) to enable the management of such applications subsequently by mapping "Behavioural" patterns and operators to a resource management system. Patterns allow the abstraction of common interactions between components, thereby enabling reuse of interactions that have proven useful in similar domains. A user may build an application in a structured fashion by selecting the most appropriate set of patterns, and by combining them according to operator semantics. Users may also define new patterns found to be useful, and add these as components for use by others. Patterns and Operators also provide additional capability that is not easily representable via visual components. Our approach treats patterns as first class entities but differs from other works [7, 8] in that the user may explicitly define structural constraints between components, separately from the behavioural constraints. Our approach is somewhat similar to that of van der Aalst et al. [17] – although they do not make a distinction between structural and behavioural patterns. They also focus on Petri net models of their patterns, whereas our concern is to link patterns with particular resource managers and composition tools. The approach presented here is primarily aimed at computational scientists and developers, who have some understanding of the computational needs of their application domain. A scientist should be aware about the likely co-ordination and interaction types between components of the application (such as a database or numeric solver etc). The structural and behavioural patterns presented here will enable such scientists and developers to utilise common usage scenarios within a domain (either the use of particular components, such as database systems, or interactions between components, such as the use of streaming). Section 2 introduces our concept of Patterns and Operators, and Section 3 demonstrates how these are implemented in the Triana (the workflow system for the European GridLab project [9]) – and describe theme (1) mentioned above. Theme (2) is then explained in Section 4.

2 Structured Composition of Applications in PSEs Based on Grids

Structural Pattern Templates encode component connectivity, representing topologies like a ring, a star or a pipeline, or design patterns like Facade, Proxy or Adapter [15]. The possibility of representing these structural constraints allows, for example, the representation of common software architectures in high-performance computing applications. For example, the pipeline pattern may be used in a signal processing application where the first stage may consist of a signal generator service producing data to a set of intermediate stages for filtering. Frequently, the last stage consists of a visualisation service for observing results. The proxy pattern, for instance, allows the local presence of an entity's surrogate, allowing access to the remote entity. Grid services, for example, are usually accessed through a proxy (or gatekeeper).

Behavioural Pattern Templates capture recurring themes in component interactions, and define the temporal and the (control and data) flow dependencies between the components. Generally, these applications involve distribution of code from a master, the replication of a code segment (such as within a loop), or parameter sweeps over one or more indices. We provide several behavioural patterns such as Master-Slave, Client-Server, Streaming, Peer-to-Peer, Mobile Agents/Itinerary, Remote Evaluation, Code-on-Demand, Contract, Observer/Publish-Subscriber, Parameter sweep, Service Adapter, and so on. For example, the Service Adapter pattern "attaches additional properties or behaviours to an existing application to enable it to be invoked as a service" [16]. The Master-Slave pattern, in turn, can be mapped to many parallel programming libraries, and represents the division of a task into multiple (usually independent) sub-units – and shares some similarities with the Client-Server pattern – although the control flow in the latter is more complex.

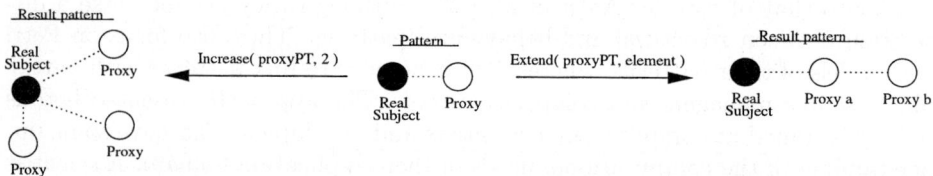

Fig. 1. The increase and extend structural operators.

Structural Operators support the composition of structural patterns, without modifying the structural constraints imposed on the pattern. This provides a user with a simple and flexible way to refine structural patterns. There are several structural operators such as increase, decrease, extend, reduce, rename, replace, replicate, embed, etc. For example in figure 1 it is possible to observe the result of applying the increase and extend operators to the Proxy pattern.

Behavioural Operators are applied over the structural operator templates combined with the behavioural patterns after instantiating the templates with specific runnable components. Behavioural operators act upon pattern instances for execution control and reconfiguration purposes. Behavioural operators include: Start (starts the execution of a specific pattern instance), Stop (stops the execution of a pattern instance saving its current state), Resume (resumes the execution of a pattern instance from the point where it was stopped), Terminate (terminates the execution of a specific pattern instance), Restart (allows the periodic execution of a pattern instance), Limit (limits the execution of a specific pattern instance to a certain amount of time; when the time expires the execution is terminated), Repeat (allows the repetition of the execution of a specific pattern a certain number of times), etc. Both structural operators and behavioural operators can be combined into scripts which may be later reused in similar applications.

3 Implementation over the Triana GCE

A prototype has been implemented by extending Triana [14], and allows developers to utilise a collection of pre-defined patterns from a library. Triana comes with components (called *units*) for signal processing, mathematical calculations, audio and image processing, etc, and provides a wizard for the creation of new components, which can then be added to the toolbox. Structural Patterns appear as standard components that can be combined with other patterns or executable units. Triana provides both a composition editor, and a deployment mechanism to support this. The Pattern library provided within Triana treats patterns as "group units" (i.e. units made up of others). Each element within such group units is a "dummy" component (or a place holder) and can subsequently be instantiated with executables from the Triana toolbox. Hence, structural pattern templates are collections of dummy components that can be instantiated with other structural pattern templates or with executables.

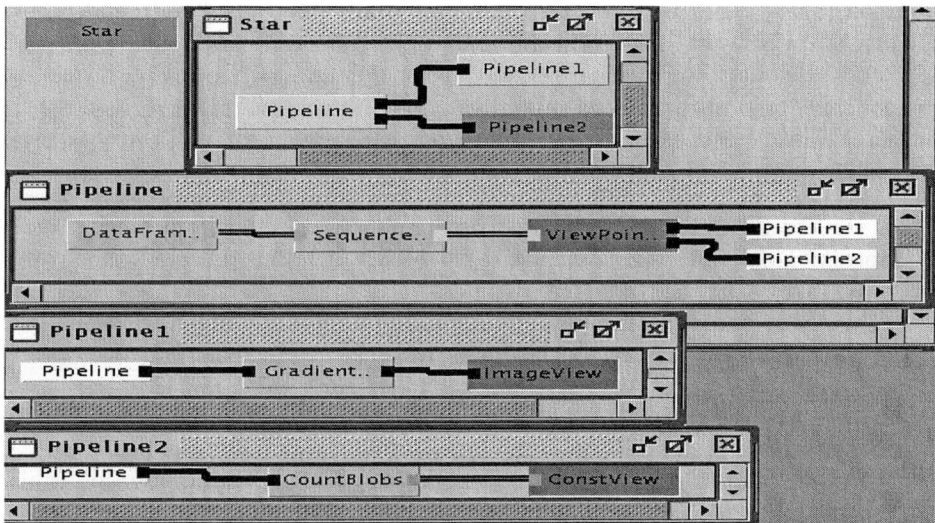

Fig. 2. A possible final configuration for the image processing of the "Galaxy Formation example".

A Galaxy simulation application with Triana is illustrated in figure 2. The Galaxy formation example may be represented by a star pattern template, where the nucleus contains the actions necessary to generate and control the animation execution, and the satellites represent image processing and analysis actions. Both the actions at the nucleus and at the satellites are supported by pipeline templates. As such, the *Pipeline* pattern instance shown in the figure, represents the actions at the nucleus, and consists of three stages. *Pipeline* is connected to the two satellites, namely, *Pipeline1* and *Pipeline2*, and produce data to these two pattern instances. *Pipeline1* is embedded in one of the satellites and connects two units for image processing. *Pipeline2* binds two units that together support analysis of the data produced at the nucleus. See [12] for details.

4 Mapping to the DRMAA API

Behavioural patterns are implemented over the run-time system used to execute the components. There is no visual representation of these, as they are provided as a collection of scripts that need to be configured by a user prior to execution. We map behavioural patterns over DRMAA [10]. Pattern execution essentially involves coordination between modules. Execution is therefore delegated via DRMAA to third party resource management systems (DRMAA provides a generalised API to execute jobs over Distributed Resource Management Systems (DRMSs)). DRMAA includes common operations on jobs like termination or suspension. A job is a running application on a DRMS and it is identified by a *job_id* attribute that is passed back by the DRMS upon job submission. This attribute is used by the functions that support job control and monitoring. DRMAA API uses an IDL-like definition (with IN defining an input parameter, OUT defining an output parameter, and INOUT defining a parameter that may be changed), and also provides support for handling errors (via error codes).

To configure and execute an application using the patterns library, a user (developer) needs to undertake the following:

- A structural pattern – such as a "pipeline" – is selected from the patterns library. This appears as a standard Triana (group) unit. Figure 2 illustrates a number of different pipeline instances.
- A user may add or remove elements from the structural pattern chosen. This is achieved by using structural operators such as "increase" or "decrease" respectively.
- A user may now select a behavioural pattern – such as "dataflow" – to indicate how interaction between the units/elements is to take place.
- An entity at the pattern level is defined, the *pattern executor*, responsible for enforcing the selected behavioural pattern at each element.
- All component place-holders are instantiated with components (*Applications*) that may represent a unit in Triana or a group of units organized in a workflow.
- A user may now wish to use a behavioural operator – such as "start" or "stop" – on the behavioural pattern. These operators are supported by functions in the DRMAA API that manage the execution of the *Applications* by a resource manager. The execution of each *Application* is supported by a job (running executable) in the resource manager.

A user therefore may select structural patterns/operators followed by behavioural patterns/operators – all of which are implemented in Java. It is important to note that behavioural patterns/operators can only be applied to structural patterns – and not to arbitrary Triana units. A user does not need to know the actual implementation of any of these patterns/operators to make use of them – as they are primarily pre-defined group units in Triana or scripts. We therefore do not expect the user to be familiar with any particular programming language or scripting tool. Experienced developers, however, may add their own operators or patterns to our library.

Application execution using DRMAA requires the definition of attributes like the application's name, its initial input parameters, the necessary remote environment that has to be set up for the application to run, and so forth. These attributes are used to explicitly configure the task to be run via a particular resource manager. Although DRMAA has the notion of sessions, only one session can be active at a time. A single DRMAA session for all the operators is assumed. Hence drmaa_init and drmaa_exit routines are called, respectively, after the pattern instance is created and in the end of the script program. As an example, we show how a pipeline pattern can be mapped to DRMAA: Element pattern_elements[MAX_ELEMS] – contains the *Elements* that compose a specific pattern instance. Similarly, job_identifiers[MAX_ELEMS] represents the identifiers returned by the drmaa_run_job routine for jobs created to support pattern_elements. The order of the activities is preserved. DRMAA variables frequently used: INOUT jt is a job template (opaque handle), and INOUT drmaa_context_error_buf contains a context-sensitive error upon failed return. The examples are illustrated in a Java-like notation.

Start Operator – to initiate execution of Pipeline Elements.

```
/* launch all activities in the pipeline */
for( int index = Pipeline.pattern_elements.length -1 ; index >= 0;
index -- ) {
   int ret = drmaa_allocate_job_template( jt, drmaa_context_error_buf );
   process_error( ret, drmaa_context_error_buf );
   define_attributes( jt, Pipeline.pattern_elements[index] );
   /* Pipeline.startTime defines the time at which all
   elements in the pipeline instance should start running.  */
   ret = drmaa_set_attribute( jt, drmaa_start_time,
                              Pipeline.startTime,
                              drmaa_context_error_buf );
   process_error( ret, drmaa_context_error_buf );
   /* run one job at the specified time */
   ret = drmaa_run_job( job_id, jt, drmaa_context_error_buf );
   process_error( ret, drmaa_context_error_buf ); }
```

Repeat Operator – in this instance a single operator is used to re-execute an entire pattern instance a certain number of times ("*n*" in the code).

```
for( int count = 0; count < n; count++ ) {
   Start( Pipeline );
   /* wait for all the jobs that compose the pipeline to terminate */
   drmaa_synchronize( Pipeline.job_identifiers, timeout, 0,
                      drmaa_context_error_buf );
   /* timeout is bigger than all jobs' execution times */ }
```

5 Conclusions and Future Work

The extension of a Problem Solving Environment (Triana) with Patterns and Operators is described. Composition is achieved using a pattern extended graphical interface provided with Triana – whereas execution is managed by mapping Operators to the DRMAA API. We believe a Pattern based approach is particularly useful for reuse of component libraries in PSEs, and for mapping applications constructed in PSEs to a range of different execution environments. The DRMAA API was selected because of the significant focus it initially received within the Grid community – and the availability of commercial resource management systems (such as Grid Engine from Sun Microsystems) that make use of it. We are also investigating alternatives to DRMAA (such as Java CoG) [11] – primarily as current versions of DRMAA are aimed at executing batch jobs. With the emerging focus on Web Services in the Grid community, the DRMAA API has also lagged behind other equivalent developments (such as the Java CoG kit).

Patterns provide a useful extension to existing PSEs, as they enable the capture of common software usage styles across different application communities. The pipeline and star structural patterns, for instance, are commonly found in scientific applications (such as integrating a data source with a mesh generator, followed by a visualiser). Describing such compositions in a more formal way (as we have attempted to do here), will enable practitioners in the community identify common software libraries and tools. This is particularly important as software that performs similar functionality is available from a variety of different vendors. Providing the right balance between tools that require users to possess programming skills, and those that are based on a visual interface is difficult to achieve. By combining the visual interface of Triana with more advanced patterns and operators, we are attempting to enhance the functionality offered through (a variety of) existing workflow tools. Full usage of these ideas by the applications community is still a future aim for us.

References

1. E. Gallopoulos, E. Houstis and J. Rice, "Computer as Thinker/Doer:Problem-Solving Environments for Computational Science", IEEE Computational Science and Engineering, 1(2), 1994.
2. M. Li and M. A. Baker, "A Review of Grid Portal Technology", to appear in Book, "Grid Computing: Software Environment and Tools" (ed: Jose Cunha and O.F.Rana), Springer Verlag, 2004
3. J. Novotny, M. Russell and O. Wehrens "GridSphere: A Portal Framework for Building Collaborations", 1st International Workshop on Middleware for Grid Computing (at ACM/IFIP/USENIX Middleware 2003), Rio de Janeiro, Brazil, June 2003. See Web site at: http://www.gridsphere.org/. Last visited: January 2004.
4. O. Bunin, Y. Guo, and J. Darlington, "Design of Problem-Solving Environment for Contingent Claim Valuation", Proceedings of EuroPar, LNCS 2150, Springer Verlag, 2001.

5. S. Fleeter, E. Houstis, J. Rice, C. Zhou, and A. Catlin, "GasTurbnLab: A Problem Solving Environment for Simulating Gas Turbines", Proceedings of 16th IMACS World Congress, 104-5, 2000.
6. G. Fox, D. Gannon and M. Thomas, "A Summary of Grid Computing Environments", Concurrency and Computation: Practice and Experience (Special Issue), 2003. Available at:http://communitygrids.iu.edu/cglpubs.htm
7. B. Wydaeghe, W. Vanderperren, "Visual Composition Using Composition Patterns", Proc. Tools 2001, Santa Barbara, USA, July 2001.
8. ObjectVenture, The ObjectAssembler Visual Development Environment. See Web site at: http://www.objectventure.com/objectassembler.html. Last visited: March 2003.
9. The GridLab project. See Web site at: http://www.gridlab.org/. Last visited: January 2004.
10. Habri Rajic, Roger Brobst et al., "Distributed Resource Management Application API Specification 1.0". Global Grid Forum DRMAA Working Group. See Web site at: http://www.drmaa.org/. Last visited: September 2003.
11. Gregor von Laszewski, Ian Foster, Jarek Gawor, and Peter Lane, "A Java Commodity Grid Kit," Concurrency and Computation: Practice and Experience, vol. 13, no. 8-9, pp. 643-662, 2001, http:/www.cogkits.org/.
12. M.C.Gomes, O.F.Rana, J.C.Cunha "Pattern Operators for Grid Environments", Scientific Programming Journal, Volume 11, Number 3, 2003, IOS Press, Editors: R. Perrot and B. Szymanski.
13. M.C.Gomes, J.C.Cunha, O.F.Rana, "A Pattern-based Software Engineering Tool for Grid Environments", Concurrent Information Processing and Computing proceedings, NATO Advanced Research Workshop, Sinaia, Romenia, June 2003, IOS Press.
14. I. Taylor et al., "Triana" (http://www.trianacode.org/). Triana is the workflow engine for the EU GridLab project (http://www.gridlab.org/). Last Visited: January 2004.
15. E. Gamma, R. Helm, R. Johnson, J. Vlissides, "Design Patterns: Elements of Reusable Object-Oriented Software", Addison-Wesley, 1994.
16. O. F. Rana, D. W. Walker, "Service Design Patterns for Computational Grids", in "Patterns and Skeletons for Parallel and Distributed Computing", F. Rabhi and S. Gorlatch(Eds), Springer, 2002.
17. W.M.P. van der Aalst, A.H.M. ter Hofstede, B. Kiepuszewski, and A.P. Barros. Workflow Patterns. Distributed and Parallel Databases, 14(3), pages 5-51, July 2003

Databases, Workflows and the Grid in a Service Oriented Environment

Zhuoan Jiao, Jasmin Wason, Wenbin Song, Fenglian Xu, Hakki Eres,
Andy J. Keane, and Simon J. Cox

School of Engineering Sciences, University of Southampton, UK
{z.jiao,j.l.wason,w.song,f.xu,hakki.eres,ajk,sjc}@soton.ac.uk

Abstract. As the Grid moves towards adopting a service-oriented architecture built on Web services, coupling between processes will rely on secure, reliable, and transacted messages and be supported by databases. We have built a generic toolkit targeted at design engineers, which provides convenient methods to access a grid-enabled repository. In this paper we report how we have developed it further, and integrated it into a workflow toolkit to support a range of activities that design engineers have previously attempted to perform by multiple ad-hoc methods in the workflows used to improve designs. It also presents opportunities for improving the process of design search in a variety of ways that would have been otherwise hard to implement. We show the potential of our grid-enabled data repository in the context of workflow management, engineering optimisation process monitoring and steering.

1 Introduction

Engineering design search and optimisation (EDSO) is a computationally and data intensive process. Its aim is to achieve improved designs by exploiting engineering modelling and analysis. The quality of a particular design is measured by the value of an objective function. The design search process systematically modifies the variables which describe the design to increase, or reduce, this quality measure, whilst ensuring that the design variables satisfy various constraints. At the heart of this process lie workflows which, through a series of computational experiments, can build a repository containing samples of the design variables with their corresponding objective function values. In previous work [1] we focused on making grid-enabled databases more accessible to engineering designers so that they could be used routinely as a repository. Files and data structures are associated with metadata and services are provided to archive, locate (by querying the metadata), and retrieve them.

In this paper, we now demonstrate that this repository may in fact be exploited in a much wider variety of ways throughout the whole process that engineers follow to locate improved designs efficiently. We give some additional background in section 2 and describe our repository architecture and workflow construction environment in section 3. We then use case studies drawn from engineering design practice to illustrate, in sections 4-6, scenarios in which the database is used to store, monitor, and steer optimisation workflows. Section 7 summarizes our current and future work.

2 Background

The aspirations of the Grid to allow resources to be shared seamlessly and securely address many of the challenges encountered in engineering design search, which requires coupling of computing power, data resources and software applications. Many commercial engineering design search packages exist, for example iSIGHT [2] and ModelCenter [3]. However, they suffer from a limitation that data is managed internally in proprietary formats making it difficult to share with processes running outside of the package environments. Another common practice is to use the file-system as a medium for transferring data between various processes, which works fine in a local environment, but falls short in the Grid environment where processes can be dispatched to run on different machines with varied hardware and software configurations: in such an environment secure, reliable, transacted messages coupled to databases offer a better way to provide resilience and quality of service.

Our approach is to bring Grid technologies into an environment and mode of working familiar to engineers. In particular the routine use of a database-driven repository allows data to be passed from process to process on the Grid, and retrieved and exploited by other tools or packages. To achieve this we have developed the Geodise [4] database toolkit [1] which provides a set of Web services and a Java client API to allow higher level applications to manage data on the Grid. The toolkit is used by engineers at the University of Southampton working on a variety of design optimisation problems and by the GENIE [5] climate modelling project. In [6] we described in detail how the toolkit was used successfully in the Matlab [7] environment together with the Geodise Computational Toolkit. Matlab is a scripting language popular in the engineering community for its ease of use and rich functionality. We provide a number of Matlab functions to access the repository through our API, which users can incorporate into their scripts to manage data. It also provides a powerful execution engine for our workflows, and is available on a wide variety of platforms and operating systems. Other environments, including existing design packages, could be equally supported by writing routines in the appropriate language to access the API.

Whilst an expert can simply use their favourite editor to write a script, incorporating calls to our grid toolkits, and then run it from a Matlab command prompt, we also provide a workflow construction tool to assist users with the whole sequence of tasks required to develop and execute their design search. In this paper, we will focus on demonstrating how the repository is exploited within our workflow tool environment throughout each step in the design process, particularly application monitoring.

3 Architecture

The Geodise Database Toolkit is designed to help engineers manage the large amounts of data produced by their applications. We provide the means to store and share files and variables (primitive values and data structures) in a Grid-enabled repository. Each file or variable has some standard metadata associated with it (e.g. archive date, file size), which may be supplemented with additional user-defined custom metadata. Related data, such as that referring to a whole design job, may be logically grouped together as a datagroup and metadata can be added so that the entire collection can be described. Data may belong to multiple datagroups, which may also contain sub-datagroups, so users can describe and exploit relationships or hierarchies.

A data file is archived to a Globus [8] server using GridFTP [9] and upon archiving it is assigned a Universally Unique Identifier (UUID) which allows it to be physically located and retrieved later. Its associated metadata is stored in a relational database (Oracle 9i [10]): standard metadata is stored in tables for efficient access, whilst custom metadata is stored as native XML for flexibility and this can be transparently generated from user-defined Matlab structures with our XML Toolbox for Matlab [11]. Variables are stored in a similar way and the database is queried using a simple syntax to locate files/variables based on their characteristics. Scalability of the system is dependent on the underlying database software, particularly the efficiency of querying large quantities of unstructured XML. The concept and benefits of using a relational database containing metadata on top of a transacted file system is also at the heart of the WinFS file-system in Longhorn, the next generation of Windows [12]. In other work we demonstrate how this metadata becomes even more powerful in knowledge reuse when described in a semantically consistent way to allow key concepts and relationships to be defined and shared using ontological language [13].

All database access is supplied through secure Web services for storage and query of metadata, file location and authorisation, which can be invoked with platform independent client code anywhere on the Grid. Web service communication is secure, allowing certificate based authentication and authorisation using the same delegated credentials that provide a point of single sign-on for computational jobs on the Grid [14]. Secure requests are signed with the user's proxy certificate which the Web service verifies and uses to authorise database access, record data ownership and allow users to grant access permissions to others. The architecture described permits data archiving, querying and retrieval from any location with web access, Java support, the required APIs and a valid proxy certificate. In addition to its use from scripts written in Matlab, the database toolkit has also been integrated with the Geodise workflow construction environment (WCE). The full WCE architecture is described in [15].

Figure 1 shows the main roles of the WCE in assisting engineers to build and reuse workflows, verify resources, execute workflow scripts, and monitor or control the workflow as it runs locally or on the Grid. A user's first task in the WCE is to load-up or visually construct a series of connected tasks (a workflow), shown in Figure 2. The component view on the left pane of the WCE GUI contains a hierarchy of tasks and related computational components (e.g. Matlab functions). The hierarchy is created from an XML document describing the available named components, their inputs and outputs. A component can be dragged onto the main workflow view and may be linked to other components. The default initial values of function inputs and outputs can be altered in the component property window. Dataflow is configured by making associations between the output and input parameters of different components. The complete workflow description is saved as XML for reuse in the graphical editor, but more importantly as a Matlab script (e.g. optim.m in Figure 1) which can be stored in the Geodise repository, reused and edited outside the WCE without engineers learning new workflow formats, and run on compute resources with Matlab support. The available compute resources are pre-described in a user-editable configuration file and displayed by the WCE, which is able to verify the resources are accessible to the user and configured properly before execution. The WCE also assists in constructing scripts to monitor and steer workflow execution (e.g. monitor.m) which may be stored in the repository for later use. We now give examples of how workflows can be managed, monitored and steered.

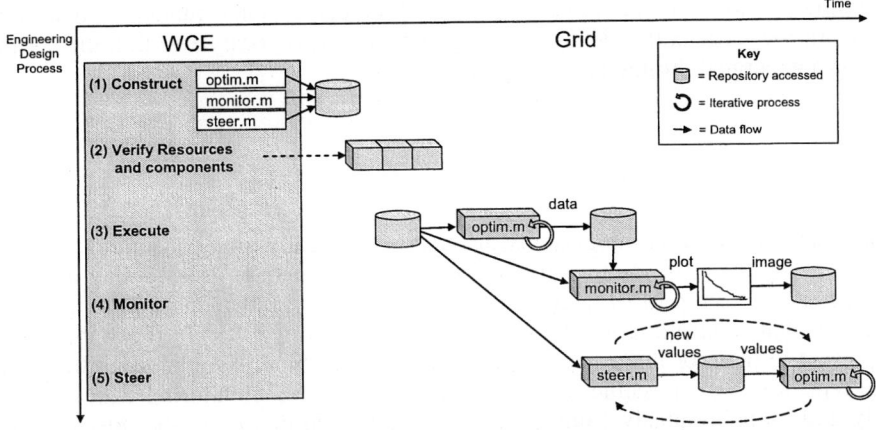

Fig. 1. Roles of Workflow Construction Environment (WCE) in workflow construction, resource verification, execution, monitoring and steering.

4 Storing Workflows

An example of constructing a workflow for a two-dimensional airfoil design optimisation problem is shown in Figure 2.

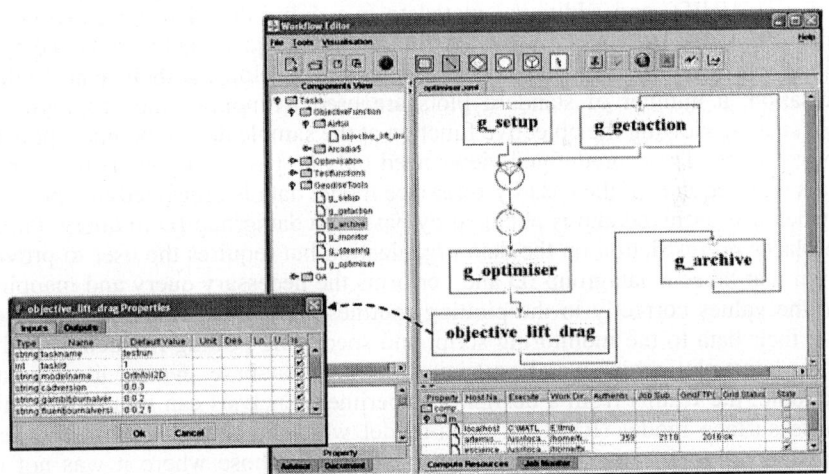

Fig. 2. The workflow construction environment allows users to select and configure workflow components, run workflows locally or on the Grid and monitor their progress.

The goal of the optimisation is to minimize the drag/lift ratio by varying the weights of six basis functions and the thickness-to-chord ratio [16]. After construction a user may save the workflow locally, or archive it to the repository for sharing and reuse. The WCE calls the database API to store and group the workflow script, its XML description, and some additional metadata which can be later queried to help locate the workflow efficiently. By using datagroups it is also possible to link the script to any data generated during its execution.

5 Monitoring

Optimisations based on high fidelity analysis are typically time consuming, therefore it may be desirable to monitor the progress of the search as it is running and make changes if necessary. Most integrated packages [2], [3] implement monitoring on the local machine controlling the optimisation. However, it is difficult, if not impossible, at present to monitor the process from a separate machine. By exploiting the central data repository it is possible to retrieve and post-process monitoring data independently and asynchronously from the running script. Furthermore, the post-processed information (e.g. a plot or image) may itself be returned to the database, from where it can be accessed and retrieved from any location, or even sent to a mobile device.

Whilst technologies exist that allow monitoring of Grids for scheduling, performance analysis and fault detection, e.g. the NetLogger Toolkit [17] and Gridscape [18], we focus on monitoring the progress of the user's application. This progress is described by the ongoing deposition of variables and data into the repository as transactions within and between grid processes occur. The user can supply a script (monitor.m in Figure 1) which periodically performs a query to retrieve the data, render it, and then return the image to a predefined location, e.g. the database. Data deposited for monitoring purposes are aggregated together into a datagroup assigned with an identifier (a UUID) and additional metadata (e.g. job index, job name) to help the monitor script, or a user operating away from the WCE, query and retrieve the data.

Along with bespoke visualisations a user might develop for their data, in design optimisation, a number of standard plots are used to monitor the progress of the search, such as viewing the objective function at the sample design points explored, or its convergence. These predefined views need only be passed (by value or reference) the data they require. In the pass-by-reference model data is organized in a prescribed way when it is archived and is obtained by passing a datagroup ID to query. Pass-by-value places no restriction on the data organization but requires the user to provide a function that takes a datagroup ID and performs the necessary query and mapping to return the values correctly to the plotting routines. The WCE can assist the user in linking their data to the monitoring script and specifying polling frequency, and displays the latest plot, which may be retrieved from the database, in a monitoring panel.

Example: the results from a design of experiment analysis can be used to build a response surface model (RSM), a meta-model which interpolates from the design points where the objective function was calculated to those where it was not [16]. Searching this meta-model for potentially better designs is then rapid, and further expensive simulations need only be performed to validate areas which the meta-model identifies as promising. Whilst this cycle of building, searching and tuning the RSM is automated as far as possible, monitoring the evolution of the response surface as the design search progresses can sometimes reveal additional ways in which the search can be accelerated. For example, the sampling process for the initial build can simply be terminated if the important features of the response surface remain the same as more design points are added. Figure 3 shows a response surface model built from

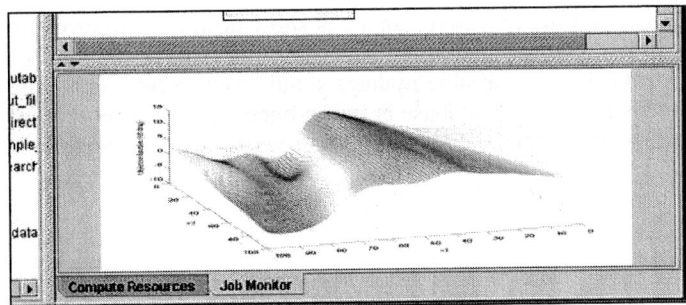

Fig. 3. Response Surface Modelling monitoring view screen shot.

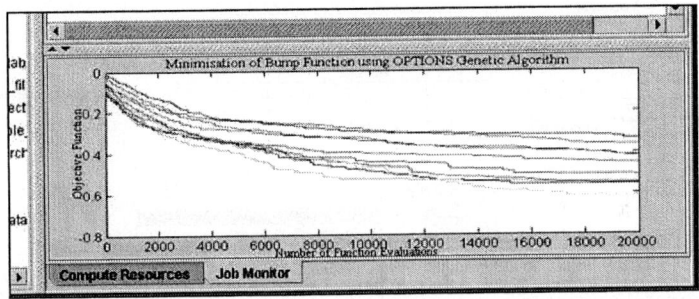

Fig. 4. Improvement in objective function as more design points are searched using a genetic algorithm optimiser. Multiple curves are shown for searches using different start points.

results with 16 runs of the airfoil problem with two design variables. The model is updated as more data becomes available in the database or at user specified intervals.

Designers may wish to view detailed information about each design point, such as geometry or analysis results. These might reveal unforeseen problems with the analysis procedure or geometry generation in certain regions of the design space, which could yield misleading results from the design search and significant wasted time and resource if only discovered at the end. As the picture of the design space emerges, a skilled designer may be able to use physical insight to explain features or patterns in the good (or bad) designs to derive a more efficient formulation of their problem.

Another unknown at the start of the design process is how long it is worth continuing to search. If analyzing further designs is unlikely to yield any further improvement then the job can be terminated. If the converse is true it may be desirable to extend the search. Figure 4 shows how the minimum value of the objective function improves as more design points are analysed during a genetic algorithm optimisation.

6 Steering

We have described how the repository provides a convenient location to deposit and then monitor variables as the search progresses. Steering is achieved by updating values in the database which are retrieved by the running script to control the design search. A number of uses for steering have been identified [19], e.g. to modify the design variables when the objective functions show no further improvements early in the search. In practical implementation from our scripts, two important issues arise: 1) Side-effects from arbitrary variable updates should be carefully understood and controlled; 2) Information in our database is 'write-once', so data is not modified or overwritten, instead scripts query to locate the latest version of the variable. Thus as variables are updated, a complete log of all intermediate values is recorded, which is important for provenance or repeatability.

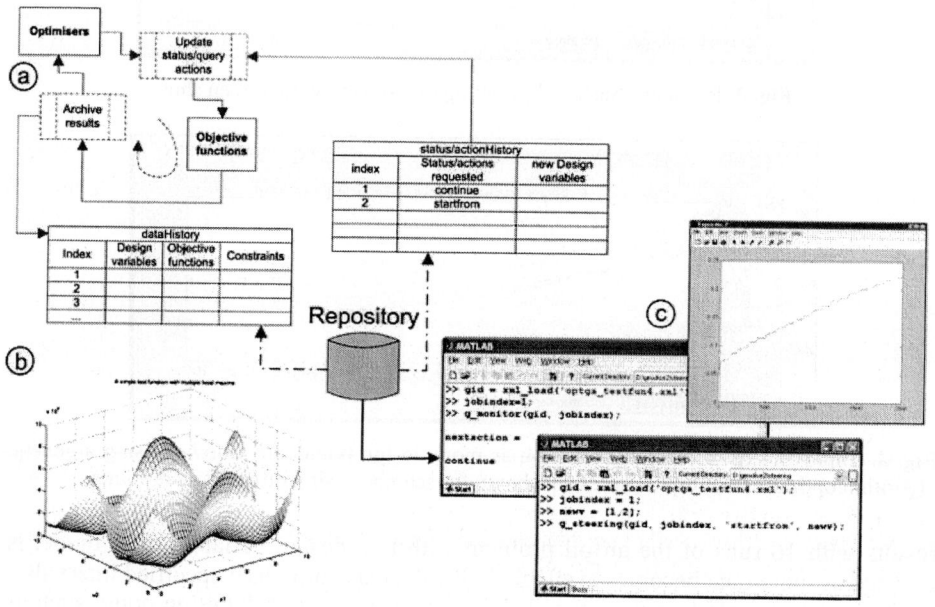

Fig. 5. Illustration of optimisation monitoring and steering using a grid-enabled database.

Traditional optimisation workflows are augmented with a set of grid-enabled query and archive functionalities that users can plug into their existing scripts at various points (Figure 5a). In each iteration, the optimisation process retrieves the current user action request from the action history in the database, and takes the appropriate action to achieve steering. To demonstrate how the system works we provide a simple example using a typical test function with multiple local optima. During optimisation, a user can monitor progress and interactively steer the process. The landscape of the test objective function is shown in Figure 5b. The search progress is monitored from another Matlab session based on the datagroup ID which provides the entry point to the database. The search history is shown in Figure 5c along with the effect of the user's steering activity.

7 Conclusions and Future Work

In this paper we have shown a number of roles for our repository throughout the process of EDSO. The toolkit enables engineers to write scripts in a familiar language and archive them along with data produced at various stages of the design optimisation process to a grid-enabled database. This makes the data available outside the optimisation package boundary, and accessible from various locations by authorised users. This in turn allows monitoring and steering of the optimisation, and thus enables new ways to accelerate, improve and reduce the cost of the design process.

An important issue for future consideration is provision of a redundant failover system for the metadata services and database. Future work will also include investigating data lifetime management and cleanup strategies, and integrating Geodise semantic and knowledge tools into the WCE.

References

1. Wason, J.L., Molinari, M., Jiao, Z., Cox, S.J.: Delivering Data Management for Engineers on the Grid. Proc. Euro-Par 2003, LNCS (2003) 413-416
2. iSIGHT, http://www.engineous.com/index.htm (2004)
3. ModelCenter, http://www.phoenix-int.com (2004)
4. Geodise Project, http://www.geodise.org
5. GENIE (Grid ENabled Integrated Earth system model) project, http://www.genie.ac.uk
6. Eres, M.H., Pound, G.E., Jiao, Z., Wason, J.L., Xu, F., Keane, A.J., Cox, S.J.: Implementation and utilisation of a Grid-enabled Problem Solving Environment in Matlab. Future Generation Computer Systems (in press).
7. Matlab 6.5. http://www.mathworks.com
8. The Globus Toolkit 2.4, http://www.globus.org
9. GridFTP, http://www.globus.org/datagrid/gridftp.html
10. Oracle 9i Database, http://otn.oracle.com/products/oracle9i
11. Molinari, M.: XML Toolbox for Matlab, GEM/Geodise (2004) http://www.soton.ac.uk/~gridem/xml_toolbox
12. Longhorn Developer Center, http://msdn.microsoft.com/longhorn
13. Chen, L., Shadbolt, N.R., Goble, C., Tao, F., Cox, S.J., Puleston C.: Managing Semantic Metadata for the Semantic Grid. Submitted to ISWC2004, Japan (2004)
14. Foster, I., Kesselman, C., Tsudik, G., Tuecke, S.: A Security Architecture for Computational Grids. Proc. 5th ACM Conference on Computer and Communications Security Conference (1998) 83-92
15. Xu, F., Cox, S.J.: Workflow Tool for Engineers in a Grid-Enabled Matlab Environment. Proc. UK e-Science All Hands Meeting (2003) 212-215
16. Song, W., Keane, A.J., Eres, M.H., Pound, G.E., Cox, S.J.: Two Dimensional Airfoil Optimisation using CFD in a Grid Computing Environment. Proc. Euro-Par 2003, LNCS (2003) 525-532
17. Gunter, D., Tierney, B., Jackson, K., Lee, J., Stoufer, M.: Dynamic Monitoring of High-Performance Distributed Applications, Proc. IEEE HPDC-11 (2002) 163-170
18. Gibbins, H., Buyya, R.: Gridscape: A Tool for the Creation of Interactive and Dynamic Grid Testbed Web Portals. Proc. 5th IWDC International Workshop (2003) 131-142
19. Shahroudi, K. E.: Design by Continuous Collaboration Between Manual and Automatic Optimization, Technical Report (1997) http://ftp.cwi.nl/CWIreports/SEN/SEN-R9701.pdf

A Parallel Programming Tool for SAR Processors*

M. Lettere, D. Guerri, and R. Fontanelli

Synapsis Srl, Livorno, Italy

Abstract. In the context of Italian Space Agency COSMO SkyMed project a quantitative and qualitative study of a set of image processing algorithms for *SAR Processors* has been carried out. The algorithms showed some interesting patterns in terms of structure and parallelism exploitation. During the activity of prototyping and analysis, an abstraction (**SPE Chain Model**) of the algorithmic behaviour has been defined in order to simplify performance modeling, design and implementation of parallel image processing algorithms. According to the defined abstraction, a parallel programming tool (**SPE**- Sar Parallel Executor) has been developed. SPE enables the implementation of efficient, structured and object oriented parallel image processing algorithms conforming to the **SPE Chain Model** and reuse of pre-existing sequential code. A set of image processing algorithms belonging to different classes of applications have been tested to validate both the *SPE Chain Model* and the *SPE* programming tool. The results show that no significant difficulties arise in the porting of already existing code to *SPE* and that writing new parallel algorithms is intuitive and productive and provides, at the same time, concrete high performance solutions required in real-time industry environments.

1 Introduction

Earth observation is based on the application of computationally challenging image processing algorithms on image data. Images are acquired at fine geometric resolutions and raw data is quite huge (26500*5600 double precision complex pixel values for a raw image [2]). This well known fact, along with requirements related to real-time industry production, led to a first study which focused on quantitative aspects (*flops*, memory usage) of a large set of algorithms [2]. The target of this study was to show how parallelism could be employed to reduce the intrinsic weight of some data and computation intensive operations.

A second study focused on *qualitative* aspects such as logical, functional and data dependencies among computational steps of an algorithm. The target was to define a parallel programming model, called **SPE Chain Model**, to help

---

* Big thanks goes to Gustavo Ovando, J.M.Moreno and M.J.Stefanini from Argentinian Space Agency (CONAE). This work has been partially supported by the CINECA Institute.

developers, not necessarily (*hpc*) experts, to design parallel image processing algorithms.

According to the *SPE Chain Model*, the considered algorithms have been split into sequences of **Macro Phases (MP)** which represent aggregations of logically and functionally related computation steps. For each MP, *quantitative* aspects and *qualitative* aspects have been analyzed in order to establish *analytical performance models* used for predicting performance.

The main requirements for the *SPE Chain Model* were similar to those of similar tools [8] [7] [6] [9]: strong *object oriented (OO)* design [1], modularity, reusability and adherence to the analytical performance models. However, the *SPE Chain Model* was born from a generalization activity based on the study of a set of SAR algorithms. Thus it misses the richness and the complexity of other general purpose models (Active Objects, Distributed Shared Memory, Dynamic Load Balancing, wide area distribution and mobile agents).

A programming tool called **SPE** has been developed to implement algorithms designed according to the *SPE Chain Model*. *SPE* is based on two class libraries: **SPEAPI** used for writing parallel image processing algorithms and **SPEENG** a set of runtime support classes.

A set of different case studies using *SPE* shows that the model is very stable and general. In a scenario where domain experts cooperate with hpc developers, the design of new parallel image processing algorithms, turned out to be very intuitive and productive. There are also no significant difficulties porting already existing code to *SPE*. Tested algorithms show that the performance of their *SPE* implementations closely matches the performance of pre-existing, low level implementations, demonstrating that the use of high level programming constructs doesn't introduce a significant overhead. Moreover, algorithms implemented from scratch with *SPE*, are very efficient despite the short time it takes to develop them.

This paper presents the *SPE Chain Model* (section 2), the implementation of *SPE* (section 3) and the results obtained with two case study algorithms (section 4).

2 The SPE Chain Model

This section shows how an algorithm is designed and executed using *SPE Chain Model* and *SPE*.

2.1 Algorithm Design

Qualitative analysis of the studied algorithms, has shown that image processing algorithms can be split into sequences of *Macro Phases (MP)*. *MP*s are aggregations of computational steps that are functionally related because they exploit the same data, share a common stencil or can be executed concurrently.

For the *SPE Chain Model*, an algorithm is a sequence of *MP*s connected by entities called **Bindings**. *Bindings* are used to exchange data among *MP*s

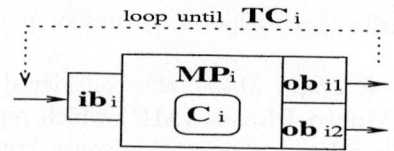

Fig. 1. Example of Macrophase Execution Loop.

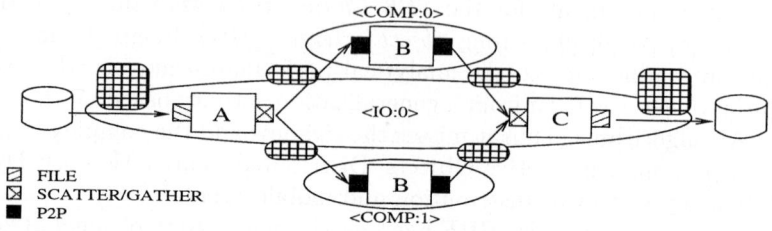

Fig. 2. Example of parallel program with 2 Node classes.

according to a specific transfer policy called the *type* of the *Binding*. A special kind of *Binding*, called **FILE Binding**, connects a *MP* to the file system.

Figure 1 shows the behaviour, or *Execution Loop* of a *Macro Phase* MP_i. MP_i receives data from a set of preceding *MPs* through its *Input Binding* ib_i. It executes its code C_i and finally it sends the output data through its *Output Bindings* (in order ob_{i1} and ob_{i2}) to the following *MPs*. If MP_i is *iterative*, its *Execution Loop* is repeated until a programmable *Termination Condition* TC_i is verified.

2.2 Algorithm Execution

For the *SPE Chain Model*, a parallel program consists of a number of **Nodes** organized in **Node Classes**. A *Node* matches the concept of *process*. A *Node Class* is a group of possibly related *Nodes*. *Bindings* match the concept of inter-process communications [5].

A *Node* is programmed by statically assigning to it a sequence of *Macro Phases* and the parallelism degree (number of *Nodes*) of a *Node Class* is set statically in the algorithm parameters.

Nodes are identified by a pair <class-name:id> composed of their class and their class internal **Node ID**.

In the program of figure 2, data is read and split in two sub-images by *Macro Phase A*. Each *Macro Phase B* receives a sub-image through its *P2P Binding* and executes the algorithm specific sequential code on the sub-image. Finally the output sub-images are sent through a *P2P Output Binding* to a *Macro Phase C* who is responsible for recollecting the output image data and write it to the filesystem.

Figure 2 shows one possible allocation of the program where the *IO Class* has only one *Node* <IO:0> who is responsible for distributing and recollecting

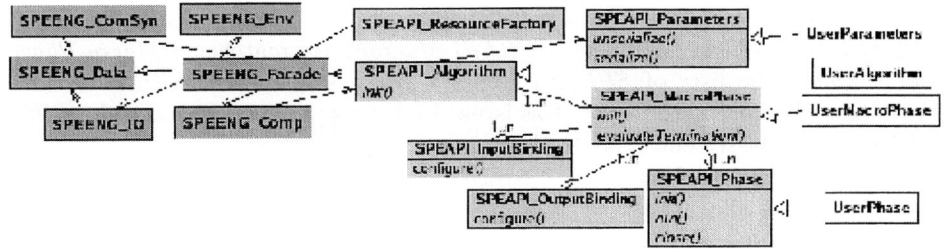

Fig. 3. Class diagram of SPE.

the data (*A* and *C*) and *COMP Class Nodes* execute in parallel (2 process) the algorithm specific code.

Available types of *Bindings* are *FILE, P2P (point-to-point), MP (multi-point), BKCH (block-change), SCATTER, GATHER, ON_DEMAND* and *ROUND_ROBIN*.

Using pairs of *Input/Output Bindings*, it is possible to implement a great variety of communication schemes with well-known semantics and many of the common parallel patterns like *Farms* or *Pipelines* [5].

There is a special type of *Binding* called **BLOCKCHANGE (BKCH)**. Image data is often seen as a two-dimensional matrix. During the computation of an image processing algorithm it is often necessary to perform some operations on whole rows or on whole columns of the matrix [3]. In *SPE* the data is split among a set of *Nodes* according to a specific storing order (*row-major* or *column-major*). A couple of *BKCH Bindings* changes the storing order of the distributed image data. Assuming the *BKCH* operation as a standard operation in *SPE*, makes it possible for the programmer to benefit from a very optimized implementation of the routine. The way *SPE* is designed makes it possible for the *SPE* development team to rapidly integrate other highly specialized operations or communication patterns. This can be achieved by subclassing an abstract *Binding* class and overriding the methods that implement the underlying communication strategy.

3 A Sample Implementation of SPE

This section describes a sample implementation of *SPE* based on a class library for writing parallel algorithms conforming to the *SPE Chain Model* (*SPEAPI*) and a collection of classes that implement the runtime support for executing *SPEAPI* algorithms (*SPEENG*).

The class library *SPEAPI* was designed to ease the development of object oriented, modular, reusable, strong structured parallel image processing algorithms. The target is to enable a programmer with no particular *hpc* expertise, to design a parallel program by simply implementing the abstract methods *init()*, *run()*, *close()* and *evaluateTermination()* of the *SPEAPI* classes as shown in figure 3.

SPEAPI_Algorithm models a parallel algorithm and *contains* the sequence of *SPEAPI_MacroPhase* instances. *SPEAPI_MacroPhase* matches a *Macro Phase* and *contains* the lists of *SPEAPI_Binding* and a sequence of *SPEAPI_Phase* instances. The *SPEAPI_Phase* class is used for wrapping sequential code. *SPEAPI_Binding* classes encapsulate the concept of *Bindings*.

SPEAPI_ResourceFactory is a *singleton factory* class used for requesting memory buffers allocation.

SPEENG implements the runtime support for algorithms written using *SPEAPI* with the idea to distribute responsibilities across a set of *systems* accessible through a singleton *facade* class called *SPEENG_Facade*.

SPEENG_ComSyn and *SPEENG_IO* are responsible for executing specific communication and file system access strategies related to *SPEAPI_Binding* instances.

For implementing these systems, *double-buffering* and *asynchronous communications* have been used to optimize performance and overlap computation and communication [5]. Moreover, *SPEENG* currently adopts *MPI* as communication software and standard *POSIX IO* for filesystem access. These two classes are *wrapper* classes that isolate *SPE* from all the implementation choices. Thus, adopting different choices implies just rewriting part of the code of *SPEENG_ComSyn* or *SPEENG_IO*.

SPEENG_Data is a *factory* class responsible for handling memory allocation and deallocation requests. *SPEENG_Env* and *SPEENG_Comp* are responsible respectively for storing environment information and managing the *Execution loop* of an *SPEAPI_Algorithm*.

As shown in figure 3, *SPEENG_Comp* accesses *SPEAPI_Algorithm* for managing its *Execution loop*. *SPEAPI_Algorithm* accesses *SPEENG_Facade* for requesting runtime support services (communication, IO, status or error notification, execution time). *SPEAPI_ResourceFactory* accesses *SPEENG_Facade* for requesting memory handling facilities.

4   SPE: Two Case Studies

This section shows two algorithms developed with the described sample implementation of *SPE*. The results, measured in terms of **generality of the model**, **usability** of *SPEAPI* and **performance**, show that there are no significant difficulties porting already existing code to *SPE* and the development of new efficient and scalable parallel algorithms, in a scenario where domain experts cooperate with hpc developers, is intuitive and productive.

The performance tests have been executed on the **CINECA** *Linux Beowulf Cluster* which has a peak performance of about 3 TeraFlops and is composed of 256 dual-processors (SMP) connected through a 2 Gbit/s network (*http://www.cineca.it/HPSystems/Resources/*).

CSA*(Chirp Scaling Algorithm)* is an image focusing algorithm for *SAR* processing. The *SPE* development team had already made an experience in porting

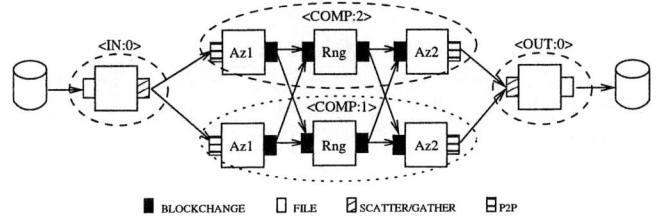

Fig. 4. SPEAPI structure of CSA.

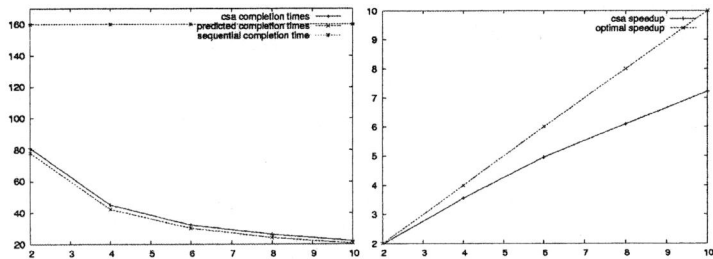

Fig. 5. Completion time, in seconds, (left) and speed up (right) for CSA.

a sequential prototype to a very optimized parallel implementation using low level tools (*C and MPI*). The porting to *SPE* validated the assessment related to the ease of integrating already existing code into a parallel program designed with *SPEAPI*. Moreover the performance of the *SPEAPI* version closely approximated the performance of the optimized low level implementation and of other similar solutions [4]. Figure 4 shows the *SPEAPI* structure of the algorithm. *CSA* first computes on the columns (*azimuth*) of the matrix representing the image. The computation is based on *Fourier Transforms (FT)* that require whole columns to be accessible locally on a *Node*. Data is stored inside *Macro Phase Az1* in column-major order. The second step of *CSA* computes *FT* on whole rows of the image (*range*). Thus the *Bindings* between *Az1 and Rng* are of type *BLOCKCHANGE* and a communication implies a change in the storing order. *Az2* is created to implement the last inverse FT in azimuth direction and another *BLOCKCHANGE* between *Rng1 and Az2* is necessary. The implementation requires two global exchanges of image data among all the *COMP Class Nodes*. Figure 5 shows the completion time with a varying number of *COMP Class Nodes* on image blocks sized 540 Mbyte. The adherence to the values provided by the analytical model can easily be seen and the *speedup*, compared to other results [3], is very encouraging.

P-FLOOD is an iterative algorithm that works on raster images and *DEM* (digital elevation model [2]) data to study the flow of water during a rainy timespan. *P-FLOOD* has been designed and developed from scratch using *SPEAPI* and its development demonstrated how *SPEAPI* can be used in a scenario where domain experts work together with programmers at the implementation

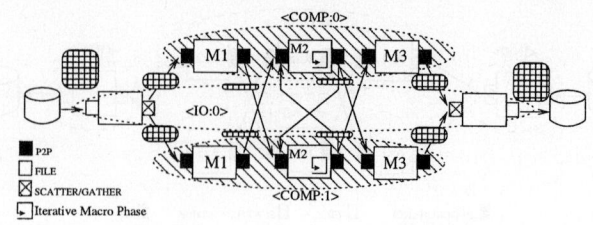

Fig. 6. SPEAPI structure of P-FLOOD.

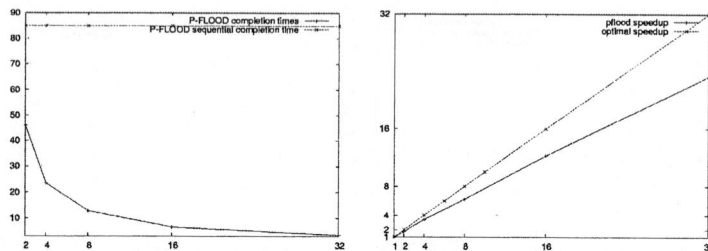

Fig. 7. Completion time, in seconds, (left) and speed up (right) for P-FLOOD.

of a parallel image processing algorithm. In *P-FLOOD* the data is read from a file by *Node* $< IO : 0 >$. The data is split in sub-images which are sets of rows and sent to the $< COMP : * >$ *Nodes*. The computational core is the iterative *Macro Phase M2* of figure 6. At each iteration the quantity of water at every pixel is computed. For this computation, information related to the nine neighbors of the pixel, in a 3x3 square stencil, have to be known. This implies that at each iteration of *M2*, $< COMP : i >$ has to exchange one row with both $< COMP : i - 1 >$ and $< COMP : i + 1 >$ to update the borders of its sub-image. This behaviour is modeled by the *P2P Input and Output Bindings* of *M2*. *M1* and *M3* are necessary to perform respectively the first and the last exchange of border rows.

The performance tests with a varying number of COMP Nodes (figure 7) have been executed on a *DEM* of 1000*1000 pixels (4Mbyte) with 50 iterations.

5 Conclusions and Future Work

This paper shows a new object oriented and strong structured abstraction called *SPE Chain Model* that simplifies design, implementation and performance modeling of parallel image processing algorithms. Moreover it presents *SPE*, a parallel programming tool that implements the *SPE Chain Model* and shows two sample algorithms that were implemented using *SPE*.

The results of this work are very positive in terms of *SPEAPI* usability, code reuse and performance. *SPE* is quite general since many new algorithms can be developed from scratch and already existing algorithms can easily be ported to it. Performance tests showed that no significant overhead is introduced by the high level programming constructs because the performance of pre-existing implementations can be matched very closely. Moreover *SPE* enables the development of new parallel image processing algorithms that are fast and scalable.

The parallel algorithm implementations written to test *SPE* show that most of the code written with *SPEAPI* can be automated. This fact was used to produce a further abstraction layer based on an XML representation of the *SPEAPI* structure of an algorithm. The idea is to create a RAD tool which enables a programmer to easily design the parallel structure of an algorithm by simply interacting with graphical widgets.

Acknowledgments

The design foundations of the platform utilized for carried out the work described in the paper has been funded by **Telespazio S.p.A.** within the ASI COSMO-SkyMed project.

References

1. Connie U. Smith, LLoyd G. Williams: Performance Solutions, A Practical Guide To Creating Responsive, Scalable Software. Addison-Wesley, Object Technology (2001)
2. J.C. Curlander, R. N. McDnough: SYNTHETIC-APERTURE RADAR-SYSTEM AND SIGNAL PROCESSING. WILEY INTERSCIENCE, 1991
3. J. J. Mallorqui, M. Barà, A. Broquetas, M. Wis, A. Martinez, L. Nogueira, V. Moreno: PARALLEL ALGORITHMS FOR HIGH SPEED SAR PROCESSING.
4. Yiming Pi, Hui Long, Shunji Huang: A SAR PARELLEL PROCESSING ALGORITHM AND ITS IMPLEMENTATION. Department of Electronic Engineering, University of Electronic Science and Technology of China.
5. K. Hwang ADVANCED COMPUTER ARCHITECTURE: Parallelism, Scalability, Programmability. McGraw-Hill, Series in Computer Science (1993)
6. M. Vanneschi: The programming model of ASSIST, an environment for parallel and distributed portable applications. Parallel Computing, Vol. 28, Issue 12 (December 2002)
7. L.V. Kale, Sanjeev Krishnan: CHARM++ : A Portable Concurrent Object Oriented System Based On C++. Object Oriented Programming Systems, Languages and Applications, Sept-Oct 1993. ACM Sigplan Notes, Vol. 28, No. 10, pp. 91-108.
8. Chialin Chang, Alan Sussman, Joel Saltz: CHAOS++: RUNTIME SUPPORT FOR DISTRIBUTED DYNAMIC DATA STRUCTURES IN C++. CRPC Vol. 3 Issue 3 - Summer 1995
9. D. Caromel, F. Belloncle and Y. Roudier: The C++// system. Parallel Programming Using C++, G.Wilson and P. Lu editors, MIT Press, 1996, ISBN 0-262-73118-5

Topic 17
High Performance Bioinformatics

Mohammed J. Zaki, David A. Bader, Johan Montagnat, and Concettina Guerra

Topic Chairs

Bioinformatics is the science of managing, mining, and interpreting information from biological sequences and structures. Genome sequencing projects have contributed to an exponential growth in complete and partial sequence databases. Similarly, the rapidly expanding structural genomics initiative aims to catalog the structure-function information for proteins. Advances in technology such as microarrays have launched the subfield of genomics and proteomics to study the genes, proteins, and the regulatory gene expression circuitry inside the cell. What characterizes the state of the field is the flood of data that exists today or that is anticipated in the future. Combined with the fact that many of the bioinformatics tasks are highly compute intensive (e.g., ad initio protein folding), it is clear that high performance computing has a fundamental role to play in various bioinformatics problems ranging from the protein folding problem to large-scale genomics to inferring pathways and regulatory networks. The goal of this session is to present the latest research in high-performance computing applied to bioinformatics tasks. We are especially interested in scalable, parallel, and distributed algorithms for mining and analyzing bioinformatics data, as well as system tools that support large-scale high performance bioinformatics.

There were six papers submitted to this session, out of which three were selected for presentation at the conference. Weiguo Liu and Bertil Schmidt, in their paper, *A Generic Parallel Pattern-based System for Bioinformatics*, present a generic programming framework for sequence alignment algorithms in bioinformatics. Alexandros Stamatakis, Thomas Ludwig, Harald Meier, in *Parallel Inference of a 10.000-taxon Phylogeny with Maximum Likelihood*, present their work on using parallelism to construct the phylogenetic or evolutionary trees for a large number of sequences. Gilles Parmentier, Denis Trystram, Jaroslaw Zola, in *Cache-based parallelization of multiple sequence alignment problem*, use their CaLi software cache library to study a parallel multiple sequence alignment algorithm. These papers represent some of the new advances in the field of high performance bioinformatics, and we hope they will stimulate further research in this exciting area of research.

A Generic Parallel Pattern-Based System for Bioinformatics

Weiguo Liu and Bertil Schmidt

School of Computer Engineering, Nanyang Technological University, Singapore 639798
liuweiguo@pmail.ntu.edu.sg, asbschmidt@ntu.edu.sg

Abstract. Parallel program design patterns provide users a new way to get parallel programs without much effort. However, it is always a serious limitation for most existing parallel pattern-based systems that there is no generic description for the structure and behavior of a pattern at application level. This limitation has so far greatly hindered the practical use of these systems. In this paper, we present a new parallel pattern-based system for bioinformatics. The underlying programming techniques are based on generic programming, a programming technique suited for the generic representation of abstract concepts. This allows the new system to be built in a generic way at application level. We show how this system efficiently addresses the shortcomings of existing systems and leads to significant runtime savings for some popular applications in bioinformatics on PC clusters.

1 Introduction

Parallel design patterns are based on sequential program design patterns used in object-oriented languages [1]. A parallel program design pattern names, abstracts, and identifies the key aspects of a common parallel structure that make it useful for creating a reusable and efficient parallel program. Parallel design patterns are similar to parallel program libraries. However, there is an important difference between them. When a library is used, a programmer must define the structure of the application and make calls to the library code. Conversely, a parallel design pattern defines the structure of a framework and the programmer supplies code for specific application-dependent routines.

In the past decade, many parallel pattern-based systems have been developed to employ design patterns related concepts in the parallel computing domain in the context of object-oriented programming techniques. Some of the systems based on similar ideas include *Code* [2], *Frameworks* [15], *Enterprise* [16], *HeNCE* [4], *Tracs* [3], and *DPnDP* [14]. However, most of these systems lack practical usability for the following reasons [1, 9]:

- Although more and more parallel patterns are supported by most of the systems, the parallel patterns are not expressed in a compact generic manner. This is also the root of other shortcomings.
- The way to implement a parallel application for a new algorithm is very complex for the user.

- These systems are not flexible enough for the user to reuse the components of existing parallel patterns at application level.
- Most of the systems are overstaffed and the code generated by them is inefficient and suffers from great loss of performance.

Generic programming is a program design technique that deals with finding abstract representations of algorithms, data structures, and other software concepts [18]. Generic programming can use both the traditional objected-oriented techniques (such as inheritance and virtual functions) and templates. This makes generic programming a practical and flexible technique in program design. In practice, the ideas of generic programming have been successfully applied to the field of high performance computing. The STL (the Standard Template Library, which later was adapted and incorporated into the C++ standard library) and Janus [6] are two examples of generic programming applications.

In this paper, we present a new system using generic programming techniques. We demonstrate how the new system can efficiently address most of the shortcomings of existing systems and generate parallel programs with substantial performance gains for some dynamic programming applications in bioinformatics.

The rest of the paper is organized as follows: Section 2 describes the characters and classification for dynamic programming algorithms which have wide applications in bioinformatics. In Section 3, we present implementation details for our system. Section 4 gives its application and performance evaluation for some popular algorithms in bioinformatics. Section 5 provides conclusions.

2 Dynamic Programming Problems in Bioinformatics

Dynamic programming is an important algorithm design technique in bioinformatics. Typical applications using this technique are highly compute intensive and suffer from long runtimes on sequential architectures. Dynamic programming algorithms can be classified according to the matrix dimension and the dependency relationship of each cell on the matrix [8]: a dynamic programming algorithm is called a tD/eD algorithm if its matrix dimension is t and each matrix cell depends on $O(n^e)$ other cells. For example, three dynamic programming algorithms are defined as follows:

Algorithm 1 (2D/0D): Given $D[i, 0]$ and $D[0, j]$ for $0 \leq i, j \leq n$, compute
$$D[i,j] = \min\{D[i-1, j] + x_i, D[i, j-1] + y_j, D[i-1, j-1] + z_{i,j}\} \quad \text{for } 1 \leq i, j \leq n,$$
where x_i, y_j and $z_{i,j}$ are computed in constant time.

Algorithm 2 (2D/1D): Given $w(i, j)$ for $1 \leq i < j \leq n$; $D[i, i] = 0$ for $1 \leq i \leq n$ compute
$$D[i,j] = w(i,j) + \min_{i < k \leq j} \{D[i, k-1] + D[k, j]\} \quad \text{for } 1 \leq i, j \leq n$$

Algorithm 3 (2D/2D): Given $w(i, j)$ for $1 \leq i < j \leq 2n$; $D[i, 0]$ and $D[0, j]$ for $0 \leq i, j \leq n$, compute
$$D[i,j] = \min_{\substack{0 \leq i' < i \\ 0 \leq j' < j}} \{D[i', j'] + w(i' + j', i + j)\} \quad \text{for } 1 \leq i, j \leq n$$

The dynamic programming formulation of a problem always yields an obvious algorithm whose time complexity is determined by the matrix dimension and the dependency relationship. If a dynamic programming algorithm is a tD/eD problem, it takes time $O(n^{t+e})$ provided that the computation of each term takes constant time. According to above classification method, we classify some popular dynamic programming algorithms in bioinformatics as shown in Table1.

Table 1. A classification of some popular dynamic programming algorithms in bioinformatics

Algorithms	Applications	Classification	References
Smith-Waterman algorithm with linear and affine gap penalty	Genome local alignment	2D/0D	[10, 17]
Syntenic alignment	Generalized genome global alignment		
Smith-Waterman algorithm with general gap penalty	Genome local alignment	2D/1D	[5, 17]
Nussinov algorithm	RNA base pair maximization		
Viterbi algorithm	Gene sequence alignment using HMMs, Multiple sequence alignment	2D/0~2D	[5]
Double dynamic programming algorithm	Protein threading	2D/2D	[12]
Spliced alignment	Gene finding	3D/0~1D	[7]
Zuker algorithm	RNA secondary structure prediction	3D/0~1D	[19]
CYK algorithm	RNA secondary structure alignment	3D/1D	[5]

According to the classification of these algorithms, although they solve wide variety of problems, all the algorithms have three aspects in common:

1. a matrix with initial values
2. dependency relationship of each cell
3. an order to compute the matrix

Because these algorithms exhibit the similar characters in these aspects, from the point of view of parallel computing they can be treated in a similar way, i.e. a similar partitioning and message-passing strategy. In the next section, we will demonstrate the implementation of a parallel pattern-based system for the algorithms in bioinformatics.

3 The Generic Parallel Pattern-Based System

The system is presently implemented using standard C++. The implementation uses the MPI library provided by MPICH 1.2.5.

3.1 System Overview

Generic programming techniques support static polymorphism through the use of templates. Compared with dynamic polymorphism using inheritance and virtual functions, there are many advantages of static polymorphism, such as type-safe and efficiency. One of the most important advantages of generic programming is that the templates can be parameterized with user-defined classes. This advantage of generic programming techniques can facilitate the development of parallel programs. For example by using generic programming techniques, we can easily separate the communication part of a parallel program from the sequential application part. This is also the main idea of parallel design pattern. The code in Fig. 1 shows the structure of the class template `GenericPattern`. It is also the generic description for the system.

```
1. template<class datatype,
2.           class AlgorithmIni,
3.           class SequentialComp,
4.           class ParallelCommu>
5. class GenericPattern{
6.   void ParComputing(){
7.     AlgorithmIni::preprocess();
8.     while the cyclic loop is not completed{
9.       ParallelCommu::prepare();
/*SequentialComp is invoked in the following method*/
10.      ParallelCommu::launch();
11.    }
12.    AlgorithmIni::postprocess();
13.  }
14.};
```

Fig. 1. The structure of the template `GenericPattern`

Concrete parallel applications can be implemented by extending and instantiating the template parameters (lines 1, 2, 3, 4) of `GenericPattern`. These parameters encapsulate the abstract structure and behavior of a set of parallel patterns in an application independent manner. The parameter `AlgorithmIni` initializes the matrix that will be computed. `SequentialComp` processes the sequential computing for each cell of the matrix. `ParallelCommu` consists of the communication behavior between all processors participating in the parallel computing. In fact, these three template parameters represent the three aspects of the dynamic programming algorithms mentioned in Section 2 respectively. By defining these three parameters, the parallel part (`ParallelCommu`) is separated from the sequential application parts (`AlgorithmIni` and `SequentialComp`). Thus, both parts of a parallel program can evolve independently. This allows for the rapid prototyping of parallel programs, and permits users to experiment with alternative communication structures quickly and easily. Currently, we have integrated the block and block-cyclic (line 8) partitioning methods and the space-saving method [13] into the system. The system also supports the two dimensional and three dimensional applications with nested recursive dependency relationships. Hence, it can be used to develop efficient parallel programs for many regular and irregular dynamic programming applications [11].

3.2 Generic Parallel Pattern: Extensibility and Flexibility

An important aspect of our system is the generic representation for a set of patterns, i.e. a generic pattern. With this generic pattern, we mainly focus on the extensibility of the system rather than how many limited patterns it can support. In order to achieve a good extensibility, the template parameters of `GenericPattern` are also defined as class templates. Thus, the user can extend the generic pattern by specifying these application-independent templates. Different specialization will lead to different policies, i.e. different implementation strategies for a concrete parallel application.

Fig.2 shows how to use the generic pattern to develop parallel programs for new algorithms. The user only needs to specify the relevant template parameters according to the characters of the algorithm so as to generate corresponding policy. A policy is yet to be further instantiated in order to generate the concrete parallel program.

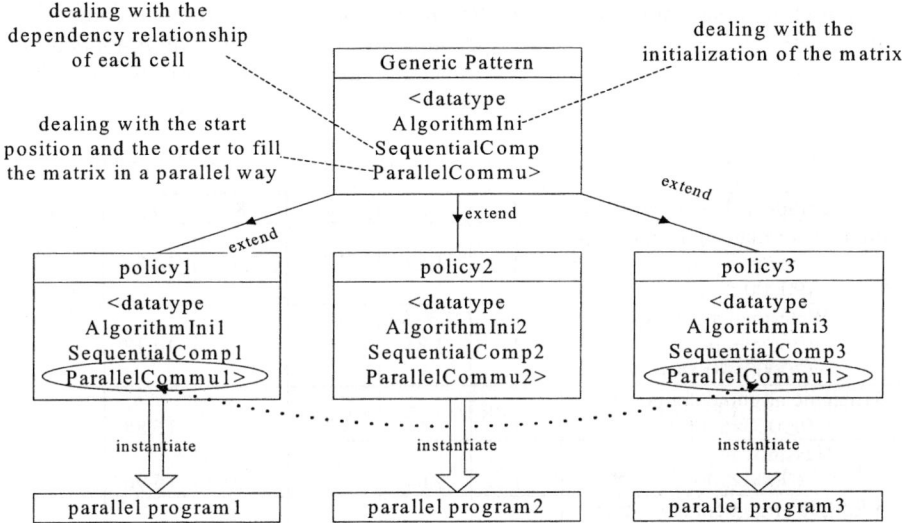

Fig. 2. Extending the generic pattern to implement application-dependent parallel programs

Because the algorithms in bioinformatics share similar characters, this procedure is not complex. The user only needs to provide the sequential application specific code while the system supplies the necessary parallel code. For example, the Smith-Waterman algorithm with linear and affine gap penalty and the syntenic alignment algorithm share analogous characters in all the three aspects. The obvious differences between them are the number of matrices they will compute and the dependency relationship of each cell. Hence, extending the generic pattern to parallelize these three algorithms is quite simple and similar.

From Fig. 2 we can also find that each template parameter is defined independent of other parameters. Yet different template parameters can interact with each other via standard interfaces. Consequently, the system has a good flexibility. For instance in Fig. 2, policy3 and policy1 share the same parallel characters. Thus we can entirely reuse the overall design of `ParallelCommu1` to develop policy3. The user can therefore reuse the components in existing patterns to develop new applications in a flexible way.

4 Experiments and Performance Evaluation

We have used the described system to develop parallel applications for some of the algorithms presented in Table 1. They are running on a distributed memory cluster which comprises 8 AlphaServer ES45 nodes. Each node contains 4 Alpha-EV68 1GHz processors with 1GB RAM for each processor. All the nodes are connected with each other by a Gbit/sec Quadrics switch.

Table 2 shows the speedups for different number of processors. It is important to note that these applications are implemented in different methods. The linear space method [13] is used to reduce the RAM needed by the Smith-Waterman algorithm (with linear gap penalty and affine gap penalty) and the Syntenic alignment algorithm for long sequences. Similar space-saving method is used for the three dimensional applications such as the multiple alignment algorithm for three sequences and the spliced alignment algorithm. As to the Nussinov algorithm, we store and compute the whole matrix. Notice the super linear speedups are observed in several applications. This is because of the effects due to better caching are different according to different implementation methods for specific applications.

Table 2. Speedups for some well-known parallel applications in bioinformatics

Number of Processors	1	2	4	8	16	32
Smith-Waterman algorithm with linear gap penalty (90000×90000)	1	1.87	3.7	7.37	14.74	28.25
Smith-Waterman algorithm with affine gap penalty (70000×70000)	1	1.85	3.68	7.31	14.13	27.65
Syntecnic alignment algorithm (60000×60000)	1	2	4.25	8.31	16.84	33.68
Nussinov algorithm (5000×5000)	1	2.04	4.07	8.03	15.96	29.87
Multiple alignment algorithm for three sequences (3000×3000×3000)	1	2	4.04	8.72	17.7	33.37
Spliced alignment algorithm (4000×4000×1000)	1	2.12	4.55	9.19	18.2	36

Since the system has been implemented using generic programming techniques, it is interesting to compare its performance with a system using object-oriented programming techniques. In order to investigate this, we have implemented the system in both ways, with generic programming techniques and using inheritance and virtual functions. Table 3 presents the performance comparison for the sequential applications between these two methods.

From Table 3 we can see that the code generated by the system using generic programming techniques is faster. This is because the generic programming relies on static polymorphism, which resolves interfaces at compile time. On the other hand in dynamic polymorphism using inheritance and virtual functions, the determination of which virtual function to use cannot be made at compile time and must instead be made during run time. Thus, more overhead is associated with the invocation of a virtual function.

Table 3. Performance comparison between the system using generic programming and virtual functions (system1: using virtual functions; system2: using generic programming techniques)

Sequential Applications	Runtimes of system1(sec)	Runtimes of system2(sec)	Runtimes reduction
Nussinov (5000×5000)	821	795	3.2%
Smith-Waterman with linear gap penalty (100000×100000)	551	503	8.7%
Smith-Waterman with affine gap penalty (100000×100000)	917	868	5.3%
Syntenic alignment algorithm(100000×100000)	1543	1319	14.5%
Multiple alignment for three sequences (3000×3000×3000)	643	624	3%
Spliced alignment (4000×4000×1000)	730	642	12.1%

5 Conclusions

In this paper we have described a new parallel pattern-based system using generic programming techniques. The generic representation for a set of algorithms based on their characters is one important feature that distinguishes this system from other pattern-based approaches in parallel computing. To achieve higher extensibility and flexibility, the template parameters of the generic pattern are defined as class templates and they can interact with each other in a flexible way. The system also allows a programmer to reuse the components of a design for a specific application. Thus, the user can develop new parallel applications quickly and easily. Moreover, because the generic patterns are also defined independently from each other, adding new patterns to the system is quite simple.

We have presented the effectiveness of this system for some popular two dimensional and three dimensional applications in bioinformatics: Smith-Waterman, Nussinov, syntenic alignment, multiple sequence alignment and spliced sequence alignment. The measurements show that the new system can generate parallel programs with substantial performance gains for these algorithms. Our future work in parallel program design patterns will include adding more patterns to the system and identifying more applications that can benefit from this technique. Now, we are working on the identification of parallel design patterns that are frequently used on popular data structures in bioinformatics such as sequences, trees and matrices. It would also be interesting to develop the cost analysis and performance prediction tools for the system.

References

1. J. Anvik, J. Schaeffer, D. Szafron, and K. Tan: Why Not Use a Pattern-Based Parallel Programming System, *Proceedings of EURO-PAR'2003*, LNCS 2790, 2003.
2. J.C. Browne, M. Azam, and S. Sobek: CODE: A unified approach to parallel programming, *IEEE Software*, pp.10-18, 1989.

3. A. Bartoli, P. Corsini, G. Dini, C.A. Prete: Graphical design of distributed applications through reusable components, *IEEE Parallel Distrib. Technol*, 3, 1995.
4. J. C. Browne, S. I. Hyder, J. Dongarra, K. Moore, P. Newton: Visual programming and debugging for parallel computing, *IEEE Parallel Distrib. Technol*, 3, 1995.
5. R. Durbin, S. Eddy, A. Krogh, G. Mitchison: Biological Sequence Analysis- Probabilistic Models of Protein and Nucleic Acids, Cambridge University Press, 1998.
6. J. Gerlach: Generic Programming of Parallel Application with JANUS, *Parallel Processing Letters*, Vol. 12, No. 2, pp. 175-190, 2002.
7. M.S. Gelfand, A.A. Mironov, and P. A. Pevzner: Gene recognition via spliced sequence alignment, *Proc. Natl. Acad. Sci*, Vol. 93, pp. 9061-9066, 1996.
8. Z. Galil, K. Park: Dynamic Programming with Convexity, Concavity and Sparsity, *Theoretical Computer Science*, 92, pp. 49-76, 1992.
9. D. Goswami, A. Singh, B.R. Preiss: From Design Patterns to Parallel Architectural Skeletons, *Journal of Parallel and Distributed Computing*, 62(4), 669-695, 2002.
10. X. Huang, K. M. Chao: A generalized global alignment algorithm, *Bioinformatics*, 19(2), pp. 228-233, 2003.
11. W. Liu and B. Schmidt: Parallel Design Pattern for Computational Biology and Scientific Computing Applications, *Proceedings of IEEE International Conference on Cluster Computing*, 2003.
12. D. W. Mount: Bioinformatics-Sequence and Genome Analysis, Cold Spring Harbor Laboratory Press, 2001.
13. W. Miller, E. Myers: Optimal alignments in linear space, *Computer Applications in the Biosciences*, 4, pp. 11-17, 1988.
14. S. Siu, A. Singh: Design patterns for parallel computing using a network of processors, *Sixth IEEE International Symposium on High Performance Distributed Computing*, pp. 293-304, 1997.
15. A. Singh, J. Schaeffer, M. Green: A template-based tool for building applications in a multi-computer network environment, *Parallel Computing*, 461-466, 1989.
16. J. Schaeffer, D. Szafron, G. Lobe, I. Parsons: The enterprise model for developing distributed applications, *IEEE Parallel Distrib. Technol*, 1, pp. 85-96, 1993.
17. T. F. Smith, M. S. Waterman: Identification of common subsequences, *Journal of Molecular Biology*, pp. 195-197, 1981.
18. D. Vandevoorde, N. M. Josuttis: C++ Template: The Complete Guide, Addison Wesley, 2002.
19. M. Zuker, P. Stiegler: Optimal computer folding of large RNA sequences using thermodynamics and auxiliary information, *Nucleic Acids Research*, 9, 1981.

Parallel Inference of a 10.000-Taxon Phylogeny with Maximum Likelihood

Alexandros Stamatakis[1], Thomas Ludwig[2], and Harald Meier[1]

[1] Technische Universität München, Department of Computer Science,
Boltzmannstr. 3, D-85748 Garching b. München, Germany
{Alexandros.Stamatakis,Harald.Meier}@in.tum.de
[2] Ruprecht-Karls-Universität, Department of Computer Science,
Im Neuenheimer Feld 348, D-69120 Heidelberg, Germany
thomas.ludwig@informatik.uni-heidelberg.de

Abstract. Inference of large phylogenetic trees with statistical methods is computationally intensive. We recently introduced simple heuristics which yield accurate trees for synthetic as well as real data and are implemented in a sequential program called RAxML. We have demonstrated that RAxML outperforms the currently fastest statistical phylogeny programs (MrBayes, PHYML) in terms of speed and likelihood values on real data. In this paper we present a non-deterministic parallel implementation of our algorithm which in some cases yields super-linear speedups for an analysis of 1.000 organisms on a LINUX cluster. In addition, we use RAxML to infer a 10.000-taxon phylogenetic tree containing representative organisms from the three domains: Eukarya, Bacteria and Archaea. Finally, we compare the sequential speed and accuracy of RAxML and PHYML on 8 synthetic alignments comprising 4.000 sequences.

1 Introduction

Within the ParBaum project at the Technische Universität München, we work on phylogenetic tree inference based on the maximum likelihood method by J. Felsenstein [2]. We intend to develop novel systems and algorithms for computation of huge phylogenetic trees based on sequence data from the ARB [6] ssu rRNA (small subunit ribosomal RiboNucleic Acid) database. In a recent paper [11] we implemented simple heuristics in RAxML (Randomized Axelerated Maximum Likelihood) which accelerate the tree optimization process and yield good results in terms of final likelihood values. In a series of experiments with 9 real data alignments containing 101 up to 1.000 organisms we demonstrate that RAxML is the currently -to the best of our knowledge- fastest and most accurate sequential program for real data under the HKY85 [4] model of nucleotide substitution. In this paper we describe the parallel non-deterministic implementation of RAxML and report speedup values on a LINUX cluster for an alignment containing 1.000 organisms. Finally, we use the sequential and parallel version of RAxML to infer the -to the best of our knowledge- first integral maximum likelihood-based tree containing 10.000 sequences from the three domains: Eukarya, Bacteria and Archaea. This large alignment has been extracted in cooperation with biologists from ARB. The source code of the sequential and parallel program including all alignment files and final trees is freely available for download at WWWBODE.IN.TUM.DE/~STAMATAK.

Related Work: A comparison of popular phylogeny programs using statistical approaches such as fastDNAml [7], MrBayes [5], treepuzzle [14], and PAUP [8] based on synthetic data may be found in [15]. MrBayes carries out bayesian phylogenetic inference and outperforms all other phylogeny programs in terms of speed and tree quality in this survey. More recently, Guidon et al. published their new maximum likelihood program PHYML [3], which seems to be able to compete with MrBayes. To the best of our knowledge apart from RAxML, MrBayes and PHYML are currently the fastest and most accurate programs for phylogenetic tree inference. In addition, results in [3] and [11] suggest that traditional maximum likelihood methods are still significantly faster than bayesian phylogenetic inference. Thus, maximum likelihood-based programs currently represent the only statistical approach for computation of trees comprising more than 500 sequences. Another important issue is that MrBayes and PHYML have high memory consumption compared to RAxML. For a 1.000 sequence alignment RAxML consumed 199MB, PHYML 880MB, and MrBayes 1.195MB of main memory. Furthermore, both MrBayes and PHYML exited with error messages due to excessive memory requirements for the 10.000 taxon alignment on a processor equipped with 4GB (!) of main memory. Therefore, we made an effort to port MrBayes and PHYML to a 64-bit Itanium2 1.3GHz processor with 8GB of memory. While MrBayes exited for unknown reasons, PHYML finally required 8.8GB of main memory in contrast to RAxML which consumed only 800MB. In what concerns parallel computing, the parallel implementations of bayesian methods are relatively closely coupled such that high performance computers with expensive communication infrastructure are required [1]. For PHYML there exists no parallel implementation yet. There also exists a popular parallel implementation for fastDNAml [13] which is however based on the old sequential algorithm from 1994.

2 Heuristics

Sequential Algorithm: The heuristics of RAxML belong to the class of algorithms, which optimize the likelihood of a starting tree which already comprises all sequences. In contrast to other programs RAxML starts by building an initial tree with the dnapars parsimony program from Felsenstein's PHYLIP package [9] for two reasons: *Firstly*, parsimony is related to maximum likelihood under simple models of evolution such that one can expect to obtain a starting tree with a relatively good likelihood value compared to random or neighbor joining starting trees. *Secondly*, dnapars uses stepwise addition [2] for tree building and is relatively fast. The stepwise addition algorithm enables the construction of distinct starting trees by using a randomized input sequence order. Thus, RAxML can be executed several times with different starting trees and thereby yields a set

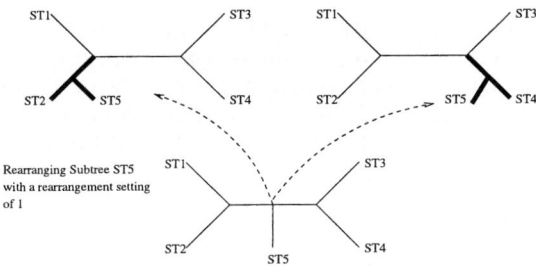

Fig. 1. Rearrangements traversing one node for subtree ST5, branches which are optimized by RAxML are indicated by bold lines.

of distinct final trees which can be used to build a consensus tree. To expedite computations, some optimization steps have been removed from dnapars.

The tree optimization process represents the second and most important part of the heuristics. RAxML performs standard subtree rearrangements by subsequently removing all possible subtrees from the currently best tree `t_best` and re-inserting them into neighboring branches up to a specified distance of nodes. RAxML inherited this optimization strategy from fastDNAml. One rearrangement step in fastDNAml consists of moving all subtrees within the currently best tree by the minimum up to the maximum distance of nodes specified (lower/upper rearrangement setting). This process is outlined for a single subtree (ST5) and a distance of 1 in Figure 1. In fastDNAml the likelihood of each thereby generated topology is evaluated by exhaustive branch length optimizations. If one of those alternative topologies improves the likelihood `t_best` is updated accordingly and once again all possible subtrees are rearranged within `t_best`. This process of rearrangement steps is repeated until no better topology is found. The rearrangement process of RAxML differs in two major points: In fastDNAml after each insertion of a subtree into an alternative branch the branch lengths of the entire tree are optimized. As depicted in Figure 1 with bold lines RAxML only optimizes the three local branches adjacent to the insertion point either analytically or by the Newton-Raphson method before computing its likelihood value. Since the likelihood of the tree strongly depends on the topology per se this fast pre-scoring can be used to establish a small list of potential alternative trees which are very likely to improve the score of `t_best`. RAxML uses a list of size 20 to store the best 20 trees obtained during one rearrangement step. This list size proves to be a practical value in terms of speed and thoroughness of the search. After completion of one rearrangement step the algorithm performs global branch length optimizations on those 20 best topologies only. Due to the capability to analyze significantly more alternative topologies in less time a higher upper rearrangements setting can be used e.g. 5 or 10 which results in significantly improved final trees. Another important change especially for the initial optimization phase, i.e. the first 3-4 rearrangement steps, consists in the subsequent application of topological improvements during one rearrangement step. If during the insertion of one specific subtree into an alternative branch a topology with a better likelihood is encountered this tree is kept immediately and all subsequent subtree rearrangements of the current step are performed on the improved topology. This enables rapid initial optimization of random starting trees [11]. The exact implementation of the RAxML algorithm is indicated in the C-like pseudocode below. The algorithm is passed the user/parsimony starting tree t, the initial rearrangement setting rStart (default: 5) and the maximum rearrangement setting rMax (default: 21). Initially the rearrangement stepwidth ranges from rL = 1 to rU = rStart. Fast analytical local branch length optimization a is turned off when functions rearr(), which actually performs the rearrangements, and optimizeList20() fail to yield an improved tree for the first time. As long as the tree does not improve the lower and upper rearrangement parameters rL, rU are incremented by rStart. The program terminates when the upper rearrangement setting is greater or equal to the maximum rearrangement setting, i.e. rU >= rMax.

```
optimize(tree t, int rStart, int rMax)
{
  int rL, rU;
  boolean a = TRUE, impr = TRUE, stop = FALSE;
  while(!stop){
    if(impr){
      rL = 1;
      rU = rStart;
      rearr(t, rL, rU, a);
    }
    else{
      if(!a){
        a = FALSE;
        rL = 1;
        rU = rStart;
      }
      else{
        rL += rStart;
        rU += rStart;
      }
      if(rU < rMax) rearr(t, rL, rU, a);
      else stop = TRUE;
    }
    impr = optimizeList20();
  }
}
```

Parallel Algorithm: The parallel implementation is based on a simple master-worker architecture and consists of two phases. In **phase I** the master distributes the alignment file to all worker processes if no common file system is available, otherwise it is read directly from the file. Thereafter, each worker independently computes a randomized parsimony starting tree and sends it to the master process. Alternatively, it is possible to start the program directly in **phase II** by specifying a tree file name in the command line. In **phase II** the master initiates the optimization process for the best parsimony or specified starting tree. Due to the high speed of a single topology evaluation, the requirement for atomicity of a specific subtree rearrangement by function rearrangeSubtree() and the high communication cost, it is not feasible to distribute work by single topologies as e.g. in parallel fastDNAml. Therefore, we distribute work by sending

Parallel Inference of a 10.000-Taxon Phylogeny with Maximum Likelihood

the subtree ID along with the currently best topology `t_best`, to each worker. The `sequential` and `parallel` implementation of RAxML on the master-side is outlined in the pseudocode of function `rearr()` which actually executes subtree rearrangements. The worker simply executes function `rearrangeSubtree()`.

```
void rearr(tree t_best, int rL, int rU, boolean a)
{
 boolean impr;
 worker w;
 for(i = 2; i < #species * 2 - 1; i++){
   if(sequential){
     impr = rearrangeSubtree(t_best, i, rL, rU, a);
     if(impr) applySubsequent(t_best, i);
   }
   if(parallel){
     if(w = workerAvailable) sendJob(w, t_best, i);
     else putInWorkQueue(i);
   }
 }
 if(parallel){
   while(notAllTreesReceived){
     w = receiveTree(w_tree);
     if(likelihood(w_tree) > likelihood(t_best)) t_best = w_tree;
     if(notAllTreesSent) sendJob(w, t_best, nextInWorkQueue());
   }
 }
}
```

In the sequential case rearrangements are applied to each individual subtree i. If the tree improves through this subtree rearrangement `t_best` is updated accordingly, i.e. subsequent topological improvements are applied. In the parallel case subtree IDs are stored in a work queue. Obviously, the subsequent application of topological improvements during 1 rearrangement step (1 invocation of `rearr()` is closely coupled. Therefore, we slightly modify the algorithm to break up this dependency according to the following observation: Subsequent improved topologies occur only during the first 3–4 rearrangement steps (initial optimization phase). Thereafter, the likelihood is improved only by function `optimizeList20()`. This phase requires the largest amount of computation time, especially with big alignments ($\approx$ 80% of execution time). Thus, during the initial optimization phase we send only one single subtree ID i=2,...,#species * 2 - 1 along with the currently best tree `t_best` to each worker for rearrangements. Each worker returns the best tree `w_tree` obtained by rearranging subtree i within `t_best` to the master. If `w_tree` has a better likelihood than `t_best` at the master, we set `t_best` = `w_tree` and distribute the updated best tree to each worker along with the following work request. The program assumes that the initial optimization **phase IIa** is terminated if no subsequent improved topology has been detected during the last three rearrangement steps. In the final optimization **phase IIb**, we reduce communication costs and increase granularity by generating only $5 * \#workers$ jobs (subtree ID spans). Finally, irrespective of the current optimization phase the best 20 topologies (or #*workers* topologies if #*workers* > 20) computed by each worker during one rearrangement step are stored in a local worker tree list. When all `#species * 2 - 3` subtree rearrangements of `rearr()` have been completed, each worker sends its tree list to the master. The master process merges the lists and redistributes the 20 (#*workers*) best tree topologies to the workers for branch length optimization, like in parallel fastDNAml. When all topologies have been globally optimized the master starts the next iteration of function `optimize()`. Due to the required changes to the algorithm the parallel program is non-deterministic, since final output depends on the number of workers and on the arrival sequence of results for runs with equal numbers of workers, during the initial optimization **phase IIa**. This is due to the altered implementation of the subsequent application of topological improvements during the initial rearrangement steps which leads to a traversal of search space on different paths.

3 Results

For our experiments we extracted alignments of 1.000 and 10.000 taxa (1000_ARB, 10000_ARB) from the ARB database containing organisms from the domains Eukarya, Bacteria and Archaea. We used the HKY85 model of sequence evolution and a transition/transversion ratio of 2.0. Furthermore, we generated 8 synthetic 4.000 taxon alignments (SIM_1,...,SIM_8) with a length of 2.000 base pairs and distinct parameter settings for comparison of PHYML and RAxML.

Synthetic Data Tests: In Table 1 we list the topological distance to the simulated "true" tree (normalized Robinson-Foulds rate) and execution time in seconds of PHYML and RAxML for the 8 synthetic 4.000 taxon alignments. Details on the generation of the simulated data sets, a discussion of results, and supplementary experiments with real-data are provided in [12].

Table 1. Topological accuracy and execution times for PHYML & RAxML on simulated data

data	PHYML	secs	RAxML	secs	data	PHYML	secs	RAxML	secs
SIM_1	0.065	18944	0.065	9152	SIM_5	0.028	24182	0.035	91178
SIM_2	0.039	22273	0.037	50609	SIM_6	0.027	32614	0.031	176686
SIM_3	0.033	24907	0.027	97962	SIM_7	0.027	34750	0.032	185454
SIM_4	0.030	30870	0.031	85080	SIM_8	0.026	18828	0.036	78061

Scalability Tests: We conducted parallel tests with a fixed starting tree for 1000_ARB. The program was executed on the 2.66GHz Xeon cluster on the RRZE [10] on 1, 4, 8, 16, and 32 processors with an initial rearrangement setting `rStart` of 5. To calculate the speedup values we only take into account the number of workers, since the master process hardly produces any load. In Figure 2 we plot "fair" and "normal" speedup values obtained for the experiments with

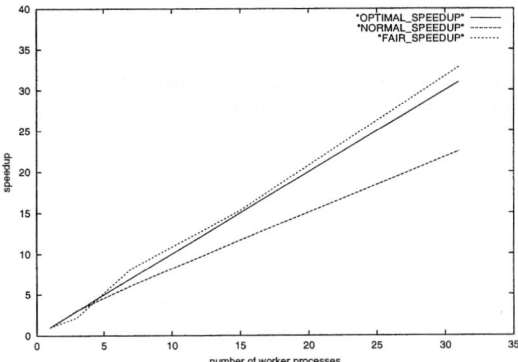

Fig. 2. Normal, fair, and optimal speedup values for 1000_ARB with 3,7,15, and 31 worker processes on the RRZE PC Cluster.

the 1000_ARB data set at the RRZE PC-cluster. "Fair" speedup values measure the first point of time at which the parallel code encounters a tree with a better likelihood than the final tree of the sequential run or vice versa. These "fair" values better correspond to real program performance. Furthermore, we also indicate "normal" speedup values which are based on the entire execution time of the parallel program, irrespective of final likelihood values. Since we intend to explore the effect of non-determinism on program performance we executed the parallel code 4 times for each job-size and calculated average "normal"/"fair" execution times and likelihood values.

Inference of a 10.000-taxon tree: The computation of the 10.000-taxon tree was conducted using the sequential, as well as the parallel version of RAxML. One of the advantages of RAxML consists in the randomized generation of starting trees. Thus, we computed 5 distinct randomized parsimony starting trees sequentially along with the first 3–4 rearrangement steps on a small cluster of Intel Xeon 2.4GHz processors at our institute. This phase required an average of 112.31 CPU hours per tree. Thereafter, we executed several subsequent parallel runs (due to job run-time limitations of 24 hrs) with the respective starting trees on either 32 or 64 processors at the RRZE 2.66GHz Xeon-cluster. The parallel computation required an average of 1689.6 accumulated CPU hours per tree. The best likelihood for 10000_ARB was -949570.16 the worst -950047.78 and the average -949867.27. PHYML reached a likelihood value of -959514.50 after 117.25 hrs on the Itanium2. Note, that the parsimony starting trees computed with RAxML had likelihood values ranging between -954579.75 and -955308.00. The average time required for computing those starting trees was 10.99 hrs. Since bootstrapping is not feasible for this large data size and in order to gain some basic information about similarities among the 5 final trees we built a consensus tree using the extended majority rule with consense from PHYLIP (consense constantly exited with a memory error message when given more than 5 trees). The consensus tree has 4777 inner nodes which appear in all 5 trees, 1046 in 4, 1394 in 3, 1323 in 2, and 1153 in only 1 tree (average: 3.72).

4 Conclusion

We presented an efficient parallel implementation of recently introduced heuristics for phylogenetic inference under simple models of site substitution which achieves optimal speedup values. Thus, RAxML provides a fast and practicable approach for sequential and parallel inference of large phylogenetic trees containing up to 10.000 organisms. We were able to compute the -to the best of our knowledge- first 10.000-taxon tree with maximum likelihood using RAxML on a medium-size commodity PC cluster. However, at this tree size, there arise new, yet unresolved problems such as, assessment of quality and visualization which require further investigation.

References

1. Feng, X. et al.: Parallel algorithms for Bayesian phylogenetic inference. J. Par. Dist. Comp. (2003) 63:707–718
2. Felsenstein, J.: Evolutionary Trees from DNA Sequences: A Maximum Likelihood Approach. J. Mol. Evol., (1981) 17:368–376,
3. Guindon, S. et al.: A Simple, Fast, and Accurate Algorithm to Estimate Large Phylogenies by Maximum Likelihood. Syst. Biol., (2003) 52(5):696–704
4. Hasegawa,M. et al.: Dating of the human-ape splitting by a molecular clock of mitochondrial DNA. J. Mol. Evol., (1985) 22:160–174
5. Huelsenbeck, J.P. et al.: MRBAYES: Bayesian inference of phylogenetic trees. Bioinf., (2001) 17(8):754–755
6. Ludwig, W. et al.: ARB: A Software Environment for Sequence Data. Nucl. Acids Res., (2004) 32(4):1363–1371
7. Olsen, G. et al.: fastdnaml: A Tool for Construction of Phylogenetic Trees of DNA Sequences using Maximum Likelihood. Comput. Appl. Biosci., (1994) 10:41–48
8. PAUP: PAUP.CSIT.FSU.EDU, visited May 2003
9. PHYLIP: EVOLUTION.GENETICS.WASHINGTON.EDU, visited Nov 2003
10. RRZE: WWW.RRZE.UNI-ERLANGEN.DE, visited Oct 2003
11. Stamatakis, A. et al.: New Fast and Accurate Heuristics for Inference of Large Phylogenetic Trees. Proc. of IPDPS2004 (2004)
12. Stamatakis, A. et al.: RAxML-III: A Fast Program for Maximum Likelihood-based Inference of Large Phylogenetic Trees. Bioinf. to be published
13. Stewart, C. et al.: Parallel Implementation and Performance of fastdnaml - a Program for Maximum Likelihood Phylogenetic Inference. Proc. of SC2001 (2001)
14. Strimmer, K. et al.: Quartet Puzzling: A Maximum-Likelihood Method for Reconstructing Tree Toologies. Mol. Biol. Evol., (1996) 13:964-969
15. Williams, T.L. et al.: An Investigation of Phylogenetic Likelihood Methods. Proc. of BIBE'03 (2003)

Cache-Based Parallelization of Multiple Sequence Alignment Problem

Gilles Parmentier[1], Denis Trystram[2], and Jaroslaw Zola[2,3],⋆

[1] Laboratoire LBMC
Ecole Normale Superieure of Lyon, France
[2] Laboratoire ID–IMAG⋆⋆, France
[3] Institute of Computer & Information Sciences
Czestochowa University of Technology, Poland
zola@imag.fr

Abstract. In this paper we present new approach to the problem of parallel multiple sequence alignment. The proposed method is based on the application of caching technique and is aimed to solve, with high precision, large alignment instances on the heterogeneous clusters. The cache is used to store partial alignment guiding trees which can be reused in future computations, and is applied to eliminate redundancy of computations in parallel environment. We describe an implementation based on the CaLi library, the software designed for caches implementation. We report preliminary experimental results and finally, we propose some extensions of our method.

1 Introduction

Today, multiple sequence alignment is a main concern in bioinformatics. It is both a comparison tool for sequences but also an input data for phylogeny inference. In this case, it is important to have a reliable multiple alignments [1]. In general formulation, for a set of sequences we want to compute the phylogeny and multiple alignment. This problem is known as the "Generalized Tree Alignment Problem" [2] and has been shown to be Max–SNP–HARD [3].

In our previous work [4] we have proposed a new heuristic to solve the problem of multiple sequence alignment based on the simultaneous reconstruction of the alignment and the tree used to guide alignment construction. This general approach allows fast approximated alignment of large datasets (the ones containing many long sequences) as well, as the more precise alignment of small ones. In this case the method, called *PhylTree* (*PT*), explores space of the partial guiding trees of some fixed size k, being an input parameter. Unfortunately, the size of the analyzed search space grows exponentially with growth of k. Thus, a *PT* is too complex to be applied for large data.

In this paper we propose parallel version of the *PT* method, which allows accurate alignment of large datasets to be computed in reasonable time. First, we describe the basic parallel algorithm, then we show its extension based on the application of the

---
⋆ The work of Jaroslaw Zola has been supported by French Government.
⋆⋆ Laboratory ID–IMAG is funded by CNRS, INRIA, INPG and UJF.

techniques similar to these used in the web-caching and distributed caching [5]. We present also results of the experiments obtained using our software for parallel multiple sequence alignment on the heterogeneous clusters.

2 Multiple Alignment with Tree Construction

The usual aim of multiple sequence alignment is to understand the relation between the involved sequences. To do this, one need to know the relationships between analyzed species, the *phylogeny*. This one is figured out to be a tree, with the common ancestor at the root, and sequences of the dataset at the leaves. Usually, when using a multiple alignment program, we do not have this knowledge and estimation of the tree must be performed.

2.1 Related Works

Multiple alignment [6–8] and multiple alignment with tree estimation have been addressed a lot in the recent years [9–11]. As we have already said, these problems are computationally intensive. The important number of proposed solutions relay on an approach named *progressive multiple alignment*. Two main parts can be distinguished in this method. The first one consists in the construction of a tree aiming to estimate the ancestral relationships of the dataset. This is important as it is the best criterion to construct the multiple alignment. Then, such a tree is used to guide the multiple alignment and here, depending on the method, different criteria may be used.

2.2 A New Approach: PhylTree Method

The *PT* method follows general scheme of multiple alignment with simultaneous building of phylogenetic tree. The principle of this algorithm is to iteratively group sequences which are supposed to be closely related from the evolutionary point of view. The algorithm proceeds in two main phases (see Fig. 1). First, the distance matrix for all sequences is computed. In our case it is done using pairwise sequence alignment, which is a typical approach. Next, the second phase starts, it consists in four iteratively repeated steps: (i) neighborhood determination, (ii) phylogenetic analysis, (iii) new taxa construction, (iv) update of distance matrix.

The distance matrix computation requires $\Theta(\frac{n^2-n}{2})$ pairwise alignments. It is easy to notice that this process can be well parallelized since all computations are independent.

After building the distance matrix we proceed with the main part of the method. First, we look for relevant groups of sequences. More precisely, for each taxon from the input dataset we search neighborhood of fixed size k. Here, neighborhood is defined as follows:

$$ng(t) = \{i \in T | \forall j \in T, D_{ti} \leq D_{tj} \text{ or } j \in ng(t)\} \tag{1}$$

where: T denotes input set of taxa, D is a distance matrix for T and $ng(t)$ is a neighborhood of taxon t, $|ng(t)| = k$. We will use this notation in the rest of the paper. The process of neighborhoods determination is very fast, and additionally number of input

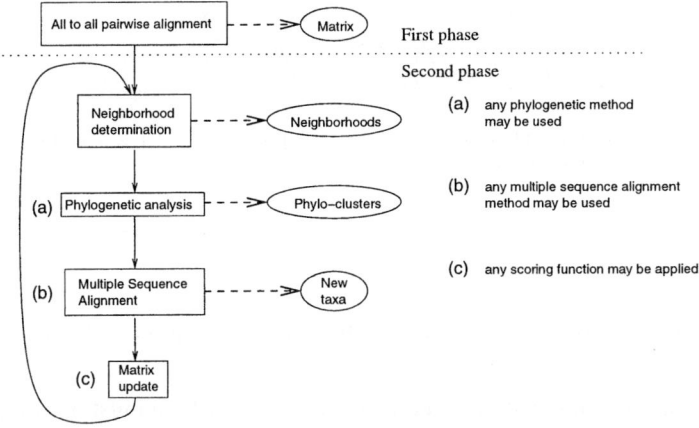

Fig. 1. General scheme of *PhylTree* method.

taxa decreases with every iteration. Thus, parallelization of this stage is useless, not to say impossible.

The crucial and the most complex part of the whole process is the phylogenetic analysis. In this step, for each neighborhood we construct all possible guiding trees. The number of trees with k leaves growths exponentially with k. This means that the process of analysis is very time consuming and requires a special treatment.

As a result of phylogenetic analysis, for each neighborhood we obtain the best tree with respect to the corresponding score of multiple alignment. We call such tree a *partial guiding tree*. In fact, we can obtain several such trees and we will discuss implications of this fact in the last section. This way we have a list of best trees which are next used to find the set of the most relevant sequences in the input dataset. These groups are called *phylo-clusters*. Formally, *phylo-cluster* is defined as a set of m taxa ($m \leq k$) which appear as leaves of a subtree of size m (see Fig. 2). Additionally, to be a *phylo-cluster* set of taxa must appear in the neighborhood of each of its components. This condition is due to the relevance of the groups. A *phylo-cluster* that does not appear in the neighborhood of its components was built from distant taxa.

The next step is the update of the distance matrix. Previously determined *phylo-clusters* (with appropriate trees) become new entries in the distance matrix and corresponding values in the matrix are modified. The way the matrix is updated depends on scoring function we use. In our implementation it is mean average of distances between *phylo-cluster* components and other taxa. The whole process is extremely fast and finishes an iteration.

3   Parallelization of the Method

In the previous section we have described the *PT* method highlighting some properties which make this approach too complex to solve the large instances of the multiple alignment problem. Parallel version of the *PT* method tries to face these issues.

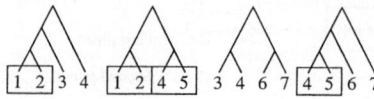

Fig. 2. Best trees for some neighborhoods (numbers represent taxa). $<1,2>$ and $<4,5>$ are the only *phylo-clusters*.

3.1 Generic Scheme

The parallel *PT* algorithm is based on the following observations:

- Distance matrix computation is an independent task which must be completed before starting the second phase.
- The atomic operation for the whole process is alignment which is used to build the distance matrix and is a basic operation during the phylogenetic analysis as well.
- Processing of different neighborhoods often requires computations which overlaps. Namely, if two neighborhoods have common subset of more than one element, alignment for such set will be computed twice. At the same time, neighborhoods can be processed concurrently.
- Phylogenetic analysis results with set of optimal partial guiding trees. Selecting only one of them, we constrain the exploration of the search space which maybe valuable for a user.
- Large input datasets can contain sequences which were previously used as an input and which may tend to form neighborhoods analyzed in the past.

In order to parallelize our algorithm we have chosen the distributed master-worker (DM/W) paradigm addressing heterogeneous computational clusters. Such architecture seems to be the most accurate to efficiently carry on computations required by *PT*. Master, which is a single, arbitrary chosen node, stores input and cached data. Thus, it is responsible for all I/O operations and is used to manage the whole system. As a computational power set of workers is used, which can be set of any machines able to run single pairwise alignment computations.

At the beginning of computations, master distributes requests for pairwise alignments among workers and gathers results to have a complete distance matrix at the end. Here, any scheduling method which guarantees good load balancing can be used. During this step, the master performs additional task. It measures performance of each computing node. This allows to approximate efficiency of nodes and optimize scheduling in the next stages. Since pairwise alignment is an atomic operation done by workers, such an approximation seems to be reasonable.

As soon, as the distance matrix is computed the main loop of the *PT* is started. The master generates neighborhoods for all input taxa. Next, it searches cache to check which neighborhoods have to be analyzed (we detail cache usage in the next subsection). These are send to workers and there processed. Master receives best guiding trees found by workers, updates cache and finally, modifies the distance matrix.

3.2 Application of the Cache

For a long time, caching has been deployed as a one of the methods to improve the efficiency of software and hardware systems. Today, the cache applications cover a wide variety of computer science areas, including hardware, Internet technologies with web-caching and distributed caching, data bases and complex computational systems [5, 12].

In our application, we can distinguish two levels of cache. Both relaying on the observations made in the previous section. The first level of cache is to keep partial guiding trees being result of phylogenetic analysis. According to formula (1), neighborhood $ng(t)$ of taxon t contains simply t and set of its $k-1$ closest elements. As it is easy to notice, the only knowledge used to determine neighborhood is distance matrix. This property gives us some possibility to compare two neighborhoods without explicit comparison of their elements, and further, to build unique key describing given neighborhood. The caching system can next associate this key with corresponding best partial guiding tree.

Definition 1. *ATAC graph (all to all comparison graph). ATAC graph, denoted $AG(ng)$, for the neighborhood ng is a labeled complete graph whose nodes are the elements of ng. Every edge of AG graph is labeled by distance between the nodes it connects.*

$AG(ng)$ describes precisely neighborhood ng. This allows us to write binary predicates to compare AG graphs and this way to compare neighborhoods. Such an approach has very important property. It permits us to use caching system as a heuristic for alignment process. More precisely, depending on the form of predicate we can have exact solution, the predicate compares if two neighborhoods are exactly the same, or approximation, predicates compare if two neighborhoods are similar. Here, "similar" is a user defined criterion. Of course, the graph comparison is a complex task by itself. However, considered graphs are small (usually $k \in [4,9]$) and lot of methods or relaxations exists which can be successfully applied here.

The second level cache is maintained in a distributed fashion. As we have already pointed, during neighborhoods processing strong redundancy of computations occurs. Hence, application of cache on each node combined with accurate method for request routing can lead to a great improvement in efficiency. The use of distributed cache requires to select carefully the scheduling method, especially because we are addressing heterogeneous architecture. Scheduling must guarantee that neighborhoods which have common subset will be assigned to workers in the way maximizing cache usage – the computations will be replaced by cache requests. Here, the knowledge about nodes efficiency collected during the distance matrix computation can be utilized. The distributed cache is currently under development and first analyzes are very promising for further research.

4 Experimental Results

To validate our approach of parallel multiple sequence alignment, we have performed several experiments with actual biological data. The main purpose of these experiments was to verify what are benefits coming from the cache usage. Currently only the first level cache application has been tested.

4.1 Testing Environment

The parallel *PT* method has been implemented using message passing interface and C++ language. To perform scheduling during the first phase, the guided self-scheduling method is used. As a communication layer LAM/MPI version 7 is utilized. The caching system has been implemented using the CaLi library [13], and is based on the classical *Least Frequently Used with Size* replace policy. This policy evicts entries which are accessed least frequently and if two entries have the same number of references the smaller one is removed. However, during experiments no evictions were performed since cache capacity was never exceeded. To compare neighborhoods *AG* graphs are tested for equality. The sequence alignment is based on the Sankoff algorithm [14].

As a hardware platform we have used *ACCORD* cluster [15]. This machine is a heterogeneous cluster containing one server (Dual AMD Athlon 1600 MHz with 1 GB RAM), and SMP nodes based on AMD Athlon 1600 MHz, AMD Athlon 1200 MHz and Intel Pentium 750 MHz processors. Each node contains 512 MB of RAM. As a physical connection layer FastEthernet is used.

4.2 Experiments

We have simulated a set of requests for multiple alignment of sequences coming from *TreeBase* [16]. Each request was processed with neighborhood size $k = 5$. The first request contained 16 DNA sequences of average length 1000 bp (reference M993). Every next request was extended with 8 new sequences. When processing of the last request had finished, we have resubmitted all requests. This allowed us to measure efficiency of the system when usage of the first level cache is maximal.

To process requests the server could use up to 10 weakly heterogeneous computational nodes. Table 1 presents result of the experiment. T_{seq} is the execution time of the sequential version of the *PT* method on dual AMD Athlon 1600 MHz node. T is the time of execution of the parallel version. T_{opt} represents the time of execution of the parallel version when cache hit ratio is maximal, that is 1. Finally, T_c is the time of execution of the parallel version but without cache support. In this case, during computations all available processors are used.

The multiple sequence alignments which we have obtained using parallel *PT* method were exactly the same as these obtained using the sequential one.

Table 1. Results of requests processing for M993 entry.

# Sequences	T_{seq} [s]	T [s]	# Cache hits	Hit ratio	# CPU used	Speedup	T_{opt} [s]	T_c [s]
16	24456	4850	19	0.40	5	5.04	2070	5334
24	32112	6245	22	0.40	6	5.14	3554	7284
32	55450	9028	30	0.32	7	6.14	4721	10631
40	71136	13382	35	0.29	7	6.24	5500	13885
48	87617	13833	48	0.35	7	6.33	5674	15643
55	103619	20559	63	0.43	7	5.03	6714	21766
63	133525	27215	71	0.41	9	4.90	7851	28485

4.3 Analysis

Based on the presented results we can state several observations. As expected, application of the cache increases the efficiency of the system. This is a natural consequence of replacing large part of the computations by caching system. However, since not every partial tree can be cached, application of the cache has some limitations. This is suggested by results of processing when the cache hit ratio is maximal. It is obvious that we can not do better than the time required to process uncachable data (time used to serve cache can be neglected). In the simplified way, this can be expressed by the following formula (we consider only second phase of the *PT* method):

$$T_N = T_1\beta + \frac{(1-\beta-\eta)T_1}{N} \quad (2)$$

where: T_N is a time of execution on N processors, T_1 is a time of execution on a single processor, β is a sequential part of the computations and η is a fraction of computations which are handled by the caching system. In our case β refers to the neighborhoods determination, multiple alignment for best tree, and the distance matrix update. It is easy to notice, that formula (2) is similar to the well known Amdhal's Law. What is more, with growth of request size the speedup remains the same but the number of the processors used increases. This can be explained by change of the β/η ratio – cache usage is more or less constant but number of required computations increases. However, we can expect that cache hit ratio will increase when more and more requests will be processed, and the introduction of second level cache will speed up computations of uncached requests.

One of the interesting observations concerns processors usage. If the processed request is not big enough not all available processors are used. This is direct consequence of use of DM/W architecture. Since the number of neighborhoods to process during a single iteration can be significantly reduced by caching system number of required processors stays small. On the other hand, this property remains very useful for large requests since it allows a better load balancing.

The presented results show, that application of the first level cache can noticeably increase efficiency of the whole system. Additionally, the second level cache has to be implemented to more effectively utilize computational power during phylogenetic analysis.

5 Conclusions

In this paper we have presented a new approach to the problem of parallel multiple sequence alignment. The originality of this method flows from two things: concurrency of alignment and guiding tree construction and application of two level caching. We have presented also preliminary experimental results based on the alignment of real biological data.

The current work concerns application of the second level cache. Additionally, our method provides several possibilities for extensions. These include experiments with different scheduling protocols, design of new cache update policies and research on different distributed caching schemes.

One of the most interesting issues is to develop support for the human-in-the-loop interaction [17]. As we have mentioned in Section 2.2, using only one from the set of best trees may strongly constrain the exploration of the search space. On the other hand, it is computationally infeasible to use several trees for each neighborhood, and the term "best tree" may be not well defined. We believe that introducing user interaction we can partially solve this problem. End-user could have possibility to arbitrary choose one of the best guiding trees and to change the value of the parameter k during processing. This way he/she could direct alignment process using expert knowledge, and dynamic change of k value should cause better global cache usage. This should lead finally to more accurate (from the user point of view) alignments.

References

1. Philips, A., Janies, D., Wheeler, W.: Multiple sequence alignment in phylogenetic analysis. Mol. Phyl. Evol. **16** (2000) 317–330
2. Sankoff, D., Morel, C., Cedergren, R.: Evolution of 5s rna and the nonrandomness of base replacement. Nature New Biology **245** (1973) 232–234
3. Jiang, T., Lawler, E., Wang, L.: Aligning sequences via an evolutionnary tree: Complexity and approximation. In: Proc. 26th Ann. ACM Symp. on Theory of Comp. (1994) 760–769
4. Guinand, F., Parmentier, G., Trystram, D.: Integration of multiple alignment and phylogeny reconstruction. In: Eur. Conf. on Comp. Biology, Poster Abstr. (2002)
5. Cao, P., Irani, S.: Cost-aware www proxy caching algorithms. In: USENIX Symposium on Internet Technologies and Systems. (1997) 193–206
6. Katoh, K., Misawa, K., Miyata, T.: MAFFT: a novel method for rapid multiple sequence alignment based on fast Fourier transform. Nucleic Acid Res. **30** (2002) 3059–3066
7. Lee, C., Grasso, C., Sharlow, M.: Multiple sequence alignment using partial order graphs. Bioinformatics **18** (2002) 452–464
8. Morgenstern, B.: DIALIGN 2: improvement of the segment-to-segment approach to multiple sequence alignment. Bioinformatics **15** (1999) 211–218
9. Higgins, D., Thompson, J., Gibson, T.: CLUSTALW: improving the sensitivity of progressive multiple sequence alignment through sequence weighting, position-specific gap penalties and weight matrix choice. Nucleic Acids Res. **22** (1994) 4673–4680
10. Pei, J., Sadreyev, R., Grishin, N.: PCMA: fast and accurate multiple alignment based on profile consistency. Bioinformatics **19** (2003) 427–428
11. Notredame, C., D.G., H., Henriga, J.: T–Coffee: A novel method for fast and accurate multiple sequence alignment. J. Mol. Biol. **302** (2000) 205–217
12. Iyer, S., Rowston, A., Druschel, P.: Squirrel: A decentralized, peer-to-peer web cache. In: 12th ACM Symp. on Princ. of Distr. Computing. (2002)
13. http://icis.pcz.pl/~zola/CaLi: CaLi – generic computational buffers library. Web Page (2004)
14. Sankoff, D.: Minimal mutation trees of sequences. SIAM J. Appl. Math. **28** (1975) 35–42
15. http://eltoro.pcz.pl: Accord cluster. Web Page (2003)
16. http://www.treebase.org: Treebase – a database of phylogenetic knowledge. Web Page (2004)
17. Licklider, J.: Man-computer symbiosis. IRE Trans. on Hum. Fact. in Elect. **HFE–1** (1960) 4–11

Topic 18
Peer-to-Peer and Web Computing

Seif Haridi, Karl Aberer, Peter Van Roy, and Michele Colajanni

Topic Chairs

Peer-to-peer computing has evolved recently from an attractive new paradigm into an exciting and vibrant research field bringing together researchers from distributed systems, networks and theory. Peer-to-peer systems are characterized as being decentralized, self-organizing distributed systems, in which all or most communication is symmetric. Peer-to-peer technologies are aimed for building large-scale distributed services. In particular peer-to-peer architectures could be used in the context of web and grid computing to provide scalable self-organizing resources/services. Because of their size, autonomy and possibly high volatility, peer-to-peer systems provide the opportunity for researchers to re-evaluate many fields of distributed computing, such as protocols, infrastructures, security, certification, fault tolerance, scheduling, performance analysis, etc.

This topic examines peer-to-peer technologies, applications, and systems, and also identifies key research issues and challenges. 19 papers were submitted, seven has been accepted as regular papers, and one as short. Here comes a short description of the accepted papers.

In "A P2P Grid Services-Based Protocol: Design and Evaluation" the authors propose to use the Gnutella-protocol for discovering grid-peers and implement an OGSA-compliant grid service for the Gnutella protocol. In "A Small World Overlay Network for Resource Discovery" the authors suggest that every peer in the system describe its resources in a XML document. A distance function is defined which takes two XML documents and returns the complement of the ratio of matches between the documents. In "A Hybrid Peer-to-Peer Network Solution to the Synchronization and Session Control in 3D Multi-User Virtual Environments" the paper is describing an alternative to a centralized maintenance of a distributed 3D environment using the Gnutella network. In "TAP: Topology-Aware Peer-to-Peer Network with Expanding-Area Lookup" the authors propose a combination of DHT mechanism with a small hierarchical structure (4 levels are proposed) to handle the locality problem in Distributed Hash Table overlay networks. In "Improving the Scalability of Logarithmic-Degree DHT-based Peer-to-Peer Networks" the authors propose an improvement of the network size for given routing-table size of logarithmically structured P2P DHT-based networks. In "The *ncast* Primitive for Peer-to-Peer Networks" the paper proposes a communication primitive similar to a multicast operation, where only the number of delivered messages is specified. The paper "Going Large-scale in P2P Experiments Using the JXTA Distributed Framework" basically presents a case study of a framework for distributed applications based on JXTA which is a middleware for building peer-to-peer applications. Finally "Using Similarity Groups to Increase Performance of P2P Computing" is about the design of new strategies based on various concepts of similarity that can be used to group together peers in a Peer-to-peer network.

The ncast Primitive for Peer-to-Peer Networks

James Pascoe and Vaidy Sunderam

Dept. of Math and Computer Science,
Emory University, Atlanta, GA 30322, USA
vss@mathcs.emory.edu

Abstract. Many emerging applications of peer-to-peer networking will benefit from new, more appropriate communication primitives at the transport level. This paper presents the design and preliminary implementation of two variants of a novel primitive termed ncast. Unlike conventional multiway communications, the ncast primitives are not based on identity; their semantics specify the *number* of message deliveries in a given *collaborative peer group*. Both variants provide a means for fulfilling message delivery *quotas* in a peer-to-peer network, but they differ in the strength of the semantics they provide. A description of this operation and its semantics is followed by examples and implementation outlines. The paper concludes with a brief discussion of issues relating to the utility of such primitives in collaborative peer group networks.

1 Introduction

This paper introduces a primitive called ncast, for use in peer-to-peer networks. Generally, peer-to-peer systems such as Tapestry [11], Pastry [6], Chord [7] and CAN [5] support two paradigms of communication (see Figure 1). *Unicast* protocols offer single sender, single receiver semantics and are analogous to point-to-point exchanges. *Propagation* schemes disseminate messages from a single origin to the whole network such as in the Gnutella search protocol [3]. In addition, the recent introduction of protocols such as Acknowledged Multicast [2] suggests the emergence of peer-to-peer *multicast* algorithms, which typically contact multiple nodes in a single operation.

The protocols proposed in this paper are members of a family of message propagation primitives designed to support *collaborative peer groups* [9]. Such collections of entities occur both in CSCW settings as well as in Grid environments, where resource sharing on peer networks is the goal. Collaborative peer groups are defined as peer-to-peer overlay networks with controlled membership and multiway communication primitives that offer well-defined semantics. Unlike conventional paradigms, collaborative peer group primitives do not use the *identity* of a message recipient as the basis for the transport destination. Rather, literally adopting the tenet that peers are equivalent in a peer-to-peer system, our primitives deliver messages based on group/peer attributes such as a given number of peers, "nearby" peers, "any willing peer" and so on. In the context of Grids, this model can also encompass addressing peer nodes based on capabilities or resources, e.g. "any peer that is willing to offer X amount of resource type Y

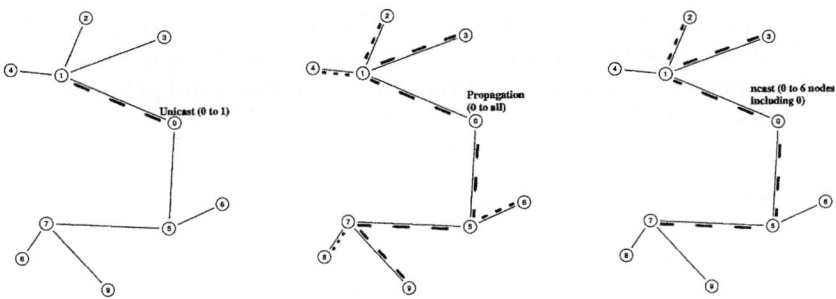

Fig. 1. Paradigms of Peer-to-Peer Communication (Unicast, Propagation and ncast).

for Z time". In these and similar situations, the actual *identity* of the peer is not important – and therefore, traditional communication primitives that address by name, or by group membership, are less appropriate. This project proposes to develop a new suite of communication primitives that are suitable to the *common needs* of peer-to-peer systems, in environments that support collaboration of multiple kinds, including resource sharing.

This paper describes two exemplar primitives, collectively termed ncast, which provide sender oriented semantics in terms of the *number* of nodes that *deliver* a message. In this context, a message has been delivered when it reaches the application. ncast differs from conventional schemes in its notion of which recipients are contacted. In operations such as Acknowledged Multicast, the list of recipients is explicitly stated [2], that is, the application submits one or more identifiers and the multicast transport ensures correct delivery of the message. Using ncast, *any* node that has not previously delivered the message is eligible as a recipient. The important consideration is the *number* of nodes which deliver the message. Such primitives are likely to be useful useful in numerous peer group applications. For example, consider scenarios including distributed voting, student group learning, emergency personnel engaged in disaster recovery and police officers coordinating a public order operation. In these and similar situations, cooperation is not based on the participants knowledge of each other's *specific identities* but on *commonalities*, either in terms of purpose or situation. Furthermore, there are a number of instances where the delivery of a message to a *number* of participants, as opposed to a group of *specific* individuals, is important. For example, it may be the case that in order for a distributed vote to be valid, at least 100 voters must be contacted. Similarly, in a command and control scenario, it is often more important to address a particular number of personnel that fulfill some criteria (e.g. proximity), as opposed to a group of specific individuals. In the Grid context, the actual identity of a resource sharing partner is less important than the type and quantity of resources they offer. In a data grid, a partitioned data set may need to be stored on a given *number* of servers, irrespective of their identities. It is these types of requirements that the ncast primitives are designed to address.

In this paper, ncast refers to two variants that differ in semantic strength. greedy_ncast is the less expensive (but semantically weaker) protocol and ensures delivery to *at least* a given number of nodes. `safe_ncast` guarantees delivery to an exact number of nodes, but is expected to be more expensive and difficult to implement. The design of the `ncast` primitives and preliminary experiences with their implementation and use are discussed in the following sections.

2 System Model and Semantic Definition

Our model assumes an asynchronous communication system in which there are no upper bounds on the time for message delivery [1]. Within such a system, a set of *nodes* N can form dynamic, self organizing *logical overlays* or *peer-to-peer networks* that support application oriented forms of *interaction* [9]. A node is defined to be capable of performing operations including message routing, object location and communication. Nodes posses unique identifiers and are symmetric in their abilities. There are no distinguished roles, centralized services or global information oracles. A variety of topologies may form and while these are not limited solely to trees, it is assumed that cycles are not present in the graph[1]. In order to operate, nodes store and maintain information about their *neighbors*, i.e. other nodes with which they share a direct connection. NE_i is used to denote the set of available nodes that are neighbors of i.

When compared to conventional networks, peer-to-peer overlays exhibit a high degree of dynamicity in terms of node *availability*. Nodes are 'available' when they are connected to at least one other node and can become unavailable as a result of a voluntary leave, a network partition, failure or non-graceful shutdown. Availability of neighbors is typically determined through the periodic exchange of explicit 'heartbeat' messages – usually at the application, transport, or data link level. When available, nodes engage in predominantly two forms of communication: unicast (one to one) and propagation (one to all available nodes). A propagation primitive could be used to address the same applications as `ncast`; however, it is expensive to contact every node where a subset will suffice, particularly in large networks. In `ncast`, the aim is to reach the requisite number of deliveries (termed a *quota*), but with a *minimal* number of messages. It is this minimization in the presence of network dynamicity, lack of global knowledge and asynchrony that makes the problem challenging. Furthermore, the recipients of the message (m) are arbitrary, and successful delivery to any n nodes will suffice. As there are no assumptions on the timeliness of message delivery, it is necessary to introduce a timeout constant δ_t to ensure termination; completion time upper bounds are derived from this constant.

In defining the semantics of `ncast`, the dynamicity of peer-to-peer networks presents a conflict of goals. On one hand, `ncast` should employ all means possible to fulfill a quota. For example, the arrival of new nodes can be considered as favorable and as a result, `ncast` may contact $> n$ nodes, thereby increasing the likelihood of success. On the other hand, a decrease in the size of the network may mean that there are *not* enough nodes to satisfy the quota. In certain scenarios, it

---

[1] This simplifying assumption has since been removed, and subsequent versions of the algorithm successfully perform `ncast` in arbitrary graphs.

is necessary to guarantee that if the protocol can not reach its quota, then strictly $< n$ nodes deliver m. By allowing ncast to proactively contact new nodes in all cases, it would be impossible to guarantee this property. Thus, two primitives are introduced: greedy_ncast is designed to exploit network dynamicism, while safe_ncast guarantees that if n nodes can not deliver m, strictly $< n$ nodes deliver m. Their semantics are:

- greedy_ncast
  - $success \equiv m$ is delivered by *at least* n nodes within $n * \delta_t$ units of time from t where t is the instant the timeout is set by the sender;
  - $failure \equiv m$ is not delivered by a minimum of n nodes within $n*\delta_t$ units of time from t. However it may be the case that at a time $t' > (t + (n * \delta_t))$, m is delivered by $\geq n$ nodes. It may also be the case that at the instant failure is returned, $n \leq \#N$ where $\#$ returns the length of a set;
- safe_ncast
  - $success \equiv m$ is delivered by exactly n nodes within $n * \delta_t$ units of time after t;
  - $failure \equiv$ strictly $< n$ nodes deliver m within $n * \delta_t$ units of time after t, but it may be the case that $\#N$ is $\geq n$ at the instant failure is returned.

3 Quotas and the ncast Algorithms

Operation of the ncast scheme is based on the notion of a quota, defined as a requisite number of deliveries for a particular message. Based on this, the ncast primitives operate through two algorithms which execute as part of a middleware library on each node. The *quota manager* is primarily responsible for processing messages associated with the ncast protocol and for coordinating the actions associated with fulfilling outstanding quotas. Conversely, the *event manager* is designed to respond to events that occur during the protocols execution, for example, servicing timeout expirations and responding to the arrival of new nodes. The reason for distinguishing event handling from protocol message processing is to provide an element of extensibility in the design.

Both the quota manager and the event manager coordinate their operation using two tables which are stored in memory at each node. The *quota table* stores entries that relate to the quotas a particular node is trying to fulfill and the *pending table* retains information reflecting the quotas that have been delegated to other nodes. The quota table is denoted as qt and stores entries of the form shown in Table 1(top). Conversely, the pending table (pt) operates with the format shown in Table 1(bottom). The roles of these fields will become evident in the ensuing discussion, however, two require prior introduction here. The *current delivery level* relates to a particular quotas progress towards fulfillment. If the *current delivery level* $= n$, then the quota is fulfilled. Each ncast message is also labeled with a sequence number. This combined with the source field uniquely identifies each ncast message and permits a node to detect messages that are 'new'. Nodes store sequence numbers *as they are received* in a list called seq. Note that this does not require the maintenance of global information. Moreover, if a node receives an ncast from a node that is not listed in seq, then the message is new and a record is established.

3.1 The Informal Operation of ncast

Suppose that the application of node s calls greedy_ncast to send a message m to n nodes. Using a NEW_QUOTA message, s submits m and n to its quota manager. The quota manager responds by making an entry in the quota table of the form: $(n, s, 0, s, greedy)$. As the message is new, it is eligible for delivery at s, however, since s originated m, message delivery is deferred until the rest of the quota is fulfilled. The quota manager then issues sub-quotas of value $(n-1)/\#NE_s$ to each available neighbor. Since s has no knowledge of the networks density and topology, any remainder from the division is added to a randomly selected sub-quota. Each sub-quota is entered into the pending table and a timeout of value $sub\_quota * \delta_t$ is set. Each neighbor ne receives the sub-quota and based on its sequence number list, determines if m is a new message. If so, it is delivered by ne and the sub-quota is decremented. If the message has been seen before, it is not delivered and the sub-quota remains unchanged. ne then enters the message into its own quota table, whereby the local quota manager issues new sub-quotas taking care not to propagate the message back to s (hence the 'previous hop' field in the quota table). If message delivery fulfilled the quota, then a QUOTA_FILLED message is returned to the previous hop. If the previous hop is unavailable, then the QUOTA_FILLED message is sent to the source of the ncast. If the source of the ncast is unavailable, then the operation is considered defunct and s discards m. However, if greedy_ncast succeeds, m is delivered to s signifying success.

During the course of the execution of the protocol, several events can occur, including the expiration of timeouts and the arrival of new nodes. If a timeout expires and a quota has not been fulfilled a QUOTA_MISSED message is sent either to the previous hop or to the source of the ncast depending on availability. The QUOTA_MISSED message includes details of the quotas target and shortfall. When new neighbors arrive, the event manager issues new sub-quotas for any 'greedy' entries in the quota table. Thus, it is possible that a shortfall in one part of the network is compensated by greater node density in another. However, it should be noted that if either primitive returns failure, it does not imply that $n > \#N$ at the time s returned failure. If a node sends a QUOTA_MISSED message and is subsequently joined by several new neighbors, there is at present, no mechanism to take advantage of this.

The safe_ncast algorithm does not exploit favorable network events as this can lead to semantic violation. At any time, there are a maximum of $n-1$ instances of the message m circulating within the network. The final instance is held by the sender and its delivery depends on the overall outcome of the operation. The reason for this is that if a message m is sent to a node, there is no way of ascertaining whether m has or has not been delivered at that node unless a QUOTA_FILLED or a QUOTA_MISSED message is returned. Consider a node i that was sent a message m and returns either a QUOTA_FILLED or a QUOTA_MISSED message. In this case, it is certain that m was delivered by the number of nodes indicated in the response. However, if i is arbitrarily slow, then the sender may timeout waiting for i to respond. If this occurs, then the sender has no guarantee that either m was delivered or discarded and there is no way of ascertaining that m will not be delivered by i in the future. If the sender sends m to another neighbor without first knowing the outcome at i, it is impossible to guarantee that strictly n nodes will deliver m. Similarly in the negative case, it is not possible to guarantee that strictly $< n$ nodes have delivered m as i may subsequently deliver the message. Thus, in order to guarantee safety, the protocol employs a QUOTA_ABORT message. If a timeout expires for a safe_ncast, the operation is aborted and the sender is informed. Since at any time $< n$ instances of m are in the network, the sender discards its copy of m thus guaranteeing that $< n$ nodes have or will deliver m. The rationale of this overview is to communicate the essence of the scheme. Before formalizing the quota and event managers as pseudo-code, two examples are provided to further illustrate the operation of the algorithm.

Table 1. Exemplar Quota (top) and Pending tables (bottom).

n	source	current delivery level	sequence num	previous hop	greedy/safe
5	0	0	0	0	greedy

issued to	sub-quota	timeout	sequence num	source	greedy/safe
1	2	$t + (2 * \delta_t)$	0	0	greedy
5	2	$t + (2 * \delta_t)$	0	0	greedy

Worked Example 1: $n \leq N$ Consider the topology depicted in Figure 1 and that an application residing at node 0 wishes to send a greedy_ncast message m to 5 other nodes. The quota manager at node 0 detects the request, adds it to the quota table and calculates the sub-quotas. Since node 0 has two neighbors, the sub-quotas are calculated as 2 and 2 since the fifth delivery will be at node 0. Delivery at the sender depends on whether the operation succeeds in delivering m to 4 other nodes. Node 0 then dispatches two NEW_QUOTA messages to nodes 1 and 5. As each sub-quota is dispatched, a record is added to the pending table and a timeout is set. A snapshot of node 0's quota and pending tables at this stage is given in Table 1. Nodes 1 and 5 receive the message, deliver m (since it is new) and decrement the quota by 1. At this point, neither quota has been fulfilled, so nodes 1 and 5 prepare and send sub-quota messages to their available neighbors (excluding node 0). In the event that no neighbor is available, a QUOTA_MISSED message is returned to the previous hop or to the source of the ncast if the previous hop is unavailable. If new neighbors arrive at a node, the event manager dispatches additional sub-quotas for all greedy_ncast entries in the quota table.

Since both nodes require only a single delivery to fulfill their quotas, they each select a random neighbor (e.g. 2 and 6), update their tables and set appropriate timeouts. Nodes 2 and 6 receive and deliver the NEW_QUOTA message, fulfilling the quota in the process. Both nodes then send QUOTA_FILLED messages to 1 and 5 which in turn send a QUOTA_FILLED message to node 0. As the QUOTA_FILLED messages are back propagated to node 0, each node cancels the appropriate timeout and prunes its tables of redundant entries. When node 0 receives a

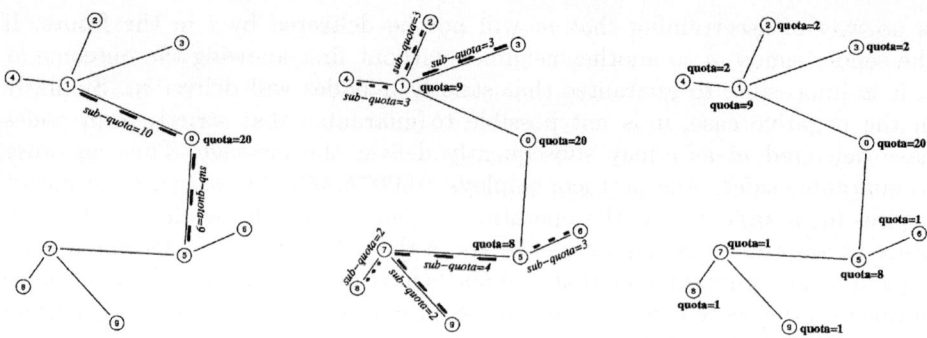

Fig. 2. Quota Values in Worked Example 2.

QUOTA_FILLED message from 5 and 1, it cancels the timeout and delivers m, fulfilling the original quota and returning success.

Worked Example 2: $n > N$ Now consider the same scenario as above, but with $n = 20$. The quota manager at node 0 enters a new record into the quota table and issues sub-quotas of 10 and 9 to nodes 1 and 5 depending on which node is randomly selected to service the remainder (see Figure 2a). Assuming for simplicity, that node 1 was selected to fulfill the quota of 10, nodes 1 and 5 receive and deliver m before issuing their own sub-quotas (Figure 2b). Each neighbor delivers m and is now required to fulfill either a quota of 1 or 2 (Figure 2c). This is not possible since the neighbor of nodes 5 and 1 (excluding 0) are leaf nodes. Nodes 2,3,4,6,7,8 and 9 now wait for either δ_t or $2 * \delta_t$ units of time for new neighbor to become available. If the number of nodes in the network does not increase, nodes 6,7,8 and 9 each send a QUOTA_MISSED message to node 5 after δ_t units of time. Node 5 then returns a QUOTA_MISSED message to node 0. After a further δ_t time units, nodes 2,3 and 4 each return a QUOTA_MISSED message to node 1. Finally, node 1 sends a QUOTA_MISSED message to the sender which then discards m and returns failure.

4 Discussion and Future Research

This paper has presented the design and preliminary implementation an ncast scheme for peer-to-peer networks. The paper began with a brief summary of existing peer-to-peer communication paradigms before suggesting that certain peer-to-peer and Grid resource-sharing collaborative scenarios would benefit from the provision of sender oriented delivery semantics in terms of the *number* of recipients as opposed to other identity based approaches. Section 3 presented a design for fulfilling this requirement and demonstrated the these informal algorithms by means of two worked examples with different parameters.

As mentioned earlier, we have already designed and implemented an enhanced version of these algorithms that do not impose the restriction that the peer graph be free of cycles. In conjunction with a separate algorithm for autonomically reconfiguring a peer-to-peer network such that it maintains certain desirable properties (small diameter, small degree, and well-connectedness), this enhanced algorithm performs very well in (simulated) random peer-to-peer networks of several thousand nodes when performing ncast to a few hundred. Although time metrics are less insightful in simulations, message count metrics were very encouraging: in numerous experiments, between 1.05 and 1.09 messages per recipient were required when performing ncast to 100 nodes in a 1000-node peer to peer network.

Based on these promising results, we intend to fully develop and implement the network organization and ncast algorithms. In the short to mid term, analysis and verification of the ncast primitives will be conducted and consolidated into the design of these schemes. Simulation studies will then determine appropriate values for δ_t and will investigate the cost of the scheme. Following this, the ncast protocols will be implemented; a prototype implementation based on JXTA [8] has already produced encouraging results. A full treatment of these topics will be given in a subsequent paper.

References

1. H. Attiya and J. Welch. *Distributed Computing: Fundamentals, Simulations and Advanced Topics.* McGraw-Hill, 1998.
2. Kirsten Hildrum, John D. Kubiatowicz, Satish Rao, and Ben Y. Zhao. Distributed Object Location in a Dynamic Network. In *Proc. of the Fourteenth ACM Symposium on Parallel Algorithms and Architectures*, pages 41–52, August 2002.
3. Andy Oram, editor. *Peer-to-peer: Harnessing the Power of Disruptive Technologies.* O'Reilly, March 2001.
4. J. S. Pascoe, V. S. Sunderam, and R. J. Loader. The Design and Semantic Verification of an ncast Algorithm for Peer-to-Peer Networks. Technical Report 030101, Emory University, Dpt. of Math & Computer Science, 2003.
5. Sylvia Ratnasamy, Paul Francis, Mark Handley, Richard Karp, and Scott Shenker. A Scalable Content Addressable Network. In *Proc. ACM SIGCOMM*, San Diego, CA, August 2001. ACM Press, New York.
6. Antony Rowstron and Peter Druschel. Pastry: Scalable, decentralized object location and routing for large-scale peer-to-peer systems. In *Proc. IFIP/ACM International Conference on Distributed Systems Platforms (Middleware)*, pages 329–350, Heidelberg, Germany, November 2001. ACM Press, New York.
7. Ion Stoica, et. al., Chord: A Scalable Peer-to-peer Lookup Protocol for Internet Applications. In *Proc. ACM SIGCOMM*, pages 149–160, San Diego, CA, August 2001. ACM Press, New York.
8. Sun Microsystems. JXTA Resolver. World Wide Web, October 2002. Available from: http://java.sun.com/features/2002/10/jxta_res.html.
9. V. S. Sunderam, J. S. Pascoe, and R. J. Loader. Towards a Framework for Collaborative Peer Groups. In *Proc. IEEE/ACM International Symposium on Cluster Computing and the Grid (CCGRID)*, Tokyo, Japan, May 2003.
10. The ns team. The Network Simulator (ns2) Homepage. 2002. Available at: http://www.isi.edu/nsnam/ns/.
11. B. Y. Zhao, J. D. Kubiatowicz, and A. D. Joseph. Tapestry: An Infrastructure for Fault-tolerant Wide-area Location and Routing. Technical Report UCB/CSD-01-1141, University of California at Berkeley, April 2001.

A P2P Grid Services-Based Protocol: Design and Evaluation

Domenico Talia and Paolo Trunfio

DEIS, University of Calabria
Via P. Bucci 41c, 87036 Rende, Italy
{talia,trunfio}@deis.unical.it

Abstract. Several aspects of today's Grids are based on centralized or hierarchical services. However, as Grid sizes increase from tens to thousands of hosts, functionalities should be decentralized to avoid bottlenecks and guarantee scalability. A way to ensure Grid scalability is to adopt Peer-to-Peer (P2P) models and techniques to implement nonhierarchical decentralized Grid services and systems. Standard P2P protocols based on a pervasive exchange of messages, such as Gnutella, appear to be inadequate for OGSA Grids, where peers communicate among them through Grid Services mechanisms. This paper proposes a modified Gnutella discovery protocol, named *Gridnut*, which makes it suitable for OGSA Grids. In particular, Gridnut uses appropriate message buffering and merging techniques to make Grid Services effective as a way to exchange messages in a P2P fashion. We present the design of Gridnut, and compare Gnutella and Gridnut performances under different network and load conditions.

1 Introduction

Many aspects of today's Grids are based on centralized or hierarchical services. However, as Grids used for complex applications increase their size from tens to thousands of nodes, we should decentralize their functionalities to avoid bottlenecks and ensure scalability. As argued in [1] and [2], a way to provide Grid scalability is to adopt *Peer-to-Peer* (*P2P*) models and techniques to implement nonhierarchical decentralized Grid systems.

Recently, the Grid community has undertaken a development effort to align Grid technologies with Web Services. The *Open Grid Services Architecture* (*OGSA*) defines *Grid Services* as an extension of Web Services and lets developers integrate services and resources across distributed, heterogeneous, dynamic environments and communities [3]. OGSA adopts the Web Services Description Language (WSDL) to define the concept of a Grid Service using principles and technologies from both the Grid and Web Services communities. Web Services and the OGSA both seek to enable interoperability between loosely coupled services, independent of implementation, location, or platform. The OGSA model provides an opportunity to integrate

P2P models in Grid environments since it offers an open cooperation model that allows Grid entities to be composed in a decentralized way.

In [4], Fox and colleagues explore the concept of a *Peer-to-Peer Grid* designed around the integration of Peer-to-Peer and OGSA models. A Peer-to-Peer Grid is built in a *service* model, where a *service* is a Web Service that accepts one or more inputs and gives one or more results. These inputs and results are the messages that characterize the system. All the entities in the Grid (i.e., users, computers, resources, and instruments) are linked by messages, whose communication forms a distributed system integrating the component parts. In a Peer-to-Peer Grid, access to services can be mediated by "servers in the core", or by direct Peer-to-Peer interactions between machines "on the edge". The server approach best scales within pre-existing hierarchical organizations, but P2P approaches best support local dynamic interactions. The Peer-to-Peer Grid architecture is a mix of structured (Grid-like) and unstructured dynamic (P2P-like) services, with peer groups managed locally and arranged into a global system supported by core servers. A key component of a Peer-to-Peer Grid is the messaging subsystem, that manages the communication among resources, Web Services and clients to achieve the highest possible system performance and reliability.

Although Grid Services are appropriate for implementing loosely coupled P2P applications, they appear to be inefficient to support an intensive exchange of messages among tightly coupled peers. In fact Grid Services operations, as other RPC-like mechanisms, are subject to an invocation overhead that can be significant both in terms of activation time and memory/processing consumption [5]. The number of Grid Service operations that a peer can efficiently manage in a given time interval depends strongly on that overhead. For this reason, standard P2P protocols based on a pervasive exchange of messages, such as Gnutella [6], are inappropriate on large OGSA Grids where a high number of communications take place among hosts.

To overcome this limitation, we propose a modified Gnutella protocol, named *Gridnut*, which uses appropriate message buffering and merging techniques that make Grid Services effective as a way for exchanging messages among Grid nodes in a P2P fashion. Although the pure Gnutella protocol is not scalable (since the load on each node grows linearly with the total number of queries), and several advancements have been proposed to improve the performance of decentralized search (see for instance [7]), we worked on it because it is a reference model for several more sophisticated systems to which our approach can be also applied.

Gnutella defines both a protocol to discover hosts on the network, based on the *Ping/Pong* mechanism, and a protocol for searching the distributed network, based on the *Query/QueryHit* mechanism. Here we discuss only the Gridnut discovery protocol, even if we are also designing the Gridnut search protocol.

The remainder of the paper is organized as follows. Section 2 presents the design of the Gridnut protocol focusing on message routing and buffering rules. Section 3 compares the performance of Gridnut and Gnutella protocols under different network and load conditions. Finally, Section 4 concludes the paper.

2 Gridnut Design

The two basic principles of the Gridnut protocol that make it different from Gnutella are

a) *Message buffering*: to reduce communication overhead, messages to be delivered to the same peer are buffered and sent in a single packet at regular time intervals.

b) *Collective Pong*: when a peer *B* must respond to a Ping message received from *A*, it waits to receive all the Pong messages from its neighbors, then merge them with its Pong response and send back the Pong collection as a single message to *A*.

Since the Gridnut protocol is derived from the Gnutella discovery protocol, we adopt here the Gnutella terminology. Each Grid node executes a *Gridnut servent*, i.e., an application that performs both client and server Gridnut tasks.

A Gridnut servent is composed of three logical components (see Figure 1):
- *Peer Service*: a Grid Service through which remote Gridnut servents can connect and deliver messages to this servent.
- *Client Interface*: an interface through which local users and applications can issue Grid nodes discovery requests and get results.
- *Network Module*: a component that interacts with remote Gridnut servents on the basis of the Peer Service and Client Interface input.

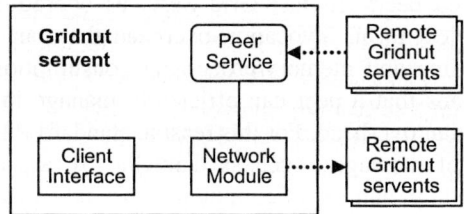

Fig. 1. Gridnut servent components

2.1 Peer Service

The Peer Service is a *persistent* Grid Service, activated at the Gridnut servent's startup and terminated when the servent leaves the network. Each Peer Service is assigned a globally unique name, the *Grid Service Handle* (*GSH*), that distinguishes a specific Grid Service instance from all other Grid Service instances. This handle is used within a Gridnut network to uniquely identify both the Peer Service and the associated Gridnut servent. For instance, a valid handle could be:

`http://p1.deis.unical.it:8080/ogsa/services/p2p/PeerService`

The Peer Service supports three main *operations*:
- `connect`: used by a remote servent to connect this servent. The operation receives the *handle* of the requesting servent and returns a *reject* response if the connection is not accepted (for instance, because the maximum number of connections has been reached).
- `disconnect`: used by a remote servent to disconnect this servent. The operation receives the *handle* of the requesting servent.

- `deliver`: used by a connected servent to deliver messages to this servent. The operation receives the *handle* of the requesting servent and an *array of messages* to be delivered to this servent.

2.2 Messages

A servent connects itself to the Gridnut network by establishing a connection with one or more servents currently in the network (a discussion of the connection and disconnection phases is outside the scope of this paper). Once a servent joined successfully the Gridnut network, it communicates with other servents by sending and receiving *Ping* and *Pong* messages:
- A *Ping* is used to discover available nodes on the Grid; a servent receiving a Ping message is expected to respond with a Pong message.
- A *Pong* is a response to a Ping; it includes the URL of a set of reachable Gridnut servents, each one representing an available Grid node.

The logical structure of Ping and Pong messages is shown in Figure 2.

Fig. 2. Structure of Gridnut messages

The meaning of fields in Figure 2 is the following:
- *GUID* (*Global Unique Identifier*): a string identifying the message on the network.
- *TTL* (*Time To Live*): the number of times the message will be forwarded by servents before it is removed from the network.
- *Hops*: the number of times the message has been forwarded by servents.
- *Handles*: an array of zero, one or more reachable Gridnut servents' URLs.

For the purposes of this paper, Pong messages do not include further information because here we use the discovery protocol to locate all the active nodes on the Grid. The search protocol we are designing (not discussed in the paper) will be used for host characterization, discovery of needed services, etc.

2.3 Data Structures

Each Gridnut servent uses a set of data structures to perform its functions.

A *connection list* (*CL*) is used to maintain a reference to all directly connected servents (i.e., references to the connected servents' Peer Services). Entries into the CL are updated by the `connect` and `disconnect` operations.

A *routing table* (*RT*) is used to properly route messages through the network. The RT contains a set of records having a structure [*GUID, Handle*], used to route messages with a given *GUID* to a servent with a given *Handle*.

The results of the discovery tasks are stored into a *result set* (*RS*), that users and applications can access for their purposes.

Finally, each Gridnut servent uses a set of internal *transmission buffers*, in which messages are stored and processed before to deliver them to the proper servent. In

particular, a servent S_o uses two separated transmission buffers for each of its neighbors:
- A *pong buffer* (B_p), in which Pong messages with an equal GUID are merged before the delivery. The notation $B_p(S_k)$ indicates the pong buffer in which S_o inserts Pong messages directed to a servent S_k.
- A *fast buffer* (B_f), used for Ping and Pong messages that are to be fast delivered to a given servent. We use the notation $B_f(S_k)$ to indicate the fast buffer in which S_o inserts messages directed to a servent S_k.

A thread T_k is associated to each couple of buffers $B_p(S_k)$ and $B_f(S_k)$. T_k periodically delivers the buffered messages to S_k, on the basis of the rules described below.

2.4 Routing Rules

In Gridnut, like in Gnutella, Ping messages are forwarded to all directly connected servents, whereas Pong messages are sent along the same path that carried the incoming Ping message.

However, there are two main differences between Gnutella and Gridnut message routing and transmission modalities:
1) In Gnutella implementations, messages are sent as a byte stream over TCP sockets, whereas Gridnut messages are sent through a Grid Service invocation (by means of the `deliver` operation).
2) In standard Gnutella implementations, each message is forwarded whenever it is received, whereas Gridnut messages, as mentioned before, are buffered and merged to reduce the number of Grid Service invocations and routing operations executed by each servent.

Let consider a servent S_o having a set of neighbors $S_1...S_n$. When a neighbor delivers an array of messages to S_o, each message is processed separately by S_o as specified below.

Let us suppose that S_o received from S_k the message *Ping*[GUID=g, TTL=t, Hops=h] (this notation means that g, t, and h are the actual values of GUID, TTL and Hops of this Ping); S_o performs the following operations:

```
t = t - 1; h = h + 1;
if (RT contains a record with GUID=g)
    insert a Pong[GUID=g, TTL=h, Hops=0, Handles=Ø] into Bf(Sk);
else if (t == 0)
    insert a Pong[GUID=g, TTL=h, Hops=0, Handles={S0}] into Bf(Sk);
else {
    insert a record [GUID=g, Handle=Sk] into RT;
    insert a Pong[GUID=g, TTL=h, Hops=0, Handles={S0}] into Bp(Sk);
    for (i:1..n; i ≠ k)
        insert a Ping[GUID=g, TTL=t, Hops=h] into Bf(Si);
}
```

In the following we use the term "dummy Pong" to refer to a Pong having Handles=Ø.

Let us suppose that S_o received from S_k the message *Pong*[GUID=g, TTL=t, Hops=h, Handles=H] (where H is a set of servents' handles); the following operations are performed by S_o:

```
t = t - 1; h = h + 1;
if (t == 0)
    insert H into RS;
else if (RT contains a record R with GUID=g) {
    S_r = value of the Handle field of R;
    insert a Pong[GUID=g, TTL=t, Hops=h, Handles=H] into B_p(S_r);
}
```

Finally, to start a new discovery task, S_0 must perform the following operations:

```
clear RS;
g = globally unique string; t = initial TTL;
insert the record [GUID=g, Handle=S_0] into RT;
for (i:1..n)
    insert a Ping[GUID=g, TTL=t, Hops=0] into B_f(S_i);
```

The discovery task is completed when the RS contains the handles of all the reachable servents in the network.

2.5 Buffering Rules

Let consider again a servent S_0 connected to a set of N servents $S_1...S_n$.

Within a pong buffer $B_p(S_k)$, a set of counters are used. A counter C_g counts the number of Pong messages with GUID=g till now inserted in $B_p(S_k)$.

When a Pong P_1 having GUID=g and containing a set H_1 of Handles is inserted into $B_p(S_k)$, the following operations are performed:

```
C_g = C_g + 1;
if (B_p(S_k) contains a Pong P_0 with GUID=g) {
    add H_1 to the current Handles set of P_0;
    if (C_g >= N)
        mark P_0 as ready;
}
else {
    insert P_1 into B_p(S_k);
    if (C_g >= N)
        mark P_1 as ready;
}
```

Whenever a Pong message is marked as *ready*, it can be delivered to the servent S_k. To avoid blocking situations due to missed Pong messages, a Pong could be marked as ready also if a *timeout* has been reached. In the following we do not consider failure situations, therefore no timeouts are used.

Differently from a pong buffer, messages inserted into a fast buffer $B_f(S_k)$ are immediately marked as *ready* to be delivered to S_k.

As we have mentioned before, a thread T_k is used to periodically deliver the buffered messages to S_k. In particular, the following operations are performed by T_k every time it is activated:

```
get the set of ready messages M from B_p(S_k) and B_f(S_k);
deliver M to S_k through a single deliver operation;
```

The time interval I_a between two consecutive activations of T_k is a system parameter. In the worst case, exactly a `deliver` operation can be invoked by S_0 for each of its N neighbors. Therefore, the maximum number of `deliver` operation invoked by S_0

during an interval of time I is equal to $(I / I_a) \times N$. Obviously, increasing the value of I_a the number of deliver operations can be reduced, but this could produce a delay in the delivery of messages. In our prototype we use $I_a = 5\ msec$.

3 Performance Evaluation

In this section we compare some experimental performance results of Gridnut and Gnutella protocols. To perform our experiments we developed a Java prototype of a Gridnut servent, which can also work as a standard Gnutella servent for comparison purposes. In our prototype the Peer Service is an object accessed through Remote Method Invocation (RMI). The goal of our tests is to verify how significantly Gridnut reduces the workload - number of Grid Service operations - of each peer. In doing this, we compared Gridnut and Gnutella by evaluating two parameters:

1) ND, the average number of deliver operations processed by a servent to complete a discovery task. In particular, $ND = P / (N \times T)$, where: P is the total number of deliver operations processed in the network, N is the number of servents in the network, and T is the overall number of discovery tasks completed.
2) $ND(d)$, the average number of deliver operations processed by servents that are at distance d from the servent S_0 that started the discovery task. For instance: $ND(0)$ represents the number of deliver operations processed by S_0; $ND(1)$ represents the number of deliver operations processed by a servent distant one hop from S_0.

Both ND and $ND(d)$ have been evaluated considering seven different network topologies. We distinguished the network topologies using a couple of numbers $\{N,C\}$, where N is the number of servents in the network, and C is the number of servents directly connected to each servent (i.e., each servent has exactly C neighbors). The network topologies we experimented are characterized by $\{N,C\}$ respectively equal to $\{10,2\}$, $\{10,4\}$, $\{30,3\}$, $\{30,4\}$, $\{50,4\}$, $\{70,4\}$ and $\{90,4\}$. Notwithstanding the limited number of used servents, the number of exchanged messages among servents was extremely high and performance trends are evident.

Resulting networks were completely connected, i.e., each servent can reach any other servent in the network in a number of steps lower or equal than TTL.

3.1 Number of Deliver Operations

For each network topology, we measured ND under four load conditions. We use R to indicate the number of discovery tasks that are initiated in the network at each given time interval. The following values for R have been used: 1, 3, 5 and 10. In particular,

- $R=1$ indicates that, at each time interval, only one discovery task is initiated, therefore only messages with a given GUID are simultaneously present in the network;
- $R=10$ indicates that, at each time interval, ten discovery tasks are initiated, therefore messages with up to ten different GUID are simultaneously present in the network.

Table 1a and Table 1b report the *ND* measured in Gnutella and Gridnut networks, respectively. *ND* values are measured for network topologies ranging from {10,2} to {90,4}, under load conditions ranging from $R=1$ to $R=10$.

Table 1a. ND in Gnutella networks

{N,C}	{10,2}	{10,4}	{30,3}	{30,4}	{50,4}	{70,4}	{90,4}
R=1	3.60	4.53	4.91	5.49	6.00	6.27	6.52
R=3	3.61	4.54	4.95	5.48	6.01	6.32	6.53
R=5	3.61	4.55	4.96	5.47	6.01	6.35	6.54
R=10	3.60	4.54	4.99	5.49	6.02	6.35	6.53

Table 1b. ND in Gridnut networks

{N,C}	{10,2}	{10,4}	{30,3}	{30,4}	{50,4}	{70,4}	{90,4}
R=1	2.12	5.91	3.86	5.74	5.75	5.72	5.73
R=3	1.96	4.54	3.48	4.81	4.76	4.70	4.89
R=5	1.85	3.98	3.11	4.28	4.22	4.16	4.03
R=10	1.70	2.93	2.52	3.19	3.22	3.10	2.91

In Gnutella (see Table 1a), *ND* is not influenced by the *R* factor, apart from little variations due to measurements errors. This is because in Gnutella no buffering strategies are adopted, and one `deliver` operation is executed to move exactly one message in the network. Obviously, the value of *ND* increases with the size of the network, ranging from an average value of 3.61 in a {10,2} network, to an average value of 6.53 in a {90,4} network.

In Gridnut (see Table 1b), *ND* depends from both network topology and load condition. For a given value of *R*, *ND* mainly depends from the value of *C* (number of connections per servent), whereas it varies a little with the value of *N* (number of servents). For instance, if we consider the value of *ND* for $R=1$, we see that it varies in a small range (from 5.72 to 5.91) for all the networks with *C*=4.

If we consider networks with the same value of *N*, we see that *ND* decreases when the value of *C* is lower. For instance, the *ND* for a network {10,2} is lower than the *ND* for a network {10,4}, with any value of *R*. Moreover, because a single `deliver` operation is performed to deliver more buffered messages, for a given topology the value of *ND* decreases when *R* increases.

Figure 3 compares the values of *ND* in Gridnut and Gnutella in five network topologies: {10,2}, {30,3}, {50,4}, {70,4} and {90,4}. For Gridnut networks the values of *ND* when $R=1$, 3, 5, and 10 are represented, whereas for Gnutella networks the average of the *ND* values measured when $R=1$, 3, 5, and 10 is represented.

We can see that the number of `deliver` operations is lower with Gridnut in all the considered configurations. In particular, when the number of discovery tasks increases, the Gridnut strategy maintains the values of ND significantly low in comparison with Gnutella.

3.2 Distribution of Deliver Operations

Table 2a and Table 2b report the value of $ND(d)$ measured in Gnutella and Gridnut networks, respectively. Notice that in the {10,4} network the maximum distance between any couple of servents is 2, therefore no values have been measured for $d > 2$. For analogous reasons, there are no values for $d > 4$ in {30,3}, {30,4} and {50,4} networks.

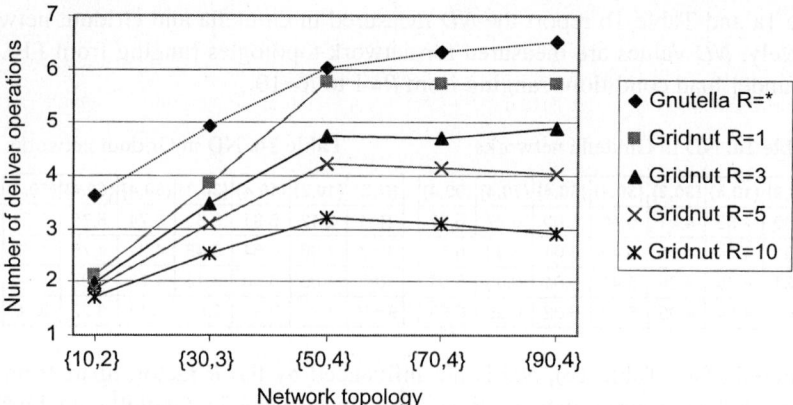

Fig. 3. Comparison between ND in Gridnut networks and ND in Gnutella networks

Table 2a. ND(d) in Gnutella networks

{N,C}	{10,2}	{10,4}	{30,3}	{30,4}	{50,4}	{70,4}	{90,4}
d=0	9.00	9.00	29.00	29.00	49.00	69.00	89.00
d=1	4.50	4.08	9.67	7.82	12.44	17.28	22.50
d=2	3.50	4.00	4.39	4.32	5.53	6.72	8.20
d=3	2.50	–	3.04	4.00	4.11	4.41	4.46
d=4	2.00	–	3.00	4.00	4.00	4.01	4.02
d=5	2.00	–	–	–	–	4.00	4.00

Table 2b. ND(d) in Gridnut networks

{N,C}	{10,2}	{10,4}	{30,3}	{30,4}	{50,4}	{70,4}	{90,4}
d=0	2.00	4.00	3.00	4.00	4.00	4.00	4.00
d=1	2.00	5.35	3.00	4.51	4.07	4.04	4.22
d=2	2.00	6.76	3.07	5.40	5.20	4.89	4.52
d=3	2.01	–	4.05	6.40	5.84	5.61	5.50
d=4	2.34	–	4.80	6.82	6.65	6.32	6.26
d=5	2.82	–	–	–	–	6.78	6.67

In Gnutella (see Table 2a) the value of *ND(0)* is always equal to *N-1*. This is because S_0 receives, through its neighbors, a Pong message from each of other servents in the network, and each of those messages are delivered to S_0 by means of a separated deliver operation. *ND(1)* is always greater or equal than *ND(0)* divided by *C*. The equality is obtained only for networks in which *C* is sufficiently little compared to *N*, as in {10,2} and {30,3} networks. In general, the value of *ND(d)* decreases when *d* increases, and it reaches the minimum value, equal to *C*, on the servents more distant from S_0.

In Gridnut (see Table 2b) the value of *ND(0)* is always equal to *C*, because S_0 must process exactly a deliver operation for each servent directly connected to it. The value of *ND(d)* increases slightly with *d*, reaching its maximum on the servents more distant from S_0. *ND(d)* increases with *d* because the number of "dummy Pong" messages increase moving away from S_0. Anyway, the value of *ND(d)* remains always of the order of *C*, even for *d* equal to TTL.

Comparing the results in Tables 2a and 2b, we can see that Gridnut implies a much better distribution of deliver operations among servents in comparison with Gnutella. In Gnutella, the servent that started the discovery task and its closest neighbors must process a number of Grid Service operations that becomes unsustainable when the size of the network increases to thousands of nodes. In Gridnut, conversely, the number of Grid Service operations processed by each servent remains always in the order of the number of connections per peer. This Gridnut behaviour results in significantly lower discovery times since communication and computation overhead due to Grid Services invocations are considerably reduced as shown in Tables 2a and 2b. For example, considering a {90,4} network with R ranging from 1 to 10, Gnutella discovery experimental times vary from 2431 to 26785 msec, whereas Gridnut times vary from 2129 to 8286 msec.

4 Conclusions

The Gridnut protocol modifies the Gnutella discovery protocol to make it suitable for OGSA Grids. It uses message buffering and merging techniques to make Grid Services effective as a way for exchanging messages among Grid nodes in a P2P mode. We compared Gridnut and Gnutella performance considering different network topologies and load conditions. Experimental results show that appropriate message buffering and merging strategies produce significant performance improvements, both in terms of number and distribution of Grid Service operations processed.

We are extending Gridnut to support also distributed search by modifying the original Query/QueryHit Gnutella mechanism. In doing this, the buffering mechanism is maintained, whereas the collection mechanism is modified since the number of responding nodes will be limited by the query constraints.

The Gridnut protocol can be an effective way to discover active nodes in a OGSA Grids. Currently we are designing a *Presence Management Service* (*PMS*) that uses Gridnut as mechanism to discover active Grid nodes in a P2P fashion. Presence management is a key aspect in large-scale Grids, in which hosts join and leave the network dynamically over the time, as in typical P2P environments. The PMS allows users and schedulers to efficiently locate active nodes and support execution of large-scale distributed applications in dynamic Grid environments.

References

1. Foster, I., Iamnitchi, A.: On Death, Taxes, and the Convergence of Peer-to-Peer and Grid Computing. 2nd International Workshop on Peer-to-Peer Systems, Berkeley (2003)
2. Talia, D., Trunfio, P.: Toward a Synergy between P2P and Grids. IEEE Internet Computing, vol. 7 n. 4 (2003) 94-96
3. Foster, I., Kesselman, C., Nick, J. M., Tuecke, S.: The Physiology of the Grid: An Open Grid Services Architecture for Distributed Systems Integration.
http://www.globus.org/ research/papers/ogsa.pdf
4. Fox, G., Gannon, D., Ko, S., Lee, S., Pallickara, S., Pierce, M., Qiu, X., Rao, X., Uyar, A., Wang, M., Wu, W.: Peer-to-Peer Grids.
http://grids.ucs.indiana.edu/ptliupages/publications/ p2pGridbook.pdf
5. The Globus Alliance: Globus Toolkit 3.0 - Performance Tuning Guide.
http://www-unix.globus.org/toolkit/3.0/ogsa/docs/performance_guide.html
6. Clip2: The Gnutella Protocol Specification v.0.4.
http://www9.limewire.com/developer/ gnutella_protocol_0.4.pdf
7. Lv, Q., Cao, P., Cohen, E., Li, K., Shenker, S.: Search and Replication in Unstructured Peer-to-Peer Networks. 16th ACM Int. Conference on Supercomputing, New York (2002)

A Synchronization and Session Control Based Protocol for 3D Multi-user Virtual Environments in a Peer-to-Peer Networks[*]

Azzedine Boukerche[1], Rafael Rabêlo[2], Marcelo Laffranchi[2], and Regina Araujo[2]

[1] SITE – University of Ottawa
Ottawa, Canada
boukerch@site.uottawa.ca

[2] DC - Universidade Federal de São Carlos
13565-905 São Carlos, SP, Brazil
{rabelo,laffranchi,regina}@dc.ufscar.br

Abstract. This paper describes a solution to support 3D Multi User Virtual Environments in a hybrid peer-to-peer Gnutella network which provides session control and distributed synchronization to collaborative applications. For that, two components compliant to the emerging multi-user extension to the MPEG-4 standard were implemented and integrated into the Gnutella network. This solution minimizes the disadvantages of hibrid solutions, based on proxies, which have to be re-configured whenever a new virtual environment is made available in the network. It also provides shorter network latency compared to content providers solutions based on the client-server model and can be deployed on devices with restricted capabilities.

1 Introduction

The synchronization of 3D virtual environments that are shared among multiple users (3D Multi-User Virtual Environments – MUVEs) in a network is a challenge for developers. Any change made to the environment by one participating user has to be broadcasted to all other participant users who share the same virtual environment - VE. For that, efficient communication models, such as the traditional client-server model, are used to support the exchange of information among users' client terminals. The advantages of the client-server model are well known and include simpler management and billing model as well as simpler protection against frauds. However, the model centered on a server can be subjected to bottlenecks and single point of failures. The peer-to-peer communication model can overcome these pitfalls by providing direct collaboration among all nodes in a network, without a central coordination and no single point of failure. However, this is a distributed solution and therefore, more complex to manage. Examples of 3DMUVEs implemented as peer-to-peer

---

[*] Dr. A. Boukerche's work was partially supported by Grants from NSERC, Canada Research Chair Program and Canada Foundation for Innovation.

applications include MIMAZE [1], MASSIVE [2] and DIVE [3]. Hybrid solutions combine the advantages of both centralized and distributed approaches. Mauve and colleagues [4], describe a hybrid solution based on proxies which are used as an extension to central servers. The proxies have some of the servers functionalities and are located close to the users´ nodes. One major problem with such a solution is that the proxies need to be reconfigured everytime a new application emerges. Moreover, the proxies have to support several architectures for different applications at the same time. In order to reduce the disadvantages of the hybrid solutions based on proxies, this paper describes the implementation and preliminary evaluation of a solution to support 3DMUVEs in a hybrid peer-to-peer network in which new applications are spontaneously made available by the different nodes of the network. The target network for this work is the Gnutella network [5], which is accessed through the Limewire client software [6]. Gnutella network has a decentralized nature, what makes it more resistent to software failures and malicious attacks, besides having a simple and basic protocol, that can be used in different experiments, and a very active development community. The Gnutella version used in this work contains special nodes, called Ultrapeers, with larger processing power and network bandwidth. As a result of this work, the Limewire client software was extended to support games search, as an example of 3D MUVEs. The Gnutella network was also extended to support session control and shared VE synchronization.The components responsible for session control and synchronization are compliant to the ongoing multiuser extension to the MPEG-4 standard to support 3DMUVEs. MPEG-4 is an ISO/IEC Standard for coding and delivery of different media formats in a wide variety of networks and computing plataforms. Preliminary results show the viability of 3D MUVEs in hybrid peer-to-peer networks.

2 The Hybrid Peer-to-Peer Gnutella Network

When the Gnutella network emerged, back in the year 2000, it was considered as the Napster next generation, the network that started the open sharing "fever" of music files and led the main music record companies to an alert state about copyright issues. It was soon discovered that Gnutella generated a lot too much network traffic. It was then reorganized, mirrowing more efficient structures, such as the FastTrack (proprietary protocol used by the KazaA application), and incorporating concepts such as the Ultrapeers. The Ultrapeers are special nodes with larger processing power and network bandwidth capacity. These nodes relieve other nodes of the network from receiving most of the messages traffic and so reducing the total traffic of the network, making it faster. With the incorporation of the ultrapeers concept to the Gnutella network, it started to be considered a hybrid peer-to-peer network. In order to use the Gnutella network, a user node needs an application to provide the connection to other nodes, forming an ad hoc network, with no central control, which acts either as a client or as a content provider. One of the best known access interface software to the Gnutella network is Limewire [6]. Since there are no centralized servers, when a user wants to enter in the Gnutella network, he initially connects to one of the several existing Ultrapeers, almost always available. These machines then forward information such as IP addresses and network port to other Gnutella nodes (also known as "peers"). Once connected to the network, the nodes interact among themselves by exchanging messages of the following types: *Ping, Pong, Query, Rich Query, Query Response* and *Get/Push*. Characteristics of the Gnutella network, such as the TTL (Time to Live) counter, prevents the re-forwarding of messages to other nodes of the network indefinitely. Although the Gnutella network has been slightly overshadowed by other efficient network protocols, such as the FastTrack, it was chosen as the target network for this work because of its conceptual simplicity and also because it is an open software, allowing the realization and evaluation of further experiments on the support to 3D MUVEs. The existence of 3DMUVEs applications support in the Gnutella network is unknown to the authors of this paper – most of the shared content in the Gnutella network is related to music files. The implementation of shared MUVEs requires session control and synchronization - next section describes the ongoing extension to multi-user in the MPEG-4 standard which specifies the support for such requirements.

3 The Ongoing Extension to Multi-user Support in the MPEG-4 Standard

MPEG-4 is an ISO/IEC (ISO/IEC JTC1/SC29/WG11) standard, developed by the MPEG *(Moving Picture Experts Group)* for coding and delivery of different media formats in a wide variety of networks and computational plataforms [7]. There are currently some versions of the MPEG-4 Player (MPEG-4 client terminal) available for interactive TV set-top-boxes. The deployment of the MPEG-4 standard in cellular telephones and mobile devices in general is already on course and it is just a matter of time before we can see complex multimedia applications in these devices, through the MPEG-4 Player. The emerging extension to the MPEG-4 which supports 3D MUVEs, MPEG-4MU, is widely based on the ***Pilot/Drone*** mechanism proposed by the ***Living Worlds*** specification [8]. According to that specification, ***Pilots*** are master copies of shared objects in a VE. Any changes in the *Pilot* state or behavior are replicated to other instances of that object which are named ***Drones*** (***Pilots*** replicas). The replication of changes in the *Pilot,* or any other object in the graphics scene, can be realized basically through the BIFS-Command Protocol. When these commands arrive at the remote users' client terminals, they are locally processed, causing the update of the scene, so that all users that receive a BIFS command have a consistent view of the shared VE. The MPEG-4MU architecture components responsible for managing multi-user sessions and scene synchronization are the MUTech Session Controller (MSC) and the MUTech Bookeeper (MBK) components, respectively. The MPEG-4MU defines a standard interface for the access to these components and suggest the set of functions these components should perform - it is then left to developers how these functions are implemented. Next section shows how the MSC and MBK components can be used to support session control and synchronization in 3DMUVEs.

4 A Hybrid Peer-to-Peer Solution to Support 3D MUVEs

In the proposed solution to support 3D MUVEs, the MSC and MBK components, compliant to the ongoing extension to the MPEG-4 standard, are implemented and integrated to the Gnutella network so that a wide range of devices can access this network through a modified version of the Limewire access interface. Figure 1 illustrates a possible scenario. When a user selects a VE, the application and the MBK component are downloaded to his machine. When the device has capabilities restrictions, the MBK can be located at an Ultrapeer node. The Ultrapeers of the Gnutella network take over the role of session controllers (MSC component). In the current state of the work, only Ultrapeers nodes can assume this role because they have larger processing power and network band capacity.

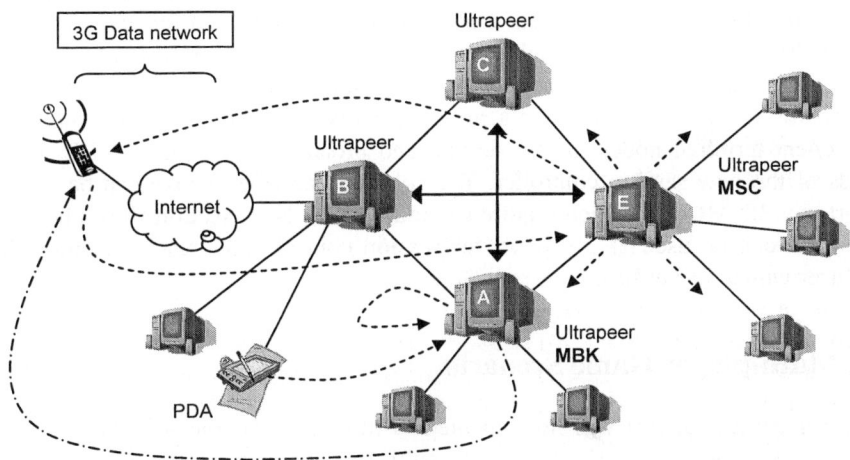

Fig. 1. Session Control and synchronization components integrated to the Gnutella Network

The main tasks of a session controller are, basically, to keep information about session, zones and users up to date, besides receiving requests for connection from the network nodes and checking if these can apply to become session controllers Ultrapeers. It would be hard to contemplate the use of Ultrapeers as session controllers if this task demanded too much processing time (after all, Ultrapeer nodes spontaneously donate their processing power and network connection to the realization of the session controlling tasks). In order to avoid the interruption of a user´s activity in the shared VE when a session controller Ultrapeer leaves the network (through a normal or abnormal disconnection), a control transfer scheme was devised which works as follows: if an Ultrapeer leaves the network for any reason, a successor Ultrapeer takes over the session so that the session does not end while there are participating users. Thus, a list of potential session controllers Ultrapeers, ready to take over a session control if the actual controller Ultrapeer leaves the scene, is generated. The list follows the connection order – the next to establish the connection and fulfil the requirements to be an Ultrapeer is the next to take over control. If by any chance the control transfer can not be realized (for absolute lack of volunteers machines that can be session controllers), the session then will be terminated (currently, there is always an existing Ultrapeer up and running in the Gnutella network for the music file sharing application). The Ultrapeers selection scheme as successors session controllers works as follows: when the current Ultrapeer session controller receives a connection request from a node, it checks first if this node has volunteered as a session controller and as an Ultrapeer. These options are offered to the user through the LimeWire client. If the node has larger processing power and network bandwidth, it is assigned as session controller number two in the session controllers hierarchy, otherwise it enters the session as an ordinary node. Thus, this node starts receiving session update information and starts managing the current session controller status (through heartbeat message). At the moment, up to four successors session controllers are maintained in latent state (ready to take over the control of a session). If any of these successors session controllers get disconnected, a new successor is searched for and configured as such automatically. As soon as the first successor session controller in the list realizes that the current session controller got disconnected (heartbeat message is not received followed by no reply for a ping), it starts establishing connections to all users´ client terminal nodes in that session and broadcasting them its URL, ie, the address of the new session controller. The solution described above can be used to support any 3D MUVEs which require session control. Next section shows how this mechanism can be used as a support for session control and synchronization in a multi-user games scenario.

5 A Multi-player Game Scenario

Multi-user games support was implemented to illustrate how the solution described above works. When a user wants to play multi-user games in the Gnutella network, he needs to have, in his machine, the modified Limewire client software to support shared games, besides the 3D Player MPEG-4 and the game application itself. If this machine is set up as an Ultrapeer node, the game application to be downloaded by the user has, embedded in it, the components responsible for the session control, and the game synchronization, besides information on existing sessions and players (current number of players, Ping response time, and if the node is a session controller Ultrapeer or not). Once the game is downloaded, an MPEG-4 Player is opened in that user´s client terminal. After downloading and selecting the desired game, another LimeWire window is opened with information on the session and zones for that game, besides the number of players. The user can then select the session he wishes to participate. The following situations may occur: *The session controler Ultrapeer leaves the network before a connection can be established* – in this case a message is shown in the Limewire window and the user will have to select the session again; *There are no opened sessions for the selected game* – the user may create a session and starts controlling it – soon friends might join the session; *The selected session is full* – by mirrowing what happens in the chats on the web, a new function will have to be implemented in the Limewire, which drives the user to another session automatically – if the user so wishes. When a session is selected successfully, the user is ready to play. His presence is broadcast (through the session controller) to all other players of that session. Any modification to the game state (caused by users´ interaction), is broadcast, by the MBK component, to all players, so that synchronization is achieved among all participating users. The performance of the control transfer mechanism is discussed in [10].

6 Conclusion

This paper describes a solution to support 3D MUVEs in a hybrid peer-to-peer Gnutella network. Session control and synchronization were implemented as specified by the ongoing multi-user extension to the MPEG-4 Standard, through the MSC and MBK components. These componentes were integrated to the Gnutella network, as well as the session control transfer mechanism described above. Preliminary tests show that the system scale and can be a viable solution to 3D MUVEs. The main advantages of this solution over more traditional solutions based either on client-server model or hybrid solutions using proxies are: (1) any user who has a 3D MUVE and wishes to start and control a session may do so without having to register to a content provider; (2) because the implemented solution uses the MPEG-4 Player and the ongoing multi-user MPEG-4 extension, it is easier to make this solution available to different devices, from cellular phones to PDAs and set-top-boxes; (3) the use of Ultrapeers as session controllers provides shorter network latencies for the nodes that connect to them, as usually the Ultrapeers are located closer to those nodes. This work has provided a better understanding of the issues involved in the implementation of 3D MUVEs in hybrid peer-to-peer networks.

References

1. C. Diot and L. Gautier. "A distributed arqchitecture for multiplayer interactive applications on the internet". IEEE Networks magazine, 13 (4), 1999.
2. Chris Greenhalgh and Steve Benford. "MASSIVE: a Distributed VR System Incorporating Spatial Trading". In Proc, IEEE 15th Intl Conf. on Distr. Computing Systems, 1995.
3. O. Hagsand. "Interactive MUVEs in the DIVE System". IEEE Comp., 3(1), 1996.
4. Mauve, M., Fischer, S. and Widmer, J. "A Generic Proxy System for Networked Computer Games", in ACM NetGames, pp 25-28, Germany, 2002.
5. Gnutella (2003). http://www.gnutella.com
6. Limewire (2003). http://www.limewire.org
7. N4264 (2001). "Overview of MPEG-4 Standard", (V.18 – Singapore Version).
8. LivingWorlds (1997) http://www.vrml.org/WorkingGroups/living-worlds/draft_2/index.htm.
9. Singla A., Rohrs C. (2003). "Ultrapeers: Another Step Towards Gnutella Scalability". http://rfd-gnutella.sourceforge.net/Proposals/Ultrapeer/ Ultrapeers.htm
10. A. Boukerche and R. Araujo, "Hybrid Peer-to-Peer Network Solution to the Synchronization and Session Control in Distributed Interactive Simulation and Collaborative Virtual Environments", Technical report in Preparation.

Large-Scale Deployment in P2P Experiments Using the JXTA Distributed Framework

Gabriel Antoniu[1], Luc Bougé[2], Mathieu Jan[1], and Sébastien Monnet[3]

[1] IRISA/INRIA, Gabriel.Antoniu@irisa.fr
[2] IRISA/ENS Cachan, Brittany Extension
[3] IRISA/University of Rennes I

Abstract. The interesting properties of P2P systems (high availability despite peer volatility, support for heterogeneous architectures, high scalability, etc.) make them attractive for distributed computing. However, conducting *large-scale experiments* with these systems arises as a major challenge. Simulation allows only to partially model the behavior of P2P prototypes. Experiments on real testbeds encounter serious difficulty with *large-scale deployment and control* of peers. This paper shows how an optimized version of the *JXTA Distributed Framework* (JDF) can help deploying, configuring and controlling P2P experiments. We report on our experience in the context of our JUXMEM JXTA-based grid data sharing service for various configurations.

1 How to Test P2P Systems at a Large Scale?

The scientific distributed systems community has recently shown a growing interest in the Peer-to-Peer (*aka* P2P) model [1]. This interest is motivated by properties exhibited by P2P systems such as high availability despite peer volatility, support of heterogeneous architectures and, most importantly, high scalability. For example, the *KaZaA* network has shown to scale up to 4,500,000 users, an unreachable scale for distributed systems based on the traditional client-server model.

However, the experimental validation phase remains a major challenge for designers and implementers of P2P systems. Validating such highly-scalable systems requires the use of large-scale experimentations, which is extremely difficult. Consider for instance popular P2P software, like *Gnutella* or *KaZaA*: workloads of these systems are not fully analyzed and modeled because the behavior of such systems cannot be precisely reproduced and tested [2]. Recently, P2P systems like *CFS* [3], *PAST* [4], *Ivy* [5] and *OceanStore* [6] based on smarter localization and routing schemes have been developed. However, most of the experiments published for these systems exhibit results obtained either by simulation, or by actual deployment on small testbeds, typically consisting of less than a few tens of *physical* nodes [7]. Even when larger scales are reached via emulation [8], no experimental methodology is discussed for automatic deployment and volatility control. For instance, failures are simulated by manually stopping the peers using the `kill` signal! There is thus a crucial need for infrastructures providing the ability to test P2P systems at a large scale. Several approaches have been considered so far.

Simulation. Simulation allows one to define a model for a P2P system, and then study its behavior through experiments with different parameters. Simulations are often executed on a single sequential machine. The main advantage of simulation is the reproducibility of the results. However, existing simulators, like *Network Simulator* [9], or *SimGrid* [10], need significant adaptations in order to meet the needs of a particular P2P system. This holds even for specific P2P simulators, like *ng-simulator* used by *NeuroGrid* [11]. Also, the technical design of the simulated prototype may be influenced by the functionalities provided by the simulator to be used, which may result in deviations from reality. Last but not least, simulators model simplified versions of real environments. Further validation by alternative techniques such as emulation or experiments in real environments is still necessary.

Emulation. Emulation allows one to configure a distributed system in order to reproduce the behavior of another distributed system. Tools like *dummynet* [12] or *NIST Net* [13] allow to configure various characteristics of a network, such as the latency, the loss rate, the number of hops between physical nodes, and sometimes the number of physical nodes (e.g., *ModelNet* [14] and *Emulab/Netbed* [15]). Thisn way, networks with various sizes and topologies can be emulated. However, the heterogeneity of a real environment (in terms of physical architecture, but also of software resources) cannot be faithfully reproduced. More importantly, deployment of P2P prototypes is essentially left to the user: it is often overlooked, but it actually remains a major limiting factor.

Experiments on real testbeds. Real testbeds such as *GridLab* [16] or *PlanetLab* [17] are large-scale, heterogeneous distributed environments, usually called *grids*. They are made of several interconnected sites, with various resources ranging from sensors to supercomputers, including clusters of PC. Such environments have proven helpful for realistic testing of P2P systems. Even though experiments are not reproducible in general on such platforms, this is a mandatory step in order to validate a prototype. Here again, deployment and configuration control have in general to be managed by the user. This is even more difficult than in the case of emulation, because of the much larger *physical* scale.

To sum up, actually *deploying and controlling* a P2P system over large-scale platforms arises as a central challenge in conducting realistic P2P experiments. The contribution of this paper is to introduce an enhanced version of *JXTA Distributed Framework* (JDF) [18] and demonstrate its use in large-scale experiments for a P2P system in a grid computing context.

2 Deploying and Controlling Large-Scale P2P Experiments in JXTA

In this paper, we focus on deploying and controlling P2P experiments on grid architectures. Let us stress that we do not consider the *actual testing strategy*, i.e., which aspects of the behavior are monitored and how. We only address the technical infrastructure needed to support specific testing activities. As seen in Section 1, just *deploying* a P2P prototype on such a scale is a challenging problem.

2.1 The Five Commandments of Large-Scale P2P Experiments

A tool aiming to facilitate large-scale P2P experiments should, at least, observe the following 5 commandments.

Commandment C1: *You shall provide the ability to easily specify the requested virtual network of peers.* Therefore, a specification language is required to define what kind of application-specific peers are needed, how they are interconnected, using which protocols, where they should be deployed, etc.

Commandment C2: *You shall allow designers of P2P prototypes to trace the behavior of their system.* Therefore, the tool should allow to retrieve the outputs of each peer, such as log files as well as result files, for off-line analysis.

Commandment C3: *You shall provide the ability to efficiently deploy peers on a large number of physical nodes.* For instance, a hierarchical deployment strategy may be useful when testing on a federation of distributed clusters.

Commandment C4: *You shall provide the ability to synchronize peers between stages of a test.* Indeed, a peer should have the possibility to wait for other peers to reach a specific state before going through the next step of a test.

Commandment C5: *You shall provide the ability to control the simulation of peers' volatility.* In typical P2P configurations, some peers may have a high probability of failure, while others may be almost stable. The tool should allow one to *artificially enforce* various volatile behaviors when testing on a (hopefully stable!) testbed.

2.2 The JXTA Project

The JXTA platform (for *juxtapose*) [19] is an open-source framework initiated by Sun Microsystems. JXTA specifies a set of language- and platform-independent, XML-based protocols. It provides a rich set of building blocks for P2P interaction, which facilitate the design and implementation of custom P2P prototypes. The basic entity is the regular peer (called *edge peer*). Before being able to communicate within the JXTA virtual network, a peer must discover other peers. In general, this is done through the use of specialized *rendezvous peers*. Communication between peers is direct, except when firewalls are in the way: in this case, *relay peers* must be used. Peers can be members of one or several *peer groups*. A peer group consists of peers that share a common set of interests, e.g., peers that share access to some resources. The reference implementation of JXTA is in Java. Its current version 2.2.1 includes around 50,000 lines of code. An implementation in C is under development.

2.3 The JXTA Distributed Framework Tool (JDF)

The purpose of JDF is to facilitate automated testing of JXTA-based systems. It provides a generic framework allowing to easily define custom tests, deploy all the required

```xml
<network analyze-class="prototype.test.Analyze">
  <profile name="Rendezvous">
      <!-- peer information -->
     <peer base-name="peerA" instances="1"/>
        <!-- rendezvous information -->
     <rdvs is-rdv="true"/>
        <!-- transport information -->
     <transports>
       <tcp enabled="true" base-port="13000"/>
     </transports>
        <!-- bootstrap information -->
     <bootstrap class=
               "prototype.test.Rendezvous"/>
  </profile>
  <profile name="Edge">
     <peer base-name="peerB" instances="1"/>
     <rdvs is-rdv="false">
       <rdv profile="Rendezvous"/>
     </rdvs>
     <transports>
       <tcp enabled="true" base-port="13000"/>
     </transports>
     <bootstrap class="prototype.test.Edge"/>
  </profile>
</network>
```

```java
public class Peer extends JxtaBootStrapper {
   public static void main(String[] args) {
      Peer peer = new Peer();
      peer.start(args);
      peer.stop();
   }

   // Start the test on the local peer
   public void start(String[] args) {
      // Start JXTA locally
      super.startJxta();
      // Get a custom P2P service
      p2pService = group.lookupService(...);
      // Use the API of the service
      p2pService.doSomething();
   }

   // Stop the test on the local peer
   public void stop() throws Exception {
      // Stop the custom P2P service
      p2pService.stopApp();
      // Stop JXTA locally
      super.stopJxta();
   }

   // Store the local results
   protected void updateProperties() {
      super.updateProperties();
      // store own results ...
      setProperty(PROPERTY_TAG,
                  property_result);
   }
}
```

Fig. 1. An example of required input files: a network description file defining 2 profiles (left), and a basic Java test class inside the original JDF's framework from Sun Microsystems (right).

resources on a distributed testbed and run the tests with various configurations of the JXTA platform.

JDF is based on a regular Java Virtual Machine (JVM), a Bourne shell and ssh or rsh. File transfers and remote control are handled using either ssh/scp or rsh/rcp. JDF assumes that all the physical nodes are visible from the control node. JDF is run through a regular shell script which launches a distributed test. This script executes a series of elementary steps: install all the needed files; initialize the JXTA network; run the specified test; collect the generated log and result files; analyze the overall results; and remove the intermediate files. Additional actions are also available, such as killing all the remaining JXTA processes. This can be very useful if the test badly failed for some reason. Finally, JDF allows one to run a sequence of such distributed tests.

A distributed test is specified by the following elements. 1) A *network description file* defining the requested JXTA-based configuration, in terms of edge peers, rendezvous peers, relay peers and their associated Java classes, and how they are linked together. This description is done through the use of profiles, as shown on the left side of Figure 1. Two profiles are described. Profile Rendezvous specifies a rendezvous peer configured to use TCP. Its associated Java class is protoype.test.Rendezvous. Profile Edge specifies an edge peer. It is configured to use a peer with Profile Rendezvous for its rendezvous peer through TCP a connection. The instance attribute of the peer tag specifies how many peers of the profile will be launched on a given physical node. 2) The set of Java classes describing the behavior of each

```
<profile name="RendezVous1" replicas="1"/>      <profile name="RendezVous1" replicas="1"/>
<profile name="RendezVous2" replicas="1"/>      <profile name="RendezVous2" replicas="1"/>
<profile name="Edge1" replicas="9"/>            <profile name="Edge1" replicas="99"/>
<profile name="Edge2" replicas="9"/>            <profile name="Edge2" replicas="99"/>
```

Fig. 2. A small JXTA network (left), and a large one (right).

peer. These Java classes must *extend* the framework provided by JDF, in order to easily start JXTA, stop JXTA and save the results into files, as shown on the right side of Figure 1. These result files are collected by JDF on each physical node and sent back to the control node to be analyzed by an additional Java class specified by the user (`prototype.test.Analyze` in the example). 3) A *node file* containing the list of physical nodes where to deploy and run the previously described JXTA network, as well as the path of the JVM used on each physical node. 4) An optional file containing the list of librairies to deploy on each physical node (a default one is provided if omitted).

3 Improving and Extending the JXTA Distributed Framework

We are currently building a prototype for a data-sharing service for the grid, based on the JXTA platform. Therefore, we need a tool to easily deploy and control our prototype for large-scale experiments. We describe how we improved the JDF tool, with the goal of making it better observe the 5 commandments stated above.

3.1 Improving JDF Functionalities

Commandment C1 requires JDF to provide the ability to easily specify the requested virtual network of peers. Developers can easily modify the number of peers hosted on each physical node, but JDF requires that a *specific* profile must be explicitly defined for each physical node, which is obviously not scalable. To facilitate testing on various large scale configurations, we introduced the ability to specify that a single profile can be shared by multiple physical nodes. This is specified by the `replicas` attribute. Let us assume we deploy JXTA on 2 clusters of 10 nodes. Each cluster runs one JXTA rendezvous peer, to which 9 edge peers are connected. Peers with profile `Edge1` are connected to the peer with profile `Rendezvous1` and similarly for peers with profiles `Edge2` and `Rendezvous2`. In the original version of JDF, this would require 2 rendezvous profiles plus 2×9 edge profiles. In contrast, our enhanced version of JDF requires only 2 rendezvous profiles plus 2 edge profiles. An excerpt of the corresponding new JDF network file is shown on the left side of Figure 2.

Now, imagine we want to run the same experiment on a larger JXTA configuration, in which 99 edge peers are connected to a rendezvous peer in each of two 100-nodes clusters. Thanks to the `replicas` attribute, it suffices to substitute 99 for 9 in the network description file. In the original JDF, this would require to define 2×99 edge profiles plus 2 rendezvous profiles!

Other enhancements mainly include improved performance for various phases of JDF execution. For instance, only the modified files are transmitted when updating the deployed environment. The ability to use NFS instead of `scp`/`rcp` has also been added when deploying a distributed test on a NFS-enabled cluster.

3.2 Enhancing JDF to Interact with Local Batch Systems

An important issue when running tests on grid infrastructures regards the necessary interaction between the deployment tool and the resource allocator available on the testbed. Many testbeds are managed by batch systems (like PBS [20], etc.), which dynamically allocate physical nodes to jobs scheduled on a given cluster. To observe commandment C3 regarding efficient deployment, we have enhanced JDF to make it interact with local batch systems via Globus, or directly with PBS.

JDF takes a static node file as input, which explicitly lists the physical nodes involved. On testbeds using dynamic resource allocators, such a list cannot be provided at the time of the job submission. Our enhanced version of JDF allows the node file to be dynamically created once the job is actually scheduled, using the actual list of physical nodes provided by the batch system.

This solves the problem on a single cluster. However, conducting P2P experiments on a federation of clusters requires to co-schedule jobs across multiple batch systems. This problem is beyond the scope of this paper, but we are confident that JDF will easily take advantage of any forthcoming progress in this domain.

3.3 Controlling the Simulation of Volatility

One of the most significant features of P2P systems is to support a high volatility. We think that a JDF-like tool should provide the designer with the possibility of running his system under various volatility conditions (Commandment C5). Using a stable testbed, the tool should provide a means to enforce node failures in a controlled way. The original version of JDF did not consider this aspect. We have therefore developed an additional facility to generate a *failure control file* from various volatility-related parameters. This file is given as an input to JDF, in order to control peer uptime. The file is statically generated before the test is launched, based on (empirical) statistical laws and on relevant parameters (e.g., the MTBF of the network peers). At runtime, the uptime of each peer is controlled by an additional service thread, which is started by JDF. This thread is used to kill the peer at the specified time. This way, the volatility conditions for a given test can be changed by simply varying the MTBF parameter. Note that using a pre-computed failure control file also enhances the reproducibility of the test.

As of now, the failure control file is generated according to a simple statistical law, where peer failures are independent. It would be interesting to consider dependencies between peers: the failure of one specific peer may induce the failure of some others. These dependencies could be associated with probabilities, for instance, the failure of peer A induces the failure of peers B and C with a probability of 50%. This should make it possible to simulate complex correlated failures, for instance, network partitions. Another direction would be to let the failure injection mechanism take into account the peer status. The JDF failure control threads could be used to monitor the state of their respective peers and regularly synchronize to generate correlated failures.

4 Case Study: Towards Large-Scale Experiments in JUXMEM

We are using this enhanced version of JDF to deploy and control our JXTA-based service called JUXMEM (for Juxtaposed Memory) [21]. We believe that these results can be applied to other JXTA-based services, such as the Meteor project [22]. Actually, we were able to deploy and run Meteor's peers using JDF without any modification in the prototype.

4.1 The JUXMEM Project: A JXTA-Based Grid Data-Sharing Service

JUXMEM is designed as a compromise between DSM systems and P2P systems: it provides *location transparency* as well as *data persistence* in a *dynamic environment*. The software architecture of the data-sharing service mirrors a hardware architecture consisting of a federation of distributed clusters. The architecture is therefore *hierarchical*. Its ultimate goal is to provide a data sharing service for grid computing environments, like DIET [23].

JUXMEM consists of a network of peer groups, called *cluster* groups, each of which generally corresponds to a physical cluster. All the *cluster* groups are enclosed by the *juxmem* group, which includes all the peers who are members of the service. Each *cluster* group consists of a set of peers which provide memory for data storage (*providers*). A *manager* monitors the providers in a given *cluster* group. Any peer can use the service to allocate, read or write to data as a *client*. All providers which host copies of the same data block make up a *data* group, uniquely identified by an ID. Clients only need to specify this ID to read/write a data block: the platform transparently locates it. JUXMEM can cope with peer volatility: each data block is replicated across a certain number of providers, according to a redundancy degree specified at allocation time. The number of replicas is dynamically monitored and maintained through dynamic replication when necessary.

4.2 Experimenting with Various Configurations

For our experiments, we used a cluster of the Distributed ASCI Supercomputer 2 (DAS-2) located in The Netherlands. This cluster is made of 72 nodes, managed by a PBS scheduler. Once it is scheduled, the JDF job deploys JXTA peers over the available cluster nodes using `ssh/scp` commands. Figure 3 reports the time needed to deploy, configure and update a JUXMEM service on a variable number of physical nodes using our optimized version of JDF.

The control node executes a sequential loop of `scp/ssh` commands in order to deploy JUXMEM and its whole environment on each physical node. The total size of all libraries for JXTA, JDF and JUXMEM is about 4 MB. As expected, the *deployment* time is thus linear with the number of physical nodes, e.g., it takes 20 s for 16 nodes, 39 s for 32 nodes.

Of course, this initial deployment step is required only once per physical node, so its cost can be "shared" by a sequence of experiments. Usually, these experiments consist of incremental modifications to some specific libraries. In this case, only the modified files need to be transmitted and updated on each physical node. For instance, the size of JUXMEM's `jar` file is around 100 kB. Modifying JUXMEM only requires the transmission of this file. This is much faster, as reported by the *update* curve.

As seen in Section 2.3, the JDF network description file defines what types of JXTA peers are requested, how they are interconnected, etc. The configuration step on a given

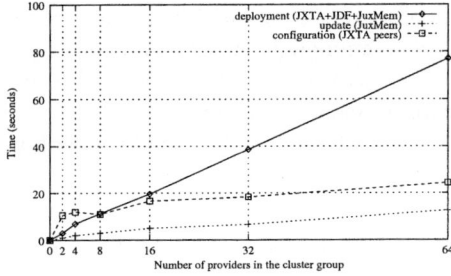 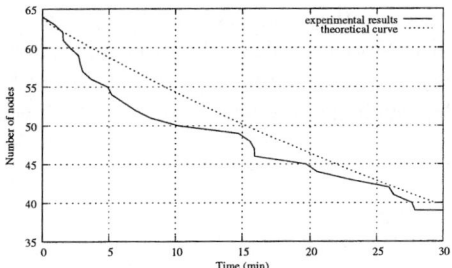

Fig. 3. Time needed to deploy, update and configure the JUXMEM service on various network sizes.

Fig. 4. Number of remaining peers in an experiment with a MTBF set to 1 minute.

physical node consists in 2 phases: 1) based on this description, generate the specific JXTA configuration file of each peer; 2) for each peer, update this file with its specific list of rendezvous peers. As expected, the *configuration* curve shows that this takes a time which increases slowly and linearly with the number of physical nodes.

Note that, in order to reach larger scales, each physical node can host several peers. This is easily specified using the `instances` attribute of the `peer` tag in the JDF network description file. For instance, we have been able to deploy 10 peers on each node without any significant overhead, amounting to a total of 640 peers without any additional effort.

Finally, once the configuration is complete, JUXMEM is started by invoking a JVM for each peer. This initialization time is dependent on the number of peers launched on each physical node. According to commandment C4, JDF should provide a way to synchronize the initialization of JXTA peers. This could be handled by a JDF control thread, through a mechanism similar to the failure control. This feature is still being implemented at this time.

4.3 Experimenting with Various Volatility Conditions

The simplest way to model architecture failures is to provide a global MTBF. Our enhanced version of JDF allows the programmer to generate a failure schedule according to a given MTBF. We carried out a sample experiment on 64 physical nodes with a failure control file generated to provide a MTBF of 1 minute: the uptime in minutes for each physical node follows an exponential distribution with a rate parameter of $\frac{1}{64}$. The files automatically collected by JDF allow to check which peer went down, and when. Figure 4 reports the results of a specific single run of 30 minutes: 25 peers were killed. Observe that these results are somewhat biased as peers do not start exactly at the same time, and the clocks are not perfectly well-synchronized.

5 Conclusion

Validating P2P systems at a large scale is currently a major challenge. Simulation is nowadays most-widely used, since it leads to reproducible results. However, the significance of these results is limited, because simulators rely on a simplified model of reality. More complex validation approaches based on emulation and execution on "real" large-scale testbeds (e.g., grids) do not have this drawback. However, they leave the deployment at the user's charge, which is a major obstacle in practice.

In this paper, we state *five commandments* which should be observed by a deployment and control tool to successfully support large-scale P2P experiments. Our contribution consists in enhancing the *JXTA Distributed Framework* (JDF) to fulfill some of these requirements. This enhancement mainly includes a more precise and concise specification language describing the virtual network of JXTA peers, the ability to use various batch systems, and also to control the volatility conditions during large-scale tests. Some preliminary performance measurements for the basic operations are reported.

Further enhancements are needed for JDF to fully observe the five commandments. A hierarchical, tree-like scheme for the ssh/rsh commands could be used to balance the load of copying files from the control node to other physical nodes, along the lines of [24]. We plan to integrate a synchronization mechanism for peers to support more complex distributed tests. The final goal is to have a rich generic tool allowing to deploy, configure, control and analyze large-scale distributed experiments on a *federation of clusters* from a single control node. We intend to further develop this tool, in order to use it for the validation of JUXMEM, our JXTA-based grid data sharing service. JDF could also be very helpful for other JXTA-based services, and the approach can be easily generalized to other P2P environments.

Acknowledgments

The authors thank Thilo Kielmann's group from the Vrije Universiteit for making the DAS-2 testbed available to us. We also thank Bernard Traversat, Senior Architect of Project JXTA at Sun Microsystems, for his useful comments, as well as the reviewers, for their suggestions.

References

1. Milojicic, D.S., Kalogeraki, V., Lukose, R., Nagaraja, K., Pruyne, J., Richard, B., Rollins, S., Xu, Z.: Peer-to-peer computing. Technical Report HPL-2002-57, HP Labs (2002) available at http://www.hpl.hp.com/techreports/2002/HPL-2002-57.pdf.
2. Gummadi, K.P., Dunn, R.J., Saroiu, S., Gribble, S.D., Levy, H.M., Zahorjan, J.: Measurement, modeling, and analysis of a peer-to-peer file-sharing workload. In: 19th ACM Symposium on Operating Systems Principles (SOSP '03), Bolton Landing, NY, ACM Press (2003) 314–329
3. Dabek, F., Kaashoek, F., Karger, D., Morris, R., Stoica, I.: Wide-area cooperative storage with CFS. In: 18th ACM Symposium on Operating Systems Principles (SOSP '01), Chateau Lake Louise, Banff, Alberta, Canada (2001) 202–215
4. Rowstron, A., Druschel, P.: Storage management and caching in PAST, a large-scale, persistent peer-to-peer storage utility. In: 18th ACM Symposium on Operating Systems Principles (SOSP '01), Chateau Lake Louise, Banff, Alberta, Canada (2001) 188–201
5. Muthitacharoen, A., Morris, R., Gil, T.M., Chen, B.: Ivy a read/write peer-to-peer file system. In: 5th Symposium on Operating Systems Design and Implementation (OSDI '02), Boston, MA (2002) 31–44

6. Kubiatowicz, J., Bindel, D., Chen, Y., Eaton, P., Geels, D., Gummadi, R., Rhea, S., Weatherspoon, H., Weimer, W., Wells, C., Zhao, B.: OceanStore: An architecture for global-scale persistent storage. In: 9th International Conference on Architecture Support for Programming Languages and Operating Systems (ASPLOS 2000). Number 2218 in Lecture Notes in Computer Science, Cambridge, MA, Springer-Verlag (2000) 190–201
7. Rhea, S., Eaton, P., Geels, D., Weatherspoon, H., Zhao, B., Kubiatowicz, J.: Pond: the OceanStore prototype. In: 2nd USENIX Conference on File and Storage Technologies (FAST '03), San Francisco, CA (2003) 14
8. Rhea, S., Geels, D., Roscoe, T., Kubiatowicz, J.: Handling churn in a DHT. Technical Report CSD-03-1299, UC Berkeley (2003) Available at http://oceanstore.cs.berkeley.edu/publications/papers/.
9. The ns manual (formerly ns notes and documentation). http://www.isi.edu/nsnam/ns/doc/ns_doc.pdf
10. Casanova, H.: SimGrid: A toolkit for the simulation of application scheduling. In: First IEEE/ACM International Symposium on Cluster Computing and the Grid (CCGrid 2001), Brisbane, Australia (2001) 430–441
11. Joseph, S.: P2P metadata search layers. In: Second International Workshop on Agents and Peer-to-Peer Computing (AP2PC 2003). Number 2872 in Lecture Notes in Computer Science, Bologna, Italy, Springer-Verlag (2003)
12. Rizzo, L.: Dummynet and forward error correction. In: 1998 USENIX Annual Technical Conference, New Orleans, LA (1998) 129–137
13. Carson, M., Santay, D.: NIST Net - a Linux-based network emulation tool. (2004) To appear in special issue of Computer Communication Review.
14. Vahdat, A., Yocum, K., Walsh, K., Mahadevan, P., Kostic, D., Chase, J., Becker, D.: Scalability and accuracy in a large-scale network emulator. In: 5th Symposium on Operating Systems Design and Implementation (OSDI '02), Boston, MA (2002) 271–284
15. White, B., Lepreau, J., Stoller, L., Ricci, R., Newbold, S.G.M., Hibler, M., Barb, C., Joglekar, A.: An integrated experimental environment for distributed systems and networks. In: 5th Symposium on Operating Systems Design and Implementation (OSDI '02), Boston, MA (2002) 255–270
16. Allen, G., Davis, K., Dolkas, K.N., Doulamis, N.D., Goodale, T., Kielmann, T., Merzky, A., Nabrzyski, J., Pukacki, J., Radke, T., Russell, M., Seidel, E., Shalf, J., Taylor, I.: Enabling applications on the grid: A GridLab overview. International Journal of High Performance Computing Applications 17 (2003) 449–466
17. PlanetLab: an open community testbed for planetary-scale services. http://www.planet-lab.org/pubs/2003-04-24-IntelITPlanetLab.pdf
18. JXTA Distributed Framework. http://jdf.jxta.org/
19. The JXTA project. http://www.jxta.org/
20. The Portable Batch System. http://www.openpbs.org/
21. Antoniu, G., Bougé, L., Jan, M.: JuxMem: Weaving together the P2P and DSM paradigms to enable a Grid Data-sharing Service. Kluwer Journal of Supercomputing (2004) To appear. Preliminary electronic version available at URL http://www.inria.fr/rrrt/rr-5082.html.
22. Project Meteor. http://meteor.jxta.org/
23. The DIET project: Distributed interactive engineering toolbox. http://graal.ens-lyon.fr/~diet/
24. Martin, C., Richard, O.: Parallel launcher for clusters of PC. In Imperial College Press, L., ed.: Parallel Computing (ParCo 2001), Naples, Italy, World Scientific (2001) 473–480

Experimental Study of Multi-criteria Scheduling Heuristics for GridRPC Systems

Yves Caniou* and Emmanuel Jeannot

[1] LORIA, INRIA-Lorraine
[2] LORIA, Université Henri Poincaré

Abstract. We study in this paper several scheduling heuristics for GridRPC middlewares. When dealing with performance issue, the scheduling strategy is one of the most important feature. However, many heuristics implemented in available middlewares were not designed for this context (for instance MCT in NetSolve). Therefore, schedulers are not necessarily effective. We propose to use heuristics based on a non-intrusive module which is able to estimate the duration of all tasks in the system. Three criteria are examined among which the execution time of the application, e.g. the makespan. Experiments on a real platform show that the proposed heuristics outperform MCT for at least two of these three criteria.

Keywords: time-shared and heterogeneous resources, dynamic scheduling heuristics, historical trace manager, completion dates estimations, DAGs

1 Introduction

GridRPC [1] is an emerging standard promoted by the global grid forum (GGF).This standard defines both an API and an architecture. A GridRPC architecture is heterogeneous and composed of three parts: a set of clients, a set of servers and an agent (also called a registry). The agent has in charge to map a client request to a server. In order for a GridRPC system to be efficient, the mapping function must choose a server that fulfills several criteria. First, the total execution time of the client application, e.g. the makespan, has to be as short as possible. Second, each request of every clients must be served as fast as possible. Finally, the resource utilization must be optimized.

Several middlewares instantiate the GridRPC model (NetSolve [2], Ninf [3], DIET [4], etc.). In these systems, a server executes each request as soon as it has been received: it never delays the start of the execution. In this case, we say that the execution is *time-shared* (in opposition to *space-shared* when a server executes at most one task at a given moment). In NetSolve, the scheduling module uses MCT (Minimum Completion Time) [5] to schedule requests on the servers. MCT was designed for scheduling an application for space-shared servers. The goal was to minimize the makespan of a set of independent tasks. This leads to the following drawbacks:

- *mono-criterion and mono-client.* MCT was designed to minimize the makespan of an application. It is not able to give a schedule that optimizes other criteria such as the response time of each request. Furthermore, optimizing the last task completion

---

* This work is partially supported by the Région Lorraine, the French ministry of research ACI GRID.

date does not lead to minimize each client application makespan. However, in the context of GridRPC, the agent has to schedule requests from more than one client.
- *load balancing*. MCT tries to minimize the execution time of the last request. This leads to over-use the fastest servers. In a time-shared environments, this implies to delay previously mapped tasks and therefore degrades the response time of the corresponding requests.

Furthermore, MCT requires sensors that give information on the system state. It is mandatory to know the network and servers state in order to take good scheduling decisions. However, supervising the environment is intrusive and disturbs it. Moreover, the information are sent back from time to time to the agent: they can be out of date when the scheduling decision is taken.

In order to tackle these drawbacks we propose and study three scheduling heuristics designed for GridRPC systems. Our approach is based on a prediction module embedded in the agent. This module is called the Historical Trace Manager (HTM) and it records all scheduling decisions. It is not intrusive and since it runs on the agent there is no delay between the determination of the state and its availability. The HTM takes into account that servers run under the time-shared model and is able to predict the duration of a given task on a given server as well as its impact on already mapped tasks. In this paper, we show that the HTM is able to predict precise task durations when the server load is not too high. The proposed heuristics use the HTM to schedule the tasks. We have plugged the HTM and our heuristics in the NetSolve system and performed intensive series of tests on a real distributed platform (more than 50 days of continuous computations) for various experiments with several clients.

In [5], MCT was only studied in the context of independent tasks submission. In this paper, the study is more general. The tests are submissions of task graphs (1D meshes and stencil). For some of them, independent tasks are submitted during the execution of the graphs. We do not assume any knowledge of the applications graphs and thus, we schedule dynamically each request. The goal is to serve at best each client request (a task of a graph as well as an independent one). We have compared our heuristics against MCT in NetSolve on three criteria. Results show that the proposed heuristics outperform MCT on at least two of the three criteria with gain up to 20% for the makespan and 60% for the average response time.

2 MCT in GridRPC Systems

GridRPC is an emerging standard for Grid Computing. It allows the remote execution of tasks to computational servers. The architecture is composed of clients that submit requests to an agent (or a registry), the agent sends back the identity of the servers that can execute the tasks. The client sends input data to the chosen server which performs the computation and sends back output data.

NetSolve [2] instantiates this model. A NetSolve agent uses Minimum Completion Time [5] (MCT) to map tasks on servers. The MCT heuristic chooses the server that will complete the task the soonest. In order to determine the completion time of a given task on a server it assumes that the load on the network and on the server is constant during the execution. This leads to the following remarks:

1. the agent needs an evaluation of each server load;
2. the load on a given server is seldom constant. If the server is loaded by some tasks, they are likely to finish during the computation of the new mapped task;
3. The time-shared model implies that mapping a task on a loaded server delays the concurrent running tasks. In this paper, we call this delay the *perturbation*. The perturbation may make the previously decision obsolete.

The load evaluation of a network or of a machine is a difficult task and is often performed by sensors. NetSolve can use its own sensors or NWS [6]. It faces two major problems: accuracy (MCT considers that the server load is constant when mapping a task and do not consider the perturbation a new task will produce on the server) and intrusiveness (sensors require some cpu cycles on the servers to evaluate the load).

3 Historical Trace Manager

The Historical Trace Manager is an attempt to efficiently answer the three remarks exposed in the previous section. It is a prediction module that runs on the agent. It is accurate and non intrusive. It simulates the execution of the tasks on each server and therefore is able to predict load variation and to compute the perturbation of the new task to already mapped ones.

We use the following model to simulate time-shared resources. When n tasks are using the same resource (CPU, network, etc.), each one uses $1/n$ of its peak power. The HTM needs several types of information: server and network peak performances, the size of input and output data and the number of operations of each task. All these information are static and can be computed off-line Therefore, the HTM answers the three remarks exposed in Section 2:

1. the HTM is not intrusive because it uses only static information. Furthermore, since it runs on the agent, information are immediately available for the scheduling heuristic. The accuracy of the information given by the HTM is high and will be experimentally demonstrated in the results section;
2. the HTM is able to compute the completion date of every tasks on a given server by simulating their executions. Therefore, the load is not assumed constant;
3. The HTM can simulate the mapping of a task on any server. Hence, it can compute the perturbation of this task on all the concurrent running ones. The perturbation can then be used to take efficient scheduling decisions, as we will see in the next sections.

4 Heuristics

We introduce here three heuristics, HMCT, MP and MSF, which are compared to MCT in section 5. The HTM simulates the new task on each server and gives resultant information to the scheduler. Therefore, the heuristic considers the perturbation the new task will induce on each running one, and computes the 'best' server accordingly.

Historical MCT: HMCT. HMCT chooses the server that will complete the request the fastest. Unlike MCT which assumes a constant behavior of the environment during the execution of the task to predict its finishing date (see Section 2), HMCT uses HTM estimations which are far more precise.

Minimum Perturbation: MP. The HTM simulates the new task on each server. MP chooses the server that minimizes the sum of the HTM estimated delays that running tasks will suffer. It aims to minimize overall running tasks completion dates and each client can thus expect a lower *response time*.

Minimum Sum Flow: MSF. MSF relies on HTM estimations to compute the sum of the duration (also called the sumflow [7, 8]) of all the tasks. It then chooses the server which minimizes the sumflow. MSF aims at reducing the duration of all the tasks (not only the new one).

Table 1. Resources Composing the Environment and task durations on the unloaded servers.

Type	Machine	Processor	Speed	Memory	Swap	System	task 1	task 2	task 3
Server	spinnaker	xeon	2 GHz	1 Go	2 Go	Linux	15	30	43
	artimon	pentium IV	1.7 GHz	512 Mo	1024 Mo	Linux	17	33.5	49.5
	soyotte	sparc Ultra-1		64 Mo	188 Mo	SunOS	128	256	382.5
	fonck	sparc Ultra-1		64 Mo	188 Mo	SunOS	127.5	254	380.9

5 Experiments and Results

5.1 Experiments

The HTM and the heuristics HMCT, MP and MSF have been implemented in the NetSolve agent. Several experiments have been conducted onto an heterogeneous environment whose resources are given in Table 1. Three types of computing intensive tasks have been used for these experiments. The duration of each task on each unloaded server is given in Table 1. computing Table 2 gives a summary of the series of tests performed on this platform. 8 scenarii (from (a) to (h)) have been submitted to this heterogeneous environment. Two types of graphs have been used. 1D meshes and stencil graphs. In a scenario, only one client submits a stencil graph while there can be several clients submitting a 1D mesh application (5 or 10). Stencil graphs have a width of 5 or 10 and a depth of 25 or 50. In scenarii (c), (f), (g) and (h), independent tasks are submitted during the execution of the graph(s). Their arrival dates are drawn from a Poisson distribution of parameter μ. We use the GSL library [9] for all the probabilistic calls. Experiments in a scenario are randomized by a different seed and run 6 times per heuristics. 6 runs were sufficient for the average values given in Section 5.3 to be stabilized. This leads to 576 experiments and about 50 days of computations.

5.2 Accuracy of HTM Predictions

Scheduling decisions rely on the accuracy of the HTM, e.g the ratio of the HTM estimated duration by the real post-mortem duration. We show in this section that it is

Table 2. Scenarii, Modalities and Number of Experiments.

Scenario	Application(s)		Independent Tasks		Experiment	
	nbclients	width x depth	nbtasks	μ (sec)	nbseeds x nbrun	total nbtasks
(a) 1D-mesh	10	1x50	-	-	4 x 6	500
(b) 1D-mesh	10	1x variable	-	-	4 x 6	500
(c) 1D-mesh + 250 independent tasks	5	1x50	250	20	4 x 6	500
(d) stencil (task 1)	1	10x50	-	-	1 x 6	500
(e) stencil (task 3)	1	10x50	-	-	1 x 6	500
(f) stencil + 175 independent tasks	1	10x25	175	28	2 x 6	425
(g) stencil + 87 independent tasks	1	10x25	87	40	4 x 6	337
(h) stencil + 87 independent tasks	1	5x25	87	25	4 x 6	212

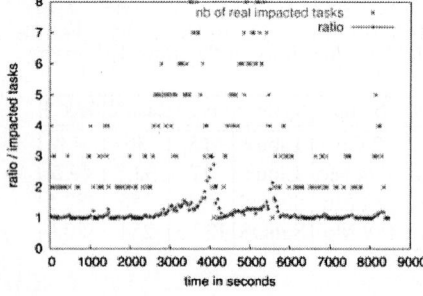

Fig. 1. 500 Independent Tasks Submitted at a Rate $\mu = 17$ Scheduled by MSF on artimon.

Fig. 2. Scenario (d) Scheduled by MP on Spinnaker.

high but may degrade with the server load and for some types of submission. Two main information appear on Figures 1(1) in dark dots: the value for each task (represented by its submission date on the abscissa) of the HTM estimated duration divided by the real post-mortem one. Hence, the closest to 1 is the ratio, the most accurate the prediction is; (2) in light dots: the number of tasks that have interfered with the considered task during its execution.

We can observe in Figure 1 the accuracy of the HTM on the 222 tasks which have been scheduled by MSF on artimon. Estimations are degrading with the load of the server. Indeed, until 2800 seconds, estimations are more than 96% accurate on the average and still 93% accurate until 3156 seconds. Then, more than 6 tasks are simultaneously executed on artimon, thus increasing prediction errors. The HTM regains a high accuracy when the load decreases. We can note that the HTM estimations accuracy is not decreasing with time.

We can observe on Figure 2 the information collected on the 301 tasks scheduled by MP on spinnaker, the fastest server, during an experiment of the scenario (d). We can observe two regions: before and after time equals 1000 seconds. Until time reaches 1000 seconds, accuracy is uneven. During that time, the server executes an average of 4 tasks running concurrently, with a peak equal to 7 tasks. After 1000 seconds, a maximum number of 3 tasks are running concurrently. The HTM is over 95% accurate.

As a conclusion, we see that the HTM accuracy is high when the number of simultaneous running jobs is lower than 5. Moreover, for some scenarii ((d) and independent tasks submissions) the HTM is able to regain a high accuracy alone and the accuracy does not degrade with time.

5.3 Comparison Between Heuristics

For each scenario we compare each heuristics on three criteria [10]. MCT is used as the comparison baseline. First, we compare the application completion time, e.g. the *makespan*. This comparison is done for all the scenarii. Second, when independent tasks are involved (scenarii (c), (f), (g) and (h)), we compare the *response time*, e.g. in our context (see section 2), the average duration of a task. The response time is only measured for each independent task. Third, when independent tasks are involved, we give the percentage of independent tasks that finish sooner when the experiment is scheduled with our heuristic than with MCT. If an heuristic is able to give a good response time and a majority of tasks finishing sooner than if scheduled with MCT, it improves the quality of service given to *each client*.

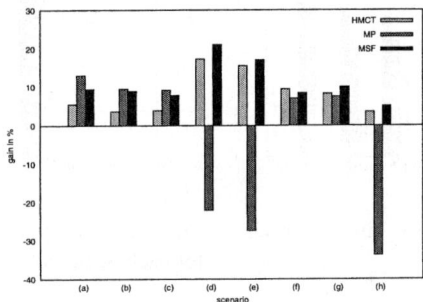

Fig. 3. Gain on the Makespan for each Client of each Scenario.

Figure 3 shows the average gain on the makespan performed by each heuristic over MCT *for each client* and for each scenario. We see that MSF and HMCT always outperform MCT up to 22%. MP outperforms HMCT and MSF when applications are 1D-mesh graphs (scenarii (a), (b) and (c)). However, MP shows negative performance for some stencil graphs scenarii. This is explained as follows. Due to its design, MP can unnecessarily map a task on a slow server, for example when it is the only idle one. This delays the completion of the graph because some critical tasks are mapped on slow servers. The client must wait for the completion of these tasks before being able to resume the execution of the application.

Figure 4 shows, for each heuristic, the average gain on the response time that *each independent task* benefits when scheduled with the considered heuristic. Scenarii (a), (b), (d) and (e) are not concerned and only appear to ease the reading. We see that HMCT, MP and MSF outperform MCT for all the scenarii. For scenarii (f) and (g) MSF is the best heuristic with a gain up to 40%. For scenarii (c) and (h) MP is the best heuristic, with more than 67% of gain in scenario (h). For this scenario HMCT shows nearly no gain.

The average percentage of tasks that finish sooner than MCT is given for each heuristic in Figure 5. On the opposite of the two former figures where a gain is observed as soon as the value is positive, the percentage here has to be superior to 50% to express a gain (note that scenarii (a), (b), (d) and (e) are still not concerned here). We see that, as for the response time, HMCT, MP and MSF always outperform MCT. MSF is the best heuristic for scenarii (f) and (g). MP is the best for scenarii (c), (f) and (h), where 90% of the tasks finish sooner than if scheduled with MCT. For this scenario HMCT has the same performance than MCT.

Figure 4 and Figure 5 show together that MP and MSF are able to offer a good quality of service to each independent task. This means that on the average a task scheduled by one of these two heuristics finishes sooner than MCT and the average gain is high.

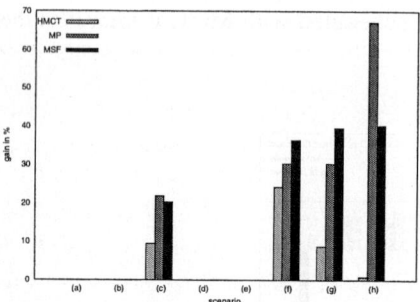

Fig. 4. Gain on the Response Time for each Client of each Scenario.

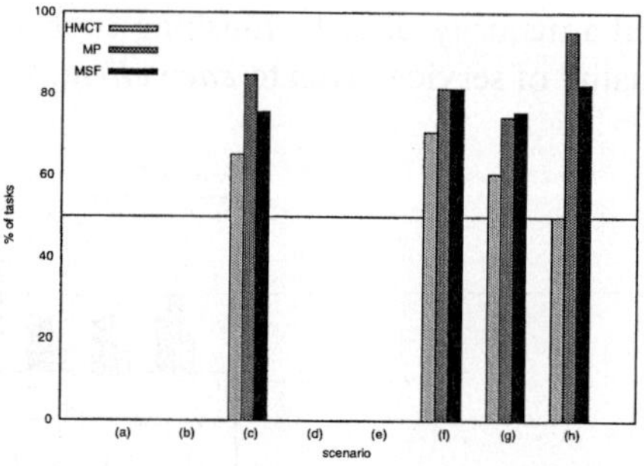

Fig. 5. Percentage of Tasks that Finish Sooner for each Scenario.

6 Conclusion

In the paper, we tackle the problem of scheduling tasks in gridRPC systems. MSF, which rely on a non-intrusive predictive module called the Historical Trace Manager. It simulates the execution of tasks on the environment with the time-shared model.

For our tests, we have used NetSolve, a gridRPC environment, in which we have implemented the HTM and all the heuristics. An extensive study has been performed on a real test platform. Numerous scenarii involving different kind of submissions, with different application graphs possibly concurrent to independent tasks, have been submitted to the modified NetSolve agent.

We have shown that the HTM is accurate when the server load is not too high. In some cases, the HTM accuracy degrades during the execution due to error accumulations.

Our heuristics were compared against Minimum Completion Time, the default NetSolve scheduling heuristic.Results show that our heuristics outperform MCT on at least two of the three observed criteria: makespan, response time and percentage of tasks finishing sooner than when scheduled with MCT. MSF appears to be the best overall heuristic as it always outperforms MCT on all the criteria.

In our future work, we will implement a mechanism to synchronize the HTM to the reality. This mechanism will be based on the task completion date. The goal is to limit the effect of error accumulation seen in some experiments. Early developments show an improvement in the HTM predictions accuracy.

References

1. Nakada, H., Matsuoka, S., Seymour, K., Dongarra, J., Lee, C., Casanova, H.: A GridRPC Model and API for End-User Applications (2003) https://forge.gridforum.org/projects/gridrpc-wg/document/GridRPC_EndUser_16dec03/en/1
2. Casanova, H., Dongarra, J.: Netsolve: A network server for solving computational science problems. In: Proceedings of Super-Computing -Pittsburg. (1996)
3. Nakada, H., Sato, M., Sekiguchi, S.: Design and implementations of ninf: towards a global computing infrastructure. Future Generation Computing Systems, Metacomputing Issue (1999) 649–658
4. Caron, E., Desprez, F., Fleury, E., Lombard, D., Nicod, J., Quinson, M., Suter, F.: Une approche hiérarchique des serveurs de calcul. to appear in Calculateurs Parall'eles, numéro spécial metacomputing (2001) http://www.ens-lyon.fr/ desprez/DIET/index.htm
5. Maheswaran, M., Ali, S., Siegel, H., Hengsen, D., Freund, R.: Dynamic matching and scheduling of a class of independent tasks onto heterogeneous computing system. In: Proceedings of the 8th Heterogeneous Computing Workshop (HCW '99). (1999)
6. Wolski, R., Spring, N., Hayes, J.: The network service: A distributed resource performance forecasting service for metacomputing. Journal of Future Generation Computing Systems (1999) 757–768
7. Baker, K.: Introduction to Sequencing and Scheduling. Wiley (1974)
8. Pinedo, M.: Scheduling: Theory, Algorithms ans Systems. Prentice Hall (2001)
9. Galassi, M., Theiler, J.: The gnu standard library (1996) http://www.gnu.org/software/gsl/gsl.html
10. Dutot, P.F., Eyraud, L., Mounier, G., Trystram, D.: Bi-Criteria Algorithm for Scheduling Jobs on Cluster Platforms. In: Proceedings of SPAA 2004. (2004) to appear.

Using Similarity Groups to Increase Performance of P2P Computing

J. Bourgeois[1], J.B. Ernst-Desmulier[1], F. Spies[1], and J. Verbeke[2]

[1] LIFC, University of Franche-Comte
[2] Lawrence Livermore National Laboratory

Abstract. This article aims to demonstrate how to build new types of groups called similarity groups into the JNGI project. This is done in order to increase the relevance of task dispatching and therefore to increase the performance of JNGI.

1 Introduction

While P2P first applications were essentially dedicated to file sharing, P2P is now seen as a possible computing paradigm for every field of computer science like parallelism.

JNGI [1] is a project including the dynamism of the P2P environment to maximize the utilization of volatile resources, redundancy for higher reliability, the division of computational resources into groups to increase scalability, and heterogeneity to make better use of all the available computing power. JNGI divides computing peers into several peer groups to limit the amount of communication seen by each peer. This in turn improves scalability.

The aim of this article is to propose other types of groups called similarity groups. These groups can be used either for qualitative (structural) or quantitative (performance) purposes. Indeed, JNGI is designed to contain thousands of highly heterogeneous nodes and needs to group them according to relevant criteria. It will be shown in this article that the use of similarity groups can dramatically increase the performance of a computation. This paper is divided into 2 parts: Section 1 describes similarity groups and how they are included into JNGI and section 2 sums up the results obtained with similarity groups.

2 Similarity Groups

At every levels, peers' heterogeneity arise problems which must be overcome [2]. For example, peers run different OSs, so P2P applications need to be cross-platform. This heterogeneity is rarely taken into account by applications which simply do not know what the capacity of the underlying hardware is, e.g. low frequency processor versus high frequency processor and low bandwidth connection versus high bandwidth connection.

Grouping Nodes. Choosing different criteria for grouping peers is the first step to the grouping strategy. There are 3 criteria used for grouping peers, distance, qualitative and quantitative.

There are two possibilities to apply a distance criterion. The first one is to group peers belonging to the same geographical area. The second possibility is to group computing nodes which are close on a networking way, for example, by using the number of routers crossed between 2 nodes. The second criterion is to group peers on a qualitative criterion like their Operating System or for example on their JVM version. The quantitative criterion is the last grouping technique. It uses physical characteristics like CPU speed, memory size, etc. The grouping strategy is a mix of the three criteria described above.

For our environment, we use some of the criteria defined by the Network Weather Services [3]. Our criteria used to group peers are the following: CPU speed, percentage of available CPU, memory size, free memory, latency and bandwidth.

Job Submission. The second step of our implementation is to adapt the grouping criteria to JNGI. When a new node is connected to JNGI, it contacts the higher-level monitor group. A monitor peer in this group replies to the node and asks for its characteristics.

Thanks to these parameters, the monitor can steer the peer to the worker group which matches its parameters.

Job submission has also to be modified in order to take the groups into account.

A job can be submitted to JNGI in three different ways : the first level of submission is when the user specify the name of the wanted group. The job is then run on this group. At the second level, the user submits a job with some constraints. The constraints are compared with the different group characteristics. If the group characteristics match the constraints, the job is sent to this group. The third level is when the user sends a job and chooses a quality of service like the computation speed. This kind of submission needs JNGI to use a performance prediction service like [4] adapted to P2P.

Adding these submission levels to JNGI increases its potential: JNGI can be used as a computing platform or as a testing structure by using the constraints.

3   Case Study: Root Finding Method

The Durand-Kerner's method [5] is used for extracting roots $\omega_1, ..., \omega_n$ from a polynomial with complex coefficients:

$$P(z) = \sum_{i=0}^{n} a_i z^{n-i} \ with \ a_0 = 1, \ a_n \neq 0 \ and \ a_i \ in \ \mathbb{C} \qquad (1)$$

where n stands for the order of the polynomial. It uses the Weierstrass operator as an iterative function which is able to converge a vector towards the polynomial root solutions.

Experiments. Our experiments have been made on one local network and on two local networks interconnected with a 512Kbps link. Two routers Cisco 2500 running an IOS 12.0 interconnect the two LANs which simulates peers connected with high bandwidth. On each local network, 4 PCs, Pentium 4 1.7Ghz, are used in order to have a total of 8 JNGI workers. The PCs are connected with a 100 Mbps Fast-Ethernet Allied Telesyn switch. With this, we have the opportunity to access a homogeneous or heterogeneous 4-computer platform. Similarity groups are interesting if there are heterogeneous characteristics e.g. network bandwidth.

On this JNGI platform, we run the Durand-Kerner application using the second level of submission with a job submitter and 4 workers. The job submitter's role is to manage the execution and to balance the roots between the workers. The number of interval roots is fixed at the beginning of the computation. This value is much greater than the number of workers to have a good load balancing.

Workers compute roots asynchronously, that is to say that they do not wait for other workers to finish their computation to receive updated roots. This iteration is repeated until convergence is reached.

Results. For these experiments, two kinds of polynomials have been solved. One has 4 monomials, curves with name ending with "(4)" and the other 16 monomials, curves with name ending with "(16)".

In figure 1(a), the 1LAN-NOGROUP(4) and 1LAN-NOGROUP(16) curves correspond to the execution of our application on the 4-computer local network without group management. The 1LAN-GROUP(4) and 1LAN-GROUP(16) curves correspond to the execution of our application on the 4-computer local network managing groups. There is only 1 group because the network is fully homogeneous in order to be able to compare the results with or without group management. 1LAN-NOGROUP and 1LAN-GROUP curves show the overhead of group management with exactly the same hardware architecture. We can see that managing groups does not have any cost during the execution. The only cost for managing groups is during the allocation of the tasks to the workers, but this is quite small.

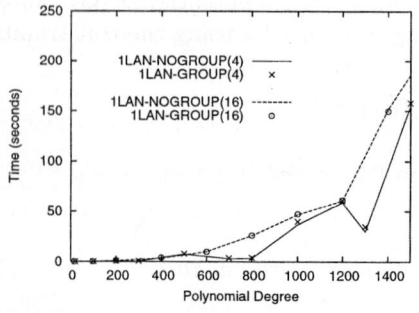
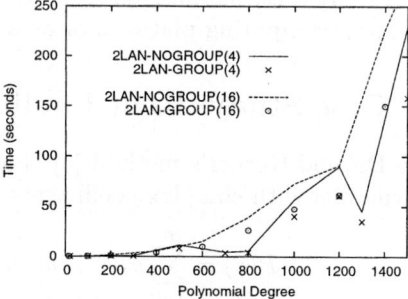

(a) Impact of the group management into JNGI

(b) Impact of similarity groups over a heterogeneous network

Fig. 1. Results of experiments.

In figure 1(b), the 2LAN-NOGROUP(4) and 2LAN-NOGROUP(16) curves correspond to the execution of our application on the 2 local networks, using 4 workers. The 2LAN-GROUP(4) and 2LAN-GROUP(16) curves show the execution of our application on the 2 local networks, using 4 workers. Each group correspond to a 4-worker LAN. 2LAN-NOGROUP and 2LAN-GROUP curves illustrate the interest of managing groups on heterogeneous network architecture. When you do not manage groups, you do not have the ability to choose your target platform and you are going to have some communications above slow bandwidth links which will act as a bottleneck for the application. A performance decrease of around 50% can be noticed. Performance cannot be achieved with such a strategy. On the contrary, managing groups according to the network bandwidth established automatically 2 groups of computers from the 2 local networks. Then, the workers' allocation takes care of the architecture avoiding bottlenecks like this slow link between the 2 LANs. Network heterogeneity must be taken into account on P2P architecture to have good performance and this can be done by using similarity groups.

4 Conclusion

Similarity groups have proved to be efficient in the JNGI project but they can be used in any other P2P computing architecture. These groups allow to take into account hardware and software heterogeneity. They can be used to classify computing platforms according to either qualitative or quantitative parameters. The efficiency of qualitative groups is difficult to measure wheareas the efficiency of quantitative groups has been easily shown. Experiments have pointed out that similarity groups allow to structure the P2P architecture. Network heterogeneity which can act as a bottleneck in P2P computing can be softened by using similarity groups.

References

1. J. Verbeke, N. Nadgir, G. Ruetsch, and I. Sharapov, "Framework for peer-to-peer distributed computing in a heterogeneous, decentralized environment," in *Proceedings of GRID 2002, Baltimore*, Sun Mic. Inc.,Palo Alto, CA 94303, USA, Jan 02.
2. S. Saroiu, P. K. Gummadi, and S. D. Gribble, "A measurement study of peer-to-peer file sharing systems," in *SPIE Multimedia Computing and Networking (MMCN2002)*, San Jose, CA, January 18-25 2002.
3. R. Wolski, N. Spring, and J. Hayes, "The network weather service: A distributed resource performance forecasting service for metacomputing," *Journal of Future Generation Computing Systems*, vol. 15, no. 5-6, pp. 757–768, October 1999.
4. J. Bourgeois, F. Spies, M. Zemerly, and T. Delaitre, "Chronos: A performance characterisation tool inside the EDPEPPS toolset," *Journal of Supercomputing*, vol. 15, no. 2, pp. 123–140, February 2000.
5. R. Couturier, P. Canalda, and F. Spies, "Iterative algorithms on heterogeneous network computing: Parallel polynomial root extracting," in *9th International Conference on High Performance Computing (HiPC)*, ser. Lecture Notes in Computer Science, vol. 2552. Bangalore, India: Springer, December 2002.

Improving the Scalability of Logarithmic-Degree DHT-Based Peer-to-Peer Networks*

Bruno Carton[1] and Valentin Mesaros[2]

[1] CETIC, rue Clément Ader 8, B-6041 Charleroi, Belgium
bruno.carton@cetic.be
[2] Université catholique de Louvain, place Sainte Barbe 2,
B-1348 Louvain-la-Neuve, Belgium
valentin@info.ucl.ac.be

Abstract. High scalability in Peer-to-Peer (P2P) systems has been achieved with the emergence of the networks based on Distributed Hash Table (DHT). Most of the DHTs can be regarded as exponential networks. Their network size evolves exponentially while the minimal distance between two nodes as well as the routing table size, i.e., the degree, at each node evolve linearly or remain constant. In this paper we present a model to better characterize most of the current logarithmic-degree DHTs. We express them in terms of *absolute* and *relative* exponential structured networks. In relative exponential networks, such as Chord, where all nodes are reachable in at most H hops, the number of paths of length inferior or equal to H between two nodes grows exponentially with the network size. We propose the `Tango` approach to reduce this redundancy and to improve other properties such as reducing the lookup path length. We analyze `Tango` and show that it is more scalable than the current logarithmic-degree DHTs. Given its scalability and structuring flexibility, we chose `Tango` to be the algorithm underlying our P2P middleware.

1 Introduction

Over the past few years, Peer-to-Peer (P2P) networks have become an important research topic due to their interesting potentials such as self-organization, decentralization and scalability. A P2P network is principally characterized by its structuring policy and the lookup protocol employed. Not long after the emergence of the first popular P2P networks, Napster and Gnutella, it was realized that scalability in these networks was an important issue. A better alternative are the P2P networks based on DHT (Distributed Hash Table). These networks are self-organized, fully distributed and highly scalable. Furthermore, given that each node has a well defined routing table, the lookup for any node/item can be accomplished within a relatively small number of hops. As the network size increases *exponentially*, the maximum lookup length as well as the routing table size at each node (i.e., the degree) increase *linearly* like in Chord [1], Pastry [2] and Tapestry [3], or even remain constant like in Koorde [4] and DH [5].

---

* This work was funded at CETIC (www.cetic.be) by the Walloon Region (DGTRE) and the E.U. (ERDF and ESF), and at UCL by the Information Society Technologies programme of the European Commission, Future and Emerging Technologies under IST-2001-33234 PEPITO.

The DHT based P2P networks are also called *structured networks*, since they follow a well defined structure. A closer look to their structure allowed us to notice that most of the logarithmic-degree DHTs fall into two main categories, depending on the nodes' view of the network (we defer the definition of node's view to Section 2). We call them *absolute* and *relative structured exponential networks*. A first contribution of this paper is the description of a model to better characterize the exponential structured networks as absolute and relative. Related to this work is the research described in [6] where a model based on the concept of k-ary search is proposed for reasoning about DHT networks. Their model addresses only relative structured exponential networks, while ours is more general, addressing the absolute networks, too.

Our model allowed us to observe that in the relative exponential structured networks the fingers of a node are not totally exploited. Hereinafter we denote the "fingers" of a node n to be the single-hop connections of n, and hence representing the entries in the routing table of node n. In Section 3 we propose an approach, that we called Tango, to structure the relative exponential networks for increasing their scalability. Tango reduces the redundancy in the multiplicity of paths between two nodes of a relative exponential network and, as such, it reduces the path length between the nodes. The Tango approach is the second and the main contribution of this paper. In Section 4 we compare Tango with DKS [7], and with the DH constant-degree network.

2 Structured Exponential Network

A structured exponential network is a network built incrementally using well-defined steps. It is composed of nodes linked together via directed edges according to structuring rules, and characterized by an exponential factor k which is the number of instances of network Net_i used to define the subsequent network Net_{i+1}. The network Net_1 is the initial network composed of one node. At step i, network Net_i is built by using k instances of network Net_{i-1} linked to one another.

We identify two methods for connecting all k instances of Net_{i-1} at the ith step: absolute and the relative connections. They lead to absolute and relative structured exponential network, respectively. We illustrate both methods for a network of size 64, built in four steps, and parameterized by an exponential factor $k = 4$. Each node is identified both numerically by using a unique identifier ranging from 0 to 63, and graphically by using k shapes (i.e., light square, light circle, bold square and bold circle). The shape organizes the nodes within the network whereas the size of the shape determines the network building step. Small shapes stand for instances of Net_1, medium shapes for instances of Net_2, and large shapes for instances of Net_3. The network instance of Net_4 regroups the four network instances of Net_3. However, for simplicity, Net_4 is not marked in the figures. In order to distinguish the fingers of the reference node from the other nodes, we represent them as non-gray numbers whereas the other nodes are in gray. Moreover, we introduce the $\oplus$ and the $\ominus$ operators. In a network of size S, we define the operators as $m \oplus n = (m+n) \mod S$, and $m \ominus n = (m-n+S) \mod S$.

An absolute structured exponential network is represented in Figure 1 (left). In such a network, each node has the same view of the network. For instance, all nodes see that nodes ranging from 0 to 15 are sitting in the large light square. That is, if a node sees that a node m is sitting in a given shape then all the nodes see that m is sitting

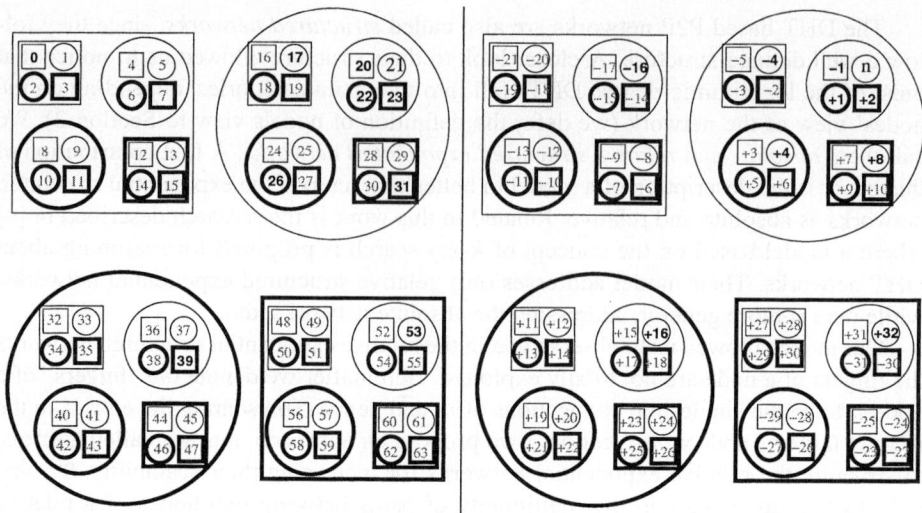

Fig. 1. (left) Absolute structured exponential network of size 64 with $k = 4$. (right) View of a node n in a relative structured exponential network of size 64 with $k = 4$.

in that given shape. In such a network, at the ith step, the $k - 1$ fingers of a node n are pointing to the $k - 1$ other instances of Net_{i-1}. Moreover, it does not matter to which node inside each Net_{i-1} n points to. For instance, in the network represented in Figure 1 (left), the fingers of node 21 at the third step can be any instance of nodes a,b,c where $a \in [16 \dots 19]$, $b \in [24 \dots 27]$, and $c \in [28 \dots 31]$.

A relative structured exponential network differs from an absolute one by the fact that the view of the network owned by a particular node is relative to its position within the network. For instance, nodes sitting in the large light square are found at distance $dist$ from the reference node, with $-21 \leq dist \leq -6$. Moreover, in a relative exponential network, a node n has to point precisely to the nodes occupying relatively the same positions in the $k - 1$ other instances of Net_{i-1}. For instance, as represented in Figure 1(right), the fingers of node n at the third step are $n \ominus 16$, $n \oplus 16$, $n \oplus 32$.

Most logarithmic-degree DHT-based P2P networks can be expressed either in terms of an absolute or in terms of a relative structured exponential network. For instance, Pastry and Tapestry can be seen as instances of the absolute structured exponential network by instantiating the employed alphabet to the shapes used in Figure 1. On the other hand, Chord and DKS can be seen as instances of the relative structured exponential network.

This model allows us to state that networks built with the relative and the absolute approaches scale at the same rate. Indeed, let S_i be the size of network Net_i and H_i be the maximum number of hops to reach any node in Net_i. Then, for both structures we have $S_i = k * S_{i-1}$ with $S_1 = 1$, $H_i = i - 1$, and a number of $(k-1) * (i-1)$ fingers at each node. Moreover, this model allows us to state that if at the ith step, a node n points to node m, then in an absolute network, the networks reachable in at most i hops by n and m, using all the fingers established in the first i steps, are identical while they differ in a relative network. This difference is at the foundation of the Tango definition and its propention to increase finger utilization.

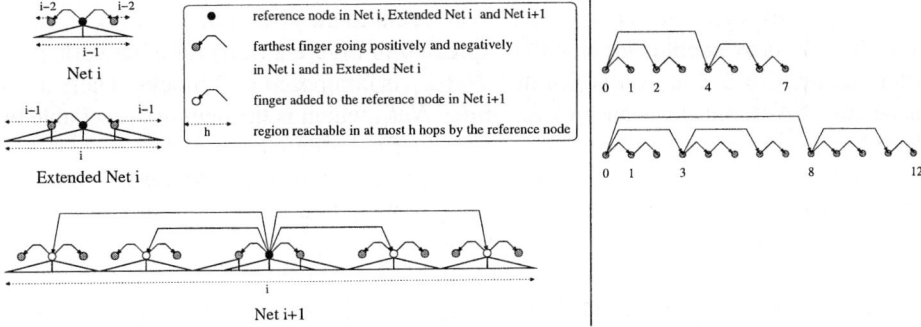

Fig. 2. (left) Network building pattern in Tango where $k = 5$. (right) Paths from node 0 to all the other nodes in a Chord network of size 8 and in a Tango network of size 13.

3 Tango: A Novel Approach for Reducing Unexploited Redundancy

In a relative exponential network we can identify two types of redundancy. The first one results from the commutative property of the addition operation and from the fact that each node owns, relatively, the same fingers. For example, in Chord, node 0 can reach node 6 via node 4 (6=0+4+2) and also via node 2 (6=0+2+4). The second type of redundancy results from the underutilization of fingers.

To have a clear explanation, we introduce the notion of positive and negative regions of a given node n. A node m is found in the positive region of node n iff $m \ominus n < n \ominus m$, otherwise, node m is found in the negative region of node n.

We propose Tango, an approach to address the second type of redundancy, and thus increasing network scalability by taking into account that the networks reachable in at most i hops by n and its fingers added at step i, using all the fingers established in the first i steps, are different in a relative network. Indeed, the region covered in at most i hops via the farthest finger added in the positive (resp. negative) region at step i and the region covered in at most i hops via the closest finger added in the positive (resp. negative) region at step $i + 1$ overlap partially. For example, the regions reachable in at most 3 hops by node 21 via node 29 (i.e., from 24 to 39) and via node 37 (i.e., from 32 to 47) overlap. Let a valid path between two nodes in a network instance Net_i be any path between these nodes whose length is at most $i - 1$ hops. In a relative network, all these overlap regions increase exponentially the number of valid paths between two nodes. Moreover, the cumulated size of the overlap, i.e., the amount of unexploited redundancy in an instance of Net_i grows exponentially with i.

3.1 Tango Definition

In order to prevent overlapping, the region comprised between the farthest finger added in the positive (resp. negative) region at step i and the closest finger added in the positive (resp. negative) region at step $i+1$ has to be equal to the size of the network instance Net_i. This improvement is graphically expressed in Figure 2 (left) for a network characterized by $k=5$. One can notice that Net_{i+1} is composed of 5 blocks. There are 4 instances of Net_i and 1 instance of $Extended\ Net_i$, which is the network reachable by the reference node in at most i hops by using the fingers defined in Net_i.

Let k_i^+ and k_i^- be the number of fingers added in the positive and, respectively, the negative regions of a node at step i. Hence, knowing that at each construction step i there are $k-1$ fingers added to a node, we obtain $k = k_i^+ + k_i^- + 1$.

Let $d_{i,j}^+$ (resp. $d_{i,j}^-$) be the distance at which the jth positive (resp. negative) finger of the ith step should be placed. Let S_i^+ (resp. S_i^-) be the size of the positive (resp. negative) region of a reference node at step i. Equations 1 establish the size growth and the fingers positioning in Tango. One can note that for $k_i^- = 0$, the Tango network corresponds to an improved version of Chord, and DKS with an arity k. The reader can refer to Section 4 for a comparison between Chord, DKS and Tango.

$$\begin{aligned}
&d_{2,j}^\pm = j & &j \in [1 \ldots k_2^\pm] & &S_1^\pm = 0 &\\
&d_{i,j}^\pm = d_{i,j-1}^\pm \pm S_{i-1} & &j \in [1 \ldots k_i^\pm],\ i > 2 & &S_i^\pm = S_{i-1}^\pm + d_{i,k_i^\pm}^\pm & i > 1 \quad (1)\\
&d_{i,0}^\pm = d_{i-1,k_{i-1}^\pm}^\pm & &i > 2 & &S_i = S_i^+ + S_i^- + 1 & i > 0
\end{aligned}$$

3.2 Key-Based Routing

The purpose of key-based routing is to route a message tagged with key Key to the node responsible of Key. Let p_n^+ (resp. p_n^-) be the first node encountered in the positive (resp. negative) region of n. The responsibility of a node n is defined in Equation 2.

Beside the node responsibility, there is the finger responsibility defining the node to which a message should be forwarded to. In Tango we split the finger responsibility of a given finger F in negative and positive sides[1]. Than, let the focused network be an instance Net_l and let $Sp_{i,j}^\pm$ (resp. $Sn_{i,j}^\pm$) be the sizes of the positive (resp. negative) finger responsibility as defined in Equations 3 and 4. The finger responsibility $R_{i,j}^\pm$ of finger located at position $P_{i,j}^\pm$ related to the distance $d_{i,j}^\pm$ are defined in Equation 5. Hence, by using its finger $F_{i,j}$, a node can cover the region $R_{i,j}$ in at most $i-1$ hops.

$$R_n = \left[n \ominus \left(\left\lfloor \frac{k_2^+}{k-1} \right\rfloor * (n \ominus p_n^- \ominus 1) \right) \ldots \left(\left\lceil \frac{k_2^-}{k-1} \right\rceil * (p_n^+ \ominus n \ominus 1) \right) \oplus n \right] \quad (2)$$

$$Sp_{i,j}^- = S_i^- \ ;\ Sp_{l,k_l^+}^+ = S_i^+ \ ;\ Sp_{i,k_i^+}^+ = S_{i+1}^+ \ ;\ Sp_{i,j}^+ = S_i^+ \quad (3)$$

$$Sn_{i,j}^+ = S_i^+ \ ;\ Sn_{l,k_l^-}^- = S_i^- \ ;\ Sn_{i,k_i^-}^- = S_{i+1}^- \ ;\ Sn_{i,j}^- = S_i^- \quad (4)$$

$$R_{i,j}^\pm = [P_{i,j}^\pm \ominus Sn_{i,j}^\pm \ldots P_{i,j}^\pm \oplus Sp_{i,j}^\pm] \quad (5)$$

---

[1] The denomination of Tango comes from its ability to have positive routing steps followed by negative routing steps and vice versa.

3.3 Tango in a Sparse and Dynamic Network

In a sparse network, the position of a finger F (i.e., P) of a node n may correspond to a missing node. In that case, n points to the node responsible of P. Hence, the nodes are playing the finger role of the missing nodes laying within their responsibility. In order to preserve the lookup efficiency, each node adapts its routing table in order to reach the same part of the network in the same number of hops as it would have been done by each missing nodes within its responsibility. That is why in Tango, we define the finger position $P(n)$ and the finger node $F(n)$ of a node n as in Equation 6, where $j \in [1 \ldots k_i^-], g \in [1 \ldots k_i^+]$ and $i \in [1..l]$.

$$\begin{aligned} P_{i,j}^-(n) &= R_n.inf \ominus d_{i,j}^- & F_{i,j}^-(n) &= m \quad s.t. \quad P_{i,j}^-(n) \in R_m \\ P_{i,g}^+(n) &= R_n.sup \oplus d_{i,g}^+ & F_{i,g}^+(n) &= m \quad s.t. \quad P_{i,g}^+(n) \in R_m \end{aligned} \quad (6)$$

To deal with the dynamics in a Tango network, the algorithms of join, fault tolerance and correction on use defined in DKS can be applied directly to Tango. Moreover, due to the symmetry provided by the Tango networks featured with $k^+ = k^-$, the correction on use can be made more efficient. For more details, the reader can refer to [8].

4 Analysis

In this section we shortly compare Tango with DKS, and with the Distance Halving constant-degree network. For more details, the reader can refer to [8].

4.1 Tango vs. DKS

DKS generalizes Chord to allow a tradeoff between the maximum lookup length in the network (i.e., the diameter) and the size of the routing table at each node (i.e., the degree). The structure of DKS characterized by $k = 2$ is the same as the one of Chord. Tango also supports the tradeoff between the diameter and the degree. Moreover, the network covered with Tango is much larger than the network covered with DKS and Chord, while keeping the same network diameter and the same degree at a node. That is, in Tango the exponential factor is bigger than in Chord and DKS. From Equation 1, one can deduce the size of the network covered with Tango at a step $i > 2$ together with the roots (i.e., z_1 and z_2) of its characteristic equation. For a step i sufficiently high, the exponential factor in Tango can be approximated to z_1, where $k < z_1 < k+1$, and thus we obtain $S_i \approx z_1^{i-1}$. Although, the search cost in Tango is $O(logN)$, for apprx. the same network size the highest search cost in Tango is 75% of the one in Chord.

$$S_i = (k+1) * S_{i-1} - S_{i-2} \quad z_1 = \frac{k+1+\sqrt{(k+1)^2-4}}{2}, z_2 = \frac{k+1-\sqrt{(k+1)^2-4}}{2} \quad (7)$$

$$S_i = k^{i-1} + \sum_{j=1}^{i-2} k^{i-j-2} * (d_{j,k_j^+}^+ + d_{j,k_j^-}^-) \quad i > 2 \quad (8)$$

In order to compare the size growth in Tango and DKS, one can define the network size covered by Tango at the ith construction step as in Equation 8. Note that the first

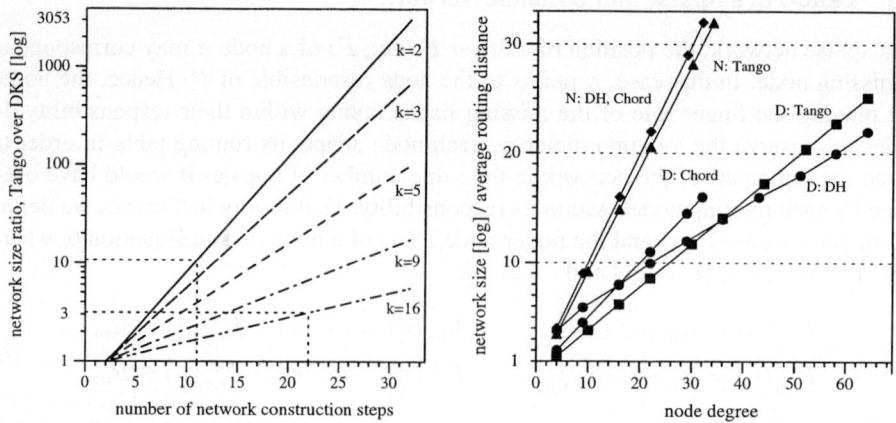

Fig. 3. (left) Ratio between the network sizes covered by Tango and DKS, with the same number of fingers at different construction steps. (right) Network size [N] and average routing distance [D] of Tango, DH and Chord, with respect to different values of node degree.

term of the equation corresponds to the network size covered by DKS at the ith step, i.e., k^{i-1}. The second term, which also increases exponentially, corresponds to the difference between the two network sizes; it actually represents the cumulated unexploited redundancy in DKS. In Figure 2 (right) we present an example of how Tango covers a larger network than Chord (DKS, $k = 2$) even at the very early building steps. With a routing table of size 3, a node in Chord can cover a network of size 8 in 3 hops, whereas in Tango, in 3 hops, a node can cover a larger network, i.e., of size 13.

To better understand the relation between Tango and DKS, in Figure 3 (left) we plotted the ratio between the network sizes covered in Tango and DKS at each construction step ranging from 1 to 32, for five different values of k. One can note that for a given k, the ratio between the network sizes is growing exponentially at each step. It is also interesting to note that the growth ratio of the ratio decreases as k increases. However, since increasing k leads to increasing the resource consuming and the maintenance cost, it is likely that relative small values of k will be employed.

4.2 Tango vs. Constant-Degree Networks

A constant-degree network is a network whose size can increase exponentially, while the node degree remains fixed and the diameter increases logarithmically. Some examples are those based on the de Bruijn graph, such as Koorde and DH. In our analysis we were interested in the average routing distance and the network size for Tango ($k = 3$) and DH with respect to different node degrees. We also plot them for Chord to have a third party reference. To compute the average routing distance for DH we used the μ_d formula for de Bruijn graphs given in [9] and doubled it to achieve load balancing, as suggested in [5]. As shown in Figure 3 (right), for the same node degrees (inferior to 34), and almost the same network size, Tango provides lower values for the average routing distance than DH.

5 Conclusion

First, in this paper we presented a model to better characterize the structure of the current logarithmic-degree P2P exponential structured networks, such as Tapestry, Pastry, Chord and DKS, in terms of absolute and relative exponential structured networks.

On the other hand, we proposed the Tango approach to better structure the relative exponential networks to increase their scalability by exploiting the redundancy in the lookup paths. We showed that Tango is more scalable than the current logarithmic-based DHTs. We analyzed the structure of Tango with respect to the one of DKS and, implicitly, to the one of Chord. Particularly, we observed that, for small values of the exponential factor k, Tango is much more scalable than DKS (and Chord), while for big values of k the scalability of the two networks is more comparable. However, since increasing k leads to increasing the resource consuming and the maintenance cost, it is likely that relative small values of k will be employed. We also analyzed Tango with respect to DH, a constant-degree network. We observed that, for networks with relative large node degrees, the average routing distance in Tango and DH are comparable.

Given its structuring flexibility and its scalability potential, we chose Tango to be the algorithm underlying our recently released P2P middleware [10], and demo applications: PostIt and Matisse [10]. As future work, we plan to address the redundancy in Tango resulting from the commutative property of the finger addition operation.

Acknowledgments

We thank Peter Van Roy and Kevin Glynn from UCL, Belgium, and Luc Onana from KTH, Sweden, for their constructive comments.

References

1. I. Stoica, R. Morris, D. Karger, F. Kaashoek, and H. Balakrishnan. Chord: A Scalable Peer-To-Peer Lookup Service for Internet Applications. In *ACM SIGCOMM*, August 2001.
2. A. Rowstron and P. Druschel. Pastry: Scalable, Decentralized Object Location, and Routing for Large-Scale Peer-to-Peer Systems. In *ICDSP*, November 2001.
3. B. Zhao, J. Kubiatowicz, and A. Joseph. Tapestry: An Infrastructure for Fault-tolerant Wide-area Location and Routing. Technical Report CSD-011141, U.C. Berkeley, April 2001.
4. F. Kaashoek and D. Karger. Koorde: A Simple Degree-optimal Hash Table. In *IPTPS*, February 2003.
5. M. Naor and U. Wieder. Novel Architectures for P2P Applications: the Continous-Discrete Approach. In *ACM SPAA*, June 2003.
6. S. El-Ansary and L. Onana et al. A Framework for Peer-to-Peer Lookup Services based on k-ary Search. Technical Report TR-2002-06, SICS, May 2002.
7. L. Onana and S. El-Ansary et al. DKS(N, k, f): A Family of Low Communication, Scalable and Fault-Tolerant Infrastructures for P2P Applications. In *CCGRID2003*, May 2003.
8. B. Carton, V. Mesaros, and P. Van Roy. Improving the Scalability of Logarithmic-Degree Peer-to-Peer Networks. Technical Report RR-2004-01, UC-Louvain, January 2004.
9. D. Loguinov and A. Kumar et al. Graph-Theoretic Analysis of Structured Peer-to-Peer Systems: Routing Distances and Fault Resilience. In *SIGCOMM*, August 2003.
10. P2PS v 1.0, Peer-to-Peer System Library, October 2003. Université catholique de Louvain, and CETIC, Belgium. www.mozart-oz.org/mogul/info/cetic_ucl/p2ps.html.

A Small World Overlay Network for Resource Discovery

Koen Vanthournout, Geert Deconinck, and Ronnie Belmans

Katholieke Universiteit Leuven,
Department Electrical Engineering (ESAT), Belgium
Koen.Vanthournout@esat.kuleuven.ac.be
http://www.esat.kuleuven.ac.be

Abstract. Interest is rising in genuinely distributed ('peer-to-peer') resource discovery systems, which all provide mechanisms for self-organization of their resource providers and requesters into an overlay network. This paper proposes such a system that uses XML description files to construct overlay networks that exhibit small world properties (i.e., a small diameter and a regular structure) and in which similar resources are grouped together. Resource users can then be linked directly to the regions with the resources of their interest, which allows exploitation of group and time locality. Simulations results validate these concepts.

1 Introduction

Resource discovery (RD) is an indispensable enabling technology for networked distributed systems. As such, a multitude of existing technologies provide the required functionalities and these can be divided into roughly two clusters: third-party systems and genuinely distributed systems. For the first type, a separate entity, independent from the resource providers and requesters, gathers references to resources and solves resource queries (e.g., LDAP [6], Matchmaking [8], etc.). Genuinely distributed systems, on the other hand, are embedded in the users (both resource providers and requesters) of the system (e.g., Gnutella [5], Freenet [4], CAN [9], etc.).

The field of third party RD systems is well established, in contrast to the fairly new genuinely distributed approach. The latter has distinct advantages, though:

- There is no dependency on a central coordinating authority.
- Every participant makes some resources available and, consequently, the available resources grow with the number of participants, which allows for scalable systems.
- The larger the number of participants, the larger the inherent redundancy in the system, which can be exploited to achieve increased fault-resilience.

All these genuinely distributed resource discovery systems or, so called by popular consensus, peer-to-peer systems compose of two main components: an overlay network and a query resolving technique. A closer look at these components of the peer-to-peer RD systems reveals a number of short-comings. Gnutella [5], by far the most spread system, deploys a random overlay structure and uses flooding to propagate queries, both being inefficient. Freenet [4] nodes are (initially) also gathered in a random structure, but the use of local caches of files with similar keys and single query forwarding to the neighbor node with the closest matching keys, allows convergence of a Freenet network to a more efficient structure. But it can only be used for files and not for other (non-copyable) resources, e.g., processing servers. A third type of systems (CAN [9, 12], Chord [11], Pastry [10], Tapestry [17] and P-Grid [2]) uses regular overlay network structures. A unique identifier is assigned to all nodes and this is then used to position the nodes on a predictable position in the network, whose structure reflects the identifier space. An ID is computed for each resource and a reference to the position of those resources is then stored at the node with the closest ID. The result is a search-optimized overlay network. But one can predict difficulties with variable resources: if the description of a resource changes, so does its identifier and, potentially, the position of its reference. This is a symptom of the fact that there is no or little correlation between the ID's and the nature and type of the resources. Functionality is not reflected in the structure of the overlay networks and, consequently, neither group locality, nor time locality [7] are exploited:

- *group locality:* Requesters can be divided into groups that tend to search for the same types of resources.
- *time locality:* The same entity may request the same or comparable resources several times within short time intervals.

Mentionable is the work presented in [15], which proposes methods to construct an overlay network that links nodes that are interested in similar documents. Their usefulness is limited, though, to file-sharing systems: it assumes that nodes interested in a certain resource (in this case, file) also provide similar resources.

Hence, there is a need for a distributed RD system that supports resources with variable properties, uses attribute naming[1] and where the nodes self-organize into a regular structure that reflects the nature of the resources [14].

The system that is proposed in this paper fills this gap. It uses a metric based on the differences between descriptions in XML of the resources, to construct an overlay network in which nodes with a comparable functionality are grouped together and in which resource users are close to the resources for which they announced an interest. The next section describes the construction of this overlay network, followed by the query forwarding strategy. Simulation results are presented at the end to illustrate the usefulness of this approach.

2 The Overlay Network

2.1 Distance Between Functionalities

A desirable property for RD overlay networks is regularity, because a regular, predictable structure will ease and optimize query forwarding. But to allow nodes

---

[1] Attribute based queries in, e.g., XML, offer a more powerful search mechanism than hash-based ID's.

to self-organize into such a regular network, some distance-value is needed: if new nodes can calculate the distance from the position they should be in the regular structure to any other node they encounter in that network, they can hop through the network, from node to node, until they arrive at their proper position, where they can form links to their surrounding neighbors. The same goes for resource queries: if a node can calculate the distance from its position to the place in the network where the query can be answered, then it can select its neighbor closest to that place and forward the query there. This is the strategy deployed by CAN, Chord, Pastry, P-Grid and Tapestry. But, as discussed before, they all deploy a regular identifier space and a unique ID per node that has no semantic correlation to the nature and type of the resources.

The system, here proposed, employs a method to calculate a distance that does reflect this: all nodes contain an XML description file with all its resources, be it files, processing power, printing services, domotica device or any other conceivable resource, and a list of their main interests; the resources they are likely to request. The distance between two nodes and, thus, between their XML description files($\mathcal{D}(xml_1, xml_2)$) is then defined as:

$$\mathcal{D}(xml_1, xml_2) = 1 - \frac{size(\mathcal{N}_{common})}{size(\mathcal{N}_{common}) + size(\mathcal{N}_{nc,xml_1}) + size(\mathcal{N}_{nc,xml_2})}$$

with $\mathcal{N}_{common}$ the set of common XML nodes, where a XML node is defined as common iff the node and all its ancestors exist in both files. Text and values can only be contained by leaf nodes and count as an separate node for the purpose of the distance metric. Attributes are ignored. *, which yields an automatic match, is allowed for text fields in interest and query XML (see Sections 2.4 and 3). $\mathcal{N}_{nc,xml_n}$ is the set of non-matching nodes for XML file n.

An example description file of a server offering processing cycles:

```
<entityDescription>
  <description>
    <IntelligentDevice>
      <static>
        <intelligentDeviceType>
          <computationDevice>
            <processor/>
          </computationDevice>
        </intelligentDeviceType>
      </static>
      <dynamic>
        ...
      </dynamic>
      <ComputationDevice>
        <Processor>
          <static>
            <flopsPerSecond>8200</flopsPerSecond>
            <OS>LINUX</OS>
            <architecture>i486</architecture>
            <memoryInGb>12</memoryInGb>
          </static>
          <dynamic>
            <loadInProcent>46</loadInProcent>
          </dynamic>
        </Processor>
      </ComputationDevice>
    </IntelligentDevice>
  </description>
</entityDescription>
```

This metric is at the same time the novelty, the strength, but also the weakness of this RD system: it guarantees the construction of a semantic overlay network, but ill-constructed XML will deteriorate that very construction. As such, semantic gossiping [1], which is designed to overcome this type of problems, would be very useful. Secondly, the above described metric is simple and lean and serves the purpose of concept-proof well, though a real-life system will require a more robust metric with more stringent syntax rules and an expanded set of masks ($<$, $>$, !=, etc.). But that is beyond this paper's scope.

2.2 Companions

The companions of a node are the nodes that offer the most similar resources: the distance metric, applied on the resource descriptions only, yields the smallest results. Every node looks for a fixed number of companions (this number defines the dimensionality of the overlay network) and these companions indicate the correct position of that node in the network.

When a new node enters the system, the only initialization needed is the address of an active node of the network. To this initial neighbor, it transmits a companion request, which contains the requester's XML description, the highest distance metric of its current neighbors (the threshold), the addresses of its current neighbors and a hop limit. A node that receives such a request checks if it knows any nodes with a lower distance metric result for the requester than the threshold, that are not in the neighbor list attached to the request. If so, it replies their addresses to the requester. If the hoplimit has not been reached, the request is forwarded to all nodes that matched the request, if any. Else a single message is send to the neighbor with the lowest metric result. If a message is received for a second time, it is dropped.

The requester listens to answers for a fixed time, after which it selects its new neighbors from the replies it received. It announces the construction of the new links to those new neighbors and then repeats the whole cycle by retransmitting a companion request. Convergence is reached when it receives no answers any more: its current neighbors are its companions. Because of the link announcements, these companions can update their links to the newcomer. Furthermore, all nodes periodically restart the convergence cycle, as to cope with changes in the network's node composition.

2.3 Small Worlds and Far Links

If all nodes would only locate their companions and solely form links to those, the result would be network partitioning and, because of the high regularity, a high network diameter and characteristic path length (see Section 4). This can be avoided by the use of far links: a fraction of the nodes constructs a link to the node that yields the worst result for the distance metric. The impact of this on the network's regularity is minor, but the network's diameter decreases significantly [3, 16]: a small world topology emerges, which combines the best of random networks (a low diameter) and regular networks (a predictable structure). And it is an effective prevention against network partitioning. Their construction is equal to the construction of companion links, but with word 'lowest' replaced by 'highest' and vice versa.

2.4 Resource Providers and Resource Users

The above-described links will result in an overlay network that reflects the functionality: nodes are clustered by functionality and, as a result, the answers to queries for comparable resources are to be found in the same region. Next optimization is to link resource requesters to the regions of their interest. Each resource requester constructs a number of links (predetermined) to the nodes whose descriptions match its expressions of interests best. This is realized by using the same convergence method as described in Section 2.2, but the metric now compares the XML expressions of interest of the requester to the description of the resources of the potential recipients of these links.

2.5 Orphans

One problem emerges after above links are formed. If a node deviates from the bulk of the nodes it resembles, then it might become an orphan, i.e., it does link to the nodes of its kind, but it is too different for other nodes to refer to it (if there happens to be no far link or resource user link to it) and thus the node becomes unreachable. The solution to this is to have each node also link to the last nodes that announced their link and to which this node has no link already (announcements are periodically repeated). Experiments show that one such anti-orphan link per node eliminates the existence of orphans, though a node can still be temporary unreachable if a new node, in the process of converging to its proper spot, traverses the network and 'steals' anti-orphan links on its way.

3 Resource Queries

A query consists of the address of the requester, the XML file that contains the query and a hoplimit. A query XML file contains the description of a resource, possibly with masks. Only fully matching resources (distance metric is zero) yield a match.

When a node receives a query, then it checks if itself or any of its neighbors matches the XML file. If so, a query answer is send to the requester. If not and if the hoplimit is not yet reached, the query is forwarded to the closest matching neighbor.

4 Results

In order to optimize and evaluate the system, it was simulated on top of an in-house developed network simulator, for various numbers of nodes. Nodes are connected in random sequential order to the network with random time slots between the additions. They are initially connected to a random active node. After all nodes have been added, the simulation continues for sufficient time to allow all nodes to converge. Then following metrics are calculated:

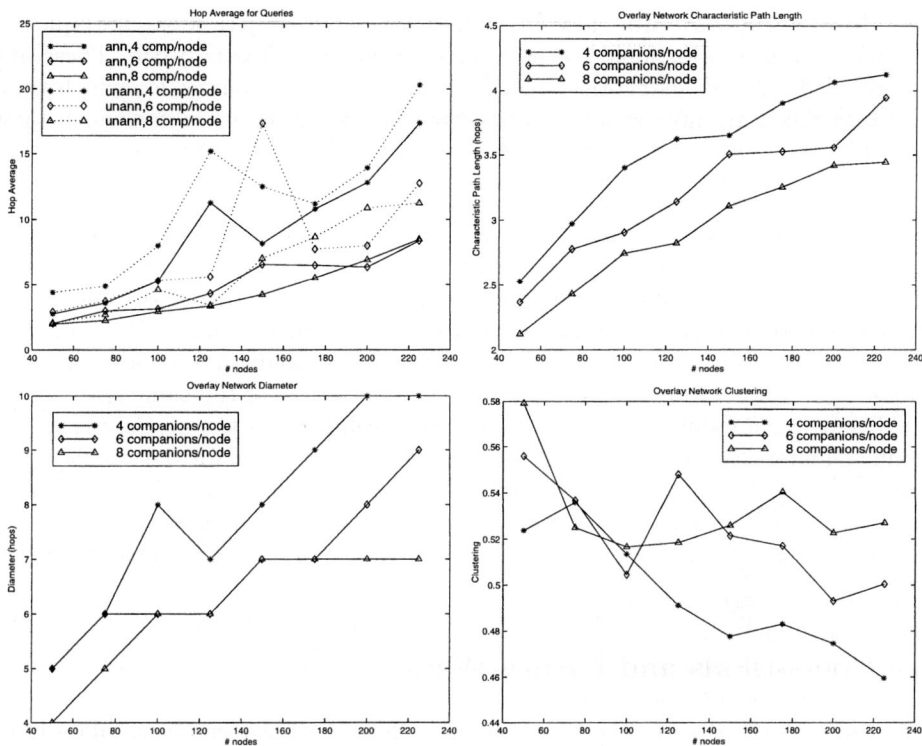

Fig. 1. The average number of hops per queries, characteristic path length, diameter and clustering for overlay networks with different numbers of companions per node (ann: the node announced an interest for the query's class of resources, unann: no interest was announced). 50% of the nodes form a far link and all nodes construct one anti-orphan link.

- **Clustering** (C) [16,3]: A node i has k_i edges to k_i other nodes. If this node and its neighbors would be part of a clique, there would be $k_i(k_i - 1)$ edges between them (for a directed graph). E_i is the number of edges that actually exist between these k_i nodes. The clustering coefficient of node i is then: $C_i = E_i/(k_i(k_i - 1))$. The clustering coefficient of the whole network is the average of all individual C_i's. It is an indication of how regular the network is.

- **Characteristic path length** (L) [16]: The characteristic path length L of a graph G is the average of the number of edges in the shortest path between all pairs of vertices. L is lowest for random networks and high for perfect regular structures. It is a property of small world networks to have a high C and at the same time a low L, which means that they possess the positive properties of both regular and random networks.

- **Diameter** (D): The diameter of a graph is equal to the longest shortest path in the graph.

- **Hop average for queries:** Every network is tested with a batch of queries with a size proportional to the number of nodes in the network. These are divided into two categories: queries from nodes for resources for which they announced an interest and queries from nodes for resources for which they did not. The average number of hops a query travels before an answer is found, evaluates the network's efficiency.

Figure 3 gives these metrics for networks with 4, 6 and 8 companions per node. Less than four companions per node results in networks with deficient properties for larger numbers of nodes. The results show that indeed small world networks are constructed: the clustering coefficient stays between 0.46 and 0.58, while the characteristic path length stays as low as 4.12 for 4 companion per node networks of size 225 and 3.44 for 8 companion per node networks of the same size.

The average number of hops a query travels illustrates the usefulness of announcing interests. Only three clusters of resources were used for the simulations, but already a clear profit is made, as can be seen in the larger average number of hops needed for queries to unannounced resource than for queries to announced resources. As the diversity of the resources rise, more distinct results can be expected.

5 Conclusions and Future Work

The use of XML description files and the differences between those, allows for the construction of small world networks in which nodes with similar files are clustered together. This can be used for distributed resource discovery systems that exploit group and time locality: resource users can be linked directly to regions where the resources of their interest are located. Simulations confirmed the validity of this concept. Further simulations with larger numbers of nodes and investigation of the convergence algorithms, used for the positioning of nodes and for query forwarding, will provide additional information on scalability. The effects of node and message failures have been studied and are presented in [13].

Acknowledgements

This work is partially supported by the K.U.Leuven Research Council (GOA/ 2001/04) and the Fund for Scientific Research - Flanders through FWO Krediet aan Navorsers 1.5.148.02.

References

1. K. Aberer, P. Cudré-Mauroux, and M. Hauswirth. A framework for semantic gossiping. *SIGMOD Rec.*, 31(4):48–53, 2002.
2. K. Aberer, A. Datta, and M. Hauswirth. Efficient, self-contained handling of identity in peer-to-peer systems. *IEEE Transactions on Knowledge and Data Engineering*, 16(7):858–869, Jul 2004.

3. Réka Albert and Albert-Lásló Barabási. Statistical mechanics of complex networks. *Reviews of Modern Physics*, 74:47–97, Jan 2002.
4. I. Clarke, O. Sandberg, et al. Freenet: A distributed anonymous information storage and retrieval system. *LNCS*, 2009:46–66, 2001.
5. Gnutella. The gnutella protocol specification. http://rfc-gnutella.sourceforge.net.
6. T. Howes and M. Smith. Rfc 1823: The ldap application program interface. http://www.faqs.org/rfcs/rfc1823.html, Aug 1995.
7. A. Iamnitchi, M. Ripeanu, and I. Foster. Locating data in (small-world?) peer-to-peer scientific collaborations. In *1st Int'l Workshop on Peer-to-Peer Systems (IPTPS'02)*, Cambridge, MA, Mar 2002.
8. Rajesh Raman, Miron Livny, and Marvin Solomon. Matchmaking: Distributed resource management for high throughput computing. In *7th IEEE Int'l Symposium on High-Performance Distributed Computing*, 1998.
9. S. Ratnasamy, P. Francis, et al. A scalable content addressable network. In *Proc. of ACM SIGCOMM*, pages 161–172, San Diego, USA, Aug 2001.
10. A. Rowstron and P. Druschel. Pastry: Scalable, decentralized object location, and routing for large-scale peer-to-peer systems. In *IFIP/ACM Int'l Conference on Distributed Systems Platforms*, pages 329–350, Heidelberg, Germany, 2001.
11. Ion Stoica, Robert Morris, David Liben-Nowell, et al. Chord: A scalable peer-to-peer lookup protocol for internet applications. *IEEE/ACM Transactions on Networking*, 11(1):17–32, Feb 2003.
12. C. Tang, Z. Xu, and S. Dwarkadas. Peer-to-peer information retrieval using self-organizing semantic overlay networks. In *Proc. of the 2003 conference on Applications, Technologies, Architectures, and Protocols for Computer Communications*, pages 175–186. ACM Press, 2003.
13. Koen Vanthournout, Geert Deconinck, and Ronnie Belmans. Building dependable peer-to-peer systems. In *Twin Workshops on Architecting Dependable Systems (WADS 2004)*, Florence, Italy, Jun 2004. Accepted for publication and presentation.
14. Koen Vanthournout, Geert Deconinck, and Ronnie Belmans. A taxonomy for resource discovery. *Lecture Notes in Computer Science*, 2981:78–91, Mar 2004.
15. S. Voulgaris, A.-M. Kermarrec, L. Massoulie, and OTHERS. Exploiting semantic proximity in peer-to-peer content searching. In *Proc. 10th IEEE Int'l Workshop on Future Trends in Distributed Computing Systems (FTDCS)*, May 2004.
16. Duncan J. Watts and Steven H. Strogatz. Collective dynamics of 'small-world' networks. *Nature*, 393:440–442, 1998.
17. Ben Y. Zhao, Ling Huang, Jeremy Stribling, et al. Tapestry: A resilient global-scale overlay for service deployment. *IEEE Journal on Selected Areas in Communications*, 22(1):41–53, Jan 2004.

Author Index

Aberer, K. 1013
Agrawal, G. 372
Akashi, Y. 644
Albert, E. 21
Aldinucci, M. 506, 596, 638
Alfaro, F.J. 873
Allard, J. 497
Almási, G. 833
Alonso, J.M. 444, 487
Alonso, M. 882
Amato, N. 803
Ammirati, P. 477
Anckaert, B. 284
Andersson, J. 263
Anshus, O.J. 163
Antoniu, G. 1038
Araki, T. 399
Araujo, R. 1032
Arbenz, P. 476
Archer, C. 833
Ashby, T.J. 654

Bader, D.A. 988
Badia, R.M. 183
Bagdanov, A.D. 752
Bal, H.E. 578
Baldo, L. 214
Bandyopadhyay, S. 856
Bastoul, C. 272
Beivide, R. 900, 908
Belmans, R. 1068
Benkner, S. 271
Bermúdez, A. 864
Berrayana, S. 846
Bilò, V. 949
Bilas, A. 832
Bjørndalen, J.M. 163
Bode, A. 38
Bongo, L.A. 163
Böszörményi, L. 776
Bougé, L. 1038
Boukerche, A. 1032
Bourgeois, J. 1056
Busch, C. 820

Cáceres, E.N. 828

Campa, S. 638
Caniou, Y. 1048
Cao, J. 916
Carton, B. 1060
Carvalho, S. 254
Casado, R. 864
Casanova, H. 230
Castaños, J.G. 220, 833
Catthoor, F. 206
Cazorla, F.J. 535
Chaudron, M. 579
Choi, J.H. 460
Choi, T.U. 957
Christos, K. 803
Chuang, P.-f. 541
Chung, K.-D. 941, 957
Clematis, A. 477
Cohen, A. 292
Colajanni, M. 1013
Cole, M. 37
Coppola, M. 381, 638
Corbalán, J. 238
Cores, F. 792
Cotronis, Y. 436
Coulaud, O. 90
Cox, S.J. 972
Cristal, A. 9
Cunha, J.C. 38, 964

Da Costa, G. 55
D'Agostino, D. 477
D'Ambra, P. 697
da Silva, F.A.B. 254
Damevski, K. 108
Das, S.K. 916
Dash, M. 363
Davis, K. 689
de Alfonso, C. 487
de Amorim, C.L. 784
De Bosschere, K. 82, 284
De Bus, B. 284
Deconinck, G. 1068
Dehne, F. 828
De La Luz, V. 507
Demirkiran, I. 304

Author Index

de Pinho, L.B. 784
DeRose, L. 39, 141
Desprez, F. 230
De Sutter, B. 284
Di Giacomo, M. 391
Di Martino, B. 271
di Serafino, D. 962
Doğan, A. 468
Donatiello, L. 117
Dongarra, J. 47
Duato, J. 832
Dussere, M. 90

Ebcioğlu, K. 506
Ekanadham, K. 141
Eltayeb, M. 468
Erciyes, K. 933
Eres, H. 972
Ericsson, M. 263
Ernst-Desmulier, J.B. 1056
Erway, C.C. 833
Esnard, A. 90
Evripidou, P. 561

Fagni, T. 347
Fang, L. 1
Faure, E. 846
Feautrier, P. 272
Fernandes, L.G. 214
Fernandez, E. 535
Fernandez, J. 689
Ferrari, G. 916
Fischer, S. 606
Fladrich, U. 726
Floros, E. 436
Fontanelli, R. 980
Frachtenberg, E. 689
Fraňa, K. 726
Franke, H. 460
Fürlinger, K. 624

Gámez, J.A. 891
Gannon, D. 1
Gao, G. 336
García, G. 487
Gaydadjiev, G. 571, 760
Genius, D. 846
Gerndt, M. 133, 155
Gianuzzi, V. 477
Giersch, A. 246

Girbal, S. 292
Glimcher, L. 372
Goldschmidt, B. 776
Gomes, C. 964
Gorlatch, S. 606, 614
Goscinski, A.M. 578
Gouranton, V. 497
Granat, R. 742
Gray, W.A. 409
Gregg, D. 318
Gregorio, J.A. 900
Grenholm, O. 551
Gropp, W. 833
Grundmann, R. 726
Guerra, C. 988
Guerri, D. 980

Hagemann, M. 624
Hagersten, E. 551
Hameurlain, A. 346
Hansen, J. 804
Haridi, S. 1013
Heidelberger, P. 833
Hempel, R. 476
Hermenegildo, M. 21
Hernández, P. 792
Hernández, V. 444, 487
Herrera, J. 429
Hipp, M. 189
Hippold, J. 718
Hong, H.-O. 941
Houstis, E. 962
Hruschka, E.R. 254
Hu, Z. 644
Huedo, E. 429

Ibañez, P. 516
Iosevich, V. 198
Irwin, M.J. 310, 507
Iwasaki, H. 644
Izu, C. 908

Jan, M. 1038
Jeannot, E. 1048
Jiao, Z. 972
Joita, L. 409
Jones, E. 632
Jonsson, I. 742
Jul, E. 578

Kadayif, I. 304

Kågström, B. 742
Kakehi, K. 644
Kandaswamy, G. 1
Kandemir, M. 304, 310, 507
Kao, O. 751
Karl, W. 149, 506
Keahey, K. 399
Keane, A.J. 972
Kelly, P.H.J. 614
Kennedy, A.D. 654
Kereku, E. 133
Kessler, C. 614
Keyes, D. 697
Knijnenburg, P.M.W. 535
Koch, A. 328
Kodeboyina, D. 1
Koelma, D. 752
Kola, G. 452
Kolcu, I. 310
Kosar, T. 452
Kosch, H. 751
Kranakis, E. 917
Kranzlmueller, D. 38
Krishnan, S. 1
Krizanc, D. 803, 917
Kruis, J. 734
Kubale, M. 804
Kuszner, Ł. 804
Kuzmanov, G. 760
Kwok, Y.-K. 925
Kyriacou, C. 561

Labarta, J. 183, 238
Laffranchi, M. 1032
Lane, P. 399
Lauwereins, R. 206
Lecointre, L. 497
Lee, C. 398
Lee, S.-W. 941
Lettere, M. 980
Li, T. 133
Li, Z. 925
Lilja, D.J. 541
Lima, A.A.B. 355
Limet, S. 497
Liu, W. 989
Livny, M. 452
Llabería, J.M. 516, 525
Llorente, I.M. 429
López, P. 882

López-García, P. 21
Löwe, W. 263
Lombardo, S. 423
Ludwig, T. 117, 997
Luque, E. 98, 220, 792
Lysne, O. 832

Machì, A. 423
Maebe, J. 82
Magdon-Ismail, M. 820
Magini, S. 638, 679
Malony, A.D. 38, 119
Margalef, T. 98
Markatos, E. 220
Martinez, M. 391
Martínez, J.M. 882
Martorell, X. 833
Matsuzaki, K. 644
Mattoso, M. 355
Mauve, M. 606
Mavronicolas, M. 820
Medek, O. 734
Meier, H. 997
Melideo, G. 949
Melin, E. 497
Mesaros, V. 1060
Miguel-Alonso, J. 900
Miles, J. 409
Mohr, B. 39, 47, 155
Moltó, G. 444
Mongelli, H. 828
Monnet, S. 1038
Montagnat, J. 988
Montero, R.S. 429
Moore, S. 47
Morajko, A. 98
Morajko, O. 98
Morancho, E. 525
Moreira, J.E. 833
Müller, J. 606
Müller-Pfefferkorn, R. 72

Nadolski, A. 804
Nagel, W.E. 72, 117, 726
Nam, Y.J. 799
Nieplocha, J. 632
Nisbet, A.P. 318
Numrich, R.W. 632

O'Boyle, M.F.P. 271, 654

Olivé, À. 525
Orlando, S. 346
Özer, E. 318
Özgüner, F. 468
Ozturk, O. 310

Pakin, S. 64
Parashar, M. 173
Park, C. 799
Park, S.-H. 941
Park, Y.-W. 941
Parker, S. 108
Parmentier, G. 1005
Pascoe, J. 1014
Pérez, C. 671
Pétrot, F. 846
Perego, R. 220, 347
Pesciullesi, P. 381, 638, 679
Petrini, F. 689, 832
Petrutiu, S. 363
Pietracaprina, A. 803
Pillon, M. 55
Pinnow, K. 833
Plale, B. 1
Plata, O. 697
Potiti, L. 638
Prencipe, G. 578
Priol, T. 398, 671
Proenca, A. 222
Puebla, G. 21
Puente, V. 900
Puppin, D. 398

Qazzaz, B. 792
Quiles, F.J. 864
Quinlan, D. 614

Rabêlo, R. 1032
Radović, Z. 551
Raffin, B. 497
Ramirez, A. 535
Rana, O.F. 409, 964
Rantakokko, J. 615
Rasmussen, C.E. 632
Ratterman, J. 833
Rauber, T. 708
Ravazzolo, R. 381, 638
Ribes, A. 671
Richard, O. 55
Ripoll, A. 792

Robert, S. 497
Robert, Y. 246
Robles, A. 891
Rocha, R. 662
Rock, M. 328
Rodriguez, G. 183
Roisenberg, P. 214
Ronsse, M. 82
Rosenstiel, W. 189
Rünger, G. 708, 718
Russello, G. 579

Saif, T. 173
Sakellariou, R. 417, 535
Sánchez, J.L. 873, 891
Sancho, J.C. 689, 891
Santana, O.J. 9
Santonja, V. 882
Santos, L.P. 222
Santos Costa, V. 662
Sbaraglia, S. 141
Schenk, O. 624
Scheuermann, P. 363
Schmidt, B. 989
Schuster, A. 198
Schwiegelshosh, U. 398
Scott, J. 391
Seelam, S. 39
Seinstra, F.J. 752
Sendag, R. 541
Seznec, A. 506
Shende, S.S. 119
Silva, F. 662
Silvestri, F. 347
Sinha, B.P. 856
Sivasubramaniam, A. 507
Skillicorn, D. 346
Slominski, A. 1
Sloot, P. 962
Smeds, N. 833
Song, S.W. 828
Song, W. 972
Sottile, M.J. 632
Spielman, D.A. 812
Spies, F. 1056
Stahl, R. 206
Stamatakis, A. 997
Steihaug, T. 698
Steinmacher-burow, B. 833
Stiller, J. 726

Stoutchinin, A.   336
Sudharsanan, S.   751
Sunderam, V.   1014
Suppi, R.   792
Suter, F.   230
Szkaliczki, T.   776
Szwarcfiter, J.L.   828

Talia, D.   962, 1022
Tao, J.   149
Temam, O.   292
Teng, S.-h.   812
Tischler, F.   768
Toonen, B.   833
Torquati, M.   596, 638
Torres, E.F.   516
Träff, J.L.   155
Trancoso, P.   561
Trenkler, B.   72
Trunfio, P.   1022
Trystram, D.   1005
Tulone, D.   587
Tvrdík, P.   734

Uhl, A.   751, 768
Üngör, A.   812
Urrutia, J.   917
Utrera, G.   238

Valduriez, P.   355
Valero, M.   9, 535
Vallejo, F.   900
Vandeputte, F.   284
Van Roy, P.   1013

van Steen, M.   579
Vanthournout, K.   1068
Vassiliadis, S.   571, 760
Velho, P.   214
Verbeke, J.   1056
Verkest, D.   206
Vetter, J.   117
Villalobos, J.E.   891
Viñals, V.   516
Vivien, F.   246

Walker, D.   476
Wason, J.   972
Watson, P.   346
Wattenhofer, R.   820
Webber, T.   214
Weidendorfer, J.   133
Wolf, F.   47
Wu, J.   916

Xu, F.   972

Yalçınkaya, Y.   698
Yang, X.Y.   792

Zaki, M.J.   988
Zanikolas, S.   417
Zapata, E.L.   697
Zeffer, H.   551
Zima, H.P.   271
Zimmermann, W.   263
Zoccolo, C.   381, 638, 679
Zola, J.   1005

Lecture Notes in Computer Science

For information about Vols. 1–3064

please contact your bookseller or Springer

Vol. 3194: R. Camacho, R. King, A. Srinivasan (Eds.), Inductive Logic Programming. XI, 361 pages. 2004. (Subseries LNAI).

Vol. 3177: Z.R. Yang, H. Yin, R. Everson (Eds.), Intelligent Data Engineering and Automated Learning – IDEAL 2004. VXIII, 852 pages. 2004.

Vol. 3174: F. Yin, J. Wang, C. Guo (Eds.), Advances in Neural Networks - ISNN 2004. XXXV, 1021 pages. 2004.

Vol. 3172: M. Dorigo, M. Birattari, C. Blum, L. M.Gambardella, F. Mondada, T. Stützle (Eds.), Ant Colony, Optimization and Swarm Intelligence. XII, 434 pages. 2004.

Vol. 3166: M. Rauterberg (Ed.), Entertainment Computing – ICEC 2004. XXIII, 617 pages. 2004.

Vol. 3158: I. Nikolaidis, M. Barbeau, E. Kranakis (Eds.), Ad-Hoc, Mobile, and Wireless Networks. IX, 344 pages. 2004.

Vol. 3157: C. Zhang, H. W. Guesgen, W.K. Yeap (Eds.), PRICAI 2004: Trends in Artificial Intelligence. XX, 1023 pages. 2004. (Subseries LNAI).

Vol. 3156: M. Joye, J.-J. Quisquater (Eds.), Cryptographic Hardware and Embedded Systems - CHES 2004. XIII, 455 pages. 2004.

Vol. 3155: P. Funk, P.A. González Calero (Eds.), Advanced in Case-Based Reasoning. XIII, 822 pages. 2004. (Subseries LNAI).

Vol. 3153: J. Fiala, V. Koubek, J. Kratochvíl (Eds.), Mathematical Foundations of Computer Science 2004. XIV, 902 pages. 2004.

Vol. 3152: M. Franklin (Ed.), Advances in Cryptology – CRYPTO 2004. XI, 579 pages. 2004.

Vol. 3150: G.-Z. Yang, T. Jiang (Eds.), Medical Imaging and Augmented Reality. XII, 378 pages. 2004.

Vol. 3149: M. Danelutto, M. Vanneschi, D. Laforenza (Eds.), Euro-Par 2004 Parallel Processing. XXXIV, 1081 pages. 2004.

Vol. 3148: R. Giacobazzi (Ed.), Static Analysis. XI, 393 pages. 2004.

Vol. 3146: P. Érdi, A. Esposito, M. Marinaro, S. Scarpetta (Eds.), Computational Neuroscience: Cortical Dynamics. XI, 161 pages. 2004.

Vol. 3144: M. Papatriantafilou, P. Hunel (Eds.), Principles of Distributed Systems. XI, 246 pages. 2004.

Vol. 3143: W. Liu, Y. Shi, Q. Li (Eds.), Advances in Web-Based Learning – ICWL 2004. XIV, 459 pages. 2004.

Vol. 3142: J. Diaz, J. Karhumäki, A. Lepistö, D. Sannella (Eds.), Automata, Languages and Programming. XIX, 1253 pages. 2004.

Vol. 3140: N. Koch, P. Fraternali, M. Wirsing (Eds.), Web Engineering. XXI, 623 pages. 2004.

Vol. 3139: F. Iida, R. Pfeifer, L. Steels, Y. Kuniyoshi (Eds.), Embodied Artificial Intelligence. IX, 331 pages. 2004. (Subseries LNAI).

Vol. 3138: A. Fred, T. Caelli, R.P.W. Duin, A. Campilho, D.d. Ridder (Eds.), Structural, Syntactic, and Statistical Pattern Recognition. XXII, 1168 pages. 2004.

Vol. 3137: P. De Bra, W. Nejdl (Eds.), Adaptive Hypermedia and Adaptive Web-Based Systems. XIV, 442 pages. 2004.

Vol. 3136: F. Meziane, E. Métais (Eds.), Natural Language Processing and Information Systems. XII, 436 pages. 2004.

Vol. 3134: C. Zannier, H. Erdogmus, L. Lindstrom (Eds.), Extreme Programming and Agile Methods - XP/Agile Universe 2004. XIV, 233 pages. 2004.

Vol. 3133: A.D. Pimentel, S. Vassiliadis (Eds.), Computer Systems: Architectures, Modeling, and Simulation. XIII, 562 pages. 2004.

Vol. 3131: V. Torra, Y. Narukawa (Eds.), Modeling Decisions for Artificial Intelligence. XI, 327 pages. 2004. (Subseries LNAI).

Vol. 3130: A. Syropoulos, K. Berry, Y. Haralambous, B. Hughes, S. Peter, J. Plaice (Eds.), TeX, XML, and Digital Typography. VIII, 265 pages. 2004.

Vol. 3129: Q. Li, G. Wang, L. Feng (Eds.), Advances in Web-Age Information Management. XVII, 753 pages. 2004.

Vol. 3128: D. Asonov (Ed.), Querying Databases Privately. IX, 115 pages. 2004.

Vol. 3127: K.E. Wolff, H.D. Pfeiffer, H.S. Delugach (Eds.), Conceptual Structures at Work. XI, 403 pages. 2004. (Subseries LNAI).

Vol. 3126: P. Dini, P. Lorenz, J.N.d. Souza (Eds.), Service Assurance with Partial and Intermittent Resources. XI, 312 pages. 2004.

Vol. 3125: D. Kozen (Ed.), Mathematics of Program Construction. X, 401 pages. 2004.

Vol. 3124: J.N. de Souza, P. Dini, P. Lorenz (Eds.), Telecommunications and Networking - ICT 2004. XXVI, 1390 pages. 2004.

Vol. 3123: A. Belz, R. Evans, P. Piwek (Eds.), Natural Language Generation. X, 219 pages. 2004. (Subseries LNAI).

Vol. 3122: K. Jansen, S. Khanna, J.D.P. Rolim, D. Ron (Eds.), Approximation, Randomization, and Combinatorial Optimization. IX, 428 pages. 2004.

Vol. 3121: S. Nikoletseas, J.D.P. Rolim (Eds.), Algorithmic Aspects of Wireless Sensor Networks. X, 201 pages. 2004.

Vol. 3120: J. Shawe-Taylor, Y. Singer (Eds.), Learning Theory. X, 648 pages. 2004. (Subseries LNAI).

Vol. 3118: K. Miesenberger, J. Klaus, W. Zagler, D. Burger (Eds.), Computer Helping People with Special Needs. XXIII, 1191 pages. 2004.

Vol. 3116: C. Rattray, S. Maharaj, C. Shankland (Eds.), Algebraic Methodology and Software Technology. XI, 569 pages. 2004.

Vol. 3114: R. Alur, D.A. Peled (Eds.), Computer Aided Verification. XII, 536 pages. 2004.

Vol. 3113: J. Karhumäki, H. Maurer, G. Paun, G. Rozenberg (Eds.), Theory Is Forever. X, 283 pages. 2004.

Vol. 3112: H. Williams, L. MacKinnon (Eds.), Key Technologies for Data Management. XII, 265 pages. 2004.

Vol. 3111: T. Hagerup, J. Katajainen (Eds.), Algorithm Theory - SWAT 2004. XI, 506 pages. 2004.

Vol. 3110: A. Juels (Ed.), Financial Cryptography. XI, 281 pages. 2004.

Vol. 3109: S.C. Sahinalp, S. Muthukrishnan, U. Dogrusoz (Eds.), Combinatorial Pattern Matching. XII, 486 pages. 2004.

Vol. 3108: H. Wang, J. Pieprzyk, V. Varadharajan (Eds.), Information Security and Privacy. XII, 494 pages. 2004.

Vol. 3107: J. Bosch, C. Krueger (Eds.), Software Reuse: Methods, Techniques and Tools. XI, 339 pages. 2004.

Vol. 3106: K.-Y. Chwa, J.I. Munro (Eds.), Computing and Combinatorics. XIII, 474 pages. 2004.

Vol. 3105: S. Göbel, U. Spierling, A. Hoffmann, I. Iurgel, O. Schneider, J. Dechau, A. Feix (Eds.), Technologies for Interactive Digital Storytelling and Entertainment. XVI, 304 pages. 2004.

Vol. 3104: R. Kralovic, O. Sykora (Eds.), Structural Information and Communication Complexity. X, 303 pages. 2004.

Vol. 3103: K. Deb, e. al. (Eds.), Genetic and Evolutionary Computation – GECCO 2004. XLIX, 1439 pages. 2004.

Vol. 3102: K. Deb, e. al. (Eds.), Genetic and Evolutionary Computation – GECCO 2004. L, 1445 pages. 2004.

Vol. 3101: M. Masoodian, S. Jones, B. Rogers (Eds.), Computer Human Interaction. XIV, 694 pages. 2004.

Vol. 3100: J.F. Peters, A. Skowron, J.W. Grzymała-Busse, B. Kostek, R.W. Świniarski, M.S. Szczuka (Eds.), Transactions on Rough Sets I. X, 405 pages. 2004.

Vol. 3099: J. Cortadella, W. Reisig (Eds.), Applications and Theory of Petri Nets 2004. XI, 505 pages. 2004.

Vol. 3098: J. Desel, W. Reisig, G. Rozenberg (Eds.), Lectures on Concurrency and Petri Nets. VIII, 849 pages. 2004.

Vol. 3097: D. Basin, M. Rusinowitch (Eds.), Automated Reasoning. XII, 493 pages. 2004. (Subseries LNAI).

Vol. 3096: G. Melnik, H. Holz (Eds.), Advances in Learning Software Organizations. X, 173 pages. 2004.

Vol. 3095: C. Bussler, D. Fensel, M.E. Orlowska, J. Yang (Eds.), Web Services, E-Business, and the Semantic Web. X, 147 pages. 2004.

Vol. 3094: A. Nürnberger, M. Detyniecki (Eds.), Adaptive Multimedia Retrieval. VIII, 229 pages. 2004.

Vol. 3093: S.K. Katsikas, S. Gritzalis, J. Lopez (Eds.), Public Key Infrastructure. XIII, 380 pages. 2004.

Vol. 3092: J. Eckstein, H. Baumeister (Eds.), Extreme Programming and Agile Processes in Software Engineering. XVI, 358 pages. 2004.

Vol. 3091: V. van Oostrom (Ed.), Rewriting Techniques and Applications. X, 313 pages. 2004.

Vol. 3089: M. Jakobsson, M. Yung, J. Zhou (Eds.), Applied Cryptography and Network Security. XIV, 510 pages. 2004.

Vol. 3087: D. Maltoni, A.K. Jain (Eds.), Biometric Authentication. XIII, 343 pages. 2004.

Vol. 3086: M. Odersky (Ed.), ECOOP 2004 – Object-Oriented Programming. XIII, 611 pages. 2004.

Vol. 3085: S. Berardi, M. Coppo, F. Damiani (Eds.), Types for Proofs and Programs. X, 409 pages. 2004.

Vol. 3084: A. Persson, J. Stirna (Eds.), Advanced Information Systems Engineering. XIV, 596 pages. 2004.

Vol. 3083: W. Emmerich, A.L. Wolf (Eds.), Component Deployment. X, 249 pages. 2004.

Vol. 3080: J. Desel, B. Pernici, M. Weske (Eds.), Business Process Management. X, 307 pages. 2004.

Vol. 3079: Z. Mammeri, P. Lorenz (Eds.), High Speed Networks and Multimedia Communications. XVIII, 1103 pages. 2004.

Vol. 3078: S. Cotin, D.N. Metaxas (Eds.), Medical Simulation. XVI, 296 pages. 2004.

Vol. 3077: F. Roli, J. Kittler, T. Windeatt (Eds.), Multiple Classifier Systems. XII, 386 pages. 2004.

Vol. 3076: D. Buell (Ed.), Algorithmic Number Theory. XI, 451 pages. 2004.

Vol. 3075: W. Lenski (Ed.), Logic versus Approximation. IX, 205 pages. 2004.

Vol. 3074: B. Kuijpers, P. Revesz (Eds.), Constraint Databases and Applications. XII, 181 pages. 2004.

Vol. 3073: H. Chen, R. Moore, D.D. Zeng, J. Leavitt (Eds.), Intelligence and Security Informatics. XV, 536 pages. 2004.

Vol. 3072: D. Zhang, A.K. Jain (Eds.), Biometric Authentication. XVII, 800 pages. 2004.

Vol. 3071: A. Omicini, P. Petta, J. Pitt (Eds.), Engineering Societies in the Agents World. XIII, 409 pages. 2004. (Subseries LNAI).

Vol. 3070: L. Rutkowski, J. Siekmann, R. Tadeusiewicz, L.A. Zadeh (Eds.), Artificial Intelligence and Soft Computing - ICAISC 2004. XXV, 1208 pages. 2004. (Subseries LNAI).

Vol. 3068: E. André, L. Dybkjær, W. Minker, P. Heisterkamp (Eds.), Affective Dialogue Systems. XII, 324 pages. 2004. (Subseries LNAI).

Vol. 3067: M. Dastani, J. Dix, A. El Fallah-Seghrouchni (Eds.), Programming Multi-Agent Systems. X, 221 pages. 2004. (Subseries LNAI).

Vol. 3066: S. Tsumoto, R. Słowiński, J. Komorowski, J.W. Grzymała-Busse (Eds.), Rough Sets and Current Trends in Computing. XX, 853 pages. 2004. (Subseries LNAI).

Vol. 3065: A. Lomuscio, D. Nute (Eds.), Deontic Logic in Computer Science. X, 275 pages. 2004. (Subseries LNAI).